CCSS Your Common Core Edition

GLENCOE MATH

ACCELERATED

A PRE-ALGEBRA PROGRAM

AUTHORS

Carter • Cuevas • Day • Malloy • Molix-Bailey • Price • Willard

McGraw Hill Education

Bothell, WA • Chicago, IL • Columbus, OH • New York, NY

connectED.mcgraw-hill.com

The McGraw·Hill Companies

McGraw Hill Education

STEM McGraw-Hill is committed to providing instructional materials in Science, Technology, Engineering, and Mathematics (STEM) that give all students a solid foundation, one that prepares them for college and careers in the 21st century.

Send all inquiries to:
McGraw-Hill Education
8787 Orion Place
Columbus, OH 43240

ISBN: 978-0-07-663798-0
MHID: 0-07-663798-0

Printed in the United States of America.

11 12 13 14 15 DOR 20 19 18 17 16

CONTENTS IN BRIEF

Units organized by CCSS clusters

Glencoe Math Accelerated is organized into units based on groups of related standards. This year, you will study and understand the content in the four units shown below.

GO digital

it's all at connectED.mcgraw-hill.com

Go to the Student Center for your eBook, Resources, Homework, and Messages.

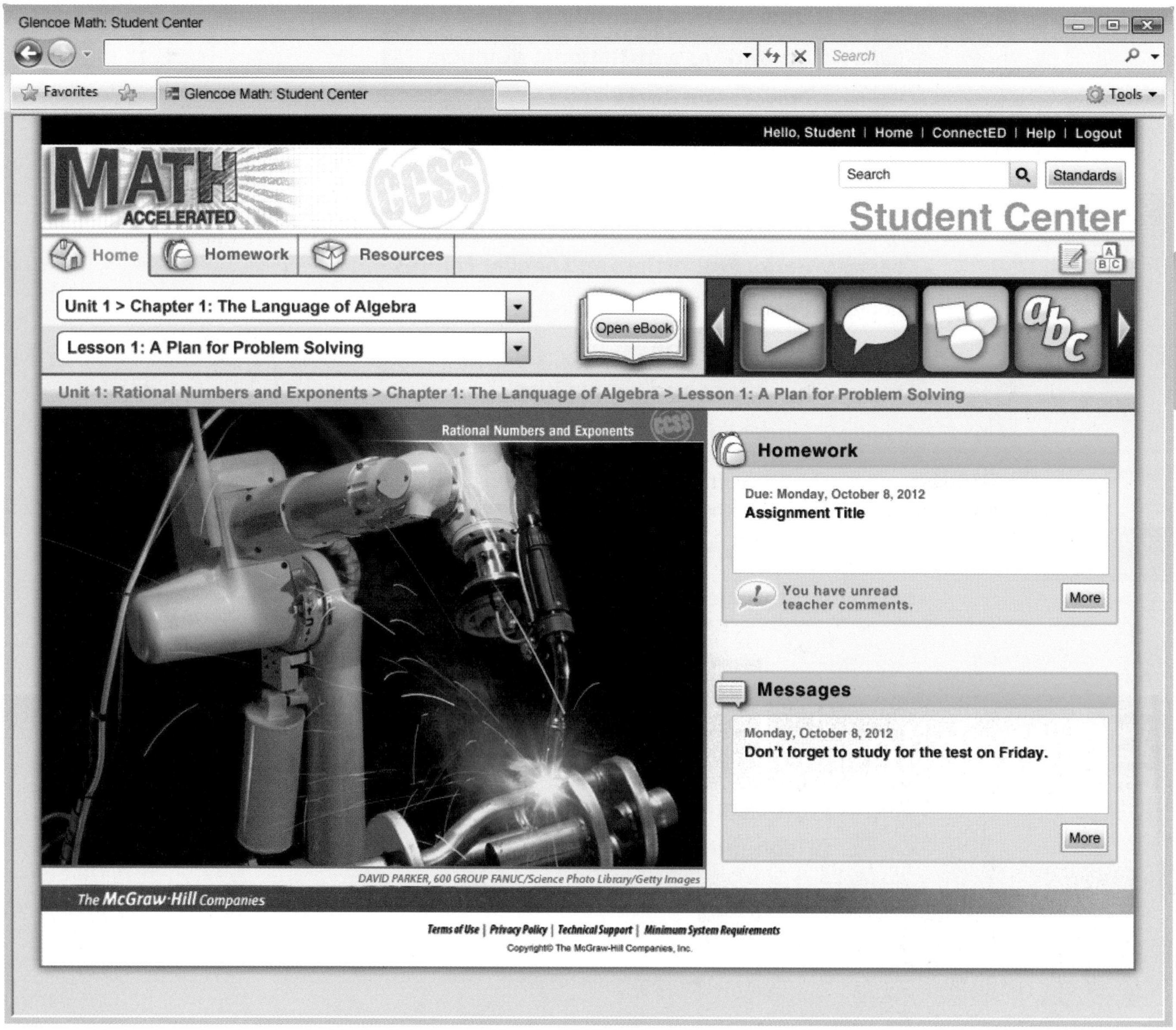

Write your Username ____________ Password ____________

Get your resources online to help you in class and at home.

Vocab
Find activities for building vocabulary.

Watch
Watch animations and videos.

Tutor
See a teacher illustrate examples and problems.

Tools
Explore concepts with virtual manipulatives.

Check
Self-assess your progress.

eHelp
Get targeted homework help.

Masters
Provides practice worksheets.

ISG
Interactive Study Guide

GO mobile

Scan this QR code with your smart phone* or visit mheonline.com/apps.

*May require quick response code reader app.

Available on the App Store

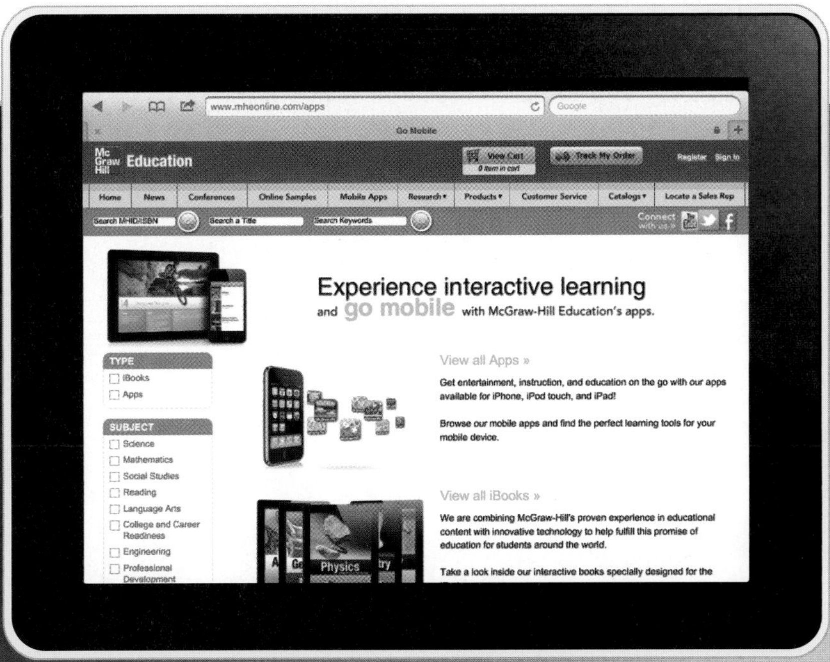

UNIT 1 Rational Numbers and Exponents

Chapter 1
The Language of Algebra

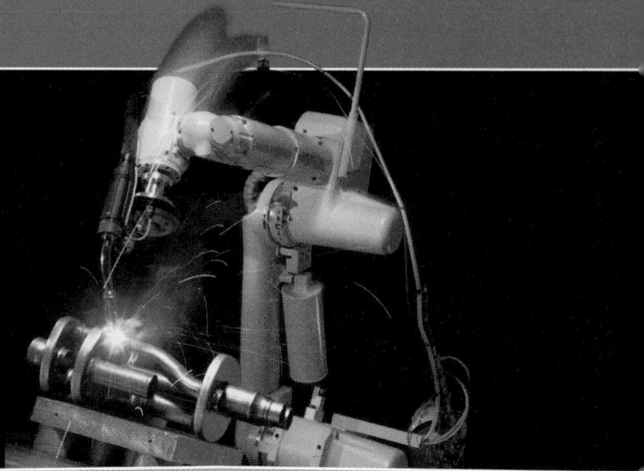

Essential Question
HOW can you use numbers and symbols to represent mathematical ideas?

Interactive Study Guide
Use pages 1-24 with this chapter.

Each chapter contains:
- Chapter Preview
- Foldables Study Organizer
- Lesson Getting Started and Notes
- Mid-Chapter Check
- 21st Century Career
- Chapter Review

Go to page 25 to learn about a 21st Century Career in **Animal Conservation!**

Chapter 2
Operations with Integers

Essential Question

WHAT happens when you add, subtract, multiply, and divide integers?

Interactive Study Guide
Use pages 25–46 with this chapter.

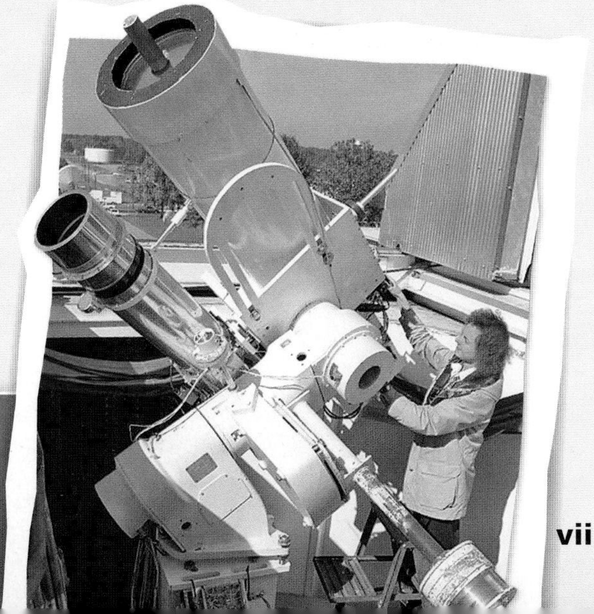

Go to page 68 to learn about a 21st Century Career in Astronomy!

Chapter 3
Operations with Rational Numbers

Essential Question

WHAT happens when you add, subtract, multiply, and divide rational numbers?

Interactive Study Guide

Use pages 47–68 with this chapter.

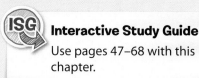

Go to page 113 to learn about a 21st Century Career in Fashion Design!

(t)Pixtal/age fotostock, (bkgd)©Roy Lawe/Alamy, (b)Justin Pumfrey/The Image Bank/Getty Images

Chapter 4
Powers and Roots

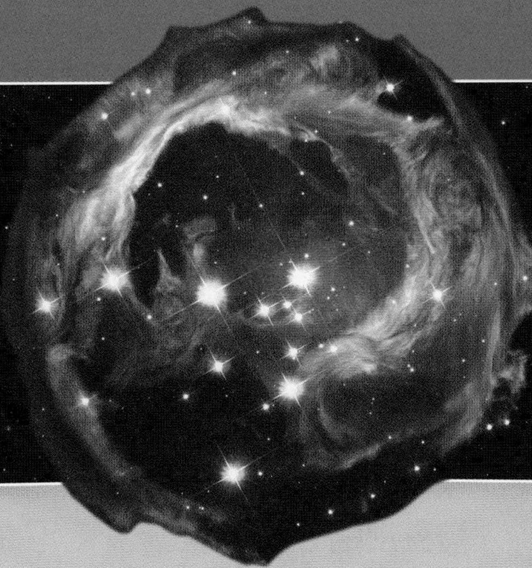

Essential Question
WHY is it useful to write numbers in different ways?

Interactive Study Guide
Use pages 69–92 with this chapter.

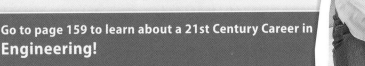

Go to page 159 to learn about a 21st Century Career in **Engineering!**

UNIT 2 Proportionality and Linear Relationships

Chapter 5
Ratio, Proportion, and Similar Figures

Essential Question

HOW can you identify and represent proportional relationships?

Interactive Study Guide

Use pages 93–122 with this chapter.

Go to page 211 to learn about a 21st Century Career in **Engineering!**

Chapter 6
Percents

Essential Question

HOW can you use proportional relationships to solve real-world percent problems?

Interactive Study Guide
Use pages 123–144 with this chapter.

Go to page 267 to learn about a 21st Century Career in **Video Game Design!**

Chapter 7 Algebraic Expressions

Essential Question
WHY are algebraic rules useful?

Interactive Study Guide
Use pages 145–164 with this chapter.

Go to page 309 to learn about a 21st Century Career in **Design Engineering!**

Chapter 8
Equations and Inequalities

Essential Question

HOW are equations and inequalities used to describe and solve multi-step problems?

Interactive Study Guide
Use pages 165–190 with this chapter.

Go to page 352 to learn about a 21st Century Career in **Veterinary Medicine!**

Chapter 9
Linear Functions

Essential Question
HOW are linear functions used to model proportional relationships?

Interactive Study Guide
Use pages 191–214 with this chapter.

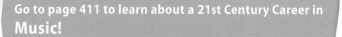

Go to page 411 to learn about a 21st Century Career in **Music!**

UNIT 3 Introduction to Sampling and Inference

Chapter 10
Statistics and Probability

Essential Question

HOW are statistics used to draw inferences about and compare populations?

Interactive Study Guide

Use pages 215–240 with this chapter.

Go to page 461 to learn about a 21st Century Career in Market Research!

CCSS **UNIT 4** Creating, Comparing, and Analyzing Geometric Figures

Chapter 11
Congruence, Similarity, and Transformations

Essential Question

HOW can you determine congruence and similarity?

ISG **Interactive Study Guide**

Use pages 241–266 with this chapter.

Go to page 527 to learn about a 21st Century Career in **Computer Animation!**

Chapter 12
Volume and Surface Area

Essential Question
HOW are two-dimensional figures used to solve problems involving three-dimensional figures?

Interactive Study Guide
Use pages 267–296 with this chapter.

Go to page 591 to learn about a 21st Century Career in **Architecture!**

Common Core State Standards for MATHEMATICS, Accelerated 7th Grade

Glencoe Math Accelerated focuses on four critical areas: (1) adding, subtracting, multiplying, and dividing rational numbers; (2) analyzing proportional relationships and using expressions and equations; (3) using sampling to draw inferences about a population; and (4) solving problems involving angle measure, area, surface area, and volume.

Content Standards

Domains 7.NS, 8.NS, 8.EE

Rational Numbers and Exponents

- Apply and extend previous understandings of operations with fractions to add, subtract, multiply, and divide rational numbers.
- Know that there are numbers that are not rational, and approximate them by rational numbers.
- Work with radicals and integer exponents.

Domains 7.RP, 7.EE, 8.EE

Proportionality and Linear Relationships

- Analyze proportional relationships and use them to solve real-world and mathematical problems.
- Use properties of operations to generate equivalent expressions.
- Solve real-life and mathematical problems using numerical and algebraic expressions and equations.
- Understand the connections between proportional relationships, lines, and linear equations.
- Analyze and solve linear equations and pairs of simultaneous linear equations.

Domain 7.SP

Introduction to Sampling and Inference

- Use random sampling to draw inferences about a population.
- Draw informal comparative inferences about two populations.
- Investigate chance processes and develop, use, and evaluate probability models.

Domains 7.G, 8.G

Creating, Comparing, and Analyzing Geometric Figures

- Draw, construct and describe geometrical figures and describe the relationships between them.
- Solve real-life and mathematical problems involving angle measure, area, surface area, and volume.
- Understand congruence and similarity using physical models, transparencies, or geometry software.
- Solve real-world and mathematical problems involving volume of cylinders, cones and spheres.

Mathematical Practices

1. Make sense of problems and persevere in solving them.
2. Reason abstractly and quantitatively.
3. Construct viable arguments and critique the reasoning of others.
4. Model with mathematics.
5. Use appropriate tools strategically.
6. Attend to precision.
7. Look for and make use of structure.
8. Look for and express regularity in repeated reasoning.

Rational Numbers and Exponents

Domains 7.NS, 8.NS, 8.EE

7.NS Apply and extend previous understandings of operations with fractions to add, subtract, multiply, and divide rational numbers.

1. Apply and extend previous understandings of addition and subtraction to add and subtract rational numbers; represent addition and subtraction on a horizontal or vertical number line diagram.

a. Describe situations in which opposite quantities combine to make 0.

b. Understand $p + q$ as the number located a distance $|q|$ from p, in the positive or negative direction depending on whether q is positive or negative. Show that a number and its opposite have a sum of 0 (are additive inverses). Interpret sums of rational numbers by describing real-world contexts.

c. Understand subtraction of rational numbers as adding the additive inverse, $p - q = p + (-q)$. Show that the distance between two rational numbers on the number line is the absolute value of their difference, and apply this principle in real-world contexts.

d. Apply properties of operations as strategies to add and subtract rational numbers.

2. Apply and extend previous understandings of multiplication and division and of fractions to multiply and divide rational numbers.

a. Understand that multiplication is extended from fractions to rational numbers by requiring that operations continue to satisfy the properties of operations, particularly the distributive property, leading to products such as $(-1)(-1) = 1$ and the rules for multiplying signed numbers. Interpret products of rational numbers by describing real-world contexts.

b. Understand that integers can be divided, provided that the divisor is not zero, and every quotient of integers (with non-zero divisor) is a rational number. If p and q are integers, then $-\frac{p}{q} = \frac{(-p)}{q} = \frac{p}{(-q)}$. Interpret quotients of rational numbers by describing real-world contexts.

c. Apply properties of operations as strategies to multiply and divide rational numbers.

d. Convert a rational number to a decimal using long division; know that the decimal form of a rational number terminates in 0s or eventually repeats.

3. Solve real-world and mathematical problems involving the four operations with rational numbers.

8.NS Know that there are numbers that are not rational, and approximate them by rational numbers.

1. Know that numbers that are not rational are called irrational. Understand informally that every number has a decimal expansion; for rational numbers show that the decimal expansion repeats eventually, and convert a decimal expansion which repeats eventually into a rational number.

2. Use rational approximations of irrational numbers to compare the size of irrational numbers, locate them approximately on a number line diagram, and estimate the value of expressions (e.g., π^2).

8.EE Work with radicals and integer exponents.

1. Know and apply the properties of integer exponents to generate equivalent numerical expressions.

2. Use square root and cube root symbols to represent solutions to equations of the form $x^2 = p$ and $x^3 = p$, where p is a positive rational number. Evaluate square roots of small perfect squares and cube roots of small perfect cubes. Know that $\sqrt{2}$ is irrational.

3. Use numbers expressed in the form of a single digit times an integer power of 10 to estimate very large or very small quantities, and to express how many times as much one is than the other.

4. Perform operations with numbers expressed in scientific notation, including problems where both decimal and scientific notation are used. Use scientific notation and choose units of appropriate size for measurements of very large or very small quantities (e.g., use millimeters per year for seafloor spreading). Interpret scientific notation that has been generated by technology.

Domains 7.RP, 7.EE, 8.EE

Proportionality and Linear Relationships

7.RP Analyze proportional relationships and use them to solve real-world and mathematical problems.

1. Compute unit rates associated with ratios of fractions, including ratios of lengths, areas and other quantities measured in like or different units.
2. Recognize and represent proportional relationships between quantities.
   a. Decide whether two quantities are in a proportional relationship, e.g., by testing for equivalent ratios in a table or graphing on a coordinate plane and observing whether the graph is a straight line through the origin.
   b. Identify the constant of proportionality (unit rate) in tables, graphs, equations, diagrams, and verbal descriptions of proportional relationships.
   c. Represent proportional relationships by equations.
   d. Explain what a point (x, y) on the graph of a proportional relationship means in terms of the situation, with special attention to the points $(0, 0)$ and $(1, r)$ where r is the unit rate.
3. Use proportional relationships to solve multistep ratio and percent problems.

7.EE Use properties of operations to generate equivalent expressions.

1. Apply properties of operations as strategies to add, subtract, factor, and expand linear expressions with rational coefficients.
2. Understand that rewriting an expression in different forms in a problem context can shed light on the problem and how the quantities in it are related.

7.EE Solve real-life and mathematical problems using numerical and algebraic expressions and equations.

3. Solve multi-step real-life and mathematical problems posed with positive and negative rational numbers in any form (whole numbers, fractions, and decimals), using tools strategically. Apply properties of operations to calculate with numbers in any form; convert between forms as appropriate; and assess the reasonableness of answers using mental computation and estimation strategies.
4. Use variables to represent quantities in a real-world or mathematical problem, and construct simple equations and inequalities to solve problems by reasoning about the quantities.
   a. Solve word problems leading to equations of the form $px + q = r$ and $p(x + q) = r$, where p, q, and r are specific rational numbers. Solve equations of these forms fluently. Compare an algebraic solution to an arithmetic solution, identifying the sequence of the operations used in each approach.
   b. Solve word problems leading to inequalities of the form $px + q > r$ or $px + q < r$, where p, q, and r are specific rational numbers. Graph the solution set of the inequality and interpret it in the context of the problem.

8.EE Understand the connections between proportional relationships, lines, and linear equations.

5. Graph proportional relationships, interpreting the unit rate as the slope of the graph. Compare two different proportional relationships represented in different ways.

6. Use similar triangles to explain why the slope m is the same between any two distinct points on a non-vertical line in the coordinate plane; derive the equation $y = mx$ for a line through the origin and the equation $y = mx + b$ for a line intercepting the vertical axis at b.

8.EE Analyze and solve linear equations and pairs of simultaneous linear equations.

7. Solve linear equations in one variable.

a. Give examples of linear equations in one variable with one solution, infinitely many solutions, or no solutions. Show which of these possibilities is the case by successively transforming the given equation into simpler forms, until an equivalent equation of the form $x = a$, $a = a$, or $a = b$ results (where a and b are different numbers).

b. Solve linear equations with rational number coefficients, including equations whose solutions require expanding expressions using the distributive property and collecting like terms.

Introduction to Sampling and Inference

Domain 7.SP

7.SP Use random sampling to draw inferences about a population.

1. Understand that statistics can be used to gain information about a population by examining a sample of the population; generalizations about a population from a sample are valid only if the sample is representative of that population. Understand that random sampling tends to produce representative samples and support valid inferences.

2. Use data from a random sample to draw inferences about a population with an unknown characteristic of interest. Generate multiple samples (or simulated samples) of the same size to gauge the variation in estimates or predictions.

7.SP Draw informal comparative inferences about two populations.

3. Informally assess the degree of visual overlap of two numerical data distributions with similar variabilities, measuring the difference between the centers by expressing it as a multiple of a measure of variability.

4. Use measures of center and measures of variability for numerical data from random samples to draw informal comparative inferences about two populations.

7.SP Investigate chance processes and develop, use, and evaluate probability models.

5. Understand that the probability of a chance event is a number between 0 and 1 that expresses the likelihood of the event occurring. Larger numbers indicate greater likelihood. A probability near 0 indicates an unlikely event, a probability around $\frac{1}{2}$ indicates an event that is neither unlikely nor likely, and a probability near 1 indicates a likely event.

6. Approximate the probability of a chance event by collecting data on the chance process that produces it and observing its long-run relative frequency, and predict the approximate relative frequency given the probability.

7. Develop a probability model and use it to find probabilities of events. Compare probabilities from a model to observed frequencies; if the agreement is not good, explain possible sources of the discrepancy.

a. Develop a uniform probability model by assigning equal probability to all outcomes, and use the model to determine probabilities of events.

b. Develop a probability model (which may not be uniform) by observing frequencies in data generated from a chance process.

8. Find probabilities of compound events using organized lists, tables, tree diagrams, and simulation.

a. Understand that, just as with simple events, the probability of a compound event is the fraction of outcomes in the sample space for which the compound event occurs.

b. Represent sample spaces for compound events using methods such as organized lists, tables and tree diagrams. For an event described in everyday language (e.g., "rolling double sixes"), identify the outcomes in the sample space which compose the event.

c. Design and use a simulation to generate frequencies for compound events.

Domains 7.G, 8.G

Creating, Comparing, and Analyzing Geometric Figures

7.G Draw, construct, and describe geometrical figures and describe the relationships between them.

1. Solve problems involving scale drawings of geometric figures, including computing actual lengths and areas from a scale drawing and reproducing a scale drawing at a different scale.
2. Draw (freehand, with ruler and protractor, and with technology) geometric shapes with given conditions. Focus on constructing triangles from three measures of angles or sides, noticing when the conditions determine a unique triangle, more than one triangle, or no triangle.
3. Describe the two-dimensional figures that result from slicing three-dimensional figures, as in plane sections of right rectangular prisms and right rectangular pyramids.

7.G Solve real-life and mathematical problems involving angle measure, area, surface area, and volume.

4. Know the formulas for the area and circumference of a circle and use them to solve problems; give an informal derivation of the relationship between the circumference and area of a circle.
5. Use facts about supplementary, complementary, vertical, and adjacent angles in a multi-step problem to write and solve simple equations for an unknown angle in a figure.
6. Solve real-world and mathematical problems involving area, volume and surface area of two- and three-dimensional objects composed of triangles, quadrilaterals, polygons, cubes, and right prisms.

8.G Understand congruence and similarity using physical models, transparencies, or geometry software.

1. Verify experimentally the properties of rotations, reflections, and translations:
   a. Lines are taken to lines, and line segments to line segments of the same length.
   b. Angles are taken to angles of the same measure.
   c. Parallel lines are taken to parallel lines.
2. Understand that a two-dimensional figure is congruent to another if the second can be obtained from the first by a sequence of rotations, reflections, and translations; given two congruent figures, describe a sequence that exhibits the congruence between them.
3. Describe the effect of dilations, translations, rotations, and reflections on two-dimensional figures using coordinates.
4. Understand that a two-dimensional figure is similar to another if the second can be obtained from the first by a sequence of rotations, reflections, translations, and dilations; given two similar two-dimensional figures, describe a sequence that exhibits the similarity between them.
5. Use informal arguments to establish facts about the angle sum and exterior angle of triangles, about the angles created when parallel lines are cut by a transversal, and the angle-angle criterion for similarity of triangles.

8.G Solve real-world and mathematical problem involving volume of cylinders, cones, and spheres.

9. Know the formulas for the volumes of cones, cylinders, and spheres and use them to solve real-world and mathematical problems.

Chapter 1

The Language of Algebra

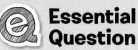

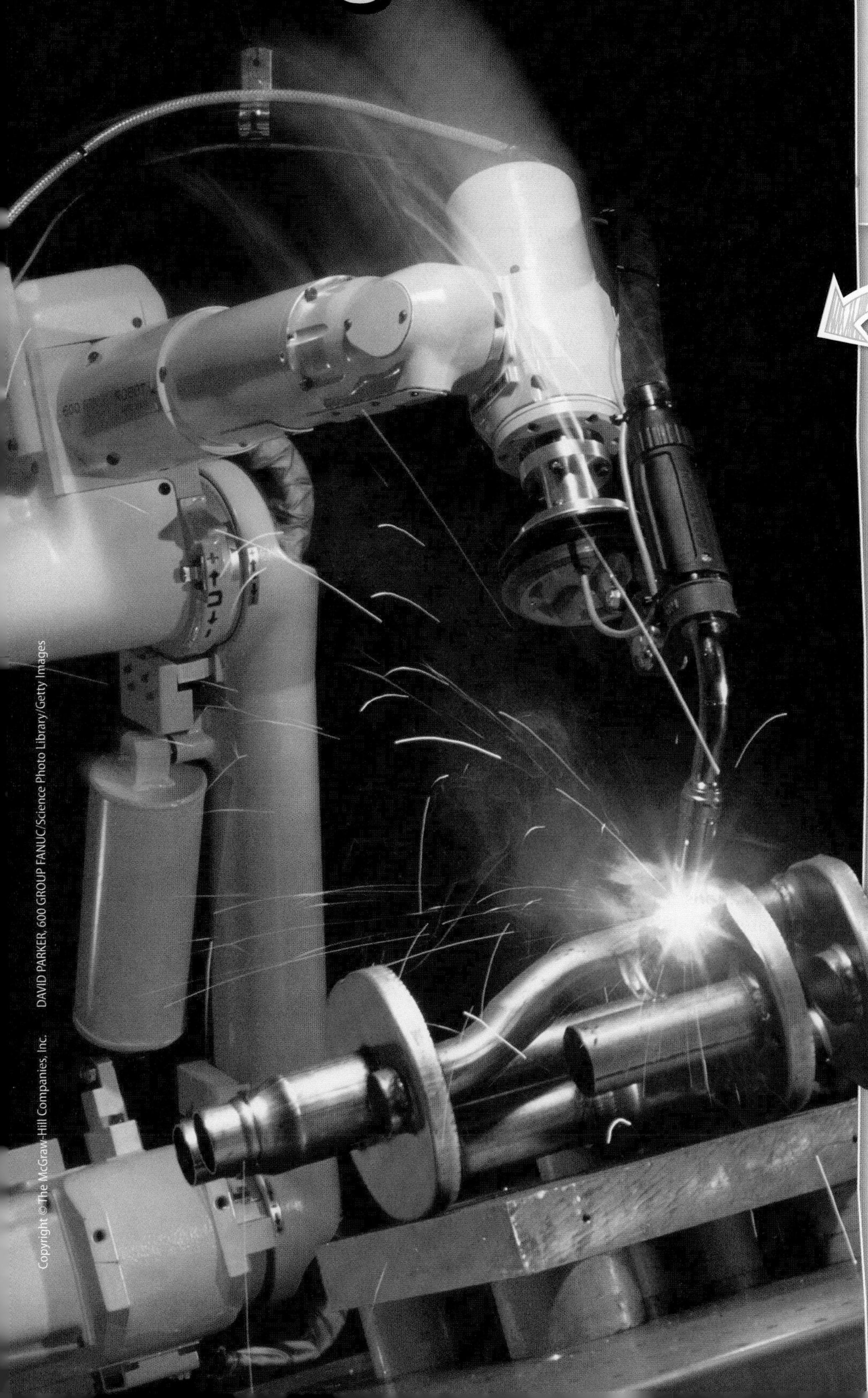

Essential Question

How can you use numbers and symbols to represent mathematical ideas?

Common Core State Standards

Content Standards
7.NS.3, 7.EE.1, 7.EE.2, 7.EE.3, 7.EE.4

Mathematical Practices
1, 3, 4, 5, 7, 8

Math in the Real World

Engineering Robotic arms can do everything from welding car parts to vacuuming a carpet. Programmers use a coordinate system to teach robots how to locate objects and move them in three-dimensional space.

Interactive Study Guide
See pages 1–4 for:

- Chapter Preview
- Are You Ready?
- Foldable Study Organizer

Lesson 1-1

A Plan for Problem Solving

Interactive Study Guide

See pages 5–6 for:
- Getting Started
- Real-World Link
- Notes

Essential Question

How can you use numbers and symbols to represent mathematical ideas?

Common Core State Standards

Content Standards
7.NS.3, 7.EE.3

Mathematical Practices
1, 3, 4, 5

Vocabulary

four-step plan

What You'll Learn

- Use the four-step plan to solve problems.
- Solve multi-step problems.

Real-World Link

Cell Phones The table shows the results of a survey about how teens use their cell phones. About how many times as great is the number of teens who use their phones to take pictures compared to the number who exchange videos?

I Use My Cell Phone to...	Number of Teens
access social network sites	46
exchange videos	61
play games	92
play music	120
take pictures	177

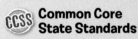

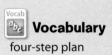

Key Concept The Four-Step Plan

Understand
- Read the problem quickly to gain a general understanding of it.
- Ask, "What facts do I know?"
- Ask, "What do I need to find out?"
- Ask, "Is there enough information to solve the problem? Is there extra information?"

Plan
- Reread the problem to identify relevant facts.
- Determine how the facts relate to one another.
- Make a plan and choose a strategy for solving it. There may be several strategies that you can use.
- Estimate what you think the answer should be.

Solve
- Use your plan to solve the problem.
- If your plan does not work, revise it or make a new plan.

Check
- Reread the problem. Is there another solution?
- Examine your answer carefully.
- Ask, "Is my answer reasonable and close to my estimate?"
- Ask, "Does my answer make sense?"
- If your answer is not reasonable, make a new plan and solve the problem another way.
- You may also want to check your answer by solving the problem again in a different way.

It is often helpful to have an organized plan for solving math problems. The **four-step plan** shown above can be used to solve any math problem.

Example 1

Tutor

Refer to the table at the beginning of the lesson. About how many times as great is the number of teens who use their cell phones to take pictures compared to the number who exchange videos? Use the four-step plan.

Understand You know that 177 teens use cell phones to take pictures and 61 use them to exchange videos. You need to find *about* how many times as great the first group of teens is than the second group.

Plan Divide the number of teens who take pictures by the number who exchange videos. Since the question asks for *about* how many times as great, you can estimate.

Solve $177 \div 61 \rightarrow 180 \div 60 = 3$ Estimate.

So, the number of teens who use their cell phones to take pictures is about 3 times as great as the number who exchange videos.

Check $61 \times 3 = 183$. Since 183 is close to 177, the answer is reasonable. ✓

Watch Out!
The phrase *how many times as great* means you should use division to compare the numbers. The phrase *how many more* means you should use subtraction.

Got It? Do this problem to find out.

1. About how many more teens in the survey use cell phones to play games than to access social network sites?

Solving Multi-Step Problems

In some problems, the *Solve* step may require two or more substeps.

Example 2

Tutor

The tower of a free-fall ride at an amusement park is 251.5 feet tall. The ride begins to free-fall from 18.9 feet below the top of the tower, and the brakes are applied 41.4 feet from the bottom. How long is the free-fall portion of the ride?

Understand You know the tower is 251.5 feet tall, and that the ride starts 18.9 feet from the top and ends 41.4 feet from the bottom. You need to find the length of the free-fall portion of the ride.

Plan Draw a diagram. Subtract the starting height of the ride from the height of the tower. Then subtract the height where the brakes are applied.

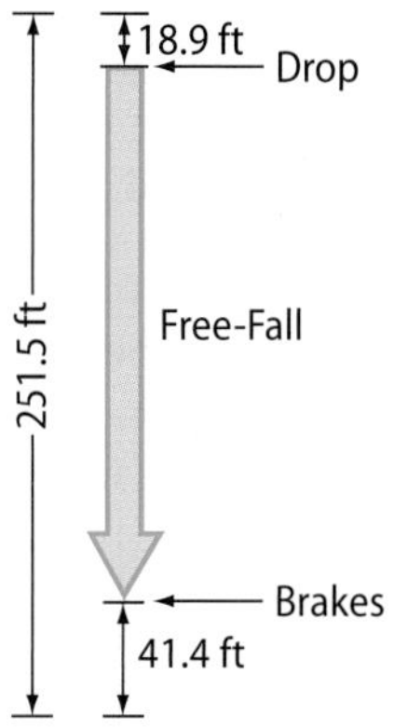

Solve $251.5 - 18.9 = 232.6$ feet
$232.6 - 41.4 = 191.2$ feet

So, the free-fall is 191.2 feet.

Check Since $20 + 40 + 190 = 250$, the answer is reasonable. ✓

Different Methods
There is often more than one way to solve a problem. Using a different method is a good way to check your answer.

Got It? Do this problem to find out.

2. A rectangular garden is 12 meters long and 9 meters wide. A border is sold in 1.5-meter sections. How many sections are needed to surround the garden?

Guided Practice

Check

Use the four-step plan to solve each problem.

1. **Financial Literacy** Antonia bought a video game system for $323.96. She paid in 12 equal installments. About how much did she pay each month? (Example 1)

2. At the 2010 Olympics in Vancouver, Yu-Na Kim scored 228.56 points in women's figure skating. Akiko Suzuki scored 181.44 points in the event. About how many more points did Yu-Na Kim score than Akiko Suzuki? (Example 1)

3. The table shows the number of Calories burned per hour for various activities. (Example 2)

   a. This week, Aiden hiked for 3 hours and bowled for 2 hours. What was the total number of Calories he burned this week doing these activities?

   b. Over the weekend, Celia walked for 3 hours and Marco bowled for 4 hours. How many more Calories did Celia burn than Marco?

Calories Burned per Hour

Activity	Calories
bowling	219
hiking	438
inline skating	548
walking	314

Independent Practice

Go online for Step-by-Step Solutions

eHelp

Use the four-step plan to solve each problem.

4. **STEM** The planet Mercury takes 56.6 Earth days to rotate once about its axis. About how many times does it rotate about its axis during one Earth year? (Example 1)

5. CCSS **Use Math Tools** The gas tank of Tyrell's truck holds 19.8 gallons. When the tank is empty, Tyrell fills it with gas that costs $3.65 per gallon. About how much will it cost to fill the tank? (Example 1)

6. The table shows the speed, in pages per minute (ppm), for three different brands of printer. (Example 2)

   a. Andrew needs to print a document that is 180 pages long. How much longer will it take to print the document on a Sure-Fire printer than on a Centroid printer?

   b. Carlos can print a document on an Apricot printer in 4.5 minutes. How long would it take him to print the same document on a Sure-Fire printer?

Printer Speeds

Brand	Speed (ppm)
Apricot	8
Sure-Fire	12
Centroid	15

7. The towns of Jackson, Bellevue, Salem, and Denton lie along a highway, in the order listed. The distance from Jackson to Salem is 84 miles. The distance from Bellevue to Denton is 61 miles. The distance from Bellevue to Salem is 18 miles. What is the distance from Jackson to Denton?

8. The sales of Café Mocha's Coffee of the Month are shown.

   a. How many more cups of this kind of coffee did Café Mocha sell in January than in April?

   b. If Café Mocha charges $1.98 for each cup of coffee, how much money did Café Mocha earn from their coffee of the month in March?

   c. In May, Café Mocha raised their price for each cup of coffee to $2.25. How much money did Café Mocha earn from their coffee of the month in May?

Coffee of the Month Sales

Month	Number of Cups Sold
January	850
February	765
March	587
April	500
May	387

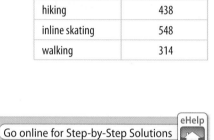

H.O.T. Problems Higher Order Thinking

9. CCSS **Model with Mathematics** Use the information in the table to write a problem that could be solved using the four-step plan. Then solve.

Video Game Scores	
Player	**High Score**
Brian	1231
Keiko	1670
Sonia	3598

10. CCSS **Persevere with Problems** Noah has a $20 gift card for an online music store. All songs at the store have the same price. After Noah buys 3 songs, the balance on the card is $16.25. What will be the remaining balance on the card after he buys 5 more songs?

11. CCSS **Justify Conclusions** Sheri bought 6 items at the grocery store. The total cost of the items before tax was $12.35. Sheri concluded that at least one of the items cost more than $2.00. Explain how Sheri could justify this conclusion.

12. **Building on the Essential Question** Explain how you could use the four-step plan to solve a problem that you might encounter in everyday life.

Standardized Test Practice

13. A 26-mile section of a highway is designated as a scenic byway. New signs will be placed at the beginning and end of this section and at every 0.5 mile in between. How many new signs are needed?

A 13 **C** 52
B 14 **D** 53

14. Five new students join the Chess Club. The club president schedules games so that each of the new members plays a game against each of the other new members. How many games does the president schedule?

F 25 **H** 10
G 20 **J** 5

15. Gypsy wants to cover the floor of a rectangular room using carpet squares with a side length of 2.5 feet. How many carpet squares does she need?

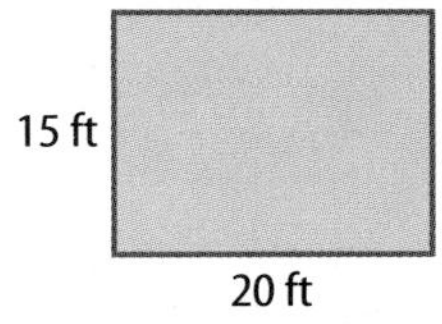

A 14 **C** 48
B 28 **D** 120

CCSS Common Core Review

Find the area of each figure. 6.G.1

16.

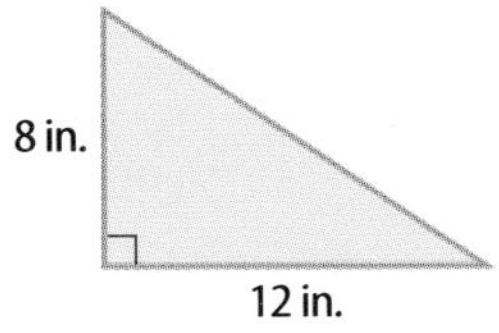

17.

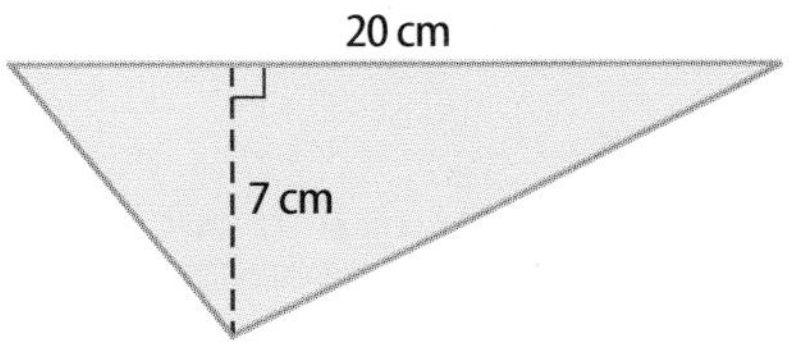

18.

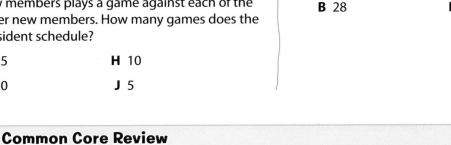

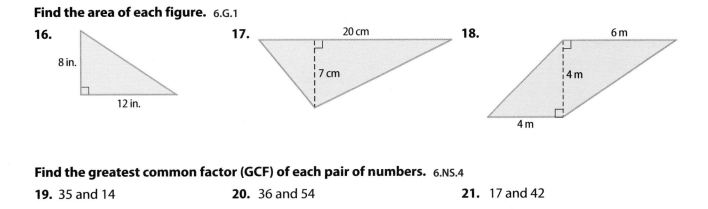

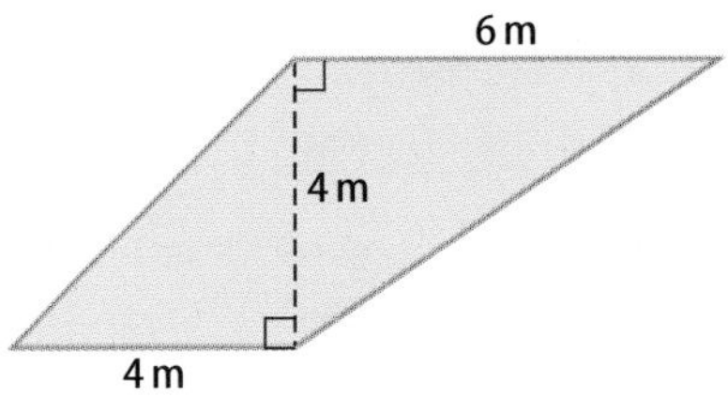

Find the greatest common factor (GCF) of each pair of numbers. 6.NS.4

19. 35 and 14 **20.** 36 and 54 **21.** 17 and 42

22. 16 and 40 **23.** 15 and 24 **24.** 45 and 35

Lesson 1-2

Words and Expressions

ISG Interactive Study Guide

See pages 7–8 for:
- Getting Started
- Vocabulary Start-Up
- Notes

Essential Question

How can you use numbers and symbols to represent mathematical ideas?

CCSS Common Core State Standards

Content Standards
7.NS.3

Mathematical Practices
1, 3, 4

Vocab Vocabulary

numerical expression
evaluate
order of operations

What You'll Learn

- Translate verbal phrases into numerical expressions.
- Use the order of operations to evaluate expressions.

Real-World Link

Inventions The first consumer digital camera was introduced in 1990, the portable music digital player was introduced in 2001, and a video sharing Website began in 2005. You can use numbers and operations to find how many years apart these events occurred.

Stockbyte/Getty Images

Translate Verbal Phrases into Expressions

To find the number of years between the introduction of digital cameras and the portable music digital player, you can use the numerical expression 2001 – 1990. **Numerical expressions** contain a combination of numbers and operations such as addition, subtraction, multiplication, and division.

Example 1

Tutor

Write a numerical expression for each verbal phrase.

a. the total amount of money if you have nine dollars and twelve dollars

Phrase the sum of nine and twelve

Expression $9 + 12$

b. the age difference between fifteen years old and ten years old

Phrase the difference of fifteen and ten

Expression $15 - 10$

Got It? **Do these problems to find out.**

1a. the cost of ten yo-yos if each costs three dollars

1b. the number of students in each group if fifteen students are divided into five equal groups

1c. the number of students on three buses if each bus holds twenty-two students

1d. the amount of money Nina earned if she mowed the lawn for fifteen dollars and walked the dog for four dollars

Grouping Symbols

Grouping symbols include:

- parentheses ( ),
- brackets [ ], and
- fraction bars, as in $\frac{6+4}{2}$, which means $(6+4) \div 2$

Key Concept Order of Operations

Step 1 Simplify the expressions inside grouping symbols.

Step 2 Multiply and/or divide in order from left to right.

Step 3 Add and/or subtract in order from left to right.

To **evaluate** an expression, you find its numerical value. If an expression has more than one operation, use the order of operations. The **order of operations** are the rules to follow when evaluating an expression with more than one operation. These rules ensure that numerical expressions have only one value.

Example 2

Tutor

Watch Out!

In Example 2b, you divide before you multiply because division comes first in order from left to right.

Evaluate each expression.

a. $20 - 3 \times 5$

$20 - 3 \times 5 = 20 - 15$ — Multiply 3 and 5 first.

$= 5$ — Subtract 15 from 20.

b. $30 \div 5 \times 3$

$30 \div 5 \times 3 = 6 \times 3$ — Divide 30 by 5.

$= 18$ — Multiply 6 and 3.

c. $4(10 - 7) + 2 \cdot 3$

$4(10 - 7) + 2 \cdot 3 = 4(3) + 2 \cdot 3$ — Evaluate (10 − 7) first.

$= 12 + 2 \cdot 3$ — 4(3) means 4 × 3 or 12.

$= 12 + 6$ — 2 • 3 means 2 × 3 or 6.

$= 18$ — Add 12 and 6.

d. $5[11 - (7 + 5) \div 4]$

$5[11 - (7 + 5) \div 4] = 5[11 - 12 \div 4]$ — Evaluate (7 + 5).

$= 5(11 - 3)$ — Divide 12 by 4.

$= 5(8)$ — Subtract 3 from 11.

$= 40$ — Multiply 5 and 8.

e. $\frac{60 - 15}{2 + 7}$

$\frac{60 - 15}{2 + 7} = (60 - 15) \div (2 + 7)$ — Rewrite as a division expression.

$= 45 \div 9$ — Evaluate (60 − 15) and (2 + 7).

$= 5$ — Divide 45 by 9.

Got It? Do these problems to find out.

2a. $6 - 3 + 5$

2b. $24 \div 3 \times 9$

2c. $2[(10 - 3) + 6(5)]$

2d. $\frac{19 - 7}{25 - 22}$

Example 3

Real World

Tutor

A cell phone company charges $20 per month and $0.15 for each call made or received. Write and evaluate an expression to find the cost for 40 calls. Then make a table showing the cost for 40, 50, 60, and 70 calls.

First, write an expression for 40 calls.

Words	$20 per month and $0.15 for each call made or received
Expression	$20 + 0.15 \cdot 40$

$$20 + 0.15 \cdot 40 = 20 + 6 \quad \text{Multiply.}$$
$$= 26 \quad \text{Add.}$$

So, 40 calls will cost $26. Make a table showing the costs for 40, 50, 60, and 70 calls.

Number of Calls	Expression	Cost ($)
40	$20 + 0.15 \cdot 40$	26.00
50	$20 + 0.15 \cdot 50$	27.50
60	$20 + 0.15 \cdot 60$	29.00
70	$20 + 0.15 \cdot 70$	30.50

As the number of calls increases by 10, the cost increases by $1.50.

Got It? **Do this problem to find out.**

3. The same phone company offers another plan where they charge $15 per month and $0.25 for each call made or received. Write and evaluate an expression to find the cost for 40 calls during one month. Then make a table showing the cost for 40, 50, 60, and 70 calls.

Guided Practice

Check

Write a numerical expression for each verbal phrase. (Example 1)

1. the cost of six electronic handheld games if each costs eight dollars

2. the cost of one box of cereal if four boxes cost twelve dollars

Evaluate each expression. (Example 2)

3. $18 + 2 \times 4$

4. $2 \times 9 \div 3$

5. $4(6) + 9$

6. $6(17 - 8)$

7. $4[6(2) - 3]$

8. $3[(20 - 7) + 1]$

9. $\frac{15 - 5}{6 - 4}$

10. $\frac{34 + 18}{27 - 14}$

11. A tour bus costs $75 plus $6 for each passenger. Write and evaluate an expression to find the total cost for 25 passengers. Then make a table showing the cost for 25, 30, 35, and 40 passengers. (Example 3)

Independent Practice

Go online for Step-by-Step Solutions eHelp

Write a numerical expression for each verbal phrase. (Example 1)

12. the height difference between fourteen and nine inches

13. the total number of fish if you had eight and bought four more

14. the number of weeks until vacation if vacation is twenty-eight days away

15. the total length of songs on a CD if each of nine songs is three minutes long

16. the total money earned if Lee sold four scarves at twenty dollars each

17. the number of people left to play if ten of fifteen have played

Evaluate each expression. (Example 2)

18. $3 \cdot 6 - 4$

19. $18 - 4 \times 2$

20. $16 \div 4 + 15$

21. $2[3 + 7(4)]$

22. $6 + 4(3)$

23. $12(11) - 56$

24. $8[(12 - 5) + 4]$

25. $4(8) \div (8 - 6)$

26. $\frac{28 + 12}{13 - 5}$

27. $7[8(3) \div (15 - 9)]$

28. $\frac{5 + 9}{10 - 3}$

29. $\frac{16 - 8}{15 - 11}$

30. A decorative floor pattern has 1 red square tile surrounded by 12 blue tiles. Write and evaluate an expression to show how many total red and blue tiles are needed if there are 15 red tiles. Then make a table showing the total number of tiles if there are 15, 20, 25, or 30 red tiles. (Example 3)

31. **Financial Literacy** To place an ad in a newspaper, it costs \$8 plus \$0.75 for each line. Write and evaluate an expression to find the total cost for an ad that has 6 lines. Then make a table showing the cost if there are 6, 10, 14, and 18 lines in the ad. (Example 3)

32. The table shows the prices of admission to the local zoo.

a. Write an expression that can be used to find the total cost of admission for 4 adults, 3 children, and 1 senior.

b. Find the total cost.

Zoo Admission	
Ticket	**Cost**
adults (12–64)	\$8
children (3–11)	\$5
seniors (65+)	\$4

33. CCSS **Multiple Representations** In this problem, you will investigate expressions using a toothpick sequence.

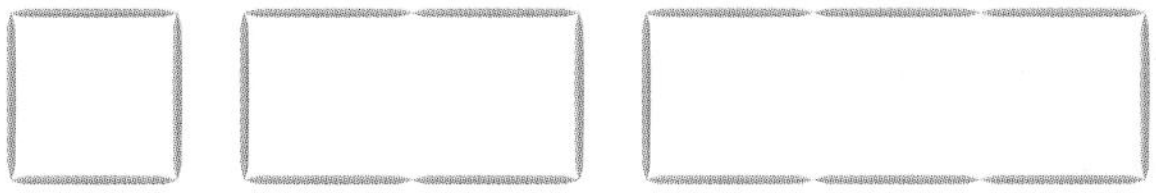

a. Table Make a table showing the term number and number of toothpicks.

b. Words Write a verbal rule to find the number of toothpicks for any term.

34. Grace has a coupon for \$2 off the regular price of each pair of jeans she buys. She buys 4 pairs of jeans with a regular price of \$23 each.

a. Write and evaluate an expression to find the total cost of the jeans.

b. Lamar also buys 4 pairs of the jeans. He has a coupon for \$2 off the total purchase price. Write and evaluate an expression for the total cost of his jeans.

H.O.T. Problems Higher Order Thinking

35. CCSS **Model with Mathematics** Write two different expressions, each with more than one operation, that have a value of 20.

36. CCSS **Persevere with Problems** Leah bought two packages of special decorative trim for a craft project. One package contains 350 centimeters of trim. The second package contains 200 inches of trim.

a. If 1 inch ≈ 2.54 centimeters, write and evaluate an expression to find the number of centimeters in 200 inches.

b. Find the total number of centimeters of trim in the two packages.

c. How many packages of trim containing 90 inches each will Leah need to buy to have about the same amount of trim from part **b**? Explain.

37. **Building on the Essential Question** Explain why the rules for the order of operations are important. Support your answer with two numerical examples.

Standardized Test Practice

38. Rusty is evaluating $96 \div 3 \times 4 + 7$ as shown below.

$$96 \div 3 \times 4 + 7$$
$$96 \div 12 + 7$$
$$8 + 7$$
$$15$$

What should Rusty have done differently in order to evaluate the expression correctly?

A multiplied $(96 \div 3)$ by $(4 + 7)$

B divided 96 by (3×11)

C multiplied $(96 \div 3)$ by 4 and added 7

D divided 96 by $(4 + 7)$

39. Evaluate $7[12 - (6 - 2) \div 4]$.

F 7 **H** 77

G 14 **J** 83

40. A man is a musician. He charges \$50 for each of the first three hours he plays and \$24.95 for each additional hour. Which expression *cannot* be used to find the total amount he charges if he plays for 7 hours?

A 4 × \$24.95 + 3 × \$50

B 7(\$50 + \$24.95)

C \$50 + \$50 + \$50 + 4(\$24.95)

D 3(\$50) + 4(\$24.95)

41. **Short Response** Max wants to buy 5 hats and 4 T-shirts. Write an expression to find the total cost of 5 hats and 4 T-shirts.

Hats	\$15
T-shirts	\$20

Common Core Review

Round each decimal to the indicated place value. 5.NBT.4

42. 0.52; tenths **43.** 3.951; hundredths **44.** 28.17; ones

45. 33.78; tenths **46.** 12.7255; thousandths **47.** 6.05; tenths

48. 7.0515; hundredths **49.** 0.09; ones **50.** 54.9806; hundredths

Inquiry Lab

Rules and Expressions

Inquiry **WHAT is an advantage of using a rule to describe a numerical pattern?**

Content Standards
Preparation for 7.EE.4

Mathematical Practices
1, 2, 3, 4

Patterns A pattern is made from toothpicks. The first term uses 3 toothpicks, the second term uses 5 toothpicks, and the third term uses 7 toothpicks, as shown below. What will be the sixth and seventh terms in the pattern?

Investigation

Watch

Step 1 Use toothpicks to create the first three terms described in the pattern.

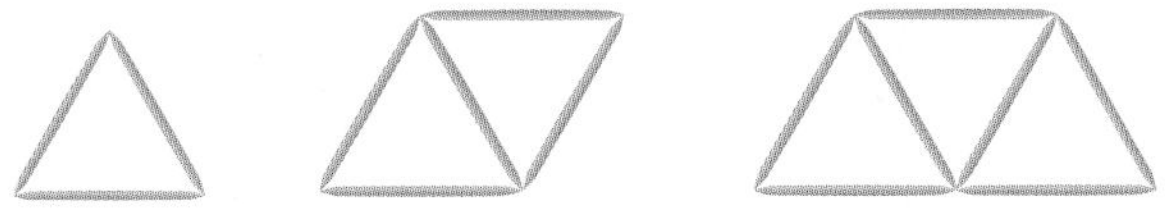

Step 2 Use toothpicks to create the next two terms in the pattern.

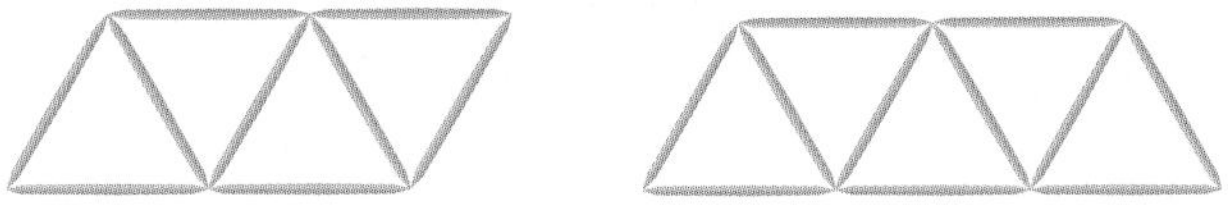

Step 3 Copy and complete the table to show the number of toothpicks needed to create the fourth and fifth terms if the pattern continues.

Term Number	1	2	3	4	5	6	7
Number of Toothpicks	3	5	7	■	■	■	■

+2 +2 +2

Step 4 Use the pattern to determine the sixth and seventh terms. Add these numbers to your table.

Collaborate

Work with a partner.

1. Look at the table for the toothpick pattern. How is the number of toothpicks related to the term number?
2. Describe a rule you could use to find the number of toothpicks needed to make any figure in the pattern.
3. Use the rule you wrote for Exercise 2 to find the number of toothpicks you would need to create the twelfth term.

Analyze

The pattern shown below is made from tiles.

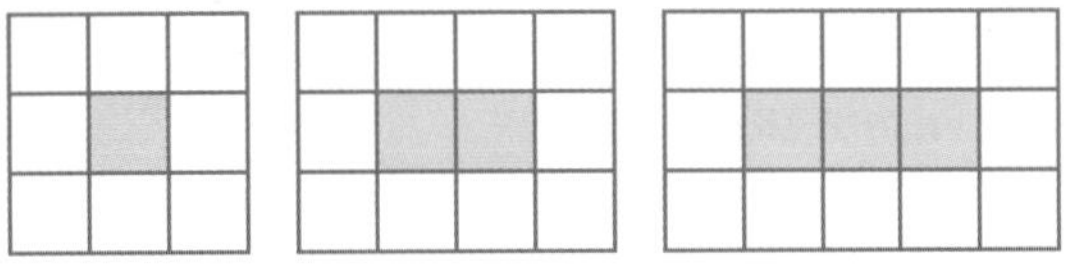

4. Make a table to show how many white tiles are needed for each blue tile.
5. Describe a rule you could use to find the number of white tiles needed to make any figure in the pattern.
6. How many white tiles would you need to make a figure that has 15 blue tiles?
7. How many white tiles would you need to make a figure that has 80 blue tiles?
8. A figure in the pattern has exactly 24 white tiles. How many blue tiles are there?
9. Do you think there is ever a figure in the pattern that requires exactly 73 white tiles? Why or why not?

The pattern below is made from pennies.

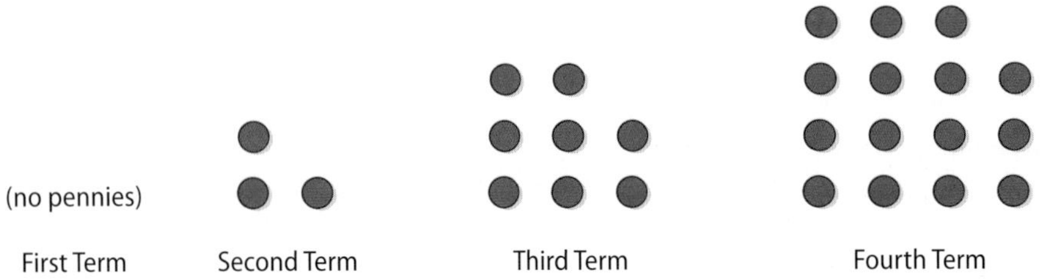

10. Describe a rule you could use to find the number of pennies needed to make any term in the pattern.
11. How many pennies would you need to make the tenth term of the pattern?
12. How many pennies would you need to make the fifty-first term of the pattern?
13. CCSS **Reason Abstractly** The figures show store display shelves of different heights that a carpenter has made for local stores. How many cubes would the carpenter need to build a display shelf that is 8 feet tall?

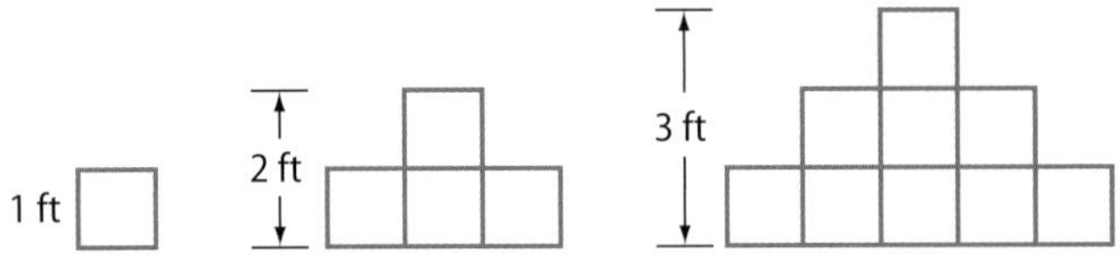

14. CCSS **Justify Conclusions** Is it possible for the carpenter to build a store display shelf using exactly 85 cubes? Explain your reasoning.

Reflect

15. Inquiry WHAT is an advantage of using a rule to describe a numerical pattern?

Lesson 1-3

Variables and Expressions

ISG Interactive Study Guide

See pages 9–10 for:
- Getting Started
- Real-World Link
- Notes

Essential Question

How can you use numbers and symbols to represent mathematical ideas?

Common Core State Standards

Content Standards
7.NS.3, 7.EE.4

Mathematical Practices
1, 3, 4

Vocabulary

algebra
variable
algebraic expression
defining a variable
Substitution Property of Equality

What You'll Learn

- Translate verbal phrases into algebraic expressions.
- Evaluate expressions containing variables.

Real-World Link

Online Games Did you know that nearly 250 million people worldwide participate in online gaming? Mathematical expressions can be used to represent the total points earned by a player in an online game.

Algebraic Expressions and Verbal Phrases

Algebra is a branch of mathematics that uses symbols. A variable is often used in algebra. A **variable** is a letter or symbol used to represent an unknown value. Any letter can be used as a variable.

The letter x is often used as a variable. → $4 - x$

$5m$ means $5 \cdot m$. ab means $a \cdot b$. → $5m + 6$, ab

$\frac{y}{3}$ means $y \div 3$. → $\frac{y}{3}$

An expression like $5m + 6$ is an **algebraic expression** because it contains at least one variable and at least one mathematical operation.

The first step in translating verbal phrases into algebraic expressions is to choose a variable and a quantity for the variable to represent. This is called **defining a variable**. All of the steps involved in writing algebraic expressions are shown below.

Words	Describe the situation. Use only the most important words.
Variable	Define a variable by choosing a variable to represent the unknown quantity.
Expression	Translate your verbal model into an algebraic expression.

Example 1

Tutor

Translate each phrase into an algebraic expression.

a. three dollars more than the cost of a sandwich

Words	three dollars more than the cost of a sandwich
Variable	Let c represent the cost of a sandwich.
Expression	$3 + c$

b. Mari had \$2 and made \$6 an hour babysitting.

Words	two more than six dollars per hour
Variable	Let n represent the number of hours.
Expression	$2 + 6n$

Products with Variables

When you write the product of a number and a variable, you should always write the number first. Write $2 + 6n$ rather than $2 + n6$.

Got It? **Do these problems to find out.**

1a. two miles less than the athlete ran

1b. five points more than the points scored by field goals if each field goal is worth 3 points

Key Concept Substitution Property of Equality

Words	If two quantities are equal, then one quantity can be replaced by the other.
Symbols	For all numbers a and b, if $a = b$, then a may be replaced by b.

To evaluate an algebraic expression, replace the variable(s) with known values and then use the order of operations. When you replace a variable with a number, you are using the **Substitution Property of Equality**.

Example 2

Tutor

Evaluate $d + 5 - f$ if $d = 16$ and $f = 18$.

$d + 5 - f = 16 + 5 - 18$	Replace d with 16 and f with 18.
$= 21 - 18$	Add 16 and 5.
$= 3$	Subtract 18 from 21.

Got It? **Do these problems to find out.**

2a. Evaluate $6 - e + f$ if $e = 3$ and $f = 9$.

2b. Evaluate $7k + h$ if $k = 4$ and $h = 10$.

Example 3

Tutor

Evaluate each expression if $r = 1$, $s = 5$, and $t = 8$.

a. $6s + 2t$

$6s + 2t = 6(5) + 2(8)$ — Replace s with 5 and t with 8.

$= 30 + 16$ or 46 — Multiply. Then add.

b. $\frac{st}{20}$

$\frac{st}{20} = st \div 20$ — Rewrite as a division expression.

$= (5 \cdot 8) \div 20$ — Replace s with 5 and t with 8.

$= 40 \div 20$ or 2 — Multiply. Then divide.

c. $r + (40 - 3t)$

$r + (40 - 3t) = 1 + (40 - 3 \cdot 8)$ — Replace r with 1 and t with 8.

$= 1 + (40 - 24)$ — Multiply 3 and 8.

$= 1 + 16$ or 17 — Subtract 24 from 40. Then add 1 and 16.

Watch Out!

In Example 3a, use parentheses when you replace the variables with numbers. This will help you remember to multiply.

Got It? Do these problems to find out.

Evaluate each expression if $a = 4$, $b = 8$, and $c = 12$.

3a. $3a + 2c$

3b. $\frac{ab}{16}$

3c. $c + (5b - 2a)$

Real World Example 4

Tutor

A company rents a houseboat for \$200 plus an extra \$30 per day.

a. Write an expression that can be used to find the total cost to rent a houseboat.

Words	two-hundred-dollar rental fee plus thirty dollars per day
Variable	Let d represent the number of days.
Expression	$200 + 30d$

The expression is $200 + 30d$.

b. Suppose the Gregoran family wants to rent a houseboat for six days. What will be the total cost?

$200 + 30d = 200 + 30(6)$ — Replace d with 6.

$= 200 + 180$ or 380 — Multiply. Then add.

The total cost will be \$380.

Got It? Do this problem to find out.

4. At a garage sale, each DVD was marked at \$5, and each CD was marked at \$3. Write an expression to find the total cost to buy some DVDs and CDs. Then find the cost of buying 4 DVDs and 7 CDs.

Guided Practice

Check

Translate each phrase into an algebraic expression. (Example 1)

1. four dollars less than the cost of the sweater
2. thirteen more students than teachers
3. money earned babysitting at \$10 per hour
4. thirty pencils divided among some students

Evaluate each expression if $g = 6$, $h = 10$, and $j = 5$. (Examples 2 and 3)

5. $h + 15$
6. $g - 3$
7. $20 - h + g$
8. $22 - 3j$
9. $\frac{gh}{j}$
10. $4g + (3h - 4j)$

11. One pint of liquid is the same as 16 fluid ounces. (Example 4)
    a. Suppose the number of pints of liquid is represented by p. Write an expression to find the number of fluid ounces.
    b. How many fluid ounces are in 5 pints?

Independent Practice

eHelp Go online for Step-by-Step Solutions

Translate each phrase into an algebraic expression. (Example 1)

12. three times as many balloons
13. twenty-four pieces of candy divided among some students
14. the number of people increased by thirteen
15. the number of inches in any number of feet
16. four more than the number of weeks in a group of days
17. four less than the amount of cents in a number of dimes

Evaluate each expression if $a = 9$, $b = 4$, and $c = 11$. (Examples 2 and 3)

18. $b + 9$
19. $13 - a$
20. $2c - 5$
21. $18 + 4b$
22. $\frac{ab}{6}$
23. $\frac{8a}{b}$
24. $5c - 4a$
25. $7b - 2c$
26. $45 - \frac{bc}{2}$
27. $\frac{ac}{3} - 15$
28. $4b + 3c - 5a$
29. $6c - 2a + 6b$

30. A studio charges a sitting fee of \$25 plus \$7 for each portrait sheet ordered. Write an expression that can be used to find the total cost to have photographs taken. Then find the cost of purchasing twelve portrait sheets. (Example 4)

31. One gallon of water is equal to 231 cubic inches. Write an expression for the number of gallons of water in any number of cubic inches of water. (Example 4)

Evaluate each expression if $x = 9$, $y = 4$, and $z = 12$.

32. $7z - (y + x)$
33. $(8y + 5) - 2z$
34. $(5z - 4x) + 3y$
35. $6x - (z - 2y)$
36. $2x + (4z - 13) - 5$
37. $(29 - 3y) + 4z - 7$

38. **Financial Literacy** A cell phone company offers two different monthly plans. Plan A costs a flat rate of \$0.10 per minute for all calls. Plan B costs \$29.99 for the first 500 minutes and \$0.08 for each additional minute. Which plan is less costly if a person uses 750 minutes per month? Explain.

39 One bushel of apples from a dwarf apple tree is equal to 42 pounds. Write an expression to find the number of pounds of apples in any number of bushels. If one tree can produce 6 bushels, how many pounds of apples will an orchard of 100 trees produce?

40. A car rental company charges a one-time fee of \$95, plus \$65 per day. Write an expression that can be used to find the cost of renting a car for any number of days. If George rents a car for 4 days, what is the total cost?

41. CCSS **Multiple Representations** In this problem, you will use algebra to describe a relationship. Jacinda used the table below to help convert measurements while she was cooking.

Number of Cups (c)	4	8	12	16
Number of Quarts (q)	1	2	3	4

a. **Words** Write an expression in words that describes the relationship between the number of quarts and the number of cups.

b. **Symbols** Write an algebraic expression that represents the number of quarts in c cups.

c. **Numbers** Use the expression in part **b** to find the number of quarts in 100 cups.

H.O.T. Problems Higher Order Thinking

42. CCSS **Model with Mathematics** Write an algebraic expression that has two different variables and two different operations: addition, subtraction, multiplication, or division. Then write a real-world problem that uses the expression.

43. CCSS **Find the Error** John is writing an algebraic expression for the phrase *five less than a number*. Find his mistake and correct it.

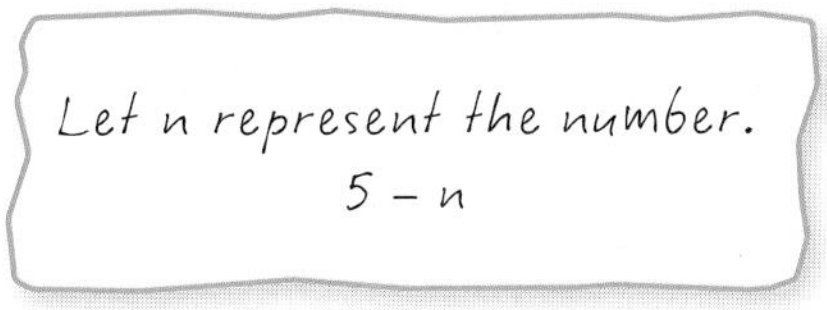
Let n represent the number.
5 − n

44. CCSS **Persevere with Problems** Franco constructed the objects below using toothpicks.

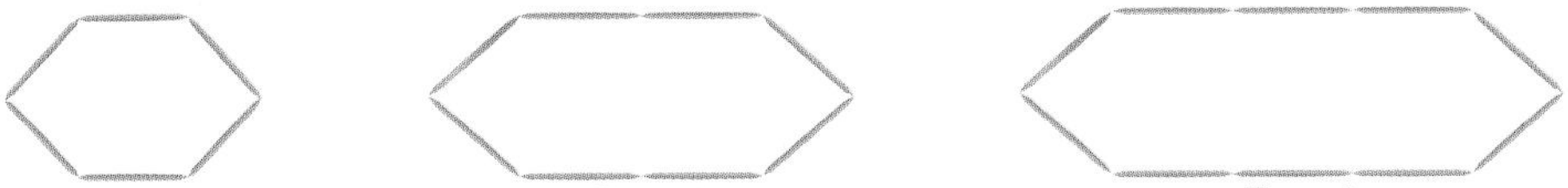

Figure 1 Figure 2 Figure 3

Write two different rules that relate the figure number to the number of toothpicks in each figure. Explain how you arrived at your answers.

45. **Building on the Essential Question** Cassandra needs to evaluate the expression $a(x + y)$. After she replaces the variables with numerical values, in which order should she perform the operations of addition and multiplication? Explain.

Standardized Test Practice

46. What word phrase is equivalent to the expression $5x + 9$?

A five cents more than nine nickels
B nine cents plus five cents
C nine cents more than five nickels
D nine cents less than five nickels

47. Which rule describes the ordered pairs in this table?

x	y
1	1
2	4
3	7
4	10

F $y = x$
G $y = 2x$
H $y = 3x - 2$
J $y = 2x + 2$

48. What is "8 more than the quotient of five and a number n" written as an algebraic expression?

A $5 + 8n$
B $8 + 5n$
C $8 + 5 \div n$
D $5 + n \div 8$

49. **Short Response** The table below shows how much Ava will pay to rent one DVD and one game for the number of days given.

Number of Days	Total Cost ($)
2	10
4	20
6	30

How much will Ava pay to rent one DVD and one game for 7 days?

CCSS Common Core Review

Find the value of each expression. 7.NS.3

50. $3 \cdot 6 - 4$

51. $12 - 3 \times 3$

52. $9 + 18 \div 3$

53. $56 \div (7 \cdot 2) \times 6$

54. $75 \div (7 + 8) - 3$

55. $70 - (16 \div 2 + 21)$

56. $\dfrac{45 - 18}{9 \div 3}$

57. $\dfrac{8 \div 8 + 11}{15 - 4(3)}$

58. $4(20 - 13) + 4 \times 5$

The final standings of a hockey league are shown. A win is worth 3 points, and a tie is worth 1 point. Zero points are given for a loss. 7.NS.3

Team	Wins	Losses	Ties
Knights	14	9	7
Huskies	11	9	10
Wildcats	10	9	11
Mustangs	9	10	11
Panthers	10	14	6

59. How many points do the Wildcats have?

60. How many points do the Huskies have?

61. How many more points do the Knights have than the Panthers?

62. How many fewer points do the Mustangs have than the Huskies?

63. At the end of the season, all teams with more than 40 points go to the playoffs. Which teams from the league will go to the playoffs?

Find the least common multiple (LCM) of each pair of numbers. 6.NS.4

64. 6 and 8

65. 5 and 7

66. 6 and 10

67. 4 and 8

68. 1 and 9

69. 3 and 12

70. 10 and 12

71. 9 and 12

Lesson 1-4

Properties of Numbers

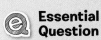

<dInteractive Study Guide

See pages 11–12 for:
- Getting Started
- Real-World Link
- Notes

Essential Question

How can you use numbers and symbols to represent mathematical ideas?

Common Core State Standards

Content Standards
7.EE.1, 7.EE.2

Mathematical Practices
1, 3, 4, 7

Vocabulary

properties
Commutative Property
Associative Property
counterexample
simplify
deductive reasoning

What You'll Learn

- Identify and use properties of addition and multiplication.
- Use properties to simplify algebraic expressions.

Real-World Link

Crafts Duct tape has been used to create everything from flip-flops and prom dresses to wallets and homemade flotation devices. Algebraic properties can be used to find the amount of duct tape needed to make an item.

Properties of Addition and Multiplication

In algebra, **properties** are statements that are true for any numbers. For example, the expressions $30 + 10$ and $10 + 30$ have the same value, 40. This illustrates the **Commutative Property of Addition**. Likewise, $30 \cdot 10$ and $10 \cdot 30$ have the same value, 300. This illustrates the **Commutative Property of Multiplication.**

Key Concept Commutative Properties

Words The order in which numbers are added or multiplied does not change the sum or product.

Symbols For any numbers a and b, $a + b = b + a$.
For any numbers a and b, $a \cdot b = b \cdot a$.

Examples

$6 + 9 = 9 + 6$
$15 = 15$

$4 \cdot 7 = 7 \cdot 4$
$28 = 28$

To evaluate the expression $16 + (14 + 58)$, use mental math by grouping the numbers as $(16 + 14) + 58$ since $4 + 6 = 10$. This illustrates the **Associative Property of Addition**. There is also an **Associative Property of Multiplication**.

Key Concept Associative Properties

Words The way in which numbers are grouped when added or multiplied does not change the sum or product.

Symbols For any numbers a, b, and c, $(a + b) + c = a + (b + c)$.
For any numbers a, b, and c, $(a \cdot b) \cdot c = a \cdot (b \cdot c)$.

Examples

$(3 + 6) + 1 = 3 + (6 + 1)$
$9 + 1 = 3 + 7$
$10 = 10$

$(5 \cdot 9) \cdot 2 = 5 \cdot (9 \cdot 2)$
$45 \cdot 2 = 5 \cdot 18$
$90 = 90$

Jeff Greenberg/Alamy

In addition to the Commutative and Associative Properties, the Identity and Zero Properties are also true for any numbers.

Identity
The word identity means *sameness of essential character*. The additive identity, 0, and multiplicative identity, 1, allow the original number to remain the same.

Key Concept Number Properties

Property	Words	Symbols	Examples
Additive Identity	When 0 is added to any number, the sum is the number.	For any number a, $a + 0 = 0 + a = a$	$5 + 0 = 5$ $0 + 5 = 5$
Multiplicative Identity	When any number is multiplied by 1, the product is the number.	For any number a, $a \cdot 1 = 1 \cdot a = a$.	$8 \cdot 1 = 8$ $1 \cdot 8 = 8$
Multiplicative Property of Zero	When any number is multiplied by 0, the product is 0.	For any number a, $a \cdot 0 = 0 \cdot a = 0$	$3 \cdot 0 = 0$ $0 \cdot 3 = 0$

Do these properties apply to subtraction or division? One way to find out is to look for a counterexample. A **counterexample** is an example that shows a statement is not true.

Example 1

Tutor

Is division of whole numbers associative? If not, give a counterexample.

The Associative Property of Multiplication states $(a \cdot b) \cdot c = a \cdot (b \cdot c)$. To determine whether the Associative Property applies to division, check $(a \div b) \div c \stackrel{?}{=} a \div (b \div c)$.

$(27 \div 9) \div 3 \stackrel{?}{=} 27 \div (9 \div 3)$ Pick values for a, b, and c.

$(3) \div 3 \stackrel{?}{=} 27 \div (3)$ Simplify.

$1 \neq 9$ Simplify.

We found a counterexample. So, division of whole numbers is not associative.

Counterexample
You can disprove a statement by finding only one counterexample.

Got It? Do this problem to find out.

1. Is subtraction of decimals associative? If not, give a counterexample.

Example 2

Tutor

Name the property shown by each statement.

a. $4 + (a + 3) = (a + 3) + 4$

The order of the numbers and variables changed. This is the Commutative Property of Addition.

b. $1 \cdot (3c) = 3c$

The expression was multiplied by 1 and remained the same. This is the Multiplicative Identity Property.

Got It? Do these problems to find out.

2a. $d + 0 = d$

2b. $8 \cdot 1 = 8$

2c. $14 + (9 + 10) = (14 + 9) + 10$

2d. $5 \times 7 \times 2 = 7 \times 2 \times 5$

Simplify Algebraic Expressions

Vocabulary Link
simplify
Everyday Use to make easier to understand
Math Use to shorten an algebraic expression by performing all operations. A simplified expression is often easier to understand.

To **simplify** an algebraic expression, perform all possible operations. You can use the properties you learned in this lesson. Using facts, properties, or rules to reach valid conclusions is called **deductive reasoning**.

Example 3

Tutor

Simplify each expression.

a. $(3 + e) + 7$

$(3 + e) + 7 = (e + 3) + 7$	Commutative Property of Addition
$= e + (3 + 7)$	Associative Property of Addition
$= e + 10$	Simplify.

b. $8 \cdot (x \cdot 5)$

$8 \cdot (x \cdot 5) = 8 \cdot (5 \cdot x)$	Commutative Property of Multiplication
$= (8 \cdot 5) \cdot x$	Associative Property of Multiplication
$= 40x$	Simplify.

c. $9 + (0 + q)$

$9 + (0 + q) = (9 + 0) + q$	Associative Property of Addition
$= 9 + q$	Additive Identity Property

d. $4 \cdot (a \cdot 0)$

$4 \cdot (a \cdot 0) = 4 \cdot 0$	Multiplicative Property of Zero
$= 0$	Multiplicative Property of Zero

Got It? **Do these problems to find out.**

3a. $12 \cdot (10 \cdot z)$

3b. $(15 + w) + 16$

3c. $11 \cdot (b \cdot 0)$

3d. $10 + (p + 18)$

Guided Practice

Check

1. Is subtraction of whole numbers commutative? If not, give a counterexample. (Example 1)

Name the property shown by each statement. (Example 2)

2. $8 \cdot 4 = 4 \cdot 8$

3. $6 \cdot 1 = 6$

4. $9 + 3 + 20 = 3 + 9 + 20$

5. $7 + 0 = 7$

6. $13 + 12 = 12 + 13$

7. $6 \times (1 \times 9) = (6 \times 1) \times 9$

Simplify each expression. (Example 3)

8. $9 + (5 + y)$

9. $3 + (k + 8)$

10. $(15 + s) + 4$

11. $(12 + m) + 4$

12. $(1 \cdot d) \cdot 14$

13. $(z \cdot 11) \cdot 3$

Independent Practice

Go online for Step-by-Step Solutions eHelp

State whether each conjecture is true. If not, give a counterexample. (Example 1)

14. The sum of two odd numbers is always odd.

15. The product of odd numbers is always even.

16. Division of whole numbers is commutative.

17. All multiples of 3 are odd.

Name the property shown by each statement. (Example 2)

18. $0 + 14 = 14$

19. $8 \cdot 1 = 8$

20. $15 + 17 = 17 + 15$

21. $(2 \cdot 8) \cdot 5 = 2 \cdot (8 \cdot 5)$

22. $14 \times 0 \times 3 = 0$

23. $4 + (9 + 2) = (4 + 9) + 2$

24. $7 + x + 11 = x + 7 + 11$

25. $5k \times 1 = 5k$

Simplify each expression. (Example 3)

26. $(d + 12) + 16$

27. $(54 + p) + 16$

28. $14 + (27 + m)$

29. $(r + 32) + 24$

30. $(8 \cdot s) \cdot 9$

31. $g \cdot (5 \cdot 7)$

32. $11 \cdot (t \cdot 4)$

33. $15b(5)$

34. $6(12c)$

35. $(7 + p) + 13$

36. $29 + (1 + t)$

37. $4 \cdot (x \cdot 2)$

38. Use the table to write an expression that shows how many total baskets the Cavaliers made during the season. Simplify the expression.

Cavaliers' Baskets	
free throws	1484
2-point field goals	*f*
3-point field goals	494

39. Moreno likes to do her social studies homework before she does her math homework. Is doing social studies homework and math homework commutative? Explain your reasoning.

40. CCSS **Justify Conclusions** Ming said, "12, 20, and 36 are all divisible by 2 and by 4. So any whole number that is divisible by 2 is also divisible by 4." Do you agree? If yes, explain your reasoning. If not, give a counterexample.

41. A ceramics class needs to buy blocks of clay for an activity. Use the table at the right to write an expression that represents the total cost of clay, in dollars, for the activity. Simplify the expression.

Ceramics Class	
number of students	19
blocks of clay per student	*b*
price per block of clay ($)	6

Translate each verbal expression into an algebraic expression. Then simplify the expression.

42. the product of seven and four times a number multiplied by three

43. the sum of two times a number and five added to six times the number

44. eight more than six times a number added to one more than nine times the number

45. the product of eleven and five times a number multiplied by four

46. the difference of twelve times a number and nine times the number

47. the sum of four times a number and seven less than two times the number

48. Denzel wants to know the total number of boys in the three clubs shown in the table.

a. What expression should Denzel evaluate in order to find the total number of boys?

b. Explain how you can use properties of numbers to make the expression easier to evaluate using mental math.

Club Membership		
Club	**Boys**	**Girls**
chess	17	19
drama	28	23
music	13	21

49. **Financial Literacy** The Center of Wonders science center has the rates shown.

a. Write an algebraic expression for the total cost for five people to get into the center, visit the planetarium, and watch a 3-D movie.

b. If the cost of admission to the center is \$12, how much will it cost for four people to get into the center and watch a 3-D movie?

Center of Wonders	
Type of Ticket	**Cost (\$)**
admission	a
planetarium	4.50
3-D movie	7.75

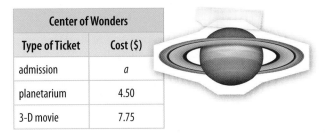

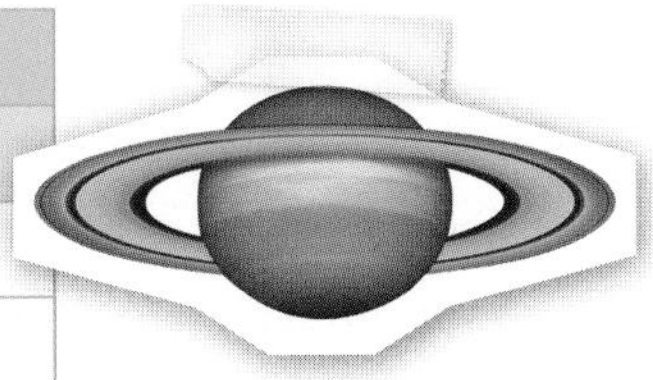

c. Children get a discount of \$2.50 on their tickets to the planetarium if they also see a 3-D movie. Write an expression to find the cost for two adults and two children to get into the center, see a 3-D movie, and visit the planetarium.

H.O.T. Problems Higher Order Thinking

50. CCSS **Identify Structure** Write an algebraic expression that can be simplified using at least two different properties. Simplify the expression showing each step and provide a justification for each step.

51. CCSS **Construct an Argument** Is the following statement *true* or *false*? Explain your reasoning.

$$15 + (4 \cdot 6) = (15 + 4) \cdot 6$$

52. CCSS **Find the Error** Meghan is simplifying the expression $8(3) \cdot 4 \cdot 2(3)$. Find her mistake and correct it.

$$8(3) \cdot 4 \cdot 2(3) = 24 \cdot 4 \cdot 2$$
$$= 96 \cdot 2$$
$$= 192$$

53. CCSS **Persevere with Problems** If you take any two whole numbers and add them together, the sum is always a whole number. This is the Closure Property for Addition. The set of whole numbers is *closed* under addition.

a. Is the set of whole numbers closed under subtraction? If not, give a counterexample.

b. Suppose you had a very small set of numbers that contained only 0 and 1. Would this set be closed under addition? If not, give a counterexample.

c. There is also a Closure Property for Multiplication of Whole Numbers. State this property using the addition property above as a guideline.

d. Is the set {0, 1} closed under multiplication? Explain.

54. **Building on the Essential Question** The number 1 is the identity for multiplication. Do you think that division has an identity? Explain your reasoning.

Standardized Test Practice

55. Which statement is an example of the Identity Property?

A $3 \cdot x \cdot 0 = 0$

B $7(4x) = (4 \cdot 7)x$

C $5 + (4 + x) = (5 + 4) + x$

D $4x + 0 = 4x$

56. Which expression can be used to find the perimeter of the rectangle below?

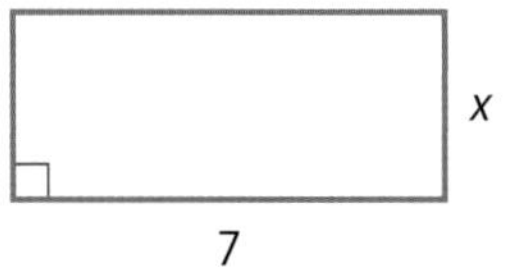

F $2x + 7$

G $2x + 14$

H $x + 7$

J $x + 14$

57. Which property is illustrated by the statement below?

$$12 \cdot (n \cdot 5) = (12 \cdot n) \cdot 5$$

A Commutative Property

B Associative Property

C Identity Property

D Zero Property

58. **Short Response** Simplify the expression. Show and justify each step.

$$10 \cdot (x \cdot 3)$$

Common Core Review

Translate each phrase into an algebraic expression. 6.EE.2a

59. Bianca's salary plus a $200 bonus

60. three more than the number of cakes baked

61. six feet shorter than the mountain's height

62. eight less than the quotient of the number of quarters and four

63. **STEM** The number of times a cricket chirps can be used to estimate the temperature in degrees Fahrenheit. The expression $c \div 4 + 37$, where c is the number of chirps in 1 minute, shows this relationship. 6.EE.2c

a. Find the approximate temperature if a cricket chirps 136 times a minute.

b. What is the temperature if a cricket chirps 100 times in a minute?

Evaluate each expression. 5.OA.1

64. $50 \div 2 \times 5$

65. $6(8 - 4) + 3 \cdot 7$

66. $16 - 2 \cdot 4$

67. $18 + 2 \cdot 3$

68. $49 - 25 + 5$

69. $3(7 \cdot 5) \cdot 2$

70. $90 \div 6 \div 3$

71. $90 \div (6 \div 3)$

72. $20 + 8 \div 4$

Convert each measurement to the given units. 5.MD.1

73. 7 centimeters to meters

74. 72 inches to feet

75. 300 milliliters to liters

76. 24 gallons to quarts

77. 12 kilograms to grams

78. 2 miles to feet

79. 18 yards to feet

80. 9 millimeters to centimeters

ISG Interactive Study Guide

See page 13 for:
• Mid-Chapter Check

21ST CENTURY CAREER in Animal Conservation

Shark Scientist

Are you fascinated by sharks, especially those that are found around the coasts of the United States? If so, you should consider a career as a shark scientist. Shark scientists use satellite-tracking devices, called tags, to study and track the movements of sharks. By analyzing the data transmitted by the tags, scientists are able to learn more about the biology and ecology of sharks. Their research is helpful in protecting shark populations around the world.

College & Career READINESS

Explore college and careers at ccr.mcgraw-hill.com

Is This the Career for You?

Are you interested in a career as a shark scientist? Take some of the following courses in high school.

- Algebra
- Calculus
- Physics
- Statistics

Find out how math relates to a career in Animal Conservation.

Interactive Study Guide
See page 14 for:
- Problem Solving
- Career Project

Lesson 1-5

Problem-Solving Strategies

Interactive Study Guide

See pages 15–16 for:
- Getting Started
- Real-World Link
- Notes

Essential Question

How can you use numbers and symbols to represent mathematical ideas?

Common Core State Standards

Content Standards
7.NS.3, 7.EE.3

Mathematical Practices
1, 3, 4, 8

Vocabulary

look for a pattern

guess, check, and revise

make a table

work backward

What You'll Learn

- Use problem-solving strategies to solve nonroutine problems.
- Select an appropriate strategy.

Real-World Link

Movie Snacks Nothing goes better with a movie than popcorn! When you buy snacks at the movie theater, you can receive your change in several different combinations. You can use different problem-solving strategies to determine how you will receive your change.

Use Problem-Solving Strategies

There are many problem-solving strategies in mathematics. One common strategy is to **look for a pattern**. To use this strategy, analyze the first few numbers in a pattern and identify a rule that is used to go from one number to the next. Then use the rule to extend the pattern and find a solution.

Real World Example 1

Tutor

Ramon got a text from Angela. After 10 seconds, he forwarded it to 2 of his friends. After 10 more seconds, those 2 friends forwarded it to 2 more friends. If the text was forwarded like this every 10 seconds, how many people received Angela's text after 40 seconds?

Look for a pattern in the data and extend the pattern.

1, 2, 4, ■, ■

×2 ×2 ×2 ×2

Time (s)	People Getting Text
0	1
10	2
20	4
30	■
40	■

To continue the pattern, multiply each term by 2.

$4 \times 2 = 8 \qquad 8 \times 2 = 16$

So, $1 + 2 + 4 + 8 + 16$ or 31 people received the text.

Got It? Do this problem to find out.

1. Tamiko makes a necklace by stringing beads with the following colors: red, yellow, blue, red, yellow, blue, ... If she continues in this way, what will be the color of the 50th bead?

To solve other problems, you can make a reasonable guess and then check it in the problem. You can then use the results to improve your guess until you find the solution. This strategy is called **guess, check, and revise**.

Example 2

Tutor

Each hand in the human body has 27 bones. There are 6 more bones in the fingers than in the wrist. There are 3 fewer bones in the palm than in the wrist. How many bones are in each part of the hand?

Make a guess to find the number of bones in each part of the hand.

Wrist	Palm (wrist − 3)	Fingers (wrist + 6)	Total Bones (27)	Correct?
5	$5-3=2$	$5+6=11$	$5+2+11=18$	This is too low.
7	$7-3=4$	$7+6=13$	$7+4+13=24$	This is too low.
9	$9-3=6$	$9+6=15$	$9+6+15=30$	This is too high.
8	$8-3=5$	$8+6=14$	$8+5+14=27$	This is correct. ✓

There are 8 bones in the wrist, 5 bones in the palm, and 14 bones in the fingers.

Watch Out!

Make sure you answer the entire question that is asked in a problem. This problem asks for the number of bones in each part of the hand, so you must give the number of bones in the wrist, the palm, and the fingers.

Got It? Do this problem to find out.

2. Colin sold student tickets and guest tickets to a school play. He sold a total of 154 tickets in all. He sold 8 more student tickets than guest tickets. How many of each type of ticket did he sell?

Another strategy for solving problems is to **make a table**. A table allows you to organize information in an understandable way.

Example 3

Tutor

A vending machine accepts dollars, and each item in the machine costs 65 cents. If the machine gives back only nickels, dimes, and quarters, what combinations of those coins are possible as change for one dollar?

The machine will give back \$1.00 − \$0.65 or 35 cents in change in a combinations of nickels, dimes, and quarters.

Make a table showing different combinations of nickels, dimes, and quarters that total 35 cents. Organize the table by starting with the combinations that include the most quarters.

quarters	dimes	nickels
1	1	0
1	0	2
0	3	1
0	2	3
0	1	5
0	0	7

The total for each combination of these coins is 35 cents. There are 6 combinations possible.

Got It? Do these problems to find out.

3a. The product of two whole numbers is 36. What are the possible values for the sum of the two numbers?

3b. Zane paid for a DVD with a \$20 bill. The price of the DVD was \$19.50. How many combinations of nickels, dimes, and quarters are possible as change for \$0.50?

In most problems, a set of conditions or facts is given and an end result must be found. However, some problems start with the result and ask for something that happened earlier. The **work backward** strategy can be used to solve problems like this.

To use the work backward strategy, start with the end result and *undo* each step.

Example 4

Kendrick spent half of the money he had this morning on lunch. After lunch, he loaned his friend a dollar. Now he has $1.50. How much money did Kendrick start with in the morning?

Start with the end result, $1.50, and work backward to find Kendrick's starting amount.

Kendrick now has $1.50. →	$1.50
Undo the $1 he loaned to his friend. →	+1.00
	$2.50
Undo the half he spent for lunch. →	× 2
	$5.00

The amount Kendrick started with was $5.00.

Check Kendrick started with $5. If he spent half of that on lunch and loaned his friend $1.00, he would have $1.50 left. The solution is correct. ✓

Strategies
Other problem-solving strategies include:
- draw a diagram
- make a model
- solve a simpler problem
- use logical reasoning

Got It? **Do this problem to find out.**

4. Some passengers boarded a bus at the bus station. At the first stop, 9 more passengers boarded the bus. At the second stop, half of the passengers exited the bus. This left 12 passengers on the bus. How many passengers boarded the bus at the bus station?

Guided Practice

Use a strategy to solve each problem.

1. The figures below are made with toothpicks. Make a table that relates the perimeter to the figure number. Then find the perimeter of the twelfth figure. (Example 1)

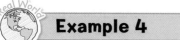

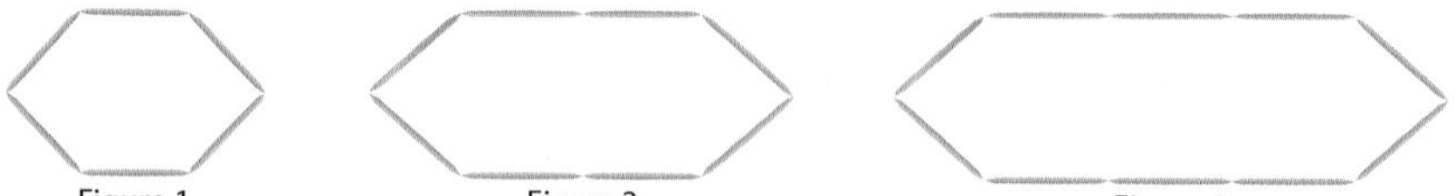

Figure 1 Perimeter = 6 | Figure 2 Perimeter = 8 | Figure 3 Perimeter = 10

2. The product of two consecutive non-negative odd integers is 783. What are the integers? (Example 2)

3. Jorge had 55 football cards. He traded 8 cards for 5 from Elise. He traded 6 more for 4 from Leon and 5 for 3 from Bret. Finally, he traded 12 cards for 9 from Ginger. How many cards does Jorge have now? (Example 3)

4. Tia used half of her allowance to buy a ticket to the class play. Then she spent $1.75 for ice cream. Now she has $2.25 left. How much is her allowance? (Example 4)

Independent Practice

eHelp Go online for Step-by-Step Solutions

Use a strategy to solve each problem.

5. **STEM** A ball bounces back 0.6 of its height on every bounce. If a ball is dropped from 200 feet, how high does it bounce on the fifth bounce? Round to the nearest tenth. (Example 1)

6. Rafael is burning a CD for Selma. The CD will hold 35 minutes of music. Which songs should he select from the list to record the maximum time on the CD without going over? (Example 2)

Song	A	B	C	D	E
Time	5 min 4 s	9 min 10 s	4 min 12 s	3 min 9 s	3 min 44 s
Song	F	G	H	I	J
Time	4 min 30 s	5 min 0 s	7 min 21 s	4 min 33 s	5 min 58 s

7 The cubes at the right are each numbered 1 through 6. During a game, both are rolled and the faces landing up are added. How many ways can a person playing the game roll a sum less than 8? (Example 3)

8. To catch a 7:30 A.M. bus, Don needs 30 minutes to shower and dress, 15 minutes for breakfast, and 10 minutes to walk to the bus stop. To catch the bus, what is his latest possible wake-up time? (Example 4)

9. **Financial Literacy** Mr. and Mrs. Delgado each own an equal number of shares of a stock. Mr. Delgado sells one-third of his shares for $2700. What was the total value of Mr. and Mrs. Delgado's stock before the sale?

10. The three counters shown in the table are used for a board game. If the counters are tossed, how many ways can at least one counter with Side A occur?

Counters	Side 1	Side 2
Counter 1	A	B
Counter 2	A	C
Counter 3	B	C

11. Odell has the same number of quarters, dimes, and nickels. In all he has $4 in change. How many of each coin does he have?

12. A certain number is multiplied by 3, and then 5 is added to the result. The final answer is 41. What is the number?

13. Lawanda put $15 of her paycheck into her savings account. Then she spent one-half of what was left on clothes. She paid $24 for a concert ticket and later spent one-half of what was then left on a book. When she got home, she had $14 left. What was the amount of Lawanda's paycheck?

14. The spinner at the right is used to play a certain game. On your turn, you must spin the spinner twice. How many different combinations of colors could you spin? List all possible combinations.

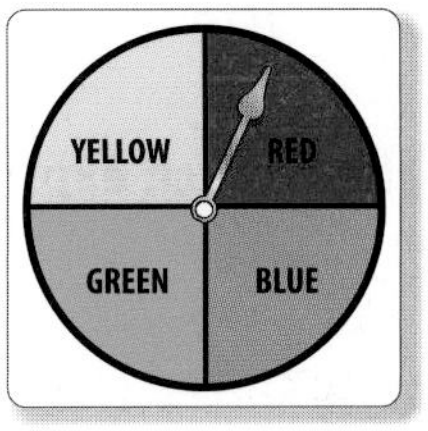

15. Brianne is three times as old as Camila. Four years from now she will be just two times as old as Camila. How old are Brianne and Camila now?

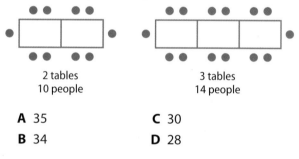

H.O.T. Problems Higher Order Thinking

16. **CCSS Model with Mathematics** Write a real-world problem that can be solved by working backward. Include the answer to your problem.

17. **CCSS Justify Conclusions** Tyler has 23 baseball cards in his collection. Each week, he buys a package with 8 cards to add to his collection. If Tyler continues in this way, will he ever have exactly 504 cards in his collection? Explain your reasoning.

18. **CCSS Identify Repeated Reasoning** Find the value of 1^2, 11^2, and 111^2. Then use your results to predict the value of $111{,}111^2$.

19. **Building on the Essential Question** Explain how to decide which strategy to use when solving a problem.

Standardized Test Practice

20. The sandwich choices at a local deli are shown in the table below. If a customer orders a sandwich with cheddar cheese and only one type of meat, how many different sandwiches are available?

Sandwich Choices	
Meats	**Cheeses**
ham	cheddar
salami	Swiss
turkey	colby

A 1 **C** 4
B 3 **D** 6

21. The product of three consecutive whole numbers is 210. What is the sum of the three whole numbers?

F 70 **H** 15
G 18 **J** 7

22. A restaurant has rectangular tables that can be placed together end-to-end. When 2 tables are placed together, 10 people can be seated. When 3 tables are placed together, 14 people can be seated. How many people can be seated when 7 tables are placed together?

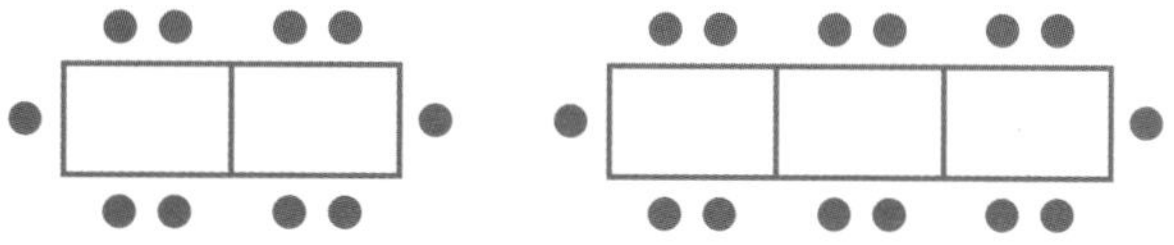

A 35 **C** 30
B 34 **D** 28

23. **Short Response** Isabelle took half of the cherries from a bowl. Then Roberto took half of the remaining cherries. Finally, Malia took 3 cherries from the bowl. This left 5 cherries in the bowl. How many cherries were in the bowl before Isabelle arrived?

CCSS Common Core Review

Evaluate each expression. 5.OA.1

24. $3[2(5) - 6]$
25. $12 + 6 \div 3$
26. $24 \div (2 + 4) + 2$
27. $9 + 2(12 - 4)$
28. $30 - 24 \div 6$
29. $19 - (7 - 4) \times 3$

Evaluate each expression if $x = 6$, $y = 4$, and $z = 9$. 7.NS.1, 7.NS.2

30. $2y - 5$
31. $\frac{xz}{6}$
32. $3z - 2x + y$
33. $\frac{y + z + 5}{x}$
34. $y \div 2 + 2x$
35. $z(x + 4)$

Lesson 1-6

Ordered Pairs and Relations

Interactive Study Guide

See pages 17–18 for:
- Getting Started
- Vocabulary Start-Up
- Notes

Essential Question

How can you use numbers and symbols to represent mathematical ideas?

Common Core State Standards

Content Standards
Preparation for 7.RP.2a, 7.RP.2b, 7.RP.2d, 8.EE.5

Mathematical Practices
1, 3, 4

Vocabulary

coordinate system
coordinate plane
y-axis
origin
x-axis
ordered pair
x-coordinate
y-coordinate
graph
relation
domain
range

What You'll Learn

- Use ordered pairs to locate points.
- Use graphs to represent relations.

Real-World Link

Bungee Jumping People bungee jump from bridges, from cliffs, and even into volcanoes! The table describes four bungee jumping sites and the approximate heights and times of the jumps. There are different ways to represent this information.

Location	Height (ft)	Time of Fall (s)
Europabrücke, Austria	630	6
Glenns Ferry Bridge, Idaho	170	3
Macau Tower, China	764	7
Navajo Bridge, Arizona	452	5

Ordered Pairs

In mathematics, a **coordinate system** or **coordinate plane** is used to locate points. The coordinate system is formed by the intersection of two number lines that meet at right angles at their zero points.

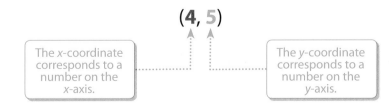

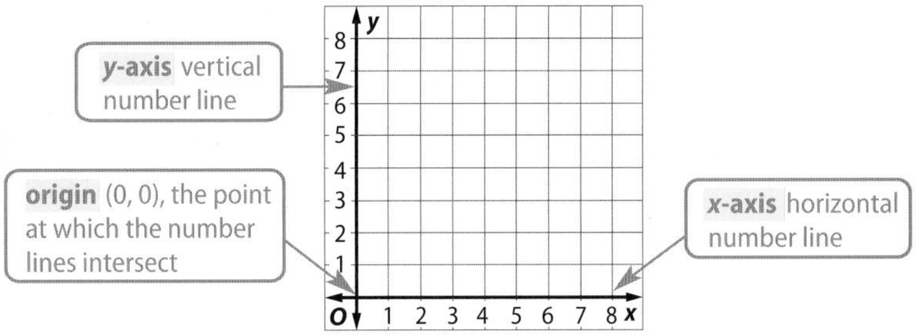

An **ordered pair** of numbers is used to locate any point on a coordinate plane. The first number is called the ***x*-coordinate**, and the second number is called the ***y*-coordinate**.

(4, 5)

The *x*-coordinate corresponds to a number on the *x*-axis.

The *y*-coordinate corresponds to a number on the *y*-axis.

To **graph** an ordered pair, draw a dot at the point that corresponds to the ordered pair. The coordinates are your directions to locate the point.

Example 1

Tools Tutor

Graph each ordered pair on a coordinate plane.

Coordinate Planes
Unless the units are marked otherwise, you can assume that each unit on the x-axis and y-axis represents 1 unit.

a. $J(5, 3)$

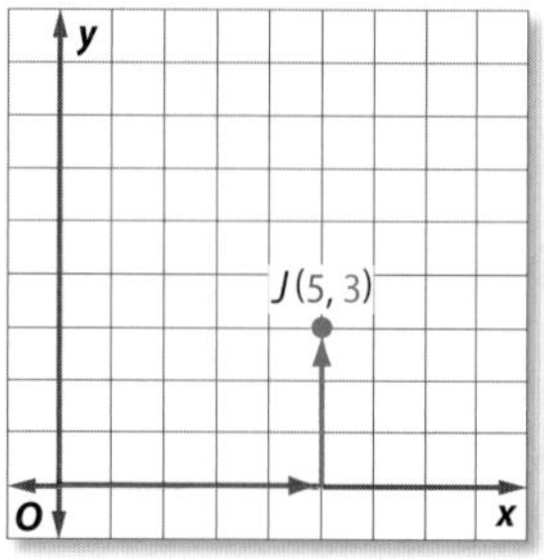

Step 1 Start at the origin.

Step 2 Since the x-coordinate is 5, move 5 units to the right.

Step 3 Since the y-coordinate is 3, move 3 units up. Draw a dot.

b. $K(0, 4)$

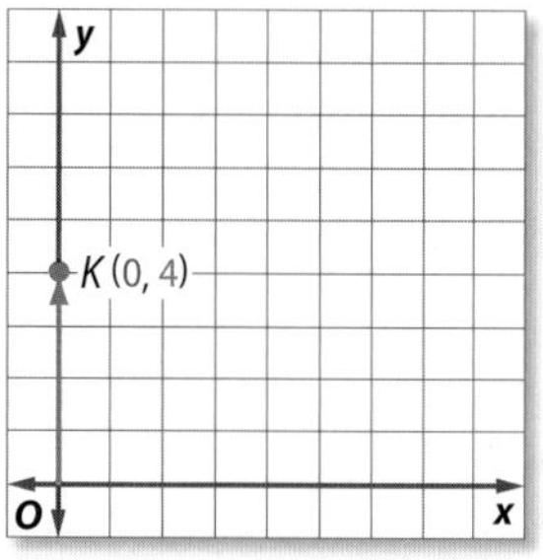

Step 1 Start at the origin.

Step 2 Since the x-coordinate is 0, you do not need to move right.

Step 3 Since the y-coordinate is 4, move 4 units up. Draw the dot on the axis.

Got It? Do these problems to find out.

1a. $Q(2, 3)$ **1b.** $R(5, 0)$ **1c.** $S\left(3, 1\frac{1}{2}\right)$ **1d.** $T\left(6\frac{1}{2}, 5\frac{1}{2}\right)$

Sometimes a point on a graph is named by using a capital letter. To identify its location, you can write the ordered pair that represents the point.

Example 2

Tutor

Write the ordered pair that names each point.

a. A

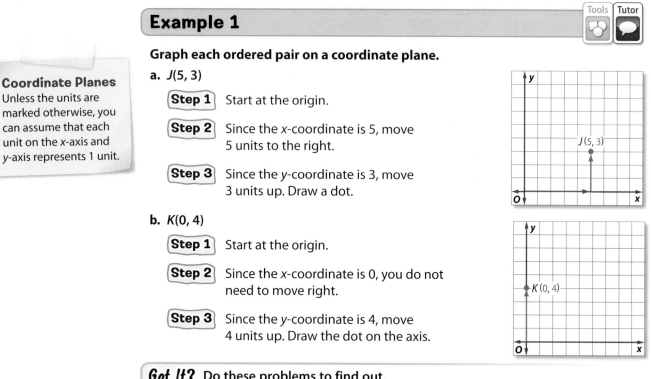

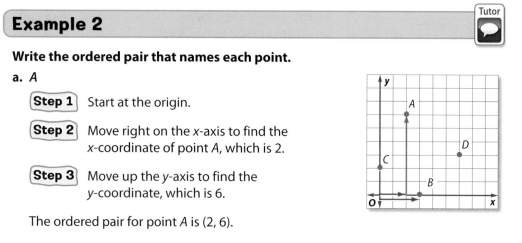

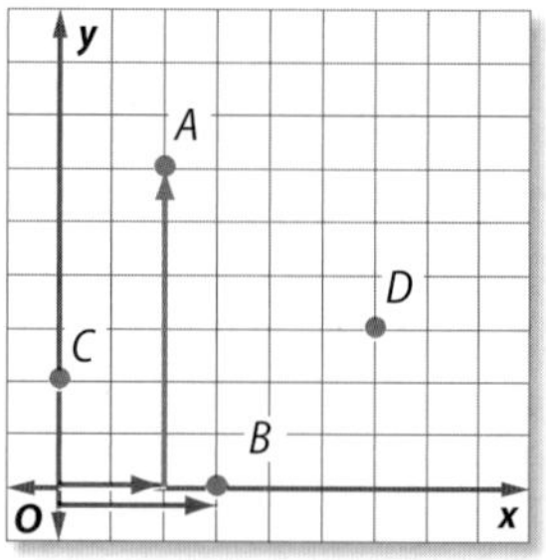

Step 1 Start at the origin.

Step 2 Move right on the x-axis to find the x-coordinate of point A, which is 2.

Step 3 Move up the y-axis to find the y-coordinate, which is 6.

The ordered pair for point A is (2, 6).

b. B

The x-coordinate of point B is 3, and the y-coordinate is 0.

The ordered pair for point B is (3, 0).

Got It? Do these problems to find out.

2a. C **2b.** D

Relations

Domain and Range
The domain of a relation is also called the *input*. The range of a relation is also called the *output*.

A set of ordered pairs such as {(2, 3), (3, 5), (4, 1)} is a **relation**. A relation can also be shown in a table or a graph. The **domain** of the relation is the set of *x*-coordinates. The **range** of the relation is the set of *y*-coordinates.

Ordered Pairs

(2, 3)
(3, 5)
(4, 1)

The domain is {2, 3, 4}.

The range is {3, 5, 1}.

Table

x	*y*
2	3
3	5
4	1

Graph

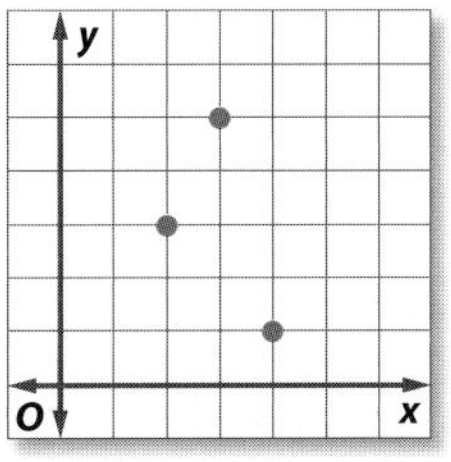

Example 3

Express the relation {(0, 2), (1, 4), (2, 5), (3, 8)} as a table. Then determine the domain and range.

x	0	1	2	3
y	2	4	5	8

The domain is {0, 1, 2, 3}, and the range is {2, 4, 5, 8}.

Got It? **Do this problem to find out.**

3. Express the relation {(2, 4), (0, 3), (1, 4), (1, 1)} as a table. Then determine the domain and range.

Real World

Example 4

A seahorse swims at a rate of about 5 feet per hour.

a. **Make a table of ordered pairs in which the *x*-coordinate represents the hours and the *y*-coordinate represents the number of feet for 1, 2, and 4 hours.**

x	*y*
1	5
2	10
4	20

The points appear to lie in a line.

b. **Graph the ordered pairs and describe the graph.**

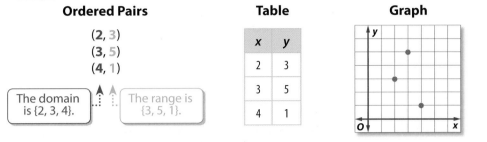

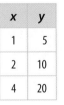

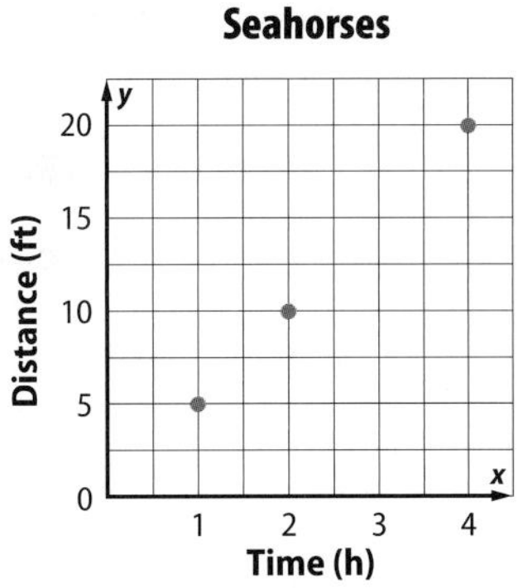

Got It? **Do these problems to find out.**

4. One square mile is equal to six hundred forty acres.

a. Make a table of ordered pairs in which the *x*-coordinate represents the number of square miles and the *y*-coordinate represents the number of acres in 1, 2, and 3 square miles.

b. Graph the ordered pairs. Then describe the graph.

Guided Practice

Check

Graph each ordered pair on a coordinate plane. (Example 1)

1. *F*(6, 0) **2.** *A*(2, 5) **3.** *W*(4, 1) **4.** *Z*(0, 1)

Refer to the coordinate plane shown at the right. Write the ordered pair that names each point. (Example 2)

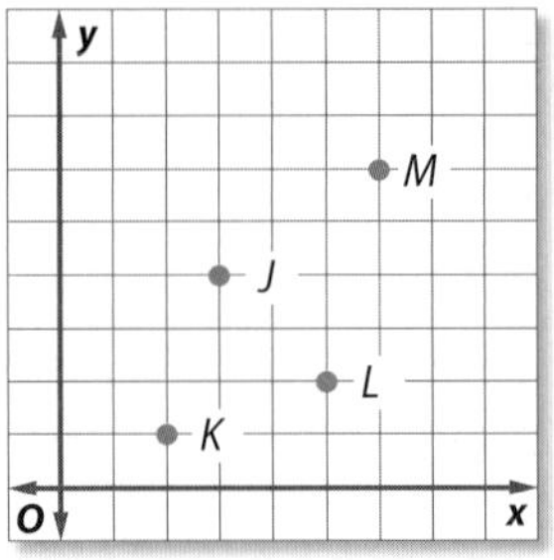

5. *J* **6.** *K*

7. *L* **8.** *M*

Express each relation as a table. Then determine the domain and range. (Example 3)

9. {(3, 4), (1, 5), (4, 2)}

10. {(1, 3), (2, 6), (3, 3), (4, 7)}

11. One quart is equal to two pints. (Example 4)

a. Make a table of ordered pairs in which the *x*-coordinate represents the number of quarts and the *y*-coordinate represents the number of pints in 1, 2, 3, and 4 quarts.

b. Graph the ordered pairs. Then describe the graph.

Independent Practice

Go online for Step-by-Step Solutions

eHelp

Graph each ordered pair on a coordinate plane. (Example 1)

12. *A*(4, 7) **13.** *B*(0, 4) **14.** *C*(7, 3) **15.** *D*(3, 4)

16. *F*(6, 1) **17.** *G*(6, 5) **18.** *H*(3, 0) **19.** *J*(2, 2)

Refer to the coordinate plane shown at the right. Write the ordered pair that names each point. (Example 2)

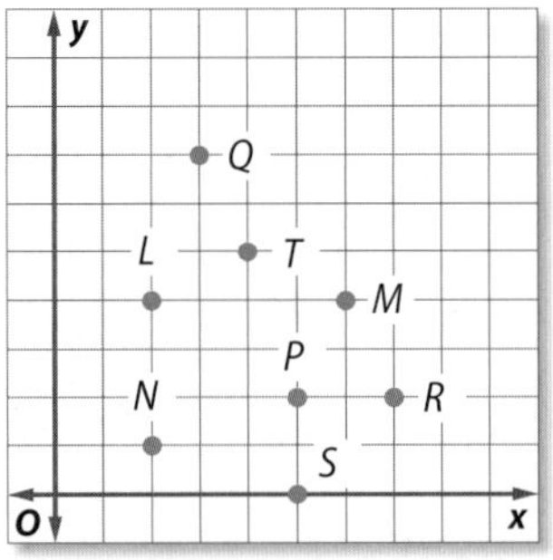

20. *L* **21.** *M*

22. *N* **23.** *P*

24. *Q* **25.** *R*

26. *S* **27.** *T*

Express each relation as a table. Then determine the domain and range. (Example 3)

28. {(4, 5), (2, 1), (5, 0), (3, 2)}

29 {(0, 2), (2, 2), (4, 1), (3, 5)}

30. {(6, 0), (4, 5), (2, 1), (3, 1)}

31. {(5, 1), (3, 7), (4, 8), (5, 7)}

32. The cost of a mini pizza is $7 at a local pizza store. (Example 4)

a. Make a table of ordered pairs in which the *x*-coordinate represents the number of mini pizzas and the *y*-coordinate represents the cost of 1, 3, 5, and 7 mini pizzas.

b. Graph the ordered pairs. Then describe the graph.

33. Aaron is hiking in a state park. He averages 3 miles per hour. (Example 4)

a. Make a table of ordered pairs in which the x-coordinate represents the number hours and the y-coordinate represents the number of miles hiked in 1, 2, 4, and 6 hours.

b. Graph the ordered pairs.

34. CCSS **Multiple Representations** In this problem, you will explore more about relations. Suppose Jamal has only 30 minutes to practice the piano and study for a science test.

a. Table Make a table of ordered pairs showing at least 6 ways Jamal can split the time between the two activities. Let the x-coordinate and the y-coordinate represent the number of minutes spent playing the piano and the number of minutes spent studying, respectively.

b. Graph Graph the ordered pairs.

c. Words Describe the general pattern of points of your graph.

d. Graph Choose a point on the graph that is *not* one of the points you plotted. Use the coordinates to predict a pair of values for the piano time and study time.

35 Six students in Mr. Maloney's class made a table of ordered pairs for their height in inches x and their shoe size y.

Height (in.)	58	56	62	60	59	61
Shoe Size	6	$5\frac{1}{2}$	$8\frac{1}{2}$	8	7	$7\frac{1}{2}$

a. Graph the ordered pairs.

b. Compare this graph to the graph in Example 4.

36. CCSS **Multiple Representations** The numbers 4, 7, 10, 13, … form an *arithmetic sequence* because each term can be found by adding the same number to the previous term.

Term Number	1	2	3	4
Term	4	7	10	13

a. Numbers Write the set of ordered pairs (term number, term).

b. Graph Graph the ordered pairs.

c. Words Describe the shape of the graph.

d. Words If possible, write a rule to find the twentieth term. Explain how you found the rule or why you cannot write a rule.

H.O.T. Problems Higher Order Thinking

37. CCSS **Reason Abstractly** Write a vocabulary term that completes the following.

A point is to a number line as a(n) __?__ is to a coordinate grid.

38. CCSS **Persevere with Problems** Describe all of the possible locations for the graph of (x, y) if $x = 2$.

39. CCSS **Justify Conclusions** Refer to Exercises 32 and 33. For which graph would it make more sense to connect the points with line segments? Explain.

40. **Building on the Essential Question** Explain why the point $M(4, 3)$ is different from the point $N(3, 4)$.

Standardized Test Practice

41. On the map of a campsite shown below, the tent is located at (3, 7). Which point represents the location of the tent?

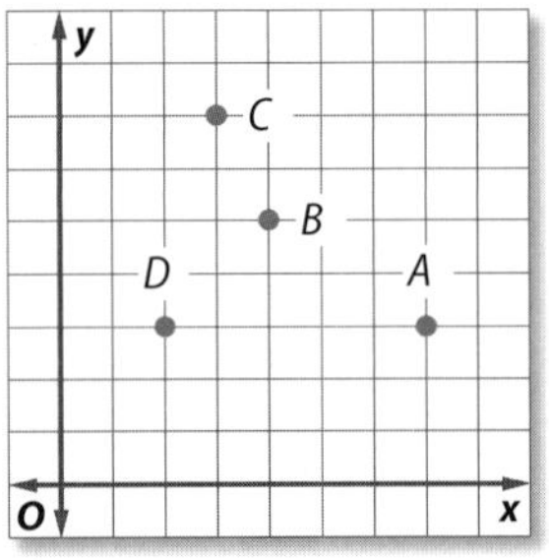

A point *A**   **C** point *C*

B point *B*   **D** point *D*

42. Rectangle *ABCD* has vertices *A*(1, 3), B(1, 6), and *C*(5, 6). What are the coordinates of point *D*?

F (6, 5)   **H** (5, 3)

G (5, 1)   **J** (6, 1)

43. What is the domain of the relation below?

x	y
1	3
2	4
4	8
6	1
7	4

A {1, 2, 4, 6, 7}

B {1, 3}

C {1, 3, 4, 8}

D {(1, 3), (2, 4), (4, 8), (6, 1), (7, 4)}

44. **Short Response** Point *Z* is located at (4, 7) on a coordinate plane. Point *T* is located 3 units to the right and 4 units down from point *Z*. What is the *x*-coordinate of point *T*?

CCSS Common Core Review

Name the property shown by each statement. 7.EE.1

45. $5 \cdot 3 = 3 \cdot 5$

46. $6 \cdot 2 \cdot 0 = 0$

47. $0 + 13 = 13$

48. $(5 + x) + 6 = 5 + (x + 6)$

State whether each conjecture is true. If not, give a counterexample. 7.EE.1

49. Division of whole numbers is associative.

50. Subtraction of whole numbers is commutative.

51. Melinda purchased the items shown in the table. 7.EE.3

a. Write an expression to show the total cost of the items.

b. Suppose the cost of the sweater is \$25. How much did she spend in all?

Item	Price (\$)
sweater	s
purse	$s + 12$
belt	$s + 4$

Find the value of each expression. 6.EE.2c

52. $x + 7$ if $x = 8$

53. $d - 5$ if $d = 12$

54. $18 + m$ if $m = 4$

55. $f - 12$ if $f = 28$

56. $s + 14$ if $s = 32$

57. $19 - b$ if $b = 5$

Find the value of each expression. 5.OA.1

58. $8 + 12 \times 4 \div 8$

59. $13 - 6 \times 2 + 1$

60. $14 - 2 \times 7 + 0$

61. $30 - 14 \times 2 + 8$

62. $11 + 4 \times (12 - 7)$

63. $(11 - 7) \times 3 - 5$

Lesson 1-7

Words, Equations, Tables, and Graphs

ISG Interactive Study Guide

See pages 19–20 for:
- Getting Started
- Real-World Link
- Notes

Essential Question

How can you use numbers and symbols to represent mathematical ideas?

CCSS Common Core State Standards

Content Standards
7.EE.4

Mathematical Practices
1, 3, 4, 8

Vocabulary

equation

What You'll Learn

- Use multiple representations to represent relations.
- Translate among different verbal, tabular, graphical, and algebraic representations of relations.

Real-World Link

Fireworks Physics can be used to calculate the path of fireworks. In general, for every 1-inch increase in shell diameter, a firework's height increases by about 100 feet. This relationship can be represented using an equation, a table, or a graph.

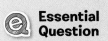

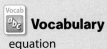

Represent Relations

You have already seen that a relation may be represented as a set of ordered pairs. You can also write a rule for the operation(s) performed on the domain value to get the range value. A table may list the *x*-coordinates (domain values), the rule, and the *y*-coordinates (range values).

Example 1

Tools Tutor

In a game of *What's My Rule?* Kinna picked the card shown at the right. Make a table for four different domain values and write an algebraic expression for the rule. Then state the domain and range of the relation.

What's My Rule?
double a number, then add three

Step 1 Create a table showing the *x*-coordinates, the rule, and the *y*-coordinates. Enter four different domain values.

Step 2 The rule "double a number, then add three" translates to $2x + 3$. Use the rule to complete the table.

Step 3 The domain is {1, 2, 3, 4}. The range is {5, 7, 9, 11}.

x	Rule: $2x + 3$	y
1	$2(1) + 3$	5
2	$2(2) + 3$	7
3	$2(3) + 3$	9
4	$2(4) + 3$	11

Got It? Do this problem to find out.

1. Jenna picked the game card shown. Make a table for four different domain values and write an algebraic expression for the rule. Then state the domain and range of the relation.

What's My Rule?
triple the number and subtract one

Equation An equation always contains an equal sign. For example, $60t$ is an expression, but $d = 60t$ is an equation.

Multiple Representations

Words, equations, tables, and graphs can be used to represent relations. An **equation** is a mathematical sentence stating that two quantities are equal. Relations are often written as equations with two variables—one to represent domain values and one to represent range values.

Concept Summary Multiple Representations

Words

Distance is equal to 60 miles per hour times the number of hours.

Equation

$d = 60t$

Table

Time (h)	Distance (mi)
1	60
2	120
3	180
4	240

Graph

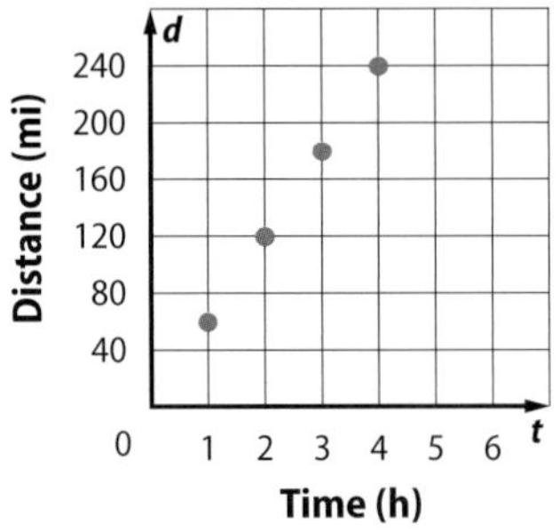

Example 2

STEM The navigation message from a satellite to a GPS in an airplane is sent once every 12 minutes.

a. Write an equation to find the number of messages sent in any number of minutes.

Let t represent the time and n represent the number of messages. The equation is $n = t \div 12$.

b. Make a table to find the number of messages in 120, 180, 240, and 300 minutes. Then graph the ordered pairs.

t	$t \div 12$	n
120	$120 \div 12$	10
180	$180 \div 12$	15
240	$240 \div 12$	20
300	$300 \div 12$	25

GPS Messages

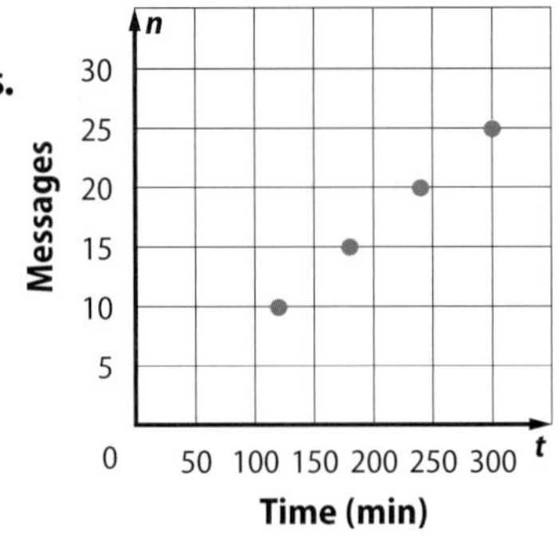

Got It? **Do this problem to find out.**

2. Sound travels at about 1088 feet per second at 32°F in dry air at sea level.

a. Write an equation to find the distance traveled by sound for any number of seconds.

b. Make a table to find the distance sound travels in 0, 1, 2, and 3 seconds. Then graph the ordered pairs.

Guided Practice

Check

Copy and complete each table. Then state the domain and range of the relation. (Example 1)

1. The team scores 6 points for each touchdown.

Number of Touchdowns	Number of Points
x	*y*
1	■
2	■
5	■
7	■

2. Bob spent 5 more than 3 times what Anna spent

Anna's Spending ($)	Bob's Spending ($)
x	*y*
2	■
4	■
6	■
8	■

3. CCSS **Multiple Representations** There are 16 ounces in 1 pound. (Example 2)

a. Symbols Write an equation that can be used to find the number of ounces in any number of pounds.

b. Table Make a table to find the number of ounces in 5, 8, 11, and 13 pounds.

c. Graph Graph the ordered pairs for the relation.

Independent Practice

Go online for Step-by-Step Solutions

eHelp

Copy and complete each table. Then state the domain and range of the relation. (Example 1)

4. Each ticket to the school musical costs $8.

Number of Tickets	Total Cost ($)
x	*y*
4	■
7	■
9	■
12	■

5. The dog weighs 4 pounds more than the cat.

Weight of Cat (lb)	Weight of Dog (lb)
x	*y*
3	■
6	■
9	■
12	■

6. Today's attendance is 4 less than half of yesterday's attendance.

Yesterday's Attendance	Today's Attendance
x	*y*
14	■
18	■
22	■
26	■

7. Casey has 5 less than 4 times as many baseball cards than Ben.

Ben's Cards	Casey's Cards
x	*y*
3	■
7	■
11	■
15	■

8. **Multiple Representations** One roll of quarters contains 40 quarters. (Example 2)

a. **Symbols** Write an equation that can be used to find the number of quarters q in any number of rolls of quarters r.

b. **Table** Make a table to find the number of quarters in 3, 4, 5, and 6 rolls.

c. **Graph** Graph the ordered pairs for the relation.

9. **Multiple Representations** Sales for a new video game offered by Technogames is shown at the right. (Example 2)

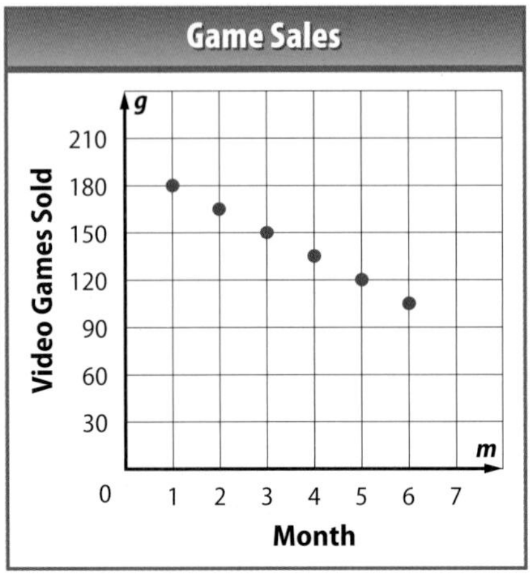

a. **Table** Make a table showing the domain (month) and the range (video games sold).

b. **Symbols** Write an equation that can be used to find the number of games sold g for any month m.

c. **Numbers** State the domain and range of the relation.

10. **Multiple Representations** Kevin's Flooring sells different sizes of square floor tiles. Carl wants to purchase 10 tiles.

a. **Symbols** Write an equation that can be used to find the area of any 10 square floor tiles. (*Hint:* area = side × side)

b. **Table** Make a table to find the area covered by 10 tiles that measure 6, 12, 15, and 24 inches on one side.

c. **Graph** Graph the ordered pairs for the relation.

11. **Multiple Representations** The table shows the temperatures at various depths in a lake.

Depth (ft)	Temperature (°F)
0	74
10	72
20	71
30	61
40	55
50	53

a. **Graph** Graph the ordered pairs on a coordinate plane.

b. **Symbols** Can you write one equation that can be used to find the temperature t based on the depth in the lake d? Explain.

c. **Analyze** State the domain and range of the relation.

H.O.T. Problems Higher Order Thinking

12. **Model with Mathematics** Tickets to an amusement park are $26 each. Write an equation to find the total cost c for t tickets. Graph four ordered pairs for the relation.

13. **Model with Mathematics** Write about a real-world situation that can be represented by the equation $y = 4x$.

14. **Identify Repeated Reasoning** Write a rule for the relation shown in the table.

x	1	2	3	4
y	10	12	14	16

15. **Building on the Essential Question** Give an example of a situation in which it is best to represent a relation by a table. Give an example of a situation in which it is best to represent a relation by an equation. Justify your choice in each case.

Standardized Test Practice

16. **Short Response** The table below follows a rule.

x	y
1	5
3	11
5	17
7	23
8	26
10	32

Write a rule for the relationship shown in the table.

17. Walik is buying CDs from an online store. Each CD costs \$12.99. There is a flat shipping charge of \$4.95. Which expression represents the cost of purchasing n CDs?

A $n(12.99 + 4.95)$
B $4.95n + 12.99$
C $12.99n + 4.95$
D $(12.99 - 4.95)n$

18. Which of the following equations would best describe the graph shown below?

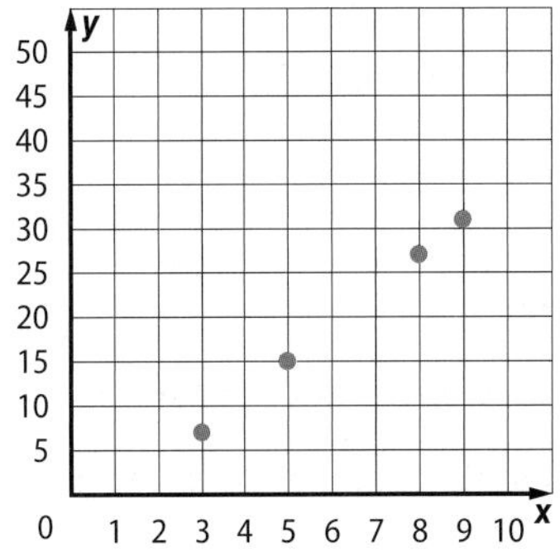

F $y = 2x + 1$
G $y = 3x - 2$
H $y = 5x - 10$
J $y = 4x - 5$

19. The relation that is represented by the equation $y = 20 - 2x$ has the domain {3, 4, 5, 6}. Which of the following sets is the range of the relation?

A {3, 4, 5, 6}
B {8, 10, 12, 14}
C {12, 13, 14, 15}
D {14, 16, 18, 20}

Common Core Review

Refer to the coordinate plane at the right. Write the ordered pair that names each point. 5.G.1

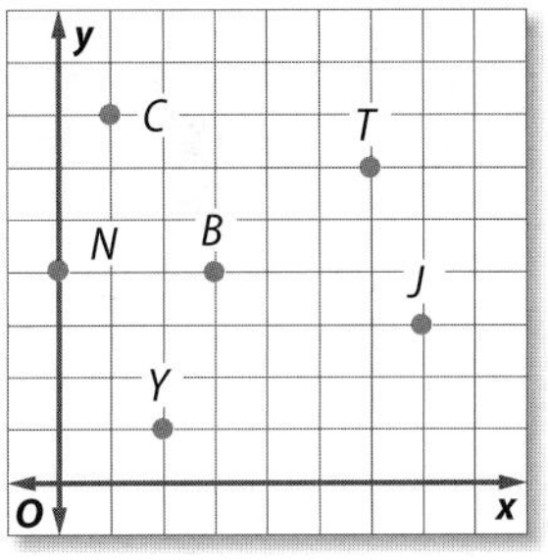

20. C **21.** J **22.** N

23. T **24.** Y **25.** B

Simplify each expression 7.EE.1

26. $(m + 8) + 4$ **27.** $(17 + p) + 9$ **28.** $21 + (k + 16)$

29. $(6 \cdot c) \cdot 8$ **30.** $8 \cdot (y \cdot 2)$ **31.** $3(25s)$

32. **Financial Literacy** There are 20 nickels in one dollar. 6.EE.2

a. Write an algebraic expression that can be used to find the number of nickels in any number of dollars n.

b. How many nickels are in \$7.00?

Find each quotient. 6.NS.2

33. $84 \div 12$ **34.** $135 \div 15$ **35.** $792 \div 24$ **36.** $1938 \div 19$

37. $1904 \div 112$ **38.** $5040 \div 80$ **39.** $11{,}045 \div 47$ **40.** $36{,}192 \div 104$

Evaluate each expression if $a = 3$, $b = 8$, and $c = 2$. 7.NS.3

41. $\frac{ab}{2}$ **42.** $5a - (b + c)$ **43.** $2a - c + b$ **44.** $23 - \frac{5b}{2}$

Check

Chapter Review

Interactive Study Guide
See pages 21–24 for:
- Vocabulary Check
- Key Concept Check
- Problem Solving
- Reflect

Lesson-by-Lesson Review

Lesson 1-1 A Plan for Problem Solving (pp. 2–5)

Use the four-step plan to solve each problem.

1. Carlos traveled 347 miles to go to his cousins' house. It took him about 5.5 hours to drive there. Estimate his average speed in miles per hour.

2. Julie walked 3 miles each day in March and April. How many miles did she walk altogether over this time?

3. A restaurant is offering selected three-course meals for one price. Customers can choose from 2 appetizers, 3 main courses, and 2 desserts. How many different dinner combinations are available?

4. There are 140 students going on a field trip. A maximum of 22 students can fit on each bus. How many buses will be needed for the field trip?

Example 1

Dana is tying 18 bows onto gift baskets. Each bow uses 4 feet of ribbon. How many feet of ribbon does Dana need to make all the bows?

Understand You know the number of bows and the length of ribbon for each bow. You need to find the total ribbon for all the bows.

Plan Multiply the number of bows by the length of ribbon for each bow.

Solve $18 \times 4 = 72$

So, Dana needs 72 feet of ribbon in all.

Check Estimate. $20 \times 4 = 80$.

Since $80 \approx 72$, the answer is reasonable. ✓

Lesson 1-2 Words and Expressions (pp. 6–10)

Write a numerical expression for each verbal phrase.

5. the total number of students, if there are nineteen in one class and thirteen in another

6. the number of soccer players on each of five teams, if there are a total of thirty-five players

Evaluate each expression.

7. $6(7) + 3$

8. $4[(12 - 4) + 2]$

9. $3[6 + (6 - 2)]$

10. $9(2) + 8(7)$

11. $10(4 \div 2)$

12. $4[25 \div (2 + 3)]$

13. **Financial Literacy** Yu, Collin, and Sydney spent \$284 to make bracelets. They sold the bracelets for \$674. If they split the profits evenly, how much did each person earn?

Example 2

Write the phrase *the total number of postcards, if eight people each buy five* as a numerical expression.

Phrase the product of 8 and 5

Expression $8 \cdot 5$

Example 3

Evaluate the expression $3[(5 - 4) + 6]$.

$3[(\mathbf{5 - 4}) + 6] = 3(\mathbf{1 + 6})$ Evaluate (5 – 4).

$= 3(7)$ 3(7) means 3×7.

$= 21$ Multiply 3 and 7.

Lesson 1-3 Variables and Expressions (pp. 13–18)

14. Translate into an algebraic expression *the quotient of the number of points and three.*

Evaluate each expression if $a = 4$, $b = 8$, and $c = 11$.

15. $3a + b$

16. $16 - c$

17. $3a + 4b - c$

18. $\frac{3b}{a} + ac$

19. $12 + 3a + b$

20. $6(a + b + c)$

21. $3a - c$

22. $2ab + ac$

23. There are 36 inches in a yard. Write an expression to find the number of yards in x inches.

Example 4

Evaluate $x - 5 + 2y$ if $x = 6$ and $y = 4$.

$x - 5 + 2y = 6 - 5 + 2(4)$	Replace x with 6 and y with 4.
$= 6 - 5 + 8$	Multiply 2 and 4.
$= 1 + 8$	Subtract 5 from 6.
$= 9$	Add 1 and 8.

Lesson 1-4 Properties of Numbers (pp. 19–24)

Name the property shown by each statement.

24. $12 \times 0 = 0$

25. $(3 \cdot 5) \cdot 2 = 3 \cdot (5 \cdot 2)$

Simplify each expression.

26. $3 \cdot (2 \cdot x)$

27. $(5 + v) + 7$

28. $10 \cdot (4 \cdot x)$

29. $(2 + 3) + (4 + x)$

30. $(7 + y) + 13$

31. $1 \cdot (25 \cdot x)$

32. Gloria has 58 dolls. If she does not add any dolls to her collection, write a number sentence that represents the situation. Then name the property that is illustrated.

Example 5

Simplify $(4 + x) + 6$.

$(4 + x) + 6 = (x + 4) + 6$	Commutative (+)
$= x + (4 + 6)$	Associative (+)
$= x + 10$	Simplify.

Lesson 1-5 Problem-Solving Strategies (pp. 26–30)

Use a strategy to solve each problem.

33. Mia has 14 coins in her pocket that total \$1.10. What are the coins in her pocket?

34. Yanira is 3 years older than Tim and twice as old as Hannah. Tim is 2 years older than Hannah. How old are Yanira, Tim, and Hannah?

35. The sum of two consecutive even integers is 150. What are the two integers?

Example 6

Tina bought twice as many pencils as Frank bought at the school store, and Joel bought 5 more pencils than Frank did. Altogether they bought 17 pencils. How many pencils did Frank buy?

Make a guess. Check your guess and revise.

Frank	Tina	Joel	Total	
2	$2(2) = 4$	$2 + 5 = 7$	13	too low
3	$2(3) = 6$	$3 + 5 = 8$	17	✓

Frank bought 3 pencils.

Lesson 1-6 Ordered Pairs and Relations (pp. 31–36)

Express each relation as a table. Then determine the domain and range.

36. {(2, 3), (2, 6), (2, 5)}

37. {(1, 4), (2, 8), (3, 12), (4, 16)}

38. {(3, 6), (4, 12), (5, 18)}

39. {(1, 1), (2, 4), (3, 9), (4, 16)}

40. It costs $2 per person to ride the Ferris wheel.

a. Make a table of ordered pairs in which the *x*-coordinate represents the number of people and the *y*-coordinate represents the cost for 4, 8, 12, and 16 people.

b. Graph the ordered pairs and then describe the graph.

Example 7

Express the relation {(2, 1), (5, 6), (2, 7), (6, 1)} as a table. Then determine the domain and range.

x	2	5	2	6
y	1	6	7	1

The domain is {2, 5, 6}, and the range is {1, 6, 7}.

Lesson 1-7 Word, Equations, Tables, and Graphs (pp. 37–41)

41. Shanna downloaded 5 more songs than videos.

a. Write an equation that can be used to find the number of songs downloaded given the number of videos downloaded.

b. Complete the table for the number of songs downloaded when the number of videos downloaded is 2, 4, 6, and 8.

Number of Videos	Number of Songs
v	*s*
2	■
4	■
6	■
8	■

c. Graph the ordered pairs for the relation.

Example 8

There are 3 apples for each horse.

a. Write an equation to find the number of apples needed for any number of horses.

Let *h* represent the number of horses and let *a* represent the number of apples. The equation is $a = 3h$.

b. Make a table for 3, 5, 7, and 11 horses.

Number of Horses	Number of Apples
h	*a*
3	9
5	15
7	21
11	33

c. Graph the ordered pairs for the relation.

Apples for Horses

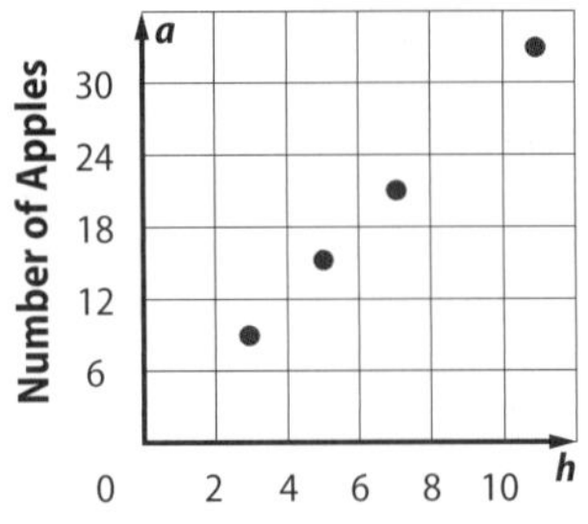

Chapter 2

Operations with Integers

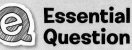

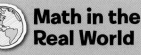

Essential Question

What happens when you add, subtract, multiply, and divide integers?

Common Core State Standards

Content Standards
7.NS.1, 7.NS.1a, 7.NS.1b, 7.NS.1c, 7.NS.1d, 7.NS.2, 7.NS.2a, 7.NS.2b, 7.NS.2c, 7.NS.3, 7.EE.3

Mathematical Practices
1, 2, 3, 4, 5, 7

Math in the Real World

Weather Extremes Desert temperatures are either high or low. Typical summer days sizzle to over 100°F and reach as high as 134°F. Winter nights can be near freezing and fall as low as −16°F. Little rain, monsoon thunderstorms, flash floods, and massive windstorms make desert living an adventure in weather extremes.

Interactive Study Guide
See pages 25–28 for:
- Chapter Preview
- Are You Ready?
- Foldable Study Organizer

Lesson 2-1

Integers and Absolute Value

Interactive Study Guide

See pages 29–30 for:
- Getting Started
- Vocabulary Start-Up
- Notes

Essential Question

What happens when you add, subtract, multiply, and divide integers?

Common Core State Standards

Content Standards
Preparation for 7.NS.1, 7.NS.1a

Mathematical Practices
1, 2, 3, 4, 7

Vocabulary

negative number
positive number
integer
opposites
coordinate
inequality
absolute value

Math Symbols

$<$ is less than
$>$ is greater than

What You'll Learn

- Compare and order integers.
- Find the absolute value of an expression.

Real-World Link

Geocaching Geocaching is an outdoor treasure hunting game. Some treasures, or geocaches, are located hundreds of feet above sea level. Others are hidden in lakes and can only be reached by snorkeling or scuba diving.

Compare and Order Integers

A **negative number** is a number less than zero. A **positive number** is a number greater than zero. Negative numbers like -3 and positive numbers like $+3$ are members of the set of integers. An **integer** is any number from the set $\{\ldots, -3, -2, -1, 0, 1, 2, 3, \ldots\}$, where … means continues indefinitely.

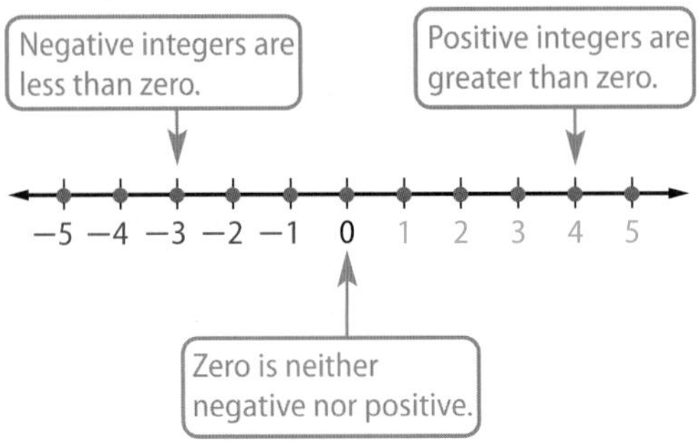
Negative integers are less than zero.
Positive integers are greater than zero.
−5 −4 −3 −2 −1 0 1 2 3 4 5
Zero is neither negative nor positive.

Integers such as $+3$ and -3 are called **opposites**, because they are the same distance from zero on the number line.

Example 1

Tutor

Write an integer for each situation. Then identify its opposite and describe what it means.

a. 23°F below zero

Because it is *below* zero, the integer is -23. Its opposite is $+23$ or 23, which means 23°F above zero.

b. 11 inches more than normal

Because it is *more than* normal, the integer is $+11$ or 11. Its opposite is -11, which means 11 inches less than normal.

Got It? **Do these problems to find out.**

1a. a loss of 8 yards

1b. a deposit of \$15

Digital Vision/Getty Images

To graph an integer, locate the point named by the integer on a number line. The **coordinate** is the number that corresponds to the point on a number line.

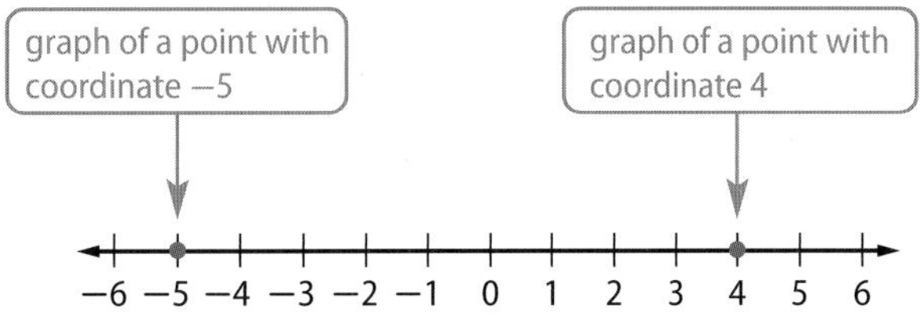

Any mathematical sentence containing < or > is called an inequality. An **inequality** compares numbers or quantities. When two numbers are graphed on a number line, the number to the left is always less than the number to the right.

Example 2

Use the integers graphed on the number line below.

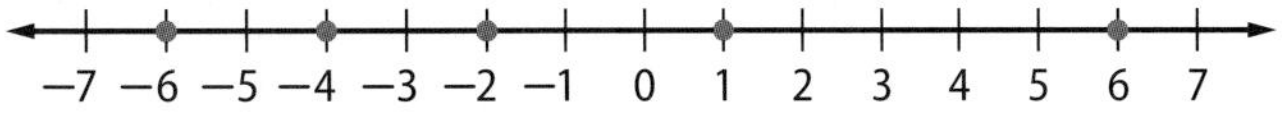

Inequalities
The inequality symbol always points to the lesser number.

a. Write two inequalities involving 1 and −2.
Since 1 is to the right of −2, 1 is greater than −2. So, $1 > -2$.
Since −2 is to the left of 1, −2 is less than 1. So, $-2 < 1$.

b. Replace the ● with <, >, or = in −4 ● −6 to make a true sentence.
Since −4 is to the right of −6, −4 is greater. So, $-4 > -6$.

Got It? **Do these problems to find out.**

2a. Write two inequalities involving −7 and −3.

2b. Replace the ● with <, >, or = in −1 ● 2 to make a true sentence.

Example 3

Bethany and her friends played a question-and-answer video game. Their scores at the end of the game were 1, −5, 0, −1, 2, and 4. Order the scores from least to greatest.

Graph each integer on a number line.

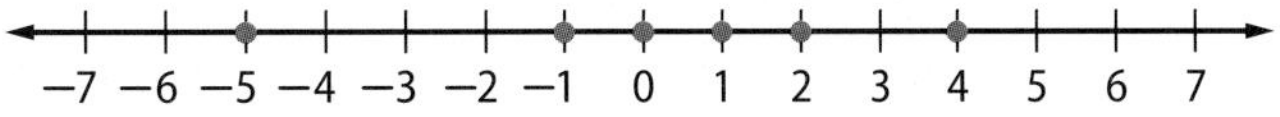

Write the numbers as they appear from left to right. The scores −5, −1, 0, 1, 2, and 4 are in order from least to greatest.

Got It? **Do this problem to find out.**

3. The recorded highs in degrees Celsius at Niagara Falls from February 21 to 28 of a recent year are 4, 2, 3, −6, −5, −1, 0, and 1. Order the temperatures from greatest to least.

Key Concept Absolute Value

Words The absolute value of a number is the distance the number is from zero on the number line. The absolute value of a number is always greater than or equal to zero.

Example $|6|$ and $|-6|$

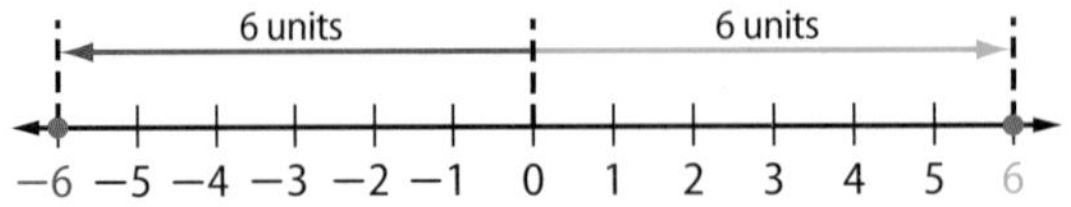

Symbols

$|6| = 6$ The absolute value of 6 is 6.

$|-6| = 6$ The absolute value of −6 is 6.

Notice on the number line that −6 and 6 are each 6 units from 0, even though they are on opposite sides of 0. The **absolute value** of a number is the distance the number is from zero on a number line. So, −6 and 6 have the same absolute value.

Example 4

Tools Tutor

Evaluate each expression.

a. $|-4|$

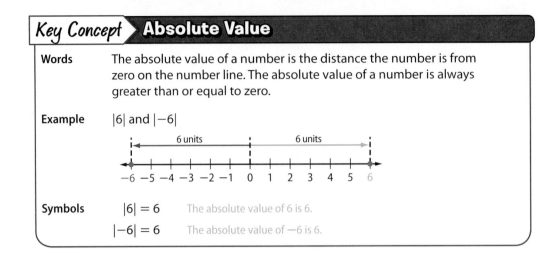

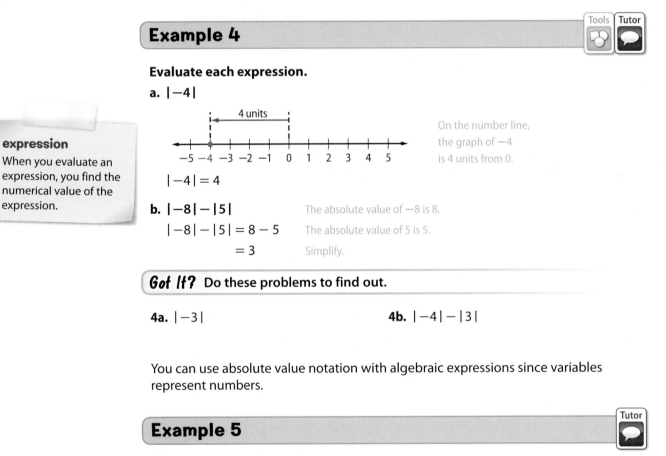

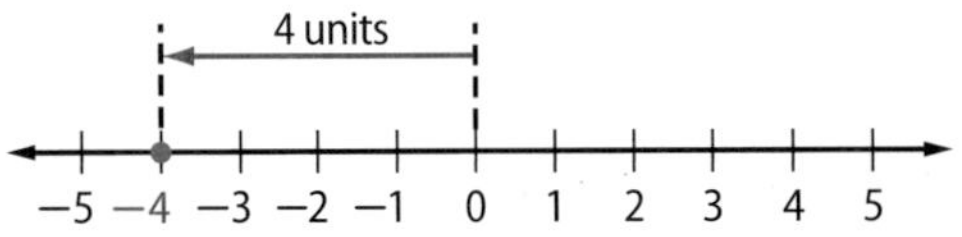

On the number line, the graph of −4 is 4 units from 0.

$|-4| = 4$

b. $|-8| - |5|$ The absolute value of −8 is 8.

$|-8| - |5| = 8 - 5$ The absolute value of 5 is 5.

$= 3$ Simplify.

expression When you evaluate an expression, you find the numerical value of the expression.

Got It? Do these problems to find out.

4a. $|-3|$

4b. $|-4| - |3|$

You can use absolute value notation with algebraic expressions since variables represent numbers.

Example 5

Tutor

Evaluate $6 + |x|$ if $x = -2$.

$6 + |x| = 6 + |-2|$ Replace x with −2.

$= 6 + 2$ The absolute value of −2 is 2.

$= 8$ Simplify.

Got It? Do these problems to find out.

5a. Evaluate $|y| + 8$ if $y = -7$.

5b. Evaluate $9 - 5|z|$ if $z = 3$.

Guided Practice

Check

Write an integer for each situation. Identify its opposite and describe its meaning. (Example 1)

1. a bank withdrawal of \$500

2. a gain of 4 pounds

Write two inequalities using the number pairs. Use the symbols < or >. (Example 2)

3. 2 and −5

4. −4 and −8

5. −1 and 1

Replace each ● with <, >, or = to make a true sentence. (Example 2)

6. −9 ● −16

7. −7 ● 7

8. −6 ● 0

9. Order the state temperatures from least to greatest. (Example 3)

State	AL	AK	CA	FL	HI	ME	NJ	OH	TX
Temperature	−27	−80	−45	−2	12	−48	−34	−39	−23

Evaluate each expression. (Example 4)

10. $|-12|$

11. $|-14| + |3|$

12. $|18| - |-5|$

Evaluate each expression if $x = 7$ and $y = -6$. (Example 5)

13. $15 - |y|$

14. $|y| + x$

15. $3|y|$

Independent Practice

Go online for Step-by-Step Solutions

eHelp

Write an integer for each situation. Identify its opposite and describe its meaning. (Example 1)

16. 5 strokes above par

17. 200 feet below sea level

18. an elevator descent of 18 floors

19. no gain on fourth down

Write two inequalities using the number pairs. Use the symbols < or >. (Example 2)

20. 5 and −11

21. −8 and 14

22. −6 and −1

23. 0 and −4

24. $|51|$ and $|50|$

25. $|-27|$ and $|-30|$

Replace each ● with <, >, or = to make a true sentence. (Example 2)

26. −11 ● −9

27. −14 ● −17

28. 15 ● −6

29. −2 ● 16

30. 21 ● 0

31. 0 ● −35

32. $|13|$ ● $|-13|$

33. $|-27|$ ● $|-27|$

34. In a recent year, Jimmy Johnson was the point leader in NASCAR's Chase to the Cup. Other drivers' standings are shown in the table. (Example 3)

a. Write an integer to describe each driver's standing with respect to the leader.

b. Order the integers from least to greatest.

Chase to the Cup

Driver	Number of Points Behind the Leader
K. Busch	40
K. Harvick	50
J. Gordon	20
T. Stewart	30

35 The top fourth-round scores of a recent PGA Championship were +4, −2, +6, +1, −4, −3, +5, −1, +2, and +3. Order the scores from least to greatest. (Example 3)

Evaluate each expression. (Example 4)

36. $|8|$

37. $|-17|$

38. $|-21|$

39. $-|15|$

40. $|0| + -|4|$

41. $-|-7| + |12|$

42. $|12| - |-2|$

43. $|-32| - |-6|$

44. $|18 - 4| - |5|$

Evaluate each expression if $x = -3$, $y = 4$, and $z = 2$. (Example 5)

45. $10 - |x|$

46. $2y - |x|$

47. $|z| + 19$

48. $3y + 3z + |x|$

49. $|4yz| - 3|x|$

50. $2(z + y) - |x|$

51. CCSS **Justify Conclusions** Movies are ranked based on ticket sales. The top movies for one week are listed in the table showing the change in position from the previous week. Which movie had the greatest absolute change in position? Explain.

Movie	A	B	C	D	E	F	G	H
Change in Position	−2	−7	+1	−3	+2	−8	−4	0

STEM The table at the right shows the freezing point of various elements.

Element	Freezing Point (°C)
chlorine	−101
helium	−272
krypton	−157
neon	−249
nitrogen	−201

52. Write two inequalities using the freezing point of neon and helium.

53. Order the temperatures from least to greatest.

54. Is the absolute value of the freezing point of chlorine greater than or less than the absolute value of the freezing point of nitrogen?

55. The average temperature of Saturn is −218°F while the average temperature of Jupiter is −162°F. Which planet has the lower average temperature? Explain.

Order the integers in each set from greatest to least.

56. $\{4, -2, -10, 3\}$

57. $\{-13, 5, 0, -5\}$

58. $\{7, -26, -15, 32, -19\}$

59. $\{-28, 62, -35, 20, -59\}$

60. $\{-42, 1, -6, 74, 0, -11\}$

61. $\{88, -72, -83, 232, -165, -94\}$

H.O.T. Problems Higher Order Thinking

62. CCSS **Model with Mathematics** Write a real-world situation in which you compare two negative integers.

CCSS **Persevere with Problems Determine whether each statement is *always*, *sometimes*, or *never* true. Explain your reasoning.**

63. $|x| = |-x|$

64. $|x| = -|x|$

65. $|-x| = -|x|$

66. CCSS **Reason Abstractly** Write a vocabulary term that completes the analogy: The symbol = is related to an equation in the same way as > is related to $a(n)$ __?__.

67. CCSS **Identify Structure** What is the least integer value of n such that $n > 0$?

68. **Building on the Essential Question** Order the integers −12, −5, −15, −10, −3 from least to greatest without using a number line. Explain your method.

Standardized Test Practice

69. Which of the following statements is false if $a = 3$ and $b = -3$?

A $|b| = a$ **C** $|b| = |a|$

B $|b| > 0$ **D** $|b| < 0$

70. If $|x| = 1$, what is the value of x?

F 1 **H** -1 and 0

G -1 **J** 1 and -1

71. **Short Response** Order $|-5|$, $-|9|$, -4, 0, $-|10|$, and 7 from least to greatest. Explain how you determined the order.

72. The table shows the number of points selected players have at the end of a game.

Player	Points
A	−10
B	−50
C	−5
D	0
E	−15

Which list shows the order of the players from greatest to least points?

A D, C, A, E, B **C** D, C, A, B, E

B B, E, A, C, D **D** B, C, A, D, E

CCSS Common Core Review

Name the property shown by each statement. 7.EE.1

73. $42 + 36 = 36 + 42$

74. $16 + 0 = 16$

75. $(19 \cdot 15) \cdot 2 = 19 \cdot (15 \cdot 2)$

76. $33 \cdot 0 = 0$

77. $(7 + 9) + 6 = 7 + (9 + 6)$

78. $25 \cdot 1 = 25$

Translate each phrase into an algebraic expression. 6.EE.2a

79. nine less than twice the number of boys

80. the distance Darren ran plus four more miles

81. three times the difference of the number of video games sold and two

82. CCSS **Multiple Representations** A roll of wrapping paper costs \$3. 7.EE.4

a. Symbols Write an equation that can be used to find the cost y of buying x number of rolls of wrapping paper.

b. Table Make a table to find the cost of 3, 4, 5, and 6 rolls.

c. Graph Graph the ordered pairs.

Find each quotient. 6.NS.2

83. $14\overline{)2730}$

84. $54\overline{)4736}$

85. $63\overline{)19{,}278}$

86. $45\overline{)20{,}250}$

87. $8\overline{)122{,}628}$

88. $103\overline{)21{,}129}$

Find the area of each figure. 6.G.1

89.

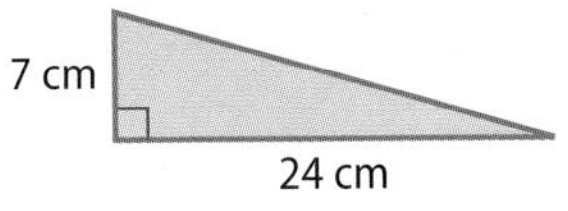

90.

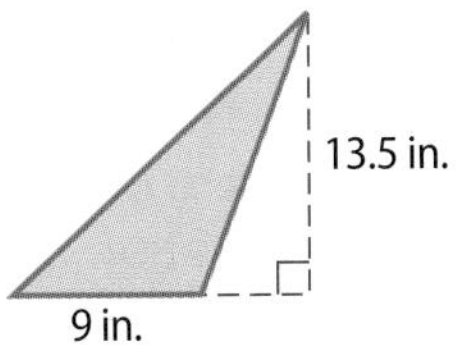

91.

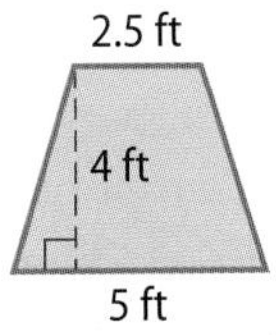

Inquiry Lab
Adding Integers

Inquiry WHAT is the rule for adding two integers?

CCSS Content Standards
7.NS.1, 7.NS.1a, 7.NS.1b

Mathematical Practices
1, 3

Entertainment On a quiz show, contestants earn points for answering a question correctly and lose points for answering a question incorrectly. You can use a positive integer to represent the points a contestant earns and a negative integer to represent the points a contestant loses. Suppose a contestant has -2 points and answers a question incorrectly for -4 points. What is the contestant's new score?

Investigation 1

Tools

Find the sum $-2 + (-4)$ using algebra tiles.

Step 1 Use algebra tiles and an integer mat to model operations with integers. Let [1] represent the integer 1, and let [−1] represent the integer -1.

Recall that addition means to *combine*. The expression $-2 + (-4)$ tells you to combine a set of 2 negative tiles with a set of 4 negative tiles.

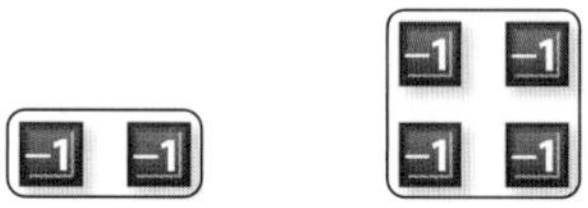

-2 $\qquad$ -4

Step 2 Combine the two sets of tiles on the mat.

Step 3 There are 6 negative tiles on the mat.

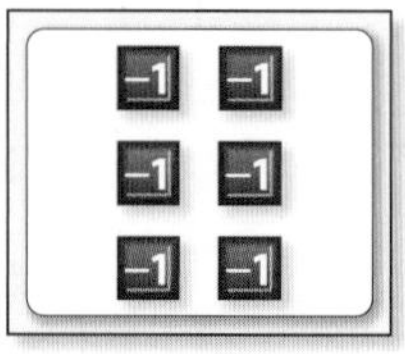

So, $-2 + (-4) = -6$. The contestant's new score is -6 points.

One positive tile paired with one negative tile is called a **zero pair**. You can add or remove zero pairs from a mat because adding or removing zero does not change the value of the tiles on the mat. You will use zero pairs in Investigation 2.

Investigation 2

Watch | Tools

Find the sum −3 + 2 using algebra tiles.

Step 1 The expression −3 + 2 tells you to combine a set of 3 negative tiles with a set of 2 positive tiles.

Step 2 Combine the two sets of tiles on the mat.

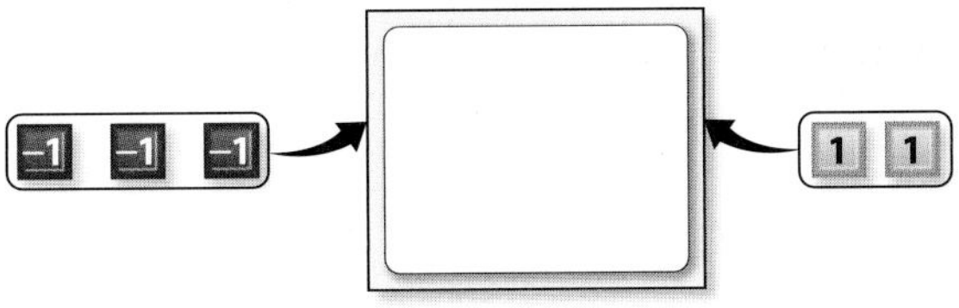

Step 3 Remove two zero pairs.

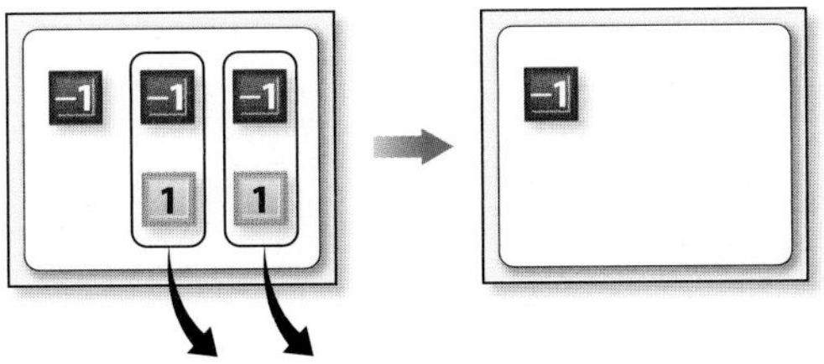

There is one negative tile remaining. So, −3 + 2 = −1.

Collaborate

Work with a partner.

1. Copy and complete the addition table below using algebra tiles. For example, to find the value in the shaded cell of the table, the addends are −4 and 3. The sum is −1. So, −4 + 3 = −1.

Addition Table

+	4	3	2	1	0	−1	−2	−3	−4
4	8	7	6	5	4	3	2	1	0
3	7	6	5	4	3	2	1	0	−1
2	6	5							
1	5	4							
0	4	3							
−1	3	2							
−2	2	1							
−3	1	0							
−4	0								

addends (top row); sums (body); addends (left column)

Analyze

CCSS Reason Inductively For Exercises 2–4, use the completed Addition Table.

2. What is true about the sum when both addends are positive?

3. What is true about the sum when both addends are negative?

4. Look at all the sums in the table that are zero. What is true about the addends that result in a sum of zero?

5. Suppose one addend is positive and one addend is negative. How can you determine whether the sum will be positive or negative?

Find each sum. Use algebra tiles if needed.

6. $6 + (-5)$ **7.** $-3 + (-3)$ **8.** $-4 + 6$ **9.** $-6 + (-3)$

10. $-5 + 3$ **11.** $6 + (-6)$ **12.** $-2 + (-5)$ **13.** $-1 + 7$

14. $-3 + 3$ **15.** $-4 + (-4)$ **16.** $8 + (-4)$ **17.** $-7 + (-6)$

For Exercises 18–23, tell what expression is modeled by the algebra tiles. Then find the sum.

18.

19.

20.
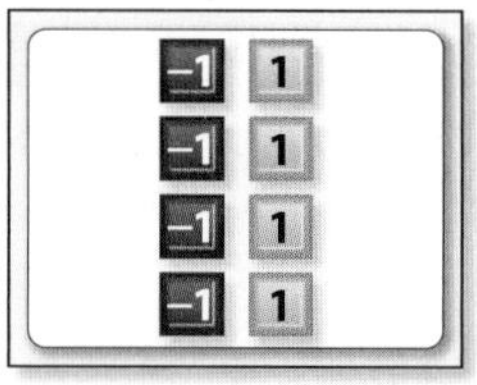

21.

22.

23.
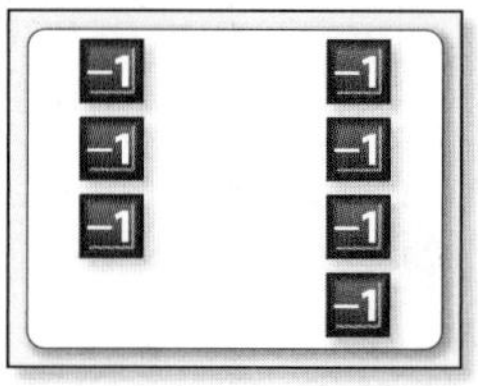

For Exercises 24–26, does it appear that the given property is true for addition of integers? If so, write two examples that illustrate the property. If not, give a counterexample.

24. Additive Identity Property

25. Commutative Property of Addition

26. Associative Property of Addition

27. CCSS **Justify Conclusions** On a quiz show, contestants earn points for answering questions correctly and lose points for answering questions incorrectly. On the first two questions, a contestant answers one question correctly and one question incorrectly. Is it possible for the contestant's score after the two questions to be 0? Explain your reasoning.

Reflect

28. Inquiry WHAT is the rule for adding two integers?

Lesson 2-2

Adding Integers

ISG Interactive Study Guide

See pages 31–32 for:
- Getting Started
- Real-World Link
- Notes

Essential Question

What happens when you add, subtract, multiply, and divide integers?

CCSS Common Core State Standards

Content Standards
7.NS.1, 7.NS.1a, 7.NS.1b, 7.NS.1d, 7.NS.3, 7.EE.3

Mathematical Practices
1, 2, 3, 4

Vocabulary

additive inverse

What You'll Learn

- Add two integers.
- Add more than two integers.

Real-World Link

Financial Literacy The money you earn can be represented by a positive integer while the money you spend or borrow can be represented by a negative integer. Every time you find the total amount of money you have after earning or spending it, you are adding integers.

Add Integers

The equation $-2 + (-3) = -5$ is an example of adding two integers with the same sign. Notice the sign of the sum is the same as the sign of the addends.

Key Concept: Add Integers with the Same Sign

Words To add integers with the same sign, add their absolute values. The sum is:
- positive if both integers are positive.
- negative if both integers are negative.

Examples $-2 + (-4) = -6$ $\quad$ $8 + 1 = 9$

Example 1

Tools Tutor

Find $-1 + (-5)$.

Use a number line.

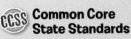

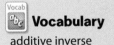

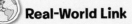

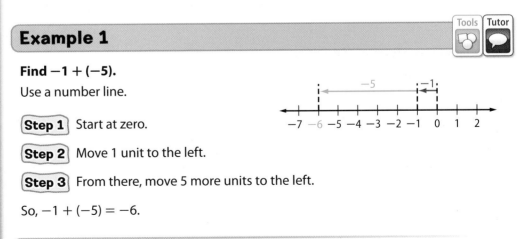

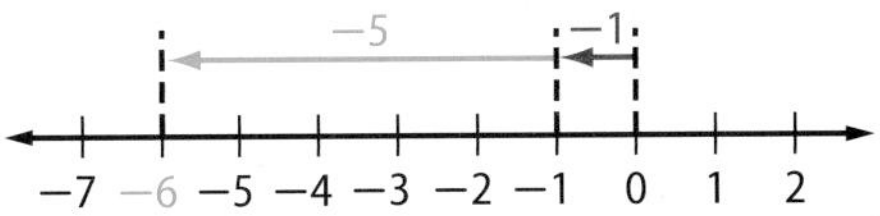

Step 1 Start at zero.

Step 2 Move 1 unit to the left.

Step 3 From there, move 5 more units to the left.

So, $-1 + (-5) = -6$.

Got It? Do these problems to find out.

Find each sum.

1a. $-3 + (-4)$ **1b.** $-6 + (-14)$ **1c.** $-7 + (-2)$

Example 2

Find $-3 + (-6)$.

$-3 + (-6) = -9$ Add $|-3|$ and $|-6|$. Both numbers are negative, so the sum is negative.

Got It? **Do these problems to find out.**

Find each sum.

2a. $-8 + (-2)$

2b. $-1 + (-12)$

Key Concept Add Integers with Different Signs

Words To add integers with different signs, subtract their absolute values. The sum is:

- positive if the positive integer's absolute value is greater.
- negative if the negative integer's absolute value is greater.

Examples $9 + (-4) = 5$ $-9 + 4 = -5$

Adding Integers on a Number Line

The sum of p and q is located a distance $|q|$ from p on a number line. The sum is left of p if q is negative and right of p if q is positive.

To add integers with different signs you can also use a number line. Start at zero. Move right to model a positive integer. Move left to model a negative integer.

Example 3

Find each sum.

a. $-3 + 6$

Use a number line.

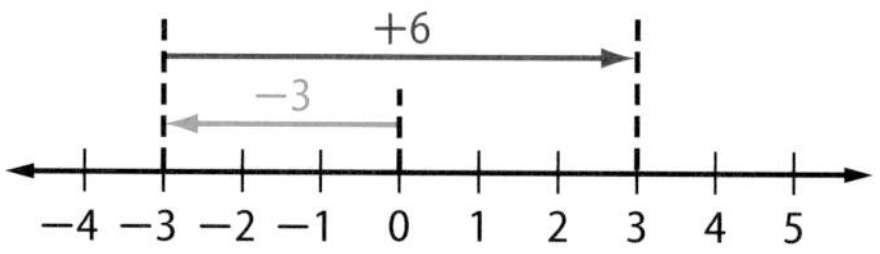

Step 1 Start at zero.

Step 2 Move 3 units to the left.

Step 3 From there, move 6 units to the right. The sum, 3, is 6 units to the right of -3.

So, $-3 + 6 = 3$.

Parentheses

Use parentheses around negative numbers that come after operation symbols. So, write $1 + (-4)$, *not* $1 + -4$.

b. $1 + (-4)$

Use a number line.

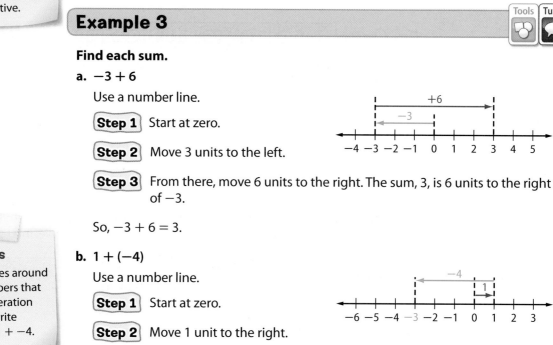

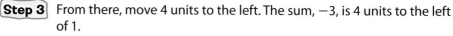

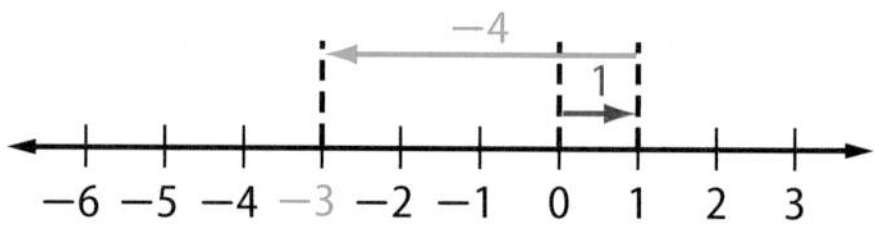

Step 1 Start at zero.

Step 2 Move 1 unit to the right.

Step 3 From there, move 4 units to the left. The sum, -3, is 4 units to the left of 1.

So, $1 + (-4) = -3$.

Got It? **Do these problems to find out.**

3a. $5 + (-2)$

3b. $4 + (-8)$

Example 4

Tutor

Find each sum.

a. $12 + (-6)$

$12 + (-6) = 6$

To find $12 + (-6)$, subtract $|-6|$ from $|12|$.
The sum is positive because $|12| > |-6|$.

b. $-7 + 5$

$-7 + 5 = -2$

To find $-7 + 5$, subtract $|5|$ from $|-7|$.
The sum is negative because $|-7| > |5|$.

Watch Out!

Check the sign of the sum. The sign should be the same as the addend with the greater absolute value. A positive sum does not need a sign.

Got It? Do these problems to find out.

4a. $-20 + 4$

4b. $16 + (-5)$

Real World Example 5

Tutor

A blue whale was at a depth of 275 feet below the surface of the water. After 10 minutes, it rose 194 feet. What is the current depth of the blue whale? Write an addition equation and then solve. Interpret the sum.

Words	Beginning depth plus increase after 10 minutes = current depth
Variable	Let d = the current depth.
Equation	$-275 + 194 = d$

Solve the equation.

$-275 + 194 = d$

$-81 = d$

Estimate $-275 + 200 = -75$.

To find the sum, subtract 194 from $|-275|$.

The sum is negative because $|-275| > |194|$.

The current depth is -81 feet or 81 feet below the surface.

Check for Reasonableness $-81 \approx -75$ ✔

Got It? Do this problem to find out.

5. A scuba diver is 120 feet below the water's surface. She then ascends 83 feet. What is her current depth? Write an addition equation and then solve. Interpret the sum.

Add More than Two Integers

The integers -4 and 4 are opposites. Opposites have the same absolute value but different signs. An integer and its opposite are also called **additive inverses**. The Additive Inverse Property is useful when adding integers.

Key Concept Additive Inverse Property

Words	The sum of any number and its additive inverse is zero.
Examples	$2 + (-2) = 0$
Symbols	$a + (-a) = 0$

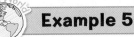

Example 6

Tutor

Find each sum.

a. $-6 + (-15) + 6$

$$-6 + (-15) + 6 = -6 + 6 + (-15) \quad \text{Commutative Property}$$
$$= 0 + (-15) \quad \text{Additive Inverse Property}$$
$$= -15 \quad \text{Identity Property of Addition}$$

b. $7 + (-1) + 26 + (-13)$

$$7 + (-1) + 26 + (-13) = 7 + 26 + (-1) + (-13) \quad \text{Commutative Property}$$
$$= (7 + 26) + [-1 + (-13)] \quad \text{Associative Property}$$
$$= 33 + (-14) \text{ or } 19 \quad \text{Simplify.}$$

Adding Mentally

- One way to add a group of integers mentally is to look for addends that are opposites.
- Another way is to group the positive addends together and the negative addends together. Then add.

Got It? **Do these problems to find out.**

6a. $4 + (-2) + (-7)$

6b. $-10 + 3 + (-7) + 12$

6c. $-9 + 12 + (-11)$

6d. $-8 + 4 + (-1) + 6$

Guided Practice

Check

Find each sum. (Examples 1–4)

1. $-5 + (-6)$

2. $14 + (-5)$

3. $-18 + 11$

4. $16 + (-13)$

CCSS **Reason Abstractly** **Write and solve an addition equation. Then interpret the sum.** (Example 5)

5. A football team gained 6 yards on a play. It then lost 10 yards on the next play. What was the total change of yardage for the team?

6. The temperature in Rockford, Illinois, was −3°F. The temperature rose 3 degrees. What is the current temperature?

Find each sum. (Example 6)

7. $-7 + 14 + 7$

8. $11 + (-2) + (-10)$

9. $-5 + 4 + (-5) + 3$

Independent Practice

Go online for Step-by-Step Solutions

eHelp

Find each sum. (Examples 1–4)

10. $-7 + (-3)$

11. $-6 + (-14)$

12. $17 + (-8)$

13. $21 + (-11)$

14. $-15 + 20$

15. $-16 + 18$

16. $-25 + (-4)$

17. $-23 + (-2)$

18. $8 + (-13)$

19. $12 + (-16)$

20. $-11 + (-5)$

21. $13 + (-2)$

CCSS **Reason Abstractly** **Write and solve an addition equation. Then interpret the sum.** (Example 5)

22. A contestant has −1500 points. He loses another 1250 points. What is his new score?

23. Ty had $450 in his savings account. He made a withdrawal of $160 to buy a bicycle. What is the balance in his account after the withdrawal?

Find each sum. (Example 6)

24. $10 + (-3) + 3$

25. $-7 + (-1) + 7$

26. $-15 + 8 + (-9)$

27. $-6 + (-2) + 14$

28. $5 + (-3) + 2 + (-3)$

29. $7 + (-6) + 4 + (-6)$

30. $-15 + 4 + (-5) + 10$

31. $8 + (-11) + (-19) + 11$

32. $13 + 20 + (-17) + (-13)$

33. **STEM** A shark rises 68 feet from the depth shown in the diagram. What is its current depth?

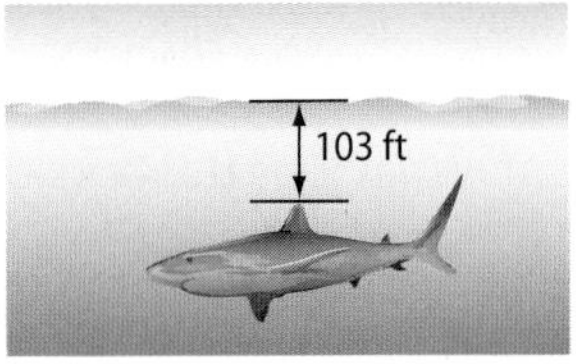

34. Sally begins hiking at an elevation of 324 feet. Then, she descends 201 feet and then climbs 55 feet higher. Finally, she descends 83 feet. Write an addition sentence to describe the situation. Then solve and interpret the sum.

35. The table below shows the change in music sales to the nearest percent after one year.

Style of Music	Percent of Music Sold	Percent Change After One Year
rock	33	+1
rap/hip hop	10	+2
pop	9	−2
country	14	−1

a. After one year, what is the percent of music sold for each of these music categories?

b. What was the total percent change in the sale of these types of music? Interpret the sum.

Find each sum.

36. $|-4 + 15|$

37. $|-2 + 11|$

38. $|17 - 25|$

H.O.T. Problems Higher Order Thinking

39. CCSS **Model with Mathematics** Write a real-world problem that can be solved by using the number line.

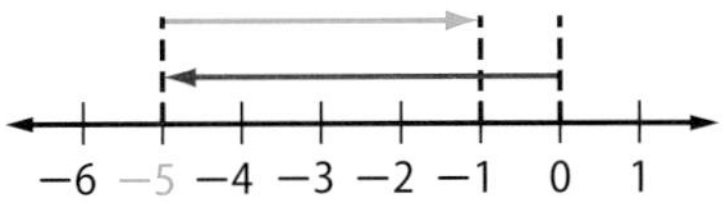

40. CCSS **Model with Mathematics** Describe two situations in which opposite quantities combine to make zero.

41. CCSS **Use a Counterexample** *True* or *false*? $-n$ always names a negative number. If *false*, give a counterexample.

CCSS **Persevere with Problems** **Name the property or properties illustrated by each equation.**

42. $a[b + (-b)] = [b + (-b)]a$

43. $a[b + (-b)] = 0$

44. **Building on the Essential Question** Explain how you know the sum of 5, 4, and −5 is positive without actually adding.

Standardized Test Practice

45. The outside temperature was 65°F. The table shows two temperature changes that occurred that day.

Change 1	+4°F
Change 2	−10°F

What was the outside temperature after the second change?

A 14°F **C** 69°F

B 59°F **D** 79°F

46. **Short Response** On a drive, a football team gained 3 yards, lost 11 yards, and gained 15 yards. What was their total yardage gained for that drive?

47. A bird is flying at an altitude of 100 feet. It descends 30 feet, and then ascends 50 feet. Which of the following expressions best represents this situation?

F $100 + 30 + (-50)$

G $100 + (-30) + (-50)$

H $100 + (-30) + 50$

J $100 + 30 + 50$

48. A team started with 4 points. They gained 3 points, lost 6 points, and gained 2 points. How many points does the team have now?

A 2 **C** 4

B 3 **D** 5

CCSS Common Core Review

49. The record low temperature for Wisconsin is 54°F below zero. Write an integer to represent this situation. Then write its opposite and describe its meaning. 6.NS.5

CCSS Model with Mathematics Refer to the coordinate plane shown. Write the ordered pair that names each point. 5.G.2

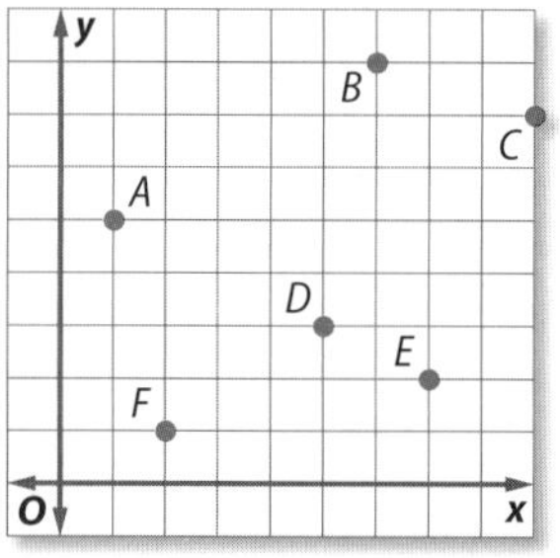

50. point *A* **51.** point *B*

52. point *C* **53.** point *D*

54. point *E* **55.** point *F*

Name the property shown by each statement. 6.EE.3

56. $12 \cdot 7 = 7 \cdot 12$ **57.** $1 \cdot 4 \cdot 0 = 0$ **58.** $4xy \cdot 1 = 4xy$

59. A soccer league ranks each team in their league using points. A team gets three points for a win, one point for a tie, and zero points for a loss. Write an expression that can be used to find the total number of points a team receives. 6.EE.2

Evaluate each expression. 7.NS.3

60. $9 \div 3 \cdot 6$ **61.** $25 - 12 \div 4$

62. $(7 \cdot 5) + (8 \cdot 2)$ **63.** $(34 \div 2) - (8 \cdot 2)$

Evaluate each expression if $a = 5$, $b = 12$, and $c = 7$. 6.EE.2

64. $2c - 11$ **65.** $4a - 16$ **66.** $bc - ac$ **67.** $(3b - c) - 4c$

68. $6a + 3b$ **69.** $5c - a + b$ **70.** $(9c - ab) + a$ **71.** $7b + (ab - c)$

Inquiry Lab

Subtracting Integers

Inquiry HOW is subtraction of integers related to addition of integers?

CCSS Content Standards
7.NS.1, 7.NS.1c
Mathematical Practices
1, 3, 5, 7

Weather On a winter day, the temperature at noon is 5°F. The temperature at 6:00 P.M. is −2°F. What is the difference in the temperatures?

This problem requires you to subtract integers to find the difference 5 − (−2). You can use algebra tiles to model subtraction. Recall that one meaning of subtraction is to *take away*. For example, to model 7 − 4, place 7 positive tiles on the mat. Then take away 4 positive tiles. The model shows that 7 − 4 = 3.

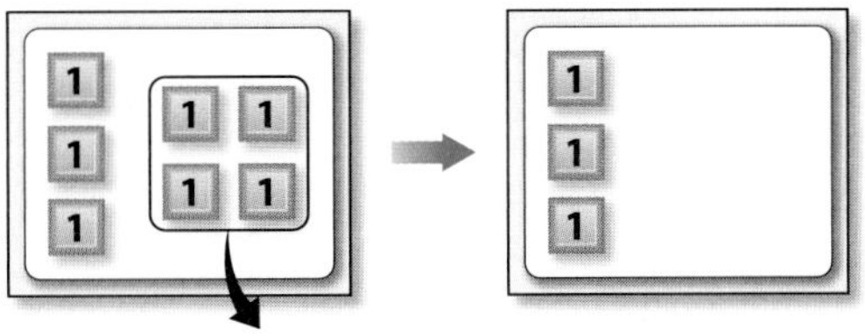

7 − 4 = 3

Investigation 1

Tools

Find the difference 5 − (−2) using algebra tiles.

Step 1 Place 5 positive tiles on the mat.

Step 2 There are no negative tiles to remove. Add two zero pairs to the mat.

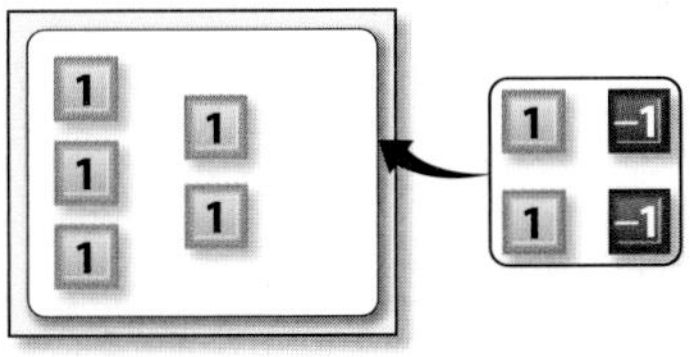

Step 3 Remove the 2 negative tiles.

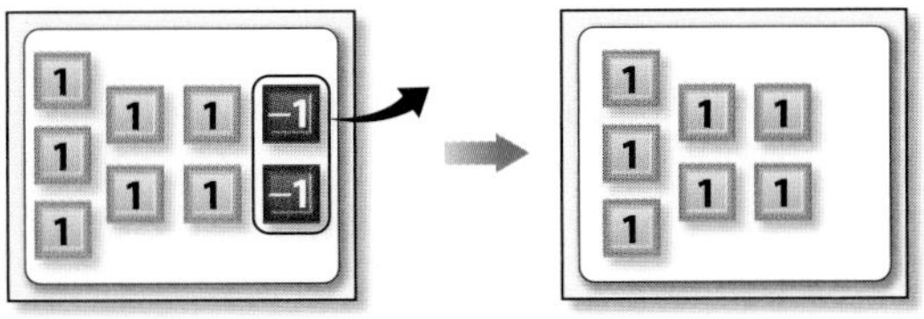

There are 7 positive tiles remaining. So, 5 − (−2) = 7.

The difference in the temperatures is 7°F.

Investigation 2

Find the difference −6 − 3 using algebra tiles.

Step 1 Place 6 negative tiles on the mat.

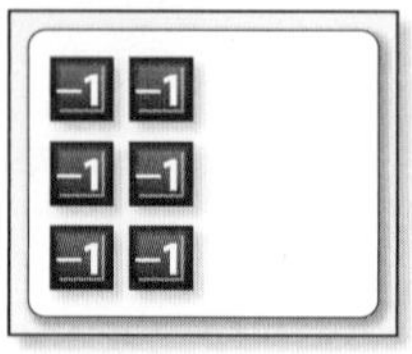

Step 2 There are no positive tiles to remove. Add three zero pairs to the mat.

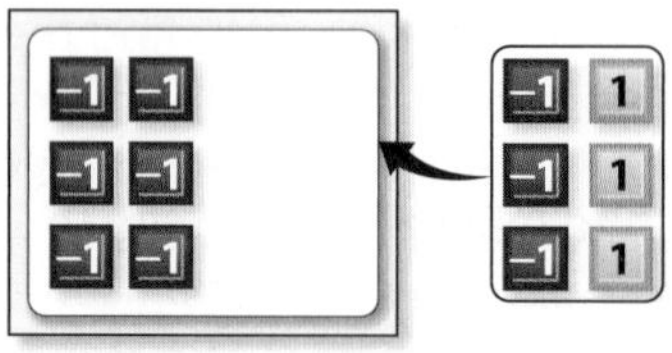

Step 3 Remove the 3 positive tiles.

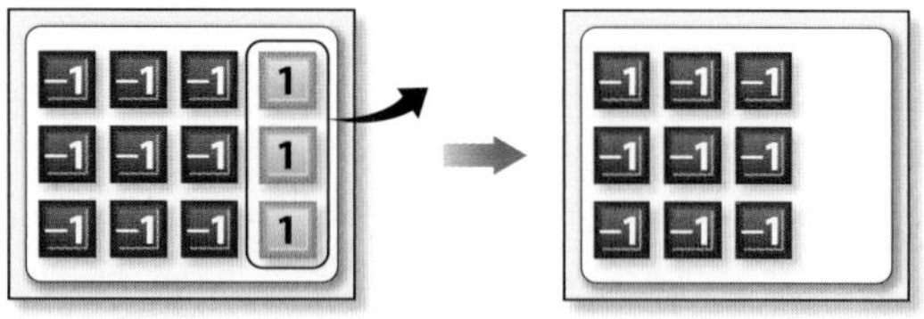

There are 9 negative tiles remaining. So, $-6 - 3 = -9$.

Collaborate

Work with a partner.

1. Copy and complete the table by finding the value of each difference or sum. Use algebra tiles to help you find the differences.

Difference	Value	Sum	Value
$2 - 5$		$2 + (-5)$	
$-4 - (-3)$		$-4 + 3$	
$-5 - 1$		$-5 + (-1)$	
$7 - (-2)$		$7 + 2$	

Analyze

2. CCSS **Identify Structure** In each row of the table, how are the difference and the sum related to each other?
3. CCSS **Reason Inductively** In each row of the table, what is true about the value of the difference and the value of the sum?
4. CCSS **Justify Conclusions** What sum do you think has the same value as $-13 - (-7)$? Explain your reasoning.

Reflect

5. Inquiry HOW is subtraction of integers related to addition of integers?

Lesson 2-3

Subtracting Integers

Interactive Study Guide

See pages 33–34 for:
- Getting Started
- Real-World Link
- Notes

Essential Question

What happens when you add, subtract, multiply, and divide integers?

Common Core State Standards

Content Standards
7.NS.1, 7.NS.1c, 7.NS.1d, 7.NS.3, 7.EE.3

Mathematical Practices
1, 3, 4, 7

Vocabulary

inductive reasoning

What You'll Learn

- Subtract integers.
- Find distance on the number line.

Real-World Link

Ants Ants are social insects that live in colonies. Ant colonies can be found both above and below ground. Most colonies build their nest below ground where they can dig to 20 feet below the surface. Other types of ants build their colonies above ground. The nests above ground can sometimes reach more than 6 feet tall!

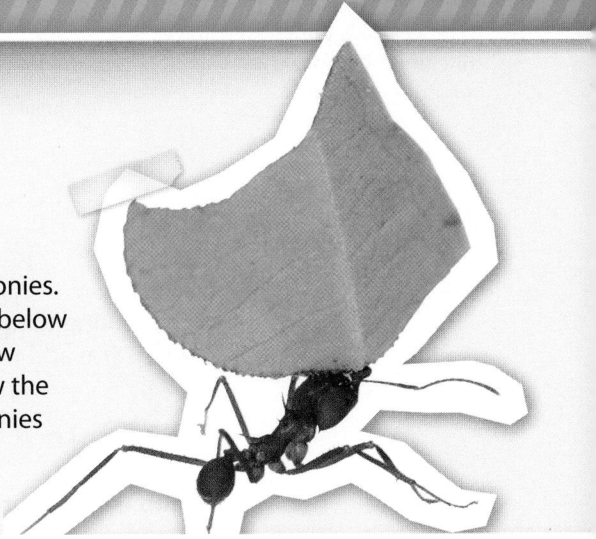

Key Concept Subtract Integers

Watch

Words	To subtract an integer, add its additive inverse.
Examples	$2 - 7 = 2 + (-7)$
Symbols	$a - b = a + (-b)$

The number line below shows $4 - 9 = -5$.

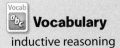

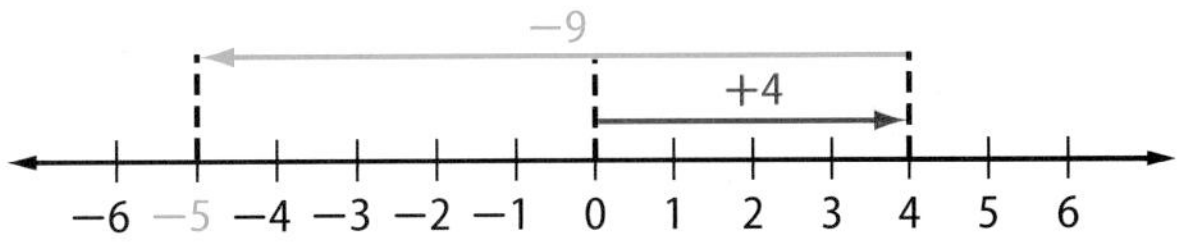

When you subtract 9 on the number line, the result is the same as adding −9.

9 and −9 are additive inverses

$4 - \mathbf{9} = -5$

$4 + (\mathbf{-9}) = -5$

same result

Example 1

Tools Tutor

Find each difference.

a. $6 - 15$

$6 - \mathbf{15} = 6 + (\mathbf{-15})$ To subtract 15 add −15.

$= -9$ Simplify.

b. $-7 - 8$

$-7 - \mathbf{8} = -7 + (\mathbf{-8})$ To subtract 8 add −8.

$= -15$ Simplify.

Got It? **Do these problems to find out.**

1a. $4 - 15$

1b. $-3 - 12$

Conjecture
A conjecture is the conclusion based on inductive reasoning.

Inductive reasoning is the process of reaching a conclusion based on a pattern of examples. In Example 1, you subtracted a positive integer by adding its additive inverse. Use inductive reasoning to see if the method also applies to subtracting a negative integer.

Adding the Additive Inverse			Subtracting an Integer		
Input	Rule: $4 + (-x)$	Output	Input	Rule: $4 - x$	Output
2	$4 + (-2)$	2	2	$4 - 2$	2
1	$4 + (-1)$	3	1	$4 - 1$	3
0	$4 + 0$	4	0	$4 - 0$	4
−1	$4 + 1$	5	−1	$4 - (-1)$	■

Continuing the pattern in the first column, $4 - (-1) = 5$. The result is the same as when you add the additive inverse.

Example 2

Tutor

Find each difference.

a. $9 - (-2)$

$9 - (-2) = 9 + 2$ To subtract −2, add 2
$= 11$

b. $3 - (-5)$

$3 - (-5) = 3 + 5$ To subtract −5, add 5
$= 8$

Got It? **Do these problems to find out.**

2a. $18 - (-2)$

2b. $-5 - (-11)$

Find Distance on a Number Line

You can find the distance between two integers by using a number line. You can also determine the distance by finding the absolute value of the difference of the numbers. The number line below shows that the distance between −4 and 5 is 9 units.

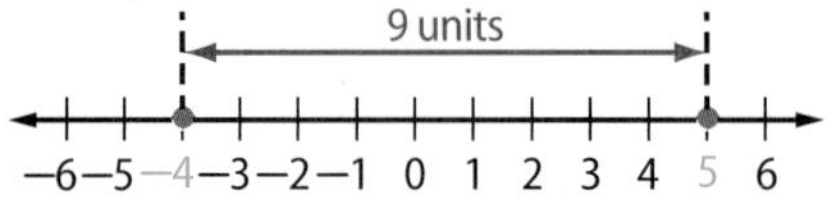

$|-4 - 5| = |-9|$ or 9

Example 3

Tutor

Find the distance between 3 and −3 on a number line.

$|3 - (-3)| = |6|$ Find the absolute value of the difference of 3 and −3.
$= 6$ Simplify.

The distance between the integers is 6 units.

Check using the number line.

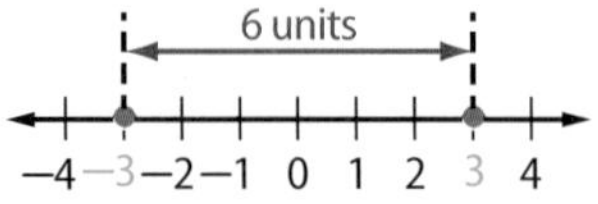

Got It? **Do these problems to find out.**

Find the distance between each pair of integers.

3a. −30 and −22

3b. −6 and 5

Example 4

Tutor

The top of an iceberg is 15 feet above sea level. The bottom of the iceberg is 105 feet below sea level. What is the distance from the top to the bottom of the iceberg?

Write an expression to find the distance, and then evaluate.

Words	15 feet above sea level minus 105 feet below sea level
Expression	$15 - (-105)$

$|15 - (-105)| = |15 + 105|$ Find the absolute value of the difference.

$= |120|$ Add.

$= 120$ Simplify.

So, the distance from the top to the bottom of the iceberg is 120 feet.

Got It? **Do this problem to find out.**

4. A gopher begins digging at 7 inches below ground level and stops digging at 16 inches below ground level. How far did the gopher dig?

Guided Practice

Check

Find each difference. (Examples 1 and 2)

1. $3 - 5$
2. $-10 - 14$
3. $17 - (-14)$
4. $-7 - (-11)$

Find the distance between each pair of integers. (Example 3)

5. -1 and -7
6. -8 and 8
7. -4 and 6

8. Moira dug a hole 2 feet deep to transplant a shrub. The shrub stands 3 feet above the ground. Write an expression to find the distance from the bottom of the hole to the top of the shrub. What is the distance? (Example 4)

Independent Practice

Go online for Step-by-Step Solutions

eHelp

Find each difference. (Examples 1 and 2)

9. $6 - 7$
10. $4 - 8$
11. $-12 - (-7)$
12. $-15 - (-6)$
13. $11 - (-14)$
14. $9 - (-8)$
15. $-5 - 2$
16. $-9 - 3$
17. $5 - (-10)$
18. $1 - (-18)$
19. $-15 - (-14)$
20. $-12 - (-11)$
21. $-20 - (-30)$
22. $-38 - (-40)$
23. $-32 - 28$
24. $-47 - 34$

Find the distance between each pair of integers. (Example 3)

25. -18 and -12
26. -3 and -11
27. -15 and 7
28. -6 and 12
29. 5 and -15
30. 12 and -7

31. At the end of the first round of a game show, Jillian had a score of 40 points and Marty had a score of −50 points. Find the difference between their two scores. (Example 4)

32. The lowest elevation in Louisiana is 8 feet below sea level. The highest elevation is 535 feet above sea level. What is the difference in elevation between the highest and lowest points? (Example 4)

33. CCSS **Model with Mathematics** A jellyfish is swimming at 4 meters below sea level. It swims to 1 meter below sea level. Draw and label a number line to find the distance the jellyfish swam. How far did the jellyfish swim?

Find each difference.

34. $125 - (-114)$

35. $-320 - (-106)$

36. $-2200 - (-3500)$

37 **Financial Literacy** The daily closing prices for a company's stock are shown.

Date	May 3	May 4	May 5	May 6	May 7
Closing Price	\$33.30	\$30.59	\$31.04	\$31.97	\$30.15
Change	—	■	■	■	■

a. Find the change in the closing price since the previous day.

b. What is the difference between the highest and lowest change values?

38. CCSS **Multiple Representations** In this problem, you will apply subtraction of integers to a real-world situation. An underwater video camera is 7 feet below the surface. It will be lowered an additional f feet.

a. **Symbols** Write an expression to show how many total feet below the surface the camera will be after it is lowered.

b. **Table** Make a table to show the depth of the camera if it is lowered 5, 8, 10, or 12 feet.

H.O.T. Problems Higher Order Thinking

39. CCSS **Identify Structure** Write a subtraction expression with a positive integer and a negative integer whose difference is positive. Then find the difference.

40. CCSS **Find the Error** Rick is finding $-6 - (-2)$. Find his mistake and correct it.

$$-6 - (-2) = 6 - 2$$
$$= 4$$

41. CCSS **Use a Counterexample** *True* or *false?* A subtraction expression with a positive integer and a negative will have a difference of zero. If *false*, give a counterexample.

42. **Building on the Essential Question** Write an expression involving the subtraction of a negative integer. Then write an equivalent addition expression. Explain why the result is the same.

Standardized Test Practice

43. The melting point of mercury is −39°C. The freezing point of alcohol is −114°C. How much warmer is the melting point of mercury than the freezing point of alcohol?

A −153°C **C** 75°C
B −75°C **D** 153°C

44. Which expression is modeled below?

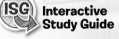

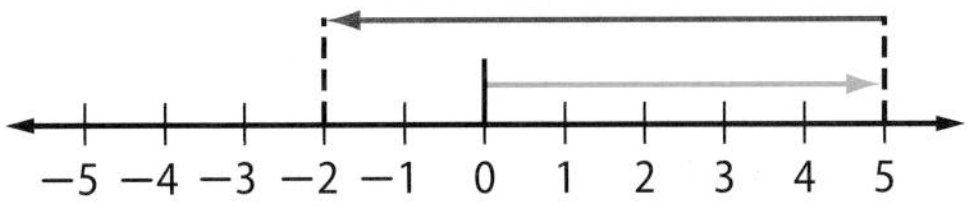

F 5 − 7 **H** 0 − 2
G 5 − 2 **J** 5 + 2

45. **Short Response** The crest of a mountain is 5740 feet above sea level. The base of the mountain is 25 feet below sea level. What is the difference between the crest and base of the mountain?

46. Which statement about subtracting integers is always true?

A positive − negative = positive
B negative − positive = positive
C negative − negative = positive
D positive − positive = positive

Common Core Review

47. A team gained 4 yards on one play. On the next play, they lost 5 yards. Write an addition equation and solve to find the change in yardage. Interpret the sum. 7.NS.1

Replace each ● with <, >, or = to make a true sentence. 6.NS.7

48. −18 ● −8 **49.** 0 ● −3 **50.** 9 ● −9

51. **Multiple Representations** It costs $6 to buy a student ticket to the movies. 7.EE.4

a. Symbols Write an expression that can be used to find the cost of any number of student tickets.

b. Table Make a table to find the cost of 2, 4, 5, and 7 tickets.

c. Graph Graph the ordered pairs.

Translate each phrase into an algebraic expression. 6.EE.2a

52. eight more than the amount Kira saved

53. five runs fewer than the Pirates scored

54. the quotient of a number and four, minus five

55. seven increased by the quotient of a number and eight

Simplify each expression. 7.EE.1

56. $(n + 4) + 7$ **57.** $24 + (s + 15)$ **58.** $(8 + r) + 12$

59. $(7 \cdot p) \cdot 8$ **60.** $(g \cdot 4) \cdot 9$ **61.** $5(12b)$

Convert each measurement to the given unit. 5.MD.1

62. 3 meters to centimeters **63.** 8 feet to inches

64. 100 grams to kilograms **65.** 12 quarts to gallons

ISG Interactive Study Guide
See page 35 for:
• Mid-Chapter Check

21ST CENTURY CAREER in Astronomy

Space Weather Forecaster

Did you know that space weather, or the conditions on the sun and in space, can directly affect communication systems and power grids here on Earth? If you enjoy learning about the mysteries of space, then you should consider a career involving space weather. A space weather forecaster uses spacecraft, telescopes, radar, and supercomputers to monitor the sun, solar winds, and the space environment in order to forecast the weather in space.

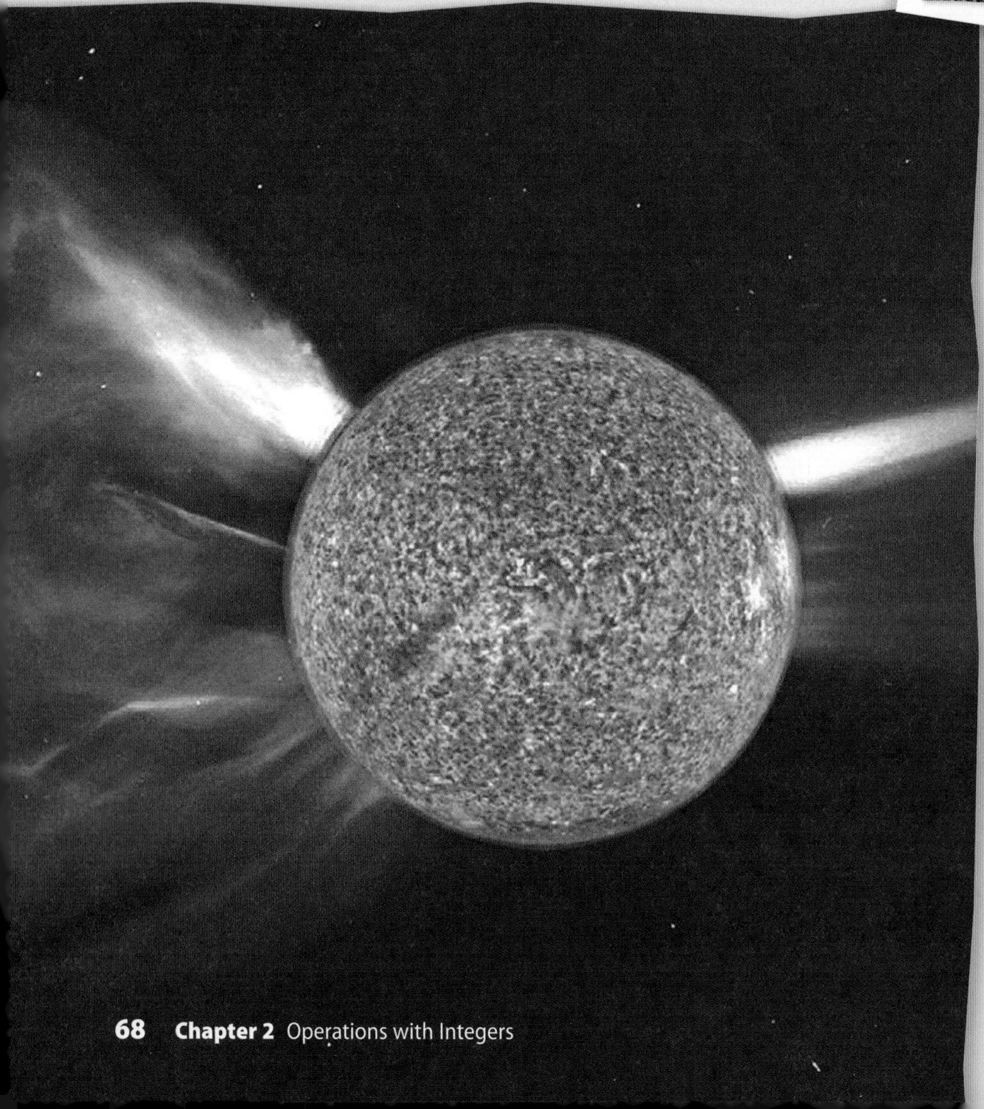

College & Career READINESS

Explore college and careers at ccr.mcgraw-hill.com

Is This the Career for You?

Are you interested in a career as a space weather forecaster? Take some of the following courses in high school.

- Astronomy
- Calculus
- Chemistry
- Earth Science
- Physics

Find out how math relates to a career in Astronomy.

Interactive Study Guide
See page 36 for:
- Problem Solving
- Career Project

Inquiry Lab
Multiplying Integers

Inquiry **HOW do you determine the sign of the product when you multiply two integers?**

CCSS Content Standards
7.NS.2, 7.NS.2a, 7.NS.2c

Mathematical Practices
1, 3, 4

Sports Drink At a volleyball tournament, the amount of sports drink in a team's cooler decreases at a rate of 3 quarts per game. What integer represents the total change in the number of quarts of sports drink in the cooler after two games?

Investigation 1

Tools

Step 1 Write an expression to represent the total change. The integer −3 represents a decrease of 3 quarts per game. After two games, the total change is 2 × (−3).

Step 2 Model 2 × (−3) by placing 2 sets of 3 negative tiles on the mat, as shown. There are 6 negative tiles.

So, after two games, the total change in the amount of sports drink in the cooler is −6 quarts.

2 × (−3) = −6

Investigation 2

Watch Tools

Find the product −2 × (−3) using algebra tiles.

Step 1 The expression −2 × (−3) means to *remove* 2 sets of 3 negative tiles. Place 2 × 3, or 6 zero pairs, on the mat.

Step 2 Remove 2 sets of 3 negative tiles from the mat. There are 6 positive tiles remaining.

So, −2 × (−3) = 6.

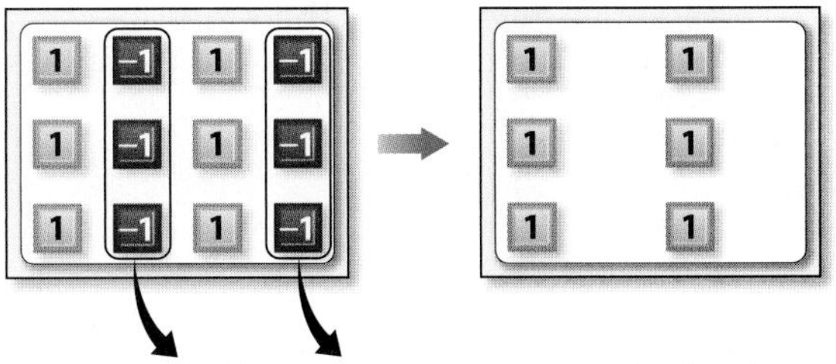

−2 × (−3)

−2 × (−3) = 6

Collaborate

Work with a partner. Use algebra tiles if necessary.

1. 3 × (−4)
2. 2 × (−5)
3. 4 × (−1)
4. −2 × (−4)
5. −5 × (−2)
6. −3 × (−3)
7. −5 × 2
8. −3 × 2
9. −1 × 4
10. 6 × 2
11. −6 × (−2)
12. 2 × (−1)

Analyze

13. CCSS **Reason Inductively** Look at your results from Exercises 1–12. What is true about the factors when the product is positive?

14. CCSS **Reason Inductively** Look at your results from Exercises 1–12. What is true about the factors when the product is negative?

15. Keri has a gift card for an online music store. Each time she buys an album, $10 is deducted from the value of the gift card. You can represent the change in the value of the gift card by the integer -10. Suppose Keri buys 4 albums. Write and evaluate a product of integers to find the total amount deducted from the value of the gift card.

For Problems 16–18, tell what product could be modeled by the algebra tiles and find the product.

16.

17.

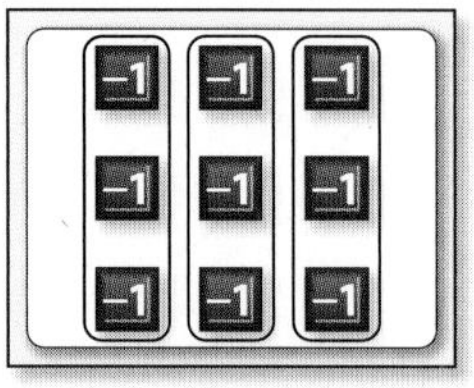

18.

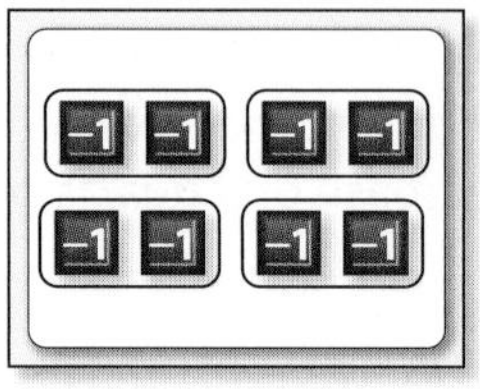

CCSS **Justify Conclusions** **Properties can be used to justify each statement you make when verifying, or proving, another statement. For Exercises 19 and 20, copy and complete the proof by writing the correct property from the list at the right.**

Mathematical Properties

- Additive Inverse Property
- Distributive Property
- Multiplicative Property of Zero
- Multiplicative Identity Property

19. Show that $2(-1) = -2$.

Statements	Properties
$0 = 2(0)$	
$0 = 2[1 + (-1)]$	
$0 = 2(1) + 2(-1)$	
$0 = 2 + 2(-1)$	

Conclusion: In the last statement, $0 = 2 + 2(-1)$. In order for this to be true, $2(-1)$ must equal the additive inverse of 2. Therefore, $2(-1) = -2$.

20. Show that $(-2)(-1) = 2$.

Statements	Properties
$0 = -2(0)$	
$0 = -2[1 + (-1)]$	
$0 = -2(1) + (-2)(-1)$	
$0 = -2 + (-2)(-1)$	

21. Write a conclusion for Exercise 20.

Reflect

22. Inquiry HOW do you determine the sign of the product when you multiply two integers?

Lesson 2-4

Multiplying Integers

ISG Interactive Study Guide

See pages 37–38 for:
- Getting Started
- Real-World Link
- Notes

Essential Question

What happens when you add, subtract, multiply, and divide integers?

CCSS Common Core State Standards

Content Standards
7.NS.2, 7.NS.2a, 7.NS.2c, 7.NS.3, 7.EE.3

Mathematical Practices
1, 2, 3, 4, 7

What You'll Learn

- Multiply integers.
- Simplify algebraic expressions.

Real-World Link

Ballooning Hot air balloon races grab the attention of people from miles around. During a balloon race, pilots must fly over a target on the ground and drop a marker as close to the target as possible. During a recent race, the winning balloon descended at a rate of about 8 meters per second.

Multiply Integers

Multiplication is repeated addition. So, $3(-40)$ means that -40 is used as an addend 3 times.

$$3(-40) = (-40) + (-40) + (-40)$$
$$= -120$$

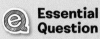

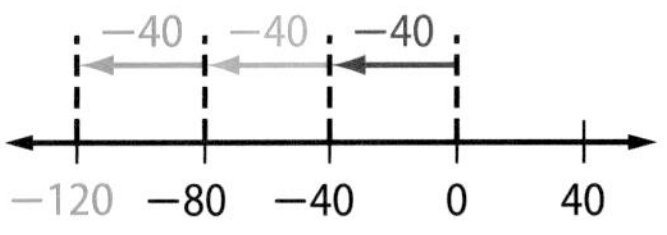

By the Commutative Property of Multiplication, $3(-40) = -40(3)$. This and other similar examples demonstrate the rule for multiplying integers with different signs.

Key Concept: Multiply Two Integers with Different Signs

Words The product of two integers with different signs is negative.

Examples $2(-6) = -12$ $-2(6) = -12$

Example 1

Tutor

Find each product.

a. $-3 \cdot 12$

$-3(12) = -36$ The factors have different signs. The product is negative.

b. $4(-7)$

$4(-7) = -28$ The factors have different signs. The product is negative.

Got It? Do these problems to find out.

1a. $7(-8)$

1b. $-6 \cdot 12$

Key Concept: Multiply Two Integers with the Same Sign

Words The product of two integers with the same sign is positive.

Examples $4 \cdot 6 = 24$ $-4(-6) = 24$

The product of two positive integers is positive. What is the sign of the product of two negative integers? Look at the pattern below.

Factor	Rule: Times −5	Product
2	−5(2)	−10
1	−5(1)	−5
0	−5(0)	0
−1	−5(−1)	5
−2	−5(−2)	10

One positive and one negative factor: Negative product

Two negative factors: Positive product

+5 +5 +5 +5

Each product is 5 more than the previous product.

Example 2

Tutor

Find each product.

a. −5(−7)

$-5(-7) = 35$ The product is positive.

b. −8(−14)

$-8(-14) = 112$ The product is positive.

Got It? **Do these problems to find out.**

2a. $-5(-11)$

2b. $-13(-4)$

Real World Example 3

Tutor

An airplane descends at a rate of 175 feet per minute. What is the airplane's change in altitude after 5 minutes? Write a multiplication expression. Then find and interpret the product.

Descends
The word *descends* means downward, so the rate per minute is represented by a negative integer.

Words	5 minutes times descent rate of 175 feet per minute
Expression	$5 \times (-175)$

$5(-175) = -875$ The product is negative.

So, the airplane descends 875 feet after 5 minutes.

Got It? **Do this problem to find out.**

3. A scuba diver descends from the surface at a rate of 7 feet per minute. What was the scuba diver's depth after 15 minutes? Write a multiplication expression. Then find and interpret the product.

Use the Commutative and Associative Properties of Multiplication to multiply more than two integers.

Example 4

Tutor

Find $-4(12)(-5)$.

Mental Math
Look for products that are multiples of ten to make the multiplication simpler.

Method 1 **Use the Associative Property**

$-4(12)(-5) = [-4(12)](-5)$	Associative Property
$= -48(-5)$	$-4(12) = -48$
$= 240$	$-48(-5) = 240$

Method 2 **Use the Commutative Property**

$-4(12)(-5) = -4(-5)(12)$	Commutative Property
$= 20(12)$	$-4(-5) = 20$
$= 240$	$20(12) = 240$

Got It? **Do these problems to find out.**

4a. $-7(9)(-6)$

4b. $-3(-4)(-5)$

Algebraic Expressions

You can use the rules for multiplying integers to simplify and evaluate algebraic expressions.

Example 5

Tutor

Simplify $-7a(4b)$.

$-7a(4b) = (-7)(a)(4)(b)$	$-7a = (-7)(a)$, $4b = (4)(b)$
$= (-7 \cdot 4)(a \cdot b)$	Commutative and Associative Properties of Multiplication
$= -28ab$	$-7 \cdot 4 = -28$, $a \cdot b = ab$

Got It? **Do these problems to find out.**

5a. $-3(6y)$

5b. $-9x(3y)$

Example 6

Tutor

Evaluate $3xy$ if $x = 2$ and $y = -6$.

$3xy = 3(2)(-6)$	Replace x with 2 and y with -6.
$= [3(2)](-6)$	Associative Property of Multiplication
$= 6(-6)$	The product of 3 and 2 is positive.
$= -36$	The product of 6 and -6 is negative.

Got It? **Do these problems to find out.**

6a. Evaluate $2rs$ if $r = 5$ and $s = -10$.

6b. Evaluate $4ab$ if $a = -8$ and $b = -4$.

Guided Practice

Check

Find each product. (Examples 1, 2, and 4)

1. $-6 \cdot 7$ **2.** $-5(-8)$ **3.** $8(-3)(-5)$ **4.** $-2(-9)(-5)$

5. **Financial Literacy** Mr. Heppner bought lunch with his debit card every day for 5 days. Each day he spent \$8. If these were his only transactions, what was the change in his account balance? Write a multiplication expression. Then find and interpret the product. (Example 3)

Simplify each expression. (Example 5)

6. $-2 \cdot 9m$ **7.** $-3a(7b)$ **8.** $-6e(-4f)$

Evaluate each expression. (Example 6)

9. $8j$, if $j = -11$ **10.** $-9cd$, if $c = -3$ and $d = -7$

Independent Practice

eHelp Go online for Step-by-Step Solutions

Find each product. (Examples 1, 2, and 4)

11 $3(-9)$ **12.** $8(-9)$ **13.** $25 \cdot 3$ **14.** $-4(-8)$

15. $-7(-7)$ **16.** $2(-11)(5)$ **17.** $-8(-7)(-6)$ **18.** $-8(-20)(5)$

19. **STEM** The temperature dropped 2°F every hour for the last 6 hours. What is the total change in temperature? Write a multiplication expression. Then find and interpret the product. (Example 3)

20. An elevator takes passengers from the ground floor down to an underground parking garage. Where will the elevator be in relation to the ground floor after 5 seconds if it travels at a rate of 3 feet per second? Write a multiplication expression. Then find and interpret the product. (Example 3)

Simplify each expression. (Example 5)

21. $5(-6m)$ **22.** $-5 \cdot 10s$ **23.** $-9m(-9n)$ **24.** $11a(7c)$

25. $4a(b)(-9)$ **26.** $-12(-j)(-3k)$ **27.** $3e(-2f)(9g)$ **28.** $3r(7s)(5t)$

Evaluate each expression. (Example 6)

29. $10n$, if $n = -10$ **30.** $8m$, if $m = -5$

31. $4xy$ if $x = -6$ and $y = 3$ **32.** $-15st$, if $s = 4$ and $t = -9$

33. CCSS **Model with Mathematics** Diego borrowed \$780 from his father to buy a new saddle for his horse. He earns \$65 each weekend cleaning stables and uses his earnings to pay off his debt. Write an expression Diego can use to determine how much he owes each week. Then make a table to show how much he owes after 2, 4, 6, and 8 weeks. Based on the pattern in the table, predict how long it will take Diego to repay his father.

34. The table shows the number of Calories burned per minute for a 120-pound person during different activities. What is the change in the number of Calories in a 120-pound person's body if he runs for 20 minutes and swims for 25 minutes? Interpret the product.

Activity	Calories per Minute
ballet dancing	6
bicycling	12
running	18
swimming	8

Replace each ● with <, >, or = to make a true sentence.

35. $-4(6)$ ● $3(8)$

36. $3(-2)$ ● $4(-1)$

37. $(-11)(-3)(6)(-2)$ ● $-24(18)$

38. $(-5)(-2)(9)$ ● $(4)(10)(-9)$

39. CCSS **Reason Abstractly** Two teams of students played a trivia game. Each team earned 5 points for every correct answer, lost 8 points for every incorrect answer, and lost 2 points for every passed question. For each team, write an expression to determine the number of points they earned. Which team won?

Team	Correct	Incorrect	Passed
1	12	3	1
2	13	2	7

40. **Financial Literacy** The price of stock fell \$2 each day for 14 consecutive days. The original price of the stock was \$41. Write an expression that you could use to find the price of the stock on any day.

41. CCSS **Multiple Representations** In this problem, you will investigate the relationship between time and altitude. A hot air balloon is at an altitude of 600 feet. It begins descending at a constant rate of 15 feet per minute.

a. Table Complete the table below that shows the altitude every 5 minutes.

Time (minutes)	$-15x + 600$	Altitude (feet)
0	$-15(0) + 600$	600
5	■	■
10	■	■
15	■	■

b. Words What does 0 minutes represent?

c. Graph Graph the ordered pairs.

d. Symbols Determine the time it will take for the balloon to reach the ground. Explain how you solved.

H.O.T. Problems Higher Order Thinking

42. CCSS **Identify Structure** Name two integers that have a product between -10 and -15.

43. CCSS **Identify Structure** Name all of the values of x if $7|x| = 63$.

44. CCSS **Persevere with Problems** Positive integers A and C satisfy $A(A - C) = 23$. What is the value of C?

45. CCSS **Identify Structure** Calculate $(-10)(5)(18)[7 + (-7)]$ mentally. Justify your answer.

CCSS **Use a Counterexample** **Determine whether each of the following is *true* or *false*. If *false*, give a counterexample.**

46. The product of three negative integers is positive.

47. The product of four negative integers is positive.

48. **Building on the Essential Question** When multiplying more than two integers, how can you determine the sign of the product?

Standardized Test Practice

49. A submarine descends at a constant rate of 300 feet per minute. Which of the following equations best represents how to find the change in altitude of the submarine after 5 minutes?

A $5(-300) = -1500$

B $5(300) = 1500$

C $-5(300) = -1500$

D $-5(-300) = 1500$

50. Which of the following equations is modeled by the number line below?

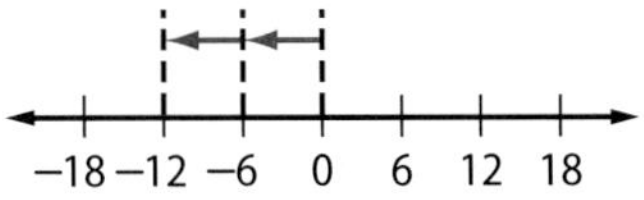

F $-2(-6) = -12$ **H** $2(6) = -12$

G $-2(6) = -12$ **J** $2(-6) = -12$

51. Simplify $-14(-2)(-12)$.

A -336 **C** 168

B -168 **D** 288

52. **Short Response** The distance from the water line to the bottom of a cargo ship changes based on the weight of the cargo.

Weight of Cargo (tons)	Depth of Ship (feet)
35	30
25	25
15	20

If the pattern in the table continues, find the depth of the ship if the ship is carrying 100 tons of cargo.

CCSS Common Core Review

53. The highest point in California is Mount Whitney, with an elevation of 14,494 feet. The lowest point is Death Valley, with an elevation of −282 feet. How much greater is the elevation of Mount Whitney than Death Valley? **7.NS.1d**

Find each sum. **7.NS.1b**

54. $6 + (-9) + 9$

55. $-7 + (-13) + 4$

56. $9 + 16 + (-10)$

57. $-12 + 18 + (-12)$

58. The table shows the lowest recorded temperatures for certain states. **6.NS.7b**

Record Lowest Temperatures by State			
State	Station	Date	Temperature (°F)
Alaska	Prospect Creek Camp	Jan. 23, 1971	−80
Montana	Rogers Pass	Jan. 20, 1954	−70
Wisconsin	Danbury	Jan. 24, 1922	−54

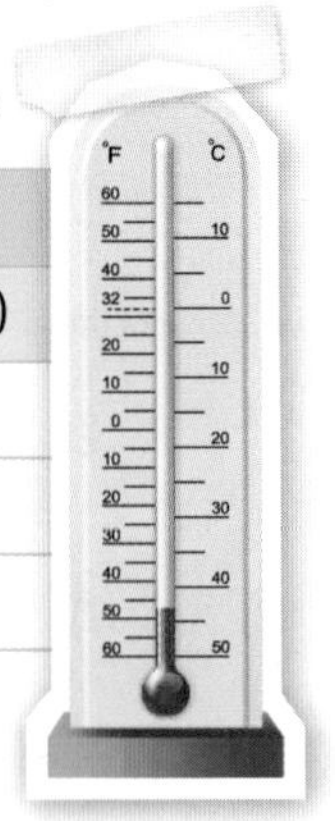

a. Compare the lowest temperature in Montana and the lowest temperature in Wisconsin using an inequality.

b. Write the temperatures in order from greatest to least.

Find each difference **7.NS.1c**

59. $-12 - 16$

60. $-25 - (-50)$

61. $8 - 10$

62. $5 - (-5)$

63. Challenger Deep is 10,924 meters below sea level. Write an integer to represent this depth of Challenger Deep. Then identify its opposite and describe the meaning. **6.NS.5**

Lesson 2-5

Dividing Integers

Interactive Study Guide

See pages 39–40 for:
- Getting Started
- Real-World Link
- Notes

Essential Question

What happens when you add, subtract, multiply, and divide integers?

Common Core State Standards

Content Standards
7.NS.2, 7.NS.2b, 7.NS.2c, 7.NS.3, 7.EE.3

Mathematical Practices
1, 2, 3, 4

What You'll Learn

- Divide integers.
- Find the mean (average) of a set of data.

Real-World Link

Hair Did you know that people with blond hair have more hairs on their head (120,000) than people with red hair (80,000)? However, no matter what color hair you have, the average person loses about 560 hairs per week to make way for new growth.

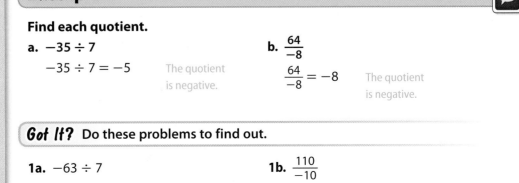

Divide Integers

The expression $-45 \div 5$ is an example of dividing integers with different signs. Since division and multiplication are inverse operations, one way to find the quotient is to use a related multiplication sentence.

Think of this factor . . . to find this quotient.

$$5 \cdot (\mathbf{-9}) = -45 \quad \longrightarrow \quad -45 \div 5 = \mathbf{-9}$$

The dividend and the divisor have different signs, so the quotient is negative.

The following division expressions are equivalent and have a quotient of -5.

$$-\left(\frac{45}{9}\right) = \frac{-45}{9} = \frac{45}{-9}$$

Key Concept: Divide Integers with Different Signs

Words The quotient of two integers with different signs is negative.

Example $-10 \div 5 = -2$ $\quad$ $10 \div (-5) = -2$

Example 1

Tutor

Find each quotient.

a. $-35 \div 7$

$-35 \div 7 = -5$ The quotient is negative.

b. $\frac{64}{-8}$

$\frac{64}{-8} = -8$ The quotient is negative.

Got It? Do these problems to find out.

1a. $-63 \div 7$

1b. $\frac{110}{-10}$

Stockbyte/Getty Images

Key Concept: Divide Integers with the Same Signs

Words	The quotient of two integers with the same sign is positive.
Example	$-10 \div (-5) = 2$ $\quad$ $10 \div 5 = 2$

You can also use multiplication and division sentences to find the quotient of integers with the same sign.

Think of this factor . . . to find this quotient.

$$-3 \cdot (\mathbf{-12}) = 36 \longrightarrow 36 \div (-3) = \mathbf{-12}$$

The quotient is always positive when both integers are positive or both integers are negative.

Watch Out!
Always check your work after finding an answer. Does $-8 \times 7 = -56$?

Example 2

Tutor

Find each quotient.

a. $48 \div 12$

$48 \div 12 = 4$ — The quotient is positive.

b. $\frac{-56}{-8}$

$\frac{-56}{-8} = 7$ — The quotient is positive.

Got It? Do these problems to find out.

2a. $-35 \div (-5)$

2b. $\frac{39}{3}$

You can use the rules for dividing integers to evaluate algebraic expressions.

Example 3

Tutor

Evaluate each expression if $x = -6$ and $y = -3$.

a. $12y \div x$

$12y \div x = 12(-3) \div (-6)$ — Replace y with -3 and x with -6.

$= -36 \div (-6)$ — The product of 12 and -3 is negative.

$= 6$ — The quotient of -36 and -6 is positive.

b. $\frac{-5x}{y}$

$\frac{-5x}{y} = \frac{-5(-6)}{-3}$ — Replace x with -6 and y with -3.

$= \frac{30}{-3}$ — The product of -5 and -6 is positive.

$= -10$ — The quotient of 30 and -3 is negative.

Got It? Do these problems to find out.

3a. $2y \div x$

3b. $3x \div (-y)$

Mean (Average)

Division is used in statistics to find the average, or mean, of a set of data. To find the mean of a set of numbers, find the sum of the numbers and then divide by the number of items in the set.

Real World

Example 4

Tutor

The wind chill temperatures in degrees Fahrenheit for 7 days are shown below. Find the mean and interpret the quotient.

−6, −5, 2, −10, 1, −9, 6

$$\frac{-6 + (-5) + 2 + (-10) + 1 + (-9) + 6}{7} = \frac{-21}{7}$$ Find the sum of the temperatures. Divide by the number of days.

$$= -3$$ Simplify.

The mean wind chill temperature is −3°F.

Got It? **Do these problems to find out.**

4a. Linda has scores of −3, −2, 1, and 0 during 4 rounds of golf. Find the mean and interpret the quotient.

4b. The last four transactions Jesse posted in her checkbook were −\$35, \$23, −\$156, and \$60. Find the mean and interpret the quotient.

Concept Summary Operations with Integers

Words	Examples	
Adding Integers		
Same Signs: Add absolute values. The sum has the same sign as the integers.	$3 + 2 = 5$ $-3 + (-2) = -5$	
Different Signs: Subtract absolute values. The sum has the same sign as the integer with the greater absolute value.	$-3 + 2 = -1$ $3 + (-2) = 1$	
Subtracting Integers		
To subtract an integer, add its additive inverse.	$3 - 5 = 3 + (-5)$ or -2 $3 - (-5) = 3 + 5$ or 8	
Multiplying and Dividing Two Integers		
Same Signs: The product or quotient is positive.	$3 \cdot 2 = 6$ $-3(-2) = 6$	$6 \div 3 = 2$ $-6 \div (-3) = 2$
Different Signs: The product or quotient is negative.	$-3 \cdot 2 = -6$ $3(-2) = -6$	$-6 \div 3 = -2$ $6 \div (-3) = -2$

Guided Practice

Check

Find each quotient. (Examples 1 and 2)

1. $40 \div (-10)$
2. $\frac{39}{13}$
3. $-26 \div (-3)$
4. $\frac{-54}{6}$
5. $-48 \div 3$
6. $\frac{72}{-18}$
7. $36 \div (-4)$
8. $\frac{-72}{-9}$

Evaluate each expression if $s = -2$ and $t = 7$. (Example 3)

9. $14s \div t$
10. $\frac{-10t}{s}$
11. $4t \div (2s)$

12. **Financial Literacy** The following are the changes of a value of a certain stock over the last 5 days: −\$7, +\$3, +\$6, −\$2, −\$5. Find the mean and interpret the quotient. (Example 4)

Independent Practice

Go online for Step-by-Step Solutions

eHelp

Find each quotient. (Examples 1 and 2)

13. $-33 \div 11$
14. $28 \div (-14)$
15. $-36 \div (-2)$
16. $-60 \div (-5)$
17. $\frac{-150}{10}$
18. $\frac{600}{-20}$
19. $126 \div 9$
20. $750 \div 15$
21. $-770 \div 7$
22. $-560 \div 8$
23. $\frac{-350}{-70}$
24. $\frac{-480}{-16}$

Evaluate each expression. (Example 3)

25. $\frac{n}{-13}$, if $n = -182$
26. $252 \div k$, if $k = 9$
27. $\frac{-6a}{b}$, if $a = -24$ and $b = -4$
28. $\frac{9y}{x}$, if $x = -21$ and $y = -35$

29. **Financial Literacy** The last 5 transactions at Mr. Brigham's ATM were \$250, −\$60, −\$94, \$300, and −\$186. Find the mean and interpret the quotient. (Example 4)

30. The table shows the extreme high and low temperatures for different states. The expression $\frac{5(F - 32)}{9}$, where F represents the temperature in degrees Fahrenheit, can be used to convert temperatures from degrees Fahrenheit to degrees Celsius.

    a. Find the extreme high and low temperatures for each state in degrees Celsius. Round to the nearest tenth.

    b. The difference between the extreme high and low temperatures is called the range. Find the range of the temperatures in degrees Celsius for each state.

    c. List the states in order from least to greatest ranges.

Extreme Temperatures

State	Extreme Low (°F)	Extreme High (°F)
Arizona	−40	128
Florida	−2	109
Kentucky	−34	114
Michigan	−51	112
New York	−57	108

31. **Financial Literacy** A small clothing company's total income was \$64,000, while its total expenses were \$67,600. Use the expression $\frac{I - E}{12}$, where I represents total income and E represents total expenses, to find the average difference between the company's income and expenses each month. Interpret the quotient.

Replace each ● with <, >, or = to make a true sentence.

32. $-80 \div (-2)$ ● $120 \div 3$
33. $\frac{240}{-80}$ ● $\frac{-150}{-50}$

Find the value of *x* that makes each statement true.

34. $-375 \div x = -15$

35. $22 = x \div (-34)$

36. $x \div (-17) = -35$

37. $-689 \div x = 53$

CCSS Reason Abstractly For Exercises 38–40, write an expression to represent each real-world situation. Then evaluate the expression and interpret the quotient.

38. Jean owes her parents $90, to be paid in 5 equal installments. How much is each installment?

39 The temperature dropped a total of 40°F over an 8 hour period. What was the mean hourly temperature drop?

40. In October, the full price of a television is $695. Over the next seven months, the price of the television drops a total of $140. How much did the price of the television drop each month on average?

41. **STEM** The table shows the deepest point of each of the Great Lakes.

Great Lake	Deepest Point (m)
Erie	−64
Huron	−229
Michigan	−281
Ontario	−244
Superior	−406

a. What is the mean of the deepest points of the Great Lakes?

b. Suppose each of the deepest points were 10 meters higher. Find the mean. Compare the new mean to the original.

42. **CCSS Multiple Representations** In this problem, you will investigate the relationships between distance and time. Suppose Joseph was tracking a submarine. The submarine descended 480 feet in 20 minutes.

a. Numbers How many feet did the submarine descend per minute?

b. Symbols Write an expression to determine the change, in feet, of the submarine's elevation after any number of minutes.

c. Table Use the expression you wrote in part **b** to create a table showing the change in the submarine's depth after 5, 10, 15, and 18 minutes.

H.O.T. Problems Higher Order Thinking

43. **CCSS Model with Mathematics** Write a division expression with a quotient between −20 and −25. Then describe a real-world situation in which the equation represents the solution.

44. **CCSS Persevere with Problems** The mean temperature during 5 days was −10°F. Give a sample set of what the temperatures might have been for the 5 days.

45. **CCSS Reason Inductively** Find the next two numbers in the pattern 1024, −256, 64, −16, ... Explain your reasoning.

46. **CCSS Persevere with Problems** Addition and multiplication are said to be *closed* for whole numbers, but subtraction and division are not. That is, when you add or multiply any two whole numbers, the result is a whole number. Which operations are closed for integers?

47. **Building on the Essential Question** Explain whether the Associative Property and Commutative Property are true for division of integers. Support your reasoning with an example. Then explain your reasoning to a classmate.

Standardized Test Practice

48. Yesterday's low temperature was 24 degrees Fahrenheit. Use the expression $\frac{5(F-32)}{9}$, where F represents the temperature in degrees Fahrenheit, to find the approximate low temperature in degrees Celsius.

A −9.8°C **C** 4.4°C

B −4.4°C **D** 9.8°C

49. The temperature dropped 30°F in a 6 hour period. What was the mean hourly temperature change?

F −5°F **H** 5°F

G −6°F **J** 6°F

50. The depth of a reservoir decreased 84 inches in two weeks. If the water depth changed by the same amount each day, how much did the depth of the water change per day?

A −7 inches **C** 6 inches

B −6 inches **D** 7 inches

51. Miss Washer recorded the low temperature each day for a week. What is the mean low temperature?

Day	M	T	W	Th	F
Temperature (°F)	−18	12	−7	9	−2

F 9.6°F **H** −9.6°F

G 1.2°F **J** −1.2°F

Common Core Review

52. STEM During low tide, in Wrightsville, North Carolina, the beachfront in some places is about 350 feet from the ocean to the homes. High tide can change the width of a beach at a rate of −17 feet an hour. It takes 6 hours for the ocean to move from low to high tide. **7.NS.3**

a. What is the change in the width of the beachfront from low to high tide?

b. What is the distance from the ocean to the homes at high tide?

Find each product. **7.NS.2a**

53. −6 • 12

54. −15(−4)

55. 8(−11)

56. 2(−5)(−3)

57. −3(4)(−9)

58. −6(−2)(−2)

Find each difference. **7.NS.1c**

59. 3 − 8

60. 4 − 5

61. 2 − 9

62. −9 − (−7)

63. −7 − (−10)

64. −11 − (−12)

65. Financial Literacy The starting balance in a checking account was $50. What was the balance after checks were written for $25 and for $32? **7.NS.1b**

Refer to the coordinate plane shown at the right. Write the ordered pair that names each point. **5.G.2**

66. *A*

67. *G*

68. *C*

69. *E*

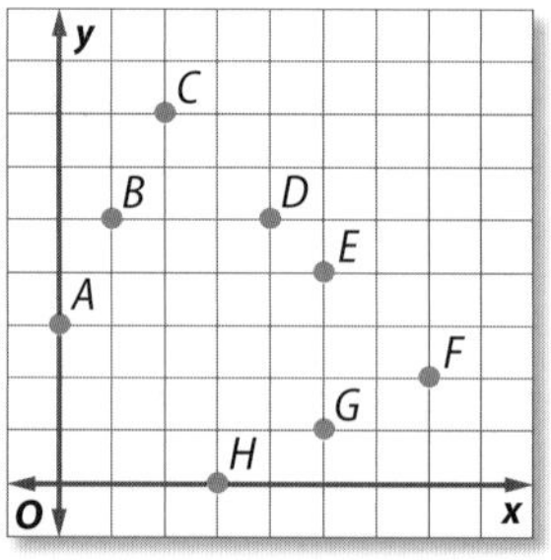

Lesson 2-6

Graphing in Four Quadrants

Interactive Study Guide

See pages 41–42 for:
- Getting Started
- Vocabulary Start-Up
- Notes

Essential Question

What happens when you add, subtract, multiply, and divide integers?

Common Core State Standards

Content Standards
Preparation for 7.RP.2a, 7.RP.2b, 7.RP.2d, 8.EE.5

Mathematical Practices
1, 3, 4, 7

Vocabulary
quadrants

What You'll Learn

- Graph points on a coordinate plane.
- Graph algebraic relationships.

Real-World Link

Video Games Programmers of 3-D video games use several coordinate systems or spaces to create a game engine. The most commonly used spaces are local space, world space, and camera space. In local space, objects are placed on a coordinate grid at the object's relative origin.

Graph Points

The coordinate system you used in Lesson 1–6 can be extended to include points below and to the left of the origin.

origin

The *x*-axis extends to the right and left of the origin. Notice that the numbers to the left of zero on the *x*-axis are negative.

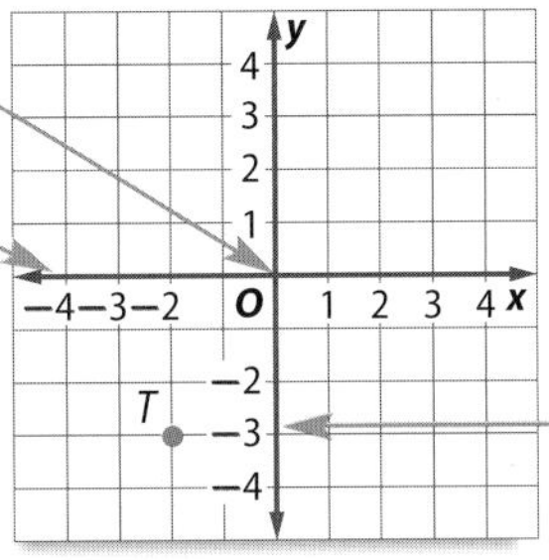

The *y*-axis extends above and below the origin. Notice that the numbers below zero on the *y*-axis are negative.

Recall that a point graphed on the coordinate system has an *x*-coordinate and a *y*-coordinate. The dot at the ordered pair (−2, −3) is the graph of point *T*.

Example 1

Tools Tutor

Write the ordered pair that names each point.

a. *J*
The *x*-coordinate is −4.
The *y*-coordinate is −3.
The ordered pair is (−4, −3).

b. *L*
The *x*-coordinate is 2.
The *y*-coordinate is −2.
The ordered pair is (2, −2).

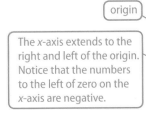

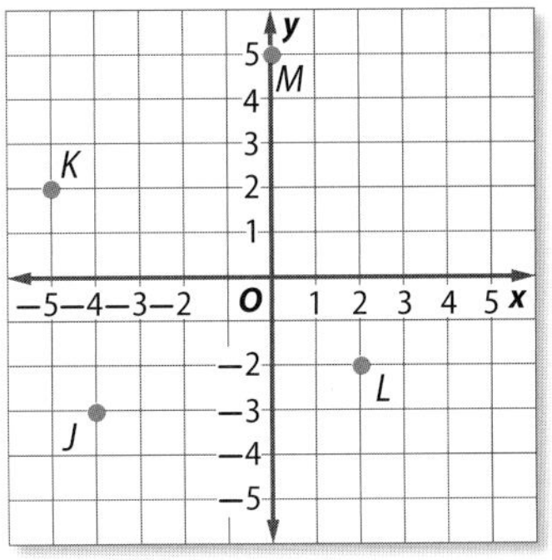

Got It? **Do these problems to find out.**

1a. *M*

1b. *K*

Nicola Tree/Stockbyte/Getty Images

The x-axis and the y-axis separate the coordinate plane into four regions, called **quadrants**. The quadrants are named I, II, III, and IV.

The axes and points on the axes are not located in any of the quadrants.

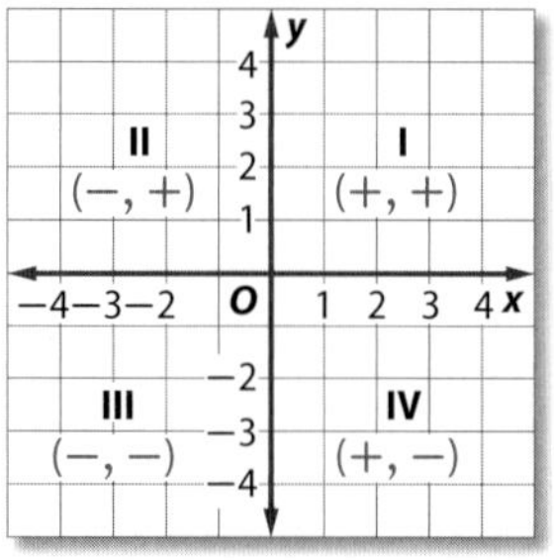

Example 2

Watch Tools Tutor

Graph and label each point on a coordinate plane. Name the quadrant in which each point lies.

a. $A(-2, -4)$

Start at the origin. Move 2 units left. Then move 4 units down and draw a dot. Point $A(-2, -4)$ is in Quadrant III.

b. $B(0, 2)$

Start at the origin. Since the x-coordinate is 0, the point will lie on the y-axis. So, move 2 units up. Point $B(0, 2)$ is not in a quadrant. It is on the y-axis.

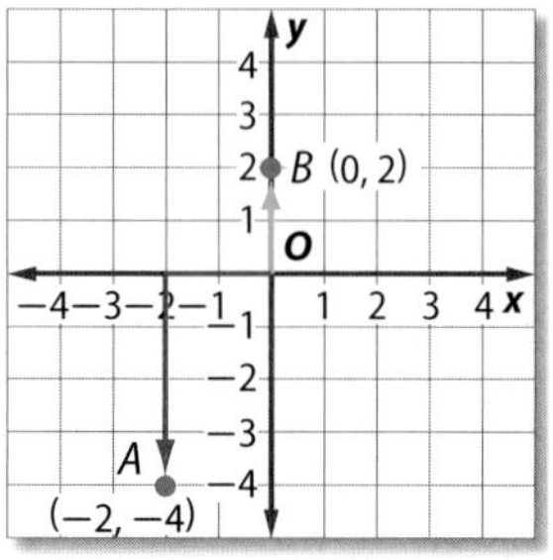

Got It? **Do these problems to find out.**

2a. $H(4, -3)$ **2b.** $I(-1, 4)$ **2c.** $J(0, -2)$

Graph Algebraic Relationships

You can use a coordinate graph to show relationships between two numbers.

Example 3

Real World

Tools Tutor

The difference between John and Tarie's golf score is 2. If x represents John's score and y represents Tarie's score, make a table of possible values for x and y. Graph the ordered pairs and describe the graph.

Choose values for x and y that have a difference of 2. Then graph the ordered pairs.

The points are along a diagonal line that crosses the x-axis at $x = 2$.

$x - y = 2$		
x	y	(x, y)
2	0	(2, 0)
1	−1	(1, −1)
0	−2	(0, −2)
−1	−3	(−1, −3)
−2	−4	(−2, −4)

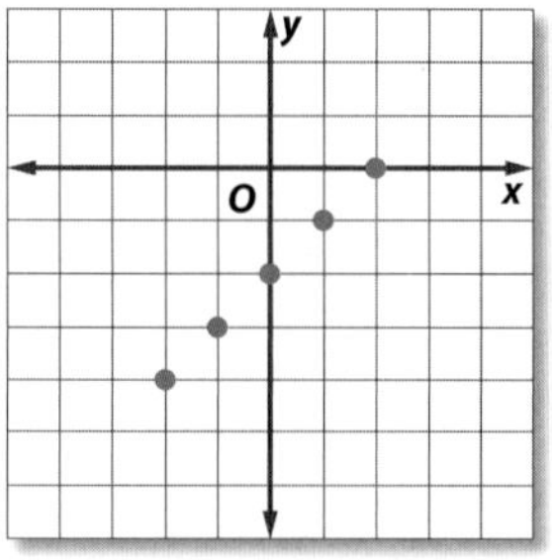

Scale

When no numbers are shown on the x- or y-axis, you can assume that each square is one unit long on each side.

Got It? **Do this problem to find out.**

3. The sum of two golf scores is 3. If x represents one score and y represents the other score, make a table of possible values for x and y. Graph the ordered pairs and describe the graph.

Guided Practice

Check

Name the ordered pair for each point graphed at the right. (Example 1)

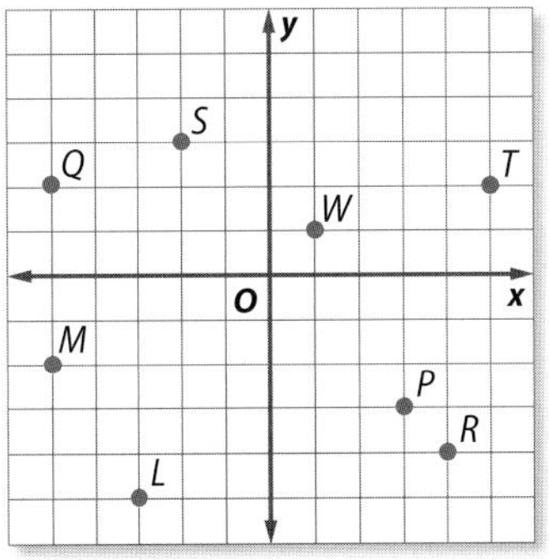

1. *Q*

2. *P*

3. *T*

4. *M*

Graph and label each point on a coordinate plane. Name the quadrant in which each point is located. (Example 2)

5. $A(-2, 3)$

6. $B(4, -1)$

7. $C(-3, -2)$

8. $D(0, -5)$

9. CCSS **Model with Mathematics** The difference of two temperatures is 4°F. If *x* represents the first temperature and *y* represents the second temperature, make a table of possible values for *x* and *y*. Graph the ordered pairs and describe the graph. (Example 3)

Independent Practice

eHelp

Go online for Step-by-Step Solutions

Name the ordered pair for each point graphed at the right. (Example 1)

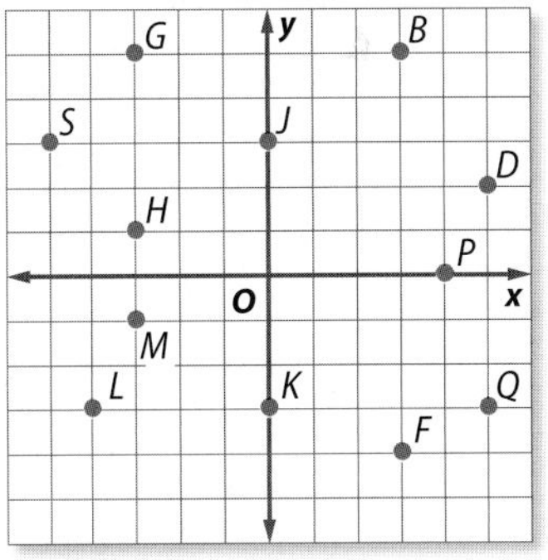

10. *S*

11. *H*

12. *D*

13. *B*

14. *M*

15. *L*

16. *F*

17. *Q*

18. *K*

19. *J*

Graph and label each point on a coordinate plane. Name the quadrant in which each point is located. (Example 2)

20. $Z(-1, 1)$

21. $Y(-2, 3)$

22. $X(5, 6)$

23. $W(6, 2)$

24. $V(-1, -6)$

25. $S(2, -1)$

26. $T(-5, 0)$

27. $R(0, -4)$

28. $P(-4, 5)$

29. $Q(-3, 3)$

30. $N(1, -1)$

31. $K(5, -3)$

32. CCSS **Model with Mathematics** After two plays, the Wildcats gained a total of 16 yards. If *x* represents the number of yards for play one, and *y* represents the number of yards for play two, make a table of possible values for *x* and *y*. Graph the ordered pairs and describe the graph. (Example 3)

33. CCSS **Model with Mathematics** The distance between two runners in a race is 10 feet. If *x* represents the position of one runner in relation to a water stop and *y* represents the position of the second runner, make a table of possible values for *x* and *y*. Graph the ordered pairs and describe the graph. (Example 3)

CCSS **Persevere with Problems** **Name the quadrant in which each point lies.**

34. $A(5, |-6|)$

35. $E(|-5|, -3)$

36. $J(x, y)$ if $x < 0, y > 0$

37. $U(x, y)$ if $x > 0, y < 0$

38. Consider the points $A(-4, 3)$, $B(1, 3)$, $C(1, 2)$, and $D(-4, 2)$.

a. Graph the points on a coordinate plane and connect them to form a rectangle.

b. Add 4 to the x-coordinate of each ordered pair and redraw the figure.

c. Compare the two rectangles.

39 **STEM** The table shows temperatures in degrees Celsius and the corresponding temperatures in degrees Fahrenheit. Graph the ordered pairs (°Celsius, °Fahrenheit) to show the relationship between degrees Celsius and degrees Fahrenheit.

Celsius	−10	−5	0	5	10
Fahrenheit	14	23	32	41	50

40. **Financial Literacy** The table shows the balance on a \$50 music card after a certain number of songs have been downloaded.

a. Make a graph to show how the number of songs downloaded and the remaining balance are related.

b. Use your graph to find the balance on the card after 25 songs have been downloaded.

Songs Downloaded	Balance (\$)
0	50
5	45
10	40
15	35

For each graph, create a table showing the rule and the values for *x* and *y*.

41.

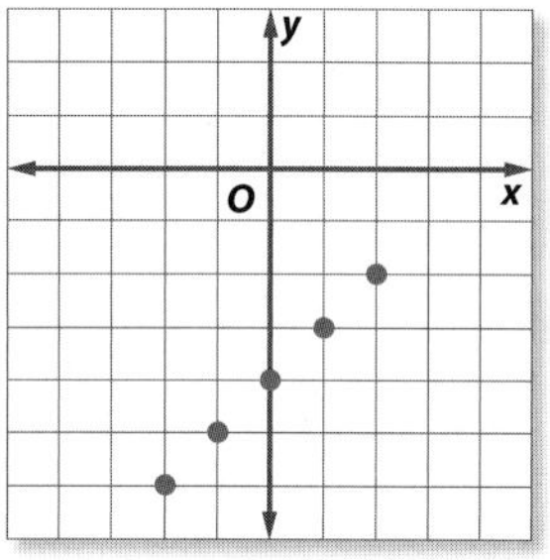

42.

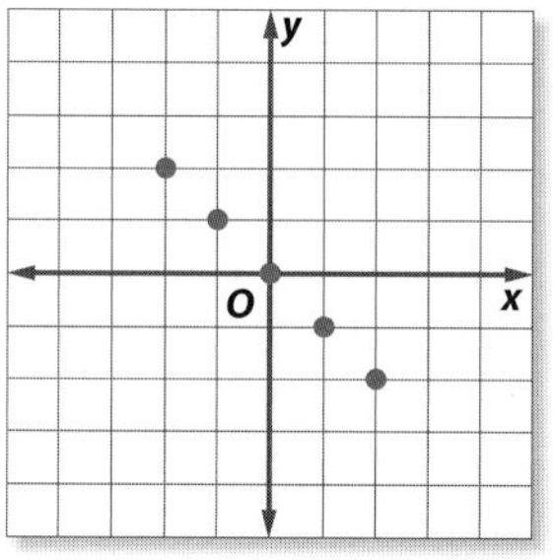

Graph and label each point on a coordinate plane.

43. $A(-6.5, 3)$

44. $B(-2, -5.75)$

45. $C(4.1, -1)$

46. $D(-3.4, 1.5)$

H.O.T. Problems Higher Order Thinking

47. CCSS **Identify Structure** Write the coordinates of a point located in quadrant II.

48. CCSS **Persevere with Problems** The product of two numbers is 12.

a. Make a table using −3, −2, −1, 1, 2, and 3 as x values.

b. Graph the ordered pairs. Compare and contrast this graph with the one in Example 3.

49. CCSS **Persevere with Problems** Determine whether each statement is *always*, *sometimes*, or *never* true. Explain or give a counterexample to support your answer.

a. Both x- and y-coordinates of a point in quadrant I are negative.

b. The x-coordinate of a point that lies on the x-axis is negative.

50. **Building on the Essential Question** How does the location of the points $(-7, 8)$ and $(8, -7)$ change if you multiply each of the coordinates in each ordered pair by -1? Explain your reasoning to a classmate.

Standardized Test Practice

51. Which point on the graph best represents the location of the library?

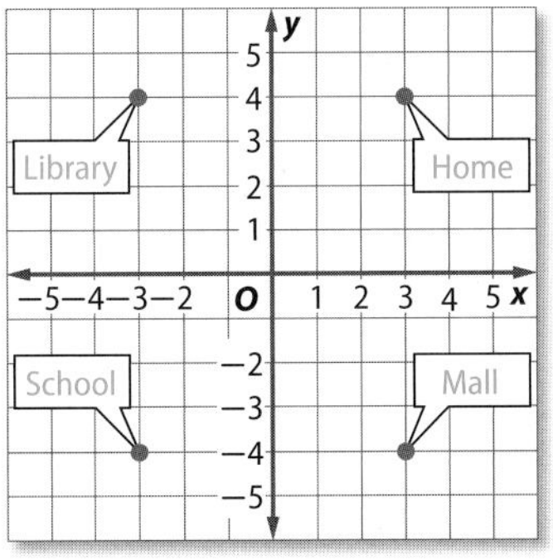

A (3, 4)
B (−3, −4)
C (−3, 4)
D (3, −4)

52. What building is located at point (−3, −4) on the graph above?

F School
G Mall
H Library
J Home

53. In which quadrant on the coordinate plane is point (2, −3)?

A quadrant I
B quadrant II
C quadrant III
D quadrant IV

54. **Short Response** Juan wants to rent 4 DVDs. Each DVD costs $3 for two days. Complete the table to show his total cost for the number of days given.

Number of Days	Total Cost ($)
2	■
4	■
6	■

CCSS Common Core Review

Find each quotient. 7.NS.2b

55. −27 ÷ (−9)

56. −77 ÷ 7

57. −300 ÷ 6

58. STEM A glacier was receding at a rate of 300 feet per day. What is the glacier's movement in 5 days? (*Hint*: The word *receding* means *moving backward*.) 7.NS.2a

59. Lincoln High School's swim team finished the 4 × 100-meter freestyle relay in 5 minutes 18 seconds. Prospect High School's swim team finished the race in 5 minutes 7 seconds. Write an integer that represents Lincoln's finish compared to Prospect's finish. 7.NS.1c

Evaluate each expression. 7.NS.1

60. $|-9-1|$

61. $|10|-|-4|$

62. $|16|+|-5|$

Find each sum. 7.NS.1

63. −85 + 15

64. −13 + (−8)

65. −10 + 12

Evaluate each expression if $a = -5$, $b = 4$, and $c = -9$. 7.NS.1, 7.NS.2

66. $4a + c$

67. $4b + c$

68. $-b + 2a$

69. $-b - 2a$

70. $a(b + c)$

71. $a(b - c)$

72. $|a - b|$

73. $4a \div (5b)$

74. $(c - a) \cdot (a - c)$

Check

Chapter Review

ISG Interactive Study Guide
See pages 43–46 for:
- Vocabulary Check
- Key Concept Check
- Problem Solving
- Reflect

Lesson-by-Lesson Review

Lesson 2-1 Integers and Absolute Value (pp. 46–51)

Write two inequalities using the number pairs. Use the symbols $<$ or $>$.

1. -20 and -18
2. 0 and -5

Replace each ● with $<$, $>$, or $=$ to make a true sentence.

3. $5 \bullet -5$
4. $7 \bullet 7$
5. $-3 \bullet 1$
6. $-14 \bullet -22$

Evaluate each expression.

7. $|-16|$
8. $|4|$
9. $|-34|$
10. $|-2| + |-11|$

11. Jamal traded away 7 shortstop cards for 5 pitcher cards. Find an integer that represents the change in the number of cards Jamal had after the trade.

Example 1

Write two inequalities comparing -5 and -4. Use the symbols $<$ or $>$.

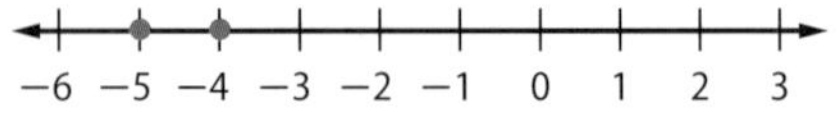

Since -4 is to the right of -5, $-4 > -5$.

Since -5 is to the left of -4, $-5 < -4$.

Example 2

Evaluate $|-5|$.

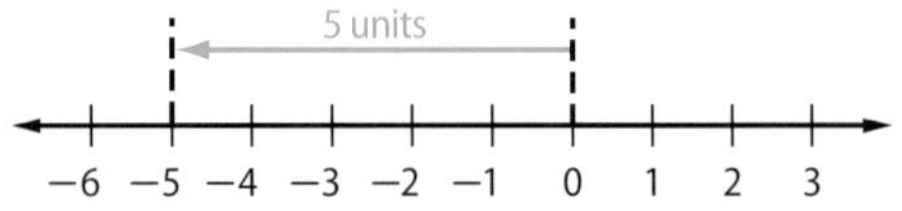

The graph of -5 is 5 units from 0.

So, $|-5| = 5$.

Lesson 2-2 Adding Integers (pp. 55–60)

Find each sum.

12. $-5 + (-1)$
13. $-3 + (-7)$
14. $-6 + 10$
15. $4 + (-9)$
16. $7 + (-2)$
17. $14 + (-5)$
18. $-12 + 5 + (-6)$
19. $2 + 8 + (-3)$
20. $-4 + 9 + (-2)$
21. $-7 + 5 + (-4)$

22. A contestant on a quiz game show has -25 points. If she loses an additional 50 points, what is her score? Write an addition equation and then solve.

23. A golfer's scores for the last five weeks are -3, $+5$, -1, -2, and $+4$. What is the sum of his scores?

Example 3

Find $-2 + (-3)$.

Use a number line.

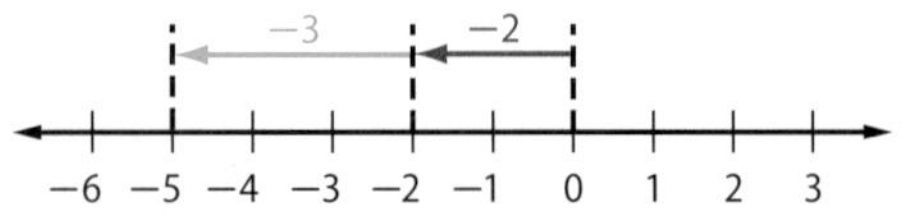

Start at zero. Move 2 units to the left. From there, move 3 more units to the left.

So, $-2 + (-3) = -5$.

Example 4

Find $9 + (-4)$.

$9 + (-4) = 5$ Subtract $|-4|$ from $|9|$. The sum is positive.

Lesson 2-3 Subtracting Integers (pp. 63–67)

Find each difference.

24. $13 - 7$

25. $-2 - 5$

26. $8 - (-3)$

27. $-1 - (-4)$

28. $-4 - 6$

29. $3 - 5$

30. $7 - (-7)$

31. $-4 - 10$ —

32. $-5 - (-9)$

33. $13 - 3$

34. $-12 - (-3)$ —

35. $-9 - (-9)$

36. The table shows the highest and lowest elevations for North America. Find the difference between the highest and lowest elevations.

Lowest Elevation (feet)	Highest Elevation (feet)
−282	20,320

37. On Mars, the temperature ranges from 68°F during the day to −220°F at night. What is the difference in temperature between day and night?

Example 5

Find −13 − 4.

$-13 - \mathbf{4} = -13 + (\mathbf{-4})$ To subtract 4, add −4.

$= -17$

So, $-13 - 4 = -17$.

Example 6

Find 10 − (−2).

$10 - (\mathbf{-2}) = 10 + \mathbf{2}$ To subtract −2, add 2.

$= 12$

So, $10 - (-2) = 12$.

Lesson 2-4 Multiplying Integers (pp. 71–76)

Find each product.

38. $-2 \cdot 3$

39. $-5 \cdot 6$

40. $-7 \cdot (-9)$

41. $-12 \cdot (-4)$

42. $11(-7)$

43. $-10(14)$

44. $5(-2)(-6)$

45. $-7(4)(3)$

46. $10(-4)(9)$

47. $-3(-2)(-8)$

48. $9(-8)(1)$

49. $-10(-2)(3)$

50. $4(-7)(-3)$

51. $-12(-12)(-2)$

52. A scuba diver starts at the surface and descends at a rate of 25 feet per minute. Write and evaluate a multiplication expression to find the depth of the scuba diver after 6 minutes.

53. For each jump she completes incorrectly in an ice-skating competition, Dawn receives −2 points. If Dawn completes six jumps incorrectly and no jumps correctly, what is her score?

54. A helicopter descends at a rate of 450 feet per minute. Write and evaluate a multiplication expression to find the change in altitude of the helicopter after 5 minutes.

Example 7

Find 3(−7).

$3(-7) = -21$ The factors have different signs. The product is negative.

Example 8

Find −5(−4).

$-5(-4) = 20$ The factors have the same sign. The product is positive.

Lesson 2-5 Dividing Integers (pp. 77–82)

Find each quotient.

55. $-16 \div (-4)$

56. $-56 \div (-8)$

57. $\frac{-30}{5}$

58. $15 \div (-3)$

59. $-88 \div -11$

60. $\frac{170}{-10}$

61. $18 \div (-9)$

62. $-144 \div (-12)$

63. $\frac{-720}{9}$

64. $350 \div (-70)$

65. For the first five legs of a bicycle race, Elena was 32 seconds, 5 seconds, 10 seconds, 8 seconds, and 12 seconds behind the leader. What was the average time she was behind the leader?

66. Yesterday's high temperature in Death Valley was 114°F. What was the temperature in degrees Celsius? Use the expression $\frac{5(F-32)}{9}$ to convert degrees Fahrenheit to degrees Celsius. Round to the nearest integer.

Example 9

Find $-24 \div (-6)$.

$-24 \div (-6) = 4$ The quotient is positive.

Example 10

Find $15 \div (-3)$.

$15 \div (-3) = -5$ The quotient is negative.

Lesson 2-6 Graphing in Four Quadrants (pp. 83–87)

Graph and label each point on a coordinate plane. Name the quadrant in which each point is located.

67. $P(7, -12)$

68. $Q(6, 9)$

69. $R(-10, 10)$

70. $S(-8, -9)$

71. $T(4, -5)$

72. $V(0, -1)$

73. $X(5, 2)$

74. $Y(-3, 3)$

75. $W(2, 0)$

76. $Z(-1, -4)$

77. The coordinate plane shown represents the position of players' pieces in a board game. Name the quadrant in which each player's game piece is located.

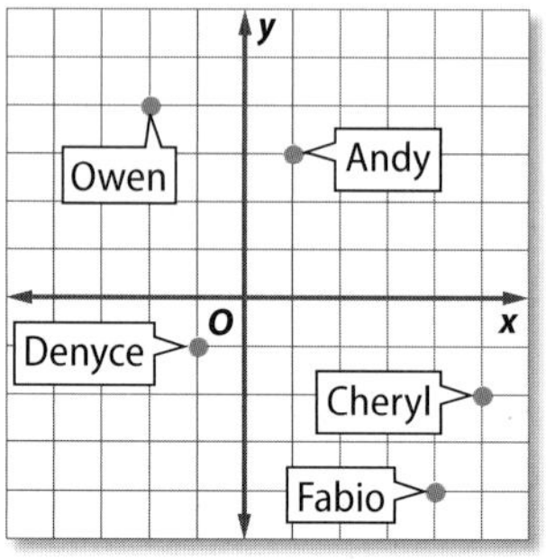

Example 11

Graph and label point $J(-3, 5)$ on a coordinate plane. Name the quadrant in which the point is located.

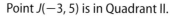

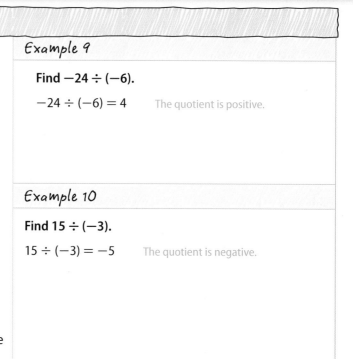

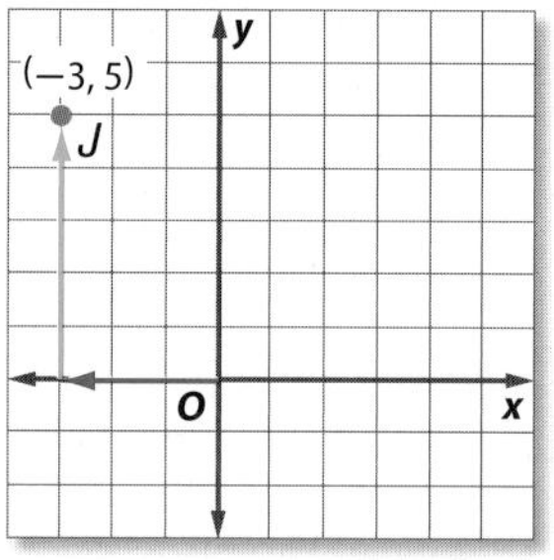

Point $J(-3, 5)$ is in Quadrant II.

Chapter 3 Operations with Rational Numbers

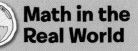

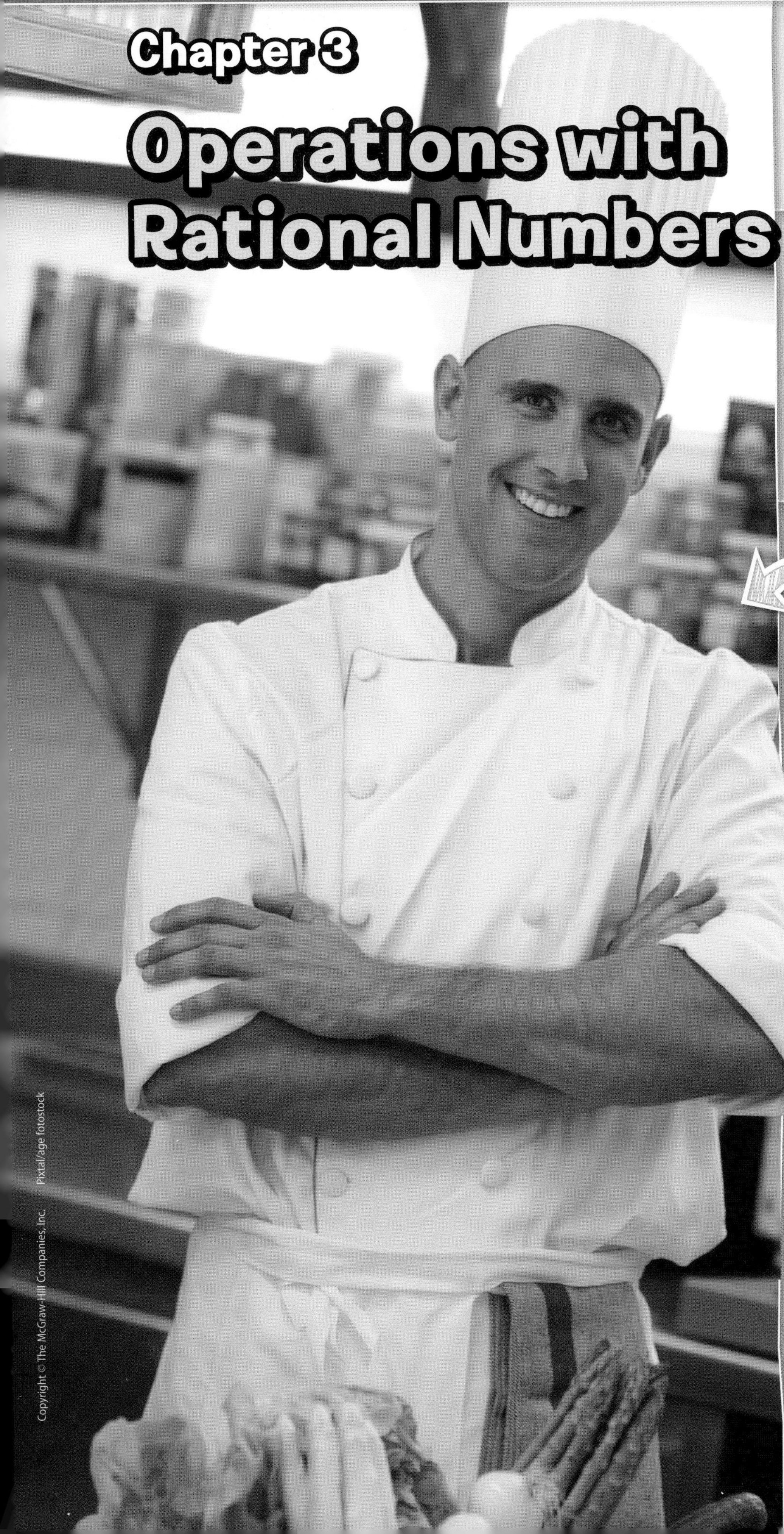

Essential Question

What happens when you add, subtract, multiply, and divide rational numbers?

Common Core State Standards

Content Standards
7.NS.1, 7.NS.1d, 7.NS.2, 7.NS.2a, 7.NS.2c, 7.NS.2d, 7.NS.3, 7.EE.3, 8.NS.1

Mathematical Practices
1, 3, 4, 5, 6, 7

Math in the Real World

Cooking Cookbooks are among the top-selling books each year. Recipes in the cookbooks give a set of instructions to make a dish. Most recipes include fractions and mixed numbers. Being able to understand and use rational numbers will help you make a more delicious meal.

Interactive Study Guide
See pages 47–50 for:
- Chapter Preview
- Are You Ready?
- Foldable Study Organizer

Inquiry Lab

Fractions and Decimals on the Number Line

Inquiry HOW can you graph a negative fraction, or a negative decimal to the tenths place, on a number line?

CCSS Content Standards
Preparation for 7.NS.1

Mathematical Practices
1, 3, 5

Finance Before 2001, the change in the value of a share of stock on the New York Stock Exchange was represented by a positive or negative fraction or mixed number. The changes in the value of one share of MathCorp during a week of trading one month are shown in the table.

Monday	Tuesday	Wednesday	Thursday	Friday
$-\frac{5}{8}$	$1\frac{3}{8}$	$-2\frac{4}{5}$	$\frac{1}{8}$	$3\frac{7}{16}$

How can you graph the fraction that represents the change in value of one share of MathCorp stock on Monday?

Investigation 1

Watch Tools

Use a number line to graph $-\frac{5}{8}$.

Step 1 Draw a number line showing −1, 0, and 1.

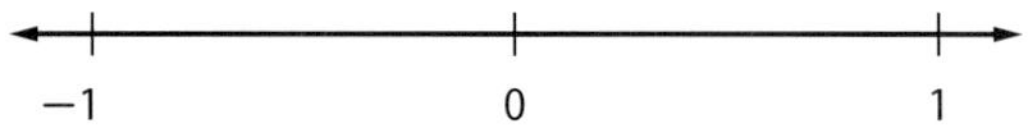

Step 2 Since the denominator is eighths, divide the number line between 0 and 1 into 8 equal parts. Then divide the number line between 0 and −1 into 8 equal parts.

Label the number line with $\frac{1}{8}$, $\frac{2}{8}$, $\frac{3}{8}$, and so on.

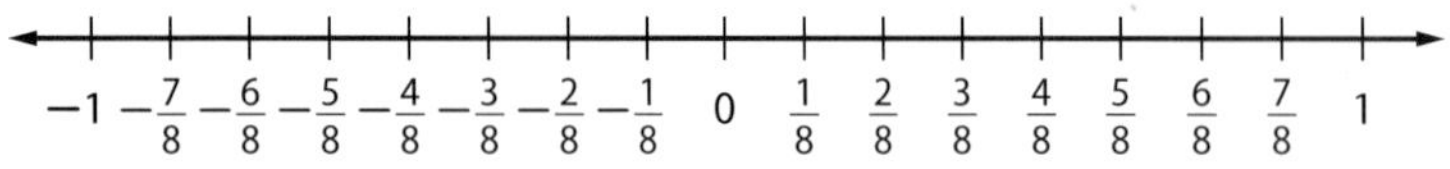

Step 3 Locate $\frac{5}{8}$ on the number line. Then find $-\frac{5}{8}$, the point opposite $\frac{5}{8}$ on the number line. Draw a dot on the number line above the $-\frac{5}{8}$ mark.

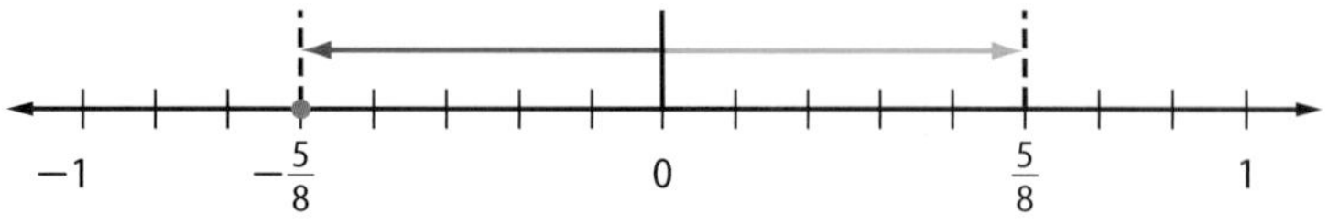

Investigation 2

Use a number line to graph −2.8.

Step 1 Draw a number line showing −2 and −3.

Step 2 The decimal is expressed in tenths. So, divide the number line between −2 and −3 into 10 equal parts.

On the number line, −2.1 lies just to the left of −2, −2.2 lies to the left of −2.1, and so on. Label the number line.

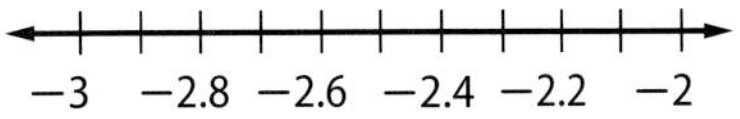

Step 3 Draw a dot on the number line above the −2.8 mark.

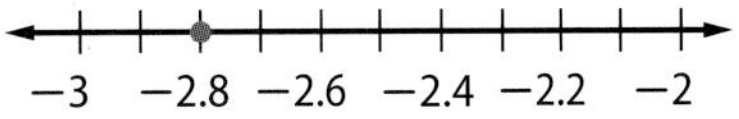

Collaborate

CCSS Model with Mathematics Work with a partner. Graph each number on a number line.

1. $-\frac{3}{8}$

2. −1.6

3. In Investigation 2, you graphed −2.8 on a number line. From that point, in what direction on the number line would you move if you wanted to graph −3.2?

4. Equivalent numbers are graphed in the same place on a number line. Use a number line to determine if the following pairs of numbers are equivalent.

   a. $\frac{2}{3}$ and 0.8

   b. −1.75 and $-1\frac{3}{4}$

Analyze

5. Use a number line to determine which number is greater in each pair of numbers. Justify your reasoning.

   a. $-2\frac{5}{8}$ or −2.75

   b. −1.4 or −1.7

   c. $\frac{1}{5}$ or $-1\frac{1}{5}$

6. **CCSS Identify Structure** Name a decimal between 1 and 2 and its opposite. Then graph the decimals on a number line.

7. **CCSS Identify Structure** Explain why $\frac{3}{4}$ is graphed to the right of $\frac{1}{4}$ but $-\frac{3}{4}$ is graphed to the left of $-\frac{1}{4}$.

Reflect

8. **Inquiry** HOW can you graph a negative fraction, or a negative decimal to the tenths place, on a number line?

Lesson 3-1

Fractions and Decimals

ISG Interactive Study Guide

See pages 51–52 for:
- Getting Started
- Vocabulary Start-Up
- Notes

Essential Question

What happens when you add, subtract, multiply, and divide rational numbers?

CCSS Common Core State Standards

Content Standards
7.NS.2, 7.NS.2d, 8.NS.1, 7.EE.3

Mathematical Practices
1, 3, 4, 5, 7

Vocabulary
repeating decimal
terminating decimal
bar notation

What You'll Learn

- Write fractions as terminating or repeating decimals.
- Compare fractions and decimals.

Real-World Link

Robotics If you could design your own robot, what would it look like? What would it be able to do? In an annual robot competition, middle school students apply math and science to design, program, and test their own robots. The goal is to make their 'bots outperform the competition!

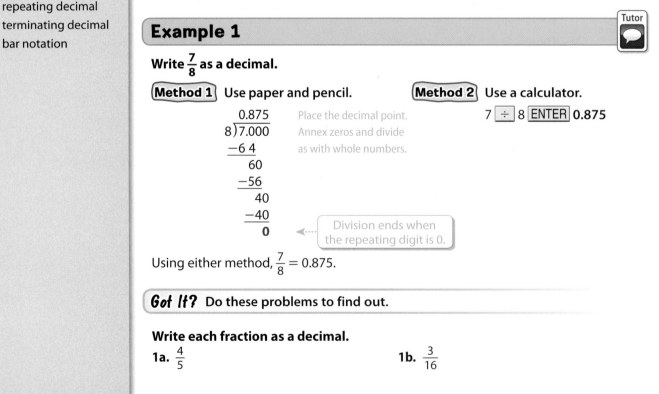

Write Fractions as Decimals

Some fractions like $\frac{1}{2}$ and $\frac{3}{4}$ can be written as a decimal by making equivalent fractions with denominators of 10, 100, or 1000. However, any fraction $\frac{a}{b}$, where $b \neq 0$, can be written as a decimal by dividing the numerator by the denominator. So, $\frac{a}{b} = a \div b$. The decimal form of a rational number is called a **repeating decimal**.

If the repeating digit is zero, then the decimal is a **terminating decimal**.

Example 1

Tutor

Write $\frac{7}{8}$ as a decimal.

Method 1 Use paper and pencil.

```
   0.875
8)7.000
 -6 4
    60
   -56
    40
   -40
     0
```

Place the decimal point. Annex zeros and divide as with whole numbers.

Division ends when the repeating digit is 0.

Method 2 Use a calculator.

7 [÷] 8 [ENTER] 0.875

Using either method, $\frac{7}{8} = 0.875$.

Got It? Do these problems to find out.

Write each fraction as a decimal.

1a. $\frac{4}{5}$

1b. $\frac{3}{16}$

Jeff Greenberg/Alamy

Vocabulary Link
Terminating
Everyday Use bringing to an end
Math Use a decimal whose digits end

Not all fractions have repeating digits that are zero. Sometimes a nonzero digit or a group of digits repeats without end in the quotient

$$\frac{1}{6} \longrightarrow 6\overline{)1.000} = 0.166$$

```
   0.166      ← The digit 6 repeats.
6)1.000
 −6
  40          ← The remainder after
 −36             each step is 4.
   40
  −36
    4
```

Check 1 [÷] 6 [ENTER] **0.1666666667** ✓ The last digit is rounded.

You can indicate that the digit 6 repeats by annexing dots. So,

$\frac{1}{6} = 0.1666666666\ldots$. This decimal is called a repeating decimal.

Repeating decimals have a pattern in their digits that repeats without end. **Bar notation** is a bar or line placed over the digit(s) that repeats. The table shows some examples of repeating decimals and their bar notations.

Decimal	Bar Notation
0.166666...	$0.1\overline{6}$
0.353535...	$0.\overline{35}$
12.6888888...	$12.6\overline{8}$
5.714285714285...	$5.\overline{714285}$

Example 2

Tutor

Write each fraction as a decimal. Use a bar to show a repeating decimal.

a. $\frac{5}{12}$

$\frac{5}{12} \rightarrow 12\overline{)5.0000\ldots} = 0.4166\ldots$ The digit 6 repeats.

So, $\frac{5}{12} = 0.41\overline{6}$.

b. $-\frac{2}{11}$

$-\frac{2}{11} \rightarrow 11\overline{)2.0000\ldots} = 0.1818\ldots$ The digits 18 repeat.

So, $-\frac{2}{11} = -0.\overline{18}$.

Got It? **Do these problems to find out.**

2a. $-\frac{5}{6}$

2b. $\frac{7}{9}$

It is helpful to memorize these fraction-decimal equivalents.

Concept Summary Fraction-Decimal Equivalents

$\frac{1}{2} = 0.5$	$\frac{1}{3} = 0.\overline{3}$	$\frac{1}{4} = 0.25$	$\frac{1}{5} = 0.2$	$\frac{1}{10} = 0.1$	$\frac{1}{100} = 0.01$
$\frac{2}{3} = 0.\overline{6}$	$\frac{3}{4} = 0.75$	$\frac{2}{5} = 0.4$	$\frac{3}{5} = 0.6$	$\frac{4}{5} = 0.8$	$\frac{5}{6} = 0.8\overline{3}$

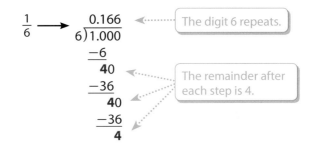

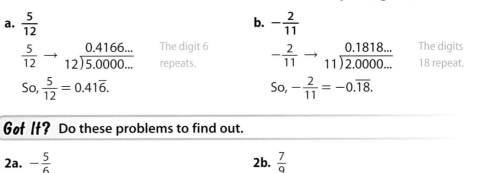

Example 3

According to the USDA, teenage boys should consume an average of 2700 Calories per day. About 360 Calories should come from milk. To the nearest hundredth, what part of a teenage boy's total Calories should come from milk?

Divide the number of Calories that should come from milk, 360, by the number of total Calories, 2700.

360 [÷] 2700 [ENTER] **0.133...** or $0.1\overline{3}$

Look at the digit to the right of the thousandths place. Round down since $3 < 5$.

Milk should be 0.13 of the daily Calories consumed by a teenage boy.

Got It? Do this problem to find out.

3. In a recent Masters Tournament, Zach Johnson's first shot landed on the fairway 45 out of 56 times. To the nearest thousandth, what part of the time did his shot land on the fairway?

Compare Fractions and Decimals

It may be easier to compare numbers when they are written as decimals.

Example 4

Comparing Decimals
When comparing two decimals, compare the digits in the same place-value position.

Replace each ● with <, >, or = to make a true sentence.

a. $\frac{1}{4}$ ● 0.2

$\frac{1}{4}$ ● 0.2	Write the sentence.
0.25 ● 0.2**0**	Write $\frac{1}{4}$ as a decimal. Annex a zero to 0.2.
$0.25 > 0.20$	In the hundredths place, $5 > 0$.

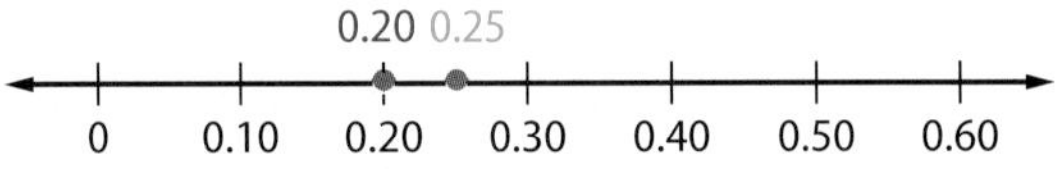

Check Since 0.20 is to the left of 0.25 on the number line, $\frac{1}{4} > 0.2$.

b. $-\frac{5}{8}$ ● $-\frac{6}{9}$

Write the fractions as decimals and then compare the decimals.

$-\frac{5}{8} = -0.625$ $\quad$ $-\frac{6}{9} = -0.666\ldots$ or $-0.\overline{6}$

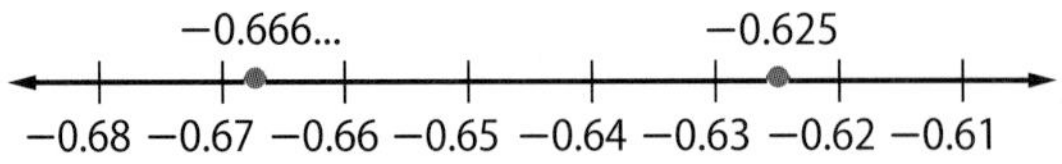

Since -0.625 is to the right of $-0.\overline{6}$ on the number line, $-\frac{5}{8} > -\frac{6}{9}$.

Got It? Do these problems to find out.

4a. $\frac{7}{8}$ ● 0.87

4b. $-\frac{7}{15}$ ● $-\frac{5}{12}$

Example 5

Real World

Tools | Tutor

> **Use a Graph**
> You can use a graph to visualize data, analyze trends, and make predictions. In this example, you can compare the decimals on a number line.

Thirty out of 36 seventh graders and 34 out of 40 eighth graders participated in a marathon for charity. Which class had a greater fraction participating in the marathon?

Write each fraction as a decimal.
Then compare the decimals.

seventh graders: $\frac{30}{36} = 0.8\overline{3}$

eighth graders: $\frac{34}{40} = 0.85$

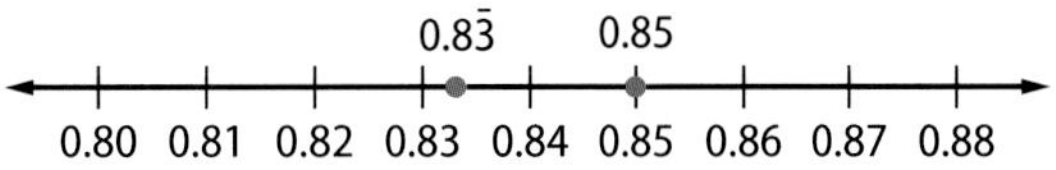

On a number line, $0.8\overline{3}$ is to the left of 0.85. Since $0.8\overline{3} < 0.85$, $\frac{30}{36} < \frac{34}{40}$.

So, a greater fraction of eighth graders participated in the marathon.

Got It? **Do this problem to find out.**

5. Over the weekend, $\frac{16}{28}$ of the eighth grade girls and $\frac{19}{30}$ of the eighth grade boys went to see a new comedy movie. Did a greater fraction of girls or boys see the movie?

Guided Practice

Check

Write each fraction as a decimal. Use a bar to show a repeating decimal. (Examples 1 and 2)

1. $\frac{3}{5}$ **2.** $\frac{5}{16}$ **3.** $-\frac{3}{20}$

4. $\frac{5}{8}$ **5.** $-\frac{2}{3}$ **6.** $-\frac{7}{9}$

7. In one season, the New England Patriots converted 16 of 20 fourth downs. What part of the time did the Patriots convert on fourth down? (Example 3)

Replace each ● with <, >, or = to make a true sentence. (Example 4)

8. $0.89 \bullet \frac{11}{13}$ **9.** $-\frac{2}{3} \bullet -\frac{3}{5}$ **10.** $-0.21 \bullet \frac{1}{5}$

11. $\frac{5}{9} \bullet \frac{6}{11}$ **12.** $-\frac{9}{15} \bullet -0.61$ **13.** $\frac{3}{4} \bullet \frac{7}{9}$

14. Of Nikki's home water usage, $\frac{7}{50}$ comes from lawn watering, and $\frac{3}{20}$ comes from cooking. Does a greater fraction of water usage come from lawn watering or from cooking? (Example 5)

15. On his first reading test, Tre answered $\frac{26}{30}$ questions correctly. On his second reading test, he answered $\frac{34}{40}$ questions correctly. On which test did Tre have the better score? (Example 5)

Independent Practice

Go online for Step-by-Step Solutions eHelp

Write each fraction as a decimal. Use a bar to show a repeating decimal. (Examples 1 and 2)

16. $\frac{3}{8}$ **17.** $\frac{7}{20}$ **18.** $-\frac{8}{25}$ **19.** $-\frac{3}{16}$

20. $\frac{4}{5}$ **21.** $\frac{9}{25}$ **22.** $-\frac{1}{8}$ **23.** $-\frac{7}{16}$

24. $\frac{3}{11}$ **25.** $\frac{33}{45}$ **26.** $-\frac{5}{11}$ **27.** $-\frac{2}{9}$

28. The customer service department resolved 106 of 120 customer complaints in a one-hour time span. To the nearest thousandth, find the resolve rate of the customer service department. (Example 3)

29 In a recent season, Niklas Backstrom of the Minnesota Wild saved 955 out of 1028 shots on goal. To the nearest thousandth, what part of the time did Backstrom save shots on goal? (Example 3)

Replace each ● with <, >, or = to make a true sentence. (Example 4)

30. $\frac{6}{15}$ ● 0.4 **31.** 0.7 ● $\frac{17}{20}$

32. $\frac{5}{6}$ ● $\frac{7}{8}$ **33.** $\frac{5}{7}$ ● $\frac{10}{14}$

34. $-\frac{2}{9}$ ● $-\frac{1}{4}$ **35.** $-\frac{1}{8}$ ● $-\frac{1}{10}$

36. $0.\overline{6}$ ● $\frac{5}{9}$ **37.** $\frac{1}{2}$ ● 0.67

The graph at the right shows the amount of rain, in inches, that fell in a 5-day period. (Example 5)

38. On which days did it rain less than one-fifth of an inch?

39. Did more or less than one-fourth inch of rain fall on Tuesday? Explain.

40. Suppose it rained $\frac{9}{10}$ inch on Saturday. How does this compare to the previous five days?

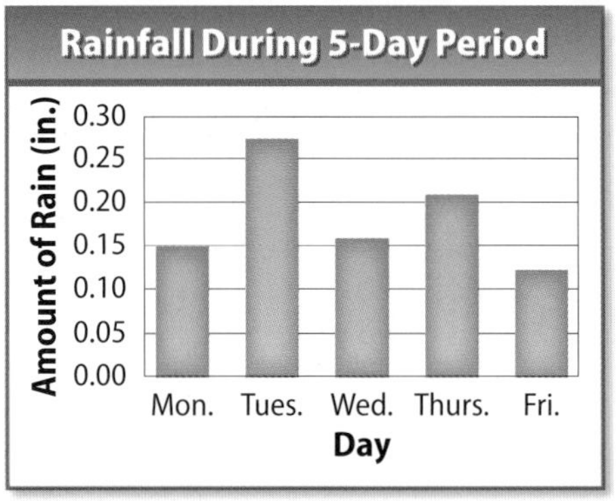

Replace each ● with <, >, or = to make a true sentence.

41. $-\frac{5}{13}$ ● $-0.\overline{36}$ **42.** $0.\overline{54}$ ● $\frac{6}{11}$

43. $-5.\overline{42}$ ● $-5\frac{3}{7}$ **44.** $-\frac{5}{16}$ ● $-\frac{8}{25}$

45. -2.2 ● $-2\frac{2}{7}$ **46.** $-5\frac{1}{3}$ ● $-5\frac{3}{10}$

47. A carpenter has some bolts that are marked $\frac{1}{2}$, $\frac{5}{16}$, $\frac{3}{32}$, $\frac{3}{4}$, and $\frac{3}{8}$. If all measurements are in inches, how should these bolts be arranged from least to greatest?

Order each group of numbers from least to greatest.

48. $-0.29, -\frac{3}{11}, -\frac{2}{7}$ **49.** $2\frac{3}{5}, 2.67, 2\frac{2}{3}$

50. $-1.\overline{1}, -1\frac{1}{8}, -1\frac{1}{10}$ **51.** $\frac{2}{25}, \frac{1}{13}, 0.089$

52. The table shows the number of times at bat and hits that players on the Rawson Middle School team had last season. Order the players based on their batting averages from greatest to least. (*Hint:* Divide the number of hits by the number of at bats.)

Player	Hits	At Bats
Kristen	35	47
Cho	51	73
Brooke	36	50
Alma	49	65
Jessica	46	60

Write each decimal using bar notation.

53. 0.99999…

54. 4.636363…

55. −10.3444…

56. −22.8151515…

57. CCSS **Multiple Representations** Use the number line shown.

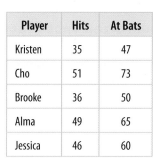

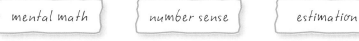

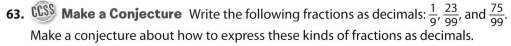

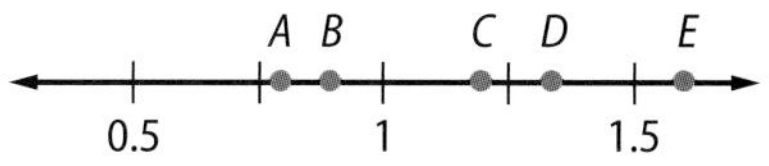

a. Numbers Find a fraction or mixed number that might represent each point on the graph.

b. Symbols Write an inequality using two of your values.

58. The table shows the number of each type of bead on 3 bracelets that Mrs. Fraser made for a craft show. Which bracelet has the greatest fraction of glass beads? the least?

Mrs. Fraser's Bracelets			
Bead Type	Bracelet 1	Bracelet 2	Bracelet 3
glass	9	10	9
clay	5	5	4
metal	12	18	14

H.O.T. Problems Higher Order Thinking

59. CCSS **Model with Mathematics** Give one example each of real-world situations where it is most appropriate to give a response in fractional form and in decimal form.

60. CCSS **Justify Conclusions** Are there any rational numbers between $0.\overline{4}$ and $\frac{4}{9}$? Explain.

61. CCSS **Identify Structure** A *unit fraction* is a fraction that has 1 as its numerator. Write the four greatest unit fractions that are repeating decimals. Write each fraction as a decimal.

62. CCSS **Use Math Tools** Luke is making lasagna that calls for $\frac{4}{5}$ pound of mozzarella cheese. The store only has packages that contain 0.75- and 0.85-pound of mozzarella cheese. Which of the following techniques might Luke use to determine which package to buy? Justify your selection(s). Then use the technique(s) to solve the problem.

mental math | number sense | estimation

63. CCSS **Make a Conjecture** Write the following fractions as decimals: $\frac{1}{9}$, $\frac{23}{99}$, and $\frac{75}{99}$. Make a conjecture about how to express these kinds of fractions as decimals.

64. **Building on the Essential Question** Explain how 0.5 and $0.\overline{5}$ are different. Which is greater?

Standardized Test Practice

65. Sherman answered $\frac{4}{5}$ of the multiple-choice questions on his science test correctly. Write this fraction as a decimal.

A 0.4 **C** 0.8

B 0.45 **D** 4.5

66. Which of the following show the fractions $\frac{2}{5}, \frac{3}{8}, \frac{1}{3}, \frac{1}{2}$, and $\frac{5}{12}$ in order from least to greatest?

F $\frac{1}{3}, \frac{3}{8}, \frac{2}{5}, \frac{1}{2}, \frac{5}{12}$ **H** $\frac{1}{3}, \frac{3}{8}, \frac{2}{5}, \frac{5}{12}, \frac{1}{2}$

G $\frac{1}{2}, \frac{1}{3}, \frac{2}{5}, \frac{3}{8}, \frac{5}{12}$ **J** $\frac{1}{2}, \frac{5}{12}, \frac{2}{5}, \frac{3}{8}, \frac{1}{3}$

67. The fraction $\frac{7}{9}$ is found between which pair of fractions on a number line?

A $\frac{3}{5}$ and $\frac{3}{4}$ **C** $\frac{7}{10}$ and $\frac{3}{4}$

B $\frac{7}{10}$ and $\frac{4}{5}$ **D** $\frac{3}{5}$ and $\frac{2}{3}$

68. **Short Response** Which item(s) shown in the table have a recycle rate less than one half?

Material	Fraction Recycled
paper	$\frac{5}{11}$
aluminum cans	$\frac{5}{8}$
glass	$\frac{2}{5}$

Common Core Review

Find each product or quotient. 7.NS.2

69. 4(−12)(−5) **70.** −2(42)(3) **71.** −54 ÷ (−6) **72.** 72 ÷ (−9)

73. A scuba diver descends from the surface of the lake at a rate of 6 meters per minute. Where will the diver be in relation to the lake's surface after 4 minutes? 6.RP.2

Name the ordered pair for each point graphed at the right. 6.NS.6b

74. *E* **75.** *G*

76. *C* **77.** *D*

78. *A* **79.** *B*

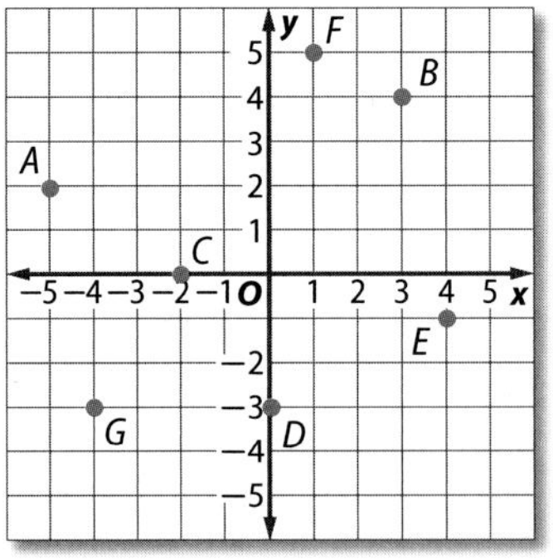

Evaluate each expression if *x* = 7, *y* = 3, and *z* = 5. 6.EE.2

80. $x + y + z$ **81.** $4x - z$

82. $6y - z$ **83.** $9x + 8y$

84. A store sells new and used CDs. Each type of CD sells for the same price. Roberta bought an equal number of new and used CDs. Find the missing values. 6.NS.3

Roberta's Purchases			
Description	Number	Unit Cost	Price
new CD	■	$9.95	$39.80
used CD	■	$5.98	$23.92
		Total:	$63.72

Write each decimal in word form. 5.NBT.3

85. 0.34 **86.** 5.836

87. 0.3 **88.** 1.6

Lesson 3-2

Rational Numbers

ISG Interactive Study Guide

See pages 53–54 for:
- Getting Started
- Vocabulary Start-Up
- Notes

Essential Question

What happens when you add, subtract, multiply, and divide rational numbers?

CCSS Common Core State Standards

Content Standards
7.NS.2, 7.NS.2d, 8.NS.1, 7.EE.3

Mathematical Practices
1, 3, 4, 7

Vocab Vocabulary

rational numbers

What You'll Learn

- Write rational numbers as fractions.
- Identify and classify rational numbers.

Real-World Link

Monkeys New World monkeys are primates that are found in Mexico, Central America, and South America. Their common characteristics include being small to mid-sized, possessing high intelligence, and being skilled in using their hands. In a similar way, numbers can also be organized into sets based on shared characteristics.

Rational Numbers

When you first learned to count using the numbers 1, 2, 3, …, you were using members of the set of *natural numbers*, N.

If you add zero to the set of natural numbers, the result is the set of *whole numbers*, W = {0, 1, 2, 3, …}.

Whole numbers and their opposites make up the set of *integers*, Z = {…, −3, −2, −1, 0, 1, 2, 3, …}.

Any number that can be written in the form $\frac{a}{b}$, where *a* and *b* are integers and $b \neq 0$ is part of the set of **rational numbers**, Q. Some examples of rational numbers are shown below.

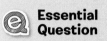

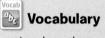

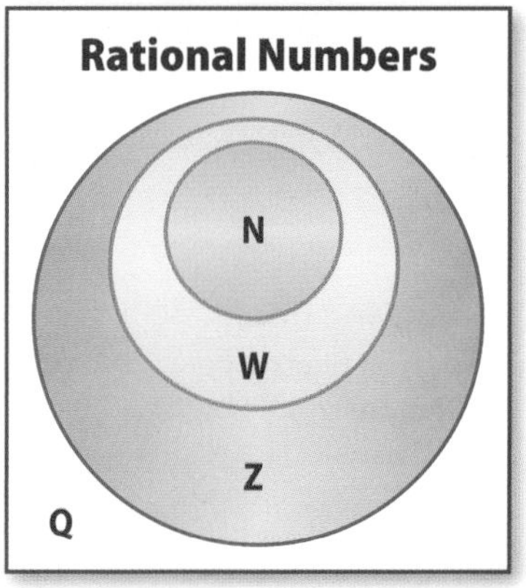

$$0.87 \qquad -23 \qquad \frac{2}{3} \qquad -2.\overline{56} \qquad 1\frac{1}{2}$$

Example 1

Write each rational number as a fraction.

a. $6\frac{1}{6}$

$6\frac{1}{6} = \frac{37}{6}$ Write $6\frac{1}{6}$ as an improper fraction.

b. −23

$-23 = \frac{-23}{1}$ or $-\frac{23}{1}$

Got It? **Do these problems to find out.**

1a $4\frac{2}{3}$

1b. 7

Terminating and repeating decimals are rational numbers because they can be written as fractions

Decimal Point

Use the word *and* to represent the decimal point.

- Read 0.625 as *six hundred twenty-five thousandths.*
- Read 20.005 as *twenty and five thousandths.*

Example 2

Tutor

a. Write 0.64 as a fraction in simplest form.

$0.64 = \frac{64}{100}$ — 0.64 is 64 hundredths.

$= \frac{16}{25}$ — The GCF of 64 and 100 is 4.

thousands	hundreds	tens	ones	tenths	hundredths	thousandths	ten-thousandths
0	0	0	0 .	6	4	0	0

b. A handheld video game system weighs 9.675 ounces. Write this decimal as a mixed number in simplest form.

$9.675 = 9\frac{675}{1000}$ — 0.675 is 675 thousandths.

$= 9\frac{27}{40}$ — The GCF of 675 and 1000 is 25.

thousands	hundreds	tens	ones	tenths	hundredths	thousandths	ten-thousandths
0	0	0	9 .	6	7	5	0

Got It? **Do these problems to find out.**

Write each decimal as a fraction in simplest form.

2a. 0.84

2b. 5.875

2c. Rock music accounted for 0.35 of the total music sales in a recent year. Write this decimal as a fraction in simplest form.

Repeating Decimals

When *two* digits repeat, multiply each side by 100. Then subtract N from $100N$ to eliminate the repeating part.

Example 3

Tutor

Write $0.\overline{6}$ as a fraction in simplest form.

$N = 0.6666\ldots$ — Let N represent the number.

$\mathbf{10}N = \mathbf{10}(0.6666\ldots)$ — Multiply each side by 10 because one digit repeats.

$10N = 6.666\ldots$

Subtract N from $10N$ to eliminate the repeating part, $0.666\ldots$.

$$\begin{aligned} 10N &= 6.666\ldots \\ -N &= 0.666\ldots \\ \hline 9N &= 6 \end{aligned}$$

$10N - N = 10N - 1N$ or $9N$

$\frac{9N}{\mathbf{9}} = \frac{6}{\mathbf{9}}$ — Divide each side by 9.

$N = \frac{6}{9}$ or $\frac{2}{3}$

Check 6 [÷] 9 [ENTER] 0.666666667 ✓

Got It? **Do this problem to find out.**

3. Write $0.\overline{42}$ as a fraction in simplest form.

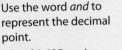

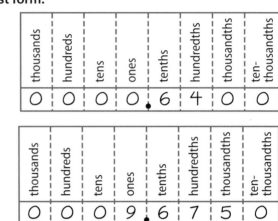

Ratios

Rational comes from the word ratio. A *ratio* is the comparison of two quantities by division. Recall that $\frac{a}{b} = a \div b$, where $b \neq 0$.

Key Concept Rational Numbers

Watch

A rational number is any number that can be expressed as the quotient $\frac{a}{b}$, where a and b are integers and $b \neq 0$.

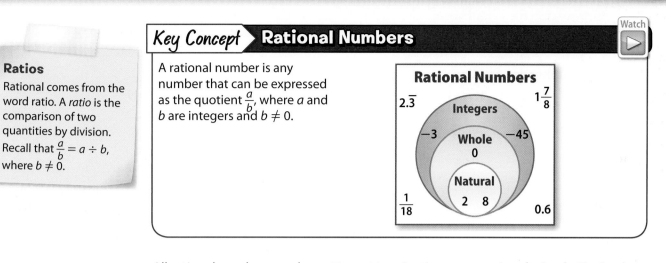

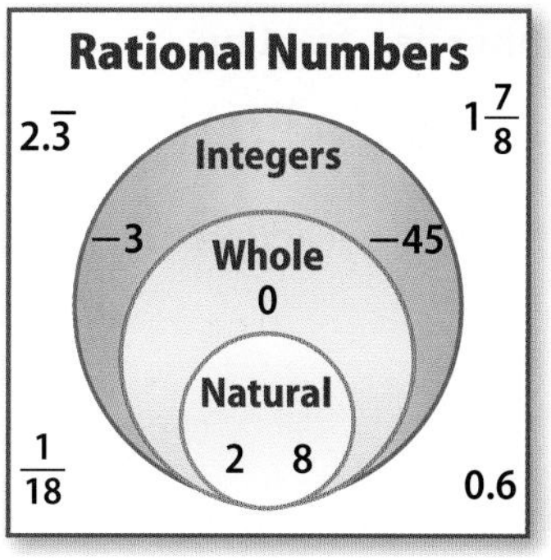

All rational numbers can be written as terminating or repeating decimals. Decimals that neither terminate nor repeat, such as the numbers below, are called *irrational numbers*. You will learn more about irrational numbers in Chapter 4.

$\pi = 3.141592\ldots$ → The digits do not repeat.

$8.787787778\ldots$ → The same block of digits does not repeat.

Example 4

Tutor

Identify all sets to which each number belongs.

a. $-2\frac{6}{11}$

Since $-2\frac{6}{11}$ can be written as $-\frac{28}{11}$, it is rational.

b. **1.313313331…**

This is a nonterminating and nonrepeating decimal. So, it is irrational.

c. **45**

45 is a natural number, a whole number, an integer, and a rational number.

Got It? **Do these problems to find out.**

4a. 0 **4b.** $1\frac{4}{5}$ **4c.** 1.414213562...

Guided Practice

Check

Write each number as a fraction. (Example 1)

1. $3\frac{3}{4}$ **2.** -9 **3.** $-1\frac{3}{4}$

Write each decimal as a fraction or mixed number in simplest form. (Examples 2 and 3)

4. 0.07 **5.** $-3.\overline{85}$ **6.** $0.\overline{78}$

7. There are approximately 2.54 centimeters in 1 inch. Express 2.54 as a mixed number.

Identify all sets to which each number belongs. (Example 4)

8. -632 **9.** $0.\overline{56}$ **10.** 21

Independent Practice

Go online for Step-by-Step Solutions — eHelp

Write each number as a fraction. (Example 1)

11. $1\frac{5}{6}$ **12.** -12 **13.** $-10\frac{7}{8}$ **14.** 49

Write each decimal as a fraction or mixed number in simplest form. (Example 2)

15. 3.625 **16.** 0.55 **17.** -5.36

18. -0.265 **19.** -1.3 **20.** 0.9

21. **Financial Literacy** Recently, one U.S. dollar was equal to 0.506 British pounds. Express 0.506 as a fraction in simplest form. (Example 2)

22. The estimated portions for various age groups of the population for 2010 are shown in the table. (Example 2)

a. Find the fraction in simplest form of the population that is 19 years of age or younger.

b. Find the fraction in simplest form of the population that is 20 to 64 years of age.

Age Group	Portion of Population
19 years and under	0.27
20 to 64 years	0.60
65 years and over	0.13

Write each decimal as a fraction or mixed number in simplest form. (Example 3)

23. $-2.\overline{5}$ **24.** $0.\overline{36}$ **25.** 0.161616…

26. $9.\overline{27}$ **27.** $-0.\overline{09}$ **28.** $-10.\overline{74}$

Identify all sets to which each number belongs. (Example 4)

29. -8 **30.** 14 **31.** 9.23

32. $1\frac{5}{9}$ **33.** 0.323322333… **34.** 3.141516…

35. Maria has a bead that is 0.6 inch long. She wants to use the bead to fill a space that is $\frac{5}{8}$ inch long. Will the bead fit? Explain.

36. All of the Calories in one cup of milk come from fat, protein, and carbohydrates. Use the table to find the fraction of Calories that comes from protein. Write the fraction in simplest form.

Nutrient	Decimal Part of Calories
fat	0.03
protein	■
carbohydrates	0.53

Replace each ● with <, >, or = to make a true sentence.

37. -0.23 ● -0.3 **38.** $\frac{8}{9}$ ● 0.888…

39. 0.714 ● $\frac{5}{7}$ **40.** $-1\frac{1}{11}$ ● -0.9

41. $4.\overline{63}$ ● $4\frac{5}{8}$ **42.** $-5.\overline{3}$ ● 5.333…

Write each decimal as a fraction or mixed number in simplest form.

43. $0.\overline{652}$ **44.** $0.1\overline{8}$

45. $0.72\overline{4}$ **46.** $3.5\overline{96}$

47. $9.2\overline{43}$ **48.** $0.24\overline{67}$

49. CCSS **Multiple Representations** Pi (π) is a nonrepeating, nonterminating decimal. Two common estimates for pi are 3.14 and $\frac{22}{7}$.

a. Graph Use a calculator to find the value of π to seven decimal places. Graph this value, 3.14, and $\frac{22}{7}$ on a number line.

b. Symbols Write an inequality comparing the values.

c. Words To find the circumference of a circle, you multiply pi by the diameter *d* of the circle. Explain when you might use 3.14 to find the circumference and when you might use $\frac{22}{7}$ to find the circumference.

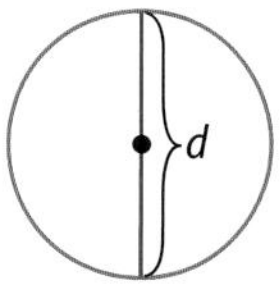

50. The mathematician Archimedes believed that π was between $3\frac{1}{7}$ and $3\frac{10}{71}$.

a. Express each mixed number as a decimal rounded to the nearest thousandth. Was Archimedes' theory correct? Explain.

b. The Rhind Papyrus records that the Egyptians used $\frac{256}{81}$ for π. Express the fraction as a decimal rounded to the nearest thousandth. Which value is closer to the actual value of π, Arcihimedes' or the Egyptians' value?

Order each set of rational numbers from least to greatest.

51. $-3.4, 3\frac{4}{11}, -3.\overline{42}, 3.38$

52. $\frac{1}{3}, 0.\overline{13}, \frac{5}{13}, 0.32$

53. $-1\frac{13}{14}, -1.9, -1\frac{9}{11}, -1.95$

54. $9\frac{4}{5}, 9.\overline{79}, 9\frac{11}{13}, 9.82$

55. A lion's speed is $\frac{5}{7}$ the speed of a cheetah. Find the least rational number with a denominator of 9 that is greater than $\frac{5}{7}$. Find the greatest rational number with a denominator of 8 that is less than $\frac{5}{7}$. Write an inequality comparing the three numbers.

H.O.T. Problems Higher Order Thinking

56. CCSS **Model with Mathmatics** Choose a repeating decimal in which three digits repeat. Write the number as a fraction or mixed number in simplest form.

57. CCSS **Construct an Argument** Explain why $0.\overline{76}$ is greater than 0.76.

58. CCSS **Persevere with Problems** Antonio stated that $0.\overline{9} = 1$. Show that he is correct.

59. CCSS **Use a Counterexample** Determine whether the following statements are *true* or *false*. If true, explain your reasoning. If false, give a counterexample.

a. All integers are rational numbers.

b. All whole numbers are integers.

c. A rational number is always an integer.

d. All natural numbers are rational.

60. **Building on the Essential Question** How do you compare and order fractions and decimals? Give an example to explain your reasoning.

Standardized Test Practice

61. Which fraction is between 0.12 and 0.15?

A $\frac{3}{25}$ **C** $\frac{3}{20}$

B $\frac{1}{8}$ **D** $\frac{1}{5}$

62. Which of the following is not a rational number?

F $\frac{4}{9}$ **H** $0.\overline{62}$

G -4.27 **J** $-3.131131113\ldots$

63. Last football season, Jason made 0.85 of his field goal attempts. Write this decimal as a fraction in simplest form.

A $\frac{85}{100}$ **C** $\frac{17}{20}$

B $\frac{20}{17}$ **D** $\frac{100}{85}$

64. **Short Response** The table shows the results of a survey about how students get to school.

Method of Transportation	Portion of Students
bus	0.40
walk	0.18
car	0.36
bicycle	0.04
other	0.02

Which method of transportation do most students use to get to school? Write the fraction of students who use this method of transportation in simplest form.

CCSS Common Core Review

Write each fraction as a decimal. Use a bar to show a repeating decimal. 8.NS.1

65. $-\frac{5}{8}$ **66.** $\frac{1}{6}$ **67.** $-\frac{2}{10}$ **68.** $\frac{4}{7}$

69. Ms. Adepoju grades students' exams by starting with a perfect score of 100, and marking points off for incorrect answers. The table shows the results of 5 students on a recent exam. 6.NS.7

a. Write an integer to describe each student's grade with respect to a perfect score of 100.

b. Order the students from highest to lowest grades.

Student	Points Taken Off
Ava	5
Brennan	3
Denny	6
Jose	0
Hao	7

Convert each measurement to the given units. 5.MD.1

70. 24 feet to inches **71.** 24 ounces to pounds **72.** 300 minutes to hours

73. Mount Kilimanjaro's elevation is 5895 meters. Lake Assal's elevation is −155 meters. Find the difference between these elevations. 7.NS.1

Find each product. 7.NS.2, 7.NS.2a

74. $-6(-12)$ **75.** $15(-3)(-4)(0)$ **76.** $-3(5)(-9)$ **77.** $14(-20)$

Find each quotient. 7.NS.2, 7.NS.2b

78. $\frac{1}{-1}$ **79.** $\frac{-16}{8}$ **80.** $\frac{-100}{-10}$ **81.** $\frac{0}{-5}$

Lesson 3-3

Multiplying Rational Numbers

ISG Interactive Study Guide

See pages 55–56 for:
- Getting Started
- Real-World Link
- Notes

Essential Question

What happens when you add, subtract, multiply, and divide rational numbers?

CCSS Common Core State Standards

Content Standards
7.NS.2, 7.NS.2a, 7.NS.2c, 7.NS.3, 7.EE.3

Mathematical Practices
1, 3, 4, 5

What You'll Learn

- Multiply positive and negative fractions.
- Evaluate algebraic expressions with fractions.

Real-World Link

Ice Cream A survey was taken to find the most popular ice cream flavors in the United States. Some of the top ten flavors contain chocolate, some contain fruit, and some contain both! You can use rational numbers to analyze the results of the survey.

Key Concept Multiply Fractions

Words	To multiply fractions, multiply the numerators and multiply the denominators.
Symbols	$\frac{a}{b} \cdot \frac{c}{d} = \frac{a \cdot c}{b \cdot d}$, where $b, d \neq 0$
Example	$\frac{3}{4} \cdot \frac{1}{2} = \frac{3 \cdot 1}{4 \cdot 2}$ or $\frac{3}{8}$

The green area in the area model below shows the product of $\frac{1}{3}$ and $\frac{1}{2}$. One sixth of the second circle is shaded green. So, $\frac{1}{3} \times \frac{1}{2} = \frac{1}{6}$.

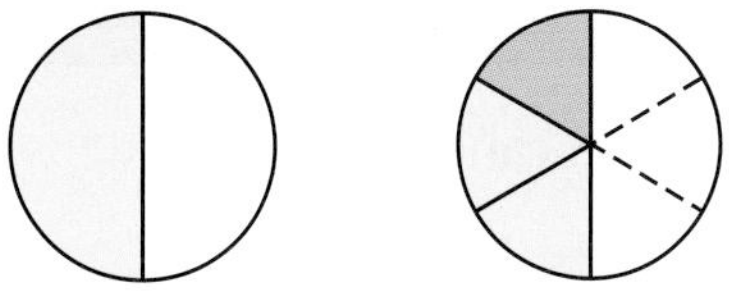

Draw a circle and shade $\frac{1}{2}$ of it yellow.

Then shade $\frac{1}{3}$ of the yellow part green.

Example 1

Tutor

Find $\frac{1}{6} \cdot \frac{2}{3}$. Write the product in simplest form.

$\frac{1}{6} \cdot \frac{2}{3} = \frac{1 \cdot 2}{6 \cdot 3}$ ← Multiply the numerators. ← Multiply the denominators.

$= \frac{2}{18}$ or $\frac{1}{9}$ Simplify. The GCF of 2 and 18 is 2.

Got It? Do these problems to find out.

Find each product. Write in simplest form.

1a. $\frac{1}{2} \cdot \frac{4}{10}$ **1b.** $\frac{5}{12} \cdot \frac{6}{10}$ **1c.** $\frac{3}{4} \cdot \frac{8}{21}$

If the fractions have common factors in the numerators and denominators, you can simplify before you multiply.

Negative Signs When working with negative fractions, assign the negative sign to the numerator. Example: Write $-\frac{7}{9}$ as $\frac{-7}{9}$.

Example 2

Tutor

Find each product. Write in simplest form.

a. $\frac{3}{4}\left(-\frac{7}{9}\right)$

$\frac{3}{4}\left(-\frac{7}{9}\right) = \frac{\overset{1}{\cancel{3}}}{4}\left(\frac{-7}{\underset{3}{\cancel{9}}}\right)$ Divide 3 and 9 by their GCF, 3.

$= \frac{1 \cdot -7}{4 \cdot 3}$ Multiply the numerators and multiply the denominators.

$= -\frac{7}{12}$ Simplify.

b. $2\frac{1}{3} \cdot 2\frac{5}{7}$

Estimate $2 \cdot 3 = 6$

$2\frac{1}{3} \cdot 2\frac{5}{7} = \frac{7}{3} \cdot \frac{19}{7}$ Rename $2\frac{1}{3}$ as $\frac{7}{3}$ and $2\frac{5}{7}$ as $\frac{19}{7}$.

$= \frac{\overset{1}{\cancel{7}}}{3} \cdot \frac{19}{\underset{1}{\cancel{7}}}$ Divide by the GCF, 7.

$= \frac{1 \cdot 19}{3 \cdot 1}$ Multiply.

$= \frac{19}{3}$ or $6\frac{1}{3}$ Simplify.

Check for Reasonableness The solution is close to the estimate. ✔

Got It? Do these problems to find out.

2a. $-\frac{9}{12} \cdot -\frac{2}{3}$

2b. $\frac{6}{9} \cdot -\frac{3}{11}$

2c. $3\frac{3}{8} \cdot 2\frac{1}{3}$

Evaluate Expressions with Fractions

Variables can represent fractions in algebraic expressions.

Example 3

Tutor

Evaluate $\frac{1}{2}ab$ if $a = \frac{6}{7}$ and $b = -\frac{4}{9}$. Write in simplest form.

$\frac{1}{2}ab = \frac{1}{2}\left(\frac{6}{7}\right)\left(-\frac{4}{9}\right)$ Replace a with $\frac{6}{7}$ and b with $-\frac{4}{9}$.

$= \frac{1}{\underset{1}{\cancel{2}}}\left(\frac{\overset{2}{\cancel{6}}}{7}\right)\left(-\frac{\overset{2}{\cancel{4}}}{\underset{3}{\cancel{9}}}\right)$ The GCF of 6 and 9 is 3. The GCF of 2 and 4 is 2.

$= -\frac{4}{21}$ Simplify.

Got It? Do these problems to find out.

Evaluate each expression if $x = \frac{3}{8}$, $y = -2\frac{2}{9}$, and $z = -\frac{7}{10}$. Write in simplest form.

3a. xy

3b. $5x$

3c. yz

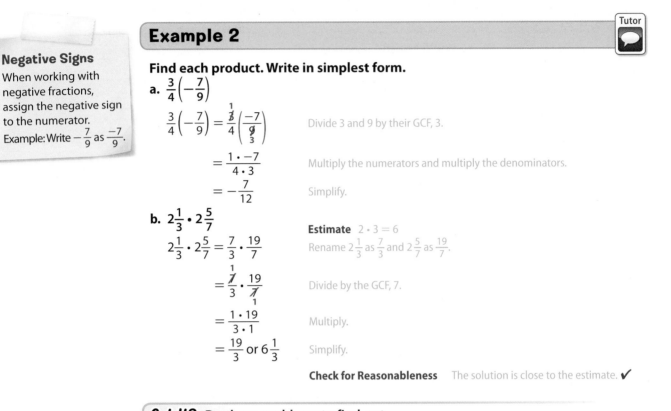

Example 4

Real World

Tutor

The first hill on a certain roller coaster is 255 feet tall. The first drop on another roller coaster is about $\frac{11}{20}$ as tall as the first coaster. Find the height of the hill on the second roller coaster.

To find the height of the hill on the second roller coaster, multiply $\frac{11}{20}$ by 255.

$\frac{11}{20} \cdot 255 = \frac{11}{20} \cdot \frac{255}{1}$ Rename 255 as $\frac{255}{1}$.

$= \frac{11}{\cancel{20}_{4}} \cdot \frac{\cancel{255}^{51}}{1}$ Divide by the GCF, 5.

$= \frac{11 \cdot 51}{4 \cdot 1}$ Multiply.

$= \frac{561}{4}$ or $140\frac{1}{4}$ Simplify.

So, the height of the drop is about 140 feet.

Got It? Do these problems to find out.

4a. The Willis Tower in Chicago is about 1450 feet tall. The Empire State Building in New York City is about $\frac{4}{5}$ as tall. About how tall is the Empire State Building?

4b. The longest suspension bridge in the United States is the 4260-foot Verrazano-Narrows Bridge in New York City. The Tacoma Narrows Bridge in Tacoma, Washington, is about $\frac{11}{12}$ of that length. About how long is the Tacoma Narrows Bridge?

Guided Practice

Check

Find each product. Write in simplest form. (Examples 1 and 2)

1. $\frac{7}{8} \cdot \frac{1}{2}$

2. $\frac{1}{3} \cdot \frac{2}{5}$

3. $-\frac{2}{3} \cdot \frac{3}{16}$

4. $-\frac{3}{5}\left(-\frac{10}{21}\right)$

5. $-4\frac{1}{2}\left(-1\frac{1}{9}\right)$

6. $-2\frac{1}{2} \cdot 5\frac{2}{3}$

Evaluate each expression if $x = \frac{14}{15}$, $y = -1\frac{2}{5}$, and $z = -\frac{3}{7}$. Write the product in simplest form. (Example 3)

7. xy

8. $z \cdot z$

9. xz

10. $\frac{3}{4}xz$

11. $4y$

12. $\frac{7}{3}z$

13. "Midway" is the name of 252 towns in the United States. "Pleasant Hill" occurs $\frac{5}{9}$ as many times. How many towns named "Pleasant Hill" are there in the United States? (Example 4)

14. Of the 480 students at Pleasantview Middle School, $\frac{13}{20}$ play a school sport. How many students play a sport? (Example 4)

Independent Practice

Go online for Step-by-Step Solutions eHelp

Find each product. Write in simplest form. (Examples 1 and 2)

15. $\frac{3}{4} \cdot \frac{1}{8}$

16. $\frac{3}{7} \cdot \frac{1}{6}$

17. $\frac{2}{3} \cdot \frac{4}{9}$

18. $\frac{1}{12} \cdot \frac{3}{8}$

19. $\frac{5}{10} \cdot \frac{2}{9}$

20. $\frac{4}{5} \cdot \frac{5}{8}$

21. $-\frac{1}{15}\left(-\frac{10}{13}\right)$

22. $-\frac{6}{10}\left(-\frac{1}{8}\right)$

23. $3\frac{1}{3}\left(-\frac{1}{5}\right)$

24. $\frac{12}{45}\left(-\frac{9}{16}\right)$

25. $-1\frac{1}{2} \cdot \frac{2}{3}$

26. $4\frac{3}{8}\left(-3\frac{3}{7}\right)$

Evaluate each expression if $a = \frac{10}{24}$, $b = -3\frac{1}{8}$, and $c = -\frac{4}{5}$. Write the product in simplest form. (Example 3)

27. bc

28. ab

29. $2c$

30. $\frac{2}{3}abc$

31. $-4bc$

32. $-\frac{19}{5}ac$

33. The average person living in Argentina consumes about 145 pounds of beef per year. The average person living in the United States consumes about $\frac{3}{5}$ as much. How many pounds of beef does the average American consume every year? (Example 4)

34. The Golden Gate Bridge in San Francisco is 4200 feet long. The Brooklyn Bridge in New York City is $\frac{19}{50}$ as long. How long is the Brooklyn Bridge? (Example 4)

Find each product. Write in simplest form.

35. $\frac{3}{5} \cdot \frac{10}{28} \cdot \frac{2}{9}$

36. $\frac{2}{3} \cdot \frac{1}{4} \cdot \frac{6}{13}$

37. $3\frac{1}{2} \cdot \left(-1\frac{1}{14}\right) \cdot \frac{4}{5}$

38. $4\frac{1}{5} \cdot \left(-1\frac{3}{7}\right) \cdot \frac{6}{11}$

39. $-\frac{6}{11} \cdot (-4) \cdot -2\frac{3}{4} \cdot \frac{1}{3}$

40. $\left(-\frac{9}{10}\right) \cdot 7 \cdot 2\frac{1}{3} \cdot \frac{1}{21}$

41. Dexter's lawn is $\frac{2}{3}$ of an acre. If $7\frac{1}{2}$ bags of fertilizer are needed for 1 acre, how much will he need to fertilize his lawn?

42. A certain hybrid car can travel $1\frac{4}{11}$ times as far as a similar nonhybrid car can, each on one gallon of gasoline. If the nonhybrid car can travel 33 miles per gallon of gasoline, how far can the hybrid travel on $\frac{4}{5}$ gallon of gasoline?

Complete each conversion.

43. ■ ounces = $\frac{3}{4}$ pound
(*Hint*: 1 pound = 16 ounces)

44. ■ feet = $\frac{2}{3}$ mile
(*Hint*: 1 mile = 5280 feet)

45. $\frac{5}{6}$ foot = ■ inches

46. $\frac{1}{4}$ minute = ■ seconds

47. ■ cups = $\frac{1}{4}$ gallon
(*Hint*: 1 gallon = 16 cups)

48. $\frac{3}{4}$ year = ■ weeks

49. Use a cookbook or the Internet to find a recipe for guacamole. Change the recipe to make $2\frac{1}{4}$ times the original amount.

50. The table shows the statistics from the last election for eighth grade class president. There are 540 students in the eighth grade.

a. How many students voted for Hector?

b. How many students voted for Nora?

c. Were there other candidates for class president? How do you know? Explain your reasoning. If there were other candidates, what fraction of the student body voted for them?

Class Elections	
Fraction of class that voted	$\frac{3}{4}$
Fraction of votes for Hector	$\frac{3}{5}$
Fraction of votes for Nora	$\frac{1}{3}$

51. CCSS **Use Math Tools** The expression $\frac{1}{2} \times 4$ means $\frac{1}{2}$ of 4. The number line shows that the product of $\frac{1}{2}$ and 4 is 2. Find each product using a number line.

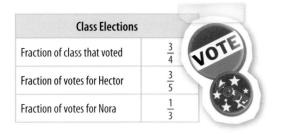

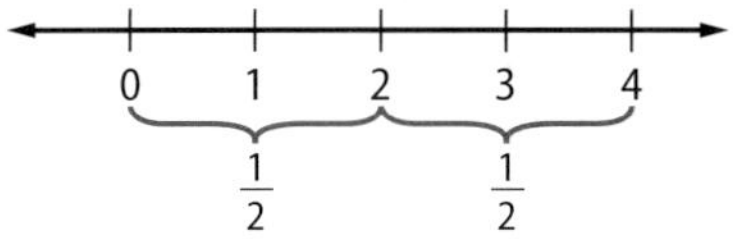

a. $\frac{2}{3}$ of 6
b. $\frac{3}{4}$ of 8
c. $\frac{1}{2}$ of $\frac{2}{3}$
d. $\frac{1}{2}$ of 2
e. $\frac{2}{3}$ of $\frac{3}{2}$
f. $\frac{3}{7}$ of $\frac{7}{3}$

g. Look back at the solutions for Exercises d–f. What pattern do you notice?

h. What is the product of $\frac{a}{b} \cdot \frac{b}{a}$ where $a, b \neq 0$?

H.O.T. Problems Higher Order Thinking

52. CCSS **Justify Conclusions** Find two rational numbers greater than $\frac{1}{3}$ whose product is less than $\frac{1}{3}$. Explain your reasoning to a classmate.

53. CCSS **Find the Error** Kelly is finding $-4\frac{1}{6} \cdot 2\frac{2}{9}$. Find her mistake and correct it.

$$-4\frac{1}{6} \cdot 2\frac{2}{9} = -4\frac{1}{\cancel{6}_3} \cdot 2\frac{\cancel{2}^1}{9}$$
$$= -8\frac{1}{27}$$

CCSS **Persevere with Problems** **Find each missing fraction.**

54. $\frac{2}{3} \cdot \frac{x}{y} = -\frac{3}{8}$

55. $\frac{a}{b} \cdot \left(-\frac{3}{4}\right) = \frac{5}{8}$

56. $\frac{8}{9} \cdot \frac{m}{n} = \frac{14}{27}$

57. $-\frac{9}{10} \cdot \frac{c}{d} = -\frac{3}{5}$

58. $\frac{r}{s} \cdot \frac{4}{5} = \frac{1}{10}$

59. $\frac{1}{3} \cdot \frac{x}{y} = -\frac{1}{5}$

60. CCSS **Make a Conjecture** Investigate the product of a fraction between 0 and 1 and a whole number or mixed number. Is the product *always*, *sometimes*, or *never* less than the whole number or mixed number? Explain.

61. **Building on the Essential Question** Estimate $3\frac{3}{5} \cdot 4\frac{2}{3}$. Then find the actual product. Explain why the estimate and the product are different. What could you do to make your estimate closer to the actual product?

Standardized Test Practice

62. Of the students in Mr. Bogg's class, $\frac{3}{5}$ participate in an after-school sport. Of these, $\frac{1}{3}$ participate in track and field. What fraction of the students participates in track and field?

A $\frac{1}{5}$ **C** $\frac{3}{5}$

B $\frac{1}{3}$ **D** $\frac{14}{15}$

63. Which statement is shown on the number line below?

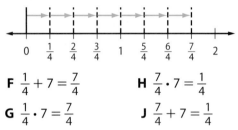

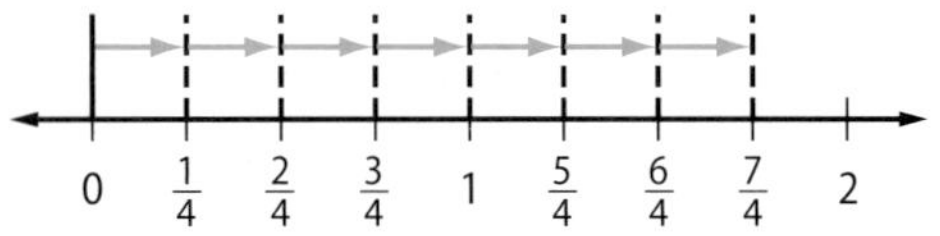

F $\frac{1}{4} + 7 = \frac{7}{4}$ **H** $\frac{7}{4} \cdot 7 = \frac{1}{4}$

G $\frac{1}{4} \cdot 7 = \frac{7}{4}$ **J** $\frac{7}{4} + 7 = \frac{1}{4}$

64. What is the value of the expression $2ab$ if $a = \frac{5}{7}$ and $b = -\frac{3}{8}$?

A $-2\frac{15}{56}$

B $-\frac{15}{28}$

C $\frac{15}{28}$

D $2\frac{15}{56}$

65. **Short Response** The length of one side of a square garden tile is $1\frac{2}{3}$ feet. Write mixed numbers to represent the perimeter and the area of the tile.

Common Core Review

Write each decimal as a fraction or mixed number in simplest form. 8.NS.1

66. 4.02 **67.** 0.215 **68.** −5.125

69. $-0.\overline{3}$ **70.** $4.\overline{5}$ **71.** $-2.\overline{05}$

Replace each ● with <, >, or = to make a true statement. 6.NS.7

72. $0.3 \bullet \frac{1}{4}$ **73.** $\frac{5}{8} \bullet 0.65$ **74.** $\frac{2}{5} \bullet 0.4$

75. $\frac{7}{8} \bullet \frac{8}{9}$ **76.** $\frac{1}{5} \bullet 0.\overline{5}$ **77.** $3\frac{4}{9} \bullet 3.\overline{4}$

78. In an online survey, about $\frac{1}{4}$ of teenagers go to sleep between 9 and 10 P.M., while $\frac{13}{50}$ of teenagers go to sleep at 12 A.M. or later. Which group is larger? **6.NS.7**

Find each product. 7.NS.2

79. 14(−5) **80.** −8(−11) **81.** −7(−8)(−3) **82.** 2(−8)(−9)(10)

83. −50(−5) **84.** (12)(−2)(8) **85.** (−1)(16)(−2) **86.** 14(−2)(−3)

Find the greatest common factor for each set of numbers. 6.NS.4

87. 12, 20 **88.** 32, 14 **89.** 64, 48

90. 35, 9 **91.** 4, 8, 10 **92.** 18, 30, 42

93. 21, 45, 51 **94.** 24, 40, 64 **95.** 42, 63, 84

ISG Interactive Study Guide

See page 57 for:
- Mid-Chapter Check

21ST CENTURY CAREER in Fashion Design

Fashion Designer

Do you enjoy reading fashion magazines, keeping up with the latest trends, and creating your own unique sense of style? You might want to consider a career in fashion design. Fashion designers create new designs for clothing, accessories, and shoes. In addition to being creative and knowledgeable about current fashion trends, fashion designers need to be able to take accurate measurements and calculate fit by adding, subtracting, multiplying, and dividing measurements.

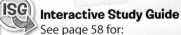

College & Career READINESS

Explore college and careers at ccr.mcgraw-hill.com

Is This the Career for You?

Are you interested in a career as a fashion designer? Take some of the following courses in high school.

- Algebra
- Art
- Digital Design
- Geometry

Find out how math relates to a career in Fashion Design.

Interactive Study Guide

See page 58 for:

- Problem Solving
- Career Project

Lesson 3-4

Dividing Rational Numbers

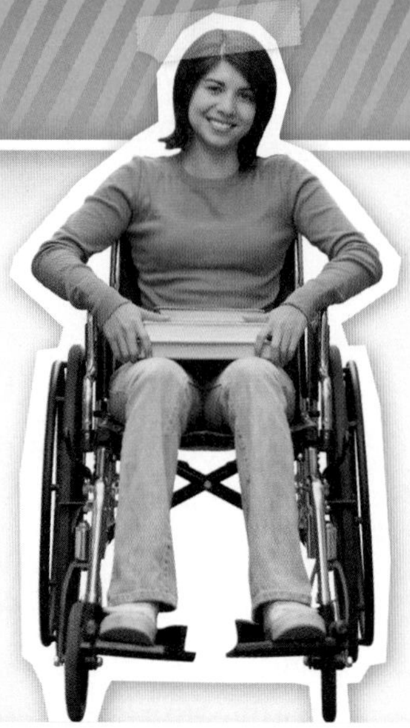

Interactive Study Guide

See pages 59–60 for:
- Getting Started
- Real-World Link
- Notes

Essential Question

What happens when you add, subtract, multiply, and divide rational numbers?

Common Core State Standards

Content Standards
7.NS.2, 7.NS.2a, 7.NS.2c, 7.NS.3, 7.EE.3

Mathematical Practices
1, 3, 4, 5, 7

Vocabulary

multiplicative inverse
reciprocal

What You'll Learn

- Divide positive and negative fractions using multiplicative inverses.
- Divide algebraic fractions.

Real-World Link

Global Literacy After learning the history of Mexico's holiday *El Día de los Muertos*, or *Day of the Dead*, students created clay containers to commemorate loved ones. They made their containers from two slabs of clay that that they cut into thirds.

Divide Fractions

All of the properties of integers also apply to rational numbers. Two numbers whose product is 1 are called **multiplicative inverses** or **reciprocals**. The statement $\frac{1}{4} \cdot 4 = 1$ demonstrates this property too.

Key Concept: Inverse Property of Multiplication

Words	The product of a number and its multiplicative inverse is 1.
Symbols	For every number $\frac{a}{b}$, where $a, b \neq 0$, there is exactly one number $\frac{b}{a}$ such that $\frac{a}{b} \cdot \frac{b}{a} = 1$.
Example	$\frac{2}{3} \cdot \frac{3}{2} = 1$

Example 1

Tutor

Find the multiplicative inverse of each number.

a. $\frac{7}{16}$

$\frac{7}{16}\left(\frac{16}{7}\right) = 1$ The product is 1.

The multiplicative inverse or reciprocal of $\frac{7}{16}$ is $\frac{16}{7}$.

b. $-6\frac{1}{3}$

$-6\frac{1}{3} = -\frac{19}{3}$ Write $-6\frac{1}{3}$ as an improper fraction.

$-\frac{19}{3}\left(-\frac{3}{19}\right) = 1$ The product is 1.

The multiplicative inverse or reciprocal of $-6\frac{1}{3}$ is $-\frac{3}{19}$.

Got It? Do these problems to find out.

1a. $-\frac{7}{9}$

1b. $2\frac{1}{12}$

Andersen Ross/Blend Images LLC

Key Concept Divide Fractions

Words To divide by a fraction, multiply by its multiplicative inverse.

Examples $\frac{4}{9} \div \frac{3}{5} = \frac{4}{9} \cdot \frac{5}{3}$ $\qquad \frac{a}{b} \div \frac{c}{d} = \frac{a}{b} \cdot \frac{d}{c}$, where b, c, and $d \neq 0$

To demonstrate this concept, consider $\frac{4}{9} \div \frac{3}{5}$ and $\frac{a}{b} \div \frac{c}{d}$.

$$\frac{\frac{4}{9}}{\frac{3}{5}} = \frac{\frac{4}{9} \cdot \frac{5}{3}}{\frac{3}{5} \cdot \frac{5}{3}}$$ Multiply the numerator and denominator by $\frac{5}{3}$, the multiplicative inverse of $\frac{3}{5}$.

$$= \frac{\frac{4}{9} \cdot \frac{5}{3}}{1}$$ $\frac{3}{5} \cdot \frac{5}{3} = 1$

$$= \frac{4}{9} \cdot \frac{5}{3}$$

$$\frac{\frac{a}{b}}{\frac{c}{d}} = \frac{\frac{a}{b} \cdot \frac{d}{c}}{\frac{c}{d} \cdot \frac{d}{c}}$$ Multiply the numerator and denominator by $\frac{d}{c}$, the multiplicative inverse of $\frac{c}{d}$.

$$= \frac{\frac{a}{b} \cdot \frac{d}{c}}{1}$$ $\frac{c}{d} \cdot \frac{d}{c} = 1$

$$= \frac{a}{b} \cdot \frac{d}{c}$$

Example 2

Tutor

Dividing By a Whole Number
When dividing by a whole number, always rename it as an improper fraction first. Then multiply by its reciprocal.

Find each quotient. Write in simplest form.

a. $\frac{1}{9} \div \frac{5}{12}$

$\frac{1}{9} \div \frac{5}{12} = \frac{1}{9} \cdot \frac{12}{5}$ Multiply by the reciprocal of $\frac{5}{12}$, $\frac{12}{5}$.

$= \frac{1}{\cancel{9}_3} \cdot \frac{\cancel{12}^4}{5}$ Divide by the GCF, 3.

$= \frac{4}{15}$ Simplify.

b. $\frac{3}{7} \div 8$

$\frac{3}{7} \div 8 = \frac{3}{7} \div \frac{8}{1}$ Write 8 as $\frac{8}{1}$.

$= \frac{3}{7} \cdot \frac{1}{8}$ Multiply by the reciprocal of $\frac{8}{1}$, $\frac{1}{8}$.

$= \frac{3}{56}$ Simplify.

Got It? Do these problems to find out.

2a. $\frac{1}{3} \div \frac{7}{15}$ **2b.** $\frac{5}{8} \div \left(-\frac{3}{4}\right)$ **2c.** $\frac{3}{4} \div 11$ **2d.** $-\frac{6}{7} \div 12$

Example 3

Tutor

Find $-4\frac{2}{3} \div 3\frac{1}{9}$.

$-4\frac{2}{3} \div 3\frac{1}{9} = -\frac{14}{3} \div \frac{28}{9}$ Rename the mixed numbers as improper fractions.

$= -\frac{14}{3} \cdot \frac{9}{28}$ Multiply by the reciprocal, $\frac{9}{28}$.

$= -\frac{\cancel{14}^1}{\cancel{3}_1} \cdot \frac{\cancel{9}^3}{\cancel{28}_2}$ Divide out common factors.

$= -\frac{3}{2}$ or $-1\frac{1}{2}$ Simplify.

Got It? Do these problems to find out.

3a. Find $6\frac{3}{8} \div \left(-4\frac{1}{4}\right)$. **3b.** Find $-6\frac{4}{5} \div \left(-2\frac{2}{5}\right)$.

Division can be used to find the number of equal size groups in a real-world situation.

Example 4

Tutor

Tessa feeds her dog Roscoe $3\frac{3}{4}$ cups of dog food per day. If she buys a bag of food that contains 165 cups, how many days will the bag of food last?

To find how many days, divide. $165 \div 3\frac{3}{4}$ THINK How many $3\frac{3}{4}$s are in 165?

$165 \div 3\frac{3}{4} = \frac{165}{1} \div \frac{15}{4}$ Rewrite 165 and $3\frac{3}{4}$ as improper fractions.

$= \frac{165}{1} \cdot \frac{4}{15}$ Multiply by the reciprocal of $\frac{15}{4}$, $\frac{4}{15}$.

$= \frac{\overset{11}{\cancel{165}}}{1} \cdot \frac{4}{\underset{1}{\cancel{15}}}$ Divide out common factors.

$= 44$ Simplify.

So, the bag of dog food will last 44 days.

Check

Tessa feeds her dog about 4 cups of food for a little longer than 40 days, so the number of cups the bag contains should be about 4(40) or 160 cups. The answer is reasonable.

Watch Out!

When working with fractions, dividing just the whole number portions may not give a good estimate for an answer. First round each fraction or mixed number to the nearest whole number, then divide.

Got It? **Do this problem to find out.**

4. A box of cereal contains $15\frac{3}{5}$ ounces. If one bowl holds $2\frac{2}{5}$ ounces of cereal, how many bowls of cereal are in one box?

Divide Algebraic Expressions

You can divide algebraic fractions in the same way that you divide numerical fractions.

Example 5

Tutor

Find $\frac{5}{3ab} \div \frac{15}{abc}$. Write the quotient in simplest form.

$\frac{5}{3ab} \div \frac{15}{abc} = \frac{5}{3ab} \cdot \frac{abc}{15}$ Multiply by the reciprocal of $\frac{15}{abc}$, $\frac{abc}{15}$.

$= \frac{\overset{1}{\cancel{5}}}{3\underset{1}{\cancel{ab}}} \cdot \frac{\overset{1}{\cancel{ab}}c}{\underset{3}{\cancel{15}}}$ Divide out common factors.

$= \frac{c}{9}$ Simplify.

Got It? **Do these problems to find out.**

Find each quotient. Write in simplest form.

5a. $\frac{5ab}{6} \div \frac{10b}{7}$

5b. $\frac{mn}{4} \div \frac{m}{8}$

Guided Practice

Check

Find the multiplicative inverse of each number. (Example 1)

1. $\frac{6}{7}$ **2.** $-5\frac{1}{2}$ **3.** -63

4. $9\frac{9}{10}$ **5.** -101 **6.** $\frac{11}{35}$

Find each quotient. Write in simplest form. (Examples 2 and 3)

7. $-\frac{4}{5} \div \frac{8}{9}$ **8.** $-\frac{5}{7} \div \frac{2}{35}$ **9.** $\frac{4}{9} \div (-2)$

10. $\frac{7}{9} \div (-14)$ **11.** $-2\frac{1}{5} \div \left(-3\frac{2}{3}\right)$ **12.** $7\frac{1}{9} \div \left(-1\frac{1}{3}\right)$

13. Sonia is making a quilted wall hanging that is 38 inches wide. If each quilt square is $4\frac{3}{4}$ inches wide, how many squares will she need to complete one row of the wall hanging? (Example 4)

Find each quotient. Write in simplest form. (Example 5)

14. $\frac{4ab}{c} \div \frac{3a}{2c}$ **15.** $\frac{mn}{6} \div \frac{3m}{p}$ **16.** $\frac{3xy}{yz} \div \frac{6y}{5}$

Independent Practice

eHelp Go online for Step-by-Step Solutions

Find the multiplicative inverse of each number. (Example 1)

17. $-\frac{4}{5}$ **18.** $6\frac{1}{8}$ **19.** $\frac{10}{19}$

20. $-4\frac{2}{7}$ **21.** 19 **22.** -54

23. $5\frac{2}{3}$ **24.** -12 **25.** $-\frac{2}{9}$

Find each quotient. Write in simplest form. (Examples 2 and 3)

26. $-\frac{1}{8} \div \frac{2}{5}$ **27.** $-\frac{5}{12} \div \frac{2}{3}$

28. $-\frac{6}{7} \div \left(-\frac{16}{21}\right)$ **29.** $-\frac{4}{9} \div (-24)$

30. $-\frac{9}{10} \div (-21)$ **31.** $-6\frac{1}{9} \div 3\frac{2}{3}$

32. $-10\frac{3}{5} \div \left(-2\frac{2}{5}\right)$ **33.** $2\frac{3}{8} \div 1\frac{1}{6}$

34. Hannah is making chocolate chip cookies. The dry ingredients are shown at the right. How many batches of cookies can she make if she has $7\frac{1}{2}$ cups of brown sugar? (Example 4)

Chocolate Chip Cookies
$\frac{1}{2}$ cup granulated sugar
$1\frac{1}{2}$ cups packed brown sugar
$2\frac{1}{2}$ cups all-purpose flour
$\frac{3}{4}$ teaspoon salt
1 teaspoon baking powder
1 teaspoon baking soda

35. How many play costumes can be made with $49\frac{1}{2}$ yards of fabric if each costume requires $4\frac{1}{8}$ yards? (Example 4)

Find each quotient. Write in simplest form. (Example 5)

36. $\frac{x}{20} \div \frac{x}{5}$ **37.** $\frac{m}{6n} \div \frac{7m}{3n}$

38. $\frac{m}{np} \div \frac{3m}{2p}$ **39.** $\frac{5a}{3bc} \div \frac{2a}{9bc}$

40. Barbara babysat for $3\frac{1}{4}$ hours and earned \$19.50. What was her hourly rate?

41. A train traveled 405 miles in $4\frac{1}{2}$ hours. How fast was the train traveling on average? (*Hint:* Distance equals the rate multiplied by the time.)

42. Sydney reduced her favorite photograph to put in a scrapbook. How many times as wide is the actual photo than the reduced photo?

4 in.

3 in.

Evaluate each expression if $m = 2\frac{2}{5}$, $n = -\frac{3}{10}$, and $p = 6$.

43. $mn \div p$

44. $\frac{m}{n}$

45. $np \div m$

46. Ms. Augello is making tie-dyed shirts with her students. Each gallon of hot water needs $\frac{2}{3}$ cup of dye. If Ms. Augello has $5\frac{1}{4}$ cups of dye, how many batches of solution will she be able to make?

47. CCSS **Make a Conjecture** The model at the left shows $\frac{3}{4} \div \frac{1}{2}$. The model at the right shows $\frac{3}{4} \div \frac{1}{4}$.

How many $\frac{1}{2}$s are in $\frac{3}{4}$?

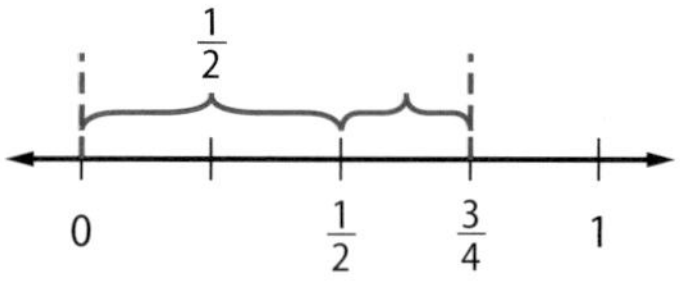

There are $1\frac{1}{2}$ $\frac{1}{2}$s in $\frac{3}{4}$.

How many $\frac{1}{4}$s are in $\frac{3}{4}$?

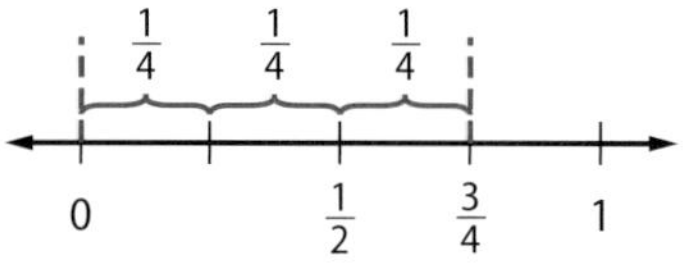

There are three $\frac{1}{4}$s in $\frac{3}{4}$.

Make a conjecture about what happens to the quotient as the value of the divisor increases. Test your conjecture.

H.O.T. Problems Higher Order Thinking

48. CCSS **Use Math Tools** Choose two fractions and use an area model or number line to show that division of rational numbers is not commutative.

49. CCSS **Persevere with Problems** Give a counterexample to this statement. *The quotient of two fractions between 0 and 1 is always a whole number.*

50. CCSS **Construct an Argument** Which is greater, $40 \cdot \frac{1}{4}$ or $40 \div \frac{1}{4}$? Explain.

51. CCSS **Identify Structure** Is a whole number divided by a proper fraction *always*, *sometimes*, or *never* greater than the whole number?

52. **Building on the Essential Question** Explain why, for a positive number n, $n \div \frac{1}{2} > n$.

Standardized Test Practice

53. Heidi is having a party. She is planning that each of her 16 guests will have $\frac{3}{4}$ cup of snack mix. She has made 12 cups of snack mix. Which expression could Heidi use to determine if she has made enough snack mix for each of her guests?

A $16 \div \frac{3}{4}$ **C** $12 \div \frac{3}{4}$

B $16 \div 12$ **D** $\frac{3}{4}(12)$

54. A bag of potting soil contains $4\frac{1}{4}$ pounds of soil. Each flower that Mr. Henderson plants will need $\frac{1}{8}$ pound of soil. How many flowers will he be able to plant?

F 16 **H** 32

G 28 **J** 34

55. A recipe for one batch of soft pretzels calls for $\frac{1}{4}$ cup of salt and $\frac{2}{3}$ cup of sugar. If Mrs. Valdez uses $\frac{7}{8}$ cup of salt and $2\frac{1}{3}$ cups of sugar, how many batches of pretzels is she making?

A $3\frac{1}{2}$ **C** $2\frac{1}{4}$

B 3 **D** 2

56. **Short Response** Popcorn is sold in a variety of sizes. Use the table to find how many times as large the regular bag of popcorn is than the snack bag.

Size	Amount (cups)
Snack	$3\frac{1}{2}$
Regular	$8\frac{3}{4}$
Large	12

Common Core Review

Find each product. Write in simplest form 7.NS.2a

57. $2 \cdot \frac{9}{16}$ **58.** $-4\frac{4}{7} \cdot 2\frac{5}{8}$ **59.** $\frac{3}{20} \cdot \left(-\frac{10}{11}\right)$

60. $-6\frac{1}{2} \cdot \left(-3\frac{1}{4}\right)$ **61.** $-\frac{5}{6} \cdot \left(-1\frac{7}{35}\right)$ **62.** $1\frac{1}{8} \cdot 1\frac{1}{3}$

63. The White House covers an area of 0.028 square mile. What fraction of a square mile is this? 8.NS.1

64. The Wildcat football team was penalized the same amount four times during the third quarter. The total of the four penalties was 60 yards. If −60 represents a loss of 60 yards, write a division sentence to represent this situation. Then express the number of yards of each penalty as an integer. 7.EE.3

Find each product. 7.NS.2

65. $12(-6)$ **66.** $-12(-11)$ **67.** $4(-2)(-6)$

Find each sum or difference. 7.NS.1

68. $23 - (-13)$ **69.** $-42 + (-26)$ **70.** $-80 - (-80)$

71. $n + 2n$ **72.** $-4x - (-3x)$ **73.** $5n - 10n$

Translate each of the following to a mathematical expression. 6.EE.2a

74. Add 7 and 5, and then multiply the result by 3.

75. Subtract the quotient of 6 and 3 from 12.

76. Divide the sum of 10 and 15 by 5.

77. Subtract the sum of 3 and 5 from 15.

Lesson 3-5

Adding and Subtracting Like Fractions

Interactive Study Guide

See pages 61–62 for:
- Getting Started
- Real-World Link
- Notes

Essential Question

What happens when you add, subtract, multiply, and divide rational numbers?

Common Core State Standards

Content Standards
7.NS.1, 7.NS.1d, 7.NS.3, 7.EE.3

Mathematical Practices
1, 3, 4, 5, 6

Vocabulary
like fractions

What You'll Learn

- Add rational numbers with common denominators.
- Subtract rational numbers with common denominators.

Real-World Link

Technology In a survey, users of E-readers were asked to describe why they prefer E-readers over books. One-eighth said that it was because they can change the font size and read faster. Five-eighths said that it was because the devices are portable—it's like having a small library with them wherever they go!

Key Concept: Add Like Fractions

Words To add fractions with like denominators, add the numerators and write the sum over the denominator.

Symbols $\frac{a}{c} + \frac{b}{c} = \frac{a+b}{c}$, where $c \neq 0$

Example $\frac{2}{8} + \frac{3}{8} = \frac{2+3}{8}$ or $\frac{5}{8}$

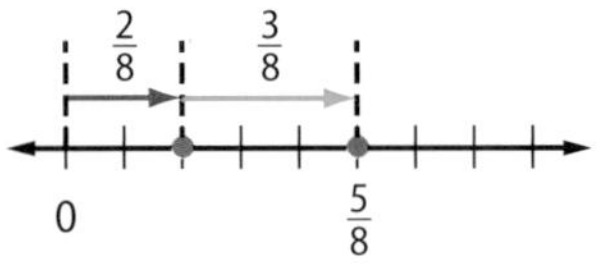

Like fractions are fractions with the same denominator.

Example 1

Tutor

Find each sum. Write in simplest form.

a. $\frac{7}{10} + \frac{6}{10}$

Estimate $1 + \frac{1}{2} = 1\frac{1}{2}$

$\frac{7}{10} + \frac{6}{10} = \frac{7+6}{10}$ — The denominators are the same. Add the numerators.

$= \frac{13}{10}$ or $1\frac{3}{10}$ — Simplify and rename as a mixed number. Is the answer reasonable?

b. $\frac{5}{8} + \left(-\frac{7}{8}\right)$

Estimate $\frac{1}{2} + (-1) = -\frac{1}{2}$

$\frac{5}{8} + \left(-\frac{7}{8}\right) = \frac{5+(-7)}{8}$ — The denominators are the same. Add the numerators.

$= \frac{-2}{8}$ or $-\frac{1}{4}$ — Simplify. Compare to the estimate. Is it reasonable?

Got It? Do these problems to find out.

1a. $\frac{5}{6} + \frac{4}{6}$

1b. $\frac{4}{7} + \left(-\frac{6}{7}\right)$

1c. $\frac{1}{5} + \frac{4}{5}$

1d. $-\frac{5}{8} + \frac{11}{8}$

Example 2

Tutor

Find $2\frac{3}{8} + 3\frac{7}{8}$. Write in simplest form.

Estimate $2 + 4 = 6$

$$2\frac{3}{8} + 3\frac{7}{8} = (2 + 3) + \left(\frac{3}{8} + \frac{7}{8}\right)$$ Add the whole numbers and fractions separately.

$$= 5 + \frac{10}{8}$$ Add the numerators.

$$= 5\frac{10}{8} \text{ or } 6\frac{1}{4}$$ Simplify. Rename $5\frac{10}{8}$ as $6\frac{2}{8}$ or $6\frac{1}{4}$.

Check for Reasonableness $6\frac{1}{4} \approx 6$ ✓

Alternative Method

When adding or subtracting mixed numbers, you can write them as improper fractions before adding or subtracting. If any of the numbers are negative, it is easier to use this method.

$$2\frac{3}{8} + 3\frac{7}{8} = \frac{19}{8} + \frac{31}{8}$$
$$= \frac{50}{8} \text{ or } 6\frac{1}{4}$$

Got It? Do these problems to find out.

Find each sum. Write in simplest form.

2a. $1\frac{3}{4} + 4\frac{3}{4}$

2b. $3\frac{2}{5} + 8\frac{1}{5}$

2c. $-2\frac{3}{7} + \left(-4\frac{5}{7}\right)$

Key Concept Subtract Like Fractions

Words To subtract fractions with like denominators, subtract the numerators and write the difference over the denominator.

Symbols $\frac{a}{c} - \frac{b}{c} = \frac{a - b}{c}$, where $c \neq 0$

Example $\frac{4}{9} - \frac{3}{9} = \frac{4 - 3}{9}$ or $\frac{1}{9}$

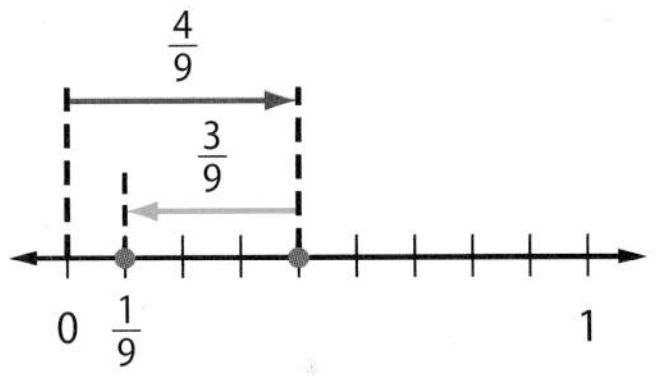

The rule for subtracting fractions with like denominators is similar to the rule for addition.

Example 3

Tutor

Find $\frac{3}{10} - \frac{9}{10}$. Write in simplest form.

Estimate $\frac{1}{2} - 1 = -\frac{1}{2}$

$$\frac{3}{10} - \frac{9}{10} = \frac{3 - 9}{10}$$ The denominators are the same. Subtract the numerators.

$$= \frac{-6}{10} \text{ or } -\frac{3}{5}$$ Simplify.

Check for Reasonableness $-\frac{3}{5} \approx -\frac{1}{2}$ ✓

Got It? Do these problems to find out.

Find each difference. Write in simplest form.

3a. $\frac{5}{15} - \frac{10}{15}$

3b. $\frac{3}{9} - \frac{4}{9}$

3c. $\frac{7}{8} - \frac{3}{8}$

Watch Out!

When you subtract a negative number, you add the opposite. If the number you are subtracting from is negative, the result can be positive, negative, or zero.

Example 4

Evaluate $x - y$ when $x = \frac{3}{5}$ and $y = -\frac{4}{5}$.

To subtract a negative number, add its additive inverse.

$x - y = \frac{3}{5} - \left(-\frac{4}{5}\right)$ — Replace x with $\frac{3}{5}$ and y with $-\frac{4}{5}$.

$= \frac{3}{5} + \frac{4}{5}$ — The additive inverse of $-\frac{4}{5}$ is $\frac{4}{5}$.

$= \frac{3+4}{5}$ — The denominators are the same. Add the numerators.

$= \frac{7}{5}$ or $1\frac{2}{5}$ — Simplify and rename as a mixed number.

Check Use a number line.

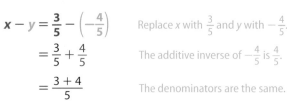

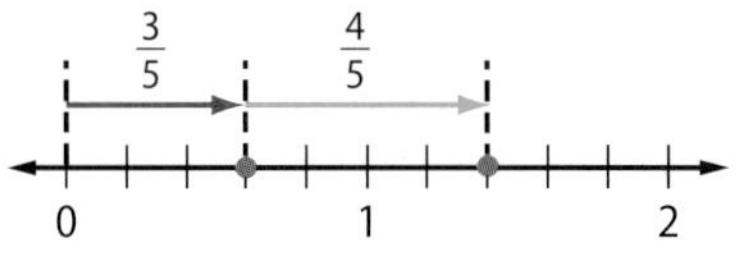

$\frac{3}{5} - \left(-\frac{4}{5}\right) = \frac{7}{5}$ or $1\frac{2}{5}$ ✓

Got It? Do these problems to find out.

Evaluate each expression if $a = \frac{3}{8}$, $b = -\frac{5}{8}$, and $c = \frac{7}{8}$.

4a. $a - b$

4b. $b - c$

4c. $c - a$

Real World Example 5

LaShaun has $5\frac{1}{8}$ yards of ribbon to border scrapbook pages. If she uses $1\frac{7}{8}$ yards on one page, how much ribbon is left?

Subtract the amount of ribbon she will use from the total amount of ribbon.

Estimate $5\frac{1}{8} - 1\frac{7}{8} \approx 5 - 2$ or 3 yards

$5\frac{1}{8} - 1\frac{7}{8} = 4\frac{9}{8} - 1\frac{7}{8}$ — Rename $5\frac{1}{8}$ as $4\frac{9}{8}$.

$= 3\frac{2}{8}$ — Subtract the whole numbers and then the fractions.

$= 3\frac{2}{8}$ or $3\frac{1}{4}$ — Simplify.

LaShaun has $3\frac{1}{4}$ yards of ribbon remaining.

Check for Reasonableness $3\frac{1}{2} \approx 3$ ✓

Got It? Do this problem to find out.

5. The Daytona International Speedway is one of the longest tracks used in NASCAR races. It is $2\frac{2}{4}$ miles long. Richmond International Speedway is $\frac{3}{4}$ mile long. How much longer is the Daytona Speedway than the Richmond Speedway?

You can use the same rules for finding the distance on the number line between two fractions as you did for finding the distance between two integers.

Example 6

Tutor

Find the distance between $-\frac{1}{7}$ and $\frac{5}{7}$ on a number line. Simplify if necessary.

Step 1 The common denominator is 7. Divide each unit between −1 and 1 into seven parts.

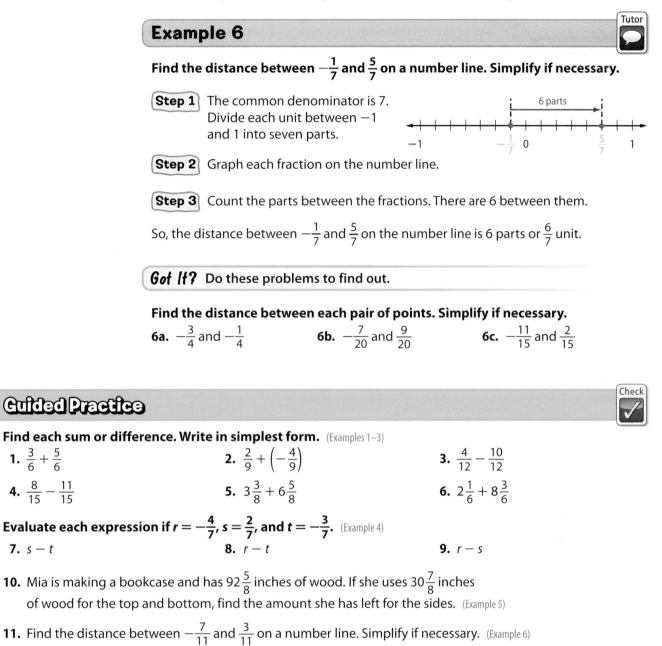

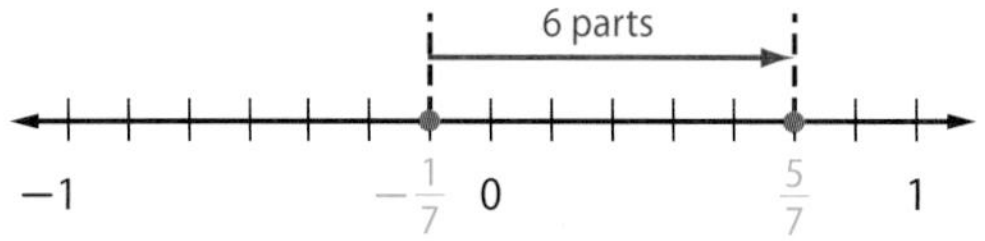

Step 2 Graph each fraction on the number line.

Step 3 Count the parts between the fractions. There are 6 between them.

So, the distance between $-\frac{1}{7}$ and $\frac{5}{7}$ on the number line is 6 parts or $\frac{6}{7}$ unit.

Got It? Do these problems to find out.

Find the distance between each pair of points. Simplify if necessary.

6a. $-\frac{3}{4}$ and $-\frac{1}{4}$ **6b.** $-\frac{7}{20}$ and $\frac{9}{20}$ **6c.** $-\frac{11}{15}$ and $\frac{2}{15}$

Guided Practice

Check

Find each sum or difference. Write in simplest form. (Examples 1–3)

1. $\frac{3}{6}+\frac{5}{6}$ **2.** $\frac{2}{9}+\left(-\frac{4}{9}\right)$ **3.** $\frac{4}{12}-\frac{10}{12}$

4. $\frac{8}{15}-\frac{11}{15}$ **5.** $3\frac{3}{8}+6\frac{5}{8}$ **6.** $2\frac{1}{6}+8\frac{3}{6}$

Evaluate each expression if $r=-\frac{4}{7}$, $s=\frac{2}{7}$, and $t=-\frac{3}{7}$. (Example 4)

7. $s-t$ **8.** $r-t$ **9.** $r-s$

10. Mia is making a bookcase and has $92\frac{5}{8}$ inches of wood. If she uses $30\frac{7}{8}$ inches of wood for the top and bottom, find the amount she has left for the sides. (Example 5)

11. Find the distance between $-\frac{7}{11}$ and $\frac{3}{11}$ on a number line. Simplify if necessary. (Example 6)

Independent Practice

Go online for Step-by-Step Solutions eHelp

Find each sum or difference. Write in simplest form. (Examples 1–3)

12. $\frac{5}{6}+\left(-\frac{4}{6}\right)$ **13.** $-\frac{11}{12}+\frac{7}{12}$ **14.** $\frac{3}{14}+\left(-\frac{5}{14}\right)$ **15.** $-\frac{3}{7}+\frac{6}{7}$

16. $5\frac{1}{4}+5\frac{1}{4}$ **17.** $12\frac{5}{9}+\left(-1\frac{1}{9}\right)$ **18.** $\frac{2}{15}-\frac{7}{15}$ **19.** $\frac{5}{11}-\frac{7}{11}$

20. $-\frac{1}{5}-\frac{4}{5}$ **21.** $-\frac{7}{20}-\frac{7}{20}$ **22.** $2\frac{12}{13}-\left(-7\frac{10}{13}\right)$ **23.** $-8\frac{3}{10}-4\frac{9}{10}$

Evaluate each expression if $x=-\frac{5}{9}$, $y=\frac{2}{9}$, and $z=-\frac{4}{9}$. (Example 4)

24. $y-z$ **25.** $y-x$ **26.** $x-z$ **27.** $z-x$

28. **CCSS Be Precise** Nan was $59\frac{7}{8}$ inches tall at the end of summer. She was $62\frac{1}{8}$ inches by March. How much did she grow during that time? (Example 5)

29. Yahto needs $3\frac{3}{4}$ cups of sugar to make cookies. He needs an additional $\frac{3}{4}$ cup for bread. Find the total amount of sugar that Yahto needs. (Example 5)

Find the distance between each pair of points. Simplify if necessary. (Example 6)

30. $-\frac{5}{8}$ and $\frac{1}{8}$

31. $-\frac{7}{15}$ and $-\frac{4}{15}$

32. $-\frac{1}{8}$ and $\frac{3}{8}$

33. $-\frac{3}{16}$ and $\frac{3}{16}$

Find each sum or difference. Write in simplest form.

34. $-2\frac{9}{10} + \left(-9\frac{9}{10}\right) + \left(-6\frac{9}{10}\right)$

35. $\frac{1}{9} - 2\frac{4}{9} - \frac{5}{9}$

36. A triathlon is a race with swimming, biking, and running. If an athlete swims for $18\frac{2}{4}$ minutes, bikes for $59\frac{1}{4}$ minutes, and runs for $37\frac{3}{4}$ minutes, what is his total time?

37. The table shows the weight of Leon's dog during its first 5 years.

Age (years)	1	2	3	4	5
Weight (pounds)	$17\frac{2}{8}$	$18\frac{5}{8}$	$19\frac{4}{8}$	$18\frac{3}{8}$	$20\frac{7}{8}$

a. How much weight did Leon's dog gain or lose between years 3 and 4? between years 1 and 5?

b. If Leon's dog gains $1\frac{3}{8}$ pounds each year between years 5 and 7, how much will his dog weigh?

38. A lasagne recipe uses $1\frac{2}{4}$ teaspoons basil, $\frac{1}{4}$ teaspoon pepper, and 4 teaspoons parsley. If you double the recipe, how many teaspoons of seasoning will you use?

H.O.T. Problems Higher Order Thinking

39. **CCSS Use Math Tools** Write a subtraction problem with a difference of $-\frac{2}{3}$.

40. **CCSS Persevere with Problems** Lopez Construction is replacing a window in a house. The window is currently 3 feet wide by 4 feet tall. The homeowner wants to add 9 inches to each side of the window. What is the new perimeter of the window in feet? Justify your reasoning.

41. **CCSS Find the Error** Xavier said the sum of $-4\frac{1}{9}$ and $1\frac{7}{9}$ is $-3\frac{8}{9}$. Is he correct? Explain your reasoning.

42. **CCSS Use Math Tools** Explain how you could use mental math to find the following sum. Then find the sum. Support your answer with a model.

$$1\frac{1}{4} + 2\frac{1}{3} + 3\frac{2}{3} + 4\frac{1}{2} + 5\frac{1}{2} + 6\frac{3}{4}$$

43. **Building on the Essential Question** Write a real-world problem about cooking that can be solved by adding or subtracting fractions. Then solve the problem.

Fuse/Getty Images

Standardized Test Practice

44. The average times it takes Miguel to cut his lawn and his neighbor's lawn are given in the table.

Lawn	Time to Cut (h)
Miguel's	$\frac{3}{4}$
Neighbor's	$1\frac{1}{4}$

Last summer, he cut his lawn 10 times and his neighbor's 6 times. How many hours did he spend cutting both lawns?

A $13\frac{1}{2}$ h **C** $14\frac{1}{2}$ h

B 14 h **D** 15 h

45. A piece of wood is $1\frac{9}{16}$ inches thick. A layer of padding $\frac{15}{16}$ inch thick is placed on top of the wood. What is the total thickness of the wood and the padding?

F $1\frac{3}{8}$ in. **H** $1\frac{15}{16}$ in.

G $1\frac{1}{2}$ in. **J** $2\frac{1}{2}$ in.

46. Ronata is putting lace around the tablecloth shown below. How much lace will she need to cover all 4 sides?

$72\frac{2}{3}$ in.

$52\frac{1}{3}$ in.

A $20\frac{1}{2}$ in. **C** 125 in.

B 41 in. **D** 250 in.

47. **Short Response** Simplify the expression below.

$$-5\frac{7}{9} - 2\frac{4}{9} + 1\frac{8}{9}$$

Common Core Review

Find each quotient. Write in simplest form. 6.NS.1

48. $\frac{2}{7} \div \frac{5}{14}$

49. $-3\frac{1}{8} \div \frac{5}{16}$

50. $4\frac{2}{3} \div \left(-3\frac{1}{9}\right)$

Find each product. Write in simplest form. 5.NF.4

51. $\frac{3}{10} \cdot \frac{4}{21}$

52. $\frac{3}{8}(-6)$

53. $\frac{5}{19} \cdot 2\frac{2}{18}$

54. A shelf $16\frac{5}{8}$ inches wide is to be placed in a space that is $16\frac{3}{4}$ inches wide. Will the shelf fit in the space? Explain. 6.NS.7b

55. The Hawks started a play on their own 31-yard line. They lost 9 yards on one play and another 5 yards on the next play. Find the team's field location after the two plays. 7.NS.1

Find the least common multiple of each pair of numbers. 6.NS.4

56. 6, 8

57. 12, 15

58. 3, 7

59. 15, 45

60. 4, 10

61. 8, 20

Evaluate each expression if $a = -2$, $b = 5$, and $c = -8$. 6.EE.2c

62. $3 - b$

63. $9 - c$

64. $a - 6$

65. $b - 5$

66. $c - b$

67. $c - a$

68. $a - b + c$

69. $a + b - c$

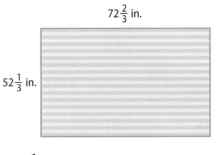

Lesson 3-6

Adding and Subtracting Unlike Fractions

ISG Interactive Study Guide

See pages 63–64 for:
- Getting Started
- Real-World Link
- Notes

Essential Question

What happens when you add, subtract, multiply, and divide rational numbers?

Common Core State Standards

Content Standards
7.NS.1, 7.NS.1d, 7.NS.3, 7.EE.3

Mathematical Practices
1, 3, 4

Vocabulary

unlike fractions

What You'll Learn

- Add unlike fractions.
- Subtract unlike fractions.

Real-World Link

Climate The Sahara Desert covers almost eleven countries, fully or partially. It is one of the driest places in the world, receiving only a couple of inches of rainfall each year.

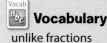

Key Concept: Add Unlike Fractions

Words To add fractions with unlike denominators, rename the fractions with a common denominator. Then add and simplify as with like fractions.

Symbols

$$\frac{2}{3}+\frac{1}{2}=\frac{2}{3}\cdot\frac{2}{2}+\frac{1}{2}\cdot\frac{3}{3}$$
$$=\frac{4}{6}+\frac{3}{6}$$
$$=\frac{7}{6} \text{ or } 1\frac{1}{6}$$

Unlike fractions are fractions with different denominators. Use the least common multiple of the denominators to rename the fractions before adding them.

Example 1

Tutor

Find $\frac{3}{5}+\frac{1}{3}$. Write in simplest form.

Estimate $1 + 0 = 1$

$\frac{3}{5}+\frac{1}{3}=\frac{3}{5}\cdot\frac{3}{3}+\frac{1}{3}\cdot\frac{5}{5}$ Use 3 • 5 or 15 as the common denominator.

$=\frac{9}{15}+\frac{5}{15}$ Rename each fraction with the common denominator.

$=\frac{14}{15}$ Add the numerators.

Check for Reasonableness $\frac{14}{15}\approx 1$ ✓

Got It? Do these problems to find out.

Find each sum. Write in simplest form.

1a. $\frac{1}{6}+\frac{3}{4}$

1b. $\frac{2}{7}+\frac{3}{14}$

1c. $\frac{2}{5}+\frac{3}{10}$

1d. $\frac{3}{8}+\frac{5}{24}$

Example 2

Tutor

Find each sum. Write in simplest form.

a. $-\frac{5}{6}+\frac{1}{8}$ **Estimate** $-1+0=-1$

$-\frac{5}{6}+\frac{1}{8}=\frac{-5}{6}\cdot\frac{4}{4}+\frac{1}{8}\cdot\frac{3}{3}$ The LCD of 6 and 8 is 24.

$=\frac{-20}{24}+\frac{3}{24}$ Rename each fraction using the LCD, 24.

$=\frac{-17}{24}$ or $-\frac{17}{24}$ Simplify.

b. $-\frac{3}{10}+\left(-5\frac{3}{4}\right)$ **Estimate** $-\frac{1}{2}+-6=-6\frac{1}{2}$

$-\frac{3}{10}+\left(-5\frac{3}{4}\right)=\frac{-3}{10}+\frac{-23}{4}$ Write $-5\frac{3}{4}$ as an improper fraction.

$=\frac{-3}{10}\cdot\frac{2}{2}+\frac{-23}{4}\cdot\frac{5}{5}$ The LCD of 10 and 4 is 20.

$=\frac{-6}{20}+\frac{-115}{20}$ Rename each fraction using the LCD, 20.

$=\frac{-121}{20}$ or $-6\frac{1}{20}$ Simplify.

LCD
You can rename fractions using any common denominator. However, using the least common denominator will make simplifying the solution easier.

Got It? Do these problems to find out.

2a. $3\frac{3}{4}+\frac{5}{14}$ **2b.** $-6\frac{8}{9}+7\frac{5}{12}$ **2c.** $3\frac{3}{5}+\left(-4\frac{5}{6}\right)$

Key Concept Subtract Unlike Fractions

To subtract fractions with unlike denominators, rename the fractions with a common denominator. Then subtract and simplify as with like fractions.

The rule for subtracting fractions with unlike denominators is similar to the rule for addition.

Example 3

Tutor

Find each difference. Write in simplest form.

a. $\frac{3}{8}-\frac{3}{4}$

$\frac{3}{8}-\frac{3}{4}=\frac{3}{8}-\frac{3}{4}\cdot\frac{2}{2}$ The LCD is 8.

$=\frac{3}{8}-\frac{6}{8}$ Rename using the LCD.

$=\frac{-3}{8}$ or $-\frac{3}{8}$ Simplify.

b. $9\frac{3}{5}-7\frac{2}{3}$

$9\frac{3}{5}-7\frac{2}{3}=\frac{48}{5}-\frac{23}{3}$ Rename $9\frac{3}{5}$.

$=\frac{48}{5}\cdot\frac{3}{3}-\frac{23}{3}\cdot\frac{5}{5}$ The LCD is 15.

$=\frac{144}{15}-\frac{115}{15}$ Rename using the LCD.

$=\frac{29}{15}$ or $1\frac{14}{15}$ Simplify.

Reasonableness
Use estimation to check whether your answer is reasonable.
$\frac{3}{8}-\frac{3}{4}\approx\frac{1}{2}-1$
$\approx-\frac{1}{2}$
$-\frac{3}{8}$ is close to $-\frac{1}{2}$.

Got It? Do these problems to find out.

3a. $\frac{3}{4}-\frac{8}{9}$ **3b.** $7\frac{1}{6}-6\frac{5}{8}$ **3c.** $5\frac{1}{3}-\left(-4\frac{5}{9}\right)$

Example 4

Tutor

STEM To set up a computer network in an office, a 100-foot cable is cut and used to connect three computers to the server as shown. How much cable is left to connect the third computer?

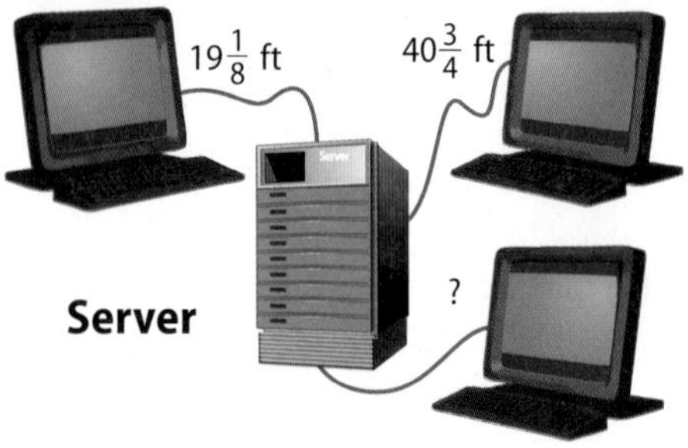

Add the measures of the cables that were already used and subtract that sum from 100, the length of the cable. **Estimate** $100 - (19 + 41) \approx 100 - 60$ or 40 feet

$19\frac{1}{8} + 40\frac{3}{4} = 19\frac{1}{8} + 40\frac{6}{8}$ Rename $40\frac{3}{4}$ using the LCD, 8.

$= 59\frac{7}{8}$ Simplify.

$100 - 59\frac{7}{8} = 99\frac{8}{8} - 59\frac{7}{8}$ Rename 100 as $99\frac{8}{8}$.

$= 40\frac{1}{8}$ Simplify.

There is $40\frac{1}{8}$ feet of cable left to connect the third computer.

Check for Reasonableness $40\frac{1}{8} \approx 40$ ✓

Got It? Do this problem to find out.

4. At a recent frog-jumping contest, the winning frog jumped $21\frac{1}{3}$ feet. The second-place frog jumped $20\frac{1}{2}$ feet. How much farther did the first-place frog jump?

Guided Practice

Check

Find each sum. Write in simplest form. (Examples 1 and 2)

1. $\frac{1}{15} + \frac{3}{5}$

2. $-\frac{5}{9} + \frac{1}{6}$

3. $\frac{7}{8} + \left(-\frac{2}{7}\right)$

4. $8\frac{5}{12} + 11\frac{1}{4}$

5. $-2\frac{1}{3} + \left(-7\frac{1}{2}\right)$

6. $4\frac{3}{8} + 10\frac{5}{12}$

Find each difference. Write in simplest form. (Example 3)

7. $-\frac{1}{4} - \frac{7}{9}$

8. $\frac{3}{5} - \frac{9}{10}$

9. $\frac{5}{8} - \frac{7}{12}$

10. $-1\frac{1}{3} - 4\frac{2}{7}$

11. $5\frac{5}{6} - \left(-2\frac{1}{4}\right)$

12. $12\frac{1}{2} - 6\frac{3}{8}$

13. Dwayne needs $1\frac{2}{3}$ cups of shredded cheese to put in his enchilada casserole and $\frac{3}{4}$ cup of cheese for the top. If he has 3 cups of cheese in all, how much cheese will he have left? (Example 4)

Independent Practice

Go online for Step-by-Step Solutions eHelp

Find each sum or difference. Write in simplest form. (Examples 1 and 3)

14. $-\frac{8}{35} + \frac{2}{5}$

15. $\frac{5}{7} + \left(-\frac{10}{21}\right)$

16. $-\frac{5}{8} + \left(-\frac{8}{9}\right)$

17. $-\frac{1}{3} + \left(-\frac{10}{11}\right)$

18. $\frac{7}{8} - \frac{2}{5}$

19. $\frac{1}{6} - \frac{5}{7}$

20. $-\frac{2}{5} - \frac{1}{3}$

21. $-\frac{3}{10} - \frac{1}{15}$

22. $-\frac{3}{4} - \frac{5}{6}$

23 During spring training, the Detroit Tigers won about $\frac{2}{3}$ of the games they played while the Cleveland Indians won $\frac{7}{15}$ of the games they played. What fraction more of the games did Detroit win than Cleveland?

24. In a college dormitory, $\frac{1}{10}$ of the residents are juniors and $\frac{2}{5}$ of the residents are sophomores. What fraction of the students at the dormitory are juniors and sophomores?

Find each sum or difference. Write in simplest form. (Examples 2 and 3)

25. $8\frac{1}{2} + 3\frac{4}{5}$

26. $-10\frac{2}{3} + 9\frac{7}{12}$

27. $16\frac{5}{6} - 12\frac{1}{3}$

28. $6\frac{6}{7} - 11\frac{7}{8}$

29. $-4\frac{1}{9} - 7\frac{2}{3}$

30. $-\frac{5}{6} + 8\frac{1}{4}$

31. $\frac{9}{16} + 3\frac{5}{6}$

32. $-5\frac{3}{5} + \left(-7\frac{1}{6}\right)$

33. $-10\frac{1}{2} - 6\frac{5}{7}$

34. Sybrina wants to make a 17-inch necklace with a $\frac{3}{4}$-inch bead, a $1\frac{1}{2}$-inch bead, and another $\frac{3}{4}$-inch bead on it. What is the length of the remaining part of the necklace? (Example 4)

35. Kenzie is making three desserts for a party. The recipes call for $\frac{2}{3}$ cup sugar, $1\frac{5}{6}$ cups sugar, and $2\frac{3}{4}$ cups sugar. If she has 6 cups of sugar, how much sugar will she have left over?

36. The length of a page in a yearbook is 10 inches. The top margin is $\frac{1}{2}$ inch, and the bottom margin is $\frac{3}{4}$ inch. What is the length of the page inside the margins?

37. CCSS **Multiple Representations** The perimeter of a geometric figure is the distance around the figure. You can find the perimeter of a rectangle by adding the measures of all four sides.

a. Table Copy and complete the table at the right by listing the lengths and widths of three additional rectangles that have a perimeter of 20.

b. Graph Write the values from the table as ordered pairs (ℓ, w). Graph the ordered pairs on the coordinate plane.

c. Numbers Use the graph to predict the length of a rectangle with a perimeter of 20 inches and a width of $5\frac{1}{2}$ inches. Check the prediction by finding the actual length of the rectangle.

Perimeter of 20	
Length (ℓ)	Width (w)
8	2
■	■
■	■
■	■

38. Use the table to find the total average precipitation for a certain city that falls in August, September, and October.

Average Precipitation	
Month	Amount (in.)
Aug.	$2\frac{47}{50}$
Sept.	$1\frac{22}{25}$
Oct.	$1\frac{1}{2}$

39. Use the Internet or another source to find out the monthly rainfall totals in your community during the past year. How much rain fell in August, September, and October?

Find each difference. Write in simplest form.

40. $-3\frac{2}{5} - \left(-2\frac{4}{7}\right)$

41. $-19\frac{3}{8} - \left(-4\frac{3}{4}\right)$

42. $8\frac{5}{12} - \left(-12\frac{13}{18}\right)$

43. $-35\frac{5}{6} - 23.\overline{3}$

44. $-17\frac{7}{8} - (-17.\overline{9})$

45. $24.\overline{56} - (-12.\overline{1})$

46. The length of a rectangle is $3\frac{1}{3}$ inches. The width is $\frac{1}{5}$ of the length. Find the width and the perimeter of the rectangle. Support your answer with a drawing.

H.O.T. Problems Higher Order Thinking

47. CCSS **Model with Mathematics** Write a subtraction problem using unlike fractions with a least common denominator of 24. Find the difference.

48. CCSS **Justify Conclusions** Is the difference between a positive mixed number and a negative mixed number *always, sometimes,* or *never* positive? Justify your answer with an example.

49. CCSS **Construct an Argument** Explain why you cannot add or subtract fractions with unlike denominators without renaming the fractions. You may use a diagram to illustrate your answer.

50. CCSS **Find the Error** Cooper is adding the fractions $\frac{1}{3}$, $\frac{7}{9}$, and $\frac{4}{15}$. His first step is to find the least common denominator of 3, 9, and 15. Find his mistake and correct it.

> The least common denominator of 5, 9, and 15 is 90 because you can divide 90 by all of those numbers without getting a remainder.

51. CCSS **Persevere with Problems** A set of measuring cups has measures of 1 cup, $\frac{3}{4}$ cup, $\frac{1}{2}$ cup, $\frac{1}{3}$ cup, and $\frac{1}{4}$ cup. How could you measure $\frac{1}{6}$ cup of milk by using these measuring cups?

52. **Building on the Essential Question** Suppose you use 24 instead of 12 as a common denominator when finding $2\frac{3}{4} - 5\frac{5}{6}$. Will you get the correct answer? Explain.

Standardized Test Practice

53. A recipe for snack mix contains $2\frac{1}{3}$ cups of mixed nuts, $3\frac{1}{2}$ cups of granola, and $\frac{3}{4}$ cup of raisins. What is the total amount of snack mix?

A $5\frac{2}{3}$ c **C** $6\frac{2}{3}$ c

B $5\frac{7}{12}$ c **D** $6\frac{7}{12}$ c

54. **Short Response** The graph shows the results of an election for class president. What fraction of the votes did Michaela receive?

Class Election Results

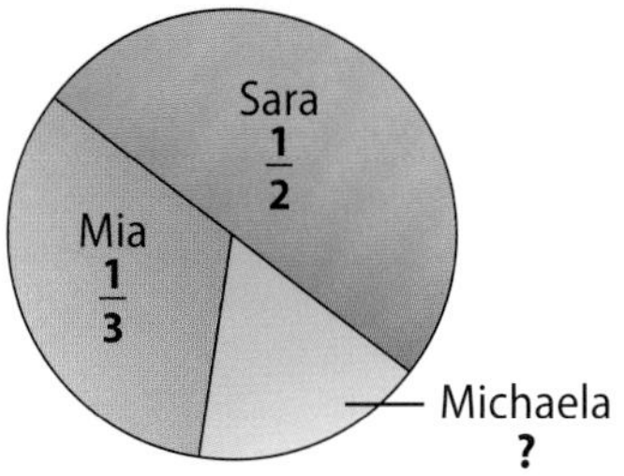

55. The results of a grocery store survey are listed in the table. Find the fraction of families who grill out 2 or more times per month.

How Often Do You Grill Out?	
Times per Month	**Fraction of People**
less than 1	$\frac{11}{50}$
1	$\frac{2}{25}$
2–3	$\frac{4}{25}$
4 or more	$\frac{27}{50}$

F $\frac{2}{25}$ **H** $\frac{7}{10}$

G $\frac{23}{100}$ **J** $\frac{39}{50}$

56. Marco spent $\frac{4}{5}$ hour doing homework on Monday. On Tuesday, he spent $1\frac{1}{3}$ hours doing homework. How long did he spend working on homework for those two days?

A $\frac{8}{15}$ h **C** $1\frac{16}{15}$ h

B $1\frac{5}{8}$ h **D** $2\frac{2}{15}$ h

CCSS Common Core Review

Find each sum or difference. Write in simplest form. 7.NS.1d

57. $6\frac{1}{12} - \left(-8\frac{5}{12}\right)$ **58.** $2\frac{5}{12} + \left(2\frac{7}{12}\right)$ **59.** $2\frac{3}{8} - 1\frac{5}{8}$

60. A 3-foot-long shelf is to be installed between two walls that are $32\frac{5}{8}$ inches apart. How much of the shelf must be cut off so that it fits between the walls? 7.NS.3

Find each quotient. Write in simplest form. 7.NS.2b

61. $\frac{8}{9} \div \frac{4}{3}$ **62.** $12 \div \frac{4}{9}$ **63.** $\frac{a}{7} \div \frac{a}{42}$

64. How many $\frac{1}{4}$-pound hamburgers can be made from $2\frac{3}{4}$ pounds of ground beef? 7.EE.3

65. STEM A *micron* is a unit of measure that is approximately 0.000039 inch. Express this decimal as a fraction. 8.NS.1

Multiply. 7.NS.2a

66. $-2 \cdot (5 \cdot 10)$ **67.** $-1 \cdot (-5 \cdot 10)$ **68.** $15 \cdot (7 - 11)$

Solve by looking for a pattern. 7.NS.2

69. A ball bounces back up 0.5 of the height from where it starts on every bounce. If a ball is dropped from 36 feet, how high does it bounce up on the fourth bounce? Round to the nearest tenth.

Chapter Review

Interactive Study Guide
See pages 65–68 for:
- Vocabulary Check
- Key Concept Check
- Problem Solving
- Reflect

Lesson-by-Lesson Review

Lesson 3-1 Fractions and Decimals (pp. 94–100)

Write each fraction or mixed number as a decimal. Use a bar to show a repeating decimal.

1. $\frac{3}{10}$
2. $\frac{2}{5}$
3. $-\frac{5}{6}$
4. $-7\frac{4}{9}$
5. $\frac{5}{8}$
6. $1\frac{4}{15}$

Replace each ● with <, >, or = to make a true sentence.

7. $\frac{3}{7}$ ● $\frac{4}{9}$
8. $-\frac{5}{8}$ ● $-\frac{3}{5}$
9. $2\frac{1}{2}$ ● $2\frac{5}{12}$
10. $\frac{5}{8}$ ● 0.625
11. $4.\overline{37}$ ● $4\frac{19}{50}$
12. −2.54 ● $-2\frac{27}{50}$
13. Antoine is cutting a $5\frac{5}{16}$-inch board for a project. Write $5\frac{5}{16}$ as a decimal.
14. A basketball player successfully made 21 out of 39 free throw attempts. To the nearest thousandth, what part of the time was he successful in making his free throws?

Example 1

Write $\frac{3}{4}$ as a decimal.

$$\begin{array}{r} 0.75 \\ 4\overline{)3.00} \\ -2\,8 \\ \hline 20 \\ -20 \\ \hline 0 \end{array}$$

Divide 3 by 4.

Divide until the remainder is zero or until a sequence of numbers repeats.

Example 2

Replace the ● with <, >, or = to make $\frac{4}{5}$ ● 0.75 a true sentence.

$\frac{4}{5}$ ● 0.75 — Write the sentence.

0.8 ● 0.75 — Write $\frac{4}{5}$ as a decimal.

0.8 > 0.75 — In the tenths place, 8 > 7.

Lesson 3-2 Rational Numbers (pp. 101–106)

Write each decimal as a fraction or mixed number in simplest form.

15. 2.08
16. −0.45
17. 0.875
18. −0.56
19. $0.\overline{1}$
20. $-2.\overline{03}$
21. $0.\overline{5}$
22. $10.\overline{27}$

Identify all sets to which each number belongs.

23. −4
24. $3\frac{1}{3}$
25. 1.151551555…
26. $-0.\overline{67}$
27. Suzanne practiced playing the piano for $1.\overline{6}$ hours after school. Write $1.\overline{6}$ as a mixed number.
28. James rode his motorbike for 10.4 miles in a competition. Write 10.4 as a mixed number.

Example 3

Write 1.25 as a fraction in simplest form.

$1.25 = 1\frac{25}{100}$ — 1.25 is *1 and 25 hundredths.*

$= 1\frac{1}{4}$ — Simplify. The GCF of 25 and 100 is 25.

Example 4

Write $0.\overline{7}$ as a fraction in simplest form.

$N = 0.777\ldots$

$10N = 10(0.777\ldots)$ — Multiply each side by 10.

$10N = 7.777\ldots$

$-N = 0.777\ldots$ — Subtract N from $10N$.

$9N = 7$ — Simplify.

$N = \frac{7}{9}$ — Divide each side by 9.

Lesson 3-3 Multiplying Rational Numbers (pp. 107–112)

Find each product. Write in simplest form.

29. $\frac{1}{5} \cdot \frac{3}{4}$

30. $-\frac{3}{7} \cdot \frac{4}{9}$

31. $-\frac{2}{3} \cdot (-5)$

32. $-3\frac{1}{2} \cdot \left(-5\frac{1}{5}\right)$

Evaluate each expression if $a = -\frac{2}{3}$ and $b = -4\frac{1}{4}$.

33. ab

34. $2a$

35. $-4b$

36. $-3ab$

37. Mireille has a piece of ribbon that is 10 inches long. Abi's ribbon is $\frac{5}{8}$ as long. How long is Abi's ribbon?

38. A liter of water weighs approximately $2\frac{1}{5}$ pounds. While backpacking, Enrique wants to carry $3\frac{1}{2}$ liters of water with him. Find the weight of the water that Enrique is taking with him.

Example 5

Find $\frac{3}{8} \cdot \frac{20}{27}$. Write in simplest form.

$\frac{3}{8} \cdot \frac{20}{27} = \frac{3 \cdot 20}{8 \cdot 27}$ Multiply the numerators. Multiply the denominators.

$= \frac{60}{216}$ or $\frac{5}{18}$ Simplify. The GCF of 60 and 216 is 12.

Example 6

Find $-4\frac{1}{6} \cdot \frac{3}{5}$. Write in simplest form.

$-4\frac{1}{6} \cdot \frac{3}{5} = -\frac{25}{6} \cdot \frac{3}{5}$ Rename $-4\frac{1}{6}$ as an improper fraction.

$= -\frac{\cancel{25}^{5}}{\cancel{6}_{2}} \cdot \frac{\cancel{3}^{1}}{\cancel{5}_{1}}$ Divide by the GCFs, 5 and 3.

$= -\frac{5}{2}$ or $-2\frac{1}{2}$ Multiply. Then simplify.

Lesson 3-4 Dividing Rational Numbers (pp. 114–119)

Find the multiplicative inverse of each number.

39. -16

40. $\frac{7}{9}$

41. $3\frac{4}{5}$

42. $-4\frac{1}{3}$

43. $-\frac{1}{11}$

44. $2\frac{9}{10}$

Find each quotient. Write in simplest form.

45. $\frac{7}{9} \div \left(-\frac{4}{15}\right)$

46. $-2\frac{2}{3} \div 2\frac{2}{7}$

47. $\frac{3}{5} \div \frac{9}{10}$

48. $3\frac{1}{9} \div \left(-1\frac{1}{6}\right)$

49. $\frac{4}{5} \div \frac{5}{6}$

50. $6\frac{2}{3} \div \left(-3\frac{1}{3}\right)$

Find each quotient. Write in simplest form.

51. $\frac{2ab}{3} \div \frac{a}{6}$

52. $\frac{pq}{5} \div \frac{p}{10}$

53. $\frac{3ab}{2} \div \frac{7b}{10}$

54. $\frac{7mn}{8} \div \frac{3m}{4}$

55. Pilar drinks $1\frac{3}{4}$ glasses of milk each day. At this rate, how many days will it take her to drink a total of 14 glasses?

56. Tahn plants $6\frac{1}{2}$ flats of tomatoes in a row. How many rows will she need to plant 52 flats?

Example 7

Find the multiplicative inverse of $2\frac{3}{4}$.

$2\frac{3}{4} = \frac{11}{4}$ Rename $2\frac{3}{4}$ as an improper fraction.

$\frac{11}{4} \cdot \frac{4}{11} = 1$ The product is 1.

The multiplicative inverse of $2\frac{3}{4}$ is $\frac{4}{11}$.

Example 8

Find $\frac{4}{9} \div \frac{2}{15}$. Write in simplest form.

$\frac{4}{9} \div \frac{2}{15} = \frac{4}{9} \cdot \frac{15}{2}$ Multiply by the reciprocal of $\frac{2}{15}$, $\frac{15}{2}$.

$= \frac{\cancel{4}^{2}}{\cancel{9}_{3}} \cdot \frac{\cancel{15}^{5}}{\cancel{2}_{1}}$ Divide out common factors.

$= \frac{10}{3}$ or $3\frac{1}{3}$ Simplify.

Example 9

Find $\frac{cd}{4} \div \frac{d}{20}$. Write in simplest form.

$\frac{cd}{4} \div \frac{d}{20} = \frac{cd}{4} \cdot \frac{20}{d}$ Multiply by the reciprocal.

$= \frac{c\cancel{d}^{1}}{\cancel{4}_{1}} \cdot \frac{\cancel{20}^{5}}{\cancel{d}_{1}}$ Divide out common factors.

$= \frac{5c}{1}$ or $5c$ Simplify.

Lesson 3-5 Adding and Subtracting Like Fractions (pp. 120–125)

Find each sum or difference. Write in simplest form.

57. $\frac{8}{15} + \left(-\frac{2}{15}\right)$

58. $\frac{6}{12} - \frac{11}{12}$

59. $\frac{3}{7} - \left(-\frac{2}{7}\right)$

60. $-\frac{1}{3} - \left(-\frac{1}{3}\right)$

61. $2\frac{5}{12} - \left(-8\frac{7}{12}\right)$

62. $5\frac{3}{7} + 2\frac{6}{7}$

63. Samantha is going to walk $3\frac{5}{16}$ miles today and $2\frac{3}{16}$ miles tomorrow. What is the total distance she will walk?

64. Last week, Douglas fed his puppy $10\frac{1}{4}$ cups of food. This week, the puppy will be fed an additional $1\frac{1}{4}$ cups of food. Find the total amount of food the puppy will be fed this week.

65. Harry's sunflowers have grown to be $8\frac{1}{4}$ feet tall. Sonya's sunflowers are $6\frac{3}{4}$ feet tall. How much taller are Harry's flowers?

66. Last month Clarissa read $41\frac{3}{8}$ books for the Read-a-thon. Mona read $27\frac{5}{8}$ books. How many more books did Clarissa read?

Example 10

Find $\frac{3}{4} - \left(-\frac{3}{4}\right)$. Write in simplest form.

$\frac{3}{4} - \left(-\frac{3}{4}\right) = \frac{3}{4} + \frac{3}{4}$ — To subtract $-\frac{3}{4}$, add $\frac{3}{4}$.

$= \frac{3+3}{4}$ — The denominators are the same. Add the numerators.

$= \frac{6}{4}$ — Simplify.

$= 1\frac{1}{2}$ — Simplify.

Example 11

Find $5\frac{7}{8} - 8\frac{3}{8}$. Write in simplest form.

$5\frac{7}{8} - 8\frac{3}{8} = \frac{47}{8} - \frac{67}{8}$ — Write the mixed numbers as improper fractions.

$= \frac{47-67}{8}$ — Subtract the numerators.

$= \frac{-20}{8}$ — Simplify the numerator.

$= -\frac{5}{2}$ or $-2\frac{1}{2}$ — Simplify.

Lesson 3-6 Adding and Subtracting Unlike Fractions (pp. 126–131)

Find each sum or difference. Write in simplest form.

67. $\frac{2}{5} + \frac{1}{15}$

68. $-3\frac{5}{6} - 2\frac{1}{2}$

69. $\frac{4}{7} + \left(-1\frac{1}{3}\right)$

70. $\frac{3}{10} - \left(-\frac{1}{8}\right)$

71. $25\frac{1}{3} - 14\frac{2}{5}$

72. $7\frac{3}{4} + 1\frac{3}{8}$

73. $-\frac{5}{9} - 3\frac{2}{3}$

74. $-4\frac{1}{6} + \frac{3}{4}$

75. Monica needs $2\frac{3}{4}$ cups of flour for a batch of cookies and $3\frac{1}{3}$ cups of flour for a dozen muffins. How many cups of flour does Monica need altogether?

76. Dane and his family drove 357.9 miles in one day. If their trip is a total of $524\frac{3}{4}$ miles, how much farther do they need to drive?

77. Ricardo swam 75.5 meters in the school pool. Helen swam $93\frac{3}{4}$ meters the same day. How much further did Helen swim that day?

Example 12

Find $-\frac{3}{8} + \frac{5}{6}$. Write in simplest form.

$-\frac{3}{8} + \frac{5}{6} = -\frac{3}{8} \cdot \frac{3}{3} + \frac{5}{6} \cdot \frac{4}{4}$ — The LCD is 24. Rename the fractions using the LCD.

$= -\frac{9}{24} + \frac{20}{24}$ — Simplify.

$= \frac{-9+20}{24}$ — Add the numerators.

$= \frac{11}{24}$ — Simplify.

Example 13

Find $6\frac{5}{9} - 4\frac{11}{12}$. Write in simplest form.

$6\frac{5}{9} - 4\frac{11}{12} = 6\frac{20}{36} - 4\frac{33}{36}$ — The LCD is 36. Rename the fractions using the LCD.

$= 5\frac{56}{36} - 4\frac{33}{36}$ — Since $\frac{20}{36}$ is less than $\frac{33}{36}$, rename $6\frac{20}{36}$.

$= 1\frac{23}{36}$ — Subtract the whole numbers and then the fractions.

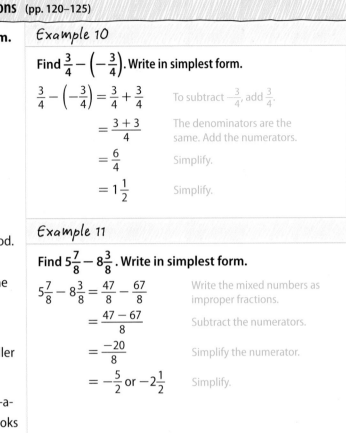

Chapter 4

Powers and Roots

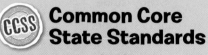

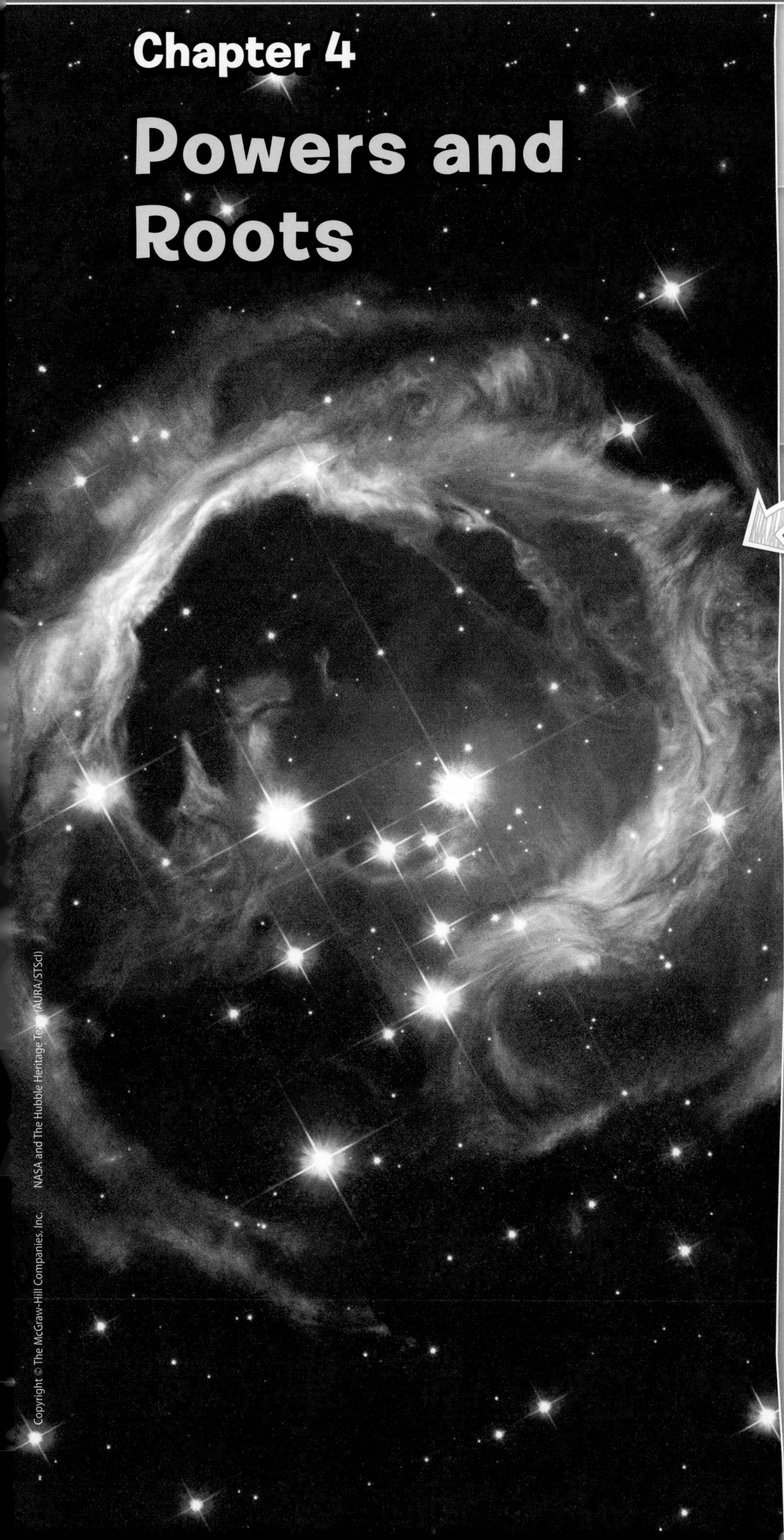

Essential Question

Why is it useful to write numbers in different ways?

Common Core State Standards

Content Standards
8.NS.1, 8.NS.2, 8.EE.1, 8.EE.2, 8.EE.3, 8.EE.4

Mathematical Practices
1, 2, 3, 4, 5, 6, 7, 8

Math in the Real World

Galaxies Most galaxies contain hundreds of billions of stars. The distance between stars and across galaxies is enormous. The Milky Way, a typical galaxy, is about 1,000,000,000,000,000,000 kilometers across. Scientific notation is a useful way to write very large numbers like this one.

Interactive Study Guide
See pages 69–72 for:
- Chapter Preview
- Are You Ready?
- Foldable Study Organizer

Lesson 4-1

Powers and Exponents

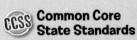

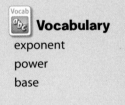

Interactive Study Guide

See pages 73–74 for:
- Getting Started
- Vocabulary Start-Up
- Notes

Essential Question

Why is it useful to write numbers in different ways?

Common Core State Standards

Content Standards
8.EE.1

Mathematical Practices
1, 3, 4, 6, 8

Vocabulary

exponent
power
base

What You'll Learn

- Write expressions using exponents.
- Evaluate expressions containing exponents.

Real-World Link

Computers Data storage capacity is measured in bytes and is based on powers of 2. The standard scientific meanings for the prefixes *mega-* and *giga-* are one million and one billion, respectively. In computer science, a megabyte equals 2^{20} bytes and a gigabyte equals 2^{30} bytes.

Use Exponents

An expression like 5 • 5 • 5 with equal factors can be written using an exponent. An **exponent** tells how many times a number is used as a factor. A number that is expressed using an exponent is called a **power**. The number that is multiplied is called the **base**. So, 5 • 5 • 5 equals the power 5^3.

base ····> 5^3 <···· exponent

power

Read and Write Powers

Power	Words	Factors
5^1	5 to the first power	5
5^2	5 to the second power or 5 squared	5 • 5
5^3	5 to the third power or 5 cubed	5 • 5 • 5
5^4	5 to the fourth power or 5 to the fourth	5 • 5 • 5 • 5
⋮	⋮	⋮
5^n	5 to the *n*th power or 5 to the *n*th	$\underbrace{5 \cdot 5 \cdot 5 \cdot \ldots \cdot 5}_{n \text{ factors}}$

Example 1

Tutor

Write each expression using exponents.

a. $(-8) \cdot (-8) \cdot (-8)$

The base −8 is a factor 3 times.

$(-8) \cdot (-8) \cdot (-8) = (-8)^3$

b. $(k + 2)(k + 2)(k + 2)(k + 2)$

The base $(k + 2)$ is a factor 4 times.

$(k + 2)(k + 2)(k + 2)(k + 2) = (k + 2)^4$

c. $5 \cdot r \cdot r \cdot s \cdot s \cdot s \cdot s$

$5 \cdot r \cdot r \cdot s \cdot s \cdot s \cdot s = 5 \cdot (r \cdot r) \cdot (s \cdot s \cdot s \cdot s)$ — Group factors with like bases.

$= 5 \cdot r^2 \cdot s^4$ or $5r^2s^4$ — $r \cdot r = r^2, s \cdot s \cdot s \cdot s = s^4$

Got It? Do these problems to find out.

1a. $\left(\frac{1}{2}\right)\left(\frac{1}{2}\right)\left(\frac{1}{2}\right)$

1b. $x \cdot x \cdot x \cdot x \cdot x$

1c. $(c - d)(c - d)$

1d. $9 \cdot f \cdot f \cdot f \cdot f \cdot g$

Vocabulary Link
Evaluate
Everyday Use Determine the significance or worth of something.
Math Use Find the value of an expression.

Evaluate Expressions

Since powers represent repeated multiplication, they need to be included in the rules for order of operations.

Concept Summary: Order of Operations

Step 1	Simplify the expressions inside grouping symbols.
Step 2	Evaluate all powers.
Step 3	Multiply and/or divide in order from left to right.
Step 4	Add and/or subtract in order from left to right.

Exponents
An exponent goes with the number, variable, or quantity in parentheses immediately preceding it.

Real World Example 2

Tutor

The playing area for beach volleyball includes the playing court and the free zone. Evaluate each expression to find the area of the playing court and the free zone.

a. The playing court is a rectangle with an area of 2^7 square meters.

$2^7 = 2 \cdot 2 \cdot 2 \cdot 2 \cdot 2 \cdot 2 \cdot 2$ — 2 is a factor 7 times.

$= 128$ — Simplify.

The area of the playing court is 128 square meters.

b. The area of the free zone is $2^2 \cdot 3^2 \cdot 5$ square meters.

$2^2 \cdot 3^2 \cdot 5 = 2 \cdot 2 \cdot 3 \cdot 3 \cdot 5$ — Evaluate powers.

$= 180$ — Multiply.

The area of the free zone is 180 square meters.

Got It? Do this problem to find out.

2. **STEM** A tennis ball is dropped from the top of a building. After 8 seconds, the tennis ball hits the ground. The distance in meters the ball traveled is represented by $4.9(8)^2$. How far did the ball drop?

Watch Out!
$(-3)^2$ is not the same as -3^2.
- $(-3)^2 = (-3)(-3) = 9$
- $-3^2 = (-1)(3^2) = -9$

Example 3

Tutor

Evaluate $x^2 + y^3$ if $x = 6$ and $y = -2$.

$x^2 + y^3 = 6^2 + (-2)^3$ — Replace x with 6 and y with -2.

$= 36 - 8$ — Evaluate powers; $6^2 = (6 \cdot 6)$ or 36; $(-2)^3 = (-2)(-2)(-2)$ or -8.

$= 28$ — Subtract.

Got It? Do these problems to find out.

Evaluate each expression if $a = 5$, $b = -2$, and $c = \frac{3}{4}$.

3a. $10 + b^2$ **3b.** $(a + b)^3$ **3c.** $2 - c^2$

Guided Practice

Check

Write each expression using exponents. (Example 1)

1. $2 \cdot 2 \cdot 2 \cdot 2 \cdot 2 \cdot 2$
2. $d \cdot d \cdot d \cdot d \cdot d \cdot d$
3. $\left(-\frac{1}{4}\right)\left(-\frac{1}{4}\right)\left(-\frac{1}{4}\right)$
4. $4 \cdot m \cdot m \cdot m \cdot q \cdot q \cdot q$
5. $(y - 3)(y - 3)(y - 3)$
6. $(a + 1)(a + 1)$
7. The longhorn beetle can have a body length of more than 2^4 centimeters. How many centimeters long is this? (Example 2)
8. **STEM** Theo sends an E-mail to three friends. Each friend forwards the E-mail to three friends. Each of those friends forwards it to three friends, and so on. Write the number of E-mails sent during the fifth stage as a power. Then find the value of the power. (Example 2)

Evaluate each expression if $a = 3$, $b = -4$, and $c = 3.5$. (Example 3)

9. $a^3 + 2$
10. $3(b - 1)^2$
11. $c^2 + b^2$
12. $4c - 7 + b^3$

Independent Practice

Go online for Step-by-Step Solutions

eHelp

Write each expression using exponents. (Example 1)

13. $11 \cdot 11 \cdot 11 \cdot 11$
14. $3 \cdot 3 \cdot 3 \cdot 3 \cdot 3$
15. $(-8)(-8)(-8)(-8)(-8)(-8)$
16. $(-14) \cdot (-14) \cdot (-14)$
17. $\left(-\frac{1}{5}\right)\left(-\frac{1}{5}\right)\left(-\frac{1}{5}\right)\left(-\frac{1}{5}\right)$
18. $(-1.5)(-1.5)(-1.5)$
19. $ab \cdot ab \cdot ab \cdot ab$
20. $5 \cdot p \cdot p \cdot p \cdot q \cdot q \cdot q$
21. $3 \cdot 7 \cdot m \cdot m \cdot n \cdot n \cdot n \cdot n$
22. $8(c + 4)(c + 4)$
23. $(n - 5)(n - 5)(n - 5)$
24. $(2x + 3y)(2x + 3y)$
25. **STEM** The longest chain of active volcanoes is in the South Pacific. This chain is more than $3 \cdot 10^4$ miles long and has approximately $3^5 \cdot 5$ volcanoes. (Example 2)
    - **a.** How long is the chain of volcanoes?
    - **b.** How many volcanoes are there?
26. A water park has a wave pool that contains about $2^6 \cdot 4^3 \cdot 10^2$ gallons of water. How many gallons of water is this? (Example 2)

Evaluate each expression if $x = -2$, $y = 3$, and $z = 2.5$. (Example 3)

27. y^4
28. z^3
29. $7x^2$
30. xy^3
31. $z^2 + x$
32. $y^4 + 9$
33. $2y + z^3$
34. $x^2 + 2y - 3$
35. $y^2 - 3x + 8$
36. $4(y + 1)^4$
37. $3(2z + 4)^2$
38. $5(x^3 + 6)$

39. CCSS **Be Precise** The table shows the minimum areas of different sports fields.

Sport	Minimum Field Area (ft^2)
field hockey	$2^6 \cdot 10^3$
men's lacrosse	$3^2 \cdot 7 \cdot 10^3$
women's soccer	$2^4 \cdot 5^2 \cdot 7 \cdot 13$

a. Find the minimum area of each playing field.

b. Order the areas from least to greatest.

c. How much greater is the area of a field hockey field than the area of a men's lacrosse field?

Evaluate each expression.

40. 9^2 **41.** 11^3 **42.** $\left(-\frac{2}{3}\right)^3$

43. $(-5)^4$ **44.** $(-2)^7$ **45.** $2 \cdot 4^4$

46. $6^3 \cdot 4$ **47.** $3^5 \cdot 10$ **48.** $2^2 \cdot 10$

49. $7^3 \cdot 2^2$ **50.** $5 \cdot 2^4$ **51.** $(4.5)^4 \cdot 2$

Replace each ● with <, >, or = to make a true statement.

52. 2^5 ● 5^2 **53.** 3^6 ● 6^3

54. 2^6 ● 8^2 **55.** 8^3 ● 4^5

56. $(-6)^4$ ● 6^4 **57.** $(-4)^6$ ● $(-4)^7$

58. CCSS **Multiple Representations** In this problem, you will explore volume of a cube. The volume of a cube equals the side length cubed.

a. **Symbols** Write an equation showing the relationship between side length *s* and volume *V* of a cube.

b. **Table** Make a table of values showing the volume of a cube with side lengths of 1, 2, 4, 8, and 16 centimeters.

c. **Analyze** Use your table to make a conjecture about the change in volume when the side length of a cube is doubled. Justify your response by writing an algebraic expression.

H.O.T. Problems Higher Order Thinking

59. CCSS **Model with Mathematics** Write a real-world problem that involves multiplying two expressions with exponents. Then solve.

60. CCSS **Persevere with Problems** Determine whether x^3 is *always*, *sometimes*, or *never* a positive number for $x \neq 0$. Explain your reasoning.

61. CCSS **Justify Conclusions** Suppose the population of the United States is about 230 million. Is this number closer to 10^7 or 10^8? Explain.

62. CCSS **Identify Repeated Reasoning** Use the pattern below to predict the value of 5^0. Explain your reasoning.

$$5^4 = 625$$
$$5^3 = 125$$
$$5^2 = 25$$
$$5^1 = 5$$

63. **Building on the Essential Question** Describe the advantages of using exponents to represent numeric values.

Standardized Test Practice

64. Marta observed that a bacterium cell doubled every 3 minutes.

Time (min)	Number of Bacteria
3	2^1
6	2^2
9	2^3
12	2^4

Which expression represents the number of cells after one-half hour?

A 2^{10} **C** 2^{20}

B 2^{15} **D** 2^{30}

65. **Short Response** Suppose a certain forest fire doubles in size every 8 hours. If the initial size of the fire was 1 acre, how many acres will the fire cover in 3 days?

66. Which of the following is equivalent to $4^3 \cdot 5^2$?

F $12 \cdot 25$

G $3 \cdot 3 \cdot 3 \cdot 3 \cdot 2 \cdot 2 \cdot 2 \cdot 2 \cdot 2$

H $4 \cdot 4 \cdot 4 \cdot 5 \cdot 5$

J $4 \cdot 4 \cdot 4 \cdot 5 \cdot 5 \cdot 5$

67. Evaluate $\left(\frac{4}{5}\right)^2$.

A $\frac{8}{25}$ **C** $\frac{8}{10}$

B $\frac{16}{25}$ **D** $1\frac{3}{5}$

Common Core Review

Find each sum or difference. 7.NS.1

68. $-12 + (-7)$

69. $25 - (-5)$

70. $-15 + 8$

71. $-9 - (-9)$

72. $3 + (-11)$

73. $-18 - 2$

Name the property shown by each statement. 7.EE.1

74. $87 + 0 = 0$

75. $19 \times 5 = 5 \times 19$

76. $12 \cdot 0 = 0$

77. Kari grew $1\frac{5}{8}$ inches last year and $2\frac{3}{4}$ inches this year. How many total inches did Kari grow in the past two years? 7.NS.3

78. A dance instructor charges a sign-up fee of \$50 plus \$8 for each group lesson. Write an expression that can be used to find the total cost of dance lessons. Then find the cost of 15 lessons. 7.EE.4

Write an integer for each situation. Then identify its opposite and explain its meaning. 7.NS.1a

79. 150 feet below sea level

80. a profit of \$75

Find the greatest common factor for each pair of numbers. 6.NS.4

81. 8 and 12

82. 18 and 24

83. 12 and 14

84. 27 and 36

85. 57 and 63

86. 45 and 80

Find each quotient. 7.NS.2b

87. $-24 \div (-6)$

88. $60 \div (-4)$

89. $-56 \div 8$

90. $-81 \div (-3)$

Lesson 4-2

Negative Exponents

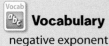

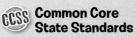

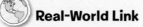

ISG Interactive Study Guide

See pages 75–76 for:
- Getting Started
- Real-World Link
- Notes

Essential Question

Why is it useful to write numbers in different ways?

Common Core State Standards

Content Standards
8.EE.1

Mathematical Practices
1, 3, 4, 6

Vocabulary

negative exponent

What You'll Learn

- Write expressions using negative exponents.
- Evaluate numerical expressions containing negative exponents.

Real-World Link

Snowflakes Have you ever heard that no two snowflakes are exactly alike? This is because they are made up of water molecules, which grow at varying patterns and rates depending on humidity, air currents, and time in the air. Negative exponents are useful for describing small measures, like the diameter of snowflakes.

Key Concept: Negative and Zero Exponents

Symbols For $a \neq 0$ and any whole number n, $a^{-n} = \frac{1}{a^n}$.
For $a \neq 0$, $a^0 = 1$.

Example $8^{-2} = \frac{1}{8^2}$ $\quad x^0 = 1, x \neq 0$

A **negative exponent** is the result of repeated division. Extending the pattern below shows that $\frac{1}{100}$ or $\frac{1}{10^2}$ can be defined as 10^{-2}.

Exponential Form	Standard Form	
$10^3 = 10 \cdot 10 \cdot 10$	1000	÷10
$10^2 = 10 \cdot 10$	100	÷10
10^1	10	÷10
10^0	1	÷10
10^{-1}	$\frac{1}{10}$	÷10
10^{-2}	$\frac{1}{100}$	÷10
10^{-3}	$\frac{1}{1000}$	

Example 1

Watch Out!
Remember 2^{-3} equals $\frac{1}{2^3}$, *not* -8.

Write each expression using a positive exponent.

a. 2^{-3}

$2^{-3} = \frac{1}{2^3}$ Definition of negative exponent

b. m^{-4}

$m^{-4} = \frac{1}{m^4}$ Definition of negative exponent

Got It? Do these problems to find out.

1a. 3^{-5} **1b.** y^{-3} **1c.** 2^0

Example 2

Division by Zero
Since division by zero is undefined, assume that no denominator equals zero. So, in Example 1b, $m \neq 0$.

Write each fraction as an expression using a negative exponent other than −1.

a. $\frac{1}{4^2}$

$\frac{1}{4^2} = 4^{-2}$ Definition of negative exponent

b. $\frac{1}{100}$

$\frac{1}{100} = \frac{1}{10^2}$ Definition of exponent

$= 10^{-2}$ Definition of negative exponent

Got It? Do these problems to find out.

2a. $\frac{1}{6^3}$ **2b.** $\frac{1}{25}$ **2c.** $\frac{1}{27}$

Negative exponents are often used in science when dealing with very small numbers. Usually the number is a power of 10.

Example 3

Real World

STEM Geckos have tiny hairs on the bottom of their feet that are about 0.000001 meter long. Write the decimal as a fraction and as a power of 10.

$0.000001 = \frac{1}{1{,}000{,}000}$ Write the decimal as a fraction.

$= \frac{1}{10^6}$ $1{,}000{,}000 = 10^6$

$= 10^{-6}$ Definition of negative exponent

Therefore, 0.000001 is $\frac{1}{1{,}000{,}000}$ as a fraction and 10^{-6} as a power of 10.

Got It? Do these problems to find out.

3a. The slowest-moving fish is a sea horse. It swims at a maximum speed of 0.0001 mile per minute. Write the decimal as a fraction and as a power of ten.

3b. The smallest species of ant has a mass of 0.00001 gram. Write the decimal as a fraction and as a power of ten.

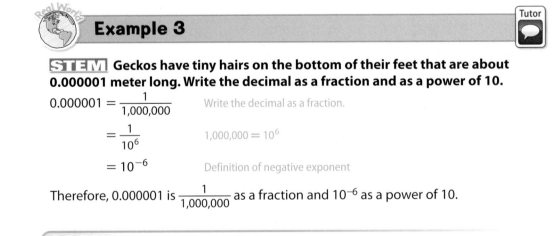

Evaluate Expressions

Algebraic expressions containing negative exponents can be written using positive exponents and then evaluated.

Order of Operations
Remember to follow the order of operations when evaluating expressions.

Example 4

Tutor

Evaluate $4a^{-5}$ if $a = -2$.

$4a^{-5} = 4 \cdot (-2)^{-5}$ Replace a with -2.

$= 4 \cdot \frac{1}{(-2)^5}$ Definition of negative exponent

$= 4 \cdot \frac{1}{-32}$ Find $(-2)^5$.

$= \overset{1}{\cancel{4}} \cdot \frac{1}{\underset{-8}{\cancel{-32}}}$ Simplify.

$= \frac{1}{-8}$ Simplify.

Got It? Do these problems to find out.

Evaluate each expression if $m = 4$ and $n = 3$.

4a. m^{-2} **4b.** $6mn^{-4}$ **4c.** $-n^{-3}$ **4d.** $-4m^{-2}$

Guided Practice

Check

Write each expression using a positive exponent. (Example 1)

1. 6^{-2} **2.** $(-2)^{-3}$ **3.** x^{-5} **4.** b^{-7}

Write each fraction as an expression using a negative exponent other than −1. (Example 2)

5. $\frac{1}{2^6}$ **6.** $\frac{1}{8^2}$ **7.** $\frac{1}{9}$ **8.** $\frac{1}{36}$

9. When a baseball is hit, it comes in contact with the bat for less than 0.001 of a second. Write 0.001 using a negative exponent other than −1. (Example 3)

Evaluate each expression if $x = -4$ and $y = 2$. (Example 4)

10. y^{-7} **11.** x^{-3} **12.** 3^x **13.** $8y^{-4}$

Independent Practice

Go online for Step-by-Step Solutions eHelp

Write each expression using a positive exponent. (Example 1)

14. 11^{-6} **15.** 7^{-1} **16.** $(-4)^{-5}$ **17.** $(-5)^{-4}$

18. a^{-2} **19.** k^{-8} **20.** b^{-15} **21.** r^{-20}

Write each fraction as an expression using a negative exponent other than −1. (Example 2)

22. $\frac{1}{9^4}$ **23.** $\frac{1}{10^3}$ **24.** $\frac{1}{7^6}$ **25.** $\frac{1}{6^5}$

26. $\frac{1}{4}$ **27.** $\frac{1}{49}$ **28.** $\frac{1}{144}$ **29.** $\frac{1}{125}$

Write each decimal using a negative exponent. (Example 3)

30. The minimum thickness of Saturn's A ring is one-tenth kilometer.

31. STEM The diameter of a typical atom is 0.00000001 centimeter. Write the decimal using a negative exponent.

Evaluate each expression if $n = 3$, $p = -2$, and $q = 6$. (Example 4)

32. n^{-5} **33.** $(pq)^{-2}$ **34.** p^{-3}

35. $-q^{-1}$ **36.** 9^p **37.** 2^{-q}

38. $6n^{-3}$ **39** $4pq^{-2}$ **40.** 7^pq^2

41. The table at the right shows the average lengths of different objects.

Object	Length (cm)
pinhead	10^{-3}
cell	10^{-4}
virus	10^{-7}
atom	10^{-10}

a. How many times as long is a virus than an atom?

b. About how many viruses would fit across a pinhead?

c. A football field is about 10^2 meters long. How many times as long is this than a cell?

42. STEM The pH of a substance is a measure of its acidity. The pH scale ranges from 0 to 14, with a pH of 7 being neutral. As the pH decreases, the substance is more acidic. The table shows the pH of several common substances.

	Substance	pH	Hydrogen Ion Concentration
Acids	coffee	5	10^{-5}
	milk	6	10^{-6}
Neutral	pure water	7	10^{-7}
Bases	egg whites	8	10^{-8}
	baking soda	9	10^{-9}

a. Which substance in the table has the greatest hydrogen ion concentration? How many times as great is that hydrogen ion concentration than that of egg whites?

b. Which substance has a hydrogen ion concentration of *one millionth*?

c. How many times as great is the hydrogen ion concentration of coffee as the hydrogen ion concentration of pure water?

43. CCSS **Be Precise** A grain of sand has a volume of about $\frac{1}{10{,}000}$ cubic millimeter.

a. Write this number using a negative exponent.

b. An empty bottle used to create sand art can hold about 10^{10} grains of sand. What is the approximate volume of the sand art bottle?

c. If one cubic centimeter is equal to 10^3 cubic millimeters, how many cubic centimeters of sand will the bottle hold?

44. The wavelengths of X-rays are between 1 and 10 nanometers. If a nanometer is equal to a billionth of a meter, express the greatest wavelength of an X-ray in meters. Write the expression using a negative exponent.

45 The shortest period of time ever measured directly was a light burst of a laser lasting about 0.000000000000001 second. Write this decimal as a fraction and as a power of ten.

46. CCSS **Multiple Representations** In this problem, you will explore negative exponents when using powers of 10, $10^{-1} = \frac{1}{10}$ or 0.1.

a. Table Copy and complete the table shown.

b. Reasoning Do you notice a pattern between the negative powers of 10 and their decimal equivalents? Explain.

c. Words Write a rule that could be used to find the decimal equivalent of any negative power of 10.

d. Numbers Use the rule from part **c** to find the value of 10^{-12}.

Power	Fraction	Decimal
10^{-1}	$\frac{1}{10}$	0.1
10^{-2}	■	■
10^{-3}	■	■
10^{-4}	■	■
10^{-5}	■	■

H.O.T. Problems Higher Order Thinking

47. CCSS **Model with Mathematics** A pizza is cut into 25 equal pieces. Write a fraction that represents one piece of the pizza. Then write this fraction as a power that has a negative exponent. Show the steps you would take to write the fraction as a power.

48. CCSS **Find the Error** Mahala is evaluating the expression $2 \cdot 4^{-2}$. Find her mistake and correct it.

$$2 \cdot 4^{-2} = (2 \cdot 4)^{-2}$$
$$= 8^{-2} \text{ or } \frac{1}{64}$$

49. CCSS **Reason Inductively** Consider the following sets of numbers:

Set 1: $2^{-2}, (-2)^{-2}, (-2)^2, 2^2$

Set 2: $2^{-3}, (-2)^{-3}, (-2)^3, 2^3$

a. Simplify each expression in Set 1. Which expressions, if any, are equal?

b. Simplify each expression in Set 2. Which expressions, if any, are equal?

c. Explain why the number of equal expressions is different for each list.

d. Make a conjecture: $2^{-x} = (-2)^{-x}$, if and only if ________.

e. Make a conjecture: $(-2)^x = 2^x$, if and only if ________.

50. CCSS **Persevere with Problems** Compare and contrast x^{-n} and x^n, where $x \neq 0$. Then give a numerical example to show the relationship.

51. CCSS **Justify Conclusions** Investigate the fraction $\frac{1}{2^n}$. Does it increase or decrease as the value of n increases? Explain.

52. **Building on the Essential Question** Explain the difference between the expressions $(-3)^4$ and 3^{-4}.

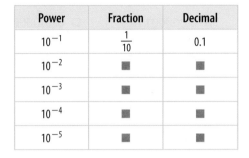

Standardized Test Practice

53. DNA contains the genetic code of an organism. The length of a DNA strand is about 10^{-7} meter. Which of the following represents the length of the DNA strand as a decimal?

A 0.00001 m **C** 0.0000001 m

B 0.000001 m **D** 0.00000001 m

54. When simplified, 2^{-5} is equal to which of the following?

F -32 **H** $\frac{1}{32}$

G $-\frac{1}{32}$ **J** 32

55. Which of the following shows the expressions $4^0, 4^{-2}, 4^2$, and 4^{-3} in order from least to greatest?

A $4^{-3}, 4^{-2}, 4^2, 4^0$

B $4^0, 4^{-2}, 4^{-3}, 4^2$

C $4^2, 4^0, 4^{-2}, 4^{-3}$

D $4^{-3}, 4^{-2}, 4^0, 4^2$

56. **Short Response** It takes light 5.3×0.000001 seconds to travel one mile. Write 0.000001 as a fraction and as a power of 10.

Common Core Review

57. The table shows the elevations of geographic places in relation to sea level. Order the elevations from least to greatest. **6.NS.7b**

Place	Elevation (ft)
Denakil	–410
Lake Assal	–509
Qattara Depression	–435
Turpan Pendi	–505

Find each sum or difference. Write in simplest form. **7.NS.1**

58. $-\frac{3}{8} + \frac{1}{4}$

59. $-\frac{5}{9} - \left(-\frac{2}{3}\right)$

60. $\frac{5}{6} + \frac{11}{12}$

61. $-1\frac{4}{5} - \left(-2\frac{3}{10}\right)$

62. $-4\frac{5}{12} + 2\frac{1}{6}$

63. $2\frac{8}{15} - \left(-4\frac{1}{3}\right)$

64. **Financial Literacy** Rami withdrew \$75 from his savings each month for 3 months. What is his total withdrawal for the 3 months? **7.NS.3**

Evaluate each expression. **7.EE.3**

65. $4n$ if $n = -12$

66. $-9p$ if $p = -7$

67. $3xy$ if $x = -4$ and $y = 5$

68. $-8st$ if $s = -2$ and $t = 6$

Write each expression using exponents. **8.EE.1**

69. $15 \cdot 15 \cdot 15 \cdot 15 \cdot 15$

70. $bc \cdot bc \cdot bc \cdot bc \cdot bc \cdot bc \cdot bc \cdot bc$

71. $4 \cdot 9 \cdot x \cdot x \cdot y \cdot y \cdot y \cdot y$

72. $(p + 2)(p + 2)(p + 2)$

Find each product. **6.NS.3**

73. 25×0.001

74. 107×0.0001

75. 3.8×0.01

76. 0.5×0.021

77. 1.5×0.003

78. 4.2×0.0005

Lesson 4-3

Multiplying and Dividing Monomials

Interactive Study Guide

See pages 77–78 for:
- Getting Started
- Real-World Link
- Notes

Essential Question

Why is it useful to write numbers in different ways?

Common Core State Standards

Content Standards
8.EE.1

Mathematical Practices
1, 2, 3, 4

Vocabulary

monomial

What You'll Learn

- Multiply monomials.
- Divide monomials.

Real-World Link

Football Did you know that football is the most watched sport on American television? The playing field in football is 120 yards long and $53\frac{1}{3}$ yards wide. Recently, a local high school cleared an area that measured 2^7 yards long and 2^6 yards wide to make room for a new football field.

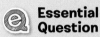

Key Concept Product of Powers Property

Words	Multiply powers with the same base by adding their exponents.		
Symbols	$a^m \cdot a^n = a^{m+n}$	Example	$2^4 \cdot 2^3 = 2^{4+3}$ or 2^7

Recall that exponents are used to show repeated multiplication.

$$10^1 \cdot 10^3 = \underbrace{\overbrace{(10)}^{\text{1 factor}} \cdot \overbrace{(10 \cdot 10 \cdot 10)}^{\text{3 factors}}}_{\text{4 factors}} = 10^4$$

A **monomial** is a number, a variable, or a product of numbers and/or variables.

Monomials	Not Monomials
$80, x, 8x$	$x + 5, x^2 - y^2$

Example 1

Tutor

Find each product. Express using positive exponents.

a. $4^3 \cdot 4^5$

$4^3 \cdot 4^5 = 4^{3+5}$ — Product of Powers Property; the common base is 4.

$= 4^8$ — Add the exponents.

b. $6 \cdot 6^{-4}$

$6 \cdot 6^{-4} = 6^1 \cdot 6^{-4}$ — $6 = 6^1$

$= 6^{1+(-4)}$ — Product of Powers Property; the common base is 6.

$= 6^{-3}$ — Add the exponents.

$= \frac{1}{6^3}$ — Definition of negative exponent

Got It? Do these problems to find out.

1a. $5^2 \cdot 5^3$

1b. $12^3 \cdot 12^{-2}$

Example 2

Tutor

Watch Out!

When multiplying powers, do not multiply the bases.

$6^1 \cdot 6^4 = 6^5$, not 36^5.

Find each product. Express using positive exponents.

a. $b^2 \cdot b^2$

$b^2 \cdot b^2 = b^{2+2}$ Product of Powers Property; the common base is *b*.

$= b^4$ Add the exponents.

b. $t^{-5} \cdot t^3$

$t^{-5} \cdot t^3 = t^{-5+3}$ Product of Powers Property; the common base is *t*.

$= t^{-2}$ Add the exponents.

$= \frac{1}{t^2}$ Definition of negative exponent

Got It? Do these problems to find out.

2a. $y^6 \cdot y^3$

2b. $r^6 \cdot r^{-5}$

2c. $a^7 \cdot a^6$

2d. $x^{-6} \cdot x^2$

Example 3

Tutor

Find each product. Express using positive exponents.

a. $2x^3 \cdot 8x^4$

$2x^3 \cdot 8x^4 = 2 \cdot 8 \cdot x^3 \cdot x^4$ Commutative Property of Multiplication

$= 2 \cdot 8 \cdot x^{3+4}$ Product of Powers Property; the common base is *x*.

$= 2 \cdot 8 \cdot x^7$ Add the exponents.

$= 16x^7$ Multiply.

b. $-3n^{-9} \cdot 2n^8$

$-3n^{-9} \cdot 2n^8 = -3 \cdot 2 \cdot n^{-9} \cdot n^8$ Commutative Property of Multiplication

$= -3 \cdot 2 \cdot n^{-9+8}$ Product of Powers Property; the common base is *n*.

$= -3 \cdot 2 \cdot n^{-1}$ Add the exponents.

$= -6n^{-1}$ Multiply.

$= -\frac{6}{n}$ Definition of negative exponent

Got It? Do these problems to find out.

3a. $(5a^2)(-3a^4)$

3b. $6b^{-4} \cdot 2b^2$

3c. $(6x^3)(-3x^5)$

3d. $10n^7 \cdot 5n^2$

Key Concept Quotient of Powers Property

Watch

Words	Divide powers with the same base by subtracting their exponents.
Symbols	$a^m \div a^n = a^{m-n}$, where $a \neq 0$
Example	$3^6 \div 3^2 = 3^{6-2}$ or 3^4

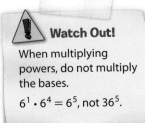

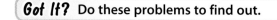

You can use repeated multiplication of exponents to demonstrate the Quotient of Powers Property.

$$\frac{4^5}{4^2} = \frac{\overbrace{4 \cdot 4 \cdot 4 \cdot 4 \cdot 4}^{\text{5 factors}}}{\underbrace{4 \cdot 4}_{\text{2 factors}}} = \frac{4 \cdot 4 \cdot 4 \cdot \overset{1}{\cancel{4}} \cdot \overset{1}{\cancel{4}}}{\underset{1}{\cancel{4}} \cdot \underset{1}{\cancel{4}}}$$

$$= \underbrace{4 \cdot 4 \cdot 4}_{\text{3 factors}} \text{ or } 4^3$$

Example 4

Tutor

Find each quotient.

Watch Out! When dividing powers, remember that the denominator cannot equal zero. So, in Example 4b, $c \neq 0$.

a. $\frac{8^5}{8^3}$

$\frac{8^5}{8^3} = 8^{5-3}$ — Quotient of Powers Property; the common base is 8.

$= 8^2$ — Subtract the exponents.

b. $\frac{c^7}{c^2}$

$\frac{c^7}{c^2} = c^{7-2}$ — Quotient of Powers Property; the common base is *c*.

$= c^5$ — Subtract the exponents.

c. $\frac{2^{-2}}{2^2}$

$\frac{2^{-2}}{2^2} = 2^{-2-2}$ — Quotient of Powers Property; the common base is 2.

$= 2^{-4}$ — Subtract the exponents.

$= \frac{1}{2^4}$ — Definition of negative exponent

d. $\frac{d^5}{d^{-2}}$

$\frac{d^5}{d^{-2}} = d^{5-(-2)}$ — Quotient of Powers Property; the common base is *d*.

$= d^7$ — Subtract the exponents.

Got It? Do these problems to find out.

4a. $\frac{3^9}{3^2}$ **4b.** $\frac{b^7}{b^6}$ **4c.** $\frac{4^5}{4^{-2}}$ **4d.** $\frac{s^{-4}}{s^1}$

Real World Example 5

Tutor

STEM The table shows the approximate diameters of Earth, Mars, and Neptune. About how many times as great is Neptune's diameter than Earth's diameter?

Planet	Approximate Diameter (mi)
Mars	2^{12}
Earth	2^{13}
Neptune	2^{15}

Write a division expression.

$\frac{2^{15}}{2^{13}} = 2^{15-13}$ — Quotient of Powers Property

$= 2^2$ or 4 — Simplify.

So, Neptune's diameter is about 4 times as great as Earth's diameter.

Got It? Do these problems to find out.

5a. About how many times as great is the diameter of Earth than the diameter of Mars?

5b. The diameter of a small asteroid is 10^{-1} kilometer. The diameter of Ceres is 10^3 kilometers. About how many times as great is the diameter of Ceres than the diameter of the smaller asteroid?

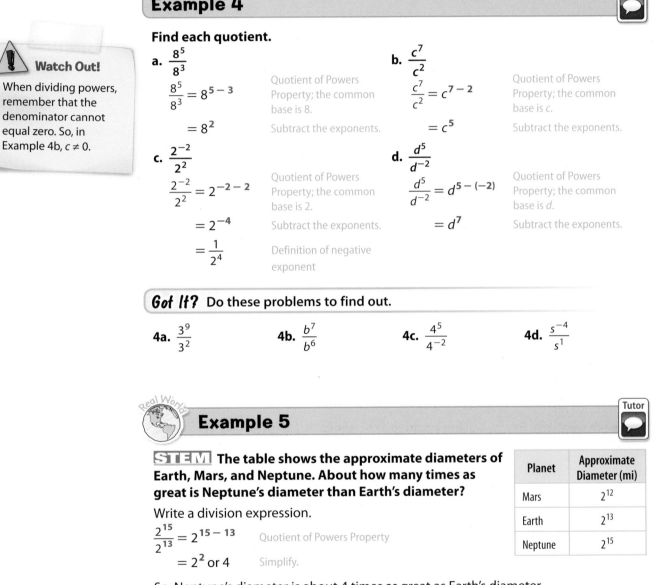

Guided Practice

Check

Find each product. Express using positive exponents. (Examples 1–3)

1. $2^4 \cdot 2^6$
2. $8^5 \cdot 8$
3. $5^{-6} \cdot 5^9$
4. $3^2 \cdot 3^{-5}$
5. $x^{10} \cdot x^6$
6. $-w^2(5w^7)$
7. $m^8 \cdot m^{-10}$
8. $y^{-4} \cdot y^{12}$

Find each quotient. Express using positive exponents. (Example 4)

9. $\frac{4^5}{4^3}$
10. $7^9 \div 7$
11. $\frac{6^7}{6^{-5}}$
12. $9^{-2} \div 9^6$
13. $\frac{r^8}{r^4}$
14. $b^{11} \div b^2$
15. $\frac{c^{-7}}{c^2}$
16. $n^5 \div n^{-4}$

17. The Grand Canyon is approximately 2^9 kilometers long. Mariner Valley is a canyon on Mars that is approximately 2^{12} kilometers long. About how many times as long is the length of Mariner Valley than that of the Grand Canyon? (Example 5)

18. A snake is 2^5 inches long. An earthworm is 2^{-1} inch long. About how many times as long is the length of the snake than the length of the earthworm? (Example 5)

Independent Practice

Go online for Step-by-Step Solutions

eHelp

Find each product. Express using positive exponents. (Examples 1–3)

19. $5^6 \cdot 5^2$
20. $(-2)^3 \cdot (-2)^2$
21. $a^7 \cdot a^2$
22. $(t^3)(t^3)$
23. $4^{-5} \cdot 4^6$
24. $6^5 \cdot 6^{-5}$
25. $c^2 \cdot c^{-3}$
26. $(w^{-4})(w^6)$
27. $(10x)(4x^{-7})$
28. $6p^7 \cdot 9p^7$
29. $m^{-5} \cdot (-4m^6)$
30. $(-8s^3)(-3s^4)$

Find each quotient. Express using positive exponents. (Example 4)

31. $\frac{5^{10}}{5^2}$
32. $\frac{7^6}{7}$
33. $\frac{a^8}{a^7}$
34. $\frac{k^{12}}{k^9}$
35. $\frac{8^{-7}}{8^4}$
36. $\frac{3^3}{3^{-1}}$
37. $\frac{b^4}{b^5}$
38. $\frac{y^{15}}{y^{-2}}$
39. $(-1.5)^8 \div (-1.5)^3$
40. $8^{15} \div 8^{-9}$
41. $r^{20} \div r^6$
42. $(-n)^{-6} \div (-n)^4$

43. Sound intensity is measured in decibels. The decibel scale is based on powers of ten, as shown. (Example 5)
    a. How many times as intense is a rock concert as a normal conversation?
    b. How many times as intense is a vacuum cleaner as a person whispering?

Sound	Decibels	Intensity
rock concert	110	10^{11}
vacuum cleaner	80	10^8
normal conversation	60	10^6
whispering	20	10^2

44. A large beetle can be 2^7 millimeters long. One of the smallest beetles can be 2^{-2} millimeter long. How many times as great is the length of the large beetle than the length of the small beetle? (Example 5)

45 A person weighing 5^3 pounds can experience forces 5 times his or her body weight while running. Find $5^3 \cdot 5$ to find the number of pounds exerted on each foot of the person while running.

46. The largest sea cucumbers are more than 10^2 times longer than the smallest sea cucumbers. If the smallest species are about 10 millimeters long, find the approximate length of the largest sea cucumbers.

47. **STEM** A nurse draws a sample of blood. A cubic millimeter of the blood contains 22^5 red blood cells and 22^3 white blood cells. Compare the number of red blood cells to the number of white blood cells as a fraction. Explain its meaning.

Find each missing exponent.

48. $(5^{\blacksquare})(5^2) = 5^3$

49. $(9^{10})(9^{\blacksquare}) = 9^8$

50. $a^{12} \cdot a^{\blacksquare} = a^{19}$

51. $\frac{6^{\blacksquare}}{6^4} = 6^8$

52. $\frac{x^7}{x^{\blacksquare}} = 1$

53. $c^{10} \div c^{\blacksquare} = c^{13}$

54. **CCSS Multiple Representations** In this problem, you will investigate area and volume. The formulas $A = s^2$ and $V = s^3$ can be used to find the area of a square and the volume of a cube, respectively, with side length s.

a. **Table** Copy and complete the table shown.

b. **Words** How are the area and volume each affected if the side length is doubled? tripled?

c. **Words** How are the area and volume each affected if the side length is squared? cubed?

Side Length (units)	Area of Square (units^2)	Volume of Cube (units^3)
s	s^2	s^3
$2s$	$\blacksquare$	$\blacksquare$
$3s$	$\blacksquare$	$\blacksquare$
s^2	$\blacksquare$	$\blacksquare$
s^3	$\blacksquare$	$\blacksquare$

Find each product or quotient. Express using exponents.

55. $ab^5 \cdot 8a^2b^5$

56. $10x^3y \cdot (-2xy^2)$

57. $\frac{n^3(n^5)}{n^2}$

58. $\frac{s^7}{s \cdot s^2}$

59. $7x^2y^5 \cdot 3y$

60. $12a^7 \cdot (-4a^2b^6)$

H.O.T. Problems Higher Order Thinking

61. **CCSS Reason Abstractly** Write two algebraic expressions whose quotient is x^5.

62. **CCSS Find the Error** Noah is multiplying $(4a^2)(4a^3)$. Find his error and correct it.

$$(4a^2)(4a^3) = 4a^{2+3}$$
$$= 4a^5$$

63. **CCSS Persevere with Problems** Use the Quotient of Powers Property and the equation $\frac{a^n}{a^n} = 1$ to show that a nonzero number raised to the zero power equals 1.

64. **CCSS Use a Counterexample** *True* or *false*. For any integer a, $(-a)^2 = -a^2$. If true, explain your reasoning. If false, give a counterexample.

65. **Building on the Essential Question** Explain how to use division of powers to divide large numbers.

Standardized Test Practice

66. In the metric system, one meter is equal to 10^2 centimeters. One kilometer is 10^3 meters. How many centimeters are in one kilometer?

A 1000

B 10,000

C 100,000

D 1,000,000

67. Which of the following expressions is equivalent to $12^5 \div 12^{-2}$?

F 12^7

G 12^3

H 12^{-3}

J 12^{-7}

68. Which of the following expressions is equivalent to the product of $5a^3$ and $3a^8$?

A $8a^{11}$

B $8a^{24}$

C $15a^{11}$

D $15a^{24}$

69. **Short Response** The formula $A = \frac{1}{2}bh$ can be used to find the area of a triangle with base b and height h. Write an expression in simplest form to represent the area of the triangle shown below. Show your work.

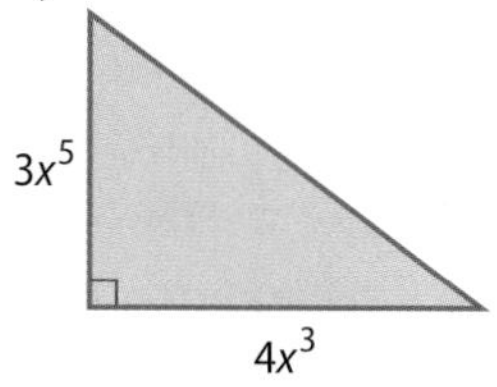

Common Core Review

Find each product. 7.NS.2a

70. $-4 \cdot -25$

71. $-14(5)$

72. $12(-6)$

73. Which numbers in the table can be expressed as whole numbers raised to a power? Name the cities and express the numbers as powers. 8.EE.1

74. The nameplate on an office door measures 16 inches long and 4 inches wide. What is the area of the nameplate? 6.G.1

75. The top of an ocean buoy is 24 inches above sea level. The bottom of the buoy is 6 inches below sea level. Write and evaluate an expression to find the distance from the top of the buoy to the bottom of the buoy. 7.NS.1c

Miles to Kentucky Dam	
City	**Miles**
Chicago	400
Lexington	250
St. Louis	225
Louisville	200
Nashville	125
Evansville	100
Paducah	25

Write each expression using a positive exponent. 8.EE.1

76. 8^{-15}

77. w^{-6}

Find each sum or difference. Write in simplest form. 7.NS.1

78. $\frac{3}{5} + \frac{3}{10}$

79. $-\frac{1}{2} + \frac{3}{8}$

80. $-6\frac{2}{3} - \frac{8}{9}$

Evaluate each expression if $a = -3$, $b = 7$, and $c = 5$. 7.EE.3

81. $\frac{1}{ab}$

82. $\frac{1}{(b)(b)}$

83. $\frac{c}{75}$

84. $\frac{ac}{5ab}$

85. $7ab + c$

86. $10a + 10b$

Lesson 4-4

Scientific Notation

ISG Interactive Study Guide

See pages 79–80 for:
- Getting Started
- Real-World Link
- Notes

Essential Question

Why is it useful to write numbers in different ways?

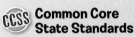

CCSS Common Core State Standards

Content Standards
8.EE.1, 8.EE.3, 8.EE.4

Mathematical Practices
1, 3, 4, 7

Vocab Vocabulary

standard form
scientific notation

What You'll Learn

- Express numbers in standard form and in scientific notation.
- Compare and order numbers written in scientific notation.

Real-World Link

Space Earth is the third planet from the Sun in our solar system. Because Earth's rotation about the Sun is not circular, the maximum distance between Earth and the Sun is about 95 million miles and the minimum distance is about 91 million miles.

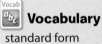

Key Concept Scientific Notation

Words	A number is expressed in scientific notation when it is written as the product of a factor and a power of 10. The factor must be greater than or equal to 1 and less than 10.
Symbols	$a \times 10^n$, where $1 \leq a < 10$ and n is an integer.
Examples	$3{,}500{,}000 = 3.5 \times 10^6$ $\quad$ $0.00004 = 4 \times 10^{-5}$

Numbers that do not contain exponents are written in **standard form**. However, when you deal with very large numbers like 12,760,000 or very small numbers like 0.00001276, it can be difficult to keep track of the place value. A number that is expressed as a product of a factor and a power of 10 is written in **scientific notation**.

When a number is expressed in scientific notation the exponent tells you how many places to move the decimal point.

Example 1

Tutor

Express each number in standard form.

a. 2×10^3

$2 \times 10^3 = 2000$ — Move the decimal point 3 places to the right.

b. 6.8×10^5

$6.8 \times 10^5 = 680{,}000$ — Move the decimal point 5 places to the right.

c. 3.25×10^{-4}

$3.25 \times 10^{-4} = 0.000325$ — Move the decimal point 4 places to the left.

Got It? **Do these problems to find out.**

1a. 4×10^2 **1b.** 5.94×10^7 **1c.** 1.3×10^{-3}

When expressing a number in scientific notation, the sign of the exponent can be determined by evaluating the number in standard form. If a number in standard form is greater than or equal to 1, then the exponent is *positive*. If a number is between 0 and 1, then the exponent is *negative*.

Example 2

Tutor

Scientific Notation
When numbers are expressed in scientific notation, no more than one digit is to the left of the decimal point.

Express each number in scientific notation.

a. 4,000,000

$4{,}000{,}000 = 4 \times 10^6$ The decimal point moves 6 places. The exponent is positive.

b. 0.072

$0.072 = 7.2 \times 10^{-2}$ The decimal point moves 2 places. The exponent is negative.

Got It? Do these problems to find out.

2a. 900 **2b.** 18,900 **2c.** 0.000064

One way to estimate a very large or a very small number is to express it in the form of a single digit times an integer power of 10. For example, the population of the United States in 2010 was 308,745,538. The number 3×10^8 is an estimate of that number.

Example 3

Tutor

The population of Kansas is 2,853,118 people. Write an estimation in scientific notation for the population.

$2{,}853{,}118 \approx 3{,}000{,}000$ Estimate.
$3{,}000{,}000 = 3 \times 10^6$ Write in scientific notation.

The population of Kansas is about 3×10^6 people.

Got It? Do these problems to find out.

Estimate each value using scientific notation.

3a. 3,612,500 cm **3b.** 0.000000251 ft **3c.** 4.215×10^{-3} kg

Real World

Example 4

Tutor

STEM **The space shuttle traveled at about 8 kilometers per second. At this rate, the shuttle would take about 4.5×10^4 seconds to fly to the moon. Is it more appropriate for a newspaper to report this time as about 4.5×10^4 seconds or about 12.5 hours? Explain your reasoning.**

The measure 12.5 hours is more appropriate. The number 4.5×10^4 seconds is very large, so choosing the larger unit of measure is more meaningful.

Got It? Do this problem to find out.

4. A dime is about 5.875×10^{-3} foot in diameter. Is it more appropriate to report that the diameter of a dime is 5.875×10^{-3} foot or 7.05×10^{-1} inch? Explain your reasoning.

Compare and Order Numbers

To compare and order numbers in scientific notation, first compare the exponents. With positive numbers, the number with a greater exponent is greater. If the exponents are the same, compare the factors.

Power of 10
When writing a number in scientific notation, the power of 10 is determined by the direction and number of places you move the decimal point.

Real World Example 5

Tutor

STEM The table shows different geologic time periods. Order the time periods from oldest to youngest.

Geologic Time Periods	
Period	Number of Years Ago
Jurassic	2.08×10^8
Silurian	4.38×10^8
Tertiary	6.64×10^7
Triassic	2.45×10^8

Step 1 Order the numbers according to their exponents.

The Tertiary period has an exponent of 7. So, it is the youngest period.

Step 2 Order the numbers with the same exponent by comparing the factors.

4.38	>	2.45	>	2.08
Silurian		Triassic		Jurassic
↓		↓		↓

So, $4.38 \times 10^8 > 2.45 \times 10^8 > 2.08 \times 10^8$.

The time periods ordered from oldest to youngest are Silurian, Triassic, Jurassic, and Tertiary.

Got It? Do this problem to find out.

5. **STEM** Approximately 1.372×10^7 square kilometers of Antarctica and about 1.834×10^6 square kilometers of Greenland are covered by an ice cap. Which land mass has a greater area covered by ice?

Guided Practice

Check

Express each number in standard form. (Example 1)

1. 4.16×10^3
2. 3.2×10^{-2}
3. 1.075×10^5

Express each number in scientific notation. (Example 2)

4. 1,600,000
5. 135,000
6. 0.008

Estimate each value using scientific notation. (Example 3)

7. 0.000007109 kg
8. 3.7085×10^{14} mL
9. 18,900,435 cm

10. If you could walk at the rate of about 1 mile every 20 minutes without stopping, it would take about 1.4×10^2 hours to walk from Columbus, Ohio, to Washington, D.C. Is it more appropriate to report the time as 1.4×10^2 hours or 8.4×10^3 minutes? Explain. (Example 4)

11. Order 3.4×10^2, 3.5×10^2, 3.7×10^{-2}, and 400 from least to greatest. (Example 5)

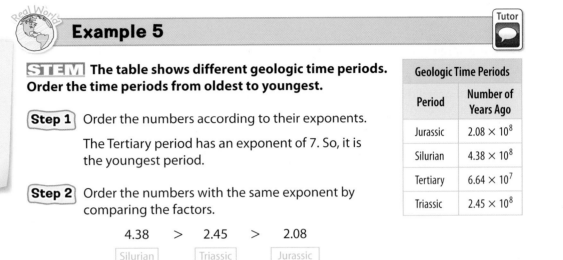

Independent Practice

Go online for Step-by-Step Solutions eHelp

Express each number in standard form. (Example 1)

12. 6.89×10^4

13. 1.5×10^{-4}

14. 2.3×10^{-5}

15. 9.51×10^{-3}

16. 3.062×10^6

17. 7.924×10^2

18. A dollar bill is approximately 1.09×10^{-2} centimeter thick. Write 1.09×10^{-2} in standard form.

19. It is estimated that more than 1.71×10^{11} E-mails are sent each day around the world. Write 1.71×10^{11} in standard form.

Express each number in scientific notation. (Example 2)

20. 700,000

21. 32,000,000

22. 0.045

23. 0.000918

24. 1,000,000

25. 0.006752

Estimate each value using scientific notation. (Example 3)

26. 0.00000095 centimeter

27. 8.375×10^{-23} pound

28. 56,300,001 miles

29. The distance between Earth and the Moon is about 3.84×10^5 kilometers. Estimate this distance using scientific notation. (Example 3)

30. The usual growth rate of human hair is 3.3×10^{-4} meter per day. Is it more appropriate to report the rate as 3.3×10^{-4} meter per day or 0.33 millimeter per day? Explain your reasoning. (Example 4)

31. One ounce of a certain cheese has 219 milligrams of calcium. Is it more appropriate to include on the nutrition label that the cheese has 2.19×10^{-4} kilogram of calcium or 219 milligrams of calcium? (Example 4)

Order each set of numbers from least to greatest. (Example 5)

32. 2.4×10^2, 2.45×10^{-2}, 2.45×10^2, 2.4×10^{-2}

33. 2.81×10^4, 2805, 2.08×10^5, 3.2×10^4, 3.024×10^2

34. 5.9×10^6, 5.9×10^4, 5.01×10^5, 5.1×10^{-3}

35. 9,562,301, 9.05×10^{-6}, 9.5×10^6, 905,000

36. List the states in the table at the right from least to greatest production of maple syrup. (Example 5)

State	Amount of Syrup Produced (L)
Maine	1.10×10^6
New Hampshire	3.14×10^5
New York	9.65×10^5
Vermont	1.89×10^6
Wisconsin	3.79×10^5

37. STEM A sheet of gold leaf is approximately 1.25×10^{-5} centimeter thick.

a. Write the value of the thickness as a decimal.

b. Use the formula $V = \ell wh$ to find the volume in cubic meters of a sheet of gold that is 2 meters wide and 5 meters long.

Replace each ● with <, >, or = to make a true statement.

38. 5.72×10^8 ● 5.8×10^8

39. 35,400 ● 35.4×10^3

40. 0.042 ● 4.2×10^{-3}

41. 5×10^5 ● 5,000,000

42. 27,000 ● 2.76×10^3

43. 6.4×10^{-5} ● 0.000649

44. **Justify Conclusions** A news article reported the population of Illinois to be 12,869,257 and the estimated population of Indiana to be 6×10^6.

a. Estimate the number of people living in Illinois. Express the number in scientific notation.

b. Which state, Illinois or Indiana, has the greater population? Explain your reasoning.

c. Estimate the combined population of the two states. Express the number in scientific notation.

45. The Moon travels around the Earth at a speed of about 3.68×10^3 kilometers per hour. If the Moon orbits the Earth every 27.3 days, about how far does it travel in one orbit around the Earth?

46. The speed of light is about 3×10^5 kilometers per second. The distance between Earth and the Moon is about 3.84×10^5 kilometers. Find how long it would take for light to travel from Earth to the Moon.

47. In a recent year, U.S. route 59 in the Houston area averaged approximately 338,510 vehicles per day. About how many vehicles was this during the entire year? Write the number in scientific notation. Verify your solution by using estimation.

H.O.T. Problems Higher Order Thinking

48. **Identify Structure** Write two numbers in scientific notation with different exponents. Then find the sum, difference, product, and quotient of the two numbers. Write the answers in scientific notation.

49. **Persevere with Problems** A *googol* is a number that is 1 followed by 100 zeros. A *centillion* is a number that is 1 followed by 303 zeros. Write each of these numbers in scientific notation.

50. **Justify Conclusions** Miami is the second most populous city in Florida.

a. Which number better describes the population of Miami: 3.8×10^4 or 3.8×10^6? Explain.

b. Express Miami's population in another form.

c. Which notation is best to use when describing population? Explain.

51. **Persevere with Problems** Which number is twice as great as 3×10^2: 6×10^2, 3×10^4, or 6×10^4? Explain.

52. **Building on the Essential Question** Your friend thinks 7.8×10^3 is greater than 6.5×10^2 because $7.8 > 6.5$. Explain why your friend's reasoning is incorrect.

Standardized Test Practice

53. The slowest land mammal is the three-toed sloth, which moves 0.07 mile per hour. Which expression represents this number?

A 7×10^{-3} **C** 7×10^{2}

B 7×10^{-2} **D** 7×10^{3}

54. The distance from Earth to the sun is about 9.5×10^{7} miles. Which of the following represents this distance?

F 9,500,000 mi

G 95,000,000 mi

H 950,000,000 mi

J 9,500,000,000 mi

55. **Short Response** The weight of a fruit fly is about 1.3×10^{-4} pound. Estimate the weight of a fruit fly. Express the weight in the form of a single digit times an integer power of 10.

56. **Short Response** A 45-acre farm produces 366,400 pounds of avocados per year. Write an expression, in scientific notation, for the number of pounds of avocados produced per year.

CCSS Common Core Review

57. **STEM** Which type of molecule in the table has a greater mass? How many times greater is it than the other type? **8.EE.1**

Molecule	Mass (kg)
penicillin	10^{-18}
insulin	10^{-23}

Find each product or quotient. Express using exponents. 8.EE.1

58. $a \cdot a^5$

59. $(n^4)(n^4)$

60. $-3x^2(4x^3)$

61. $\frac{3^8}{3^{-5}}$

62. Use the table to write an expression that shows how many bushels of green beans were picked. Simplify the expression. **7.EE.1**

Bushels of Green Beans Picked	
Acre A	40
Acre B	38
Acre C	g

63. Translate *eight times the product of six times a number, all divided by 2* into an algebraic expression. Then simplify the expression. **6.EE.2a**

64. **Financial Literacy** Of Quincy's monthly paycheck, $\frac{7}{25}$ goes to pay his mortgage, and $\frac{9}{20}$ goes to pay his college loans. Does a greater fraction of his paycheck go to paying his mortgage or his college loans? **8.NS.1**

Write each expression using a positive exponent. 8.EE.1

65. 3^{-2}

66. h^{-8}

67. $(-5)^{-1}$

68. x^{-12}

69. 5^{-3}

70. $(-7)^{-4}$

Write each expression using exponents. 8.EE.1

71. $4 \cdot 4 \cdot 4 \cdot 4 \cdot 4$

72. $(6 \cdot 6 \cdot 6) \cdot 6$

73. $3 \cdot 2 \cdot 3 \cdot 2 \cdot 2$

74. $7 \cdot 7 \cdot 8 \cdot 8 \cdot 8 \cdot 7$

75. $10 \cdot 5 \cdot 10 \cdot 5 \cdot 5 \cdot 5$

76. $(2 \cdot 2) \cdot 2 \cdot 2 \cdot 2$

ISG Interactive Study Guide

See page 81 for:
- Mid-Chapter Check

21ST CENTURY CAREER in Engineering

Robotics Engineer

Are you mechanically inclined? Do you like to find new ways to solve problems? If so, a career as a robotics engineer is something you should consider. Robotics engineers design and build robots to perform tasks that are difficult, dangerous, or tedious for humans. For example, a robotic insect was developed based on a real insect. Its purpose was to travel over water surfaces, take measurements, and monitor water quality.

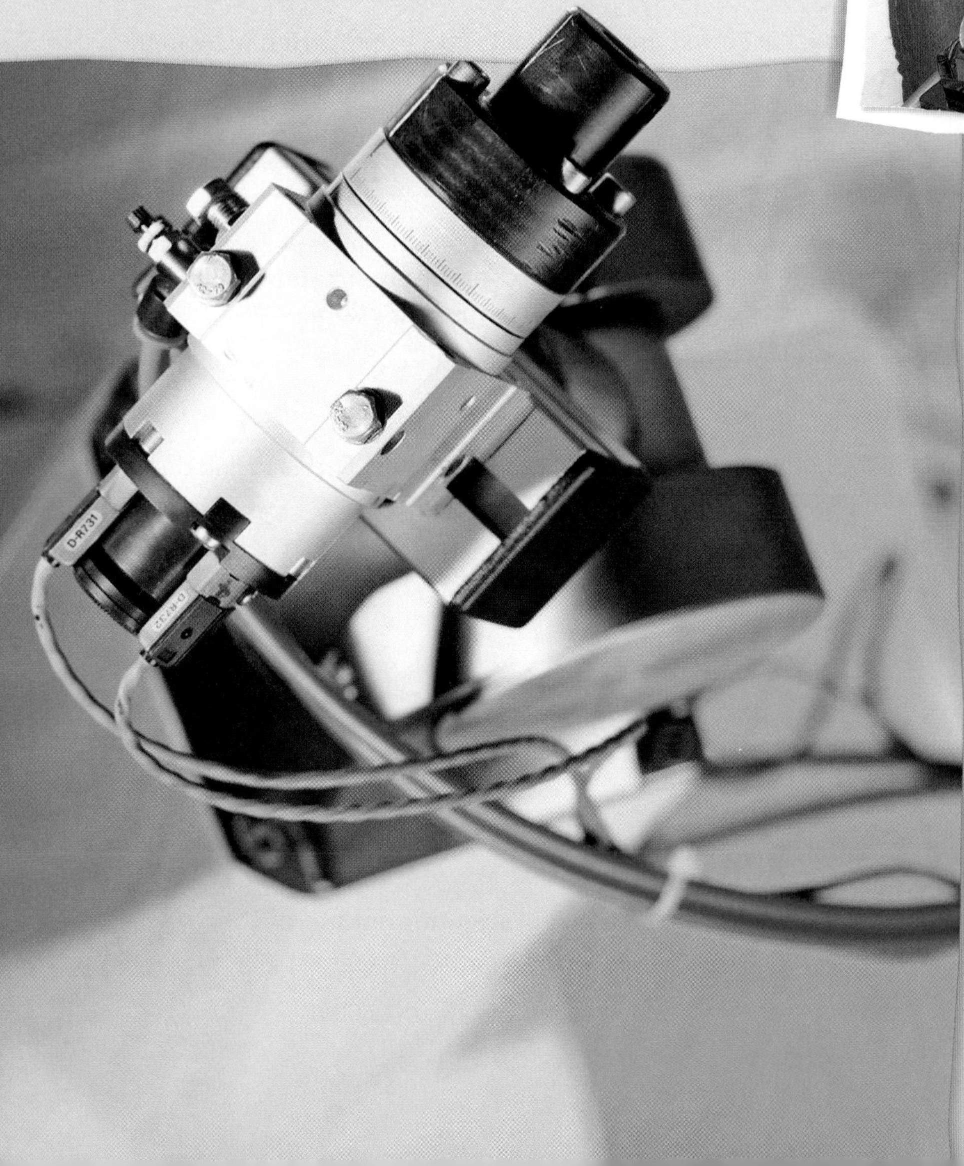

College & Career READINESS

Explore college and careers at ccr.mcgraw-hill.com

Is This the Career for You?

Are you interested in a career as a robotics engineer? Take some of the following courses in high school.

- Calculus
- Electro-Mechanical Systems
- Fundamentals of Robotics
- Physics

Find out how math relates to a career in Engineering.

Interactive Study Guide
See page 82 for:
- Problem Solving
- Career Project

Lesson 4-5

Compute with Scientific Notation

ISG Interactive Study Guide

See page 83–84 for:
- Getting Started
- Vocabulary Start-up
- Real-World Link

Essential Question

Why is it useful to write numbers in different ways?

Common Core State Standards

Content Standards
8.EE.1, 8.EE.3, 8.EE.4

Mathematical Practices
1, 3, 4, 5, 7

What You'll Learn

- Multiply and divide numbers in scientific notation.
- Add and subtract numbers in scientific notation.

Real-World Link

Aircraft The SR-71 Blackbird is one of the world's fastest airplanes. It is capable of traveling at a cruising speed of Mach 3, or three times the speed of sound. The speed of sound is approximately 760 miles per hour.

Multiplication and Division with Scientific Notation

You can apply the Product of Powers and Quotient of Powers properties to multiply and divide numbers written in scientific notation.

Real World

Example 1

Tutor

STEM Scientists estimate that there are over 3.5×10^6 ants per acre in the Amazon rain forest. If the Amazon rain forest covers approximately 1 billion acres, find the total number of ants. Write in scientific notation.

Step 1 Write the number of acres in scientific notation.

1 billion $= 1 \times 10^9$

Step 2 Multiply the number of ants per acre by the number of acres to find the total number of ants.

$(3.5 \times 10^6) \times (1 \times 10^9)$

$= (3.5 \times 1) \times (10^6 \times 10^9)$ Commutative and Associative Properties

$= (3.5) \times (10^6 \times 10^9)$ Multiply 3.5 by 1.

$= 3.5 \times 10^{6+9}$ Product of Powers

$= 3.5 \times 10^{15}$ Add the exponents.

So, there are about 3.5×10^{15} ants in the Amazon rain forest.

Got It? Do these problems to find out.

Evaluate each expression. Express the result in scientific notation.

1a. $(4.62 \times 10^5)(8.15 \times 10^9)$

1b. $(7.53 \times 10^{-8})(2.93 \times 10^{-3})$

1c. $(1.2 \times 10^7)(1500)$

1d. $(6.4 \times 10^{-5})(12{,}000)$

Example 2

Tutor

Evaluate $\frac{7.56 \times 10^8}{3.15 \times 10^3}$. Express the result in scientific notation.

$$\frac{7.56 \times 10^8}{3.15 \times 10^3} = \left(\frac{7.56}{3.15}\right)\left(\frac{10^8}{10^3}\right) \quad \text{Associative Property}$$

$$= 2.4\left(\frac{10^8}{10^3}\right) \quad \text{Divide 7.56 by 3.15.}$$

$$= 2.4 \times 10^{8-3} \quad \text{Quotient of Powers}$$

$$= 2.4 \times 10^5 \quad \text{Subtract the exponents.}$$

Got It? **Do these problems to find out.**

Evaluate each expression. Express the result in scientific notation.

2a. $\frac{4.62 \times 10^5}{1.4 \times 10^{-9}}$

2b. $\frac{2.5627 \times 10^{-9}}{5.23 \times 10^{-3}}$

Real World

Example 3

Tutor

In 2010, the population of China was about 1.3×10^9. According to census data, the population of the United States was 308,745,538. About how many times greater was the population of China than the population of the United States in 2010?

Estimate the population of the United States and write in scientific notation.

$308{,}745{,}538 \approx 300{,}000{,}000$ or 3×10^8

Find $\frac{1.3 \times 10^9}{3 \times 10^8}$.

$$\frac{1.3 \times 10^9}{3 \times 10^8} = \left(\frac{1.3}{3}\right)\left(\frac{10^9}{10^8}\right) \quad \text{Associative Property}$$

$$\approx 0.4 \times \left(\frac{10^9}{10^8}\right) \quad \text{Divide 1.3 by 3. Round to the nearest tenth.}$$

$$\approx 0.4 \times 10^{9-8} \quad \text{Quotient of Powers}$$

$$\approx 0.4 \times 10^1 \quad \text{Subtract the exponents.}$$

$$\approx 4 \times 10^0 \quad \text{Write in scientific notation.}$$

So, the population of China was about 4 times greater than the population of the United States in 2010.

Decimal Point
Since 0.4×10^1 is not written in scientific notation, move the decimal point 1 place to the right and subtract one from the exponent.

Got It? **Do this problem to find out.**

3. Until 2008, the world's largest working cattle ranch was located in Australia. It was about 6×10^6 acres. The largest ranch in the United States is 825,000 acres. About how many times larger was the ranch in Australia than the largest ranch in the United States?

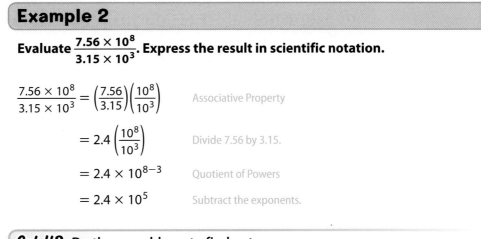

Addition and Subtraction with Scientific Notation

When adding or subtracting decimals in standard form, you line up the place values. When adding or subtracting in scientific notation, the place value is represented by the exponent. Each exponent must have the same value in order to add or subtract.

> **Watch Out!**
> When you move the decimal point to the right, the number increases, so the value of the exponent must decrease.
> $(3.12 \times 10^4) =$
> (31.2×10^3)

Example 4

Tutor

Evaluate each expression. Express the result in scientific notation.

a. $\mathbf{(5.45 \times 10^3) + (3.12 \times 10^4)}$

$(5.45 \times 10^3) + (3.12 \times 10^4)$	
$= (5.45 \times 10^3) + (31.2 \times 10^3)$	Write 3.12×10^4 as 31.2×10^3.
$= (5.45 + 31.2) \times 10^3$	Distributive Property
$= 36.65 \times 10^3$	Add 5.45 and 31.2.
$= 3.665 \times 10^4$	Write 36.65×10^3 in scientific notation.

b. $\mathbf{(2.78 \times 10^5) - (46{,}500)}$

$(2.78 \times 10^5) - (46{,}500)$	
$= (2.78 \times 10^5) - (4.65 \times 10^4)$	Write 46,500 in scientific notation.
$= (27.8 \times 10^4) - (4.65 \times 10^4)$	Write 2.78×10^5 as 27.8×10^4.
$= (27.8 - 4.65) \times 10^4$	Distributive Property
$= 23.15 \times 10^4$	Subtract 4.65 from 27.8.
$= 2.315 \times 10^5$	Write 23.15×10^4 in scientific notation.

Got It? **Do these problems to find out.**

4a. $(1.7 \times 10^7) + (6.25 \times 10^5)$

4b. $0.00864 + (5.67 \times 10^{-4})$

4c. $(2.84 \times 10^{11}) - (5.4 \times 10^9)$

4d. $0.0000321 - (4.9 \times 10^{-7})$

Guided Practice

Check

1. About 1×10^6 fruit flies weigh 1.3×10^2 pounds. How much does one fruit fly weigh? Write in scientific notation. (Example 1)

Evaluate each expression. Express the result in scientific notation. (Examples 2 and 4)

2. $(1.217 \times 10^5) - (5.25 \times 10^4)$

3. $(2.003 \times 10^4) + (7.98 \times 10^7)$

4. $\dfrac{8.25 \times 10^{10}}{2.75 \times 10^4}$

5. $(3.45 \times 10^7) - (24{,}650{,}000)$

6. $523 + (6.2 \times 10^3)$

7. $\dfrac{9.02 \times 10^3}{4.1 \times 10^5}$

8. The equatorial circumference of Earth is about 4×10^4 kilometers. The equatorial circumference of Jupiter is about 439,263.8 kilometers. About how many times greater is Jupiter's circumference than Earth's? (Example 3)

Independent Practice

Go online for Step-by-Step Solutions eHelp

9. The United States has the most miles of roads in the world at about 4×10^6 miles. Japan has about 7.3×10^5 miles. How many more miles of roads does the United States have than Japan? Write in scientific notation. (Example 1)

10. The speed of light is about 1.9×10^5 miles per second. It takes about 500 seconds for light to travel from the sun to Earth. What is the approximate distance between Earth and the sun? Write in scientific notation. (Example 1)

Evaluate each expression. Express the result in scientific notation. (Examples 1, 2, and 4)

11. $(5.32 \times 10^8)(3.54 \times 10^3)$

12. $(1.48 \times 10^{-5})(6.5 \times 10^{-6})$

13. $(9.5 \times 10^{-4})(28{,}400)$

14. $(0.042)(3.15 \times 10^4)$

15. $\frac{4.97 \times 10^6}{7.1 \times 10^{-8}}$

16. $\frac{1.86 \times 10^8}{3.1 \times 10^{-4}}$

17. $\frac{4.7 \times 10^9}{376}$

18. $\frac{99{,}500}{5 \times 10^2}$

19. $(3.205 \times 10^3) + (5.83 \times 10^5)$

20. $6{,}263{,}000 + (5.4 \times 10^8)$

21. $(2.764 \times 10^8) - (6.2 \times 10^7)$

22. $(9.518 \times 10^7) - 22{,}000$

23. $(4.21 \times 10^{-3})(56{,}200)$

24. $(8.08 \times 10^6)(3.34 \times 10^3)$

25. $(7.57 \times 10^2)(1.10 \times 10^5)$

26. $(0.0159)(5.19 \times 10^{-3})$

27. The diameter of Mars is about 7×10^6 meters. A standard table tennis ball is 0.04 meter in diameter. About how many times greater is the diameter of Mars than that of a table tennis ball? (Example 3)

28. The United States has a total area (including water) of about 9,826,630 square kilometers. Rhode Island is the smallest state with an area (including water) of about 4×10^3 square kilometers. About how many times greater is the area of the United States than the area of Rhode Island? (Example 3)

29. Earth is 1.55×10^8 kilometers from the sun. Mercury is 5.80×10^7 kilometers from the sun. Find the difference in distances and express your answer in scientific notation.

30. STEM Each minute, there are approximately 6×10^3 flashes of lightning around the world. The air around a lightning bolt is heated to about 5.4×10^4 degrees Fahrenheit, which is about five times hotter than the sun. Write each answer in scientific notation and in standard form.

a. About how many flashes of lightning are there in a day?

b. About how hot is the sun in degrees Fahrenheit?

31. A music Web site recently announced that over 4×10^9 songs have been downloaded. It also announced that it has 5×10^7 registered users. Find the average number of downloads per user and express your answer in scientific notation.

32. **Use Math Tools** The table shows the weights of various marine and land animals.

   a. Which animal is about 10 times lighter than a right whale?

   b. About how many times heavier is the blue whale than the African elephant?

   c. Estimate the combined weight of the fin whale, right whale, and white rhinoceros. Write the combined weight in scientific notation and in standard form.

Mammal	Weight (lb)
African elephant	1.44×10^4
blue whale	2.87×10^5
fin whale	9.92×10^4
right whale	8.82×10^4
white rhinoceros	7.94×10^3

33. The average width of a human hair is 4×10^{-3} centimeter. If the cross section of the average hair is round, use the formula $A = 3.14r^2$ to find the approximate area of the cross section of a hair. Write your answer in scientific notation.

34. A contractor is using a blend of two different types of sand for a new sand volleyball court. He is using 1.6×10^3 cubic feet of sand that weighs 95 pounds per cubic foot and 1.25×10^3 cubic feet of sand that weighs 88 pounds per cubic foot. How many tons of sand is being used for the vollyball court?

Evaluate each expression. Express the result in scientific notation.

35. $\dfrac{(2.8 \times 10^{-7})(14{,}000{,}000{,}000)}{3.92 \times 10^4}$

36. $\dfrac{(9.6 \times 10^{20})(3 \times 10^6)}{2 \times 10^5}$

37. $\dfrac{1.86 \times 10^8}{3.1 \times 10^{-4}} + 5.4 \times 10^{10}$

38. $\dfrac{4.5 \times 10^4}{75{,}000{,}000} \times (4.9 \times 10^6)$

39. $\left(\dfrac{6 \times 10^{-100}}{2.5 \times 10^{-60}}\right)(3.7 \times 10^{15})$

40. $\left(\dfrac{180{,}000}{5 \times 10^8}\right)(9 \times 10^2)$

41. $(8.2 \times 10^4 + 8{,}249) \times 10^8$

42. $(5.29 \times 10^4 - 52{,}000) \times 10^5$

H.O.T. Problems Higher Order Thinking

43. **Identify Structure** Write an addition expression and a subtraction expression, each with a value of 2.4×10^{-3}.

44. **Which One Doesn't Belong?** Identify the expression that does not belong with the other three. Explain your reasoning.

5.25×10^7	$(2.1 \times 10^2)(2.5 \times 10^5)$	52.5×10^8	$(2.1)(2.5) \times 10^{(2+5)}$

45. **Persevere with Problems** There are about 2.5×10^{10} red blood cells in the average adult. A googol is 1×10^{100}. About how many adults would it take to have a total of 1 googol red blood cells?

46. **Building on the Essential Question** How does writing numbers in different ways help to make it easier to compute with very large or very small numbers?

Standardized Test Practice

47. Ariana is evaluating $(8 \times 10^3) + (4 \times 10^5)$, as shown below.

$$(8 \times 10^3) + (4 \times 10^5)$$
$$(8 + 4) + (10^3 + 10^5)$$
$$12 + (10^3 + 10^5)$$
$$12 + 10^8$$

What should Ariana have done differently to evaluate the expression correctly?

A made both numbers have the same power of 10
B subtracted the exponents
C multiplied 8×4 instead of adding $8 + 4$
D made the last line 12×10^8

48. What is the value of $(2.8 \times 10^3)(1{,}600{,}000)$?

F 4.48×10^{18} **H** 44.8×10^9
G 4.48×10^6 **J** 4.48×10^9

49. After its first year in business, a movie Web site announced that over 500,000,000 movies were downloaded by 4×10^6 registered users. What is the average number of movies per user?

A about 1.25×10^{-2} movies
B about 125 movies
C 1.25×10^3 movies
D about 12.5 movies

50. **Short Response** Earth is 1.55×10^8 kilometers from the sun. Venus is 109 million kilometers from the sun. Find the difference in distances and express your answer in scientific notation.

Common Core Review

Write an integer for each situation. Then identify its opposite. 7.NS.1a

51. 58°C below zero
52. 15 gallons per mile more than usual
53. a withdrawal of $4500
54. a scuba diver's descent of 50 feet
55. a bank deposit of $68.00
56. an airplane's ascent of 300 feet

Complete each expression. 7.NS.1c

57. $18 - 5 = 18 + \blacksquare$
58. $\blacksquare - (-3) = 12$
59. $12 = 10 - \blacksquare$

60. The volume of one cube is 5^3 cubic inches. What is the volume of 3.5 of these cubes? 8.EE.1

61. The speed of sound is approximately 7.6×10^2 miles per hour. Write 7.6×10^2 in standard form. 8.EE.3

62. The SR-71 Blackbird is more than 30 years old. It can fly at altitudes above 8×10^4 feet. Is it more appropriate to report the altitude as 8×10^4 feet or as 9.6×10^5 inches? 8.EE.4

Determine whether each equation is *true or false*. If the equation is *false*, explain why. 7.NS.2c

63. $3 \times (-4) = -12$
64. $-\frac{8}{4} = \frac{-8}{-4}$
65. $-15 \div (-3) = 5$
66. $-36 \div 6 = 6$
67. $-12 \times (-11) = 132$
68. $-1 \times (-1) = 1$

Inquiry Lab

Scientific Notation Using Technology

Inquiry **WHAT is a benefit of knowing how to interpret calculator notation when working with scientific notation on your graphing calculator?**

CCSS **Content Standards**
8.EE.3, 8.EE.4
Mathematical Practices
1, 3

Solar System The table shows the mass of some planets in our solar system. What is the mass of Earth written in scientific notation?

Planet	Mass (kg)
Mars	641,850,000,000,000,000,000,000
Earth	5,973,700,000,000,000,000,000,000
Saturn	568,510,000,000,000,000,000,000,000

Investigation 1

Step 1 Press CLEAR to clear the home screen.

Step 2 Enter the value for Earth's mass. Press ENTER.

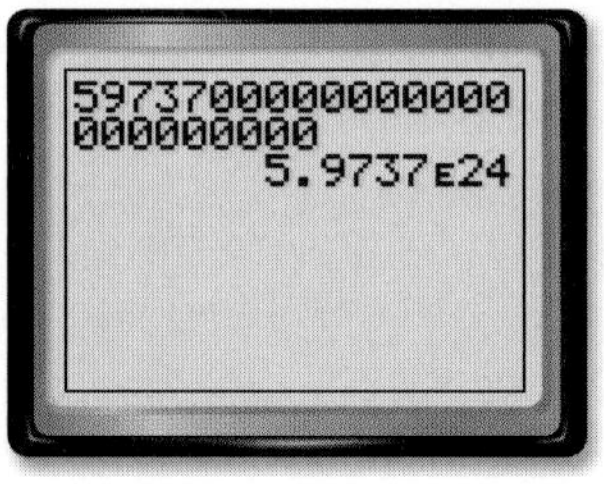

Collaborate

Work with a partner.

1. How is the number 5,973,700,000,000,000,000,000,000 written in scientific notation similar to and different from the calculator notation shown on your screen?
2. Repeat Steps 1 and 2 for the mass of Mars. What is the mass of Mars in calculator notation?
3. Based on your answer for Exercise 2, what is the mass of Mars in scientific notation?

Analyze

4. What does the E symbol represent on the calculator screen? What does the value after the E symbol represent?
5. CCSS **Reason Inductively** Without entering the value in your calculator, predict how the mass of Saturn will be displayed on the calculator screen.

Reflect

6. CCSS **Justify Conclusions** When Saturn's mass is divided by Mars's mass, suppose the calculator displays 8.857365428E2. What does this value represent? Explain.

Investigation 2

A human blood cell is about 0.000001 meter in diameter. The Moon is about 3,476,000 meters in diameter. How many times greater is the diameter of the Moon than the diameter of a blood cell?

Step 1 Press CLEAR to clear the home screen.

Step 2 Enter the values into the calculator.

3476000 ÷ 0.000001 ENTER

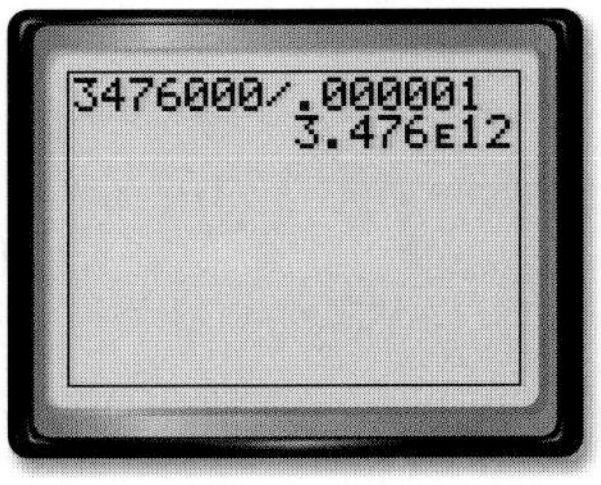

Collaborate

Work with a partner.

7. Write 3.476E12 in scientific notation. Interpret the meaning of the answer displayed on the calculator screen. Write this value in standard form.

8. The diameter of Earth is about 1.2742×10^7 meters. About how many times greater is the diameter of Earth than the diameter of the Moon? Use your calculator.

9. A *micrometer* is 0.000001 meter. Use your calculator to determine how many micrometers are in each of the following. Write your answer in both calculator and scientific notation.

   a. 5000 meters b. 4.08E14 meters c. 2.9E^{-}10 meter
   d. 63,000 meters e. 2.34E11 meters f. 9.67E^{-}8 meter

10. A *nanometer* (nm) is 0.000000001 meter. Express this in both calculator and scientific notation.

Analyze

Put your calculator in scientific mode by pressing MODE ▶ ENTER. Then press CLEAR to return to the home screen.

The approximate areas of several states are given in the table.

State	Area (mi^2)
Alaska	656,000
Texas	269,000
California	164,000
Michigan	97,000
Pennsylvania	46,000
New Jersey	9,000

11. Enter the area of Alaska on your calculator. Press ENTER. What is displayed on the screen? What does this value represent?

12. Using your calculator, find the areas of the remaining states in scientific notation.

13. CCSS **Reason Inductively** How can you use your graphing calculator to find how many times greater the area of Alaska is than the area of New Jersey? About how much greater is it?

Reflect

14. Inquiry WHAT is a benefit of knowing how to interpret calculator notation when working with scientific notation on your graphing calculator?

Lesson 4-6

Square Roots and Cube Roots

ISG Interactive Study Guide

See pages 85–86 for:
- Getting Started
- Real-World Link
- Notes

See pages 85–86 for:

Essential Question

Why is it useful to write numbers in different ways?

Common Core State Standards

Content Standards
8.NS.2, 8.EE.2

Mathematical Practices
1, 2, 3, 4, 7

Vocabulary
square root
perfect square
radical sign
cube root
perfect cube

What You'll Learn

- Find square roots.
- Find cube roots.

Real-World Link

Rain Forest Tropical rainforests contain the greatest diversity of plants and animals on Earth—and they cover less than 5 percent of Earth's land! Just a four-square-mile patch of rainforest contains thousands of species of plants and trees, and hundreds of species of mammals, birds, reptiles, and amphibians.

Key Concept Square Roots

Words	A **square root** of a number is one of its two equal factors.
Symbols	If $x^2 = y$, then x is a square root of y.

Number like 9, 25, and 144 are **perfect squares**, because they are squares of integers. The opposite of squaring a number is finding the square root.

A **radical sign**, $\sqrt{\ }$ is used to indicate a nonnegative square root. Every positive number has both a positive and a negative square root.

$$\sqrt{36} = 6 \qquad -\sqrt{36} = -6 \qquad \pm\sqrt{36} = \pm 6 \text{ or } 6, -6$$

A negative number like −36 has no real-number square root because the square of a number cannot be negative. You will learn about real numbers in the next lesson.

Example 1

Tutor

Find each square root.

a. $\sqrt{9}$

$\sqrt{9} = 3$ — Find the positive square root of 9; $3^2 = 9$.

b. $-\sqrt{64}$

$-\sqrt{64} = -8$ — Find the negative square root of 64; $8^2 = 64$.

c. $\pm\sqrt{4}$

$\pm\sqrt{4} = \pm 2$ — Find both square roots of 4; $2^2 = 4$.

d. $\sqrt{-81}$

There is no real square root because no number times itself is equal to −81.

Got It? Do these problems to find out.

1a. $\sqrt{49}$

1b. $-\sqrt{16}$

1c. $\pm\sqrt{100}$

1d. $\sqrt{-49}$

Design Pics/Corey Hochachka

You can estimate the square root of an integer that is not a perfect square by determining between which two consecutive integers the square root lies.

Example 2

Tools Tutor

Estimate each square root to the nearest integer.

a. $\sqrt{33}$

The largest perfect square less than 33 is 25. $\sqrt{25} = 5$

The smallest perfect square greater than 33 is 36. $\sqrt{36} = 6$

Plot each square root on a number line. Then estimate $\sqrt{33}$.

$25 < 33 < 36$	Write an inequality.
$5^2 < 33 < 6^2$	$25 = 5^2$ and $36 = 6^2$
$\sqrt{5^2} < 33 < \sqrt{6^2}$	Find the square root of each number.
$5 < \sqrt{33} < 6$	Simplify.

So, $\sqrt{33}$ is between 5 and 6. Since 33 is closer to 36 than to 25, the best integer estimate for $\sqrt{33}$ is 6.

Check Check using a calculator.

[2nd] [√] 33 [ENTER] **5.744562647**

$\sqrt{33} \approx 6$ ✓

b. $-\sqrt{129}$

The largest perfect square less than 129 is 121. $\sqrt{121} = -11$

The smallest perfect square greater than 129 is 144. $\sqrt{144} = -12$

The negative square root of 129 is between the integers −11 and −12. Plot each square root on a number line. Then estimate $-\sqrt{129}$.

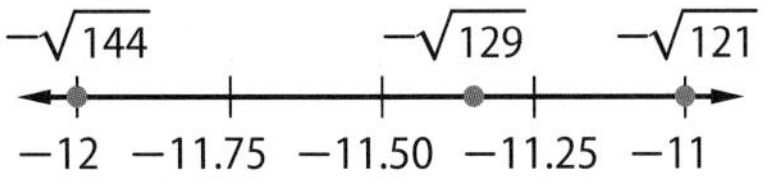

So, $-\sqrt{129}$ is between −12 and −11. Since 129 is closer to 121 than to 144, the best integer estimate for $-\sqrt{129}$ is −11.

Check Check using a calculator.

[(−)] [2nd] [√] 129 [ENTER] **−11.357816691**

$\sqrt{-129} \approx -11$ ✓

Watch Out!

When keying your calculator to find the negative of a square root, for example $-\sqrt{129}$, be sure to use the [(−)] button instead of the [−] button for the negative sign.

Got It? **Do these problems to find out.**

2a. $\sqrt{60}$

2b. $-\sqrt{23}$

2c. $\sqrt{14}$

2d. $-\sqrt{79}$

When finding square roots in real-world situations, use the positive, or *principal*, square root when a negative answer does not make sense.

Example 3

Tutor

On a clear day, the number of miles a person can see to the horizon can be found using the formula $d = 1.22 \cdot \sqrt{h}$, where d is the distance to the horizon in miles and h is the person's distance from the ground in feet. The observation deck of Seattle's Space Needle is 520 feet high. How far to the horizon can a person standing on the observation deck see? Round to the nearest tenth.

Choose a Form
Express a number as a square root if an exact answer is needed. Express a number as a decimal if an approximation is sufficient.

Estimate The distance is between $1 \cdot \sqrt{400}$ and $1 \cdot \sqrt{900}$. So, it's between 20 and 30.

$d = 1.22 \cdot \sqrt{h}$	Write the equation.
$= 1.22 \cdot \sqrt{520}$	Replace h with 520.
$\approx 1.22 \cdot 22.8$	Use a calculator.
≈ 27.8	Simplify.

The approximate distance to the horizon is 27.8 miles to the nearest tenth.

Check for Reasonableness $20 < 27.8 < 30$ ✓

Got It? Do these problems to find out.

3a. Spring Port Ledge Lighthouse in Maine is approximately 55 feet tall. Calculate about how far a person who is standing at the top of the lightbouse can see on a clear day. Round to the nearest tenth of a mile.

3b. The observation deck of the Washington Monument is 500 feet high. Calculate about how far a person on the observation deck can see on a clear day. Round to the nearest tenth of a mile.

Key Concept: Cube Roots

Words	A cube root of a number is one of its three equal factors.
Symbols	If $x^3 = y$, then $x = \sqrt[3]{y}$.
Examples	Since $2 \times 2 \times 2$ or $2^3 = 8$, 2 is a cube root of 8. Since $-6 \times (-6) \times (-6) = -216$, -6 is a cube root of -216.

A **cube root** of a number is one of three equal factors of the number. The symbol $\sqrt[3]{}$ is used to indicate the cube root of a number.

Every integer has exactly one cube root.

- The cube root of a positive number is positive.
- The cube root of zero is zero.
- The cube root of a negative number is negative.

Example 4

Tutor

Find each cube root.

a. $\sqrt[3]{343}$

$\sqrt[3]{343} = 7$ $\quad 7^3 = 7 \cdot 7 \cdot 7$ or 343

b. $\sqrt[3]{-729}$

$\sqrt[3]{-729} = -9$ $\quad (-9)^3 = (-9) \cdot (-9) \cdot (-9)$ or -729

Got It? **Do these problems to find out.**

4a. $\sqrt[3]{64}$

4b. $\sqrt[3]{-1331}$

You can also estimate cube roots mentally by using **perfect cubes**.

Example 5

Tutor

Estimate $\sqrt[3]{83}$ to the nearest integer. Do not use a calculator.

$\sqrt[3]{83}$

The first perfect cube less than 83 is 64. $\quad \sqrt[3]{64} = 4$

The first perfect cube greater than 83 is 125. $\quad \sqrt[3]{125} = 5$

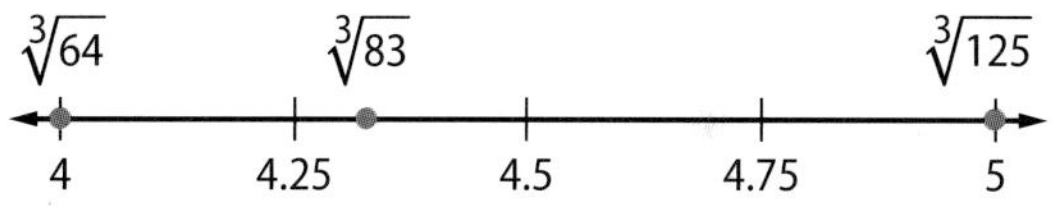

The cube root of 83 is between the integers 4 and 5. Since 83 is closer to 64 than to 125, you can expect $\sqrt[3]{83}$ to be closer to 4 than to 5.

Got It? **Do these problems to find out.**

5a. $\sqrt[3]{72}$

5b. $\sqrt[3]{-2024}$

Guided Practice

Check

Find each square root. (Example 1)

1. $\sqrt{16}$ **2.** $-\sqrt{100}$ **3.** $\pm\sqrt{81}$

Estimate each square root to the nearest integer. (Example 2)

4. $\sqrt{27}$ **5.** $-\sqrt{48}$ **6.** $\pm\sqrt{39}$

7. A baseball diamond is actually a square with an area of 8100 square feet. Most baseball teams cover their diamond with a tarp to protect it from the rain. The sides are all the same length. How long is the tarp on each side? (Example 3)

Find each cube root. (Example 4)

8. $\sqrt[3]{512}$ **9.** $\sqrt[3]{2197}$ **10.** $\sqrt[3]{-1000}$ **11.** $\sqrt[3]{-343}$

Estimate each cube root to the nearest integer. (Example 5)

12. $\sqrt[3]{74}$ **13.** $\sqrt[3]{39}$ **14.** $\sqrt[3]{-636}$ **15.** $\sqrt[3]{-879}$

Independent Practice

Go online for Step-by-Step Solutions eHelp

Find each square root. (Example 1)

16. $\sqrt{36}$

17. $\sqrt{9}$

18. $-\sqrt{169}$

19. $-\sqrt{144}$

20. $\pm\sqrt{-25}$

21. $\pm\sqrt{1}$

Estimate each square root to the nearest integer. (Example 2)

22. $\sqrt{83}$

23. $\sqrt{34}$

24. $-\sqrt{102}$

25. $-\sqrt{14}$

26. $\pm\sqrt{78}$

27. $\pm\sqrt{146}$

28. The table shows the heights of the tallest roller coasters at Cedar Point. Use the formula from Example 3 to determine how far a rider can see from the highest point of each ride. Round to the nearest tenth. (Example 3)

a. Millennium Force

b. Mean Streak

c. How much farther can a rider see on the Top Thrill Dragster than on the Magnum XL-200?

Cedar Point Attractions

Roller Coaster	Height (ft)
Mean Streak	161
Magnum XL-200	205
Millennium Force	310
Top Thrill Dragster	420

Find each cube root. (Example 4)

29. $\sqrt[3]{-1728}$

30. $\sqrt[3]{-2744}$

31. $\sqrt[3]{216}$

32. $\sqrt[3]{1331}$

Estimate each cube root to the nearest integer. Do not use a calculator. (Example 5)

33. $\sqrt[3]{499}$

34. $\sqrt[3]{576}$

35. $\sqrt[3]{-79}$

36. $\sqrt[3]{-1735}$

37. The area of a square is 215 square centimeters. Find the length of a side to the nearest tenth. Then find its approximate perimeter.

38. Order $\sqrt{77}$, -8, $-\sqrt{83}$, 9, -10, $-\sqrt{76}$, $\sqrt{65}$ from least to greatest.

H.O.T. Problems Higher Order Thinking

39. CCSS **Reason Abstractly** Write a number that completes the analogy.
x^2 is to 121 as x^3 is to ___?___ .

40. CCSS **Identify Structure** Find a square root that lies between 17 and 18.

41. CCSS **Persevere with Problems** Use inverse operations to evaluate the following.

a. $(\sqrt{246})^2$

b. $(\sqrt{811})^2$

c. $(\sqrt{732})^2$

42. **Building on the Essential Question** Describe the difference between an exact value and an approximation when finding square roots of numbers that are not perfect squares. Give an example of each.

Standardized Test Practice

43. Which point on the number line best represents $\sqrt{210}$?

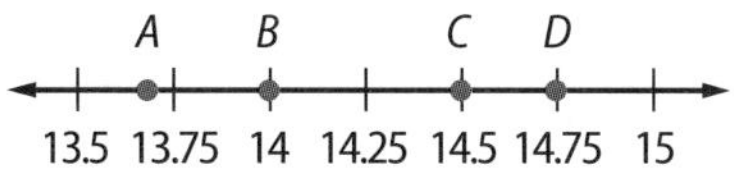

A *A*

B *B*

C *C*

D *D*

44. **Short Response** Estimate the cube root of 65 to the nearest integer.

45. The new gymnasium at Oakdale Middle School has a hardwood floor in the shape of a square. If the area of the floor is 62,500 square feet, what is the length of one side of the square floor?

F 200 ft   **H** 250 ft

G 225 ft   **J** 275 ft

46. A surveyor determined the distance across a field was $\sqrt{1568}$ feet. What is the approximate distance?

A 25.6 ft

B 30.6 ft

C 39.6 ft

D 42.6 ft

Common Core Review

Solve. 7.NS.3

47. Joseph bought four books at a book sale. Each book cost \$4.50. He paid with \$20.00. How much change did he receive?

48. Mrs. Tanner paid \$18.00 for six boxes of pencils. How much did each box of pencils cost?

49. Kathryn had \$367.50 in her bank account. She wrote a check for \$25.00, and then withdrew \$50.00 in cash. She made a deposit of \$100.00. How much money is in her bank account now?

Name the property shown by each statement. 7.EE.4

50. $3 + 6 = 6 + 3$

51. $13 + 0 = 13$

52. $(2 \cdot 5) + 6 = 6 + (2 \cdot 5)$

53. $(x + 3) + 9 = x + (3 + 9)$

54. $(y \cdot 2) \cdot 3 = y \cdot (2 \cdot 3)$

55. $28 \cdot 1 = 28$

56. $n + t = t + n$

57. $1652 \cdot 0 = 0$

58. Suppose that four-tenths of the rectangle at the right is shaded. 7.NS.1d

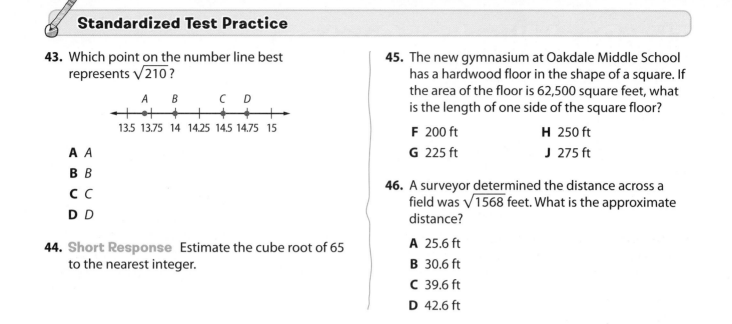

a. What fraction of the rectangle is *not* shaded?

b. What fraction of the rectangle would still need to be shaded for half of the rectangle to be shaded?

c. If an additional $\frac{7}{15}$ of the original rectangle were to be shaded, what fraction of the rectangle would be shaded?

Find each product or quotient. 7.NS.2a

59. $8 \cdot (-8)$

60. $-4 \cdot (-12)$

61. $40 \div (-5)$

62. $-150 \div (-25)$

Lesson 4-7

The Real Number System

Interactive Study Guide

See pages 87–88 for:
- Getting Started
- Real-World Link
- Notes

Essential Question

Why is it useful to write numbers in different ways?

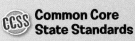

Common Core State Standards

Content Standards
8.NS.1, 8.NS.2, 8.EE.2

Mathematical Practices
1, 3, 4, 7

Vocabulary

irrational number

real numbers

What You'll Learn

- Identify and compare numbers in the real number system.
- Solve equations by finding square roots or cube roots.

Real-World Link

Weather Meteorologists use the formula $t^2 = \frac{d^3}{216}$ to predict the time t in hours a thunderstorm will last when it is d miles across.

Identify and Compare Real Numbers

Recall that *rational numbers* are numbers that can be written as fractions. Examples of rational numbers are given below.

$$1\frac{2}{5} = \frac{7}{5} \qquad -4 = -\frac{4}{1} \qquad 0.15 = \frac{15}{100}$$

$$0.\overline{3} = \frac{1}{3} \qquad \sqrt{25} = \frac{5}{1}$$

An **irrational number** is a number that cannot be written as a fraction. When written as decimals, irrational numbers neither terminate nor repeat.

Key Concept Irrational Number

Words	An irrational number is a number that cannot be expressed as $\frac{a}{b}$, where a and b are integers and $b \neq 0$.
Symbols	$\pi \approx 3.14159\ldots$ $\qquad -\sqrt{5} \approx -2.2360679\ldots$

The sets of rational numbers and irrational numbers together make up the set of **real numbers**. The diagram shows the relationship among the real numbers.

Real Numbers

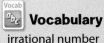

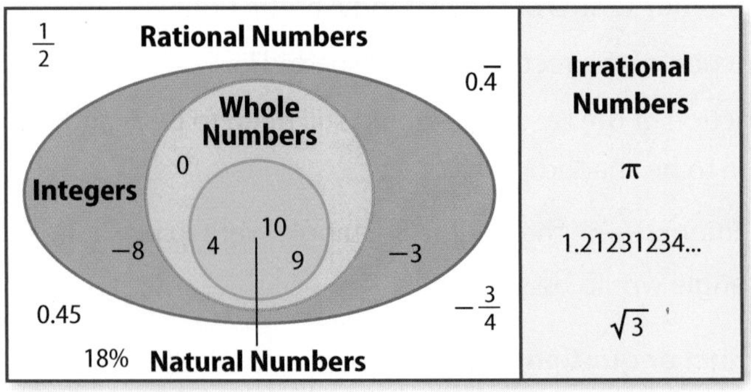

Computations with an irrational number and a rational number (other than zero) produce an irrational number.

Example 1

Integers
Integers are the whole numbers and their opposites. …, −3, −2, −1, 0, 1, 2, 3, …

Name all sets of numbers to which each real number belongs. Write *natural*, *whole*, *integer*, *rational*, or *irrational*.

a. $\frac{21}{7}$ — Since $\frac{21}{7} = 3$, this number is a natural number, a whole number, an integer, and a rational number.

b. -2.5 — Since $-2.5 = -\frac{5}{2}$, this number is a rational number.

c. $0.\overline{2}$ — Since $0.\overline{2} = 0.22222\ldots$ or $\frac{2}{9}$, this number is a rational number.

d. $\sqrt{38}$ — Since $\sqrt{38} = 6.16441400\ldots$, it is not the square root of a perfect square so it is irrational.

Got It? Do these problems to find out.

1a. 0.7

1b. $\sqrt{100}$

1c. $\frac{9}{5}$

1d. -6

Example 2

Square Roots
If an integer is not a perfect square, then its square root is an irrational number.

Replace ● with <, >, or = to make $3\frac{1}{3}$ ● $\sqrt{15}$ a true statement.

Express each number as a decimal. Then compare the decimals.

$3\frac{1}{3} = 3.33333333\ldots$

$\sqrt{15} = 3.87298334\ldots$

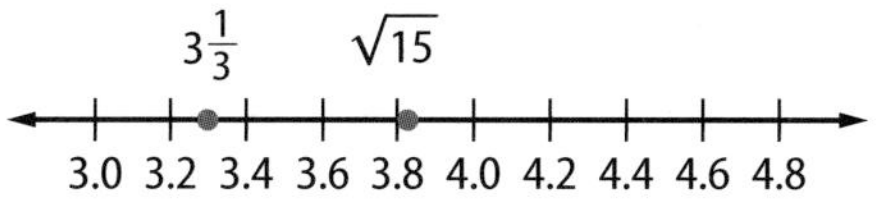

Since 3.333… is less than 3.872…, $3\frac{1}{3} < \sqrt{15}$.

Got It? Do this problem to find out.

2. Replace ● with <, >, or = to make $7\frac{2}{5}$ ● $\sqrt{57}$ a true statement.

Example 3

Order the set $\left\{8\frac{4}{5}, \sqrt{64}, 8.\overline{3}, \sqrt{76}\right\}$ from least to greatest.

Express each number as a decimal. Then order the decimals.

$8\frac{4}{5} = 8.8$

$\sqrt{64} = 8$

$8.\overline{3} = 8.33333333\ldots$

$\sqrt{76} = 8.71779788\ldots$

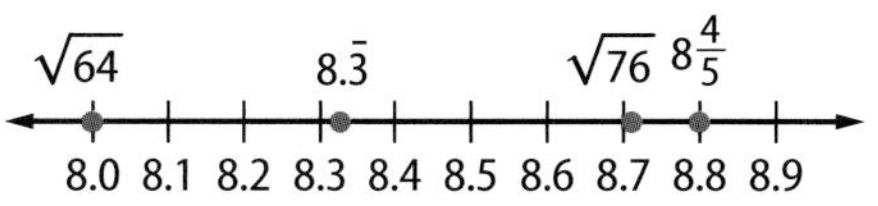

From least to greatest, the order is $\sqrt{64}$, $8.\overline{3}$, $\sqrt{76}$, and $8\frac{4}{5}$.

Got It? Do this problem to find out.

3. Order the set $\left\{\sqrt{30}, 5.6, \frac{15}{3}, 5\frac{2}{3}\right\}$ from greatest to least.

Solve Equations

By the definition of a square root, if $x^2 = y$, then $x = \pm\sqrt{y}$. By the definition of a cube root, if $x^3 = y$, then $x = \sqrt[3]{y}$. You can use these relationships to solve equations involving squares and cubes.

Example 4

Tutor

Solve each equation. Round to the nearest tenth, if necessary.

a. $a^2 = 38$

$a^2 = 38$	Write the equation.
$a = \pm\sqrt{38}$	Definition of square root
$a \approx 6.2$ and -6.2	Use a calculator.

The solutions are 6.2 and −6.2.

b. $a^3 = 125$

$a^3 = 125$	Write the equation.
$a = \sqrt[3]{125}$	Definition of cube root
$a = 5$	Check $5 \cdot 5 \cdot 5 = 125$

The solution is 5.

Study Tip

When solving for x in an equation, such as $2x^2 = 32$, first divide each side by the coefficient of the squared variable, $x^2 = 16$. Then solve using the definition of square root. $x = \pm 4$.

Got It? Do these problems to find out.

4a. $363 = 3d^2$

4b. $729 = s^3$

4c. $100 = 4n^2$

4d. $512 = x^3$

In most real-world situations, a negative square root does not make sense. Consider only the positive, or *principal*, square root.

Real World Example 5

Tutor

STEM The *aspect ratio* of a hang glider allows it to glide through the air. The formula for the aspect ratio R is $R = \frac{s^2}{A}$, where s is the wingspan and A is the area of the wing. What is the wingspan of a hang glider if its aspect ratio is 4.5 and the area of the wing is 50 square feet?

$R = \frac{s^2}{A}$	Write the formula.
$4.5 = \frac{s^2}{50}$	Replace R with 4.5 and A with 50.
$225 = s^2$	Multiply each side by 50.
$\sqrt{225} = s$	Consider the positive square root.
$15 = s$	Simplify.

The wingspan of the hang glider is 15 feet.

Got It? Do this problem to find out.

5. STEM A *tsunami* is caused by an earthquake on the ocean floor. The speed of a tsunami can be measured by the formula $\frac{s^2}{d} = 9.61$, where s is the speed of the wave in meters per second and d is the depth of the ocean in meters where the earthquake occurs. What is the speed of a tsunami if an earthquake occurs at a depth of 632 meters? Round to the nearest tenth.

Guided Practice

Check

Name all sets of numbers to which each real number belongs. Write *natural, whole, integer, rational,* or *irrational*. (Example 1)

1. 10 **2.** $\frac{1}{5}$ **3.** $\sqrt{35}$ **4.** $-\frac{14}{2}$

Replace each ● with <, >, or = to make a true statement. (Example 2)

5. $\sqrt{6}$ ● $2\frac{3}{8}$ **6.** $-5.\overline{2}$ ● $-\sqrt{29}$ **7.** $-\sqrt{42}$ ● $-6\frac{2}{3}$

Order each set of numbers from least to greatest. (Example 3)

8. $\left\{-\frac{5}{4}, -\sqrt{4}, -1.5, -1\frac{3}{4}\right\}$ **9.** $\left\{\sqrt{110}, 10\frac{1}{5}, 10.\overline{5}, 10.15\right\}$

Solve each equation. Round to the nearest tenth, if necessary. (Example 4)

10. $x^2 = 16$ **11.** $3m^2 = 222$ **12.** $4r^3 = 32$

13. The formula $A \approx 3.14r^2$ can be used to determine the area of a circle, where A is the area and r is the distance from the center of the circle to the outside edge. If the area of the largest pizza ever made was approximately 11,818 square feet, about how far is the distance from the center of the pizza to the outside edge? Round to the nearest tenth. (Example 5)

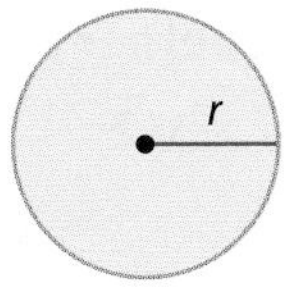

Independent Practice

Go online for Step-by-Step Solutions eHelp

Name all sets of numbers to which each real number belongs. Write *natural, whole, integer, rational,* or *irrational*. (Example 1)

14. 4 **15.** $\frac{3}{5}$ **16.** $-\frac{7}{2}$ **17.** $\sqrt{26}$

18. $-\frac{36}{4}$ **19.** 8.2 **20.** 0.55555... **21.** $-\sqrt{81}$

22. 6.01 **23.** $\frac{42}{7}$ **24.** $\sqrt{144}$ **25.** $0.\overline{18}$

Replace ● with <, >, or = to make a true statement. (Example 2)

26. $\sqrt{17}$ ● 4.2 **27.** $5.\overline{15}$ ● $\sqrt{26}$ **28.** $3\frac{5}{6}$ ● $\sqrt{10}$

29. $\sqrt{2.56}$ ● $1\frac{3}{5}$ **30.** $-\sqrt{0.25}$ ● $-\frac{1}{2}$ **31.** $-7\frac{5}{8}$ ● $-\sqrt{55}$

Order each set of numbers from least to greatest. (Example 3)

32. $\left\{2.\overline{71}, 2\frac{3}{4}, \sqrt{5}, \frac{5}{2}\right\}$ **33.** $\left\{\sqrt{64}, 8\frac{1}{7}, 8.\overline{14}, \frac{15}{2}\right\}$

34. $\left\{-\sqrt{11}, -3.\overline{3}, -3.4, -\frac{16}{5}\right\}$ **35.** $\left\{-\frac{5}{6}, -5, -\sqrt{26}, -\frac{31}{6}\right\}$

Solve each equation. Round to the nearest tenth, if necessary. (Example 4)

36. $y^2 = 64$ **37.** $130 = n^2$ **38.** $p^3 = 1331$

39. $2d^2 = 162$ **40.** $190.5 = 1.5b^2$ **41.** $1.5x^3 = 40.5$

42. The height h in feet that a pole vaulter can reach can be estimated using the formula $h = \frac{v^2}{64}$, where v is the velocity of the athlete in feet per second. If the world record height for the women's pole vault is about 16.5 feet, about how fast was the record holder running? Round to the nearest integer. (Example 5)

43. **STEM** The formula $h = 16t^2$ describes the time t in seconds that it takes for an object to fall from a height of h feet. A thrill ride has a 60-foot-tall freefall drop. How long does it take for the ride to complete its freefall? Round to the nearest tenth.

CCSS Justify Conclusions Determine whether each statement is *always, sometimes,* or *never* true. Explain your reasoning.

44. An integer is a rational number.

45. A real number can be written as a repeating decimal.

46. An irrational number can be written as a terminating decimal.

47. A whole number is an integer.

Tell whether each expression is *rational* or *irrational.* Explain.

48. $4 \times \sqrt{2}$

49. $\sqrt{49} - 15$

50. $\sqrt{10} \div 2$

51. $9 \cdot \pi$

52. In the formula $s = \sqrt{30fd}$, s is the speed of a car in miles per hour, d is the distance the car skidded in feet, and f is friction. The table shows different values of f. At an accident scene, a car made 100-foot skid marks before stopping. If the speed limit was 55 miles per hour, was the car speeding before applying the brakes on a dry, concrete road? Explain.

Road Conditions	Type of Surface	
	Concrete	Asphalt
Wet	0.4	0.5
Dry	0.8	1.0

H.O.T. Problems Higher Order Thinking

53. **CCSS Identify Structure** Find a rational number and an irrational number that are between 6.2 and 6.5. Include the decimal approximation of the irrational number to the nearest hundredth.

CCSS Use a Counterexample Tell whether each expression is *true* or *false*. If *false*, give a counterexample.

54. All whole numbers are integers.

55. All square roots are irrational numbers.

56. All rational numbers are integers.

57. **CCSS Which One Doesn't Belong?** Identify the number that does not belong with the other three. Explain your reasoning to a classmate.

$50.\overline{1}$	$-\frac{50}{2}$	-50.1	$\sqrt{50}$

58. **Building on the Essential Question** Explain the relationship between the area of a square and the length of its sides. Give an example of a square whose side length is rational and an example of a square whose side length is irrational.

Standardized Test Practice

59. For what value of x is $\frac{1}{\sqrt{x}} < \sqrt{x} < x$?

A -2 **C** $\frac{1}{2}$

B $\frac{1}{4}$ **D** 2

60. The formula $h = 16t^2$ describes the time t in seconds that it takes for an object to fall from a height of h feet. How long would it take a basketball to hit the ground from a height of 50 feet?

F 1.77 seconds
G 10.36 seconds
H 28.28 seconds
J 200 seconds

61. Which of the following is an example of an irrational number?

A -8 **C** $\sqrt{10}$

B $\frac{3}{4}$ **D** $\sqrt{16}$

62. **Short Response** The area of a triangle that has three equal sides can be found using the expression $\frac{s^2\sqrt{3}}{4}$, where s is the length of one side. What is the area in square inches of the triangle below to the nearest tenth?

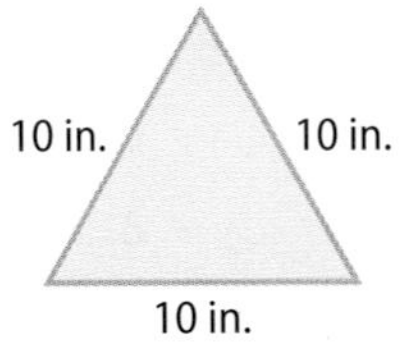

Common Core Review

Estimate each square root to the nearest integer. 8.NS.2

63. $\sqrt{79}$ **64.** $\sqrt{95}$ **65.** $-\sqrt{54}$

66. $-\sqrt{125}$ **67.** $\pm\sqrt{200}$ **68.** $\pm\sqrt{396}$

Express each number in standard form. 8.EE.1

69. 3.08×10^{-4} **70.** 1.4×10^2 **71.** 8.495×10^5

Write each expression using exponents. 8.EE.1

72. $4 \cdot 4 \cdot 4 \cdot 4$ **73.** $2 \cdot 2 \cdot 2 \cdot 2 \cdot 2$ **74.** $3 \cdot 3 \cdot 3$

Evaluate each expression. 8.EE.1

75. 2^4 **76.** 6^3 **77.** $3 \cdot 4^2$

Find each product or quotient. Express using exponents. 8.EE.1

78. $\frac{4^5}{4^3}$ **79.** $(x)(x)(x^2)(x^2)$ **80.** $(-4)^6(-4)^2$

Write the ordered pair that names each point. 6.G.3

81. *A* **82.** *B* **83.** *C*

84. *D* **85.** *E* **86.** *F*

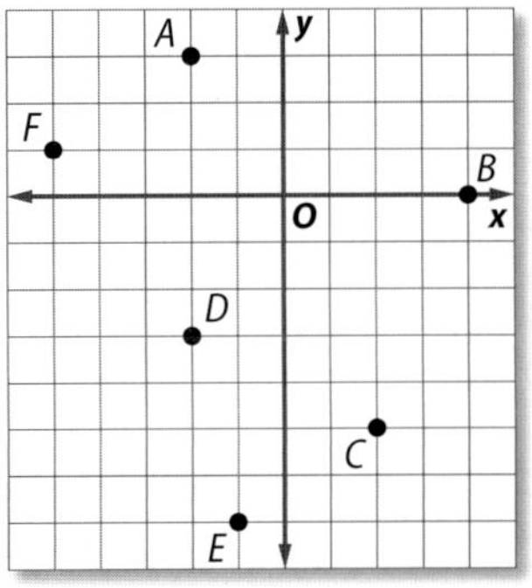

87. The reported highest point in Michigan is Mount Arvon, with an elevation of 1979 feet. The lowest point is Lake Erie, with an elevation of 571 feet. How much greater is the elevation of Mount Arvon than that of Lake Erie? 7.NS.1c

88. The temperature in Hamilton, Ohio, was –2°F. The temperature rose 2 degrees. What is the current temperature? 7.NS.1b

Check ✓

Chapter Review

ISG **Interactive Study Guide**
See pages 89–92 for:
• Vocabulary Check
• Key Concept Check
• Problem Solving
• Reflect

Lesson-by-Lesson Review

Lesson 4-1 Powers and Exponents (pp. 136–140)

Write each expression using exponents.

1. $6 \cdot 6 \cdot 6 \cdot 6 \cdot 6$ 2. 4

3. $x \cdot x \cdot x$ 4. $f \cdot f \cdot g \cdot g \cdot g \cdot g$

Evaluate each expression.

5. 3^5 6. $2 \cdot 4^3$

7. $(-4)^3$ 8. $4^0 \cdot 5$

Evaluate each expression if $w = -\dfrac{3}{4}$, $x = 4$, $y = 1$, and $z = -5$.

9. $x^2 - 6$ 10. $w^3 + y^2$

11. $2(y + z^3)$ 12. $w^4 x^2 yz$

13. Adult humans have 2^5 teeth. How many teeth do adults have?

14. Xander ran a total of 5^3 kilometers last month. How many kilometers did he run?

Example 1

Write $a \cdot a \cdot b \cdot b \cdot b \cdot b \cdot b$ using exponents.

Group the factors with like bases. Then write using exponents.

$$a \cdot a \cdot b \cdot b \cdot b \cdot b \cdot b = (a \cdot a) \cdot (b \cdot b \cdot b \cdot b \cdot b)$$
$$= a^2 b^5$$

Example 2

Evaluate $(a + 2b)^2$ if $a = 3$ and $b = -2$.

$(a + 2b)^2 = [3 + 2(-2)]^2$ $a = 3$ and $b = -2$

$= (-1)^2$ Simplify inside the brackets.

$= 1$ Simplify.

Lesson 4-2 Negative Exponents (pp. 141–146)

Write each expression using a positive exponent.

15. 9^{-4} 16. $(-10)^{-2}$

17. m^{-5} 18. c^{-5}

19. $(-4)^{-3}$ 20. y^{-9}

Write each fraction as an expression using a negative exponent other than -1.

21. $\dfrac{1}{6^3}$ 22. $\dfrac{1}{64}$

23. $\dfrac{1}{125}$ 24. $\dfrac{1}{243}$

25. $\dfrac{1}{16}$ 26. $\dfrac{1}{27}$

27. One millimeter equals 0.001 meter. Write the decimal using a negative exponent.

Example 3

Write the expression z^{-3} using a positive exponent.

$z^{-3} = \dfrac{1}{z^3}$ Definition of negative exponent

Example 4

Write $\dfrac{1}{32}$ as an expression using a negative exponent other than -1.

$\dfrac{1}{32} = \dfrac{1}{2 \cdot 2 \cdot 2 \cdot 2 \cdot 2}$ Find the prime factorization of 32.

$= \dfrac{1}{2^5}$ Definition of exponent

$= 2^{-5}$ Definition of negative exponent

Lesson 4-3 Multiplying and Dividing Monomials (pp. 147–152)

Find each product or quotient. Express using positive exponents.

28. $3^5 \cdot 3^{-2}$

29. $(-7) \cdot (-7)^4$

30. $m^3 \cdot m^6$

31. $x^8 \cdot x$

32. $(2h^7)(6h)$

33. $(5a^{-3})(-6a^4)$

34. $\frac{9^4}{9^5}$

35. $\frac{k^{10}}{k^4}$

36. Venus is about 10^8 kilometers from the Sun. Saturn is about 10^9 kilometers from the Sun. About how many times farther from the Sun is Saturn than Venus?

Example 5

Find each product or quotient.

a. $\mathbf{4t^5 \cdot 2t^8}$

$4t^5 \cdot 2t^8 = (4 \cdot 2)(t^5 \cdot t^8)$ Commutative Property of Multiplication

$= (8)(t^{5+8})$ The common base is t.

$= 8t^{13}$ Add exponents.

b. $\mathbf{\frac{n^{15}}{n^9}}$

$\frac{n^{15}}{n^9} = n^{15-9}$ The common base is n.

$= n^6$ Subtract exponents.

Lesson 4-4 Scientific Notation (pp. 153–158)

Express each number in standard form.

37. 5.82×10^3

38. 9×10^{-2}

39. 1.1×10^{-4}

40. 2.52×10^5

Express each number in scientific notation.

41. 379

42. 0.000561

43. 47,000

44. 0.0072

45. **STEM** The mass of the Sun is 1.98892×10^{15} exagrams. Express in standard form.

Example 6

Express 0.0049 in scientific notation.

$0.0049 = 4.9 \times 0.001$ The decimal point moves 3 places.

$= 4.9 \times 10^{-3}$ The exponent is negative.

Lesson 4-5 Compute with Scientific Notation (pp. 160–165)

Evaluate each expression. Express the result in scientific notation.

46. $(4.45 \times 10^9)(1.3 \times 10^6)$

47. $\frac{5.85 \times 10^5}{3.9 \times 10^{-6}}$

48. $(7.4 \times 10^4) + (3.56 \times 10^5)$

49. $(3.6 \times 10^7) - (2.85 \times 10^5)$

50. A fin whale weighs 9.92×10^4 pounds. A blue whale weighs 2.87×10^5 pounds. Estimate how many more pounds the blue whale weighs than the fin whale.

51. A male elephant weighs 1.5×10^4 pounds. A female elephant weighs 7.9×10^3 pounds. How much more does the male elephant weigh than the female elephant? Express your result in scientific notation.

Example 7

Evaluate each expression.

a. $\mathbf{(2.4 \times 10^3)(2 \times 10^5)}$

$= (2.4 \times 2)(10^3 \times 10^5)$ Associative and Commutative Properties

$= (4.8)(10^3 \times 10^5)$ Multiply 2.4 by 2.

$= 4.8 \times 10^{3+5}$ Product of Powers

$= 4.8 \times 10^8$ Add the exponents.

b. $\mathbf{(9.5 \times 10^{11}) + (6.3 \times 10^9)}$

$= (950 \times 10^9) + (6.3 \times 10^9)$ Write 9.5×10^{11} as 950×10^9.

$= (950 + 6.3) \times 10^9$ Distributive Property

$= 956.3 \times 10^9$ Add 950 and 6.3.

$= 9.563 \times 10^{11}$ Write 956.3×10^{11} in scientific notation.

Lesson 4-6 Square Roots and Cube Roots (pp. 168–173)

Find each square root or cube root.

52. $\sqrt{169}$ **53.** $-\sqrt{25}$

54. $\sqrt[3]{-64}$ **55.** $\sqrt[3]{729}$

Estimate each square root or cube root to the nearest integer.

56. $\sqrt{15}$ **57.** $-\sqrt{52}$

58. $\sqrt[3]{90}$ **59.** $\sqrt[3]{415}$

60. The *period* of a pendulum is the time in seconds it takes to make one complete swing. The period P of a pendulum is given by the formula $P = 6.28\sqrt{\frac{\ell}{32}}$, where ℓ is the length of the pendulum. If a clock's pendulum is 8 feet long, find the period.

Example 8

Find each square or cube root.

a. $\pm\sqrt{256} = \pm 16$ Find both square roots of 256; $16^2 = 256$.

b. $\sqrt[3]{125} = 5$ $5^3 = 5 \cdot 5 \cdot 5$ or 125

Example 9

Estimate $\sqrt{70}$ to the nearest integer.

The first perfect square less than 70 is 64. $\sqrt{64} = 8$

The first perfect square greater than 70 is 81. $\sqrt{81} = 9$

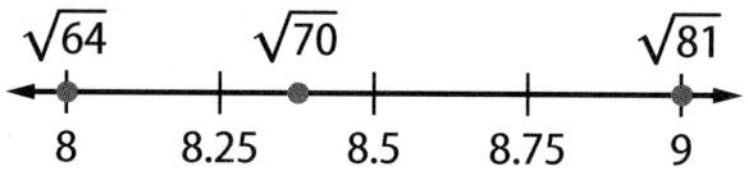

Since 70 is closer to 64 than 81, $\sqrt{70}$ is closer to 8 than to 9.

Lesson 4-7 The Real Number System (pp. 174–179)

Name all of the sets of numbers to which each real number belongs. Write *natural*, *whole*, *integer*, *rational*, or *irrational*.

61. 18 **62.** $\frac{6}{11}$

63. $\sqrt{74}$ **64.** $4.\overline{5}$

Replace each ● with <, >, or = to make a true statement.

65. $6.\overline{25}$ ● $\sqrt{39}$ **66.** $-\sqrt{70}$ ● $-8\frac{1}{5}$

67. $-11\frac{1}{9}$ ● $-\sqrt{124}$ **68.** $\sqrt{68}$ ● $8.\overline{4}$

Solve each equation. Round to the nearest tenth, if necessary.

69. $m^3 = 512$ **70.** $4y^2 = 5.76$

71. The formula $A = 3.14r^2$ can be used to determine the area of a circle, where A is the area and r is the distance from the center of the circle to the outside edge. If the area of a circular garden is 700 square feet, about how far is the distance from the center of the garden to the outside edge? Round to the nearest tenth.

Example 10

Replace ● with <, >, or = to make $\sqrt{12}$ ● $\frac{10}{3}$ a true statement.

$\sqrt{12} = 3.46410162\ldots$ $\frac{10}{3} = 3.333\ldots$

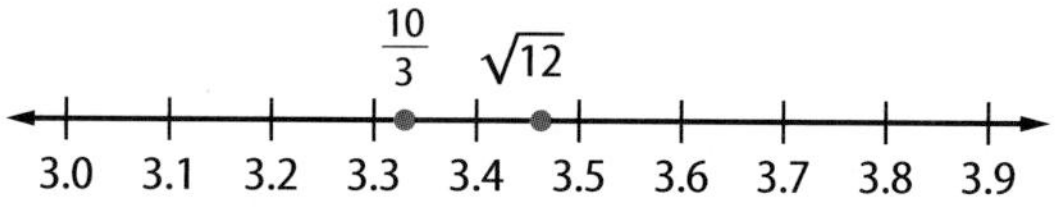

Since $\sqrt{12}$ is to the right of $\frac{10}{3}$, $\sqrt{12} > \frac{10}{3}$.

Example 11

Solve $4n^2 = 44$.

$4n^2 = 44$	Write the equation.
$n^2 = 11$	Divide each side by 4.
$n = \pm\sqrt{11}$	Definition of square root
$n \approx 3.3$ and -3.3	Use a calculator.

The solutions are approximately 3.3 and −3.3.

Chapter 5

Ratio, Proportion, and Similar Figures

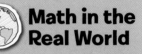

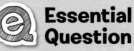

Essential Question

How can you identify and represent proportional relationships?

Common Core State Standards

Content Standards
7.RP.1, 7.RP.2, 7.RP.2a, 7.RP.2b, 7.RP.2c, 7.RP.2d, 7.RP.3, 7.NS.3, 7.G.1, 8.EE.5

Mathematical Practices
1, 2, 3, 4, 5, 6, 7

Math in the Real World

Images As an object moves farther from a mirror, its image gets smaller. Regardless of the distance that the object is from the mirror, the height of the object remains proportional to the height of its image. Proportional relationships can be described using numbers and equations.

Interactive Study Guide
See pages 93–96 for:
- Chapter Preview
- Are You Ready?
- Foldable Study Organizer

Lesson 5-1
Ratios

ISG Interactive Study Guide

See pages 97–98 for:
- Getting Started
- Real-World Link
- Notes

Essential Question

How can you identify and represent proportional relationships?

CCSS Common Core State Standards

Content Standards
7.RP.1

Mathematical Practices
1, 3, 4, 5

Vocabulary

ratio

What You'll Learn

- Write ratios as fractions in simplest form.
- Simplify ratios involving measurements.

Real-World Link

Surfing A surfer who rides inside the curl of a wave is said to be *tube riding*. The shape of the tube is described as square, round, or almond, depending on how the length of the wave compares to its width.

Key Concept Ratios

Words A **ratio** is a comparison of two quantities by division.

Examples

Numbers			Algebra		
3 to 9	3:9	$\frac{3}{9}$	a to b	$a:b$	$\frac{a}{b}$

A ratio is a comparison of two quantities by division. If the first number being compared is less than the second, the ratio is usually written as a fraction in simplest form.

Ratios can express part to part, part to whole, or whole to part relationships.

Example 1

Tutor

Express the ratio *12 baskets in 18 attempts* as a fraction in simplest form. Explain its meaning.

$$\frac{12}{18} = \frac{2}{3} \quad (\div 6)$$

Divide the numerator and denominator by the GCF, 6.

The ratio of baskets to shots attempted is 2 to 3. This means that for every 3 shots attempted, 2 were made. Also, $\frac{2}{3}$ of the shots attempted were baskets.

Got It? Do this problem to find out.

1. Refer to the table. Express the ratio of the life span of a bottlenose dolphin to the life span of a mouse as a ratio in simplest form. Explain its meaning.

Animal	Life Span (years)
bottlenose dolphin	30
mouse	3

Example 2

Ten out of every 30 Americans own a portable MP3 player. Express this ratio as a fraction in simplest form. Explain its meaning.

$$\frac{10}{30} = \frac{1}{3}$$ (÷ 10 above and below)

Divide the numerator and denominator by the GCF, 10.

The ratio of Americans who own a portable MP3 player is 1 to 3. This means that for every 3 Americans, 1 owns a portable MP3 player. Also, $\frac{1}{3}$ of Americans own a portable MP3 player.

Got It? **Do this problem to find out.**

2. Fifteen out of 100 campsites at a campground are reserved for campers with pets. Express this ratio as a fraction in simplest form. Explain its meaning.

Simplify Ratios Involving Measurements

When writing a ratio involving measurements, both quantities should have the same unit of measure.

Example 3

Express the ratio *8 ounces to 3 pounds* as a fraction in simplest form.

$$\frac{8 \text{ ounces}}{3 \text{ pounds}} = \frac{8 \text{ ounces}}{48 \text{ ounces}}$$ Convert 3 pounds to ounces.

$$= \frac{1 \text{ ounce}}{6 \text{ ounces}}$$ Divide the numerator and denominator by the GCF, 8.

Written in simplest form, the ratio is $\frac{1}{6}$.

Watch Out!
Always double check that the units in a ratio match. If the units do not match, you will not get the correct comparison.

Got It? **Do this problem to find out.**

3. Express the ratio *15 inches to 1 foot* as a fraction in simplest form.

Guided Practice

Express each ratio as a fraction in simplest form. (Example 1)

1. 12 boys to 16 girls

2. 24 out of 60 light bulbs

3. 36 DVDs out of 84 DVDs

4. 50 tiles to 25 tiles

5. In Mr. Blackwell's class, 15 out of 24 students play sports. Express this ratio as a fraction in simplest form. Explain its meaning. (Example 2)

Express each ratio as a fraction in simplest form. (Example 3)

6. 3 pints to 4 quarts

7. 2 pounds to 6 ounces

8. 9 inches to 1 yard

9. 6 gallons to 3 quarts

Independent Practice

Go online for Step-by-Step Solutions eHelp

Express each ratio as a fraction in simplest form. (Example 1)

10. 9 out of 15 pets

11. 20 wins out of 36 games

12. 4 players to 52 cards

13. 45 out of 60 days

14. 16 pens to 10 pencils

15. 96 people to 3 buses

16. On a full-sized piano, there are 36 black keys and 52 white keys. Express the ratio of black keys to white keys as a fraction in simplest form. Explain its meaning. (Example 2)

17. In a restaurant, 72 out of 108 tables are booths. Express the ratio of tables that are booths to the total number of tables as a fraction in simplest form. Explain its meaning. (Example 2)

Express each ratio as a fraction in simplest form. (Example 3)

18. 10 yards to 10 feet

19. 4 ounces to 2 pounds

20. 18 quarts to 4 gallons

21. 6 feet to 14 inches

22. A department store conducted a study to determine what age groups shop in its store.

Age Group	Number
0–17	25
18–30	75
31–45	54
46+	26

a. Express the ratio of people ages 0–17 to people ages 18–30 as a fraction in simplest form.

b. Express the ratio of people 30 or under to people over the age of 30 as a fraction in simplest form.

c. Express the ratio of people ages 18–30 to the total number of people as a fraction in simplest form.

23. A water park has 14 body slides, 8 tube slides, 2 types of swimming pools, and 6 water play areas. Use this information to write each ratio as a fraction in simplest form.

a. body slides : tube slides

b. play areas : tube slides

c. slides : all attractions

d. slides : not slides

24. A cell phone store displayed the phone at the right on a poster. The length of the phone on the poster is 3 feet 4 inches. Write a ratio comparing the length of the actual cell phone to the length of the cell phone on the poster as a fraction in simplest form.

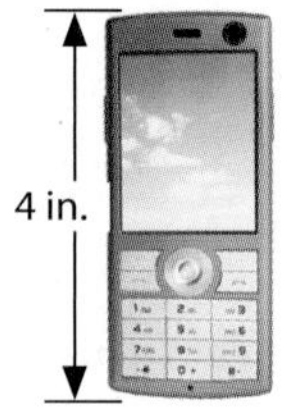

25. CCSS **Justify Conclusions** The table shows the heart rates of different animals.

Animal	Heart Rate (beats/min)	Mass (g)
cat	150	2000
cow	65	800,000
hamster	450	60
horse	44	1,200,000

a. What is the ratio of a cat's mass to its heart rate? Express the ratio as a fraction in simplest form.

b. Order the animals from greatest mass to heart rate ratio to least mass to heart rate ratio.

c. Which animal had the greatest ratio? Explain your reasoning.

26. Bonnie's Boutique had a T-shirt sale to make room for new inventory. At the end of the sale, 16 T-shirts were left. Of these, 6 were blue. Write the ratio of blue T-shirts to total T-shirts left as a fraction in simplest form.

27 When cooking a turkey, you should bake it for about 1 hour for every four pounds of meat. If an 18 pound turkey is cooked for 4 hours, was it cooked long enough? If not, how long should the turkey have been cooked?

28. The table shows the number of touchdowns and interceptions each NFL quarterback had in a recent season. Which quarterback had the best touchdown to interception ratio? Explain its meaning.

Player	Touchdowns	Interceptions
Drew Brees	26	11
Carson Palmer	28	13
Tom Brady	24	12
Philip Rivers	22	9

Compare each pair of ratios using <, >, or =.

29. \$27 for 9 key chains,
\$45 for 15 key chains

30. 8 girls out of 18 students,
12 girls out of 22 students

31. 6 cases for \$48,
14 cases for \$88

32. 24 thriller movies out of 36 DVDs,
10 thriller movies out of 15 DVDs

H.O.T. Problems Higher Order Thinking

33. CCSS **Model with Mathematics** Give three different examples of ratios that might occur in a real-world situation.

34. CCSS **Use Math Tools** Which of the technique(s) listed below might you use to determine which of the following ratios is the greatest: 440:1200, 1200:3750, or 350:450? Justify your selection(s). Then use the technique(s) to solve the problem.

mental math | a calculator | estimation

35. CCSS **Persevere with Problems** The ratio of Mieko's age to his brother Sado's age is 2:3. In ten years, the ratio will be 4:5. How old is Mieko?

36. CCSS **Which One Doesn't Belong?** Select the ratio that does not have the same value as the other three. Explain your reasoning to a classmate.

2 boys:3 girls | 2 qt:3 gal | 2 spoons:3 utensils | 2 ft:3 ft

37. CCSS **Persevere with Problems** Students in Mrs. Miller's class are measuring the length and width of a table using nonstandard materials like pencils and pieces of paper. No matter which tools they use, will the ratio of the length to the width *always*, *sometimes*, or *never* be the same? Explain.

38. **Building on the Essential Question** In a recent year, the Jacksonville Jaguars took the ball away from their opponents 21 times. They gave the ball up to their opponents 22 times. A sports writer claims the takeaway/giveaway ratio is 21 – 22, or −1. Is this statement correct? Explain.

Standardized Test Practice

39. Of 48 orchestra members, 30 are girls. What ratio compares the number of boys to girls in the orchestra?

A 3:8 **C** 3:5

B 8:3 **D** 5:3

40. Which of the following ratios does *not* describe a relationship between the squares shown?

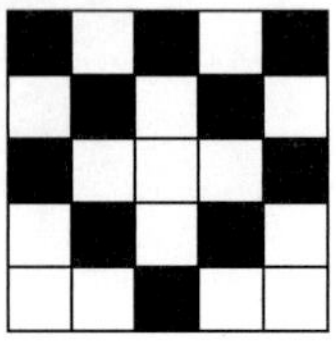

F 2 black:3 white **H** 2 black:5 total

G 2 black:5 white **J** 3 white:5 total

41. Of newly manufactured volleyballs, 12 were defective and 56 passed inspection. What ratio compares the number of defective volleyballs to the total number of volleyballs manufactured?

A 3:14 **C** 4:14

B 3:17 **D** 4:17

42. **Short Response** Marisol counted the number of coins she had in her piggy bank. The table shows her results.

Pennies	Nickels	Dimes	Quarters
47	14	18	21

Write a ratio that compares the number of dimes to the number of total coins.

Common Core Review

Write each rational number as a decimal. 7.NS.2d

43. $-12\frac{1}{3}$ **44.** $\frac{35}{8}$ **45.** $2\frac{5}{6}$

46. $-1\frac{3}{4}$ **47.** $\frac{37}{100}$ **48.** $\frac{42}{5}$

49. There are approximately 30.48 centimeters in 1 foot. Express 30.48 as a mixed number. 7.EE.3

Name the ordered pair for each point graphed at the right. 6.NS.6c

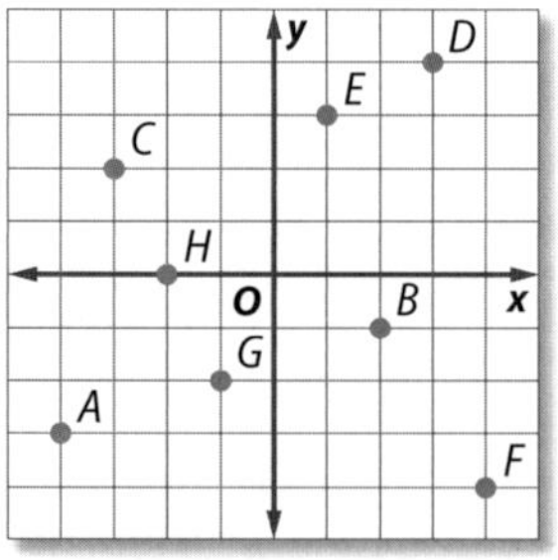

50. *A* **51.** *B*

52. *C* **53.** *D*

54. *E* **55.** *F*

56. **STEM** To simulate space travel, NASA's Lewis Research Center in Cleveland uses a 430-foot shaft. The free fall of an object in the shaft takes 5 seconds to travel the 430 feet. 7.NS.2b

a. Write an integer to represent the change in the height of the object.

b. On average, how far does the object travel each second?

Divide. Round to the nearest cent. 6.NS.3

57. \$5.75 ÷ 4 **58.** \$8.30 ÷ 6

59. \$2.27 ÷ 8 **60.** \$11.50 ÷ 5

Find each difference. 7.NS.1c

61. 9 − (−5) **62.** 3 − 8 **63.** 15 − (−4)

64. −12 − 4 **65.** −1 − 14 **66.** 11 − 3

Lesson 5-2

Unit Rates

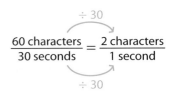

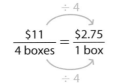

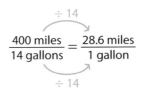

ISG Interactive Study Guide

See pages 99–100 for:
- Getting Started
- Vocabulary Start-Up
- Notes

Essential Question

How can you identify and represent proportional relationships?

CCSS Common Core State Standards

Content Standards
7.RP.1

Mathematical Practices
1, 3, 4

Vocabulary

rate
unit rate

What You'll Learn

- Find unit rates.
- Compare and use unit rates to solve problems.

Real-World Link

World Records *The Guinness Book of Records* was first published in 1955. Recent editions have featured feats such as the world's fastest human and the most hot dogs eaten in ten minutes. In 2011, the world record for texting speed was 264 characters in 1 minute, 17 seconds.

Find Unit Rates

A **rate** is a ratio of two quantities having different kinds of units. Suppose you text 60 characters in 30 seconds. This can be written as a rate.

$$\frac{60 \text{ characters}}{30 \text{ seconds}}$$ ← different kinds of units

When a rate is simplified so that it has a denominator of 1, it is called a **unit rate**. To write a rate as a unit rate, divide the numerator and denominator of the rate by the denominator.

$$\frac{60 \text{ characters}}{30 \text{ seconds}} = \frac{2 \text{ characters}}{1 \text{ second}} \quad (\div 30 \text{ top and bottom})$$

Example 1

Tutor

Express each rate as a unit rate. Round to the nearest tenth, if necessary.

a. \$11 for 4 boxes of cereal

Write the rate that compares the cost to the number of boxes.

$$\frac{\$11}{4 \text{ boxes}} = \frac{\$2.75}{1 \text{ box}} \quad (\div 4 \text{ top and bottom})$$

Divide the numerator and denominator by 4.

So, the cost is \$2.75 per box of cereal.

b. 400 miles on 14 gallons of gasoline

Write the rate that compares the number of miles to the number of gallons.

$$\frac{400 \text{ miles}}{14 \text{ gallons}} = \frac{28.6 \text{ miles}}{1 \text{ gallon}} \quad (\div 14 \text{ top and bottom})$$

Divide the numerator and denominator by 14.

So, the car traveled 28.6 miles per gallon of gasoline.

Got It? Do these problems to find out.

1a. \$18.50 for 5 pounds

1b. 100 meters in 14 seconds

Compare Unit Rates

You can also compare unit rates to solve problems.

Example 2

Tutor

Financial Literacy An online music store sells 15 songs for $12. Another online music store sells 10 songs for $9. Which online store has the lower cost per song?

Step 1 Find the unit rates of the two stores.

$$\frac{12 \text{ dollars}}{15 \text{ songs}} \overset{\div 15}{=} \frac{0.8 \text{ dollar}}{1 \text{ song}}$$

Divide the numerator and denominator by 15.

For the 15 songs, the unit rate is $0.80 per song.

$$\frac{9 \text{ dollars}}{10 \text{ songs}} \overset{\div 10}{=} \frac{0.9 \text{ dollar}}{1 \text{ song}}$$

Divide the numerator and denominator by 10.

For the 10 songs, the unit rate is $0.90 per song.

Step 2 Compare the rates. Since $0.80 < $0.90, the first store has a better rate per song.

Alternative Method
Use the LCM. In the first case, 30 songs cost $12 × 2 or $24. In the second, 30 songs cost $9 × 3 or $27. The first has the lower cost per song.

Got It? Do this problem to find out.

2. Financial Literacy A store sells two different sizes of the same brand of sunscreen, an 8-ounce bottle for $5.76 and a 12-ounce bottle for $8.88. Which size bottle is the better buy per ounce? Explain.

Once you know the unit rate, you can use it to solve problems involving any amount.

Example 3

Tutor

A typical bottlenose dolphin will take about 34 breaths in 4 hours. How many breaths will a bottlenose dolphin take in 7 hours?

Step 1 Find the unit rate.

$$34 \text{ breaths in 4 hours} = \frac{34 \text{ breaths} \div 4}{4 \text{ hours} \div 4} \text{ or } \frac{8.5 \text{ breaths}}{1 \text{ hour}}$$

Step 2 Multiply this unit rate by 7 to find the number of breaths a dolphin will take in 7 hours.

Step 3 $\frac{8.5 \text{ breaths}}{1 \cancel{\text{hour}}} \cdot 7 \cancel{\text{hours}} = 59.5 \text{ breaths}$ Divide out the common units.

A bottlenose dolphin will take 59.5 breaths in 7 hours.

Got It? Do this problem to find out.

3. A bakery can make 195 doughnuts in 3 hours. At this rate, how many doughnuts can the bakery make in 8 hours?

Guided Practice

Express each rate as a unit rate. Round to the nearest tenth or to the nearest cent, if necessary. (Example 1)

1. $120 for 5 days of work

2. 275 miles on 14 gallons

3. 338 points in 16 games

4. $19.49 for 6 pounds

5. 17 gallons in 4 minutes

6. 180 feet in 19 seconds

7. **Financial Literacy** Jamal is comparing prices of several different brands of peanuts. Which brand is the best buy? Explain. (Example 2)

Peanuts		
Brand	**Size (oz)**	**Price**
Barrel	10	$3.39
Mr. Nut	14	$4.54
Chip's	18	$6.26

8. Aisha drove 170 miles in 2.5 hours. At this same rate, how far will she drive in 4 hours? (Example 3)

Independent Practice

Go online for Step-by-Step Solutions

Express each rate as a unit rate. Round to the nearest tenth or to the nearest cent, if necessary. (Example 1)

9. 156 students in 6 classes

10. 424 Calories in 3 servings

11. 147.5 miles in 2.5 hours

12. $29.95 for 4 DVDs

13. $231 for 3 game tickets

14. 5 tablespoons in 4 quarts

15. $97.50 for 15 pizzas

16. 400 meters in 58 seconds

17. 352 miles on 16 gallons

18. $60 for 8 hours of work

19. **Financial Literacy** The Party Planner sells 10 paper plates for $2.50. Use the table to determine which company sells paper plates for the same price per plate. Explain. (Example 2)

Store	Number of Plates	Price
Party Time	15	$3.75
Good Times	20	$6.00
Birthday, Inc.	25	$7.50

20. **STEM** Building A has 7500 square feet of office space for 320 employees. Building B has 9500 square feet of office space for 370 employees. Which building has more square feet of space per employee? Explain. (Example 2)

21. A recipe that makes 2 dozen cookies calls for $\frac{3}{4}$ cup sugar. How much sugar is needed to make $4\frac{1}{2}$ dozen cookies? (Example 3)

22. A farmers' market sells ears of sweet corn. At this same rate, how much will it cost to buy 28 ears of sweet corn? (Example 3)

23. Which swimmer has the fastest rate?

Swimmer	Jenny	Dana	Kaitlin
Event	50 m	100 m	200 m
Time (s)	25.02	119.2	248.07

24. **Financial Literacy** A gallon of milk sells for \$3.18, and a quart of milk sells for \$0.76. Which item has the better unit price? Justify your answer using two different methods.

25. The line graph shows Alicia's and Jermaine's average rates in a race.

   a. Express each person's speed as a unit rate.

   b. How long would it take Alicia and Jermaine each to run 1 mile? (*Hint*: One mile = 5280 feet)

   c. Suppose Max runs at a rate of 5 feet per second. Predict where the line representing his speed would be graphed. Explain.

Racing Rates

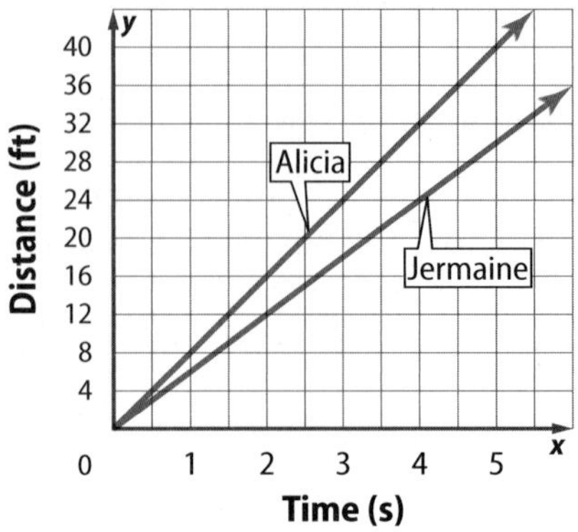

26. **CCSS Multiple Representations** In this problem, you will explore rates. Abigail left Charlotte, North Carolina, 30 minutes before Juan did, to travel to Greenville, South Carolina. Two and a half hours after Abigail left, she had traveled 155 miles, and Juan had traveled 144 miles.

   a. Table Make a function table to show how far each person has traveled after Abigail has traveled 1, 2, 3, and 4 hours.

   b. Graph Make a graph of the data. Do the two lines intersect? If so, what does this intersection point represent?

27. **STEM** The term *salinity* refers to the amount of salt in water. The salinity of Earth's oceans is about 525 grams of dissolved salt for every 15 kilograms of water. Find the grams of salt for every gram of ocean water.

H.O.T. Problems Higher Order Thinking

28. **CCSS Model with Mathematics** Give a real-world example of a rate for a unit rate of 40 miles per hour.

29. **CCSS Justify Conclusions** In which situation will the rate $\frac{x \text{ miles}}{y \text{ hours}}$ decrease? Give an example to explain your reasoning.

   a. x increases; y is unchanged **b.** x is unchanged; y increases

30. **CCSS Find the Error** Nadia writes the rate \$15.75 for 4 pounds as a unit rate. Find her mistake and correct it.

$$\frac{\$15.75}{4 \text{ pounds}} = \frac{\$7.88}{2 \text{ pounds}}$$

31. **CCSS Persevere with Problems** A 96-ounce container of orange juice costs \$4.80. At what price should a 128-ounce container be sold in order for the unit rate for both containers to be the same? Explain your reasoning.

32. **Building on the Essential Question** Explain why a horse that runs $\frac{3}{4}$ mile in 1 minute and 9 seconds is faster than a sprinter who runs a 100-yard dash in 12 seconds.

Standardized Test Practice

33. On Monday, Ms. Moseley drove 340 miles in 5 hours. On Tuesday, she drove 198 miles in 3 hours. Based on these rates, which statement is true?

A Her rate on Monday was 2 miles per hour slower than her rate on Tuesday.

B Her rate on Tuesday was 2 miles per hour slower than her rate on Monday.

C Her rate on Monday was the same as her rate on Tuesday.

D Her rate on Tuesday was 2 miles per hour faster than her rate on Monday.

34. Kalyin keyboards at a rate of 60 words per minute for 5 minutes and 45 words per minute for 10 minutes. How many words did she type in all?

F 105 **H** 650

G 510 **J** 750

35. Short Response The table shows the costs of different-sized bags of snack mix.

Snack Mix	
Size	**Price**
x-small	\$1.96 for 12 oz
small	\$2.20 for 16 oz
large	\$4.00 for 36 oz
x-large	\$5.40 for 48 oz

Pilar wants to buy the one that costs the least per ounce. What size should she buy? Explain.

36. Sal paid \$2.79 for a gallon of milk. Find the cost per quart of milk at this rate.

A \$0.55

B \$0.70

C \$0.93

D \$1.40

CCSS Common Core Review

Express each ratio as a fraction in simplest form. 6.RP.3

37. 155 apples to 75 oranges

38. 7 cups to 9 pints

39. 11 gallons to 11 quarts

40. 900 pounds to 16 tons

41. Samuel wants to buy 6 benches and 4 picnic tables. Write an expression to find the total cost of 6 benches and 4 picnic tables. 6.EE.1

Benches	\$20
Picnic tables	\$100

42. The research submarine *Alvin*, which was used to locate the wreck of the *Titanic*, descended at a rate of about 100 feet per minute. Write an integer to represent the distance *Alvin* traveled in 5 minutes. 7.NS.2a

43. Evaluate $a - b + c$ if $a = 5$, $b = 7$, and $c = 10$. 7.EE.3

Complete. 5.MD.1

44. 36 in. = ■ ft

45. 4 yd = ■ in.

46. ■ fl oz = 6 c

47. ■ m = 4 cm

48. 476 mL = ■ L

49. 3.5 km = ■ m

Find each sum or difference. Write in simplest form. 7.NS.1

50. $-\frac{5}{6} + 3\frac{1}{12}$

51. $2\frac{1}{2} - 3\frac{5}{8}$

52. $-2\frac{3}{4} + 1\frac{7}{12}$

53. $-5\frac{1}{6} + \left(-3\frac{1}{5}\right)$

54. $-10\frac{1}{2} - 7\frac{3}{5}$

55. $\frac{7}{16} + 2\frac{1}{5}$

Lesson 5-3

Complex Fractions and Unit Rates

Interactive Study Guide

See pages 101–102 for:
- Getting Started
- Real-World Link
- Notes

Essential Question

How can you identify and represent proportional relationships?

Common Core State Standards

Content Standards
7.RP.1, 7.NS.3

Mathematical Practices
1, 3, 4, 7

Vocabulary

complex fraction

What You'll Learn

- Simplify complex fractions.
- Find unit rates.

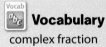

Real-World Link

Music Knowing the beats per minute, or BPM, of songs is useful for both professional DJs and casual music lovers. DJs use BPM to blend together songs that have similar tempos. Anyone can use an app to calculate the BPM of songs, and then create playlists for different occasions based on BPM values.

Simplify Complex Fractions

Fractions like $\frac{\frac{1}{3}}{\frac{1}{2}}$ are called complex fractions. **Complex fractions** are fractions with one or more fractions in the numerator, denominator, or both. Complex fractions are simplified when both the numerator and denominator are integers.

Example 1

Tutor

Simplify $\frac{\frac{1}{3}}{\frac{1}{2}}$.

Recall that a fraction can also be written as a division problem.

$\frac{\frac{1}{3}}{\frac{1}{2}} = \frac{1}{3} \div \frac{1}{2}$ Write the complex fraction as a division problem.

$= \frac{1}{3} \times \frac{2}{1}$ Multiply by the reciprocal of $\frac{1}{2}$, which is $\frac{2}{1}$.

$= \frac{2}{3}$ Simplify.

So, $\frac{\frac{1}{3}}{\frac{1}{2}}$ is equal to $\frac{2}{3}$.

Got It? **Do these problems to find out.**

Simplify.

1a. $\frac{\frac{1}{3}}{\frac{1}{4}}$

1b. $\frac{\frac{2}{3}}{\frac{3}{4}}$

1c. $\frac{\frac{1}{5}}{\frac{6}{7}}$

Example 2

Tutor

Simplify $\frac{\frac{1}{4}}{2}$.

$$\frac{\frac{1}{4}}{2} = \frac{1}{4} \div 2$$ Write the complex fraction as a division problem.

$$= \frac{1}{4} \times \frac{1}{2}$$ Multiply by the reciprocal of 2, which is $\frac{1}{2}$.

$$= \frac{1}{8}$$ Simplify.

So, $\frac{\frac{1}{4}}{2}$ is equal to $\frac{1}{8}$.

Watch Out!

To divide by a whole number, first write it as a fraction with a denominator of 1. Then multiply by the reciprocal. So, $\frac{\frac{1}{4}}{2}$ can be written as $\frac{1}{4} \div \frac{2}{1}$.

Got It? Do these problems to find out.

2a. $\frac{\frac{2}{3}}{7}$

2b. $\frac{\frac{2}{4}}{2}$

2c. $\frac{14}{\frac{7}{9}}$

Find Unit Rates

When the fractions of a complex fraction represents different units, you can find the unit rate.

Real World Example 3

Tutor

Josiah can jog $1\frac{1}{3}$ miles in $\frac{1}{4}$ hour. Find his average speed in miles per hour.

Write a rate that compares the number of miles to hours.

$$\frac{1\frac{1}{3}\text{ mi}}{\frac{1}{4}\text{ h}} = 1\frac{1}{3} \div \frac{1}{4}$$ Write the complex fraction as a division problem.

$$= \frac{4}{3} \div \frac{1}{4}$$ Write the mixed number as an improper fraction.

$$= \frac{4}{3} \times \frac{4}{1}$$ Multiply by the reciprocal of $\frac{1}{4}$, which is $\frac{4}{1}$.

$$= \frac{16}{3} \text{ or } 5\frac{1}{3}$$ Simplify.

So, Josiah jogs at an average speed of $5\frac{1}{3}$ miles per hour.

Got It? Do these problems to find out.

3a. A truck driver drove 350 miles in $8\frac{3}{4}$ hours. What is the speed of the truck in miles per hour?

3b. Aubrey can walk $4\frac{1}{2}$ miles in $1\frac{1}{2}$ hours. Find her average speed in miles per hour.

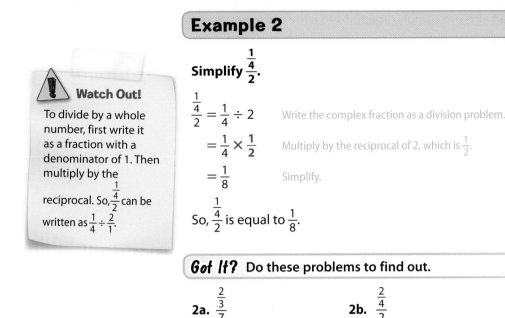

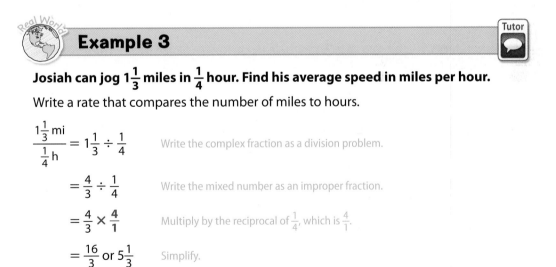

You can also use complex fractions to write some percents as fractions.

Example 4

Tutor

Percent
Percent is a ratio that compares a number to 100.

Write $33\frac{1}{3}\%$ as a fraction in simplest form.

$33\frac{1}{3}\% = \dfrac{33\frac{1}{3}}{100}$ Definition of percent

$= 33\frac{1}{3} \div 100$ Write the complex fraction as a division problem.

$= \frac{100}{3} \div 100$ Write $33\frac{1}{3}$ as an improper fraction.

$= \frac{\overset{1}{\cancel{100}}}{3} \times \frac{1}{\underset{1}{\cancel{100}}}$ Multiply by the reciprocal of 100, which is $\frac{1}{100}$.

$= \frac{1}{3}$ Simplify.

So, $33\frac{1}{3}\% = \frac{1}{3}$.

Got It? **Do these problems to find out.**

4a. $4\frac{1}{2}\%$

4b. $12\frac{1}{2}\%$

4c. $10\frac{2}{3}\%$

4d. $3\frac{1}{3}\%$

Guided Practice

Check

Simplify. (Examples 1 and 2)

1. $\dfrac{18}{\frac{3}{4}}$

2. $\dfrac{\frac{3}{6}}{4}$

3. $\dfrac{\frac{1}{3}}{\frac{1}{4}}$

4. $\dfrac{\frac{1}{3}}{\frac{3}{5}}$

5. $\dfrac{11}{\frac{1}{2}}$

6. $\dfrac{\frac{4}{7}}{2}$

7. Monica reads $7\frac{1}{2}$ pages of a mystery book in 9 minutes. What is her average reading rate in pages per minute? (Example 3)

8. Patrick drove 220 miles to his grandmother's house. The trip took him $4\frac{2}{5}$ hours. What is his average speed in miles per hour? (Example 3)

Write each percent as a fraction in simplest form. (Example 4)

9. $6\frac{2}{3}\%$

10. $7\frac{1}{2}\%$

11. $10\frac{1}{2}\%$

12. $15\frac{1}{3}\%$

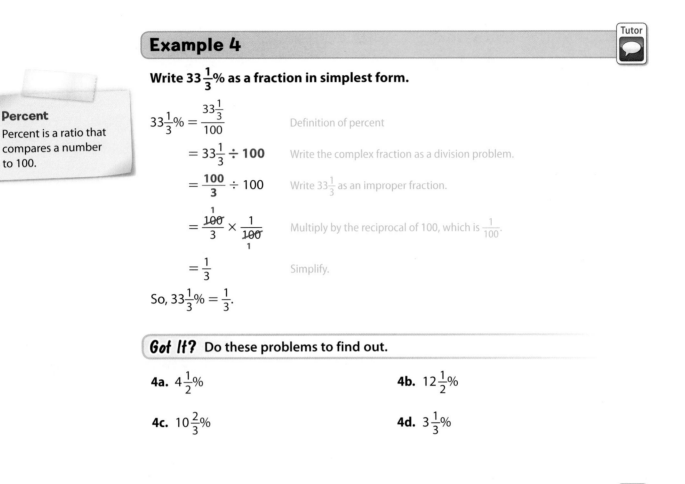

Independent Practice

eHelp Go online for Step-by-Step Solutions

Simplify. (Examples 1 and 2)

13. $\dfrac{1}{\frac{2}{3}}$

14. $\dfrac{2}{\frac{3}{11}}$

15. $\dfrac{12}{\frac{3}{5}}$

16. $\dfrac{1}{\frac{1}{4}}$

17. $\dfrac{\frac{8}{9}}{6}$

18. $\dfrac{\frac{2}{5}}{9}$

19. $\dfrac{\frac{9}{10}}{9}$

20. $\dfrac{\frac{4}{5}}{10}$

21. $\dfrac{\frac{1}{2}}{\frac{1}{4}}$

22. $\dfrac{\frac{1}{4}}{\frac{7}{10}}$

23. $\dfrac{\frac{1}{12}}{\frac{5}{6}}$

24. $\dfrac{\frac{5}{6}}{\frac{5}{9}}$

25. Richard rowed a canoe $3\frac{1}{2}$ miles in $\frac{1}{2}$ hour. What is his average speed in miles per hour? (Example 3)

26. A small airplane used $4\frac{3}{8}$ gallons of fuel to fly a $1\frac{1}{4}$-hour trip. How many gallons were used each hour? (Example 3)

27. Sally wants to make cookies for her little sister's tea party. Sally is cutting a roll of cookie dough into pieces that are $\frac{1}{2}$ inch thick. If the roll is $11\frac{1}{2}$ inches long, how many cookies can she make? (Example 3)

28. Mary Alice is making a curtain for her kitchen window. She bought $2\frac{1}{2}$ yards of fabric. Her total cost was $15. What was the cost per yard? (Example 3)

Write each percent as a fraction in simplest form. (Example 4)

29. $56\frac{1}{4}\%$

30. $15\frac{3}{5}\%$

31. $13\frac{1}{3}\%$

32. $2\frac{2}{5}\%$

33. $7\frac{3}{4}\%$

34. $8\frac{1}{3}\%$

35. Mrs. Frasier is making costumes for the school play. The table shows the amount of material needed for each complete costume. She bought $14\frac{3}{4}$ yards of material.

Play Costumes	
top	$1\frac{1}{4}$ yards
bottom	$\frac{1}{2}$ yard

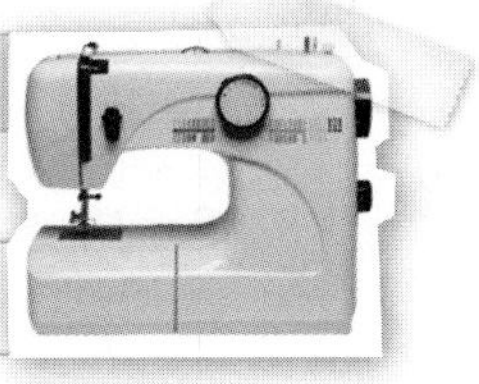

a. How many complete costumes can she make? How much material will be left over?

b. If she spent a total of $44.25 on fabric, what was the cost per yard? Explain how you solved.

36. **Financial Literacy** The value of a certain stock increased by $1\frac{1}{4}\%$. Write the percent as a fraction in simplest form.

37. A high school Family and Consumer Science class has $12\frac{1}{2}$ pounds of flour with which to make soft taco shells. There are $3\frac{3}{4}$ cups of flour in a pound, and it takes about $\frac{1}{3}$ cup of flour per shell. How many soft taco shells can they make?

38. Emma runs $\frac{3}{4}$ mile in 6 minutes. Joanie runs $1\frac{1}{2}$ miles in 11 minutes. Whose speed is greater? Explain.

39. **Financial Literacy** A bank is offering home loans at an interest rate of $5\frac{1}{2}$%. Write the percent as a fraction in simplest form.

40. CCSS **Justify Conclusions** For a project, Karl measured the wingspan of a butterfly and a moth. His measurements are shown below. How many times larger is the moth than the butterfly? Justify your answer.

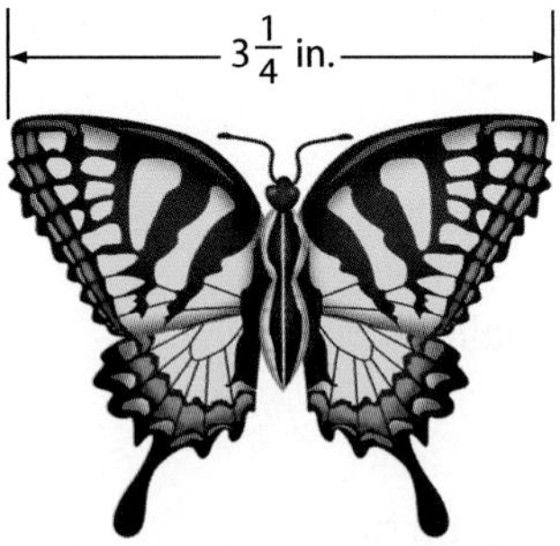

Black Swallowtail Butterfly

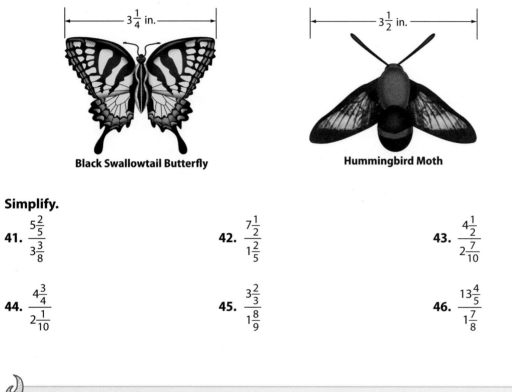

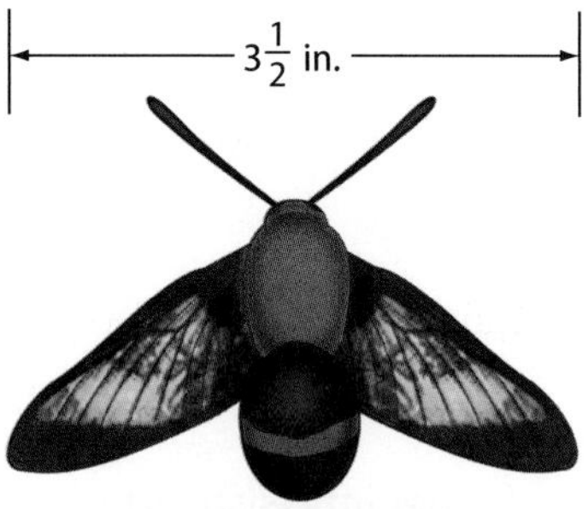

Hummingbird Moth

Simplify.

41. $\dfrac{5\frac{2}{5}}{3\frac{3}{8}}$

42. $\dfrac{7\frac{1}{2}}{1\frac{2}{5}}$

43. $\dfrac{4\frac{1}{2}}{2\frac{7}{10}}$

44. $\dfrac{4\frac{3}{4}}{2\frac{1}{10}}$

45. $\dfrac{3\frac{2}{3}}{1\frac{8}{9}}$

46. $\dfrac{13\frac{4}{5}}{1\frac{7}{8}}$

H.O.T. Problems Higher Order Thinking

47. CCSS **Identify Structure** Write three different complex fractions that can be simplified to $\frac{1}{4}$.

48. CCSS **Persevere with Problems** A motorized scooter has tires with a circumference of 22 inches. The tires make one revolution every $\frac{1}{10}$ second. Find the speed of the scooter in inches per second. (*Hint:* The speed of an object spinning in a circle is equal to the circumference divided by the time it takes to complete one revolution.)

49. **Building on the Essential Question** Explain how complex fractions can be used to solve problems involving ratios.

Standardized Test Practice

50. Debra can run $20\frac{1}{2}$ miles in $2\frac{1}{4}$ hours. How many miles per hour can she run?

A $46\frac{1}{8}$ miles per hour

B $22\frac{3}{4}$ miles per hour

C $18\frac{1}{4}$ miles per hour

D $9\frac{1}{9}$ miles per hour

51. Tina wants to give away 6 bundles of thyme from her herb garden. If she has $\frac{1}{2}$ pound of thyme, how much will each bundle weigh?

F $\frac{1}{2}$ lb **H** $\frac{1}{12}$ lb

G 3 lb **J** 12 lb

52. LaShondra is using a model to simplify the complex fraction below.

$$\frac{\frac{2}{3}}{\frac{1}{12}}$$

Which statement shows how to use the model?

A The figure is divided into twelfths. Count the twelfths that fit within $\frac{2}{3}$ of the figure.

B The figure is divided into twelfths. Remove $\frac{2}{3}$ of the twelfths, and count those remaining.

C Count the number of thirds in the figure. Multiply this number by 12.

D Count the number of rectangles in the figure. Divide this number by 3.

CCSS Common Core Review

Write each expression using exponents. 8.EE.1

53. $3 \cdot 3 \cdot 3 \cdot 3 \cdot 3 \cdot 3 \cdot 3$

54. $(-4) \cdot (-4) \cdot (-4) \cdot (-4)$

55. $\left(\frac{3}{4}\right)\left(\frac{3}{4}\right)$

56. $m \cdot m \cdot m \cdot g \cdot g$

57. $(x + 1)(x + 1)(x + 1)$

58. $5 \cdot 5 \cdot d \cdot d \cdot s \cdot s \cdot s \cdot s$

Express each number in scientific notation. 8.EE.3

59. 80,000

60. 3200

61. 0.0054

62. 2300

63. 0.0000000098

64. 47

65. Financial Literacy A warehouse store sells two different sizes of the same brand of ketchup, a 114-ounce bottle and a 44-ounce bottle. Which size bottle is the better buy per ounce? Explain. 7.RP.1

Ketchup	
114 ounces	$3.48
44 ounces	$1.90

Classify each polygon with the name that best describes it. 5.G.3

66. 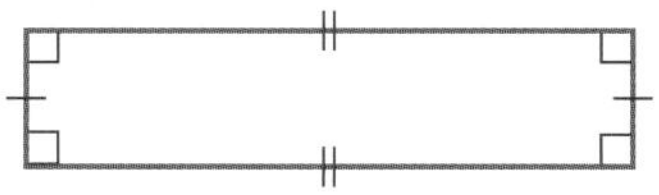

67. 

68. 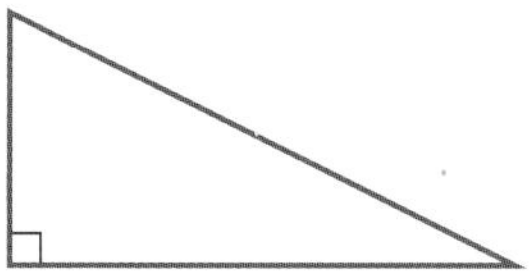

Replace each ● with <, >, or = to make a true sentence. 5.NBT.3b

69. 0.925 ● 1.023

70. 0.15 ● 0.099

71. 7.3 ● 7.30

Lesson 5-4

Converting Rates

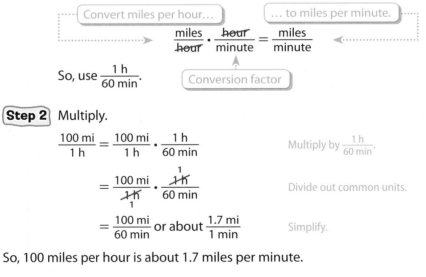

ISG Interactive Study Guide

See pages 103–104 for:
- Getting Started
- Real-World Link
- Notes

Essential Question

How can you identify and represent proportional relationships?

CCSS Common Core State Standards

Content Standards
7.RP.1, 7.RP.3

Mathematical Practices
1, 3, 4, 6, 7

Vocab Vocabulary

dimensional analysis

What You'll Learn

- Convert rates using dimensional analysis.
- Convert between systems of measurement.

Real-World Link

Sea Creature Leafy seadragons get their name from the leaflike appendages that camouflage them. They move very slowly using tiny fins on their necks and backs. You can use fractions involving units to calculate their average speed in inches per minute.

Dimensional Analysis

Dimensional analysis is the process of including units of measurement as factors when you compute. For example, you know that 1 hour = 60 minutes. You can write conversion factors $\frac{1 \text{ hour}}{60 \text{ minutes}}$ or $\frac{60 \text{ minutes}}{1 \text{ hour}}$. Each ratio is equivalent to 1 because the numerator and denominator represent the same amount.

Real-World Example 1

Tutor

Convert 100 miles per hour to miles per minute.

Step 1 You need to convert miles per hour to miles per minute. Choose a conversion factor that converts hours to minutes, with minutes in the denominator.

Convert miles per hour… … to miles per minute.

$$\frac{\text{miles}}{\cancel{\text{hour}}} \cdot \frac{\cancel{\text{hour}}}{\text{minute}} = \frac{\text{miles}}{\text{minute}}$$

Conversion factor

So, use $\frac{1 \text{ h}}{60 \text{ min}}$.

Step 2 Multiply.

$$\frac{100 \text{ mi}}{1 \text{ h}} = \frac{100 \text{ mi}}{1 \text{ h}} \cdot \frac{1 \text{ h}}{60 \text{ min}}$$ Multiply by $\frac{1 \text{ h}}{60 \text{ min}}$.

$$= \frac{100 \text{ mi}}{\cancel{1 \text{ h}}_1} \cdot \frac{\cancel{1 \text{ h}}^1}{60 \text{ min}}$$ Divide out common units.

$$= \frac{100 \text{ mi}}{60 \text{ min}} \text{ or about } \frac{1.7 \text{ mi}}{1 \text{ min}}$$ Simplify.

So, 100 miles per hour is about 1.7 miles per minute.

Got It? Do this problem to find out.

1. **Financial Literacy** The average teenager spends \$1742 per year on fashion-related items. How much is this per week?

Example 2

Tutor

Watch Out!
Make sure that you choose conversion factors that allow you to divide out the common units.

Tyree and three friends attend skydiving class before their first jump. The instructor tells them they will travel at about 176 feet per second. How many miles per hour is this?

You need to convert feet per second to miles per hour.

Use 1 mile = 5280 feet and 1 hour = 3600 seconds.

$$\frac{176 \text{ ft}}{1 \text{ s}} = \frac{176 \text{ ft}}{1 \text{ s}} \cdot \frac{1 \text{ mi}}{5280 \text{ ft}} \cdot \frac{3600 \text{ s}}{1 \text{ h}}$$ Multiply by $\frac{1 \text{ mi}}{5280 \text{ ft}}$ and $\frac{3600 \text{ s}}{1 \text{ h}}$.

$$= \frac{\overset{1}{\cancel{176 \text{ ft}}}}{1 \cancel{\text{s}}} \cdot \frac{1 \text{ mi}}{\underset{\underset{1}{\cancel{30}}}{\cancel{5280 \text{ ft}}}} \cdot \frac{\overset{120}{\cancel{3600 \text{ s}}}}{1 \text{ h}}$$ Divide the common factors and units.

$$= \frac{120 \text{ mi}}{1 \text{ h}}$$ Simplify.

So, 176 feet per second is equivalent to 120 miles per hour.

Got It? Do these problems to find out.

2a. The TGV is a high speed rail train in France. At top speed, it runs at an average of 320 kilometers per hour. How many meters per second is this?

2b. An adult cheetah can reach a speed of about 70 miles per hour. How fast is this in feet per second?

Key Concept Measurement Conversions

Length	
Customary to Metric	**Metric to Customary**
1 in. ≈ 2.540 cm	1 cm ≈ 0.394 in.
1 ft ≈ 0.305 m	1 m ≈ 3.279 ft
1 yd ≈ 0.914 m	1 m ≈ 1.094 yd
1 mi ≈ 1.609 km	1 km ≈ 0.621 mi
Capacity	
Customary to Metric	**Metric to Customary**
1 fl oz ≈ 29.574 mL	1 mL ≈ 0.034 fl oz
1 pt ≈ 0.473 L	1 L ≈ 2.114 pt
1 qt ≈ 0.946 L	1 L ≈ 1.057 qt
1 gal ≈ 3.785 L	1 L ≈ 0.264 gal
Mass or Weight	
Customary to Metric	**Metric to Customary**
1 oz ≈ 28.350 g	1 g ≈ 0.035 oz
1 lb ≈ 0.454 kg	1 kg ≈ 2.203 lb

The table above shows conversion factors between the Customary and Metric systems for units of length, capacity, and mass or weight.

You can also use dimensional analysis to convert between measurement systems. The two conversion factors $\frac{1 \text{ ft}}{0.305 \text{ m}}$ and $\frac{0.305 \text{ m}}{1 \text{ ft}}$ use the same conversion. Use the factor that will correctly divide out the appropriate common unit.

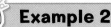

Example 3

Tutor

Complete each conversion. Round to the nearest hundredth.

a. 12 centimeters to inches

Use 1 inch ≈ 2.54 centimeters.

$12 \text{ cm} \approx 12 \text{ cm} \cdot \frac{1 \text{ in.}}{2.54 \text{ cm}}$ Multiply by $\frac{1 \text{ in.}}{2.54 \text{ cm}}$.

$\approx 12 \cancel{\text{cm}} \cdot \frac{1 \text{ in.}}{2.54 \cancel{\text{cm}}}$ Divide out common units, leaving the desired unit, inch.

$\approx \frac{12 \text{ in.}}{2.54}$ or 4.72 in. Simplify.

So, 12 centimeters is approximately 4.72 inches.

b. 4 quarts to liters

Use 1 quart ≈ 0.946 liter.

$4 \text{ qt} \approx 4 \text{ qt} \cdot \frac{0.946 \text{ L}}{1 \text{ qt}}$ Multiply by $\frac{0.946 \text{ L}}{1 \text{ qt}}$.

$\approx 4 \cancel{\text{qt}} \cdot \frac{0.946 \text{ L}}{1 \cancel{\text{qt}}}$ Divide out common units, leaving the desired unit, quart.

$\approx 4 \cdot 0.946 \text{ L}$ or 3.78 L Simplify.

So, 4 quarts is approximately 3.78 liters.

Got It? Do these problems to find out.

3a. 6 mi ≈ ■ km **3b.** 12 oz ≈ ■ g **3c.** 11 yd ≈ ■ m

Real World Example 4

Tutor

Dimensions

When converting between systems, ask yourself how many dimensions are involved. In Example 4, there are two dimensions involved, length and time.

At top speed, a giant tortoise can travel about 900 feet per hour. How many centimeters per second can a giant tortoise travel at top speed?

To convert feet to centimeters, use 1 foot = 12 inches and 1 inch ≈ 2.54 centimeters.

To convert hours to seconds, use 1 hour = 60 minutes and 1 minute = 60 seconds.

$\frac{900 \text{ ft}}{1 \text{ h}} \cdot \frac{12 \text{ in.}}{1 \text{ ft}} \cdot \frac{2.54 \text{ cm}}{1 \text{ in.}} \cdot \frac{1 \text{ h}}{60 \text{ min}} \cdot \frac{1 \text{ min}}{60 \text{ s}}$

$= \frac{900 \cancel{\text{ft}}}{1 \cancel{\text{h}}} \cdot \frac{12 \cancel{\text{in.}}}{1 \cancel{\text{ft}}} \cdot \frac{2.54 \text{ cm}}{1 \cancel{\text{in.}}} \cdot \frac{1 \cancel{\text{h}}}{60 \cancel{\text{min}}} \cdot \frac{1 \cancel{\text{min}}}{60 \text{ s}}$ Divide out common units.

$= \frac{27{,}432 \text{ cm}}{3600 \text{ s}}$ Multiply.

$= \frac{7.62 \text{ cm}}{1 \text{ s}}$ Divide.

At top speed, a giant tortoise will travel 7.62 centimeters per second.

Got It? Do this problem to find out.

4. At a recent Winter Olympics, USA short track speed skater Apolo Ohno won a gold medal by skating about 12 meters per second. Rounded to the nearest hundredth, how many miles per hour is this?

Guided Practice

Check

1. In Brazil, about 20 acres of rain forest are destroyed each minute. At this rate, how much rain forest is destroyed per day? (Example 1)

2. Lexi can paint 5 yards of fencing in one hour. At this rate, how many inches does she paint per minute? (Example 2)

Complete each conversion. Round to the nearest hundredth, if necessary. (Example 3)

3. 8 in. ≈ ■ cm
4. 5 L ≈ ■ gal
5. 15 oz ≈ ■ g
6. 24 cm ≈ ■ in.
7. 9 pt ≈ ■ L
8. 3 m ≈ ■ ft

9. An elephant can eat up to 440 pounds of vegetation every day. How many grams per minute is this? Round to the nearest hundredth. (Example 4)

Independent Practice

Go online for Step-by-Step Solutions

eHelp

10. A candy company can produce 4800 sour lemon candies per minute. How many candies can they produce each hour? (Example 1)

11. In a recent year, 51.9 billion aluminum cans were recycled. About how many cans per week is this? (Example 1)

12. The average American student spends almost 1500 hours per year watching television. To the nearest hundredth, how many minutes per day is this? (Example 2)

13. A thrill ride at an amusement park travels 55 miles per hour. To the nearest hundredth, how many feet per second is this? (Example 2)

Complete each conversion. Round to the nearest hundredth, if necessary. (Example 3)

14. 4 L ≈ ■ qt
15. 16 in. ≈ ■ cm
16. 13 m ≈ ■ ft
17. 8 yd ≈ ■ m
18. 18 lb ≈ ■ kg
19. 7 L ≈ ■ gal
20. 1500 g ≈ ■ oz
21. 15 ft ≈ ■ m
22. 28 fl oz ≈ ■ mL

23. STEM The velocity of sound through wood at 0° Celsius is 1454 meters per second. How many miles is this per hour? Round to the nearest hundredth. (Example 4)

24. A certain car in Canada can travel 15 kilometers per 1 liter of gasoline. How many miles per gallon is this? Round to the nearest hundredth. (Example 4)

Complete each conversion. Round to the nearest hundredth, if necessary.

25. 8 in. ≈ ■ mm
26. 16 L ≈ ■ c
27. 2 km ≈ ■ yd
28. 250 fl oz ≈ ■ L
29. 2750 g ≈ ■ lb
30. 5 gal ≈ ■ mL

31. Crystal's times for each portion of a triathlon are shown in the table. Round to the nearest hundredth.

	Swim	Bike	Run
Distance (km)	1.5	40	10
Time (min)	40	86	64

a. How many meters per second did she run?

b. What was her speed in miles per hour for the aquabike portion (swimming and biking)?

Order each group of rates from least to greatest.

32. 100 oz/min, 2500 g/min, 10 lb/min

33. 500 m/h, 7 yd/min, 6 in./s

34. 32 mi/gal, 15 m/mL, 6600 yd/qt

35. 500 kg/h, 5 oz/s, 18 lb/min

36. STEM The sprinkler system in the Willis Tower pumps up to 1500 gallons of water per minute. How many liters of water can the system pump in $\frac{1}{4}$ minute? Round to the nearest hundredth.

37 The average American consumes 20 gallons of ice cream in one year. At this rate, how many liters of ice cream will 50 Americans consume in one week? Round to the nearest hundredth.

Replace each ● with <, >, or = to make a true sentence.

38. 10 m ● 390 in.

39. 520 oz ● 15 kg

40. 14 pt ● 6622 mL

Financial Literacy Use dimensional analysis and data in the table to make each conversion. Round to the nearest hundredth, if necessary.

41. 150 dollars to euros

42. 275 dollars to pounds

43. 570 yuan to dollars

44. 500 pesos to dollars

Exchange Rates Per 1 U.S. Dollar		
Country	**Currency**	**Rate**
European Union	euro	0.790
Mexico	peso	13.444
China	yuan	6.375
United Kingdom	pound	0.637

45. CCSS **Model with Mathematics** Write and solve a real-world problem in which dimensional analysis is used to convert square feet to square yards. *Hint:* 1 square yard = 9 square feet

H.O.T. Problems Higher Order Thinking

46. CCSS **Be Precise** Give two examples of different measurements that are equivalent to 10 centimeters per second.

47. CCSS **Which One Doesn't Belong?** Select the rate that does not have the same value as the other three. Explain your reasoning.

60 mi/h	88 ft/s	500 ft/min	1440 mi/day

48. CCSS **Persevere with Problems** A recipe for fruit punch uses the ingredients shown in the table. About how many cups of each ingredient are needed? Round to the nearest tenth.

Fruit Punch	
900 mL	cranberry juice
700 mL	apple juice
300 mL	pineapple juice
150 mL	lemon juice
900 mL	club soda

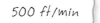

49. CCSS **Identify Structure** What property of multiplication allows you to multiply a rate by a conversion factor without changing its value? Explain.

50. **Building on the Essential Question** Explain how you would convert 10 miles per hour to meters per second.

Standardized Test Practice

51. A speed of 55 miles per hour is the same rate as which of the following?

A 34 kilometers per hour
B 50 kilometers per hour
C 88 kilometers per hour
D 98 kilometers per hour

52. A piece of notebook paper measures $8\frac{1}{2}$ inches by 11 inches. Which of the following metric approximations is the same?

F 2 m by 2.8 m
G 3 cm by 4 cm
H 22 cm by 28 cm
J 30 m by 40 m

53. A car's mileage is registered at 29,345.5 miles. The driver sees a sign that warns of road work in 1000 feet. What will be the car's mileage when the road work begins?

A 29,345.7
B 29,345.9
C 29,356.2
D 29,356.5

54. **Short Response** Convert 565 miles per hour to feet per second. Show the procedure you used.

Common Core Review

Express each rate as a unit rate. Round to the nearest tenth or to the nearest cent, if necessary. 7.RP.1

55. \$183 for 4 concert tickets

56. 100 feet in 14.5 seconds

57. 254.1 miles on 10.5 gallons

58. 9 inches of snow in 12 hours

59. **Financial Literacy** Mrs. Gallagher wants to buy the package of soda that is less expensive per can. Which pack of sodas shown should she buy? Explain your reasoning. 7.RP.1

Express each ratio as a fraction in simplest form. 7.RP.1

60. 12 cars out of 30 vehicles

61. 5 cups to 5 quarts

62. 15 soccer balls out of 35 balls

63. 8 pencils to 20 crayons

Simplify each expression. 6.EE.3

64. $(x - 3) + 2$

65. $(8 \cdot y) \cdot (-4)$

66. $25 + (d - 8)$

67. $9(5m)$

68. $(x + 1) - 9$

69. $5(3 \cdot r)$

70. Clive is making hamburgers for a cookout. How many $\frac{1}{4}$-pound hamburgers can he make from $2\frac{3}{4}$ pounds of ground beef? 7.NS.3

Find each product or quotient. 7.NS.2

71. $-12 \cdot (-10)$

72. $-18 \div 3$

73. $9 \cdot (-14)$

74. $54 \div (-6)$

75. $-14 \cdot 2$

76. $-72 \div (-4)$

Write each fraction as a decimal. 7.NS.2d

77. $\frac{4}{5}$

78. $\frac{3}{4}$

79. $\frac{3}{8}$

80. $\frac{9}{25}$

Lesson 5-5

Proportional and Nonproportional Relationships

ISG Interactive Study Guide

See pages 105–106 for:
- Getting Started
- Real-World Link
- Notes

Essential Question

How can you identify and represent proportional relationships?

Common Core State Standards

Content Standards
7.RP.2, 7.RP.2a, 7.RP.2b, 7.RP.2c

Mathematical Practices
1, 3, 4

Vocabulary

proportional

constant of proportionality

nonproportional

What You'll Learn

- Identify proportional and nonproportional relationships in tables.
- Describe a proportional relationship using an equation.

Real-World Link

Dances A middle school student advisory council is planning a spring dance with a Glow-in-the-Dark theme. The "tickets" will be glow-in-the-dark wristbands. Students on the council found the cost of the wristbands from two companies. Proportions can be used to compare the costs.

Identify Proportional Relationships

Two quantities are **proportional** if they have a constant ratio or rate. The constant ratio is called the **constant of proportionality**.

$$\frac{\text{Total Cost (\$)}}{\text{Number of Wristbands}} = \frac{6}{4} = \frac{9}{6} = \frac{12}{8} = \frac{3}{2} \qquad \frac{\text{Total Cost (\$)}}{\text{Number of Necklaces}} = \frac{6}{4} \neq \frac{10}{5} \neq \frac{15}{6}$$

The total cost is proportional to the number of wristbands purchased. The constant of proportionality is $\frac{3}{2}$. However, the rates for the necklaces are not constant. For relationships in which the ratios or rates are *not* constant, the two quantities are said to be **nonproportional**.

Example 1

Tutor

Determine whether the cost of coffee is proportional to the number of pounds. If the relationship is proportional, identify the constant of proportionality. Explain your reasoning.

Coffee (pounds)	Cost (dollars)
1	3
2	6
3	9
4	12

Write the rate of cost to pounds of coffee for each column in the table. Simplify each fraction.

$\frac{3}{1}$ $\quad \frac{6}{2} = \frac{3}{1}$ $\quad \frac{9}{3} = \frac{3}{1}$ $\quad \frac{12}{4} = \frac{3}{1}$ All the rates are equal.

The rates are equal, so the cost is proportional to the number of pounds of coffee. The constant of proportionality is 3.

Got It? Do this problem to find out.

1. Determine whether the number of legs are proportional to the number of spiders. If the relationship is proportional, identify the constant of proportionality. Explain your reasoning.

Number of Spiders	1	2	3	4
Number of Legs	8	16	24	32

Example 2

Tutor

Determine whether the distance is proportional to the time traveled. If the relationship is proportional, identify the constant of proportionality. Explain your reasoning.

Time (hours)	Distance (miles)
1	50
2	70
3	90

Write the rate of distance to time for each hour in simplest form.

$\frac{50}{1}$ $\frac{70}{2} = \frac{35}{1}$ $\frac{90}{3} = \frac{30}{1}$ The rates are not equal.

The distance is *not* proportional to the time.

Got It? Do this problem to find out.

2. Determine whether the number of ice cubes is proportional to the number of drinks. If the relationship is proportional, identify the constant of proportionality. Explain your reasoning.

Drinks	1	2	3
Ice Cubes	6	14	22

Use Proportional Relationships

Constant of Proportionality
The constant of proportionality is also called the unit rate.

Proportional relationships can also be described using equations of the form $y = kx$, where k is the constant ratio or the constant of proportionality.

Real World Example 3

Tutor

A circle's circumference is proportional to its diameter. Use the figure to find the constant of proportionality. Then write an equation relating the circumference of the circle to its diameter. What is the circumference of a circle with a 6-inch diameter?

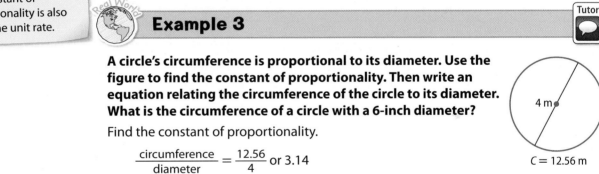

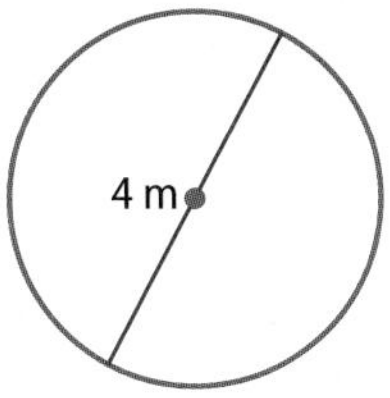

$C = 12.56$ m

Find the constant of proportionality.

$\frac{\text{circumference}}{\text{diameter}} = \frac{12.56}{4}$ or 3.14

Words	The circumference is about 3.14 times the diameter.
Variable	Let C = circumference and d = diameter.
Equation	$C = 3.14d$

$C = 3.14\mathbf{d}$ Write the equation.
$= 3.14(\mathbf{6})$ Replace d with 6.
$= 18.84$ Multiply.

The circumference is about 18.84 inches.

Got It? Do this problem to find out.

3. The cost for $2\frac{1}{2}$ pounds of meat is \$7.20. Find the constant of proportionality. Then write an equation relating cost to pounds. How much will 4 pounds cost?

Guided Practice

Check

Determine whether the set of numbers in each table is proportional. If the relationship is proportional, identify the constant of proportionality. Explain your reasoning. (Examples 1 and 2)

1.

Blue Paint (quarts)	1	2	3	4
Yellow Paint (quarts)	5	6	7	8

2.

Ice Tea Mix (cups)	1	2	3	4
Sugar (cups)	2	4	6	8

3. The cost of 13 gallons of gasoline is \$41.47. Find the constant of proportionality. Then write an equation relating cost to the number of gallons of gasoline. How much does 18.5 gallons of gasoline cost? (Example 3)

Independent Practice

eHelp Go online for Step-by-Step Solutions

Determine whether the set of numbers in each table is proportional. If the relationship is proportional, identify the constant of proportionality. Explain your reasoning. (Examples 1 and 2)

4.

Cans of Concentrate	1	2	3	4
Cans of Water	4	8	12	16

5.

Shaded Squares	1	2	3	4
Total Squares	8	15	30	42

6.

Junk E-mails	10	20	30	40
Total E-mails	15	30	45	60

7.

Weeks	5	6	7	8
Days	35	42	49	56

8. **Financial Literacy** A store is having a sale where all jeans are $\frac{1}{4}$ off the regular price. Find the constant of proportionality, then write an equation relating the sale price to the regular price. How much would a pair of \$29 jeans cost on sale? (Example 3)

9. Luke earned \$54 after mowing 3 lawns. Find the constant of proportionality, and then write an equation comparing earnings to lawns mowed. How much would Luke earn after mowing 7 lawns? (Example 3)

Copy and complete each table. Determine whether the relationship is proportional. If so, identify the constant of proportionality.

10. Ms. Rollins had an end-of-year pizza party for the chess team. At the party, every 2 students had 5 slices of pizza.

Number of Students	2	4	6	8	10
Slices of Pizza	■	■	■	■	■

11. Admission to an amusement park is \$4 plus \$1.50 per ride.

Number of Rides	1	2	3	4	5
Cost (\$)	■	■	■	■	■

12. It will cost \$7 per person to hold a birthday party at the recreation center.

Number of Guests	6	7	8	9	10
Cost (\$)	■	■	■	■	■

13 Eight hot dogs and ten hot dog buns come in separate packages.

a. Is the number of packages of hot dogs proportional to the number of hot dogs? If the relationship is proportional, identify the constant of proportionality. Explain your reasoning.

b. Suppose you buy the same number of packages of hot dogs and hot dog buns. Is the number of hot dogs proportional to the number of hot dog buns? If the relationship is proportional, identify the constant of proportionality. Explain your reasoning.

14. **CCSS Multiple Representations** Suppose Isabel decides to save \$20 each week for her family vacation. Her sister already has \$10 and wants to save an additional \$20 each week for the vacation. These situations are modeled in the graphs below.

Graph A

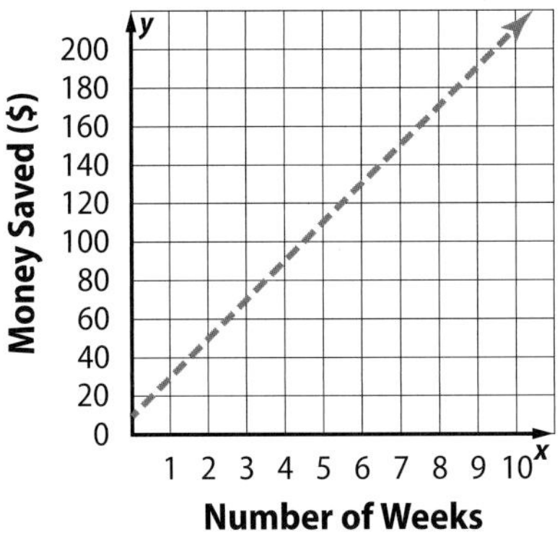

Graph B

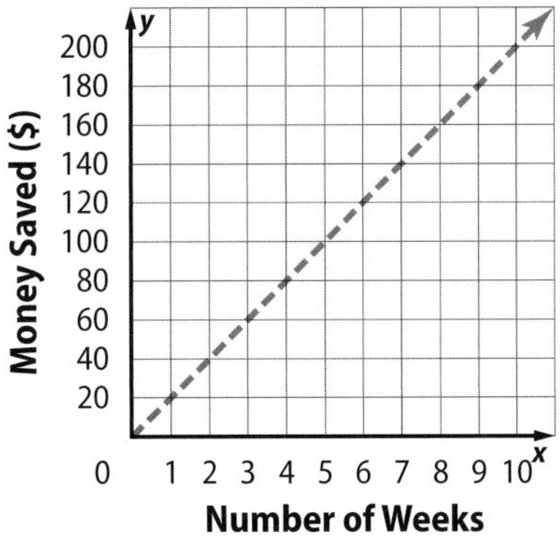

a. **Table** Make a table showing the first six weeks of savings for each girl. Which situation is proportional? Explain your reasoning.

b. **Symbols** Write an equation to represent each situation.

c. **Graph** Compare and contrast the graphs.

H.O.T. Problems Higher Order Thinking

15. **CCSS Model with Mathematics** Give examples of two similar situations in which one is a proportional relationship and the second one is nonproportional. Identify the constant of proportionality. Then write equations that describe them.

16. **CCSS Justify Conclusions** A recipe for paper maché paste includes $\frac{1}{4}$ cup of flour for every cup of water. If there are 6 cups of flour, how many gallons of water are needed? Identify the constant of proportionality. Explain your reasoning.

17. **CCSS Persevere with Problems** Many objects, such as credit cards, are shaped like golden rectangles. A *golden rectangle* is a rectangle in which the ratio of the length to the width is approximately 1.618 to 1. This ratio is called the *golden ratio*.

a. Find three different objects that are close to a golden rectangle. Make a table to display the dimensions and the ratio found in each object.

b. Describe how each ratio compares to the golden ratio.

c. Use the Internet or another source to find three examples of where the golden rectangle is used in architecture.

18. **Building on the Essential Question** This year Monica is 12 years old, and her little sister Patrice is 6 years old. Is Monica's age proportional to Patrice's age? If the relationship is proportional, identify the constant of proportionality. Explain your reasoning using a table of values.

Standardized Test Practice

19. A bicycle wheel makes 30 revolutions in 45 feet. Which of these represents an equivalent rate of bicycle wheel revolutions?

A 10 revolutions in 15 feet
B 60 revolutions in 100 feet
C 15 revolutions in 10 feet
D 100 revolutions in 60 feet

20. The cost of renting a boat for 4 hours is $51. If the cost of renting a boat is proportional to the number of rental hours, which of the following is *not* an equivalent rate?

F 6 hours for $76.50
G 3 hours for $38.25
H 7 hours for $89.25
J 5 hours for $63.00

21. The amount of sales tax paid on a purchase is proportional to the price of the item. Suppose the sales tax rate is 6.25%. If p is the price of the item and t is the sales tax, which equation represents this?

A $p = 0.0625t$
B $p = 6.25t$
C $t = 6.25p$
D $t = 0.0625p$

22. **Short Response** The prices of different-sized smoothies at an ice cream shop are shown below. Is the cost proportional to the size of the cone? If so, identify the constant of proportionality. Explain your reasoning.

Size (oz)	16	20	24
Cost ($)	3.25	3.75	4.25

Common Core Review

Complete each conversion. Round to the nearest hundredth, if necessary. 7.RP.1

23. 4 in. ≈ ■ cm

24. 5 L ≈ ■ gal

25. 1500 lb ≈ ■ kg

26. 14 yd ≈ ■ m

Express each rate as a unit rate. Round to the nearest tenth, if necessary. 7.RP.1

27. 140 miles on 6 gallons

28. 19 yards in 2.5 minutes

29. 236.7 miles in 4.5 days

30. 331.5 pages in 8.5 weeks

31. $5\frac{1}{2}$ feet in 3 minutes

32. 352 beats in $4\frac{2}{5}$ minutes

33. The table shows the record high and low temperatures for selected states. What is the difference between the highest and lowest temperatures for Kentucky? Massachusetts? 7.NS.1c

State	Lowest Temperature (°F)	Highest Temperature (°F)
Kentucky	−37	114
Massachusetts	−35	107

For Exercises 34–35, write an expression to represent each real-world situation. Then evaluate the expression and interpret the meaning of the solution. 7.NS.3

34. Nikos paid $22,500 for a new car. Over the next 3 years the value of his dropped $3750. How much did the value of the car drop each year on average?

35. The temperature dropped a total of 27°F over a 9-hour period. What was the mean hourly temperature drop?

ISG Interactive Study Guide

See page 107 for:
• Mid-Chapter Check

21ST CENTURY CAREER in Engineering

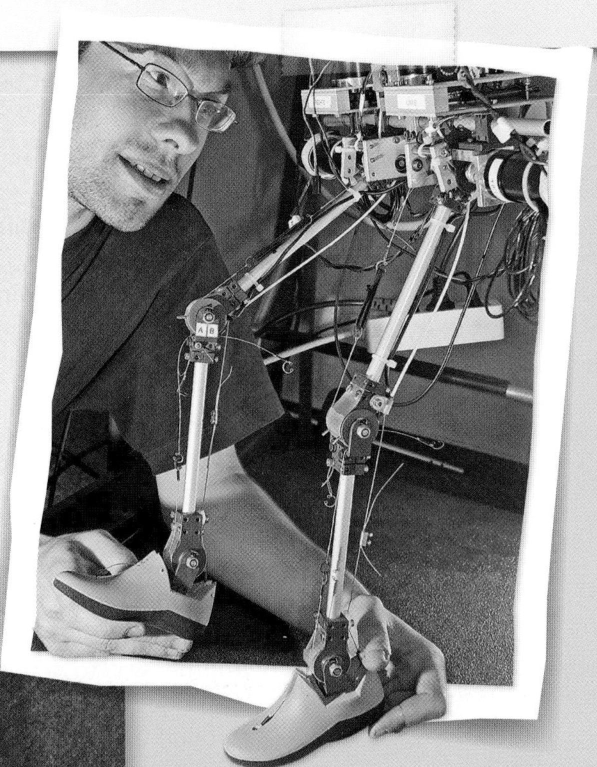

Biomechanical Engineering

Did you know that more than 700 pounds of force are exerted on a 140-pound long-jumper during the landing? Biomechanical engineers understand how forces travel through the shoe to an athlete's foot and how the shoes can help reduce the impact of those forces on the legs. If you are curious about how engineering can be applied to the human body, a career in biomechanical engineering might be a great fit for you.

College & Career READINESS

Explore college and careers at ccr.mcgraw-hill.com

Is This the Career for You?

Are you interested in a career as a biomechanical engineer? Take some of the following courses in high school.

- Biology
- Calculus
- Physics
- Trigonometry

Find out how math relates to a career in Biomechanical Engineering.

Interactive Study Guide
See page 108 for:
- Problem Solving
- Career Project

Lesson 5-6

Graphing Proportional Relationships

ISG Interactive Study Guide

See pages 109–110 for:
- Getting Started
- Real-World Link
- Notes

Essential Question

How can you identify and represent proportional relationships?

Common Core State Standards

Content Standards
7.RP.2, 7.RP.2a, 7.RP.2b, 7.RP.2d, 8.EE.5

Mathematical Practices
1, 3, 4

What You'll Learn
- Identify proportional relationships.
- Analyze proportional relationships.

Real-World Link

Parties Some birthday traditions, like having parties and giving gifts, are similar throughout the world. But many cultures celebrate significant birthdays in other ways, such as by flying flags or having a dance. The age at which a child moves into adulthood, or *coming of age*, varies by culture. The celebrated age can be 12, 13, 15, 16, or even 18.

Identify Proportional Relationships

Another way to determine whether two quantities are proportional is to graph the quantities on the coordinate plane. If the graph of the two quantities is a straight line through the origin, then the two quantities are proportional.

The cost of renting Center A and Center B for a party is shown in the graph below.

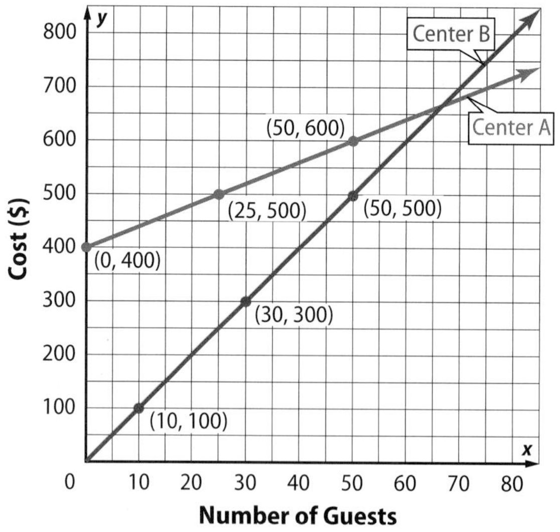

For Center A, the rate is not constant, so the relationship between the cost and the number of guests is nonproportional. Notice that the graph for Center A is a *straight line* that *does not* pass through the origin.

For Center B, the rate is constant, so the relationship between the cost and the number of guests for Center B is proportional. Notice that the graph for Center B is a *straight line* that *does* pass through the origin.

Example 1

Tutor

Determine whether each relationship is proportional by graphing on the coordinate plane. Explain your reasoning.

a. The black mamba is the fastest snake in the world. The table shows the distance the snake travels for several different times. Is the distance the snake travels proportional to the time?

> **Watch Out!**
> Since $\frac{0}{0}$ is undefined, it is not included in the list of rates.

Time (s)	0	1	2	3	4
Distance (m)	0	5	10	15	20

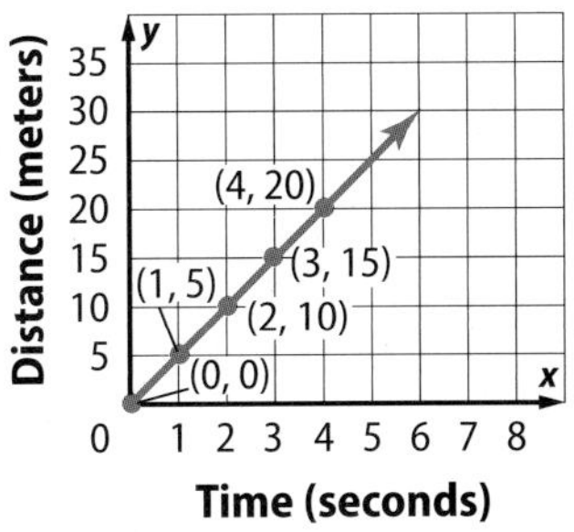

Graph the ordered pairs on the coordinate plane. Then connect the ordered pairs.

The line passes through the origin and is a straight line. So, the distance traveled in meters is proportional to the time in seconds.

Check The ratios are constant. $\frac{5}{1}, \frac{10}{2} = \frac{5}{1}, \frac{15}{3} = \frac{5}{1}, \frac{20}{4} = \frac{5}{1}$

The relationship is proportional. ✓

b. A candle is 20 centimeters tall. It burns at a rate of 2 centimeters per hour. Is the height of the candle proportional to the number of hours it burns?

Make a table to find the height of the candle after 0, 1, 2, 3, and 4 minutes.

Time (h)	0	1	2	3	4
Height (cm)	20	18	16	14	12

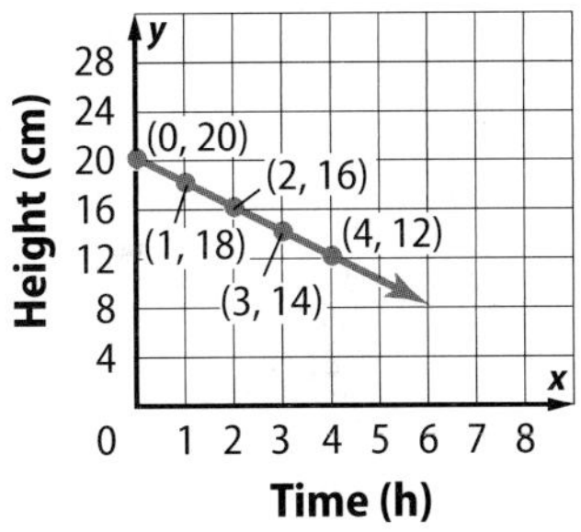

Graph the ordered pairs on the coordinate plane. Then connect the ordered pairs.

The graph is a straight line but does not pass through the origin. So, the height of the candle is nonproportional to the number of hours it burns.

Check The ratios are not constant. $\frac{18}{1}, \frac{16}{2} = \frac{8}{1}, \frac{14}{3} = 4\frac{2}{3}, \frac{12}{4} = \frac{3}{1}$

The relationship is nonproportional. ✓

Got It? Do these problems to find out.

Determine whether the cost is proportional to the number of items in each relationship by graphing on the coordinate plane. Explain.

1a.

Number of Tickets	2	4	6	8	10
Cost ($)	6	8	10	12	14

1b.

Number of Hotdogs	2	4	6	8	10
Cost ($)	6	12	18	24	30

Analyze Proportional Relationships

When two quantities are proportional, you can use a graph of the quantities to find the constant of proportionality and to analyze points on the graph. The graph of every proportional relationship passes through the point (0, 0). The point (1, *r*) tells you the constant of proportionality, or the unit rate *r*.

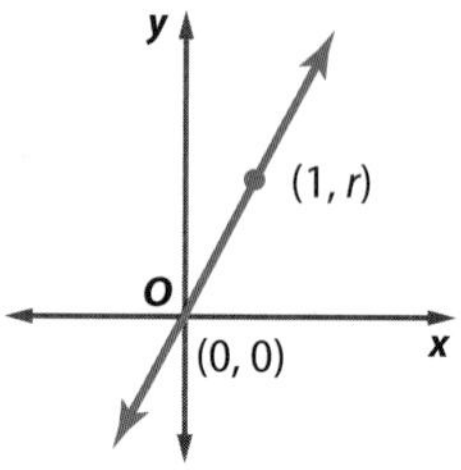

Real World

Example 2

Tutor

The length of the stretch (in millimeters) of a spring is proportional to the weight (in grams) attached to the end of the spring as shown in the graph.

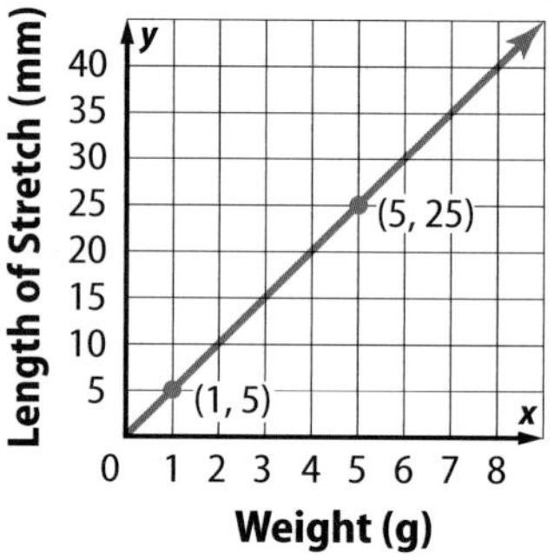

a. Find and interpret the constant of proportionality.

Use the point (5, 25) on the graph.

$$\frac{\text{length of stretch (mm)}}{\text{weight (g)}} = \frac{25}{5}$$
$$= \frac{5}{1} \text{ or } 5$$

The constant of proportionality, or unit rate, is 5 millimeters of stretch per gram of weight attached.

b. Explain what the points (0, 0), (1, 5), and (5, 25) represent.

The point (0, 0) represents the length of the stretch of the spring, 0 millimeters, when no weight is attached.

The point (1, 5) represents the length of the stretch of the spring, 5 millimeters, when a one-gram weight is attached.

The point (5, 25) represents the length of the stretch of the spring, 25 millimeters, when a five-gram weight is attached.

Got It? Do this problem to find out.

2. Keith plants a seed. Every three days after the seed sprouts he measures the height of the plant. The graph shows his results.

   a. Find and interpret the constant of proportionality.

   b. Explain what the points (0, 0), (1, 2), and (6, 12) represent.

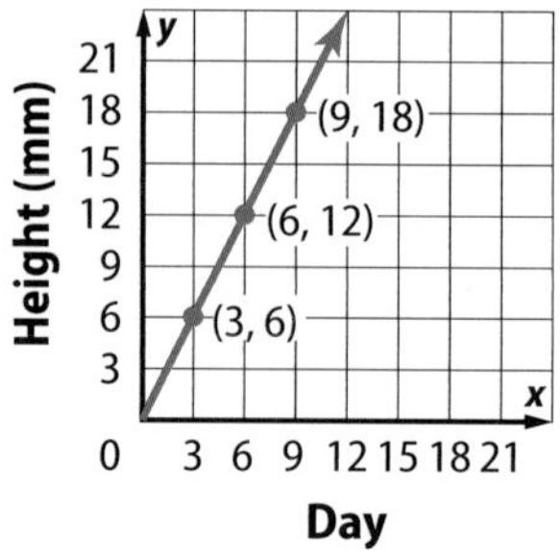

Guided Practice

Check

Determine whether each relationship is proportional by graphing on the coordinate plane. Explain your reasoning. (Example 1)

1.

Time (min)	4	6	8	10
Distance (ft)	5	10	15	20

2.

Number of Gallons	1	2	3	4
Number of Quarts	4	8	12	16

3. The formula for the area A of all parallelograms with a base of 2 centimeters is $A = 2h$, where h is the height in centimeters. Determine whether the area of all parallelograms with a base of 2 centimeters is proportional to the height in centimeters. Explain your reasoning. (Example1)

4. The number of students on a school trip is proportional to the number of teachers as shown in the graph. (Example 2)

   a. Find and interpret the constant of proportionality.

   b. Explain what the points (0, 0), (1, 25) and (4, 100) represent.

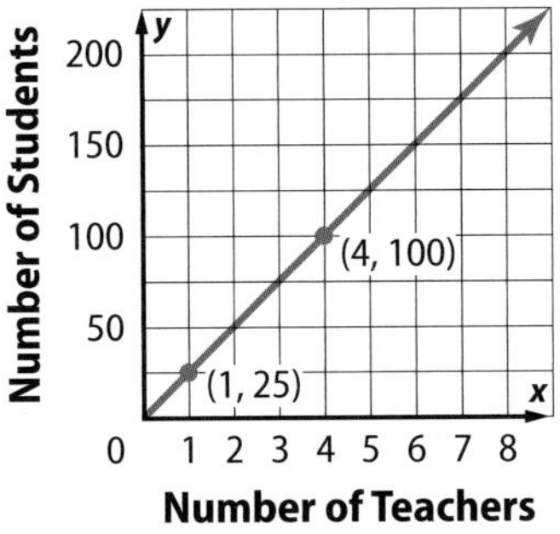

Independent Practice

Go online for Step-by-Step Solutions

eHelp

Determine whether each relationship is proportional by graphing on the coordinate plane. Explain your reasoning. (Example 1)

5.

Soap (mL)	3	6	9	12
Water (L)	2	4	6	8

6.

Number of Pizzas	2	4	6	8
Total Cost ($)	22	42	62	82

7. The cost of 3-D movie tickets is $12.50 per ticket. Determine whether the cost is proportional to the number of tickets by graphing the relationship on the coordinate plane. Explain your reasoning. (Example 1)

8. The number of pounds of walnuts in a nut mix is proportional to the number of pounds of peanuts as shown in the graph. (Example 2)

   a. Find and interpret the constant of proportionality.

   b. Explain what the points (0, 0), (1, 0.4) and (7.5, 3) represent.

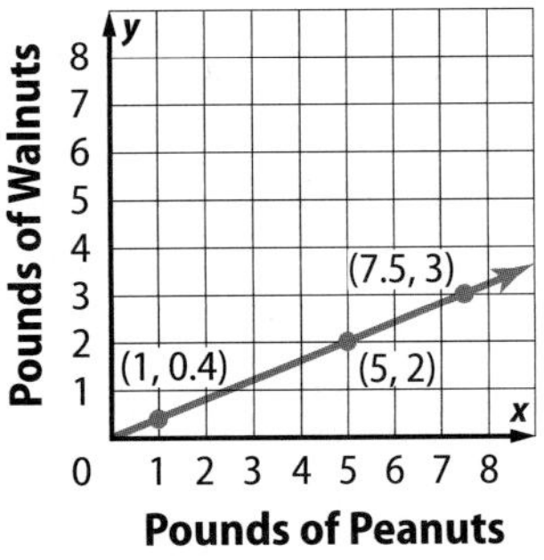

9. The perimeter of a square is proportional to the length of one side as shown in the graph. (Example 2)

   a. Find and interpret the constant of proportionality.

   b. Explain what the points (0, 0), (1, 4), and (3, 12) represent.

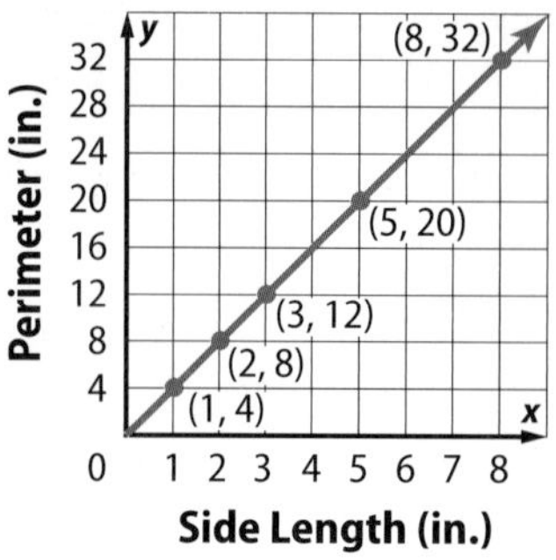

10. **Financial Literacy** The cost C to place an ad in a newspaper can be found using the formula, $C = \$5 + \$2L$, where L is the number of lines of text. Find the value of C when $L = 0$. Then explain why this shows that the cost of an ad is not proportional to the number of lines of text.

11. CCSS **Make a Prediction** On sale, boxes of cereal cost \$11 for 4 boxes, \$24.75 for 9 boxes, and \$33 for 12 boxes. Predict the cost of 10 boxes of cereal. Use a graph to explain your reasoning.

12. Sonya walks on a treadmill at a constant rate of 3.5 miles per hour. A graph showing the relationship between the time she walks and the distance she travels passes through the origin and through the point (1, 3.5). Name three other points that lie on the graph.

Complete each table so that the relationship between the two quantities is proportional. Check your work by graphing on the coordinate plane.

13.

Length of Ribbon (yd)	5	7.5	■	25
Total Cost (\$)	2	■	6	■

14.

Number of Tickets	1	3	■	■
Total Cost (\$)	6	■	24	42

15.

Boxes of Cat Food	■	6	9	15
Weight (lb)	8	16	■	■

16.

Side Length (ft)	■	4	7	9
Perimeter of a Regular Pentagon (ft)	10	20	■	■

H.O.T. Problems Higher Order Thinking

17. CCSS **Model with Mathematics** Write a real-world problem that describes a proportional relationship. Make a table of values and graph the ordered pairs on the coordinate plane.

18. CCSS **Persevere with Problems** Explain how the relationship of x and y in the graphs of $y = 3x + 1$ and $y = 3x$ differ.

19. CCSS **Justify Conclusions** The graph of $V = s^3$ passes through (0, 0). Is there a proportional relationship between V and s? Explain your reasoning.

20. **Building on the Essential Question** Explain two ways that you could determine if the following statement represents a proportional relationship.

A muffin recipe calls for 2 cups of flour per batch.

Standardized Test Practice

21. Refer to the graph below.

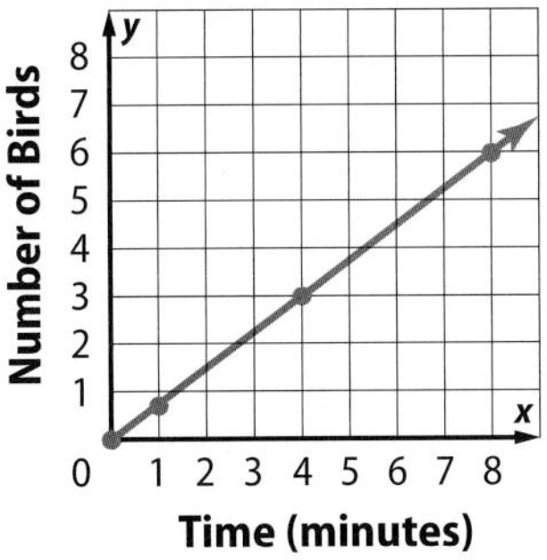

Which of the following ordered pairs represents the unit rate?

A (0, 0)
B (1, 0.75)
C (4, 3)
D (8, 6)

22. **Short Response** The number of books Jennifer reads is proportional to the number of weeks. Every 20 weeks she reads 6 books. Find the constant of proportionality.

23. Which graph of which line represents a proportional relationship?

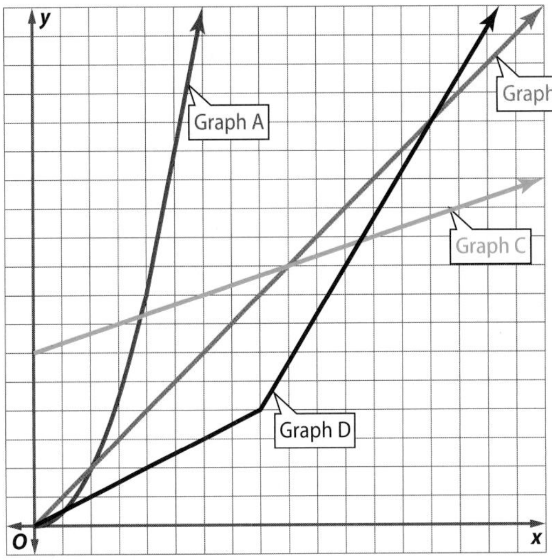

F Graph A
G Graph B
H Graph C
J Graph D

Common Core Review

Write each ratio as a fraction in simplest form. 6.RP.1

24. A bag of marbles has 20 red marbles and 45 blue marbles. What is the ratio of red marbles to blue marbles?

25. A drawer has 12 pair of white socks and 9 pair of black socks. What is the ratio of white socks to black socks?

26. A garden has 8 tomato plants and 4 pepper plants. What is the ratio of pepper plants to total plants in the garden?

Evaluate each expression. 7.NS.1d

27. $2 + (-2)$

28. $(-y) + y$

29. $-13 + 13$

Evaluate each expression. 7.NS.3

30. $6 \div 2 \cdot 3$

31. $22 - 4 \cdot 7$

32. $15 \cdot 1 + 6$

33. $(100 \div 5) - (5 \cdot 2)$

Evaluate each expression. Express the result in scientific notation. 8.EE.4

34. $(2.62 \times 10^5)(6.15 \times 10^9)$

35. $(7.12 \times 10^{-8})(2.58 \times 10^{-3})$

Order each set of numbers from least to greatest. 8.EE.1

36. $3.5 \times 10^5, 4 \times 10^5, 3.6 \times 10^{-3}, 0.004$

37. $7.1 \times 10^3, 7.15 \times 10^5, 7.01 \times 10^4, 7.5 \times 10^{-2}$

Lesson 5-7
Solving Proportions

ISG Interactive Study Guide

See pages 111–112 for:
- Getting Started
- Real-World Link
- Notes

Essential Question

How can you identify and represent proportional relationships?

CCSS Common Core State Standards

Content Standards
7.RP.2, 7.RP.2b, 7.RP.2c, 7.RP.3

Mathematical Practices
1, 3, 4

Vocab Vocabulary

proportion
cross products

What You'll Learn

- Use cross products to solve proportions.
- Use the constant of proportionality to solve proportions.

Real-World Link

Recipes You can use a recipe to make anything when you are cooking—cookies, cakes, muffins, and even punch! A great party punch is made by mixing one gallon of fruit punch and 8 ounces of frozen sherbet.

Key Concept Property of Proportions

Words	The cross products of a proportion are equal.
Symbols	If $\frac{a}{b} = \frac{c}{d}$, then $ad = cb$. If $ad = cb$, then $\frac{a}{b} = \frac{c}{d}$ if $b \neq 0$ and $d \neq 0$.

A **proportion** is an equation stating that two ratios or rates are equal. You can use the Multiplication Property of Equality to illustrate an important property of proportions. The ratios $\frac{4}{2}$ and $\frac{8}{4}$ are equal. They both simplify to $\frac{2}{1}$.

$$\frac{4}{2} = \frac{2}{1} \qquad \frac{8}{4} = \frac{2}{1}$$

$$\frac{4}{2} = \frac{8}{4}$$

Multiply each side by 2 • 4.

$$\frac{4}{2} \cdot (2 \cdot 4) = \frac{8}{4} \cdot (2 \cdot 4)$$

Simplify.

$$4 \cdot 4 = 8 \cdot 2$$

$$\frac{a}{b} = \frac{c}{d}$$

Multiply each side by bd.

$$\frac{a}{b} \cdot bd = \frac{c}{d} \cdot bd$$

Simplify.

$$ad = cb$$

The products 4 • 4 and 8 • 2, and ad and cb are called **cross products** of the proportion. The cross products of any proportion are equal.

$$\frac{4}{2} = \frac{8}{4} \quad 4 \cdot 4 = 8 \cdot 2 \qquad\qquad \frac{a}{b} = \frac{c}{d} \quad ad = bc$$

Cross products will help when you use proportional reasoning to solve a problem.

Just as in solving an equation, solving a proportion means finding the value of the variable that makes a true statement. You can use cross products to solve a proportion in which one of the quantities is not known.

Example 1

Tutor

Solve each proportion.

a. $\frac{b}{15} = \frac{66}{90}$

$\frac{b}{15} = \frac{66}{90}$

$b \cdot 90 = 15 \cdot 66$ Cross products

$90b = 990$ Multiply.

$\frac{90b}{90} = \frac{990}{90}$ Divide.

$b = 11$ Simplify.

b. $\frac{3.2}{9} = \frac{n}{36}$

$\frac{3.2}{9} = \frac{n}{36}$

$3.2 \cdot 36 = 9 \cdot n$ Cross products

$115.2 = 9n$ Multiply.

$\frac{115.2}{9} = \frac{9n}{9}$ Divide.

$12.8 = n$ Simplify.

Solve Mentally

You can also solve Example 1a mentally. Think: $90 \div 15 = 6$ and $66 \div 6 = 11$.

Got It? Do these problems to find out.

1a. $\frac{x}{4} = \frac{7}{20}$

1b. $\frac{7}{14} = \frac{c}{12}$

1c. $\frac{6.8}{t} = \frac{34}{50}$

1d. $\frac{m}{8.5} = \frac{42}{51}$

Example 2

Tutor

The wait time to ride a roller coaster is 20 minutes when 160 people are in line. At this rate, how long is the wait time when 220 people are in line?

Write and solve a proportion using ratios that compare people to wait time. Let w represent the wait time for 220 people.

$\frac{20}{160} = \frac{w}{220}$ ← wait time / ← number of people

$20 \cdot 220 = 160 \cdot w$ Cross products

$4400 = 160w$ Multiply.

$\frac{4400}{160} = \frac{160w}{160}$ Divide each side by 160.

$27.5 = w$ Simplify.

Check Check the cross products. Since $20 \cdot 220 = 4400$ and $160 \cdot 27.5 = 4400$, the answer is correct. ✓

The wait time is 27.5 minutes.

Watch Out!

When you solve a real-world problem using a proportion, be sure to compare the quantities in the same order.

Got It? Do this problem to find out.

2. Alicia's class is making care packages for a local shelter. They can make 8 care packages with 240 food items. How many care packages can they make with 500 food items?

Use the Constant of Proportionality

You can also solve problems involving proportional relationships by first finding the constant of proportionality.

Example 3

Tutor

Mrs. Hidalgo paid $30 for 4 students to visit an art museum. Find the cost for 20 students.

Write and solve an equation.

Find the constant of proportionality, or unit cost, for each student.

$\frac{\text{cost in dollars}}{\text{number of students}} = \frac{30}{4}$ or 7.50 The cost is $7.50 per student.

Words	The cost is $7.50 times the number of students.
Variable	Let *c* represent the cost. Let *s* represent the number of students.
Equation	$c = 7.5s$

Use this equation to find the cost for 20 students at the same rate.

$c = 7.5s$ Write the equation.

$c = 7.5(20)$ Replace *s* with 20.

$c = 150$ Multiply.

So, the cost for 20 students to visit the art museum is $150.

Proportions
You can set up a proportion in more than one way. In Example 3, the proportion can also be set up as $\frac{4}{20} = \frac{30}{x}$.

Check Write and solve a proportion.

Let *x* represent the cost for 20 students.

$\frac{30}{4} = \frac{x}{20}$ ← cost ← number of students

$30 \cdot 20 = 4 \cdot x$ Cross products

$600 = 4x$ Multiply.

$\frac{600}{4} = \frac{4x}{4}$ Divide each side by 4.

$150 = x$ Simplify.

Got It? Do these problems to find out.

3a. Matthew paid $49.45 for 5 DVDs at a sale. How much would it cost for 11 DVDs at the same rate?

3b. Cindy paid $34.60 for 4 tickets to a movie. How much would it cost for 7 tickets?

3c. Mrs. Jameson paid $202.50 for a group of 9 students to visit an amusement park. What would the total cost be if 4 more students wanted to join the group?

Guided Practice

Check

Solve each proportion. (Example 1)

1. $\frac{18}{m} = \frac{27}{36}$

2. $\frac{t}{21} = \frac{9}{15}$

3. $\frac{8}{17} = \frac{16}{x}$

4. $\frac{12}{7.2} = \frac{4}{p}$

5. $\frac{n}{13} = \frac{5.8}{2.6}$

6. $\frac{4.4}{2} = \frac{c}{25}$

7. *Aspect ratio* is the ratio of width to height of a television screen. A widescreen television screen has an aspect ratio of 16 inches wide to 9 inches high. If a television screen is 48 inches wide, what is the height of the screen? (Example 2)

8. Joaquin has a total of 12.5 hours of football practice after school five days a week. How many hours of practice will he have for 15 school days? For 22 school days? (Example 3)

Independent Practice

Go online for Step-by-Step Solutions

eHelp

Solve each proportion. (Example 1)

9. $\frac{6}{8} = \frac{z}{48}$

10. $\frac{8}{12} = \frac{28}{m}$

11. $\frac{b}{30} = \frac{4}{5}$

12. $\frac{18}{15} = \frac{9}{s}$

13. $\frac{7}{c} = \frac{35}{60}$

14. $\frac{k}{56} = \frac{12}{7}$

15. $\frac{32}{a} = \frac{12.8}{5.6}$

16. $\frac{11.5}{6} = \frac{n}{22.8}$

17. $\frac{9.6}{3} = \frac{p}{0.3}$

18. $\frac{14}{w} = \frac{8.4}{4.5}$

19. $\frac{v}{6} = \frac{20.7}{5.4}$

20. $\frac{10.2}{4} = \frac{h}{12}$

21. The classrooms at Lincoln Middle School are painted every summer. If 7 gallons of paint are needed to paint 4 classrooms, how many gallons of paint are needed to paint 16 classrooms? (Example 2)

22. A principal is ordering pizzas for a school pizza party. He knows that 9 pizzas will feed 25 students. If there are 300 students in the school, how many pizzas will he need to order?

23. A boat traveled 150 feet in 9.7 seconds. How far would the boat travel in 1 minute? In 1 minute 30 seconds? Round to the nearest tenth. (Example 3)

24. The record for the most amount of rain in the shortest period of time in the United States was 12 inches in 42 minutes. How much rain fell in 15 minutes? In 28 minutes?

25. The table shows the amount of each ingredient in 52 ounces of punch.

    a. If you have 130 ounces of punch, how much lime juice does the punch contain?

    b. If there are 54 ounces of sparkling lemon water in the punch, how many ounces of cranberry concentrate does the punch contain?

    c. If the punch contains 44 ounces of water, how many ounces of punch do you have?

Ingredient	Amount (ounces)
lime juice	4
water	12
cranberry concentrate	12
sparkling lemon water	24

Write a proportion that could be used to solve for each variable. Then solve.

26. 6 goals in 14 games
9 goals in *g* games

27. *s* inches in 0.54 hour
4.55 inches in 1.89 hours

28. 14 gallons for *d* dollars
8 gallons for $24.72

29. 20 boxes on 4 shelves
b boxes on 20 shelves

30. A 14-inch-wide by 20-inch-long print of the Eiffel Tower is also available as a postcard 6 inches long. What is the width of the postcard?

Solve each proportion.

31. $\frac{a}{0.28} = \frac{4}{1.4}$

32. $\frac{3}{14} = \frac{15}{m-3}$

33. $\frac{16}{x+5} = \frac{4}{5}$

34. $\frac{x-18}{24} = \frac{15}{8}$

35. $\frac{9}{7} = \frac{b+11}{14}$

36. $\frac{15-d}{12} = \frac{37.5}{75}$

37 CCSS **Multiple Representations** In this problem, you will explore proportions. A craft store is offering the specials shown for different materials.

Craft Store Sale	
ribbon	$2.67 for 3 yards
fleece	$9.50 for 2 yards
satin fabric	$10.35 for 3 yards
quilting fabric	$11.96 for 4 yards

a. Symbols Write an equation relating the cost *c* to the number of yards *n* for each material.

b. Graph Graph the cost of each material per yard on a coordinate plane. Which item costs the least per yard? The most? How is this shown on the graph?

c. Numbers How much would it cost to buy 18 inches of ribbon? How much would it cost to buy 10 meters of fleece?

H.O.T. Problems Higher Order Thinking

38. CCSS **Model with Mathematics** Give two examples that are proportional to $\frac{6 \text{ hits}}{8 \text{ at bats}}$.

39. CCSS **Persevere with Problems** Solve each proportion.

a. $\frac{4}{x} = \frac{x}{9}$ **b.** $\frac{2}{x} = \frac{x}{8}$ **c.** $\frac{4}{x} = \frac{x}{25}$ **d.** $\frac{9}{x} = \frac{x}{16}$

40. CCSS **Find the Error** Morgan is solving the proportion $\frac{x}{36} = \frac{4}{9}$. Find her mistake and correct it.

$$\frac{x}{36} = \frac{4}{9}$$
$$x(4) = 36(9)$$
$$x = 81$$

41. CCSS **Justify Conclusions** Rectangle *ABCD* has a fixed area. As the length ℓ and the width *w* change, what do you know about their product? Is the length proportional to the width? Justify your reasoning.

42. **Building on the Essential Question** Describe a situation in which it may be easier to solve a proportion using the constant of proportionality rather than cross products. Explain your reasoning.

Standardized Test Practice

43. Which equation could be used to find the total cost *c* if Fernando wanted to buy 8 pencils from the school store?

School Store Sale	
pencils	3 for \$0.45
pens	2 for \$0.75
highlighters	4 for \$1.25

A $c = 0.15 \cdot 8$

B $c = 0.45 \cdot 8$

C $c = 24 \cdot 0.45$

D $c = 0.15 \cdot 24$

44. A line to purchase concert tickets is moving at a rate of 5 feet every 20 minutes. At this rate, how long will a group of friends have been in line if they moved 30 feet?

F 30 min

G 1 h

H 1 h 30 min

J 2 h

45. Luisa typed 145 words in three minutes during her keyboarding test. Which of the following could *not* be used to determine the number of minutes it would take her to type 550 words?

A $\frac{145}{3} = \frac{550}{w}$

B $\frac{3}{145} = \frac{w}{550}$

C $\frac{145}{550} = \frac{3}{w}$

D $\frac{145}{550} = \frac{w}{3}$

46. **Short Response** There are 258 eighth graders at Henderson Middle School. The graph shows how many of the eighth graders participate in each sport.

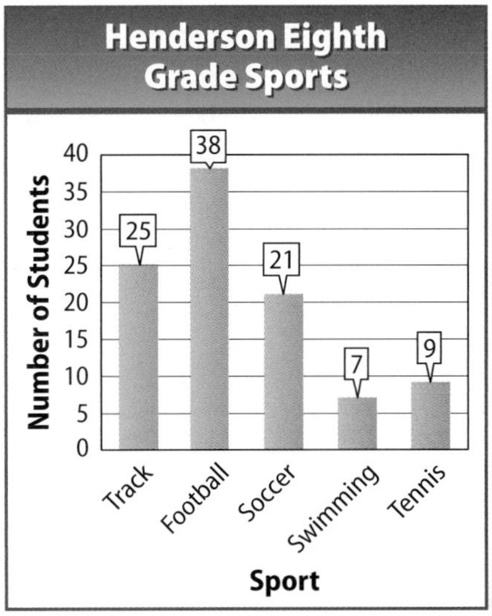

If each student participates in only one sport, write and solve a proportion you could use to predict how many students at Henderson Middle School participate in sports if the school has 645 total students.

Common Core Review

47. Determine whether the number of jars is proportional to the number of jelly beans. If the relationship is proportional, identify the constant of proportionality. **7.RP.2a, 7.RP.2b**

Jars	3	9	12	15
Jelly Beans	18	54	72	90

48. Determine whether the rental charge is proportional to the time. If the relationship is proportional, identify the constant of proportionality. Explain your reasoning. **7.RP.2a, 7.RP.2b**

Time (hours)	1	2	3	4
Rental Charge	\$13	\$23	\$33	\$43

Use dimensional analysis to complete each conversion. Round to the nearest hundredth. 6.RP.3d

49. 5 in. ≈ ■ cm (*Hint:* 1 in. ≈ 2.54 cm)

50. 10 km ≈ ■ mi (*Hint:* 1 km ≈ 0.621 mi)

51. 26.3 cm ≈ ■ in. (*Hint:* 1 cm ≈ 0.394 in.)

52. Write $5 \times 1000 + 6 \times 100 + 3 \times 1 + 8 \times \frac{1}{10} + 1 \times \frac{1}{100}$ in standard form. **5.NBT.3a**

Find each product. 7.NS.2

53. $-\frac{4}{9} \cdot \frac{2}{3}$

54. $\frac{1}{5} \cdot \frac{1}{8}$

55. $\frac{3}{4} \cdot \left(-\frac{3}{5}\right)$

56. $\frac{2}{5} \cdot \frac{5}{6}$

57. Write two inequalities involving −8 and 6. **6.NS.7a**

58. Evaluate $|-8| + |8|$. **7.NS.1b**

Lesson 5-8

Scale Drawings and Models

ISG Interactive Study Guide

See pages 113–114 for:
- Getting Started
- Vocabulary Start-Up
- Notes

Essential Question

How can you identify and represent proportional relationships?

CCSS Common Core State Standards

Content Standards
7.G.1

Mathematical Practices
1, 3, 4

Vocabulary

scale drawing
scale model
scale
scale factor

What You'll Learn

- Use scale drawings.
- Construct scale drawings.

Real-World Link

Baseball Among other attractions in Louisville, Kentucky, the town is home to the world's largest baseball bat. The bat is 120 feet long and about 9 feet in diameter. It is modeled after the 34-inch long bat that Babe Ruth used in the early 1920s.

Use Scale Drawings and Models

A **scale drawing** or a **scale model** is used to represent an object that is too large or too small to be drawn or built at actual size. The lengths and widths of objects on a scale drawing or model are proportional to the lengths and widths of the actual object.

The **scale** is determined by the ratio of a given length on the drawing or model to its corresponding length on the actual object. Consider the following scales.

1 in. = 3 ft — 1 inch represents an actual distance of 3 feet.

1 cm = 2 mm — 1 centimeter represents an actual distance of 2 millimeters.

Scales are written so that a unit length on the drawing or model is listed first.

Real World Example 1

Tutor

Suppose a model of a dragonfly has a wing length of 4 centimeters. If the length of the insect's actual wing is 6 centimeters, what is the scale of the model?

Let x represent the actual length.

Write and solve a proportion.

model length → / actual length →

$$\frac{4 \text{ cm}}{6 \text{ cm}} = \frac{1 \text{ cm}}{x \text{ cm}}$$

← model length / ← actual length

$4 \cdot x = 6 \cdot 1$ — Find the cross products.

$4x = 6$ — Simplify.

$x = 1.5$ — Divide each side by 4.

So, the scale is 1 centimeter = 1.5 centimeters.

Got It? Do these problems to find out.

1a. The pillars of the World War II memorial in Washington, D.C., are 17 feet tall. A scale model of the memorial has pillars that are 5 inches tall. What is the scale of the model?

1b. The length of a model of a bridge is 16 inches. The actual length of the bridge is 50 yards. What is the scale of the model?

If the scale drawing and model have the same unit of measure, the scale can be written without units. This is called the **scale factor**. Suppose a scale model has a scale of 1 inch = 2 feet.

scale → 1 inch = 2 feet → $\frac{1 \text{ inch}}{2 \text{ feet}}$ → $\frac{1 \text{ inch}}{24 \text{ inches}}$ → 1:24 ← scale factor

One unit on the model represents an actual distance of 24 units. So, the model's length is $\frac{1}{24}$ the size of the actual object's length.

Real World

Example 2

Tutor

The blueprint of a skateboard ramp shows that its length is 11.4 inches. If the scale on the blueprint is 1 inch = 6 feet, what is the length of the actual skateboard ramp?

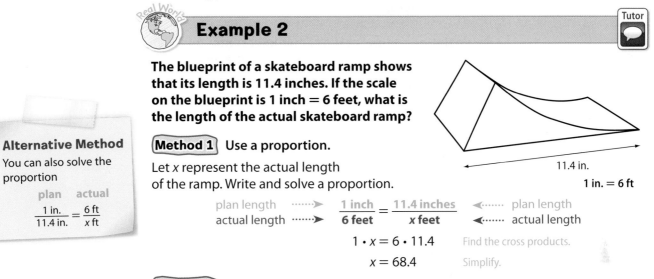

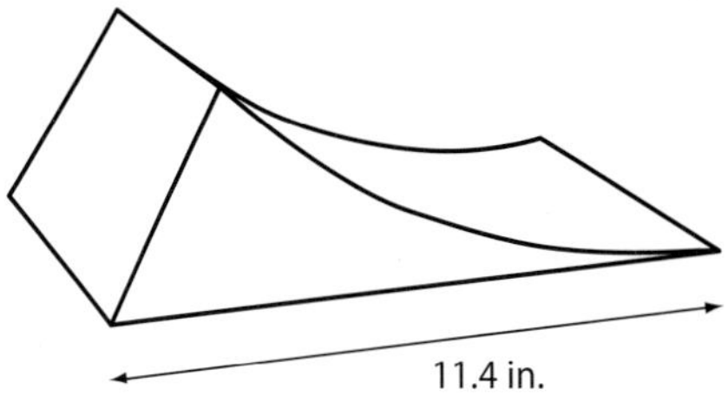

Alternative Method

You can also solve the proportion

plan actual

$\frac{1 \text{ in.}}{11.4 \text{ in.}} = \frac{6 \text{ ft}}{x \text{ ft}}$

Method 1 **Use a proportion.**

Let x represent the actual length of the ramp. Write and solve a proportion.

plan length → $\frac{1 \text{ inch}}{6 \text{ feet}} = \frac{11.4 \text{ inches}}{x \text{ feet}}$ ← plan length

actual length → ← actual length

$1 \cdot x = 6 \cdot 11.4$ Find the cross products.

$x = 68.4$ Simplify.

Method 2 **Use the scale factor.**

The actual length is proportional to the length on the scale drawing with a ratio of $\frac{1 \text{ inch}}{6 \text{ feet}}$. Find the scale factor.

$\frac{1 \text{ inch}}{6 \text{ feet}} = \frac{1 \text{ inch}}{72 \text{ inches}}$ or $\frac{1}{72}$ Convert 6 feet to inches and divide out units.

The scale factor is $\frac{1}{72}$.

So, the actual length is 72 times the blueprint length.

Words	The actual length equals 72 times the blueprint length.
Variable	Let a represent the actual length. Let b represent the blueprint length.
Equation	$a = 72b$

$a = 72\mathbf{b}$ Write the equation.

$= 72(\mathbf{11.4})$ Replace b with 11.4.

$= 820.8$ Simplify.

The actual length of the ramp is 820.8 inches or 68.4 feet.

Got It? **Do this problem to find out.**

2. A map of a natural history museum shows that the dinosaur exhibit room is 7.25 inches wide. If the scale on the map is 1 inch = 8 feet, what is the width of the actual exhibit room?

Construct Scale Drawings

To construct a scale drawing of an object, use the actual measurements of the object and the scale to which the object is to be drawn.

Vocabulary Link
Scale
Everyday Use a weighing machine
Math Use the ratio of the length in a drawing to the length of the actual object.

Real World

Example 3

Tools Tutor

Lila is painting a mural at the community center on a wall that measures 18 feet long and 12 feet tall. Make a scale drawing of the mural. Use a scale of $\frac{1}{4}$ inch = 3 feet. Use $\frac{1}{4}$-inch grid paper.

Step 1 Find the measure of the wall's length on the drawing. Let ℓ represent the length.

drawing length / actual length → $\frac{\frac{1}{4}\text{ inch}}{3\text{ feet}} = \frac{\ell\text{ inches}}{18\text{ feet}}$ ← drawing length / actual length

$\frac{1}{4} \cdot 18 = 3 \cdot \ell$ Find the cross products.

$4.5 = 3\ell$ Simplify.

$\frac{4.5}{3} = \frac{3\ell}{3}$ Divide each side by 3.

$1.5 = \ell$ Simplify.

On the drawing, the length is 1.5 or $1\frac{1}{2}$ inches.

Step 2 Find the measure of the wall's height on the drawing. Let *w* represent the width.

drawing width / actual width → $\frac{\frac{1}{4}\text{ inch}}{3\text{ feet}} = \frac{w\text{ inches}}{12\text{ feet}}$ ← drawing width / actual width

$\frac{1}{4} \cdot 12 = 3 \cdot w$ Find the cross products.

$3 = 3w$ Simplify.

$\frac{3}{3} = \frac{3w}{3}$ Divide each side by 3.

$1 = w$ Simplify.

On the drawing, the height is 1 inch.

Step 3 Make the scale drawing.
Use $\frac{1}{4}$-inch grid paper.
Since $1\frac{1}{2}$ inches = 6 squares and 1 inch = 4 squares, draw a rectangle that is 4 squares by 6 squares.

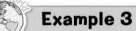

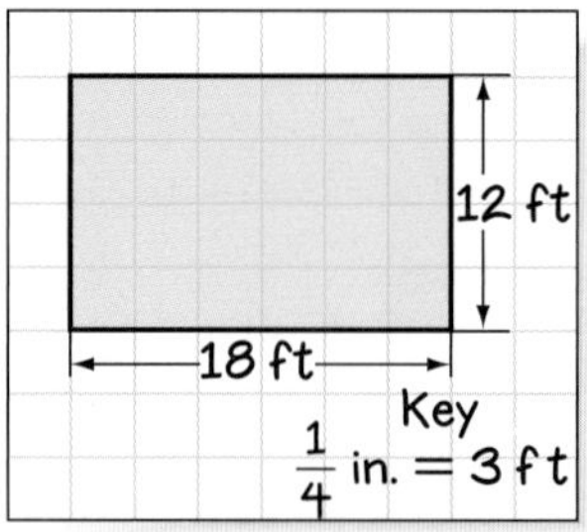

Got It? **Do this problem to find out.**

3. An architect is designing a school courtyard that is 45 feet long and 30 feet wide. Make a scale drawing of the courtyard. Use a scale of 0.5 inch = 10 feet. Use $\frac{1}{4}$-inch grid paper.

Guided Practice

Check

1. The model of a car is shown at the right. The actual car is $14\frac{1}{2}$ feet long. What is the scale of the model car? (Example 1)

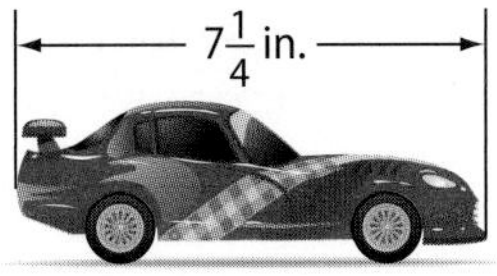

2. On the map, the scale is 1 inch = 20 miles. What is the actual distance between Kansas City and St. Louis? (Example 2)

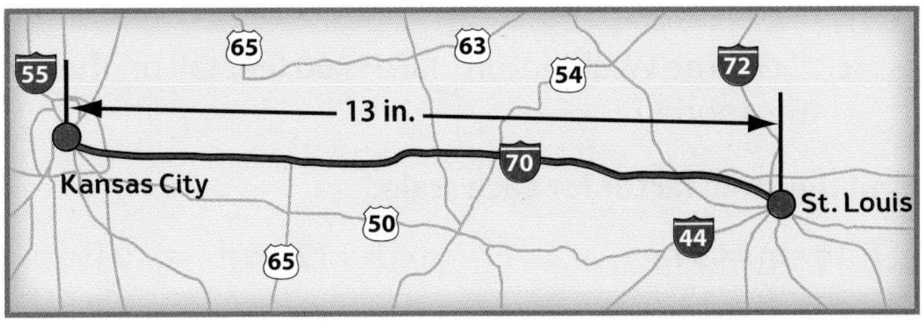

3. Marco is designing a flower garden in his backyard that is 12 feet long and 10 feet wide. Make a scale drawing of the garden. Use a scale of 0.5 inch = 2 feet. Use $\frac{1}{4}$-inch grid paper. (Example 3)

Independent Practice

Go online for Step-by-Step Solutions

eHelp

4. A model airplane is built with a wing span of 23 inches. The actual wing span is 92 feet. What is the scale of the model. (Example 1)

5. An area rug is 9 feet wide. In a photograph, the image of the rug is 3 inches wide. What is the scale? (Example 1)

6. A large American flag measures 255 feet wide. In an advertisement for renting this flag, the image of the flag is 4 inches wide. What is the scale of the flag?

7. A floor plan is shown for the first floor of a new house. If one inch represents 24 feet, what are the actual dimensions of each of the rooms listed? (Example 2)

   a. living room
   b. deck
   c. kitchen

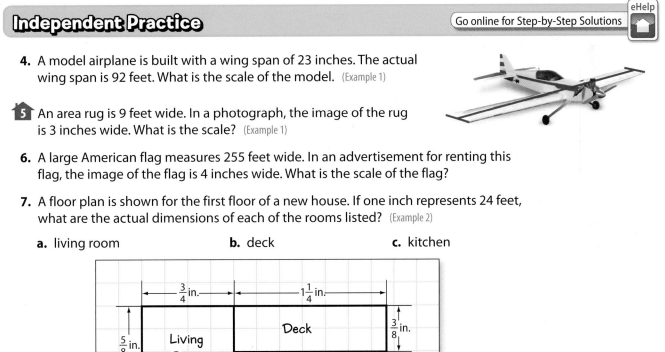

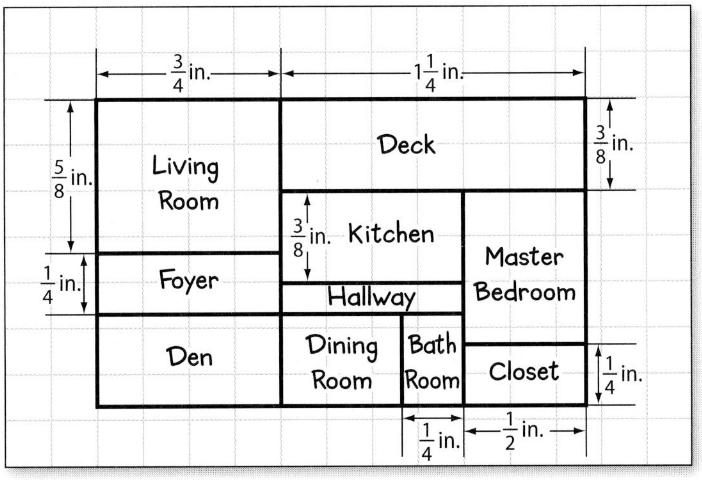

8. The actual measurements for rooms are given. Using the floor plan and scale above, find the measurements on the floor plan.

   a. master bedroom 12 feet by 15 feet
   b. den 18 feet by 9 feet
   c. dining room 12 feet by 9 feet

9. A skatepark is 24 yards wide by 48 yards long. Make a scale drawing of the skatepark that has a scale of $\frac{1}{4}$ inch = 8 yards. (Example 3)

10. The actual Mount Rushmore carving was made from a scale model with a scale of 1 inch = 1 foot. On the model, Teddy Roosevelt's moustache was 1 foot 8 inches long.

a. Find the length of Roosevelt's moustache on the monument.

b. What is the scale factor?

c. If George Washington's face is 60 feet tall on the monument, how tall was his face on the model?

Find the scale factor for each scale.

11. 10 cm = 5 m   **12.** 6 in. = 10 ft   **13.** 0.5 in. = 3 ft

14. 5 ft = 15 yd   **15.** 4 cm = 2.5 mm   **16.** 8 in. = 200 mi

17. **STEM** The 8 planets' distance from the sun is shown.

a. What scale would you use to create a scale drawing?

b. Use grid paper and your scale from part **a** to create a map of the four planets closest to the sun.

18. You build a model with a scale of 1:25. Your friend builds a model of the same object with a scale of 1:50. Which model is bigger? Explain.

Planet	Distance from the Sun (mi)
Mercury	3.6×10^7
Venus	6.7×10^7
Earth	9.3×10^7
Mars	1.42×10^8
Jupiter	4.84×10^8
Saturn	8.87×10^8
Uranus	1.8×10^9
Neptune	2.8×10^9

H.O.T. Problems Higher Order Thinking

19. CCSS **Model with Mathematics** Find a small rectangular item you use on a daily basis. Make a scale drawing of that item. Then write a problem based on your scale drawing.

20. CCSS **Persevere with Problems** Rectangle *ABCD* is reduced by a scale factor $\frac{1}{2}$. What is the area of the new rectangle?

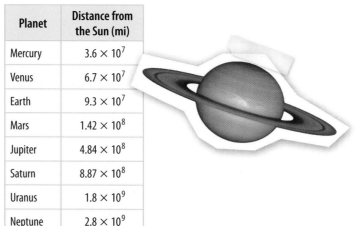

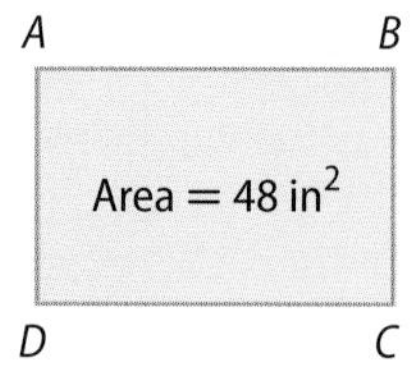

21. CCSS **Reason Inductively** Determine whether the following statement is *always*, *sometimes*, or *never* true. Justify your reasoning.
If the scale factor of a scale drawing is greater than one, the scale drawing is larger than the actual object.

22. CCSS **Which One Doesn't Belong?** Identify the scale that does not have the same scale factor. Explain your reasoning.

5 cm = 1 m	10 mm = 20 cm	10 cm = 10 m	25 mm = 0.5 m

23. CCSS **Persevere with Problems** A model of an insect has a scale of 0.25 centimeter = 1 millimeter. Is the model *smaller* or *larger* than the actual insect? Justify your reasoning by using the scale factor.

24. **Building on the Essential Question** Compare and contrast *scale* and *scale factor*.

Standardized Test Practice

25. Desiree is drawing a model of the Washington Monument which has an actual height of 555.5 feet.

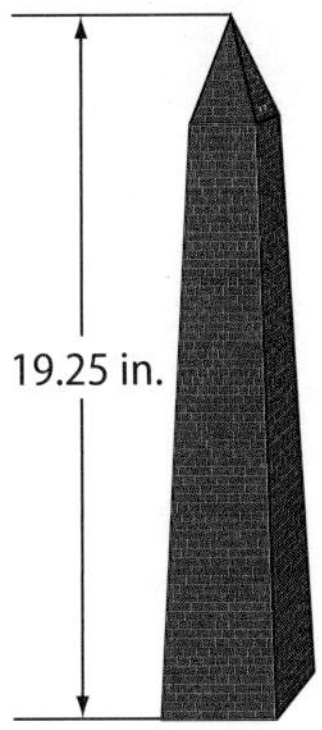

What other information is needed to find the length of the model's sides on the square base?

A the height of the top of the tower

B the length of the actual base side

C the age of the tower

D the height of the tower's first story

26. **Short Response** A blueprint has a scale of 2 inches = 2 feet. What is the scale factor of the blueprint?

27. A scale drawing of a swimming pool is shown.

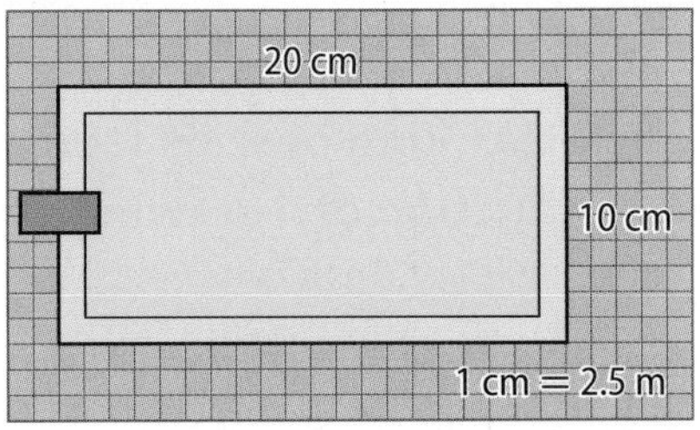

What are the actual dimensions of the swimming pool?

F 8 meters by 4 meters

G 20 meters by 10 meters

H 50 meters by 25 meters

J 80 meters by 40 meters

28. A map has a scale of 1.5 inches = 500 miles. How many inches on the map would represent 850 miles? Round to the nearest tenth.

A 2.2 inches　　**C** 2.6 inches

B 2.4 inches　　**D** 2.8 inches

CCSS Common Core Review

Solve each proportion. 7.RP.2

29. $\frac{p}{6} = \frac{24}{36}$　　**30.** $\frac{4}{10} = \frac{8}{a}$　　**31.** $\frac{18}{12} = \frac{24}{q}$

32. $\frac{5}{h} = \frac{10}{30}$　　**33.** $\frac{7}{45} = \frac{x}{9}$　　**34.** $\frac{7}{5} = \frac{10.5}{b}$

35. Write a list of the first ten terms in the numerical patterns described below. 5.OA.3

Pattern A Use the starting number 0 and the rule "Add 4."

Pattern B Use the starting number 0 and the rule "Add 8."

a. Compare the terms in pattern A to the terms in pattern B. How are they related?

b. Explain why the two patterns are related as they are.

36. Some species of bamboo can grow 245 inches in a week. Identify the constant of proportionality and then write an equation relating the height of the bamboo to the number of days. How much would a bamboo plant have grown after 3 days? 7.RP.2c

37. The conversion factor for changing meters to feet is 1 meter ≈ 3.28 feet. Find the approximate distance in feet of the 110-meter dash. 6.RP.3d

Name the property shown by each statement. 7.EE.1

38. $3 + (x + 4) = (x + 4) + 3$　　**39.** $(4y) \cdot 1 = 4y$　　**40.** $(6 \cdot 4) \cdot 10 = 10 \cdot (6 \cdot 4)$

41. $8 \cdot 0 \cdot 2 = 0$　　**42.** $0 + (3x) = 3x$　　**43.** $(5 \cdot 2) \cdot 4 = 5 \cdot (2 \cdot 4)$

Simplify each expression. 7.EE.1

44. $5 + (e + 2)$　　**45.** $(13 \cdot 24) \cdot 0$　　**46.** $(6 \cdot d) \cdot 5$

Inquiry Lab

Similar Figures

Inquiry **WHAT kind of scale factor produces an enlargement or reduction of a drawing?**

CCSS Content Standards
7.G.1

Mathematical Practices
1, 3, 6, 7

Vocab Vocabulary
similar figures

Quilts A quilt maker uses a pattern of rectangles and triangles to make certain types of quilts. How can a quilter reproduce the pattern pieces on a different scale to enlarge or reduce the pattern pieces?

Investigation 1

Step 1 On grid paper, draw a rectangle with a length of 5 units and a width of 2 units. This is the original figure.

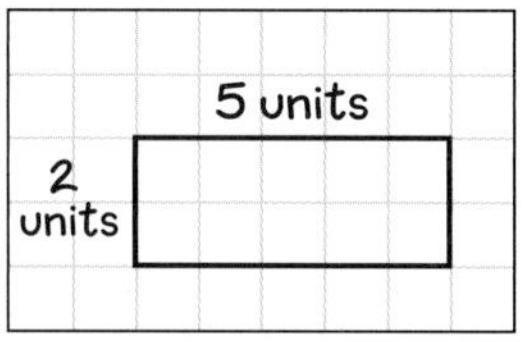

Step 2 Use a scale factor of 1.5. Draw a new rectangle with a length that is 1.5×5 units and a width that is 1.5×2 units.

The rectangle you drew in Step 2 is an enlargement of the rectangle you drew in Step 1.

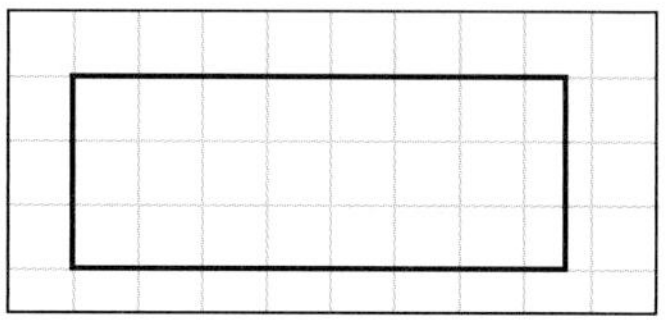

Investigation 2

Step 1 On grid paper, draw a right triangle with legs that measure 3 units and 4 units. This is the original figure.

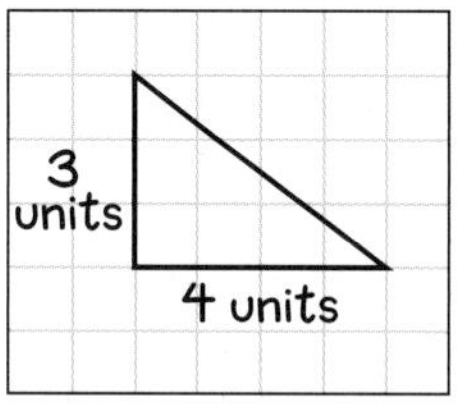

Step 2 Use a scale factor of $\frac{1}{2}$. Draw a new triangle with legs that measure $\frac{1}{2} \times 3$ units and $\frac{1}{2} \times 4$ units.

The triangle you drew in Step 2 is a reduction of the triangle you drew in Step 1.

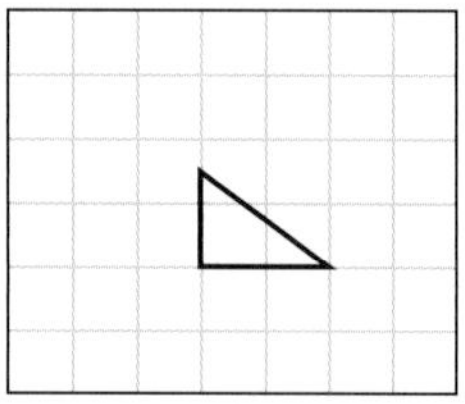

When figures have the same shape but not necessarily the same size, they are called **similar figures**.

Collaborate

Work with a partner.

1. Draw a 12-unit by 8-unit rectangle on grid paper. This is the original figure. Draw new figures using the scale factors given below.
   a. 4
   b. 3
   c. $\frac{3}{4}$
   d. 0.5

Analyze

Refer to Exercise 1.

2. Look at the scaled figures you drew. How do the angles appear to compare to the angles in the original figure?
3. Look at the scaled figures you drew. How do the lengths of the sides of the figures compare to the original figure?
4. Find the area of the original figure.
5. Copy and complete the table for the scale drawings.

Scale Drawing	Scale Factor	Area of Scale Drawing	$\frac{\text{Area of Scale Drawing}}{\text{Area of Original Figure}}$
a.	4		
b.	3		
c.	$\frac{3}{4}$		
d.	0.5		

6. How does the ratio $\frac{\text{area of scale drawing}}{\text{area of original figure}}$ compare to the scale factor?
7. CCSS **Make a Conjecture** Refer to Exercise 6. How could you find the area of a scale drawing if you know the area of the original figure and the scale factor?
8. CCSS **Be Precise** Suppose you draw a 12-inch square. If you draw a scale drawing of this square with a scale factor of $\frac{2}{3}$, what is the length of the sides of the scale drawing? Would the scale drawing be an enlargement or reduction of the original figure?
9. CCSS **Make a Conjecture** Repeat Investigations 1 and 2 using different original figures and different scale factors. How can you change the scale factor you originally used in Investigation 1 to produce an even larger figure? How can you change the scale factor you used in Investigation 2 to produce an even smaller figure?

Reflect

10. Inquiry WHAT kind of scale factor produces an enlargement or reduction of a drawing?

Lesson 5-9
Similar Figures

ISG Interactive Study Guide

See pages 115–116 for:
- Getting Started
- Real-World Link
- Notes

Essential Question

How can you identify and represent proportional relationships?

CCSS Common Core State Standards

Content Standards
7.RP.2, 7.RP.2c

Mathematical Practices
1, 3, 4

Vocabulary
similar figures
congruent
corresponding parts

Math Symbols
≅ is read *is congruent to*
~ is read *is similar to*

What You'll Learn

- Find missing measures of similar figures.
- Use scale factors to solve problems.

Real-World Link

Kites It is believed that kites were first flown in China more than 2500 years ago. Since then, kites have been used to transport materials, to pull carriages, and as tools for scientific research. However, a typical kite that you might fly in your backyard is much smaller and usually comes in the shape of a triangle or quadrilateral.

Key Concept Similar Figures

Words **Similar figures** are figures that have the same shape but not necessarily the same size. If two figures are similar, then

- the corresponding angles are **congruent**, or have the same measure, and
- the corresponding sides are proportional and opposite corresponding angles.

Model

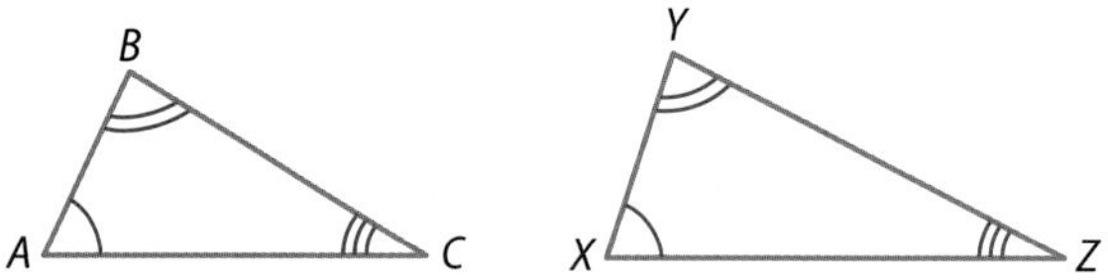

Symbols $\triangle ABC \sim \triangle XYZ$

$\angle A \cong \angle X, \angle B \cong \angle Y, \angle C \cong \angle Z$ and $\frac{AB}{XY} = \frac{BC}{YZ} = \frac{AC}{XZ}$

Figure *ABCD* is similar to figure *EFGH*. In symbols, $ABCD \sim EFGH$.

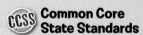

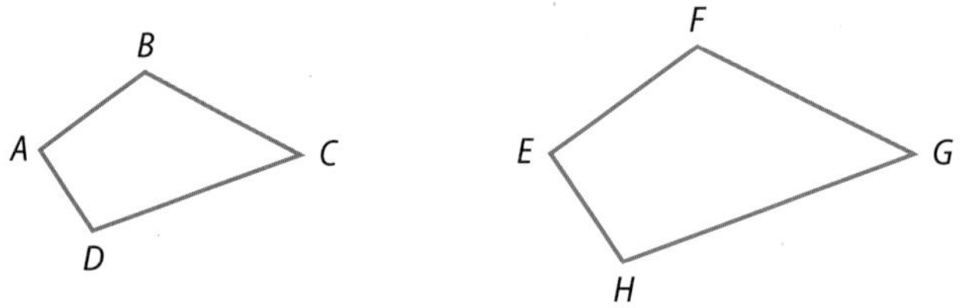

Similar figures have **corresponding parts**. These are angles and sides in the same position.

Corresponding Angles

$\angle A \leftrightarrow \angle E$ $\angle C \leftrightarrow \angle G$

$\angle B \leftrightarrow \angle F$ $\angle D \leftrightarrow \angle H$

Corresponding Sides

$\overline{AB} \leftrightarrow \overline{EF}$ $\overline{CD} \leftrightarrow \overline{GH}$

$\overline{BC} \leftrightarrow \overline{FG}$ $\overline{DA} \leftrightarrow \overline{HE}$

Since corresponding sides are proportional, you can use a proportion or the scale factor to determine the measures of the sides of similar figures when some measures are known.

Example 1

Tutor

The figures are similar. Find each missing measure.

a.

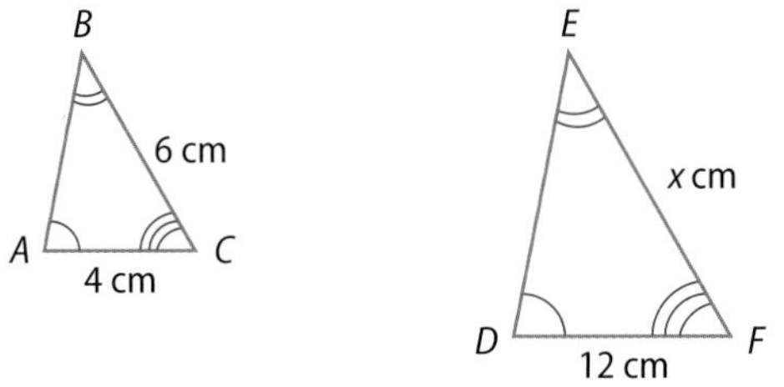

Since $\triangle ABC \sim \triangle DEF$, the corresponding angles are congruent and the corresponding sides are proportional.

$\frac{BC}{EF} = \frac{AC}{DF}$	Write a proportion.
$\frac{6}{x} = \frac{4}{12}$	Replace *BC* with 6, *EF* with *x*, *AC* with 4, and *DF* with 12.
$6 \cdot 12 = x \cdot 4$	Find the cross products.
$72 = 4x$	Simplify.
$18 = x$	Division Property of Equality

The length of $\overline{EF}$ is 18 centimeters.

Alternative Method
You can also use ratios that compare sides within the figures. In Example 1a, you can use $\frac{6}{4} = \frac{x}{12}$.

b.

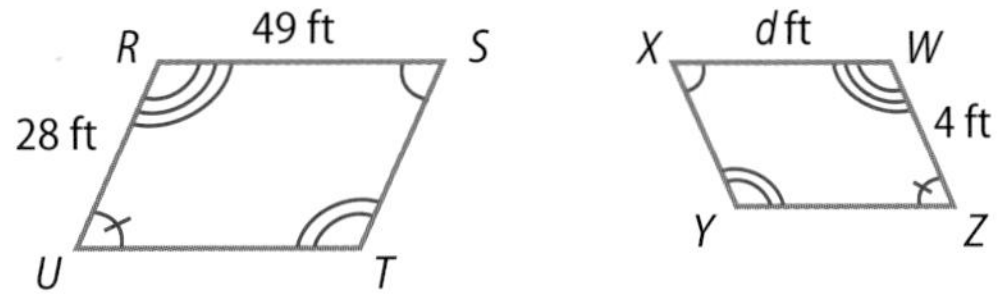

Figure *RSTU* ~ figure *WXYZ*. The corresponding sides are proportional.

$\frac{RS}{WX} = \frac{RU}{WZ}$	Write a proportion.
$\frac{49}{d} = \frac{28}{4}$	Replace *RS* with 49, *WX* with *d*, *RU* with 28, and *WZ* with 4.
$49 \cdot 4 = d \cdot 28$	Find the cross products.
$196 = 28d$	Simplify.
$7 = d$	Division Property of Equality

The length of $\overline{WX}$ is 7 feet.

Got It? Do these problems to find out.

1a.

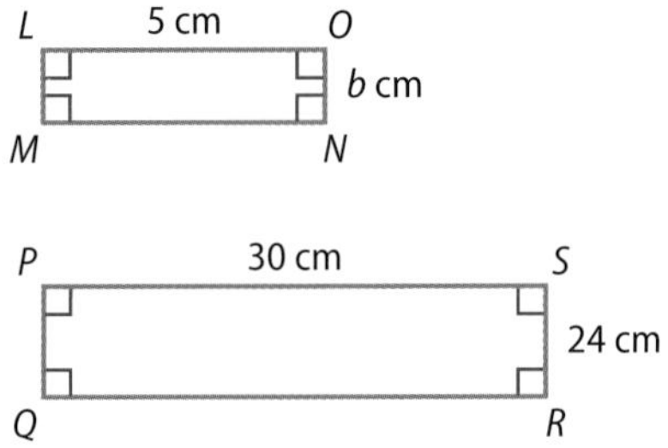

1b.

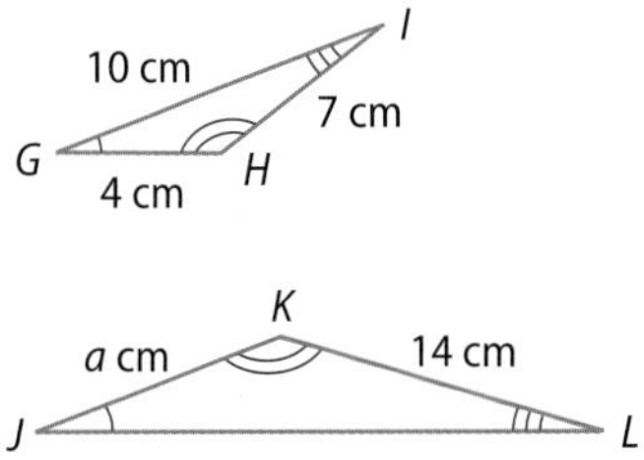

Scale Factor
The scale factor is the constant of proportionality. In Lesson 5–5, you learned that proportional relationships can be described using equations of the form $y = kx$, where k is the constant of proportionality.

Scale Factor

Recall that the scale factor is the ratio of a length on a scale drawing to the corresponding length on the real object. It is also the ratio of corresponding sides in similar figures.

Real World

Example 2

Tutor

An architect is using similar triangles to design a decorative window for the entrance of a new office building. If $\triangle ABC \sim \triangle DEF$, find the length of segment DF.

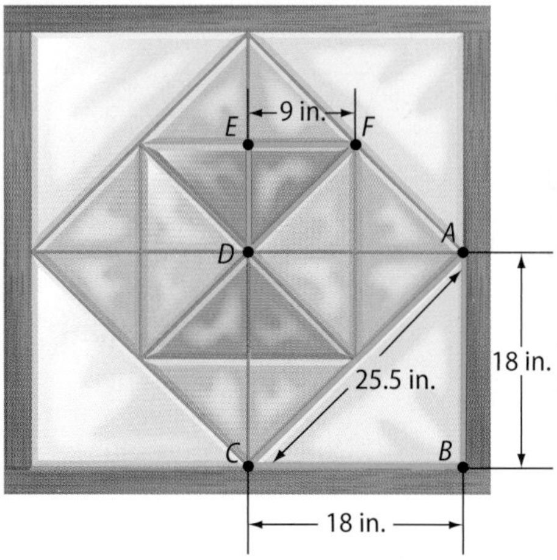

Find the scale factor from $\triangle DEF$ to $\triangle ABC$ by finding the ratio of corresponding sides with known lengths.

scale factor: $\frac{BC}{EF} = \frac{18}{9}$ or 2

Words	Tw times a length on triangle *DEF* is a corresponding length on triangle *ABC*.
Variable	Let *m* represent the measure of $\overline{DF}$.
Equation	$2m = 25.5$

$2m = 25.5$ Write the equation.
$m = 12.75$ Divide each side by 2.

So, the length of $\overline{DF}$ is 12.75 inches.

Got It? **Do this problem to find out.**

2. A rectangular blue tile has a length of 4.25 inches and a width of 6.75 inches. A similar red tile has a length of 12.75 inches. What is the width of the red tile?

Guided Practice

Check

The figures are similar. Find each missing measure. (Example 1)

1.

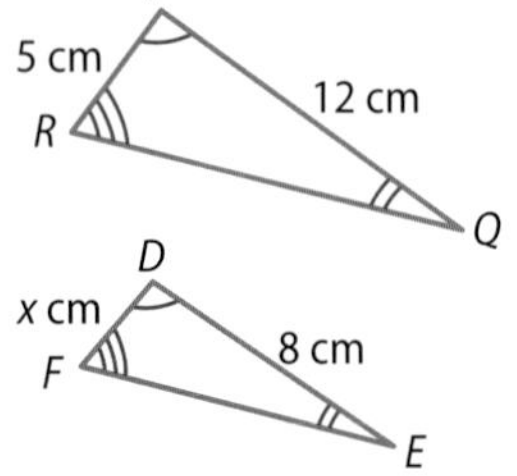

2.

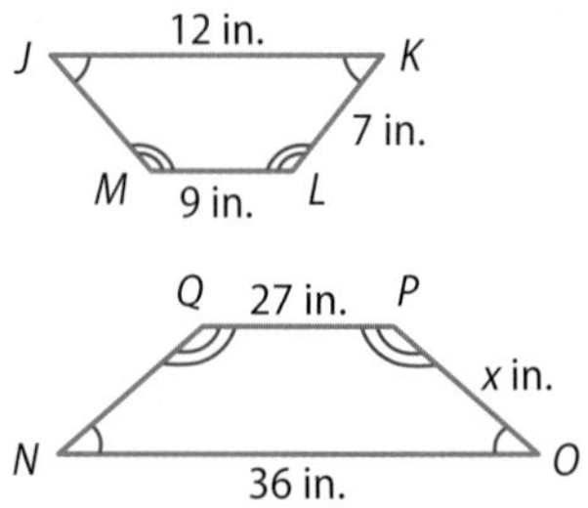

3. The logo for an electronics store is made from similar trapezoids as shown. What is the length of the missing measure? (Example 2)

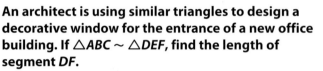

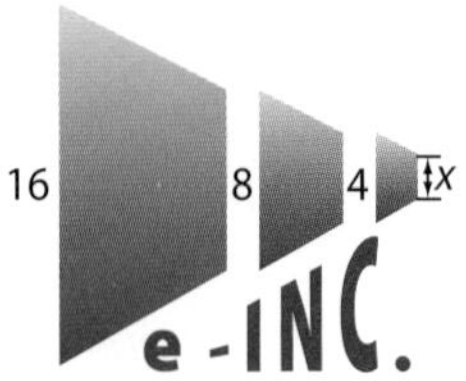

Independent Practice

Go online for Step-by-Step Solutions eHelp

The figures are similar. Find each missing measure. (Example 1)

4.

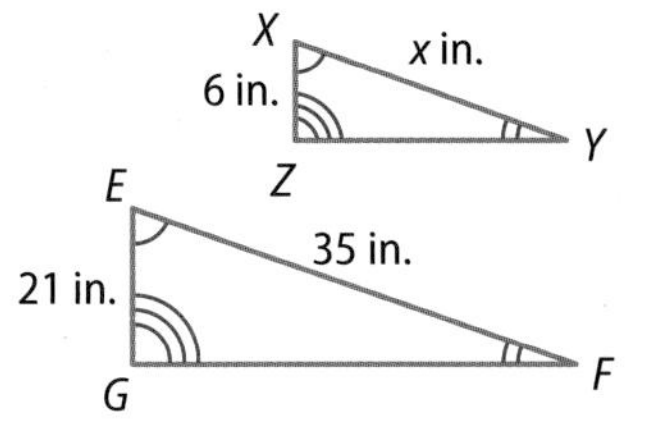

5

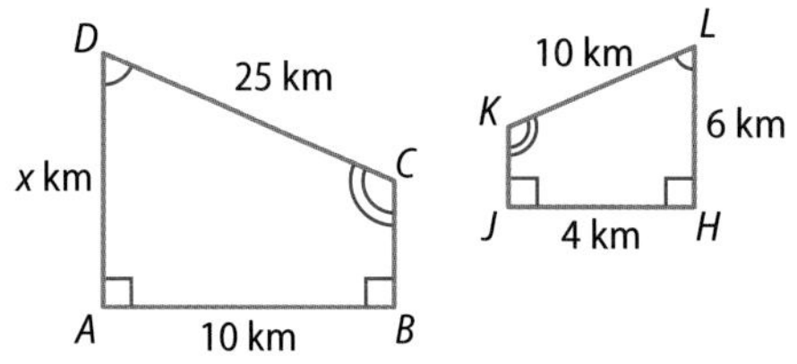

6.

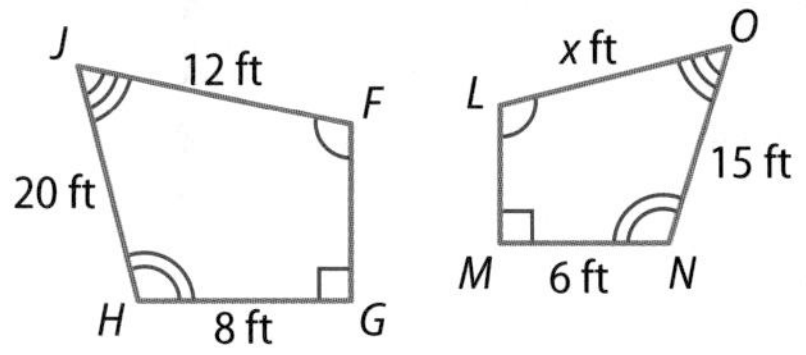

7.

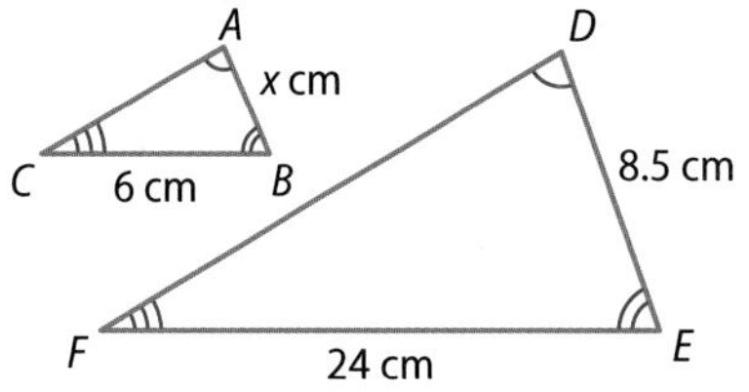

8. The design shown is made using similar triangles and quadrilaterals. Triangle A ~ triangle B and quadrilateral C ~ quadrilateral D. (Example 2)

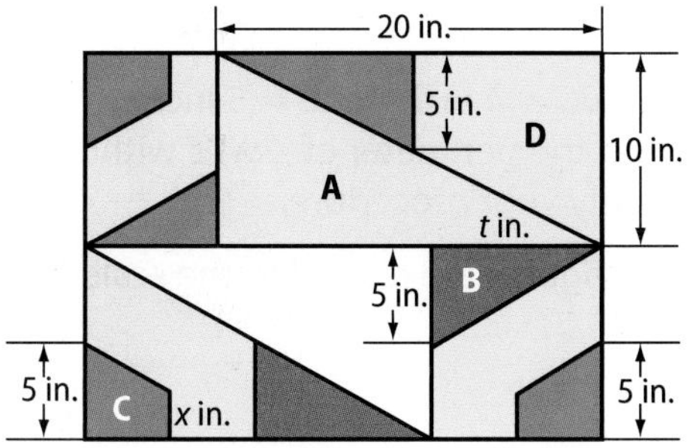

a. Find the missing measure in triangle B.

b. Find the missing measure in quadrilateral C.

9. Triangle *LMN* is similar to △*RST*. What is the value of *LN* if *RT* is 9 inches, *MN* is 21 inches, and *ST* is 7 inches? (Example 2)

10. Quadrilateral *ABCD* is similar to quadrilateral *WXYZ*. What is the value of *WZ* if *AD* is 18 feet, *CD* is 27 feet, and *YZ* is 10.8 feet?

11. Using a scale factor of $\frac{2}{3}$, draw and label a rectangle similar to rectangle *ABCD*.

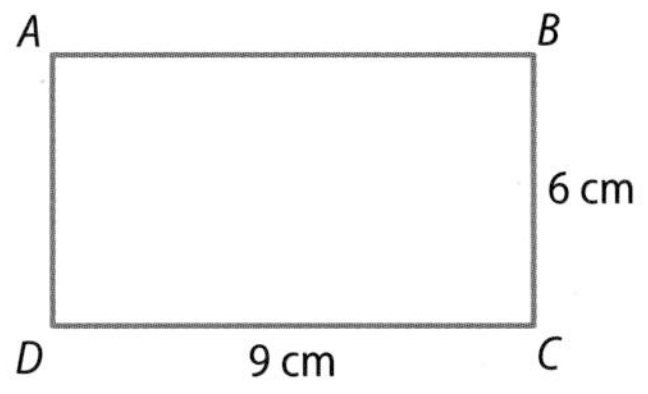

12. Using a scale factor of $\frac{4}{3}$, draw and label a triangle similar to △*XYZ*.

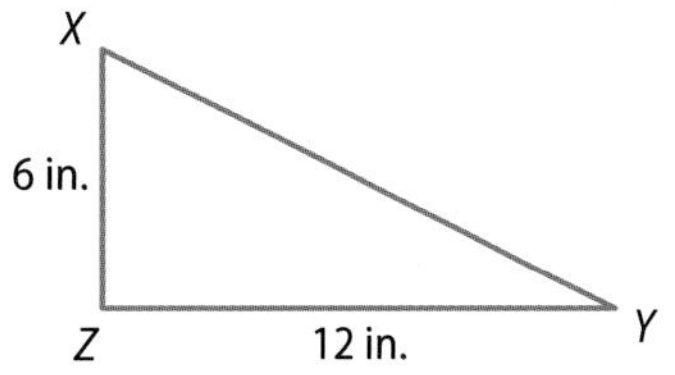

13. Figure *FGHJK* ~ figure *LMNPQ*. The scale factor from figure *FGHJK* to figure *LMNPQ* is $\frac{2}{3}$. What is the perimeter of figure *LMNPQ*?

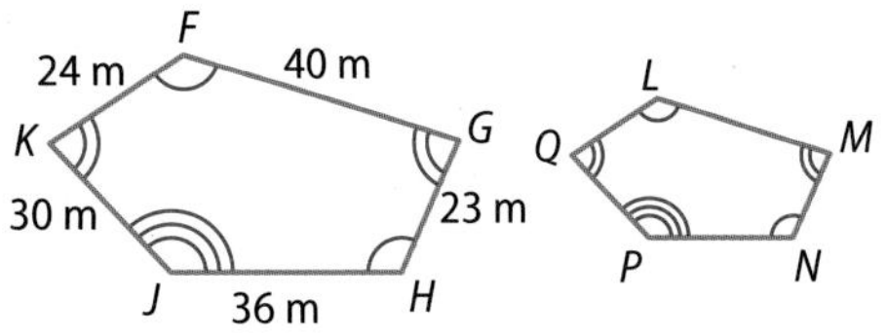

14. An image that is projected onto a movie screen measures 8.8 meters by 6.4 meters. The projection is similar to the individual frame on the movie reel. If the projection has a scale factor of 400, what are the original dimensions of a frame on a movie reel?

15 CCSS **Multiple Representations** In this problem, you will investigate the relationship between the perimeters of similar figures. In the figures below, $\triangle ABC \sim \triangle XYZ$.

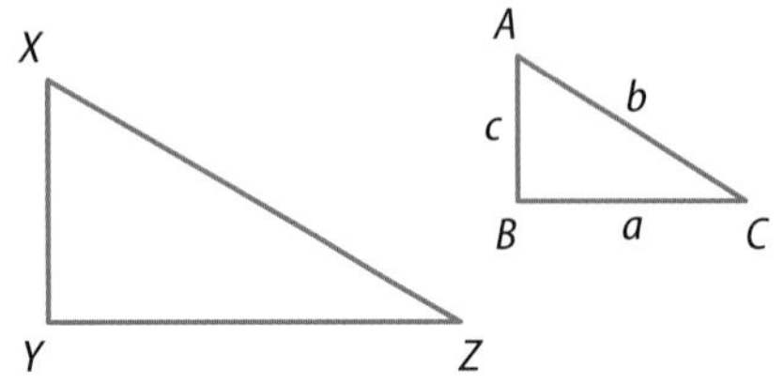

a. Geometry Write an expression for the perimeter of $\triangle ABC$.

b. Symbols If the scale factor is represented by *d*, write algebraic expressions for the measures of the sides of $\triangle XYZ$.

c. Geometry Write an expression for the perimeter of $\triangle XYZ$.

d. Symbols Use the Distributive Property to factor the expression from part **c.** Explain the meaning of the expression.

e. Make a Prediction Suppose $AB = 3$ inches, $BC = 4$ inches, $AC = 5$ inches, and the scale factor from $\triangle ABC$ to $\triangle XYZ$ is 2. Find the perimeter of $\triangle XYZ$ without calculating the lengths of $\overline{XY}$, $\overline{YZ}$, and $\overline{XZ}$. Justify your procedure.

f. Words Explain how the perimeters of similar figures are related to the scale factor.

H.O.T. Problems Higher Order Thinking

16. CCSS **Justify Conclusions** Draw two similar triangles whose scale factor is 1:3. Justify your answer.

CCSS **Reason Inductively** **Determine whether each statement is *always, sometimes,* or *never* true. Explain your reasoning.**

17. All rectangles are similar.

18. All squares are similar.

19. CCSS **Find the Error** Tony is finding the length of $\overline{AB}$ where $\triangle ABC \sim \triangle DEF$, $BC = 16$ feet, $EF = 12$ feet, and $DE = 18$ feet. Find his mistake and correct it.

$$\frac{16}{18} = \frac{12}{x}$$

$$x = 13.5 \text{ ft}$$

20. CCSS **Persevere with Problems** True or false? If $\triangle XYZ \sim \triangle RST$, then $\frac{x}{z} = \frac{r}{t}$, where *x* is the side opposite $\angle X$, *z* is the side opposite $\angle Z$, and so on. Justify your answer. If false, provide a counterexample.

21. **Building on the Essential Question** Suppose you have two triangles. Triangle *A* is similar to triangle *B*, and the measures of the sides of triangle *A* are less than the measures of the sides of triangle *B*. The scale factor is 0.25. Which is the original triangle? Explain.

Standardized Test Practice

22. Quadrilateral *ABCD* is similar to quadrilateral *EFGH*. What is the length of $\overline{FG}$?

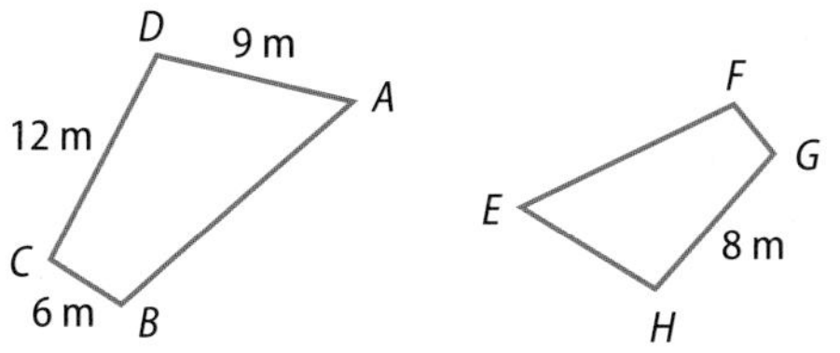

A 1.5 m **C** 4 m

B 3.8 m **D** 5.3 m

23. If polygon *ABCDE* is similar to polygon *FGHIJ*, which of the following is *not* true?

F $\angle ABC \cong \angle FGH$

G $\angle EDC \cong \angle JIH$

H $\overline{AE}$ corresponds to $\overline{FJ}$

J $\overline{BC}$ corresponds to $\overline{HI}$

24. Triangle *RST* is similar to triangle *XYZ*.

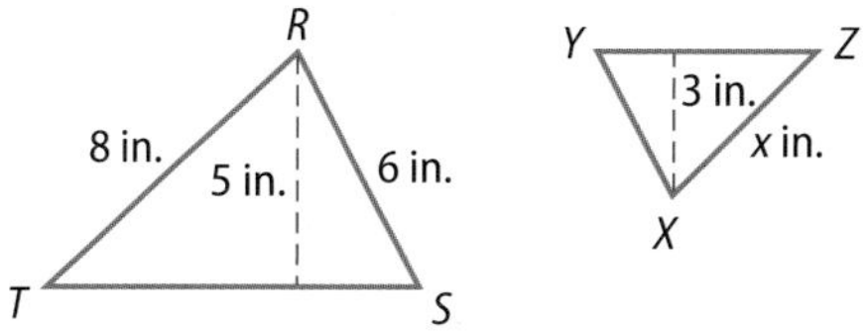

What is the length of $\overline{XZ}$?

A 1.875 in. **C** 4.8 in.

B 3.75 in. **D** 9.6 in.

25. **Short Response** Triangle *LMN* has a perimeter of 24 centimeters and is similar to triangle *DEF*. If the scale factor relating $\triangle LMN$ to $\triangle DEF$ is $\frac{1}{3}$, what is the perimeter, in centimeters, of $\triangle DEF$?

CCSS Common Core Review

26. The Statue of Zeus at Olympia is one of the Seven Wonders of the World. On a scale model of the statue, the height of Zeus is 8 inches. 7.G.1

a. If the actual height of the statue is 40 feet, what is the scale?

b. What is the scale factor?

27. Flor is making bracelets. She knows that 4 bags of beads will make 14 bracelets. If she wants to make 56 bracelets, how many bags of beads will she need? 7.RP.2b

28. A bushel of apples will make approximately 3 gallons of apple cider. The table shows the relationship between the number of bushels of apples and the number of gallons of apple cider. 7.RP.2

Apple Bushels, *b*	Gallons of Apple Cider, *g*
1	3
2	6
5	15
8	24

a. Given *b*, the number of bushels needed, write an equation that can be used to find *g*, the number of gallons of apple cider.

b. How many bushels are needed to make 54 gallons of cider?

Multiply. 7.NS.2c

29. $\frac{4}{5} \times 5$ **30.** $\frac{7}{4} \times 3$ **31.** $\frac{2}{3} \times 6$ **32.** $\frac{8}{5} \times 9$

Divide. 7.NS.2c

33. $\frac{4}{5} \div 5$ **34.** $\frac{7}{4} \div 3$ **35.** $\frac{2}{3} \div 6$ **36.** $\frac{8}{5} \div 9$

Find each quotient. Express using exponents. 8.EE.1

37. $\frac{4^{13}}{4^6}$ **38.** $\frac{8^5}{8}$ **39.** $\frac{x^9}{x^8}$ **40.** $\frac{k^{21}}{k^{14}}$

Lesson 5-10

Indirect Measurement

ISG Interactive Study Guide

See pages 117–118 for:
- Getting Started
- Real-World Link
- Notes

Essential Question

How can you identify and represent proportional relationships?

CCSS Common Core State Standards

Content Standards
7.RP.2, 7.RP.2c

Mathematical Practices
1, 3, 4

Vocabulary
indirect measurement

What You'll Learn

- Solve problems involving indirect measurement using shadow reckoning.
- Solve problems using surveying methods.

Real-World Link

Shadows Have you ever looked down and found that your shadow and the shadow of the person next to you were different lengths? This is because the length of your shadow is proportional to your height. So, unless you and the person next to you are the exact same height, your shadows will be different lengths.

Indirect Measurement

Indirect measurement allows you to use the properties of similar triangles to find measurements that are difficult to measure directly. With *shadow reckoning*, two objects and their shadows form two sides of similar triangles.

Real World Example 1

Watch Tutor

The lead statue of the Korean War Memorial in Washington, D.C., casts a 43.5-inch shadow at the same time a nearby tourist casts a 32-inch shadow. If the tourist is 64 inches tall, how tall is the lead statue?

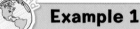

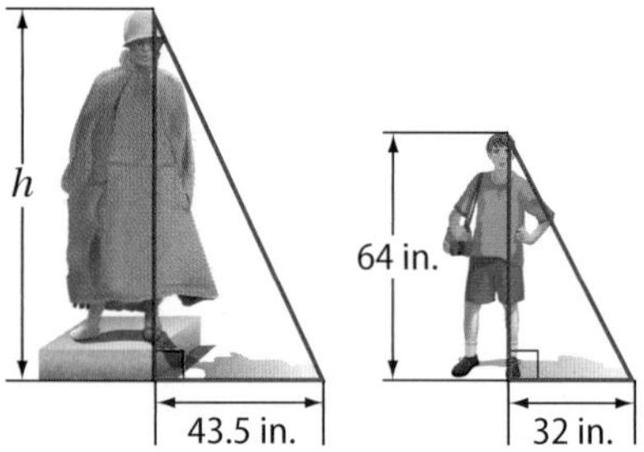

You need to find the statue's height h.

Set up a proportion comparing the tourist's shadow to the statue's shadow.

tourist's height → 64; statue's height → h; tourist's shadow → 32; statue's shadow → 43.5

$$\frac{64}{h} = \frac{32}{43.5}$$

$64 \cdot 43.5 = 32 \cdot h$ Find the cross products.

$2784 = 32h$ Multiply.

$87 = h$ Divide each side by 32.

The statue is 87 inches tall.

Got It? Do this problem to find out.

1. Suppose a bell tower casts a 27.6-foot shadow at the same time a nearby tourist casts a 1.2-foot shadow. If the tourist is 6 feet tall, how tall is the tower?

Surveying Methods

Surveyors also use similar triangles, but their method does not involve shadows. Notice in Example 2 that it is possible to measure three sides of the triangles directly.

Real World Example 2

Tutor

In the figure, $\triangle STU \sim \triangle VQU$.
Find the distance across the pond.

Since the figures are similar, corresponding sides are proportional.

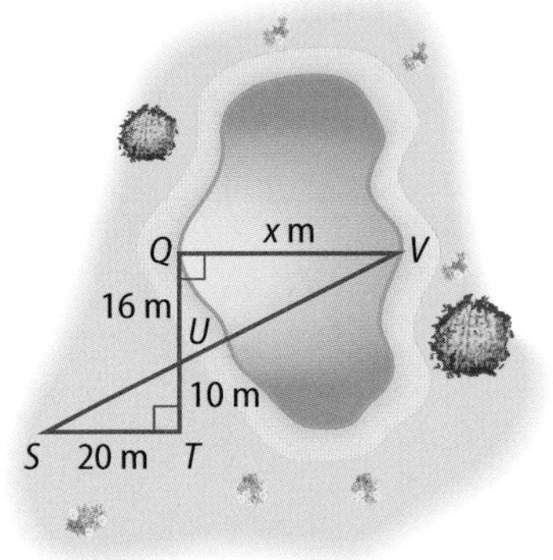

$\frac{UQ}{UT} = \frac{VQ}{ST}$	Write a proportion.
$\frac{16}{10} = \frac{x}{20}$	$UQ = 16$, $UT = 10$, $VQ = x$, and $ST = 20$
$10 \cdot x = 16 \cdot 20$	Cross products
$10x = 320$	Multiply.
$\frac{10x}{10} = \frac{320}{10}$	Divide each side by 10.
$x = 32$	Simplify.

So, the distance across the pond is 32 meters.

Got It? Do this problem to find out.

2. In the figure, $\triangle QRS \sim \triangle URT$. Find the distance from the cabin to the Mess Hall.

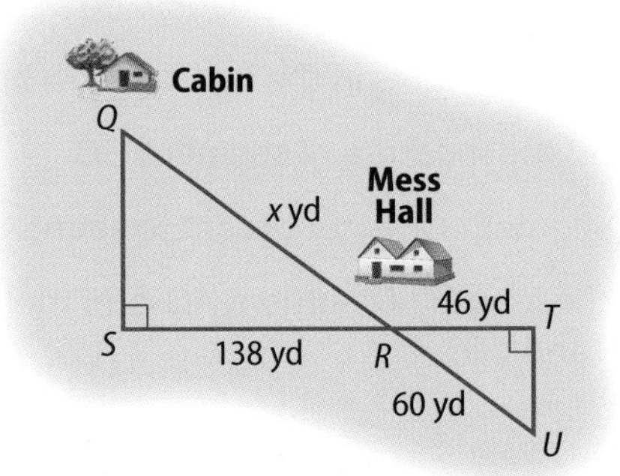

Guided Practice

Check

1. A basketball hoop in Miguel's backyard casts a shadow that is 8 feet long. At the same time, Miguel casts a shadow that is 4.5 feet long. If Miguel is 5.5 feet tall, how tall is the basketball hoop? Round to the nearest tenth. (Example 1)

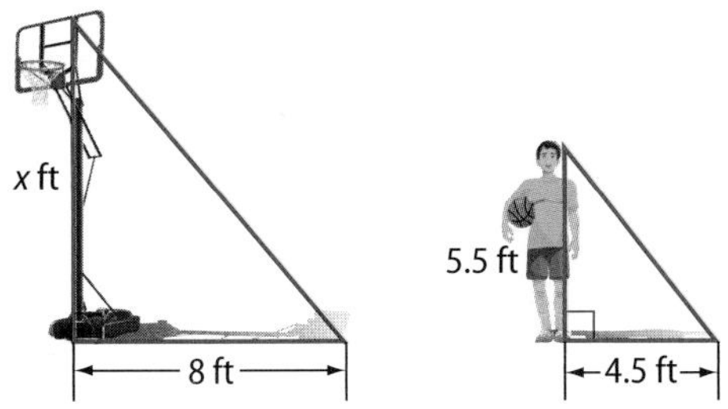

2. In the figure, $\triangle ABC \sim \triangle EBD$. Find the distance across Stallion Ravine. (Example 2)

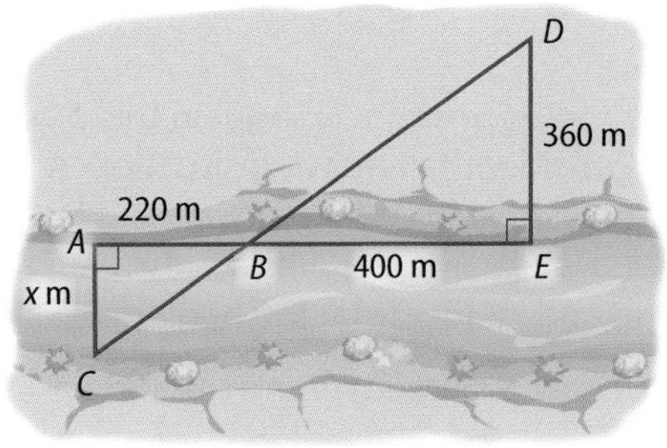

Independent Practice

Go online for Step-by-Step Solutions eHelp

3. A flagpole is 30 feet high, and a mailbox is 3.5 feet high. The mailbox casts a shadow that is 5.25 feet long. How long is the flagpole's shadow at the same time? (Example 1)

4. The height of Medina Middle School is 25 feet tall. A mail service drop box outside the school is 4 feet tall. The drop box casts a shadow that is 6 feet long. At the same time, what is the length of the shadow of the school building? (Example 1)

5. **STEM** The triangles below are similar. Find *x*. (Example 2)

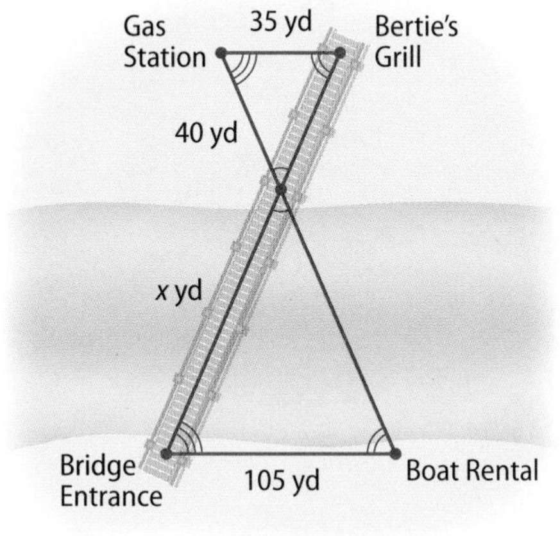

6. The triangles below are similar. How far is it from Athens to Yukon? (Example 2)

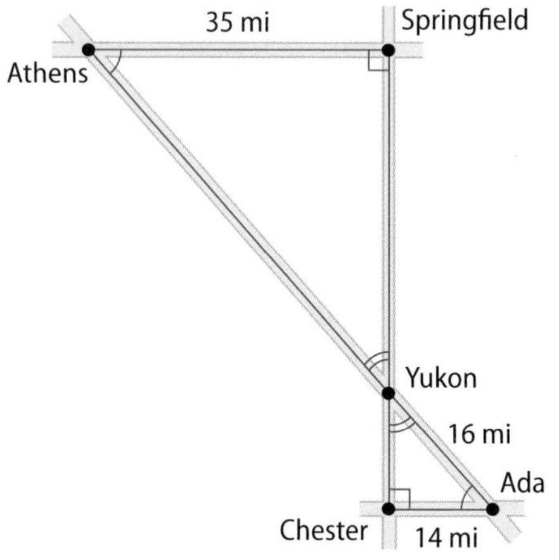

7. The height of a roller coaster is 157.5 feet. If the roller coaster's shadow is 60 feet long, how long will a person's shadow be if the person is 5 feet 3 inches tall?

8. All of the triangles in the figure below are similar.

   a. Find the measure of segment *GD*.

   b. If segment *GF* is congruent to segment *FE*, find the measure of segment *BF*.

   c. If the length of segment *AD* is 15 meters, what are the lengths of segments *BC* and *CD*?

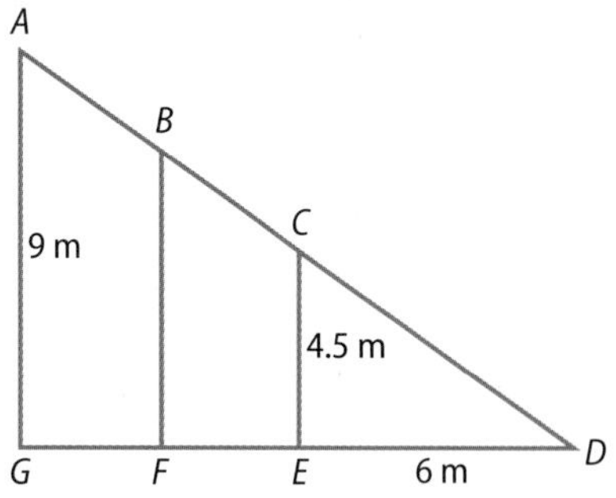

9. **CCSS Multiple Representations** In this problem, you will investigate similar triangles. Consider the following situation. A biplane starts to take off from the beginning of a runway. When the plane is level with the end of the runway, it is 500 feet above the ground. A bird is flying in the same direction. It is 8 feet above the ground and 15 feet from the beginning of the runway.

   a. **Model** Draw a diagram of the situation.

   b. **Symbols** Write and solve a proportion to find how far the plane is from the beginning of the runway.

10. Electrical poles that carry electrical wire seem to get smaller the farther away they are from you. Find the apparent height of each pole if the tallest pole is 50 feet, and the distance between each pole is 100 feet.

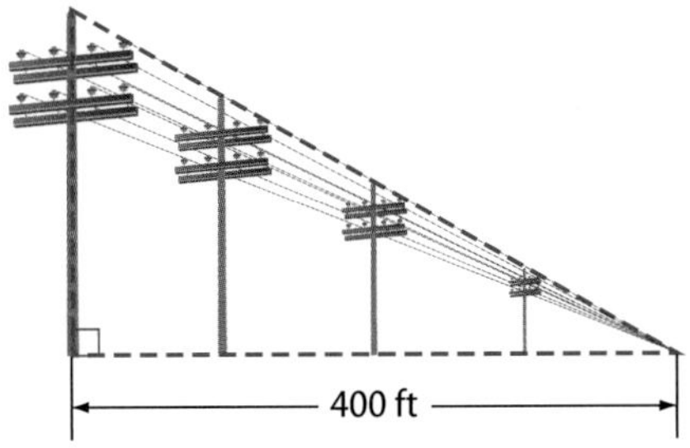

11. During a basketball game, Josh, Devon, and Marco are in the following positions. Josh is 16 feet from Devon, and Devon is $5\frac{1}{3}$ feet from Marco. If Marco is 4 feet from both *A* and *B*, how wide is the key?

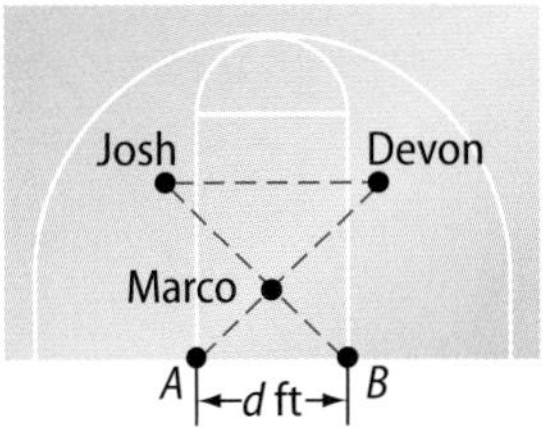

12. Use the figure at the right.

a. Write two different proportions that could be used to determine the height of the lighthouse.

b. How tall is the lighthouse?

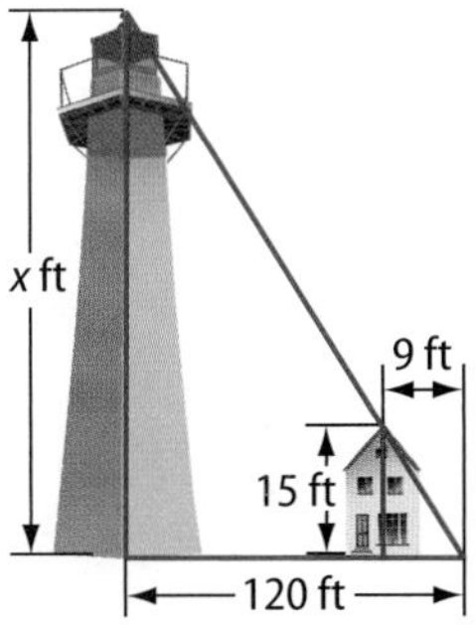

13. The Navy Pier Ferris Wheel in Chicago is 150-feet tall. If the Ferris wheel casts a $37\frac{1}{2}$-foot shadow, write and solve a proportion to find the height of a woman standing nearby who casts a $1\frac{1}{2}$-foot shadow.

14. A tree house casts a shadow of 18 feet while Jenet casts a shadow 9 feet. If Jenet is 5 feet tall, how tall is the tree house?

H.O.T. Problems Higher Order Thinking

15. CCSS **Model with Mathematics** Write a real-world problem to describe how you could determine the height of a local landmark or statue in your community using shadow reckoning.

16. CCSS **Persevere with Problems** In the diagram shown at the right, $\triangle ABC \sim \triangle EDC$.

a. Write a proportion that could be used to solve for the height *h* of the flag pole.

b. What information would you need to know in order to solve this proportion?

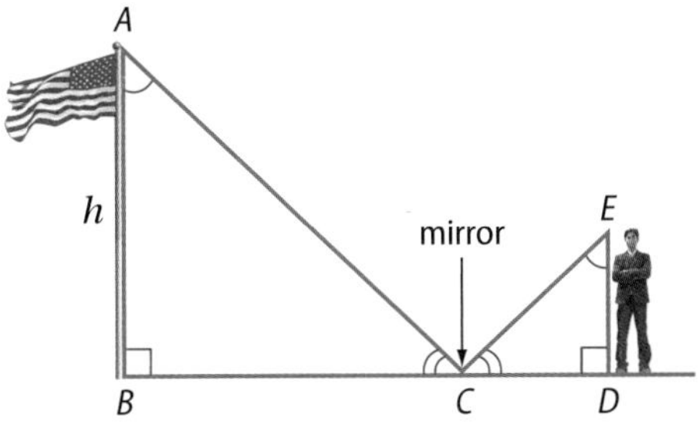

17. CCSS **Justify Conclusions** *True* or *false?* If two pairs of corresponding sides of two triangles are proportional, then you can use indirect measurement to determine the length of a missing side. Explain your reasoning.

18. **Building on the Essential Question** Give a real-world example of when you might need to use indirect measurement. Explain how you would solve the problem.

Standardized Test Practice

19. A bell tower casts a 60-inch shadow. At the same time, a statue that is 4.5 feet tall casts a 15-inch shadow. How tall is the bell tower?

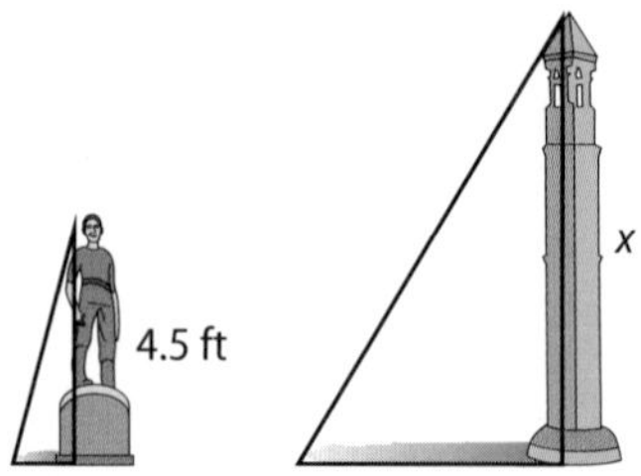

A 200 ft
B 30 ft
C 27 ft
D 18 ft

20. In the figure, $\triangle PRT \sim \triangle SRQ$. Find the distance across the golf green.

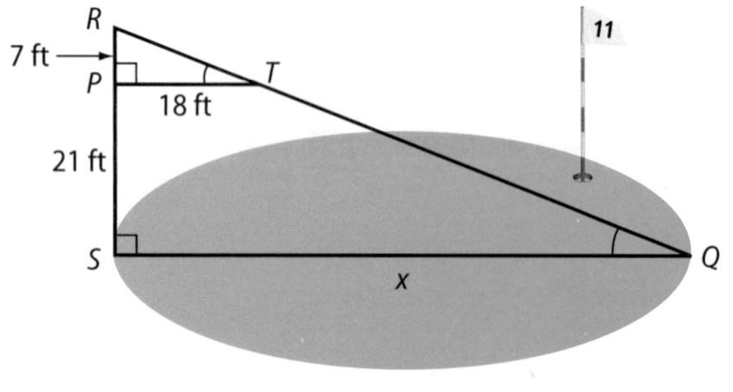

F 4.5 ft
G 35 ft
H 48 ft
J 72 ft

21. Short Response Find the length in kilometers of Beechwold Boulevard.

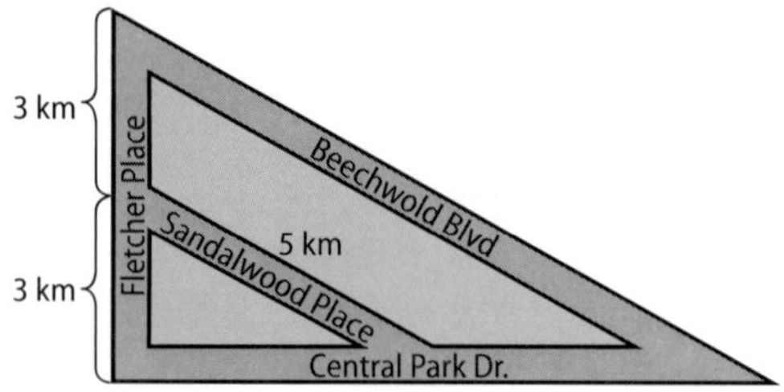

22. How tall is the street sign?

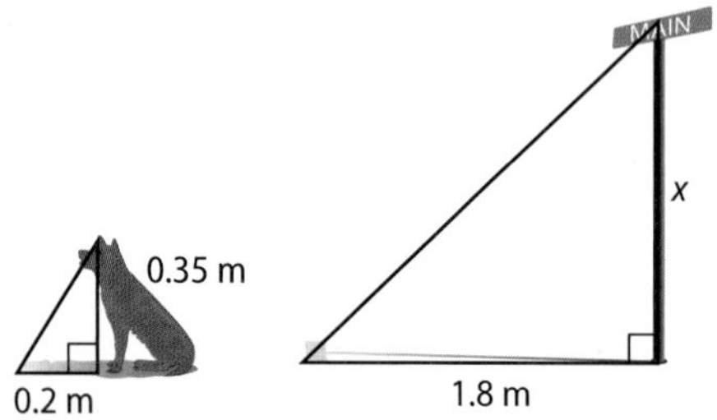

A 3.15 m
B 3.5 m
C 3.75 m
D 3.8 m

Common Core Review

23. Triangle *JMK* is similar to triangle *PRO*. What is the value of *x*? **7.RP.2**

24. How many 6-inch long pieces of ribbon can be cut from a piece that is 3 yards long? **7.RP.1**

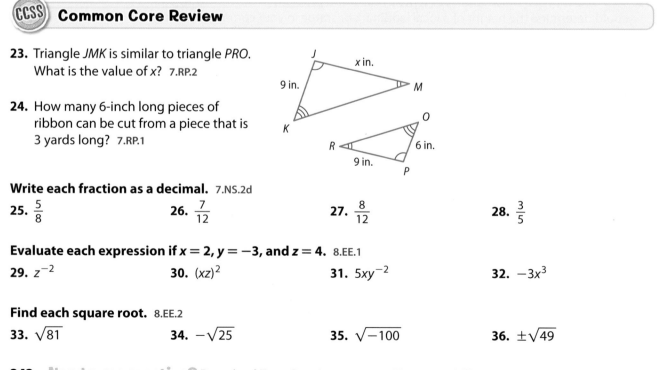

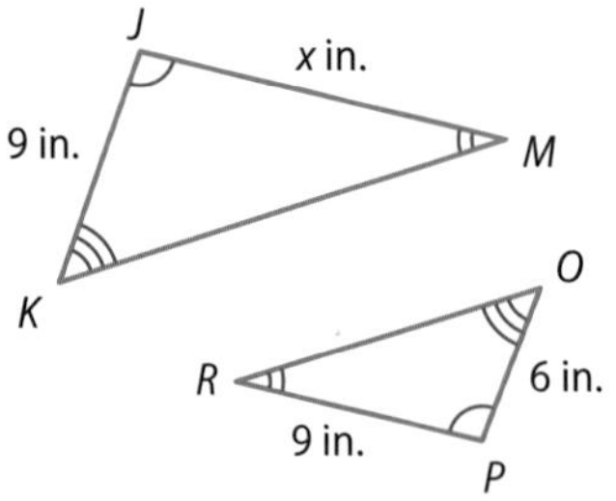

Write each fraction as a decimal. 7.NS.2d

25. $\frac{5}{8}$ **26.** $\frac{7}{12}$ **27.** $\frac{8}{12}$ **28.** $\frac{3}{5}$

Evaluate each expression if $x = 2$, $y = -3$, and $z = 4$. 8.EE.1

29. z^{-2} **30.** $(xz)^2$ **31.** $5xy^{-2}$ **32.** $-3x^3$

Find each square root. 8.EE.2

33. $\sqrt{81}$ **34.** $-\sqrt{25}$ **35.** $\sqrt{-100}$ **36.** $\pm\sqrt{49}$

Check Chapter Review

ISG Interactive Study Guide
See pages 119–122 for:
- Vocabulary Check
- Key Concept Check
- Problem Solving
- Reflect

Lesson-by-Lesson Review

Lesson 5-1 Ratios (pp. 184–188)

Express each ratio as a fraction in simplest form.

1. 10 girls out of 24 students
2. 6 red cars to 4 blue cars
3. 10 yards to 8 inches
4. 18 ounces to 3 cups
5. Jean got 12 hits out of 16 times at bat. Express this rate as a fraction in simplest form. Explain its meaning.

Example 1

Express the ratio *2 feet to 18 inches* as a fraction in simplest form.

First, convert feet to inches.

$$\frac{2\text{ ft}}{18\text{ in.}} = \frac{24\text{ in.}}{18\text{ in.}}$$

Next, divide the numerator and denominator by the GCF, 6.

$$\frac{24\text{ in.} \div 6}{18\text{ in.} \div 6} = \frac{4\text{ in.}}{3\text{ in.}} \text{ or } \frac{4}{3}$$

Lesson 5-2 Unit Rates (pp. 189–193)

Express each rate as a unit rate. Round to the nearest tenth or to the nearest cent, if necessary.

6. \$25.97 for 8 boxes
7. 400 meters in 5 minutes
8. \$175 for 4 concert tickets
9. 125 miles in 200 minutes
10. **Financial Literacy** An eight pack of juice boxes costs \$4.79, and a twelve pack of juice boxes costs \$6.59. Which is a better buy? Explain.

Example 2

Express *274 miles in 14 gallons of gasoline* as a unit rate. Round to the nearest tenth of a mile if necessary.

Write the rate that compares the miles to the number of gallons. Then divide to find the unit rate.

$$\frac{274\text{ miles}}{14\text{ gallons}} \overset{\div 14}{=} \frac{19.6\text{ miles}}{1\text{ gallon}}$$

So, the car traveled 19.6 miles on 1 gallon of gasoline.

Lesson 5-3 Complex Fractions and Unit Rates (pp. 194–199)

Simplify.

11. $\dfrac{6}{\frac{2}{5}}$

12. $\dfrac{\frac{5}{3}}{10}$

13. Noreen can walk $1\frac{1}{10}$ miles in $\frac{1}{3}$ hour. Find her average speed in miles per hour.
14. Write $66\frac{2}{3}\%$ as a fraction in simplest form.
15. Write $6\frac{1}{2}\%$ as a fraction in simplest form.
16. Write $11\frac{1}{3}\%$ as a fraction in simplest form.

Example 3

Simplify $\dfrac{\frac{5}{3}}{\frac{3}{4}}$.

$\dfrac{\frac{5}{3}}{\frac{3}{4}} = \frac{5}{1} \div \frac{3}{4}$ Write the complex fraction as a division problem.

$= \frac{5}{1} \times \frac{4}{3}$ Multiply by the reciprocal of $\frac{3}{4}$, which is $\frac{4}{3}$.

$= \frac{20}{3}$ or $6\frac{2}{3}$ Simplify.

Lesson 5-4 Converting Rates (pp. 200–205)

Complete each conversion. Round to the nearest hundredth, if necessary.

17. 7 in. ≈ ■ cm

18. 20 m ≈ ■ yd

19. 25 fl oz ≈ ■ mL

20. 4 L ≈ ■ gal

21. 18 pt ≈ ■ L

22. 12 oz ≈ ■ g

23. 26 cm ≈ ■ in.

24. 3 qt ≈ ■ L

25. 4 m ≈ ■ ft

26. 68 g ≈ ■ oz

27. **STEM** A plane is flying at a speed of 425 miles per hour. How far will the plane travel in 0.75 hour?

28. A swimming pool is being drained at a rate of 50 gallons per hour. How many milliliters per second is this? Round to the nearest tenth.

29. A runner runs 2 miles in 9.56 minutes. How many meters per second is this?

30. A family drives their car 135 miles in 3 hours. How many kilometers per hour is this?

Example 4

Complete the conversion. Round to the nearest hundredth.
18 centimeters to inches

Use 1 inch ≈ 2.54 centimeters.

$18 \text{ cm} \approx 18 \text{ cm} \cdot \frac{1 \text{ in.}}{2.54 \text{ cm}}$ Multiply by $\frac{1 \text{ in.}}{2.54 \text{ cm}}$.

$\approx 18 \cancel{\text{cm}} \cdot \frac{1 \text{ in.}}{2.54 \cancel{\text{cm}}}$ Divide out common units.

$\approx \frac{18 \text{ in.}}{2.54}$ or 7.09 in. Simplify.

Example 5

A peregrine falcon can fly at a top speed of 200 miles per hour. How many feet per second is this?

First, convert miles to feet and hours to seconds.

$$\frac{200 \text{ mi}}{1 \text{ h}} = \frac{200 \text{ mi}}{1 \text{ h}} \cdot \frac{5280 \text{ ft}}{1 \text{ mi}} \cdot \frac{1 \text{ h}}{3600 \text{ s}}$$

Next, divide out the common factors.

$$= \frac{\overset{1}{\cancel{200}} \cancel{\text{mi}}}{1 \cancel{\text{h}}} \cdot \frac{\overset{176}{\cancel{5280}} \text{ ft}}{1 \cancel{\text{mi}}} \cdot \frac{1 \cancel{\text{h}}}{\underset{\underset{1}{\cancel{30}}}{\cancel{3600}} \text{ s}}$$

$$= \frac{293.3 \text{ ft}}{1 \text{ s}}$$

The falcon can fly about 293 feet per second.

Lesson 5-5 Proportional and Nonproportional Relationships (pp. 206–210)

Determine whether the cost is proportional to the number of books purchased. If the relationship is proportional, find the constant of proportionality. Explain your reasoning.

31.

Books	1	2	3	4
Cost ($)	8	16	24	32

32.

Books	2	4	6	8
Cost ($)	2	5	7	10

33. A customer at the ring toss booth gets 8 rings for $2. Find the constant of proportionality. Write an equation relating the cost to the number of rings. At this same rate, how much would a customer pay for 11 rings? for 20 rings?

34. Mrs. Tebon buys 25 party favors for $5. At this same rate, how much would she pay for 40 party favors? for 60 party favors?

Example 6

Determine whether the distance is proportional to the time. If the relationship is proportional, find the constant of proportionality. Explain your reasoning.

Distance (meters)	30	56	69	80
Time (minutes)	1	2	3	4

Write the rate of distance to time for each minute in simplest form.

$\frac{30}{1}$ $\frac{56}{2} = \frac{28}{1}$ $\frac{69}{3} = \frac{23}{1}$ $\frac{80}{4} = \frac{20}{1}$

Since the rates are not equal, the distance is not proportional to the time, and there is no constant of proportionality.

Lesson 5-6 Graphing Proportional Relationships (pp. 212–217)

35. Determine whether the relationship is proportional by graphing on the coordinate plane. Explain your reasoning.

Time (min)	4	7	8	10
Distance (ft)	8	14	16	20

36. The cost of dance lessons is \$12 for 1 lesson, \$22 for 2 lessons, and \$32 for 3 lessons. Determine whether the cost is proportional to the number of lessons by graphing the ordered pairs on the coordinate plane. Explain your reasoning.

37. The number of squirrels is proportional to the number of trees. A graph of the relationship includes the points (0, 0), (3, 9), and (5, 15).

a. Find and interpret the constant of proportionality.

b. Explain what the points (0, 0), (3, 9), and (5, 15) represent.

Example 7

The number of slices of pizza purchased is proportional to the number of students eating. The graph below shows the relationship (students, slices). Determine the constant of proportionality. Explain what it means.

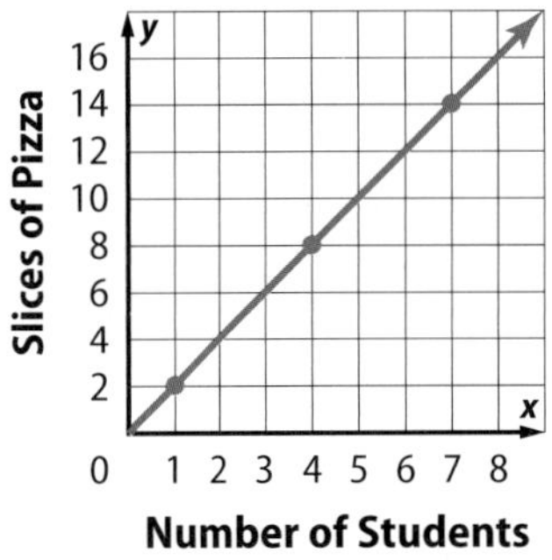

Use the point (4, 8).

$$\frac{\text{slices of pizza}}{\text{number of students}} = \frac{8}{4} = \frac{2}{1} \text{ or } 2$$

The constant of proportionality is the unit rate, 2 slices per student. It describes the number of slices of pizza purchased for every 1 student eating.

Lesson 5-7 Solving Proportions (pp. 218–223)

Solve each proportion.

38. $\frac{15}{a} = \frac{5}{4}$

39. $\frac{m}{6} = \frac{18}{15}$

40. $\frac{28}{24} = \frac{d}{12}$

41. $\frac{16.5}{21} = \frac{5.5}{t}$

42. **Financial Literacy** A homeowner whose house is assessed for \$120,000 pays \$1800 in taxes. At the same rate, what is the tax on a house assessed at \$135,000?

Example 8

Solve $\frac{4}{9} = \frac{9}{x}$.

$\frac{4}{9} = \frac{9}{x}$	Write the proportion.
$4 \cdot x = 9 \cdot 9$	Cross products
$4x = 81$	Multiply.
$\frac{4x}{4} = \frac{81}{4}$	Divide each side by 4.
$x = 20.25$	Simplify.

Lesson 5-8 Scale Drawing and Models (pp. 224–229)

On the scale drawing of a museum, the scale is 0.5 inch = 10 feet. Find the actual length of each gallery.

	Gallery	Drawing Length
43.	Modern Art	6 in.
44.	Renaissance	4.25 in.
45.	Egypt	7.5 in.

46. The length of a highway is 900 miles. If 0.5 inch on a map represents 50 miles, what is the length of the highway on the map?

Example 9

A scale model of a car has a bumper that is 3.5 inches long. The scale on the model is 1 inch = 2 feet. What is the length of the actual car bumper?

model length ···➤ $\frac{\textbf{1 in.}}{\textbf{2 ft}} = \frac{\textbf{3.5 in.}}{\textbf{x ft}}$ ◄··· model length
actual length ···➤ ◄··· actual length

$1 \cdot x = 2 \cdot 3.5$

$x = 7$

The actual length of the car bumper is 7 feet.

Lesson 5-9 Similar Figures (pp. 232–237)

The figures are similar. Determine each missing measure.

47.

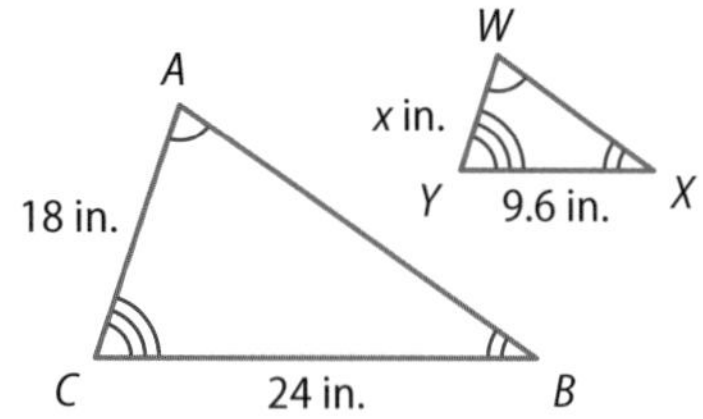

48.

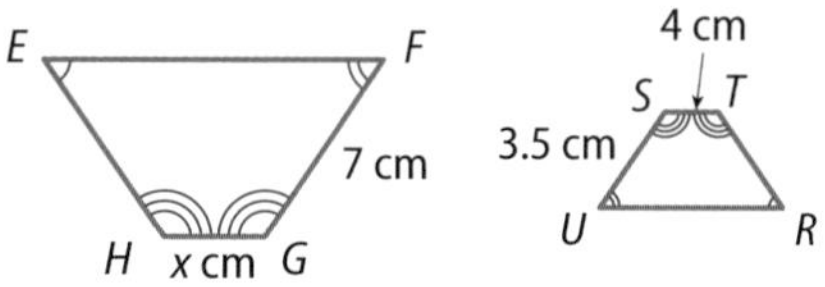

49. A mosaic is created using rectangular blocks. Block *A* has a length of 5 centimeters and a width of 2.5 centimeters. Block *B* is similar to block *A* and has a length of 7 centimeters. What is the width of block *B*?

50. Keshawn enlarges a rectangular photograph to make a poster that is similar to the photograph. The photograph is 4 inches wide and 6 inches long. The poster is 51 inches long. What is the width of the poster?

Example 10

If $\triangle LMN \sim \triangle PQR$, what is the value of x?

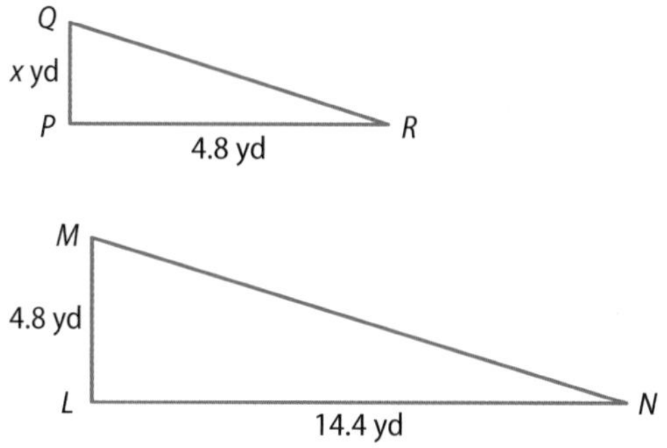

$\frac{LM}{PQ} = \frac{LN}{PR}$ Write a proportion.

$\frac{4.8}{x} = \frac{14.4}{4.8}$ Replace $\overline{LM}$ with 4.8, $\overline{PQ}$ with x, $\overline{LN}$ with 14.4, and $\overline{PR}$ with 4.8.

$x = 1.6$ Find cross products and simplify.

Lesson 5-10 Indirect Measurement (pp. 238–242)

51. At 7 feet 8 inches, the world's tallest woman casts a 46-inch shadow. At the same time, the world's shortest woman casts a 15.5-inch shadow. How tall is the world's shortest woman?

52. The largest known pyramid is the Pyramid of Khufu. At a certain time of day, a vertical yard stick casts a shadow 1.5 feet long, and the pyramid casts a shadow 241 feet long. How tall is the pyramid?

53. Mylie's house is 9 meters tall and casts a shadow 1.5 meters long. At the same time of day, a nearby doghouse casts a shadow that is 0.2 meter long. How tall is the dog house?

54. In the figure below, $\triangle ABE \approx \triangle ACD$. What is the distance across the pond?

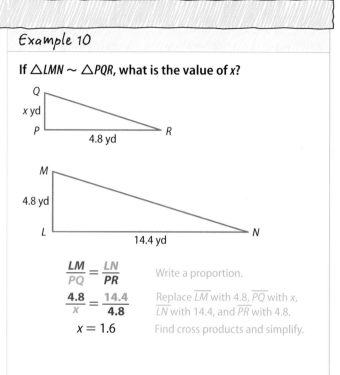

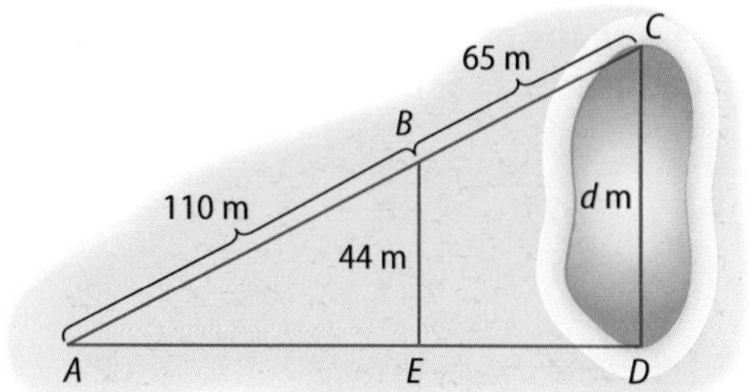

Example 11

The Washington Monument casts a 185-foot shadow at the same time as a nearby flagpole casts a 3-foot shadow. If the flagpole is 9 feet tall, how tall is the Washington Monument?

Write and solve a proportion.

flagpole height → 9 ft, monument height → x ft; flagpole's shadow → 3 ft, monument's shadow → 185 ft

$$\frac{9 \text{ ft}}{x \text{ ft}} = \frac{3 \text{ ft}}{185 \text{ ft}}$$

$9 \cdot 185 = x \cdot 3$ Cross products

$1665 = 3x$ Multiply.

$\frac{1665}{3} = \frac{3x}{3}$ Divide each side by 3.

$555 = x$

The Washington Monument is 555 feet tall.

Chapter 6

Percents

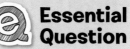

Essential Question

How can you use proportional relationships to solve real-world percent problems?

Common Core State Standards

Content Standards
7.RP.2, 7.RP.2c, 7.RP.3, 7.EE.2, 7.EE.3

Mathematical Practices
1, 3, 4, 5, 6, 7, 8

Math in the Real World

Sports In baseball, batting average, fielding percentage, and on-base percentage are used to represent player performance. For example, the highest on-base percentage of all time is .482. This means a player was on base 48.2% of the times he was at bat.

Interactive Study Guide
See pages 123–126 for:
- Chapter Preview
- Are You Ready?
- Foldable Study Organizer

Inquiry Lab

Percent Models

Inquiry HOW can you use a percent model to model a percent or part of a whole?

CCSS Content Standards
Preparation for 7.RP.3

Mathematical Practices
1, 3

Investigation 1

Advertising An Internet advertisement claims that 3 out of 5 dentists prefer a certain toothpaste. What percent does this represent?

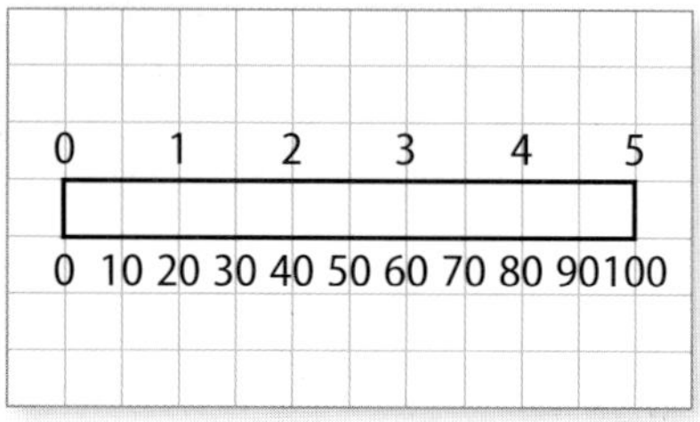

Step 1 Draw a 10-unit by 1-unit rectangle on grid paper. Label the units on the bottom from 0 to 100, because percent is a ratio that compares a number to 100.

Step 2 On the top, mark equal units from 0 to 5, because 5 represents the whole quantity.

Step 3 Draw a horizontal line from 3 on the top to the bottom of the model. The number on the bottom is the percent. Label the model as shown.

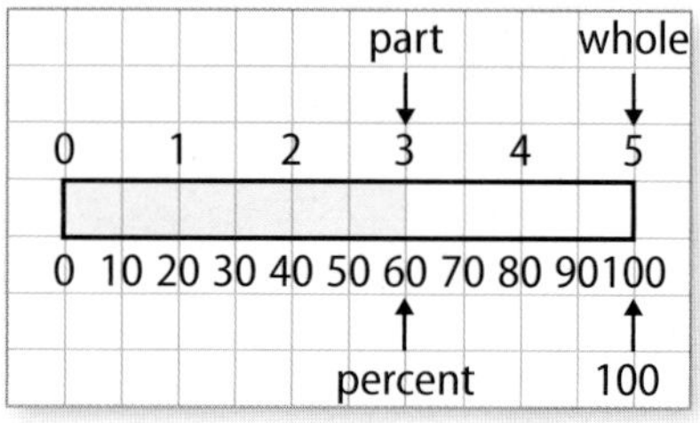

The ratio *3 out of 5* is the same as 60%. So, according to this claim, 60% of dentists prefer this toothpaste.

Investigation 2

Financial Literacy Suppose a store advertises a sale in which all merchandise is 30% off its price. If the price of a cell phone case is \$20, how much will you save?

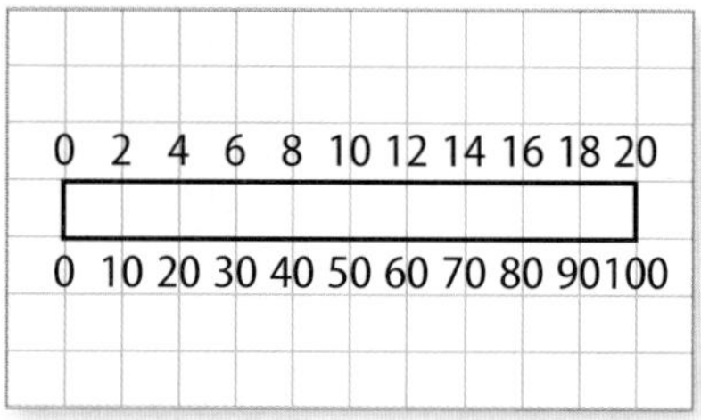

Step 1 Draw a 10-unit by 1-unit rectangle on grid paper. Label the units on the bottom from 0 to 100.

Step 2 On the top, mark equal units from 0 to 20, because 20 represents the whole quantity.

Step 3 Draw a horizontal line from 30% on the bottom to the top of the model. This number is the part. Label the model as shown.

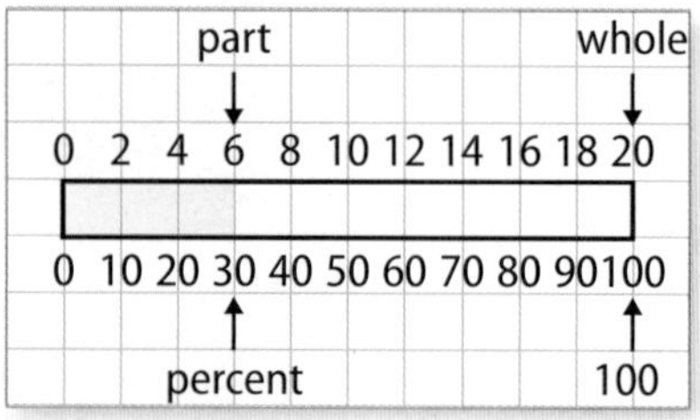

You can see that 30% of 20 is 6. So, you will save \$6 if you buy the cell phone case.

Collaborate

Draw a model and find the percent that is represented by each ratio. If it is not possible to find the exact percent using the model, estimate.

1. 4 out of 10

2. 7 out of 10

3. 4 out of 5

4. 2 out of 4

5. 6 out of 20

6. 5 out of 25

7. 8 out of 40

8. 6 out of 24

9. 1 out of 3

10. 4 out of 9

Draw a model and find the part that is represented. If it is not possible to find an exact answer from the model, estimate.

11. 20% of 80

12. 60% of 15

13. 80% of 40

14. 30% of 50

15. 25% of 50

16. 75% of 60

17. 10% of 150

18. 45% of 500

19. $33\frac{1}{3}$% of 20

20. 90% of 20

Analyze

21. CCSS **Justify Conclusions** Suppose you know that 60% of some number is 18. Use the model below to find the number *x*. Explain your reasoning.

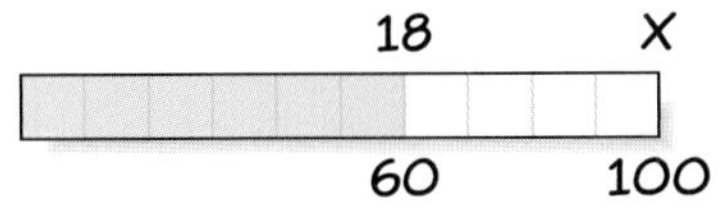

22. CCSS **Make a Conjecture** How would you use a percent model to show 200% of 10? Explain your reasoning.

What does each model show? Give the percent and the part.

23.

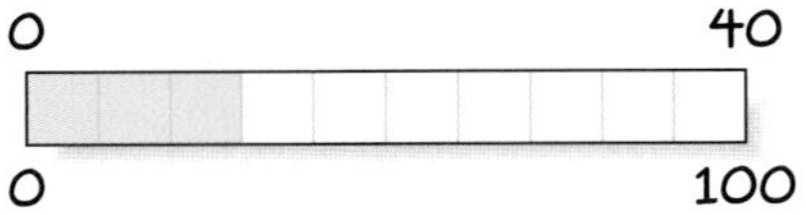

24.

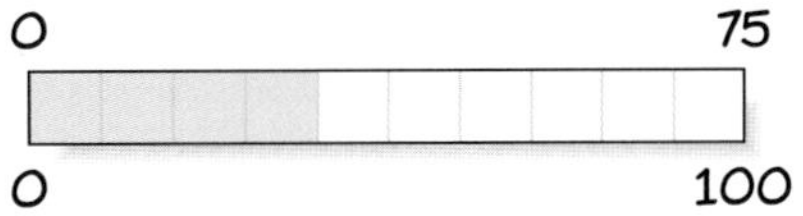

25.

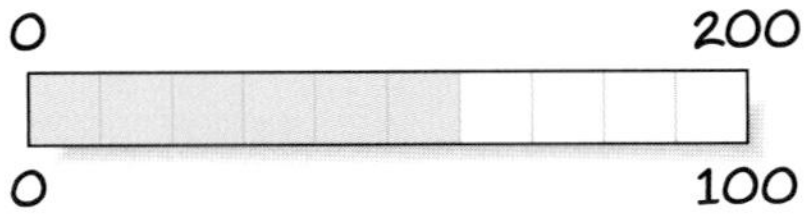

26.

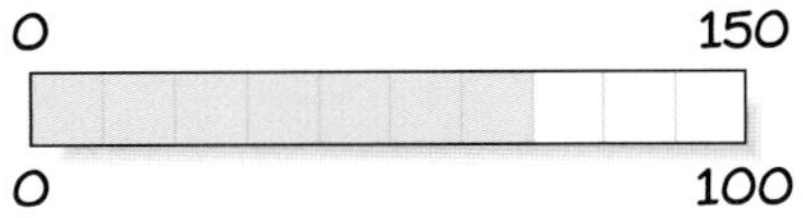

Reflect

27. Inquiry HOW can you use a percent model to model a percent or part of a whole?

Lesson 6-1

Using the Percent Proportion

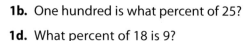

ISG Interactive Study Guide

See pages 127–128 for:
- Getting Started
- Real-World Link
- Notes

Essential Question

How can you use proportional relationships to solve real-world percent problems?

CCSS Common Core State Standards

Content Standards
7.RP.2, 7.RP.2c, 7.RP.3, 7.EE.3

Mathematical Practices
1, 3, 4, 8

Vocabulary

percent proportion

What You'll Learn

- Use the percent proportion to solve problems.
- Apply the percent proportion to real-world problems.

Real-World Link

Snacks With four different kinds of fruit, this healthy fruit salad recipe is the perfect lunch box or after school snack! You can adjust the recipe to make more or less by keeping the proportions the same.

Fruit Salad
2 cups pineapple
1 cup blueberries
3 cups grapes
2 cups strawberries

Key Concept Percent Proportion

Words $\frac{\text{part}}{\text{whole}} = \frac{\text{percent}}{100}$

Symbols $\frac{a}{b} = \frac{p}{100}$, where a is the part, b is the whole and p is the percent.

In a **percent proportion**, one ratio compares *part* of a quantity to the *whole* quantity. The other ratio is the equivalent percent written as a fraction with a denominator of 100.

Example 1

Tutor

Twelve is what percent of 32?

Words Twelve is what percent of 32?

Variable Let p represent the percent.

Proportion part → 12, whole → 32: $\frac{12}{32} = \frac{p}{100}$ } percent

Twelve is being compared to 32. So, 12 is the part and 32 is the whole.

$\frac{12}{32} = \frac{p}{100}$	Write the percent proportion.
$12 \cdot 100 = 32 \cdot p$	Find the cross products.
$1200 = 32p$	Multiply.
$\frac{1200}{32} = \frac{32p}{32}$	Divide each side by 32.
$37.5 = p$	Simplify.

So, 12 is 37.5% of 32.

Got It? Do these problems to find out.

1a. Fifteen is what percent of 20?

1b. One hundred is what percent of 25?

1c. What percent of 5 is 12?

1d. What percent of 18 is 9?

Example 2

Tutor

Check for Reasonableness
Ten percent of 450 is 45. So, the answer should be more than 45.

What number is 15.5% of 450?

The percent is 15.5, and the base is 450. Let a represent the part.

$\frac{a}{450} = \frac{15.5}{100}$ Write the percent proportion.

$a \cdot 100 = 450 \cdot 15.5$ Find the cross products.

$100a = 6975$ Multiply.

$a = 69.75$ Mentally divide each side by 100.

So, 69.75 is 15.5% of 450.

Got It? Do these problems to find out.

2a. What number is 11.4% of 330?

2b. Find 10.5% of 30.

2c. Find 15.3% of 425.

2d. What number is 63.4% of 12?

Example 3

Tutor

Whole
In percent problems, the whole usually follows the word *of*.

Seventy-eight is 60% of what number?

The percent is 60%, and the part is 78. Let b represent the whole.

$\frac{78}{b} = \frac{60}{100}$ Write the percent proportion.

$78 \cdot 100 = b \cdot 60$ Find the cross products.

$7800 = 60b$ Multiply.

$\frac{7800}{\mathbf{60}} = \frac{60b}{\mathbf{60}}$ Divide each side by 60.

$130 = b$ Simplify.

So, 78 is 60% of 130.

Got It? Do these problems to find out.

3a. Thirty percent of what number is 63?

3b. 3000 is 60% of what number?

3c. Forty-five is 3% of what number?

3d. Eighteen percent of what number is 126?

Key words and phrases indicate which type of percent problem you need to solve.

Concept Summary Types of Percent Problems

Type	Example	Proportion
Find the Percent	1 is what percent of 5? or what percent of 5 is 1?	$\frac{1}{5} = \frac{p}{100}$
Find the Part	What number is 20% of 5?	$\frac{a}{5} = \frac{20}{100}$
Find the Whole	1 is 20% of what number?	$\frac{1}{b} = \frac{20}{100}$

Real World Example 4

Tutor

The table shows the batting statistics for one season for Derek Jeter of the New York Yankees. If he had 639 at-bats, what percent of his at-bats were singles? Round to the nearest tenth.

Stat	Number
single	151
double	39
triple	4
home run	12
walk	56
strikeout	100

Compare the number of singles, 151, to the total number of at-bats, 639. Let p represent the percent.

$\frac{151}{639} = \frac{p}{100}$ Write the percent proportion.

$151 \cdot 100 = 639 \cdot p$ Find the cross products.

$15{,}100 = 639p$ Simplify.

$\frac{15{,}100}{\mathbf{639}} = \frac{639p}{\mathbf{639}}$ Divide each side by 639.

$23.6 \approx p$ Simplify.

So, about 23.6% of time Derek's at-bats were singles.

Check for Reasonableness

The number of at-bats is about 640. The fraction $\frac{160}{640}$ is equal to $\frac{1}{4}$, which is 25%. So, $\frac{151}{640}$ would be slightly less than 25%.

The answer is reasonable. ✓

Got It? Do these problems to find out.

4a. To the nearest tenth, what percent of Derek Jeter's at-bats were strikeouts?

4b. Financial Literacy The number of mutual funds that increased in value during the third quarter of the fiscal year was 15. If this is 60% of the total number of mutual funds offered, how many mutual funds were offered?

Guided Practice

Check

Use the percent proportion to solve each problem. (Examples 1-3)

1. 18 is what percent of 72?

2. What percent of 8 is 20?

3. What is 74% of 56?

4. 9 is 20% of what number?

5. What percent of 2 is 8?

6. Find 6% of 300.

7. Find 9% of 255.

8. Find 97% of 900.

9. 16 is 80% of what number?

10. 18 is 5% of what number?

11. Of the 120 math tests, 47 were Bs. To the nearest tenth, what percent of the math tests were Bs? (Example 4)

12. An Aztec Salad gets 28 of its Calories from protein. If this is 20% of the total number of Calories, how many Calories does the salad have? (Example 4)

Independent Practice

Go online for Step-by-Step Solutions eHelp

Use the percent proportion to solve each problem. (Examples 1-3)

13. 4% of what number is 10?

14. 55% of what number is 22?

15 21 is what percent of 50?

16. 16 is what percent of 64?

17. What percent of 145 is 52.2?

18. What percent of 36 is 19.8?

19. What is 60% of 120?

20. What is 80% of 125?

21. Find 65% of 440.

22. Find 83% of 200.

23. 12 is 40% of what number?

24. 34 is 20% of what number?

25. 80% of what number is 12?

26. 4% of what number is 15?

27. Sixteen of the 80 dogs at a kennel are golden retrievers. What percent of the dogs at the kennel are golden retrievers? (Example 4)

28. The number of lime-flavored gumballs in a gumball machine is 85. If this is 17% of the number of gumballs in the machine, how many gumballs are in the machine? (Example 4)

29 Use the circle graph that shows the results of a survey about New Year's resolutions.

a. Determine about how many of the 2947 people surveyed said their most important New Year's resolution was to get organized.

b. About how many said their most important New Year's resolution was to do more reading or to declutter?

New Year's Resolution

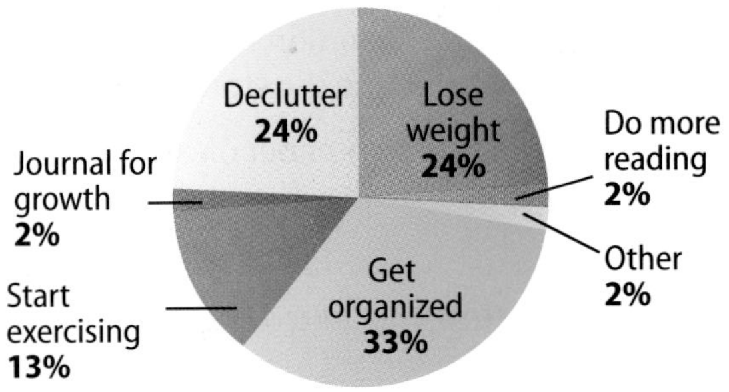

30. **Financial Literacy** Accessories Central is having a summer clearance on sunglasses. Maria wants to buy a pair of sunglasses that cost \$48 with a 65% discount. The same pair of sunglasses costs \$38 with a 55% discount at Shades, Inc. Which store has the better price for the pair of sunglasses? Explain your reasoning.

31. CCSS **Identify Repeated Reasoning** A pattern of equations is shown.

2% of 100 = 2   4% of 50 = 2   8% of 25 = 2   16% of 12.5 = 2

a. Describe the pattern.

b. Find the next equation in the pattern.

32. The bar graph shows the results of an online survey of 1242 people aged 15–25 about their political involvement over the last 12 months.

a. What percent of the people boycotted? Round to the nearest percent.

b. Of the people who signed an e-mail petition, 20% were 18 years old. How many 18-year-olds signed an e-mail petition?

c. Based on the results of the survey, predict about how many people out of 2000 would contact an official.

Political Involvement

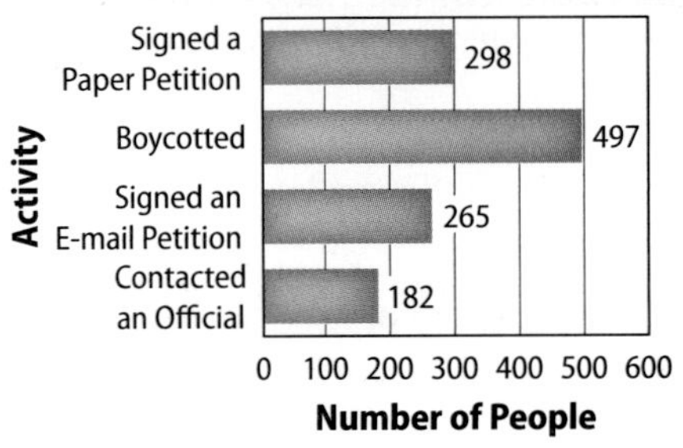

Use the percent proportion to solve each problem. Round to the nearest tenth if necessary.

33. 45 is what percent of 15?

34. 13 is 25% of what number?

35. What is 58% of 7?

36. 8 is what percent of 2000?

37. What is 0.6% of 360?

38. 41 is $5\frac{1}{3}\%$ of what number?

39. The table shows the results of a student survey.

a. What percent of the students surveyed said their favorite act was playing instruments?

b. If 10 more students vote for stand-up comedy, will the percent be greater than, less than, or equal to the current percent? Explain.

Favorite Type of Talent Show Act	
Act	**Number of Students**
stand-up comedy	198
singing	150
dancing	212
playing instruments	80

40. Use the Internet or another source to find the percent of states that begin with *A*, *I*, *O*, or *U*.

41. A cell phone store has 120 cell phones in stock. Of these, 45 have keyboards. The manager of the store wants to add more cell phones with keyboards so that 40% of the stock has keyboards.

a. Write and solve a proportion to find the number of cell phones with keyboards that should be added to the store's inventory.

b. What will be the total number of cell phones in stock?

H.O.T. Problems Higher Order Thinking

42. **Model with Mathematics** Write and solve a real-world problem involving percents.

43. **Persevere with Problems** Without calculating, arrange the following from least to greatest value. Justify your reasoning.

25% of 160, 5% of 80, 25% of 80

44. **Justify Conclusions** Sabrina spelled 82% of her spelling words correctly on her spelling tests this year. If she spells 13 out of 15 words correctly on her next test, will this help or hurt her average? Explain.

45. **Persevere with Problems** Find the value of *y* so that $y\% = \frac{3y + 9}{600}$.

46. **Find the Error** Bethany is finding what percent of 32 is 18. Find her mistake and correct it.

$$\frac{18}{32} = \frac{x}{100}$$

$$x \approx 56$$

So, 32 is about 56% of 18.

47. **Make a Conjecture** Is *x*% of *y* and *y*% of *x* *always*, *sometimes*, or *never* equivalent? Explain your reasoning.

48. **Building on the Essential Question** Kenji states that 5.9 is 125% of 47. Is his answer reasonable? Explain your reasoning.

Standardized Test Practice

49. The table shows the results of a survey of middle school students about their favorite school mascots.

Mascot	Number of Students
Falcon	60
Stallion	123
Ram	86
Tiger	131

Based on the data, predict how many out of 2000 students would vote for the falcon.

A 600
B 400
C 300
D 200

50. **Short Response** If 68 is 25% of a number, what is 600% of the number?

51. A place kicker expects to make 75% of his field goal attempts this season. If he attempts 36 field goals this season, which of the following statements does *not* represent the place kicker's expectation?

F The place kicker will make 27 field goals.
G The place kicker will miss 9 field goals.
H Less than $\frac{1}{4}$ of the field goal attempts will be missed.
J The place kicker will make more than $\frac{1}{2}$ of his field goal attempts.

52. If 40% of a number is 32, what is 35% of the number?

A 8
B 20
C 24
D 28

CCSS Common Core Review

53. In a survey, 35% of those surveyed said that they get the news from their local television station while three-fifths said that they get the news from a daily newspaper. From which source do more people get their news? **6.NS.7b**

Write each percent as a fraction in simplest form. 7.EE.3

54. 45% **55.** 120% **56.** 0.5% **57.** $83\frac{1}{3}\%$

58. In a drawing of a honeybee, the bee is 4.8 centimeters long. The actual size of the honeybee is 1.2 centimeters. What is the scale of the drawing? **7.G.1**

Find each sum or difference. Write in simplest form. 7.NS.1d

59. $\frac{17}{18} - \frac{5}{18}$ **60.** $\frac{3}{10} + \frac{7}{10}$ **61.** $\frac{1}{2} - \frac{4}{5}$

62. $\frac{7}{15} + \frac{1}{6}$ **63.** $\frac{3}{4} - \frac{4}{9}$ **64.** $\frac{1}{2} - \frac{7}{8}$

Find each product. 7.NS.2a

65. $\frac{1}{4} \times 12$ **66.** $\frac{3}{4} \times 24$ **67.** $38 \times \frac{1}{2}$ **68.** $15 \times \frac{1}{3}$

69. A police officer notes that a car traveled 200 feet in 2.2 seconds. What was the speed of the car in miles per hour? Round to the nearest tenth. **7.RP.3**

Express each number in scientific notation. 8.EE.1

70. 506,000 **71.** 10,300,000 **72.** 0.0208

Lesson 6-2

Find Percent of a Number Mentally

ISG Interactive Study Guide

See pages 129–130 for:
- Getting Started
- Real-World Link
- Notes

Essential Question

How can you use proportional relationships to solve real-world percent problems?

CCSS Common Core State Standards

Content Standards
7.RP.3, 7.EE.3

Mathematical Practices
1, 3, 4, 5

What You'll Learn

- Compute mentally with percents.
- Estimate with percents.

Real-World Link

Thrill Rides Do you enjoy thrill rides? Percents can be used to describe the experience, whether you are traveling seventy miles per hour through twists, turns, and loops, or being blasted up and down 300-foot tall towers.

Purestock/Getty Images

Find Percent of a Number Mentally

The number line shows some common percent-fraction equivalents.

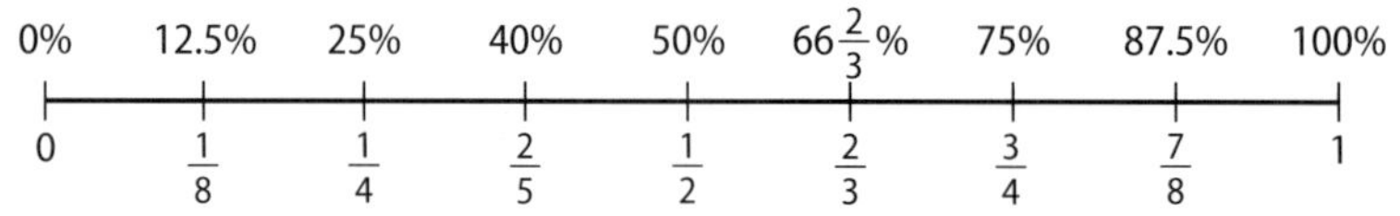

Concept Summary Percent-Fraction Equivalents

$25\% = \frac{1}{4}$	$20\% = \frac{1}{5}$	$10\% = \frac{1}{10}$	$12\frac{1}{2}\% = \frac{1}{8}$	$16\frac{2}{3}\% = \frac{1}{6}$
$50\% = \frac{1}{2}$	$40\% = \frac{2}{5}$	$30\% = \frac{3}{10}$	$37\frac{1}{2}\% = \frac{3}{8}$	$33\frac{1}{3}\% = \frac{1}{3}$
$75\% = \frac{3}{4}$	$60\% = \frac{3}{5}$	$70\% = \frac{7}{10}$	$62\frac{1}{2}\% = \frac{5}{8}$	$66\frac{2}{3}\% = \frac{2}{3}$
$100\% = \frac{1}{1}$	$80\% = \frac{4}{5}$	$90\% = \frac{9}{10}$	$87\frac{1}{2}\% = \frac{7}{8}$	$83\frac{1}{3}\% = \frac{5}{6}$

When you compute with common percents like 40% or 50%, it may be easier to use the fraction form of the percent.

Example 1

Tutor

Find the percent of each number mentally.

a. 75% of 24

$75\% \text{ of } 24 = \frac{3}{4} \text{ of } 24$ Think: $75\% = \frac{3}{4}$

$= 18$ Think: $\frac{3}{4}$ of 24 is 18.

b. 80% of 60

$80\% \text{ of } 60 = \frac{4}{5} \text{ of } 60$ Think: $80\% = \frac{4}{5}$

$= 48$ Think: $\frac{4}{5}$ of 60 is 48.

Got It? **Do these problems to find out.**

1a. 40% of 50

1b. 30% of 70

Example 2

Tutor

Compute mentally.

a. **10% of 76**

10% of 76 = 0.1 • 76 or 7.6

b. **1% of 122**

1% of 122 = 0.01 • 122 or 1.22

Got It? **Do these problems to find out.**

2a. 10% of 42

2b. 1% of 264

Real World

Example 3

Tutor

Hannah has a coupon for 20% off her entire clothing purchase. If the items she buys cost $110 originally, how much will she save with her coupon?

You need to find 20% of the total cost. First, find 10% of 110.

10% of 110 = 0.1 • 110 — Move the decimal point one place to the left.

= 11

20% is the same as 2 • 10%.

2 • 10% of 110 = 2 • 11 = 22 — Replace 10% of 110 with 11.

So, Hannah will save $22 on her purchase.

Got It? **Do this problem to find out.**

3. A $750 television is on sale for 15% off. What is the total discount?

Estimate With Percents

You can estimate when an exact answer is not needed.

Determine Reasonable Answers
Deciding whether an answer is reasonable is useful when an exact answer is not necessary.

Example 4

Tutor

Estimate.

a. **26% of 64**

26% is about 25% or $\frac{1}{4}$.

$\frac{1}{4}$ of 64 is 16.

So, 26% of 64 is about 16.

b. **$\frac{2}{3}$% of 891**

$\frac{2}{3}\% = \frac{2}{3} \times 1\%$. 1% of 900 is 9.

891 is almost 900.

So, $\frac{2}{3}$% of 891 is about $\frac{2}{3} \times 9$ or 6.

c. **39% of 81**

39% is about 40% or $\frac{2}{5}$.

81 is about 80. $\frac{2}{5}$ of 80 is 32.

So, 39% of 81 is about 32.

d. **120% of 51**

100% of 50 is 50. 20% of 50 is 10.

So, 120% of 51 is about 50 + 10 or 60.

Got It? **Do these problems to find out.**

4a. 92% of 50 **4b.** 63% of 205 **4c.** 75% of 84 **4d.** 130% of 91

Example 5

Mr. Williams ordered 4 pizzas for a birthday party. The cost of the pizzas was \$57.96. He wants to tip the delivery person about 15%. What is a reasonable amount for the tip?

Estimate the price of the pizzas. Then find 15% of the estimated price.

\$57.96 is about \$60, and 15% = 10% + 5%.

10% of \$60 is \$6.00. Move the decimal point 1 place to the left.

5% of \$60 is \$3.00 5% is one half of 10%.

So, 15% is about \$6.00 + \$3.00 or \$9.00.

A reasonable amount for the tip is \$9.

Check for Reasonableness

10% of \$58 is \$5.80 and 20% of \$58 is \$11.60.
Since $\$5.80 < \$9 < \$11.60$, the answer is reasonable. ✓

Got It? Do this problem to find out.

5. Haley went to dinner with her friends. Their bill was \$48.61. They want to leave their server a 15% tip. What would be a reasonable amount for the tip? Explain your reasoning.

Guided Practice

Find the percent of each number mentally. (Examples 1 and 2)

1. 75% of 16
2. 25% of 32
3. 10% of 37
4. 10% of 115
5. 1% of 72
6. 1% of 231
7. 80% of 200
8. $33\frac{1}{3}$% of 15
9. $62\frac{1}{2}$% of 40

10. Jasmine has finished 30% of the exercises on her homework. If there are 40 exercises in all, how many has Jasmine completed? (Example 3)

Estimate. (Example 4)

11. 11% of 70
12. 53% of 20
13. 40% of 19
14. 87% of 42
15. $\frac{1}{3}$% of 598
16. 110% of 39
17. 24% of 359
18. 91% of 1989
19. $37\frac{1}{2}$% of 81

20. Last basketball season, Carlos made 38% of the baskets he attempted. At this rate, about how many baskets will he make if he attempts 30 baskets? (Example 5)

21. STEM Lea bought a package of seeds to grow flowers. There were 72 seeds in the package. Approximately 60% of the seeds will germinate. About how many of the seeds that Lea plants will germinate? (Example 5)

Independent Practice

Go online for Step-by-Step Solutions eHelp

Find the percent of each number mentally. (Examples 1 and 2)

22. 40% of 80
23. 20% of 50
24. 25% of 280
25. 75% of 96
26. $33\frac{1}{3}$% of 27
27. $12\frac{1}{2}$% of 48
28. $8\frac{1}{3}$% of 72
29. $87\frac{1}{2}$% of 32
30. 10% of 125
31. 10% of 259
32. 1% of 30
33. 1% of 400

34. A store is having a sale where everything is 15% off. If Jeremy wants to buy items that originally cost \$50, how much will he save? (Example 3)

Estimate. (Example 4)

35. 16% of 20
36. 73% of 84
37. 46% of 88
38. 25% of 49
39. $\frac{1}{2}$% of 507
40. $\frac{1}{6}$% of 295
41 148% of 30
42. 276% of 8
43. $\frac{3}{4}$% of 801
44. $\frac{4}{5}$% of 30
45. 117% of 50
46. 194% of 15

47. **Financial Literacy** The total cost for Soledad's manicure was \$32.99. She wants to give the manicurist a 20% tip. What would be a reasonable amount for the tip? (Example 5)

48. CCSS **Justify Conclusions** In a national survey of 6700 teens, 81% of teens between the ages of 12 and 17 said they use the Internet to e-mail friends or relatives. About how many teens is this? Explain your answer. (Example 5)

49. STEM The bar graph shows the percent of each age group that has a personal profile page on a social network. Suppose there are 796 students in the 12–17-year-old age range in a school district. About how many of them are likely to have a profile page on a social network?

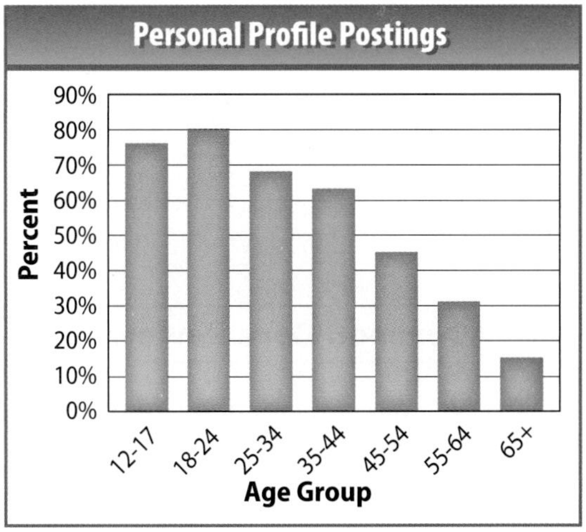

50. On a family trip, Jenna's family drove 310 miles from Gainesville to Ft. Lauderdale.

a. Her dad drove 52% of the way. About how many miles did he drive?

b. Jenna's mom drove 58% of the distance her dad drove. About how far did she drive?

c. Her older brother drove the remaining miles. About how many miles did he drive?

51 About 41% of twelfth graders participated in school performing arts last year. A high school had 1800 students, one-fourth of which were twelfth graders. About how many twelfth graders participated in school performing arts?

H.O.T. Problems Higher Order Thinking

52. **Justify Conclusions** Suppose you want to find $66\frac{2}{3}\%$ of a. List two values of a for which you could do the computation mentally. Explain your reasoning.

53. **Persevere with Problems** Find two numbers, x and y, such that 10% of x is the same as 40% of y. Explain your reasoning.

54. **Model with Mathematics** Find any receipt from a store or restaurant. Estimate a 15% tip to add to the total shown. Then find the amount of the tip and the total amount.

55. **Use Math Tools** Tobie estimated 62% of 403 as 240, while Janine estimated the same quantity as 250. How did each student obtain her estimate?

56. **Building on the Essential Question** Describe two different ways you could find 20% of 60 mentally.

Standardized Test Practice

57. Lorena, Julian, and Cho completed a group assignment that had 84 questions. Lorena answered $\frac{1}{3}$ of the questions, Julian answered 25% of the questions, and Cho answered the rest. How many questions were answered by the person who answered the greatest number of questions?

A 18 **C** 28

B 21 **D** 35

58. Which fraction is between 85% and 90%?

F $\frac{5}{6}$ **H** $\frac{9}{10}$

G $\frac{7}{8}$ **J** $\frac{10}{11}$

59. Jerome bought the items listed with the original prices shown on the receipt. If he saved 20% on each item, what is the *best* estimate of how much he saved?

RECEIPT

Qty.	Item	Amount
1	Network Cable	$50.00
1	CD 10-pack	$12.00
1	Mouse	$24.00
1	Memory	$55.00

— Thank you, come again. —

A $28 **C** $36

B $32 **D** $40

60. **Short Response** The price of Jamila's haircut is $28. There is also a 7% tax added to the bill and Jamila wants to tip the stylist 15% of the total bill. Find the approximate cost including tax and tip. Justify your answer.

Common Core Review

61. According to a survey about family activities, 35% of people said they enjoy playing games, while three-fifths enjoy watching movies, and $\frac{3}{8}$ enjoy sports. Which activity is the favorite? Explain. 6.NS.7b

Solve each proportion. 7.NS.3

62. $\frac{k}{35} = \frac{3}{7}$ **63.** $\frac{3}{t} = \frac{18}{24}$ **64.** $\frac{10}{8.4} = \frac{5}{m}$

Solve each equation. Check your solution. 6.EE.7

65. $5 = 30b$ **66.** $40n = 10$ **67.** $20 = 100k$

68. $34g = 1.7$ **69.** $3.6 = 90a$ **70.** $200j = 70$

Lesson 6-3

Using the Percent Equation

ISG Interactive Study Guide

See pages 131–132 for:
- Getting Started
- Vocabulary Start-Up
- Notes

Essential Question

How can you use proportional relationships to solve real-world percent problems?

CCSS Common Core State Standards

Content Standards
7.RP.2, 7.RP.2c, 7.RP.3, 7.EE.3

Mathematical Practices
1, 3, 4, 7

Vocabulary

percent equation

What You'll Learn

- Solve percent problems using percent equations.
- Solve real-world problems involving taxes.

Real-World Link

Elephants An African elephant can walk forward and backward, travel up to 25 miles per hour, and climb mountainous terrain—while weighing up to 15,000 pounds! Asian elephants are the African elephants' smaller cousins.

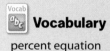

Percent Equation

A **percent equation** is an equivalent form of the percent proportion in which the percent is written as a decimal.

$\frac{\text{Part}}{\text{Whole}} = \text{Percent}$ — The percent is written as a decimal.

$\frac{\text{Part}}{\cancel{\text{Whole}}} \cdot \cancel{\text{Whole}} = \text{Percent} \cdot \text{Whole}$ — Multiply each side by the whole.

$\text{Part} = \text{Percent} \cdot \text{Whole}$ — This form is called the *percent equation.*

Example 1

Tutor

Find 62% of 75.

Estimate $\frac{3}{5}$ of 75 is 45.

The percent is 62 and the whole is 75. You need to find the part.

Words	What number is 62% of 75?
Variable	Let a represent the part.
Equation	part = percent • whole $a = 0.62 \cdot 75$

$a = 0.62 \cdot 75$ — Write the percent equation.

$= 46.5$ — Multiply.

Check for Reasonableness $46.5 \approx 45$ ✓

Got It? **Do these problems to find out.**

1a. Find 60% of 96.

1b. Find 45% of 70.

Estimation
To determine whether your answer is reasonable, estimate before finding the exact answer.

Example 2

287 is what percent of 410? **Estimate** $\frac{287}{410} \approx \frac{300}{400}$ or $\frac{3}{4}$, which is 75%.

The whole is 410 and the part is 287. Let p represent the percent.

$$\underbrace{\text{part}} = \underbrace{\text{percent}} \cdot \underbrace{\text{whole}}$$

$287 = p \cdot 410$ Write the percent equation.

$\frac{287}{\mathbf{410}} = \frac{p \cdot 410}{\mathbf{410}}$ Division Property of Equality

$0.7 = p$ Simplify.

By definition, the percent is expressed as a decimal. Convert 0.7 to a percent.

Since 0.7 = 70%, 287 is 70% of 410.

Check for Reasonableness $70 \approx 75\%$ ✓

Got It? Do these problems to find out.

2a. 15 is what percent of 125?

2b. 20 is what percent of 400?

Watch Out!
Remember that the percent is written as a decimal in the percent equation. So, use 0.55, not 55.

Example 3

33 is 55% of what number? **Estimate** 33 is 50% of 66.

The part is 33, and the percent is 55%. Let b represent the whole.

$$\underbrace{\text{part}} = \underbrace{\text{percent}} \cdot \underbrace{\text{whole}}$$

$33 = 0.55 \cdot b$ Write the percent equation.

$\frac{33}{\mathbf{0.55}} = \frac{0.55b}{\mathbf{0.55}}$ Division Property of Equality

$60 = b$ Simplify.

So, 33 is 55% of 60.

Check for Reasonableness $60 \approx 66$ ✓

Got It? Do these problems to find out.

3a. 18 is 30% of what number?

3b. 79 is 80% of what number?

The table summarizes the three types of percent problems.

Concept Summary: The Percent Equation

Type	Example	Equation
Find the Percent	15 is what percent of 60?	$15 = p(60)$
Find the Part	What number is 25% of 60?	$a = 0.25(60)$
Find the Whole	15 is 25% of what number?	$15 = 0.25b$

Solve Real-World Problems Involving Taxes

The percent equation can be used to solve real-world problems involving taxes.

Real World Example 4

Tutor

A camera costs $250. If a 6% sales tax is added, what is the total cost?

Method 1 **Find the tax first. Then add.**

Find the amount of the tax, or the part. Let t represent the amount of tax.

$t = 0.06 \cdot 250$ Write the percent equation, writing 6% as a decimal.

$= 15$ Multiply.

The tax is \$15. The total cost is \$250 + \$15 or \$265.

Method 2 **Find the total percent first.**

Find 100% + 6% or 106% of \$250. Let T represent the total cost, including tax.

$T = 1.06 \cdot 250$ Write the percent equation, writing 106% as a decimal.

$= 265$ Multiply.

Using either method, the total cost is \$265.

Percent Increase
For a value of x, an increase of 6% means $x + 0.06x$. It is the same as $1.06x$.

Got It? **Do this problem to find out.**

4. Financial Literacy Mr. Potter bought a house for \$175,000. Five years later, he sold it for a 24% profit. What was the sale price of the house?

Real World Example 5

Tutor

Mr. Li bought a memory card for \$138.89 including tax. The card had a sticker price of \$129.20. What percent sales tax did he pay?

Method 1 **Use the percent equation to find the percent of sales tax.**

The tax is \$138.89 − \$129.20 or \$9.69.

$9.69 = p \cdot 129.20$ Write the percent equation.

$\frac{9.69}{\mathbf{129.20}} = \frac{p \cdot 129.20}{\mathbf{129.20}}$ Division Property of Equality

$0.075 = p$ Simplify.

So, since 0.075 = 7.5%, Mr. Li paid 7.5% sales tax.

Method 2 **Divide the total cost of the memory card by the sticker price.**

$T = \frac{138.89}{129.20}$ Divide.

$= 1.075$

The total cost is 1.075 or 107.5% of the sticker price, so the tax is 7.5%.

Got It? **Do this problem to find out.**

5. A \$45.00 mixer sold for \$47.70 with tax. What is the percent of sales tax?

Guided Practice

Check

Solve each problem using a percent equation. (Examples 1-3)

1. What is 40% of 75?

2. Find 13% of 27.

3. 30 is what percent of 90?

4. 15 is what percent of 300?

5. 55 is 20% of what number?

6. 24 is 80% of what number?

7. Last year, Kimberly sold 95 boxes of cookies. This year she wants to sell 20% more boxes than she sold last year. How many boxes will Kimberly have to sell this year to reach her goal? (Example 4)

8. Martin wants to buy a motor scooter. The cost of a motor scooter is \$4968. If the total, including tax, is \$5290.92, what is the percent of sales tax? (Example 5)

Independent Practice

Go online for Step-by-Step Solutions

eHelp

Solve each problem using a percent equation. (Examples 1-3)

9. Find 16% of 64.

10. What is 36% of 50?

11. 8 is what percent of 40?

12. 54 is what percent of 60?

13. 16 is 25% of what number?

14. 64 is 32% of what number?

15. 39 is 50% of what number?

16. 27 is 10% of what number?

17. A commission is a fee paid to a salesperson based on a percent of sales. Suppose a salesperson at a jewelry store earns a 6% commission. What commission would be earned for selling a ring that costs \$1300 dollars? (Example 4)

18. Roberto wants to buy a new ski jacket that costs \$96. If the total cost, including tax, is \$101.28, what is the percent of sales tax? (Example 5)

Solve each problem using a percent equation.

19. Find 52.5% of 76.

20. Find 23.6% of 90.

21. 33.8 is what percent of 130?

22. 79.8 is what percent of 114?

23. **Financial Literacy** The cost, including a 6.75% sales tax, of a digital home theater system with a 40-inch high-definition television is \$2668.75. What is the original cost of the television and theater system?

24. The results of a Wimbledon Women's Championship match is shown in the table.

a. What was Bartoli's percent of receiving points won?

b. Which player had a greater percent of their first serves in?

c. Suppose in Williams' next match she has 16 break point opportunities. Based on this match, how many times will she convert on break point opportunities?

	Marion Bartoli	Venus Williams
1st Serves In	40 of 63	35 of 50
Receiving Points Won	16 of 50	30 of 63
Break Point Conversions	1 of 2	4 of 10
Net Approaches	3 of 6	12 of 17

25 A car museum wants to increase its collection by 20% over the next year. Currently, the museum has 120 cars in its collection.

a. Write and solve a multiplication equation to find how many cars the museum will have in the next year. How many cars will the museum need to add over the next year to meet its goal?

b. Make a table to find the number of cars in the museum collection if it increases by 5%, 15%, 25%, and 35%.

26. The table shows the area of the Great Lakes.

a. About what percent of the Great Lakes is covered by Lake Erie?

b. About what percent of the Great Lakes is covered by Lake Huron?

c. Suppose the area of Lake Michigan was decreased by 8%. Find its new area.

Lake	Area (square miles)
Ontario	7320
Erie	9922
Michigan	22,316
Huron	23,011
Superior	31,698

27. CCSS **Multiple Representations** In this problem, you will investigate percent relationships. In 2010, Aida saved $500. She plans to save 6% more than her previous year's savings for the next several years.

a. Symbols Write and solve a multiplication equation to find how much money she will save next year.

b. Table Let x represent the year and y represent the amount of money she has saved. Make a table using the x-values for 2010–2015.

c. Analyze Does Aida's savings increase by a constant amount each year? Explain.

Use the percent equation to solve each problem if $x = 10$.

28. $(2x)$ is 4% of what number?

29. Find $(4x)$% of 240.

H.O.T. Problems Higher Order Thinking

30. CCSS **Identify Structure** Write two real-world percent problems in which the solution is 30%.

31. CCSS **Persevere with Problems** If you found the percent of a number and the part is greater than the number, what do you know about the percent? Explain.

32. CCSS **Find the Error** Todd is finding what percent of 80 is 28. Find his mistake and correct it.

$p = 80 \cdot 0.28$
$p = 22.4$
So, 28 is 22.4% of 80.

33. CCSS **Justify Conclusions** Does taking a 10% discount on an item then adding a 10% sales tax result in the original price of the item? Support your answer with an example.

34. **Building on the Essential Question** Write two different expressions to find the total cost of an item with a price $\$y$ if the sales tax is 8%. Explain why they give the same result.

Standardized Test Practice

35. Interest on a savings account is calculated every quarter of a year. During the first quarter, Alejandra earned $54.84 in interest. This was 2% of her savings. How much was Alejandra's savings?

A $274.20 **C** $5484.00
B $2742.00 **D** $5593.68

36. A lawyer earns an annual salary of $65,490 and receives a raise. The lawyer's new annual salary is $68,109.60. About what percent of a raise did the lawyer receive?

F 3% **H** 5%
G 4% **J** 6%

37. The cost of Nate's dinner including a 15% tip was $43.70. What was the cost of dinner alone?

A $38.00 **C** $5.70
B $37.50 **D** $4.30

38. **Short Response** The table shows the capacity of two football stadiums. Suppose 75% of Ben Hill Griffin Stadium is filled and 73% of L.A. Coliseum is filled. Which stadium has more people in it? How many more?

Stadium	Capacity
L.A. Coliseum	91,000
Ben-Hill Griffin	88,548

CCSS Common Core Review

Find the percent of each number mentally. 7.RP.3

39. 75% of 64 **40.** 25% of 52 **41.** $33\frac{1}{3}$% of 27

42. STEM Carbon makes up 18.5% of the human body by weight. Determine the amount of carbon in a person who weighs 145 pounds. Round to the nearest tenth. 7.RP.3

43. The Skyway Snack Company makes a snack mix that contains raisins, peanuts, and chocolate pieces as shown in the table at the right. Suppose the company wants to sell a larger-sized bag that contains 6 cups of raisins. How many cups of chocolate pieces and peanuts should be added? 7.NS.3

Skyway's Snack Mix	
Ingredient	**Amount (cups)**
raisins	1
peanuts	$\frac{1}{2}$
chocolate pieces	$\frac{1}{3}$

Convert each rate using dimensional analysis. 6.RP.3d

44. 45 mi/h = ■ ft/s **45.** 18 mi/h = ■ ft/s

46. 26 cm/s = ■ m/min **47.** 32 cm/s = ■ m/min

Write each expression using exponents. 8.EE.1

48. 33 • 33 • 33 • 33 **49.** $xy \cdot xy \cdot xy \cdot xy \cdot xy$

50. $4 \cdot 25 \cdot a \cdot a \cdot a \cdot a \cdot b \cdot b$ **51.** $3(z - 8)(z - 8)(z - 8)(z - 8)$

Find the constant of proportionality for each table. 7.RP.2b

52.

Spiders	5	10	15	20
Legs	40	80	120	160

53.

Targets hit	2	3	4	5
Points	90	135	180	225

54. The record high temperature in Kentucky was 114°F in Greensburg in 1930. The record low temperature was −37°F in Shelbyville in 1994. What is the difference in these temperatures? 7.NS.1b

ISG Interactive Study Guide

See page 133 for:
• Mid-Chapter Check

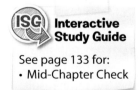

21st Century Career in Video Game Design

Video Game Designer

Are you passionate about computer gaming? You might want to explore a career in video game design. A video game designer is responsible for a game's concept, layout, character development, and game-play. Game designers use math and logic to compute how different parts of a game will work.

College & Career Readiness

Explore college and careers at ccr.mcgraw-hill.com

Is This the Career for You?

Are you interested in a career as a video game designer? Take some of the following courses in high school.

- 3-D Digital Animation
- Introduction to Computer Literacy
- Introduction to Game Development

Find out how math relates to a career in Video Game Design.

Interactive Study Guide
See page 134 for:
- Problem Solving
- Career Project

Inquiry Lab

Percent of Change

Inquiry HOW can you use a percent model to model percent of change?

CCSS **Content Standards**
7.RP.3, 7.EE.3
Mathematical Practices
1, 3, 4

Investigation 1

Sales A sleeping bag sells for $60 and then the price increases by 25%. What is the new price of the sleeping bag?

Step 1 Draw a bar diagram to represent 100% and the original price of $60. Leave room on the right side to extend the bar.

Step 2 Divide the bar into four equal parts so that each part represents the percent of increase. Label the percent of increase on the bottom of the diagram. Label the corresponding dollar amount on the top.

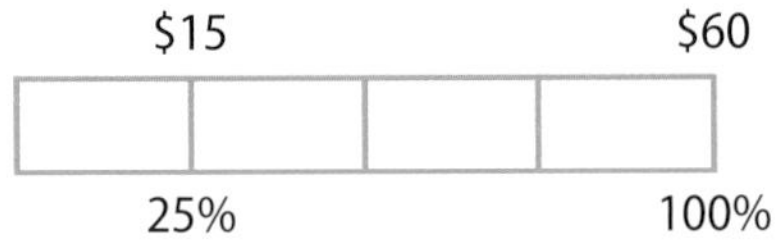

Step 3 The price of the sleeping bag increases by 25%, so add another equal part to the bar to represent the increase. Label the top and bottom of the diagram.

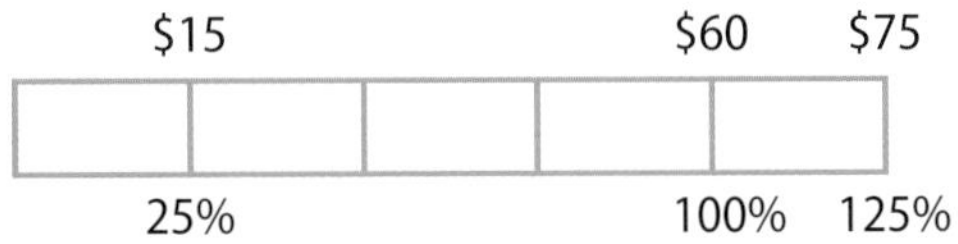

The bar diagram shows that the price of the sleeping bag after a 25% increase is $75.

You can use percents to describe a change when a number increases or decreases.

Investigation 2

Sales Suppose a tent sells for $80 and the price increases to $96. What percent does the amount of increase represent?

Step 1 Draw a bar diagram to represent the original price. Leave room on the right side to extend the bar. Label the bar to represent the current price of the tent.

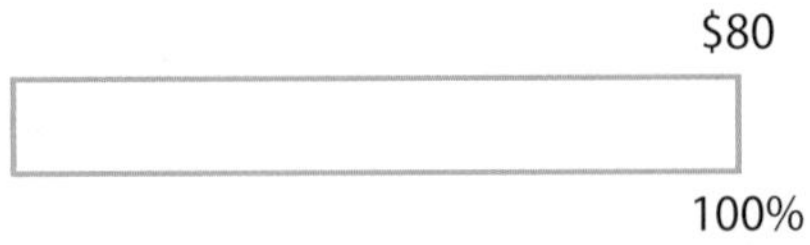

Step 2 Divide the bar diagram into equal parts so that the amount of increase is represented by one or more of the parts.

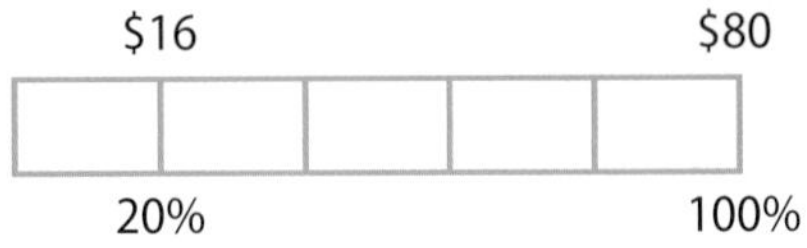

Since $\$96 - \$80 = \$16$ and $\$80 \div \$16 = 5$, divide the bar into 5 equal parts, each of which represents $16. Label the diagram.

Step 3 Since the price of the tent increases to $96, add another equal part to the bar diagram and label it.

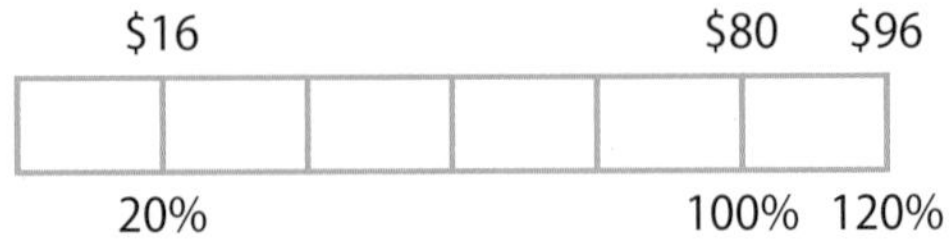

The bar diagram shows that the price of the tent increased by 20%.

Collaborate

1. The bar diagram represents the cost of a skateboard with an original price of \$120 and a discount of 40%.

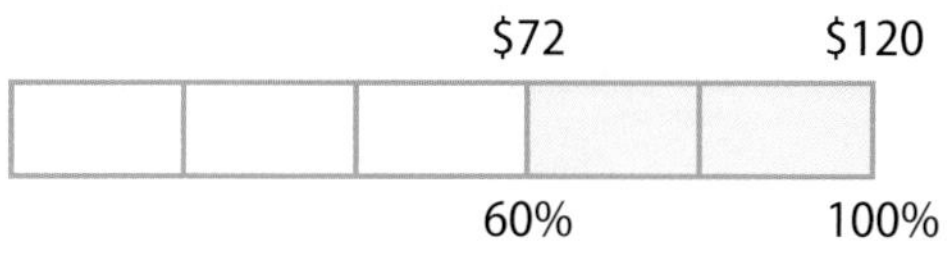

   a. What percent does each part of the bar diagram represent?

   b. What dollar amount does each part of the bar diagram represent?

   c. What is the discounted price of the skateboard?

2. Genevieve has a coupon for 25% off at her family's favorite take-out restaurant. Use a bar diagram to represent the total cost of a \$60 meal.

   a. Draw and shade a bar diagram to represent the situation.

   b. What percent does each part of the bar diagram represent?

   c. What dollar amount does each part of the bar diagram represent?

   d. What is the total cost of the meal?

3. CCSS **Model with Mathematics** Give a real-world scenario that could be represented by this bar diagram.

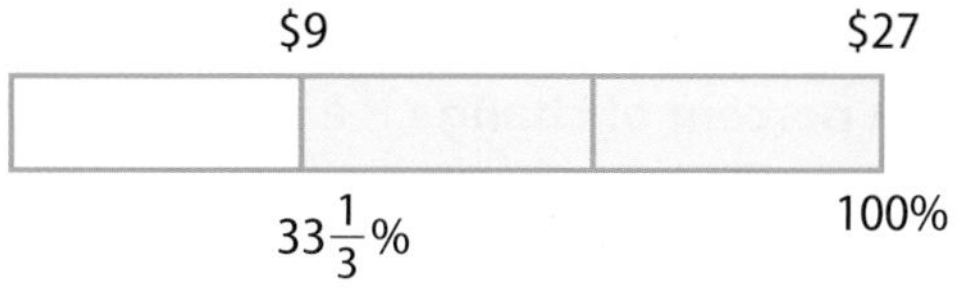

Analyze

Draw a model to find the new price. If it is not possible to find an exact answer from the model, estimate.

4. \$80 increased by 40%.
5. \$25 increased by 50%.
6. \$30 increased by 33%.
7. \$40 increased by 90%.
8. \$15 decreased by 20%.
9. \$150 decreased by 30%.
10. \$600 decreased by 75%.
11. \$20 decreased by 15%.

Draw a model to find the percent of increase or decrease. If it is not possible to find an exact answer from the model, estimate.

12. old price: \$25; new price: \$30
13. old price: \$50; new price: \$20
14. old price: \$750; new price: \$1000
15. old price: \$14; new price: \$12

16. CCSS **Reason Inductively** What would a percent model extended to more than twice its length mean with respect to percent of increase or decrease? Explain your reasoning.

Reflect

17. Inquiry HOW can you use a percent model to model percent of change?

Lesson 6-4

Percent of Change

ISG Interactive Study Guide

See pages 135–136 for:
- Getting Started
- Real-World Link
- Notes

Essential Question

How can you use proportional relationships to solve real-world percent problems?

Common Core State Standards

Content Standards
7.RP.3, 7.EE.2, 7.EE.3

Mathematical Practices
1, 3, 4, 6

Vocabulary

percent of change
percent of increase
percent of decrease
percent error

What You'll Learn

- Find percent of increase and decrease.
- Find percent error.

Real-World Link

Movies Movies sure have come a long way! The first known motion picture was filmed in 1888 and lasted for only 2.11 seconds. Today, we watch motion pictures that last an average of about two hours. The price to go to a movie has also gone up. In 1948, the average ticket price was \$0.36. In 2012, the average went up to \$7.93.

Key Concept: Percent of Change

Words	A **percent of change** is a ratio that compares the change in quantity to the original amount.
Symbols	$\text{percent of change} = \dfrac{\text{amount of change}}{\text{original amount}}$

If the percent is positive, the percent of change is a **percent of increase**. If the percent is negative, the percent of change is called a **percent of decrease**.

Example 1

Tools Tutor

Find the percent of change from 60°F to 84°F. Then state whether the percent of change is an *increase* or a *decrease*.

Step 1 Subtract to find the amount of change.

$84 - 60 = 24$ final amount – original amount

Step 2

$\text{percent of change} = \dfrac{\text{amount of change}}{\text{original amount}}$ Write a ratio that compares the amount of change to the original amount.

$= \dfrac{24}{60}$ Substitution

$= \dfrac{2}{5}$ or 0.4 Simplify.

Step 3 The decimal 0.4 is written as 40%. So, the percent of change is 40%. Since the percent of change is positive, it is a percent of increase.

Got It? Do this problem to find out.

1. Ty had 52 comic books. Now he has 61 books. Find the percent of change. Round to the nearest tenth, if necessary. Then state whether the percent of change is an *increase* or a *decrease*.

Watch Out!
When finding percent of change, don't assume the smaller number is the whole. When the percent of change is a decrease, the original amount will be larger than the new amount.

Example 2

Tools Tutor

McKenna had 318 stamps. Now she has 273 stamps. Find the percent of change. Round to the nearest tenth, if necessary. Then state whether the percent of change is an *increase* or a *decrease*.

$$\text{percent of change} = \frac{\text{amount of change}}{\text{original amount}}$$

$$= \frac{273 - 318}{318} \quad \frac{\text{final amount} - \text{original amount}}{\text{original amount}}$$

$$= -\frac{45}{318} \quad \text{Simplify.}$$

$$\approx -0.141509 \quad \text{Divide. Use a calculator.}$$

To the nearest tenth, the percent of change is −14.2%. Since the percent of change is negative, it is a percent of decrease.

Got It? Do this problem to find out.

2. Find the percent of change from 24 points to 18 points. Then state whether the percent of change is an *increase* or a *decrease*.

Key Concept: Percent Error

Words	The **percent error** is a measure of the difference between an estimate, prediction, or measurement and the actual value.
Symbols	$\text{percent error} = \frac{\text{amount of error}}{\text{actual value}} \times 100$

The amount of error is nonnegative when calculating percent error.

Example 3

Tutor

Alyssa estimates that her school auditorium has 660 seats. It actually has 750 seats. What is the percent error of her estimate?

Step 1 Find the amount of error.

$660 - 750 = -90$ Subtract the actual value from the estimate.

$|-90| = 90$ Find the absolute value of the difference.

Step 2 Find the percent error.

$\frac{90}{750} \times 100 = 0.12$ $\quad \frac{\text{amount of error}}{\text{actual value}} \times 100$

The percent error is 12%.

Got It? Do these problems to find out.

Find the percent error. Round to the nearest tenth, if necessary.

3a. estimated weight: 8 pounds, actual weight: 6.4 pounds

3b. measured length: 2.5 centimeters, actual length: 2.54 centimeters

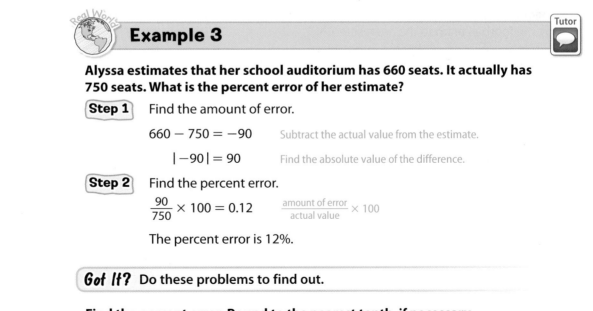

Guided Practice

Check

Find the percent of change. Round to the nearest tenth, if necessary. Then state whether the percent of change is an *increase* or a *decrease*. (Example 1)

1. From \$40 to \$32
2. From 56 inches to 63 inches
3. **Financial Literacy** On Saturday, Smoothie Central made \$1300 in sales. On Sunday, they made \$900 in sales. What is the percent of change from Saturday to Sunday, and is it an increase or decrease? (Example 2)

Find the percent error. (Example 3)

4. estimated distance: 60 miles, actual distance: 75 miles
5. measured area: 24 square inches, actual area: 22.5 square inches
6. The estimate for the amount of rain in May in one part of Texas was 5.6 inches. The actual rainfall was 2.4 inches. What was the percent error of the estimate to the nearest percent?

Independent Practice

Go online for Step-by-Step Solutions — eHelp

Find the percent of change. Round to the nearest tenth, if necessary. Then state whether the percent of change is an *increase* or a *decrease*. (Example 1)

7. From 14 inches to 26 inches
8. From \$36 to \$48
9. From 82 feet to 74 feet
10. From 16 kilograms to 5 kilograms
11. From \$128 to \$112
12. From 90 yards to 72 yards
13. From 191 ounces to 270 ounces
14. From 150 minutes to 172 minutes
15. A survey of gas prices in January showed that the cost per gallon one year was \$2.649. The following January, the cost per gallon was \$2.999. Find the percent change in gas prices from one year to the next to the nearest tenth. (Example 2)
16. Jerome High School's football team scored 38 points in their first game. The next week, they only scored 17 points. Find the percent change in the number of points scored by the football team to the nearest tenth. (Example 2)

Find the percent error. (Example 3)

17. actual height: 180 meters, estimated height: 200 meters
18. estimated time: 40 workdays, actual time: 80 workdays
19. projected cost: \$1250, actual cost: \$2000
20. actual number: 384, calculated number: 385
21. **STEM** A megabyte is 1024 kilobytes of data. Kevin incorrectly used 1000 instead of 1024. What was the percent error of his calculation to the nearest hundredth? (Example 3)
22. A bottle of vitamins should have 60 vitamins. The actual number is 62. What is the percent error to the nearest hundredth? (Example 3)
23. For a local telethon 3860 viewers called in and donated money on the first night. The next night, there was a 20% decrease in the number of calls from the first night. How many calls did the telethon receive on the second night?
24. There were 10,651 athletes who participated in the Summer Olympics one year. Four years later, 11,099 athletes participated. What was the percent of change in the number of athletes participating from one Summer Olympics to the next? Round to the nearest tenth.

25. The number of people who participated in bicycle riding one year was 40.3 million. Two years later, the number was 35.6 million. Find the percent of change in the number of bicycle riders. Round to the nearest tenth.

26. CCSS **Be Precise** A company replaced its half-gallon container of orange juice with a 59-ounce container. If a customer does not notice the change in size, what is the percent error to the nearest tenth?

27 Torie's 400 meter dash time is 74 seconds. Juliette's time is 15% faster than Torie's. What is Juliette's 400 meter dash time? Write an inequality comparing the two times. Use the symbol $<$ or $>$.

28. Christopher Columbus calculated the distance from the Canary Islands to Japan to be 3700 kilometers. The actual distance is about 19,600 kilometers. What was the percent error of his calculation to the nearest percent?

29. **Financial Literacy** The first- and second-quarter earnings of two restaurants are shown at the right. Which restaurant had the greater percent of change in the second quarter?

	Earnings ($)	
Quarter	A	B
1	17,821	8112
2	18,331	9920

30. Kyla's boxer weighs about 62 pounds. When it was a puppy it weighed 23 pounds. What is the percent of change in weight? Round to the nearest tenth.

31. CCSS **Multiple Representation** In this problem, you will compare percents of change. The table shows the population of capital cities for four different states.

City	Population 2000	Population 2006	Amount of Change	% of change
Raleigh, NC	276,093	356,321	■	■
Columbia, SC	116,278	119,961	■	■
Frankfort, KY	27,741	27,077	■	■
Columbus, OH	711,470	733,203	■	■

a. **Table** Copy and complete the table. Round to the nearest whole percent.

b. **Analyze** Compare the amounts of change and the percents of change for Columbia and Columbus. Explain the differences and similarities between the two.

H.O.T. Problems Higher Order Thinking

32. CCSS **Model with Mathematics** Give a real-world example of a percent of increase.

33. CCSS **Model with Mathematics** Find an actual measurement and a calculated measurement where the percent error of the calculation is 5%.

34. CCSS **Make a Conjecture** What happens when you find the percent error when the actual value is 0? Explain your reasoning to a classmate.

35. CCSS **Persevere with Problems** A tool is manufactured so that the mean diameter is 250 millimeters with a percent of error of no more than 2%. What are the smallest and largest diameters that are acceptable?

36. **Building on the Essential Question** The student population was 1600 and was predicted to rise 10% in five years. It rose 15% instead. Find the predicted and actual populations and describe the percent error.

Standardized Test Practice

37. If each dimension of the rectangle is tripled, what is the percent of increase in the area?

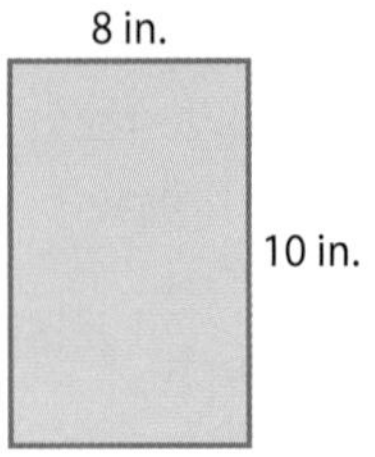

A 300% **C** 800%

B 600% **D** 900%

38. Which of the following represents the greatest percent of change?

F Boots that were originally priced at $90 are on sale for $63.

G A baby that weighed 7 pounds at birth now weighs 10 pounds.

H A bracelet that costs $12 to make is sold for $28.

J A savings account increased from $500 to $600 in 1 year.

39. The table shows the budget of a city.

Annual Budget	
Year	**Budget (millions of $)**
2009	45.6
2010	48.3
2011	45.9
2012	55.1

Which statement is supported by the table?

A The budget decreased and then increased.

B The greatest percent of change occurred from 2009 to 2010.

C The budget increased 20% from 2011 to 2012.

D The percent of change from 2009 to 2010 was the same as from 2010 to 2011.

40. **Short Response** A football team was predicted to win 5 games this year. The team won 12 games. What was the percent error of the prediction?

CCSS Common Core Review

Solve each problem using the percent equation. 7.RP.2

41. Find 12% of 72.

42. Find 42% of 150.

43. What is 37.5% of 89?

44. What is 24.2% of 60?

45. Suppose fifty-six percent of the Calories in corn chips are from fat. If one serving contains 160 Calories, estimate the number of Calories from fat in one serving of corn chips. 7.RP.3

46. Of the fish in an aquarium, 26% are angelfish. If the aquarium contains 50 fish, how many are angelfish? 7.RP.3

Find each sum. Write in simplest form. 7.NS.1d

47. $\frac{1}{10} + \frac{1}{3}$

48. $-\frac{1}{6} + \frac{7}{18}$

49. $6\frac{4}{5} + \left(-1\frac{3}{4}\right)$

Write each fraction as a decimal. 7.NS.2d

50. $\frac{7}{8}$

51. $\frac{1}{5}$

52. $\frac{6}{20}$

53. $\frac{45}{50}$

Find each sum or difference. 7.NS.1

54. $-12 - (-15)$

55. $-25 + 15$

56. $-2 - 20$

Lesson 6-5

Discount and Markup

ISG Interactive Study Guide

See pages 137–138 for:
- Getting Started
- Real-World Link
- Notes

Essential Question

How can you use proportional relationships to solve real-world percent problems?

Common Core State Standards

Content Standards
7.RP.3, 7.EE.2, 7.EE.3

Mathematical Practices
1, 3, 4

Vocabulary

markup
selling price
discount

What You'll Learn

- Solve real-world problems involving markup.
- Solve real-world problems involving discount.

Real-World Link

School Supplies Some states have sales tax holidays. Depending on where you live, clothes, computers, and school supplies may be tax free for two to seven days during the summer. This "holiday" and the back-to-school sales make it a perfect time to shop!

Using Markup

A store sells items for more than it pays for those items. The amount of increase is called the **markup**. The percent of markup is a percent of increase. The **selling price** is the amount the customer pays for an item.

Real World Example 1

Tutor

Find the selling price if a store pays $42 for a pair of in-line skates and the markup is 25%.

Method 1 **Find the amount of the markup first.**

The whole is $42. The percent is 25. You need to find the amount of the markup, or the part. Let m represent the amount of the markup.

$m = 0.25 \cdot 42$ part = percent • whole

$m = 10.5$ Multiply.

Then add the markup to the cost. So, $42 + $10.50 = $52.50.

Method 2 **Find the total percent first.**

Use the percent equation to find 100% + 25% or 125% of the price. Let p represent the price.

$p = 1.25(42)$ part = percent • whole

$p = 52.50$ Multiply.

Using either method, the selling price is $52.50.

Got It? **Do this problem to find out.**

1. Find the selling price if a store pays $75 for a bike and the markup is 40%.

Using Discount

A **discount** is the amount by which the regular price is reduced. The percent of discount is a percent of decrease.

Example 2

Real World — Tutor

Summer Sports is having a sale. A volleyball has an original price of $59. It is on sale for 65% off the original price. Find the sale price of the volleyball.

Method 1 **Find the amount of the discount.**

The percent is 65 and the whole is 59. Let *d* represent the amount of the discount.

$d = 0.65 \cdot 59$ — part = percent • whole

$d = 38.35$ — Multiply.

Subtract the discount from the original cost to find the sale price.
So, $59 − $38.35 = $20.65.

Method 2 **Find the total percent first.**

If the amount of the discount is 65%, the percent the customer will pay is 100% − 65% or 35%. Find 35% of $59.

Let *s* represent the sale price.

$s = 0.35(59)$ — part = percent • whole

$s = 20.65$ — Multiply.

Using either method, the sale price of the volleyball is $20.65.

> **Watch Out!**
> Remember to convert the percent to a decimal when finding a discount or a markup.

Got It? **Do this problem to find out.**

2. A magazine subscription has a cover price of $35. It is on sale for 67% off the original price. Find the sale price of the magazine subscription.

Example 3

Real World — Tutor

Financial Literacy **Henrik had a 25% discount on hockey equipment. The selling price was $172.50. What was the original price?**

If the amount of the discount is 25%, the selling price was 100% − 25% or 75%. 75% of the original price is $172.50.

Let *r* represent the original price.

$\frac{172.5}{r} = \frac{75}{100}$ — Write the percent proportion.

$172.5 \cdot 100 = 75r$ — Find the cross products.

$17{,}250 = 75r$ — Multiply.

$230 = r$ — Divide each side by 75.

The original price was $230.

Got It? **Do this problem to find out.**

3. Luisa got a 75% discount on a sofa. She paid a total of $225. What was the original price?

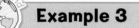

Example 4

Tutor

Cody is buying a ring that had an original price of $295 but is advertised at 30% off. Sales tax of 8.25% is applied to the discounted price. How much will Cody pay for the ring?

Discount and Markup
Remember that for markup, you can use $x + 0.0825x$ or $1.0825x$. For discount, you can use $x - 0.3x$ or $0.7x$.

Step 1 Find the discounted price.

If the amount of the discount is 30%, the discounted price is 100% − 30% or 70%. Find 70% of $295.

Let d represent the discounted price.

$d = 0.7(295)$ part = percent • whole

$d = 206.5$ Multiply.

The discounted price is $206.50.

Step 2 Add the sales tax.

Use the percent equation to find 100% + 8.25%, or 108.25% of the discounted price.

Let s represent the selling price, or the amount Cody paid.

$s = 1.0825(206.5)$ part = percent • whole

$s = 223.53625$ Multiply.

Rounded to the nearest cent, Cody paid $223.54

Got It? Do this problem to find out.

4. A CD with an original price of $11.95 is discounted 20%. Sales tax of 5.5% is added to the discounted price. How much does it cost to purchase the CD?

Guided Practice

Check

Find the selling price for each item given the cost and the percent of markup. (Example 1)

1. shoes: $30; 25% markup
2. CD player: $45; 31% markup
3. jeans: $22; 20% markup
4. guitar: $100; 34% markup
5. swim suit: $36: 28% markup
6. flash drive: $12: 35% markup

7. Find the sale price of a bike that is regularly $110 and is on sale for 45% off the original price. (Example 2)

8. An art supply store has a sale advertising 40% off all canvases. Shelly buys four large canvases and pays a total of $141.60. How much would she have paid without the discount? (Example 3)

9. Danisha picks up a takeout meal at a local restaurant that is discounted 25%. The price is $24.60 without the discount, and sales tax of 4.5% is added. How much does Danisha pay? (Example 4)

Independent Practice

Go online for Step-by-Step Solutions — eHelp

Find the selling price for each item given the cost and the percent of markup. (Example 1)

10. video game: $60; 28% markup
11. bracelet: $26.50; 35% markup
12. jacket: $25; 32% markup
13. stereo: $55; 40% markup
14. wallet: $14.50; 30% markup
15. phone: $34; 36% markup
16. television: $499; 20% markup
17. mountain bike: $255; 34% markup

18. A salon is having a sale on their hair products. Find the sale price of the shampoo and conditioner set shown below that regularly costs $30. (Example 2)

19. The unlimited rental plan at a video store costs $30 a month. It is on sale for 35% off the original price. What is the sale price of the plan? (Example 2)

Find the selling price for each item given the cost and the percent of the markup or discount. (Examples 1- 2)

20. shirt: $7; 50% discount
21. jeans: $32; 40% markup
22. sweater: $35; 28% discount
23. DVD: $15; 33% markup
24. coat $43; 40% discount
25. boots $60; 25% discount

26. Wayne buys an HD television set for $466 that was marked down $33\frac{1}{3}$% from the list price. What is the list price of the television set? (Example 3)

27. **Financial Literacy** Geoff marked down the prices of several items by 20%. Find the original price of each item and copy and complete the table. Round to the nearest cent. (Example 3)

Marked-down Price	$2.99	$4.99	$6.99	$8.99	$10.99
Original Price					

28. A hat originally priced at $28 is on sale for $15% off. What is the total cost of the hat if the tax is 6.75%? (Example 4)

29. **Financial Literacy** A store has ergonomic chairs on sale for 10% off of the regular price. Cayley finds a chair she wants to buy that has an original price of $325. How much will Cayley pay for the chair if she pays 8.5% tax? (Example 4)

State whether the percent of change is a *markup* or a *discount*. Then find the percent to the nearest tenth.

30. from $159 to $153
31. from $227 to $285
32. from $140.75 to $379.99
33. from $84.65 to $41.95
34. from $28 to $26
35. from $41.50 to $45

36. A bike shop marks up a hybrid bike from \$200 to \$330 and discounts a road bike from \$330 down to \$200.

a. By what percent is the hybrid bike marked up?

b. By what percent is the road bike discounted?

c. Are the percents equal?

37 **Financial Literacy** Two electronics stores sell a game system that has a regular price of \$299. Lassen Toys discounts the game system by \$75 and then adds 7.5% sales tax. Pineapple Systems adds the tax first and then takes the discount. Are the selling prices the same? Justify your answer.

38. CCSS **Justify Conclusions** Two electronics stores sell a game system that has a regular price of \$195. Calby Electronics discounts the game system by 35% and then adds 6.75% sales tax. Game Zinger adds the tax first and then takes the discount. Are the selling prices the same? Justify your answer.

39. A store buys shirts from a distributor for \$10 each and marks them up 50%. After a few weeks, they discount the shirts by 50%. Is the discounted price \$10? Explain your reasoning.

H.O.T. Problems Higher Order Thinking

40. CCSS **Persevere with Problems** An item at a consignment shop is marked down 10% each week until it sells. A dress originally priced at \$150 sells during the seventh week. How much did the dress sell for? What percent discount does this represent from the original price?

41. CCSS **Model with Mathematics** Describe a real-world situation and a discount price where an item was marked down 95%. Then find the regular price.

42. CCSS **Find the Error** Quinn used a credit card that discounts her purchases by 3%. The discounted price she paid for a skirt was \$76.24. Quinn is finding the regular price of the skirt. Find her error and correct it.

$r = 1.03(76.24)$
$r = 78.5272$
The regular price was \$78.53

43. CCSS **Use a Counterexample** Determine whether each statement is *true* or *false*. If false, provide a counterexample.

a. It is impossible to increase the cost of an item by more than 100%.

b. It is possible to decrease the cost of an item by less than 1%.

44. CCSS **Persevere with Problems** Suppose a store has an item on sale for 25% off the original amount. By what percent does the store have to increase the price of the item in order to sell it for the original amount? Explain.

45. **Building on the Essential Question** Write and solve a real-world problem involving a discount of an item.

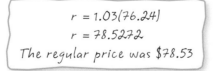

Standardized Test Practice

46. Mrs. Olsen wants to buy a DVD player that regularly costs $120 and is on sale for 30% off the original price. What is the sale price of the DVD player?

A $210 **C** $84

B $90 **D** $36

47. A store held a going-out-of-business sale where everything was 70% off. Jackson spent a total of $115.74. What would he have spent on the same items if there had been no sale?

F $385.80 **H** $165.34

G $196.76 **J** $150.46

48. A home is on the market for $315,000. After 90 days, the seller reduces the price by 8%. What is the new price of the home?

A $252,000

B $258,300

C $283,500

D $289,800

49. **Short Response** A $350 camera is marked down 50% for a 2-day sale. At the end of the sale, the price is marked up from the sales price. What percent markup is needed so that the price of the camera is $350 again?

CCSS Common Core Review

Write each rational number as a fraction. 7.EE.3

50. -12

51. 4.7

52. $3\frac{2}{3}$

53. $-9\frac{1}{4}$

54. 0.17

55. -1.03

56. Every 1000 feet above Earth's surface, air temperature decreases 5°F. How much would the temperature change from sea level to the hightest point on each mountain? 7.NS.3

Mountain	Highest Point (ft)
Mt. McKinley	20,320
Mt. St. Elias	18,008
Mt. Foraker	17,400

Find each difference. 7.NS.1d

57. $2 - (-7)$

58. $11 - 111$

59. $-15 - (-20)$

60. $-3 - 19$

61. $-17 - 5$

62. $48 - (-47)$

Determine how the next term in each sequence can be found. Then find the next two terms in the sequence. 5.OA.3

63. 16, 18, 20, 22, ...

64. 66, 57, 48, 39, ...

65. 1024, 256, 64, 16, ...

66. 1.6, 4, 10, 25, ...

Simplify each expression. 6.EE.1

67. $(-3)^4$

68. $(-2)^3$

69. -5^2

70. 0^{20}

Lesson 6-6

Simple and Compound Interest

ISG Interactive Study Guide

See pages 139–140 for:
- Getting Started
- Vocabulary Start-Up
- Notes

Essential Question

How can you use proportional relationships to solve real-world percent problems?

CCSS Common Core State Standards

Content Standards
7.RP.3

Mathematical Practices
1, 3, 4

Vocabulary

interest
simple interest
principal
compound interest

What You'll Learn

- Solve simple interest problems and apply the simple interest equation to real-world problems.
- Solve compound interest problems.

Real-World Link

Cars Have you ever dreamed of buying your first car? Do you already know what color and style you would like to have? Typically when you buy a car, you pay a certain amount up front and get a loan to cover the rest.

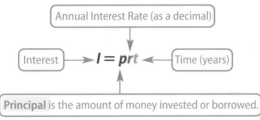

Simple Interest

Interest is the amount of money paid or earned for the use of money by a bank or other financial institution. **Simple interest** is paid only on the initial principal of a savings account or a loan. To solve simple interest problems involving an annual interest rate, use the following formula.

$$I = prt$$

- Interest → I
- Annual Interest Rate (as a decimal) → r
- Time (years) → t
- **Principal** is the amount of money invested or borrowed. → p

Example 1

Tutor

Find the simple interest. Round to the nearest cent, if necessary.

a. $1000 at 4.5% for 2 years

$I = prt$	Write the simple interest formula.
$I = 1000 \cdot 0.045 \cdot 2$	Replace p with 1000, r with 0.045, and t with 2.
$I = 90$	Simplify.

The simple interest is $90.

b. $2500 at 6.75% for 3 years

$I = prt$	Write the simple interest formula.
$I = 2500 \cdot 0.0675 \cdot 3$	Replace p with 2500, r with 0.0675, and t with 3.
$I = 506.25$	Simplify.

The simple interest is $506.25.

Got It? **Do these problems to find out.**

1a. $2250 at 6% for 4 years

1b. $4000 at 4.25% for 1 year

Example 2

Real World

Tutor

Mr. Gabel borrowed $1860 to buy a computer. He will pay $71.30 per month for 30 months. Find the simple interest rate for his loan.

Use the formula $I = prt$. First find the amount of interest he will pay.

$71.30 • 30 = $2139 — Multiply to find total amount.
$2139 − $1860 = $279 — Subtract to find the interest.
So, I = $279.

The principal is $1860. So, $p = 1860$.

The loan will be for 30 months or 2.5 years. So, $t = 2.5$.

$I = prt$ — Write the simple interest formula.

$279 = 1860 \cdot r \cdot 2.5$ — Replace I with 279, p with 1860, and t with 2.5.

$279 = 4650r$ — Simplify.

$\frac{279}{4650} = \frac{4650r}{4650}$ — Divide each side by 4650.

$0.06 = r$

The simple interest rate is 0.06 or 6%.

> **Watch Out!**
> **Converting Units** When using the formula $I = prt$, remember the time is expressed in years. For example, 6 months is 0.5 year.

Got It? Do this problem to find out.

2. **Financial Literacy** Suppose Nantai placed $2400 in the bank for 5 years. He makes $9.20 in interest each month. Find the annual interest rate.

Compound Interest

Compound interest is paid on the initial principal and on interest earned in the past.

Example 3

Tutor

What is the total amount of money in an account where $600 is invested at an interest rate of 8.75% compounded annually for 2 years?

Step 1 Find the amount of money in the account at the end of the first year.

$I = prt$ — Write the simple interest formula.

$I = 600 \cdot 0.0875 \cdot 1$ — Replace p with 600, r with 0.0875, and t with 1.

$I = 52.5$ — Simplify.

$600 + 52.5 = 652.50$ — Add the amount invested and the interest.

At the end of the first year, there is $652.50 in the account.

Step 2 Find the amount of money in the account at the end of the second year.

$I = prt$ — Write the simple interest formula.

$I = 652.50 \cdot 0.0875 \cdot 1$ — Replace p with 652.50, r with 0.0875, and t with 1.

$I = 57.09$ — Simplify.

So, the amount in the account after 2 years is $652.50 + $57.09 or $709.59.

> **Compound Interest** Because you are finding the interest after the first year, substitute 1 for t instead of 2.

Got It? Do this problem to find out.

3. What is the total amount of money in an account where $5000 is invested at an interest rate of 5% compounded annually after 3 years?

Guided Practice

Check

Find the simple interest. Round to the nearest cent, if necessary. (Example 1)

1. \$1350 at 6% for 7 years

2. \$240 at 8% for 9 months

3. \$725 at 3.25% for 5 years

4. \$3750 at 5.75% for 42 months

5. Mateo's sister paid off her student loan of \$5000 in 3 years. If she made a payment of \$152.35 each month, what was the simple interest rate for her loan? Round to the nearest hundredth. (Example 2)

Find the total amount in each account to the nearest cent if the interest is compounded annually. (Example 3)

6. \$480 at 5% for 3 years

7. \$515 at 11.8% for 2 years

8. \$6525 at 6.25% for 4 years

9. \$2750 at 8.5% for 3 years

Independent Practice

Go online for Step-by-Step Solutions eHelp

Find the simple interest. Round to the nearest cent, if necessary. (Example 1)

10. \$275 at 7.5% for 4 years

11 \$620 at 6.25% for 5 years

12. \$734 at 12% for 3 months

13. \$2020 at 8% for 18 months

14. \$1200 at 6% for 36 months

15. \$4380 at 10.5% for 2 years

16. Thomas borrowed \$4800 to buy a new car. He will be paying \$96 each month for the next 60 months. Find the simple interest rate for his car loan. (Example 2)

Find the total amount in each account to the nearest cent if the interest is compounded annually. (Example 3)

17. \$3850 at 5.25% for 2 years

18. \$4025 at 6.8% for 6 years

19. \$595 at 4.75% for 3 years

20. \$840 at 7% for 4 years

21. \$12,000 at 6.95% for 4 years

22. \$8750 at 12.25% for 2 years

23. Denise has a car loan of \$8000. Over the course of the loan, she paid a total of \$1680 in interest at a simple interest rate of 6%. How many months was the loan?

24. A certificate of deposit has an annual simple interest rate of 5.25%. If \$567 in interest is earned over a 6-year period, how much was invested?

25. **Financial Literacy** A bank offers the options shown for interest rates on their savings accounts. Which option will yield more money after 3 years with an initial deposit of \$1500? Explain.

Kingman Bank		
Option	**Rate**	**Type of Interest**
A	6.25%	simple
B	5.75%	compounded annually

Find the total amount in each account to the nearest cent if the interest is compounded twice a year.

26. \$2500 at 6.75% for 1 year

27. \$14,750 at 5% for 1 year

28. \$3750 at 10.25% for 2 years

29. \$975 at 7.2% for 2 years

30. Mrs. Glover placed \$15,000 in a certificate of deposit for 18 months for her children's college funds. Each month she makes \$56.50 in interest. Find the annual simple interest rate for the certificate of deposit.

31. Jameson received his first credit card bill for a total of \$325.42. Each month he makes a \$50 payment and the remaining balance is charged an interest rate of 1.5%. The table at the right shows his first three monthly bills. If he does not make any more charges, what will be the amount of the fifth bill? the seventh bill?

Bill Number	Bill Amount	Payment	New Balance
1	\$325.42	\$50	\$275.42
2	\$279.55	\$50	\$229.55
3	\$232.99	\$50	\$182.99

32. CCSS **Multiple Representations** In this problem, you will compare simple and compound interest. Consider the following situation. Ben deposits \$550 at a 6% simple interest rate, and Anica deposits \$550 at a 6% interest rate that is compounded annually.

a. **Table** Copy and complete the table.

b. **Graph** Graph the data on the coordinate plane. Show the time in years on the *x*-axis and the total interest earned in dollars on the *y*-axis. Plot Ben's interest balance in blue and Anica's interest in red. Then connect the points.

c. **Analyze** Compare the graphs of the two functions.

	Total Interest Earned (\$)	
Years	Ben	Anica
2	■	■
4	■	■
6	■	■
8	■	■
10	■	■

H.O.T. Problems Higher Order Thinking

33. CCSS **Model with Mathematics** Give a principal and interest rate where the amount of simple interest earned in four years would be \$80. Justify your answer.

34. CCSS **Justify Conclusions** Kai-Yo deposits \$500 into an account that earns 2% simple interest. Marcos deposits \$250 into an account that earns 4% simple interest. How much money does each have after 10 years? Who will have more money over the long run? Explain your reasoning.

35. CCSS **Find the Error** Sabino is finding the simple interest on a \$2500 investment at a simple interest rate of 5.75% for 18 months. Find his mistake and correct it.

$$I = prt$$
$$I = 2500 \cdot 0.0575 \cdot 18$$
$$I = \$2587.50$$

36. CCSS **Persevere with Problems** Determine the length of time it will take to double a principal of \$100 if deposited into an account that earns 10% simple annual interest.

37. **Building On the Essential Question** Compare simple and compound interest.

Standardized Test Practice

38. A $500 certificate of deposit has a simple interest rate of 7.25%. What is the value of the certificate after 8 years?

A $290 **C** $790
B $500 **D** $2900

39. Beatriz borrowed $1500 in student loans. She will make 30 equal monthly payments of $62.50 to pay off the loan. What is the simple interest rate for the loan?

F 4% **H** 8.5%
G 7% **J** 10%

40. A savings account with $2250 has an interest rate of 5%. If the interest is compounded annually, how much will be in the account after 2 years?

A $230.63 **C** $2480.63
B $337.50 **D** $2587.50

41. **Short Response** Which of the following plans will produce the greater earnings for an investment of $500 over 5 years? How much more will that plan earn?

Plan A	simple interest rate of 6.75%
Plan B	rate of 6.5% compounded annually

Common Core Review

42. In 2000, there were 356 endangered mammals. Nine years later, 360 species were considered endangered. What was the percent of change? Round to the nearest tenth. **7.RP.3**

Solve each problem using the percent equation. **7.RP.3**

43. 12 is what percent of 400?

44. 30 is 60% of what number?

45. In a recent year, the number of $1 bills in circulation in the United States was about 7 billion. **7.RP.3**

a. Suppose the number of $5 bills in circulation was 25% of the number of $1 bills. About how many $5 bills were in circulation?

b. If the number of $10 bills was 20% of the number of $1 bills, about how many $10 bills were in circulation?

Find each product. Write in simplest form. **7.EE.2**

46. $\frac{2}{x} \cdot \frac{3x}{7}$

47. $\frac{a}{b} \cdot \frac{5b}{c}$

48. $\frac{4t}{9r} \cdot \frac{18r}{t^2}$

49. The table shows the amount of time Craig spends jogging every day. He increases the time he jogs every day. **7.EE.4**

a. Write an equation to show the number of minutes spent jogging m for each day d.

b. How many minutes will Craig jog during day 9?

Day	Time Jogging (minutes)
1	7
2	15
3	23
4	31
5	39

Solve each problem. **7.RP.2**

50. Find 66% of 90.

51. What is 0.2% of 735?

52. Find 250% of 7000.

53. **STEM** A meteorologist predicted that the downtown area would get 16 inches of snow in a snowstorm. The downtown area ended up getting 13 inches of snow. What was the percent error of the prediction? **7.RP.3**

Inquiry Lab

Compound Interest

Inquiry **HOW can you use a calculator or spreadsheet to find balances of accounts that use compound interest?**

CCSS Content Standards
7.RP.3, 7.EE.3
Mathematical Practices
1, 3

Financial Literacy Lainey deposits $800 into an account that earns 4% interest compounded quarterly. What is the value of the account after 1 year?

Investigation 1

You can use a calculator to investigate compound interest.

Step 1 Find the interest rate for each compounding period.

A 4% interest compounded quarterly means that the interest is paid four times a year, or every 3 months. The interest rate each quarter is 4% ÷ 4 or 1%.

Step 2 Enter the principal into the calculator.

Step 3 Multiply by 1 plus the interest rate to find the amount after 1 compounding period.

Press [×] 1.01. This will multiply the previous answer, 800, by 1.01.

Step 4 Press Enter again for each remaining compounding period.

This will multiply each calculation by 1.01.

The value of the account after 1 year is about $832.48.

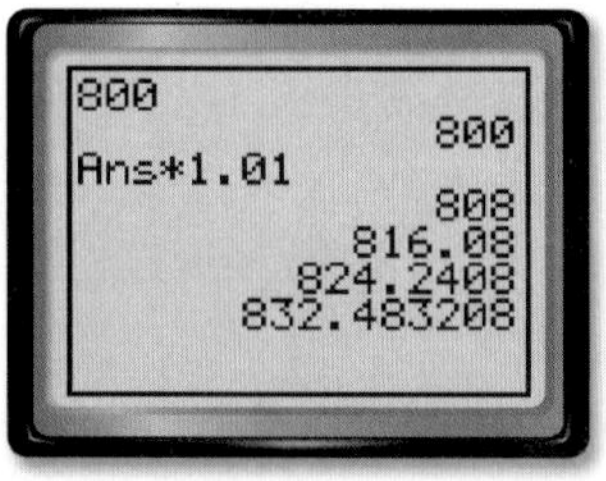

Investigation 2

Watch Tutor

Financial Literacy George deposits $1600 into an account that earns 8% interest compounded semiannually. What is the value of the account after 5 years?

An 8% interest compounded semiannually means that the interest is paid twice a year, or every 6 months. The interest rate is 8% ÷ 2 or 4%.

Compound Interest

The rate is entered as a decimal.

The spreadsheet evaluates the formula A4*B1.

The interest is added to the principal every 6 months. The spreadsheet evaluates the formula A4+B4.

◇	A	B	C	D
1	Rate	0.04		
2				
3	**Principal**	**Interest**	**New Principal**	**Time (YR)**
4	1600.00	64.00	1664.00	0.5
5	1664.00	66.56	1730.56	1.0
6	1730.56	69.22	1799.78	1.5
7	1799.78	71.99	1871.77	2.0
8	1871.77	74.87	1946.64	2.5
9	1946.64	77.87	2024.51	3.0
10	2024.51	80.98	2105.49	3.5
11	2105.49	84.22	2189.71	4.0
12	2189.71	87.59	2277.30	4.5
13	2277.30	91.09	2368.39	5.0
14				

Sheet 1 | Sheet 2 | Sheet 3

The value of the savings account after five years is $2368.39.

Analyze

1. Suppose you invest $1600 for five years at 8% simple interest. How does the simple interest compare to the compound interest shown in Investigation 2?

2. CCSS **Justify Conclusions** Use a calculator or spreadsheet to find the amount of money in a savings account if $1600 is invested for five years at 8% interest compounded quarterly. Why is a spreadsheet a better method? Explain your reasoning.

3. Suppose you leave $150 in each of three bank accounts paying 6% interest per year. One account pays simple interest, one pays interest compounded semiannually, and one pays interest compounded quarterly. Use a spreadsheet to find the amount of money in each account after three years.

4. CCSS **Reason Inductively** If the compounding occurs more frequently, how does the amount of interest change?

Reflect

5. Inquiry HOW can you use a calculator or spreadsheet to find balances of accounts that use compound interest?

Check

Chapter Review

Interactive Study Guide ISG
See pages 141–144 for:
- Vocabulary Check
- Key Concept Check
- Problem Solving
- Reflect

Lesson-by-Lesson Review

Lesson 6-1 Using the Percent Proportion (pp. 250–255)

Use the percent proportion to solve each problem.

1. 12 is what percent of 60?
2. What is 63% of 130?
3. 28 is 80% of what number?
4. 8 hours is what percent of 24 hours?
5. What distance is 72% of 120 miles?
6. 36 pounds is 15% of what weight?
7. Thirty percent of the CDs that Monique owns are classical. If Monique owns 120 CDs, how many are classical?
8. At Marie's school, 65% of the students are learning a second language. There are 143 students learning a second language. How many students are in Marie's school?
9. In a dance class, 70% of the students wear black ballet shoes. There are 30 students that wear black shoes. How many students are in the class?

Example 1

42 is what percent of 60?

$\frac{42}{60} = \frac{p}{100}$ Write the percent proportion.

$42 \cdot 100 = 60p$ Find the cross products.

$4200 = 60r$ Multiply.

$\frac{4200}{60} = \frac{60p}{60}$ Divide each side by 60.

$70 = r$ Simplify.

So, 42 is 70% of 60.

Example 2

Thirty-six is 24% of what number?

$\frac{36}{b} = \frac{24}{100}$ Write the percent proportion.

$36 \cdot 100 = b \cdot 24$ Find the cross products.

$3600 = 24b$ Multiply.

$150 = b$ Divide each side by 24.

So, 36 is 24% of 150.

Lesson 6-2 Find Percent of a Number Mentally (pp. 256–260)

Find the percent of each number mentally.

10. 50% of 36
11. 40% of 55
12. $33\frac{1}{3}$% of 27
13. 1% of 167

Estimate.

14. 24% of 40
15. 62% of 90
16. $\frac{1}{6}$% of 298
17. 130% of 250
18. Tito had 244 free throw attempts in his high school career. If he was successful 77% of the time, about how many free throws did he make?
19. There are 38 students in Mr. Raymond's science class. If 76% of them get an *A* on the final exam, about how many students got *A*'s?

Example 3

Find 40% of 90 mentally.

40% of 90 $= \frac{2}{5}$ of 90 Think: 40% $= \frac{2}{5}$.

$= 36$ Think: $\frac{2}{5}$ of 90 is 36.

So, 40% of 90 is 36.

Example 4

Estimate 78% of 112.

78% is about 75% or $\frac{3}{4}$.

$\frac{3}{4}$ of 112 is 84.

So, 78% of 112 is about 84.

Lesson 6-3 Using the Percent Equation (pp. 261–266)

Solve each problem using a percent equation.

20. 17 is what percent of 68?

21. What is $16\frac{2}{3}\%$ of 24?

22. 55 is 20% of what number?

23. 48 is what percent of 32?

24. 24 is what percent of 48?

25. 49 is what percent of 140?

26. What is 75% of 200?

27. What is 30% of 90?

28. The items in a souvenir shop are on sale for the prices shown. What percent of the original price is the sale price for each item?

Item	Original Price	Sale Price
hat	\$14.00	\$10.50
beach towel	\$17.50	\$14.00
tote bag	\$9.00	\$6.30

29. A jersey is on sale for 50% off the original price. A week later, the manager takes another 50% off. Is the jersey now free? Explain.

Example 5

84 is 60% of what number?

The part is 84 and the percent is 60%.
Let w represent the whole.

$\underbrace{\text{part}} = \underbrace{\text{percent}} \cdot \underbrace{\text{whole}}$

$84 = 0.6 \cdot w$ — Write the percent equation.

$\frac{84}{\mathbf{0.6}} = \frac{0.6w}{\mathbf{0.6}}$ — Divide each side by 0.6.

$140 = w$ — Simplify.

So, 84 is 60% of 140.

Example 6

18 is what percent of 25?

The part is 18 and the whole is 25.
Let p represent the percent.

$\underbrace{\text{part}} = \underbrace{\text{percent}} \cdot \underbrace{\text{whole}}$

$18 = p \cdot 25$ — Write the percent equation.

$\frac{18}{\mathbf{25}} = \frac{25p}{\mathbf{25}}$ — Divide each side by 25.

$0.72 = p$ — Simplify.

Since 0.72 = 72%, 18 is 72% of 25.

Lesson 6-4 Percent of Change (pp. 270–274)

Find the percent of change. Round to the nearest tenth, if necessary. Then state whether the percent of change is an *increase* or a *decrease*.

30. From 55 lb to 24 lb

31. From \$55.75 to \$75.00

Find the percent error.

32. actual distance: 3.2 m, estimated distance: 3.4 m

33. estimated time: 50 min, actual time: 90 min

34. The number of pints of mint chocolate chip sold last week was 88. If this week 110 pints are sold, what is the percent of increase?

35. A project estimated to take 30 days was completed in 75 days. What was the percent error of the estimate?

Example 7

Find the percent of change from 64 minutes to 16 minutes.

$$\text{percent of change} = \frac{\text{amount of change}}{\text{original amount}}$$

$$= \frac{16 - 64}{64}$$

$$= \frac{-48}{64}$$

$$= -\frac{3}{4} \text{ or } -0.75$$

The decimal −0.75 is written as −75%. So, the percent of change is −75%.

Since the percent of change is negative, it is a percent of decrease.

Lesson 6-5 Discount and Markup (pp. 275–280)

Find the selling price for each item given the cost and the percent of markup or discount.

36. tennis shoes: \$85; 24% discount

37. portable MP3 player: \$150; 36% markup

38. pants: \$75; 85% discount

39. amplifier: \$100; 135% markup

40. A surfboard has an original price of \$259. It is on sale for 55% off the original price. Find the sale price of the surfboard.

41. A jacket with an original price of \$49.95 is discounted 33%. Sales tax of 7% is added to the discounted price. How much does it cost to purchase the jacket?

42. A laptop case has an original price of \$45. Ellen has a coupon for 35% off the original price. Find how much Ellen paid for the laptop case.

43. Nathan bought a bicycle for \$230 at an auction. He fixed it up and sold it at a 30% markup. How much did Nathan sell the bike for?

44. Nan bought an \$85 dress on sale at 25% off the original price. She paid 5% sales tax on the sale. What was her total bill?

Example 8

Find the selling price if a store pays \$37 dollars for a video game and the markup is 25%.

$m = 0.25 \cdot 37$ part = percent • whole

$m = 9.25$ Multiply.

Add the markup and the cost. The selling price is \$37 + \$9.25 or \$46.25.

Example 9

Felicia got a 20% discount at a spa. She paid \$92. What was the regular price?

She paid 80% of the regular price.

$\frac{92}{r} = \frac{80}{100}$ Write the percent proportion.

$92 \cdot 100 = 80r$ Find the cross products.

$9200 = 80r$ Multiply.

$115 = r$ Divide each side by 80.

The regular price was \$115.

Lesson 6-6 Simple and Compound Interest (pp. 281–285)

Find the simple interest to the nearest cent.

45. \$575 at 6.25% for 7 years

46. \$12,750 at 5% for 10 years

Find the total amount in each account to the nearest cent if the interest is compounded annually.

47. \$2750 at 8% for 3 years

48. \$1500 at 12.5% for 2 years

49. Lucas borrowed \$10,500 to buy a boat. He will pay \$276.50 each month for the next 48 months. Find the simple interest rate for his loan.

50. What is the total amount of money in an account where \$4000 is invested at an interest rate of 3.5% compounded annually after 3 years?

Example 10

Find the simple interest for \$2500 invested at 3.85% for 4 years.

$I = prt$ Write the simple interest formula.

$I = 2500 \cdot 0.0385 \cdot 4$ Substitute

$I = 385$ Simplify.

The simple interest is \$385.

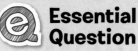

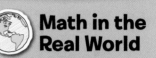

Chapter 7
Algebraic Expressions

Essential Question

Why are algebraic rules useful?

Common Core State Standards

Content Standards
7.NS.2, 7.NS.2c, 7.EE.1, 7.EE.2

Mathematical Practices
1, 2, 3, 4, 5, 7

Math in the Real World

Adventure Trips Travel agencies book adventure trips to exotic and exciting places all over the world. The destination might be snorkeling in Australia, white-water rafting in Idaho, or hiking to Machu Picchu. Algebraic expressions can help agents plan the details of these trips—costs, number of people, length of trip, and even unforeseen expenses.

Interactive Study Guide
See pages 145–148 for:
- Chapter Preview
- Are You Ready?
- Foldable Study Organizer

Lesson 7-1
The Distributive Property

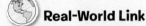

ISG Interactive Study Guide

See pages 149–150 for:
- Getting Started
- Real-World Link
- Notes

Essential Question

Why are algebraic rules useful?

Common Core State Standards

Content Standards
7.NS.2, 7.NS.2c, 7.EE.1, 7.EE.2

Mathematical Practices
1, 3, 4, 5

Vocabulary

equivalent expressions
Distributive Property

What You'll Learn

- Use the Distributive Property to write equivalent numerical expressions.
- Use the Distributive Property to write equivalent algebraic expressions.

Real-World Link

Entertainment The Newport Aquarium in Kentucky has acrylic tunnels that allow visitors to walk underneath aquatic life. The cost of admission is \$23 per person. The aquarium also offers Behind-the-Scenes Tours for \$15 per person and Penguin Encounters for \$25 per person.

Numerical Expressions

Expressions are **equivalent expressions** when they have the same value. Example 1 shows how the **Distributive Property** relates to equivalent expressions.

Key Concept Distributive Property

Watch

Words	To multiply a sum or difference by a number, multiply each term inside the parentheses by the number outside the parentheses.	
Symbols	$a(b + c) = ab + ac$	$a(b - c) = ab - ac$
Examples	$5(6 + 7) = 5 \cdot 6 + 5 \cdot 7$	$(9 - 3)8 = 9 \cdot 8 - 3 \cdot 8$

Example 1

Tutor

Use the Distributive Property to write each expression as an equivalent numeric expression. Then evaluate the expression.

a. $5(12 + 4)$

$5(12 + 4) = 5 \cdot 12 + 5 \cdot 4$

$= 60 + 20$ Multiply.

$= 80$ Add.

b. $(20 - 3)8.2$

$(20 - 3)8.2 = 20 \cdot 8.2 - 3 \cdot 8.2$

$= 164 - 24.6$ Multiply.

$= 139.4$ Subtract.

Got It? Do these problems to find out.

1a. $(6 + 3)4$

1b. $\frac{3}{4}(9 - 2)$

The Distributive Property allows you to find some products mentally. For example, you can find $7 \cdot 34$ mentally by evaluating $7 \cdot (30 + 4)$.

$$\begin{aligned} \mathbf{7} \cdot (30 + 4) &= \mathbf{7} \cdot 30 + \mathbf{7} \cdot 4 \\ &= 210 + 28 && \text{Think: } 7 \cdot 30 = 210 \\ &= 238 && \text{Think: } 210 + 28 = 238 \end{aligned}$$

Example 2

Tutor

Financial Literacy On a school visit to Washington, D.C., Dichali and his class visited the Smithsonian National Air and Space Museum. Tickets to the IMAX movie cost $8.99. Find the total cost for 20 students to see the IMAX movie.

You can use the Distributive Property and mental math to find the total cost for the movie. To find the total cost mentally, find 20($9.00 − $0.01).

$$\begin{aligned} 20(\$9.00 - \$0.01) &= 20(\$9.00) - 20(\$0.01) && \text{Distributive Property} \\ &= \$180 - \$0.20 && \text{Multiply.} \\ &= \$179.80 && \text{Subtract.} \end{aligned}$$

So, the total cost is $179.80.

You can check your result by multiplying 20 • $9 to get $180. Since $180 is close to $179.80, the answer is reasonable.

Got It? **Do these problems to find out.**

2a. A spaghetti dinner at the Italian Village restaurant costs $10.25. Use the Distributive Property and mental math to find the total cost of the dinner for Sherita, her brother, and her parents.

2b. After dinner, they each order gelato for $1.50. What is the new total?

Algebraic Expressions

Vocabulary Link
Distribute
Everyday Use to deliver to each member of a group
Math Use a property that allows you to multiply each member of a sum by a number

You can model the Distributive Property by using algebra tiles and variables.

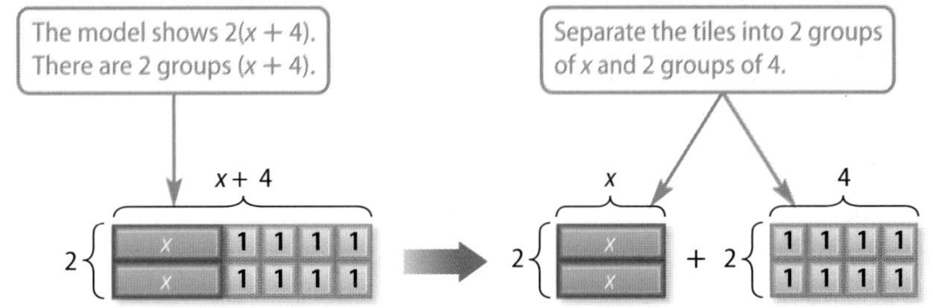

$$\mathbf{2}(x + 4) = \mathbf{2}x + \mathbf{2} \cdot \mathbf{4}$$
$$= 2x + 8$$

The expressions $2(x + 4)$ and $2x + 8$ are equivalent expressions because no matter what the value of x is, these expressions have the same value.

Example 3

Use the Distributive Property to write each expression as an equivalent algebraic expression.

a. $4(x + 5)$

$4(x + 5) = 4x + 4 \cdot 5$

$= 4x + 20$ Simplify.

b. $(y + 10)6$

$(y + 10)6 = y \cdot 6 + 10 \cdot 6$

$= 6y + 60$ Simplify.

Watch Out!

Distributive Property
In Example 3a, remember to distribute the 4 to both values inside the parentheses.

Got It? Do these problems to find out.

3a. $2.4(a + 5)$

3b. $(b + 6)3$

Example 4

Use the Distributive Property to write each expression as an equivalent algebraic expression.

a. $3(m - 4)$

$3(m - 4) = 3[m + (-4)]$ Rewrite $m - 4$ as $m + (-4)$.

$= 3 \cdot m + 3 \cdot (-4)$ Distributive Property

$= 3m + (-12)$ Simplify.

$= 3m - 12$ Definition of subtraction

b. $-9.5(n - 7)$

$-9.5(n - 7) = -9.5[n + (-7)]$ Rewrite $n - 7$ as $n + (-7)$.

$= -9.5 \cdot n + (-9.5)(-7)$ Distributive Property

$= -9.5n + 66.5$ Simplify.

Got It? Do these problems to find out.

4a. $\frac{2}{3}(d - 3)$

4b. $-7(e - 4)$

Guided Practice

Use the Distributive Property to write each expression as an equivalent numeric expression. Then evaluate the expression. (Example 1)

1. $7(9 + 3)$

2. $\frac{2}{5}(3 + 5)$

3. $(7 + 8)2.2$

4. $(5 + 6)8$

5. You purchase 3 blue notebooks and 2 red notebooks. Each notebook costs $1.30. Use mental math to find the total cost of the notebooks. Justify your answer by using the Distributive Property. (Example 2)

Use the Distributive Property to write each expression as an equivalent algebraic expression. (Examples 3 and 4)

6. $\frac{3}{4}(m + 4)$

7. $(p + 4)5$

8. $-6(b - 5)$

9. $9.5(a - 10)$

Independent Practice

eHelp Go online for Step-by-Step Solutions

Use the Distributive Property to write each expression as an equivalent numeric expression. Then evaluate the expression. (Example 1)

10. $-4(8 - 5)$ **11.** $4.5(16 + 8)$ **12.** $(23 - 7)6$

13. $12.3(9 + 4)$ **14.** $(6 + 18)\frac{2}{3}$ **15.** $\frac{5}{8}(20 - 4)$

16. Martine bought two pairs of jeans that are on sale for $32.85 each. Use mental math to find the total cost of the jeans. Justify your answer by using the Distributive Property. (Example 2)

17. **Financial Literacy** Sarah charges $6.50 per hour to babysit. She babysat for 3 hours on Friday and 5 hours on Saturday. Write two equivalent expressions for her total wages. Then find her total wages. (Example 2)

Use the Distributive Property to write each expression as an equivalent algebraic expression. (Examples 3 and 4)

18. $-7.4(10 + a)$ **19** $\frac{4}{5}(t - 15)$ **20.** $3.7(r - 1)$

21. $(b + 4)12$ **22.** $5(x - 9)$ **23.** $-\frac{1}{2}(n + 4)$

24. CCSS **Multiple Representations** The volume of seed in a bird feeder is represented by the equation $V = 12(24 - h)$.

a. Symbols Use the Distributive Property to write $12(24 - h)$ as an equivalent algebraic expression.

b. Table Make a table of ordered pairs (h, V).

c. Graph Graph the ordered pairs on the coordinate plane.

d. Words Explain what happens to the volume as the height increases.

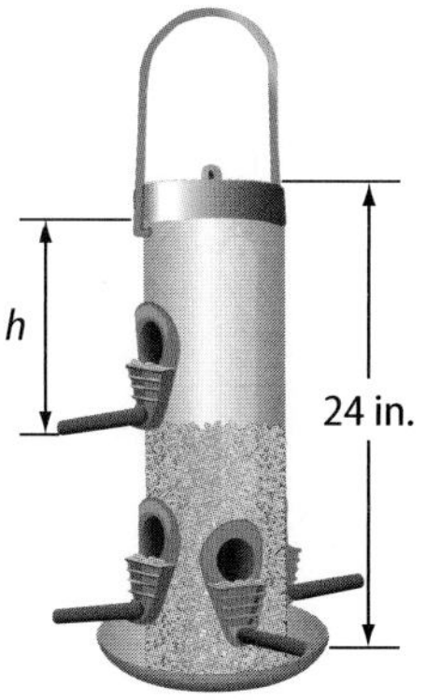

CCSS **Use Math Tools Find each product mentally. Justify your answer.**

25. $8 \cdot 22$ **26.** $13 \cdot 39$ **27.** $19 \cdot 41$

28. $29 \cdot 13$ **29.** $75 \cdot 40$ **30.** $95 \cdot 38$

Use the Distributive Property to write each expression as an equivalent numeric expression. Then evaluate the expression. $\left(\textit{Hint}: 3\frac{1}{4} \text{ can be written as the sum } 3 + \frac{1}{4}.\right)$

31. $4\frac{1}{5} \cdot 5$ **32.** $10 \cdot 5\frac{1}{2}$

33. $6 \cdot 4\frac{2}{3}$ **34.** $2\frac{2}{7} \cdot 14$

35 Aiko uses $2\frac{1}{3}$ yards of fabric to make costumes for a play. Use the Distributive Property to find how much fabric she will need if she makes 9 costumes.

36. Admission to Hersheypark Theme Park in Hershey, Pennsylvania, is $52.95 for an adult and $31.95 for children. The Diego family has a coupon for $10 off each admission ticket. Write an expression to find the cost for *x* adults and *y* children.

H.O.T. Problems Higher Order Thinking

37. **Find the Error** Julia is using the Distributive Property to simplify $3(x + 2)$. Find her mistake and correct it.

$$3(x + 2) = 3x + 2$$

38. **Persevere with Problems** Is $3 + (x \cdot y) = (3 + x) \cdot (3 + y)$ a true statement? If so, explain your reasoning. If not, give a counterexample.

39. **Building on the Essential Question** Explain how you can use the Distributive Property and mental math to simplify $2\frac{1}{2} \cdot 4\frac{1}{2}$.

Standardized Test Practice

40. Admission to a science museum is d dollars and a ticket for the 3-D movie is t dollars. Which expression represents the total cost of admission and a movie for p people?

A dtp

B $p + (dt)$

C $p(d + t)$

D $d(p + t)$

41. Which expression represents the total area of the rectangles?

F $2 + x + 7 + 2 + x + 7$

G $2x + 14$

H $2x + 7$

J $14x$

42. Which expression can be written as $7(c + d)$?

A $7c \cdot 7d$

B $(7 + c) \cdot (7 + d)$

C $7c + 7d$

D $(7 + c) + (7 + d)$

43. **Short Response** A car rental company charges \$45 per day to rent a car. If you rent the car for more than seven days, the cost will be reduced by \$10 per day.

Write an equation to show the total cost c of renting a car for d days if you rent the car for more than seven days.

Common Core Review

Find each sum or difference. Write in simplest form. 7.NS.1

44. $-\frac{5}{8} + \frac{3}{4}$

45. $-2\frac{1}{2} - \frac{2}{3}$

46. $\frac{2}{5} + \frac{1}{6}$

47. $-5\frac{6}{7} + \frac{1}{9}$

48. Jessica needs $5\frac{5}{8}$ yards of fabric to make a skirt and $14\frac{1}{2}$ yards to make a coat. How much fabric does she need in all? 7.NS.3

49. Tate's flower garden has a perimeter of 25 feet. He plans to add 2 feet 9 inches to the width and 3 feet 9 inches to the length. What is the new perimeter in feet? 7.NS.3

Evaluate each expression. 7.NS.2

50. $-6h$, if $h = -20$

51. $-4st$, if $s = -9$ and $t = 3$

52. $\frac{x}{-5}$, if $x = -85$

53. $\frac{108}{m}$, if $m = -9$

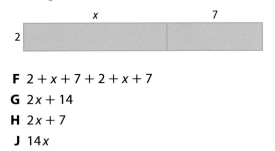

Inquiry Lab

Simplifying Algebraic Expressions

Inquiry **WHAT properties can you use to help simplify algebraic expressions?**

CCSS Content Standards 7.EE.1

Mathematical Practices 1, 3

Online Shopping Mr. Adams is ordering art supplies from an online store. Paint brushes cost \$2 per brush plus \$4 for shipping. Canvas panels cost \$4 per panel plus \$1 for shipping. Write and simplify an expression that represents the total cost to buy paint brushes and canvas panels for a class with x students.

An algebraic expression is in *simplest form* when there are no like terms and no parentheses.

Investigation 1

The expression $2x + 4 + 4x + 1$ represents the situation above. Use algebra tiles to simplify the expression.

Step 1 Model the expression.

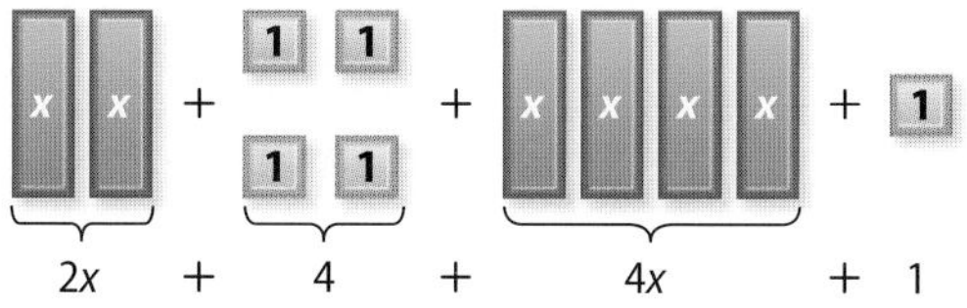

Step 2 Group like tiles together.

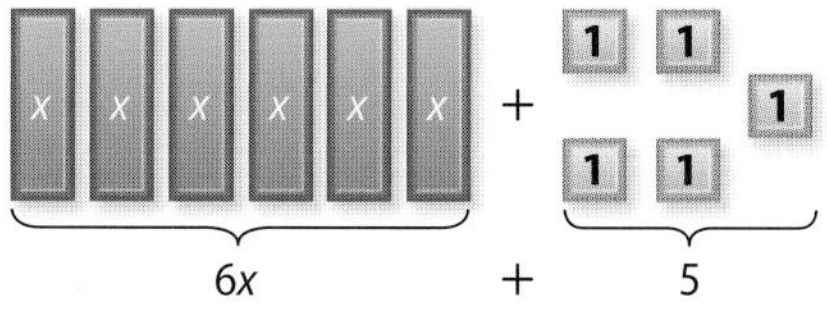

So, the expression $6x + 5$ represents the total cost to buy art supplies.

Collaborate

Work with a partner. Model and simplify each expression using algebra tiles.

1. $3x + 4 + x + 3$

2. $2x + 3 + 2x + 3$

3. $x + 9 + 5x$

4. $4x + 1 + 2x + 2 + x$

Analyze

5. Anna is training to run a marathon. She ran x miles on Monday and Tuesday, $x + 3$ miles on Wednesday, and twice as far on Thursday as she did on Wednesday. Write an expression that represents the total distance Anna ran. Then model and simplify the expression using algebra tiles.

Investigation 2

Use algebra tiles to simplify $x + 6 + 3x - 3$.

Step 1 Model the expression.

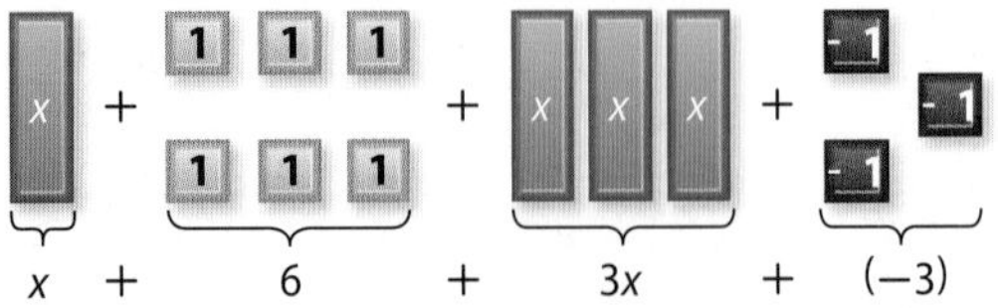

Step 2 Group like tiles together. Then remove zero pairs.

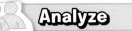

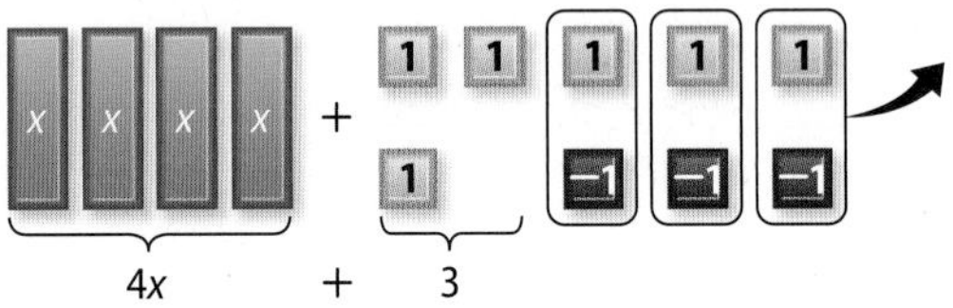

So, $x + 6 + 3x - 3 = 4x + 3$.

Collaborate

Work with a partner. Model and simplify each expression using algebra tiles.

6. $4x - 1 + 2x + 5$

7. $3x + 2x - 4$

8. $2x + 2 + 2x - 2$

9. $x - 4 + 2x + 3$

10. $7x + 4 + 3x + 19$

11. $9x + 9x - 12$

Analyze

12. A rectangle has side lengths $x + 2$ and $3x - 1$. Write an expression that represents the perimeter of the rectangle. Then model and simplify the expression using algebra tiles.

13. CCSS **Reason Inductively** When you simplify an expression using algebra tiles, what mathematical properties allow you to sort the algebra tiles by shape? Explain your reasoning.

14. CCSS **Justify Conclusions** Explain the role of zero pairs when simplifying an algebraic expression using algebra tiles. Use a mathematical property to justify your answer.

Reflect

15. Inquiry WHAT properties can you use to help simplify algebraic expressions?

Lesson 7-2

Simplifying Algebraic Expressions

Interactive Study Guide

See pages 151–152 for:
- Getting Started
- Vocabulary Start-Up
- Notes

Essential Question

Why are algebraic rules useful?

Common Core State Standards

Content Standards
7.EE.1, 7.EE.2

Mathematical Practices
1, 3, 4, 5, 7

Vocabulary

term
coefficient
like terms
constant
simplest form
simplifying the expression

What You'll Learn

- Identify parts of an algebraic expression.
- Use the Distributive Property to simplify algebraic expressions.

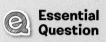

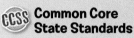

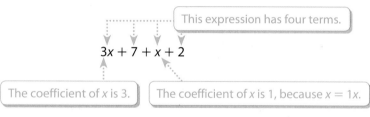

Real-World Link

Recycling Two middle school classes are having a competitive week-long recycling drive. At the end of the week, whichever class collects more recyclables wins and is treated to a pizza party! Algebraic expressions can be used to represent the results of the drive.

Parts of Algebraic Expressions

When addition or subtraction signs separate an algebraic expression into parts, each part is a **term**. The numerical part of a term that contains a variable is called the **coefficient** of the variable.

This expression has four terms.

$3x + 7 + x + 2$

The coefficient of x is 3. The coefficient of x is 1, because $x = 1x$.

In this chapter, we will work only with terms with an exponent of 1. In this case, **like terms** are terms that contain the same variables, such as $2n$ and $5n$ or $6xy$ and $4xy$. A term without a variable is called a **constant**.

Like terms

$\mathbf{5y} + 3 + \mathbf{2y}$

Constant

Example 1

Tutor

Identify the like terms in the following expressions.

a. $3x + 4y + 4x$
$3x$ and $4x$ are like terms since the variables are the same.

b. $5x + 3 + 7x + 4$
$5x$ and $7x$ are like terms since the variables are the same. Constant terms 3 and 4 are also like terms.

Got It? **Do this problem to find out.**

1. Identify the like terms in the expression $-4x + 2y + 3y + 2x$.

Rewriting a subtraction expression using addition will help you identify the terms of an expression.

Example 2

Tutor

Identify the terms, like terms, coefficients, and constants in the expression $6x - 2y + x - 5$.

$6x - 2y + x - 5 = 6x + (-2y) + x + (-5)$	Definition of subtraction
$= 6x + (-2y) + 1x + (-5)$	Identity Property

The terms are $6x$, $-2y$, x, and -5. The like terms are $6x$ and x. The coefficients are 6, −2, and 1. The constant is −5.

Got It? Do this problem to find out.

2. Identify the terms, like terms, coefficients, and constants in the expression $3n + 5m - 6m + 2$.

Simplify Algebraic Expressions

An algebraic expression is in **simplest form** if it has no like terms and no parentheses. When you use the Distributive Property to combine like terms, you are **simplifying the expression**.

Example 3

Tools Tutor

Simplify each expression.

a. $4x + 6 + 2x$

$4x + 6 + 2x = 4x + 2x + 6$	Commutative Property
$= (4 + 2)x + 6$	Distributive Property
$= 6x + 6$	Simplify.

b. $5n + 2 - n - 6$

$5n + 2 - n - 6 = 5n + 2 + (-n) + (-6)$	Definition of Subtraction
$= 5n + 2 + (-1n) + (-6)$	Identity Property
$= 5n + (-1n) + 2 + (-6)$	Commutative Property
$= [5 + (-1)]n + 2 + (-6)$	Distributive Property
$= 4n + (-4)$ or $4n - 4$	Simplify.

c. $6y - 3(x - 2y)$

$6y - 3(x - 2y) = 6y + (-3)[x + (-2y)]$	Definition of Subtraction
$= 6y + (-3x) + (-3 \cdot -2)y$	Distributive Property
$= 6y + (-3x) + 6y$	Simplify.
$= 6y + 6y + (-3x)$	Commutative Property
$= (6 + 6)y + (-3x)$	Distributive Property
$= 12y + (-3x)$ or $12y - 3x$	Simplify.

Watch Out!

Distributive Property In Example 3c, remember to distribute −3, not +3, to the terms in the parentheses.

Got It? Do these problems to find out.

3a. $4x + 6 - 3x$ **3b.** $2m + 3 - 7m - 4$ **3c.** $4(q + 8p) + p$

Real World Example 4

Tutor

Financial Literacy **You have some money in a savings account. Your sister has \$25 more than you have in her account. Write an expression in simplest form that represents the total amount of money in both accounts.**

Words	amount of your money plus amount of your sister's money
Variable	Let x = amount of your money. Let $x + 25$ = amount of your sister's money.
Expression	$x + (x + 25)$

$x + (x + 25) = (x + x) + 25$	Associative Property
$= (1\mathbf{x} + 1\mathbf{x}) + 25$	Identity Property
$= (1 + 1)\mathbf{x} + 25$	Distributive Property
$= 2x + 25$	Simplify.

The expression $2x + 25$ represents the total amount of money you and your sister have in your accounts.

Got It? Do these problems to find out.

4a. Mato and Lola both collect stamps. Lola has 16 more stamps in her collection than Mato. Write an expression in simplest form that represents the total number of stamps in both collections.

4b. Derek has as many stamps as Mato. Write an expression to represent the total of all 3 collections.

Guided Practice

Check

Identify the terms, like terms, coefficients, and constants in each expression. (Examples 1 and 2)

1. $-2a + 3a + 5b$

2. $2x + 3x + 4 + 4x$

3. $mn + 4m + 6n + 2mn$

4. $3a + 5b + 4 + 6a$

5. $3x + 4x + 5y$

6. $-4p - 6q - 5$

Simplify each expression. (Example 3)

7. $6x + 2x + 3$

8. $-2a + 3a + 6$

9. $7x + 4 - 5x - 8$

10. $5a - 2 - 3a + 7$

11. $-3(m - 1) + 4m + 2$

12. $4a - 6 - 2(a - 1)$

13. Marena is using a certain number of blue beads in a bracelet design. She will use 7 more red beads than blue beads. Write an expression in simplest form that represents the total number of beads in her bracelet design. (Example 4)

14. Kyung bought 3 CDs that cost x dollars each, 2 DVDs that cost \$10 each; and a book that cost \$15. Write an expression in simplest form that represents the total amount that Kyung spent. (Example 4)

Independent Practice

Go online for Step-by-Step Solutions — eHelp

Identify the terms, like terms, coefficients, and constants in each expression. (Examples 1 and 2)

15. $3a + 2 + 3a + 7$

16. $4m + 3 + m + 1$

17. $3c + 4d + 5c + 8$

18. $7j + 11jk + k + 9$

19. $4x + 4y + 4z + 4$

20. $3m + 3n + 2p + 4r$

Simplify each expression. (Example 3)

21. $4a + 3a$

22. $9x + 2x$

23. $-5m + m + 5$

24. $6x - x + 3$

25. $7p + 3 + 4p + 5$

26. $2a + 4 + 2a + 9$

27. $4a - 3b - 7a - 3b$

28. $-x - 2y - 8x - 2y$

29. $x + 5(6 + x)$

30. $2a + 3(2 + a)$

31. $-3(6 - 2r) - 3r$

32. $-2(2x - 5) - 4x$

For each situation, write an expression in simplest form that represents the total amount. (Example 4)

33. Mateo has y pairs of shoes. His brother has 5 fewer pairs.

34. You used p minutes one month on your cell phone. The next month you used 75 fewer minutes.

35. Nathan scored x points in his first basketball game. He scored three times as many points in his second game. In his third game, he scored 6 more than the second game.

36. On Monday, Rebekah spent d dollars on lunch. She spent \$0.50 more on Tuesday than she did on Monday. On Wednesday, she spent twice as much as she did on Tuesday.

Simplify each expression.

37. $2(x - y) + 3x$

38. $-3(a - 2b) - 4b$

39. $-4(3m + 2n) - 5m + y$

40. $\frac{2}{3}(6a + 3b) - \frac{1}{2}(a - 2b)$

41. $\frac{1}{4}(m + 2n) - \frac{1}{3}(3m - 3n)$

42. $2(x - y) - (x + y)$

43. $\frac{2}{5}(2a - b) + \frac{2}{3}(a + 2b)$

44. $-\frac{3}{4}(3x + 2y) - \frac{3}{8}(x - 3y)$

45. CCSS **Use Math Tools** Write an expression to represent each model. Then simplify the expression using algebra tiles.

a.

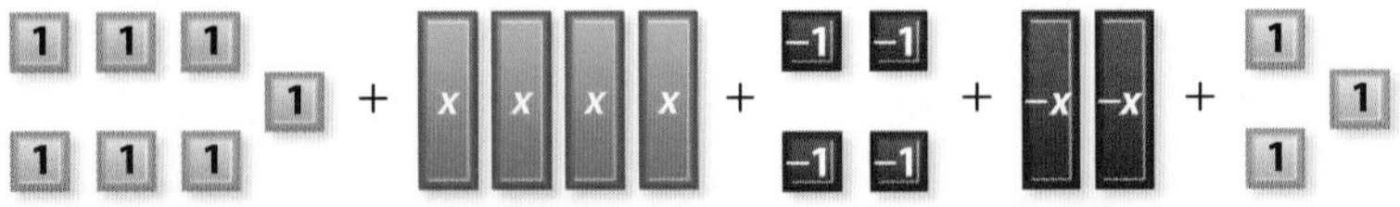

b.

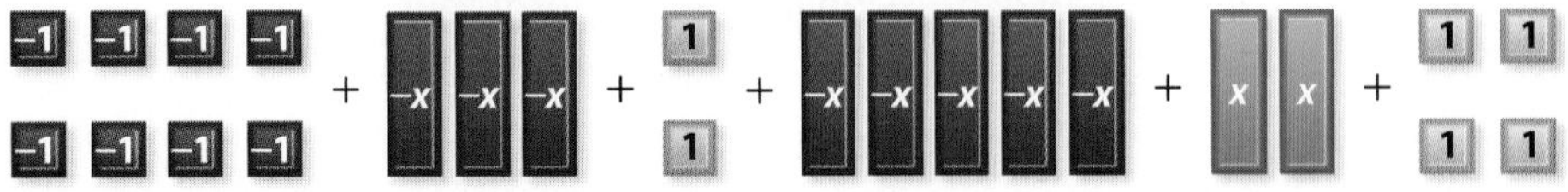

c. Use algebra tiles to write and simplify your own expression.

46. **Multiple Representations** In this problem, you will investigate the perimeter of a rectangle. Consider a rectangle that has a length that is twice its width.

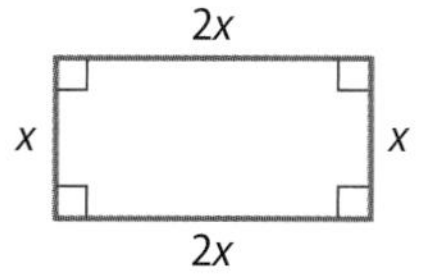

a. **Table** Make a table that shows the width of a rectangle and its perimeter for widths of 1, 2, 3, 4, 5, and 6 units.

b. **Graph** Graph the ordered pairs (width, perimeter).

c. **Symbols** Write an expression in simplest form for the perimeter of the rectangle.

d. **Words** If you double the value of x, what happens to the perimeter? Justify your reasoning.

Write an expression in simplest form for the perimeter of each rectangle.

47.

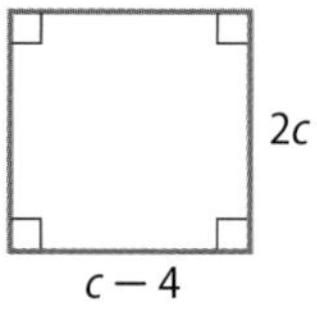

48.

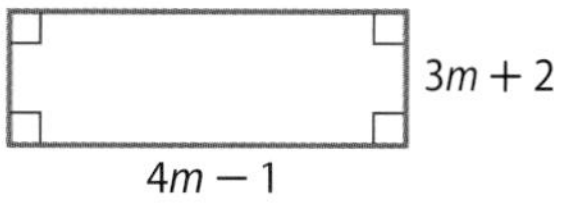

Simplify. Identify the properties you used in each step of your calculation.

49. $16 \cdot (-31) + 16 \cdot 32$

50. $72(38) + (-72)(18)$

51. $24 \cdot (-15) + 36 \cdot 15$

52. $22(-18) - 22(24)$

53. This year Ana's mother is 2 years more than 3 times Ana's age. Write an expression in simplest form for the total of their ages.

H.O.T. Problems Higher Order Thinking

54. **Identify Structure** Write an expression containing at least 2 unlike terms. Then simplify the expression.

55. **Persevere with Problems** Simplify $(2 + x)(y + 5)$.

56. **Justify Conclusions** Classify the following statement as *sometimes*, *always*, or *never* true. Explain your reasoning to a classmate.

When using the Distributive Property, if the term outside the parentheses is negative, then the sign of each term inside the parentheses will change.

57. **Which One Doesn't Belong?** Identify the algebraic expression that does not belong with the other three. Explain your reasoning.

$-6(x - 2)$	$x + 12 - 7x$	$-x - 5x + 12$	$-6x - 12$

58. **Persevere with Problems** In a three-digit number, the second and third digits are the same. The first digit is 4 more than the sum of the second and third digits. Write an expression in simplest form for the total sum of all three digits.

59. **Building on the Essential Question** Suppose your friend simplifies $4x - 2(x + 5)$ as $2x + 10$. Identify the error and correct it.

Standardized Test Practice

60. The perimeter of $\triangle DEF$ is $4x + 3y$. What is the measure of the third side of the triangle?

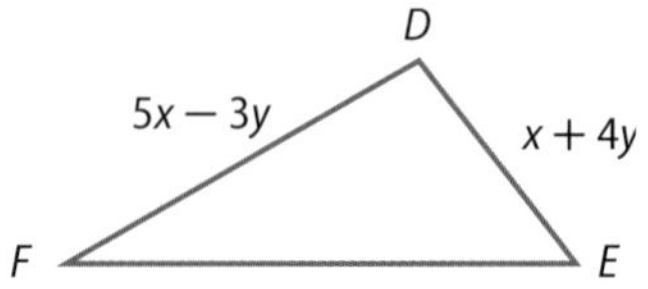

A $-2x + 2y$ **C** $x - y$
B $2x + 2y$ **D** $-x + 2y$

61. Which of the following expressions is equivalent to $4x + 4y$?

F $4xy$ **H** $4x + y$
G $4(x + y)$ **J** $x + 4y$

62. Adriana spent m minutes on her homework on Monday. She spent 45 more minutes doing her homework on Tuesday than on Monday. Which expression represents the total amount of time she spent on her homework on Monday and Tuesday?

A $m + 45$
B $m - 45$
C $2m + 45$
D $2m - 45$

63. **Short Response** Simplify the following expression.

$$7(3a - 2b) + 5b - 3(4a + 2)$$

CCSS Common Core Review

Use the Distributive Property to write each expression as an equivalent expression. 7.EE.1

64. $8(z - 3)$ **65.** $(a - 6)(-5)$ **66.** $15(s + 2)$

67. The table shows the cost of different items at a movie theater. Write two equivalent expressions for the total cost of four movie tickets and four boxes of popcorn. Then find the total cost. 7.EE.1

Item	Cost ($)
ticket	7.00
small popcorn	3.00
small drink	2.50
candy bar	1.75

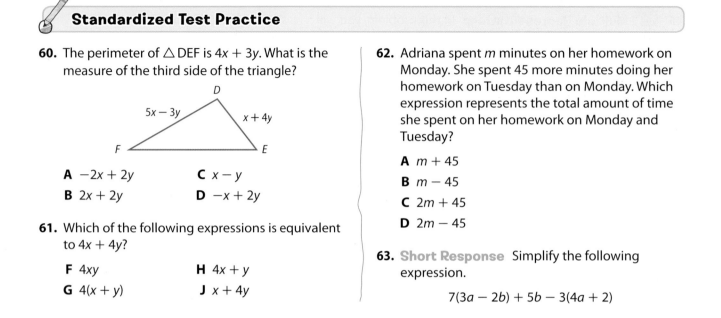

68. Simon has $1\frac{1}{4}$ cups of margarine. He needs $\frac{1}{2}$ cup for a cake and another $\frac{1}{3}$ cup for the icing. How much margarine will he have left? 7.NS.3

Write two inequalities using the number pairs. Use the symbols $<$ or $>$. 6.NS.7

69. -6 and -2 **70.** -10 and -13
71. 0 and -9 **72.** $|-11|$ and $|-7|$
73. $|15|$ and $|18|$ **74.** $|-12|$ and $|14|$

Find the value of each expression if $a = 6$ and $b = 7$. 7.EE.3

75. $\frac{4b + 3a}{b - 5}$ **76.** $\frac{6a - 2ab}{a + 2}$ **77.** $\frac{3(4a - 3b)}{b - 4}$

Find each sum or difference. 7.NS.1

78. $-21 - 6$ **79.** $62 - (-12)$ **80.** $-32 + 26$

Find each product or quotient. 7.NS.2

81. $-4 \cdot 18$ **82.** $-98 \div (-7)$ **83.** $7 \cdot (-8)$

Lesson 7-3

Adding Linear Expressions

Interactive Study Guide

See pages 153–154 for:
- Getting Started
- Real-World Link
- Notes

Essential Question

Why are algebraic rules useful?

Common Core State Standards

Content Standards
7.EE.1

Mathematical Practices
1, 2, 3, 4, 7

Vocabulary

linear expression

What You'll Learn

- Add linear expressions.
- Find perimeter by adding linear expressions.

Real-World Link

Engineering A *trebuchet* is a medieval catapult that was used to hurl large stones and other projectiles at castle walls. Building a model trebuchet requires knowledge of science, math, and engineering. If done successfully, a model can launch a clay ball thirty feet or farther!

Add Linear Expressions

A **linear expression** is an algebraic expression in which the variable is raised to the first power. You can use models to add linear expressions.

Example 1

Tools Tutor

Add. Use models if needed.

a. $(3x + 4) + (2x + 1)$

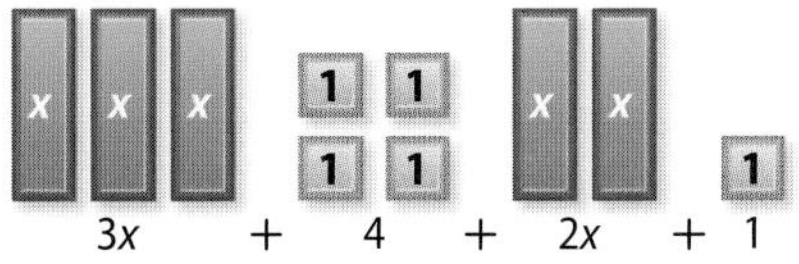

Model each linear expression.

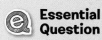

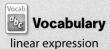

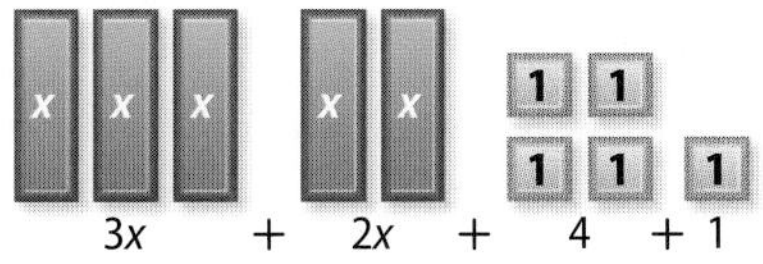

Combine the tiles that have the same shape.

$(3x + 4) + (2x + 1) = 5x + 5$

b. $(-4x + 2) + (-2x + 2)$

$$\begin{array}{r} -4x + 2 \\ + -2x + 2 \\ \hline -6x + 4 \end{array}$$

Arrange like terms in columns.

Add.

So, $(-4x + 2) + (-2x + 2) = -6x + 4$.

Got It? **Do these problems to find out.**

1a. $(x - 3) + (x - 4)$

1b. $(-x + 1) + (-3x)$

Example 2

Zero Pairs
Remember that a zero pair is one positive and one negative tile of the same unit. Since $1 + (-1) = 0$, you can remove zero pairs without affecting the value of the expression.

Add $(3x + 2) + (-x + 4)$.

Model the linear expressions.

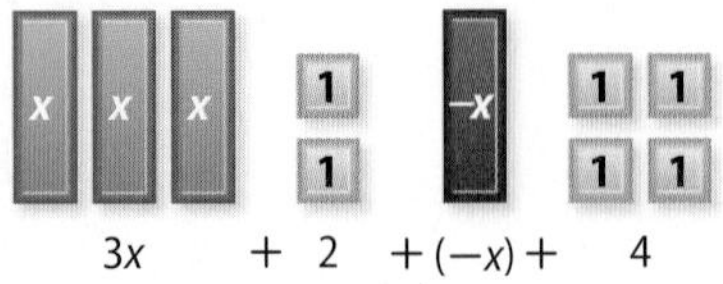

Group tiles with the same shape. Then remove any zero pairs.

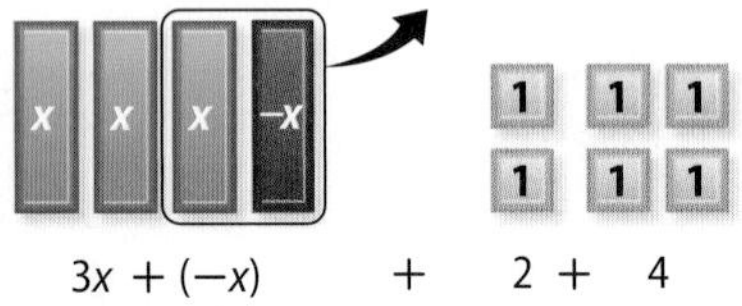

So, $(3x + 2) + (-x + 4) = 2x + 6$.

Got It? **Do these problems to find out.**

Add. Use models if needed.

2a. $(-2x + 4) + (8x - 4)$

2b. $(-4x - 1) + (5x - 3)$

Find Perimeter

Linear expressions can be used to find perimeter.

Example 3

The lengths of the sides of golden rectangles are in the ratio 1:1.62. So, the length of a golden rectangle is approximately 1.62 times greater than its width.

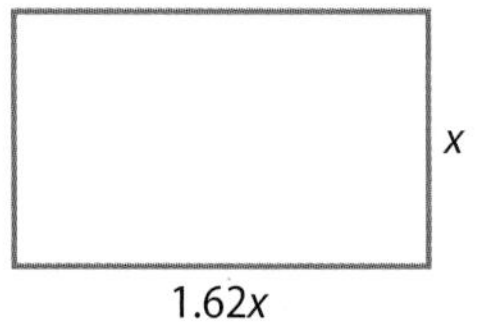

a. Write and simplify a linear expression for the perimeter of a golden rectangle.

$P = 2\ell + 2w$	Formula for the perimeter of a rectangle
$P = 2(\mathbf{1.62x}) + 2x$	Replace ℓ with $1.62x$ and w with x.
$P = 3.24x + 2x$ or $5.24x$	Simplify.

The formula is $P = 5.24x$, where x is the measure of the width.

b. Find the perimeter of a golden rectangle if its width is 8.3 centimeters.

$P = 5.24\mathbf{x}$	Perimeter of a golden rectangle
$= 5.24(\mathbf{8.3})$ or 43.492	Replace x with 8.3 and simplify.

The perimeter of the golden rectangle is 43.492 centimeters.

Got It? **Do these problems to find out.**

3. A rectangle has side lengths of $(5x - 1)$ units and $(2x + 1)$ units.

a. Write and simplify a linear expression for the perimeter of the rectangle.

b. Find the perimeter of the rectangle if the value of x is 5.4 units.

Guided Practice

Check

Add. Use models if needed. (Examples 1 and 2)

1. $(x + 5) + (2x + 3)$

2. $(-4x + 3) + (-5x + 2)$

3. $(x + 6) + (-2x - 4)$

4. $(-7x + 2) + (x + 4)$

5. Use the figure at the right. (Example 3)

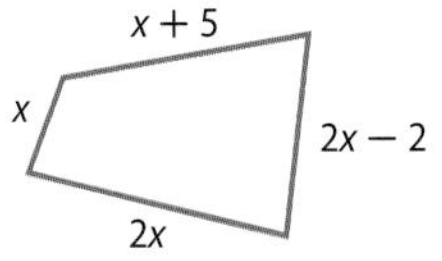

a. Write and simplify a linear expression for the perimeter of the figure.

b. Find the perimeter of the figure if $x = 4$.

Independent Practice

Go online for Step-by-Step Solutions

eHelp

Add. Use models if needed. (Examples 1 and 2)

6. $(7x + 5) + (x + 2)$

7 $(-x + 3) + (-5x + 6)$

8. $(-7x + 1) + (-x + 2)$

9. $(5x + 4) + (-9x + 5)$

10. $(-2x + 1) + (2x + 10)$

11. $(x - 1) + (x + 1)$

For each of the figures, write and simplify a linear expression for the perimeter of the figure. Then find the perimeter of each figure if $x = 0.8$. (Example 3)

12.

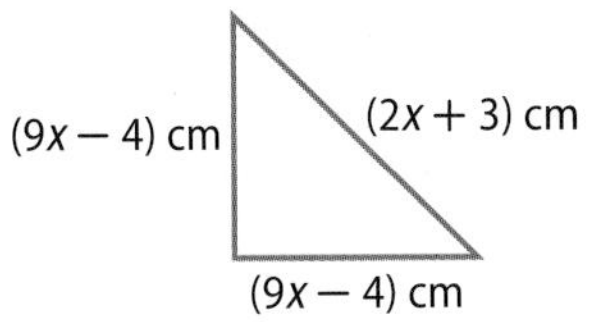

13.

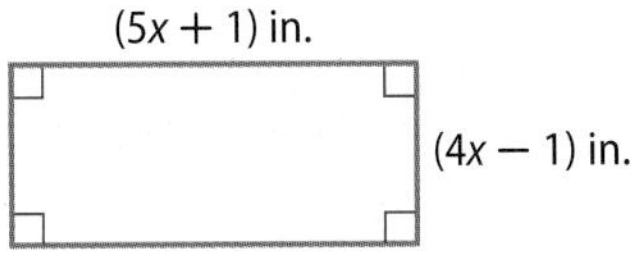

14. The angle measures of a triangle are $(x + 15)°$, $(2x - 20)°$, and $2x°$. What are the actual angle measures of the triangle?

15 CCSS **Reason Abstractly** Anna and Cole each earn x cents per newspaper that they deliver, plus tips. Anna delivered 55 newspapers and earned \$12 in tips. Cole delivered 68 newspapers and earned \$15 in tips.

a. Write a linear expression to represent Anna's total earnings.

b. Write a linear expression to represent Cole's total earnings.

c. Write a linear expression to represent their total earnings.

Add.

16. $\left(3\frac{1}{2}x - \frac{2}{3}\right) + \left(-\frac{1}{4}x + 1\frac{1}{2}\right) + \left(-1\frac{3}{4}x + \frac{5}{6}\right) + (4x)$

17. $(2a - b + 4) + (-a + 3b - 6)$

H.O.T. Problems Higher Order Thinking

18. CCSS **Identify Structure** Write two linear expressions that have a sum of $3x - 8$.

19. CCSS **Persevere with Problems** What linear expression would you add to $-4y + 2$ to have a sum of y?

20. CCSS **Justify Conclusions** Explain how algebra tiles represent like terms and zero pairs.

21. **Building on the Essential Question** Explain how to add linear expressions without using numbers in your explanation.

Standardized Test Practice

22. Add.

$(-2x - 3) + (-3x + 8)$

A $-x + 5$ **C** $-5x + 5$
B $-x + 11$ **D** $-5x + 11$

23. Keisha makes and sells x baskets at her shop. The expression $9x - 2$ represents the profit she made on Monday, and the expression $6x + 5$ represents the profit she made on Tuesday. Which expression represents the total profit Keisha made on Monday and Tuesday?

F $15x + 3$ **H** $3x + 3$
G $15x - 3$ **J** $3x - 3$

24. **Short Response** Write and simplify a linear expression for the perimeter of the triangle.

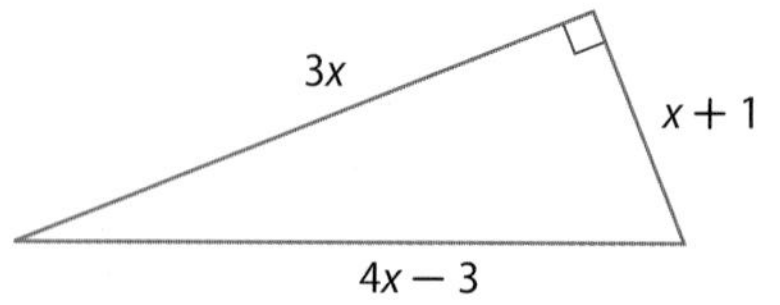

25. Jin and Henry earn x dollars plus tips for each lawn they mow. The expression $4x + 15$ represents Jin's earnings, and $3x + 10$ represents Henry's earnings. How much did they earn in all if $x = \$20$?

A $140 **C** $155
B $150 **D** $165

Common Core Review

Express each ratio as a fraction in simplest form. 7.RP.1

26. 12 cheetahs to 18 lions

27. 15 apples to 30 oranges

28. 14 out of 30 whales

29. 5 lilies to 25 daffodils

30. 9 sheep to 15 goats

31. 4 boats to 16 cars

Find each square root or cube root. 8.EE.2

32. $\sqrt{121}$

33. $\pm\sqrt{49}$

34. $\sqrt[3]{216}$

35. $\sqrt[3]{64}$

36. $-\sqrt{225}$

37. $\sqrt[3]{-512}$

38. Tara is making salt dough for a craft project. She uses 3 cups of salt to 4 cups of flour to make the dough. If she uses 9 cups of flour, how many cups of salt should she use? 7.RP.3

39. Joaquin bought a $120 jacket for a discount of 15%. If tax is 7%, what is the total cost of the jacket? 7.RP.3

Simplify each expression. 7.EE.1

40. $-3a + 4b + 9a - 6b$

41. $2x + 5(3 - x)$

42. $2.3y + 8 - 3y - 10$

43. $\frac{2}{3}z - 12\left(\frac{1}{2}\right) - \frac{4}{9}z$

44. $2 + 0.7z - 2(-5) + 0.3z$

45. $\frac{3}{16}m + 1\frac{1}{2} - \frac{1}{4}m - \frac{3}{4}$

46. $4y + 5(5 - y) - 20 + 2y$

47. $5(3 - x) - (6x + 11)$

ISG Interactive Study Guide

See page 155 for:
• Mid-Chapter Check

21ST CENTURY CAREER in Design Engineering

Roller Coaster Designer

If you have a passion for amusement parks, a great imagination, and enjoy building things, you might want to consider a career in roller coaster design. Roller coaster designers combine creativity, engineering, mathematics, and physics to develop rides that are both exciting and safe. In order to analyze data and make precise calculations, a roller coaster designer must have a solid background in high school math and science.

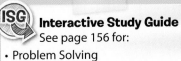

College & Career READINESS

Explore college and careers at ccr.mcgraw-hill.com

Is This the Career for You?

Are you interested in a career as a roller coaster designer? Take some of the following courses in high school to get you started in the right direction.

- Algebra
- Calculus
- Geometry
- Physics
- Trigonometry

Find out how math relates to a career in Design Engineering.

Interactive Study Guide

See page 156 for:

- Problem Solving
- Career Project

Lesson 7-4

Subtracting Linear Expressions

ISG Interactive Study Guide

See pages 157–158 for:
- Getting Started
- Real-World Link
- Notes

Essential Question

Why are algebraic rules useful?

CCSS Common Core State Standards

Content Standards
7.EE.1

Mathematical Practices
1, 3, 4, 7

What You'll Learn

- Subtract linear expressions.
- Solve real-world problems by subtracting linear expressions.

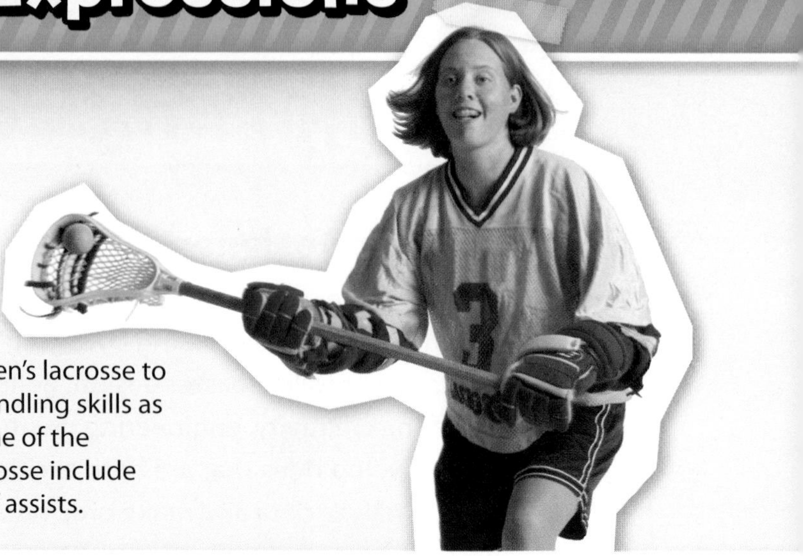

Real-World Link

Lacrosse Middle school girls play a modified version of women's lacrosse to help them acquire good ball-handling skills as they are learning the sport. Some of the statistics that are tracked in lacrosse include number of goals and number of assists.

Subtract Linear Expressions

When subtracting linear expressions, subtract like terms. As with adding linear expressions, you can use models and zero pairs if needed.

Example 1

Tools Tutor

Subtract. Use models if needed.

a. $(5x + 4) - (3x + 2)$

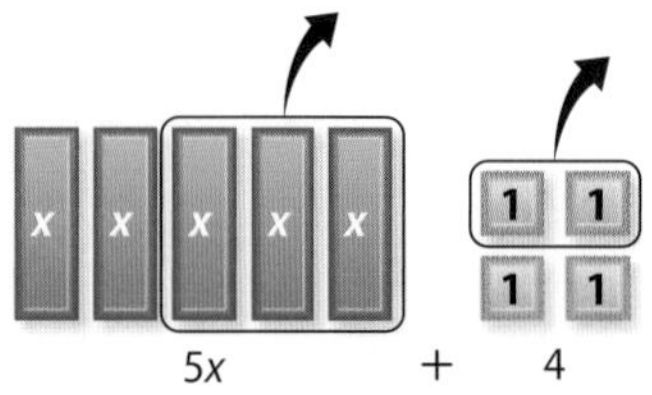

Model the linear expression $5x + 4$.

To subtract $3x + 2$, remove three x-tiles and two 1-tiles.

Then write the linear expression for the remaining tiles.

So, $(5x + 4) - (3x + 2) = 2x + 2$.

b. $-4x - 6 - (-x - 3)$

Arrange like terms in columns.
Each term is subtracted.

$$\begin{array}{r} -4x - 6 \\ -\quad -x - 3 \\ \hline \end{array} \quad \rightarrow \quad \begin{array}{r} -4x - 6 \\ +\quad x + 3 \\ \hline -3x - 3 \end{array}$$

So, $(-4x - 6) - (-x - 3) = -3x - 3$.

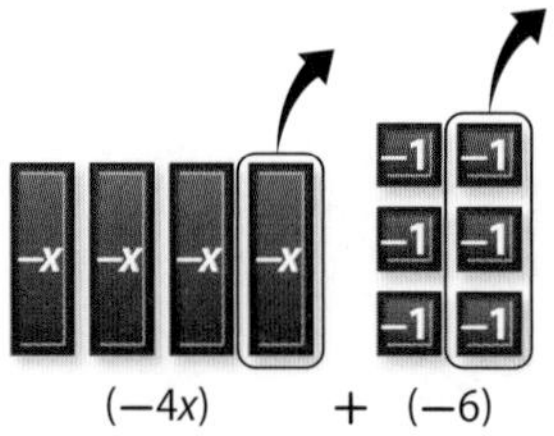

Got It? **Do these problems to find out.**

1a. $(7x - 5) - (2x - 1)$

1b. $(6x - 4) - (2x - 4)$

Alan Bailey/Rubberball/Getty Images

Example 2

Watch Tools Tutor

Find $(3x + 2) - (-2x + 1)$.

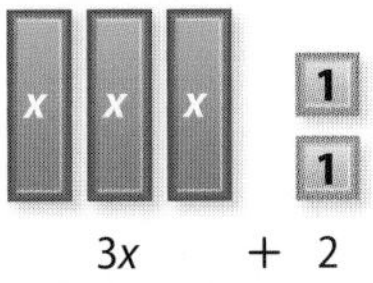

Model the linear expression $3x + 2$.

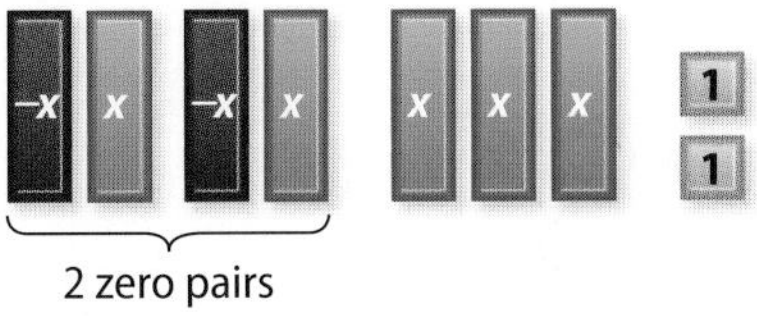

Since there are no negative x-tiles to remove, add 2 zero pairs of x-tiles.

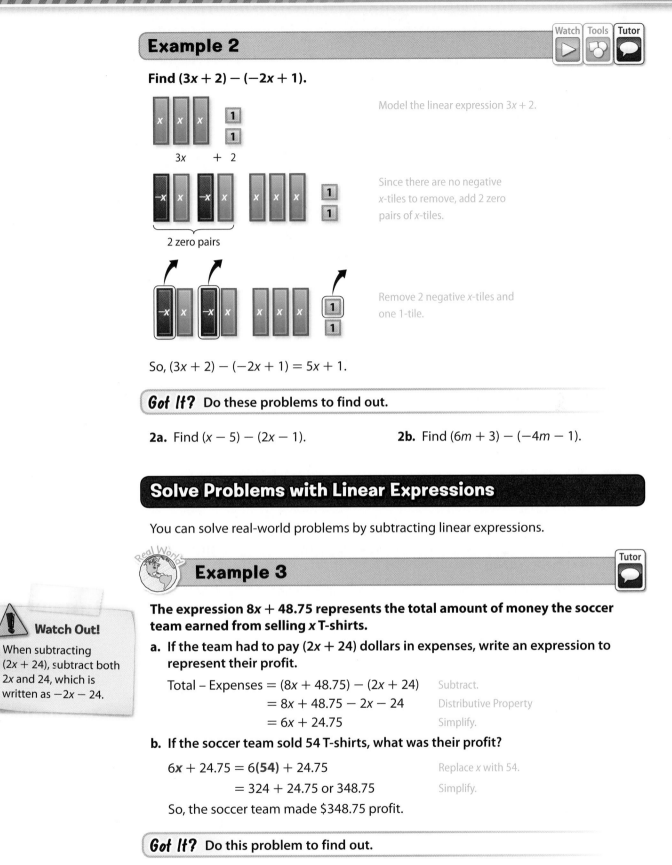

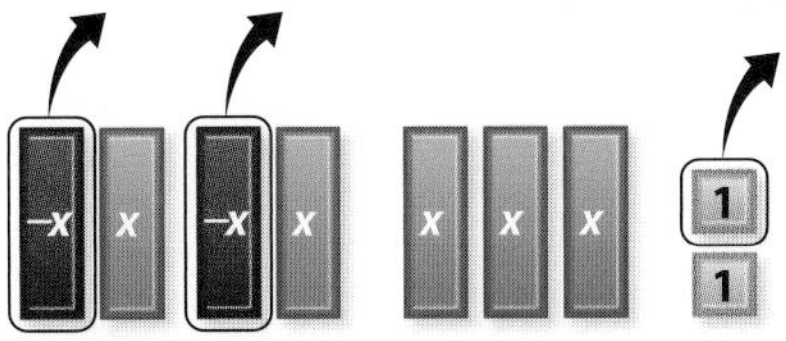

Remove 2 negative x-tiles and one 1-tile.

So, $(3x + 2) - (-2x + 1) = 5x + 1$.

Got It? Do these problems to find out.

2a. Find $(x - 5) - (2x - 1)$.

2b. Find $(6m + 3) - (-4m - 1)$.

Solve Problems with Linear Expressions

You can solve real-world problems by subtracting linear expressions.

Real World Example 3

Tutor

The expression $8x + 48.75$ represents the total amount of money the soccer team earned from selling x T-shirts.

a. If the team had to pay $(2x + 24)$ dollars in expenses, write an expression to represent their profit.

Total – Expenses	$= (8x + 48.75) - (2x + 24)$	Subtract.
	$= 8x + 48.75 - 2x - 24$	Distributive Property
	$= 6x + 24.75$	Simplify.

b. If the soccer team sold 54 T-shirts, what was their profit?

$6x + 24.75$	$= 6(\mathbf{54}) + 24.75$	Replace x with 54.
	$= 324 + 24.75$ or 348.75	Simplify.

So, the soccer team made $348.75 profit.

> **Watch Out!**
> When subtracting $(2x + 24)$, subtract both $2x$ and 24, which is written as $-2x - 24$.

Got It? Do this problem to find out.

3. After working x hours on Monday, Kay earns $9x$ dollars. On Tuesday, she earns $(7x + 3)$ dollars.

a. Write an expression to represent how much more she earned on Monday.

b. If she worked for 5 hours each day, how much more did she earn on Monday?

Guided Practice

Subtract. Use models if needed. (Examples 1 and 2)

1. $(6x + 5) - (3x + 1)$

2. $(-4x + 2) - (-2x + 1)$

3. $(9x - 4) - (-2x + 1)$

4. $(2x + 7) - (x + 1)$

5. The cost of shipping an item that weighs x pounds from Charlotte to Chicago is shown in the table. (Example 3)

Shipping Company	Cost ($)
Atlas Service	$4x + 2.80$
Mid-Atlantic Service	$3x + 1.25$

a. Write an expression to represent how much more Atlas charges than Mid-Atlantic for shipping an item.

b. If an item weighs 2 pounds, how much more does Atlas charge for shipping it?

Independent Practice

Go online for Step-by-Step Solutions

Subtract. Use models if needed. (Examples 1 and 2)

6. $(3x + 7) - (x + 5)$

7. $(-4x + 3) - (-x - 4)$

8. $(8x - 9) - (3x - 1)$

9 $(3x + 7) - (x - 2)$

10. $(5x + 6) - (2x + 5)$

11. $(x + 5) - (2x + 3)$

12. **Model with Mathematics** The expression $5.5x + 2$ represents the number of miles Celeste rode her bike, and $10x$ represents the number of miles that Kimiko rode her bike, in x hours. (Example 3)

a. Write an expression to show how many more miles Kimiko rode than Celeste.

b. If they each rode for 2 hours, how many more miles did Kimiko ride?

13 Evan plans to download x songs from a music site on the Internet. The expression $1.29x$ represents the cost at Web site A, and $0.25x + 25$ represents the cost at Web site B. How much more will Evan pay at Web site A than Web site B if he downloads an average of 30 songs per month?

14. The expression $5\frac{1}{2}x + 6$ represents the perimeter of the rectangle shown. Write an expression that represents the length of the rectangle.

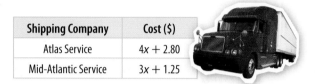

$\frac{1}{4}x + 2$

H.O.T. Problems Higher Order Thinking

15. **Identify Structure** Write two linear expressions that have a difference of $4x + 1$.

16. **Persevere with Problems** Suppose A and B represent linear expressions. If $A + B = 2x - 2$ and $A - B = 4x - 8$, find A and B.

17. **Building on the Essential Question** Explain how you can use a rule for subtracting integers to help subtract linear expressions.

Standardized Test Practice

18. Subtract.

$(-3x + 4) - (-7x - 6)$

A $-10x - 2$

B $-10x + 10$

C $4x - 2$

D $4x + 10$

19. Jorge bought x tickets to attend a football game and a baseball game. The expression $8x + 62$ represents the total cost of the football game, and $9x + 34$ represents the total cost of the baseball game. How much more did the football game cost if Jorge bought 7 tickets for each game?

F \$35 **H** \$11

G \$21 **J** \$7

20. The length of a rectangle is $7x - 4$. The width of the rectangle is $5x + 1$. Which expression represents the difference between the length and the width of the rectangle?

A $2x - 3$

B $2x + 3$

C $2x - 5$

D $2x + 5$

21. **Short Response** The expression $3x + 2$ represents the number of miles Emma walked in x hours. Lea walked $4x - 1$ miles in x hours. Write an expression that represents how much farther Lea walked than Emma.

CCSS Common Core Review

Solve each problem. 7.RP.2

22. What is 15% of 80?

23. 46 is what percent of 115?

24. 5 is 4% of what number?

25. Find 15% of 325.

26. 17 is what percent of 20?

27. 14 is 20% of what number?

Use the Distributive Property to write each expression as an equivalent expression. 7.EE.1

28. $6(n - 3)$

29. $(w + 9)8$

30. $-7(a + 5)$

31. $-4(-b - 2)$

32. There are 21 birds at a bird sanctuary, 9 of which are parrots. Write the ratio of parrots to total birds as a fraction in simplest form. 7.RP.1

33. In a survey about favorite movies, 54 out of 120 people preferred comedies. What percent of the people in the survey preferred comedies? 7.RP.3

34. The temperature in Bismarck, North Dakota, is 13°F at 9 A.M. It is −3°F at 1 P.M. What is the difference in temperature between 9 A.M. and 1 P.M.? 7.NS.3

Evaluate each expression if $a = 8$, $b = -4$, and $c = -15$. 6.EE.2a

35. $a + c$

36. bc

37. $2a + 5b$

38. abc

39. $3ac - b$

40. $b(a + c)$

41. $ab + c$

42. $4(a - b)$

43. $3b - 5c$

Inquiry Lab

Factoring Linear Expressions

Inquiry WHAT is an advantage of using algebra tiles to factor linear expressions?

CCSS **Content Standards**
7.EE.1

Mathematical Practices
1, 3

Design Joaquin is designing a rectangular banner with an area of $(2x + 8)$ feet. What are possible dimensions of the banner?

You can *factor* the expression to find possible dimensions of the banner. To factor an expression means to write it as a product of its factors.

Investigation 1

Use algebra tiles to factor $2x + 8$.

Step 1 Model the expression $2x + 8$.

Step 2 Arrange the tiles into a rectangle.

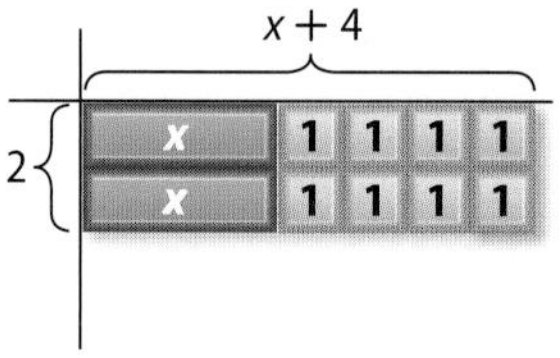

The total area of the tiles represents the product, $2x + 8$. The length and width of the rectangle represent the factors, 2 and $x + 4$.

So, $2x + 8 = 2(x + 4)$.

The possible dimensions of the banner are 2 feet by $(x + 4)$ feet.

Collaborate

Work with a partner. Factor each expression using algebra tiles.

1. $3x + 6$
2. $5x + 15$
3. $4x + 10$
4. $6x + 14$

Analyze

5. Nabuko says $4(x + 5)$ and $4x + 5$ are equivalent expressions. Explain whether Nabuko is correct.
6. The total cost to play miniature golf can be represented by the expression $(9x + 15)$ dollars. If Elliot and two friends split the cost equally, what is the cost for each friend? Explain how you can use algebra tiles to show that your answer is correct.
7. **CCSS Justify Conclusions** A painting has an area of $(4x + 12)$ inches. Possible dimensions are 4 inches by $(x + 3)$ inches. Are there other possible dimensions? Justify your answer using algebra tiles.

Investigation 2

Use algebra tiles to factor $3x - 9$.

Step 1 Model the expression $3x - 9$.

Step 2 Arrange the tiles into a rectangle with equal rows and columns.

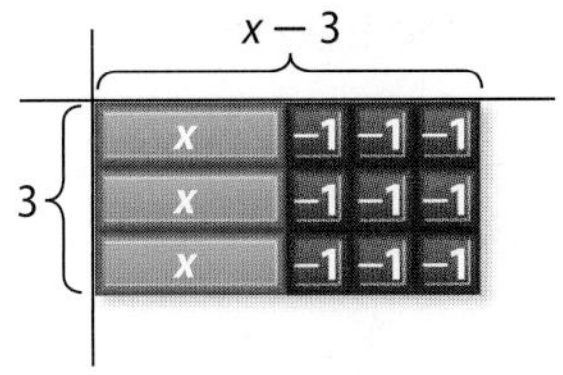

The rectangle has a width of 3 and a length of $x - 3$.

So, $3x - 9 = 3(x - 3)$

Collaborate

Work with a partner. Factor each expression using algebra tiles.

8. $2x - 4$

9. $5x - 10$

10. $4x - 14$

11. $6x - 15$

Analyze

Factor each expression.

12. $9x - 27$

13. $10x - 35$

14. Jacqui and three friends went to the movies. The total cost of their admission and snacks can be represented by the expression $(8y + 52)$. If they split the cost equally, what does each one pay?

15. Explain how you can use the Distributive Property to check that $3(x - 3)$ is the *factored form* of $3x - 9$.

16. CCSS **Justify Conclusions** Can the linear expression $2x - 3$ be factored? Support your answer using algebra tiles.

17. CCSS **Reason Inductively** Explain why algebra tiles can be used to determine whether an expression can be factored. Give examples.

Reflect

18. Inquiry WHAT is an advantage of using algebra tiles to factor linear expressions?

Lesson 7-5

Factoring Linear Expressions

ISG Interactive Study Guide

See pages 159–160 for:
- Getting Started
- Real-World Link
- Notes

Essential Question

Why are algebraic rules useful?

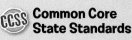

CCSS Common Core State Standards

Content Standards
7.EE.1

Mathematical Practices
1, 2, 3, 4, 7

Vocabulary

factor

factored form

What You'll Learn

- Find the greatest common factor of two monomials.
- Use properties to factor linear expressions.

Real-World Link

Marching Band Band directors create geometrical formations that are eye-catching and exciting but still follow the rhythm and feel of the music. Graph paper is used to draw formations, with different colored ink representing different sections of the band.

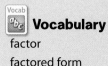

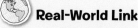

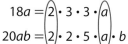

Find the GCF of Monomials

To **factor** a number means to write it as a product of its factors. A monomial can be factored using the same method you would use to factor a number. The greatest common factor (GCF) of two monomials is the greatest monomial that is a factor of both.

Example 1

Tutor

Find the GCF of each pair of monomials.

a. $4x$, $12x$

$4x = 2 \cdot 2 \cdot x$ — Write the prime factorization of $4x$ and $12x$.

$12x = 2 \cdot 2 \cdot 3 \cdot x$ — Circle the common factors.

The GCF of $4x$ and $12x$ is $2 \cdot 2 \cdot x$ or $4x$.

b. $18a$, $20ab$

$18a = 2 \cdot 3 \cdot 3 \cdot a$ — Write the prime factorization of $18a$ and $20ab$.

$20ab = 2 \cdot 2 \cdot 5 \cdot a \cdot b$ — Circle the common factors.

The GCF of $18a$ and $20ab$ is $2 \cdot a$ or $2a$.

Got It? Do these problems to find out.

Find the GCF of each pair of monomials.

1a. 12, $28c$ **1b.** $25x$, $15xy$ **1c.** $42mn$, $14mn$

Factor Linear Expressions

You can use the Distributive Property and the work backward strategy to express an algebraic expression as a product of its factors. An algebraic expression is in **factored form** when it is expressed as the product of its factors.

$8x + 4 = 4(2x) + 4(1)$ The GCF of $8x$ and 4 is 4.

$= 4(2x + 1)$ Distributive Property

Example 2

Tutor

Factor each expression.

a. $3x + 9$

Method 1 **Use a model.**

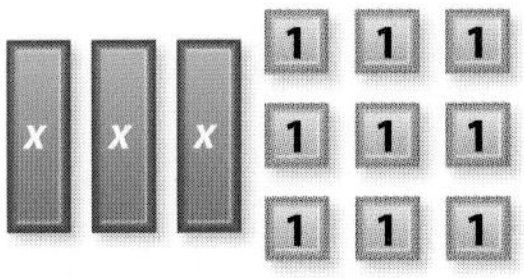

Model $3x + 9$.

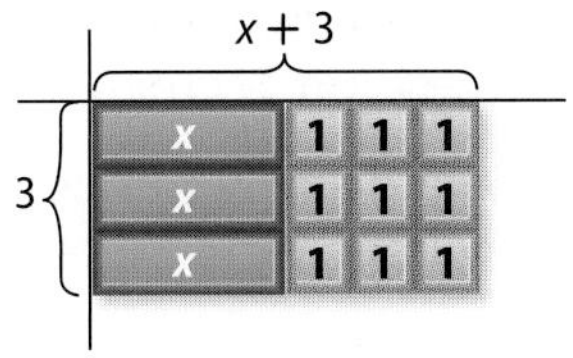

Arrange the tiles into equal rows and columns. The rectangle has a width of three 1-tiles, or 3, and a length of one x-tile and three 1-tiles, or $x + 3$.

So, $3x + 9 = 3(x + 3)$.

Method 2 **Use the GCF.**

$3x = 3 \cdot x$ Write the prime factorization of $3x$ and 9.

$9 = 3 \cdot 3$ Circle the common factors.

The GCF of $3x$ and 9 is 3. Write each term as a product of the GCF and its remaining factors.

$3x + 9 = \mathbf{3}(x) + \mathbf{3}(3)$

$= \mathbf{3}(x + 3)$ Distributive Property

So, $3x + 9 = 3(x + 3)$.

Factoring Expressions

Use algebra tiles to model the expression in Example 2b. Since you cannot rearrange the tiles to make a rectangle, the expression cannot be factored.

b. $12x + 7$

Find the GCF of $12x$ and 7.

$12x = 2 \cdot 2 \cdot 3 \cdot x$

$7 = 1 \cdot 7$

There are no common factors, so $12x + 7$ *cannot be factored*.

Got It? **Do these problems to find out.**

Factor each expression. If the expression cannot be factored, write *cannot be factored*. Use algebra tiles if needed.

2a. $4x + 28$ **2b.** $3 + 33x$ **2c.** $4x + 35$

Example 3

The garden at the right has a total area of (15x + 18) square feet. Find possible dimensions of the garden.

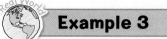

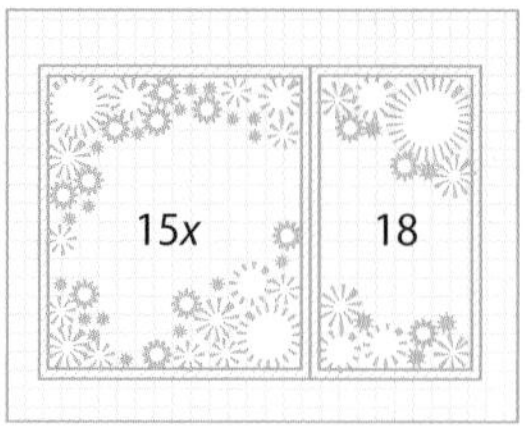

Factor 15x + 18.

$15x = 3 \cdot 5 \cdot x$ Write the prime factorization of 15x and 18.

$18 = 2 \cdot 3 \cdot 3$ Circle the common factors.

The GCF of 15x and 18 is 3. Write each term as a product of the GCF and its remaining factors.

$15x + 18 = \mathbf{3}(5x) + \mathbf{3}(6)$

$= \mathbf{3}(5x + 6)$ Distributive Property

So, the dimensions of the garden are 3 feet and (5x + 6) feet.

Check Find the product of 3 and 5x + 6. $3(5x + 6) = 15x + 18$ ✓

Got It? **Do these problems to find out.**

3a. **Financial Literacy** The Reyes family has saved $480 as a down payment for a new television. If x is the monthly payment for one year, the expression $12x + $480 represents the total cost of the television. Factor $12x + $480.

3b. Jesse wants to put down $100 toward a new computer and will pay it off in six months. If y is the monthly payment, what expression represents the total price?

Guided Practice

Find the GCF of each pair of monomials. (Example 1)

1. 32x, 18 **2.** 15y, 25 **3.** 45a, 20a

4. 16b, 12b **5.** 42s, 28s **6.** 56g, 84gh

7. 27s, 54st **8.** 18cd, 30cd **9.** 22mn, 11kmn

Factor each expression. If the expression cannot be factored, write *cannot be factored*. Use algebra tiles if needed. (Example 2)

10. 36x + 24 **11.** 6 + 3x **12.** 4x + 9

13. 13x + 21 **14.** 2x − 4 **15.** 14x − 16

16. 12 + 18x **17.** 24 + 32x **18.** 15x + 8

19. Mr. Phen's monthly income can be represented by the expression 25x + 120, where x is the number of hours worked. Factor the expression 25x + 120. (Example 3)

20. The area of a high school basketball court is (50x − 300) square feet. Factor 50x − 300 to find possible dimensions of the basketball court. (Example 3)

Independent Practice

Go online for Step-by-Step Solutions eHelp

Find the GCF of each pair of monomials. (Example 1)

21. 24, 48*m*
22. 63*p*, 84
23. 40*x*, 60*x*
24. 32*a*, 48*b*
25. 30*rs*, 42*rs*
26. 54*gh*, 72*g*
27. 36*k*, 144*km*
28. 60*jk*, 45*jkm*
29. 100*xy*, 75*xyz*

Factor each expression. If the expression cannot be factored, write *cannot be factored*. Use algebra tiles if needed. (Example 2)

30. $3x + 9$
31. $5x + 5$
32. $10x - 35$
33. $2x - 15$
34. $4x - 7$
35. $32 + 24x$
36. $12 + 30x$
37. $18x + 6$
38. $30x - 40$

39. The area of a rectangle is $(4x - 8)$ square units. Factor $4x - 8$ to find possible dimensions of the rectangle. (Example 3)

40. James has \$120 in his savings account and plans to save \$*x* each month for 6 months. The expression $\$6x + \$120$ represents the total amount in the account after 6 months. Factor the expression $6x + 120$. (Example 3)

Write an expression in factored form to represent the total area of each rectangle.

41.
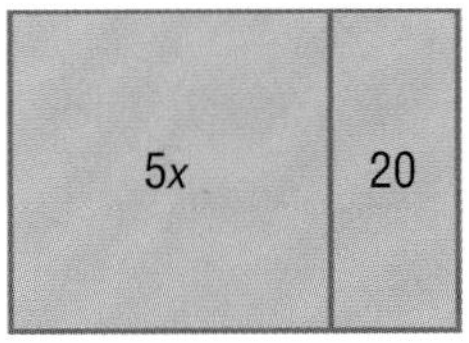

42.
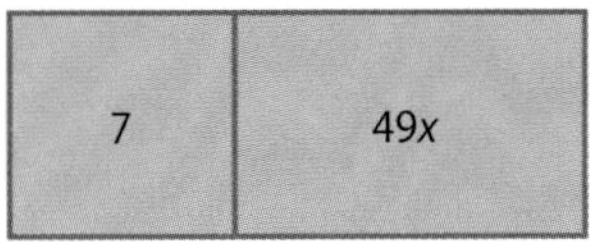

43.
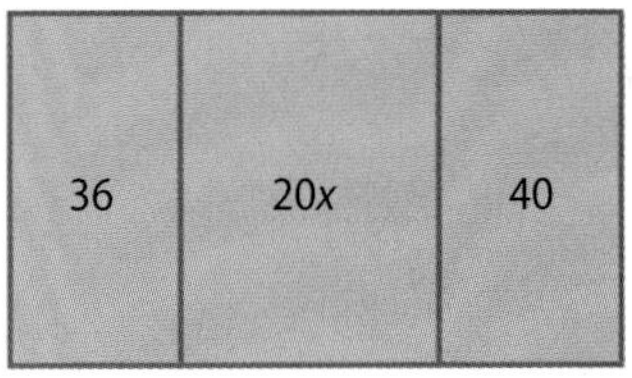

44. A square scrapbooking page has a perimeter of $(8x + 20)$ inches. What is the length of one side of the page?

45. Six friends visited a museum to see the new holograms exhibit. The group paid for admission to the museum and \$12 for parking. The total cost of the visit can be represented by the expression $\$6x + \$12$. What was the cost of the visit for one person?

46. CCSS **Reason Abstractly** The diagram represents a flower border that is 3 feet wide surrounding a rectangular sitting area. Write an expression in factored form that represents the area of the flower border.

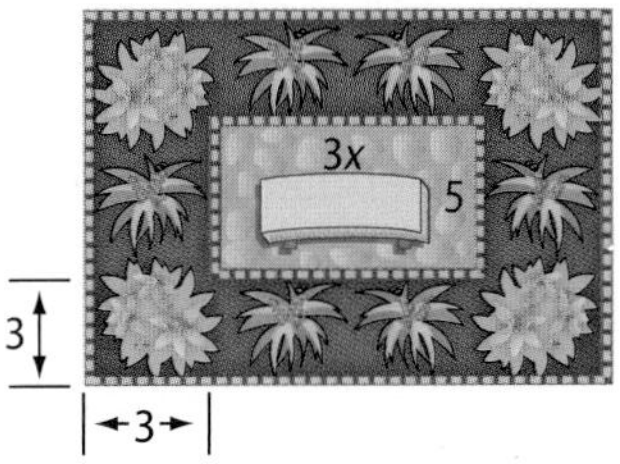

Write an expression in factored form that is equivalent to the given expression.

47. $\frac{1}{2}x + 4$
48. $\frac{2}{3}x + 6$
49. $\frac{3}{4}x - 24$
50. $\frac{5}{6}x - 30$

🔥 H.O.T. Problems Higher Order Thinking

51. CCSS **Identify Structure** Write two monomials whose greatest common factor is $4m$.

52. CCSS **Find the Error** Enrique is factoring $90x - 15$. Find his mistake and correct it.

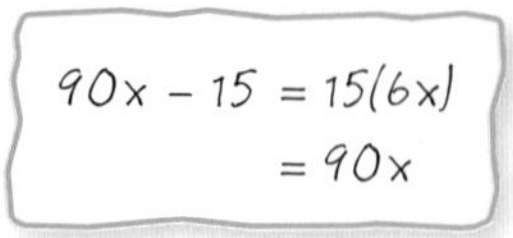

> $90x - 15 = 15(6x)$
> $\qquad\quad = 90x$

53. e **Building on the Essential Question** Explain how the GCF is used to factor an expression. Use the term *Distributive Property* in your response.

✏️ Standardized Test Practice

54. *Short Response* Factor the expression $40x + 15$.

55. Which of the following expressions cannot be factored?

 A $6 + 3x$

 B $7x + 3$

 C $15x + 10$

 D $30x + 40$

56. The Venn diagram shows the factors of 12 and $18x$.

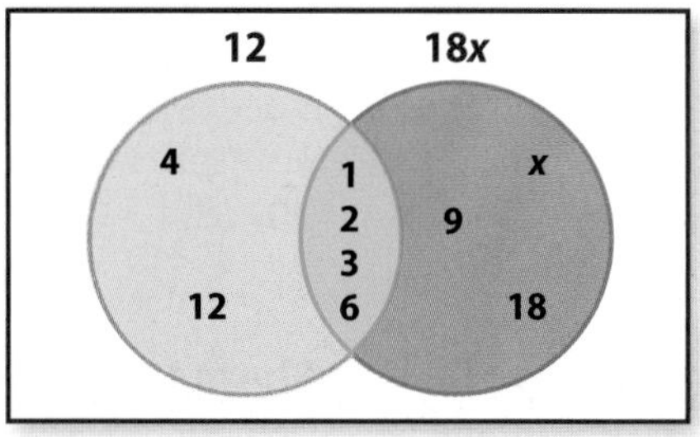

What is the greatest common factor of the two monomials?

 F 2 **H** 6

 G 3 **J** 36

CCSS Common Core Review

Find each product or quotient. 8.EE.1

57. $2^4 \cdot 2^6$

58. $\dfrac{a^3}{a^{-3}}$

59. $4x^{-2} \cdot 3x^9$

60. $\dfrac{c^5}{c^9}$

61. $(-4)s^{-8} \cdot (-4)s^7$

62. $\dfrac{12y^8}{6y^{10}}$

63. Tionne can ride 6 miles on her bike in one hour. If she rode for 1.5 hours on Saturday and 2 hours on Sunday, use mental math to find the total distance she rode that weekend. Justify your answer by using the Distributive Property. 7.NS.2c

64. A commission is a fee paid to a salesperson based on a percent of sales. Suppose a real estate agent earns a 3% commission. What commission would be earned for selling a house for $230,000? 7.RP.3

Add or subtract. 7.EE.1

65. $(-4x + 7) + (5x - 9)$

66. $(4.3x - 2) - (2.2x - 4)$

67. $\left(-\dfrac{5}{8}x + 3\right) + \left(\dfrac{3}{4}x - 8\right)$

68. $(6x - 4) - (6x + 1)$

Check

Chapter Review

ISG **Interactive Study Guide**
See pages 161–164 for:
- Vocabulary Check
- Key Concept Check
- Problem Solving
- Reflect

Lesson-by-Lesson Review

Lesson 7-1 The Distributive Property (pp. 292–296)

Use the Distributive Property to write each expression as an equivalent numeric expression. Then evaluate the expression.

1. $\frac{7}{8}(8 + 5)$
2. $(-10 + 9)3$

Use the Distributive Property to write each expression as an equivalent algebraic expression.

3. $(y + 3)7$
4. $-2(a - 7)$
5. $-\frac{2}{3}(b - 9)$
6. $(8m - 4)(-5.5)$
7. The Stuart family has 5 members. They each purchase a soda at \$2.50 each and a hot dog at \$3.50 each. Use mental math to find the total cost of the food. Justify your answer by using the Distributive Property.
8. Admission to the state fair is \$8 for adults and \$7 for students. Write two equivalent expressions if two adults and two students go to the fair. Then find the total admission cost.

Example 1

Use the Distributive Property to write 6(3 + 9) as an equivalent numberic expression. Then evaluate the expression.

$6(3 + 9) = \mathbf{6} \cdot 3 + \mathbf{6} \cdot 9$

$= 18 + 54$ Multiply.

$= 72$ Add.

Example 2

Use the Distributive Property to write 3(x − 6) as an equivalent algebraic expression.

$3(x - 6) = \mathbf{3}x - \mathbf{3} \cdot 6$

$= 3x - 18$ Simplify.

Lesson 7-2 Simplifying Algebraic Expressions (pp. 299–304)

Simplify each expression.

9. $6a + 5a$
10. $3x + 6x$
11. $7m - 2m + 3$
12. $6x - 3 + 2x + 5$
13. $a + 6(a + 3)$
14. $2(b + 3) + 3b$
15. Karen made 5 less than 4 times the number of free throws that Kimi made. Write an expression in simplest form that represents the total number of free throws made by both players.
16. Taylor jogged x miles after school. Seth jogged twice the distance that Taylor jogged. Rashida jogged 4 miles. Write an expression in simplest form to represent the total number of miles that the three students jogged.

Example 3

Simplify $-6x + 5 + x$.

$-6x + 5 + x$

$= -6x + x + 5$ Commutative Property

$= [(-6) + 1]x + 5$ Distributive Property

$= -5x + 5$ Simplify.

Lesson 7-3 Adding Linear Expressions (pp. 305–308)

Add. Use models if needed.

17. $(3x - 5) + (-5x + 12)$

18. $(-3x - 8) + (2x + 9)$

19. $(8x - 6) + (5x - 3)$

20. The angle measures of a triangle are $(x - 9)°$, $x°$, and $(2x - 5)°$. Find the measures of the angles.

Example 4

Add $(4x + 2) + (-5x + 3)$.

Arrange the linear expressions vertically, lining up like terms in columns. Then add.

$$\begin{array}{r} 4x + 2 \\ + \ -5x + 3 \\ \hline -x + 5 \end{array}$$

Lesson 7-4 Subtracting Linear Expressions (pp. 310–313)

Subtract. Use models if needed.

21. $(3x + 2) - (5x + 6)$

22. $(-3x + 4) - (8x + 4)$

23. $(5x - 7) - (-x - 9)$

24. One week, Ty made $(3x + 4)$ dollars. The following week he made $(x + 8)$ dollars. How much more did he make the first week?

Example 5

Subtract $(3x - 5) - (-8x + 1)$.

$(3x - 5) - (-8x + 1)$ Write the linear expression.

$= 3x - 5 + 8x - 1$ Definition of subtraction.

$= 3x + 8x - 5 - 1$ Group like terms.

$= 11x - 6$ Simplify.

Lesson 7-5 Factoring Linear Expressions (pp. 316–320)

Factor each expression. If the expression cannot be factored, write *cannot* be *factored*. Use algebra tiles if needed.

25. $4x + 12$

26. $9x - 54$

27. $3x - 15$

28. $5x + 12$

29. $14x + 10$

30. $28x - 42$

31. $35 + 18x$

32. $45 + 60x$

33. Dekentra made a collage on a square sheet of poster board. The perimeter of the collage is $(20x + 12)$ centimeters. What is the length of one side of the collage?

34. April has saved \$144 to buy a new car. If m is the monthly payment for one year, the expression $\$12m + \$144$ represents the total cost of the car. Factor $\$12m + \$144$.

35. The area of a name card is $(8x - 2)$ square inches. Factor $8x - 2$ to find possible dimensions of the name card.

Example 6

Factor $2x + 6$.

Method 1 Use a Model.

Model $2x + 6$.

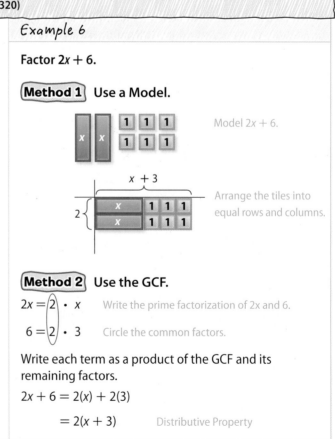

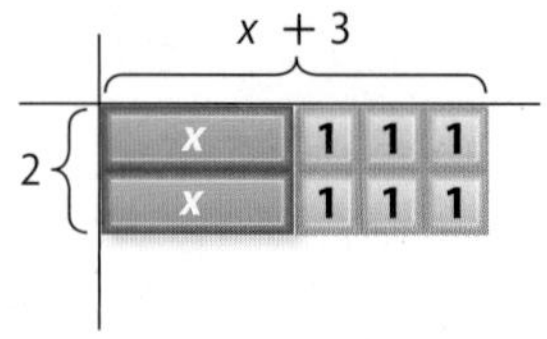

Arrange the tiles into equal rows and columns.

Method 2 Use the GCF.

$2x = \textcircled{2} \cdot x$ Write the prime factorization of 2x and 6.

$6 = \textcircled{2} \cdot 3$ Circle the common factors.

Write each term as a product of the GCF and its remaining factors.

$2x + 6 = 2(x) + 2(3)$

$= 2(x + 3)$ Distributive Property

Chapter 8
Equations and Inequalities

Essential Question

How are equations and inequalities used to describe and solve multi-step problems?

Common Core State Standards

Content Standards
7.EE.4, 7.EE.4a, 7.EE.4b, 8.EE.7, 8.EE.7a, 8.EE.7b

Mathematical Practices
1, 2, 3, 4, 5, 7

Math in the Real World

Geology Stalactites are formations that hang from the ceiling of a cave. A typical stalactite grows at a rate of 1 inch every 200 years. You can solve an inequality to find out how long it will take a stalactite to reach a length of at least 3 feet.

Interactive Study Guide
See pages 165–168 for:
- Chapter Preview
- Are You Ready?
- Foldable Study Organizer

Lesson 8-1

Solving Equations with Rational Coefficients

Interactive Study Guide

See pages 169–170 for:
- Getting Started
- Vocabulary Start-Up
- Notes

Essential Question

How are equations and inequalities used to describe and solve multi-step problems?

Common Core State Standards

Content Standards
7.EE.4, 8.EE.7, 8.EE.7b

Mathematical Practices
1, 3, 4, 5, 7

Vocabulary

solution
inverse operations
equivalent equations

What You'll Learn

- Solve equations by using the Division Property of Equality.
- Solve equations by using the Multiplication Property of Equality.

Real-World Link

Social Networks Do you have a social network profile? While gaming, blogging, and watching videos continue to be popular online activities, more preteens and teens are participating in social networking than ever before. Three-fourths of teens surveyed said they belonged to a social network, compared to 40% of adults surveyed.

Scott Quinn Photography/Stockbyte/Getty Images

Solve Equations by Dividing

An *equation* is a mathematical sentence stating that two expressions are equal. If 72 teens in a survey say they have a social network profile, and these teens are three fourths of the teens surveyed, then the equation $\frac{3}{4}x = 72$ can be used to find the total number of teens surveyed. A value for the variable that makes an equation true is called a **solution**. For $\frac{3}{4}x = 72$, the solution is 96, since $\frac{3}{4}(96) = 72$ is a true statement.

Key Concept: Properties of Equality

Words	Symbols
Multiplication Property of Equality If you multiply each side of an equation by the same nonzero number, the two sides remain equal.	For any numbers a, b, and c, if $a = b$, then $ca = cb$.
Division Property of Equality If you divide each side of an equation by the same nonzero number, the two sides remain equal.	For any numbers a, b, and c, where $c \neq 0$, if $a = b$, then $\frac{a}{c} = \frac{b}{c}$.

You can use the above properties and inverse operations to solve an equation. **Inverse operations** "undo" each other. To undo the multiplication of $\frac{3}{4}$ in $\frac{3}{4}x = 72$, you can apply the Division Property of Equality to divide each side of the equation by $\frac{3}{4}$. Applying the above properties creates **equivalent equations**, which are equations that have the same solution.

Example 1

Tools Tutor

Solve $4.2x = -52.5$. Check your solution.

$4.2x = -52.5$	Write the equation.
$\frac{4.2x}{\mathbf{4.2}} = \frac{-52.5}{\mathbf{4.2}}$	Division Property of Equality
$1x = -12.5$	$4.2 \div 4.2 = 1$; $-52.5 \div 4.2 = -12.5$
$x = -12.5$	Identity Property; $1x = x$

To check your solution, replace x with -12.5 in the original equation.

Check

$4.2\mathbf{x} = -52.5$	Write the equation.
$4.2(\mathbf{-12.5}) \stackrel{?}{=} -52.5$	Replace x with -12.5.
$-52.5 = -52.5$ ✓	The sentence is true.

The solution is -12.5.

Vocabulary Link
solution
Everyday Use the answer to a problem, such as the solution to a mystery
Math Use a value of the variable that makes an equation true

Got It? Do these problems to find out.

1a. Solve $-19.6 = 5.6x$.

1b. Solve $-20.9y = -250.8$.

1c. Solve $-13.6a = 217.6$.

1d. Solve $323.4 = -13.2z$.

Real World

Example 2

Tutor

A drive-through safari zoo charges \$12.50 per person for admission. In one hour, the park raised \$675 in admission fees. Write and solve an equation to find how many people visited that hour.

Words	Admission fee times the number of visitors equals the total money raised.
Variable	Let v = the number of zoo visitors.
Equation	$12.50 \cdot v = 675$

$12.5v = 675$	Write the equation.
$\frac{12.5v}{\mathbf{12.5}} = \frac{675}{\mathbf{12.5}}$	Division Property of Equality
$v = 54$	Simplify. Check this solution.

The zoo admitted 54 people in one hour.

Got It? Do this problem to find out.

2. A one-year, in-state camping permit for New Mexico State Parks costs \$180. If the total income from the camping permits is \$8280 during the first day of sales, write and solve an equation to find how many permits were purchased.

Solve Equations by Multiplying

Equations in which a variable is divided can be solved by multiplying each side by the same number. This method is also useful when the coefficient of the variable is a fraction.

Example 3

Tutor

Solve each equation. Check your solution.

a. $\frac{1}{4}y = -8$

$\frac{1}{4}y = -8$	Write the equation.
$\mathbf{4} \cdot \frac{1}{4}y = \mathbf{4} \cdot (-8)$	Multiplication Property of Equality
$1y = -32$	Multiplicative Inverse Property; $4 \cdot \frac{1}{4} = 1$
$y = -32$	Identity Property. Check your solution.

Multiplicative Inverse

Remember that the product of a number and its multiplicative inverse is 1. Use this property when the coefficient of x is a fraction.

b. $-\frac{3}{5}x = -\frac{6}{25}$

$-\frac{3}{5}x = -\frac{6}{25}$	Write the equation.
$-\frac{\mathbf{5}}{\mathbf{3}}\left(-\frac{3}{5}\right)x = -\frac{\mathbf{5}}{\mathbf{3}}\left(-\frac{6}{25}\right)$	Multiply each side by $-\frac{5}{3}$.
$1x = \frac{2}{5}$	Multiplicative Inverse Property; $-\frac{5}{3}\left(-\frac{3}{5}\right) = 1$
$x = \frac{2}{5}$	Identity Property. Check your solution.

Got It? **Do these problems to find out.**

3a. $7 = -\frac{1}{2}x$

3b. $\frac{6}{7}m = -\frac{3}{14}$

3c. $-9 = \frac{1}{3}n$

3d. $-\frac{1}{4} = \frac{2}{7}y$

Guided Practice

Check

Solve each equation. Check your solution. (Examples 1 and 3)

1. $1.3c = -65$

2. $-4.2 = -7m$

3. $0.8p = 9.6$

4. $\frac{1}{12}n = 12$

5. $18 = \frac{-1}{2}t$

6. $0.6h = 1.8$

7. $-3.4 = 0.4j$

8. $-\frac{3}{4}k = \frac{2}{3}$

9. $\frac{1}{25} = \frac{3}{5}m$

10. A forest preserve rents canoes for $22.50 per hour. Corey has $90 to spend. Write and solve an equation to find how many hours he can rent a canoe. (Example 2)

11. **STEM** The weight of an object on the Moon is one-sixth its weight on Earth. If an object weighs 54 pounds on the Moon, write and solve an equation to find how much it weighs on Earth. (Example 2)

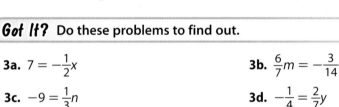

Independent Practice

Go online for Step-by-Step Solutions eHelp

Solve each equation. Check your solution. (Examples 1 and 3)

12. $9x = 5.4$

13. $0.5s = -60$

14. $6.4 = -0.4r$

15. $-7.2 = 3y$

16. $0.3x = -4.5$

17. $4.95 = 0.3t$

18. $-8.4 = -6g$

19. $-28 = -\frac{1}{14}d$

20. $\frac{1}{9}b = -108$

21. $16 = -\frac{1}{4}b$

22. $-\frac{1}{8}x = -4$

23. $-\frac{4}{9} = -\frac{4}{3}s$

24. $-\frac{25}{36} = -\frac{5}{6}r$

25. $-\frac{9}{10}k = 72$

26. $\frac{2}{3}n = -22$

27. Rashid picked a total of 420 strawberries in $\frac{5}{6}$ hour. Write and solve an equation to find how many strawberries Rashid could pick in 1 hour. (Example 2)

28. **Financial Literacy** Marcus wants to save \$378 in order to buy a new electronic keyboard. He plans to save \$15.75 every week from his paycheck. Write and solve an equation to find how many weeks Marcus will need to save. (Example 2)

Solve each equation. Check your solution.

29. $5p - 2p = -12$

30. $42 = 4x + 3x$

31. $-2(6y) = 144$

32. $72 = -12(-3x)$

33. $\frac{r}{4} = -25 + 9$

34. $\frac{m}{-3} = -5 - 18$

35. $\frac{1}{3}n = \frac{2}{9}$

36. $\frac{5}{8} = -\frac{1}{2}x$

37. $-0.7 = -\frac{7}{9}z$

38. $1\frac{7}{8}y = 4\frac{1}{2}$

39. $2\frac{1}{3} = -9m$

40. $-\frac{7}{9}t = -\frac{28}{36}$

41. The sleeping heart rate of a black bear during hibernation is about $\frac{2}{5}$ of its summer rate. If the sleeping heart rate of a bear is 28 beats per minute during hibernation, find the summer sleeping heart rate.

42. **Use the table to help you write and solve an equation.**

a. Jenna hikes at a steady rate of 2.5 miles per hour. How long will it take her to hike the Lone Star Trail?

b. It takes Ming 4 hours to hike the Mallard Lake Trail. What is his average hiking rate?

c. The round-trip distance of the Observation Point Trail is 56% of the round-trip distance of the Mystic Falls Trail. What is the round-trip distance of the Mystic Falls Trail?

Yellowstone National Park, Wyoming	
Name of Trail	**Round-Trip Distance (mi)**
Mallard Lake	6.8
Howard Eaton	5.8
Lone Star	4.8
Observation Point	1.4

43. CCSS **Use Math Tools** A student solved the equation $\frac{9}{10}m = 5\frac{1}{10}$ and found the solution $m = 459$. Use estimation to explain why the student's solution must be incorrect.

44. CCSS **Identify Structure** Explain how to use the Multiplication Property of Equality to solve $\frac{2}{5}y = 6$. Then explain how to solve the equation using the Division Property of Equality.

45. CCSS **Multiple Representations** Every autumn, the North American Monarch butterfly migrates up to 3000 miles to California and Mexico, where it hibernates until early spring. Suppose a particular butterfly travels on average 52.5 miles per day.

a. Symbols Write an equation that represents the distance *d* the butterfly will travel in *t* days.

b. Table Use the equation to complete the table.

Time (days)	1	2	3	4	5	6
Distance (miles)	■	■	■	■	■	■

c. Graph Graph the points from the table on the coordinate plane. Graph time on the *x*-axis and distance on the *y*-axis.

d. Graph Using the graph, estimate the number of days it will take the butterfly to travel 475 miles.

e. Words How many days will it take the butterfly to travel 2100 miles? Which method did you use to solve the problem?

46. The formula for finding the area of a triangle is $A = \frac{1}{2}bh$, where *A* represents the area, *b* represents the length of the base of the triangle, and *h* represents the height of the triangle. Write and solve equations to complete the table of values.

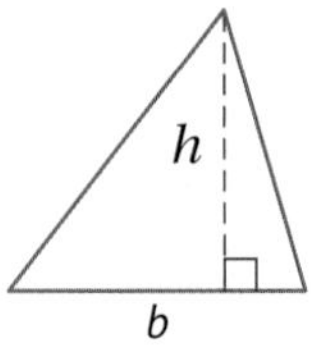

Area (*A*)	15	15	15	15	15
Base (*b*)	1	2	3	4	5
Height (*h*)	■	■	■	■	■

H.O.T. Problems Higher Order Thinking

47. CCSS **Model with Mathematics** Write a real-world example that uses an equation containing a decimal and a fraction. Then find the solution.

48. CCSS **Find the Error** Sam is solving $\frac{1}{4}x = -20$. Find his mistake and correct it.

$$\frac{1}{4}x \div 4 = -20 \div 4$$
$$x = -5$$

49. CCSS **Justify Conclusions** When you solve an equation of the form $px = q$, where *p* and *q* are rational numbers and $p \neq 0$, is it ever possible for the solution to be 0? If so, when? If not, why not? Justify your conclusion.

50. CCSS **Persevere with Problems** If $\frac{3}{10}x = 3$, what is the value of $7x + 13$?

51. **Building on the Essential Question** Suppose your friend says he can solve $3x = 15$ by using the Multiplication Property of Equality. Is he correct? Justify your response.

Standardized Test Practice

52. During a vacation, the Morales family drove 63.2 miles in 1 hour. If they averaged the same speed during their trip, which equation can be used to find how far the Morales family drove in 6 hours?

A $\frac{63.2}{x} = 6$

B $\frac{1}{6}x = 63.2$

C $6x = 63.2$

D $63.2x = 6$

53. The solution of which equation is *not* graphed on the number line below?

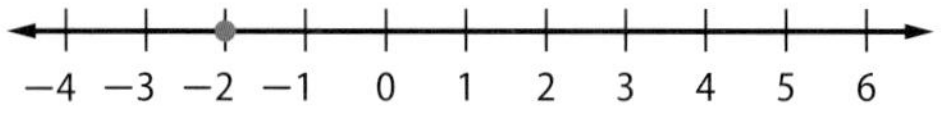

F $12.8 = -6.4x$

G $8.1x = -16.2$

H $-15 = 7.5x$

J $-18.3x = -36.6$

54. Ella paid $11.85 for 3 magazines. If each magazine was the same price, how much did each magazine cost?

A $4.59

B $4.00

C $3.95

D $3.59

55. **Short Response** Stanley paid half of what Royce paid for his baseball glove. Royce paid $64.98 for his glove. Write and solve an equation to find how much Stanley paid for his glove.

CCSS Common Core Review

Solve each equation. Check your solution. 8.EE.7

56. $x - 5 = -22$

57. $4 = 7 + p$

58. $-40 = y - 9$

59. $2.3 + r = 1.6$

60. $d - 2.7 = -1.4$

61. $t + (-16) = -24$

62. $p + \frac{1}{10} = -\frac{3}{4}$

63. $\frac{2}{3} + k = \frac{1}{6}$

64. $d - \frac{4}{9} = -\frac{1}{12}$

Simplify each expression. 6.EE.3

65. $5(t + 3)$

66. $7x - 12x$

67. $9p + 4 + 3p$

68. $3w + 4s - w + 5s$

69. $7 - 4(x + 3)$

70. $3(2 + 3x) + 21x$

71. $12x + 6 + x$

72. $4n - 9 + 5n$

73. $10(y + 8)$

74. Find the values that complete the table at the right for $y = -4x$. 7.NS.2

75. Gabriel is 12 years old, and his younger brother Elias is 2 years old. How old will each of them be when Gabriel is twice as old as Elias? 7.EE.3

x	−2	−1	0	1
y	■	■	■	■

Find the value of each expression. 6.EE.1

76. $3^3 + 7(4)$

77. $\frac{2^2 - 9}{10 + 15}$

78. $5 - 3(6 + 2^3)$

79. $5^2 - 6 + 6 \cdot 8$

80. $20 \cdot 7 - 2^4 \cdot 5$

81. $3^2[15 - (-9)]$

82. $6^2 - 4^2 \cdot 3$

83. $5(9^2 - 10^2)$

84. $-4^3 + 2 \cdot 9$

85. $10 - 5(10 - 8)^2$

86. $(4 + 3^3)(-2)$

87. $\frac{(1 - 3)^2}{(4 - 2)^2}$

Inquiry Lab

Solving Two-Step Equations

Inquiry **HOW does the arithmetic way to solve a two-step equation compare to the algebraic way?**

CCSS Content Standards
7.EE.4, 7.EE.4a, 8.EE.7

Mathematical Practices
1, 3

Cooking David is making vegetable soup. The recipe says to start by pouring three cans of vegetable broth into a pot and then add 9 fluid ounces of tomato juice. According to the recipe, the total amount of liquid should be 51 fluid ounces. How many fluid ounces of vegetable broth should each can contain?

Investigation 1

You can use a bar diagram to represent the situation.

Step 1 Draw a bar diagram that represents 3 cans of vegetable broth and 9 additional fluid ounces of tomato juice. The bar diagram should show that the total amount of liquid is 51 fluid ounces.

vegetable broth	vegetable broth	vegetable broth	tomato juice
?	**?**	**?**	**9 fl oz**

|---------------------- 51 fl oz ----------------------|

Step 2 Let x represent the number of fluid ounces of vegetable broth in each can. Then the bar diagram represents the equation $3x + 9 = 51$.

Step 3 Use the *work backward* strategy to find the number of fluid ounces of vegetable broth in each can.

Since $3x + 9 = 51$, $3x$ must equal $51 - 9$ or 42. This is shown by removing the 9 fluid ounces at the end of the bar diagram.

vegetable broth	vegetable broth	vegetable broth
?	**?**	**?**

|----------------- 42 fl oz -----------------|

Step 4 Since $3x = 42$, x must equal $42 \div 3 = 14$. This is shown by dividing the 42 fluid ounces in the bar diagram equally among the three remaining bars.

vegetable broth	vegetable broth	vegetable broth
14 fl oz	**14 fl oz**	**14 fl oz**

|----------------- 42 fl oz -----------------|

So, each can should contain 14 fluid ounces of vegetable broth.

Collaborate

Work with a partner. Use a bar diagram to solve each equation.

1. $4x + 5 = 17$
2. $5y + 2 = 22$
3. $6p + 5 = 23$

The method you used to solve an equation in Investigation 1 is an arithmetic method. You can also use algebra to solve this type of equation.

Recall that when you use algebra tiles, one $+1$ tile and one -1 tile make a zero pair. Since $1 + (-1) = 0$, you can add or subtract zero pairs from either side of an equation mat without changing the value.

Investigation 2

Model and solve $2x + 4 = 2$ using algebra tiles.

Step 1 Model the equation. Notice that it is not possible to remove four positive 1-tiles from the right side of the equation mat.

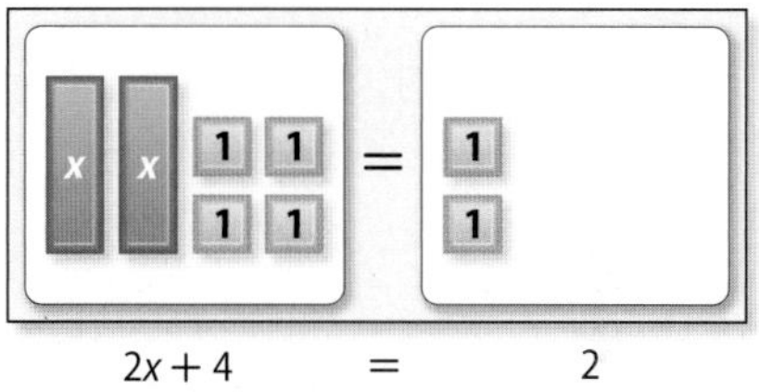

Step 2 Add 2 zero pairs to the right side of the mat so you have enough positive 1-tiles.

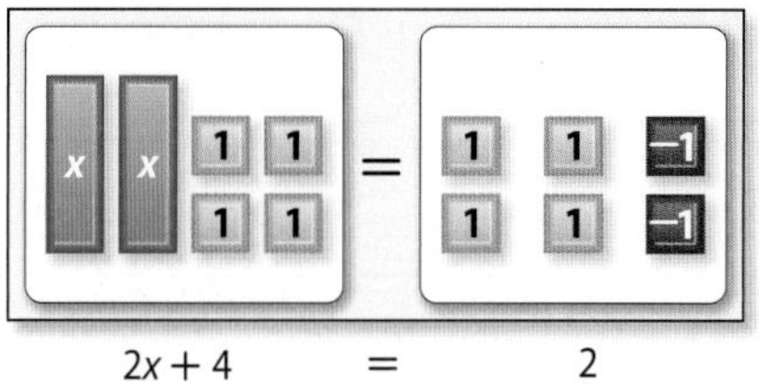

Step 3 Remove the same number of 1-tiles from each side of the mat until the *x*-tiles are alone on one side.

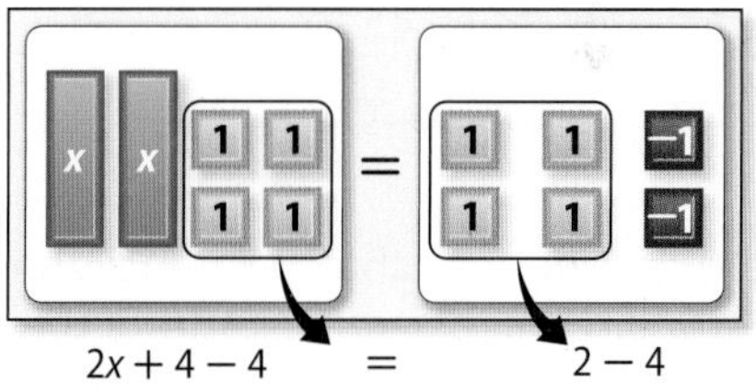

Step 4 Separate the remaining tiles into two equal groups.

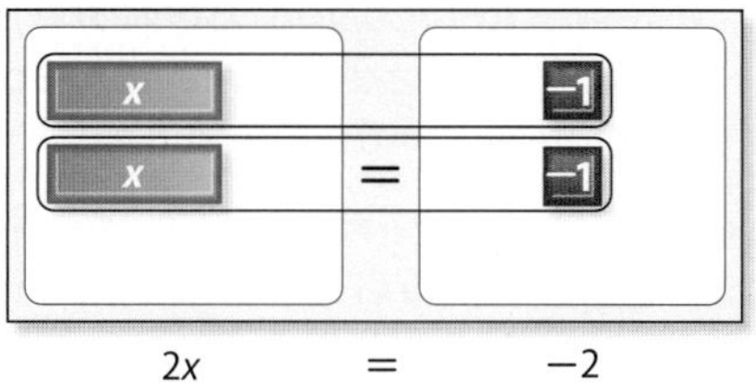

So, $x = -1$.

Collaborate

Work with a partner. Use algebra tiles to solve each equation.

4. $2x + 3 = 7$
5. $4m + 1 = 13$
6. $2 = 3g - 4$
7. $4b + 3 = -9$
8. $3x - 3 = 6$
9. $-6 = 2y + 4$
10. $-1 = 2w - 11$
11. $-4 = 4z - 8$
12. $2n - 5 = -5$

Analyze

13. What property is shown by removing tiles from each side?

14. What property is shown by separating the tiles into equal groups?

Solve each equation using either method.

15. $7m + 5 = 19$ **16.** $5 = 4x - 15$ **17.** $-3 = 4y - 11$

18. $3x - 7 = -1$ **19.** $8p - 12 = 20$ **20.** $12 = 9y - 15$

21. $7g + 10 = -4$ **22.** $-5 = 3x + 7$ **23.** $3z - 2 = -8$

24. $4n + 7 = 19$ **25.** $2m - 6 = 4$ **26.** $17 = -5w + 2$

For Problems 27–32, tell what equation is modeled by the algebra tiles and solve the equation.

27.

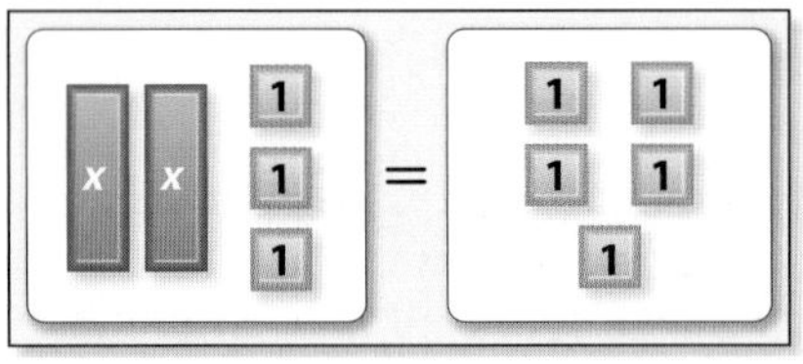

28.

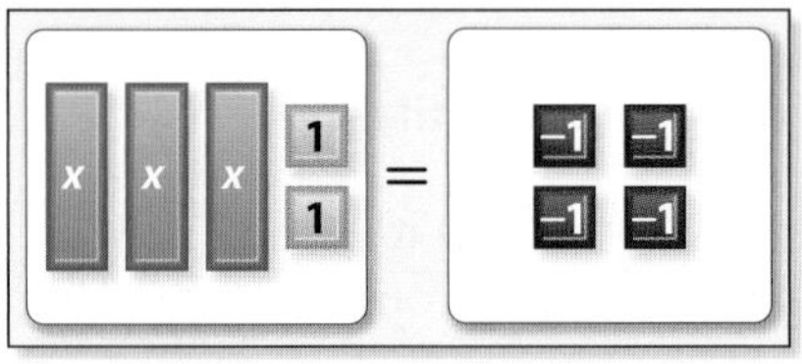

29.

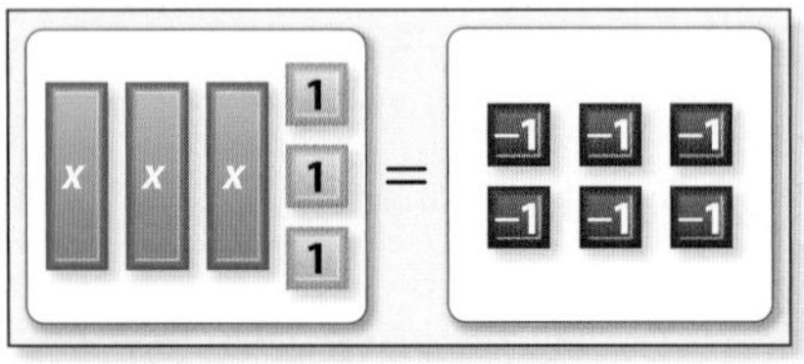

30.

31.

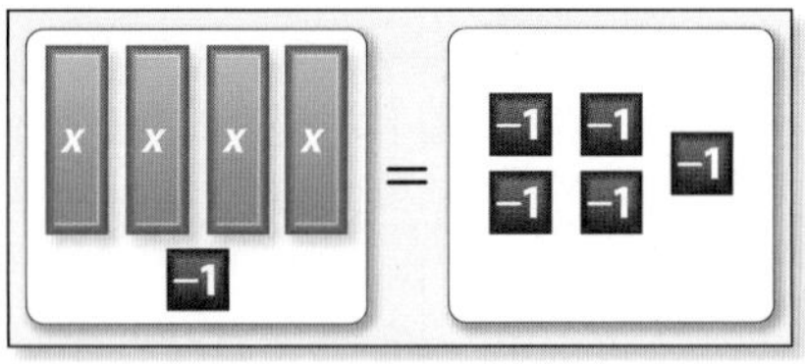

32.

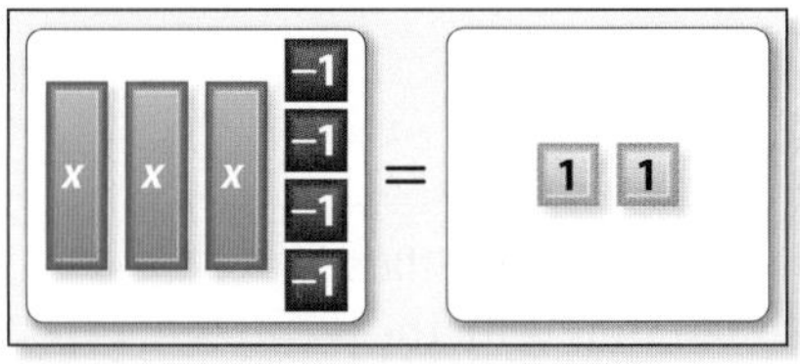

33. Mr. Jeffries bought charcoal pencils for his art class. He bought 3 full boxes, plus 4 additional charcoal pencils. Altogether, Mr. Jeffries bought 28 charcoal pencils. Solve the equation $3p + 4 = 28$ to find the number of charcoal pencils p in each box.

34. CCSS **Reason Inductively** Solve the equations $2x + 1 = 1$, $4x - 3 = -3$, and $3x + 6 = 6$. Then make a conjecture about when an equation of the form $ax + b = c$ has the solution $x = 0$.

35. CCSS **Justify Conclusions** What can you conclude about all of the equations below? Explain why your conclusion makes sense.

$2x + 1 = 7$ $2x + 2 = 8$ $2x + 3 = 9$ $2x + 4 = 10$

Reflect

36. Inquiry HOW does the arithmetic way to solve a two-step equation compare to the algebraic way?

Lesson 8-2

Solving Two-Step Equations

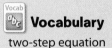

Interactive Study Guide

See pages 171–172 for:
- Getting Started
- Real-World Link
- Notes

Essential Question

How are equations and inequalities used to describe and solve multi-step problems?

Common Core State Standards

Content Standards
7.EE.4, 7.EE.4a, 8.EE.7, 8.EE.7b

Mathematical Practices
1, 3, 4, 7

Vocabulary

two-step equation

What You'll Learn

- Solve two-step equations.
- Solve real-world problems involving two-step equations.

Real-World Link

Cheerleading Cheerleaders on a middle school squad must purchase cheer shoes for \$35, plus several pairs of white ankle socks. An equation involving two operations can be used to find the total cost.

Solve Two-Step Equations

A **two-step equation** contains two operations. To solve a two-step equation, use inverse operations to undo each operation in reverse order of the order of operations.

Example 1

Tutor

Solve $3a + 9 = 33$. Check your solution.

Method 1 The Vertical Method

$3a + 9 = 33$	Write the equation.
$3a + 9 = 33$	
$\underline{\quad - \mathbf{9} = -\mathbf{9}}$	Subtraction Property of Equality
$3a \quad = 24$	Simplify.
$\frac{3a}{\mathbf{3}} = \frac{24}{\mathbf{3}}$	Division Property of Equality
$a = 8$	Simplify.

Method 2 The Horizontal Method

$3a + 9 = 33$	Write the equation.
$3a + 9 - \mathbf{9} = 33 - \mathbf{9}$	Subtraction Property of Equality
$3a = 24$	Simplify.
$\frac{3a}{\mathbf{3}} = \frac{24}{\mathbf{3}}$	Division Property of Equality
$a = 8$	Simplify.

Using either method, the solution is 8.

Check	$3\mathbf{a} + 9 = 33$	Write the equation.
	$3(\mathbf{8}) + 9 \stackrel{?}{=} 33$	Replace *a* with 8.
	$24 + 9 \stackrel{?}{=} 33$	Multiply.
	$33 = 33$ ✓	The sentence is true.

Got It? Do these problems to find out.

Solve each equation. Check your solution.

1a. $6x + 1 = 25$

1b. $4x - 5 = -33$

Example 2

Watch Tutor

Solve $\frac{1}{5}p - 12 = 20$.

$\frac{1}{5}p - 12 = 20$	Write the equation.
$\frac{1}{5}p - 12 + \mathbf{12} = 20 + \mathbf{12}$	Addition Property of Equality
$\frac{1}{5}p = 32$	Simplify.
$\mathbf{5} \cdot \frac{1}{5}p = \mathbf{5} \cdot 32$	Multiplication Property of Equality
$p = 160$	Simplify. Check your solution.

Properties of Equality
Recall that the Addition and Subtraction Properties of Equality state that the same number can be added to or subtracted from each side of an equation.

Got It? Do these problems to find out.

2a. $8 = 15 + \frac{1}{3}n$

2b. $-\frac{1}{6}x - 3 = 2$

Example 3

Tutor

Solve $9 - t = -34$.

$9 - t = -34$	Write the equation.
$9 - \mathbf{1}t = -34$	Identity Property: $t = 1t$
$9 + (-1t) = -34$	Definition of Subtraction
$\mathbf{-9} + 9 + (-1t) = \mathbf{-9} + (-34)$	Addition Property of Equality
$-1t = -43$	Simplify.
$\frac{-1t}{\mathbf{-1}} = \frac{-43}{\mathbf{-1}}$	Division Property of Equality
$t = 43$	Simplify. Check your solution.

Got It? Do these problems to find out.

3a. $-15 - b = 44$

3b. $-6.5 = -4.3 - n$

Example 4

Tutor

Solve $2x + x - 27 = 3$.

$2x + x - 27 = 3$	Write the equation.
$2x + \mathbf{1}x - 27 = 3$	Identity Property; $x = 1x$
$3x - 27 = 3$	Distributive Property; $2x + 1x = (2 + 1)x$ or $3x$
$3x - 27 + \mathbf{27} = 3 + \mathbf{27}$	Addition Property of Equality
$3x = 30$	Simplify.
$\frac{3x}{\mathbf{3}} = \frac{30}{\mathbf{3}}$	Division Property of Equality
$x = 10$	Simplify. Check your solution.

Distributive Property
You use the Distributive Property to mentally simplify $2x + x$.

$2x + 1x = (2 + 1)x$
$= 3x$

Got It? Do these problems to find out.

4a. $4 - 9c + 3c = 58$

4b. $3.4 = 0.4m - 2 + 0.2m$

Solve Real-World Problems

You can write and solve two-step equations to solve many real-world problems.

Example 5

Tutor

Deon wants to go on a camping trip with his hiking club. The trip costs \$185.75. He paid a deposit of \$45.75 and will save an additional \$17.50 per week to pay for the trip. Solve $45.75 + 17.50w = 185.75$ to find the number of weeks Deon will need to save money for the trip.

$45.75 + 17.50w = 185.75$	Write the equation.
$45.75 - \mathbf{45.75} + 17.50w = 185.75 - \mathbf{45.75}$	Subtraction Property of Equality
$17.50w = 140$	Simplify.
$\frac{17.50w}{\mathbf{17.50}} = \frac{140}{\mathbf{17.50}}$	Division Property of Equality
$w = 8$	Simplify. Check your solution.

Deon will need to save for 8 weeks.

Got It? **Do this problem to find out.**

5. Salvatore purchased a computer for \$682.20. He paid \$105.40 initially and will pay \$20.60 per month until the computer is paid off. Solve $105.40 + 20.60x = 682.20$ to find the number of months Salvatore will make payments for the computer.

Guided Practice

Check

Solve each equation. Check your solution. (Examples 1 and 2)

1. $4p + 9 = 25$

2. $-2x + 1 = 7$

3. $5y - 3 = -23$

4. $17 = 7x - 4$

5. $-4 = 8m - 12$

6. $-13 = 5 - 3z$

7. $\frac{1}{4}p - 6 = -8$

8. $-\frac{1}{6}t + 1 = 3$

9. $-\frac{1}{2}r - 12 = -27$

10. $\frac{1}{2}g + 6 = 4$

11. $-\frac{1}{8}x - 5 = -1$

12. $9 = 4 + \frac{1}{5}q$

Solve each equation. Check your solution. (Examples 3 and 4)

13. $-7 - 8d = 17$

14. $23 - 2c = 41$

15. $1 - 2k = -9$

16. $12 - m = -7$

17. $14 = 6 - x$

18. $-6 = 4 - 5b$

19. $-4 = 8y - 9y + 6$

20. $-1.3j + 0.4 = -1.16$

21. $1.1 - t + 2.2t = 5.9$

22. $5m + 4 - 7m = 10$

23. $\frac{1}{3}p + 6 - \frac{2}{3}p = 0$

24. $7.8 = 3 + 0.1n + 0.7n$

25. Kaleigh has \$25. She plans to save \$7.50 each week. Solve $25 + 7.50w = 250$ to find the number of weeks it will take Kaleigh to save \$250. (Example 5)

26. A caterer is preparing a dinner for a party. She charges a flat fee of \$16 plus \$8.25 per person. Solve $16 + 8.25p = 131.50$ to find the number of people at a dinner that costs \$131.50. (Example 5)

Independent Practice

Go online for Step-by-Step Solutions eHelp

Solve each equation. Check your solution. (Examples 1 and 2)

27. $5a + 3 = 28$

28. $3b + 15 = 27$

29. $4d - 18 = -34$

30. $25 = 2c - 9$

31. $\frac{1}{3}g + 4 = 2$

32. $\frac{1}{9}h - 3 = 2$

33. $-16 = \frac{1}{2}k - 7$

34. $20 = \frac{1}{5}m + 12$

35. $\frac{1}{4}n - 20 = -1$

36. $3.6 = 2x + 1.8$

37. $\frac{1}{8}y - \frac{1}{2} = \frac{7}{8}$

38. $\frac{1}{4}t + 1 = 2\frac{1}{4}$

Solve each equation. Check your solution. (Examples 3 and 4)

39. $46 - 8x = -18$

40. $y - 7y + 6 = 30$

41. $-7 = -\frac{1}{5}p - 1$

42. $14 = -\frac{1}{3}s - 8$

43. $x + 7 - 2x = 18$

44. $46 - 3n = -23$

45. $5.5 - 5x = 4$

46. $6 = 8.1 - 3x$

47. $8.4 - 3x - x = 2$

48. $m - 5 - 6m = 0$

49. $19 = 3 - 3d - 5d$

50. $0 = t + 4 - 9t$

51. **Financial Literacy** The cost of a family membership at a health club is shown at the right. The Johnson family budgets \$800 to use the health club. Solve $125 + 45f = 800$ to find the number of months the family can use the club. (Example 5)

HEALTH
CLUB
Family
Membership
Only $125 to join
and $45 per month!

52. The second book in a fantasy series is 112 pages longer than the first book. The total number of pages in both books is 524. Solve the equation $b + b + 112 = 524$ to find the number of pages b in the first book. (Example 5)

53. **STEM** Draven's computer downloads files at a rate of 220 kilobytes per second. The computer has already downloaded the first 550 kilobytes of a 2310-kilobyte file. Solve the equation $550 + 220s = 2310$ to find the number of seconds it will take to download the rest of the file. (Example 5)

54. The perimeter of the triangle in the figure is 22 inches. Solve the equation $x + x + 3 + 9 = 22$ to find the length x of the shortest side of the triangle. (Example 5)

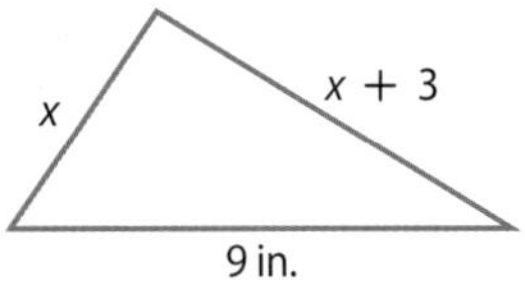

x
x + 3
9 in.

55. Tenisha bought some gel pens that cost \$1.29 each. She also bought a notebook for \$3.59. She spent a total of \$10.04 on these items. Solve the equation $1.29g + 3.59 = 10.04$ to find the number of gel pens she bought. (Example 5)

56. Aaron has a piece of yarn that is 15 inches long. For an art project, he cut off 3 pieces of yarn of equal length. This left him with $4\frac{1}{2}$ inches of yarn. Solve the equation $3p + 4\frac{1}{2} = 15$ to find the length of each piece of yarn that Aaron will use in the art project. (Example 5)

Solve each equation. Check your solution.

57. $6.1e + 1.07 = 9$

58. $-2.5c + 6.7 = -1.3$

59. $\frac{2}{3} - 6y = -1\frac{5}{6}$

60. $\frac{3}{4}x + 1.5 = 2.7$

61. $-\frac{1}{4}f + 20.5 = 12.9$

62. $54.8 - \frac{1}{5}d = 60.1$

63 Janelle and some of her friends went to the movies. Tickets cost \$6 per person, and they each received a \$1.50 student discount. Each girl also purchased a snack for \$2.25. The total cost was \$40.50. Solve the equation $6s - 1.5s + 2.25s = 40.50$ to find how many girls went to the movies.

Solve each equation. Check your solution.

64. $\frac{3x}{2} + 4x = 22$

65. $40.77 = \frac{y}{5} + 2.4y + \frac{y}{10}$

66. $\frac{x}{2} + \frac{5x}{6} + \frac{x}{4} = 380$

67. $\frac{-2x + 5}{2} = 17$

68. CCSS **Multiple Representations** In this problem, you will investigate a function. Tia's family is installing a fence around three sides of her backyard as shown at the right. The equation $2w + 24 = f$ represents the relationship between the width of the fenced area and the total amount of fencing needed.

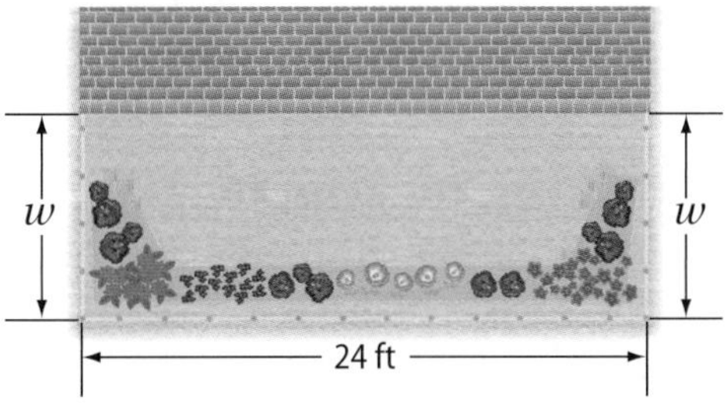

a. **Table** Make a function table to show the amount of fencing needed for widths of 12, 15, and 18 feet.

b. **Symbols** Find the width of the fenced area if Tia has 92 feet of fencing.

H.O.T. Problems Higher Order Thinking

69. CCSS **Model with Mathematics** Write a real-world example that could be solved by using the equation $2x + 7 = 15$. Then solve the equation.

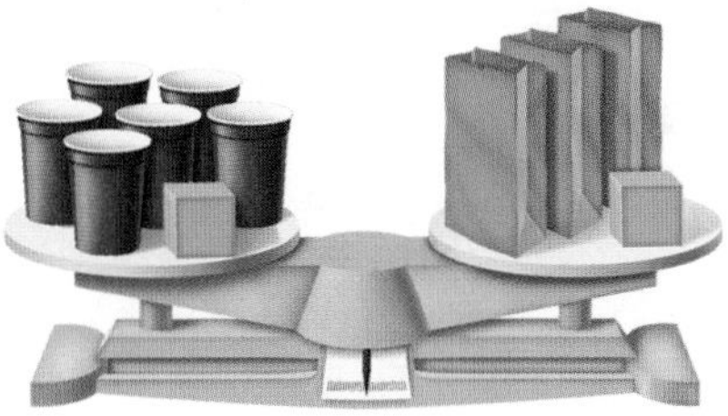

70. CCSS **Persevere with Problems** The model at the right represents the equation $6y + 1 = 3x + 1$. What is the value of x?

71. CCSS **Identify Structure** Write a two-step equation that can be solved using the Subtraction Property of Equality and the Multiplication Property of Equality. Show how to use these properties to solve the equation.

72. CCSS **Find the Error** Toshiro is solving the equation $7 - 2x = -51$. Find his mistake and correct it.

$$7 - 2x = -51$$
$$7 + 7 - 2x = -51 + 7$$
$$2x = -44$$
$$\frac{2x}{2} = \frac{-44}{2}$$
$$x = -22$$

73. **Building on the Essential Question** Evaluate $3(2) + 5$. Then solve the equation $3x + 5 = 11$. How are the problems and solutions similar? How are they different?

Standardized Test Practice

74. The results of a student council fundraiser are shown in the table.

Purchase Price for 144 Pens	Profit for 144 Pens
\$309.60	\$50.40

Use the equation below to find the selling price p of one pen.

$$144p - 309.60 = 50.40$$

A \$1.80 **C** \$2.50
B \$2.15 **D** \$2.72

75. Ms. Fraser's total monthly cell phone bill b can be found using the equation $b = 45.60 + 0.10t$, where t represents the number of text messages she made. Find the number of text messages she made in a month in which the total charge was \$56.70.

F 101 **H** 125
G 111 **J** 131

76. The distance d that Maxie can run in her first training run is represented by the equation $d = \frac{1}{2}m - 2$. What is the maximum distance m that she can run if her first training run is 3 miles?

A 10 miles
B 8 miles
C 6 miles
D 4 miles

77. Short Response Jody bought two pairs of jeans. The first pair costs \$12 less than 3 times the cost c of the second pair. The first pair of jeans costs \$45. The equation below can be used to find the cost of the second pair of jeans.

$$3c - 12 = 45$$

Solve the equation to find the cost in dollars of the second pair of jeans.

CCSS Common Core Review

Solve each equation. 8.EE.7

78. $36 = -12y$
79. $4 = \frac{x}{14}$
80. $5y = \frac{3}{2}$
81. $x - 13 = -45$
82. $\frac{2}{3} + p = 1$
83. $t + 12.4 = 16.23$

84. The difference between the record high and low temperatures in Columbus, Ohio, is 128° F. The record high temperature is 106° F. Write and solve an equation to find the record low temperature. 7.EE.4a

85. Financial Literacy You have saved some money. Your friend has saved \$40 more than you. Write an expression in simplest form that represents the total amount of money you and your friend have saved. 6.EE.2a

86. Zane has a collection of CDs. Sage's collection has 24 more CDs than Zane's. Write an expression in simplest form that represents the total number of CDs Zane and Sage have altogether. 6.EE.2a

Evaluate each expression if $x = 4$, $y = -10$, and $z = 14$. 6.EE.2c

87. xy
88. $y + z$
89. $2x - y$
90. $2z + 2y$
91. xyz
92. $z - 3x + y$
93. $4z - x - y$
94. $\frac{1}{3}(x + y)$
95. $z(3 - x)$
96. $xy - y$
97. $z - (1 - y)$
98. $\frac{1}{2}x + y$

Lesson 8-3

Writing Equations

ISG Interactive Study Guide

See pages 173–174 for:
- Getting Started
- Real-World Link
- Notes

Essential Question

How are equations and inequalities used to describe and solve multi-step problems?

CCSS Common Core State Standards

Content Standards
7.EE.4, 7.EE.4a, 8.EE.7, 8.EE.7b

Mathematical Practices
1, 3, 4, 7

What You'll Learn

- Write two-step equations.
- Solve verbal problems by writing and solving two-step equations.

Real-World Link

Tablet Computers Touch-screen computers have been around for twenty years. Their ease of use and speed are two reasons that more and more consumers are purchasing them. Accessories, such as docking stations, power adapters, and connection kits, help users get the most out of their tablets.

Write Two-Step Equations

A power adapter costs \$5 more than a docking station. The total cost of the accessories is \$63. You can summarize this information by writing an equation.

Words	Docking station plus power adapter costs \$63
Variable	Let d = the cost of the docking station. So, $d + 5$ = the cost of the power adapter.
Equation	$d + d + 5 = \$63$

Example 1

Tutor

Translate each sentence into an equation.

a. Zack has 6 shirts. This is 4 less than twice the number of shirts *n* that Xavier has.

$6 = 2n - 4$

b. Eight more than the quotient of a number *y* and −3 is −24.

$8 + \frac{y}{-3} = -24$

c. Jeremy has 13 baseball cards, which is 7 more than one-fifth the number *m* Michael has.

$13 = 7 + \frac{1}{5}m$

Got It? Do these problems to find out.

1a. Four more than 0.3 times a number x is -26.

1b. Hannah has 24 stickers. This is 6 less than twice the number of stickers n Molly has.

1c. The quotient of a number n and 7, increased by 6, is equal to 12.

Example 2

Juan's father was 29 years old when Juan was born. This year, the sum of their ages is 53. Find their ages.

Let x = Juan's age. Then, $x + 29$ = Juan's father's age.

$x + x + 29 = 53$	Write the equation.
$2x + 29 = 53$	Distributive Property
$2x + 29 - \mathbf{29} = 53 - \mathbf{29}$	Subtraction Property of Equality
$2x = 24$	Simplify.
$x = 12$	Mentally divide each side by 2.

Juan is 12 years old. His father is 12 + 29 or 41 years old.

Equations
Look for the words *is, total, equals,* or *is equal to* when you translate sentences into equations.

Got It? **Do this problem to find out.**

2. Deisha saved d dollars last month. This month she saved \$8 more than 3 times the amount she saved last month. She saved a total of \$141. Write and solve an equation to find how much she saved last month.

Two-Step Verbal Problems

In some real-world situations, you start with a given amount and then increase the amount at a constant rate.

Real World Example 3

Logan collected pledges for the charity walk-a-thon. He will receive total contributions of \$65.50 plus \$21.75 for every mile that he walks. How many miles will he need to walk to raise \$370?

First, write an equation to model the situation.

Words	\$65.50 plus \$21.75 per mile equals \$370.
Variable	Let m = the number of miles Logan walks. So, $21.75m$ = contributions for walking m miles.
Equation	$65.50 + 21.75m = 370$

$65.50 + 21.75m = 370$	Write the equation.
$65.50 - \mathbf{65.50} + 21.75m = 370 - \mathbf{65.50}$	Subtraction Property of Equality
$21.75m = 304.5$	Simplify.
$\frac{21.75m}{\mathbf{21.75}} = \frac{304.5}{\mathbf{21.75}}$	Division Property of Equality
$m = 14$	Simplify.

Logan needs to walk 14 miles to raise \$370.

Got It? **Do this problem to find out.**

3. Jasmine bought 6 DVDs, all at the same price. The tax on her purchase was \$5.04, and the total was \$85.74. What was the price of each DVD?

Guided Practice

Check

Translate each sentence into an equation. (Example 1)

1. The quotient of a number and 3, less 8, is 16.
2. Tiffani spent $95 for clothes. This is $15.80 more than 4 times the amount her sister spent for school supplies.
3. Morgan has 98 baseball cards in his collection, which is twelve less than the product of $\frac{2}{3}$ and the number of cards Tyler has.

Solve each problem by writing and solving an equation. (Examples 2 and 3)

4. Kendra pays $132 for shoes and clothes. The clothes cost $54 more than the shoes. How much do the shoes cost?
5. During the spring car wash, the Activities Club washed 14 fewer cars than during the summer car wash. They washed a total of 96 cars during both car washes. How many cars did they wash during the spring car wash?
6. **Financial Literacy** A gym charges a $49.95 activation fee and $17.50 per month for a membership. If you spend $364.95, for how many months do you have a gym membership?

Independent Practice

eHelp

Go online for Step-by-Step Solutions

Translate each sentence into an equation. (Example 1)

7. Eighteen more than half a number is 8.
8. The product of a number and 9, less 20, is 7.
9. There are 48 soccer teams in the Springtown Association. This is three less than three times the number of teams in the Lyon Association.
10. Eileen drove for 85 minutes. This is 21 more minutes than one-third the number of minutes Ethan drove.

Solve each problem by writing and solving an equation. (Examples 2 and 3)

11. In 2007, Candace Parker, from the University of Tennessee, made 37 more field goals than she did in 2006. She had a total of 497 field goals for those years. How many field goals did she make in 2006?
12. The Marsh family took a vacation that covered a total distance of 1356 miles. The return trip was 284 miles shorter than the first part of the trip. How long was the return trip?
13. Three friends share the cost of renting a game system. Each person also rents one game for $8.50. If each person pays $13.25, what is the cost of renting the system?
14. Suppose you purchase 3 identical T-shirts and a hat. The hat costs $19.75 and you spend $56.50 in all. How much does each T-shirt cost?

15. You return a book that was 6 days overdue. Including a previous unpaid balance of \$0.90, your new balance is \$2.40. How much is the daily fine for an overdue book?

16. **STEM** The main span of a suspension bridge is the roadway between the bridge's towers. The main span of the Walt Whitman Bridge in Philadelphia is 2000 feet long. This is 600 feet longer than two-fifths of the length of the main span of the George Washington Bridge in New York City. Write and solve an equation to find the length of the George Washington Bridge.

17. In his DVD collection, Domingo has eight more than twice as many animated movies as action movies. If he has 24 animated movies, write and solve an equation to find how many action movies are in his collection.

18. At the start of the school trip to Washington D.C., the tour bus has 40 gallons of gasoline in the fuel tank. Each hour, the bus uses 7 gallons of gasoline. The bus will stop for gas when there are 10 gallons left.

    a. Make a table to show how many gallons of gasoline are remaining in the tank after each hour.

    b. Write and solve an equation to find how many hours will pass before the bus will have to stop for gasoline.

19. CCSS **Multiple Representations** In this problem, you will use tables, graphs, and equations to solve a problem. Misty is saving money to buy an MP3 player that costs \$212. She has already saved \$47 and plans to save an additional \$15 per week.

Number of Weeks	Amount of Savings (\$)
1	62
2	■
3	■
4	■

    a. **Symbols** Write a variable expression to represent the amount of money saved after w weeks. Then use the expression to complete the table at the right.

    b. **Graph** Make a line graph of the data in the table. How can you use the graph to find the number of weeks it will take her to save enough money for the MP3 player?

    c. **Symbols** Write and solve an equation to find the number of weeks it will take her to save the money.

    d. **Words** Compare the methods for finding the solution that you used in parts **b** and **c**.

H.O.T. Problems Higher Order Thinking

20. CCSS **Identify Structure** Write a two-step equation with a solution of 6. Write the equation using both words and symbols.

21. CCSS **Reason Inductively** An example of two consecutive even numbers is 4 and 6. They can be represented by n and $n + 2$. Find 3 consecutive even numbers whose sum is 30.

22. CCSS **Identify Structure** The equations $\frac{x+4}{5} = 20$ and $\frac{x}{5} + 4 = 20$ are both two-step equations. Compare and contrast how to solve them.

23. CCSS **Persevere with Problems** Emelia discovered that if she takes three-fourths of her age and adds 9, it produces the same result as when she takes one-fourth of her age and adds 21. How old is Emelia?

24. **Building on the Essential Question** Explain how two-step equations are used to represent real-world problems.

Standardized Test Practice

25. An electrician charges \$35 for a house call and \$80 per hour for each hour worked. If the total charge was \$915, which equation would you use to find the number of hours n that the electrician worked?

A $35n + 2n(80) = 915$

B $80 + 35n = 915$

C $35 + (80 - n) = 915$

D $35 + 80n = 915$

26. Belinda scored 16 goals this season. This is 4 more than three times the number she scored last season. Which equation could you use to find how many goals she scored last season?

F $4n + 3 = 16$ **H** $4n - 3 = 16$

G $3n + 4 = 16$ **J** $3n - 4 = 16$

27. A hot air balloon is at an altitude of 113.2 meters. The balloon's altitude decreases by 10.8 meters every minute. Which equation can you use to find the number of minutes m until the balloon's altitude is 70 meters?

A $113.2m - 10.8 = 70$

B $113.2m + 10.8 = 70$

C $113.2 - 10.8m = 70$

D $113.2 + 10.8m = 70$

28. You and your friend spent a total of \$15 for lunch. Your friend's lunch cost \$3 more than yours did. How much did you spend for lunch?

F \$6 **H** \$8

G \$7 **J** \$9

CCSS Common Core Review

Solve each equation. Check your solution. 8.EE.7

29. $x + 12 = -10$

30. $-\frac{y}{6} = -2$

31. $2p + 13 = -7$

32. $7y + 3 = -11$

33. $-8t - 9 = -41$

34. $8z = 14$

35. $\frac{1}{5}g - 5 = -3$

36. $7 = \frac{1}{3}m + 20$

37. $18.3 = 2.5c - 1.3$

38. $2.9y + 6 = -2.7$

39. $\frac{1}{8}t + 3 = 0$

40. $\frac{2}{5} + \frac{1}{2}d = \frac{3}{5}$

41. A concert ticket costs t dollars, a hamburger costs h dollars, and soda costs s dollars. Write an expression that represents the total cost of a ticket, hamburger, and soda for n people. 6.EE.2a

42. **STEM** The air pressure decreases as the distance from Earth increases. The table shows the air pressure for certain distances. 7.RP.2a

a. Write a set of ordered pairs for the data.

b. Graph the data.

c. Is the relationship a proportional relationship? Why or why not?

Air Pressure

Height (mi)	Pressure (lb/in^2)
0 (sea level)	14.7
1	10.2
2	6.4
3	4.3
4	2.7
5	1.6

43. In 2011, the state of Illinois produced about 2×10^9 bushels of corn. This was about 4 times the amount of corn produced in Wisconsin. About how many bushels of corn were produced in Wisconsin? Write in scientific notation. 8.EE.4

Use the Distributive Property to write each expression as an equivalent algebraic expression. 6.EE.3

44. $4(x + 3)$

45. $8(y - 2)$

46. $-6(z - 7)$

47. $-2(-9 - p)$

Inquiry Lab

More Two-Step Equations

Inquiry **HOW does the arithmetic way to solve a two-step equation compare to the algebraic way?**

CCSS Content Standards
7.EE.4, 7.EE.4a, 8.EE.7

Mathematical Practices
1, 3

Fitness Marisol is planning to exercise at the gym four times this week. Each of her four workouts will be identical. During each workout, she plans to jog on a treadmill for 25 minutes and then use a rowing machine. Her goal is to exercise for a total of 220 minutes this week. How long should she use the rowing machine during each workout?

Investigation 1

You can use a bar diagram to represent the situation.

Step 1 Draw a bar diagram that represents Marisol's 4 workouts. Each day should show 25 minutes of jogging. The bar diagram should also show that the total workout time for the week is 220 minutes.

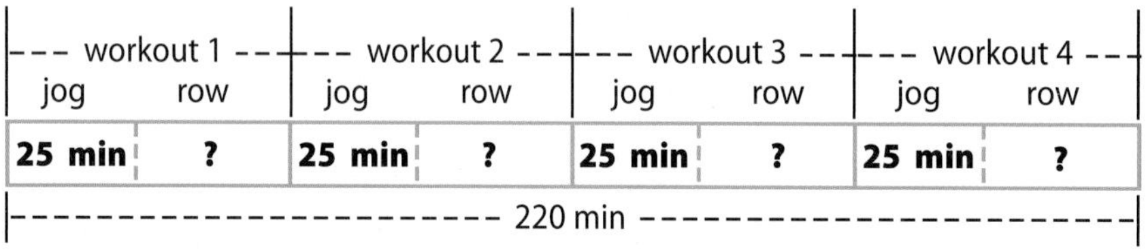

Step 2 Let x represent the number of minutes Marisol uses the rowing machine during each workout. Then the bar diagram represents the equation $4(25 + x) = 220$.

Step 3 Use the *work backward* strategy to find the number of minutes Marisol should use the rowing machine during each workout.

Since $4(25 + x) = 220$, $25 + x$ must equal $220 \div 4$ or 55. This is shown by dividing the 220 minutes equally among the four workouts in the bar diagram.

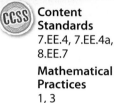

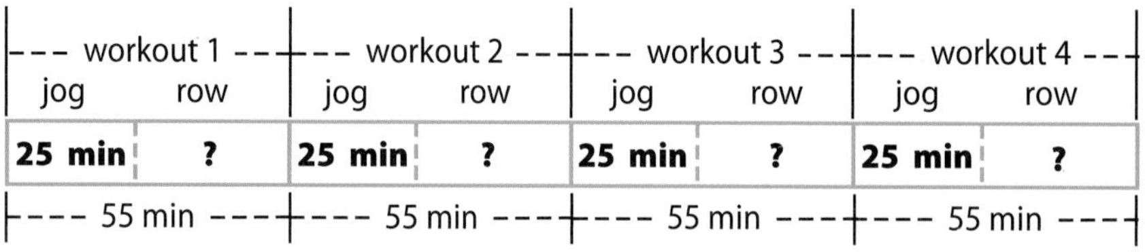

Step 4 Since $25 + x = 55$, x must equal $55 - 25 = 30$.

So, Marisol should use the rowing machine for 30 minutes during each workout

Collaborate

Work with a partner. Use a bar diagram to solve each equation.

1. $6(x + 7) = 54$ **2.** $3(12 + y) = 51$ **3.** $30 = 5(g + 6)$

The method you used to solve an equation in Investigation 1 is an arithmetic method. You can also use algebra to solve this type of equation.

Recall that when you use algebra tiles, you can remove the same number of equal groups of tiles from each side of the mat. This is often helpful when solving equations that involve parentheses.

Investigation 2

Model and solve $3(x - 2) = 15$ using algebra tiles.

Step 1 Model the equation. To model $3(x - 2)$, on the left side of the mat, make three groups that each have one x tile and two -1 tiles.

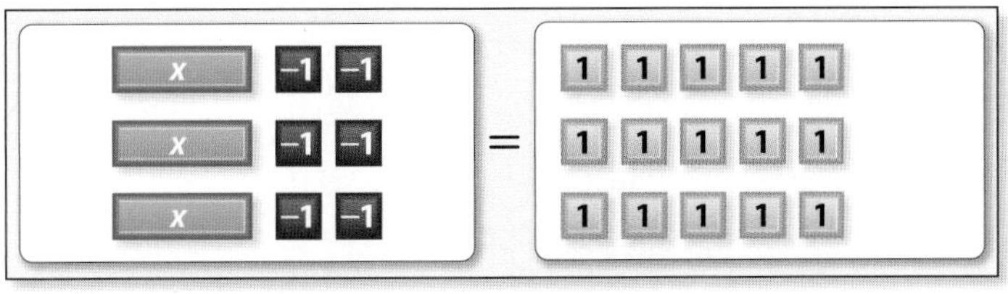

Step 2 Divide the tiles into three equal groups.

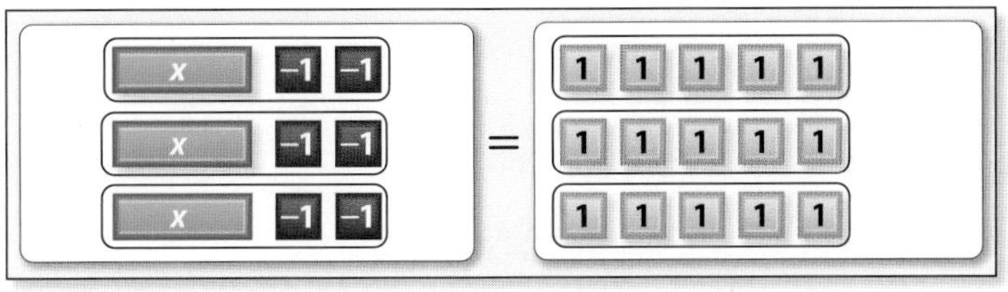

Step 3 To simplify, remove two groups from each side. You are left with $x - 2 = 5$.

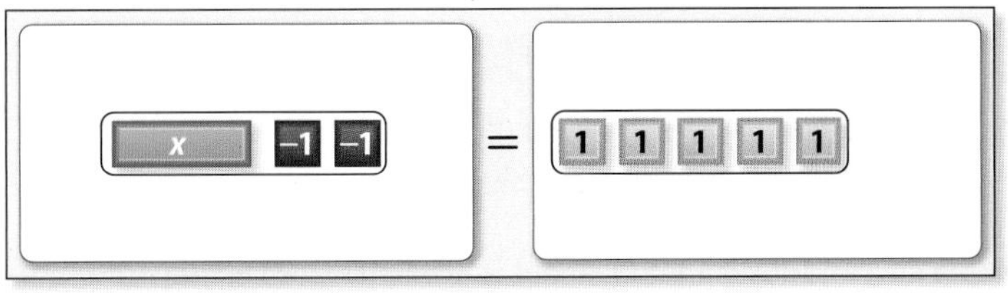

Step 4 In order to remove two -1 tiles from each side, first add two zero pairs to the right side of the mat.

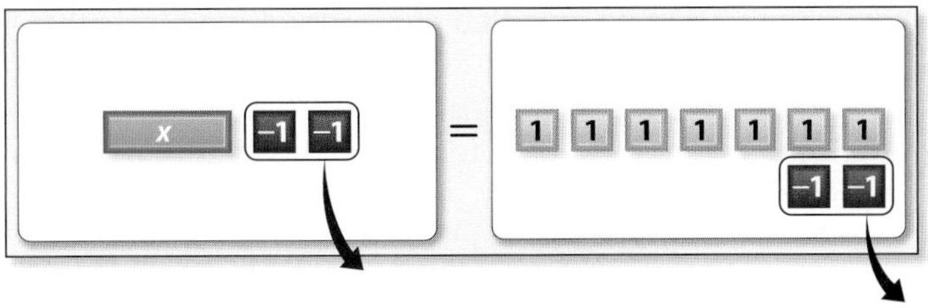

So, $x = 7$.

Collaborate

Work with a partner. Use algebra tiles to solve each equation.

4. $2(x + 4) = 12$
5. $4(z + 4) = 20$
6. $24 = 3(5 + m)$
7. $-6 = 2(n - 5)$
8. $3(x - 2) = -15$
9. $4(x + 2) = -4$

Analyze

10. CCSS **Reason Inductively** When is it useful to solve an equation of the form $a(x + b) = c$ using a bar diagram instead of using algebra tiles?

Solve each equation using either method.

11. $5(w + 1) = 25$

12. $4(n - 3) = 8$

13. $64 = 4(11 + x)$

14. $-6 = 3(m - 4)$

15. $5(c - 2) = -20$

16. $2(b + 3) = -10$

17. $3(x - 7) = -27$

18. $18 = 3(6 + x)$

19. $-12 = 3(d - 4)$

20. $2(n + 9) = 20$

21. $5(t - 5) = 25$

22. $-16 = 4(s + 2)$

For Problems 23–25, tell what equation is modeled by the algebra tiles and solve the equation.

23.

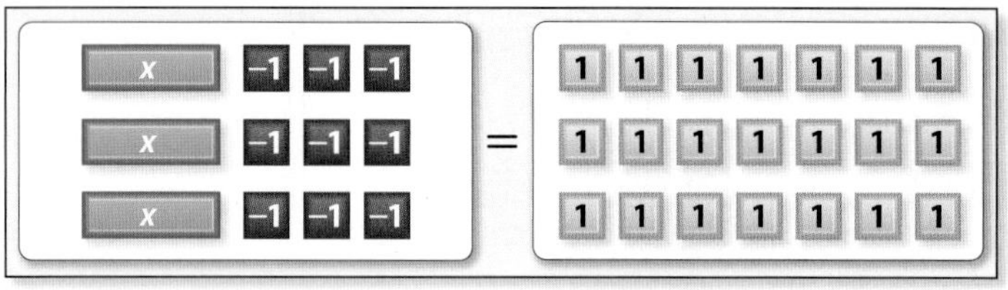

24.

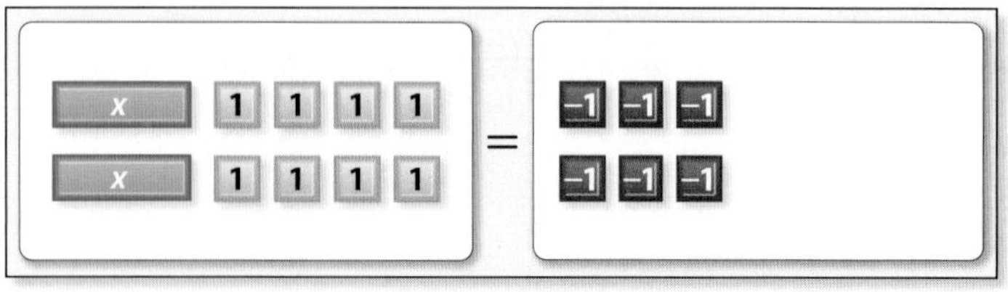

25.

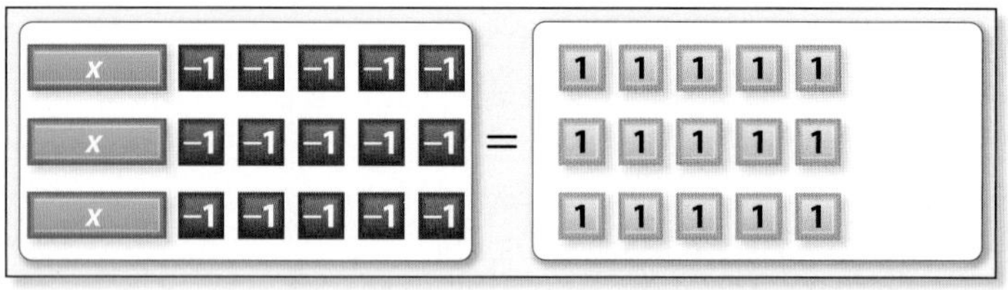

26. Mikiko bought three pairs of jeans. Each pair of jeans had the same price. She used a coupon for \$5 off the price of each pair of jeans, and the final cost, before tax, came to \$57. Solve the equation $3(x - 5) = 57$ to find the original price x of each pair of jeans.

27. Write a real-world problem that can be modeled by the equation $5(x + 2) = 45$. Then solve the problem.

28. CCSS **Justify Conclusions** In Kendall's math textbook, the equation shown below is partially covered by a drop of ink.

$$4(x - \bullet) = 24$$

According to the textbook, the solution of the equation is $x = 11$. What number is covered by the drop of ink? Justify your answer.

Reflect

29. Inquiry HOW does the arithmetic way to solve a two-step equation compare to the algebraic way?

Lesson 8-4

More Two-Step Equations

ISG Interactive Study Guide

See pages 175–176 for:
- Getting Started
- Real-World Link
- Notes

Essential Question

How are equations and inequalities used to describe and solve multi-step problems?

CCSS Common Core State Standards

Content Standards
7.EE.4, 7.EE4a, 8.EE.7, 8.EE.7b

Mathematical Practices
1, 3, 4, 5

What You'll Learn

- Solve equations of the form $p(x + q) = r$.
- Solve verbal problems by writing and solving equations of the form $p(x + q) = r$.

Real-World Link

Bowling Did you know that bowling has increased in popularity in the last decade? Forty-eight states now recognize bowling as a high school or club sport, compared to twenty states just 10 years ago. Bowling alleys typically charge for the number of games played, as well as for shoe rental.

Solve Two-Step Equations

A bowling alley charges \$3 to rent shoes. Four friends each pay for one game and for renting shoes. The total cost is \$28. You can solve the two-step equation $4(x + 3) = 28$ to find the cost of a game.

Example 1

Tutor

Solve each equation.

a. $4(x + 3) = 28$

$4(x + 3) = 28$ Write the equation.

$\frac{4(x + 3)}{\mathbf{4}} = \frac{28}{\mathbf{4}}$ Division Property of Equality

$x + 3 = 7$ Simplify.

$x + 3 \mathbf{- 3} = 7 \mathbf{- 3}$ Subtraction Property of Equality.

$x = 4$ Simplify.

b. $\frac{3}{5}(m - 6) = -9$

$\frac{3}{5}(m - 6) = -9$ Write the equation.

$\mathbf{\frac{5}{3}} \cdot \frac{3}{5}(m - 6) = \mathbf{\frac{5}{3}} \cdot (-9)$ Multiplication Property of Equality

$m - 6 = -15$ Simplify.

$m - 6 \mathbf{+ 6} = -15 \mathbf{+ 6}$ Addition Property of Equality.

$m = -9$ Simplify.

Got It? Do these problems to find out.

1a. $22 = -2(y - 3)$

1b. $3.2(x + 3.7) = 4.8$

Image Source/Getty Images

Example 2

DeAndre is ordering tickets to a concert. He buys 3 tickets that all have the same price. There is a service charge of \$4.75 per ticket. The total cost of his order is \$111.75. What is the price of each ticket?

First, write an equation to model the situation.

Words	3 times the total cost of each ticket equals \$111.75.
Variable	Let t = the price of a ticket. So, $t + 4.75$ is the total cost of a ticket including the service charge.
Equation	$3(t + 4.75) = \$111.75$

$3(t + 4.75) = \$111.75$	Write the equation.
$\frac{3(t + 4.75)}{\mathbf{3}} = \frac{\$111.75}{\mathbf{3}}$	Division Property of Equality
$t + 4.75 = 37.25$	Simplify.
$t + 4.75 - \mathbf{4.75} = 37.25 - \mathbf{4.75}$	Subtraction Property of Equality
$t = 32.50$	Simplify.

Each ticket costs \$32.50.

Got It? **Do this problem to find out.**

2. Natasha buys 5 bottles of orange juice. She has coupons for \$0.65 off the regular price of each bottle of the juice. After using the coupons, the total cost of the orange juice is \$6.20. What is the regular price of a bottle of orange juice?

Use the Distributive Property

You can also solve a two-step equation of the form $p(x + q) = r$ by using the Distributive Property. This is useful when the number outside the parentheses is not a factor of the constant term on the other side of the equation.

Example 3

Solve $6(y + 4) = -15$.

$6(y + 4) = -15$	Write the equation.
$6y + 24 = -15$	Distributive Property
$6y + 24 - \mathbf{24} = -15 - \mathbf{24}$	Subtraction Property of Equality
$6y = -39$	Simplify.
$\frac{6y}{\mathbf{6}} = \frac{-39}{\mathbf{6}}$	Division Property of Equality
$y = -6\frac{1}{2}$	Simplify.

Watch Out!
When you use the Distributive Property, make sure you multiply each term inside the parentheses by the number outside the parentheses.

Got It? **Do these problems to find out.**

3a. $11 = 5(g - 3)$

3b. $9(-3 + x) = -2$

Guided Practice

Check

Solve each equation. (Examples 1 and 3)

1. $2(n-4)=16$

2. $6(y+9)=24$

3. $35=7(m+7)$

4. $32=8(x-1)$

5. $\frac{1}{3}(c-4)=12$

6. $\frac{2}{5}(b+4)=6$

7. $15=\frac{3}{8}(z-5)$

8. $18=\frac{9}{10}(x-1)$

9. $0.2(v-5)=-1$

10. $4.5(x+3)=9.9$

11. $3(p-1.5)=13.5$

12. $4(p+9)=24.8$

Solve each problem by writing and solving an equation. (Example 2)

13. Mr. Vargas takes his class of 24 students ice skating. Each student pays an entrance fee to enter the rink and a \$4 fee to rent skates. The total cost for the students to enter the rink and rent skates is \$216. What is the ice-skating rink's entrance fee?

14. Vanessa makes 7 identical flower arrangements for the tables at a banquet. Each arrangement contains some roses and 9 tulips. Vanessa uses a total of 147 flowers to make the arrangements. How many roses are in each arrangement?

Independent Practice

Go online for Step-by-Step Solutions

eHelp

Solve each equation. (Examples 1 and 3)

15 $15(x-6)=-45$

16. $7(n-14)=-42$

17. $-2(b+5)=12$

18. $18=-3(m-6)$

19. $\frac{2}{7}(4+c)=6$

20. $\frac{4}{5}(2+m)=24$

21. $-0.5(x-2.4)=6$

22. $34=-0.4(p+22)$

23. $\frac{2}{3}(n-9)=\frac{8}{9}$

24. $-\frac{1}{3}(p+1)=\frac{1}{6}$

25. $1.5(d+3.5)=-2.4$

26. $2.6=-5.2(3.4+m)$

Solve each problem by writing and solving an equation. (Example 2)

27 Brody drives the same distance to and from work each day. He also drives an additional 1.5 miles each day to go to the gym. During a 5-day workweek, Brody drives a total of 71.25 miles. What is the distance to and from work?

28. Denisha and 4 of her friends all have discount passes to a movie theater. The passes are good for \$2.95 off the regular price of admission. The 5 friends go to the theater together and spend a total of \$47.75 on admission. What is the regular price of admission?

29. **Financial Literacy** Aiden has saved \$85 per month for the past 7 months. He plans to save the same amount each month in the future until he has saved a total of \$1955 for a new computer. For how many additional months will Aiden need to save?

30. An airplane has two sections, business class and economy class. In each section, there are 6 seats in a row. There are 24 rows in economy class, and there are 174 seats on the plane. How many rows of seats are there in business class?

31. The hexagon shown here is a regular hexagon, so each side has the same length. The perimeter of the hexagon is 20.4 centimeters. What is the value of x?

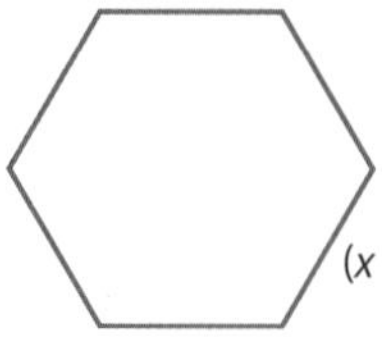

32. **STEM** Each minute, Carly's computer uploads 1.8 megabytes of data and downloads d megabytes of data. Over a period of 15 minutes, the computer uploads and downloads a total of 136.5 megabytes of data. How many megabytes of data does Carly's computer download each minute?

33. **CCSS Use Math Tools** A student solved the equation $7.9(x + 3.2) = 32.39$ and found the solution $x = 21.19$. Explain how you can use estimation to show that the student's solution is incorrect.

Solve each equation.

34. $4(x - 5) + 3 = 31$

35. $19 = -2(4 + z) - 3$

36. $3.5(c - 8) + c = 12.5$

37. $4.1(m + 4) - 2.1m = 6$

38. $5(2x - 1) = 25$

39. $23 = 9(4x - 1)$

40. $18 = \frac{3}{4}(2w - 1)$

41. $\frac{4}{5}(10y - 15) = 20$

42. $\frac{2}{3}(3p - 5) = \frac{4}{9}$

43. **Use the table to write and solve an equation for each problem.**

a. A group of 6 friends each bought a sandwich and a cup of soup. The total cost was \$50.70. Find the price of a cup of soup.

b. Kevin bought a ticket to the museum and lunch for himself and 3 friends. He bought a sandwich, a side salad, and a drinks for lunch for each person. He spent a total of \$76. Find how much money Kevin spent on each ticket.

Museum Cafeteria

Item	Price
sandwich	\$4.95
side salad	\$3.95
drink	\$1.25

H.O.T. Problems Higher Order Thinking

44. **CCSS Model with Mathematics** Write and solve a real-world problem that can be represented by the equation $4(x - 2.5) = 24$.

45. **CCSS Find the Error** Ella is solving the equation $-3(p - 4) = 14$. Find her mistake and correct it.

$$-3(p - 4) = 14$$
$$-3p - 4 = 14$$
$$-3p - 4 + 4 = 14 + 4$$
$$-3p = 18$$
$$p = -16$$

46. **CCSS Justify Conclusions** Suppose for some value of k the solution of the equation $3.1(x - k) = 0$ is $x = 7$. What must be true about k? Justify your conclusion.

47. **CCSS Persevere with Problems** Joaquin is 4 years older than Becky. Three times the sum of their ages is 114. Write and solve an equation to find Joaquin's age.

48. **Building on the Essential Question** Explain how to decide whether to use the Distributive Property or the Division Property of Equality as your first step when you solve an equation of the form $p(x + q) = r$.

Standardized Test Practice

49. Gavin bought 6 boxes of cereal during a special sale. During the sale, the regular price of each box of cereal was reduced by \$1.25. Gavin paid a total of \$14.64 for the cereal. Which equation can you use to find the regular price p of the cereal?

A $6(p + 1.25) = 14.64$

B $6(p - 1.25) = 14.64$

C $6p + 1.25 = 14.64$

D $6p - 1.25 = 14.64$

50. **Short Response** The formula $C = \frac{5}{9}(F - 32)$ describes the relationship between degrees Fahrenheit, F, and degrees Celsius, C. Suppose the current temperature is 35°C. What is the temperature in degrees Fahrenheit?

51. For which of the following is -2.5 a solution?

F $32 = 4(x + 5.5)$

G $32 = 4(x - 5.5)$

H $32 = 8(x + 6.5)$

J $32 = 8(x - 6.5)$

52. Jamila is solving the equation $\frac{3}{8}(x - 4) = 24$. If she solves the equation correctly, which of the following could be one step of her solution?

A $x - 4 = 9$

B $x - 4 = 64$

C $\frac{3}{8}x - 4 = 24$

D $\frac{3}{8}x - \frac{3}{2} = 9$

Common Core Review

Find the percent of change. Round to the nearest tenth, if necessary. Then state whether the percent of change is an increase or decrease. 7.RP.3

53. From 8 gallons to 12 gallons

54. From 82 meters to 75 meters

55. From \$216 to \$200

56. From 60 minutes to 62 minutes

57. From 95 kilograms to 150 kilograms

58. From \$1300 to \$1150

59. A ticket to the movie costs \$8.50. A small bucket of popcorn costs p dollars, and a large drink costs twice as much as the popcorn. Write an expression that represents the total cost of a movie ticket, a small popcorn, and a large drink. 6.EE.2a

60. The table shows the time it took three friends to walk different distances. Which of the three friends walked at the fastest rate? 7.RP.1

Name	Distance (mi)	Time (h)
Jonathan	$\frac{1}{4}$	$\frac{1}{8}$
Sumitra	$\frac{2}{3}$	$\frac{1}{6}$
Leah	$\frac{1}{2}$	$\frac{1}{6}$

61. A contestant's score on a game show is -7. The contestant then answers two questions correctly, each worth 4 points, and three questions incorrectly, with each error causing a loss of 5 points. What is the contestant's final score? 7.NS.3

Find each product or quotient. 7.NS.2

62. $-4(-9)$

63. $-3(-2)(-6)$

64. $-42 \div (-21)$

65. $-140 \div 5$

66. $6 \cdot (-7)$

67. $5(-3)(-4)$

68. $\frac{-400}{-25}$

69. $\frac{320}{-80}$

70. $\frac{-169}{13}$

ISG Interactive Study Guide

See page 177 for:
• Mid-Chapter Check

21ST CENTURY CAREER in Veterinary Medicine

Veterinary Technician

If you love being around animals, enjoy working with your hands, and are good at analyzing problems, a challenging career in veterinary medicine might be a perfect fit for you. Veterinary technicians help veterinarians by helping to diagnose and treat medical conditions. They may work in private clinics, animal hospitals, zoos, aquariums, or wildlife rehabilitation centers.

College & Career READINESS

Explore college and careers at ccr.mcgraw-hill.com

Is This the Career for You?

Are you interested in a career as a veterinary technician? Take some of the following courses in high school.

- Algebra
- Animal Science
- Biology
- Chemistry
- Veterinary Assisting

Find out how math relates to a career in Veterinary Medicine.

Interactive Study Guide
See page 178 for:
- Problem Solving
- Career Project

Inquiry Lab

Solving Equations with Variables on Each Side

Inquiry HOW do you solve an equation with variables on each side?

CCSS Content Standards 7.EE.4, 8.EE.7, 8.EE.7b

Mathematical Practices 1, 3

Crafts Naomi is researching Web sites that make custom stickers. Site A charges \$5 per sheet of stickers plus a \$4 setup fee. Site B charges \$3 per sheet of stickers plus a \$10 setup fee. For how many sheets of stickers is the total cost the same at both sites?

Investigation 1

Let x represent the number of sheets of stickers. Then $5x + 4$ represents the total cost of ordering stickers from Site A, and $3x + 10$ represents the total cost of ordering stickers from Site B. To find the number of sheets for which the costs are equal, Naomi needs to solve $5x + 4 = 3x + 10$.

You can use algebra tiles to solve this equation.

Step 1 Model the equation.

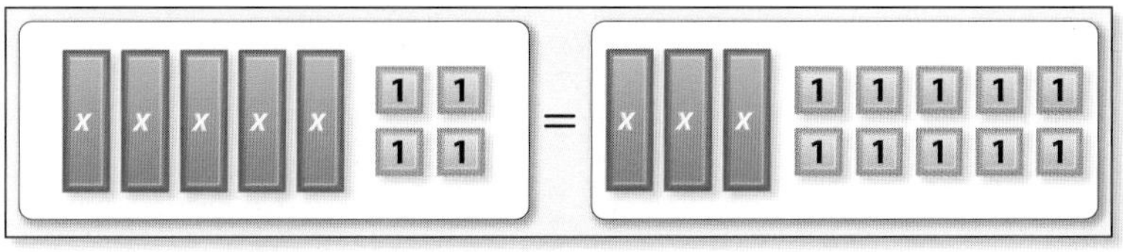

$5x + 4 \quad = \quad 3x + 10$

Step 2 Remove three x-tiles from each side of the mat so that all of the x-tiles are on one side of the mat.

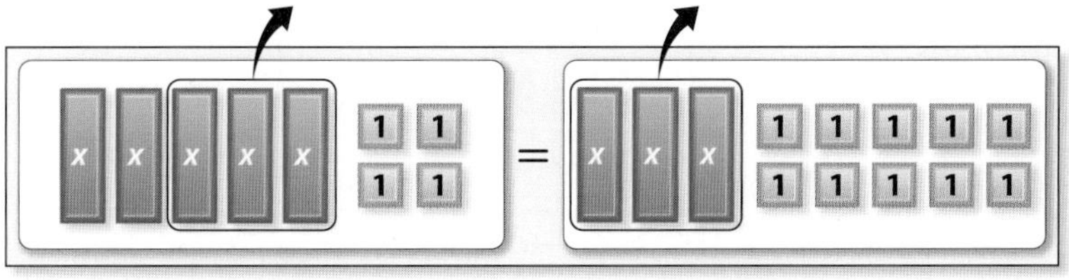

$5x - 3x + 4 \quad = \quad 3x - 3x + 10$

Step 3 Remove four 1-tiles from each side.

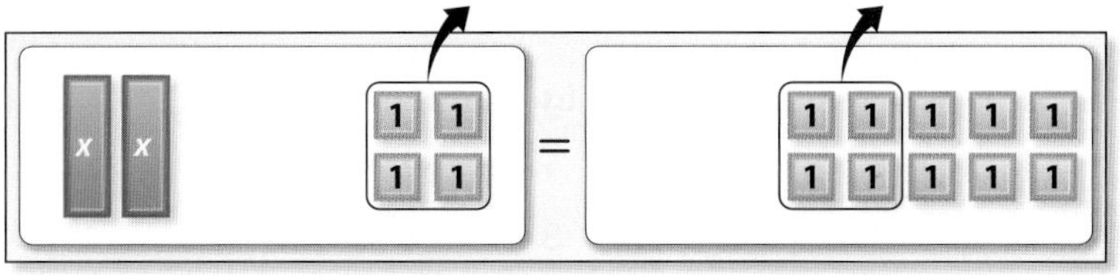

$2x + 4 - 4 \quad = \quad 10 - 4$

Step 4 Separate the remaining tiles into two equal groups. This shows that $x = 3$

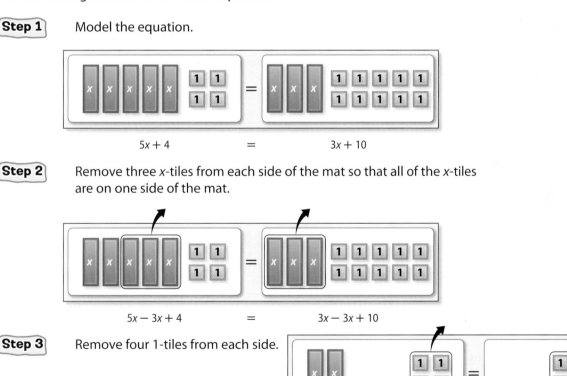

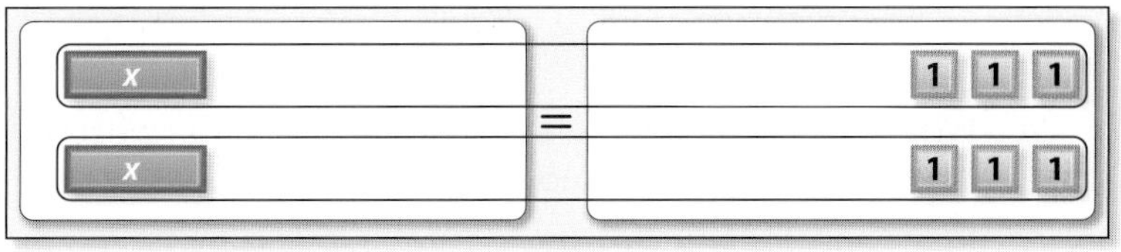

$2x \quad = \quad 6$

So, the total cost is the same at both Web sites when you order 3 sheets of stickers.

Collaborate

Work with a partner. Use algebra tiles to solve each equation.

1. $2y + 8 = 5y + 2$ **2.** $5m + 1 = m + 13$ **3.** $4p - 3 = 2p - 7$

When you solve an equation using algebra tiles, your goal is to isolate the *x*-tiles on one side of the mat and the 1-tiles on the other side of the mat. In some cases, you may need to add zero pairs to each side of the mat in order to do this.

Investigation 2

Solve $2x - 5 = 3x + 3$ using algebra tiles.

Step 1 Model the equation.

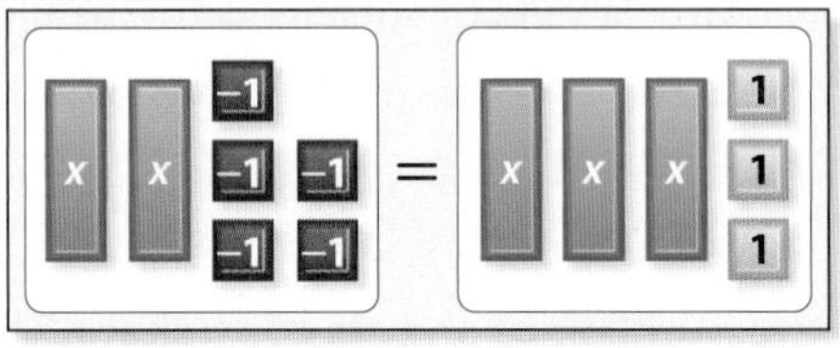

$2x - 5 = 3x + 3$

Step 2 Remove two x-tiles from each side of the mat. Now all of the *x*-tiles are on one side of the mat.

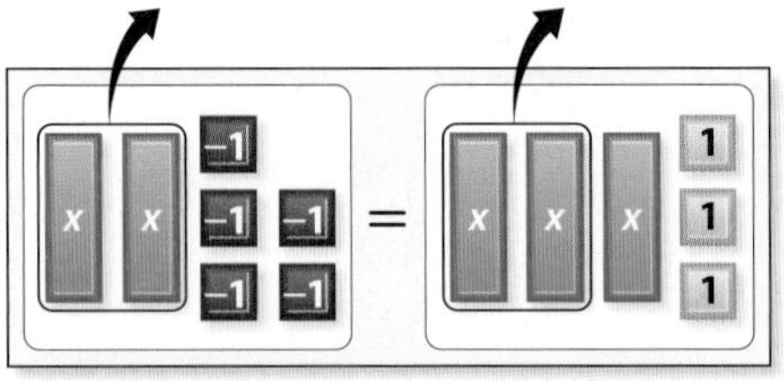

$2x - 2x - 5 = 3x - 2x + 3$

Step 3 It is not possible to remove three 1-tiles from each side of the mat. So, add three zero pairs to the left side of the mat.

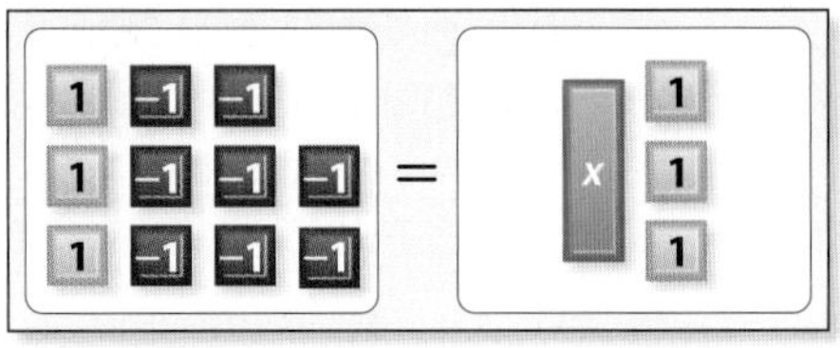

$3 + (-3) + (-5) = x + 3$

Step 4 Now you can remove three 1-tiles from each side. There are 8 negative tiles on the left side of the mat.

So, $x = -8$.

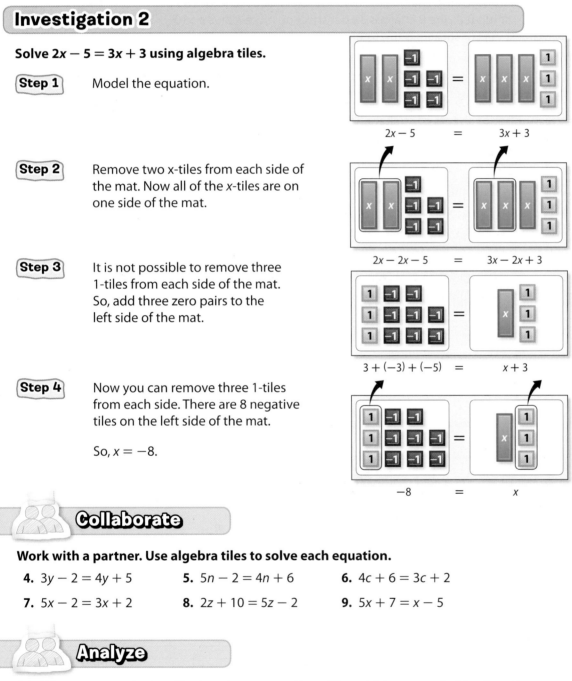

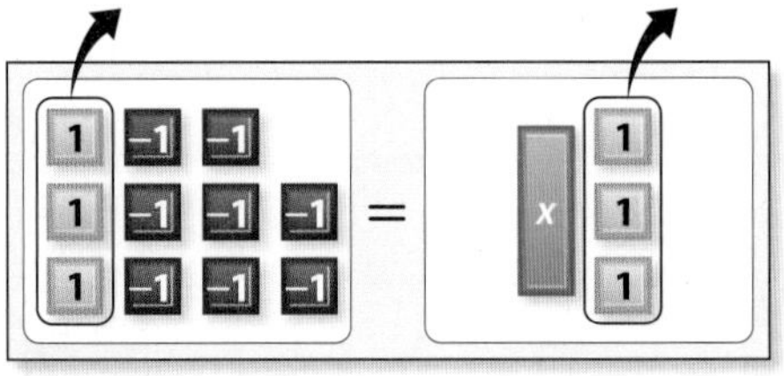

$-8 = x$

Collaborate

Work with a partner. Use algebra tiles to solve each equation.

4. $3y - 2 = 4y + 5$ **5.** $5n - 2 = 4n + 6$ **6.** $4c + 6 = 3c + 2$

7. $5x - 2 = 3x + 2$ **8.** $2z + 10 = 5z - 2$ **9.** $5x + 7 = x - 5$

Analyze

10. When you use algebra tiles to solve an equation with variables on each side, does it matter whether you remove *x*-tiles or 1-tiles first?

11. What property of equality allows you to remove the same number of *x*-tiles from each side of the mat?

For Problems 12–14, tell what equation is modeled by the algebra tiles and solve the equation.

12.

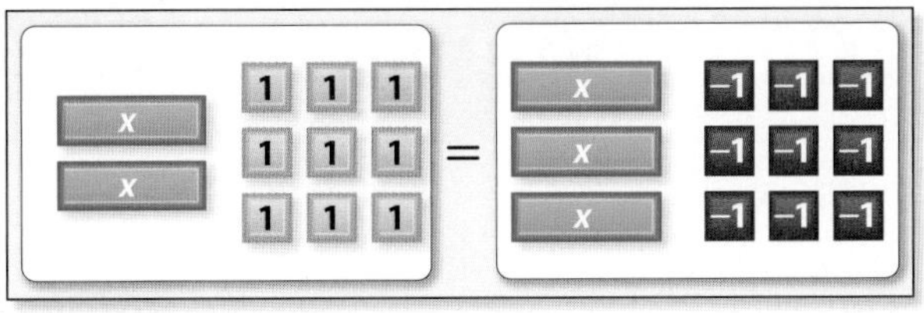

13.

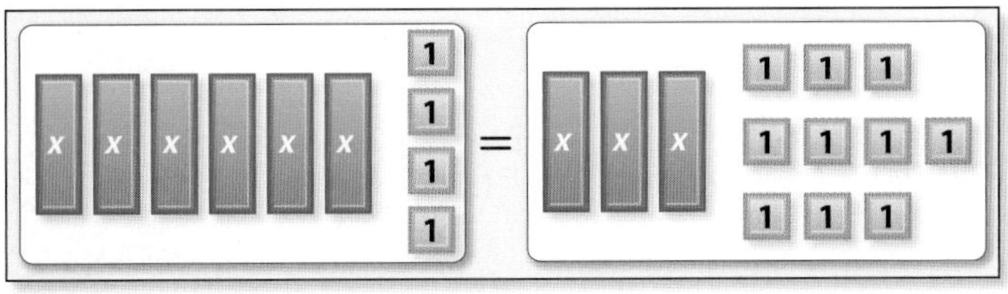

14.

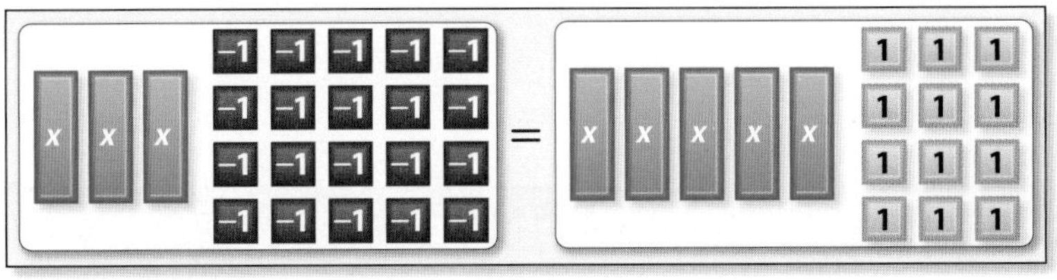

15. Drew wants to buy packages of blank DVDs. At OfficePro, the packages cost \$5 each, and Drew has a coupon for \$8 off the total purchase price. At Best Deal, the packages cost \$4 each. Write and solve an equation to find the number of packages for which the total cost is the same at both stores.

16. Bob's Bike Rentals charges an initial fee of \$6 to rent a bike, plus \$5 an hour. Boardwalk Bikes charges an \$8 initial fee, plus \$4 an hour.

a. Write and solve an equation to find the number of hours for which the total cost of bike rental is the same at both rental agencies.

b. If you want to rent a bike for 4 hours, which service will be less expensive?

17. Kaitlyn is arranging photographs on a wall. If she arranges the photographs in 4 rows of equal length, she has 20 photographs left over. If she arranges the photographs in 6 rows, each with the same length as before, she has 2 photographs left over.

a. Write and solve an equation to find the number of photographs in each row.

b. How many photographs does Kaitlyn have?

18. CCSS **Reason Inductively** Solve each of the equations below. What do you notice? Why do you think this is the case?

$4x + 9 = 6x + 9$ $\quad$ $3x - 5 = 2x - 5$ $\quad$ $5x + 4 = x + 4$ $\quad$ $2x - 6 = 5x - 6$

19. CCSS **Justify Conclusions** Rashid is comparing the cost of two courier services that deliver packages in his area. Service A costs \$4 plus a fee of \$5 per mile. Service B costs \$12 plus a fee of \$3 per mile. Rashid wants to know which service will be less expensive. Explain how the answer to this question depends upon the distance the package will travel. Justify your conclusions.

Reflect

20. Inquiry HOW do you solve an equation with variables on each side?

Lesson 8-5

Solving Equations with Variables on Each Side

Interactive Study Guide

See pages 179–180 for:
- Getting Started
- Real-World Link
- Notes

Essential Question

How are equations and inequalities used to describe and solve multi-step problems?

Common Core State Standards

Content Standards
7.EE.4, 7.EE.4a, 8.EE.7, 8.EE.7b

Mathematical Practices
1, 3, 4, 7

What You'll Learn

- Solve equations with variables on each side.

Real-World Link

Camping Camping is an activity that many families like to do together. Many campsites offer rentals for equipment, like kayaks and bicycles. The table shows the rental fees for a certain campground.

Item	Deposit ($)	Cost per Day ($)
bicycle	3.00	5.50
kayak	6.00	5.00

Equations with Variables on Each Side

The equation $5.5x + 3 = 5x + 6$ can be used to find the number of days for which renting a bike costs the same as renting a kayak. To solve equations with variables on each side, use the Addition or Subtraction Property of Equality to write an equivalent equation with the variables on one side. Then solve the equation.

Example 1

Tools Tutor

Solve $3x = x + 8$.

$3x = x + 8$ Write the equation.

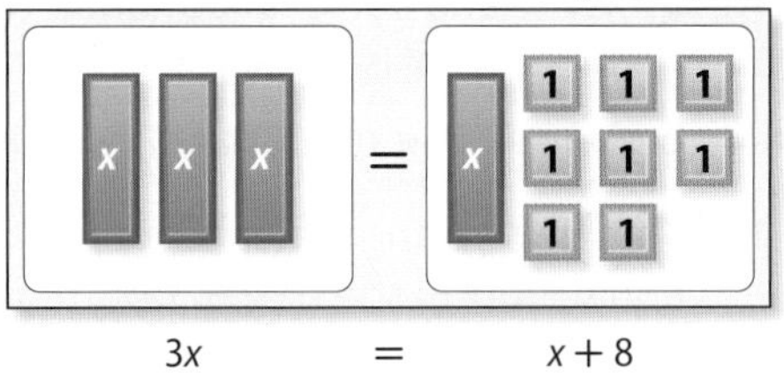

$$\begin{aligned} 3x &= x + 8 \\ \underline{-x} &= \underline{-x} \\ 2x &= 8 \end{aligned}$$

Since $3x$ is already alone on one side, use the Subtraction Property of Equality to subtract x from each side. Simplify.

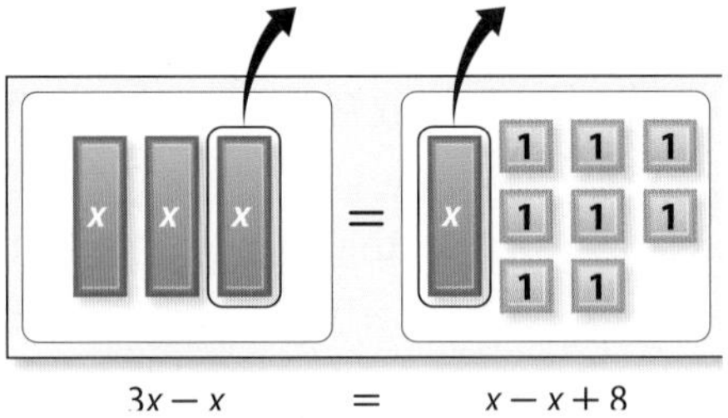

$\frac{2x}{2} = \frac{8}{2}$ Use the Division Property of Equality to divide each side by 2.

$x = 4$ Simplify.

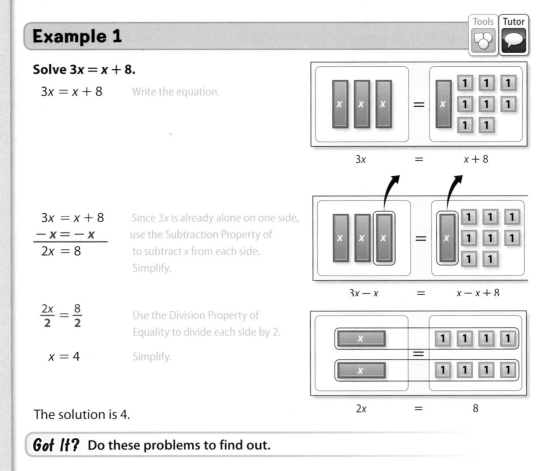

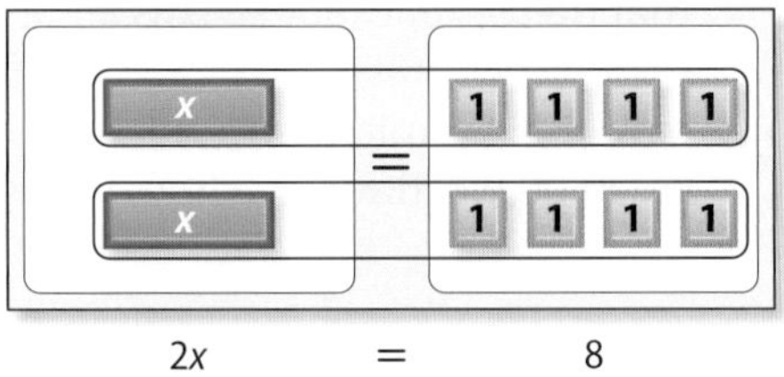

The solution is 4.

Got It? Do these problems to find out.

1a. $7x = 5x + 4$

1b. $3x - 2 = x$

Example 2

Tutor

Solve a Simpler Equation

Notice that the equations in the third step of each solution are similar to the equations that you solved in Lesson 8-2.

Solve $\frac{6}{5}y + 8 = \frac{4}{5}y - 10$.

$\frac{6}{5}y + 8 = \frac{4}{5}y - 10$	Write the equation.
$\frac{6}{5}y - \mathbf{\frac{4}{5}y} + 8 = \frac{4}{5}y - \mathbf{\frac{4}{5}y} - 10$	Subtraction Property of Equality
$\frac{2}{5}y + 8 = -10$	Simplify.
$\frac{2}{5}y + 8 - \mathbf{8} = -10 - \mathbf{8}$	Subtraction Property of Equality
$\frac{2}{5}y = -18$	Simplify.
$\mathbf{\frac{5}{2}} \cdot \frac{2}{5}y = \mathbf{\frac{5}{2}} \cdot (-18)$	Multiplication Property of Equality
$y = -45$	Simplify.

Got It? **Do these problems to find out.**

2a. $2.1x + 3 = 3.1x - 2$

2b. $\frac{1}{2}p - 15 = \frac{3}{4}p - 3$

Solve Verbal Problems

In some real-world situations you are asked to determine when the cost of two different products or services will be equal. This often results in an equation with variables on each side.

Real World Example 3

Tutor

A personal trainer charges a one-time fee of $60 plus $25 for each individual session. A fitness club charges a yearly fee of $450 plus $10 for each session with a personal trainer. Write and solve an equation to determine for what number of sessions the costs will be equal.

Let s = number of sessions.

$60 + 25s = 450 + 10s$	Write the equation.
$60 + 25s - \mathbf{10s} = 450 + 10s - \mathbf{10s}$	Subtraction Property of Equality.
$60 + 15s = 450$	Simplify.
$60 - \mathbf{60} + 15s = 450 - \mathbf{60}$	Subtraction Property of Equality.
$15s = 390$	Simplify.
$\frac{15s}{\mathbf{15}} = \frac{390}{\mathbf{15}}$	Division Property of Equality.
$s = 26$	Simplify.

You would need to have 26 training sessions in order for the costs to be equal.

Got It? **Do this problem to find out.**

3. Red Bird Cruises charges $85 per day plus a one-time fee of $75. King Cruises charges $100 per day plus a fee of $30. Write and solve an equation to determine for what number of days the charge for the cruises will be the same.

Guided Practice

Check

Solve each equation. Check your solution. (Examples 1 and 2)

1. $x + 6 = 3x$

2. $4y = 2y - 10$

3. $1.4z - 6 = 2.9z + 9$

4. $-3.8t + 4 = 4.4t - 37$

5. $\frac{4}{9}x - 1 = \frac{5}{9}x + 2$

6. $\frac{3}{8}p - 3 = \frac{1}{8}p - 5$

7. An Internet movie rental company charges a yearly membership fee of \$50 plus \$1.99 per DVD rental. Your neighborhood rental store has no membership fee and charges \$3.99 per DVD rental. Write and solve an equation to find the number of DVDs for which the cost of each will be the same. (Example 3)

Independent Practice

Go online for Step-by-Step Solutions eHelp

Solve each equation. Check your solution. (Examples 1 and 2)

8. $2x + 3 = x$

9 $8 - v = 7v$

10. $3.2 + 0.3x = 0.2x + 1.4$

11. $0.4x = 2x + 1.2$

12. $7.2 - 3c = 2c - 2$

13. $3 - 3.7b = 10.3b + 10$

14. $-\frac{1}{4}x + 6 = \frac{2}{3}x + 28$

15. $\frac{2}{5}x - 8 = 20 + \frac{3}{4}x$

16. Use the table at the right to write and solve an equation to find the number of miles a rental car must be driven for each option to cost the same for one day. (Example 3)

ABC Auto Rental		
Option	Cost per Day	Cost per Mile
A	\$25	\$0.45
B	\$40	\$0.25

17. Denzel is comparing Web sites for downloading music. One charges a \$5 membership fee plus \$0.50 per song. Another charges \$1.00 per song but has no membership fee. Write and solve an equation to find how many songs Denzel would have to buy to spend the same amount at both Web sites. (Example 3)

Solve each equation. Check your solution.

18. $5.3x + 2 - 4.1x = 3.6x - 1.6$

19. $-2.2 + 0.3z = 3 - 0.5z - 0.8z$

20. $\frac{2}{5}c + \frac{4}{5}c - 6 = c + 7$

21. $\frac{1}{8}m + \frac{2}{5} + \frac{1}{2}m = \frac{3}{8}m + \frac{3}{4}m$

22. $-\frac{1}{3}x + \frac{5}{6}x + 2 = -\frac{1}{2} + \frac{1}{6}x$

23. $\frac{4}{5}c - \frac{1}{10}c = \frac{9}{10}c$

24. Gabriella bought some school supplies for \$48, a jacket for \$56, and then bought 3 CDs. Min did not buy any school supplies but bought 11 CDs. All the CDs cost the same amount, and both students spent the same amount of money. Write and solve an equation to find the cost of one CD.

25. **Financial Literacy** One cell phone company charges \$19.95 a month and a \$2.15 tax plus \$0.21 per text message, and a second company charges \$24.95 per month plus \$0.16 per text message. For how many text messages is the cost of the plans the same?

26. Five years ago Ang was $\frac{1}{2}$ as old as Paula. Now he is $\frac{3}{5}$ as old as Paula.

Complete the table shown below. Then use the table to write and solve an equation to find their current ages.

Student	5 years ago	Now
Ang	$\frac{1}{2}a$	■
Paula	a	$a + 5$

27 Florida's coastline is 118 miles shorter than four times the coastline of Texas. It is also 983 miles longer than the coastline of Texas. Find the lengths of the coastlines of Florida and Texas.

28. Use the square shown at the right.

a. What is the value of x?

b. Find the length of each side of the square.

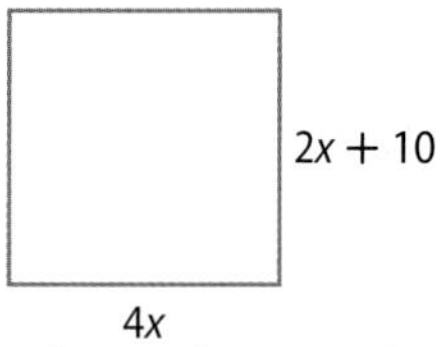

29. **CCSS Justify Conclusions** Jamie is going to fence the rectangular and triangular sections of grass shown below. The perimeters of the two sections are now equal. If w represents the width of the rectangle, how could you find the lengths of the sides of the rectangle and of the triangle? Justify your response and use your method to solve the problem.

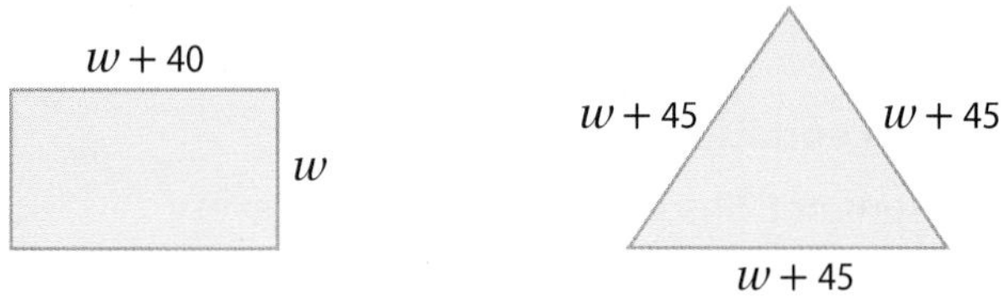

H.O.T. Problems Higher Order Thinking

30. **CCSS Identify Structure** Write an equation that has variables on each side and has a solution of -2.

31. **CCSS Persevere with Problems** The formula $F = 1.8C + 32$ can be used to find the temperature in degrees Fahrenheit F when the temperature is given in degrees Celsius C. For what value is the temperature in degrees Fahrenheit equal to the temperature in degrees Celsius? Justify your reasoning by writing and solving an equation. (*Hint:* If Fahrenheit and Celsius are equal, they can be assigned the same variable.)

32. **CCSS Find the Error** Mykia is solving the equation $10x + 6 = 8x - 4$. Find her mistake and correct it.

$$10x + 6 = 8x - 4$$
$$10x - 10x + 6 = 8x - 4 - 10x$$
$$6 = 4 - 2x$$
$$2 = -2x$$
$$-1 = x$$

33. **Building on the Essential Question** Write a real-world problem that could be solved by using the equation $54 + 3.5x = 8x$. Then solve the equation and interpret your solution.

Standardized Test Practice

34. Yesterday, the math club had 1 less than 3 times their average attendance. Last week they had 3 more than their average attendance. If the attendance for both weeks was equal, what is the average attendance?

A 1 **C** 3
B 2 **D** 4

35. For which of the following is -8 a solution?

F $-2c + 18 = 10c + 12$
G $6m - 15 = 9m + 9$
H $4 + 7s = 5s + 20$
J $5d - 13 = 19 - 3d$

36. **Short Response** What is the solution of the equation $12x + 4 = 2x - 16$?

37. A cellular company has the following options for text messaging plans.

Text Plans	Monthly Fee	Cost per Message
Plan A	\$10	\$0.15
Plan B	\$20	\$0.05

Which equation shows how many text messages would need to be sent in order for the costs for one month to be the same?

A $10 + 0.05m = 20 + 0.15m$
B $10m + 0.15 = 20m + 0.05$
C $10 + 0.15m = 20 + 0.05m$
D $10(m + 0.15) = 20(m + 0.05)$

CCSS Common Core Review

Solve each problem by writing and solving an equation. 7.EE.4a

38. Carla paid \$45 to join a golf camp for the summer. She will also pay \$15 for every private lesson that she takes. If she has budgeted \$225 for the camp, how many private lessons can she take?

39. An accountant charges \$22.50 plus \$17.50 per hour for a consultation. The Chen family paid \$83.75 for a tax consultation. How long did their consultation last?

40. STEM An atom of chlorine has 36 fewer protons than an atom of iodine. Together, an atom of chlorine and an atom of iodine have a total of 70 protons. How many protons does an atom of iodine have?

Simplify each expression. 7.EE.1

41. $2x + 5x$ **42.** $7b + 2b$ **43.** $y + 10y$

44. $5m + 4 + 7m$ **45.** $6s + 10 - 4s - 3$ **46.** $\frac{2}{3}n + \frac{2}{3} + \frac{1}{3}n$

47. $3y - 2(y + 1)$ **48.** $0.2p + 0.5(2p + 7)$ **49.** $-2(x - 2) + 4x$

50. The table shows the number of tornadoes that occurred in Nebraska in June and the total number of tornadoes for selected years. 7.EE.3

a. What decimal part, rounded to the nearest hundredth, of the annual tornadoes occurred in June for each year?

b. In which years did more than two-fifths of the tornadoes occur in June?

Annual Nebraska Tornadoes

Year	June	Total
2009	24	39
2006	8	22
2003	43	81
2000	21	61

Evaluate each expression. 6.EE.2c

51. $8c + 5$, if $c = 6$ **52.** $22 - 3h$, if $h = 4$ **53.** $36 - (-6g)$, if $g = -2$

Lesson 8-6

Inequalities

ISG Interactive Study Guide

See pages 181–182 for:
- Getting Started
- Real-World Link
- Notes

Essential Question

How are equations and inequalities used to describe and solve multi-step problems?

CCSS Common Core State Standards

Content Standards
7.EE.4

Mathematical Practices
1, 3, 4, 5

What You'll Learn

- Write inequalities.
- Graph inequalities on a number line.

Real-World Link

Water Parks Wisconsin Dells, Wisconsin, is known as the Water Park Capital of the World. The town has 20 water parks with more than 200 waterslides and 16 million gallons of water. Most of the water parks offer three different prices of tickets – adults, children, and seniors.

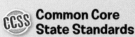

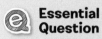

Write Inequalities

Recall that an inequality is a mathematical sentence that compares quantities that are not equal. Inequalities contain the symbols $<$, $>$, $\leq$, or $\geq$. The inequality $w > 200$ represents the number of waterslides in Wisconsin Dells.

Example 1

Tutor

Write an inequality for each sentence.

a. The DVD costs more than $15.

Words	The DVD costs more than $15.
Variable	Let d = the cost of the DVD in dollars.
Inequality	$d > 15$

b. A dog weighs less than 50 pounds.

Words	A dog weighs less than 50 pounds.
Variable	Let d = the weight of the dog in pounds.
Inequality	$d < 50$

Got It? **Do these problems to find out.**

1a. Amelia sold more than 40 magazines.

1b. Gino sent less than 35 texts yesterday.

The table below shows some common verbal phrases and the corresponding mathematical inequalities.

Inequalities
Notice that $\leq$ and $\geq$ are a combination of the $<$ or $>$ symbol with part of the symbol for equals, $=$.

Concept Summary: Inequalities

$<$	$>$	$\leq$	$\geq$
• is less than • is fewer than	• is greater than • is more than • exceeds	• is less than or equal to • is no more than • is at most	• is greater than or equal to • is no less than • is at least

Example 2

Real World

Tutor

You must be at least 18 years old to vote. Write an inequality to describe this situation.

Words	Your age is at least 18 years.
Variable	Let a = your age.
Inequality	$a \geq 18$

The inequality is $a \geq 18$.

Got It? Do this problem to find out.

2. A student must have at least 10 hours of instructor-assisted driving time to pass the course. Write an inequality to describe this situation.

Inequalities with variables are open sentences. When the variable in an open sentence is replaced with a number, the inequality may be true or false.

Example 3

Tutor

For the given value, state whether each inequality is *true* or *false*.

a. $2t + 8 > 7; t = -1$

$2t + 8 > 7$	Write the inequality.
$2(-1) + 8 \overset{?}{>} 7$	Replace t with -1.
$6 \not> 7$	Simplify.

This sentence is false.

b. $p - 42 \leq -2; p = 40$

$p - 42 \leq -2$	Write the inequality.
$40 - 42 \overset{?}{\leq} -2$	Replace p with 40.
$-2 \leq -2$	Simplify.

Although the inequality $-2 < -2$ is false, the equation $-2 = -2$ is true. So, this sentence is true.

Got It? Do these problems to find out.

3a. $3.9 + x \leq 12; x = 6$

3b. $y - \frac{1}{3} < 1; y = \frac{4}{3}$

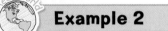

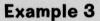

Graph Inequalities

Inequalities can be graphed on a number line. The graph helps you visualize the values that make the inequality true.

> **Graphing Inequalities**
> When inequalities are graphed, an open dot means the number is not included ($<$ or $>$) and a closed dot means it is included ($\leq$ or $\geq$).

Example 4

Tools Tutor

Graph each inequality on a number line.

a. $a > 6$

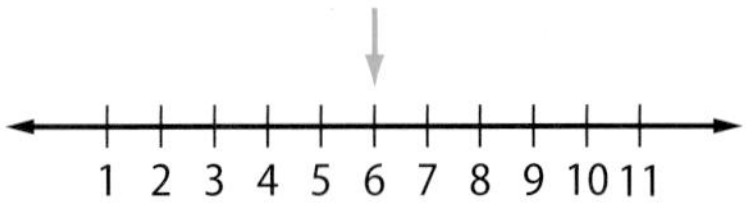

Locate 6 on the number line. It is a key point in the inequality.

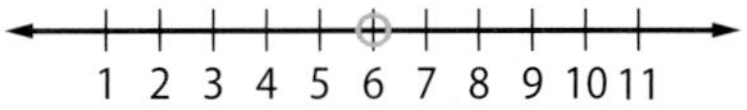

Draw an *open* circle on 6 because 6 is *not* included.

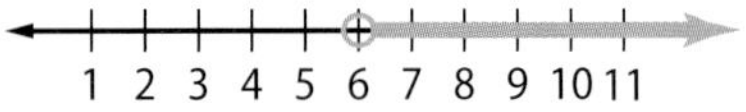

The inequality $a > 6$ means that all numbers *greater than* 6 will make the sentence true. Draw an arrow from the dot pointing to the right.

b. $x \leq -1$

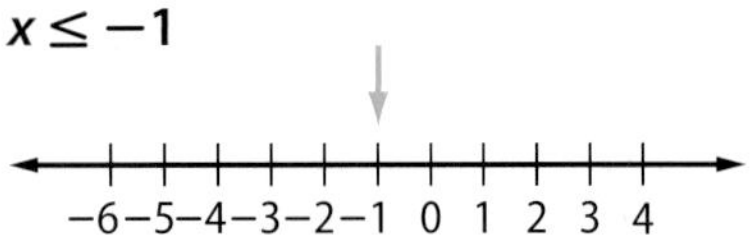

Locate −1 on the number line. It is a key point in the inequality.

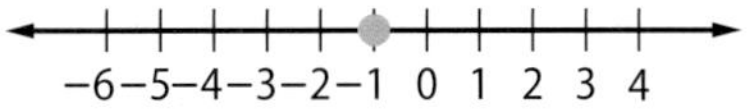

Draw a *closed* circle on −1 because −1 *is* included.

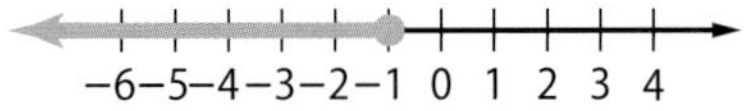

The inequality $x \leq -1$ means that all numbers *less than or equal to* −1 will make the sentence true. Draw an arrow from the circle pointing to the left.

Got It? Do these problems to find out.

4a. $x < 5$ **4b.** $x \geq -2$ **4c.** $x > 0$

Example 5

Tools Tutor

Write an inequality for the graph.

An open circle is on 2, so the point 2 is *not* included in the graph. The arrow points to the right, so the graph includes all numbers greater than 2. The inequality is $x > 2$.

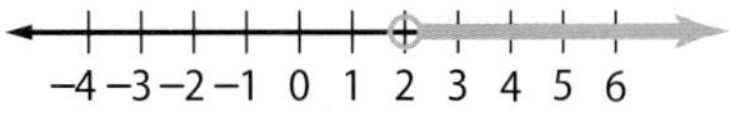

Got It? Do these problems to find out.

Write an inequality for each graph.

5a.

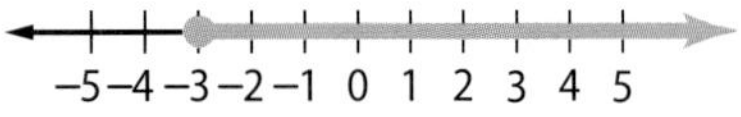

5b.

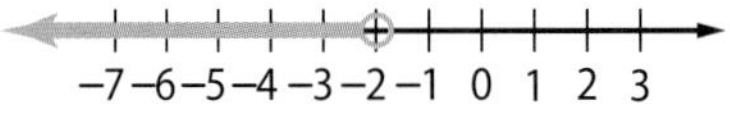

Guided Practice

Check

Write an inequality for each sentence. (Example 1)

1. Lacrosse practice will be no more than 45 minutes.
2. Mario is more than 60 inches tall.
3. More than 8000 fans attended the Wizards' opening soccer game. Write an inequality to describe the attendance. (Example 2)

For the given value, state whether the inequality is *true* or *false*. (Example 3)

4. $13 - x < 4; x = 9$

5. $45 > \frac{3}{4}x + 25; x = 20$

Graph each inequality on a number line. (Example 4)

6. $x < -1$

7. $y \geq 5$

8. $w > 9$

9. $z \leq 2$

Write an inequality for each graph. (Example 5)

10. 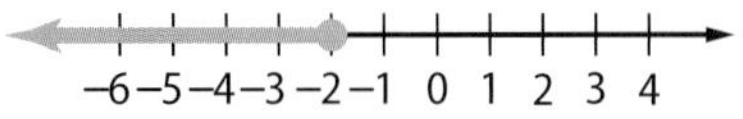

11. 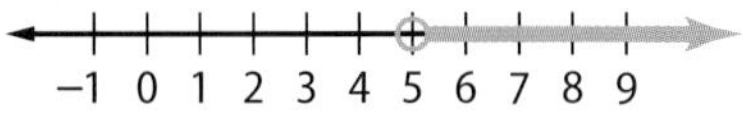

Independent Practice

Go online for Step-by-Step Solutions

eHelp

Write an inequality for each sentence. (Example 1)

12. The elevators in an office building have been approved for a maximum load of 3600 pounds.

13. Children under the age of 2 fly free.

14. An assignment requires at least 45 minutes.

15. While shopping, Abby spent no more than $50.

STEM The graph shows the average life span of various animals. (Example 2)

16. The average life span of a Galapagos tortoise is at least 4 times that of a chimpanzee. Write an inequality for the life span of a tortoise.

17. The average life span of a lobster is at most the life span of a cat. Write an inequality for the life span of a lobster.

18. The average life span of a giraffe is at least that of a horse. Write an inequality for the life span of a giraffe.

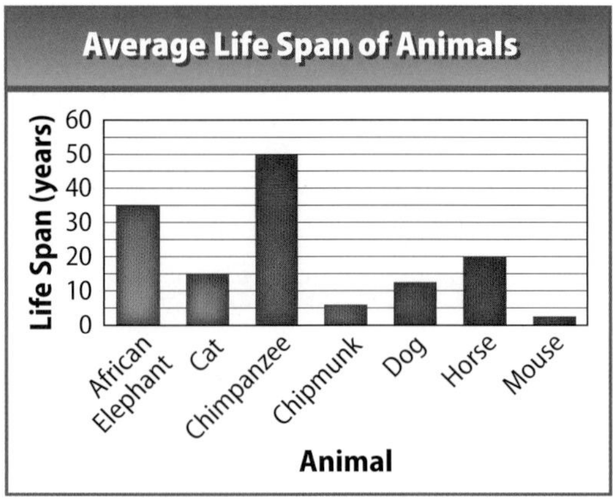

For the given value, state whether the inequality is *true* or *false*. (Example 3)

19. $13 - a < 29; a = -30$

20. $4.5b \geq -12; b = -4$

21. $\frac{2}{5}c + 18 \leq 25; c = 15$

22. $\frac{120}{d} > 40; d = 3$

23. $\frac{55}{f} > -22; f = -5$

24. $c + 19 < 2c; c = 20$

Graph each inequality on a number line. (Example 4)

25. $x < 0$

26. $y \geq 3$

27. $p > -4$

28. $t > 6$

29. $s \geq -2$

30. $r \leq -4$

Write an inequality for each graph. (Example 5)

31. 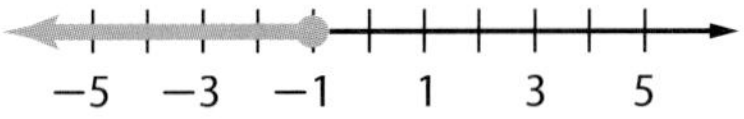

32. 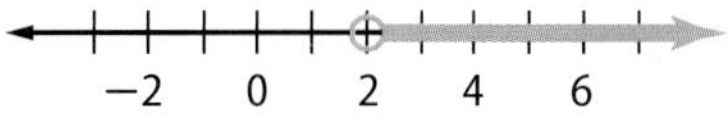

33. 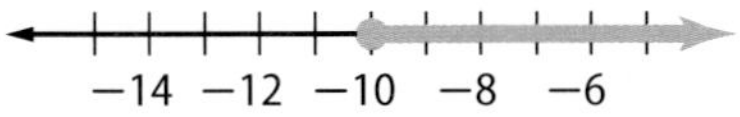

34. 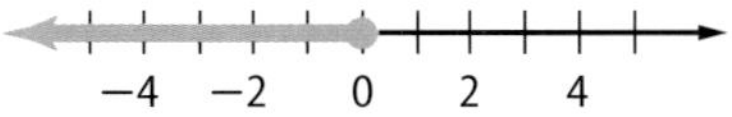

35. 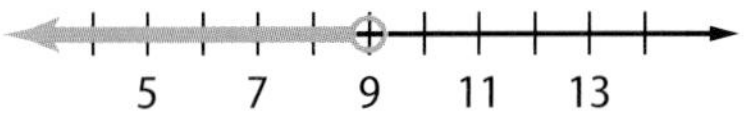

36. 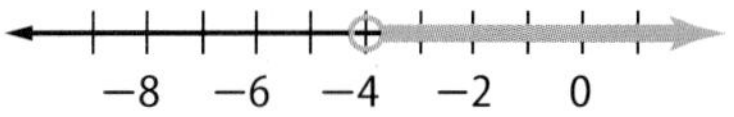

37. **Financial Literacy** Madison Middle School spends \$750 per year on club activities. It spends at least twice that amount on after-school activities. Write an inequality that represents how much it spends on after-school activities.

38. In a recent year, a professional football player threw for 4418 yards. This is at most 500 yards more than a second player's passing yards. Write an inequality that represents the second player's passing yards.

39. CCSS **Use Math Tools** Use the Internet or another source to find the state or national spending limits on certain government branches, organizations, or projects. Write an inequality to express one or more of these limits.

H.O.T. Problems Higher Order Thinking

40. CCSS **Justify Conclusions** State three numbers that could be solutions of the inequality $h \leq -12$. Then justify your response by using a number line.

41. CCSS **Model with Mathematics** Write a real-world example for the inequality below.

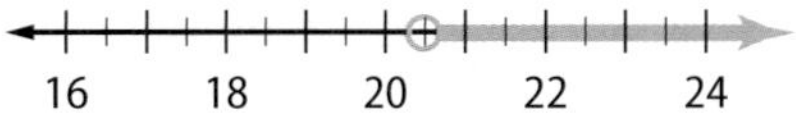

42. CCSS **Use a Counterexample** Provide a counterexample to the statement *All numbers less than 0 are negative integers.*

43. CCSS **Find the Error** Alex is graphing the inequality $p < 14$. Find his mistake and correct it.

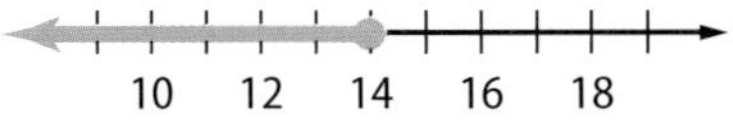

44. **Building on the Essential Question** Explain how to tell the difference between graphing an inequality with a closed circle and one with an open circle. Use examples to clarify your explanation.

Standardized Test Practice

45. Which of the following best represents the sign shown?

72"
60"
48"
36"
24"
12"
Must be over 48 inches tall to ride.

A $t > 48$

B $t < 48$

C $t \geq 48$

D $t \leq 48$

46. **Short Response** Graph the following inequality on a number line.

$$p < 3$$

47. An elevator's maximum load is 3400 pounds. Which of the following best represents that sentence?

F $\ell > 3400$ **H** $\ell \geq 3400$

G $\ell < 3400$ **J** $\ell \leq 3400$

48. The Chess Club is having a bake sale to raise money for a tournament. The club must raise at least three times what they raised at the last bake sale. What is the minimum amount they must raise if the last bake sale raised $45?

A $15 **C** $90

B $45 **D** $135

Common Core Review

Solve each equation. 8.EE.7

49. $3t - 6 = 6t + 30$

50. $2y + 14 = 42 - 5y$

51. $3x = 12 - 3x$

52. $4p = -p - 20$

53. $\frac{2}{3}m + 3 = \frac{4}{3}m - 5$

54. $-\frac{3}{4}x + \frac{1}{4} = \frac{1}{2}x - \frac{3}{4}$

55. $-8.3y - 8 = 10 + 0.7y$

56. $9 - 2.4x = -4.5 + 3x$

Solve each problem by writing and solving an equation. 7.EE.4a

57. **Financial Literacy** Jordan has $234 in his savings account. He plans to deposit $45.50 into the account each week. If his goal is to save a total of $780, for how many weeks will he need to make deposits into the account?

58. In a recent season of major league baseball, the Philadelphia Phillies won 13 more games than the Atlanta Braves. Together, the two teams won a total of 191 games. How many games did the Phillies win?

59. Four friends eat dinner at a restaurant. They decide to share the cost of the meal evenly, and each person also leaves $2.00 for the tip. If each person pays a total of $14.75, what is the total cost of the meal?

60. A cell phone plan charges $7 per month and $0.10 per minute. If the monthly cost is $25, find the number of minutes you can talk that month.

Find the area of each figure. 6.G.1

61.

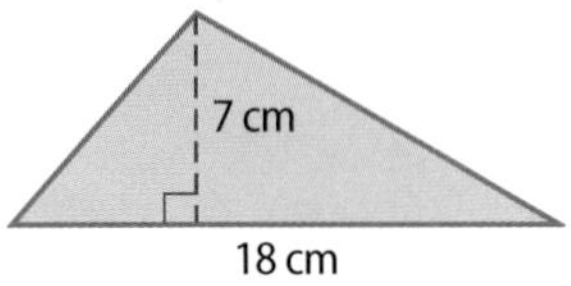

62.

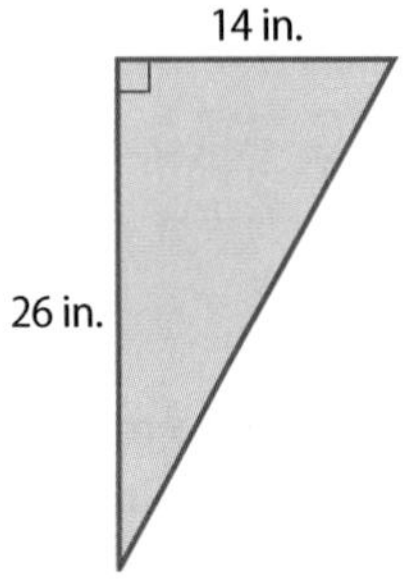

63.

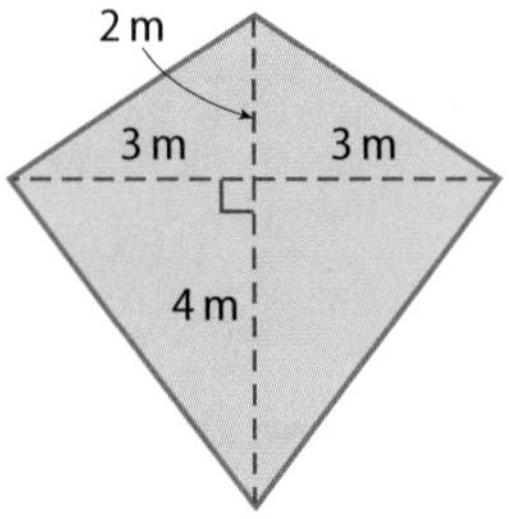

Lesson 8-7

Solving Inequalities

ISG Interactive Study Guide

See pages 183–184 for:
- Getting Started
- Real-World Link
- Notes

Essential Question

How are equations and inequalities used to describe and solve multi-step problems?

CCSS Common Core State Standards

Content Standards
7.EE.4, 7.EE.4b

Mathematical Practices
1, 3, 4

What You'll Learn

- Solve inequalities by using the Addition and Subtraction Properties of Inequality.
- Solve inequalities by multiplying or dividing by a positive or negative number.

Real-World Link

Pets Did you know that 39 percent of U.S. households own at least one dog? The amount of food that you feed your dog should be based on the dog's weight. For example, an adult dog that weighs 10 pounds or less needs about $\frac{3}{4}$ cup of food per day. A dog that weighs more than 50 pounds needs 2 to 4 cups of food each day.

Key Concept: Addition and Subtraction Properties

Words When you add or subtract the same number from each side of an inequality, the inequality remains true.

Symbols For all numbers a, b, and c,

1. If $a < b$, then $a + c < b + c$ and $a - c < b - c$.
2. If $a > b$, then $a + c > b + c$ and $a - c > b - c$.

Examples

$$5 < 9 \qquad\qquad 11 > 6$$
$$5 + \mathbf{4} < 9 + \mathbf{4} \qquad\qquad 11 - \mathbf{3} > 6 - \mathbf{3}$$
$$9 < 13 \qquad\qquad 8 > 3$$

The above properties are useful for solving inequalities. Note that these properties are also true for $a \leq b$ and $a \geq b$.

Example 1

Tutor

Solve $x + 5 > 12$. Check your solution.

	$x + 5 > 12$	Write the inequality.
	$\mathbf{-5} \quad \mathbf{-5}$	Subtraction Property of Inequality
	$x > 7$	Simplify.
Check	$x + 5 > 12$	Write the inequality.
	$\mathbf{9} + 5 \stackrel{?}{>} 12$	Replace x with any number greater than 7.
	$14 > 12$ ✓	The statement is true.

Got It? Do these problems to find out.

1a. $y + 10 < 3$

1b. $x + 7 \geq 10$

Digital Vision Ltd./SuperStock

When graphing inequalities, it is often easier to visualize the solution when the variable is on the left side of the inequality symbol.

Equations and Inequalities

In an equation, if $a = b$, then $b = a$. In an inequality, if $a < b$, then $b > a$.

Example: $7 = 2 + 5$ and $2 + 5 = 7$

$2 + 8 > 7$

but $7 < 2 + 8$

Example 2

Solve $3 \leq b - 1\frac{1}{3}$. Graph the solution on a number line.

$3 \leq b - 1\frac{1}{3}$	Write the inequality.
$3 + \mathbf{1\frac{1}{3}} \leq b - 1\frac{1}{3} + \mathbf{1\frac{1}{3}}$	Addition Property of Inequality
$4\frac{1}{3} \leq b$ or $b \geq 4\frac{1}{3}$	Simplify.

The solution is $b \geq 4\frac{1}{3}$.

Check	$3 \leq \mathbf{b} - \frac{1}{3}$	Write the inequality.
	$3 \stackrel{?}{\leq} \mathbf{4\frac{1}{3}} - 1\frac{1}{3}$	Replace b with $4\frac{1}{3}$.
	$3 \leq 3$ ✓	The statement is true.

Graph the solution.

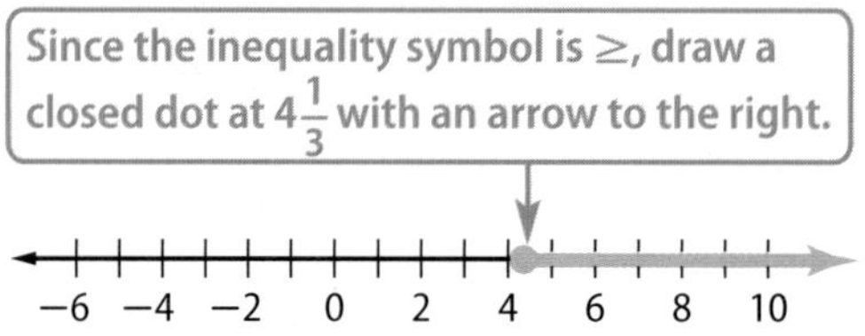

Since the inequality symbol is ≥, draw a closed dot at $4\frac{1}{3}$ with an arrow to the right.
−6 −4 −2 0 2 4 6 8 10

Got It? Do these problems to find out.

Solve each equation. Graph the solution on a number line.

2a. $3 \geq g + 7$

2b. $b + \frac{5}{7} > 2$

Key Concept Multiplication and Division Properties

Words When you multiply or divide each side of an inequality by the same *positive* number, the inequality remains true.

Symbols For all numbers a, b, and c, where $c > 0$,

1. If $a < b$, then $ac < bc$ and $\frac{a}{c} < \frac{b}{c}$.
2. If $a > b$, then $ac > bc$ and $\frac{a}{c} > \frac{b}{c}$.

Examples

$-6 < 10$	$20 > 16$
$-6 \cdot \mathbf{2} < 10 \cdot \mathbf{2}$	$\frac{20}{\mathbf{4}} > \frac{16}{\mathbf{4}}$
$-12 < 20$	$5 > 4$

These properties are also true for $a \leq b$ and $a \geq b$.

Positive Number

The statement $c > 0$ means that c is a positive number.

Some inequalities, like $4x > 8$, are solved by multiplication or division. You can multiply or divide each side of an inequality by a positive number and the inequality is still true.

Example 3

Tutor

Words to Symbols
Remember, *at most* translates to $\leq$, while *at least* translates to $\geq$.

Macy is making each of her 7 friends a bracelet. In addition to charms, she needs one package of wire for all of the bracelets. The wire is \$6. She does not want to spend more than \$45 on the bracelets. Find the maximum cost for each bracelet.

Since Macy wants to spend at most \$45, write and solve an inequality using the symbol $\leq$.

Words	Number of friends times cost of each bracelet plus \$6 must be less than or equal to \$45.
Variable	Let c = the cost of each bracelet.
Inequality	$7c + \$6 \leq \$45$

$7c + 6 \leq 45$	Write the inequality.
$7c + 6 - 6 \leq 45 - 6$	Subtraction Property of Inequality
$7c \leq 39$	Simplify.
$\frac{7c}{7} \leq \frac{39}{7}$	Division Property of Inequality
$c \leq 5\frac{4}{7}$ or $5.\overline{571428}$	Simplify.

Macy can spend no more than \$5.57 per bracelet.

Got It? Do these problems to find out.

3a. Alfonzo works for a lawn service company. It takes Alfonzo $\frac{3}{4}$ hour to mow a lawn. If he works more than 8 hours, he gets a $\frac{1}{2}$ hour lunch. Write and solve an inequality to find the number of lawns he can mow if he works at least 14 hours.

3b. Keiko prepared 28 bags of granola to sell at a school fundraiser. She also received a \$10 donation. Write and solve an inequality to find the price she should charge for each bag of granola if she wants to raise at least \$115.

Negative Number
The statement $c < 0$ means that c is a negative number.

Key Concept Multiplication and Division Properties

Words When you multiply or divide each side of an inequality by the same *negative* number, the inequality symbol must be reversed for the inequality to remain true.

Symbols For all numbers a, b, and c, where $c < 0$,

1. If $a < b$, then $ac > bc$ and $\frac{a}{c} > \frac{b}{c}$.
2. If $a > b$, then $ac < bc$ and $\frac{a}{c} < \frac{b}{c}$.

Examples

$-4 < 5$	$18 > -12$
$-4 \cdot (-3) > 5 \cdot (-3)$	$\frac{18}{-3} < \frac{-12}{-3}$
$12 > -15$	$-6 < 4$

These properties are also true for $a \leq b$ and $a \geq b$.

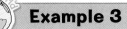

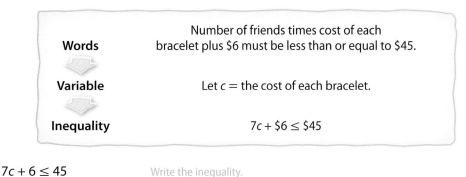

To understand the properties shown on the previous page, consider what happens when each side of an inequality is multiplied or divided by a negative number.

Graph 2 and 4 on a number line.

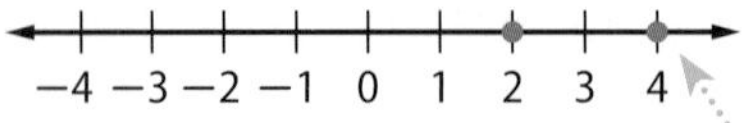

Since 4 is to the right of 2, $2 < 4$.

Now, multiply each number by −1.

Since −2 is to the right of −4, $-2 > -4$.

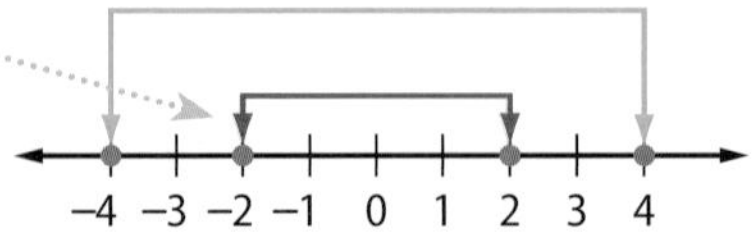

Notice that the numbers being compared switched positions as a result of being multiplied by a negative number. In other words, their order reversed. This suggests the properties shown on the previous page.

Real-World

Example 4

Tools Tutor

Solve each inequality. Then graph the solution on a number line.

a. $-5x < 45$

$-5x < 45$	Write the inequality.
$\frac{-5x}{-5} > \frac{45}{-5}$	Division Property of Inequality
$x > -9$	Simplify.

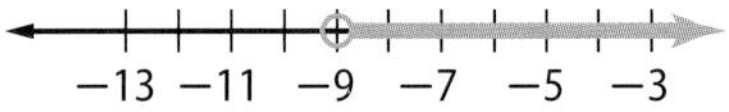

b. $\frac{b}{-8} - 2 \geq -8$

$\frac{b}{-8} - 2 \geq -8$	Write the inequality.
$\frac{b}{-8} - 2 + 2 \geq -8 + 2$	Addition Property of Inequality
$\frac{b}{-8} \geq -6$	Simplify.
$(-8)\,\frac{b}{-8} \leq -6\,(-8)$	Multiplication Property of Inequality
$b \leq 48$	Simplify.

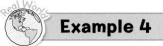

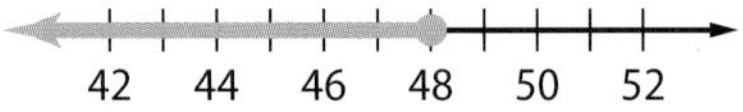

Checking Solutions

For Example 4a, try a number that is less than −9 to show it is *not* a solution.

$$-5x < 45$$
$$-5(-10) \overset{?}{<} 45$$
$$50 \not< 45$$

Got It? **Do these problems to find out.**

4a. $-\frac{y}{4} < 3$

4b. $7 \geq -2f$

Guided Practice

Check

Solve each inequality. Check your solutions. (Example 1)

1. $y + 7 \leq 12$

2. $b + 20 > -13$

3. $-7 < x + (-3)$

Solve each inequality. Graph each solution on a number line. (Examples 2 and 4)

4. $d - 9.3 \geq 12.5$

5. $3\frac{1}{5} > f - \frac{4}{5}$

6. $g - 22 \leq -40$

7. $-12 \geq -3q - 18$

8. $-8z \leq -24$

9. $18 > -\frac{2}{3}g$

10. Isabel earns \$50 plus \$2.50 for each table she cleans. Write and solve an inequality to find how many tables she must clean to earn at least \$120. (Example 3)

Independent Practice

Go online for Step-by-Step Solutions eHelp

Solve each inequality. Check your solutions. (Examples 1 and 2)

11. $a + 18 < 40$

12. $h - 12 > 52$

13. $y - 4.2 \leq 6.5$

14. $g + 5.9 \geq 10$

15 $p - 14 > 12$

16. $x + 3.75 < 5$

17. $n - 0.1 \leq 1.4$

18. $7 > z + \frac{2}{3}$

19. $22 \geq c - 2.1$

20. $13 \geq 9 + b$

21. $14\frac{1}{2} < b - 1\frac{1}{4}$

22. $t + \frac{1}{5} < 2\frac{7}{10}$

23. Montel pays a \$2 entrance fee and \$0.75 every time he plays his favorite video game. If he has \$10, write and solve an inequality to find how many video games he can play. (Example 3)

24. The dance committee has \$75 to spend on centerpieces. They spent \$30 on flowers. If each vase costs \$3, write and solve an inequality to find how many vases they can buy. (Example 3)

Solve each inequality. Graph each solution on a number line. (Example 4)

25. $4x > -36$

26. $7y \leq -49$

27. $13 - 3n < -8$

28. $3b + 15 \leq 8b - 5$

29. $45 \leq -10r$

30. $15 \leq -\frac{k}{4} - 9$

31. $\frac{7}{2}y > 63$

32. $\frac{3}{4} \leq -\frac{5}{7}m$

33. $-\frac{3}{24}b \geq -\frac{1}{4}$

34. $-\frac{3}{4}d - 6 \geq 9$

35. $\frac{4}{9}c \leq -\frac{4}{5}$

36. $-\frac{3}{4} \geq -\frac{6}{10}a$

37 Khadijah has at most three hours to work on a math assignment and a history project. If the math assignment will take $\frac{3}{4}$ hour, how much time can Khadijah spend working on her history project?

38. The 2008 attendance at the Ohio State Fair was at least 16,700 less than the attendance in 2009. If the attendance in 2008 was 809,300, write and solve an inequality to find the 2009 attendance.

Solve each inequality. Check your solutions.

39. $a - 3.5 < \frac{2}{5}$

40. $b + 4\frac{1}{2} \geq 0.4$

41. $26 \leq 5 - c$

42. $12 - d > 40$

43. $4.5 \geq \frac{3}{2}r$

44. $f - 8 < -1.1$

45. $t - 6 \leq 2.5$

46. $-9.6 \geq \frac{1}{3}y$

47. $-\frac{7}{11} \geq -\frac{14}{33}g$

Write an inequality to represent each situation. Then solve each inequality.

48. Seven more than a number is at most 24.

49. The quotient of a number and −3 is greater than the quotient of 5 and 6.

50. 18 is at least the product of −6 and a number.

51. The difference of a number and 15 is no more than −8.

52. Twelve less than a number is at most 20.

53. 28 is at least the product of 7 and a number.

54. **Financial Literacy** Brian is saving money to buy a new mountain bike. The bike that he likes costs \$375.95, and he has already saved \$285.50. Write and solve an inequality to find the amount he must still save.

55. **Multiple Representations** Consider the inequalities $b \geq 4$ and $b \leq 13$.

   a. **Graph** Graph each inequality on the same number line.

   b. **Words** Do the solution sets of the two inequalities overlap? If so, what does this overlapping area represent?

   c. **Symbols** A *compound inequality* is an inequality that combines two inequalities. Write a compound inequality for the situation.

   d. **Graph** Look back at the graph of the solutions for both inequalities. Make another graph that shows only the solution of the compound inequality.

Graph each compound inequality on a number line.

56. $-3 < n < 5$

57. $4 \geq m > -2$

58. $8 \leq g < 14$

59. **STEM** Use the body temperature scale shown.

   a. Suppose Malia has a temperature of 99.2°F. Write and solve an inequality to find how much her temperature must increase before she is considered to have a high fever.

   b. Hypothermia occurs when a person's body temperature falls below 95°F. Write and solve an inequality that describes how much lower the body temperature of a person with hypothermia will be than the body temperature of a person with a normal temperature of 98.6°F.

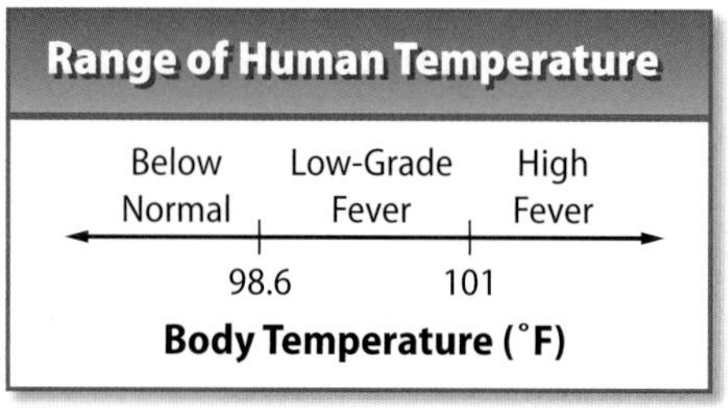

H.O.T. Problems Higher Order Thinking

60. **Persevere with Problems** Write an inequality for the following sentence.

The quotient of a number and −5 increased by 4 is at most 8.

Name three numbers that are possible solutions. Explain.

61. **Model with Mathematics** Write a real-world problem involving an inequality and negative numbers where the inequality symbol would *not* be reversed when finding the solution.

62. **Persevere with Problems** Twenty more than half a number is at least 45. Find the least number that meets this condition.

63. **Use a Counterexample** Is the following statement *true* or *false*? If false, provide a counterexample.

For all values of x, two times x is greater than x.

64. **Justify Conclusions** Is it *always*, *sometimes*, or *never* true that if $x \leq y$, then $y > x$? Explain your reasoning to a classmate.

65. **Building on the Essential Question** Explain how to solve inequalities that involve multiplication and division.

Standardized Test Practice

66. The length of the rectangle is greater than its width. Which inequality represents the possible values of x?

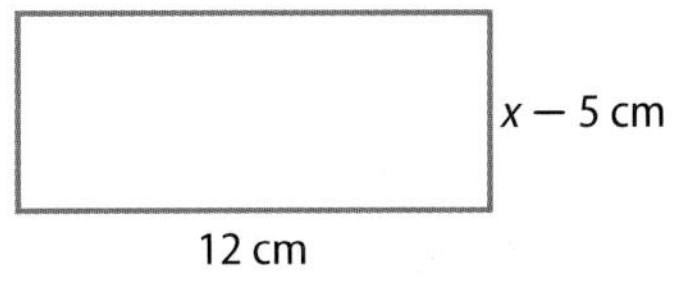

A $x \le 17$ **C** $x \ge 17$

B $x < 17$ **D** $x > 17$

67. If $n + 15 > 4$, then n could be which of the following values?

F −13 **H** −11

G −12 **J** −10

68. The solutions for which inequality are represented by the following graph?

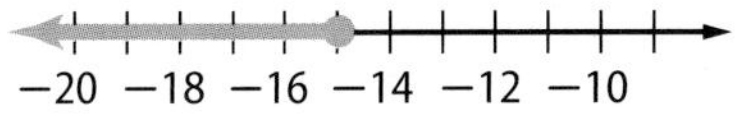

A $\frac{x}{-3} < 5$ **C** $\frac{x}{3} > -5$

B $\frac{x}{-3} \le 5$ **D** $\frac{x}{-3} \ge 5$

69. The product of a number n and four is at most thirty. Which inequality represents the possible values of n?

F $n \le 7\frac{1}{2}$ **H** $n \ge 120$

G $n \ge 7\frac{1}{2}$ **J** $n \le 120$

Common Core Review

Write an inequality for each sentence. 7.EE.4

70. Leticia made at least \$45 babysitting last weekend.

71. Marc could pay no more than \$8500 for his car.

72. Adrienne needs an 86% or better to get a B in the class.

73. There are fewer than 625 students at Everett Middle School.

74. The car's tank contains no less than 8.5 gallons of gasoline.

75. Tyrell has more than 200 foreign coins in his collection.

Solve each equation. Check your solutions. 8.EE.7

76. $3x + 1 = 7$

77. $5x - 4 = 11$

78. $4h + 6 = 22$

79. $8n + 3 = -5$

80. $37 = 4d + 5$

81. $9 = 15 + 2p$

82. $4h + 5 = -6h - 19$

83. $7d - 13 = 17 + 3d$

84. $n - 14 = 3n$

85. $2g + 12 = 3g - 1$

86. $8y + 5 = 5y - 5 + 2y$

87. $4t = 2t - 26$

88. A standard showerhead uses about 6 gallons of water per minute. The table shows the relationship between time in minutes and the number of gallons of water used. 7.RP.2

a. Given m, the number of minutes, write an equation that can be used to find g, the number of gallons used.

b. How many minutes elapsed if 72 gallons of water were used?

Taking a Shower

Time m (minutes)	Water Used g (gallons)
1	6
2	12
4	24
7	42

Lesson 8-8

Solving Multi-Step Equations and Inequalities

ISG Interactive Study Guide

See pages 185–186 for:
- Getting Started
- Real-World Link
- Notes

Essential Question

How are equations and inequalities used to describe and solve multi-step problems?

Common Core State Standards

Content Standards
7.EE.4, 7.EE.4a, 7.EE.4b, 8.EE.7, 8.EE.7a, 8.EE.7b

Mathematical Practices
1, 2, 3, 4, 7

Vocabulary

null or empty set
identity

What You'll Learn

- Solve multi-step equations.
- Solve multi-step inequalities.

Real-World Link

Field Trip A field trip is a fun and an exciting way to learn about something in a totally different way. Mr. Murphy's class of 20 students is going on a field trip to the science center. The total cost of admission and tickets for the 3-D movie for all of the students is \$270.

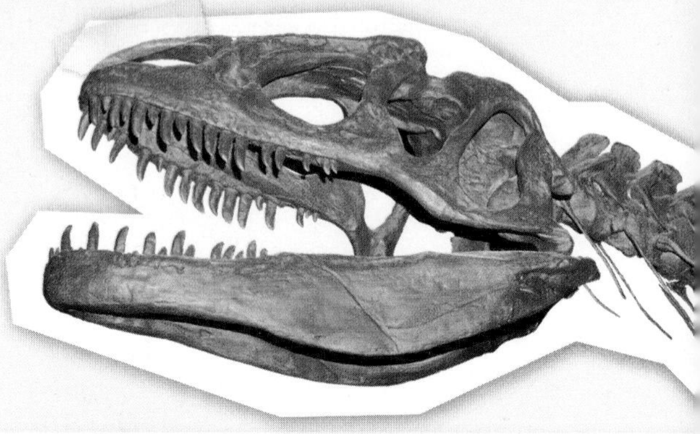

Solve Multi-Step Equations

If x represents the science center's entry fee, the expression $20(x + 2.50)$ represents the total cost for Mr. Murphy's students. Suppose 15 of the students go on a field trip to an art museum where the entry fee is twice that of the science center's fee, and there is a fee of \$1 for the audio tour. If the total cost for Mr. Murphy's students is the same at both museums, you can find the science center's entry fee by solving $20(x + 2.50) = 15(2x + 1)$.

Example 1

Tutor

Solve $20(x + 2.50) = 15(2x + 1)$. Check your solution.

$20(x + 2.50) = 15(2x + 1)$	Write the equation.
$20x + 50 = 30x + 15$	Distributive Property
$20x - \mathbf{20x} + 50 = 30x - \mathbf{20x} + 15$	Subtraction Property of Equality
$50 = 10x + 15$	Simplify.
$50 - \mathbf{15} = 10x + 15 - \mathbf{15}$	Subtraction Property of Equality
$35 = 10x$	Simplify.
$3.5 = x$	Division Property of Equality; check your solution.

Got It? **Do these problems to find out.**

Solve each equation. Check your solution.

1a. $12m + 12 = 6(3m + 3)$

1b. $5(n - 3) = 3(n + 7)$

Some equations have *no* solution. When this occurs, the solution is the **null or empty set**, shown by the symbol Ø or {}. Other equations may have every number as their solution. An equation that is true for every value of the variable is called an **identity**.

Example 2

Tutor

Identities

An identity is an equation that shows that a number or expression is equivalent to itself.

Additive identity

$a + 0 = a$

Multiplicative Identity

$a \cdot 1 = a$

Solve each equation.

a. $3(y - 5) + 25 = 3y + 10$

$3(y - 5) + 25 = 3y + 10$	Write the equation.
$3y - 15 + 25 = 3y + 10$	Use the Distributive Property.
$3y + 10 = 3y + 10$	Simplify.
$3y + 10 - \mathbf{3y} = 3y + 10 - \mathbf{3y}$	Subtraction Property of Equality
$10 = 10$	Simplify.

The statement $10 = 10$ is *always* true. The equation is an identity and the solution set is all numbers.

b. $-5s - 14 = 2(2s + 3) - 9s$

$-5s - 14 = 2(2s + 3) - 9s$	Write the equation.
$-5s - 14 = 4s + 6 - 9s$	Use the Distributive Property.
$-5s - 14 = 6 - 5s$	Simplify.
$-5s - 14 + \mathbf{5s} = 6 - 5s + \mathbf{5s}$	Addition Property of Equality
$-14 = 6$	Simplify.

The statement $-14 = 6$ is *never* true. The equation has no solutions and the solution set is Ø.

Got It? **Do these problems to find out.**

2a. $-2(3r + 4) = -5r - 8 - r$

2b. $14 + 8w = 4(8 + 2w)$

Solve Multi-Step Inequalities

Solving a multi-step inequality is similar to solving a multi-step equation. As a first step, you can use the Distributive Property to remove grouping symbols.

Example 3

Tutor

Solve $4(x - 3) > 6$. Check your solution.

$4(x - 3) > 6$	Write the inequality.
$4x - 12 > 6$	Distributive Property
$\underline{+12 \quad +12}$	Addition Property of Inequality
$4x > 18$	Simplify.
$\frac{4x}{4} > \frac{18}{4}$	Division Property of Inequality
$x > 4.5$	Simplify. Check your solution.

Got It? **Do these problems to find out.**

Solve. Check your solution.

3a. $3 < 4(x + 2)$

3b. $4(b - 3) \leq 72$

Example 4

Mariella's parents have budgeted at most \$575 for her Quinceañera celebration. The cost of the party room is \$75. How much can the family spend per guest on food if each of the 40 guests receives a \$5 favor?

First write an inequality to represent the situation.

Words	party cost ≥ room cost + the number of guests × the cost per guest
Variable	Let c = the food cost per guest so $c + 5$ = the total cost per guest.
Inequality	$575 \geq 75 + 40(c + 5)$

$575 \geq 75 + 40(c + 5)$	Write the Inequality.
$575 \geq 75 + 40c + 200$	Distributive Property
$575 \geq 40c + 275$	Simplify.
$575 - \mathbf{275} \geq 40c + 275 - \mathbf{275}$	Subtraction Property of Inequality
$300 \geq 40c$	Simplify.
$7.5 \geq c$	Division Property of Inequality

The family can spend at most \$7.50 per guest on food.

Got It? **Do this problem to find out.**

4. Sofia recycled 3 pounds less than the amount that James recycled. Hannah recycled 3 times the amount that Sofia recycled. If they recycled a total of 53 pounds, how many pounds did Sofia recycle?

Watch Out!

Inequality Signs Do not reverse the inequality sign just because there is a negative sign in the inequality. Only reverse the sign when you multiply or divide by a negative number.

Example 5

Solve $5a - 8 \geq 4(a - 3)$. Graph the solution on a number line.

$5a - 8 \geq 4(a - 3)$	Write the inequality.
$5a - 8 \geq 4a - 12$	Distributive Property.
$5a - 8 - \mathbf{4a} \geq 4a - 12 - \mathbf{4a}$	Subtraction Property of Inequality
$a - 8 \geq -12$	Simplify.
$a - 8 + \mathbf{8} \geq -12 + \mathbf{8}$	Addition Property of Inequality
$a \geq -4$	Simplify.

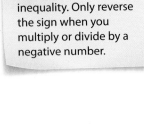

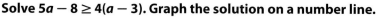

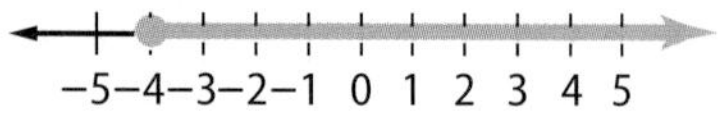

Graph the solution on a number line.

Got It? **Do these problems to find out.**

Solve. Graph each solution on a number line.

5a. $-2(k + 1) > -16 + 5k$

5b. $2p + 5 \geq 3(p - 6)$

Guided Practice

Check

Solve. Check your solution. (Examples 1 and 2)

1. $4(x + 1) + 3 = 31$

2. $33 = 7(2p - 1) - 2$

3. $2(a - 2) = 3(a - 5)$

4. $16(z + 3) = 4(z + 9)$

5. $7(x + 2) = 2(x + 2)$

6. $3(d - 2) = 5(d + 8)$

7. $6x + 4 = 2(3x - 5)$

8. $20f + (-8f - 15) = 3(4f - 5)$

9. $3(1 + 2f) - 5 = 6f - 2$

10. $7x + 5 = 10(x - 7) - 3x$

Solve. Graph each solution on a number line. (Examples 3 and 5)

11. $-2(k - 2) \geq -20$

12. $(3r + 7)2 \leq -34$

13. $-2(g - 1) > g - 4$

14. $5p + 8 \geq 3(p + 6)$

15. $6(-2z + 5) < -19z + 16$

16. $10p \leq 7(2p - 4)$

17. You and three friends are going to the fair. The cost for parking is \$5 per car and admission to the fair is \$19 per person. If you have a total of \$113, what is the maximum amount each person can spend on food? (Example 4)

Independent Practice

Go online for Step-by-Step Solutions

eHelp

Solve. Check your solution. (Examples 1 and 2)

18. $6n - 18 = 4(n + 2)$

19 $12y + 5(y - 6) = 4$

20. $12z + 4 = 2(5z + 8) - 12$

21. $d - 12 = 4(d - 6)$

22. $3x + 2 = 2(2x - 7)$

23. $6(y - 5) = 2(10 + 3y)$

24. $4(2c + 8) = 5(c + 4)$

25. $10 + 12p = 3(3 + 4p)$

26. $3x + 2 + 5(x - 1) = 8x + 17$

27. $10z + 4 = 2(5z + 8) - 12$

Solve. Graph each solution on a number line. (Examples 3 and 5)

28. $20 > 5(w + 3)$

29. $-32 \leq 9(3h + 2) + 4$

30. $3(6m - 4) \geq 24$

31. $10(3 + s) < 4s$

32. $3y - 6 > 4(y - 3)$

33. $8(2h + 6) \leq 12h + 20$

34. $3(3r + 5) \geq 24 + 10r$

35. $14t - 28 < 7(t + 6)$

36. The perimeter of a rectangle is at least 50 centimeters. The length of the rectangle is one more than 3 times the width of the rectangle. What are the minimum dimensions of the rectangle? (Example 4)

37. **Financial Literacy** Tim is taking the train to Seattle to visit his grandparents. He has at most \$15.00 to spend on snacks and reading material. Granola bars cost \$1.15 each, and magazines cost \$1.25. If Tim buys the same number of granola bars and magazines, how many can he buy? (Example 4)

38. CCSS **Reason Abstractly** Sumi is considering two gyms. Gym A has a one-time membership fee of \$45 and costs \$18 per month. Gym B has a one-time membership fee of \$60 and costs \$18 per month. Write and solve an equation to find the number of months for which the two gyms have the same cost. Explain how your solution relates to the problem situation.

39. Nomar has earned scores of 73, 85, 91, and 82 on the first four of five math tests for the grading period. He would like to finish the grading period with a test average of at least 82. What score does Nomar need to earn on the fifth test in order to achieve his goal?

Solve.

40. $-0.2(3c + 15) = 3(0.8c - 8)$

41. $2(t + 12) - 6(2t - 3) = 14$

42. $5 - \frac{1}{2}(x - 6) < 4$

43. $6n - 18 \geq 4(n + 2.1)$

44. $2.01c - 6 = -0.15c + 6.96$

45. $\frac{1}{4}x + 13 > 0.25(2x - 32)$

46. $0.5(4x + 24) = 22x - 2$

47. $6(n + 2) + 3(3n - 5) = 57$

48. Cole is having his car repaired. The mechanic said it would cost at least \$375 for parts and labor. If the cost of the parts was \$150, and the mechanic charges \$60 an hour, how many hours is the mechanic planning to work on the car?

49. A good rule to know when training for a marathon is that you will generally have enough endurance to finish a race that is 3 times your average daily distance. Tammy wants to be able to run *at least* the standard marathon distance of 26.2 miles. The length of her current daily run is about 4 miles. To the nearest hundredth, by how many miles should she increase her daily run to meet her goal?

Solve. Justify each step in the solution. Use a Property of Equality or Inequality when necessary.

50. $4(y - 3) = 2(3y + 10)$

51. $5(2f - 1) = 3(f + 3)$

52. $-1.2(w + 1.1) \leq 6.18$

53. $p > \frac{2}{3}\left(p - \frac{1}{2}\right)$

H.O.T. Problems Higher Order Thinking

54. **CCSS Identify Structure** Write a multi-step inequality that can be solved by first adding 3 to each side.

55. **CCSS Identify Structure** Explain how you can solve $45 > -6x + 3$ without multiplying or dividing by a negative number.

56. **CCSS Find the Error** Jada is solving $3x - 9 \leq 5(x + 10)$. Find her mistake and correct it.

$$3x - 9 \leq 5(x + 10)$$
$$3x - 9 \leq 5x + 10$$
$$-2x \leq 19$$
$$x \geq -9.5$$

57. **CCSS Persevere with Problems** Use the information in Example 2 about equations that have no solutions or those that are identities to solve the following inequalities. Justify each step in the solution.

a. $5x - 6 \geq 3(x - 2) + 2x$

b. $12p + 17 \leq 3(4p - 8)$

58. **Building on the Essential Question** Explain how to determine if an equation has no solution, one solution, or if all numbers are solutions. Use examples with your explanation.

Standardized Test Practice

59. Find the value of x so that the polygons have the same perimeter.

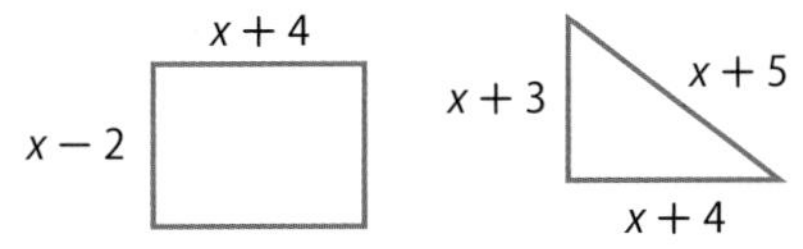

A 3
B 6
C 8
D 12

60. Short Response Damon can spend at most \$150 on supplies for a school party. He has already spent \$78.80 on food and \$29.95 on decorations. He plans to buy cases of juice that cost \$8.25 per case. Write and solve an inequality to find the number of cases of juice Damon can buy.

61. What is the solution of this inequality?

$$-4x + 16 \geq -4$$

F $x \geq 3$
G $x \leq 3$
H $x \geq 5$
J $x \leq 5$

62. Sandra's scores on the first five science tests are shown. Which inequality represents the score she must receive on the sixth test to have an average score of more than 88?

Test	Score
1	85
2	84
3	90
4	95
5	88

A $s \geq 86$
B $s \leq 88$
C $s < 88$
D $s > 86$

CCSS Common Core Review

Solve each inequality. Graph each solution on a number line. 7.EE.4

63. $-25t \leq 400$
64. $8 > \frac{q}{3}$
65. $14 \geq 7 + a$
66. $-13 \geq x - 8$
67. $-\frac{3}{4} < w - 1$
68. $3 \leq \frac{1}{2} + a$

Write an inequality for each sentence. 7.EE.4

69. Kyle's earnings were no more than \$60.

70. The 10 kilometer race time of 86 minutes was greater than the winner's time.

Write each fraction or mixed number as a decimal. Use a bar to show a repeating decimal. 7.NS.2d

71. $\frac{1}{5}$
72. $-\frac{5}{8}$
73. $7\frac{3}{10}$
74. $\frac{1}{9}$
75. $-3\frac{3}{4}$
76. $-\frac{5}{11}$

77. Financial Literacy The formula $P = I - E$ is used to find the profit P when income I and expenses E are known. One month a small business has an income of \$19,592 and expenses of \$20,345. 7.NS.1

a. What is the profit for the month?

b. What does a negative profit mean?

78. STEM The diameter of Earth is approximately 1.28×10^4 kilometers. (8.EE.4)

a. The diameter of Jupiter is about 11.1 times the diameter of Earth. Write the diameter of Jupiter in standard notation.

b. Write the diameter of Jupiter in scientific notation.

Divide. 6.NS.3

79. $7.2 \div 2$
80. \$3.75 ÷ 5
81. \$25.90 ÷ 3.5
82. $29.14 \div 4.7$

Check

Chapter Review

ISG Interactive Study Guide
See pages 187–190 for:
- Vocabulary Check
- Key Concept Check
- Problem Solving
- Reflect

Lesson-by-Lesson Review

Lesson 8-1 Solving Equations with Rational Coefficients (pp. 324–329)

Solve each equation. Check your solutions.

1. $1.2m = 2.4$
2. $\frac{1}{5}x = 4$
3. $-2x = 2.2$
4. $0.5x = 25$
5. $-\frac{1}{4}x = 16$
6. $\frac{1}{6}x = -4$

7. Rosa is baking granola bars for her friends. Each bar requires 4.8 ounces of dough. Write and solve an equation to find how many granola bars Rosa can make if she has 336 ounces of dough.

8. Samantha has 15 packets of beads. She wants to make bracelets that use 1.5 packets of beads each. Write and solve an equation to find how many bracelets Samantha can make.

Example 1

Solve $-4x = -3.2$.

$-4x = -3.2$	Write the equation.
$\frac{-4x}{\mathbf{-4}} = \frac{-3.2}{\mathbf{-4}}$	Division Property of Equality
$x = 0.8$	Simplify.

Example 2

Solve $-\frac{1}{2}a = 5$.

$-\frac{1}{2}a = 5$	Write the equation.
$\mathbf{-2}\left(-\frac{1}{2}a\right) = \mathbf{-2}(5)$	Multiplication Property of Equality
$a = -10$	Simplify.

Lesson 8-2 Solving Two-Step Equations (pp. 333–338)

Solve each equation. Check your solutions.

9. $3 + 4c = 15$
10. $2.1n - 5.31 = 18$
11. $\frac{1}{3}a + 2 = 5$
12. $\frac{1}{5}x - 3 = 7$
13. $\frac{4}{7} + 2p = \frac{2}{7}$
14. $0.12t - 0.6 = -0.06$

15. Nate read 10 more books than Maren for the summer reading program. The total number of books they read is 60. Solve $x + x + 10 = 60$ to find the number of books Nate read.

Example 3

Solve $3x + 5 = 29$.

$3x + 5 = 29$	Write the equation.
$3x + 5 \mathbf{- 5} = 29 \mathbf{- 5}$	Subtraction Property of Equality
$3x = 24$	Simplify.
$\frac{3x}{\mathbf{3}} = \frac{24}{\mathbf{3}}$	Division Property of Equality
$x = 8$	Simplify.

Lesson 8-3 Writing Equations (pp. 339–343)

Translate each sentence into an equation. Then solve each equation.

16. Toya bought some fruit for \$5 and 3 boxes of cereal and spent a total of \$17.

17. Six less than twice a number is -22.

18. Noelle spent \$36 on books and pens. She spent \$12 more on books than she did on pens. How much did she spend on books?

Example 4

The product of a number and 6 is -36. Write and solve an equation to find the number.

$6n = -36$	Write the equation.
$\frac{6n}{\mathbf{6}} = -\frac{36}{\mathbf{6}}$	Division Property of Equality
$n = -6$	Simplify.

Lesson 8-4 More Two-Step Equations (pp. 347–351)

Solve each equation.

19. $3(x - 5) = 24$

20. $\frac{4}{5}(c + 12) = -8$

21. $0.3(5 + y) = 12$

22. $2.4(b + 3.1) = 14.4$

23. $\frac{3}{8}(v - 6) = \frac{1}{8}$

24. $4.1(x - 2.7) = 254.2$

25. Talia buys 4 boxes of cereal. She has coupons for \$1.15 off the regular price of each box of cereal. After using the coupons, the total cost of the cereal is \$8.40. Write and solve an equation to find the regular price of each box of cereal.

26. Brady buys five T-shirts on sale for \$4 off the regular price. His total cost is \$55, before tax. Write and solve an equation to find the regular price of each T-shirt.

Example 5

Solve $\frac{2}{3}(p - 5) = -8$. Check your solution.

$\frac{2}{3}(p - 5) = -8$ — Write the equation.

$\frac{3}{2} \cdot \frac{2}{3}(p - 5) = \frac{3}{2} \cdot (-8)$ — Multiplication Property of Equality

$p - 5 = -12$ — Simplify.

$p - 5 + 5 = -12 + 5$ — Addition Property of Equality

$p = -7$ — Simplify.

Check $\frac{2}{3}(p - 5) = -8$

$\frac{2}{3}(-7 - 5) \stackrel{?}{=} -8$

$\frac{2}{3}(-12) \stackrel{?}{=} -8$

$-8 = -8$ ✓

Lesson 8-5 Solving Equations with Variables on Each Side (pp. 356–360)

Solve each equation. Check your solutions.

27. $3a + 6 = 2a$

28. $10 - x = 9x$

29. $8.7w - 3 = 5.5w + 3.4$

30. $\frac{2}{3}x + 6 = \frac{1}{3}x - 13$

31. $-5.2k + 2.9 = -12k + 30.1$

32. $5x + \frac{2}{5} = x - \frac{1}{10}$

33. An online DVD rental club has two membership plans as shown. Write and solve an equation to find how many months it would take for the total cost of the two plans to be the same.

Plan	Membership Fee	Cost Per Month
A	\$20	\$5
B	\$30	\$3

34. An online video game site charges a \$10 membership fee and a \$6.00 monthly cost. Another game site charges \$6 membership fee and a \$8.00 monthly cost. Write and solve an equation to find how many months it would take for the total cost of the two sites to be the same.

Example 6

Solve $8x + 6 = 4x - 10$. Check your solution.

$8x + 6 = 4x - 10$ — Write the equation.

$8x - 4x + 6 = 4x - 4x - 10$ — Subtraction Property of Equality

$4x + 6 = -10$ — Simplify.

$4x + 6 - 6 = -10 - 6$ — Subtraction Property of Equality

$4x = -16$ — Simplify.

$x = -4$ — Division Property of Equality.

Check $8x + 6 = 4x - 10$

$8(-4) + 6 \stackrel{?}{=} 4(-4) - 10$

$-32 + 6 \stackrel{?}{=} -16 - 10$

$-26 = -26$ ✓

Lesson 8-6 Inequalities (pp. 361–366)

Write an inequality for each sentence.

35. Jeremiah can spend at most \$15 at the store.

36. There are more than 35 students in the band.

For the given value, state whether each inequality is *true* or *false*.

37. $x + 6 > 7, x = 2$

38. $13 - a < 9, a = 10$

39. $16 \leq 4a, a = 4$

40. $3m + 4 \geq 12, m = 2$

41. $6x > 18, x = 3$

42. $6b + 4 > 12, b = 2$

Example 7

State whether $x - 6 > 12$ is *true* or *false* for $x = 15$.

$x - 6 > 12$ — Write the inequality.

$15 - 6 \overset{?}{>} 12$ — Replace x with 15.

$9 \overset{?}{>} 12$ — Simplify.

The sentence is false. So, $9 \not> 12$.

Lesson 8-7 Solving Inequalities (pp. 367–373)

Solve each inequality. Graph the solution on a number line.

43. $a - 15 \leq 3$

44. $x + 13 > -22$

45. $-0.3z \leq -2.4$

46. $6h - 4 > 38$

47. $-5x < -13$

48. $\frac{2}{3}x + 7 > 13$

49. Jose is buying books and CDs. He can spend at most \$150. He spends \$30 on books . If each CD costs \$20, write and solve an inequality to show the maximum number of CDs Jose can buy.

50. Tim can spend at most \$36 for cupcakes. If each cupcake costs \$3, write and solve an inequality to show the maximum number of cupcakes Tim can buy.

Example 8

Solve $x - 3 < 10$. Then graph the solution on a number line.

$x - 3 < 10$ — Write the inequality.

$x - 3 + \mathbf{3} < 10 + \mathbf{3}$ — Addition Property of Inequality

$x < 13$ — Simplify.

Graph the solution.

Lesson 8-8 Solving Multi-Step Equations and Inequalities (pp. 374–379)

Solve. Check your solutions.

51. $5(x - 3) + 1 = 16$

52. $3(c - 4) = 4(c - 6)$

53. $2(x + 2) - 4 = 2x$

54. $5 + 3y = 3(2 + y)$

55. $-4g - 5 \geq -17$

56. $18 > -12 + 6m$

57. $24 - 3c \leq 15$

58. $\frac{2}{3}k + 9 < 5$

59. A car sales associate receives a monthly salary of \$1700 a month plus \$140 for every car he sells. How many cars must he sell monthly to earn at least \$4500?

60. At the car wash, Xander receives \$200 a week plus \$10 for every car he details. How many cars must he detail weekly to earn at least \$350 a week?

Example 9

Solve $-5m + 8 \geq 23$.

$-5m + 8 \geq 23$ — Write the inequality.

$-5m + 8 - \mathbf{8} \geq 23 - \mathbf{8}$ — Subtraction Property of Inequality

$-5m \geq 15$ — Simplify.

$\frac{-5m}{\mathbf{-5}} \leq \frac{15}{\mathbf{-5}}$ — Division Property of Inequality

$m \leq -3$ — Simplify.

Chapter 9
Linear Functions

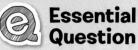

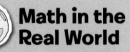

Essential Question

How are linear functions used to model proportional relationships?

Common Core State Standards

Content Standards
7.RP.2, 7.RP.2a, 7.RP.2b, 7.RP.2d, 7.EE.4, 8.EE.5, 8.EE.6, 8.EE.8, 8.EE.8a, 8.EE.8b, 8.EE.8c

Mathematical Practices
1, 2, 3, 4, 6, 7, 8

Math in the Real World

Gorillas Some gorillas consume more than 40 pounds of vegetation per day, such as leaves, stems, roots, vines, herbs, trees, and grasses. This type of vegetation is nearly half water, so gorillas rarely need to drink water in the wild. The amount of vegetation a gorilla eats each day can be represented by a function.

Interactive Study Guide
See pages 191–194 for:
- Chapter Preview
- Are You Ready?
- Foldable Study Organizer

Lesson 9-1

Functions

ISG Interactive Study Guide

See pages 195–196 for:
- Getting Started
- Vocabulary Start-Up
- Notes

Essential Question

How are linear functions used to model proportional relationships?

CCSS Common Core State Standards

Content Standards
Preparation for 7.EE.4

Mathematical Practices
1, 3, 4, 6, 8

Vocabulary

function
independent variable
dependent variable
vertical line test
function rule
function notation

What You'll Learn

- Determine whether a relation is a function.
- Write a function using function notation.

Real-World Link

Meerkats Meerkats are small mammals that live in the southern African plains. They have sharp claws that they use to forage for food and to dig underground burrows. Because meerkats work well as a family unit, they have been the focus of scientific studies, television shows, and a movie!

Relations and Functions

A **function** is a relation in which each element of the domain is paired with *exactly* one element of the range. The domain contains all values of the **independent variable**. These values are chosen and do not depend upon the other variable. The range contains all values of the **dependent variable**. These values depend on the values in the domain.

Function

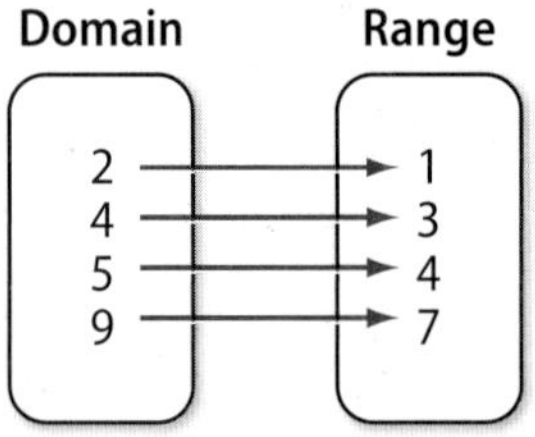

This is a function because each domain value is paired with exactly one range value.

Not a Function

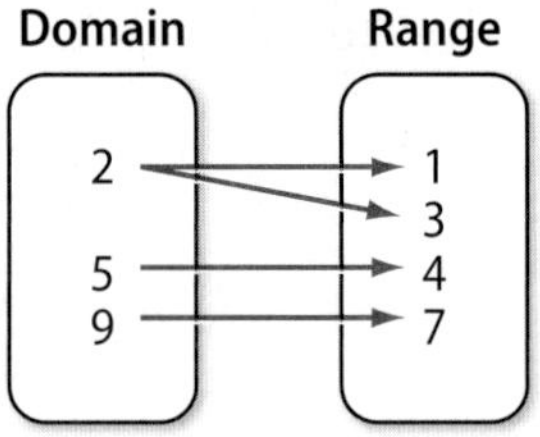

This is *not* a function because 2 in the domain is paired with two range values, 1 and 3.

Example 1

Tools Tutor

Determine whether each relation is a function. Explain.

a. {(54, 112), (56, 130), (55, 145), (54, 123), (56, 128)}

This is not a function because 54 and 56 in the domain are paired with two range values.

b.

x	4	7	6	3	5	2
y	6	10	11	8	4	9

This a function. Each domain value is paired with one range value.

Got It? Do these problems to find out.

1a. {(5, 1), (6, 3), (7, 5), (8, 0)}

1b.

x	1	6	3	1	5	2
y	7	6	2	8	2	1

Vertical Line Test
Use a pencil or straightedge to represent a vertical line. Place the pencil at the left of the graph. Move it to the right across the graph.

Another way to determine whether a relation is a function is to apply the **vertical line test** to the graph of the relation. If, for each value of x in the domain, a vertical line passes through no more than one point on the graph, then the graph represents a function. If the line passes through more than one point on the graph, it is *not* a function.

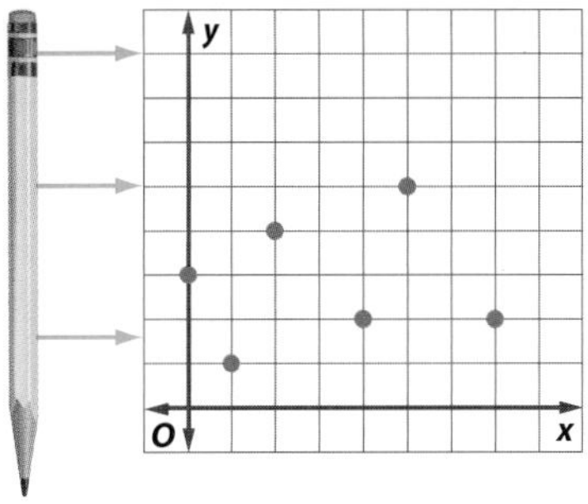

Example 2

Tutor

Determine whether the graph at the right is a function. Explain your answer.

The graph is a function because the vertical line test shows that it passes through no more than one point on the graph for each value of x.

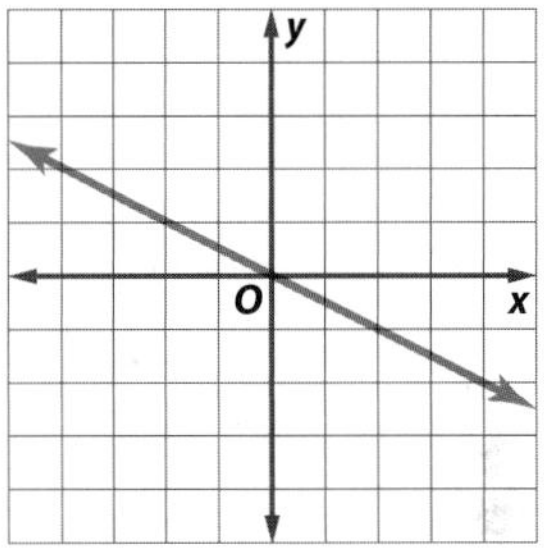

Got It? **Do this problem to find out.**

2. Determine whether the graph at the right is a function. Explain your answer.

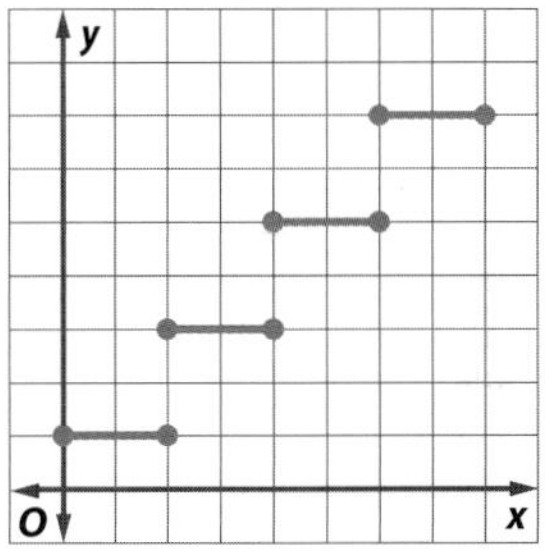

Describe Relationships

Function Notation
$f(4)$ means to find the value of the function when $x = 4$.

A **function rule** describes the operation(s) performed on the domain value (input) to get the range value (output) and can be used to write an equation that relates the variables. A function that is written as an equation can also be written in a form called **function notation**. Consider the equation $y = 2x + 3$.

Equation	**Function Notation**
$y = 2x + 3$	$f(x) = 2x + 3$

The variable y and $f(x)$ both represent the dependent variable. In the example above, when $x = 4$, $f(x) = 11$. In function notation, $f(x)$ is read "f of x" and is equal to the value of the function at x.

Example 3

If $f(x) = 4x - 7$, find each function value.

a. $f(5)$

$f(x) = 4x - 7$ Write the function.

$f(5) = 4(\mathbf{5}) - 7$ Replace x with 5.

$f(5) = 13$ Simplify.

b. $f(-6)$

$f(x) = 4x - 7$ Write the function.

$f(-6) = 4(\mathbf{-6}) - 7$ Replace x with -6.

$f(-6) = -31$ Simplify.

Got It? Do these problems to find out.

If $f(x) = 14 + 3x$, find each function value.

3a. $f(4)$

3b. $f(-7)$

A function can also describe the relationship between two quantities. For example, the distance you travel in a car depends on how long you are driving in the car. Time is the independent variable, and distance is the dependent variable. In other words, *distance is a function of time* or $d(t)$.

Real World Example 4

Functions
A function does not always have to be in the form of $f(x)$. If a function is representing a real-world situation, any variables representing the independent and dependent variables can be used.

Mariah spent $22.50 downloading songs to her digital music player from an online music store for $0.90 each.

a. Identify the independent and dependent variables. Then write a function to represent the total cost of any number of songs.

Since the total cost depends on the number of songs purchased, the total cost c is the dependent variable and the number of songs purchased s is the independent variable.

Words	total cost = cost per song times the number of songs		
Function	$c(s)$	$=$	$0.9s$

The function is $c(s) = 0.9s$.

b. Use the function to determine the number of songs that Mariah downloaded.

$c(s) = 0.9s$ Write the function.

$22.5 = 0.9s$ Substitute 22.5 for $c(s)$.

$25 = s$ Divide each side by 0.9.

So, Mariah downloaded 25 songs.

Got It? Do these problems to find out.

4. A whale-watching boat traveled at a speed of 5.5 miles per hour.

- **a.** Identify the independent and dependent variables. Then write a function to represent the total distance traveled in any number of hours spent whale watching.
- **b.** Use the function to find how long it took to travel 25 miles. Round to the nearest tenth.

Guided Practice

Check

Determine whether each relation is a function. Explain. (Examples 1 and 2)

1. {(8, 2), (4, 3), (6, 5), (1, 5)}

2.

x	1	3	8	7	3
y	4	2	9	6	4

3.

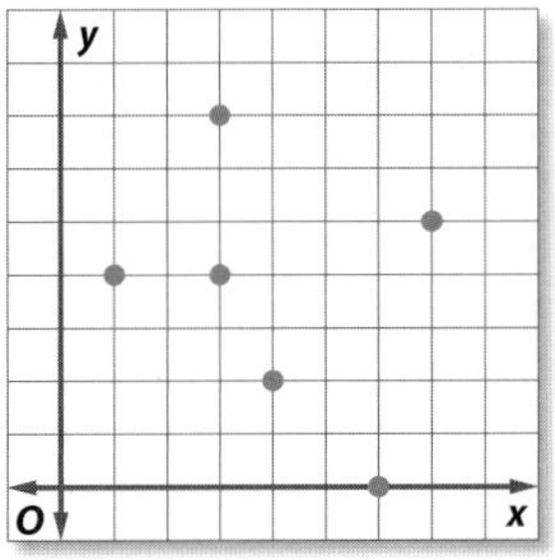

4.

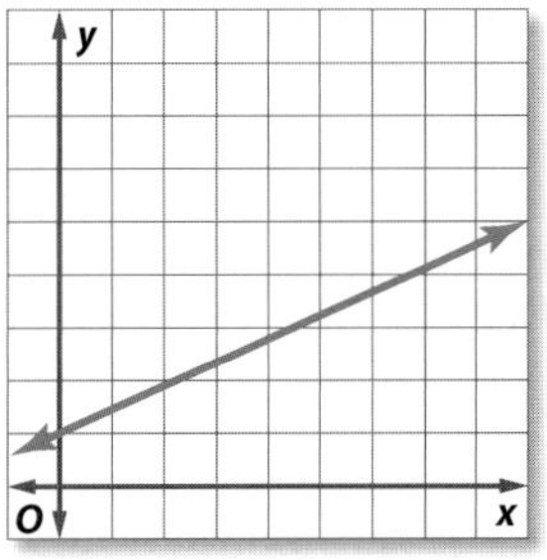

If $f(x) = 6x - 4$, find each function value. (Example 3)

5. $f(3)$

6. $f(-5)$

7. $f(8)$

8. $f(-9)$

9. The Milligan family spent \$215 to have their family portrait taken. The portrait package they would like to purchase costs \$125. In addition, the photographer charges a \$15 sitting fee per person in the portrait. (Example 4)

   a. Identify the independent and dependent variables. Then write a function to represent the total cost of any number of people in the portrait.

   b. Use the equation to find the number of people in the portrait.

Independent Practice

eHelp

Go online for Step-by-Step Solutions

Determine whether each relation is a function. Explain. (Examples 1 and 2)

10. {(10, 8), (12, 4), (15, 15), (9, 4)}

11. {(24, 16), (25, 16), (24, 17), (26, 17)}

12. {(8, 9), (9, 10), (10, 11), (11, 12)}

13. {(3, 1), (3, 3), (3, 7), (3, 9)}

14.

x	5	8	9	4	5
y	37	42	24	37	29

15.

x	50	75	100	125	150
y	8	12	16	20	24

16.

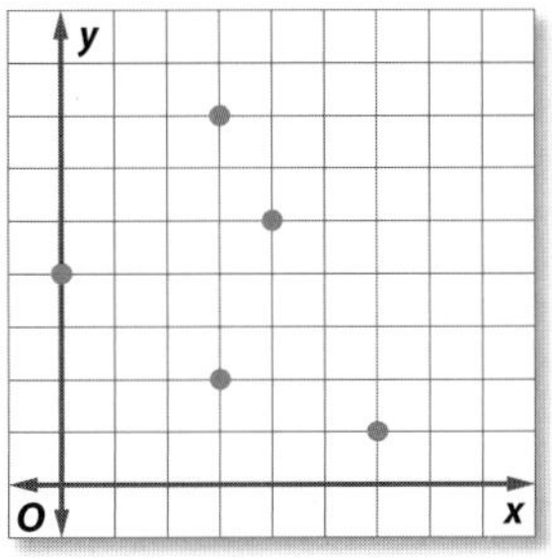

17.

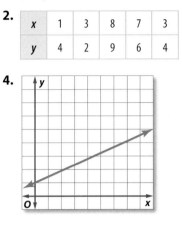

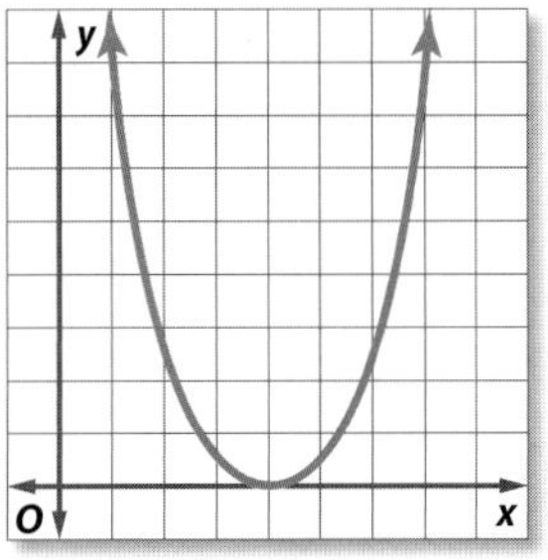

If $f(x) = 3x - 9$, find each function value. (Example 3)

18. $f(12)$

19. $f(9)$

20. $f(-8)$

21. $f(-15)$

If $g(x) = 5x + 4$, find each function value. (Example 3)

22. $g(7)$

23. $g(14)$

24. $g(-10)$

25. $g(-18)$

26. **Be Precise** A submersible descends to a depth of 1500 feet at a rate of 40 feet per minute to explore an underwater shipwreck. (Example 4)

a. Identify the independent and dependent variables. Then write a function to represent the total depth of the submersible in any number of minutes.

b. Use the equation to find the total amount of time it took to reach 1500 feet.

27. **Financial Literacy** Logan spends $128.10 to rent a moving truck from a rental company. The rental company charges him $50 for the truck plus an additional fee of $0.55 per mile that the truck is driven. (Example 4)

a. Identify the independent and dependent variables. Then write a function to represent the total cost of any number of miles driven.

b. Use the equation to find the number of miles that he drove the truck.

28. The table shows the average number of hours worked weekly by U.S. production workers for various years.

Year	Hours
1970	37
1980	35
1990	34
2000	34

a. Write the values in the table as a set of ordered pairs. Do the data represent a function? Explain.

b. The *inverse* of a relation is obtained by switching the order of the numbers in each ordered pair. For example, (1970, 37) would be (37, 1970). Write the inverses of the ordered pairs of the data. Do the data represent a function? Explain.

H.O.T. Problems Higher Order Thinking

29. **Model with Mathematics** Draw a graph to represent a real-world situation, either one that represents a relation that is a function or one that represents a relation that is not a function. Explain why the graph is or is not a function.

30. **Identify Repeated Reasoning** In an arithmetic sequence, a term is found by adding a constant value to the previous term. Consider the terms in this arithmetic sequence: 36, 33, 30, 27, …

Term Number	Term
1	36
2	33
3	30
⋮	⋮
7	■

a. What value was added to each term in the sequence?

b. In a sequence, each number is assigned a term number. Copy and complete the table. Is the set of ordered pairs (term number, term) a function? Explain.

c. Graph the set of ordered pairs. Describe the relationship between term number and term shown in the graph.

Persevere with Problems If $f(x) = 4x - 3$ and $g(x) = 8x + 2$, find each function value.

31. $f[g(3)]$

32. $g[f(5)]$

33. $g\{f[g(-4)]\}$

34. **Building on the Essential Question** How can the relationship between water depth and time to ascend to the water's surface be a function? Explain how the two variables are related. Discuss whether water depth can ever correspond to two different times.

Standardized Test Practice

35. The equation $c(d) = 3.50d$ represents the total cost of renting DVDs as a function of the number of DVDs rented. Which of the following tables contains values that satisfy this function?

A

Cost of Renting DVDs				
d	1	2	3	4
$c(d)$	3.50	7.00	10.50	14.00

B

Cost of Renting DVDs				
d	1	2	3	4
$c(d)$	3.50	6.00	9.00	12.50

C

Cost of Renting DVDs				
d	1	2	3	4
$c(d)$	3.50	7.50	10.50	14.50

D

Cost of Renting DVDs				
d	1	2	3	4
$c(d)$	7.00	10.50	14.00	17.50

36. What is the value of $f(3)$ if $f(x) = -4x + 2$?

F -14

G -10

H -6

J 14

37. **Short Response** For song downloads, Kaylee pays \$15 per month plus \$1.75 per song. Write an equation using function notation that gives the total cost as a function of the number of songs downloaded. How much would her total bill be if she downloaded 15 songs?

38. Which of the following sets of ordered pairs does *not* represent a function?

A {(1, 2), (2, 3), (3, 4), (4, 5)}

B {(2, 3), (2, 4), (2, 5), (2, 6)}

C {(1, 6), (2, 6), (3, 6), (4, 6)}

D {(5, 4), (4, 3), (3, 2), (2, 1)}

CCSS Common Core Review

Write an inequality for each sentence. 7.EE.4

39. Ronald made at least \$55 delivering mulch yesterday.

40. Abigail would pay no more than \$1500 for her airline ticket.

41. Landyn wants 50% or more of his yard reseeded.

Solve each equation. Check your solution. 8.EE.7

42. $3x + 2 = -x - 10$

43. $y - 12 = 3y - 2$

44. $4z + 1 = 15 - 3z$

45. $2 - 5x = -2x + 11$

46. $5y - 5 = -5y + 5$

47. $24 + 6z = 24 - 6z$

48. **Financial Literacy** Suppose Marcus invests \$750 at an annual rate of 6.25%. How long will it take until Marcus earns \$125 in simple interest? 7.RP.3

Find each product or quotient. 7.NS.2c

49. $\frac{8}{9} \cdot \frac{27}{28}$

50. $\frac{3}{4}\left(-\frac{1}{3}\right)$

51. $-\frac{7}{8} \cdot \frac{2}{5}$

52. $2 \cdot \frac{7}{12}$

53. $\frac{2}{9} \div \frac{1}{4}$

54. $-\frac{1}{2} \div \frac{5}{6}$

55. $-\frac{3}{5} \div -\frac{5}{9}$

56. $6\frac{2}{3} \div 5$

Evaluate each expression if $a = 12$ and $b = 8$. 6.EE.2

57. $4a - b$

58. $3b - 2a$

59. $3a + 2b$

60. $a + 5b$

Lesson 9-2
Representing Linear Functions

ISG Interactive Study Guide

See pages 197–198 for:
- Getting Started
- Real-World Link
- Notes

Essential Question

How are linear functions used to model proportional relationships?

CCSS Common Core State Standards

Content Standards
7.EE.4

Mathematical Practices
1, 2, 3, 4, 7, 8

Vocabulary

linear equation
linear function
function table
x-intercept
y-intercept

What You'll Learn

- Solve linear functions with two variables.
- Graph linear functions using ordered pairs.

Real-World Link

Racing Do you like to watch stock car racing? The Daytona 500 is the most important race on the stock car racing calendar. It was first run in 1959, and the average speed of the winner was about 135 miles per hour. However, the record for the fastest average speed is 177.6 miles per hour, which is about 3 miles per minute.

Solve Linear Functions

An equation such as $y = 60x$ is called a linear equation. A **linear equation** is an equation whose graph is a line. The solution of a linear equation consists of two numbers, one for each variable, that make the equation true.

A **linear function** is a function in which the graph of the solutions forms a straight line. One way to find solutions is to make a **function table**. A function table shows the values for *x* and *y*, and the function rule.

Example 1

Tutor

Find four solutions of $y = 7x$. Write the solutions as ordered pairs.

Step 1 Choose four values for *x* and substitute each value into the equation. We chose the values −1, 0, 1, and 2.

Step 2 Evaluate the expression to find the value of *y*.

Step 3 Write the solutions as ordered pairs.

	Step 1	Step 2	Step 3
x	$y = 7x$	y	(x, y)
−1	$y = 7(-1)$	−7	(−1, −7)
0	$y = 7(0)$	0	(0, 0)
1	$y = 7(1)$	7	(1, 7)
2	$y = 7(2)$	14	(2, 14)

Four solutions of $y = 7x$ are (−1, −7), (0, 0), (1, 7), and (2, 14).

Got It? Do these problems to find out.

Find four solutions of each function. Write the solutions as ordered pairs.

1a. $y = x + 2$

1b. $y = 3x - 1$

1c. $y = -2x + 5$

1d. $y = -4x - 6$

When solving real-world problems, check that the solution makes sense in the context of the original problem.

Real World

Example 2

Tutor

Games cost \$8 to download onto a cell phone. Ring tones cost \$1. Find four solutions of $8x + y = 20$ in terms of the numbers of games x and ring tones y Darcy can buy with \$20. Explain each solution.

Step 1 Rewrite the equation by solving for y.

$8x + y = 20$ Write the equation.

$8x - \mathbf{8x} + y = 20 - \mathbf{8x}$ Subtract $8x$ from each side.

$y = 20 - 8x$ Simplify.

Step 2 Choose four x values and substitute them into $y = 20 - 8x$.

x	$y = 20 - 8x$	y	(x, y)
1	$y = 20 - 8(\mathbf{1})$	12	(1, 12)
2	$y = 20 - 8(\mathbf{2})$	4	(2, 4)
$\frac{1}{4}$	$y = 20 - 8\left(\mathbf{\frac{1}{4}}\right)$	18	$\left(\frac{1}{4}, 18\right)$
5	$y = 20 - 8(\mathbf{5})$	−20	(5, −20)

Solution
Each solution of a two-variable equation is an ordered pair, not a single number.

Step 3 Explain each solution.

(1, 12) → She can buy 1 game and 12 ring tones.

(2, 4) → She can buy 2 games and 4 ring tones.

$\left(\frac{1}{4}, 18\right)$ → This solution does not make sense in this situation, because you cannot have a fractional number of games.

(5, −20) → This solution does not make sense in this situation, because you cannot have a negative number of ring tones.

Got It? **Do this problem to find out.**

2. **Financial Literacy** Michael and his two friends have a total of \$9 to spend on tacos and burritos for lunch. A burrito x costs \$2 and a taco y costs \$1. Find two solutions of the equation $2x + y = 9$ to find how much food they can buy with \$9. Explain each solution.

Graph Linear Functions

The x-coordinate of the point at which the graph crosses the x-axis is the **x-intercept**. The y-coordinate of the point at which the graph crosses the y-axis is the **y-intercept**.

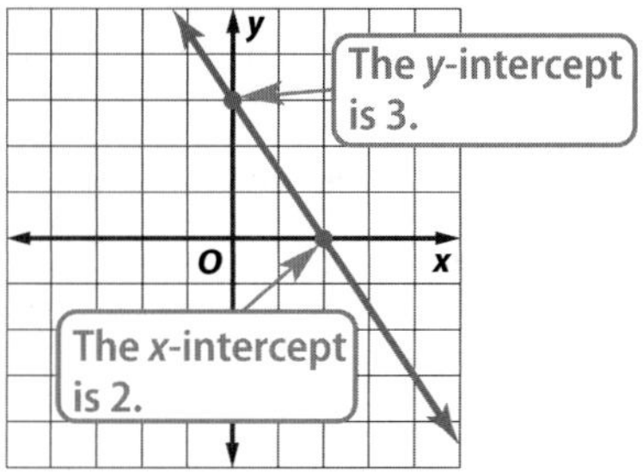

Since two points determine a straight line, a simple method of graphing a linear function is to find the points where the graph crosses the x-axis and the y-axis and then connect them.

Example 3

Graph $-2x + y = 4$.

Find the intercepts.

To find the x-intercept, let $y = 0$.

$-2x + y = 4$ — Write the equation.

$-2x + (0) = 4$ — Replace y with 0.

$-2x = 4$ — Divide each side by -2.

$x = -2$

Since $x = -2$ when $y = 0$, graph the ordered pair $(-2, 0)$.

To find the y-intercept, let $x = 0$.

$-2x + y = 4$ — Write the equation.

$-2(\mathbf{0}) + y = 4$ — Replace x with 0.

$y = 4$

Since $y = 4$ when $x = 0$, graph the ordered pair $(0, 4)$.

Standard Form
An equation written in the form $Ax + By = C$, is written in standard form.

Connect the points with a line.

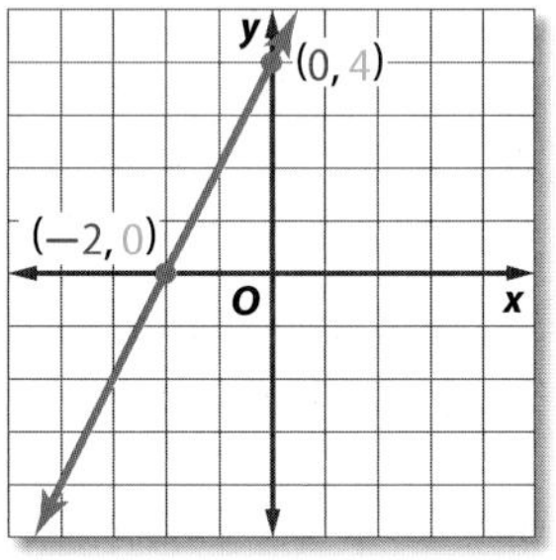

Check Check by solving the equation for y.

$-2x + y = 4$ — Write the equation.

$-2x + \mathbf{2x} + y = 4 + \mathbf{2x}$ — Add $2x$ to each side.

$y = 4 + 2x$ — Simplify.

If $x = -1$, $y = 4 + 2(-1)$ or 2. $(-1, 2)$ is on the line. ✓

Graphing Equations
Although only two points are needed to determine a straight line, always graph at least three to make certain you are correct.

Got It? Do these problems to find out.

Graph each function.

3a. $y = x - 4$

3b. $x + y = 2$

Example 4

Real World

After x hours of driving, there are y gallons of gasoline remaining in a car. For Becca's car, this situation can be represented by $1.8x + y = 18$.

a. State the x- and y-intercepts and then graph the equation.

To find the x-intercept, let $y = 0$.

$1.8x + y = 18$

$1.8x + 0 = 18$

$1.8x = 18$

$x = 10$

To find the y-intercept, let $x = 0$.

$1.8x + y = 18$

$1.8(\mathbf{0}) + y = 18$

$y = 18$

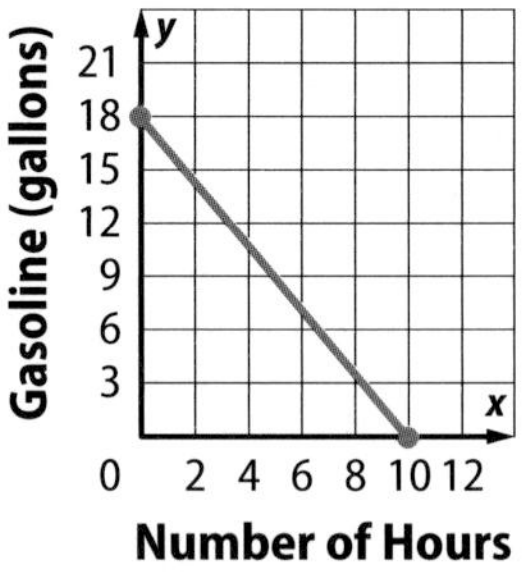

b. Interpret the x- and y-intercepts.

The x-intercept is at the point $(10, 0)$. This means that after 10 hours Becca's gas tank will be empty. The y-intercept is at the point $(0, 18)$. This means that before she begins driving, Becca's gas tank has 18 gallons of gas in it.

Got It? Do these problems to find out.

4. A cleaning service charges \$80 plus \$40 per room cleaned. The total cost y can be represented by $y = 40x + 80$, where x is the number of rooms cleaned.

a. State the x- and y-intercepts and then graph the function.

b. Interpret the x- and y-intercepts.

Guided Practice

Check

Copy and complete each table. Use the results to write four ordered pair solutions of the given function. (Example 1)

1. $y = x + 7$

x	$y = x + 7$	y
−1	■	■
0	■	■
1	■	■
2	■	■

2. $y = 2x - 3$

x	$y = 2x - 3$	y
−2	■	■
0	■	■
2	■	■
4	■	■

Find four solutions of each function. Write the solutions as ordered pairs. (Example 1)

3. $y = x + 5$
4. $y = -4x$
5. $y = 3x + 6$
6. $-x + y = 7$

7. **Financial Literacy** The amount of money Toshiko earns y for working x hours at the library is given by the linear function $y = 10x$. Find two solutions of this function. Explain each solution. (Example 2)

Graph each function by plotting ordered pairs. (Example 3)

8. $y = x - 1$
9. $y = 3x$
10. $y = 2x + 4$
11. $x + y = 6$

12. Yosune walked her dog 5 miles today and plans to walk her dog 2.5 miles every day from now on. The total number of miles walked y can be represented by $y = 2.5x + 5$, where x is the number days walked after today. (Example 4)

a. State the x- and y-intercepts and then graph the equation.

b. Interpret the x- and y-intercepts.

Independent Practice

Go online for Step-by-Step Solutions

eHelp

Copy and complete each table. Use the results to write four ordered pair solutions of the given function. (Example 1)

13. $y = x - 2$

x	$y = x - 2$	y
−1	■	■
0	■	■
1	■	■
2	■	■

14. $y = -2x$

x	$y = -2x$	y
−1	■	■
0	■	■
1	■	■
2	■	■

15. $y = 5x + 1$

x	$y = 5x + 1$	y
−2	■	■
−1	■	■
0	■	■
1	■	■

16. $y = -2x + 8$

x	$y = -2x + 8$	y
−1	■	■
0	■	■
2	■	■
4	■	■

Find four solutions of each function. Write the solutions as ordered pairs. (Example 1)

17 $y = 8x$
18. $y = -6x$
19. $y = x + 7$
20. $y = -x + 3$
21. $y = 2x + 5$
22. $y = -3x - 4$
23. $x + y = -3$
24. $2x + y = 9$

25. The circumference of a circle C with a radius of r units is approximately given by the linear equation $C \approx 6.3r$. Find two solutions of this function. Explain each solution. (Example 2)

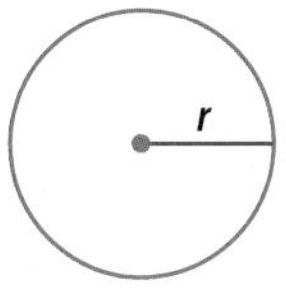

26. **CCSS Reason Abstractly** Regular rides cost 3 tickets and children's rides cost 1 ticket. Find three solutions of $3x + y = 12$ to find the number of regular rides x and children's rides y a family can go on with 12 tickets. Explain each solution. (Example 2)

Graph each function. (Example 3)

27. $y = 5x$
28. $y = -3x$
29. $y = x + 4$
30. $y = -x + 5$
31. $y = 2x + 2$
32. $y = 4x - 1$
33. $x + y = -6$
34. $x - y = 1$

35. A rectangle is x inches long and y inches wide and has a perimeter of 16 inches. The equation $16 = 2x + 2y$ represents this situation. (Example 4)

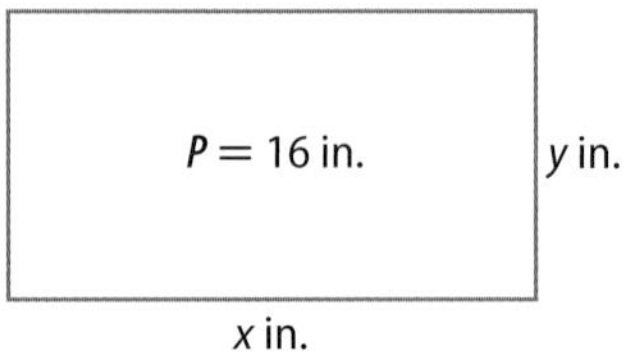

a. State the x- and y-intercepts and then graph the function.
b. Interpret the x- and y-intercepts.

36. **Financial Literacy** Mara has \$440 to pay a painter to paint her bedroom. The painter charges \$55 per hour. The equation $y = 440 - 55x$ represents the amount of money left after x number of hours worked by the painter. (Example 4)

a. State the x- and y-intercepts and then graph the function.
b. Interpret the x- and y-intercepts.

37. **CCSS Multiple Representations** A whale can swim half a mile per minute.

a. **Table** Make a table to find the number of miles a whale can swim in 5, 10, 15, and 20 minutes.
b. **Words** As the input values increase by 5, does the difference between the output values increase by the same amount? If not, what is the difference between the output values?
c. **Graph** Graph the ordered pairs (time, distance). Then draw a line through the points.
d. **Symbols** Write an equation to represent the relationship between time x and number of miles y. Explain your reasoning.

H.O.T. Problems Higher Order Thinking

38. **CCSS Identify Structure** Write a linear equation that has $(3, -2)$ as a solution. Then find another solution of the equation.

39. **CCSS Persevere with Problems** Name a linear equation that is *not* a function.

40. **CCSS Identify Repeated Reasoning** Consider the arithmetic sequence 3, 10, 17, 24, 31, 38, … .

a. Graph the ordered pairs (term number, term). Do the points seem to lie on a line?
b. In which quadrant(s) would this graph make sense? Explain.
c. If you connect the points, you are including points where the x–value is 1.5, 2.7, or 5.9. Can an arithmetic sequence have these values for x? Explain your reasoning.

41. **CCSS Model with Mathematics** Explain how a linear equation represents a function. Then describe four different real-world representations of a linear function that can be used to express the same relationship.

42. **Building on the Essential Question** Explain why a linear function has infinitely many solutions. Then determine which representation shows all the solutions of a linear function: a table, a graph, or an equation.

Standardized Test Practice

43. Which equation represents the table of values for the ordered pairs shown?

x	-1	0	1	2
y	12	9	6	3

A $x - y = -13$
B $3x - y = 9$
C $x + y = 11$
D $3x + y = 9$

44. Which of the following equations represents the line graphed below?

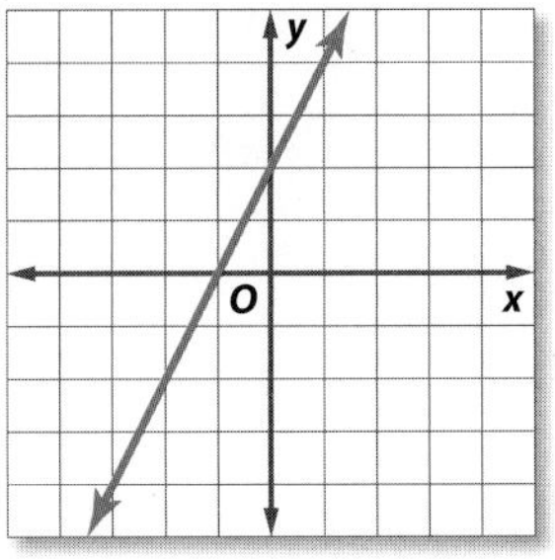

F $2x - y = -2$
G $x - y = -2$
H $x - y = 0$
J $x + 3y = -5$

45. Which of the following sets of ordered pairs represents a linear function?

A $\{(2, 2), (1, -1), (0, 0), (-1, -1)\}$

B $\{(4, -4), (3, -3), (2, -2), (1, -1)\}$

C $\{(-4, -4), (0, 0), (4, -4), (-8, -8)\}$

D $\{(-3, -5), (-1, 0), (0, 1), (3, -5)\}$

46. **Short Response** The equation $y = 64x$ describes the weight y of x cubic feet of water. Make a table of values that describe the relationship. Choose two ordered pairs from your table and describe what they mean.

Common Core Review

47. **Financial Literacy** Last month, Vikas sold 75 calendars. This year he wants to sell 40% more calendars than he sold last year. How many calendars will Vikas have to sell this year to reach his goal? **7.RP.3**

Write an expression in factored form to represent the total area for each figure. **7.EE.1**

48.

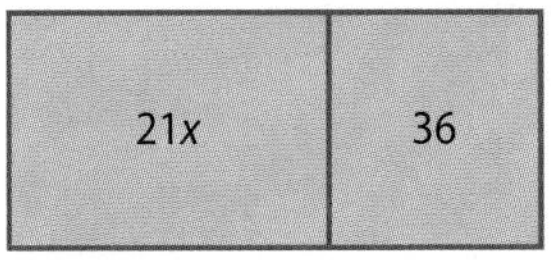

49.

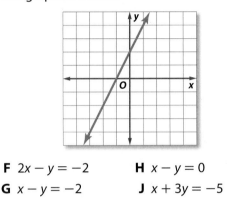

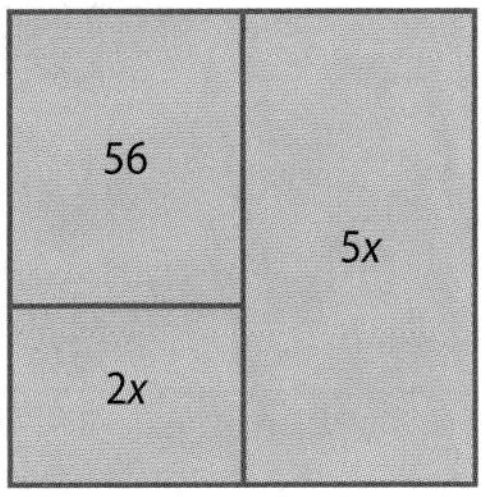

Evaluate each expression. **7.NS.3**

50. $\frac{18 - 10}{8 - 4}$

51. $\frac{16 - 7}{1 - 3}$

52. $\frac{46 - 22}{2008 - 2004}$

53. $\frac{31 - 25}{46 - 21}$

Simplify each expression. **7.EE.1**

54. $3x + 2x - 10$

55. $4y - 12 - 3y + 2$

56. $4z + 1 + 15 + 3z$

57. $2(5x - 2) + x + 11$

58. $5y - 5 - 2(y + 1)$

59. $-2(4 + 6z) - 3z$

Lesson 9-3

Constant Rate of Change and Slope

ISG Interactive Study Guide

See pages 199–200 for:
- Getting Started
- Vocabulary Start-Up
- Notes

Essential Question

How are linear functions used to model proportional relationships?

CCSS Common Core State Standards

Content Standards
7.RP.2, 7.RP.2b, 7.RP.2d, 8.EE.5

Mathematical Practices
1, 3, 4, 7

Vocabulary

rate of change

linear relationship

constant rate of change

slope

What You'll Learn

- Find the constant rate of change for a linear relationship.
- Find the slope of a line.

Real-World Link

Exercise Routines Do you work out? Personal trainers recommend a combination of cardio and strength training for an overall fitness routine. Treadmills, stationary bikes, elliptical trainers, and free weights are among the most popular pieces of exercise equipment.

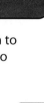

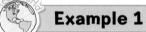

Constant Rate of Change

A **rate of change** is a rate that describes how one quantity changes in relation to another quantity. In a **linear relationship**, the rate of change between any two quantities is the same, or constant. This is called a **constant rate of change**.

Real World Example 1

Tutor

The table shows the distance traveled on a zipline tour. Find the constant rate of change between the quantities.

+2 +2 +2 +2 +2

Time (s)	2	4	6	8	10	12
Distance (ft)	12	24	36	48	60	72

+12 +12 +12 +12 +12

Find the unit rate to determine the constant rate of change.

$$\frac{\text{change in distance}}{\text{change in time}} = \frac{12 \text{ feet}}{2 \text{ seconds}}$$ The distance increases by 12 feet for every 2 seconds.

$$= \frac{6 \text{ feet}}{1 \text{ second}}$$ Write as a unit rate.

The constant rate of change is 6 feet per second.

Got It? Do this problem to find out.

1.

Days	1	2	3	4
Money ($)	100	80	60	40

You can also use a graph to find a constant rate of change.

Example 2

Tutor

A circular design on an Internet advertisement has two circles, one that is decreasing in size and one that is increasing in size. Find the constant rate of change for the radius of circle 1 in the graph shown. Then interpret its meaning.

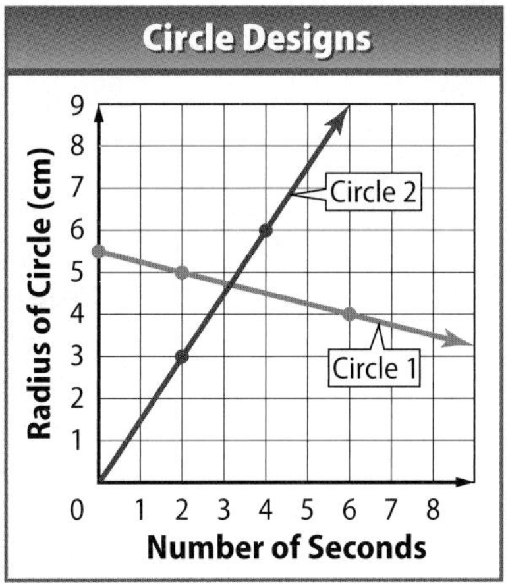

Step 1 Choose any two points on the green line, such as (2, 5) and (6, 4).

(2, 5) ⟶ 2 seconds, radius 5 cm
(6, 4) ⟶ 6 seconds, radius 4 cm

Step 2 Find the rate of change between the points.

$$\text{rate of change} = \frac{\text{change in radius}}{\text{change in time}}$$

$$= \frac{4\text{ cm} - 5\text{ cm}}{6\text{ s} - 2\text{ s}}$$ The radius goes from 5 cm to 4 cm. The time goes from 2 s to 6 s.

$$= -\frac{1\text{ cm}}{4\text{ s}}$$ Simplify.

$$= -0.25\text{ cm/s}$$ Write as a unit rate.

Watch Out! It does not matter which point you use first to find the difference in a rate of change problem. But it is important that you *always* begin with that point in both the numerator and the denominator.

The rate of change −0.25 cm/s means that the radius of the circle is decreasing at a rate of 0.25 centimeter per second.

Got It? Do this problem to find out.

2. Use the graph above. Find the constant rate of change for Circle 2. Then interpret its meaning.

Concept Summary: Rate of Change

Rate of Change	positive	negative	zero
Real-Life Meaning	increase	decrease	no change
Graph	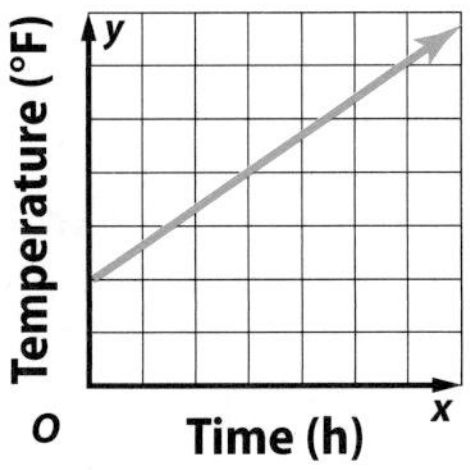	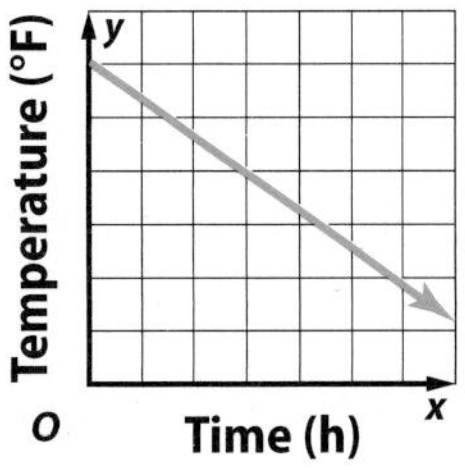	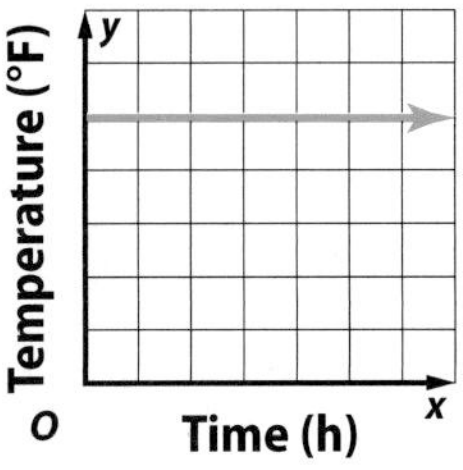

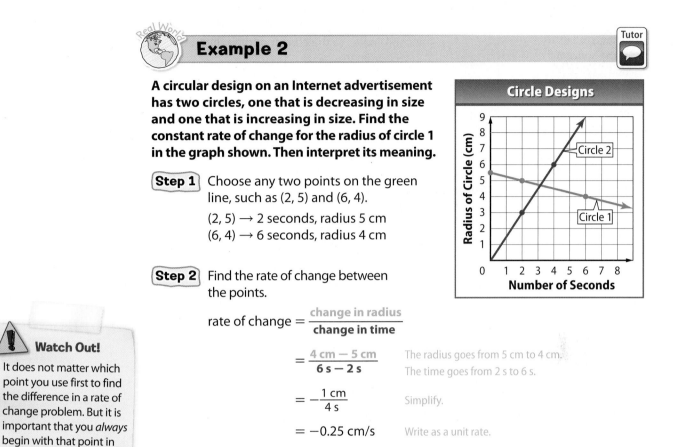

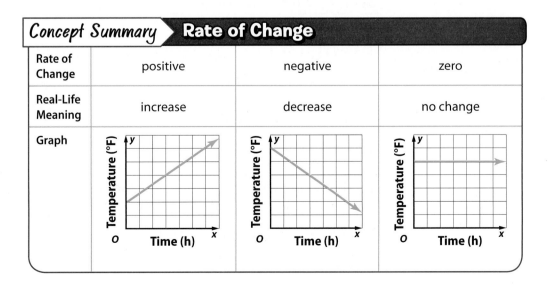

Slope
The vertical change is sometimes called "rise" while the horizontal change is called "run." You can say that slope = $\frac{\text{rise}}{\text{run}}$.

Key Concept Slope

Words The slope m of a line passing through points at (x_1, y_1) and (x_2, y_2) is the ratio of the difference in y-coordinates to the corresponding difference in x-coordinates.

Symbols $m = \frac{y_2 - y_1}{x_2 - x_1}$, where $x_2 \neq x_1$

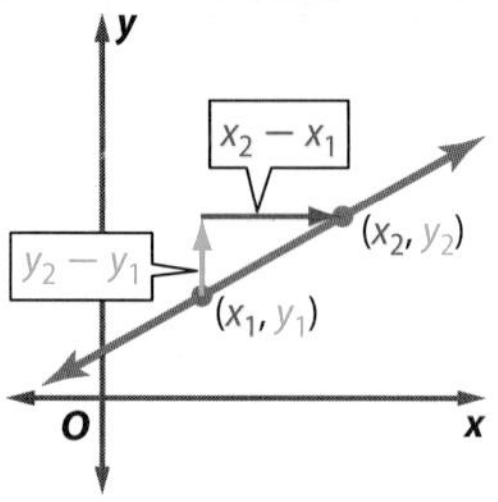

Slope is the ratio of the vertical change to the horizontal change of a line. It describes the steepness of the line. In linear functions, the slope, or rate of change, of the line is always constant. Steeper slopes represent greater rates of change. Less steep slopes represent lesser rates of change.

Example 3

Tutor

Find the slope of each line.

Choosing Points
- Any two points in a line can be chosen as (x_1, y_1) and (x_2, y_2).
- The coordinates of both points must be used in the same order.

a.

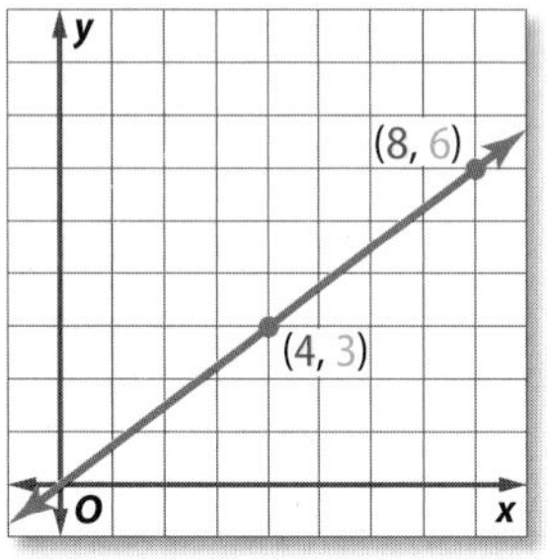

$m = \frac{y_2 - y_1}{x_2 - x_1}$ Definition of slope

$m = \frac{6 - 3}{8 - 4}$ $(x_1, y_1) = (4, 3)$, $(x_2, y_2) = (8, 6)$

$m = \frac{3}{4}$ Simplify.

b.

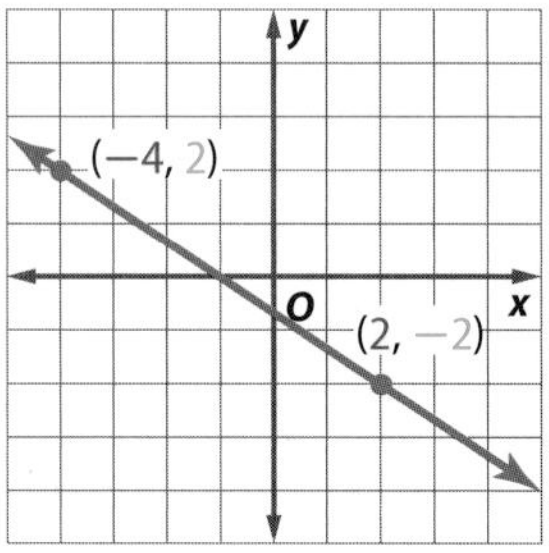

$m = \frac{y_2 - y_1}{x_2 - x_1}$ Definition of slope

$m = \frac{2 - (-2)}{-4 - 2}$ $(x_1, y_1) = (2, -2)$, $(x_2, y_2) = (-4, 2)$

$m = \frac{4}{-6} = -\frac{2}{3}$ Simplify.

Got It? Do these problems to find out.

3a.

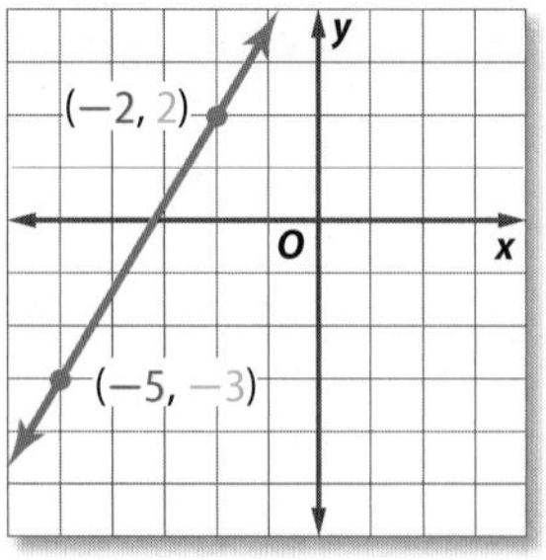

3b.

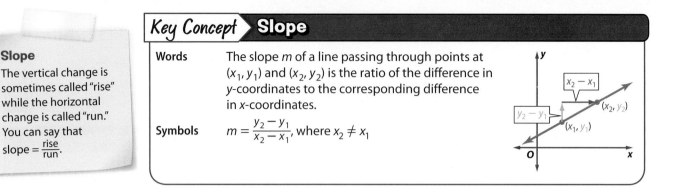

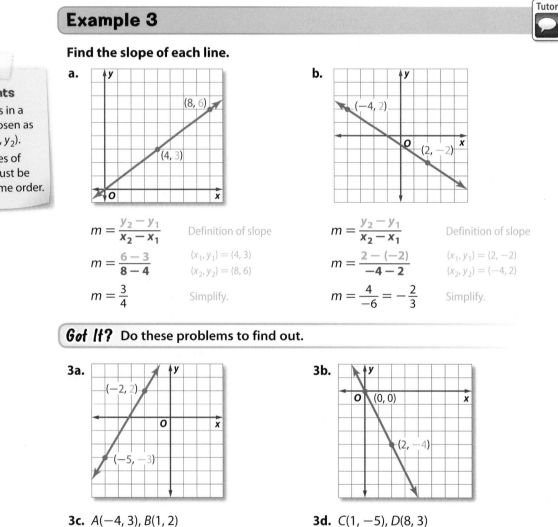

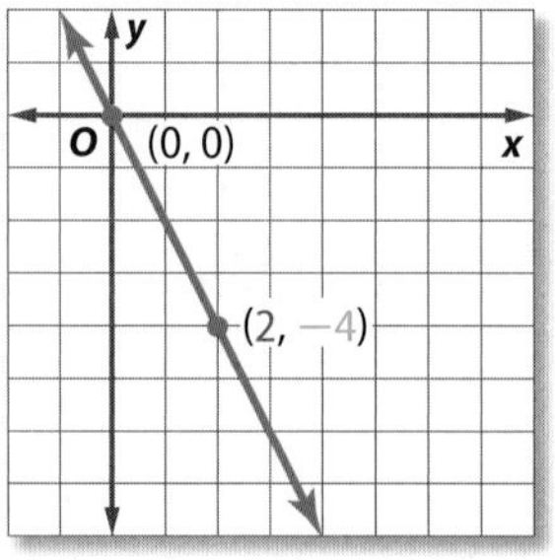

3c. $A(-4, 3)$, $B(1, 2)$

3d. $C(1, -5)$, $D(8, 3)$

Since slope is a rate of change, it can be positive (a line slanting upward), negative (a line slanting downward), or zero (a horizontal line). Vertical lines have undefined slope.

Tutor

Example 4

Find the slope of the line that passes through each pair of points.

a. $A(-3, 4), B(2, 4)$

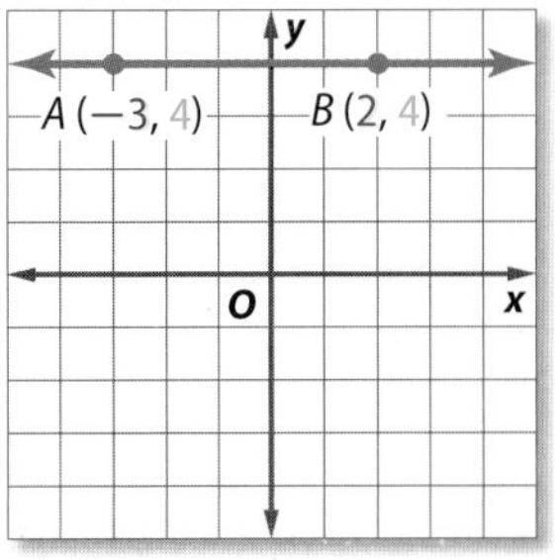

$$m = \frac{y_2 - y_1}{x_2 - x_1}$$ Definition of slope

$$m = \frac{4 - 4}{-3 - 2}$$ $(x_1, y_1) = (2, 4)$, $(x_2, y_2) = (-3, 4)$

$$m = \frac{0}{-5} = 0$$ Simplify.

b. $T(1, 3), U(1, 0)$

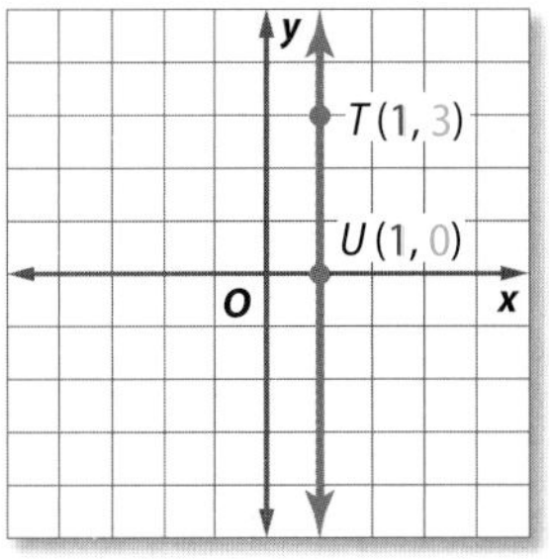

$$m = \frac{y_2 - y_1}{x_2 - x_1}$$ Definition of slope

$$m = \frac{0 - 3}{1 - 1}$$ $(x_1, y_1) = (1, 3)$, $(x_2, y_2) = (1, 0)$

$$m = \frac{-3}{0}$$ The slope is undefined.

Got It? Do these problems to find out.

4a. $E(-1, 7), F(5, 7)$

4b. $G(2, 4), H(2, -1)$

Guided Practice

Check

Find the constant rate of change between the quantities in each table. (Example 1)

1.

Items	5	10	15	20
Cost ($)	12	24	36	48

2.

Time (min)	4	6	8	10
Altitude (ft)	160	120	80	40

Find the constant rate of change for each linear function and interpret its meaning. (Example 2)

3.

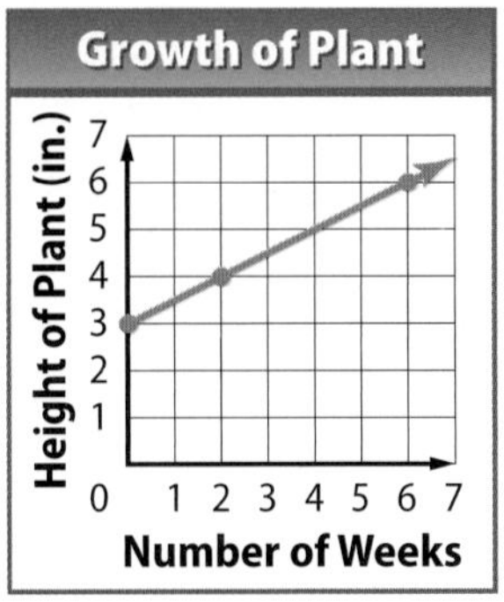

4.

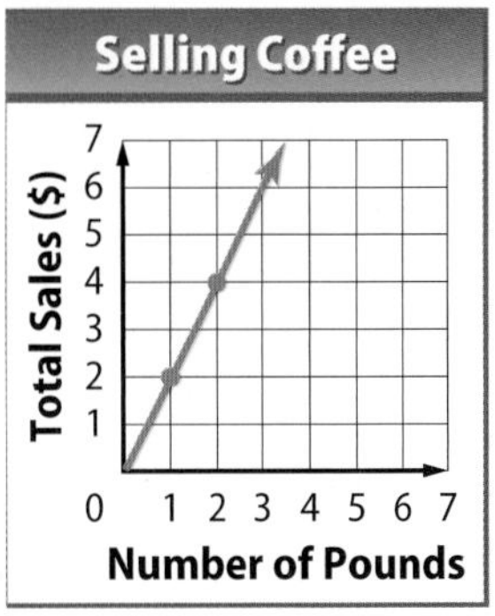

5. Find the slope of the line in the graph at the right. (Example 3)

6. Find the slope of the line that passes through the points $A(4, -5)$ and $B(9, -5)$. (Example 4)

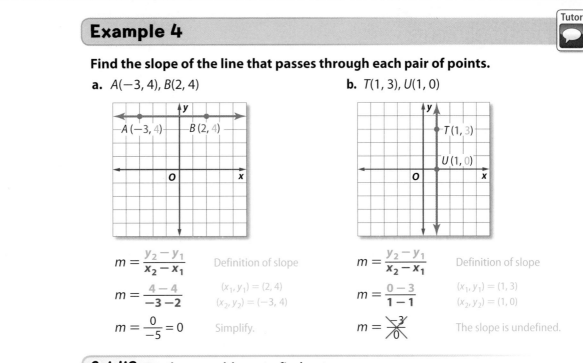

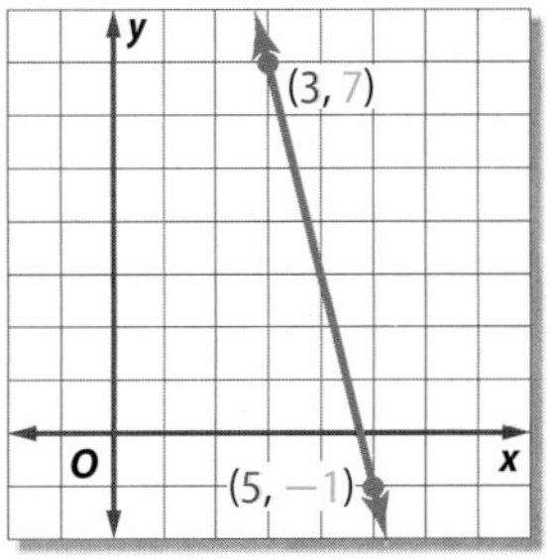

Independent Practice

Go online for Step-by-Step Solutions eHelp

Find the constant rate of change between the quantities in each table. (Example 1)

7.

Photos	4	6	8	10
Cost ($)	1.00	1.50	2.00	2.50

8.

Time (h)	3	5	7	9
Depth (ft)	864	672	480	288

9.

Time (s)	7	9	11	13
Distance (cm)	7	6	5	4

10.

Banners	6	9	12	15
Fabric (yd)	8	13	18	23

Find the constant rate of change for each linear function and interpret its meaning. (Example 2)

11.

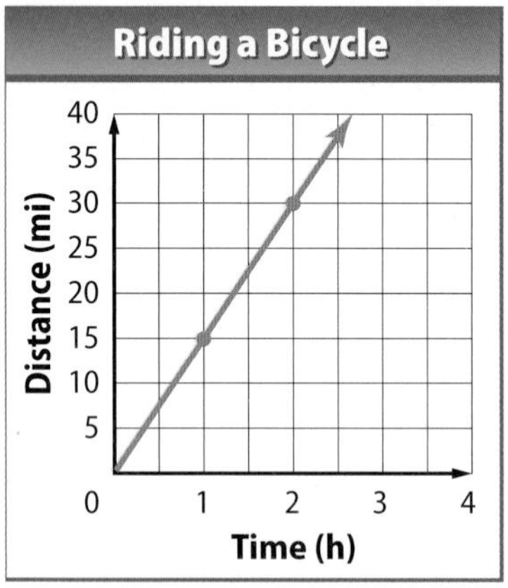

12.

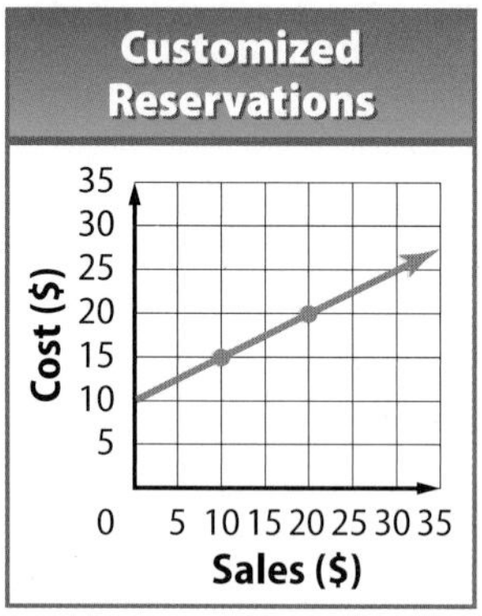

Find the slope of each line. (Example 3)

13.

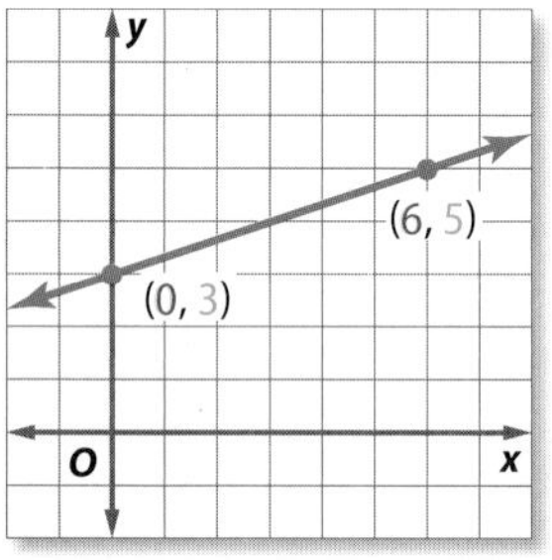

14.

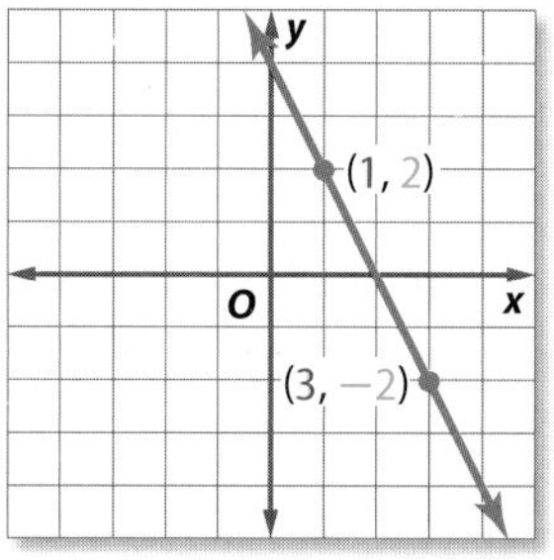

15.

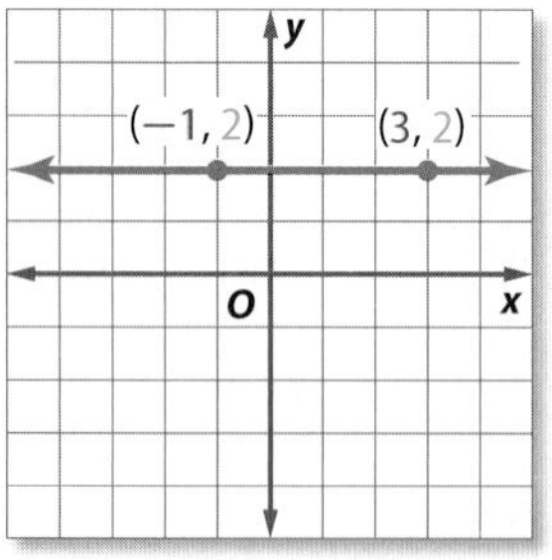

16.

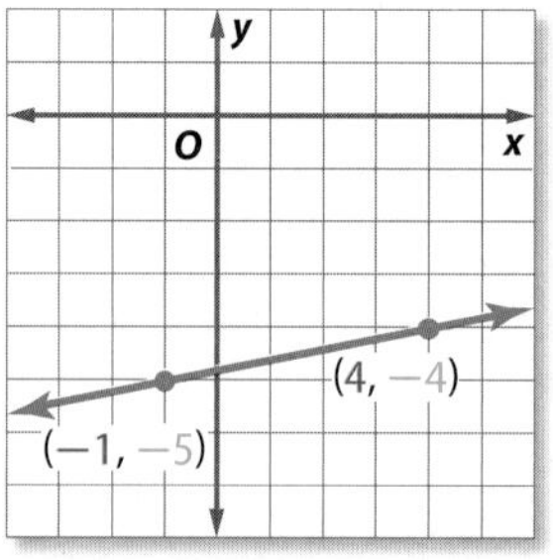

Find the slope of the line that passes through each pair of points. (Example 4)

17. $A(3, 2)$, $B(10, 8)$

18. $R(5, 1)$, $S(0, 4)$

19. $L(5, -6)$, $M(9, 6)$

20. $J(-1, 3)$, $K(-1, 7)$

21. $C(-8, 6)$, $D(1, 6)$

22. $V(5, -7)$, $W(-3, 9)$

23. **Multiple Representations** In this problem, you will investigate ordered pairs. Use the table shown.

x	y
−1	−6
0	−8
1	−10
2	−12

   a. **Numbers** What is the slope of the line represented by the data in the table?

   b. **Graph** Graph the points on a coordinate plane. Connect the points with a line.

   c. **Words** What does the point (0, −8) represent?

24. Libby is driving from St. Louis to Chicago, as shown in the graph.

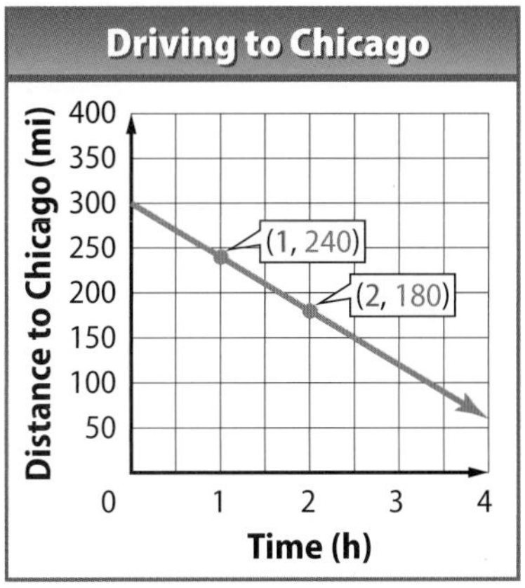

   a. Find the constant rate of change for the linear function and interpret its meaning.

   b. Find the slope of the line.

   c. How does the slope of the line compare to the rate of change you found in part a?

   d. How long does it take Libby to drive from St. Louis to Chicago?

25. The point (2, 3) lies on a line with a slope of $\frac{1}{2}$. Name two additional points that line on the line.

26. The points (0, p) and (1, 4) lie on a line with a slope of 3. What is the value of p? Explain.

H.O.T. Problems Higher Order Thinking

27. **Identify Structure** Graph a line that shows a 3-unit increase in y for every 1-unit increase in x. State the rate of change.

28. **Identify Structure** Name two points on a line that has a slope of $\frac{5}{8}$.

29. **Persevere with Problems** The terms in arithmetic sequence A have a common difference of 3. The terms in arithmetic sequence B have a common difference of 8. In which sequence do the terms form a steeper line when graphed as points on a coordinate plane? Justify your reasoning.

30. **Persevere with Problems** Refer to the graph at the right.

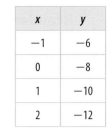

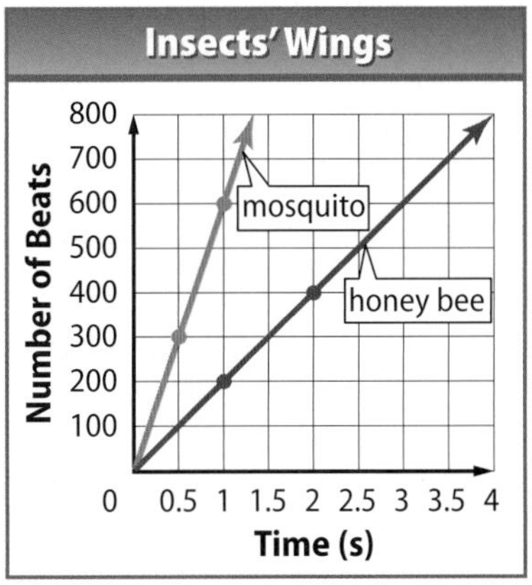

   a. What is the connection between the steepness of the lines and the rates of change?

   b. What is the connection between the unit rates and the slopes?

31. **Model with Mathematics** A person starts walking, then runs, and then sits down to rest. Sketch a graph of the situation to represent the different rates of change. Label the x-axis *Time* and the y-axis *Distance*.

32. **Building on the Essential Question** Determine whether the following statement is *always, sometimes,* or *never* true. Justify your reasoning. *A linear relationship that has a constant rate of change is a proportional relationship.*

Standardized Test Practice

33. Which of the following is true concerning the slope of the line below?

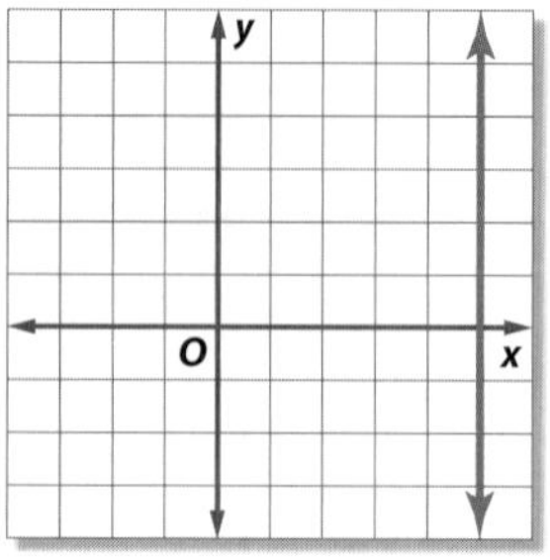

A The slope is −1.
B The slope is zero.
C The slope is 1.
D The slope is undefined.

34. A horizontal line passes through the point (2, 3). What is the slope of the line?

F 0
G $\frac{2}{3}$
H $\frac{3}{2}$
J 3

35. What is the slope of the line that passes through the points (−3, 5) and (6, −1)?

A $\frac{2}{5}$
B $\frac{2}{3}$
C $-\frac{2}{3}$
D $-\frac{3}{2}$

36. **Short Response** Find the constant rate of change for the linear function in the graph.

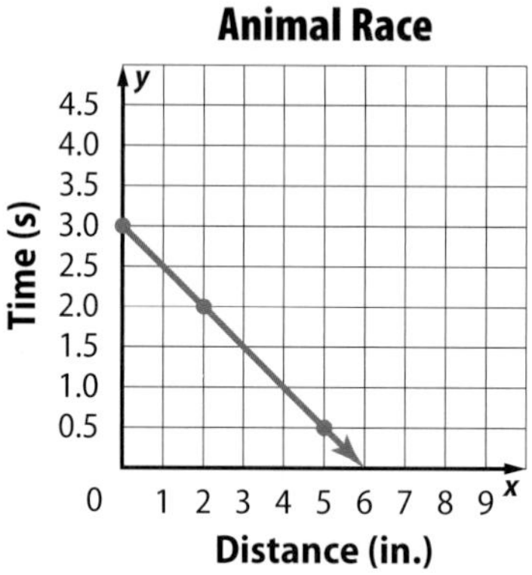

Common Core Review

Find four solutions of each equation. Write the solutions as ordered pairs. 7.EE.4

37. $y = 12x$

38. $y = \frac{1}{2}x$

39. $y = -3x + 5$

Solve each problem using the percent equation. 7.RP.3

40. 9 is what percent of 25?

41. 48 is 64% of what number?

42. 82% of 45 is what number?

Solve each equation. 8.EE.7

43. $18 + 57 + x = 180$

44. $x + 27 + 54 = 180$

45. $85 + x + 24 = 180$

46. $x + x + x = 180$

47. $2x + 3x + 4x = 180$

48. $2x + 3x + 5x = 180$

49. The table shows the percent of forest land in different states. 7.RP.3

a. For each state in the table, how many square miles of land are covered by forests? Round to the nearest square mile.

b. Which state has the greatest amount of forest land?

State	Percent of Land Covered by Forest	Area of State (Square Miles)
Illinois	11.0%	55,584
Kentucky	49.1%	39,728
Michigan	44.7%	56,804
New York	56.1%	47,214
Ohio	27.3%	40,948

 Need more practice? Download Extra Practice at **connectED.mcgraw-hill.com.**

Inquiry Lab

Slope Triangles

Inquiry **HOW are slope triangles on a given line related to the slope of the line?**

CCSS Content Standards 8.EE.6

Mathematical Practices 1, 3, 4

Financial Literacy Miranda started with $1 in her piggy bank. She continued to put money into her bank on a regular schedule. The amount of money y Miranda has in her bank after x days can be represented by the equation $y = x + 1$.

You can use the graph of the line and $\triangle ABC$ formed by the rise, run, and section of the line between points A and B to find her rate of savings. The right triangle ABC is called the *slope triangle* for the line.

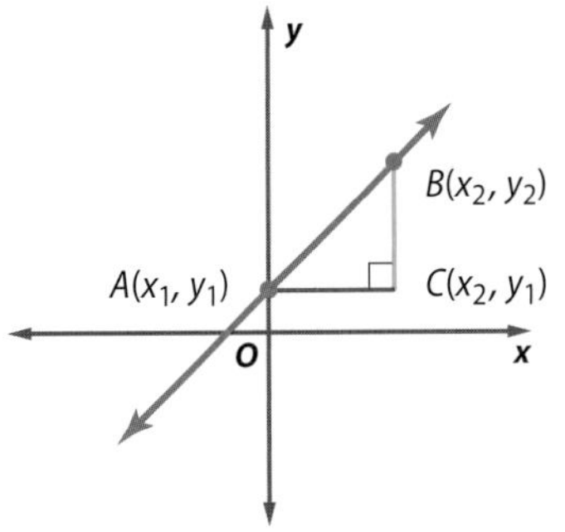

Investigation

Step 1 Graph $y = x + 1$ on grid paper. Make the graph as large as possible.

Step 2 Select two points on the line and create a right triangle.

The rate at which Miranda's savings are increasing in dollars per day is the ratio of the rise to the run of the triangle between points A and B.

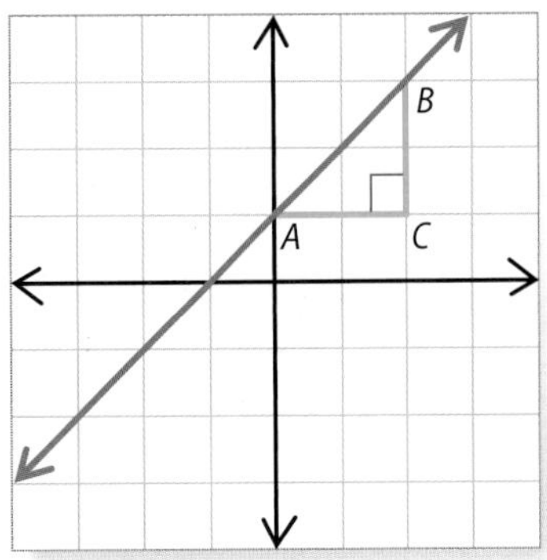

Collaborate

Work with a partner. Use your graph from the Investigation.

1. What is the slope of the line?
2. Select any two points on the line, different from the points you used in the Investigation. Make another right triangle from these two points. Label it $\triangle FGH$. Is the slope of $\triangle FGH$ the same as the slope of $\triangle ABC$? Explain.
3. What do you notice about the two triangles?
4. Select two new points on the line, different from the pairs of points you have used so far. Make another right triangle from these two points. What do you notice about this triangle?
5. Repeat the Investigation and Exercises 1–4 using several other linear equations.

Analyze

6. Write a definition for the term slope triangle.

7. **CCSS Justify Conclusions** What do you think is true of any two slope triangles formed on the same line? Explain.

8. How does the slope change when the slope triangles on the same line increase in size?

9. How does the slope change when the slope triangles on the same line decrease in size?

10. Explain the meaning of the slope for Miranda's situation.

11. Tess has $10 and is saving money regularly each week for a new cell phone. The equation $y = 5x + 10$ represents her savings after x weeks.
    a. Name two points on the graph of this line.
    b. Use a slope triangle to find the slope of the line.
    c. Explain the meaning of the slope.

12. **CCSS Model with Mathematics** Write a real-world problem that can be modeled by the equation $y = \frac{1}{2}x + 3$. Then solve the problem.

13. The graph below shows the distance Quincy traveled while driving on a country road.

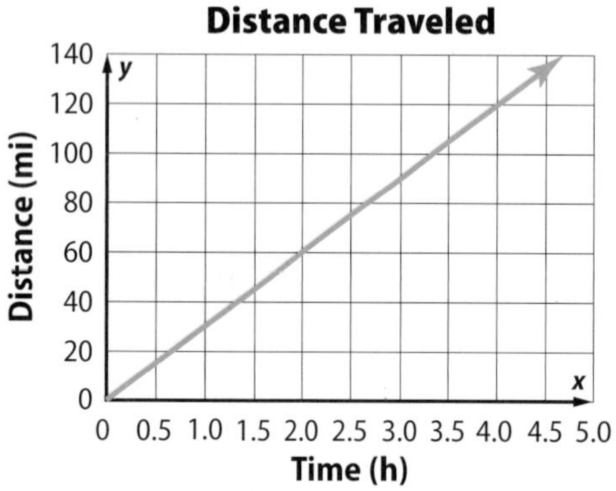

    a. Select two points on the line and create a right triangle. Use the slope triangle to find the slope of the line.
    b. The students in Mr. Jenison's class made slope triangles to find the rate at which Quincy traveled. Students could choose any two points on the line. Describe how the triangles will be alike and different.
    c. What was Quincy's rate in miles per hour on the country road?

14. **CCSS Reason Inductively** Graph each pair of points on the same coordinate plane. Use slope triangles to find the slope of line PQ and line RS. What do you notice?

    $P(1, 4)$ and $Q(3, 12)$ $R(5, 20)$ and $S(8, 32)$

15. **CCSS Persevere with Problems** In $\triangle ABC$, the points at $A(x_1, y_1)$, $B(x_2, y_2)$, and $C(x_2, y_1)$ form the slope triangle for a line. If $x_1 = 0$, $y_1 = 25$, $x_2 = 20$, and $y_2 = 75$, what is the location of point C? What is the slope of every slope triangle on this line?

Reflect

16. **Inquiry** HOW are slope triangles on a given line related to the slope of the line?

Lesson 9-4

Direct Variation

Interactive Study Guide

See pages 201–202 for:
- Getting Started
- Real-World Link
- Notes

Essential Question

How are linear functions used to model proportional relationships?

Common Core State Standards

Content Standards
7.RP.2, 7.RP.2a, 7.RP.2b, 8.EE.5

Mathematical Practices
1, 3, 4

Vocabulary

direct variation
constant of variation

What You'll Learn

- Identify direct variation.
- Use direct variation to solve problems.

Real-World Link

Video Games According to a recent survey, about 56% of U.S. households own a current video game system. Games for these systems can get expensive, so many stores offer sales on pre-owned video games. The prices for pre-owned games at a local store are shown in the table below. A *direct variation* equation can be used to represent this situation.

Number of Games	Total Cost ($)
2	16.50
4	33.00
6	49.50
8	66.00

Key Concept Direct Variation

Words A direct variation is a relationship in which the ratio of y to x is a constant, m. We say y varies directly with x.

Symbols $\frac{y}{x} = m$ or $y = mx$, where $m \neq 0$

Example $y = 2x$

Graph

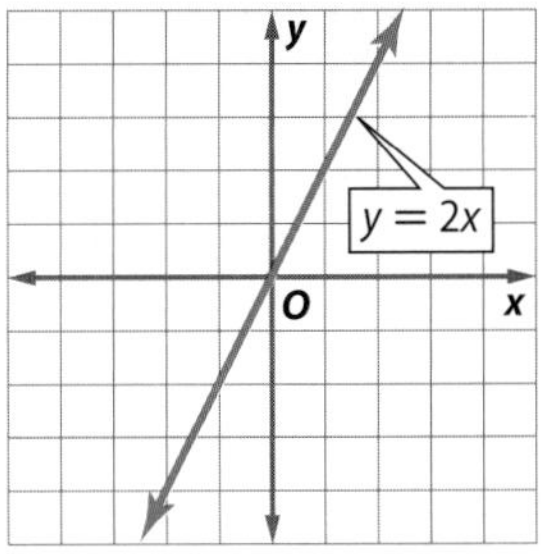

When the ratio of two variable quantities is constant, their relationship is a **direct variation**. Since (0, 0) is a solution of $y = mx$, the graph of a direct variation always passes through the origin and represents a proportional linear relationship.

In the equation $y = mx$, m is called the **constant of variation** or constant of proportionality. It is the slope of the graph of $y = mx$.

Example 1

Tutor

The graph shows the cost of different amounts of trail mix. Determine if the relationship between the cost and the weight of the trail mix is a direct variation.

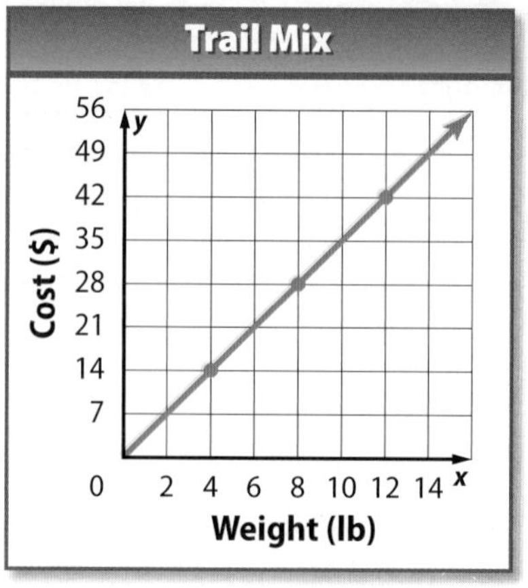

To determine if the relationship is a direct variation, find $\frac{\text{cost } y}{\text{weight } x}$ for points on the graph.

$\frac{\$14}{4 \text{ lb}} = \$3.50/\text{lb}$

$\frac{\$28}{8 \text{ lb}} = \$3.50/\text{lb}$

$\frac{\$42}{12 \text{ lb}} = \$3.50/\text{lb}$

The ratio, \$3.50/lb, is a constant rate. Since the graph passes through the origin and the ratios are constant, the cost of the trail mix varies directly with the weight of the trail mix.

Directly Proportional

Since k is a constant rate of change in a direct variation, we can say the following:

- y varies directly with x.
- y is directly proportional to x.

Got It? **Do this problem to find out.**

1. Tyler charges his customers \$10 per week plus \$5 every time he walks their dogs. Determine whether the relationship between total weekly cost and the number of times the dog is walked is a constant variation.

Example 2

Tutor

The equation $y = 40x$ represents the distance y in miles an ostrich can travel in x hours. Determine whether there is a constant of variation. If so, explain what it represents in this situation.

$y = mx$ Compare the equation to $y = mx$, where m is the constant of variation.

$y = 40x$

$40 = \frac{y}{x}$

The constant of variation is 40. This means that an ostrich can travel 40 miles per hour.

Got It? **Do these problems to find out.**

2a. The equation $P = 6s$ relates the perimeter P of a regular hexagon to the length of a side s. Determine if there is a constant of variation. If so, explain what it represents in this situation.

2b. STEM The equation $K = C + 273$ relates the Kelvin temperature K to Celsius temperature C. Determine if there is a constant of variation. If so, explain what it represents in this situation.

Example 3

Tutor

STEM The time it takes to burn amounts of information on a CD varies directly with the amount of information. Write and solve an equation to find how long it will take to fill a 700-megabyte CD.

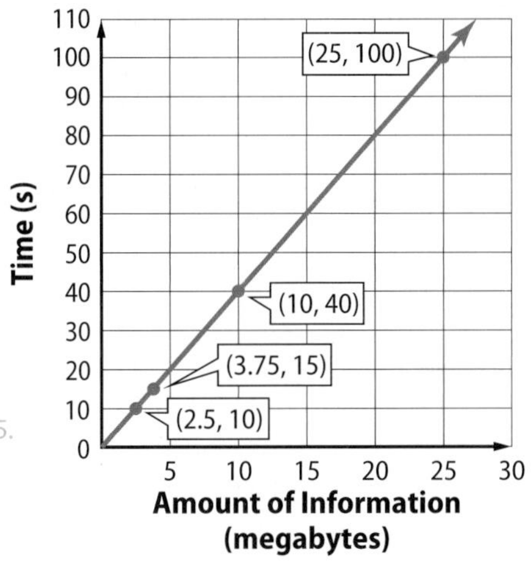

Graphs
The graph of a direct variation is always a line that goes through the origin.

Step 1 Use the equation $y = mx$. Choose any point on the graph. Then solve for m.

$y = mx$	Direct variation equation
$10 = m(2.5)$	Replace y with 10 and x with 2.5.
$4 = m$	Simplify.

Step 2 Use m to write an equation.

$y = mx$	Direct variation equation
$y = 4x$	Replace m with 4.

Step 3 Find how long it will take to fill the CD.

$y = 4x$	Write the direct variation equation.
$y = 4(700)$ or 2800	Replace x with 700 and simplify.

It will take 2800 seconds to fill the CD.

Got It? **Do these problems to find out.**

3. The cost of bulk peanuts varies directly with the weight of the peanuts. At a local grocery store, 2 pounds of peanuts costs $5.80. Write and solve an equation to find how much 5 pounds of peanuts would cost.

Compare Direct Variations

Example 4

Tutor

The equation $y = 1.5x$ represents the relationship between the number of heartbeats y and time in seconds x for a dog. The graph shows the heartbeats for a cat. Which animal has a faster heart rate? Explain.

In the equation $y = 1.5x$, the slope or unit rate is 1.5 beats per second. In the graph, the point (1, 2) represents the unit rate, which is 2 beats per second. Since $2 > 1.5$, the cat has a faster heart rate.

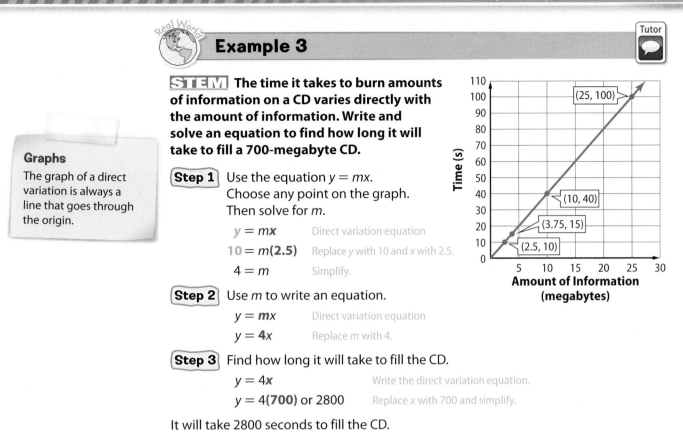

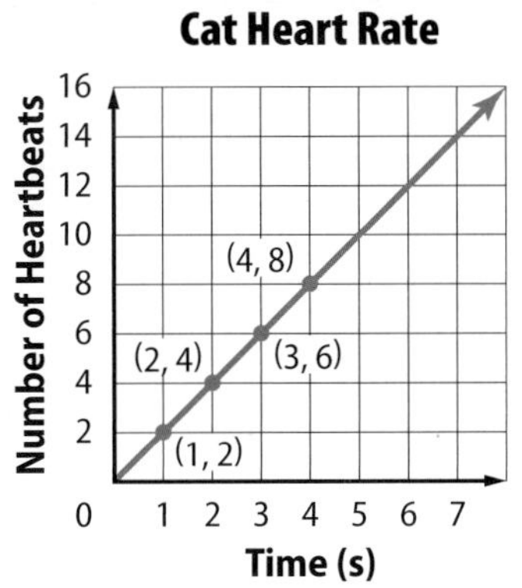

Got It? **Do this problem to find out.**

4. The equation $y = 3x$ represents the heart rate of a rabbit, where x is the time in seconds and y is the number of heartbeats. Does the rabbit or cat have a faster heart rate? Explain.

Guided Practice

Check

1. Recall the graph from Lesson 9–3 of the circular design on an Internet advertisement, shown at the right. The design has two circles, one that is decreasing in size and one that is increasing in size. (Example 1)

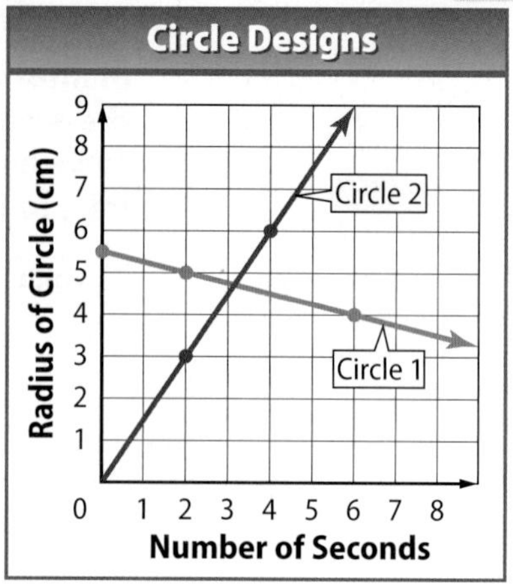

   a. Determine if the relationship between the number of seconds and the radius of circle 1 is a direct variation.

   b. Determine if the relationship between the number of seconds and the radius of circle 2 is a direct variation.

2. **Financial Literacy** The equation $y = 750x$ represents the number of dollars y Olivia earns in x weeks. Determine if there is a constant of variation. If so, explain what it represents. (Example 2)

3. **STEM** The length that a spring stretches varies directly with the amount of weight attached to it. When an 8-ounce weight is attached, a spring stretches 2 inches. (Example 3)

   a. Write an equation relating the weight x and the amount of stretch y.

   b. Predict the stretch of the spring when it has a 20-ounce weight attached.

4. The table below shows the changes in height for a kitesurfer. Assume that the height varies directly with the number of seconds. The height of a second kitesurfer increases 3.5 feet per second. Does the height of the first or second kitesurfer increase faster? Explain your reasoning. (Example 4)

Time (s)	4	6	8	10
Height (ft)	8	12	16	20

Independent Practice

eHelp Go online for Step-by-Step Solutions

Determine if the relationship between the two quantities is a direct variation. (Example 1)

5.

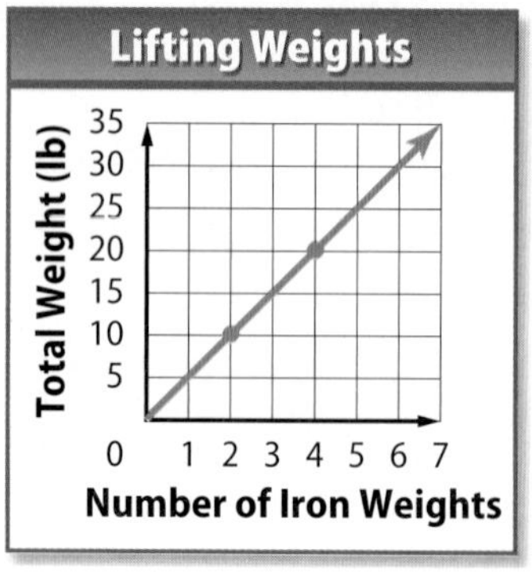

6.

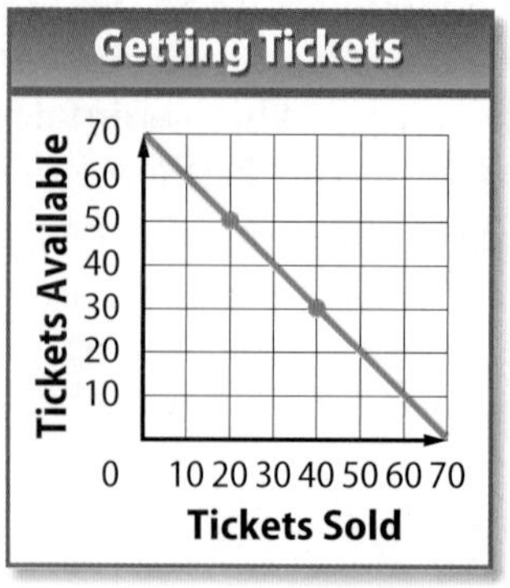

7. The equation $y = 26.2x$ represents the number of miles y Conrad runs in x marathons. Determine if there is a constant of variation. If so, explain what it represents. (Example 2)

8. The equation $y = 3.50x + 5$ represents the number of dollars y Kristin charges for driving you x miles in her taxi. Determine if there is a constant of variation. If so, explain what it represents. (Example 2)

9 **STEM** Water pressure is measured in pounds per square inch (psi), which varies directly with the depth of the water. (Example 3)

Depth (ft)	Pressure (psi)
x	y
33	14.7
66	29.4
99	44.1
132	58.8

a. Write an equation that relates the depth and the water pressure. Round to the nearest tenth.

b. The deepest dive ever recorded by an orca whale is 900 feet. What is the approximate water pressure at this depth?

10. CCSS **Model with Mathematics** The cost of cheese varies directly with the number of pounds bought. Suppose 2 pounds cost \$8.40. Write and solve an equation to find the cost of 3.5 pounds of cheese. (Example 3)

11. Erica's earnings y varies directly with the number of hours she works x. The relationship is shown in the table below. Javier's earning can be represented by the equation $y = 9.2x$. Who earns more money per hour? Explain. (Example 4)

Hours	4	5	6	7
Money Earned (\$)	40.60	50.75	60.90	71.05

12. The graph shows the cost y of x pounds of apples. The cost of x pounds of pears can be represented by the equation $y = 0.95x$. Which fruit has the lower price? Explain your reasoning. (Example 4)

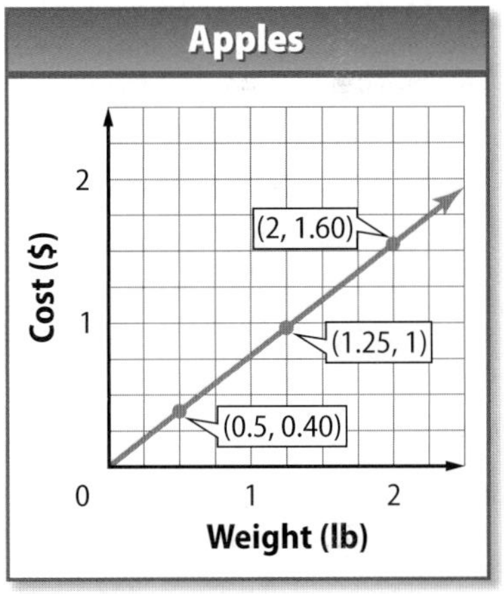

13. Use the graph at the right to determine whether each statement is *true* or *false*. Explain your reasoning.

a. There is a direct variation.

b. There is a linear relationship.

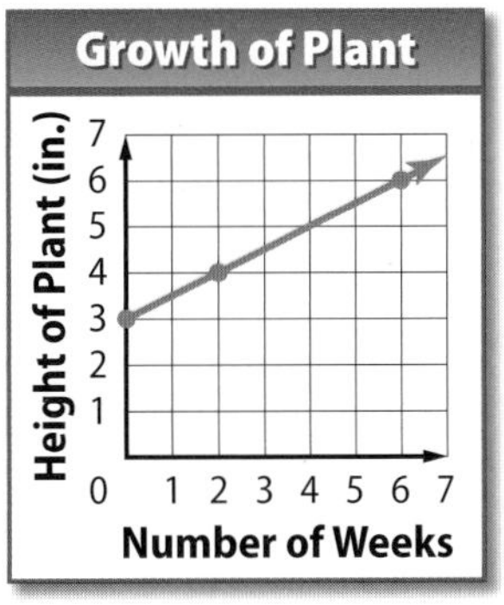

H.O.T. Problems Higher Order Thinking

14. CCSS **Find the Error** Ramiro is determining whether a line through the points with coordinates (2, 12) and (6, 18) represents a direct variation relationship. Find his mistake and correct it.

$$\frac{\text{change in } y}{\text{change in } x} = \frac{18 - 12}{6 - 2}$$
$$= \frac{6}{4} \text{ or } 1.5$$

The relationship is a direct variation.

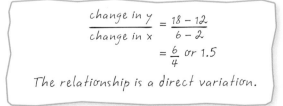

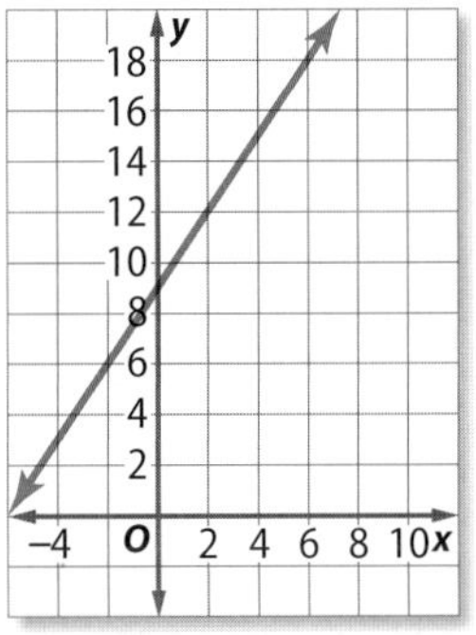

15. **Building on the Essential Question** Describe the steps you take to determine whether an equation has a constant of variation. Give an example of an equation that has a constant of variation and one that does not.

Standardized Test Practice

16. Which is *not* a true statement about the graph shown below?

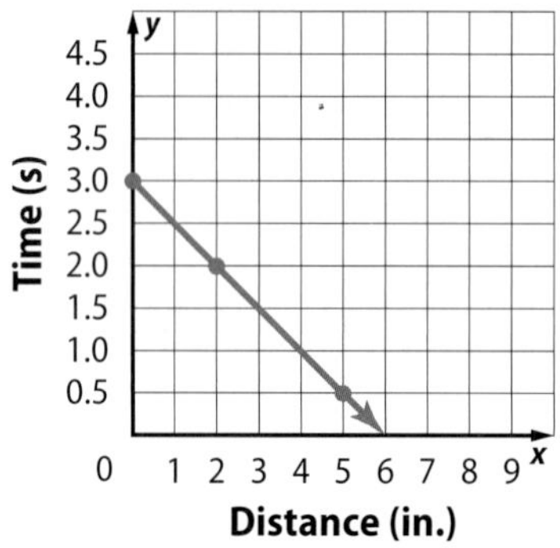

A It is a linear relationship.

B The constant of variation is 1.

C The slope is $-\frac{1}{2}$.

D The x-intercept is (6, 0).

17. The equation $y = 32x$ describes the number of miles per gallon that Melanie's car gets. Predict how many miles Melanie can drive on 12 gallons of gas.

F 44 mi **H** 384 mi

G 352 mi **J** 416 mi

18. The cost of cookies at a bake sale is shown below.

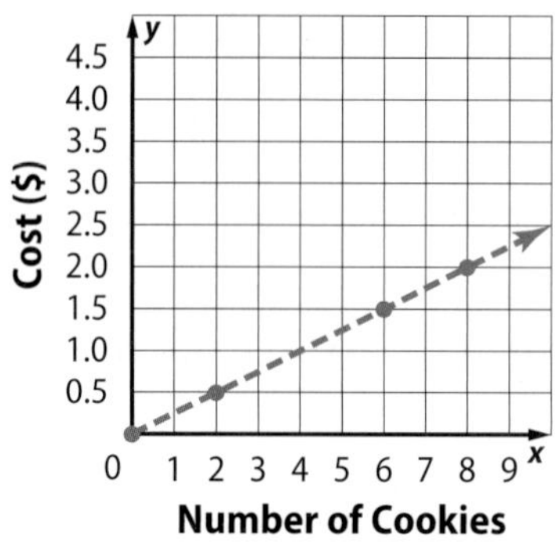

Which of the following is the best prediction for the cost of 21 cookies?

A \$5.00 **C** \$5.50

B \$5.25 **D** \$5.75

19. **Short Response** At an amusement park, the cost of admission varies directly with the number of tickets purchased. One ticket costs \$12.75. Write an equation that could be used to find the cost of any number of admission tickets.

CCSS Common Core Review

20. The costs of admission to a water park are shown in the table at the right. 7.RP.2b, 7.RP.2d, 8.EE.5

a. Find the constant rate of change between the quantities in the table.

b. For this situation, what is the meaning of the constant rate of change?

c. What is the slope of a line that connects the ordered pairs for this relationship?

Water Park Costs	
Number of People	Total Cost ($)
x	y
3	36
4	48
5	60

Find the slope of the line that passes through each pair of points. 8.EE.5

21. $R(-1, 5)$ and $S(0, 5)$

22. $A(-2, 3)$ and $B(-2, 8)$

23. $C(6, 4)$ and $D(4, 6)$

Solve each problem using the percent equation. 7.RP.3

24. 13.64 is what percent of 62?

25. 31.45 is 18.5% of what number?

ISG **Interactive Study Guide**

See page 203 for:
- Mid-Chapter Check

21ST CENTURY CAREER in Music

Mastering Engineer

Do you love listening to music? Are you interested in the technical aspects of music-making? If so, a career creating digital masters might be something to think about! A mastering engineer produces digital masters and is responsible for making songs sound better, having the proper spacing between songs, removing extra noises, and assuring that all the songs have consistent levels of tone and balance. Having a great-sounding master helps increase radio airplay and sales for recording artists.

College & Career READINESS

Explore college and careers at ccr.mcgraw-hill.com

Is This the Career for You?

Are you interested in a career as a mastering engineer? Take some of the following courses in high school.

- Algebra
- Music Appreciation
- Recording Techniques
- Sound Engineering

Find out how math relates to a career in music.

Interactive Study Guide
See page 204 for:
- Problem Solving
- Career Project

Lesson 9-5

Slope-Intercept Form

Interactive Study Guide

See pages 205–206 for:
- Getting Started
- Real-World Link
- Notes

Essential Question

How are linear functions used to model proportional relationships?

Common Core State Standards

Content Standards
7.EE.4

Mathematical Practices
1, 2, 3, 4, 7

Vocabulary

slope-intercept form

What You'll Learn

- Determine slopes and y-intercepts of lines.
- Graph linear equations using the slope and y-intercept.

Real-World Link

Space Camp A weeklong space camp costs \$800. Marissa's parents paid an initial \$400 deposit and then paid the rest in monthly payments of \$100, as shown in the table. You can use a graph to determine whether the total amount paid is proportional to the number of months.

Number of Months	Total Amount Paid (\$)
0	400
1	500
2	600
3	700

Key Concept: Slope-Intercept Form

Words The slope-intercept form of an equation is $y = mx + b$, where m is the slope and b is the y-intercept.

Symbols $y = mx + b$

Example $y = 2x + 1$ (2: slope; 1: y-intercept)

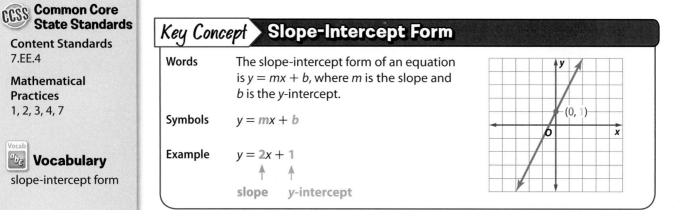

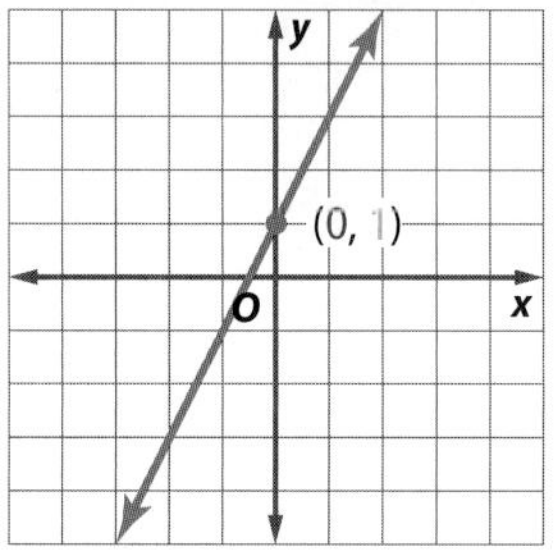

An equation with a y-intercept that is *not* 0 represents a nonproportional relationship. An equation of the form $y = mx + b$, where m is the slope and b is the y-intercept, is in **slope-intercept form**.

Example 1

Tutor

State the slope and the y-intercept of the graph of $y = \frac{1}{4}x - 6$.

$y = \frac{1}{4}x - 6$ Write the equation.

$y = \frac{1}{4}x + (-6)$ Write the equation in the form $y = mx + b$.

The slope is $\frac{1}{4}$ and the y-intercept is -6.

Got It? Do these problems to find out.

State the slope and the y-intercept of the graph of each equation.

1a. $y = 3x + 1$

1b. $y = -2x$

Tutor

Example 2

State the slope and the y-intercept of the graph of $-4x + y = 2$.

First, write the equation in slope-intercept form.

$-4x + y = 2$	Write the original equation.
$\mathbf{+4x \quad = \quad +4x}$	Add $4x$ to each side.
$y = 2 + 4x$	Simplify.
$y = 4x + 2$	Write in slope-intercept form.
$y = mx + b$	$m = 4, b = 2$

The slope of the graph is 4 and the y-intercept is 2.

Got It? **Do these problems to find out.**

State the slope and the y-intercept of the graph of each equation.

2a. $6x + y = -3$

2b. $y - 5 = -x$

2c. $y - 5x = 10$

2d. $x = 2 + y$

Graph Equations

You can use the slope-intercept form of an equation to graph a line.

Tools Tutor

Example 3

Graph $y = -\frac{2}{3}x + 3$ using the slope and y-intercept.

Constant of Proportionality

The direct variation equation $y = kx$ is a special case of a linear equation of the form $y = mx + b$. The constant of proportionality k is the slope and the y-intercept b is 0.

Step 1 Find the slope and y-intercept.

slope $= -\frac{2}{3}$

y-intercept $= 3$

Step 2 Graph the y-intercept point at (0, 3).

Step 3 Write the slope as $\frac{-2}{3}$. Use it to locate a second point on the line.

$m = \frac{-2}{3}$ ← Change in y: down 2 units; ← Change in x: right 3 units

Another point on the line is at (3, 1).

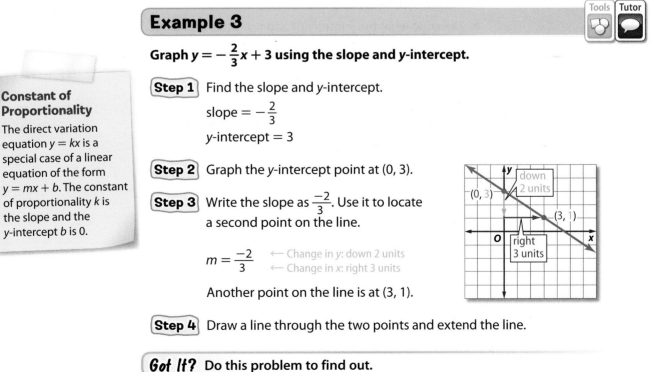

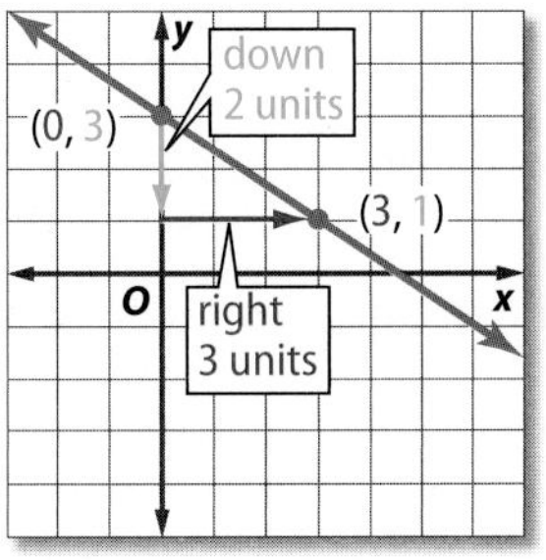

Step 4 Draw a line through the two points and extend the line.

Got It? **Do this problem to find out.**

3. Graph $y = -x - 2$ using the slope and y-intercept.

Example 4

A typical leopard gecko is 3 inches long at birth and grows at a rate of about $\frac{1}{3}$ inch per week for the first few months. The total length of a leopard gecko y after x weeks can be represented by $y = \frac{1}{3}x + 3$.

a. Graph the equation.

Step 1 First, find the slope and the y-intercept.

slope $= \frac{1}{3}$

y-intercept $= 3$

Step 2 Plot the point at (0, 3). Then go up 1 and right 3 and plot another point.

Step 3 Connect these points and extend the line.

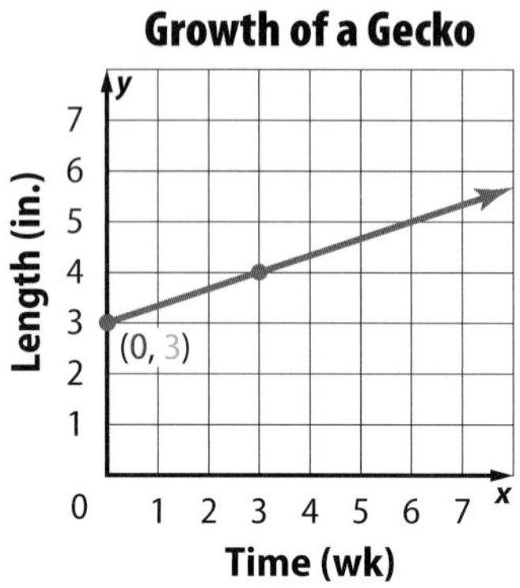

Watch Out!

It only takes two points to determine a line, but it is always wise to plot three points to confirm the line.

b. Describe what the y-intercept and the slope represent.

The y-intercept 3 represents the length of the gecko at birth. The slope $\frac{1}{3}$ represents the growth rate in inches per week, which is the rate of change.

Got It? **Do these problems to find out.**

4. Jack has written 30 pages of his novel. He plans to write 12 pages per week until he has completed his novel. The total number of pages written y can be represented by $y = 12x + 30$, where x is the number of weeks.

a. Graph the equation.

b. Describe what the y-intercept and the slope represent.

Guided Practice

State the slope and the y-intercept of the graph of each equation. (Examples 1 and 2)

1. $y = 2x + 6$

2. $y = \frac{3}{4}x - 1$

3. $7x + y = 0$

4. $4x + y = 3$

Graph each equation using the slope and y-intercept. (Example 3)

5. $y = \frac{1}{3}x + 1$

6. $y = -x + 2$

7. $y = 2x - 4$

8. $y = -0.75x - 3$

9. A kite flying 60 feet in the air is falling. The altitude of the kite can be represented by $y = -x + 60$, where x is the time in seconds. (Example 4)

a. Graph the equation.

b. Describe what the y-intercept and the slope represent.

Independent Practice

Go online for Step-by-Step Solutions eHelp

State the slope and the *y*-intercept of the graph of each equation. (Examples 1 and 2)

10. $y = x + 8$

11. $y = -\frac{5}{2}x - 2$

12. $y = \frac{1}{3}x$

13. $y = -9x$

14. $-x + y = 5$

15. $4x + y = 0$

16. $-9x + y = -5$

17. $y - 6 = \frac{1}{2}x$

Graph each equation using the slope and *y*-intercept. (Example 3)

18. $y = x - 2$

19. $y = 3x + 4$

20. $y = \frac{1}{4}x + 1$

21. $y = \frac{3}{2}x - 3$

22. $y = -2x - 6$

23. $y = -\frac{4}{3}x + 5$

24. **Financial Literacy** To replace a set of brakes, an auto mechanic charges \$40 for parts plus \$50 per hour. The total cost y can be given by $y = 50x + 40$ for x hours. (Example 4)

a. Graph the equation using the slope and *y*-intercept.

b. State the slope and *y*-intercept of the graph of the equation and describe what they represent.

25 The altitude in feet y of an albatross who is slowly landing can be given by $y = 300 - 50x$, where x represents the time in minutes. (Example 4)

a. Graph the equation using the slope and *y*-intercept.

b. State the slope and *y*-intercept of the graph of the equation and describe what they represent.

26. Sam has 15 teaspoons of chopped nuts. She uses $1\frac{1}{2}$ teaspoons for each muffin. The total amount of nuts that she has left y after making x muffins can be given by $y = -\frac{3}{2}x + 15$, as shown in the graph.

a. State the slope and *y*-intercept of the graph of the equation and describe what they represent.

b. Name the *x*-intercept and describe what it represents.

Baking

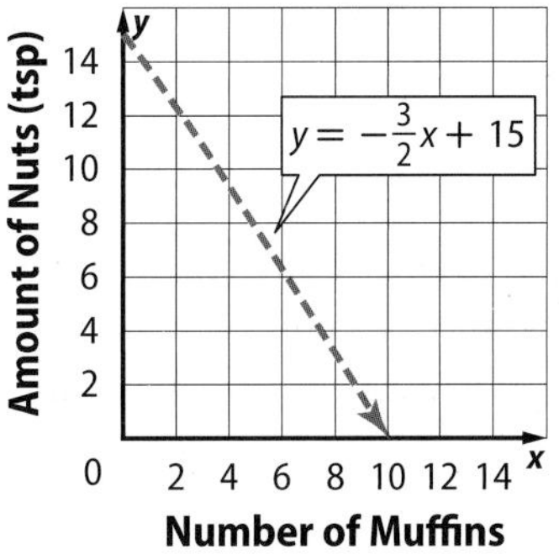

27. **CCSS Reason Abstractly** The table shows the cost of the sitting fee and each 5 × 7 portrait at two photography studios.

Studio	Sitting Fee (\$)	5 × 7 Price (\$)
Lifetime Photos	18	15
Family Photos	12	15

a. Write an equation to represent y the total cost of having your photo taken and x the number of 5 × 7s you get from each studio.

b. Graph each equation on the same coordinate plane.

c. Will the lines ever intersect? Explain.

d. Compare the slopes of each line.

Graph each equation using the slope and y-intercept.

28. $x - 2y = 8$

29. $3x + 4y = 12$

30. $y = 6$

31. $x + 4y = 0$

32. CCSS **Multiple Representations** In this problem, you will investigate graphs of equations. The Math Club is planning a trip to an amusement park. The table shows the bus prices to travel to each park and the admission price per student.

Park	Bus Fee ($)	Admission Price Per Student ($)
Wild Waves	200	16.50
Coaster Haven	250	14.50

a. Symbols Write an equation to represent the total cost *y* for *x* students at each park.

b. Graph Graph the two equations on the same coordinate plane. For how many students is the cost of both trips the same? Explain.

c. Numbers If 20 students decide to take the trip, which trip will cost less? If 28 students decide to take the trip, which trip will cost less?

d. Words Is it possible to determine the answers to part **c** by examining the graph you made in part **b**? Explain your reasoning.

For Exercises 33–35, use the graph shown at the right.

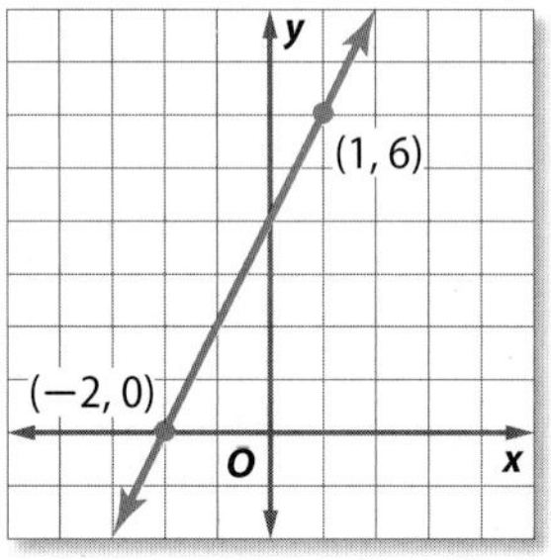

33 What is the slope of the line shown?

34. Identify the *x*-intercept and *y*-intercept of the graph.

35. What is the equation of the line? Write in slope-intercept form.

H.O.T. Problems Higher Order Thinking

36. CCSS **Identify Structure** Describe a line that has a *y*-intercept but no *x*-intercept. Identify the slope of the line.

37. CCSS **Model with Mathematics** Write a real-world problem that can be represented by an equation in the form $y = mx + b$. Solve by graphing.

38. CCSS **Persevere with Problems** Suppose the graph of a line has a negative slope and a positive *y*-intercept. Through which quadrants does the line pass? Justify your reasoning.

39. CCSS **Make a Conjecture** Describe what happens to the graph of $y = 3x + 4$ when the slope is changed to $\frac{1}{3}$.

40. CCSS **Find the Error** Maricruz is finding the slope and *y*-intercept of $x - 2y = 3$. Find her mistake and correct it.

slope = 1
y-intercept = 3

41. **Building on the Essential Question** Describe the steps you take to graph an equation using the slope and *y*-intercept.

Standardized Test Practice

42. A line has a slope of $\frac{4}{5}$ and a *y*-intercept of 10. Which of the following represents the equation of the line?

A $4x - 5y = 10$

B $5x - 4y = 10$

C $4x - 5y = -50$

D $5x - 4y = -50$

43. Which best represents the graph of $y = 3x - 1$?

F

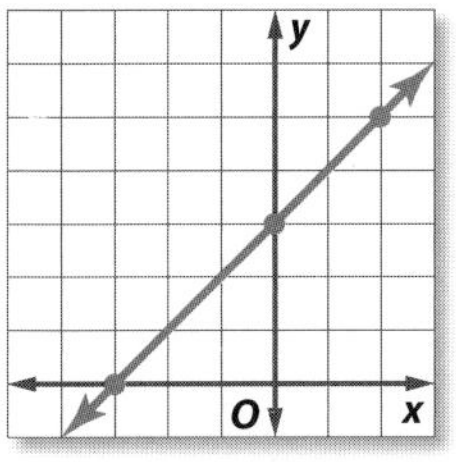

H

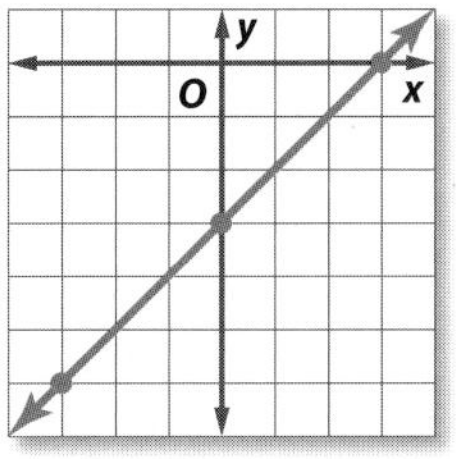

G

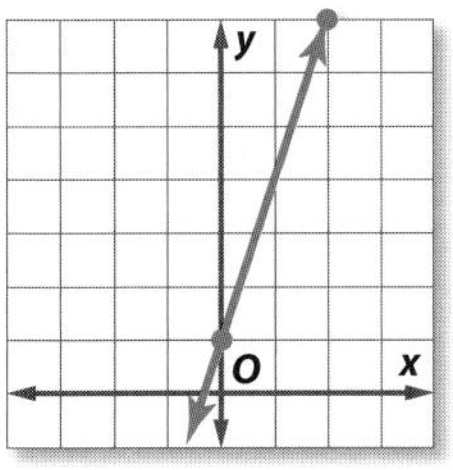

J

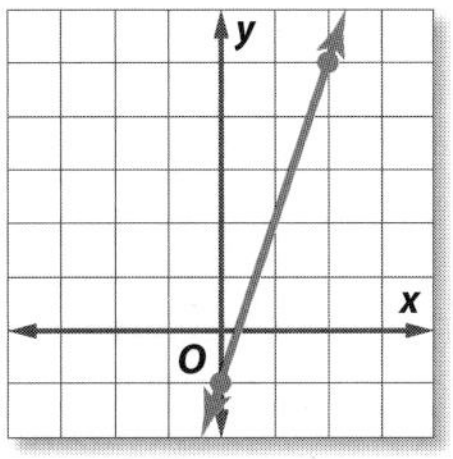

44. The cost of renting a cotton candy machine is \$30 plus \$5 for each hour. The total cost of renting a cotton candy machine is represented by $C(h) = 5h + 30$, where *h* is number of hours. What does the slope represent?

A number of hours

B cost of each hour

C cost per bag of cotton candy

D cost of renting the machine for no hours

45. **Short Response** Snow Mountain Ski Resort offers a special season pass at the beginning of each ski season. The pass costs \$35, and an additional \$25 is charged each time you ski. The total cost can be represented by $y = 35 + 25x$, where *x* is number of times you ski. Explain what the *y*-intercept and the slope represent.

CCSS Common Core Review

46. The cost of gas varies directly with the number of gallons bought. Marty bought 18 gallons of gas for \$49.50. Write an equation that could be used to find the unit cost of a gallon of gas. Then find the unit cost. **7.RP.2c**

47. Find the constant rate of change for the linear function in the table and interpret its meaning **7.RP.2b**

Amount of Birdseed (lb)	*x*	4	8	12
Total Cost (\$)	*y*	11.20	22.40	33.60

Find each sum in simplest form. **7.NS.1**

48. $\frac{1}{2} + \frac{7}{10}$

49. $\frac{1}{6} + \frac{3}{12}$

50. $\frac{7}{8} + \frac{2}{10}$

Simplify each expression. **7.NS.3**

51. $-3(5) - 7$

52. $4 + (-5)(6)$

53. $(-10 - 8) \div (-9)$

54. $(32 - 12) \div 4 \times 7$

55. $(14 - 6) \times 8 \div 2$

56. $(56 - 28) \div (7 \times 2)$

Estimate. Express in the form of a single digit times an integer power of 10. **8.EE.3**

57. 0.00000035 cm

58. 8.5×10^{-3} lb

59. 24,300,099 mi

60. 0.00075 mm

61. 1.375×10^{-43} mg

62. 91,800 m

Inquiry Lab

Families of Linear Graphs

Inquiry **HOW does changing the slope or *y*-intercept affect the graph of a linear function?**

CCSS Mathematical Practices 1, 3

Taxi Rates The taxi service Mario uses charges \$6 plus an additional \$2 per mile x. The equation $y = 2x + 6$ represents the total cost y of a taxi ride.

A **family of functions** is a set of functions that is related in some way. The family of linear functions has the parent function $y = x$.

A TI-Nspire calculator allows you to enter a function and manipulate the graph. This is useful for investigating families of linear functions because you can easily compare characteristics such as slopes and *y*-intercepts.

Investigation

Watch

Graph $y = 2x + 6$ in the default viewing window and move the line to see how the equation relates to the graph.

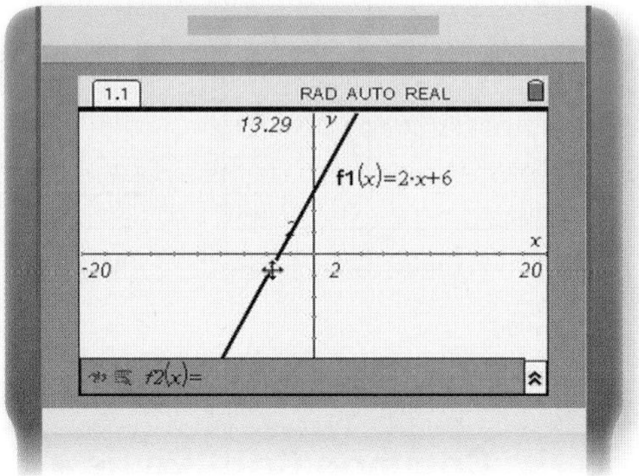

Step 1 Graph $y = 2x + 6$ in the default viewing window.

- Open a new **Graphs & Geometry** window by using the following keystrokes. (home) 2
- Enter the function $2x + 6$ in ***f1(x)*** and press (enter). When a new window is open, the cursor is automatically in the ***f(x)*** box. The graph and its equation will appear on the coordinate plane.
- Press (tab) twice to move the cursor to the graph screen.

Step 2 Grab the line and translate the line up and down.

- Use the NavPad to move the cursor over the line in the third quadrant near the *x*-axis until it blinks with a ✣ sign on it. Press (ctrl)(click) to grab the line. Use the NavPad to translate the line up and down.

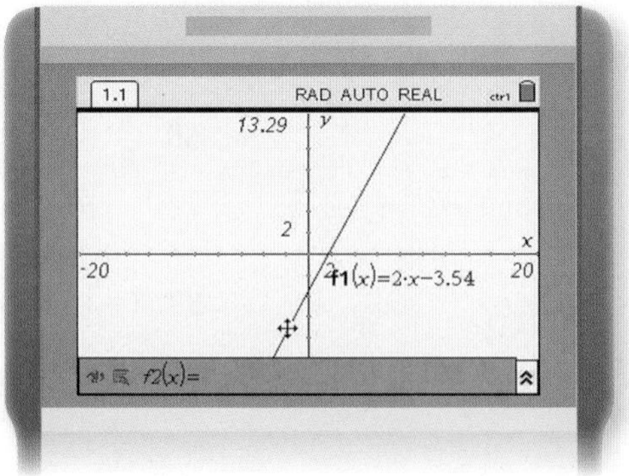

When graphing, be sure to use a viewing window that shows both the *x*- and *y*-intercepts of the graph of a function. If you need to set your own minimum and maximum values for the axes and the scale factor, use the WINDOW option from the menu button.

Collaborate

Work with a partner.

Graph $y = 3x - 4$. Grab the line and move it up and down as instructed in the Investigation.

1. What changes in the equation? What remains the same?
2. How does adding or subtracting a constant *c* to a linear function affect its graph?
3. Move the line until it crosses the *y*-axis as near to 5 as possible. Write the equation of the line.
4. Write an equation of the line that is parallel to $y = 3x - 4$ and passes through the origin.

Open a new Graphs & Geometry window. Graph $y = 3x - 4$. Use the NavPad to move over the line until a blinking ↻ appears on it. Grab the line as instructed in the Investigation and rotate the line.

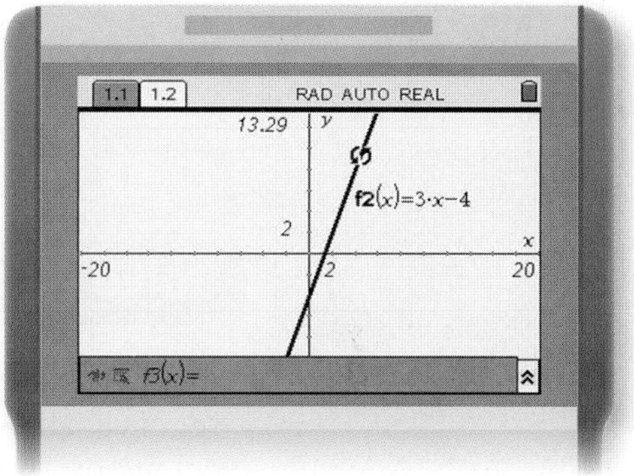

5. What changes in the equation? What remains the same?
6. How does changing the coefficient of *x* affect the graph of a linear function?
7. CCSS **Reason Inductively** Without graphing, determine whether the graph of $y = 0.5x$ or the graph of $y = 1.5x$ has a steeper slope. Explain.
8. Rotate the graph until the *x* coefficient is negative. How does changing the sign of the coefficient of *x* affect the graph of a linear function?

Analyze

Open a new Graphs & Geometry window. Graph the following lines.

$f1(x) = -2x$ $\quad$ $f2(x) = -2x + 1$ $\quad$ $f3(x) = \frac{1}{2}x + 1$

9. Describe the similarities and differences between the graphs.
10. A garden center charges \$75 per cubic yard for topsoil. The delivery fee is \$25.
    a. Write the equation that represents the charges.
    b. Describe the change in the graph if the delivery fee is changed to \$35.
    c. How does the graph change if the price of a cubic yard of topsoil is increased to \$80 when the delivery fee is \$35?
    d. What are the prices of a cubic yard of topsoil and delivery if the graph has slope 70 and *y*-intercept 40?
11. CCSS **Persevere with Problems** If the graph of $y = -3x + 5$ is parallel to the graph of $y = ax + 9$, find the value of *a*.

Reflect

12. Inquiry HOW does changing the slope or *y*-intercept affect the graph of a linear function?

Lesson 9-6

Solve Systems of Equations by Graphing

ISG Interactive Study Guide

See pages 207–208 for:
- Getting Started
- Real-World Link
- Notes

Essential Question

How are linear functions used to model proportional relationships?

CCSS Common Core State Standards

Content Standards
8.EE.8, 8.EE.8a, 8.EE.8b, 8.EE.8c

Mathematical Practices
1, 2, 3, 4, 7

Vocab Vocabulary

system of equations

What You'll Learn

- Solve systems of linear equations by graphing.
- Determine the number of solutions of a system of linear equations.

Real-World Link

Lizards It costs about \$8 per month to buy food for a pet lizard. An online store charges an initial fee of \$10 and then \$6 per month for lizard food. You can write and solve a set of equations to find when these two plans cost the same amount.

Solve Systems of Equations by Graphing

A **system of equations** is a collection of two or more equations with the same set of variables. The equations $y = 5x$ and $y = 30 + 2x$ together are a system of equations. The solution of this system is (10, 50) because the ordered pair is a solution of both equations.

$y = 5x$	Write the equations.	$y = 30 + 2x$
$50 \stackrel{?}{=} 5(10)$	Replace (x, y) with (10, 50).	$50 \stackrel{?}{=} 30 + 2(10)$
$50 = 50$ ✓	Simplify.	$50 \stackrel{?}{=} 30 + 20$
		$50 = 50$ ✓

One way to solve a system of equations is to graph the equations on the same coordinate plane. The coordinates of the point where the graphs intersect is the solution of the system of equations.

Example 1

Tools Tutor

Solve the system of equations by graphing.
$y = x$
$y = -3x + 4$

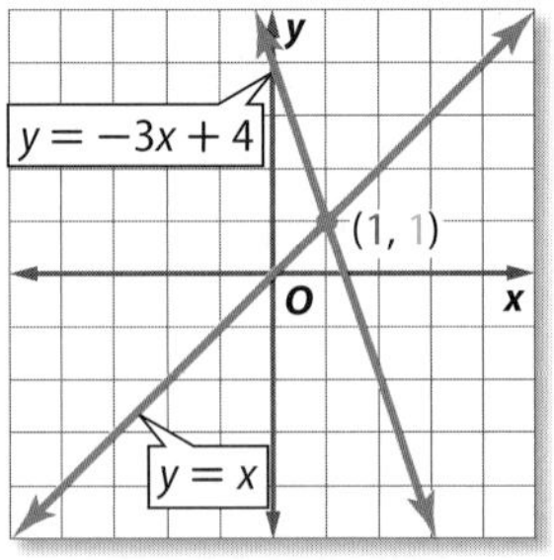

The graphs appear to intersect at (1, 1). Check this estimate by replacing x with 1 and y with 1.

Check

$y = x$	$y = -3x + 4$
$1 \stackrel{?}{=} 1$	$1 \stackrel{?}{=} -3(1) + 4$
$1 = 1$ ✓	$1 = 1$ ✓

The solution of the system of equations is (1, 1).

Got It? **Do this problem to find out.**

1. Solve the system of equations by graphing.
$y = x + 2$
$y = -2x - 4$

Example 2

Tools Tutor

Answer All Parts
When answering questions with more than one part, make sure your solutions answer all parts of the question.

Marjorie and Bryan are selling magazine subscriptions. Marjorie sells 3 times as many subscriptions as Bryan. Bryan sells 12 fewer subscriptions than Marjorie.

a. Write a system of equations to represent this situation.

Let y represent Marjorie's sales and x represent Bryan's sales.

$y = 3x$ Marjorie sells 3 times as many subscriptions as Bryan.

$y = x + 12$ Bryan sells 12 fewer subscriptions than Marjorie.

b. Solve the system by graphing. Explain what the solution means.

Graph each equation on the same coordinate plane. The equations intersect at (6, 18).

So, the solution to the system is $x = 6$ and $y = 18$. This means that Marjorie sells 18 subscriptions and Bryan sells 6 subscriptions.

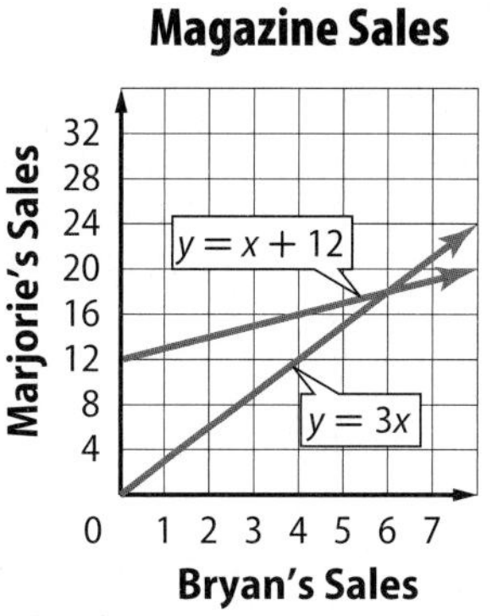

Got It? **Do these problems to find out.**

2. A doctor's office has twice as many new patients as existing patients. The number of new patients is 22 more than the number of existing patients.

a. Write a system of equations to represent this situation.

b. Solve the system by graphing. Explain what the solution means.

Example 3

Tools Tutor

Solve the system of equations by graphing.

$\mathbf{y = x - 2}$

$\mathbf{x + 3y = 6}$

Step 1 Write the equations in the form $y = mx + b$.

$y = x - 2 \rightarrow y = x - 2$

$x + 3y = 6 \rightarrow y = -\frac{1}{3}x + 2$

Step 2 Graph each equation on the same coordinate plane.

Step 3 Find the intersection of the graphs. (3, 1)

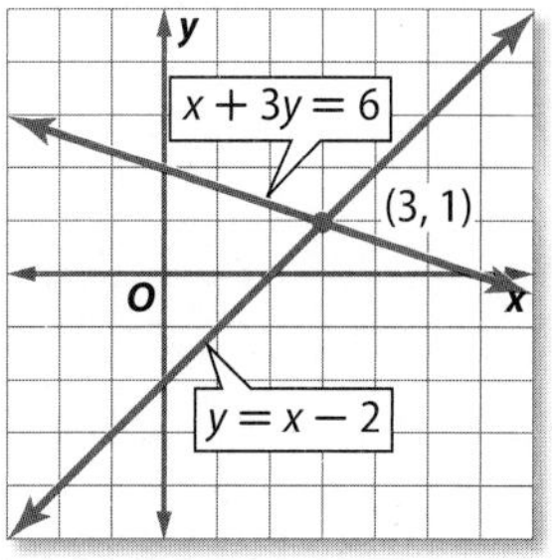

Check Use the original equations.

$y = x - 2$	$x + 3y = 6$
$1 \stackrel{?}{=} 3 - 2$	$3 + 3(1) \stackrel{?}{=} 6$
$1 = 1$ ✓	$6 = 6$ ✓

The solution of the system of equations is (3, 1).

Got It? **Do this problem to find out.**

3. $y = 3x - 3$

$x + 2y = 8$

Number of Solutions

A system of equations can have one solution, no solution, or infinitely many solutions.

Example 4

Tools Tutor

Slopes/Intercepts

When the graphs of a system of equations have

- different slopes → exactly 1 solution
- same slope, different y-intercepts → no solution
- same slope, same y-intercept → infinitely many solutions

Solve each system of equations by graphing.

a. $y = -x + 1$
$y = -x - 3$

The graphs appear to be parallel lines. Since there is no coordinate pair that is a solution to both equations, there is no solution for this system of equations.

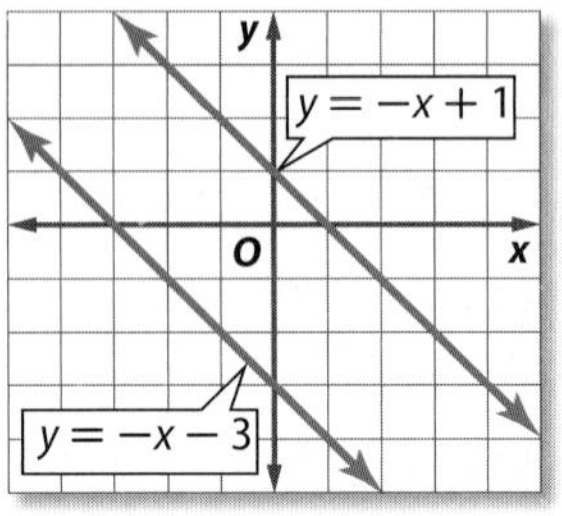

b. $y = 2x + 4$
$\frac{1}{2}y - x = 2$

Both equations have the same graph. Any ordered pair on the graph will satisfy both equations. Therefore, there are infinitely many solutions for this system of equations.

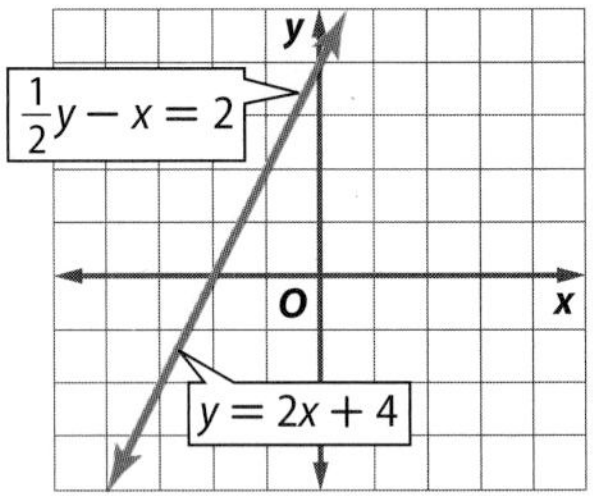

Got It? **Do these problems to find out.**

4a. $y = x + 4$
$y = x$

4b. $y = \frac{1}{2}x - 1$
$x - 2y = 2$

Concept Summary Systems of Equations

Graph	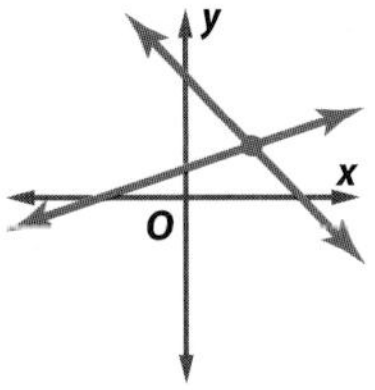 Intersecting Lines	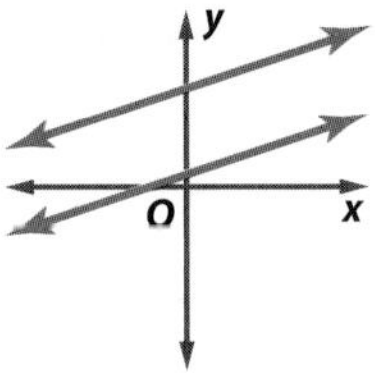 Parallel Lines	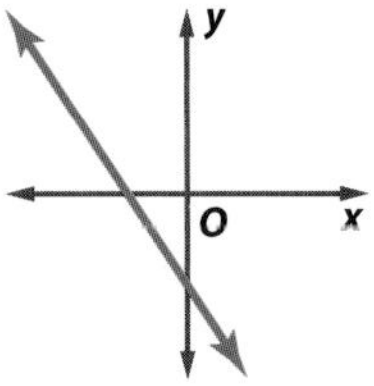 Same Line
Number of Solutions	one solution	no solutions	infinitely many

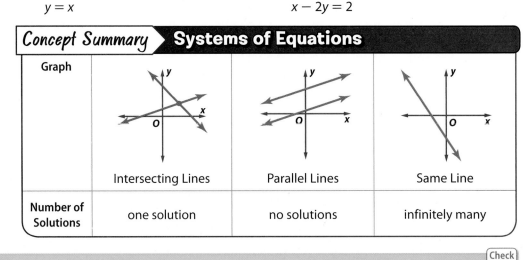

Guided Practice

Check

Solve each system of equations by graphing. (Examples 1, 3, and 4)

1. $y = -x$
$y = x - 4$

2. $y = x + 1$
$x + y = 7$

3. $y = \frac{3}{2}x - 1$
$3x - 2y = 2$

4. **Financial Literacy** Amanda pays an one-time fee of \$100 and a monthly fee of \$10 to belong to a gym. Maria pays only a monthly fee of \$20 to belong to her gym. (Example 2)

a. Write a system of equations to represent this situation.

b. Solve the system by graphing. Explain what the solution means.

Independent Practice

Go online for Step-by-Step Solutions eHelp

Solve each system of equations by graphing. (Examples 1, 3, and 4)

5. $y = 2x$
$y = x + 1$

6. $y = x$
$y = -x + 4$

7 $y = -x + 1$
$y = x - 5$

8. $y = \frac{3}{4}x$
$3x - 4y = 0$

9. $y = \frac{1}{2}x + 1$
$y = \frac{1}{2}x - 2$

10. $x + y = -3$
$x - y = -3$

11. CCSS **Reason Abstractly** Ling starts out with 50 baseball cards and plans to collect 5 per month. Jonathan starts out with 90 baseball cards and plans to sell 5 per month. (Example 2)

a. Write a system of equations to represent this situation.

b. Solve the system of equations by graphing. Explain what the solution means.

12. The sum of two numbers is 5, and the difference of the numbers is 3. (Example 2)

a. Write a system of equations to represent this situation.

b. Solve the system of equations by graphing. Explain what the solution means.

Solve each system of equations by graphing.

13. $y = x + 2$
$x = 1$

14. $y = x + 4$
$y = 0$

15. $y = 2x - 3$
$y = 5$

16. $y = -x - 4$
$x = 2$

17. $x + y = 2$
$x = -3$

18. $x - y = 6$
$y = -1$

19 **Financial Literacy** The cost of 2 bagels and 1 can of orange juice is $8. The cost of 3 bagels and 2 cans of orange juice is $13.

a. Write a system of equations to represent this situation.

b. Solve the system of equations. Explain what the solution means.

H.O.T. Problems Higher Order Thinking

20. CCSS **Identify Structure** Write a system of equations that has the solution (1, 7). Write the equations in slope-intercept form.

21. CCSS **Identify Structure** Describe the graph of a system of equations if the system has one solution, no solution, or infinitely many solutions.

22. CCSS **Reason Abstractly** Complete the following analogies.

a. Intersecting lines are related to one solution as __?__ lines are related to no solutions.

b. Lines with the same slope and different *y*-intercepts are related to no solution as lines with the same slope and same *y*-intercept are related to __?__.

23. **Building on the Essential Question** Describe the three ways that two lines can be related. Can a system of linear equations have exactly two solutions? Explain.

Standardized Test Practice

24. Amy took three times as many pictures as Jennifer. Jennifer has 16 fewer pictures than Amy. Which system of equations can be used to find the number of pictures each person took?

A $a = 3j$
 $a = j + 16$

B $a = 3j$
 $a = j - 16$

C $j = 3a$
 $j = a + 16$

D $j = 3a$
 $j = a - 16$

25. Refer to Exercise 24. How many pictures did each person take?

F Amy took 8 pictures and Jennifer took 24 pictures.

G Amy took 24 pictures and Jennifer took 8 pictures.

H Amy took 16 pictures and Jennifer took 6 pictures.

J Amy took 6 pictures and Jennifer took 16 pictures.

26. Which is the solution of the system of equations graphed below?

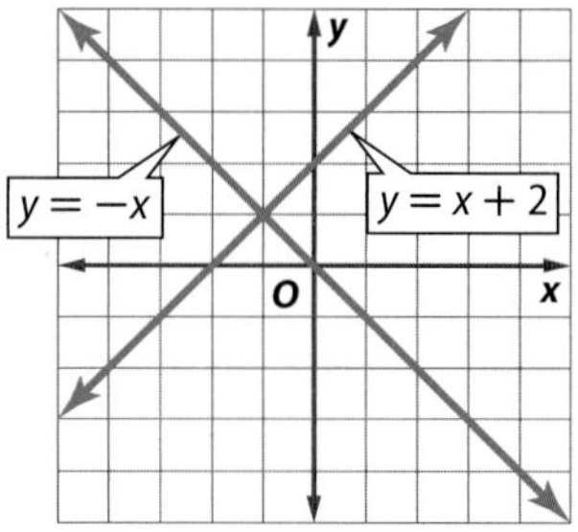

A (1, 1)

B (1, −1)

C (−1, 1)

D (−1, −1)

27. **Short Response** In one season, Tyra made 4 times as many goals as Kailey. Kailey made 9 fewer goals than Tyra. How many goals did Tyra make?

Common Core Review

Evaluate each expression for $x = 7, y = 3$, and $z = 9$. 6.EE.2c

28. $\frac{xy}{3} + 2$

29. $2x + 3z + 5y$

30. $5z - 3x - 2y$

Find the simple interest to the nearest cent. 7.RP.3

31. \$480 at 9% for 3 months

32. \$2400 at 7% for $1\frac{1}{2}$ years

33. \$350 at $3\frac{1}{2}$% for 3 years

34. \$260 at $2\frac{1}{2}$% for 2 years

35. A raindrop falls from the sky at about 17 miles per hour. How many feet per second is this? Round to the nearest foot per second. 7.RP.1

36. Use the table at the right to determine whether the cleaning fee is proportional to the time. Explain your reasoning. 7.RP.3

Time (hours)	1	2	3	4
Cleaning Fee	\$30	\$60	\$90	\$115

37. How long will it take an Air Force fighter jet to fly 5200 miles at 650 miles per hour? 7.EE.4a

Determine whether each relation is a function. Explain. 7.EE.4

38.

x	−1	0	1	2
y	−7	−3	1	5

39.

x	−3	−1	3	3
y	7	5	3	1

Multiply. 8.EE.1

40. $4^2 \cdot 4$

41. $9^2 \cdot 9^2$

42. $2^3 \cdot 2^2$

43. $5 \cdot 5^2$

Lesson 9-7

Solve Systems of Equations Algebraically

ISG Interactive Study Guide

See pages 209–210 for:
- Getting Started
- Real-World Link
- Notes

Essential Question

How are linear functions used to model proportional relationships?

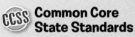

CCSS Common Core State Standards

Content Standards
8.EE.8, 8.EE.8a, 8.EE.8b, 8.EE.8c

Mathematical Practices
1, 2, 3, 4

Vocab Vocabulary
substitution

What You'll Learn

- Solve a system of equations algebraically.
- Interpret the meaning of the solutions of a system of equations.

Real-World Link

Volleyball Alisha scored a total of 12 points in two volleyball games. In the second game, she scored 2 times as many points as in the first game. You can write and solve a system of equations to find the number of points she scored in each game.

Solve Systems of Equations Algebraically

You can also use algebraic methods to solve a system of equations. One method is called **substitution**.

Example 1

Tutor

Solve the system of equations algebraically.

$y = x + 2$

$y = 5$

Since $y = 5$, replace y with 5 in the first equation.

$y = x + 2$	Write the first equation.
$5 = x + 2$	Replace y with 5.
$3 = x$	Solve for x.

The solution of this system of equations is (3, 5).

Check Check by graphing the equations on the same coordinate plane. The graphs appear to intersect at (3, 5).

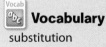

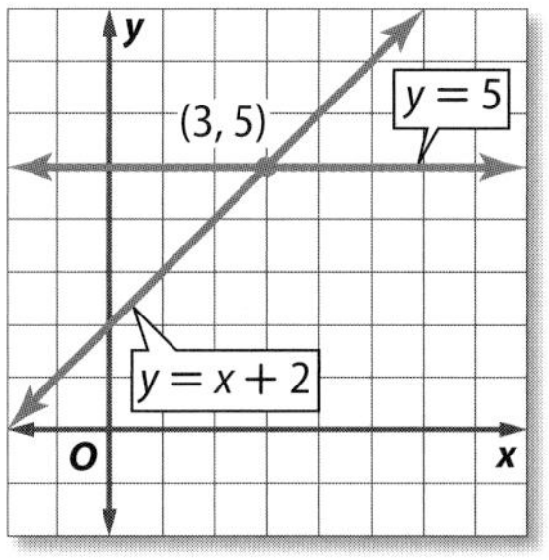

Got It? Do these problems to find out.

1a. $y = 5$
$y = x + 4$

1b. $y = 7 - x$
$x = 3$

1c. $x = y + 4$
$x = -3$

Sometimes one or both equations in a system are written in standard form. You can solve either equation in the system for either x or y, and then substitute into the other equation to find a third equation having only one variable.

Vocabulary Link
Substitution
Everyday Use the act of replacing
Math Use For all numbers a and b, if $a = b$, then a may be replaced with b.

Example 2

Tutor

Solve the system of equations algebraically.

$y - 3x = -13$

$4x + 5y = 11$

Solve the first equation for y since its coefficient is 1.

$y - 3x = -13$	Write the first equation.
$y - 3x + \mathbf{3x} = -13 + \mathbf{3x}$	Add 3x to each side.
$y = -13 + 3x$	Simplify.

Find the value for x by substituting $-13 + 3x$ for y in the second equation.

$4x + 5y = 11$	Write the second equation.
$4x + 5(-13 + 3x) = 11$	$y = -13 + 3x$
$4x - 65 + 15x = 11$	Distributive Property
$19x - 65 = 11$	Combine like terms.
$19x - 65 + \mathbf{65} = 11 + \mathbf{65}$	Add 65 to each side.
$19x = 76$	Simplify.
$\frac{19x}{\mathbf{19}} = \frac{76}{\mathbf{19}}$	Divide both sides by 19.
$x = 4$	Simplify.

Substitute 4 for x in either equation to find the value of y.

$y - 3x = -13$	Write the first equation.
$y - 3(4) = -13$	$x = 4$
$y - 12 = -13$	Simplify.
$y = -1$	Add 12 to each side.

The solution of this system of equations is (4, –1).

Got It? Do these problems to find out.

2a. $2x - 3y = -10$
$1 = 4x + y$

2b. $2x - 4y = 6$
$3x - 5y = 3$

If you solve a system of equations and the result is a false statement, the system has no solutions. If the result is a true statement, the system has infinitely many solutions.

Example 3

Tutor

Solve the system $y = x + 5$ and $2y - 2x = 10$ algebraically.

$2y - 2x = 10$	Write the second equation.
$2(x + 5) - 2x = 10$	Replace y with $x + 5$.
$2x + 10 - 2x = 10$	Distributive Property
$10 = 10$	Simplify.

$10 = 10$ is a true statement. There are infinitely many solutions.

Got It? Do these problems to find out.

3a. $4x + 2y = 8$
$y = -2x + 4$

3b. $x - y = -4$
$x - y = 1$

Interpret Solutions

You can write systems of equations to represent real-world situations and then solve algebraically. As with solving systems by graphing, you can analyze solutions and interpret their meaning in the context of the problems.

Real World

Example 4

Tutor

Karim scored a total of 16 points in two ball games. In the second game, he scored 3 times as many points as in the first game. Write and solve a system of equations to represent this situation. Interpret the solution.

Let x = points in the first game, and let y = points in the second game.

Points in first game	plus	Points in second game	is	16
x	$+$	y	$=$	16

Points in second game	is	3	times	Points in first game
y	$=$	3	$\times$	x

Use substitution to solve the system of equations $x + y = 16$ and $y = 3x$.

Substitute $3x$ for y in the first equation.

$x + y = 16$ Write the first equation.

$x + 3x = 16$ $y = 3x$

$4x = 16$ Combine like terms.

$x = 4$ Simplify.

Substitute 4 for x in either equation.

$y = 3x$ Write the second equation.

$y = 3(4)$ $x = 4$

$y = 12$ Simplify.

The solution is (4, 12). This means that Karim scored 4 points in the first game and 12 points in the second game.

Got It? Do this problem to find out.

4. The length of a rectangular garden is 4 times its width. The garden has a perimeter of 280 feet. Write and solve a system of equations to represent this situation. Interpret the solution.

Guided Practice

Check

Solve each system of equations algebraically. (Examples 1–3)

1. $y = x + 8$
   $y = 2$

2. $y = 2 - x$
   $x = 5$

3. $x + 4y = 3$
   $2x - 3y = 17$

4. $3x + 2y = 30$
   $x - 2y = -6$

5. $8x + 4y = -5$
   $8x + 4y = 12$

6. $3x + y = 10$
   $3x = -10 - y$

7. Lola paid \$13 for 6 greeting cards. Some cost \$2 each, and some cost \$3 each. Write and solve a system of equations to represent this situation. Interpret the solution. (Example 4)

Independent Practice

Go online for Step-by-Step Solutions eHelp

Solve each system of equations algebraically. (Examples 1–3)

8. $y = -x + 10$
$y = 3$

9. $y = 16 + 5x$
$x = -2$

10. $7x - 5y = 3$
$x = 4$

11. $x + \frac{1}{4}y = 9$
$x + y = 21$

12. $\frac{1}{2}x - y = 2$
$x - 2y = 4$

13. $2x = \frac{7}{2} + -\frac{3}{2}y$
$4x + 3y = 7$

14. $10x + 3y = 12$
$10x = -12 - 3y$

15. $6x + 9y = 12$
$2x + 3y = 6$

16. $5x + y = 8$
$-x + y = -4$

17. $x + 4y = 33$
$2y + 6x = 0$

18. $8x + 10y = -4$
$-4x = 2 + 5y$

19. $7x + y = 9$
$y + 7x = -2$

20. Wil paid \$35 for 5 T-shirts. Some of the T-shirts cost \$5 each and the rest cost \$10 each. Write and solve a system of equations to represent this situation. Interpret the solution. (Example 4)

21 One number added to twice another number is 30. The sum of the numbers is 19. Write and solve a system of equations to represent this situation. Interpret the solution. (Example 4)

22. CCSS **Reason Abstractly** The owners of a catering business need to seat 110 people for a banquet. Their tables seat either 6 or 8 people. They want to use twice as many tables that seat 6 as tables that seat 8. Each table will seat its maximum number of people. Write and solve a system of equations to represent this situation. Interpret the solution.

23. The sum of the digits in a two-digit number is 15. Three times the second digit equals two times the first digit. Write and solve a system of equations to represent this situation. Interpret the solution.

H.O.T. Problems Higher Order Thinking

24. CCSS **Model with Mathematics** Write a real-world problem that can be solved by writing and solving a system of equations. Then solve and interpret the solution.

25. CCSS **Persevere with Problems** The solution to the system of equations $2x + 3y = 8$ and $x - 2y = m$ is $(1, p)$. Find the values of m and p. Explain your reasoning to a classmate.

26. **Building on the Essential Question** Describe, both graphically and algebraically, a system of equations in two variables that has infinitely many solutions.

Standardized Test Practice

27. The length x of a rectangle is 8 inches more than the width y. The perimeter is 49 inches. Which system of equations represents this situation?

A $x + y = 8$ and $4x + 4y = 49$

B $x - 8 = y$ and $2x + 2y = 49$

C $x = y - 8$ and $x + y = 49$

D $x - y = 8$ and $x + y = 49$

28. If $x + y = 12$ and $3x + y = 30$, what is the value of $-2y$?

F 3

G −1

H 1

J −6

29. Solve this system of equations.

$x + 3y = -6$

$3x - y = 2$

A (0, −2)

B (−6, 0)

C infinitely many solutions

D no solution

30. **Short Response** Jileen has four times as many points as Kevin. Kevin has 15 fewer points than Jileen. Write a system of equations that can be used to find each player's points.

Common Core Review

Solve each equation. 8.EE.7a

31. $3(x - 4) - 12 = 2x - 9$

32. $-3x - 10 = 2(2x + 2) - 7x$

33. $5(x - 1) + 10 = 5(x + 1)$

34. $-2x - 11 = -3(x + 3) - 2 + x$

35. $-(x + 7) - 16 = x - 3$

36. $8x - 14 = 3(2x + 2) + 2x$

37. As a writer you are paid \$20 per column of text plus \$600 per week. This week you want your pay to be at least \$1000. Write and solve an inequality for the number of columns of text you need to write. 7.EE.4b

38. Gabe's age y is 3 years more than twelve times Briana's age x. 7.EE.4a

a. Write an equation that represents this situation.

b. Find four solutions of your equation. Write the solutions as ordered pairs.

Estimate each square root to the nearest integer. Do not use a calculator. 8.NS.2

39. $\sqrt{80}$

40. $\sqrt{98}$

41. $\sqrt{48}$

42. $-\sqrt{35}$

43. $\pm\sqrt{26}$

44. $-\sqrt{198}$

Find each cube root. 8.EE.2

45. $\sqrt[3]{27}$

46. $\sqrt[3]{125}$

47. $\sqrt[3]{-512}$

48. $\sqrt[3]{729}$

Solve each equation. Do not use a calculator. 8.EE.2

49. $x^3 = 1000$

50. $2x^3 = -2000$

51. $-5x^3 = 135$

52. $6x^3 = 48$

53. $-x^3 + 100 = 108$

54. $-2x^3 - 100 = -532$

Check Chapter Review

ISG Interactive Study Guide
See pages 211–214 for:
- Vocabulary Check
- Key Concept Check
- Problem Solving
- Reflect

Lesson-by-Lesson Review

Lesson 9-1 Functions (pp. 384–389)

1. Determine whether the relation {(5, 3), (−5, 4), (4, 2), (4, 1)} is a function. Explain.

2. Use the table below that shows the cost of gas in different years. Is the relation a function? Explain.

Year	2002	2004	2006
Cost ($)	1.36	1.82	2.26

Example 1

Determine whether the relation shown in the table below is a function. Explain.

x	9	11	13	17	21
y	7	3	−1	−5	−7

Yes; it is a function since each domain value is paired with only one range value.

Lesson 9-2 Representing Linear Functions (pp. 390–395)

Find four solutions of each equation. Write the solution as ordered pairs.

3. $y = -5x$
4. $y = 4x$
5. $y = x + 9$
6. $x + y = -1$

Graph each equation.

7. $y = -2x$
8. $y = x + 5$

9. Each small smoothie x costs \$1.50, and each large smoothie y costs \$3. Find two solutions of $1.5x + 3y = 12$ to determine how many of each type of smoothie Lisa can buy with \$12.

Example 2

Find four solutions of $y = -x + 1$. Write the solutions as ordered pairs.

Choose four values for x and then solve for y.

x	$y = -x + 1$	y	(x, y)
−1	$y = -(-1) + 1$	2	(−1, 2)
0	$y = -0 + 1$	1	(0, 1)
1	$y = -1 + 1$	0	(1, 0)
2	$y = -2 + 1$	−1	(2, −1)

Four solutions: (−1, 2), (0, 1), (1, 0), and (2, −1).

Lesson 9-3 Constant Rate of Change and Slope (pp. 396–402)

10. Find the constant rate of change between the quantities in the table below.

Time (h)	0	4	8
Money Earned ($)	0	31	62

Find the slope of the line that passes through each pair of points.

11. $F(0, 1), G(6, 4)$
12. $A(-3, 7), G(5, -1)$

13. A lizard is crawling up a hill that rises 5 feet for every horizontal change of 30 feet. Find the slope.

Example 3

Find the constant rate of change in the water level.

Time (min)	0	4	8
Water Level (ft)	5	4	3

$$\text{rate of change} = \frac{\text{change in water level}}{\text{change in time}} = \frac{5\text{ ft} - 4\text{ ft}}{0\text{ min} - 4\text{ min}} = \frac{1\text{ ft}}{-4\text{ min}} \text{ or } -\frac{1}{4}\text{ ft/min}$$

The rate of change is $-\frac{1}{4}$ feet per minute.

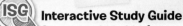

Lesson 9-4 Direct Variation (pp. 405–410)

The cost of renting a paddle boat varies directly with the number of hours, as shown in the table.

Hours	2	3	4	5
Cost ($)	18	27	36	45

14. Write an equation that relates the number of hours with the cost.

15. Find the cost of renting a paddle boat for 7 hours.

16. The temperature for one day is shown in the graph. Determine whether the relationship between the temperature and the time is a direct variation.

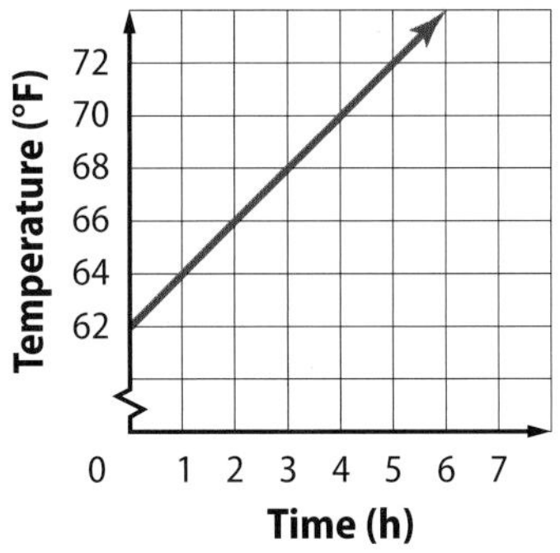

Example 4

The term varies directly with the term number. Write an equation that relates the term number with the term. Then find the 16th term.

Term Number	1	2	3	4
Term	9	18	27	36

Choose any point in the table. Then solve for m in $y = mx$.

$y = mx$ Direct variation equation.

$9 = m(1)$ Replace y with 9 and x with 1.

$9 = m$ Solve for m.

Use m to write an equation.

$y = 9x$ Direct variation equation.

$y = 9(16)$ Replace x with 16.

$y = 144$ Solve for y.

The 16th term is 144.

Lesson 9-5 Slope-Intercept Form (pp. 412–417)

State the slope and the *y*-intercept of the graph of each equation.

17. $y = 4x + 7$

18. $y = -\frac{4}{3}x$

19. $5x + y = 0$

20. $-x + y = -8$

21. $y = -8x - 7$

22. $4x - y = 6$

Graph each equation using the slope and *y*-intercept.

23. $y = -x + 4$

24. $y = 2x - 6$

25. $y = \frac{3}{2}x - 3$

26. $y = -\frac{1}{4}x + 5$

27. A balloon is rising above the ground. The height in feet y of the balloon can be given by $y = 7 + 2x$, where x represents the time in seconds. State the slope and y-intercept of the graph of the equation. Describe what they represent.

28. Jacob is ordering DVDs from a Web site. The site charges a flat rate for shipping, no matter how many DVDs he buys. The total cost y of Jacob's order is given by $y = 9x + 5$, where x represents the number of DVDs he buys. State the slope and y-intercept of the equation. Describe what they represent.

Example 5

State the slope and *y*-intercept of the graph of $y = 4x - 1$. Then graph the equation.

$y = 4x - 1$ Write the original equation.

$y = 4x + (-1)$ Write the equation in the form $y = mx + b$.

$y = mx + b$ $m = 4, b = -1$

The slope of the graph is 4 and the y-intercept is -1.

To graph the equation, first write the slope as $\frac{4}{1}$. Plot the point at $(0, -1)$. Then go up 4 and right 1. Connect the points and extend the line.

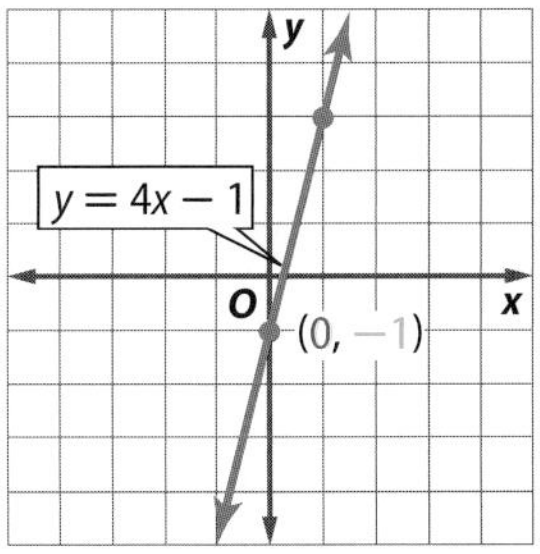

Lesson 9-6 Solve Systems of Equations by Graphing (pp. 420–424)

Solve each system of equations by graphing.

29. $y = x$
$y = \frac{1}{2}x - 1$

30. $y = x + 2$
$y = 3x$

31. $y = 2x + 1$
$x + y = -2$

32. $5x - 3y = -3$
$y = -x + 1$

33. The sum of two numbers is 9, and the difference of the numbers is 1. Write a system of equations to represent this situation. Then solve the system to find the numbers.

34. A café sells strawberry smoothies and vanilla smoothies. On Monday, the café sold twice as many strawberry smoothies as vanilla smoothies. The total number of smoothies sold was 24. Write a system of equations to represent this situation. Then solve the system and explain what the solution means.

Example 6

Solve the system of equations by graphing.

$\mathbf{y = x - 1}$

$\mathbf{y = -\frac{2}{3}x + 4}$

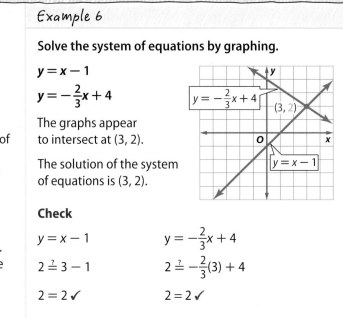

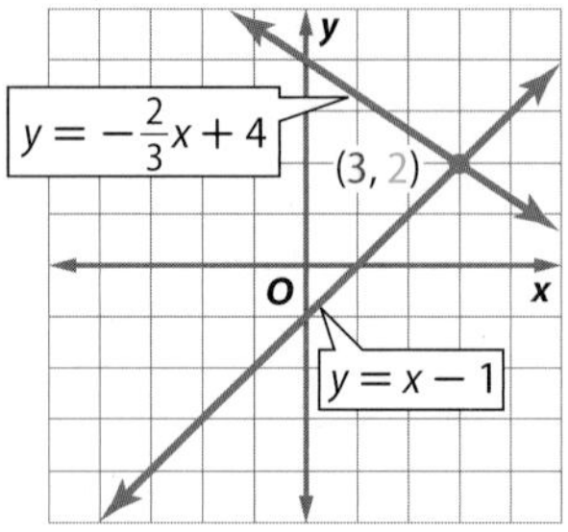

The graphs appear to intersect at (3, 2).

The solution of the system of equations is (3, 2).

Check

$y = x - 1$

$2 \stackrel{?}{=} 3 - 1$

$2 = 2$ ✓

$y = -\frac{2}{3}x + 4$

$2 \stackrel{?}{=} -\frac{2}{3}(3) + 4$

$2 = 2$ ✓

Lesson 9-7 Solve Systems of Equations Algebraically (pp. 425–429)

Solve each system algebraically.

35. $y = 4$
$y = 3x - 11$

36. $y = 6 - x$
$x = -1$

37. $-5x + y = 2$
$-3x + 6y = 12$

38. $-4x + y = 6$
$-5x - y = 21$

39. $2x - 4y = 6$
$3x - 5y = 11$

40. $8y = 6 - 2x$
$x = 3 - 4y$

41. $7x - 3y = -4$
$7x = -2 + 3y$

42. $2x + y = 3$
$y = -3x + 7$

43. One number subtracted from three times another number is 11. The sum of the numbers is 1. Write and solve a system of equations to represent this situation. Interpret the solution.

44. Tickets to a museum cost $3 for children and $8 for adults. A group of four visitors to the museum spent a total of $22 on tickets. Write and solve a system of equations to represent this situation. Interpret the solution.

Example 7

Solve the system of equations algebraically.

$y = 2x + 5$

$3y - x = 20$

Since $y = 2x + 5$, replace y with $2x + 5$ in the second equation.

$3y - x = 20$	Write the second equation.
$3(2x + 5) - x = 20$	Replace y with $2x + 5$.
$6x + 15 - x = 20$	Distributive Property
$5x + 15 = 20$	Combine like terms.
$5x = 5$	Subtract 15 from each side.
$\frac{5x}{\mathbf{5}} = \frac{5}{\mathbf{5}}$	Divide each side by 5.
$x = 1$	Simplify.

Substitute 1 for x in either equation to find the value of y.

$y = 2x + 5$	Write the first equation.
$y = 2(1) + 5$	$x = 1$
$y = 2 + 5$	Multiply.
$y = 7$	Simplify.

The solution is (1, 7).

Chapter 10
Statistics and Probability

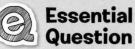

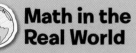

Essential Question

Essential Question: How are statistics used to draw inferences about and compare populations?

Common Core State Standards

Content Standards
7.SP.1, 7.SP.2, 7.SP.3, 7.SP.4, 7.SP.5, 7.SP.6, 7.SP.7, 7.SP.7a, 7.SP.7b, 7.SP.8, 7.SP.8a, 7.SP.8b, 7.SP.8c

Mathematical Practices
1, 2, 3, 4, 5, 6, 7

Math in the Real World

Games Statistics and statistical displays are frequently used to describe the results of the Olympic Games. In the 2010 Winter Olympics, the United States won 9 gold medals, 15 silver medals, and 13 bronze medals.

Interactive Study Guide
See pages 215–218 for:
- Chapter Preview
- Are You Ready?
- Foldable Study Organizer

Lesson 10-1

Measures of Center

ISG Interactive Study Guide

See pages 219–220 for:
- Getting Started
- Real-World Link
- Notes

Essential Question

How are statistics used to draw inferences about and compare populations?

CCSS Common Core State Standards

Content Standards
7.SP.4

Mathematical Practices
1, 3, 4

Vocab Vocabulary

statistics

measures of center

What You'll Learn

- Use the mean, median, and mode as measures of center.
- Choose an appropriate measure of center and recognize measures of statistics.

Real-World Link

Softball On average, softball players in warm weather states have faster pitching speeds than players in colder weather regions. This is because athletes are able to practice and play their sport outdoors all year.

Key Concept Measures of Center

mean	sum of the data divided by the number of items in the data set
median	middle number of the data ordered from least to greatest, or the mean of the middle two numbers
mode	number or numbers that occur most often

Statistics is the study of collecting, organizing, and interpreting information, or data. Mean, median, and mode are called **measures of center** because they are statistics that describe the center of a set of *data*.

Example 1

Real World

Tutor

Find the mean, median, and mode of the data.

House of Representatives			
California	53	New York	27
Florida	27	North Carolina	13
Illinois	18	Ohio	16
Michigan	14	Pennsylvania	18
New Jersey	12	Texas	36

$$\text{mean} = \frac{\text{total number of representatives}}{\text{number of states}}$$

$$= \frac{53 + 27 + \ldots + 36}{10}$$

$$= \frac{234}{10} \text{ or } 23.4$$ The mean is 23.4.

To find the median, order the numbers from least to greatest.

12, 13, 14, 16, 18, 18, 27, 27, 36, 53

$\frac{18 + 18}{2} = 18$ There is an even number of items. Find the mean of the two middle numbers. The median is 18.

Since both 18 and 27 appear twice, the modes are 18 and 27.

Got It? Do this problem to find out.

1. The snakes at the zoo are 62, 48, 37, 45, 50, 65, 48, 54, 48, 52, 40, and 51 centimeters long. Find the mean, median, and mode of the data.

Example 2

Tutor

The table shows the water usage of a household during the same eight months for two years. Compare the mean usage and the median usage.

Water Usage (hundreds of gallons)	
2010	2011
12	14
21	20
18	15
29	21
44	27
49	30
38	19
32	18

Step 1 Find the mean of each data set.

2010: $\frac{12 + 21 + 18 + 29 + 44 + 49 + 38 + 32}{8} = 30.375$

2011: $\frac{14 + 20 + 15 + 21 + 27 + 30 + 19 + 18}{8} = 20.5$

Step 2 Find the median of each data set.

Order each set of numbers from least to greatest.

2010: 12, 18, 21, 29, 32, 38, 44, 49

$\frac{29 + 32}{2} = 30.5$ The median is 30.5.

2011: 14, 15, 18, 19, 20, 21, 27, 30

$\frac{19 + 20}{2} = 19.5$ The median is 19.5.

Step 3 Compare each measure.

The mean usage in 2011, 20.5 hundred gallons or 2050 gallons, is less than the mean usage in 2010, 3037.5 gallons. The median usage in 2011, 1950 gallons, is less than the median usage in 2010, 3050 gallons.

Using either measure, water usage was less in 2011 than in 2010.

> **Median**
> When there is an odd number of data values, the median is a value in the data set. When there is an even number of values, the median is often *not* a value in the data set.

Got It? **Do this problem to find out.**

2. The Fitzgeralds and the Corderos are neighbors. The ages of the Fitzgerald family members are 47, 50, 22, 20, and 18. The ages of the Cordero family members are 42, 35, 9, 5, 4, 78, and 2. Compare the mean and the median ages of the two families.

Choose Appropriate Measures

Different circumstances determine which measures of center are most appropriate.

Concept Summary Using Mean, Median, and Mode

Measure	Most Useful When...
mean	• the data have no extreme values (values that are much greater or much less than the rest of the data)
median	• the data have extreme values • there are no big gaps in the middle of the data
mode	• data have many repeated numbers

Example 3

Tutor

STEM The table shows the high temperatures for the previous week. Which measure of center best represents the data? Justify your selection and then find the measure of center.

Day	Temperature (°F)
Sunday	84
Monday	83
Tuesday	89
Wednesday	90
Thursday	91
Friday	85
Saturday	80

Since the set of data has no extreme values or numbers that are identical, the mean would best represent the data.

Mean: $\frac{84 + 83 + 89 + 90 + 91 + 85 + 80}{7} = \frac{602}{7}$ or 86

The temperature 86°F is the measure of center that best represents the data.

Got It? Do this problem to find out.

3. The table shows the number of sit-ups Pablo had done in one minute for the past 7 days. Which measure of center best represents the data? Justify your selection and then find the measure of center.

Day	Number of Sit-Ups
Sunday	40
Monday	37
Tuesday	45
Wednesday	19
Thursday	49
Friday	50
Saturday	46

You can also use measures of center to show different points of view.

Example 4

Tutor

The average wait times for 10 different rides at an amusement park are 65, 21, 17, 52, 25, 17, 11, 22, 60, and 44 minutes. Would the amusement park advertise the mean or the median to show that the wait times for its rides are short? Explain.

Mean: $\frac{65 + 21 + \ldots + 44}{10} = \frac{334}{10}$ or 33.4

Median: 11, 17, 17, 21, 22, 25, 44, 52, 60, 65

$\frac{22 + 25}{2}$ or 23.5

The amusement park would want to advertise a short wait time. So, the amusement park would want to use the median, 23.5 minutes.

Got It? Do this problem to find out.

4. Maggie ran the 100-meter dash in 11.6, 11.8, 12.7, 12.6, 11.9, and 12.0 seconds. Would she want to use the mean or the median to describe her times? Explain.

Guided Practice

Check

1. The table shows the winning speeds of the Indianapolis 500 in recent years. Find the mean, median, and mode of the data. Round to the nearest tenth if necessary. (Example 1)

Year	Driver	Speed (mph)
2006	Sam Hornish Jr	157
2007	Dario Franchitti	152
2008	Scott Dixon	144
2009	Helio Castroneves	150
2010	Dario Franchitti	162
2011	Dan Wheldon	170
2012	Dario Franchitti	168

2. Last week, Sarah spent 34, 30, 45, 30, 40, 38, and 28 minutes exercising. Find the mean, median, and mode. Round to the nearest whole number. (Example 1)

3. The best performances in the Men's Outdoor Pole Vault are shown for 1991 to1995 and for 2007 to 2011. Compare the mean heights and the median heights. (Example 2)

Best Outdoor Pole Vault Performance					
1991–1995 (meters)	6.10	6.13	6.05	6.14	6.03
2007–2011 (meters)	5.95	6.04	6.01	5.95	5.90

4. The students of different eighth-grade homerooms read 38, 45, 26, 51, 42, 38, 50, and 58 books for a reading competition. Which measure of center best represents the data? Justify your selection and then find the measure of center. (Example 3)

5. Refer to the data in Exercise 2. Should Sarah use the mean or the median to show that she exercises for large amounts of time each day? Explain. (Example 4)

Independent Practice

Go online for Step-by-Step Solutions

eHelp

Find the mean, median, and mode for each set of data. If necessary, round to the nearest whole number. (Example 1)

6. The minutes spent biking each day in one week: 45, 30, 65, 90, 74, 60, 35

7. The price, in dollars, of digital cameras: 250, 200, 320, 235, 265, 200

8. The prices of the same items at two stores are shown in the table below. Compare the mean price and the median price of the items. (Example 2)

Grocery Store Prices ($)					
Store A	1.49	0.59	1.07	0.89	0.75
Store B	1.44	0.63	1.25	0.98	0.79

9. The table shows the football teams with the most all-time wins. Which measure of center best represents the data? Justify your selection and then find the measure of center. (Example 3)

University	All-Time Football Wins
Michigan	895
Texas	858
Notre Dame	853
Nebraska	846
Ohio State	825

10. The table shows the attendance at an art museum. Which measure of center would the museum use to show it has a large number of visitors? Explain. (Example 4)

Day	Attendance
Tuesday	214
Wednesday	189
Thursday	214
Friday	248
Saturday	220
Sunday	253

11. Matthew's math test scores this semester were 80, 76, 94, 90, 88, 92, 88, and 96. Which measure of center might Matthew want to use to describe his test scores? Explain. (Example 4)

12. **Financial Literacy** The hourly wages of employees in a small store are \$7, \$24, \$8, \$10, \$6, \$8, and \$8. Which measure of center might the store use to attract people to work there? Explain.

13. CCSS **Justify Conclusions** The graph shows the number of siblings that Ms. Delgado's students have. Based on the graph, which measure of center best represents the data? Explain.

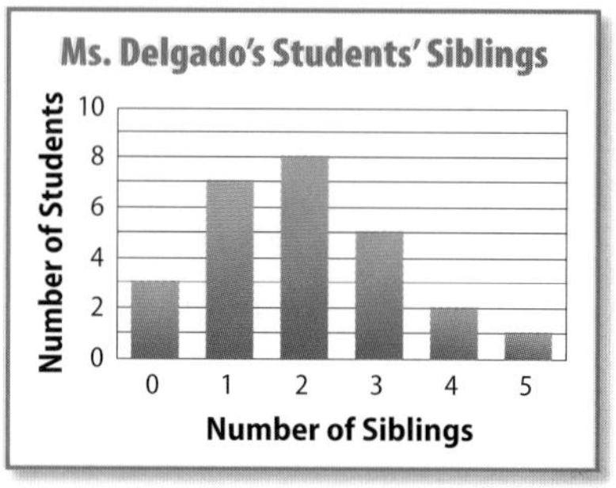

14. Winona needs to average 5.8 points from 14 judges to win the competition. The mean score of 13 judges was 5.9. What is the lowest score she can receive from the fourteenth judge and still win?

15. Survey your classmates to find their heights. Display the results of your survey in a line plot. Then find the mean, median, and mode of your data. Which measure of center would you use to represent the overall height?

H.O.T. Problems Higher Order Thinking

16. CCSS **Model with Mathematics** A data set showing the number of siblings of several students has a mean of 4 and a median that is *not* 4. Construct a data set that models the situation.

17. CCSS **Reason Inductively** Is it *always, sometimes,* or *never* possible for the mean, median, and mode to be equal? Justify your reasoning.

18. CCSS **Persevere with Problems** The ages of the players on an intramural volleyball team are 29, 25, 26, 31, 28, 23, 21, and 25.

    a. Suppose another player joins the team. What must the age of the new player be so that the mean age is 27? Explain your reasoning.

    b. What must the age of the new player be so that the median age is 25? Explain your reasoning.

19. CCSS **Justify Conclusions** A real-estate guide lists the average home prices for counties in your state. Do you think the mean, median, or mode would be the most useful average for homebuyers? Explain.

20. CCSS **Construct an Argument** Can a data set have more than one mode? median? Explain.

21. **Building on the Essential Question** Use the Internet to find some real-world data. Which measure of center best represents the data you found? Justify your selection and then find the measure of center.

Standardized Test Practice

22. The number of books read by the students in each reading class this year is shown in the table. Which measure of center would the school use to show that their students read a lot of books?

104	90
162	134
110	97
145	126

A mode **C** mean
B median **D** cannot be determined

23. The high temperatures, in degrees Fahrenheit, for one week are 79°, 81°, 77°, 81°, 82°, 75°, and 76°. If the temperature on the eighth day is 80°, which of the following would be true?

F The mode will change.
G The mean will increase and the median will remain the same.
H The median will increase and the mean will remain the same.
J Both the mean and the median will increase.

24. Jamal said that the number that best represented the following set of data is 27. Which measure of center is he referring to?

28, 32, 21, 25, 33, 32, 20, 26

A mean **C** mode
B median **D** all of the above

25. **Short Response** Shane and Cho have the bowling scores shown.

Game #	Shane	Cho
1	124	125
2	135	132
3	109	128
4	116	130
5	141	125

Shane's score for his sixth game is 155. What must Cho bowl in the sixth game in order to have the same average as Shane?

CCSS Common Core Review

Write an expression to represent each real-world situation. Then evaluate the expression and interpret the meaning of the solution. 7.NS.3

26. Collette owed the library $3.75. Then she returned a library book that was overdue by 9 days. The library charges $0.25 a day for overdue books. How much did Collette owe the library when she returned the book?

27. The Chess Club bought two pizzas that cost $13.86 each. Twelve members of the club shared the cost. How much did each member pay?

Find the constant of proportionality for each table. 7.RP.2b

28.

Dodecahedrons	1	2	3	4
Sides	12	24	36	48

29.

Containers	3	6	9	12
Weight (lb)	1.5	3	4.5	6

Solve each equation. Check your solution. 8.EE.7b

30. $9 = -\frac{1}{4}x$

31. $\frac{8}{9}y = -\frac{2}{9}$

Find the unit rate. Round to the nearest hundredth, if necessary. 6.RP.3b

32. 216 miles in 4.5 hours

33. 1650 milliliters in 25 seconds

34. $1.44 for 0.25 ounce

35. 16 errors in 157 games

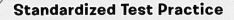

Lesson 10-2

Measures of Variability

ISG Interactive Study Guide

See pages 221–222 for:
- Getting Started
- Vocabulary Start-Up
- Notes

Essential Question

How are statistics used to draw inferences about and compare populations?

CCSS Common Core State Standards

Content Standards
7.SP.4

Mathematical Practices
1, 3, 4, 7

Vocabulary

measures of variability
range
quartiles
first quartile
third quartile
interquartile range
outlier

What You'll Learn

- Find measures of variability.
- Use measures of variability to interpret and analyze data.

Real-World Link

Spiders Spiders come in many different sizes, shapes, and colors. Some of the most dangerous spiders in the world are less than an inch long.

Measures of Variability

Measures of variability are used to describe the distribution of the data. One measure of variability is the range. The **range** of a set of data is the difference between the greatest and the least values of the set. It describes whether the data are spread out or clustered together.

Example 1

Tutor

Find the range for each set of data.

a. **Study Time (minutes)**

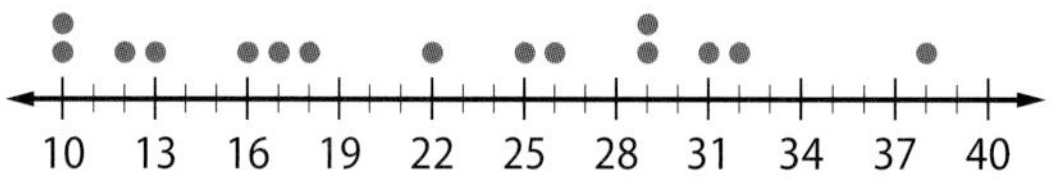

The greatest value is 38 minutes. The least value is 10 minutes. The range is $38 - 10$ or 28 minutes.

b. The age in years of Mrs. Tyznik's grandchildren: 27, 8, 5, 19, 21, 10, 4, and 21
The range is $27 - 4$ or 23 years.

Got It? **Do these problems to find out.**

1a. **Mass of Letters (grams)**

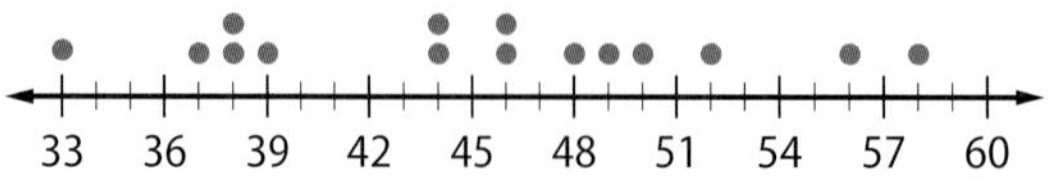

1b. The cost in dollars of DVDs: 20, 25, 15, 16, 10, and 9

In a set of data, the **quartiles** are the values that divide the data into four equal parts. Recall that the median of a set of data separates the set in half.

Quartiles

One half of the data lies between the first quartile and the third quartile.

The median of the lower half of a set of data is the **first quartile**, or Q_1.

The median of the upper half of a set of data is the **third quartile**, or Q_3.

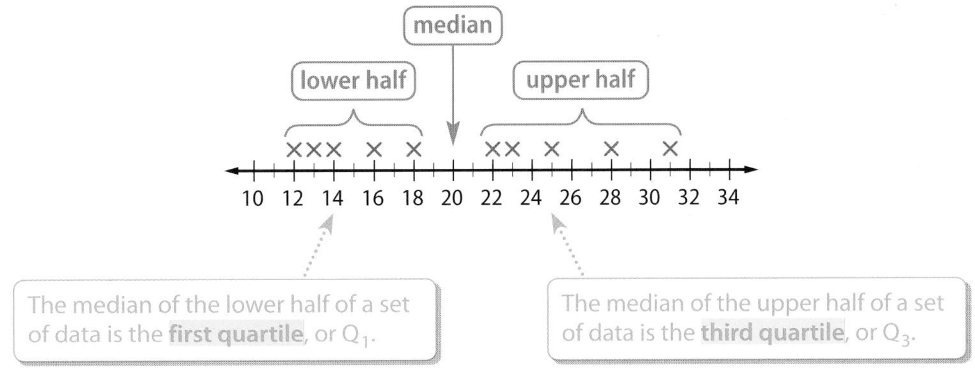

Key Concept ▶ **Interquartile Range**

Words The **interquartile range** is the range of the middle half of a set of data. It is the difference between the third quartile and the first quartile.

Symbols Interquartile range = $Q_3 - Q_1$

Example 2

Tutor

Find the measures of variability for the data in the table.

U.S. Summer Olympic Silver Medals 1976–2008		
35	0	61
31	34	32
24	39	38

Step 1 Range: $61 - 0$ or 61 medals

Step 2 Median, First Quartile, and Third Quartile:

Order the data from least to greatest.

first quartile median third quartile

0 24 31 32 34 35 38 39 61

$\frac{24 + 31}{2} = 27.5$ $\frac{38 + 39}{2} = 38.5$

The median is 34, the first quartile is 27.5, and the third quartile is 38.5. The interquartile range is $38.5 - 27.5$, or 11.

Statistics

A small interquartile range means that the data in the middle of the set are close in value. A large interquartile range means that the data in the middle are spread out, or vary.

Got It? Do this problem to find out.

2. Find the measures of variability for the data in the table.

Marley's Paper Airplane Tosses (ft)			
40	45	49	25
44	39	53	38

Outliers
A data value that is much larger or much smaller than the median is an outlier.

Data that are more than 1.5 times the value of the interquartile range beyond the quartiles are called **outliers**.

Example 3

Tutor

Find any outliers in the data set.

Step 1 Find the interquartile range.
$35 - 15 = 20$

Step 2 Multiply the interquartile range, 20, by 1.5.
$20 \times 1.5 = 30$

Step 3 Subtract 30 from the first quartile and add 30 to the third quartile.
$15 - \mathbf{30} = -15$ $\quad$ $35 + \mathbf{30} = 65$

The only outlier is 70 because it is greater than 65.

Animal Speeds		
	Animal	Speed (mph)
	squirrel	12
first quartile ···➤	turkey	15
	elephant	25
median ···➤	cat	30
	reindeer	32
third quartile ···➤	rabbit	35
	cheetah	70

Got It? **Do this problem to find out.**

3. Find any outliers in the data set.

Movie Running Time (min)				
	105	120	155	115
	96	100	110	120

Use Measures of Variability

Measures of variability can be used to interpret and compare data.

Example 4

Tutor

The table shows the Calories burned by each activity. Use measures of variability to describe the data.

Find the measures of variability.

The range is $261 - 84$ or 177.

The median is $\frac{222 + 210}{2}$ or 216.

The first quartile is 166.5.

The third quartile is 231.

The interquartile range is $231 - 166.5$ or 64.5.

In one-fourth of the activities, you will burn 166.5 Calories or less. In one-fourth of the activities, you will burn 231 Calories or more. The number of Calories burned for half of the activities is in the interval 166.5 to 231.

Number of Calories Burned for 30 Minutes	
swimming (fast crawl)	261
soccer	234
racquetball	228
football	222
basketball	210
tennis	183
downhill skiing	150
volleyball	84

Got It? **Do this problem to find out.**

4. Use measures of variability to describe the data at the right.

Cost of Video Games ($)				
	22.79	44.99	34.00	59.99
	44.76	32.50	29.25	24.95

Example 5

Tutor

The number of counties for certain western and northeastern states are shown.

Counties by Region										
Western States	15	17	23	29	33	36	39	44	53	56
Northeastern States	3	5	8	10	14	14	16	21	24	62

a. Compare the western states' range with the northeastern states' range.

The range for the western region is 56 − 15, or 41 counties, and the range for the northeastern region is 62 − 3, or 59. So, the number of counties in the Northeast vary more than in the West.

b. Do the data for either region contain an outlier?

	Western States	**Northeastern States**
First Quartile:	23	8
Third Quartile:	44	21
Interquartile Range:	$44 - 23 = 21$	$21 - 8 = 13$
Multiply by 1.5:	$21 \cdot 1.5 = 31.5$	$13 \cdot 1.5 = 19.5$
Determine Outliers:	$23 - 31.5 = -8.5$ ✗	$8 - 19.5 = -11.5$ ✗
	$44 + 31.5 = 75.5$ ✗	$21 + 19.5 = 40.5$ ✓

Since 62 is greater than 40.5, 62 is an outlier for the northeastern states' data.

c. How does the outlier affect the measures of center for the northeast region?

Calculate the mean, median, and mode without the outlier, 62.

	without the outlier	**with the outlier**
Mean:	$\frac{3 + 5 + \ldots + 24}{9} \approx 12.78$	$\frac{3 + 5 + \ldots + 62}{10} = 17.7$
Median:	14	14
Mode:	14	14

When the outlier is not included, the mean decreased by 17.7 − 12.78, or 4.92, while the median and mode did not change.

Got It? **Do this problem to find out.**

5. The number of magazines sold by each student in two homerooms is shown.

Magazines Sold										
Homeroom 102	2	5	7	7	8	9	13	14	14
	16	16	23	24	28	33	35	35	40
Homeroom 104	1	2	4	8	9	9	10	10	11
	11	12	15	16	16	20	25	26	51

a. Compare Homeroom 104's range with Homeroom 102's range.

b. Does either homeroom have an outlier in the data set?

c. How does the outlier affect the range for the number of magazines sold in that homeroom?

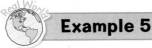

Guided Practice

Check

Find the measures of variability and any outliers for each set of data. (Examples 1–3)

1. The number of minutes spent bike riding are the following: 120, 80, 170, 100, 120, 110, 180, and 35.

2. **Animal Life Spans (years)**

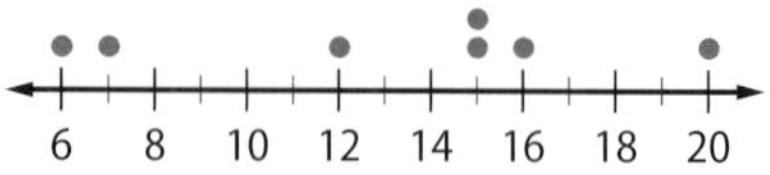

3. The number of Calories in a serving of certain fruits and vegetables is shown in the table below. (Examples 4 and 5)

Calories Per Serving								
Fruits	50	50	50	50	60	60	60	80
Vegetables	30	35	35	40	40	40	45	50

   a. Compare the fruits' range with the vegetables' range.

   b. Determine any outliers. How do the outliers affect the measures of variation for the number of Calories in fruits?

Independent Practice

Go online for Step-by-Step Solutions

eHelp

Find the measures of variability and any outliers for each set of data. (Examples 1–3)

4.

Computer Game Sales	
Day	**Number Sold**
Monday	89
Tuesday	90
Wednesday	80
Thursday	100
Friday	92
Saturday	104
Sunday	150

5.

Popcorn Sales at Movie Time Theater	
Year	**Sales ($ thousands)**
2007	0.66
2008	0.43
2009	1.25
2010	0.2
2011	0.53
2012	0.6
2013	0.58
2014	0.48

6. The number of volunteer hours per month at the community center are: 38, 43, 36, 37, 32, 37, 29, and 51.

7. **Sunny Days Per Month**

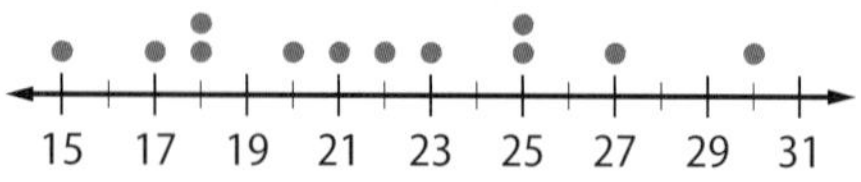

8. CCSS **Identify Structure** The table shows the scores on a recent science test. Use measures of variability to describe the data. (Example 4)

Science Test Scores					
94	64	88	100	91	106
82	79	85	91	97	109
97	88	97	103	82	100

9. The table shows the number of points Cami scored per basketball game. (Example 5)

a. Compare the range for the 2011 stats to the range for the 2012 stats.

b. Do the data for either year contain an outlier?

c. How does the outlier for the number of points scored affect the range for that year?

Cami's Basketball Stats			
2011		2012	
6	12	14	22
7	14	14	22
8	16	14	24
9	17	16	24
9	18	16	26
10	21	18	26
10	27	18	40

10. Rondell's golf scores at the end of the high school golf tournament were −1, −2, 4, −6, 3, −1, and −3. Find the measures of variability of his golf scores. Explain what the measures of variability tell you about the data.

11 The table shows average monthly temperatures for two different cities.

City	Average Monthly High Temperatures (°F)											
	Month											
	Jan.	Feb.	Mar.	Apr.	May	Jun.	Jul.	Aug.	Sep.	Oct.	Nov.	Dec.
Antelope, MT	21	30	42	58	70	79	84	84	72	58	37	24
Augusta, ME	28	32	41	53	66	75	80	79	70	58	46	34

a. Which city has a greater range of temperatures?

b. Find the measures of variability for each city.

c. Compare the medians and the interquartile ranges of the average temperatures.

d. Select the appropriate measure of center to describe the average high temperature for Augusta. Justify your response.

e. Describe the average temperatures of Antelope and Augusta, using both the measures of center and variability.

H.O.T. Problems Higher Order Thinking

12. CCSS **Model with Mathematics** Find a real-world data set with at least eight values that include one or more outliers. Display the data in a table. Find the mean and the interquartile range of the data set. Then remove the outlier(s) from the data set and find the mean and the interquartile range. Describe any differences in the values.

13. CCSS **Persevere with Problems** Write two lists of data that each have at least 15 values and a median of 10. One should have a range of 25 and the other a range of 6.

14. CCSS **Justify Conclusions** The range is *always*, *sometimes*, or *never* affected by outliers. Justify your reasoning.

15. CCSS **Construct an Argument** *True* or *false*. The interquartile range is affected by the outliers of a data set. Explain your reasoning.

16. CCSS **Use a Counterexample** *True* or *false*. Of mean, median, and mode, the mean will always be most affected by the outliers. If false, give a counterexample.

17. **Building on the Essential Question** Explain how outliers affect the calculation of measures of variability in a data set.

Standardized Test Practice

18. The table shows the total number of wins by the team that won the women's NCAA basketball tournament for the past 15 years.

Total Wins				
34	39	29	31	34
34	34	32	37	39
33	36	35	33	31

Which of the following statements is *not* supported by these data?

A Less than half of the teams won more than 34 games and half won less than 34 games.

B The range of the data is 10 games.

C About one fourth of the teams won 32 or fewer games.

D An outlier of the data is 29 games.

19. Refer to the table in Exercise 18. What is the interquartile range of the data?

F 4   **H** 8

G 5   **J** 10

20. What is the first quartile of the following set of data?

37, 12, 7, 8, 10, 5, 14, 19, 7, 15, 11

A 7   **C** 11

B 7.5   **D** 15

21. **Short Response** Find the measures of variability and any outliers for the set of data.

30, 62, 35, 80, 12, 24, 30, 39, 53, 38

Common Core Review

Find the mean, median, and mode for each set of data. Round to the nearest tenth, if necessary. 6.SP.5c

22. **Ages of People in a Play (years)**

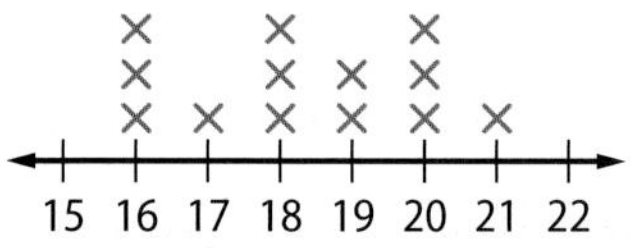

23. **Lengths of Tadpoles (cm)**

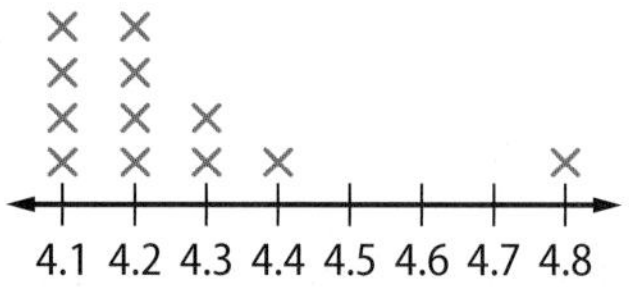

Order each set of decimals from least to greatest. 6.NS.7

24. 1.0, 1.1, 0.9, 0.5, 1.9, 10.9, 0.1

25. 7.8, 8.7, 6.7, 6.8, 7.0, 6.9

Solve each equation. 8.EE.7b

26. $2(x - 8) = 24$

27. $\frac{3}{5}(x + 10) = 45$

28. $-4\left(\frac{x}{2}\right) + 1 = 12$

29. $-200(1 - 3x) = 50$

Solve each system of equations. 8.EE.8a

30.

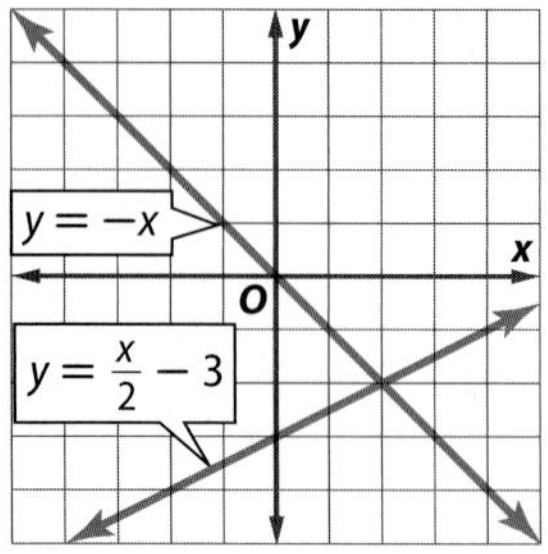

31.

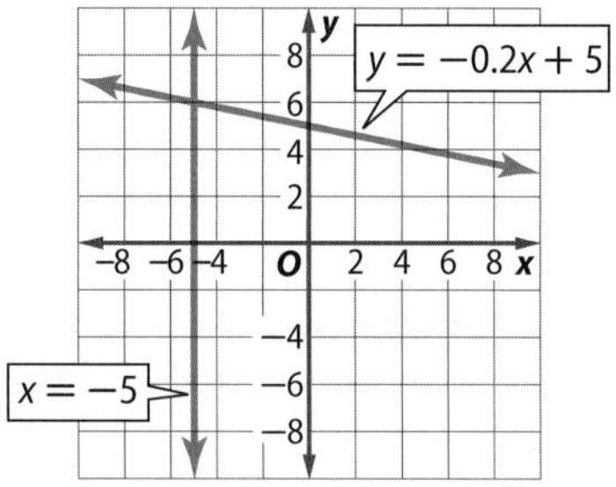

Solve each system of equations algebraically. 8.EE.8b

32. $y = x + 15$
$x = -5$

33. $y = -2x + 25$
$y = -3$

Lesson 10-3

Mean Absolute Deviation

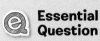

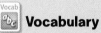

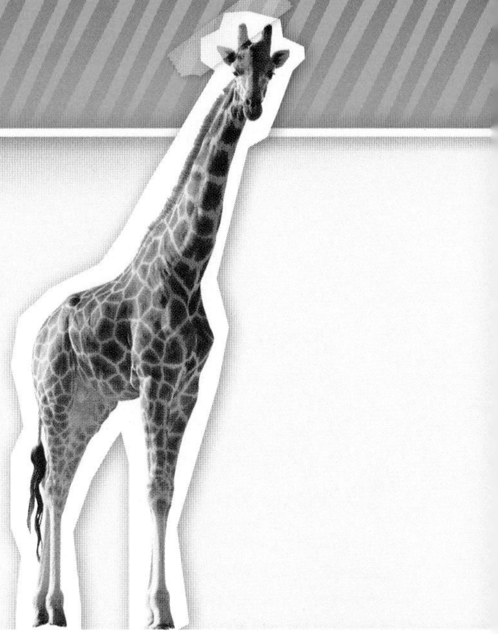

Interactive Study Guide

See pages 223–224 for:
- Getting Started
- Real-World Link
- Notes

Essential Question

How are statistics used to draw inferences about and compare populations?

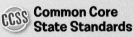

Common Core State Standards

Content Standards
7.SP.4

Mathematical Practices
1, 3, 4, 6

Vocabulary

mean absolute deviation

What You'll Learn

- Find the mean absolute deviation of a set of data.
- Compare the mean absolute deviations for two data sets.

Real-World Link

Mammals Mammals vary in size from the very small to the very large. The smallest land animal is the Hog-nosed bat from Thailand, which is about 3 centimeters long. The tallest land animal is the giraffe, which can reach a height of 20 feet.

Find Mean Absolute Deviation

You have found measures of center to describe the middle of a set of data, and you have used the interquartile range to describe the spread of a set of data. The **mean absolute deviation** is the average distance between each data value and the mean.

Example 1

Real World — Tutor

The table shows points scored by a basketball player in his last seven games. Find the mean absolute deviation. Describe what the mean absolute deviation represents.

Points Scored			
17	22	17	30
15	17	8	

Step 1 Find the mean.

$$\frac{17 + 22 + 17 + 30 + 15 + 17 + 8}{7} = 18$$

Step 2 Find the absolute value of the differences between each value in the data set and the mean.

$|17 - 18| = 1$ $|22 - 18| = 4$ $|17 - 18| = 1$ $|30 - 18| = 12$

$|15 - 18| = 3$ $|17 - 18| = 1$ $|8 - 18| = 10$

Step 3 Find the average of the absolute values.

$$\frac{1 + 4 + 1 + 12 + 3 + 1 + 10}{7} \approx 4.57$$

The mean absolute deviation is about 4.57. This means that the average distance between the mean points per game and the actual points per game is 4.57 points.

Got It? **Do this problem to find out.**

1. The number of tomatoes in five boxes was 18, 21, 16, 19, and 20. Find the mean absolute deviation. Describe what the mean absolute deviation represents.

Compare Variation

You can compare the mean absolute deviations for two data sets. A data set with a greater mean absolute deviation has data values that are more spread out from the mean than a data set with a smaller mean absolute deviation.

Example 2

Tutor

The dot plots show the heights of the girls and boys in the glee club.

Girls' Heights (inches)

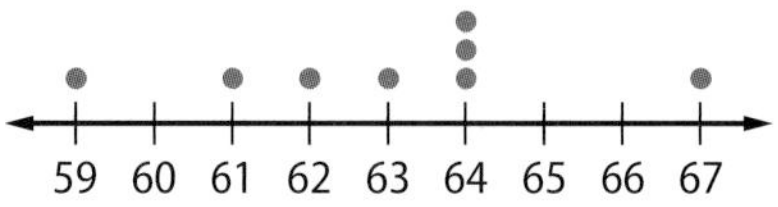

Boys' Heights (inches)

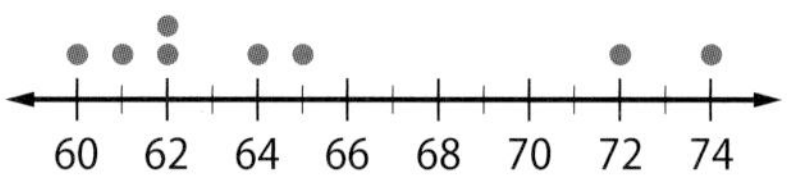

> **Watch Out!**
> The data set with the greater range may not have the greater mean absolute deviation.

a. Find the mean absolute deviation for each set of data.

Find the mean of the girls' heights.

$\frac{59 + 61 + 62 + 63 + 64 + 64 + 64 + 67}{8} = 63$ The mean is 63 inches.

Find the mean absolute deviation of the girls' heights.

$|59 - 63| = 4$ $|61 - 63| = 2$ $|62 - 63| = 1$ $|63 - 63| = 0$

$|64 - 63| = 1$ $|64 - 63| = 1$ $|64 - 63| = 1$ $|67 - 63| = 4$

$\frac{4 + 2 + 1 + 0 + 1 + 1 + 1 + 4}{8} = 1.75$ The mean absolute deviation is 1.75 inches.

Find the mean of the boys' heights.

$\frac{60 + 61 + 62 + 62 + 64 + 65 + 72 + 74}{8} = 65$ The mean is 65 inches.

Find the mean absolute deviation of the boys' heights.

$|60 - 65| = 5$ $|61 - 65| = 4$ $|62 - 65| = 3$ $|62 - 65| = 3$

$|64 - 65| = 1$ $|65 - 65| = 0$ $|72 - 65| = 7$ $|74 - 65| = 9$

$\frac{5 + 4 + 3 + 3 + 1 + 0 + 7 + 9}{8} = 4$ The mean absolute deviation is 4 inches.

b. Write a few sentences comparing their variation.

The mean absolute deviation for the girls' heights is much less than that of the boys' heights. The data for the girls are closer together than the data for the boys.

Got It? **Do this problem to find out.**

2. The dot plots show the high and low temperatures in Denver, Colorado, during one week. Find the mean absolute deviation for each set of data. Round to the nearest hundredth. Then write a few sentences comparing their variation.

High Temperature (°F)

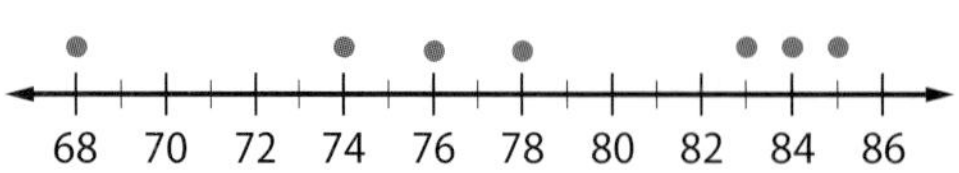

Low Temperature (°F)

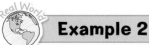

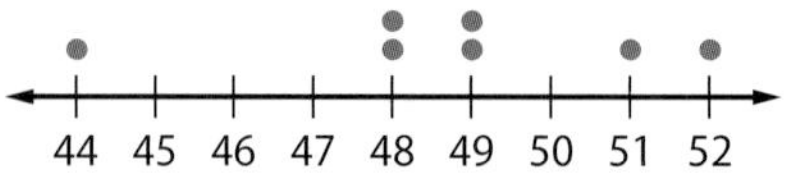

Guided Practice

Check

1. **STEM** The table shows the number of rose petals on several types of roses. Find the mean absolute deviation. Describe what the mean absolute deviation represents. (Example 1)

Rose Petals				
22	16	35	44	17
26	33	35	20	52

2. The dot plots show the number of pull-ups Joe did each day during two different weeks. Find the mean absolute deviation for each set of data. Round to the nearest hundredth. Then write a few sentences comparing their variation. (Example 2)

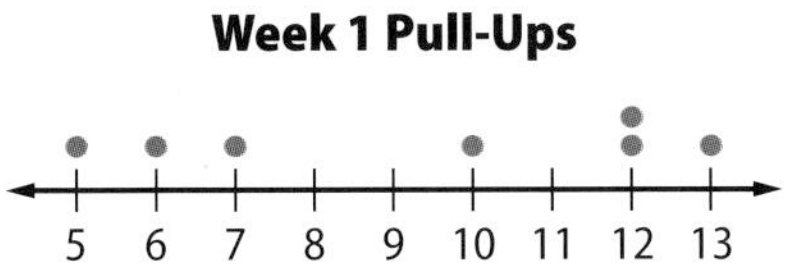

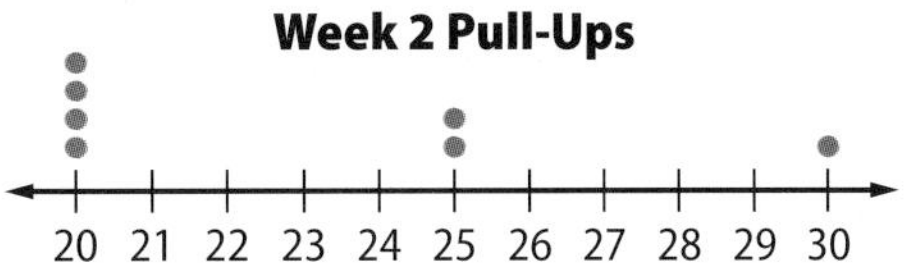

Independent Practice

Go online for Step-by-Step Solutions

eHelp

3. The weights of several adult guinea pigs in a pet store are 33, 39, 42, 33, 39, 31, 32, and 37 ounces. Find the mean absolute deviation. Describe what the mean absolute deviation represents. (Example 1)

4. The table below shows Leland's bowling scores during two different months of his bowling league. Find the mean absolute deviation for each set of data. Round to the nearest hundredth. Then write a few sentences comparing their variation. (Example 2)

Leland's Bowling Scores					
September			April		
98	103	116	112	118	145
95	90	118	130	125	105
101	121	94	118	122	150

Find the mean and the mean absolute deviation of each data set.

5. Ages of children at a family reunion: 0, 5, 7, 3, 9, 12, 5, 2, 4, 3

6. Cost of books in dollars: 4, 15, 8, 4, 6, 11, 1, 7, 3, 1

7. The high temperatures for the first eight days of January and of July are shown in the table. Predict which data set has a greater mean absolute deviation. Justify your answer.

High Temperatures (°F)			
January		July	
16	4	87	88
13	23	90	92
21	45	86	85
27	19	83	83

8. Find the mean absolute deviation for each of the following data sets.

   1, 2   1, 2, 3, 4   1, 2, 3, 4, 5, 6

   Based on the pattern, what will be the mean absolute deviation of the set {1, 2, 3, 4, 5, 6, 7, 8, 9, 10}?

9 CCSS **Be Precise** A carpenter has wood planks that are 4 yards, 8 yards, 7 yards, and 3 yards long.

a. What is the mean absolute deviation of the measurements?

b. Suppose the lengths of the boards were given in feet. What is the mean absolute deviation in feet?

c. How does the mean absolute deviation in yards compare to the mean absolute deviation in feet?

10. CCSS **Justify Conclusions** The table shows the yards gained by a football team on their first ten plays.

Yards Gained				
−9.5	12.5	17.5	−4	−6
12	8.5	−2.5	21.5	6.5

a. Find the mean and the mean absolute deviation of the data set.

b. Explain why the mean absolute deviation is greater than the mean.

H.O.T. Problems Higher Order Thinking

11. CCSS **Reason Abstractly** Describe a real-world situation in which a data set has two data values with a mean absolute deviation equal to 5 units.

12. CCSS **Find the Error** Alicia is finding the mean absolute deviation of the data set 5, 7, 8, 10, 12, 48. Find her error and correct it.

5, 7, 8, 10, 12, 48

mean: $\frac{8 + 10}{2} = 9$

$|5 - 9| = 4$ $|7 - 9| = 2$ $|8 - 9| = 1$

$|10 - 9| = 1$ $|12 - 9| = 3$ $|48 - 9| = 39$

mean absolute deviation: $\frac{4 + 2 + 1 + 1 + 3 + 39}{6} = 8.\overline{3}$

13. CCSS **Make a Conjecture** The prices of 5 items are \$5, \$9, \$16, \$27, and \$43.

a. What will happen to the mean absolute deviation if each price increases by \$3?

b. What will happen to the mean absolute deviation if each price is doubled?

14. **Building on the Essential Question** Explain why the mean absolute deviation might be useful when considering two real-world distributions with the same mean and median.

Standardized Test Practice

15. Which describes the mean absolute deviation of the data?

Distance (miles)						
60	40	15	25	30	35	40

A The average distance between each data value and the mean absolute deviation is 10 miles.

B The average distance between each data value and the mean is 10 miles.

C The average distance between the mean and the mean absolute deviation is 10 miles.

D The average distance between each data value is 10 miles.

16. Which data set has a mean absolute deviation of 4?

F 1, 6, 11

G 0, 12, 24

H 3, 10, 14

J 4, 5, 9

17. What is the mean absolute deviation of Hailey's science scores?

70 82 84 94

A 82.5

B 80

C 6.5

D 6

18. **Short Response** Matt wants to invest in the fund with the smaller mean absolute deviation of profits. Which fund should he choose? What is the mean absolute deviation for that fund?

Fund A	5%	11%	8%	2%	10%
Fund B	9%	7%	5%	3%	7%

Common Core Review

19. The cost of four popular games at a store are \$53, \$61, \$57, \$46. During a sale, the same games cost \$45, \$52, \$49, and \$39. Compare the range of the prices before the sale to the range of the prices during the sale. 7.SP.4

20. The ages of children at Jamie's party were 11, 10, 9, 12, 12, and 13. The ages of children at Katrin's party were 8, 9, 13, 12, and 11. Compare the mean and median ages of the children at the two parties. 7.SP.4

21. Baseball data presentations may show the decimals 0.1 and 0.2 to represent $\frac{1}{3}$ and $\frac{2}{3}$, respectively. For example, $7\frac{2}{3}$ innings would be displayed as 7.2. 7.RP.3

a. What is the percent error in using 0.1 in a calculation instead of the correct value of $\frac{1}{3}$?

b. What is the percent error in using 6.2 in a calculation instead of the correct value of $6\frac{2}{3}$?

c. Compare the percent errors.

State the slope and the *y*-intercept of the graph of each equation. 7.EE.4

22. $y = -3x$

23. $y = -0.5$

24. $y = \frac{x}{5} + 5$

25. $2x + 3y = 60$

Inquiry Lab

Visual Overlap of Data Distributions

Inquiry WHAT does the ratio $\frac{\text{difference in means}}{\text{mean absolute deviation}}$ tell you about how much visual overlap there is between two distributions with similar variation?

CCSS Content Standards
7.SP.3

Mathematical Practices
1, 3

Vocab

Vocabulary
distribution
visual overlap

Text Messages Ricardo is investigating the number of text messages sent and received by different age groups.

Investigation

The **distribution** of a set of data shows the arrangement of data values. You can compare two numerical data sets by comparing the shape of their distributions. The **visual overlap** of two distributions with similar variation is a visual demonstration that compares their centers to their variation, or spread.

The tables show the number of text messages sent and received daily for two different age groups. You can organize the data in a double dot plot to compare the data.

Text Messages Ages 12–15			
70	90	80	90
85	75	85	80
90	80	75	95
100	85	95	85

Text Messages Ages 16–19			
85	75	80	70
75	80	65	75
85	70	90	80
70	75	60	65

Step 1 Use a double dot plot to display the data.

Text Messages Sent and Received

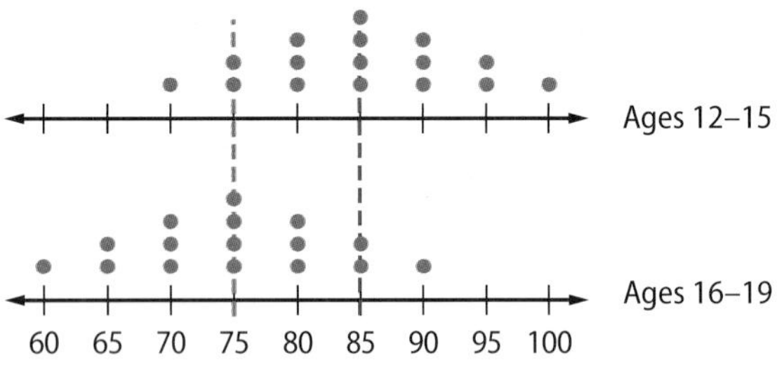

Step 2 Find the mean number of text messages for each age group.

Ages 12–15

$\frac{1360}{16} = 85$

Ages 16–19

$\frac{1200}{16} = 75$

Step 3 Draw a red vertical dotted line through both dot plots that corresponds to the mean for the age group, 12–15 years. Draw a blue vertical dotted line through both dot plots that corresponds to the mean for the age group, 16–19 years. The dotted lines show the visual overlap between the centers.

Collaborate

Work with a partner. The double dot plot below compares the number of text messages sent and received by a third age group to the age group, 12–15 years.

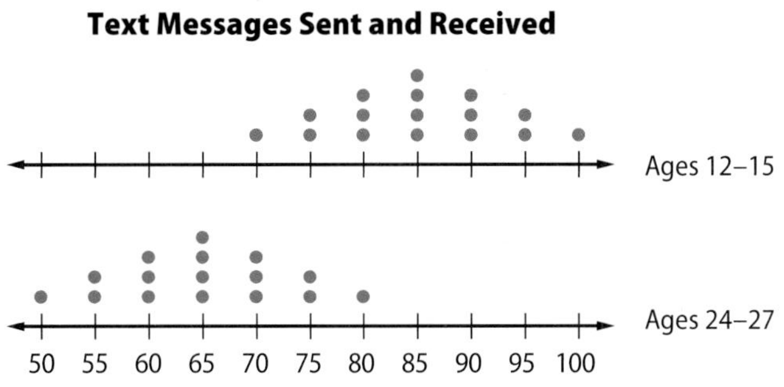

1. What is the mean number of texts for the age group, 24–27 years?
2. In the art above, where would you draw the vertical lines that are used to represent the mean of each data set?

Analyze

Work with a partner.

3. What is the difference between the means of the distributions for the Investigation?
4. What is the difference between the means of the distributions for Exercise 1?
5. The mean absolute deviation of each distribution is 6.25 text messages. Write the difference between the means and the mean absolute deviation for the Investigation as a ratio. Express the ratio as a decimal.
6. CCSS **Reason Inductively** Predict how many times as large the ratio of the difference between the means to the mean absolute deviation in Exercise 1 is, compared to the ratio in the Investigation. Explain your reasoning.
7. Write the ratio of the difference between the means and the mean absolute deviation for Exercise 1. Express the ratio as a decimal. Compare it to your prediction in Exercise 6.

Reflect

8. Inquiry WHAT does the ratio $\frac{\text{difference in means}}{\text{mean absolute deviation}}$ tell you about how much visual overlap there is between two distributions with similar variation?

Lesson 10-4

Compare Populations

ISG Interactive Study Guide

See pages 225–226 for:
- Getting Started
- Real-World Link
- Notes

Essential Question

How are statistics used to draw inferences about and compare populations?

CCSS Common Core State Standards

Content Standards
7.SP.4

Mathematical Practices
1, 2, 3, 4

Vocabulary

box plot

double box plot

What You'll Learn

- Compare two populations using the measures of center and variability.
- Compare two populations when only one is symmetric.

Real-World Link

Fundraising Every year, Boy Scouts sell gourmet popcorn to raise money for their packs and troops. In just a couple of days, Boy Scouts can earn enough money from popcorn sales to cover uniforms, equipment, and summer camp fees.

Compare Two Populations

A **box plot** uses a number line to show the distribution of a set of data. It divides a set of data into four parts using the median and quartiles. A *box* is drawn around the quartile values, and *whiskers* extend from each quartile to the minimum and maximum values that are not outliers. A **double box plot** consists of two box plots graphed on the same number line. You can draw inferences about two populations in a double box plot by comparing their centers and variability.

Example 1

Real-World

Tutor

Kacey surveyed a group of students in her science class and in her math class. The double box plot shows the results for both classes. Compare their centers and variations. Write an inference you can draw about the two populations.

How Many Times Have You Posted a Blog This Month?

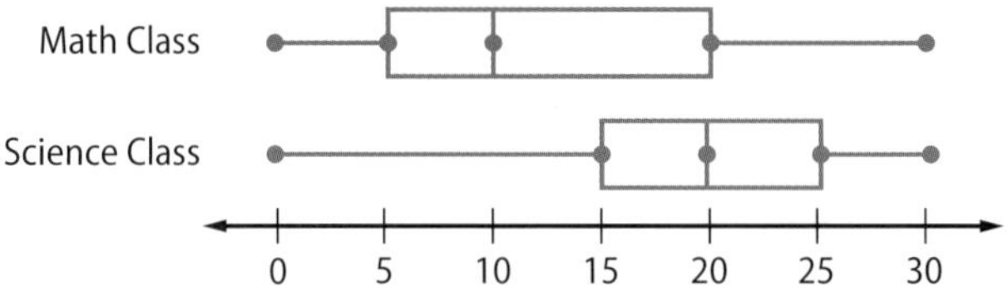

Neither box plot is symmetric. Use the median to compare the centers and the interquartile range to compare the variations.

	Math Class	Science Class
Median	10	20
Interquartile Range	20 − 5, or 15	25 − 15, or 10

Overall, the science students posted more blogs than the math students. The median for the science class is twice the median for the math class. There is a greater spread of data around the median for the math class than the science class.

Got It? Do this problem to find out.

1. The double box plot shows the costs of MP3 players at two different stores. Compare the centers and variability of the two populations. Write an inference you can draw about the two populations.

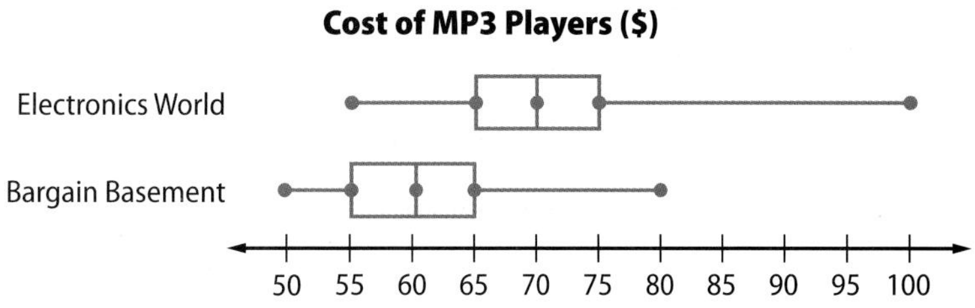

You may recall that a line plot is a way to organize and represent the data in a data set. Another name for a line plot is a *dot plot*. A dot plot uses dots to show the frequency of data values in a distribution. A *double dot plot* consists of two dot plots that are drawn on the same number line.

Real World Example 2

Tutor

The double dot plot below shows the daily high temperatures for two cities for thirteen days. Compare the centers and variations of the two populations. Write an inference you can draw about the two populations.

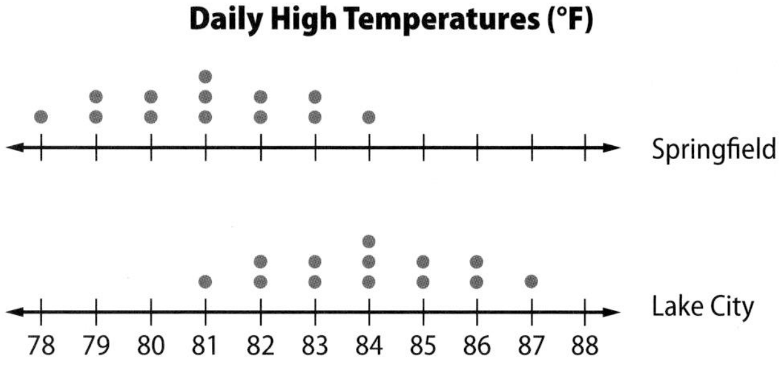

Most Appropriate Measures

To compare the centers and variations when both sets of data are symmetric, use the mean and the mean absolute deviation. When neither set is symmetric or only one is symmetric, use the median and the interquartile range to compare.

Examine the shape of the data. Both dot plots are symmetric. Use the mean to compare the centers and use the mean absolute deviation, rounded to the nearest tenth, to compare the variations.

	Springfield	Lake City
Mean	81	84
Mean Absolute Deviation	1.4	1.4

While both cities have the same variation, or spread of data about each of their means, Lake City has a greater mean temperature than Springfield. Overall, Lake City had a warmer climate than Springfield during the period the data was recorded.

Got It? Do this problem to find out.

2. The double dot plot shows the number of new e-mails in each of Pedro's and Annika's inboxes for sixteen days. Use the dot plots to compare the centers and variations of the two populations. Write an inference you can draw about the two populations.

Number of E-mails in Inbox

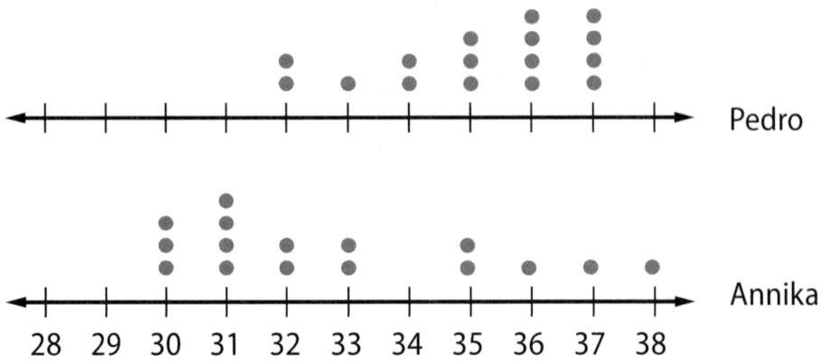

Only One Symmetric Population

You can compare two populations when only one is symmetric. The mean and median of a symmetric distribution are very similar, if not the same. So, you can use the median and interquartile range for both populations.

Real World Example 3

Tutor

Double Box Plots
Remember to compare median to median, first quartile to first quartile, and so on. It can be helpful to sketch arrows from each measure of the upper box plot to the corresponding measure in the lower box plot.

The double box plot shows the daily participants for two zip line companies for one month. Compare the centers and variations of the two populations. Which company has the greater number of daily participants?

Number of Daily Participants

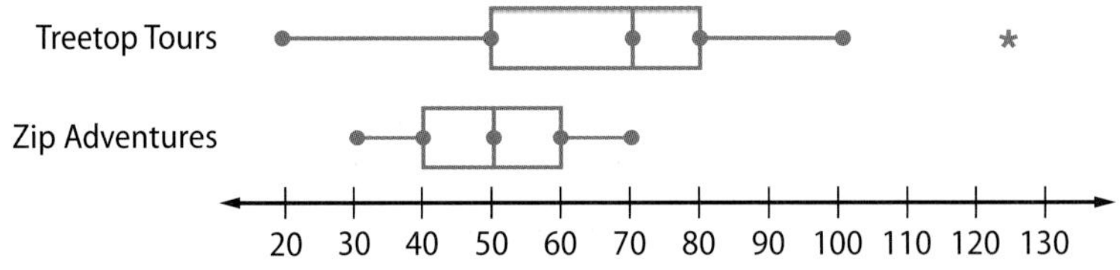

The distribution for Zip Adventures is symmetric, while the distribution for Treetop Tours is not symmetric. Use the median and the interquartile range to compare the populations.

	Treetop Tours	Zip Adventures
Median	70	50
Interquartile Range	30	20

Overall, Treetop Tours has a greater number of daily participants. However, Treetop Tours also has a greater variation, so it is more difficult to predict how many participants they may have each day. Zip Adventures has a greater consistency in their distribution.

Got It? Do this problem to find out.

3. The double dot plot shows Kareem's and Martin's race times for different three-mile races. Compare the centers and variations of the two populations. Which runner is more likely to run a faster race?

Race Times (min)

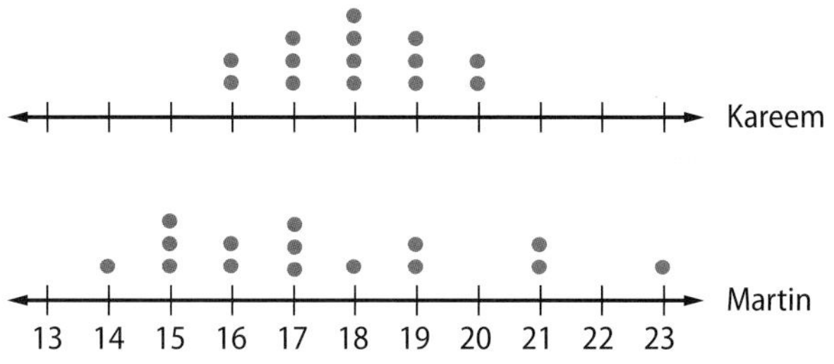

Guided Practice

1. The double dot plot below below shows the quiz scores out of 20 points for two different class periods. Compare the centers and variations of the two populations. Round to the nearest tenth. Write an inference you can draw about the two populations. (Examples 1 and 2)

Quiz Scores (points)

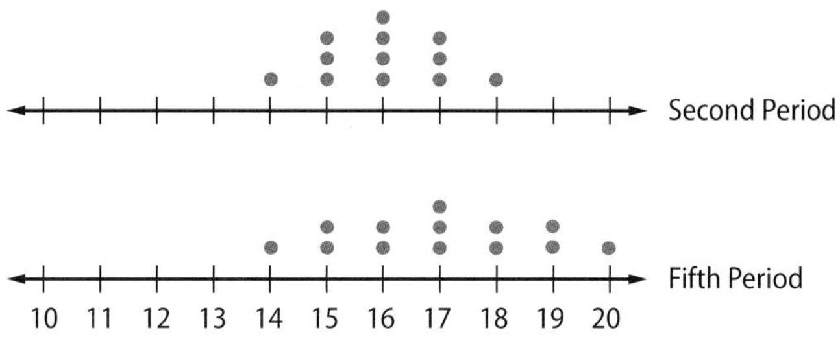

2. For a study on traffic safety, scientists recorded car speeds at different locations in Hamilton County. The double box plot below shows the speeds of cars recorded on two different roads, Hayes Road and Jefferson Road. Compare the centers and variations of the two populations. On which road are the speeds greater? (Example 3)

Speed of Cars (mph)

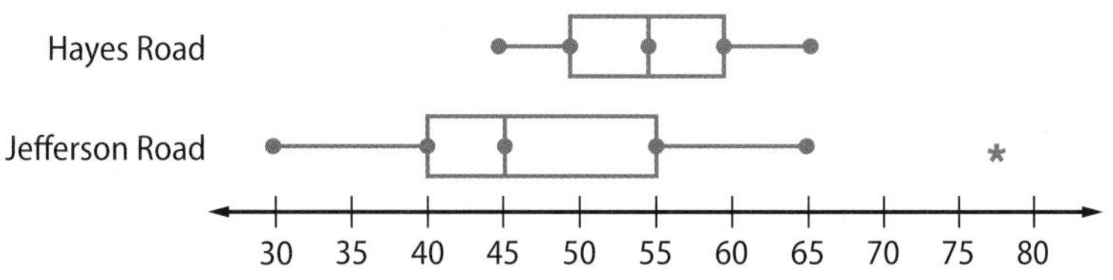

Independent Practice

Go online for Step-by-Step Solutions

eHelp

3. Jordan randomly asked customers at two different restaurants how long they waited for a table before they were seated. The double box plot shows the results. Compare their centers and variations. Write an inference you can draw about the two populations. (Examples 1 and 2)

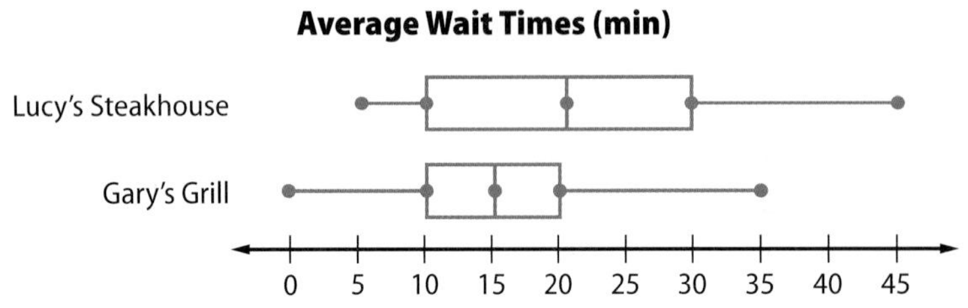

4. The double dot plot below shows the heights in inches for the girls and boys in Franklin's math class. Compare the centers and variations of the two populations. Round to the nearest tenth. Write an inference you can draw about the two populations. (Examples 1 and 2)

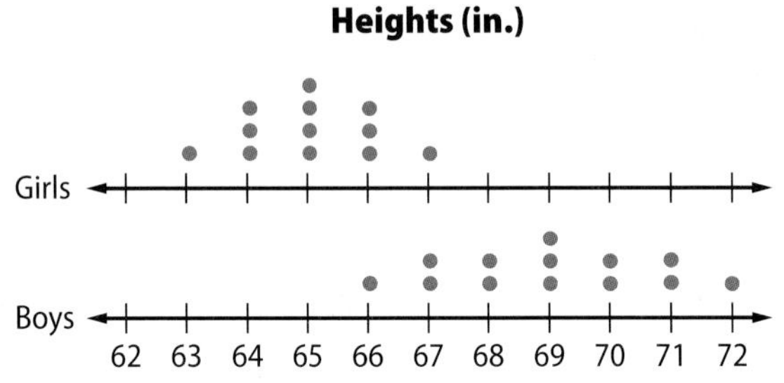

5 The double box plot shows the number of points scored by the football team for two seasons. Compare the centers and variations of the two populations. During which season was the team's performance more consistent? (Example 3)

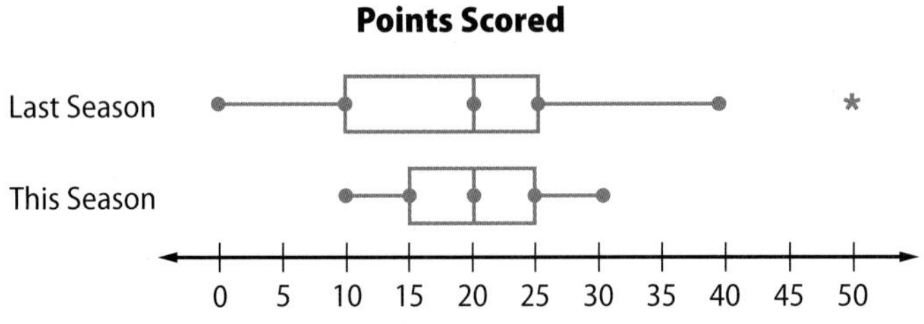

6. The double dot plot below shows the times, in hours, for flights of two different airlines flying out of the same airport. Compare the centers and variations of the two populations. Which airline's flights had shorter flight times? (Example 3)

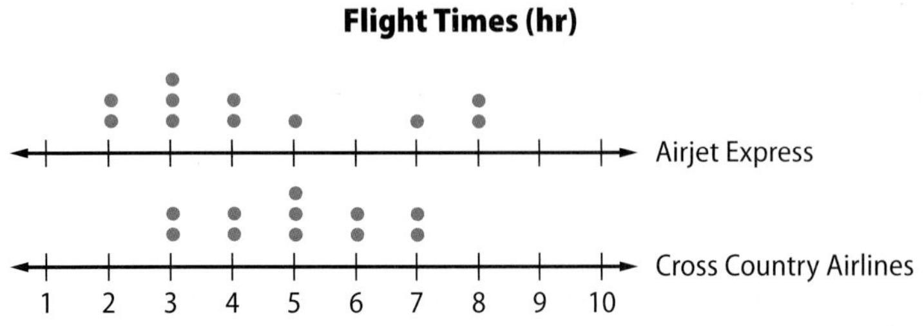

Justify Conclusions **The double box plot below shows the number of daily visitors to two different parks.**

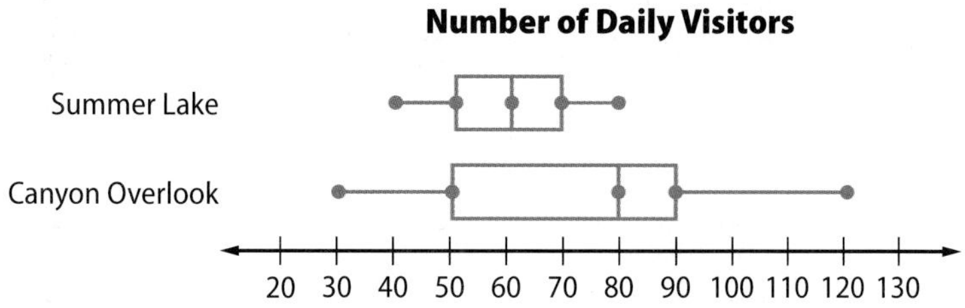

7. Use the medians to compare the centers of the data sets.
8. Which park has a greater variability in the number of daily visitors?
9. In general, which park has more daily visitors? Justify your response.
10. For which park could you more accurately predict the number of daily visitors on any given day? Explain your reasoning.

H.O.T. Problems Higher Order Thinking

11. **Persevere with Problems** The histograms below show the number of tall buildings for two cities. Explain why you cannot describe the specific location of the centers and spreads of the histograms.

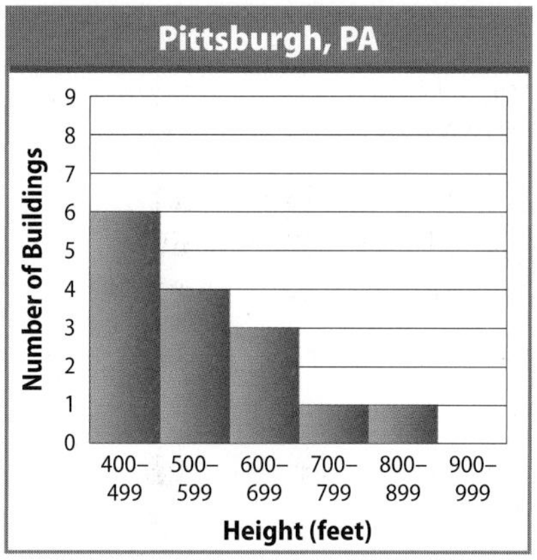

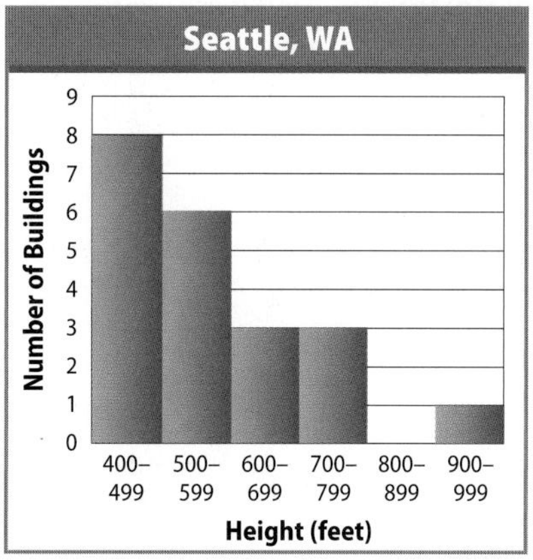

12. **Model with Mathematics** Two hockey teams, the Warriors and the Bulldogs, played 15 games each during a month. Both scored a minimum of 0 goals and a maximum of 8 goals. The Bulldogs generally scored fewer goals than the Warriors. Draw a possible double box plot for the situation.

13. **Reason Abstractly** Write a vocabulary term that completes the following analogy.

    A double bar graph is related to a bar graph in the same way as a(n) ____?____ is related to a box plot.

14. **Building on the Essential Question** Marcia recorded the daily temperatures for two cities for 30 days. The two populations have similar centers, but City A has a greater variation than City B. For which city can you more accurately predict the daily temperature? Explain.

Standardized Test Practice

15. A data distribution has a minimum value of 16, a maximum value of 45, a median of 35, a first quartile value of 22, and a third quartile value of 39. Which box plot shows data values that are generally greater than the ones given?

A

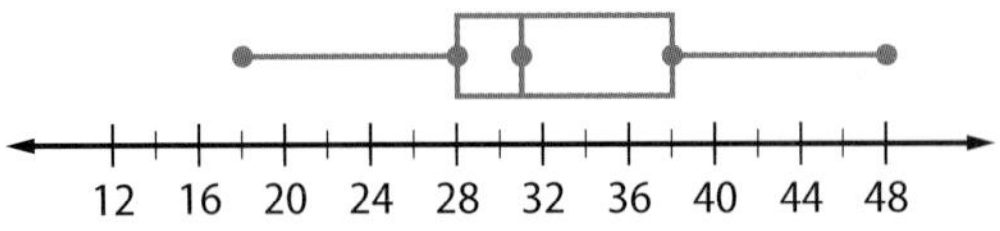

B

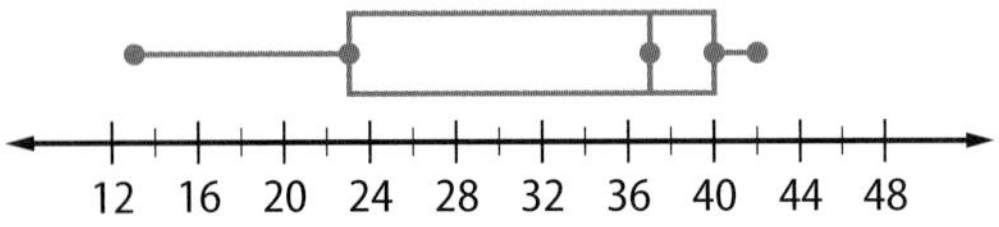

C

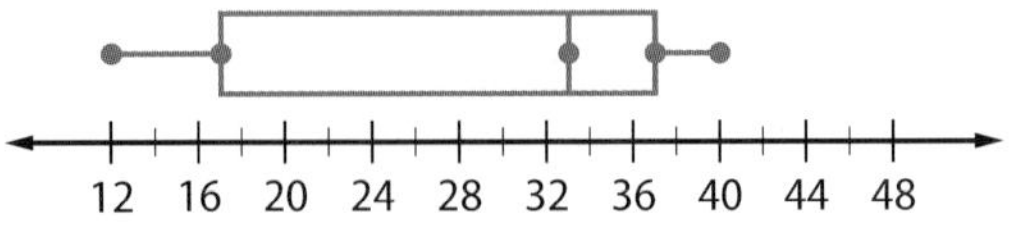

D

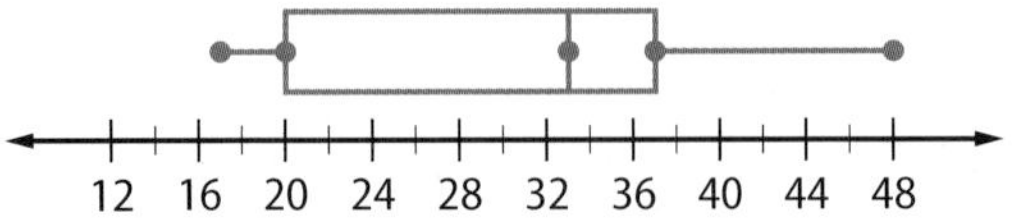

16. Which of the following is not true about the double box plot?

Speed of Roller Coasters (miles per hour)

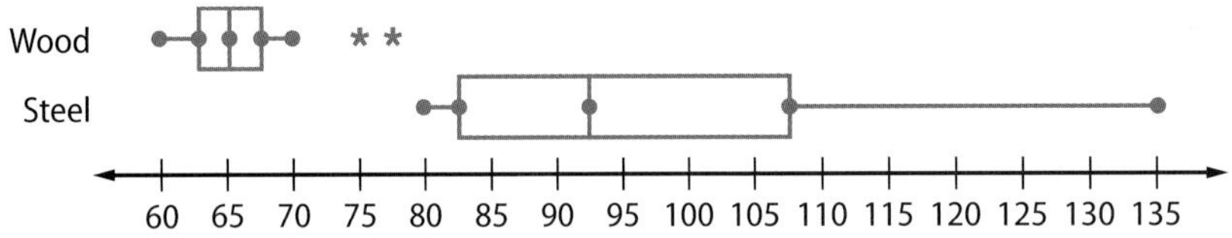

F The data for the steel coasters is symmetric.

G The data for the steel coasters is not symmetric.

H The fastest steel coaster travels 135 miles per hour.

J The slowest wooden coaster travels 60 miles per hour.

Common Core Review

Evaluate each expression for $x = -7$. 6.EE.2c

17. $\frac{3 - x}{2}$

18. $-(x - 1)$

19. $-\frac{56}{2x}$

20. $-\frac{1}{4}(x + 11)$

21. $5(x + 7)$

22. $-10\left(\frac{2.8}{x}\right)$

23. A movie reviewer rates films from 0 stars to 4 stars. The table shows the breakdown of the reviewer's ratings for the year. Compare the median rating to the mean rating. **6.SP.5c**

Movie Ratings				
4 stars	3 stars	2 stars	1 stars	0 stars
52	85	48	11	2

Solve each equation. 8.EE.7, 8.EE.7a

24. $-6x + 8 = 6x + 8$

25. $2x + 3 = -10$

26. $2x - 4 = 2x + 4$

27. $\frac{4x + 8}{4} - 2 = x$

ISG Interactive Study Guide
See page 227 for:
- Mid-Chapter Check

21ST CENTURY CAREER in Market Research

Market Research Analyst

Do you think that gathering and analyzing information about people's opinions, tastes, likes, and dislikes sounds interesting? If so, then you should consider a career in market research. Market research analysts help companies understand what types of products and services consumers want. They design Internet, telephone, or mail response surveys and then analyze the data, identify trends, and present their conclusions and recommendations. Market research analysts must be analytical, creative problem-solvers, have strong backgrounds in mathematics, and have good written and verbal communication skills.

College & Career READINESS

Explore college and careers at ccr.mcgraw-hill.com

Is This the Career for You?

Are you interested in a career as a market research analyst? Take some of the following courses in high school.

- Algebra
- Calculus
- Computer Science
- English
- Statistics

Find out how math relates to a career in Market Research.

Interactive Study Guide
See page 228 for:
- Problem Solving
- Career Project

Lesson 10-5

Using Sampling to Predict

Interactive Study Guide

See pages 229–230 for:
- Getting Started
- Vocabulary Start-Up
- Notes

Essential Question

How are statistics used to draw inferences about and compare populations?

Common Core State Standards

Content Standards
7.SP.1, 7.SP.2

Mathematical Practices
1, 3, 4

Vocabulary

sample
population
unbiased sample
random
simple random sample
stratified random sample
systematic random sample
biased sample
convenience sample
voluntary response sample

What You'll Learn

- Identify various sampling techniques.
- Determine the validity of a sample and predict the actions of a larger group.

Real-World Link

School Clubs Many schools offer a variety of clubs for students to join, like the ones listed below. Does your school have any of these clubs? If you could start a club at your school, what would it be?

Board Games Club	Environmental Club
Cooking Club	Guitar Club
Engineering Club	Sign Language Club

Identify Sampling Techniques

Suppose a school conducted a survey about new clubs. They cannot survey every student, so a randomly selected smaller group called a **sample** is chosen from the larger group, or **population**. The best sample is an **unbiased sample**, or one that is

- representative of the larger population,
- selected at **random** or without preference, and
- large enough to provide accurate data. If a sample is too small, data accurately representing the larger population may not be available.

Concept Summary: Unbiased Samples

Words	Definition	Example
simple random sample	Each item or person in a population is as likely to be chosen as any other.	Thirty student ID numbers are randomly selected by a computer.
stratified random sample	The population is divided into similar, nonoverlapping groups. A simple random sample is then selected from each group.	A population of election districts can be separated into urban, suburban, and rural strata.
systematic random sample	The items or people are selected according to a specific time or item interval.	Every 20 minutes a customer is chosen, or every 10th customer in line is chosen.

A sample that is not representative of the population is called a **biased sample**. A biased sample usually favors certain parts of the population over others.

Concept Summary Biased Samples

Words	Definition	Example
convenience sample	Includes members of the population that are easily accessed.	The first 10 students in the cafeteria line.
voluntary response sample	Involves only those who want to or can participate in the sampling.	The principal sent an e-mail to graduating seniors asking them where to hold commencement. Seniors are asked to vote through an online poll.

Example 1

Tutor

To determine the types of music a concert hall's customers like best, all the people attending a country music concert are surveyed. Identify the sample as *biased* or *unbiased* and describe its type. Explain your reasoning.

Since the attendees at a country music concert probably prefer country music, the sample is biased. The sample is a convenience sample since all of the people surveyed are in one location.

Got It? Do this problem to find out.

1. To determine which passengers' carry-on bags are to be inspected, every eighth person to check in will have his or her bags inspected. Identify the sample as *biased* or *unbiased* and describe its type. Explain your reasoning.

Making Predictions
To make a valid prediction or inference about a population, the sample would need to be random, and either simple, stratified, or systematic.

Validating and Predicting Samples

Depending on the sampling method used, you can make predictions about larger populations. When you make a prediction about a population from a sample of data, you are drawing an *inference* about that population.

Real World

Example 2

Tutor

A pet store mailed a survey to residents to determine their favorite pets. Fifty people responded and the results are shown in the table. Is this sampling method valid? If so, how many people can you expect to choose dogs as their favorite pet in a city with 1585 people? Explain.

Pet	Number
dog	20
cat	16
fish	9
gerbil	5
no pets	0

This is a biased and voluntary response sample since it involves only those who want to participate in the survey. Therefore, this sampling method will not produce an accurate and valid prediction of the total number of dogs in the city.

Got It? Do this problem to find out.

2. Of the 2000 watches made, the manufacturer tests the first 15 watches produced for defects. Of the watches, 3 were defective. Is this sampling method valid? If so, about how many of the 2000 are defective? Explain.

Example 3

Tutor

STEM From a batch of 7500 computer chips produced, the manufacturer sampled every 150th chip at random for defects and found that 2 were defective. Is this sampling method valid? If so, find how many of the 7500 computer chips you can expect to be defective. Explain.

This is a systematic random sample because the samples are selected according to a specific interval. So, this sampling method is reasonable and will produce a valid prediction.

Since every 150 chips were sampled, there were a total of $7500 \div 150$ or 50 chips sampled, and 2 were defective. Two out of 50, or 4%, were defective. So, find 4% of 7500.

Alternate Method

Set up a proportion.

$$\frac{2}{50} = \frac{n}{7500}$$

$$2 \cdot 7500 = 50 \cdot n$$

$$300 = n$$

Words	What number is 4% of 7500?
Variable	Let n = the number of defective chips.
Equation	$n = 0.04 \times 7500$

$n = 0.04 \times 7500$ — Write the equation.

$= 300$ — Multiply.

So, you would expect approximately 300 defective chips.

Got It? Do this problem to find out.

3. After finishing their meal, every fifth person that left a restaurant was surveyed about whether they ordered dessert after their meals. Out of 20 people, 12 said yes. Is this sampling method valid? If so, about how many of the 100 people who had dinner at the restaurant ate dessert?

Guided Practice

Check

Identify each sample as *biased* or *unbiased* and describe its type. Explain your reasoning. (Example 1)

1. To determine how many students at a middle school bring their lunch from home, all the students on one school bus are surveyed.
2. To determine the theme for the homecoming dance, the homecoming committee surveys one classroom.
3. To determine shopping habits at a department store, one male and one female shopper are randomly selected and surveyed from each of their 75 stores.
4. The theater group took a survey about the type of popcorn they should sell during plays. They randomly surveyed 52 students at lunch. Their results are shown in the table. Is this sampling method valid? If so, how many of the boxes of popcorn should be caramel if they order 600 boxes? (Examples 2 and 3)

Survey on Types of Popcorn to Sell

Flavor	Number of Students
butter	18
cheese	14
caramel	13
plain	7

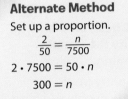

Independent Practice

Go online for Step-by-Step Solutions eHelp

Identify each sample as *biased* or *unbiased* and describe its type. Explain your reasoning. (Example 1)

5. To determine whether a new university library would be useful, all students whose student ID number ends in 2 are surveyed.

6. To determine the quality of cell phones coming off an assembly line, the manager chooses one cell phone every 20 minutes and checks it.

7 To determine the popularity of a musician, a magazine asks teenagers to log on to their Web site and participate in the survey.

8. To determine whether a candidate for governor is popular with the voters, 30% of citizens in each of the 254 counties are surveyed.

9. A school committee wanted to find out if students will recycle at school. The committee randomly surveyed 25% of the teenagers at a mall on a Saturday afternoon. The results are in the graph. Is this sampling method valid? If so, about how many of the 576 students at the school will participate in the program? (Examples 2 and 3)

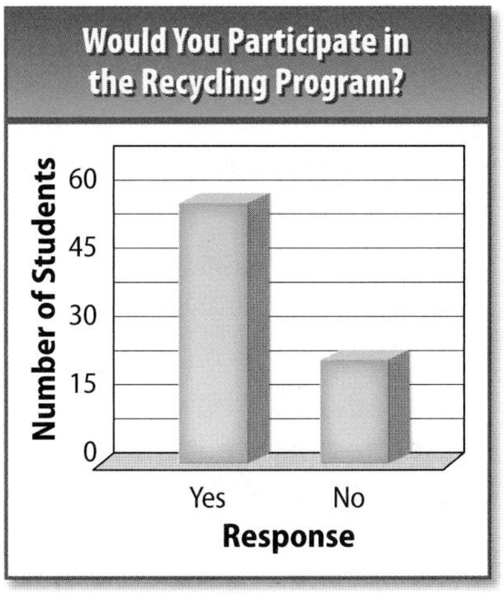

10. Seven of the 28 students in math class have the flu. Is this sampling of the students who have the flu representative of the entire school? If so, how many of the 464 students who attend the school have the flu? (Examples 2 and 3)

11. As people leave a concert, every 10th person is surveyed. They are asked if they would buy a T-shirt. One hundred forty of 800 people surveyed said yes. Is this sampling method valid? If so, how many people would you expect to buy T-shirts at the next concert if 7000 attend? Explain your reasoning.

12. Every hour, twenty customers in a grocery store are randomly selected and surveyed on their milk preference. The results are shown in the table. After reviewing the data, the store manager decided that 40% of his total milk stock should be low-fat milk. Is this a valid conclusion? If not, what information should the store manager review to make a better conclusion?

Milk Preference

Milk	Number
skim	88
low-fat	92
whole	60

13. A movie rental store is considering adding an international section. They randomly selected 300 customers, and 80 customers agree the international movie section is a good idea. Should the store add this section? Explain.

14. Ellis is planning on opening a restaurant in his community. He is conducting a survey to determine what type of food people in the community like.

a. Describe an unbiased population sample that Ellis could survey to get unbiased results.

b. Write two questions that Ellis could ask.

c. After the survey is completed, how could Ellis use the results of the survey to determine what types of food he should serve in his restaurant?

15. The student council is conducting a survey about what types of games and activites to have at the school carnival.

a. Describe an unbiased population sample they could survey to determine the types of games and activities to have at the carnival.

b. Write three questions the student council could ask their sample population.

c. Describe how the student council could use the results of the survey to determine what types of games and activities should be included at the carnival.

16. CCSS **Justify Conclusions** An online gaming site conducted a survey to determine the types of games people play online. The results are shown in the circle graph.

a. If 2500 people participated in the study, how many of them would play arcade or board games?

b. An article said 30% of Americans play card games online. Is this statement valid? Explain your reasoning.

c. Describe how the study could have been conducted so that it represented all Americans and not just online gamers.

Games People Play Online

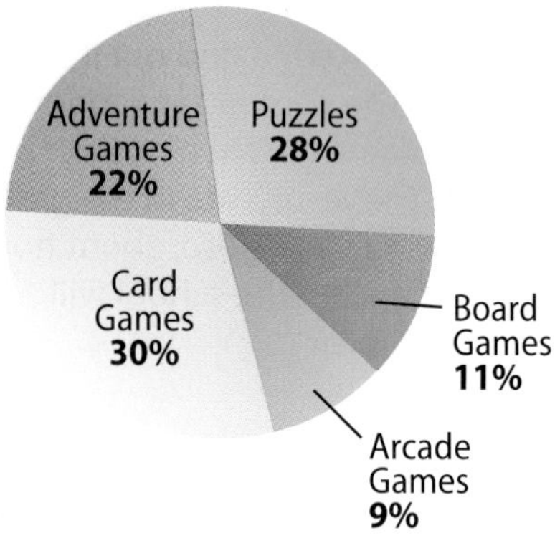

H.O.T. Problems Higher Order Thinking

17. CCSS **Model With Mathematics** Give an example of a unbiased survey. Then conduct the survey. Display the your results in a graph.

18. CCSS **Persevere with Problems** Suppose you are a farmer and want to know if your corn crop is ready to be harvested. Describe an unbiased way to determine whether the crop is ready to harvest.

19. CCSS **Justify Conclusions** If someone were to conduct a survey in person, could the surveyor's tone of voice or how they ask the questions alter the response to the question? Explain.

20. **Building on the Essential Question** Why is sampling an important part of the manufacturing process? Illustrate your answer with an unbiased and biased sampling method you can use to check the quality of DVDs.

Standardized Test Practice

21. A real estate agent surveys people about their housing preferences at an open house for a luxury townhouse. Which is the best explanation for why the results of this survey might *not* be valid?

A The survey is biased because the agent should have conducted the survey by telephone.

B The survey is biased because the sample consisted of only people who already are interested in townhouses.

C The survey is biased because the sample was a voluntary response sample.

D The survey is biased because the agent should have conducted the survey at a single-family home.

22. **Short Response** One hundred people in a music store were surveyed about what type of music they prefer. If 35% of them said they prefer rock music, how many people out of 1500 can be expected to prefer rock music?

An online survey of about 38,000 children produced the results shown in the circle graph.

How Many Cans of Soda Do You Drink in a Day?

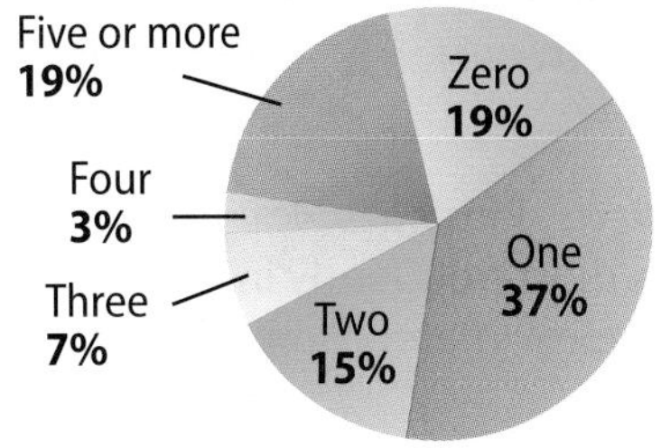

23. About how many of the children surveyed drink two cans of soda or fewer per day?

F 5700 **H** 14,060

G 8993 **J** 26,980

24. Based on the results, about how many children in a class of 30 would drink two or more cans of soda per day?

A 5 **C** 13

B 8 **D** 21

Common Core Review

25. A rectangle has a width of $2x$ inches. The length is 5 inches longer than the width. Write an expression for the perimeter of the rectangle. 7.EE.1

26. The table shows the number of points earned by each player during an online trivia game. Display the data in a histogram. 6.SP.4

Total Points		
Points	Tally	Frequency
80–96	卌 卌	10
97–113	卌	5
114–130	||||	4
131–147	||	2
148–164		0
165–181	|	1

27. Find the percent of change from 32 feet to 79 feet. Round to the nearest tenth, if necessary. Then state whether the percent of change is a *percent of increase* or a *percent of decrease*. 7.RP.3

Solve using the percent equation. Round to the nearest tenth. 7.RP.3

28. 7 is what percent of 32?

29. What is 28.5% of 84?

Factor each expression. 7.EE.1

30. $56x + 16$

31. $250t - 100$

32. $27 + 108n$

Evaluate each expression if $r = -\frac{1}{3}$, $s = \frac{1}{2}$ and $t = -\frac{3}{8}$. 6.EE.2c

33. $s - t$

34. $r + s$

35. $t - r$

Inquiry Lab
Multiple Samples of Data

Inquiry WHY is it useful to collect multiple samples of data to make an inference about a population?

CCSS Content Standards 7.SP.2, 7.SP.6, 7.SP.7
Mathematical Practices 1, 3

Language Arts Tamara is working on a project where she needs to determine which letter occurs most frequently in words in the English language. She also needs to determine approximately how often this letter occurs in words.

Investigation

When you draw a conclusion about a population from a sample of data, you are drawing an inference about that population. Sometimes drawing inferences about a population from only one sample is not as accurate as using multiple samples of data.

The table at the right contains fifteen randomly selected words from an English language dictionary.

Randomly Selected Words from an English Language Dictionary				
airport	doorstop	lemon	print	thread
blueberry	instrument	mileage	sewer	vacuum
costume	juggle	percentage	standard	whale

Step 1 Copy and complete the tables below. Find the frequency of each letter. Record the frequencies in Sample 1 rows of the tables.

Letter	a	b	c	d	e	f	g	h	i	j	k	l	m
Sample 1 Frequency	8	2	3	4	15	0	4	2	4	1	0	5	5
Sample 2 Frequency													
Sample 3 Frequency													

Letter	n	o	p	q	r	s	t	u	v	w	x	y	z
Sample 1 Frequency	6	6	4	0	11	5	9	6	1	2	0	1	0
Sample 2 Frequency													
Sample 3 Frequency													

Step 2 Randomly select another 15 words from the dictionary. Record the frequency of the letters in the rows labeled Sample 2 in the tables above.

Step 3 Repeat Step 2. Record the frequency of the letters in the rows labeled Sample 3.

Collaborate

Work with a partner.

1. The *relative frequency* of a letter occurring is the ratio of the number of times the letter occurred to the total number of times all letters occurred. Find the relative frequency for the letter *e* for each sample.
2. What is the mean relative frequency of the letter *e* for the three samples? The median relative frequency? Round to the nearest tenth if necessary.
3. Use your results to predict the relative frequency of the letter *e* for words in the English language.
4. CCSS **Use Math Tools** Research on the Internet to find the actual relative frequency of the letter *e* for words in the English language. How do you sample results compare to the actual relative frequency?
5. A restaurant randomly hands out prizes with each kids meal. There are three different prizes: a book, a game, and a free pass for a movie. The restaurant gives out the book 40% of the time, the game 40% of the time, and the movie pass 20% of the time.
   a. Design a method to simulate how many times each prize will be distributed.
   b. Use the method you described in part **a** to simulate the prize distribution 20 times. Record the frequency of each prize in a table like the one shown at the right.
   c. Repeat the process described in part b two more times. Record the frequencies of each prize selection in the Sample 2 and Sample 3 columns.
   d. Which prize was selected most often in each sample?

Prize	Sample 1 Frequency	Sample 2 Frequency	Sample 3 Frequency
Book			
Game			
Movie			

Analyze

Work with a partner. Refer to Exercise 5.

6. What is the relative frequency for the book for each sample? Round to the nearest hundredth.
7. What is the mean relative frequency of the book for the three samples? the median relative frequency? Round to the nearest tenth if necessary.
8. CCSS **Justify Conclusions** Sonia wanted to know the average length of last names in the United States. She chose a random sample of ten last names from a telephone directory. The table shows her sample.
   a. Based on this sample, what conclusion might Sonia make about the average length of last names in the United States?
   b. Do you think this is a valid conclusion? Why or why not?
   c. Explain how Sonia could improve her results.

Randomly Selected Last Names	
Bell	Lee
Chan	Mays
Diaz	Ruiz
Ford	Smith
Lamb	Zinn

Reflect

9. Inquiry WHY is it useful to collect multiple samples of data to make an inference about a population?

Lesson 10-6

Probability of Simple Events

Interactive Study Guide

See pages 231–232 for:
- Getting Started
- Vocabulary Start-Up
- Notes

Essential Question

How are statistics used to draw inferences about and compare populations?

Common Core State Standards

Content Standards
7.SP.5

Mathematical Practices
1, 3, 4,

Vocabulary

outcome
simple event
probability
sample space
complement

What You'll Learn

- Find the probability of simple events.
- Find the probability of the complement of an event.

Real-World Link

Basketball A free throw is an unguarded shot awarded when an opponent commits a foul or rule infraction. Mark Price of the NBA and Cynthia Cooper of the WNBA are among the best career free throw shooters in their leagues with percentages of 0.904 and 0.871.

Probability

When you attempt a free throw, there are two possible results: making or missing the shot. Each results is an **outcome**. A **simple event** is one outcome or a collection of outcomes The **probability** of an event is a ratio that compares the number of favorable outcomes to the number of possible outcomes.

$$P(\text{event}) = \frac{\text{number of favorable outcomes}}{\text{number of possible outcomes}}$$

The probability of an event is a number between 0 and 1 that expresses the likelihood of the event occuring. The closer a probability is to 1, the more likely it is to occur.

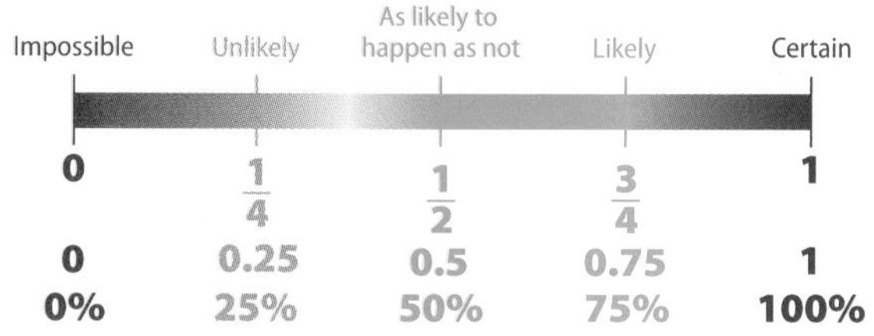

Example 1

Watch Tutor

A bag contains 7 pink, 2 white, and 6 blue marbles. One marble is selected without looking. Find $P(\text{blue})$. Then describe the likelihood of the event. Write *impossible*, *unlikely*, *equally likely*, *likely*, or *certain*.

There are 15 marbles in all, and 6 of them are blue.

$$P(\text{blue}) = \frac{\text{number of blue marbles}}{\text{number of marbles in all}}$$

$$= \frac{6}{15} \text{ or } \frac{2}{5}$$

The probability of a blue marble being chosen is $\frac{2}{5}$ or 40%. Because the probability is less than $\frac{1}{2}$ or 50%, the event is unlikely to occur.

Got It? Do this problem to find out.

1. Find $P(\text{white})$. Then describe the likelihood of the event.

The set of all possible outcomes is called the **sample space**. In Example 1, the sample space is {pink, white, blue}. When you roll a number cube, the sample space is {1, 2, 3, 4, 5, 6}. Outcomes occur at random if each outcome is equally likely to occur.

Math
P(even) is read as the *probability of rolling an even number.*

Example 2

Tutor

Katy rolls a six-sided number cube whose sides are numbered 1 through 6. Find *P*(even). Then describe the likelihood of the event. Write *impossible, unlikely, equally likely, likely,* or *certain.*

The sample space is {1, 2, 3, 4, 5, 6}.

The favorable outcomes are the even numbers: 2, 4, and 6.

$$P(\text{event}) = \frac{\text{number of favorable outcomes}}{\text{number of possible outcomes}}$$

$$= \frac{3}{6} \text{ or } \frac{1}{2} \qquad \text{Simplify.}$$

The probability of rolling an even number is $\frac{1}{2}$, 0.5, or 50%. Because the probability of the event is 50%, the event is equally likely to occur.

Got It? **Do this problem to find out.**

2. Find *P*(3, 4, 5, or 6). Then describe the likelihood of the event. Write *impossible, unlikely, equally likely, likely,* or *certain.*

Vocabulary Link
Complement
Everyday Use something that fills up, or makes complete or perfect
Math Use the set of objects that do not belong to a given set

Find Probability of the Complement

The **complement** of a set is the set of all objects that do not belong to the given set. Two events are complementary if one or the other must happen, but they cannot happen at the same time. For example, when you roll a number cube, the number cube must land on an even number or an odd number, but not both. The sum of the probability of an event and its complement is 1 or 100%.

Real World

Example 3

Tutor

Find *P*(not blue) in Example 1. Then describe the likelihood of the event. Write *impossible, unlikely, equally likely, likely,* or *certain.*

Selecting a blue marble and not selecting a blue marble are complementary events. So the sum of the probabilities is 1.

$$P(\text{blue}) + P(\text{not blue}) = 1 \qquad \text{The sum of the probabilities is 1.}$$

$$\frac{2}{5} + P(\text{not blue}) = 1 \qquad \text{Replace } P(\text{blue}) \text{ with } \frac{2}{5}.$$

$$\frac{2}{5} - \mathbf{\frac{2}{5}} + P(\text{not blue}) = 1 - \mathbf{\frac{2}{5}} \qquad \text{Subtraction Property of Equality}$$

$$P(\text{not blue}) = \frac{3}{5} \qquad \text{Simplify.}$$

The probability of selecting a marble that is not blue is $\frac{3}{5}$ or 60%. Because the probability of the event is greater than 50%, the event is likely to occur.

Got It? **Do this problem to find out.**

3. Find *P*(not 6) when you roll a standard number cube. Then describe the likelihood of the event. Write *impossible, unlikely, equally likely, likely,* or *certain.*

Guided Practice

Check

The spinner shown at right is spun once. Determine the probability of each outcome. Express each probability as a fraction and as a percent. Then describe the likelihood of the event. Write *impossible, unlikely, equally likely, likely,* or *certain.* (Examples 1–3)

1. *P*(6)

2. *P*(even)

3. *P*(greater than 6)

4. *P*(less than 5)

5. *P*(prime)

6. *P*(5 or 8)

7. *P*(not even)

8. *P*(not prime)

9. *P*(not 4)

10. *P*(not 11)

11. A dodecahedron is a three-dimensional solid with 12 identical faces. Stephanie has a number cube in the shape of a dodecahedron with faces numbered 1 through 12. What is the probability that she rolls the number cube and rolls neither a 1 nor a 2? Express the probability as a fraction and as a percent. Then describe the likelihood of the event. Write *impossible, unlikely, equally likely, likely,* or *certain.* (Examples 1–3)

Independent Practice

Go online for Step-by-Step Solutions

eHelp

A dartboard like the one shown is divided into 20 equal sections. Determine the probability of each outcome if a dart is equally likely to land anywhere on the dartboard. Express each probability as a fraction and as a percent. Then describe the likelihood of the event. Write *impossible, unlikely, equally likely, likely,* or *certain.* (Examples 1–3)

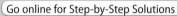

12. *P*(20)

13. *P*(less than 8)

14. *P*(odd)

15. *P*(prime)

16. *P*(greater than 16)

17. *P*(multiple of 3)

18. *P*(less than 21)

19. *P*(25)

20. *P*(even)

21. *P*(1 or 2)

22. *P*(greater than 20)

23. *P*(greater than 0)

24. *P*(not 4, 5, or 6)

25. *P*(not a multiple of 2)

26. *P*(not 20)

27. *P*(not prime)

28. *P*(not odd)

29. *P*(not 25)

30. *P*(composite)

31. *P*(not 1)

32. A bag contains 3 red tiles, 5 black tiles, and 4 white tiles. You choose a tile without looking.

a. What is the probability that you choose a white tile? Express the probability as a fraction and as a percent. Then describe the likelihood of the event. Write *impossible, unlikely, equally likely, likely,* or *certain.* (Examples 1–3)

b. What is the probability that you do not choose a white tile? Express the probability as a fraction and as a percent. Then describe the likelihood of the event. Write *impossible, unlikely, equally likely, likely,* or *certain.*

33. STEM The blood type O+ is the most common blood type in the United States. Approximately 37.4% of the population has this blood type. If you choose a person in the United States at random, what is the probability that his or her blood type is not O+?

34. Aaron has a bag of marbles, as shown in the figure. He chooses one of the marbles without looking.

a. Is Aaron more likely to choose a blue marble or a green marble? Why?

b. What is the probability that Aaron chooses a red marble or a yellow marble?

c. What is the probability that Aaron does not choose a green marble?

35. In a raffle, one ticket is randomly chosen from 80 tickets to receive free haircuts for a year. If Wakim entered 6 tickets, what is the probability that he is *not* chosen to receive the free haircuts?

36. CCSS **Justify Conclusions** The table shows the heights of the roller coasters at an amusement park. Elissa wants to ride just one roller coaster, so she chooses one of them at random.

a. Is she likely to choose a roller coaster taller than 250 feet? Why or why not?

b. Is she likely to choose a roller coaster taller than 350 feet? Why or why not?

Roller Coasters

Name	Height (ft)
Cobra	345
Screamer	410
Zipster	185
Maniac	230
Flyer	255

37. A bag contains 26 tiles, each with a different letter of the alphabet written on it. A tile is selected at random. What is the probability that the tile contains a letter in the word MATHEMATICS?

38. A bucket of ice at a picnic contains 24 juice drinks. The probability of choosing a drink at random and choosing an orange-mango drink is $\frac{2}{3}$. How many of the drinks are orange-mango drinks?

39. A jar contains red, white, and blue marbles. A marble is selected at random. $P(\text{red}) = \frac{1}{10}$ and $P(\text{white}) = \frac{2}{5}$. What is $P(\text{blue})$? Explain.

40. One piece of candy is picked, without looking, from a dish that contains ten red candies, five blue candies, three green candies, and two yellow candies. Explain how likely it is for each event to happen.

a. brown

b. red, blue, or green

H.O.T. Problems Higher Order Thinking

41. CCSS **Model with Mathematics** Give an example of a real-world situation in which the probability of an event is 25%.

42. CCSS **Construct an Argument** A refrigerator case at a cafeteria contains some containers of yogurt. You choose a container at random. The probability of choosing a peach yogurt is $\frac{1}{5}$. Construct an argument to explain why the refrigerator case cannot contain exactly 18 containers of yogurt.

43. CCSS **Persevere with Problems** A school committee consists of 4 girls and 6 boys. How many girls should be added to the committee so that the probability of choosing a student from the committee at random and choosing a girl is 70%?

44. **Building on the Essential Question** Explain how to find the probability of an event by first calculating the probability of the complement of the event.

Standardized Test Practice

45. The production records of a toy manufacturing company show that 5 out of every 75 toys have a defect. What is the probability that a randomly selected toy manufactured at the company will *not* have a defect?

A 7%
B 27%
C 70%
D $93.\overline{3}\%$

46. If Luke spins a spinner like the one shown, what is the probability that it lands on the space with the square or the space with the triangle?

F $\frac{1}{4}$
G $\frac{1}{2}$
H $\frac{2}{3}$
J $\frac{3}{4}$

47. Short Response Casey has a bag containing 3 white, 8 red, 2 blue, and 2 yellow marbles. She randomly picks a marble from the bag. What is the probability that the marble Casey picks will be white?

48. A bag contains some tiles. Each tile has the number 1, 2, 3, 4, 5, or 6 written on it. The table shows the frequency of each number in the bag. You choose a tile at random. Which is closest to the probability of choosing a 3?

Number	1	2	3	4	5	6
Frequency	13	9	20	8	10	15

A 13%
B 17%
C 20%
D 27%

Common Core Review

49. Use the graph shown. 6.SP.4

a. Describe the data.

b. Why is there a jagged line in the vertical axis?

c. How many states have no roller coasters? Explain.

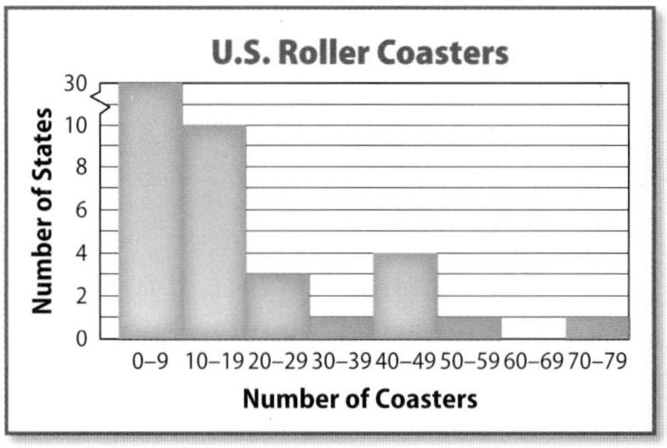

50. The double box plot below shows the average gas mileage for some cars. 7.SP.4

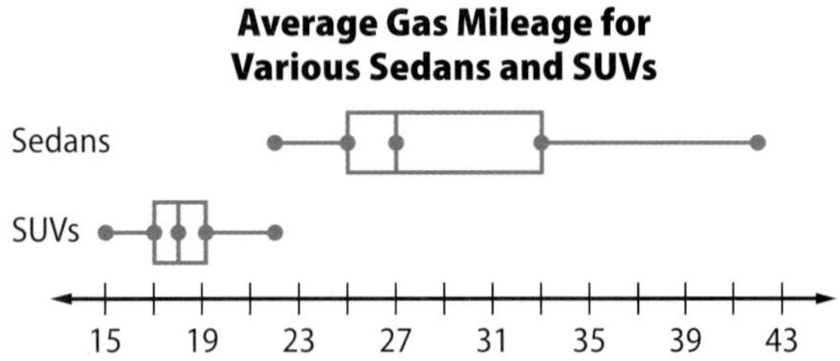

a. Which types of vehicles tend to be less fuel-efficient?

b. Compare the most fuel-efficient SUV to the least fuel-efficient sedan.

Solve each system of equations algebraically. 8.EE.8b

51. $y = x + 4$
$x = 5$

52. $y = 5x - 2$
$x = 0$

53. $y = x - 3$
$y = 0$

54. $y = 2x$
$x = 3.5$

55. $y = x - 1$
$y = 2x$

56. $x + y = 4$
$y = -1$

57. $y = 2x + 2$
$y = 4x$

58. $y = 3x + 2$
$x + y = 6$

59. $y = x$
$y + x = 12$

Inquiry Lab

Relative Frequency and Probability

Inquiry HOW is probability related to relative frequency?

Content Standards
7.SP.6, 7.SP.7, 7.SP.7a

Mathematical Practices
1, 3

Vocabulary
relative frequency

Games Jonah is playing a board game with his sister. The game uses two number cubes. Jonah will win the game if he rolls the number cubes and the sum of the numbers he rolls is 7. Jonah wants to know approximately how often the sum of the numbers rolled will be 7.

You can conduct an experiment to find the relative frequency of rolling a sum of 7. The **relative frequency** is the ratio of the number of experimental successes to the number of experimental attempts.

Investigation

Step 1 Determine all possible outcomes of the experiment. The possible outcomes are all possible sums of two number cubes: 2, 3, 4, 5, 6, 7, 8, 9, 10, 11, and 12.

Step 2 Two number cubes were rolled 50 times. The frequency of the sum of the numbers rolled was recorded in the frequency table.

Sum	2	3	4	5	6	7	8	9	10	11	12
Frequency	1	3	5	6	8	9	7	7	1	2	1

Step 3 Find the relative frequency of rolling a sum of 7.

$$\frac{\text{number of times a sum of 7 was rolled}}{\text{total number of rolls}} = \frac{9}{50} = 0.18$$

So, based on the experiment, Jonah can expect to roll a sum of 7 about 18% of the time.

Collaborate

Work with a partner.

1. Repeat the above experiment and make your own table to keep track of the results. Based on your experiment, what is the relative frequency of rolling a sum of 7?
2. If you repeat the experiment but roll the number cubes 100 times, would you expect to find the same relative frequency of rolling a sum of 7? Explain.
3. Copy and complete the table to show all possible sums when you roll two number cubes.

		Second Number Cube					
		1	2	3	4	5	6
First Number Cube	1	2	3				
	2	3	4				
	3						
	4						
	5						
	6						

4. Use your table to find the probability of rolling a sum of 7.

5. How does the probability of rolling a sum of 7 compare to the relative frequency of rolling a sum of 7.

6. What are the possible outcomes when you flip two pennies at the same time?

7. Flip two pennies 40 times and record the results in a table like the one shown below.

Result	Two Heads	One Head, One Tail	Two Tails
Frequency			

8. Find the relative frequency of flipping two heads and the relative frequency of flipping two tails.

9. Based on your results, do you think both pennies are as likely to both land on heads as they are to both land on tails? Explain.

10. The table at the right shows the sample space for flipping two coins. Find the probability that both pennies will land on heads and the probability that both pennies will land on tails.

		Second Penny	
		Heads	Tails
First Penny	Heads	H, H	H, T
	Tails	T, H	T, T

Analyze

11. In the experiment with two number cubes, why do you think you were asked to collect data by rolling the number cubes 50 times rather than 3 times?

12. CCSS **Reason Inductively** What do you think would happen if you rolled the number cubes 1000 times and found the relative frequency of rolling a sum of 7? What if you rolled the number cubes 10,000 times?

13. CCSS **Justify Conclusions** A spinner has 10 equal sections that are numbered from 1 to 10. Suppose you spin the spinner 80 times and record the results. What do you think would be the approximate relative frequency of landing on 1? Explain.

14. CCSS **Reason Inductively** Dontrell flipped a coin 100 times. He kept track of the results, as shown in the table at right.

Flipping a Coin		
Result	Heads	Tails
Frequency	98	2

   a. Based on Dontrell's experiment, what is the relative frequency of landing on heads?

   b. A fair coin is a coin that is equally likely to land on heads or tails. What is the probability that a fair coin lands on heads?

   c. What do you think might explain Dontrell's results?

15. A box of a certain brand of cereal contains one of six prizes. Each prize is equally likely to occur. You want to know how likely you are to buy two boxes at the same time and get the same prize in each box. Explain how you could use number cubes to conduct an experiment that addresses this question. Describe the data you would record and any calculations you would make.

Reflect

16. Inquiry HOW is probability related to relative frequency?

Lesson 10-7

Theoretical and Experimental Probability

ISG Interactive Study Guide

See pages 233–234 for:
- Getting Started
- Real-World Link
- Notes

Essential Question

How are statistics used to draw inferences about and compare populations?

CCSS Common Core State Standards

Content Standards
7.SP.7, 7.SP.7a, 7.SP.7b

Mathematical Practices
1, 3, 4

Vocab abc Vocabulary

uniform probability model

theoretical probability

experimental probability

What You'll Learn

- Find and compare experimental and theoretical probabilites.
- Predict the actions of a larger group.

Real-World Link

Games Redemption games are games that you play to earn tickets or points to trade for prizes. Some redemption games require varying amounts of skill to earn tickets, but others are purely games of chance.

Theoretical and Experimental Probability

In a **uniform probability model**, each outcome has an equal probability of occurring. For example, some carnival games involve spinning a wheel to win a prize. If each prize is equally likely, you can use a uniform probability model to calculate your chances of winning a particular prize.

Theoretical probability is based on uniform probability — what *should* occur when conducting a probability experiment. **Experimental probability** is based on relative frequency — what *actually* occurs during such an experiment. If the number of trials is very small, there can be a wide variation in results.

Example 1

Tools Tutor

The table shows the results of an experiment in which a number cube was rolled. Find the experimental probability of rolling a six for this experiment. Then compare the experimental probability with the theoretical probability.

Outcome	Tally	Frequency
1	IIII	4
2	𝍸 I	6
3	II	2
4	𝍸 III	8
5	III	3
6	𝍸 I	6

$$\frac{\text{number of times six is rolled}}{\text{number of possible outcomes}} = \frac{6}{4+6+2+8+3+6}$$

$$= \frac{6}{29}$$

The experimental probability of rolling a six in this case is $\frac{6}{29}$ or about 21%.

The theoretical probability of rolling a six is $\frac{1}{6}$ or about 17%.

So, rolling a six in the experiment occurred more often than expected.

Got It? Do this problem to find out.

1. Find the experimental probability of rolling a four for the experiment above. Then compare it to the theoretical probability.

Make Predictions

You can use theoretical or experimental probability to predict future events.

Example 2

Tutor

The table shows the results of a survey that asked 200 teens what they would be doing if they were not online. Out of a similar group of 450 teens, predict how many would listen to music.

Teen Hobbies (Other Than Internet)	
Activity	**Number of Responses**
watch TV	50
listen to music	52
physical activity/sports	54
read	18
write/draw	10
other	16

Find the experimental probability that a student listens to music.

$$P(\text{music}) = \frac{\text{number who said "listen to music"}}{\text{total number surveyed}}$$

$$= \frac{52}{200}$$

$$= 0.26 \text{ or } 26\%$$

Use the percent proportion to find 26% of 450.

The percent is 26%, and 450 is the whole. Let n represent the part.

Percent Proportion
compares part of a quantity to the *whole* quantity as a fraction

part ⟶ whole ⟶ $\frac{n}{450} = \frac{26}{100}$ } percent

$100 \cdot n = 26 \cdot 450$

$100n = 11{,}700$ Find the cross products.

$= 117$ Mentally divide each side by 100.

Check Estimate: 25% of 440 is 110. So, 117 is reasonable. ✓

Got It? **Do this problem to find out.**

2. Out of the 450 teens surveyed, how many would you expect to say they would be writing or drawing?

Guided Practice

Check

1. The table shows the results of an experiment in which a number cube was rolled. Find the experimental probability of rolling a 4. Then compare it to the theoretical probability. (Example 1)

Number	Frequency
1	II
2	I
3	IIII
4	III
5	III
6	II

2. Isaiah flipped a coin 40 times. He found that the coin landed heads up 18 times. Find the experimental probability of the coin landing heads up. Then compare it to the theoretical probability. (Example 1)

3. Without looking, Delores took a handful of multicolored candies from a bag and found that 40% of the candies were yellow. Suppose that there were 375 candies in the bag. How many can she expect to be yellow? (Example 2)

4. The table shows the results of a survey about the most popular pets among seventh-grade students at Stanyon Middle School. Out of a similar group of 320 seventh-graders, predict how many would choose cats as their favorite pet. (Example 2)

Most Popular Pets					
Pet	dog	cat	bird	fish	other
Number of Responses	12	11	6	5	6

Independent Practice

Go online for Step-by-Step Solutions eHelp

5 The table shows the results of an experiment in which Alexis spun the spinner shown 20 times. (Example 1)

Result	Frequency
1	IIII
2	卌
3	卌
4	卌 I

a. What is the experimental probability of the spinner landing on 4?

b. What is the experimental probability of the spinner landing on 3?

c. What is the theoretical probability of the spinner landing on 3? Compare it to the experimental probability.

6. The table shows the results of an experiment in which two spinners like the one shown above were each spun 50 times. (Example 1)

Sum	Frequency
2	II
3	III
4	卌 III
5	卌 I
6	卌 卌 卌
7	卌 III
8	卌 III

a. What is the experimental probability of spinning a sum of 6?

b. What is the experimental probability of spinning a sum of 8?

c. What is the theoretical probability of spinning a sum of 8? Compare it to the experimental probability.

7. The table shows the results of a survey of visitors to a science museum. They were asked to name their favorite exhibit at the museum. (Example 2)

Favorite Exhibits	
Exhibit	**Number of Responses**
aquarium	9
3-D cinema	12
planetarium	8
rain forest	15
dinosaurs	6

a. Out of a similar group of 425 museum visitors, predict how many would choose the planetarium as their favorite exhibit.

b. Out of a similar group of 600 museum visitors, predict how many would choose the aquarium as their favorite exhibit.

8. The table shows the results of a survey of students at Montez Middle School. The students were asked who they plan to vote for in an upcoming election for student body president. Suppose all 520 students at the school vote. (Example 2)

Student Body President Survey	
Candidate	**Number of Responses**
Abbi	5
Bryan	6
Carlos	12
Jack	9
Jasmine	16
Ming	12

a. Predict how many votes Carlos would receive.

b. Predict how many more votes Ming would receive than Jack.

9. **Financial Literacy** A group of adults were surveyed about their preferred method of paying bills. The results of the survey are shown in the table. (Example 2)

Preferred Method of Paying Bills	
Method	**Number of Responses**
check	52
cash	15
debit card	18
credit card	60
online	45
other	10

   a. Based on the survey, if you choose an adult at random, what would be the probability that he or she prefers to pay bills online? Why?

   b. A new business is expecting 3500 customers in its first month. How many of the customers should the business expect will pay their bill by check?

10. A jar contains 40 pennies, 18 nickels, 20 dimes, and 12 quarters. Sophia chooses a coin at random, notes the result, and replaces the coin. She repeats this until she has done it a total of 45 times. Predict the number of times she will choose a penny. Explain.

11. In an assembly line, 56 out of 60 randomly selected parts have no faults. Predict the number of parts that will have no faults in a similar batch of 12,000 parts.

12. The table shows the results of a survey that asked 200 youths about what is important to their personal success.

   a. If 1200 youths were surveyed, how many would you expect to say friendships or immediate family are factors in their success?

   b. Suppose 1500 youths were surveyed. How many would you expect to say work and career or friendships are a factor in their success?

State of Our Nation's Youth	
Item of Importance	**Number of Responses**
work and career	18
personal development and satisfaction	38
friendships	45
immediate family	52
make a contribution to society	47

13. **CCSS Justify Conclusions** Based on the past season, the experimental probability that a basketball player will make a free throw is 0.76. The player needs to make 5 out of 7 free throws to win a free throw tournament. Do you think the player is likely to win the tournament? Justify your answer.

14. The table shows the results of a class survey about students' favorite X Games sport.

   a. What is the probability of skateboarding being someone's favorite sport?

   b. Suppose 800 students in the school are surveyed. At this rate, predict how many will choose inline skating as their favorite sport.

   c. Of the 800 surveyed, predict how many more students will prefer BMX than MotoX. Explain.

X Games	
Game	**Number of Responses**
BMX	8
Inline	6
MotoX	4
Skate-boarding	15
Wake-boarding	7

H.O.T. Problems Higher Order Thinking

15. **CCSS Persevere with Problems** The experimental probability of a penny landing on tails is $\frac{9}{16}$. If the penny landed on heads 21 times, how many times was the coin tossed?

16. **CCSS Persevere with Problems** A survey found that 95 households out of 150 have high-speed Internet and that 125 out of 200 have cable television. What is the probability that a household has both?

17. **Building on the Essential Question** Jasmine's sock drawer has 8 white pairs, 5 black pairs, 7 navy pairs, 3 khaki pairs, and 12 other pairs. She randomly selected a pair of socks 20 times and selected a navy pair 6 times. Compare and contrast the theoretical probability with the experimental probability.

Standardized Test Practice

18. Nikki has a set of five cards numbered 1 through 5. She shuffles the cards and chooses one card at random. She repeats the process several times and records the results in the table below. Based on her results, what is the experimental probability of choosing a 1?

Number	1	2	3	4	5
Frequency	6	4	5	3	7

A 20%
B 24%
C 25%
D 30%

19. Short Response Esteban flips a coin several times. The table shows his results. What is the experimental probability that the coin lands tails-up? Express the probability as a fraction and as a percent.

Result	Heads	Tails
Frequency	31	19

20. In a survey, 60 students were asked to choose their favorite type of film, and 27 of the students chose action films. Out of a similar group of 480 students, which is the best prediction of the number of students who would choose action films as their favorite type of film?

F 45
G 130
H 216
J 288

21. Sushila has a bag containing red, blue, and green marbles. She chooses a marble at random, notes the color, and replaces it. She does this 12 times and finds that she chose a red marble *r* times. Which of the following is the experimental probability of choosing a red marble?

A $\frac{r}{12}$
B $\frac{12}{r}$
C $\frac{r}{3}$
D $\frac{3}{r}$

Common Core Review

The spinner shown at the right is spun once. Determine the probability of each outcome. Express each probability as a fraction and as a percent. Then describe the likelihood of the event. Write *impossible, unlikely, equally likely, likely,* or *certain*. 7.SP.5

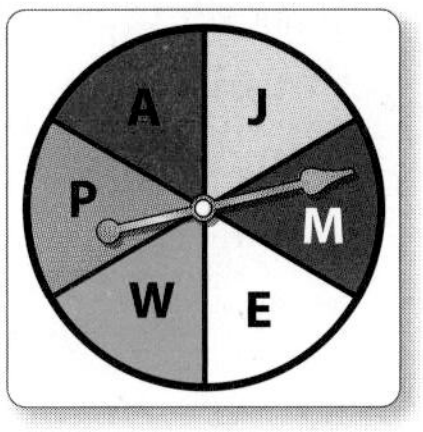

22. $P(A)$
23. $P(\text{not } A)$
24. $P(J \text{ or } W)$
25. $P(\text{vowel})$
26. $P(\text{not a vowel})$
27. $P(M, P, \text{or } J)$
28. $P(T)$
29. $P(\text{consonant})$
30. $P(\text{not } Z)$
31. $P(A, J, M, E, W, \text{or } P)$

Solve each equation. 8.EE.7a

32. $4(n - 4) = 4n$
33. $3x + 6 = 3(x + 2)$
34. $3y + 8 - y = 2(y + 4)$
35. $32 + x = 8(x - 1) - 7x$
36. $\frac{1}{2}(2m + 10) = 5 + m$
37. $-2(x - 2) - 4 = 1 - 2x$

38. STEM The table shows the relationship between the number of aluminum atoms and the total number of protons in the atoms. Write an equation that gives the number of protons *p* in any number of aluminum atoms *a*. 7.RP.2c

Aluminum Atoms	
Number of Atoms	Number of Protons
3	39
4	52
5	65

Lesson 10-8

Probability of Compound Events

Interactive Study Guide

See pages 235–236 for:
- Getting Started
- Real-World Link
- Notes

Essential Question

How are statistics used to draw inferences about and compare populations?

Common Core State Standards

Content Standards
7.SP.8, 7.SP.8a, 7.SP.8b

Mathematical Practices
1, 3, 4, 5

Vocabulary

compound event
tree diagram

What You'll Learn

- Find the number of outcomes for an event.
- Find the probability of a compound event.

Real-World Link

Pizza Pizza is a favorite food in many American households because you can get just about anything on a pizza—it satisfies almost everyone! Most pizzerias offer different sizes of pizzas, as well as different crusts and toppings.

Outcomes of Compound Events

A **compound event** consists of two or more simple events. The probability of a compound event is the fraction of outcomes in the sample space for which the compound event occurs. When you work with compound events, you can make a table, draw a **tree diagram**, or make a list to display and count all of the outcomes in the sample space.

Example 1

Tutor

A pet store sells aquariums in three shapes, hexagon, pentagon, and rectangle, and two sizes, 10 gallons and 20 gallons. How many different fish tanks can be made from the different shapes and sizes?

You can use a table or a tree diagram to find the number of outcomes. List each shape choice. Then pair each size choice with each shape.

Shape	Size
hexagon	10 gallons
pentagon	10 gallons
rectangle	10 gallons
hexagon	20 gallons
pentagon	20 gallons
rectangle	20 gallons

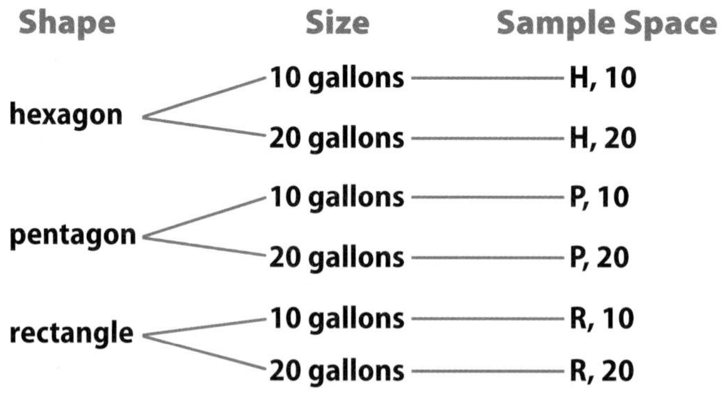

Using either method, the sample space consists of 6 outcomes.

Got It? Do this problem to find out.

1. Miranda has 5 shirts and 4 pairs of pants that she uses for her work clothes. How many different outfits can be assembled using 5 shirts and 4 pairs of pants?

Brand X Pictures/PunchStock

Example 2

Tutor

Students are assigned a temporary password the first time they visit the computer lab. Temporary passwords consist of a letter (A, B, or C), followed by a number (1 or 2), followed by a letter (X, Y, or Z). How many different temporary passwords are there?

Make a list of all the possible passwords.

A1X	A1Y	A1Z	A2X	A2Y	A2Z
B1X	B1Y	B1Z	B2X	B2Y	B2Z
C1X	C1Y	C1Z	C2X	C2Y	C2Z

So, there are 18 different temporary passwords.

> **Watch Out!**
> **Making a List**
> When you make a list, be sure to do so in an organized way. For example, first write all the passwords that begin with A1, then write all the passwords that begin with A2, and so on.

Got It? Do this problem to find out.

2. A pizza shop has regular, deep-dish, and thin pizza crusts, two different cheeses, and five toppings. How many different one-cheese and one-topping pizzas can be ordered?

Probability of Compound Events

When you know the number of outcomes, you can find the probability that a compound event will occur.

Real World Example 3

Tutor

Emilio has 2 counters. Each counter has one side marked with an E and the other side marked with a J, for Jacob. Both counters are tossed. If one counter lands with E up and the other lands with J up, Emilio wins. Otherwise, Jacob wins. What is the probability that Emilio will win?

First find the number of outcomes. The possible outcomes are EE, EJ, JE, and JJ.

There are four equally likely outcomes, with two favoring Emilio. So, the probability of Emilio winning is $\frac{1}{2}$ or 50%.

Got It? Do this problem to find out.

3. What is the probability of randomly choosing a 5-letter password that consists of only vowels if each vowel can be used more than once?

When you work with compound events, you can use the **Fundamental Counting Principle** to find the number of outcomes.

Key Concept: Fundamental Counting Principle

Words	If event M can occur in m ways and is followed by event N that can occur in n ways, then the event M followed by N can occur in $m \cdot n$ ways.
Example	If there are 4 possible sizes for fish tanks and 3 possible shapes, then there are $4 \cdot 3$ or 12 possible fish tanks.

Example 4

Lamar is going to spin each spinner once. What is the probability that he will spin red and the number 9?

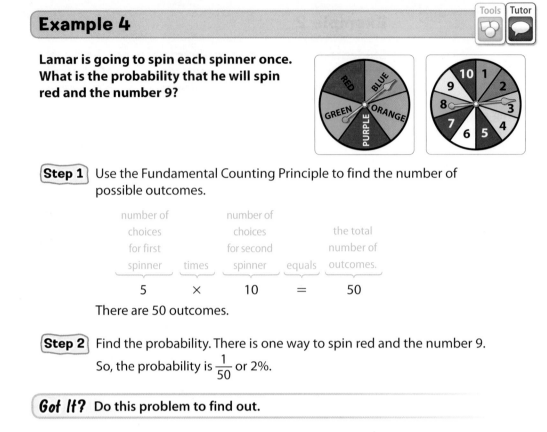

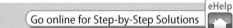

Step 1 Use the Fundamental Counting Principle to find the number of possible outcomes.

number of choices for first spinner	times	number of choices for second spinner	equals	the total number of outcomes.
5	×	10	=	50

There are 50 outcomes.

Step 2 Find the probability. There is one way to spin red and the number 9. So, the probability is $\frac{1}{50}$ or 2%.

Got It? **Do this problem to find out.**

4. Three number cubes are rolled. What is the probability of rolling three 5s?

Guided Practice

Use any method to find the total number of outcomes in each situation. (Examples 1 and 2)

1. Find the number of tennis shoes available if they come in gray or white and are available in size 6, 7, or 8.

2. The table shows the options a dealership offers for a model of a car.

Doors	Gears	Color
2-door 4-door	automatic 5-speed 4-speed	black tan white red

3. Elisa can take 6 different classes first period, 2 different classes second period, and 3 different classes third period.

4. Amber has a denim and a black purse. Ebony has a black, a red, a denim, and a brown purse. Each girl picks a purse at random to bring to the mall. What is the probability the girls will bring the same color purse? (Examples 3 and 4)

5. A spinner with 8 equal sections labeled 1–8 is spun twice. What is the probability that it will land on 6 after the first spin and on 6 after the second spin? (Examples 3 and 4)

Independent Practice

Go online for Step-by-Step Solutions

Use any method to find the total number of outcomes in each situation. (Examples 1 and 2)

6. Nathan has 4 t-shirts, 4 pairs of shorts, and 2 pairs of flip-flops.

7. Frozen yogurt is available in 2 sizes and 6 flavors.

8. Malik flips a penny, a nickel, and a dime at the same time.

Use any method to find the total number of outcomes in each situation. (Examples 1 and 2)

9. A pet store has male and female huskies with blue, green, and amber eyes.

10. A coin is tossed and a 12-sided solid with identical faces is rolled.

11 There are 3 true-false questions on a quiz.

12. Kane rolls a number cube and flips a quarter at the same time. What is the probability he rolls a 1 and the quarter lands heads up? (Examples 3 and 4)

13. STEM A newborn baby is equally likely to be male or female. What is the probability that a mother's first three children will all be girls? (Examples 3 and 4)

14. The table shows the sandwich choices for lunch at a cafe. If a one-bread, one-meat, one-cheese sandwich is chosen at random, what is the probability that it will be turkey and Swiss on wheat bread? (Examples 3 and 4)

Sandwich Choices		
Breads	**Meat**	**Cheese**
wheat rye white	roast beef turkey ham pepperoni	swiss cheddar american

15 A game requires you to toss a 10-sided numbered solid and a 6-sided numbered solid to determine how to move on a game board. Find the following probabilities.

a. *P*(same number on both)

b. *P*(odd, even) or *P*(even, odd)

16. Every U.S. citizen has a nine-digit Social Security number. The first three digits are a geographic code, the next two digits are determined by the year and the state where the number is issued, and the last four digits are random numbers. If you are less than 100 years old, what is the probability that the last two numbers are your age?

17. CCSS **Use Math Tools** Each of the spinners at the right is spun once. Use a tree diagram to find the following probabilities.

a. *P*(at least one 2)

b. *P*(at least one 3)

H.O.T. Problems Higher Order Thinking

18. CCSS **Model with Mathematics** Give an example of a real-world situation that has 16 outcomes.

19. CCSS **Find the Error** Cameron is finding the number of different outfits he can make with 4 pairs of shoes, 4 pairs of pants, and 5 shirts. Find his mistake and correct it.

> 4 + 4 + 5 = 13, so there are 13 different outfits.

20. CCSS **Justify Conclusions** Marcus has a choice of a white, gray, or black shirt to wear with a choice of tan, black, brown, or denim pants. Without calculating the number of possible outcomes, how many more outfits can he make if he buys a green shirt? Explain your reasoning to a classmate.

21. CCSS **Persevere with Problems** Write an algebraic expression for the number of possible outcomes if the spinner at the right is spun *x* number of times.

22. **Building on the Essential Question** Explain how a tree diagram might be more useful than the Fundamental Counting Principle when finding probability.

Standardized Test Practice

23. The spinner is spun twice. What is the probability that it will land on 2 after the first spin and on 5 after the second spin?

A $\frac{1}{64}$

B $\frac{1}{16}$

C $\frac{1}{8}$

D $\frac{5}{8}$

24. A bicycle lock has 4 rotating discs and each contains the digits 0–9. How many different lock combinations are possible?

F 3024

G 5040

H 6561

J 10,000

25. A restaurant offers a combo special with 3 different sandwiches, 2 different salads, and 5 different drinks. From how many different combos could Keisha choose?

A 10

B 11

C 30

D 50

26. **Short Response** Pilan can make outfits out of the clothes shown in the table. Draw a tree diagram to show all of the possible outcomes

Shoes	Shirts	Pants
white black	blue red green black	tan navy denim

Common Core Review

27. Sixty-three of the 105 students in the band said that their favorite class was music. Is this sampling representative of the entire school? If so, how many of the 848 students who attend the school would say music is their favorite class? 7.SP.1

28. Maresha took a random sample from a package of jelly beans without looking and found that 30% of the beans were red. Suppose there are 250 jelly beans in the package. How many can she expect to be red? 7.SP.2

29. Use the food data shown in the table. 7.SP.4

Fat (g) of Various Chicken Sandwiches and Burgers								
Chicken	8	13	13	15	15	18	19	20
Burgers	10	14	20	20	26	31	32	36

a. What is the median number of fat grams in each sandwich?

b. In general, which type of sandwich has a lower amount of fat? Explain.

c. Find the measures of variability and any outliers for the data.

30. Population density is a unit rate that gives the number of people per square mile. If the area of North Carolina is 48,711 square miles and its population is 8,856,505 people, what is the population density of North Carolina? Round to the nearest tenth. 6.RP.2

Add. 7.EE.1

31. $(x + 3) + (2x + 4)$

32. $(9x + 2) + (2x + 9)$

33. $(3x + 1) + (3x - 7)$

34. $(4x - 3) + (-2x + 2)$

35. $(-4x - 2) + (-5x - 1)$

36. $(3x - 8) + (-10x - 2)$

Inquiry Lab

Simulate Compound Events

Inquiry HOW can you use a simulation to find the probability of a compound event?

CCSS Content Standards
7.SP.8c

Mathematical Practices
1, 3

Vocabulary
simulation

Restaurants A restaurant randomly gives coupons to 3 out of every 8 customers. The owner of the restaurant wants to know the probability that a customer receives a coupon two days in a row.

A **simulation** is a way of modeling a problem situation. Simulations often model events that would be difficult or impractical to perform. For example, you can use a spinner to model the compound event that a customer receives a coupon two days in a row.

Investigation 1

Step 1 Create a spinner with eight equal sections. Label three of the sections C for "receives a coupon." Label the other sections D for "does not receive a coupon."

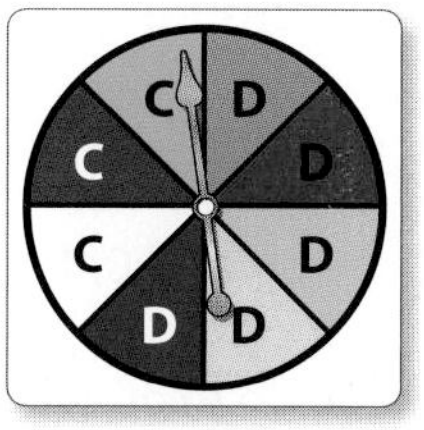

Step 2 Every two spins of the spinner represent one trial. Spin the spinner and record whether or not the customer receives a coupon on each day.

Step 3 The table shows the results after 20 trials. The circled data represents the times a coupon was given to a customer on both days.

Trial	1	2	3	4	5	6	7	8	9	10
Day 1	C	D	C	C	D	D	D	D	D	C
Day 2	D	D	C	D	D	D	C	C	D	C
Trial	**11**	**12**	**13**	**14**	**15**	**16**	**17**	**18**	**19**	**20**
Day 1	D	D	D	C	D	C	C	D	D	D
Day 2	C	D	D	D	C	C	D	D	D	D

Step 4 Calculate the experimental probability of receiving a coupon two days in a row.

$$\frac{\text{number of times customer received coupon two days in a row}}{\text{total numbers of trials}} = \frac{3}{20} \text{ or } 0.15$$

So, the probability that a customer receives a coupon two days in a row is about 15%.

Collaborate

Work with a partner.

1. Repeat the above simulation and make your own table to keep track of the results. Based on your simulation, what is the experimental probability that a customer receives a coupon two days in a row?

2. One out of every ten coupons has a red star. If a customer receives a coupon with a red star, their meal is free. Create a second spinner to simulate the probability that a customer receives a coupon and gets a free meal. Perform 60 trials.

Analyze

3. How might your results change if you simulated the results for 100 customers?
4. The owner of the restaurant randomly gives coupons to 1 out of every 8 customers.
   a. How could you change the simulation to find the experimental probability that a customer receives a coupon two days in a row?
   b. CCSS **Justify Conclusions** Would you expect the experimental probability to be greater than or less than the experimental probability you calculated in the Investigation? Why?

Investigation 2

There is a 20% chance of rain on Sunday and a 50% chance on Monday. Use a random number generator to estimate the probability that it will rain both days.

Step 1 Set up the simulation. Use a two-digit number to represent the weather on the two days. Assign 0 and 1 in the tens place to represent rain on Sunday and 0, 1, 2, 3, and 4 in the ones place to represent rain on Monday.

Step 2 To access the random number generator of your calculator, press **MATH** and then use the right arrow key to move over to the **PRB** menu. Then use the down arrow to choose **5:randInt(** and press **ENTER**. Key in **0, 99)** to complete the argument of the random integer function and press **ENTER**. The calculator will return a random integer between 0 and 99.

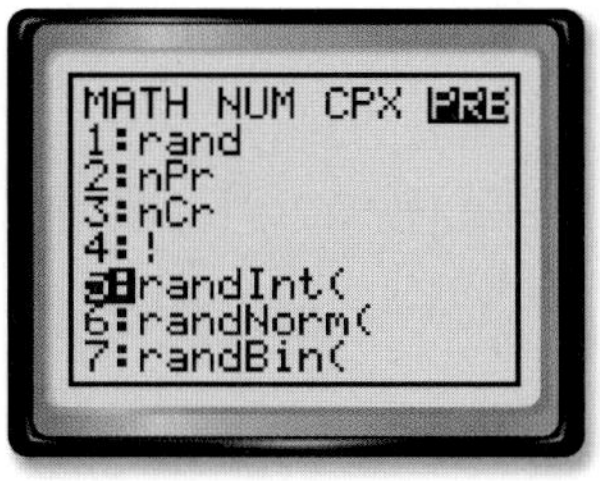

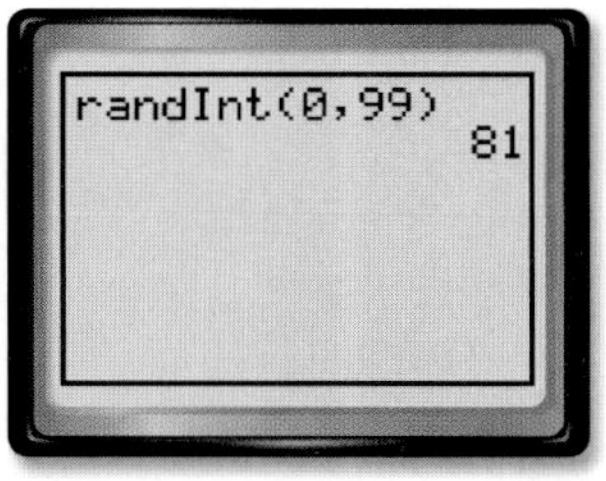

Step 3 Each time you press ENTER, the calculator will return a new random integer between 0 and 99. The table shows the results after 20 trials. Identify the trials in which it rains on both days. The circled data represents the times it will rain both days.

Trial	1	2	3	4	5	6	7	8	9	10
Result	81	28	87	52	67	57	27	41	36	93

Trial	11	12	13	14	15	16	17	18	19	20
Result	87	72	24	(10)	68	(4)	32	55	40	99

Step 4 Calculate the experimental probability that it will rain on both days.

$$\frac{\text{number of times it rains on the both days}}{\text{total numbers of trials}} = \frac{2}{20} \text{ or } 0.1$$

So, the probability that it will rain on both days is about 10%.

Collaborate

Work with a partner.

5. Repeat the above simulation and make your own table to keep track of the results. Based on your simulation, what is the experimental probability that it will rain on both days?
6. Describe how you could change the simulation so that the probability of rain on Sunday is 10% rather than 20%.

Analyze

7. What does the number 13 represent in the simulation? the number 18? the number 31? the number 96?
8. Which numbers in the simulation represent rain on both days?
9. CCSS **Reason Inductively** What do you think would happen to the experimental probability of rain on both days if you conducted 500 trials using the random number generator?
10. Explain how you could you use the simulation to estimate each probability.
    - **a.** the probability that it rains on at least one of the two days
    - **b.** the probability that it does not rain on either day
11. The probability of the Perry Panthers winning a basketball game is 60%. Design a simulation using a random number generator to find the simulated probability that the team wins four games in a row. Explain how you used the random number generator. State the number of trials you performed.
12. One out of every six boxes of Crunch Munch cereal contains a prize. Design and conduct a simulation using a number cube to estimate the probability of buying two boxes of cereal and winning at least one prize.
13. According to the Web site of a company that organizes whale-watching tours, the probability of seeing a humpback whale on any given tour is $\frac{5}{6}$. The probability of seeing a killer whale is $\frac{1}{4}$. The probability of seeing a gray whale is $\frac{3}{10}$. Explain how you can use the spinners shown below to approximate the probability that you will go on a tour and see all three types of whales.

14. A baseball player's batting average is 0.275, which means he gets a hit in 27.5% of his at-bats. Describe how you could use a simulation to find the probability that the player gets a hit in three consecutive at-bats. Be sure to describe the tools you would use and what the results of each trial of the simulation would represent.

Reflect

15. Inquiry HOW can you use a simulation to find the probability of a compound event?

Chapter Review

Check

ISG Interactive Study Guide
See pages 237–240 for:
- Vocabulary Check
- Key Concept Check
- Problem Solving
- Reflect

Lesson-by-Lesson Review

Lesson 10-1 Measures of Center (pp. 434–439)

Find the mean, median, and mode for each set of data. Round to the nearest tenth if necessary.

1. number of students in each math class: 22, 23, 24, 22, 21
2. grams of fat per serving: 2, 7, 4, 5, 6, 4, 5, 6, 3, 5
3. inches of rain last week: 1.5, 2, 2.5, 2, 1.5, 2.5, 3
4. At the movie theater, six movies are playing and their lengths are 138, 117, 158, 145, 135, and 120 minutes. Which measure of center best represent the data? Justify your selection and then find the measure of center.

Example 1

Find the mean, median, and mode of 2, 3, 2, 4, 4, 6, 4, and 7.

Mean: $\frac{2+3+2+4+4+6+4+7}{8} = \frac{32}{8}$ or 4

Median:

2, 2, 3, 4, 4, 4, 6, 7 — Arrange the numbers from least to greatest.

$\frac{4+4}{2} = 4$ — Find the middle number or the mean of the two middle numbers.

Mode: 4 — Find the data value(s) that occur most often.

Lesson 10-2 Measures of Variability (pp. 440–446)

Find the measures of variability and any outliers for each set of data.

5. the number of minutes spent reading each night: 31, 33, 32, 34, 35, 33
6. the number of fish in each fish tank: 6, 5, 7, 8, 5, 6, 7, 9, 8, 6
7. Claire earned \$5, \$7, \$10, \$6, and \$8 doing errands for her neighbors. Find the measures of variability and any outliers for the set of data.
8. The scores Mr. Han's students earned on their last test are shown in the table. Use measures of variability to describe the data in the table.

Test Scores			
99	88	81	89
77	58	92	80
83	82	74	84
76	73	99	74
82	87	82	74
86	76	85	92

Example 2

During a baking contest, bakers sampled 26, 20, 21, 24, 23, 22, 21, 27, 23, 24, and 25 cookies. Find the measures of variability and any outliers for the data.

Range: 27 − 20 or 7 cookies

Median, First Quartile, Third Quartile:

List the data from the least to greatest.

first quartile, median, third quartile

{20, 21, 21, 22, 23, 23, 24, 24, 25, 26, 27}

The median is 23, the first quartile is 21, and the third quartile is 25.

Interquartile Range: 25 − 21 or 4.

Outliers:

Multiply the interquartile range by 1.5.

$4 \times 1.5 = 6$

Subtract 6 from the lower quartile and add 6 to the upper quartile.

$21 - \mathbf{6} = 15$ $\qquad 25 + \mathbf{6} = 31$

Since there are no values less than 15 or greater than 31, there are no outliers.

Lesson 10-3 Mean Absolute Deviation (pp. 447–451)

Find the mean absolute deviation of each data set.

9. 25, 70, 75, 100

10. 0, 20, 175, 190, 175

11. 35, 50, 40, 55, 45

12. 250, 240, 300, 295, 280

13. 144, 137, 156, 149

14. The table shows the numbers of two colors of candies in several bags. Which data set has a greater mean absolute deviation? Justify your answer.

Brown Candies	18	14	20	13	17
Red Candies	10	8	5	12	8

Example 3

The number of defective light bulbs in each shipment of bulbs is given. Find the mean absolute deviation. Describe what it represents.

8, 6, 14, 3, 11, 12

The mean is 9. Find the absolute value of the differences between each data value and the mean.

$|8 - 9| = 1$ $|6 - 9| = 3$ $|14 - 9| = 5$

$|3 - 9| = 6$ $|11 - 9| = 2$ $|12 - 9| = 3$

Find the average of these differences.

$$\frac{1 + 3 + 5 + 6 + 2 + 3}{6} = 3.\overline{3}$$

The average distance between each data value and the mean is $3.\overline{3}$.

Lesson 10-4 Compare Populations (pp. 454–460)

15. The double box plot shows the numbers of floors for buildings on two streets. Compare the centers and variations. Write an inference you can draw about the two populations.

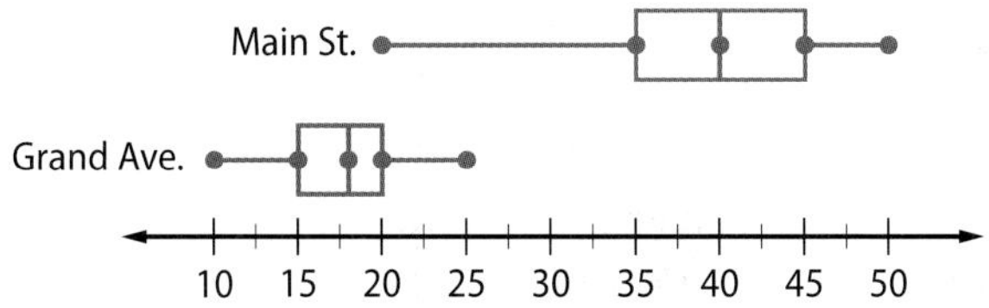

Example 4

The double dot plot shows the quiz scores of students on two topics. Compare their centers and variability. Write an inference you can draw about the two populations.

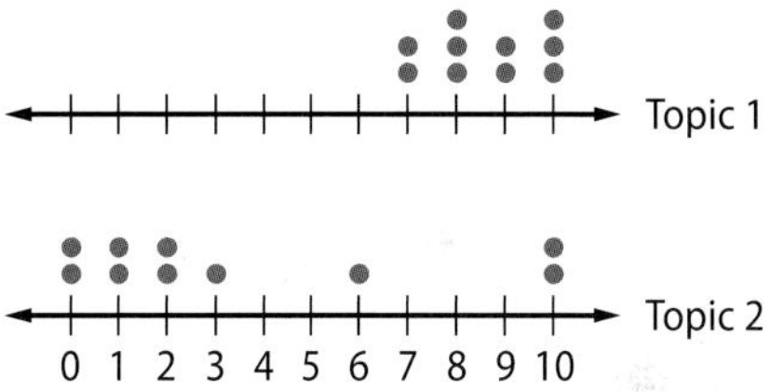

The median score for Topic 1 is 8.5 with an interquartile range of 2. The median score for Topic 2 is 2 with an interquartile range of 5. Topic 1 centers around a higher score than Topic 2, while Topic 2 has a greater variability than Topic 1.

Lesson 10-5 Using Sampling to Predict (pp. 462–467)

16. To determine the weekly top ten songs, the local radio station asks people to log on to their Web site and vote for their favorite song. Identify the sample as *biased* or *unbiased* and describe its type. Explain.

17. Forty-five out of 60 people at a steakhouse said their favorite meal was steak. Is this sampling representative of the entire town? If so, how many of the 13,000 residents would say steak was their favorite meal?

Example 5

Is polling students on the football team about their favorite sports a biased or unbiased sample? Then describe the type of sample.

biased, convenience sample

Students who play football are more likely to choose football as their favorite sport.

Lesson 10-6 Probability of Simple Events (pp. 470–474)

Mike rolls a ten-sided solid whose identical faces are numbered with the first ten square numbers. Find each probability. Then describe the likelihood of the event as *impossible, unlikely, equally likely, likely,* or *certain*.

18. P(ones digit is 2)

19. P(two-digit number)

20. P(multiple of 8)

21. P(ones digit not 3)

22. P(odd)

Example 6

Pia rolls a seven-sided solid whose identical faces are numbered 4 through 10. Find P(odd). Then describe the likelihood of the event as *impossible, unlikely, equally likely, likely,* or *certain*.

sample space: {4, 5, 6, 7, 8, 9, 10}

favorable outcomes: odd numbers, 5, 7, and 9

$$P(\text{odd}) = \frac{\text{number of favorable outcomes}}{\text{number of possible outcomes}} = \frac{3}{7}$$

The probability is about 43%. This is less than 50%, so the event is unlikely.

Lesson 10-7 Theoretical and Experimental Probability (pp. 477–481)

In a survey of randomly selected students, students chose their preference from among three meal options for a school event.

Meal	Pizza	Burgers	Vegetarian
Frequency	47	31	22

23. What was the experimental probability that a student chose burgers?

24. What was the experimental probability that a student chose the vegetarian option?

25. Out of a similar group of 450 students, predict how many would choose the vegetarian option.

Example 7

A spinner has five equal-sized sections. Two are red, one is green, one is blue, and one is orange. Lee spins the spinner 60 times. Find the theoretical and experimental probability of red.

Result	Red	Green	Blue	Orange
Frequency	22	15	11	12

theoretical: $P(\text{red}) = \frac{\text{favorable outcomes}}{\text{possible outcomes}} = \frac{2}{5}$ or 40%

experimental: $P(\text{red}) = \frac{\text{red results}}{\text{all results}} = \frac{22}{60} = \frac{11}{30}$ or $36.\overline{6}\%$

Lesson 10-8 Probability of Compound Events (pp. 482–486)

A penny is tossed and a number cube is rolled. Find each probability.

26. P(heads and 4)

27. P(tails and 1)

28. P(tails and odd number)

29. P(heads and a number greater than 2)

30. P(tails and a number less than or equal to 5)

Example 8

A card is drawn from a deck of eight cards numbered from 1 to 8 and a number cube is rolled. Find the probability of drawing a 3 and then rolling a 6.

Use the Fundamental Counting Principle to find the number of possible outcomes.

$P(3, 6) = P(\text{drawing a 3}) \cdot P(\text{rolling a 6})$

number of cards	the total number of times	sides on number cube	equals	number of outcomes
8	×	6	=	48

There are 48 outcomes.

Find the probability. There is one way draw a 3 and roll a 6. So, the probability is $\frac{1}{48}$ or 2%.

Chapter 11

Congruence, Similarity, and Transformations

Essential Question

How can you determine congruence and similarity?

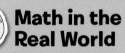

Common Core State Standards

Content Standards
7.G.2, 7.G.5, 8.G.1, 8.G.1a, 8.G.1b, 8.G.1c, 8.G.2, 8.G.3, 8.G.4, 8.G.5

Mathematical Practices
1, 2, 3, 4, 5, 6, 7

Math in the Real World

NATURE There are about 24,000 different species of butterflies in the world. They range in size from $\frac{1}{8}$ inch to 12 inches. Did you know that butterflies taste with their feet? Their taste sensors are located in the feet. By standing on their food, they can taste it!

Interactive Study Guide
See pages 241–244 for:
- Chapter Preview
- Are You Ready?
- Foldable Study Organizer

Lesson 11-1

Angle and Line Relationships

ISG Interactive Study Guide

See pages 245–246 for:
- Getting Started
- Vocabulary Start-Up
- Notes

Essential Question

How can you determine congruence and similarity?

CCSS Common Core State Standards

Content Standards
7.G.5, 8.G.5

Mathematical Practices
1, 3, 4, 5

Vocabulary

vertical angles
adjacent angles
complementary angles
supplementary angles
perpendicular lines
parallel lines
transversal
alternate interior angles
alternate exterior angles
corresponding angles

Math Symbols

|| is read *is parallel to*

⊥ is read *is perpendicular to*

$m\angle ABC$ is read *measure of angle ABC*

What You'll Learn

- Examine relationships between pairs of angles.
- Examine relationships of angles formed by parallel lines and a transversal.

Real-World Link

Roller Coasters Did you know that the tallest roller coaster in the world is 456 feet tall? When designing taller and faster coasters, designers need to know about geometry and physics. In the structure of a roller coaster, designers use different types of angles and lines to produce safe yet thrilling rides.

Key Concept Pairs of Angles

Words	Models	Symbols
When two lines intersect, they form two pairs of opposite angles, called **vertical angles**. Vertical angles are congruent.	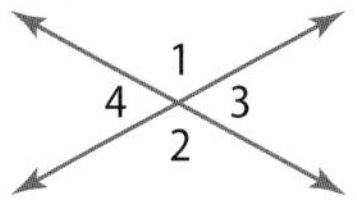 $\angle 1$ and $\angle 2$, $\angle 3$ and $\angle 4$	$\angle 1 \cong \angle 2$ $\angle 3 \cong \angle 4$
Two angles that have the same vertex between them, share a common side, and do not overlap are called **adjacent angles**.	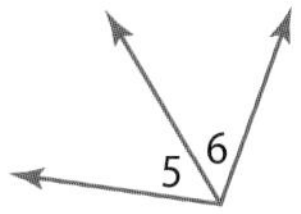	$\angle 5$ and $\angle 6$ are adjacent angles.
If the sum of the measures of two angles is 90°, the angles are called **complementary angles**.	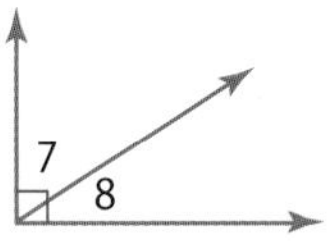	$m\angle 7 + m\angle 8 = 90°$
If the sum of the measures of two angles is 180°, the angles are called **supplementary angles**.	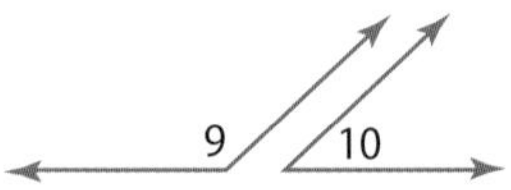	$m\angle 9 + m\angle 10 = 180°$

Pairs of angles can be classified by their relationship to each other. A special case occurs when two lines intersect to form a right angle. These lines are **perpendicular lines**.

The right angle symbol is used to indicate perpendicular lines.

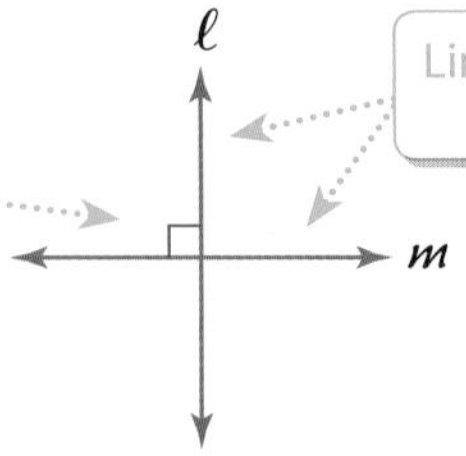

Lines ℓ and m are perpendicular. Symbols: $\ell \perp m$

Example 1

Tutor

Jun is cutting a corner off a piece of rectangular tile.

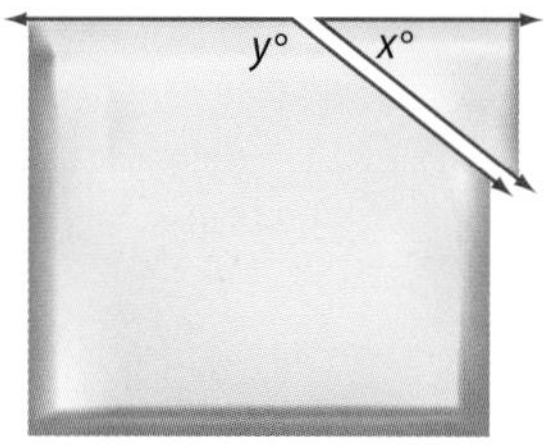

a. Classify the relationship between $\angle x$ and $\angle y$.

The angles are supplementary. The sum of their measures is 180°.

b. If $m\angle y = 135°$, what is the measure of $\angle x$?

$m\angle x + 135 = 180$ Write the equation.

$m\angle x + 135 - \mathbf{135} = 180 - \mathbf{135}$ Subtract 135 from each side.

$m\angle x = 45$ Simplify.

So, $m\angle x = 45°$.

Angles
Complementary angles and supplementary angles can either be adjacent angles or separate angles.

Got It? Do this problem to find out.

1. Angles R and S are complementary. If $m\angle R = 65.3°$, what is the measure of $\angle S$?

Parallel Lines

Two lines in a plane that never intersect are called **parallel lines**. A line that intersects two or more other lines in a plane is called a **transversal**.

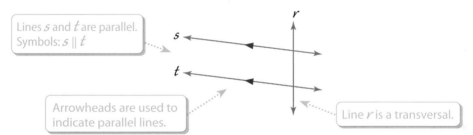

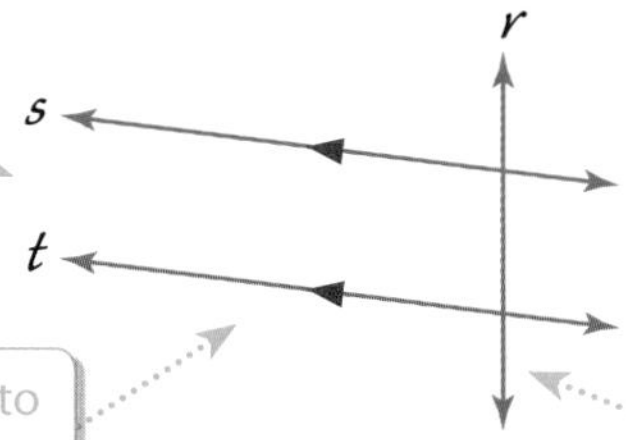

When parallel lines are cut by a transversal, special pairs of angles are congruent.

Key Concept Special Angle Relationships

When a transversal intersects two parallel lines, eight angles are formed.

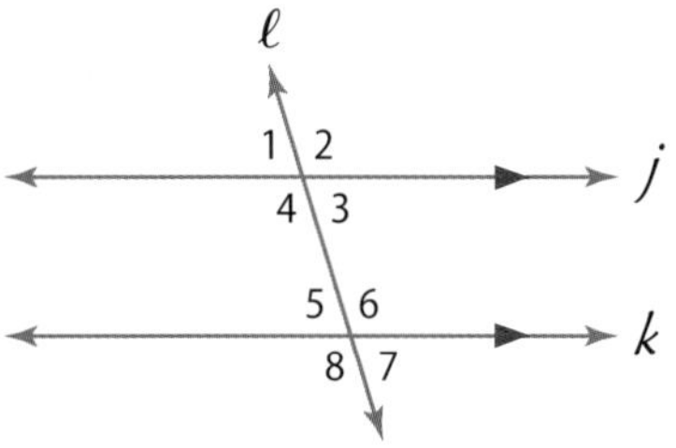

- *Interior angles* lie inside the parallel lines. $\angle 3, \angle 4, \angle 5, \angle 6$
- *Exterior angles* lie outside the parallel lines. $\angle 1, \angle 2, \angle 7, \angle 8$

The following pairs of angles are congruent.

- **Alternate interior angles** are on opposite sides of the transversal and inside the parallel lines. $\angle 3 \cong \angle 5, \angle 4 \cong \angle 6$
- **Alternate exterior angles** are on opposite sides of the transversal and outside the parallel lines. $\angle 1 \cong \angle 7, \angle 2 \cong \angle 8$
- **Corresponding angles** are in the same position on the parallel lines in relation to the transversal. $\angle 1 \cong \angle 5, \angle 2 \cong \angle 6, \angle 3 \cong \angle 7, \angle 4 \cong \angle 8$

Symbols
$\angle 3 \cong \angle 5$ is read *angle 3 is congruent to angle 5*.

Example 2

Tutor

In the figure at the right, $a \parallel b$ and q and r are transversals.

a. Classify the relationship between ∠3 and ∠5.

Since ∠3 and ∠5 are alternate interior angles, they are congruent.

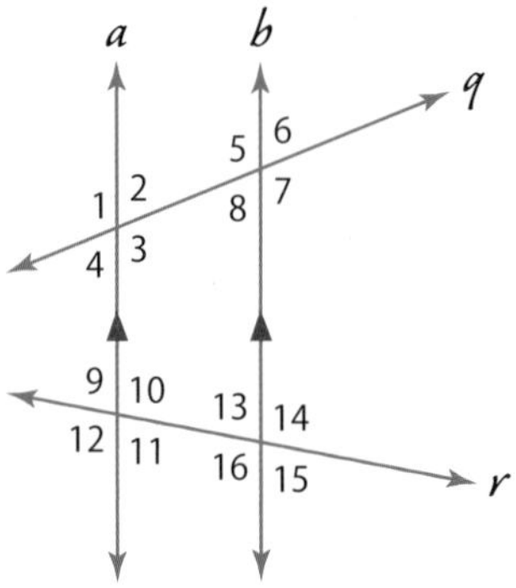

b. If $m\angle 1 = 120°$, find $m\angle 5$ and $m\angle 3$.

Since ∠1 and ∠5 are corresponding angles, they are congruent. So, $m\angle 5 = 120°$.

Since $m\angle 5$ and $m\angle 3$ are congruent, $m\angle 3 = 120°$.

Got It? Do these problems to find out.

2a. Classify the relationship between ∠11 and ∠15.

2b. If $m\angle 15 = 78.5°$, find $m\angle 11$ and $m\angle 9$.

Example 3

Tutor

In the figure at the right, $m\angle ABD = 164°$. Find the measures of ∠ABC and ∠CBD.

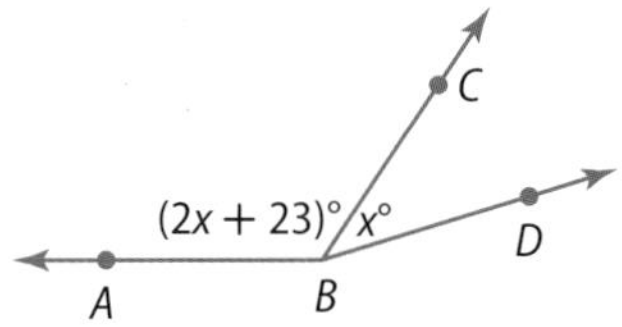

Angles *ABC* and *CBD* are adjacent angles that have a total measure of 164°.

> **Ray**
> A ray is a part of a line that extends indefinitely in one direction. In the figure at the right, $\overrightarrow{BA}$, $\overrightarrow{BC}$, and $\overrightarrow{BD}$ are rays.

Step 1 Find the value of *x*.

$m\angle ABC + m\angle CBD = 164$	Adjacent angles
$(2x + 23) + x = 164$	Replace $m\angle ABC$ with $2x + 23$ and replace $m\angle CBD$ with x.
$3x + 23 = 164$	Combine like terms.
$-23 = -23$	Subtraction Property of Equality
$3x = 141$	Simplify.
$\frac{3x}{3} = \frac{141}{3}$	Division Property of Equality
$x = 47$	Simplify.

Step 2 Replace *x* with 47 to find the measure of each angle.

$m\angle ABC = 2x + 23$ $\quad\quad$ $m\angle CBD = x$

$= 2(47) + 23 \text{ or } 117$ $\quad\quad$ $= 47$

So, $m\angle ABC = 117°$ and $m\angle CBD = 47°$.

Got It? Do these problems to find out.

Angles *RVT* and *UVW* are vertical angles, with $m\angle RVT = 3x°$ and $m\angle UVW = (5x - 36)°$.

3a. Find the value of *x*.

3b. Find $m\angle UVW$.

Guided Practice

Check

Classify the pairs of angles shown. Then find the value of *x* in each figure. (Example 1)

1.

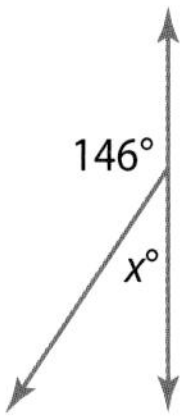

2.

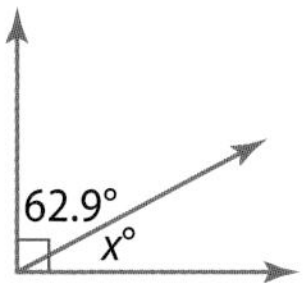

3.

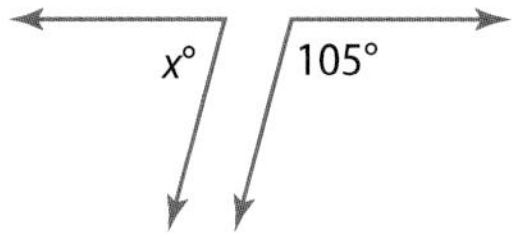

In the figure at the right, *r* || *s* and *w* is a transversal. If $m\angle 1 = 128°$, find the measure of each angle. Explain your reasoning. (Example 2)

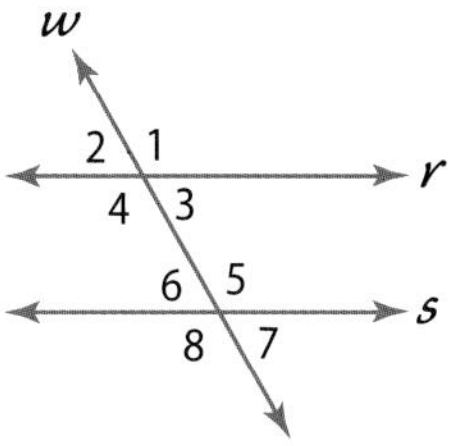

4. $\angle 5$

5. $\angle 8$

6. $\angle 2$

7. The balance beam shown below is parallel to the floor.

a. Classify the relationship between $\angle x$ and $\angle y$.

b. If $m\angle y = 117°$, find the value of *x*.

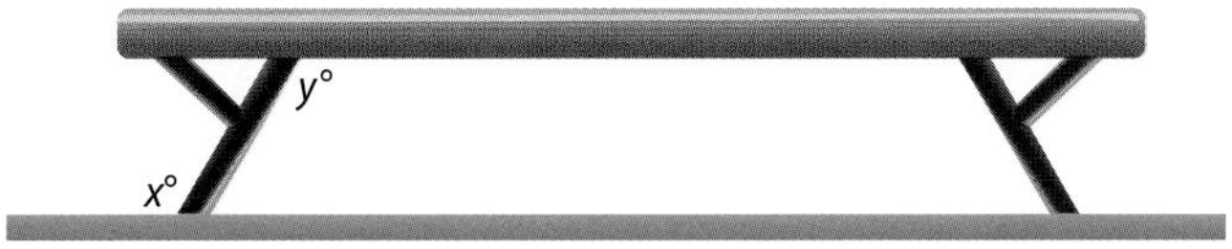

8. Angles *ABC* and *DBF* are vertical angles, with $m\angle ABC = 4x°$ and $m\angle DBF = (6x - 12)°$. (Example 3)

a. Find the value of *x*.

b. Find $m\angle DBF$.

The measure of $\angle Q$ is $(6x + 16.8)°$ and the measure of $\angle R$ is $2x°$. (Example 3)

9. If $\angle Q$ and $\angle R$ are supplementary, what is the value of *x*? What is the measure of each angle?

10. If $\angle Q$ and $\angle R$ are complementary, what is the value of *x*? What is the measure of each angle?

Independent Practice

Go online for Step-by-Step Solutions

eHelp

Classify the pairs of angles shown. Then find the value of *x* in each figure. (Example 1)

11.

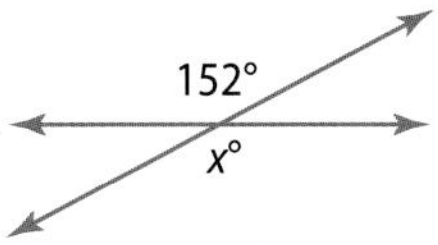

12.

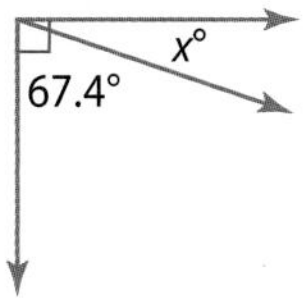

13

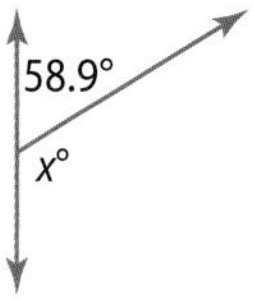

14. Look at the semicircular window.

a. Classify the relationship between $\angle 1$ and $\angle 2$.

b. If $m\angle 2$ is 24°, find $m\angle 1$.

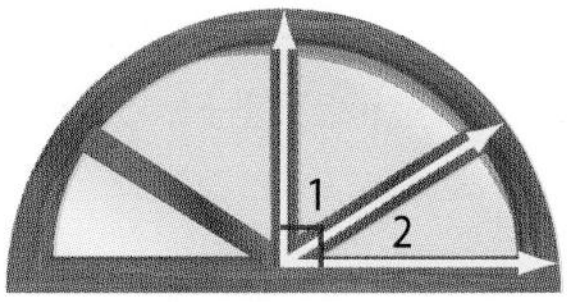

In the figure at the right, $f \parallel g$ and t is a transversal. If $m\angle 6 = 52.6°$, find the measure of each angle. Explain your reasoning. (Example 2)

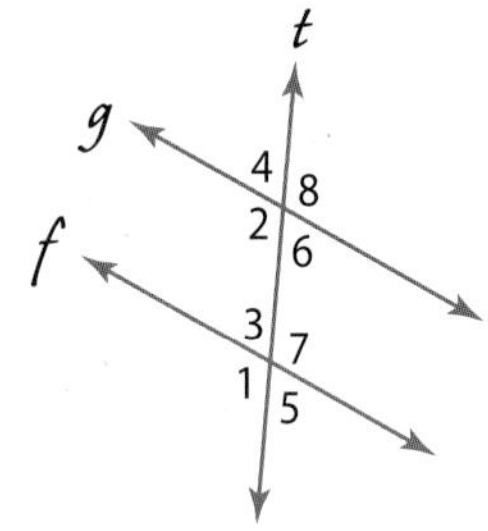

15. $\angle 3$

16. $\angle 5$

17. $\angle 2$

18. $\angle 8$

19. $\angle 4$

20. $\angle 1$

21. If $m\angle FGH = 165°$, find $m\angle 1$. (Example 3)

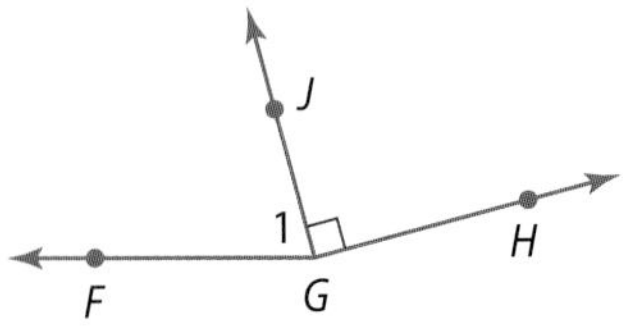

22. If $m\angle QRS = 158°$, find $m\angle TRS$. (Example 3)

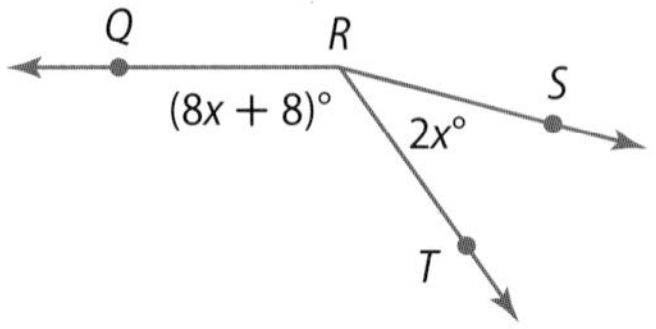

23. Angles M and P are complementary. If $m\angle M = (3x - 45)°$ and $m\angle P = (2x + 15)°$, what is the value of x? What is the measure of each angle?

24. The measure of $\angle C$ is $(4x - 24.6)°$ and the measure of $\angle D$ is $(x + 11.3)°$. If the angles are supplementary, find the value of x and $m\angle D$.

25 Use the pair of angles shown.

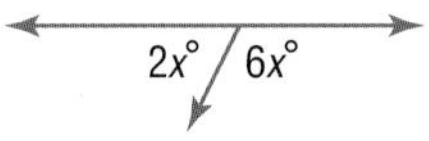

a. Write an equation to find the value of x. Then find the value of x.

b. What are the measures of the two angles?

In the figure below, $a \parallel b$ and z is a transversal.

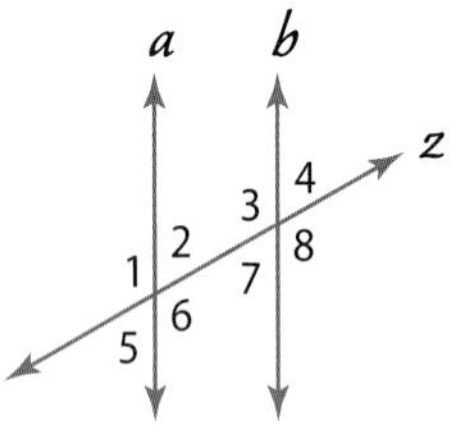

26. If $m\angle 1 = 5x°$ and $m\angle 5 = (2x - 2)°$, find the value of x, $m\angle 1$, and $m\angle 5$.

27. If $m\angle 4 = 15x°$ and $m\angle 2 = (11x + 20)°$, find the value of x, $m\angle 4$, and $m\angle 2$.

28. In the roller coaster tower at the right, the measure of $\angle 1$ is 6° less than twice the measure of $\angle 2$. Find the measures of angles 1 and 2.

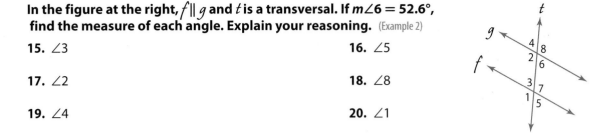

29. The sum of the measures of $\angle A$, $\angle B$, and $\angle C$ is 180°, and $m\angle B = m\angle C$. If $m\angle A = 110°$, find $m\angle B$ and $m\angle C$.

30. Angles U and V are supplementary angles. The ratio of their measures is 7:13. Find the measure of each angle.

31. Use the pair of angles shown.

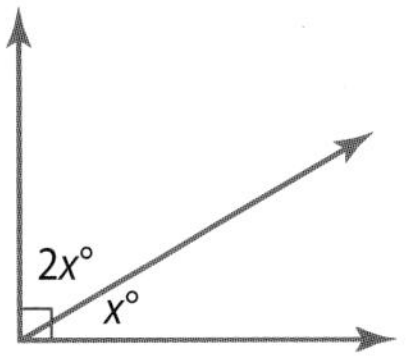

a. Write an equation to find the value of x. Then find the value of x.

b. What are the measures of the two angles?

32. Lines p, q, r, and s form the quadrilateral shown at the right. Can you conclude that both pairs of opposite sides of the quadrilateral are parallel? Explain your reasoning.

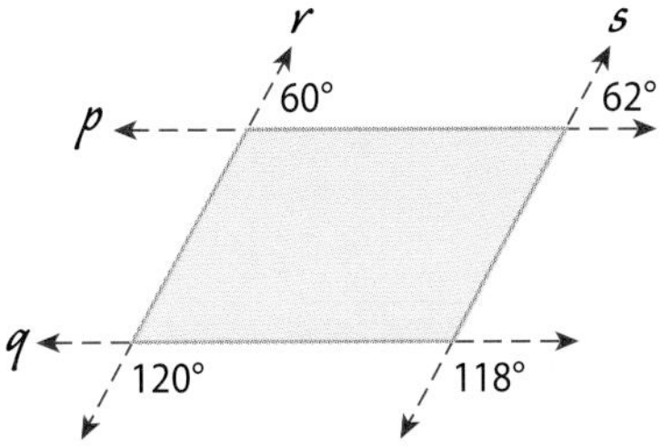

33 Use the clock at the right that shows 6 o'clock and 10 seconds.

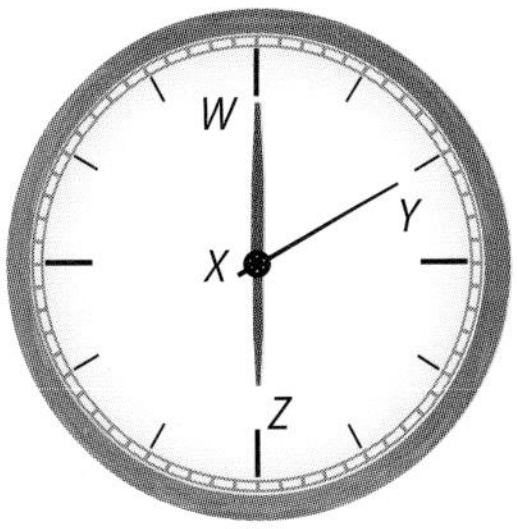

a. Find $m\angle WXY$ and $m\angle YXZ$.

b. Find the time that will show when $m\angle WXY = m\angle YXZ = 90°$.

c. When it is 25 seconds after 6 o'clock, what will be the approximate measure of $\angle WXY$?

34. CCSS **Multiple Representations** In this problem, you will investigate parallel lines on the coordinate plane. Line f passes through points at (0, 2) and (2, 3). Line g passes through points at (0, −3) and (2, −2). Line h passes through points at (1, 0) and (2, −2).

a. Graph Graph the three lines on the same coordinate plane. Label each line.

b. Words Describe the angles that are formed by the lines.

c. Analyze Describe the relationship between the slopes of parallel lines.

H.O.T. Problems Higher Order Thinking

35. CCSS **Use Math Tools** Draw a pair of complementary adjacent angles. Label the measures of the angles.

36. CCSS **Find the Error** Taylor is calculating the value of x for the missing angle shown at the right. Find her mistake and correct it.

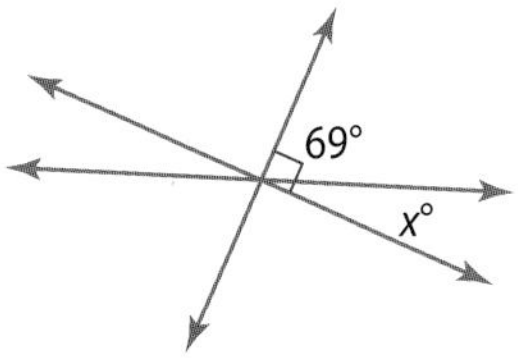

$$69 + x = 180$$
$$x = 111$$

37. CCSS **Persevere with Problems** Lines ℓ and m shown at the right are parallel and are cut by transversals j and k. Describe how the following pairs of angles are related: 1 and 2, 3 and 4, 5 and 6. Then make a conjecture about how the interior angles on the same side of a transversal are related.

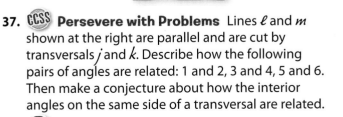

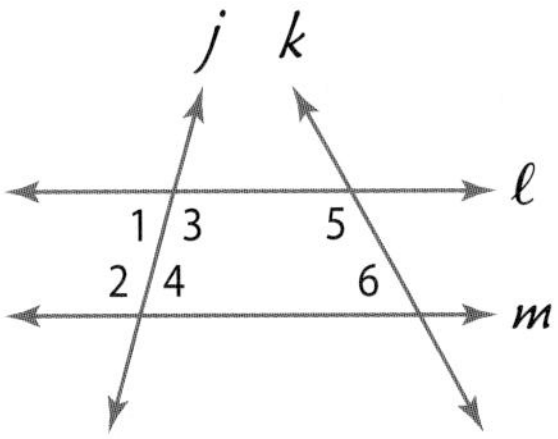

38. **Building on the Essential Question** Suppose two parallel lines are cut by a transversal and an exterior angle measures 90°. What can you conclude about the measures of the other seven angles that are formed? Explain your reasoning.

Standardized Test Practice

39. Lines a and b are parallel in the figure below. Find the value of x.

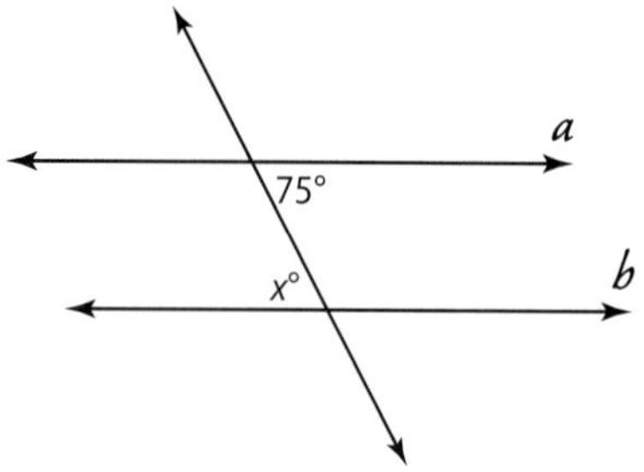

A 110 **C** 75
B 105 **D** 15

40. Which angles are not supplementary?

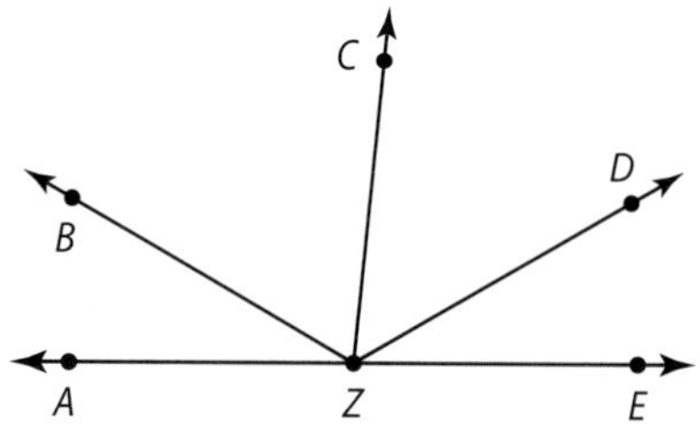

F $\angle EZC$ and $\angle CZA$ **H** $\angle AZB$ and $\angle BZE$
G $\angle BZC$ and $\angle CZD$ **J** $\angle DZE$ and $\angle AZD$

41. In the figure below, the two angles are congruent. Find the value of x.

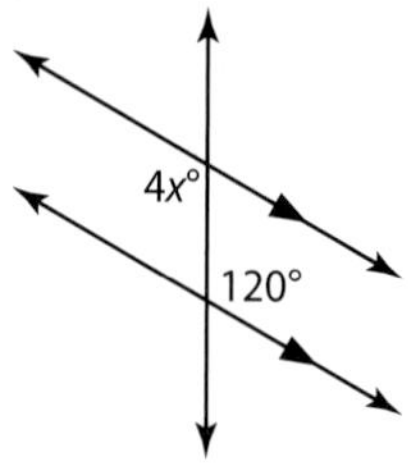

A 30
B 40
C 116
D 124

42. **Short Response** The hedge shears have two different angles shown. Find the value of x.

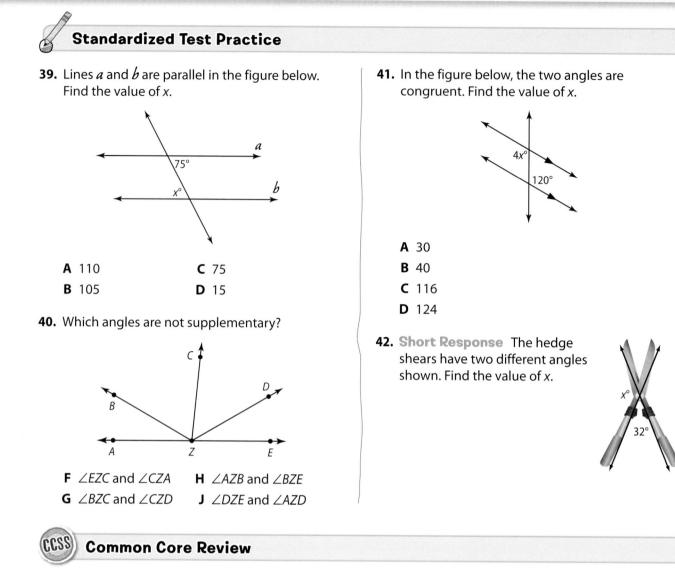

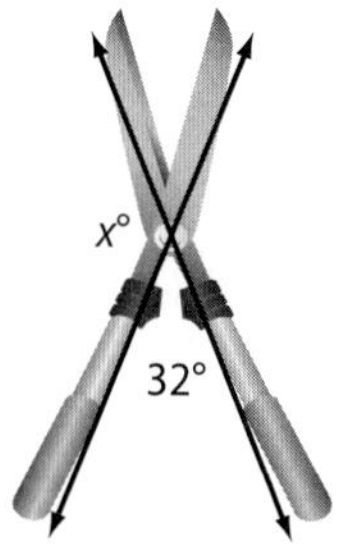

Common Core Review

43. Tobias' average for five quizzes is 86. He wants to have an average of at least 88 for six quizzes. What is the lowest score he can receive on his sixth quiz to obtain this average? **6.SP.5**

44. Cecilia has a bag containing 4 green marbles, 6 red marbles, and 6 yellow marbles. She randomly picks a marble from the bag and replaces it. She repeats the experiment 250 times. Is it likely or unlikely that the marble Cecilia picks will be green? Explain your answer. **7.SP.5**

45. In 2000, there were 356 endangered species in the U.S. Five years later, 389 species were considered endangered. What was the percent of change? State whether the percent of change is an increase or decrease. Round to the nearest tenth. **7.RP.3**

Solve each inequality. **7.EE.4**

46. $5m < 5$ **47.** $\frac{a}{-2} > 3$ **48.** $-4x \geq -16$

Evaluate each expression if $y = -4$ and $z = 6$. **6.EE.2**

49. y^2z **50.** $-6z \div (-3y)$ **51.** $2z^3 \div (4y)$ **52.** $-z + (5y^2 - 15)$

Simplify each expression. **6.EE.1**

53. $(10 - 3)^2 + (4 - 8)^2$ **54.** $[7 - (-2)]^2 + (5 - 3)^2$ **55.** $(-4 - 3)^2 + (-1 - 5)^2$

56. $(9 + 1)^2 + (2 - 12)^2$ **57.** $6 - (-1)^2 + [3 - (4 - 6)]^2$ **58.** $[(7 - 5)^2 + (5 - 4)]^2$

Inquiry Lab

Triangles

Inquiry **HOW are the measures of the interior angles of a triangle related to the measures of the exterior angles of the triangle?**

CCSS **Content Standards**
8.G.5
Mathematical Practices
1, 3, 6

Bicycles Structures made from triangles are very strong. If a force is applied to any one of the three sides of a triangle, there is at least one other side that opposes the force. A bicycle frame is a truss made of beams that meet to form angles and triangles. How are the angles in the triangular part of the truss related?

Investigation 1

Step 1 Draw a pair of parallel lines.
Draw a transversal as shown.
Label $\angle 1$ and $\angle 2$.

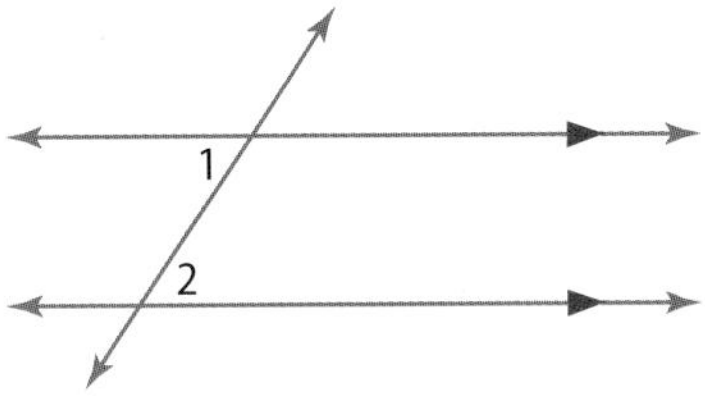

Step 2 Draw a second transversal as shown.
Label $\angle 3$ and $\angle 4$. Label the triangle formed by these lines *ABC*.

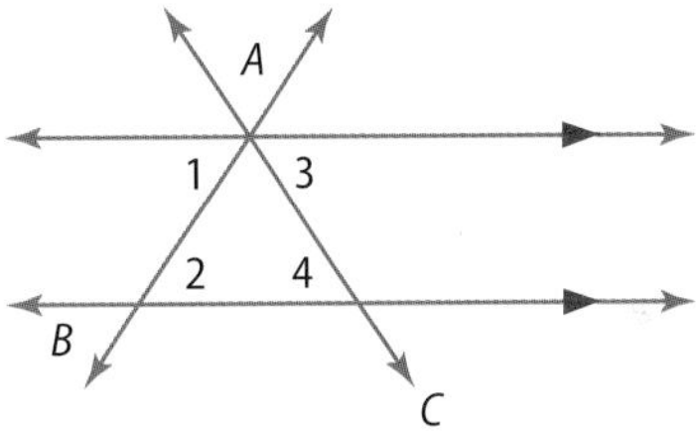

Step 3 Notice that $\angle 1$ and $\angle 2$ are alternate interior angles, so they are congruent. Similarly, $\angle 3$ and $\angle 4$ are alternate interior angles, so they are congruent.

Step 4 $\angle 1$, $\angle BAC$, and $\angle 3$ form a straight angle, so $m\angle 1 + m\angle BAC + m\angle 3 = 180°$.

Step 5 Since $m\angle 1 = m\angle 2$ and $m\angle 3 = m\angle 4$, $m\angle 2 + m\angle BAC + m\angle 4 = 180°$.

So, the sum of the measures of the interior angles of $\triangle ABC$ is 180°.

Collaborate

Work with a partner.

1. Repeat the steps of the investigation by drawing a new set of parallel lines. Then draw two transversals that intersect on one of the parallel lines, as shown above. This will form a triangle. What is the sum of the measures of its interior angles?

In Investigation 1, you worked with the interior angles of a triangle. In Investigation 2, you will work with the exterior angles of a triangle.

Investigation 2

Step 1 Draw a large triangle. Extend its sides to make three exterior angles. Label the angles as shown.

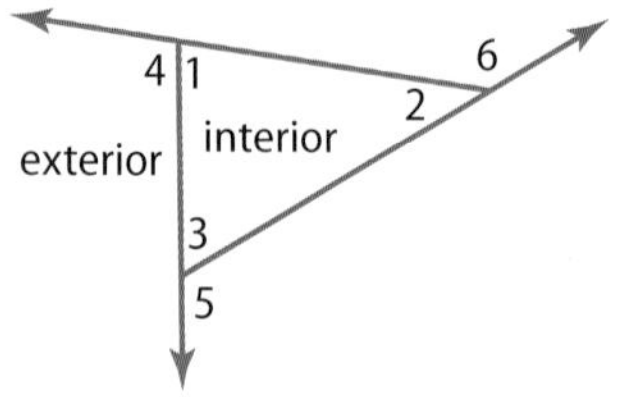

Step 2 Use a protractor to measure and label each angle to the nearest degree.

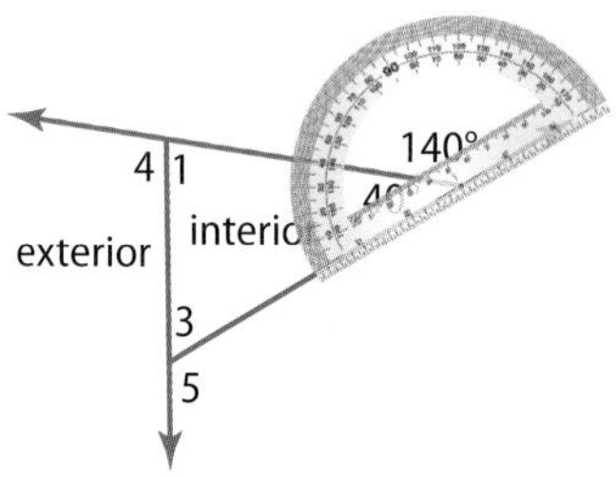

Collaborate

2. For your figure in Investigation 2, which angles are the interior angles? Which are the exterior angles?

3. What is the sum of $m\angle 1$ and $m\angle 4$?

4. What is the sum of $m\angle 2$ and $m\angle 6$? the sum of $m\angle 3$ and $m\angle 5$?

5. CCSS **Be Precise** Describe the relationship between an interior angle of a triangle and its adjacent exterior angle.

Analyze

6. CCSS **Reason Inductively** Based on your work, what is the sum of the measures of the interior angles of any triangle?

7. CCSS **Justify Conclusions** Suppose all three interior angles of a triangle are equal in measure. What is the measure of each of the triangle's exterior angles? Explain your reasoning.

Reflect

8. Inquiry HOW are the measures of the interior angles of a triangle related to the measures of the exterior angles of the triangle?

Lesson 11-2

Triangles

ISG Interactive Study Guide

See pages 247–248 for:
- Getting Started
- Vocabulary Start-up
- Real-World Link

Essential Question

How can you determine congruence and similarity?

CCSS Common Core State Standards

Content Standards
8.G.5

Mathematical Practices
1, 2, 3, 4, 7

Vocabulary

line segment
triangle
vertex
interior angle
exterior angle
congruent

Math Symbols

$m\angle 2$ is read *measure of angle 2*

$\triangle XYZ$ is read *triangle XYZ*

$\overline{AB}$ is read *segment AB*

What You'll Learn

- Find the missing angle measure of a triangle.
- Classify a triangle by its angles and by its sides.

Real-World Link

Architecture Triangles are used in the design of many structures, including buildings, bridges, and amusement park rides. The triangles used in these structures provide strength and stability.

Concept Summary — Angle Sum of a Triangle

Words	The sum of the measures of the interior angles of a triangle is 180°.	Model 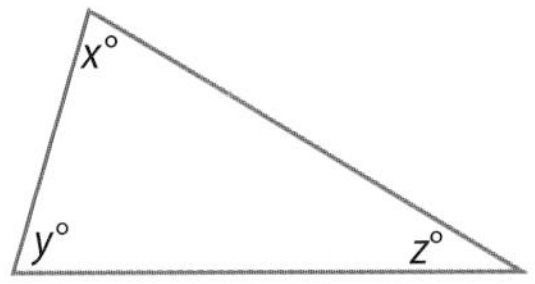
Symbols	$x + y + z = 180$	

A **line segment** is part of a line containing two endpoints and all of the points between them. A **triangle** is formed by three line segments that intersect only at their endpoints. A **vertex** is the point where the segments of a triangle intersect. The three angles that lie inside a triangle, formed by the segments and the vertices, are called **interior angles**.

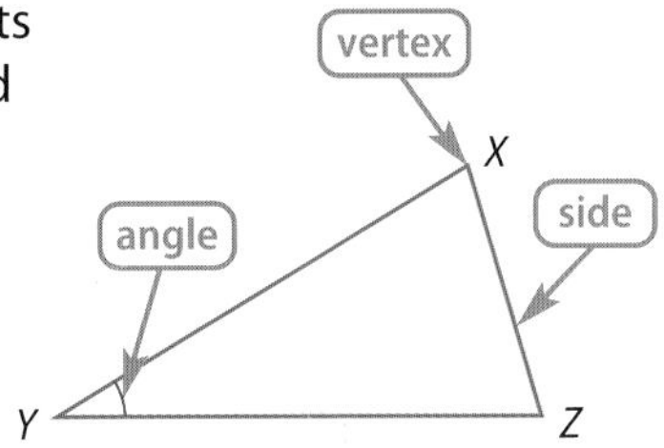

sides: $\overline{XY}$, $\overline{YZ}$, $\overline{XZ}$
vertices: X, Y, Z
angles: $\angle X$, $\angle Y$, $\angle Z$

Triangle *XYZ*, written $\triangle XYZ$, is shown at the right. When finding the measures of angles in a triangle, the notation $m\angle X$ means 'the measure of angle *X*'.

Example 1

Tutor

Find the value of *x* in $\triangle PQR$.

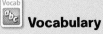

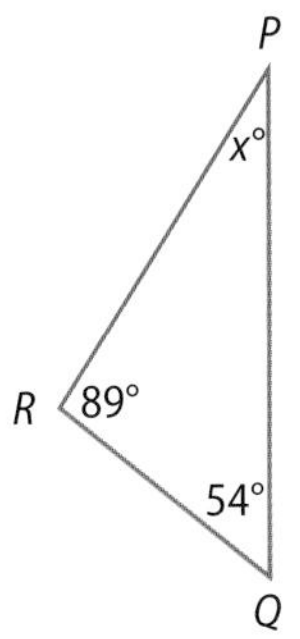

$m\angle P + m\angle Q + m\angle R = 180$	Write an equation.
$x + 54 + 89 = 180$	Substitution Property
$x + 143 = 180$	Simplify.
$x + 143 - 143 = 180 - 143$	Subtract 143 from each side.
$x = 37$	Simplify.

So, $x = 37$.

Got It? Do this problem to find out.

1. Find $m\angle E$ in $\triangle DEF$ if $m\angle D = 62°$ and $m\angle F = 39°$.

Example 2

The measures of the angles of $\triangle ABC$ are in the ratio 1:3:8. What are the measures of the angles?

Words	The sum of the measures of the angles is 180°.
Variable	Let x represent the measure of the first angle, 3*x* the measure of a second angle, and 8*x* the measure of the third angle.
Equation	$x + 3x + 8x = 180$

$x + 3x + 8x = 180$ Write the equation.

$12x = 180$ Combine like terms.

$\frac{12x}{12} = \frac{180}{12}$ Divide each side by 12.

$x = 15$ Simplify.

Check Solutions

$15 + 45 + 120 = 180$.
So, the answer is correct.

Since $x = 15$, $3x = 3(15)$ or 45, and $8x = 8(15)$ or 120.

The measures of the angles are 15°, 45°, and 120°.

Got It? Do this problem to find out.

2. The measures of the angles of a triangle are in the ratio 1:3:6. What are the measures of the angles?

Each **exterior angle** of a triangle is formed by one side of the triangle and the extension of the adjacent side. Angles 4, 5, and 6 are exterior angles.

An exterior angle is supplementary to its adjacent interior angle.

$m\angle 4 + m\angle 1 = 180°$

$m\angle 6 + m\angle 2 = 180°$

$m\angle 5 + m\angle 3 = 180°$

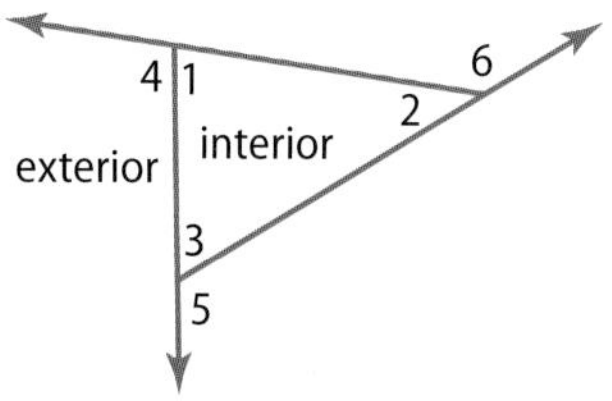

Real World Example 3

Find the measure of $\angle 2$ formed by the legs of the director's chair when the measure of $\angle 5$ is 168°.

Angle 5 is an exterior angle. It is supplementary to $\angle 2$.

$m\angle 5 + m\angle 2 = 180°$ Write an equation.

$168° + x° = 180°$ Replace $m\angle 5$ with 168°. Let $m\angle 2 = x°$.

$x = 12$ Subtraction Property of Equality

So, $m\angle 2 = 12°$.

Got It? Do this problem to find out.

3. Suppose $m\angle 4$ in the chair above is 102°. Find the measure of $\angle 1$.

Concept Summary Classify Triangles

	Acute Triangle	Obtuse Triangle	Right Triangle
By Angles	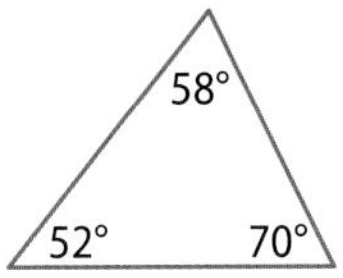 all acute angles	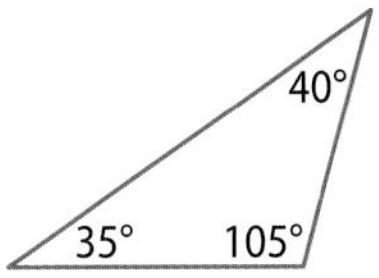 one obtuse angle	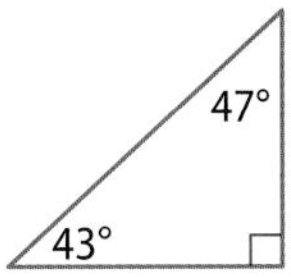 one right angle
	Scalene Triangle	**Isosceles Triangle**	**Equilateral Triangle**
By Sides	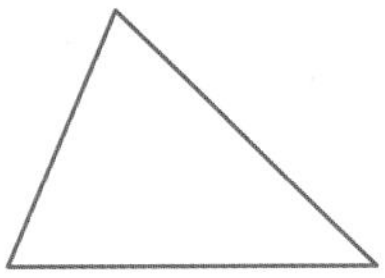 no congruent sides	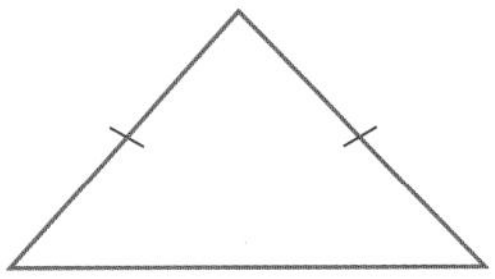 at least two sides congruent	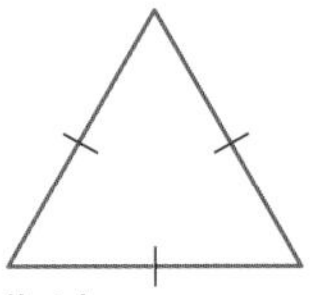 all sides congruent

Congruent Segments
Tick marks on the sides of a triangle indicate that those sides are congruent.

Triangles can be classified by their angles and by their sides. **Congruent** sides are sides that have the same length.

Example 4

Tutor

Classify each triangle by its angles and by its sides.

a.

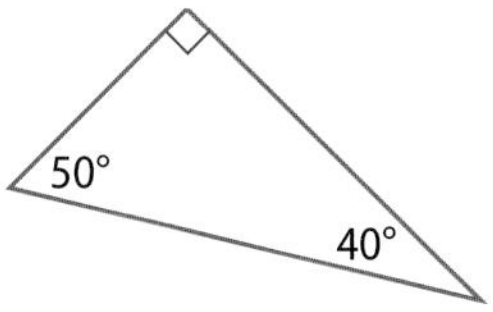

Angles: The triangle has a right angle.
Sides: The triangle has no congruent sides.

The triangle is a right scalene triangle.

b.

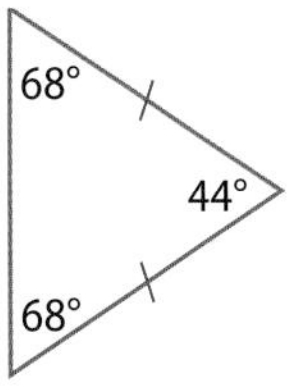

Angles: The triangle has three acute angles.
Sides: The triangle has two congruent sides.

The triangle is an acute isosceles triangle.

Got It? **Do these problems to find out.**

4a.

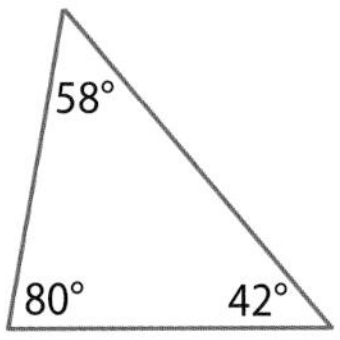

4b.

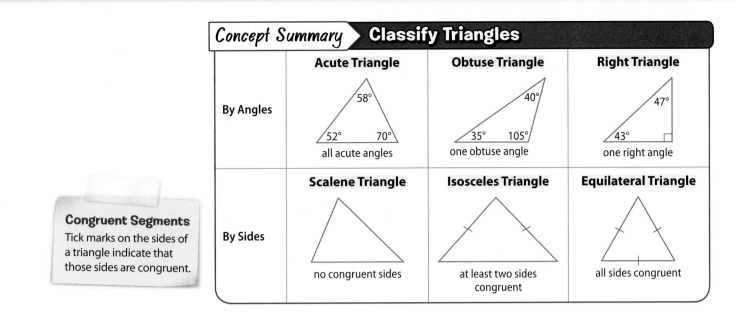

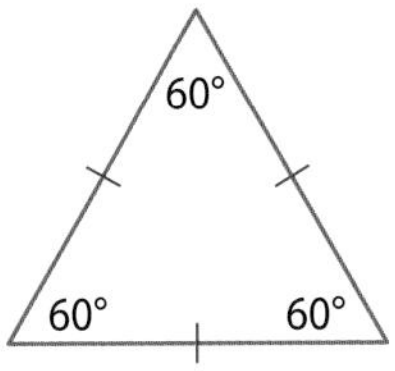

Guided Practice

Check

Find the value of *x* in each triangle. Then classify each triangle by its angles and by its sides. (Examples 1 and 4)

1.
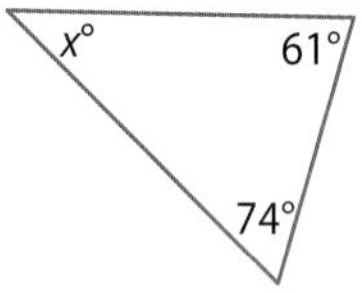

2.
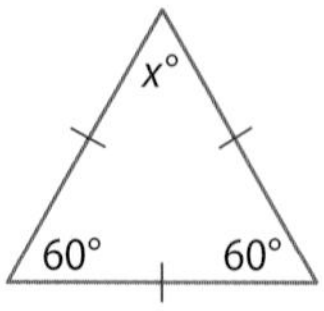

3.
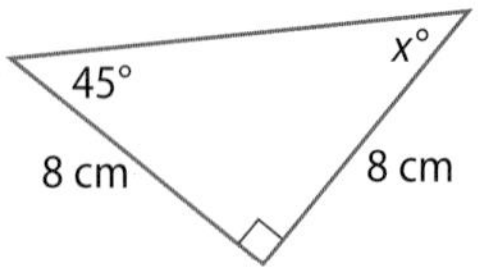

4. The measures of the angles of a triangle are in the ratio 2:3:5. What are the measures of the angles? (Example 2)

5. The truss of a house is shown at the right. If the measure of $\angle 4$ is 150°, what is the measure of $\angle 1$? (Example 3)

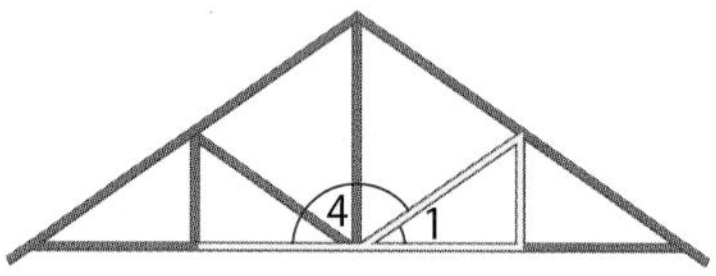

Independent Practice

Go online for Step-by-Step Solutions

eHelp

Find the value of *x* in each triangle. Then classify each triangle by its angles and by its sides. (Examples 1 and 4)

6.
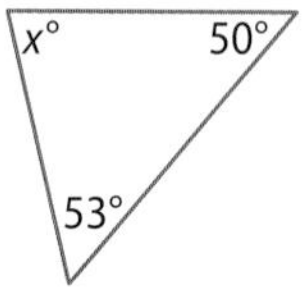

7.
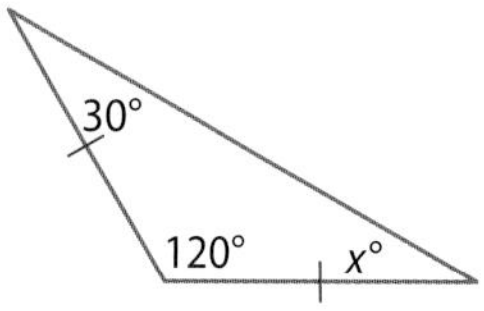

8.
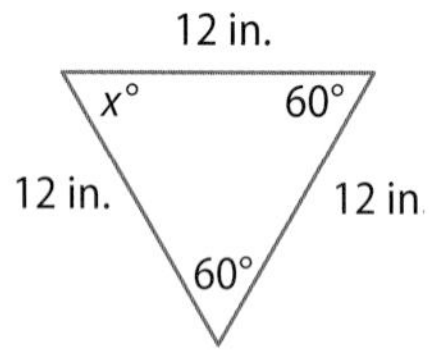

9.
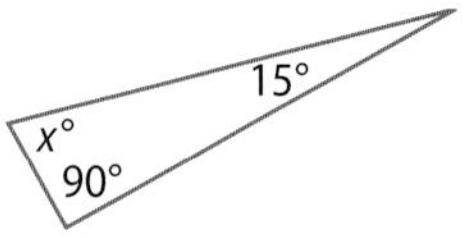

10.
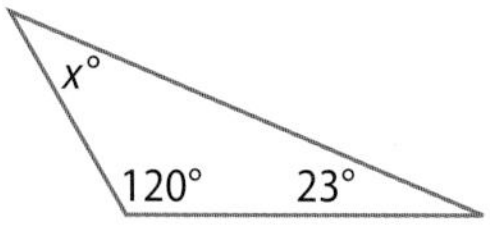

11.
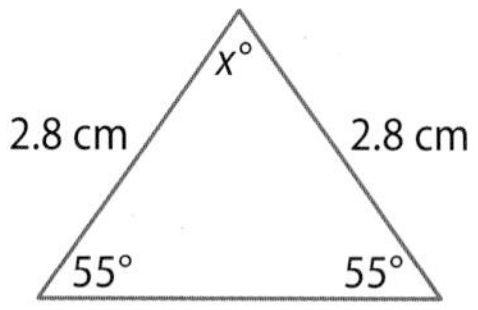

12. Determine the measures of the angles of $\triangle ABC$ if the measures of the angles are in the ratio 1:1:16. (Example 2)

13. Determine the measures of the angles of $\triangle TUV$ if the measures of the angles are in the ratio 1:6:8. (Example 2)

14. A ramp to assist wheelchairs is shown at the right. If the measure of $\angle 1$ is 16°, what is the measure of $\angle 4$? (Example 3)

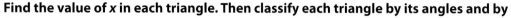

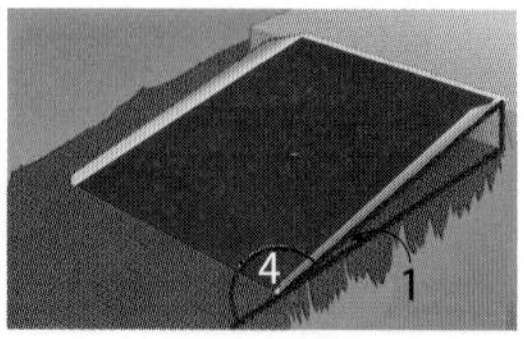

15. What are the measures of the angles in a right triangle if one exterior angle measures 120°? (Example 3)

16. What is the measure of the third angle in a triangle if the sum of the other two angle measures is 80°?

17. What is the measure of the third angle in a triangle if two of the angles measure 45° and 50°?

Find the value of *x* in each triangle.

18.

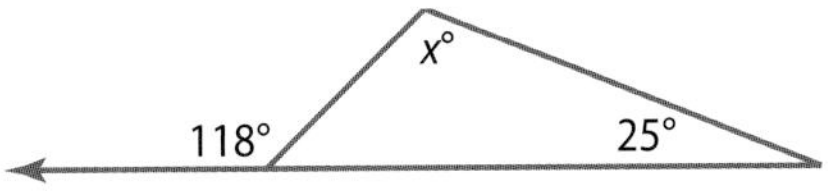

19.

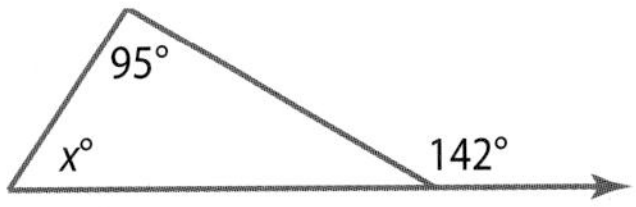

20.

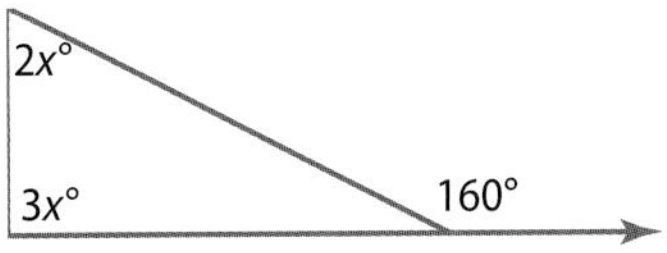

21.

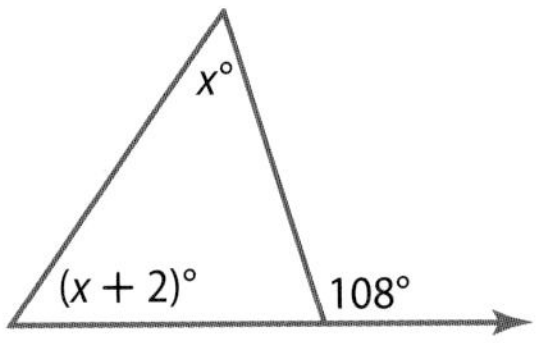

Find the measures of the angles in each triangle.

22.

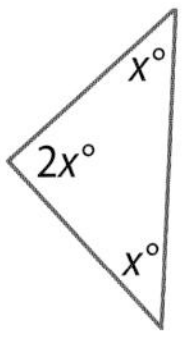

23.

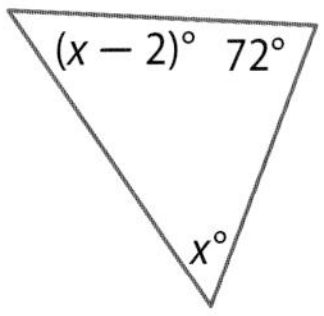

24.

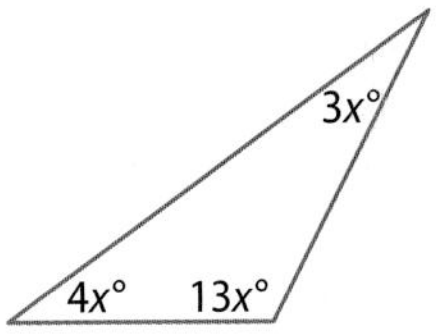

CCSS **Reason Abstractly** **The measures of the sides of a triangular flower bed are given. Classify each triangular flower bed by its sides.**

25. $2x$, $3x$, $4x$

26. y, y, y

27. $3x$, $3x$, $2x$

H.O.T. Problems Higher Order Thinking

28. CCSS **Identify Structure** Sketch each triangle. If it is not possible to sketch the triangle, write *not possible.*

a. acute scalene

b. obtuse and *not* scalene

c. right equilateral

d. obtuse equilateral

29. CCSS **Find the Error** Miguel says that an equilateral triangle is sometimes an obtuse triangle. Find his mistake and correct it.

30. CCSS **Persevere with Problems** Find the value of *x* and *y* in the figure.

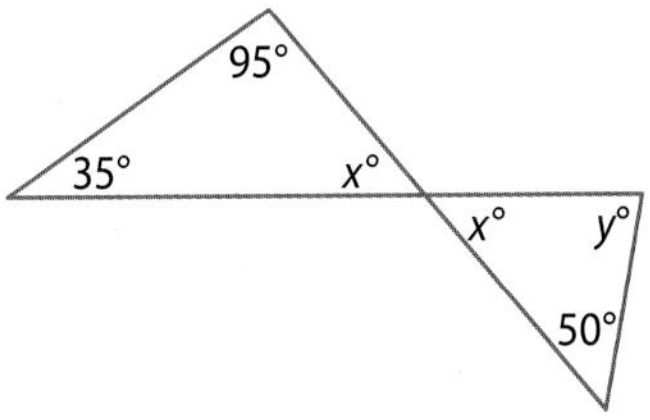

31. CCSS **Construct an Argument** What is the relationship between the measures of two acute angles of any right triangle? Explain.

32. **Building on the Essential Question** *True* or *false*? Every triangle has at least 2 acute angles. Justify your reasoning.

Standardized Test Practice

33. How would you find the value of x?

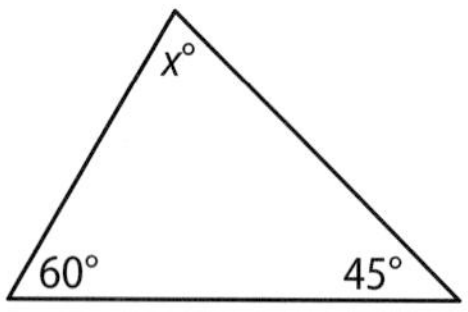

A Subtract 60 from 105.

B Add 180 to 105.

C Add 45 to 105.

D Subtract 105 from 180.

34. Which of the following best describes the triangle with the given measures?

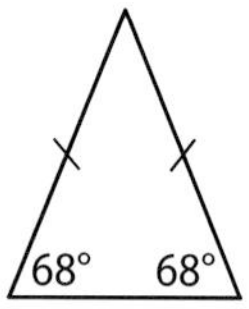

F acute isosceles triangle

G acute scalene triangle

H obtuse isosceles triangle

J acute equilateral triangle

35. Which of the following is an obtuse triangle?

A

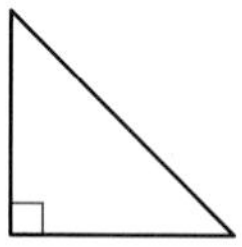

C

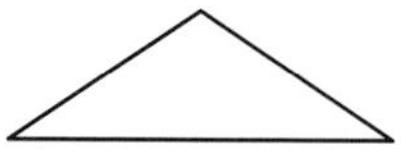

B

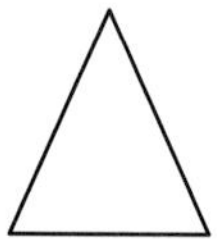

D

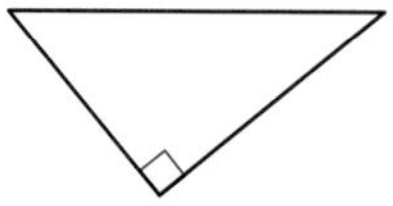

36. **Short Response** Use the triangle shown below. Find the value of x.

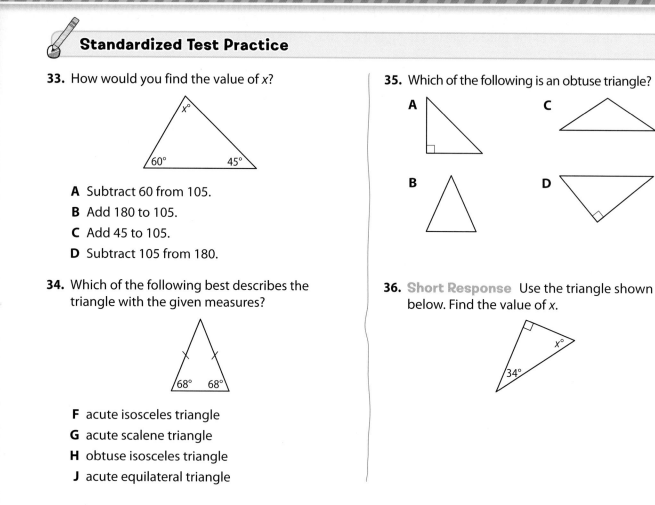

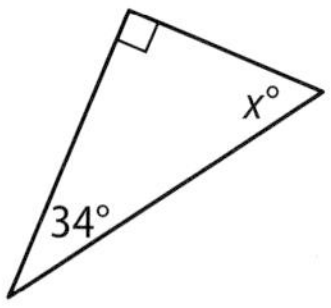

Common Core Review

Solve each equation. Round to the nearest tenth, if necessary. 8.EE.2

37. $m^2 = 81$

38. $196 = y^2$

39. $168 = 2p^2$

40. $\frac{f^2}{2} = 51$

Estimate each square root to the nearest integer. Do not use a calculator. 8.EE.2

41. $-\sqrt{52}$

42. $-\sqrt{17}$

43. $\sqrt{38}$

44. $\sqrt{140}$

45. **STEM** The time a storm will hit an area can be predicted using $d \div s = t$ where d is the distance in miles an area is from the storm, s is the speed in miles per hour of the storm, and t is the travel time in hours of the storm. Suppose it is 11:00 A.M. and a storm is heading toward a town at a speed of 30 miles per hour. If the storm is about 150 miles from the town, what time will the storm hit? 7.NS.3

Find the value of each expression. 6.EE.1

46. 11^2

47. 3^2

48. 16^2

49. 17^2

50. Angles ABC and DBE are vertical angles, with $m\angle ABC = 4x°$ and $m\angle DBE = (2x + 18)°$. 7.G.5

a. Find the value of x.

b. Find $m\angle DBE$.

Inquiry Lab
Create Triangles

Inquiry WHAT qualities determine unique triangles?

CCSS 7.G.2, 8.G.5 **Mathematical Practices** 1, 3, 5

Earrings A jewelry maker is designing earrings in the shape of a triangle. The design specifies the angle measures in the triangles. Can she use this design to create earrings with a variety of sizes?

Investigation 1

Step 1 Draw a triangle on a piece of patty paper. Copy two angles of the triangle onto separate pieces of patty paper. Extend each ray of each angle to the edge of the patty paper.

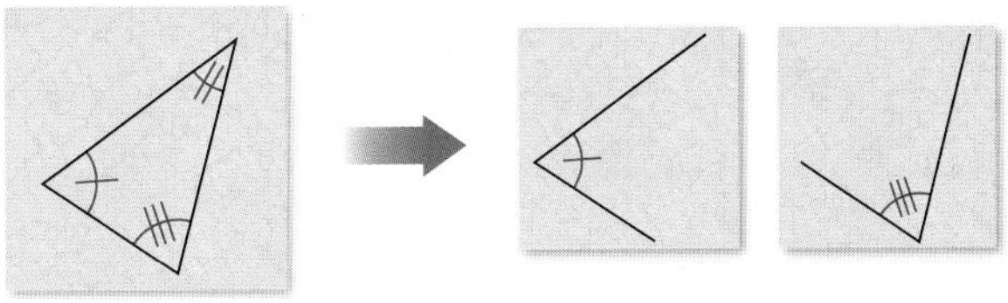

Step 2 Arrange and tape the pieces together onto a larger piece of construction paper. Extend the sides until they meet to form a triangle.

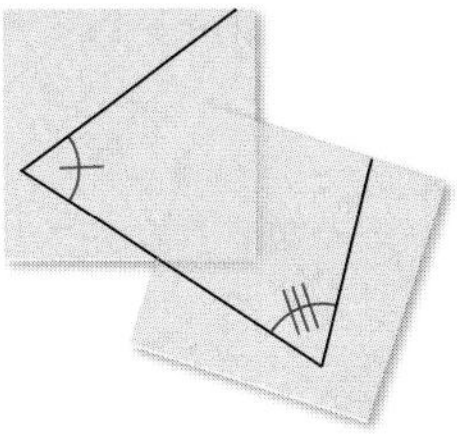

Collaborate

1. CCSS **Use Math Tools** Use a ruler and a protractor to measure the angles and side lengths of the original triangle and the new triangle created. Which measures remained the same, and which measures changed?

2. Write the lengths of the matching sides of the triangles as the ratio $\frac{\text{original}}{\text{new}}$. What do you notice about the ratios of the matching sides of the triangles?

3. Repeat Investigation 1 and form a smaller triangle. Then repeat Exercises 1 and 2. What do you notice?

4. CCSS **Make a Conjecture** Recall that triangles that have congruent angles and proportional sides are called *similar*. Are the triangles in this Investigation similar? Explain your reasoning.

Investigation 2

Step 1 Draw a triangle on a piece of patty paper. Copy two sides of the triangle and the angle between them onto separate pieces of patty paper and cut them out.

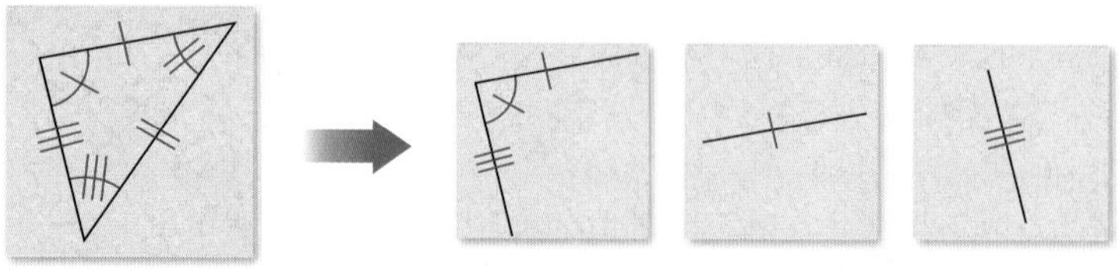

Step 2 Arrange and tape the pieces together so that the two sides are joined to form the rays of the angle. Then tape these joined pieces onto a piece of construction paper and connect the two rays to form a triangle.

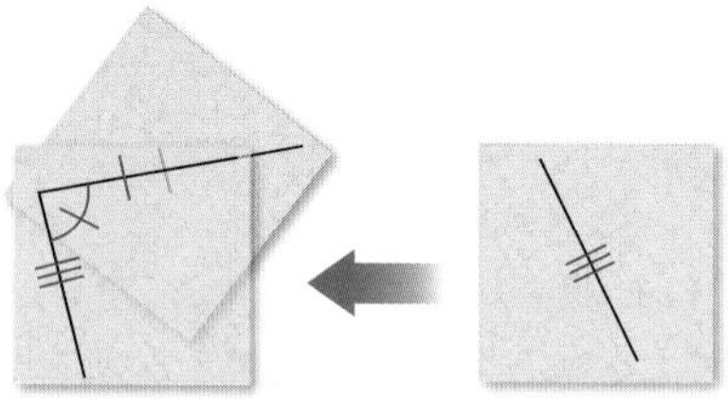

Collaborate

5. Use a ruler and a protractor to measure the angles and side lengths of your original triangle and the triangle you created. Which measures remained the same, and which measures changed?

6. Repeat Investigation 2. Can you create a different triangle than the original one drawn? Explain why or why not.

7. CCSS **Make a Conjecture** When one and only one figure can be created, the figure can be described as *unique*. Is the triangle created given the lengths of two sides and the measure of the angle between them unique? Explain

8. CCSS **Use Math Tools** Use patty paper to investigate and determine how many triangles can be created given two angles and the side between them.

Analyze

Given the following information, determine if *a unique triangle, more than one triangle,* or *no triangle* can be created.

9. 3 angles that measure 25°, 75°, and 80°

10. 2 angles that measure 90° and 40°

11. 2 angles that measure 40° and 50° and a side between them that measures 3 inches

Investigation 3

You can use dynamic geometry software to investigate triangle relationships.

Step 1 Click on **Edit**. Go to **Preferences**. Check that the distance precision is set to hundredths. On the toolbar, use the **Straightedge (segment)** tool. Click and drag to create a line segment with endpoints *A* and *B*. Click on **Measure** and **Length**.

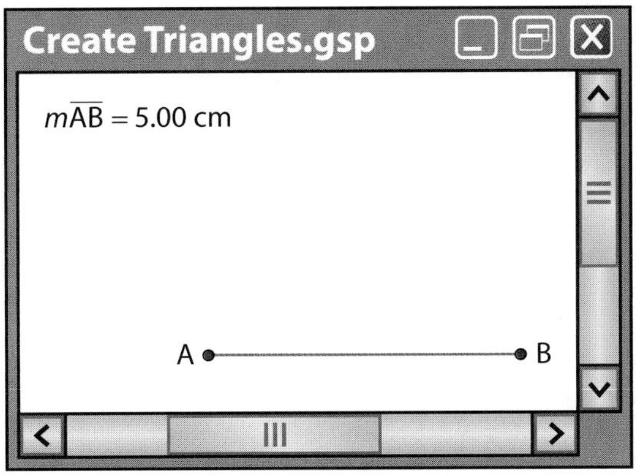

Step 2 Next, using the **Straightedge (segment)** tool, create a line segment from point *A* with endpoint *C* that is not as long as $\overline{AB}$. Click on **Measure** and **Length**. Click on points *A*, *B*, and *C*. Click on **Measure** and **Angle**.

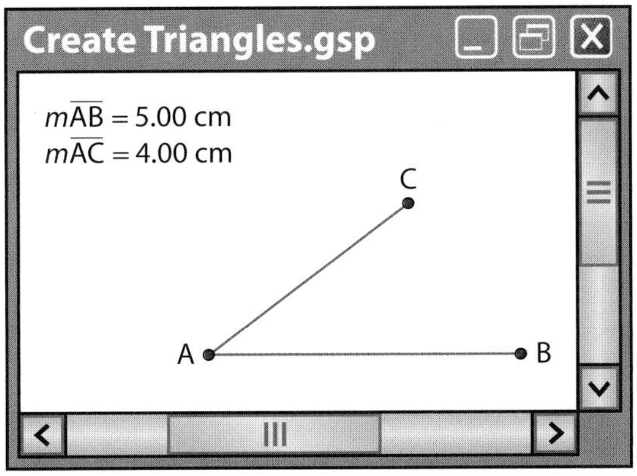

Step 3 Finally, connect points *C* and *B* with line segment $\overline{CB}$. Use the **Measure** command to measure the length of the third side and the other two angles.

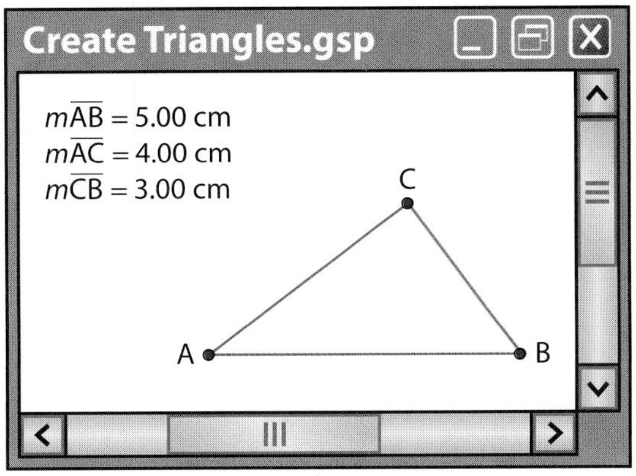

Collaborate

Work with a partner.

12. Create new triangles using the same side measures as the triangle you created in Investigation 3. How many triangles can you create?

13. Repeat Investigation 3 using different lengths for the sides of the triangle you create. Try to create a different triangle using the same side measures as the triangle you created. How many triangles can you create?

14. CCSS **Make a Conjecture** How many triangles can be created given three side lengths?

Analyze

CCSS **Persevere with Problems For Exercises 15–19, draw the triangle described below. Then determine if the triangle is unique. If it is not, draw another triangle with the same given information.**

15. triangle *GHJ* with $m\angle G = 40°$, $m\angle H = 30°$, and $m\angle J = 110°$

16. triangle *PQR* with $PQ = 4$ cm and $QR = 6$ cm.

17. triangle *LMN* with $LM = 3$ inches, $MN = 4$ inches, and $LN = 5$ inches.

18. triangle *JKL* with $m\angle J = 45°$ and $JK = 5$ cm.

19. triangle *RST* with $RS = 3$ inches, $RT = 4$ inches, and $m\angle R = 60°$.

20. Two students drew triangles *ABC* with $m\angle A = 80°$, $m\angle B = 35°$, and $m\angle C = 65°$. What must be true about the two triangles?

21. Two students drew triangles *XYZ* with $XY = 6$ inches, $YZ = 8$ inches, and $XZ = 10$ inches. What must be true about the two triangles?

22. Two students drew triangles *FGH* with $FG = 3$ cm and $GH = 5$ cm. What must be true about the two triangles?

For Exercises 23–24, refer back to the earrings the jewelry maker is designing on the first page of this Inquiry Lab.

23. How many pairs of earrings can be made from triangles with angles that measure 30°, 60°, and 90°?

24. How many pairs of earrings can be made from triangles with sides that measure 6 mm, 8 mm and 10 mm?

Reflect

25. Inquiry WHAT qualities determine unique triangles?

Lesson 11-3
Polygons

ISG Interactive Study Guide

See pages 249–250 for:
- Getting Started
- Vocabulary Start-up
- Real-World Link

Essential Question

How can you determine congruence and similarity?

Common Core State Standards

Content Standards
Extension of 8.G.5

Mathematical Practices
1, 3, 4, 7

Vocabulary

polygon
diagonal
regular polygon
tessellation

What You'll Learn

- Classify polygons.
- Determine the sum of the measures of the interior angles of a polygon.

Real-World Link

Road Signs While riding in a car, you might see several different sizes and shapes of road signs. Different shapes and colors have many different meanings. For example, a triangle sign that "points" to the right refers to a no-passing zone, and a square tilted so that it is on one of its corners is a warning sign (curves ahead, low clearance).

Classify Polygons

A **polygon** is a simple, closed figure formed by three or more line segments called *sides*. The figures below are *not* polygons.

Non Polygons

The figure has a curve. The sides overlap. The figure is not closed.

Polygons can be classified by the number of sides they have, as shown in the table.

Polygon	Sides	Polygon	Sides	Polygon	Sides	Polygon	Sides
triangle	3	pentagon	5	heptagon	7	nonagon	9
quadrilateral	4	hexagon	6	octagon	8	decagon	10

Example 1

Watch Tutor

Determine whether the figure is a polygon. If it is, classify the polygon. If it is not a polygon, explain why.

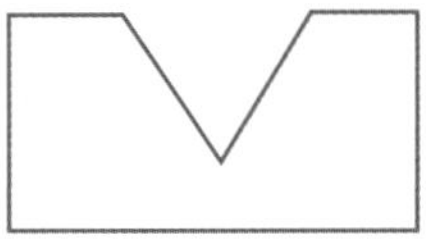

The figure has 7 sides that only intersect at their endpoints. It is a heptagon.

Got It? **Do these problems to find out.**

1a.

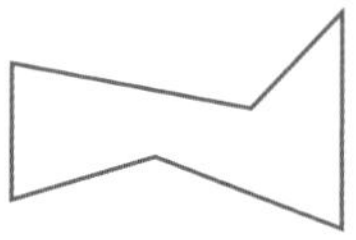

1b.

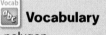

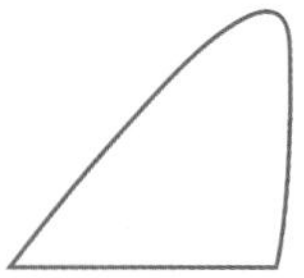

n-gon
A polygon with n sides is called an n-gon. For example, an octagon can also be called an 8-gon.

Key Concept Interior Angles of a Polygon

Words The sum of the degree measures of the interior angles of a polygon with n sides is $(n - 2)180$.

Symbols $(n - 2)180$.

Model

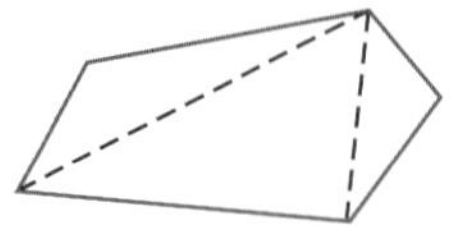

A **diagonal** is a line segment that joins two nonconsecutive vertices in a polygon. In the figures below, all possible diagonals from one vertex are shown.

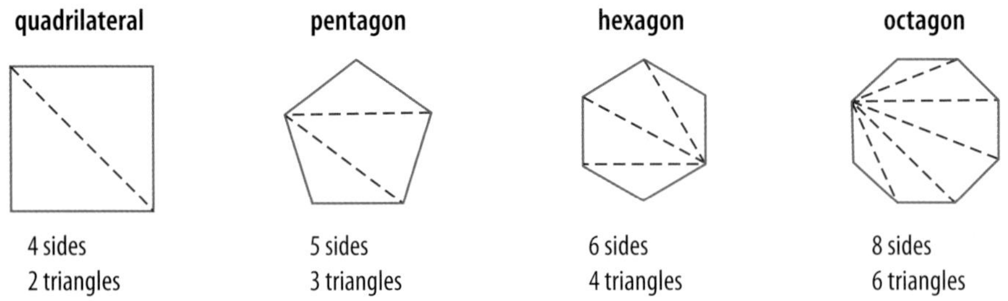

Notice that the number of triangles is 2 less than the number of sides. You can use this relationship to find the sum of the interior angle measures of a polygon. Recall that an interior angle is an angle formed at a vertex of a polygon.

Example 2

Tutor

Find the sum of the measures of the interior angles of a nonagon.

The sum of the measures of the interior angles is $(n - 2)180$. Since a nonagon has 9 sides, $n = 9$.

$(\mathbf{n} - 2)180 = (\mathbf{9} - 2)180$ Replace n with 9.

$= 7(180)$ or 1260 Simplify.

The sum of the measures of the interior angles of a nonagon is 1260°.

CHECK All possible diagonals from one vertex of a nonagon are shown below.

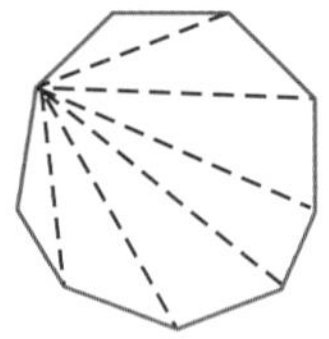

You can see that 7 triangles are formed. So, $7 \cdot 180 = 1260$ is correct. ✓

Got It? **Do these problems to find out.**

2a. Find the sum of the measures of the interior angles of a 13-gon.

2b. Find the sum of the measures of the interior angles of an 18-gon.

A **regular polygon** is a polygon that has all sides congruent and all angles congruent. Since the angles are congruent, their measures are equal.

Real World

Example 3

Tutor

The surface of a soccer ball contains 12 regular pentagons and 20 regular hexagons. What is the measure of one interior angle of a hexagon?

Step 1 Find the sum of the measures of the interior angles of a hexagon.

$(n - 2)180 = (6 - 2)180$ — A hexagon has 6 sides. Replace n with 6.

$= 4(180)$ or 720 — Simplify.

The sum of the measures of the interior angles is 720°.

Step 2 Divide the sum of the measures by 6 to find the measure of one angle.

$720 \div 6 = 120$

The measure of one interior angle of a hexagon is 120°.

Got It? **Do this problem to find out.**

3. What is the measure of an interior angle of one of the pentagonal panels in a soccer ball?

Vertex, Vertices

A *vertex* of a figure is a point where two sides of the figure meet. *Vertices* is the plural of *vertex*.

A repetitive pattern of polygons that fit together with no overlaps or holes is called a **tessellation**. The sum of the measures of the angles where the vertices meet in a tessellation is 360°.

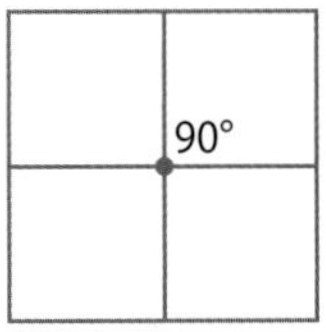

$4 \times 90° = 360°$

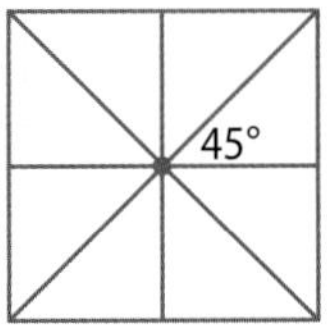

$8 \times 45° = 360°$

Example 4

Tutor

Determine whether or not a tessellation can be created using only regular hexagons. If not, explain.

The measure of each angle in a regular hexagon is 120°.

The sum of the measures of the angles where the vertices meet must be 360°. So, solve $120n = 360$.

$120n = 360$ — Write the equation.

$\frac{120n}{\mathbf{120}} = \frac{360}{\mathbf{120}}$ — Divide each side by 120.

$n = 3$ — Simplify.

Since 120° divides evenly into 360°, a regular hexagon can be used to make a tessellation.

Got It? **Do these problems to find out.**

Determine whether or not a tessellation can be created using only each regular polygon. If not, explain.

4a. pentagon

4b. octagon

Guided Practice

Check

Determine whether the figure is a polygon. If it is, classify the polygon. If it is not a polygon, explain why. (Example 1)

1.

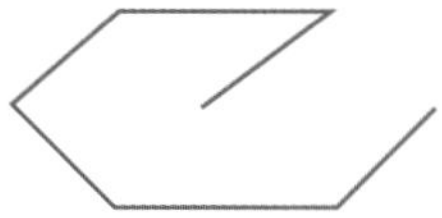

2.

3.

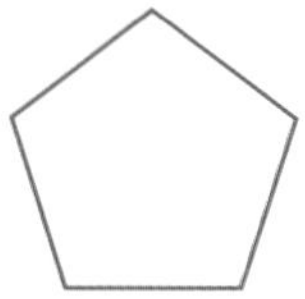

4. The sum of the measures of the interior angles of a certain regular polygon is 1800°. How many sides does this polygon have? (Example 2)

5. The kaleidoscope image at the right is a regular polygon with 14 sides. What is the measure of one interior angle of the polygon? Round to the nearest tenth. (Example 3)

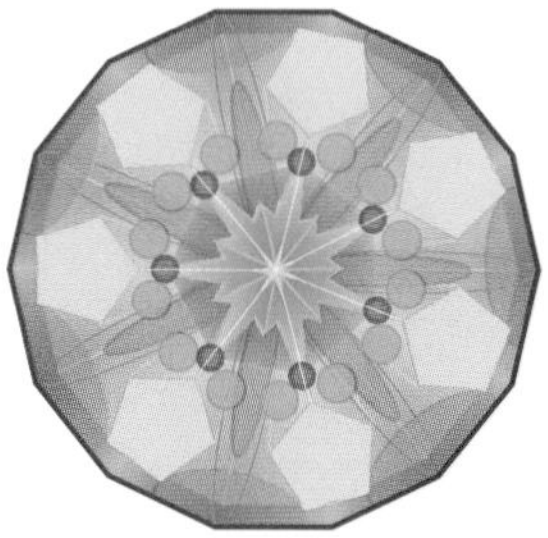

6. Find the measure of one interior angle in each regular polygon. Round to the nearest tenth if necessary.

a. nonagon **b.** heptagon **c.** 20-gon

7. Determine whether or not a tessellation can be created using only equilateral triangles. If not, explain. (Example 4)

Independent Practice

Go online for Step-by-Step Solutions

eHelp

Determine whether the figure is a polygon. If it is, classify the polygon. If it is not a polygon, explain why. (Example 1)

8.

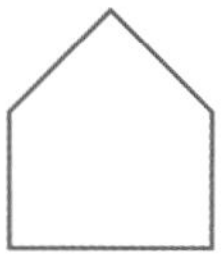

9.

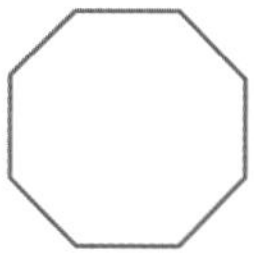

10.

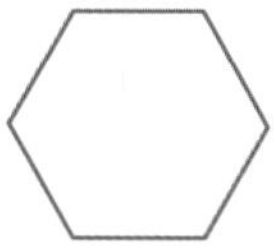

11.

12.

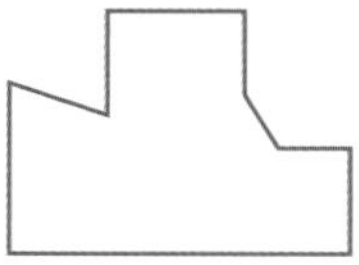

13.

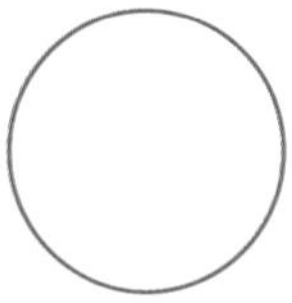

Find the sum of the measures of the interior angles of each polygon. (Example 2)

14. decagon

15. 11-gon

16. 16-gon

17. 24-gon

18. The dome in a state capitol building is octagonal. What is the measure of an interior angle of a regular octagon? (Example 3)

19 **STEM** The individual cells of a honeycomb are regular hexagons. What is the measure of an interior angle of a honeycomb cell? (Example 3)

Determine whether or not a tessellation can be created using each regular polygon. If not, explain. (Example 4)

20. quadrilateral

21. 12-gon

22. 15-gon

23. 20-gon

CCSS Identify Structure Identify the polygon given the sum of the interior angle measures.

24. 1080° **25.** 2340° **26.** 3240° **27.** 5040°

28. CCSS **Identify Structure** Refer to the mosaic at the right. How are tessellations used to create the image?

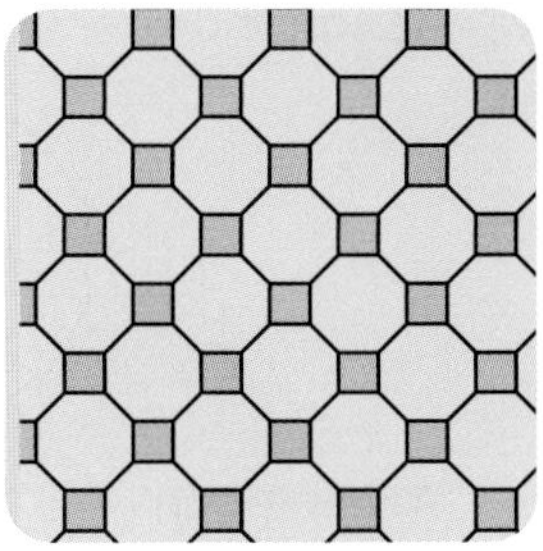

You can create a tessellation using a square and a triangle.

a. Draw a square. Then draw a triangle inside the top of the square.

b. Slide the triangle from the top to the bottom of the square.

c. Repeat this pattern unit to create a tessellation.

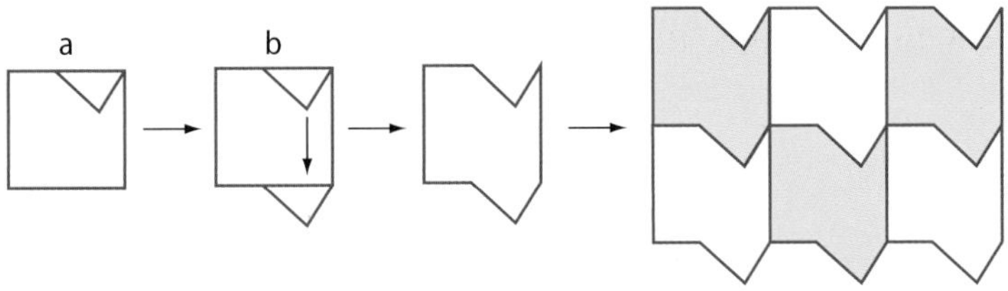

Create a tessellation for each pattern shown.

29. 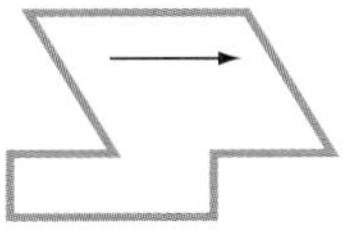

30. 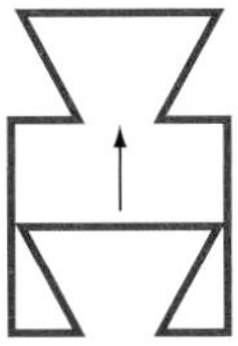

31. 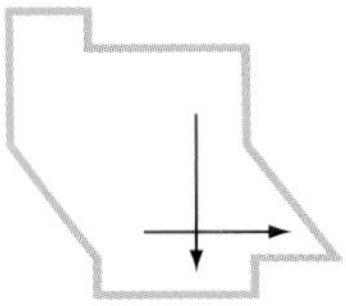

When a side of a polygon is extended, an *exterior angle* is formed. In any polygon, the sum of the measures of the exterior angles, one at each vertex, is 360°. For each regular polygon, find the measure of an exterior angle.

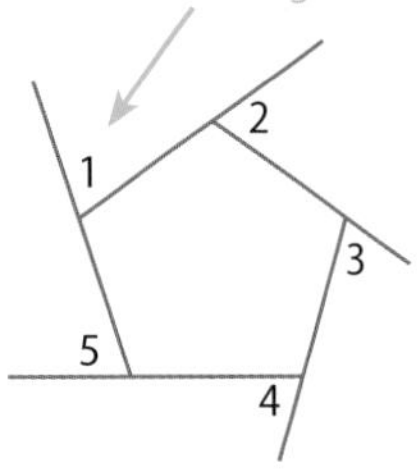

32. octagon **33** triangle

34. decagon **35.** 15-gon

H.O.T. Problems Higher Order Thinking

36. CCSS **Model with Mathematics** Use two types of polygons to create a tessellation that is different from the tessellations shown in this lesson. Identify the polygons that you used to create your tesselation. Then explain where you might see your tessellation in the real world.

37. CCSS **Reason Inductively** If the number of sides of a regular polygon increases by 1, what happens to the sum of the measures of the interior angles?

38. CCSS **Justify Conclusions** Is it possible to use a trapezoid to create a tessellation? Explain your reasoning.

39. CCSS **Persevere With Problems** Create a tessellation using regular hexagons and equilateral triangles.

40. **Building on the Essential Question** Describe the difference between a regular polygon and a polygon that is irregular. Then explain the process used to find the interior angle measure of a regular polygon.

Standardized Test Practice

41. Which term identifies the shaded part of the design shown?

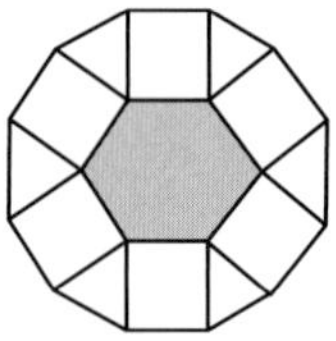

A heptagon
B hexagon
C octagon
D pentagon

42. The sum of the measures of the interior angles of a polygon is 2160°. Find the number of sides of the polygon.

F 10
G 14
H 16
J 18

43. A landscape architect is looking for a brick paver shape that will tessellate. Which shape by itself will allow her to tessellate a patio area?

A
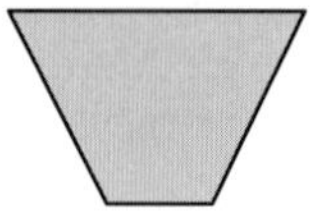

C

B
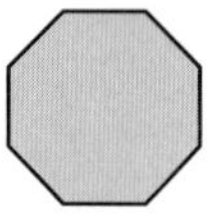

D
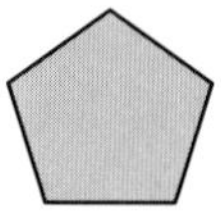

44. **Short Response** What is the measure in degrees of an interior angle of a regular polygon with 20 sides?

Common Core Review

The figures are similar. Find each missing measure. 7.RP.2

45.

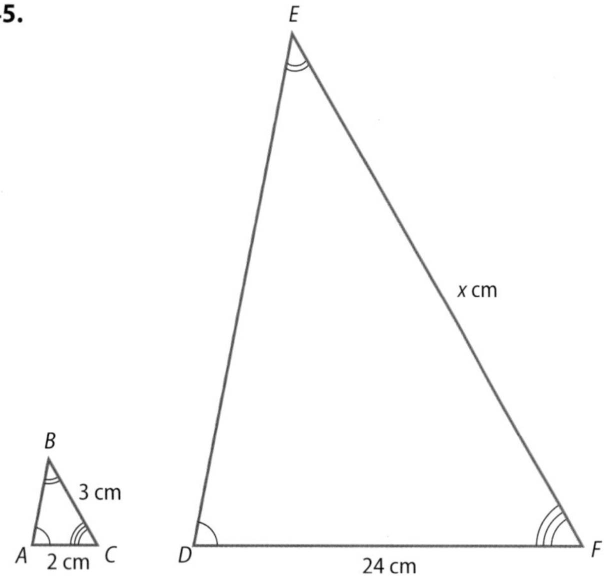

46.

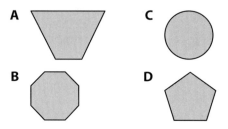

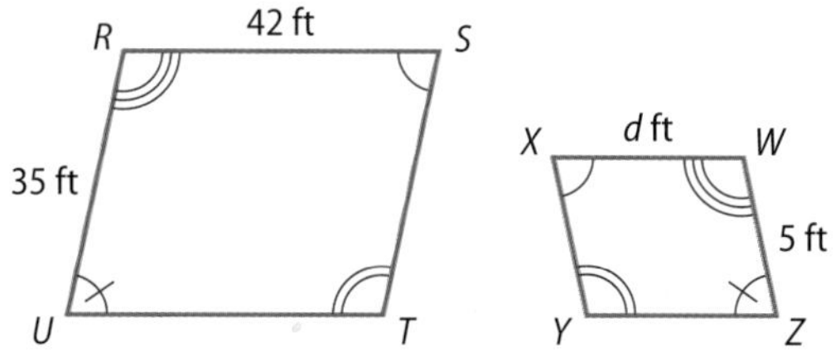

47. Clancy is making necklaces. He knows that 6 bags of beads will make 15 necklaces. If he wants to make 25 necklaces, how many bags of beads will he need? 7.RP.2

Solve each equation. Check your solutions. 8.EE.7

48. $3.5x + 1 = -2.5$

49. $\frac{2}{3}y - 8 = -2$

50. $\frac{3}{5}z + 1 = -14$

51. $33.6 - 5.6m = -11.2$

52. Solve $x - 3.4 \geq 6.2$. Graph the solution on a number line. 7.EE.4b

Find each quotient. 7.NS.2b

53. $-69 \div 23$

54. $48 \div (-8)$

55. $-24 \div (-12)$

56. $-50 \div 5$

Inquiry Lab

Transformations

Inquiry **HOW do transformations in a plane affect the size, shape, and orientation of the figure?**

CCSS 8.G.1a, 8.G.1b, 8.G.1c

Mathematical Practices
1, 3, 5

Windows Our homes have windows that open in many different ways. The motion of windows opening and closing is related to the motion of figures in a plane. What properties of a window are preserved when it opens and closes by sliding?

Investigation 1

Step 1 Draw angle *ABC* on a piece of paper. Place a dashed line on the paper as shown.

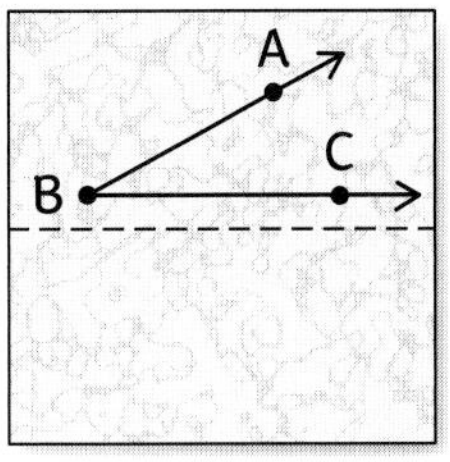

Step 2 Trace $\angle ABC$ onto a separate piece of tracing paper. Label the angle *DEF* so that *D* matches up with *A*, *E* matches up with *B*, and *F* matches up with *C*.

Step 3 Slide the tracing paper down. Tape $\angle DEF$ on the same piece of paper as $\angle ABC$, so that $\angle DEF$ is below the dashed line.

Step 4 Measure $\angle ABC$ and $\angle DEF$. $m\angle ABC = m\angle DEF$

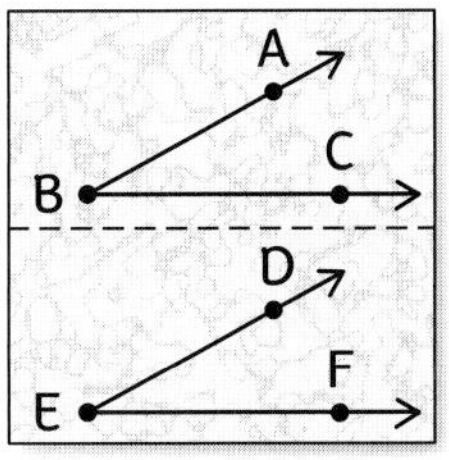

Just like the figure in Investigation 1, a window that opens and closes by sliding preserves angle measures. This motion also preserves the position, or orientation, of a figure.

Investigation 2

Step 1 Draw trapezoid *WXYZ* on a piece of tracing paper. Place a dashed line on the paper as shown.

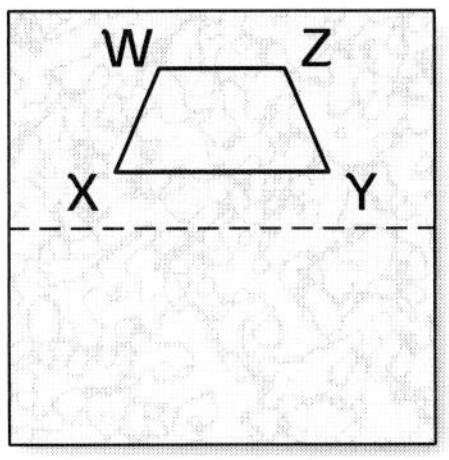

Step 2 Trace trapezoid *WXYZ* and the dashed line onto a separate piece of tracing paper. Fold the tracing paper along the dashed line, and trace the trapezoid on the folded half of the paper. Label the new trapezoid *QRST*.

Step 3 Cut out trapezoid *QRST*. Tape it beneath the dashed line on the piece of paper from Step 1.

Analyze

1. **CCSS Use Math Tools** Use a centimeter ruler to measure the lengths of the sides of both trapezoids in Investigation 2 to the nearest millimeter. How do the lengths of the following corresponding pairs of sides compare: $\overline{WX}$ and $\overline{QR}$, $\overline{XY}$ and $\overline{RS}$, $\overline{YZ}$ and $\overline{ST}$, and $\overline{WZ}$ and $\overline{QT}$?

2. Which two sides are parallel in trapezoid *WXYZ*? Are the sides that correspond to those sides in trapezoid *QRST* also parallel?

3. Did flipping the figure change the figure's orientation? If so, explain how.

Investigation 3

Step 1 On a piece of paper draw triangle *ABC*.

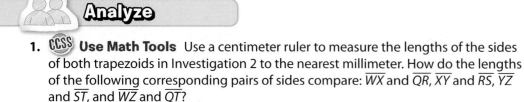

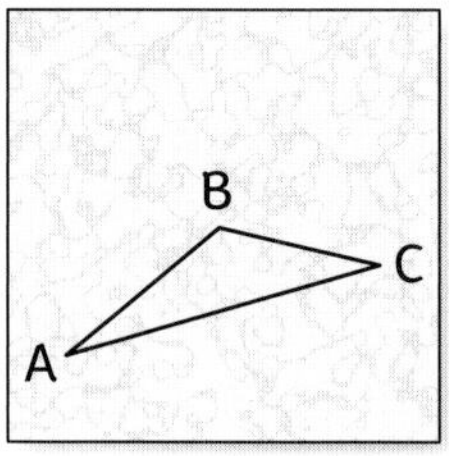

Step 2 Trace triangle *ABC* onto a piece of tracing paper. Place the eraser end of your pencil on your traced triangle at point *B*. Turn your traced triangle about point *B*. Tape your turned triangle to your paper from Step 1. Label it triangle *DEF*.

Analyze

4. Does the turned triangle *DEF* in Investigation 3 have the same side lengths and angle measures as triangle *ABC*?

5. Did turning the figure change the figure's orientation? If so, explain how.

6. For all three investigations, did sliding, flipping, or turning the figures alter the size or shape of the original figure?

7. In geometry, slides, flips and turns are called *transformations*. Their mathematical names are *translations*, *reflections*, and *rotations*. Repeat Investigations 1–3 for any quadrilateral you draw on a piece of paper.
   a. How do the measures of the angles and sides of your original figure compare to the images you got by translating, reflecting, and rotating the figure?
   b. Did translating, reflecting, or rotating your figure cause any change to your figure's orientation?

8. **CCSS Reason Inductively** How is the movement used to open or close a window related to the movement of figures on a plane?

9. Name another example of a real-world situation that involves translation, rotation, or reflection of a figure in a plane.

Reflect

10. **Inquiry** HOW do transformations of a figure in a plane affect the size, shape, and orientation of the figure?

Lesson 11-4

Translations and Reflections on the Coordinate Plane

ISG Interactive Study Guide

See pages 251–252 for:
- Getting Started
- Vocabulary Start-Up
- Notes

Essential Question

How can you determine congruence and similarity?

Common Core State Standards

Content Standards
8.G.3

Mathematical Practices
1, 3, 4, 5

Vocabulary

transformation
image
congruent
translation
reflection
line of reflection

What You'll Learn

- Define and identify transformations.
- Draw translations and reflections on a coordinate plane.

Real-World Link

Reflections Have you ever heard the term *mirror image*? A mirror image is something that appears identical but reversed. It can also be called a *reflection*. You can find reflections just about anywhere, like on your smartphone screen or even in tiles on the floor.

Translations

A **transformation** is an operation that maps an original geometric figure onto a new figure called the **image**. For certain types of transforamations, the original figure and its image are **congruent** figures. Congruent figures have the same shape and same size. Two common transformations on the coordinate plane are shown.

Key Concept Translations and Reflections

Translation	• called a *slide* • translated image and original figure are congruent • orientation is the *same* as the original figure
Reflection	• called a *flip* • figures are mirror images of each other • reflected image and original figure are congruent • orientation is *different* from the original figure

A **translation** is when you slide a figure from one position to another without turning it.

Translation

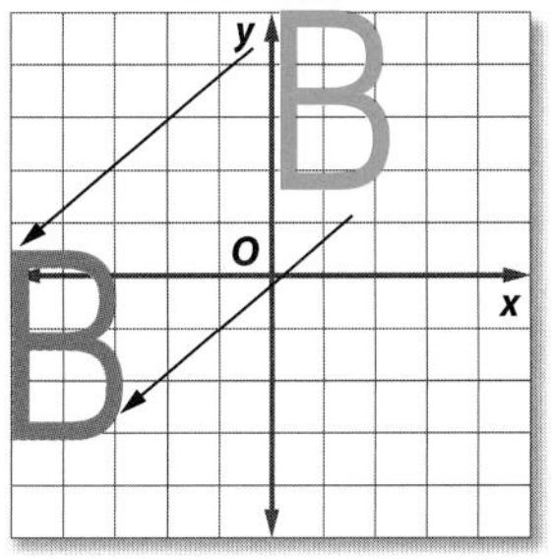

A **reflection** is when you flip a figure over a line. This line is called the **line of reflection**.

Reflection

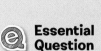

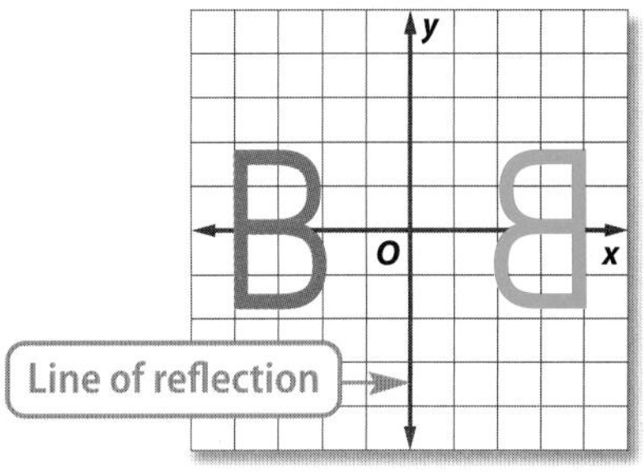

When translating a figure, every point of the original figure is moved the same distance and in the same direction.

> **Notation**
> The notation A' is read *A prime*. It corresponds to point *A*.

Translation
5 units right

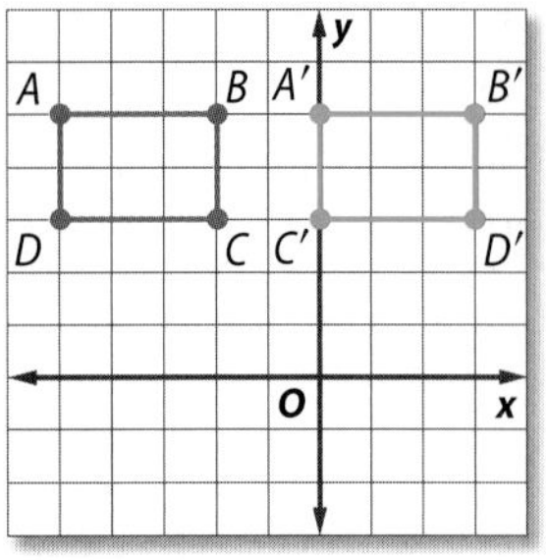

Translation
4 units up

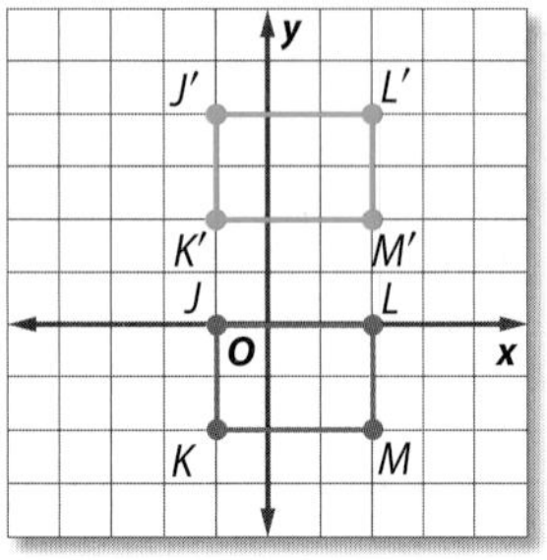

Translation
3 units left, 5 units down

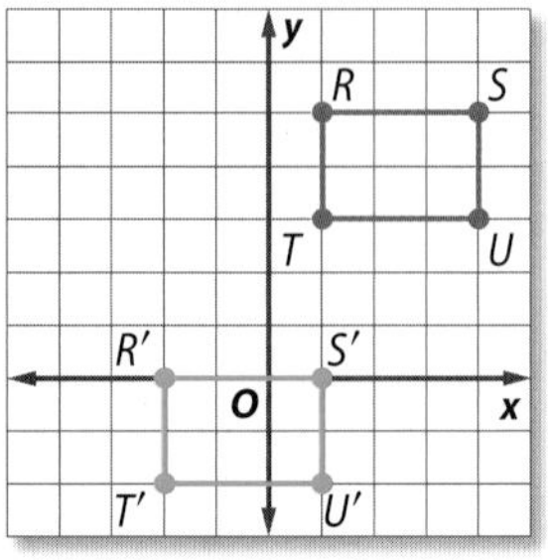

You can describe the translation using an ordered pair (a, b). For example, a translation of 3 units left and 5 units down corresponds to $(-3, -5)$. A translation moves every point $P(x, y)$ to an image $P'(x + a, y + b)$.

Example 1

> **Vocabulary Link**
> **Vertices**
> *Everyday Use* the highest points
> *Math Use* plural form of vertex, a point common to the two sides of an angle or a polygon, or the point common to the edges of a polyhedron.

Rectangle *JKLM* on the coordinate grid at the right represents a rectangular table on a tiled floor. If the table is moved 2 units to the right and 1 unit down, find the coordinates of the vertices of the table's image.

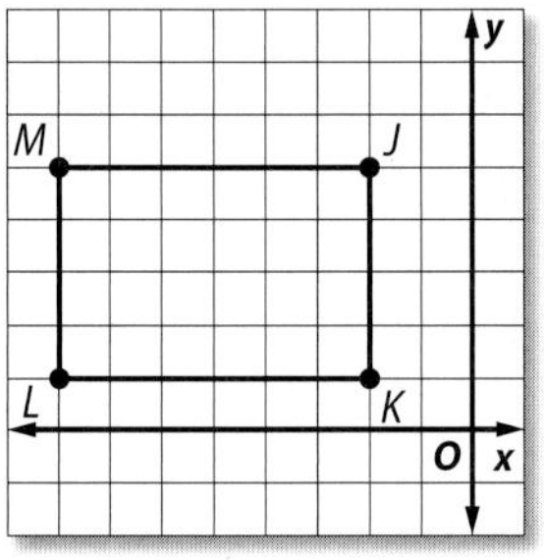

Write the translation as an ordered pair.

$(2, -1)$

To find the coordinates of the translated image, add 2 to each *x*-coordinate and add −1 to each *y*-coordinate.

original		translation		image
$J(-2, 5)$	+	$(2, -1)$	→	$J'(0, 4)$
$K(-2, 1)$	+	$(2, -1)$	→	$K'(0, 0)$
$L(-8, 1)$	+	$(2, -1)$	→	$L'(-6, 0)$
$M(-8, 5)$	+	$(2, -1)$	→	$M'(-6, 4)$

The coordinates of the vertices of rectangle $J'K'L'M'$ are (0, 4), (0, 0), (−6, 0), and (−6, 4).

Got It? Do these problems to find out.

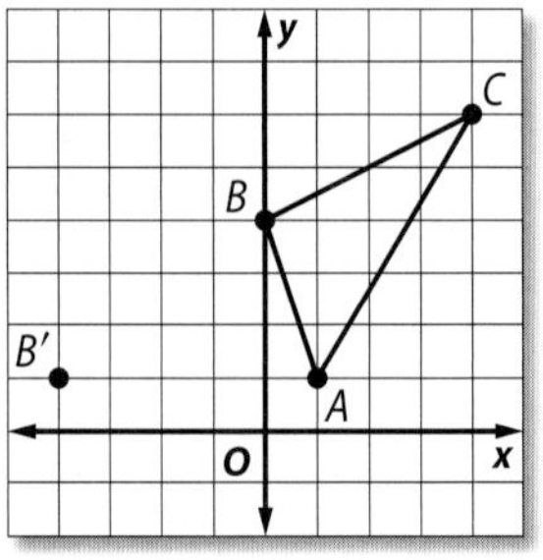

1a. Triangle *ABC* is translated so that B' is a vertex of the image. What are the coordinates of A' and C'?

1b. Suppose that triangle *ABC* is translated 8 units right and 6 units down. Find the coordinates of the vertices of its image.

Reflections

When reflecting a figure, every point of the original figure has a corresponding point on the other side of the line of reflection. Corresponding points are the same distance from the line of reflection.

Opposites
Opposites are two numbers with the same absolute value but different signs.

Reflection over *x*-axis

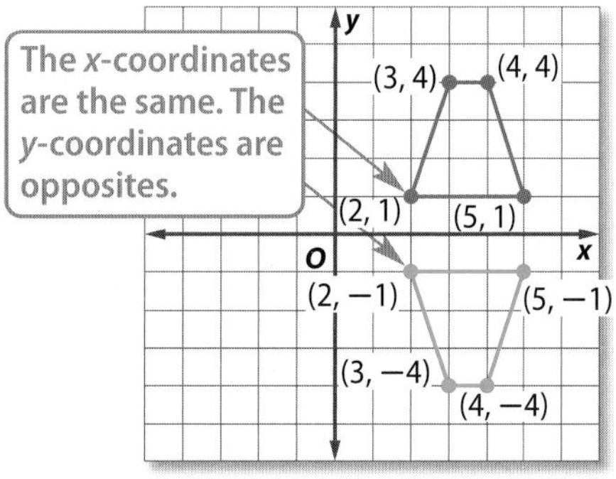

Reflection over *y*-axis

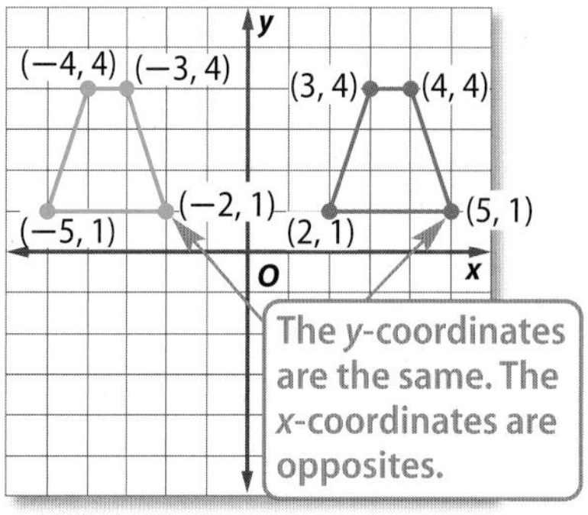

To reflect a point over the *x*-axis, use the same *x*-coordinate and multiply the *y*-coordinate by −1. To reflect a point over the *y*-axis, use the same *y*-coordinate and multiply the *x*-coordinate by −1.

Example 2

The vertices of figure *DEFG* are *D*(4, −2), *E*(5, −5), *F*(2, −4), and *G*(1, −1). Graph the figure and its image after a reflection over the *y*-axis.

To find the coordinates of a vertex of the image after a reflection over the *y*-axis, use the same *y*-coordinate. Replace the *x*-coordinate with its opposite.

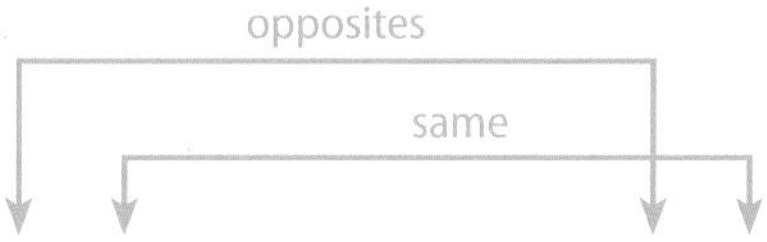

D(4, −2)	→	*D*′(−4, −2)
E(5, −5)	→	*E*′(−5, −5)
F(2, −4)	→	*F*′(−2, −4)
G(1, −1)	→	*G*′(−1, −1)

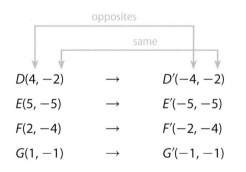

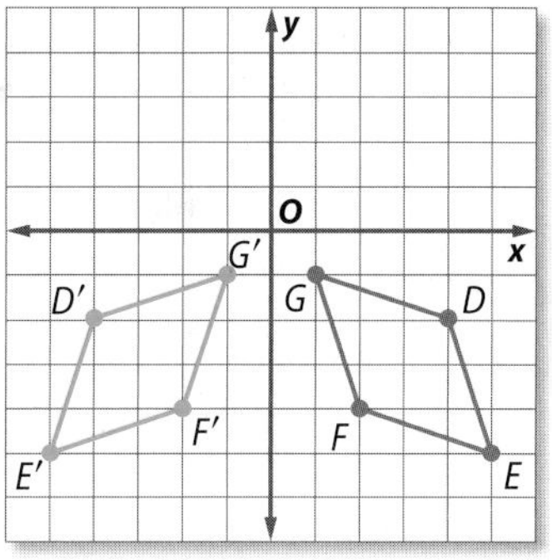

The coordinates of the vertices of the image are *D*′(−4, −2), *E*′(−5, −5), *F*′(−2, −4), and *G*′(−1, −1).

Got It? Do these problems to find out.

2a. The vertices of △*ABC* are *A*(4, −2), *B*(0, 2), and *C*(5, 2). Graph the triangle and its image after a reflection over the *x*-axis.

2b. The vertices of rectangle *WXYZ* are *W*(−3,−3), *X*(−3, 4), *Y*(2, 4) and *Z*(2, −3). Graph the rectangle and its image after a reflection over the *y*-axis.

Guided Practice

Check

1. Triangle *MNP* is shown on the coordinate plane. Find the coordinates of the vertices of the image of the triangle *MNP* translated 5 units to the right and 3 units up. (Example 1)

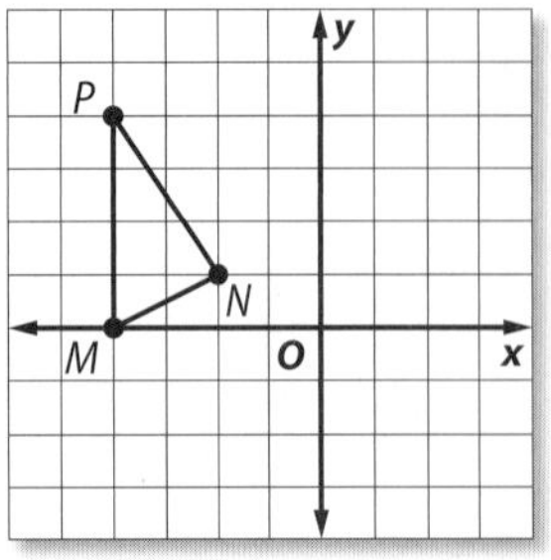

2. The vertices of $\triangle LMN$ are $L(2, 1)$, $M(5, 2)$, and $N(-1, 4)$. Graph the triangle and its image after a reflection over the *x*-axis. (Example 2)

Independent Practice

Go online for Step-by-Step Solutions eHelp

For Exercises 3 and 4, use the coordinate plane at the right. Triangle *XYZ* is shown.

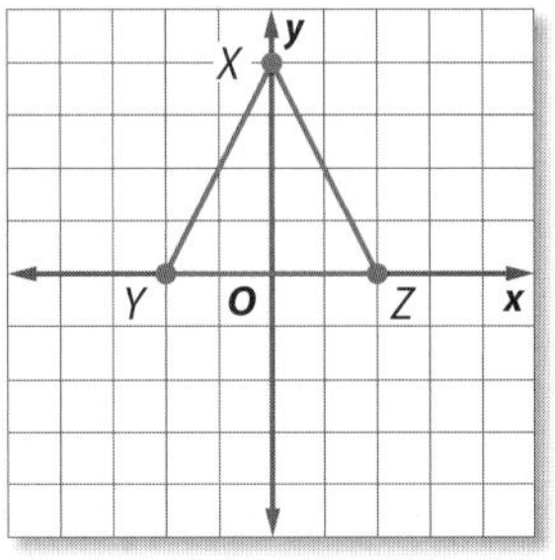

3. Find the coordinates of the vertices of the image of $\triangle XYZ$ translated 4 units to the right and 5 units down. (Example 1)

4. Find the coordinates of the vertices of the image of $\triangle XYZ$ translated 2 units to the left and 3 units up. (Example 1)

5. The vertices of $\triangle FGH$ are $F(-3, 4)$, $G(0, 5)$, and $H(3, 2)$. Graph the triangle and its image after a reflection over the *y*-axis. (Example 2)

6. The vertices of figure *RSTV* are $R(2, 4)$, $S(4, 3)$, $T(4, -2)$, and $V(1, -2)$. Graph the figure and its image after a reflection over the *y*-axis. (Example 2)

7. The vertices of figure *ABCD* are $A(-3, -1)$, $B(-5, -1)$, $C(-5, -6)$, and $D(-3, -3)$. Graph the figure and its image after a reflection over the *x*-axis. (Example 2)

8. Reflect the figure below over the *x*-axis. Sketch the figure and its image on grid paper. What is the animal? (Example 2)

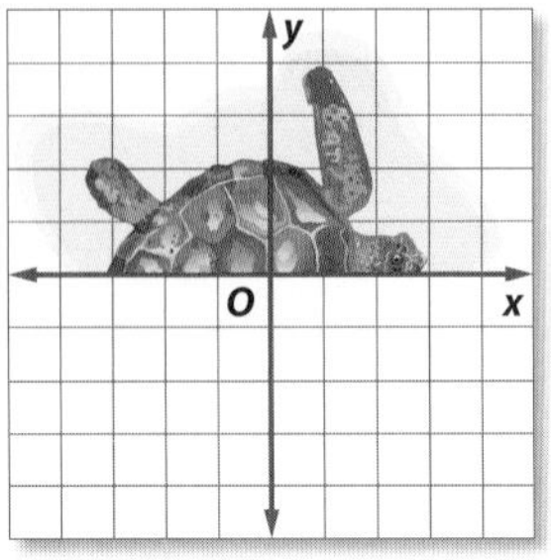

9. In chess, the rook can only move vertically or horizontally across the board. The chessboard below shows the movement of a rook after two turns. Describe this translation in words. (Example 1)

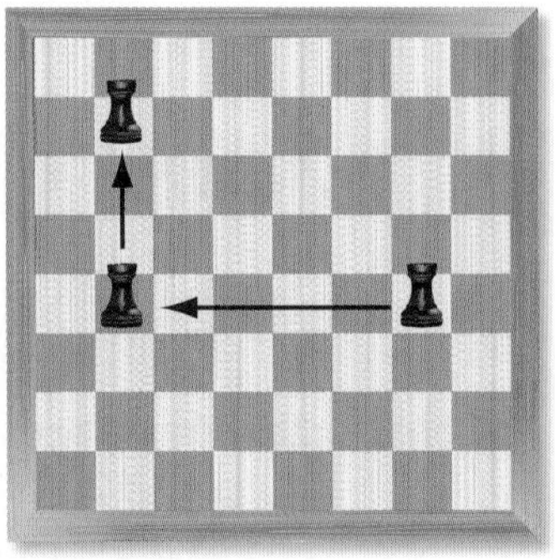

For Exercises 10–14, identify each transformation as a *translation* or a *reflection* and describe the transformation. For Exercises10–12, the green image is the original.

10. 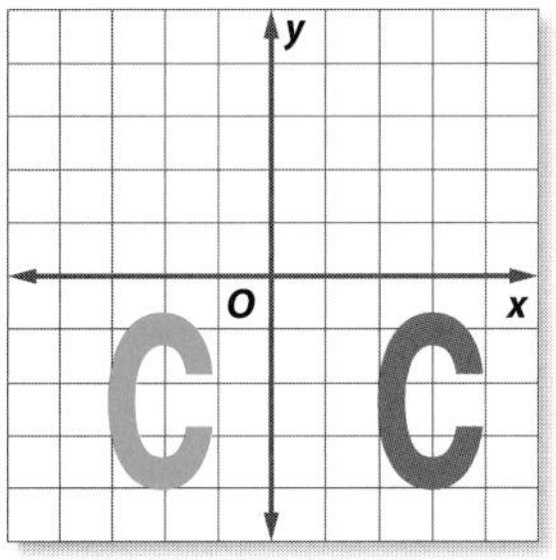

11. 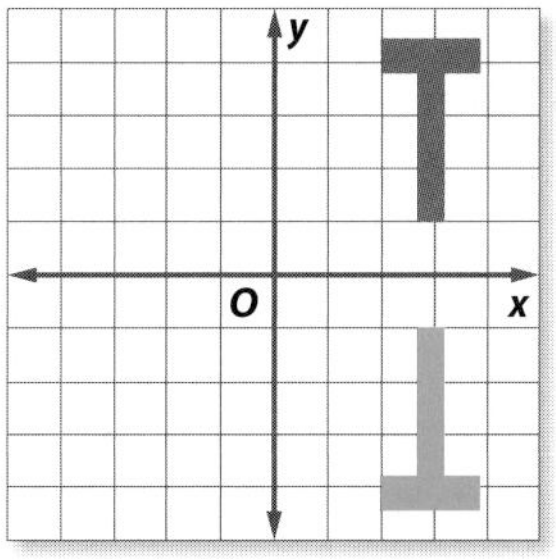

12. 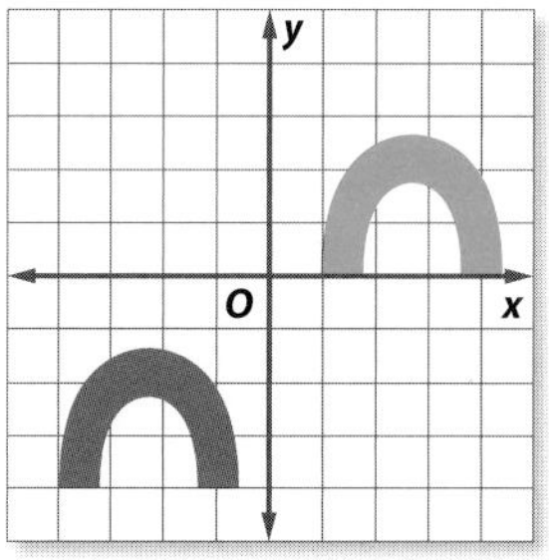

13. A figure has vertices $H(1, -1)$, $J(1, 5)$, $K(3, -1)$, and $L(3, 5)$. The image's vertices are $H'(-4, -5)$, $J'(-4, 1)$, $K'(-2, -5)$, and $L'(-2, 1)$.

14. Triangle QRS has vertices $Q(1, -1)$, $R(5, -3)$, and $S(3, 2)$. The vertices of the image are $Q'(1, 1)$, $R'(5, 3)$, and $S'(3, -2)$.

15. Triangle RST has vertices $R(4, 2)$, $S(-8, 0)$, and $T(6, 7)$. When translated, R' has coordinates $(-2, 4)$. Find the coordinates of S' and T'. Then describe the translation of triangle RST onto triangle $R'S'T'$.

16. A mosaic is a type of art created using glass, stone, tile, or other materials. Describe the transformation that maps the red outlined tile to the purple outlined tile.

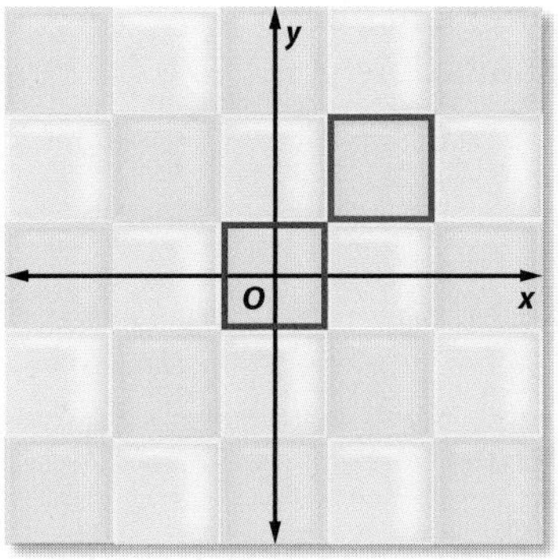

17. CCSS **Model with Mathematics** When using a rubber stamp, the image the ink makes on a page is a reflection of the stamp. Suppose you create a stamp that would print the word MATH. Draw the stamp. Is the image a reflection over the x-axis or y-axis?

H.O.T. Problems Higher Order Thinking

18. CCSS **Use Math Tools** Draw a figure on the coordinate plane. Then reflect the figure over the y-axis.

19. CCSS **Justify Conclusions** Suppose you reflect a figure over the x-axis and then you reflect the image over the y-axis. Is there a single transformation using reflections or translations that maps the original figure to its final image? If so, name it. Explain your reasoning.

20. CCSS **Which One Doesn't Belong?** Without graphing, identify the pair of points that does not represent a reflection over the y-axis. Justify your reasoning.

$E(0, 1)$ $E'(0, 1)$	$F(-2, 5)$ $F'(2, 5)$	$G(-3, -4)$ $G'(-3, 4)$	$H(5, 0)$ $H'(-5, 0)$

21. CCSS **Persevere with Problems** Discuss how an image compares to the original figure if you reflect a triangle in Quadrant I over the x-axis, then translate the image 4 units right and 3 units up. Determine if a single transformation can map the original figure to the final image.

22. **Building on the Essential Question** A figure is translated by $(2, -3)$ and then the image is translated by $(-2, 3)$. Without graphing, describe the final position of the figure. Explain your reasoning.

Standardized Test Practice

23. Which of the following is a vertex of the figure shown below after a translation of 2 units right and 2 units up?

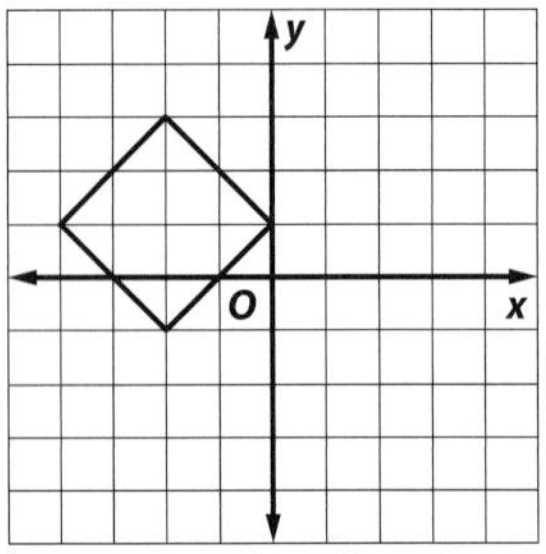

A (5, 0) **C** (2, 0)

B (0, 2) **D** (0, 5)

24. **Short Response** The coordinates of a triangle are $A(0, -1)$, $B(-2, -1)$, and $C(3, 5)$. What are the coordinates of the triangle after it has been translated 3 units left and 4 units down?

25. Which of the following *best* represents a reflection over the vertical line segment in the center of the rectangle?

F
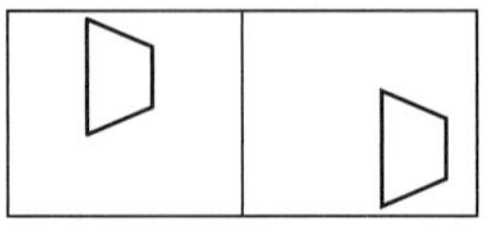

G
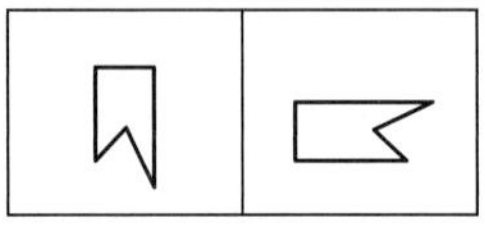

H
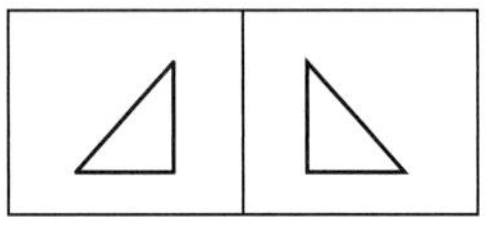

J
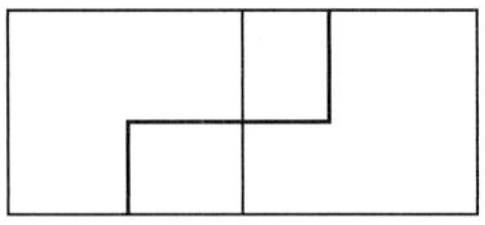

26. **Short Response** What are the coordinates of the point $(-3, 5)$ after it has been reflected over the y-axis?

Common Core Review

Name the ordered pair for each point graphed at the right. 6.NS.6c

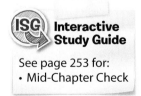

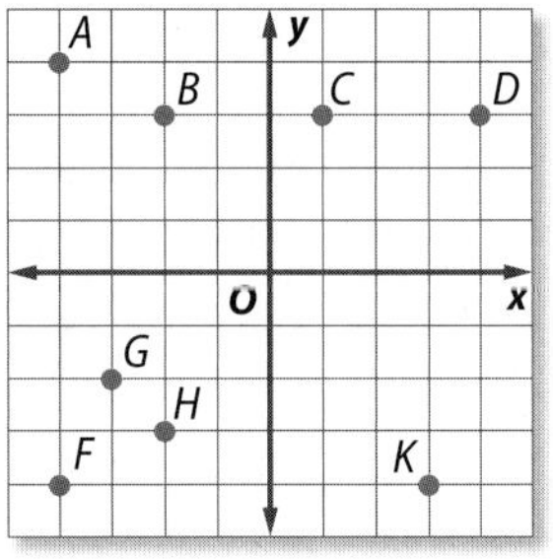

27. *A*

28. *C*

29. *G*

30. *K*

31. In their first five games, the Jefferson Middle School basketball team scored 46, 52, 49, 53, and 45 points. What was their average number of points per game? **6.SP.5c**

32. **Financial Literacy** The starting balance in a checking account was −\$50. What was the balance after a \$100 deposit was made and checks were written for \$25 and for \$32? **7.NS.1**

33. Translate the phrase *three times as many cards as Neville has* into an algebraic expression. **6.EE.2**

34. Translate the sentence below into an equation. **6.EE.2**

The sum of three times a number and five is 20.

Factor each expression. If the expression cannot be factored, write *cannot be factored*. Use algebra tiles if needed. 7.EE.1

35. $3x + 27$

36. $40 + 8x$

37. $3x + 41$

38. $30x + 40$

39. $25 + 12x$

40. $135x + 180$

ISG Interactive Study Guide

See page 253 for:
- Mid-Chapter Check

21ST CENTURY CAREER in Computer Animation

Computer Animator

Have you ever wondered how they make animated movies look so realistic? Computer animators use computer technology and apply their artistic skills to make inanimate objects come alive. If you are interested in computer animation, you should practice drawing, study human and animal movement, and take math classes every year in high school. Tony DeRose, a computer scientist at an animation studio said, "Trigonometry helps rotate and move characters, algebra creates the special effects that make images shine and sparkle, and calculus helps light up a scene."

College & Career READINESS

Explore college and careers at ccr.mcgraw-hill.com

Is This the Career for You?

Are you interested in a career as a computer animator? Take some of the following courses in high school.

- 2-D Animation
- Algebra
- Calculus
- Trigonometry

Find out how math relates to a career in Computer Animation.

Interactive Study Guide

See page 254 for:

- Problem Solving
- Career Project

Lesson 11-5

Rotations on the Coordinate Plane

Interactive Study Guide

See pages 255–256 for:
- Getting Started
- Real-World Link
- Notes

Essential Question

How can you determine congruence and similarity?

Common Core State Standards

Content Standards
8.G.3

Mathematical Practices
1, 3, 4, 7

Vocabulary

rotation
center of rotation
rotational symmetry

What You'll Learn

- Define, identify, and draw rotations.
- Determine if a figure has rotational symmetry.

Real-World Link

Fairs There are many things to see and do at the fair. You can look at the animals, browse arts and crafts projects, and ride the carnival rides. Some rides move up and down, while others move around in a circle. On a carousel, you sit on a carousel horse and ride around in a circle many times.

Rotations

A **rotation** is a transformation in which a figure is turned around a fixed point. This point is called the **center of rotation**. Rotations are also called *turns*. Since a rotated image and its orginal figure have the same size and shape, they are congruent.

The images below show clockwise rotations of a figure with a center of rotation at point *A*.

Original Figure	Clockwise Angle of Rotation		
	90°	180°	270°
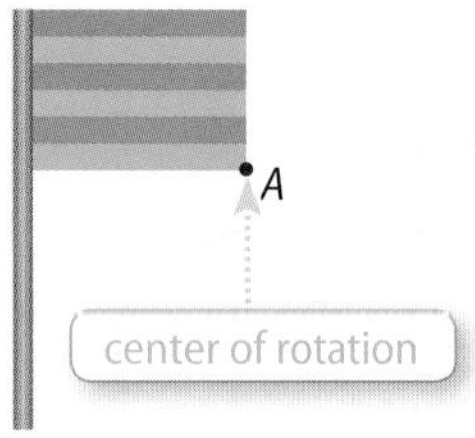	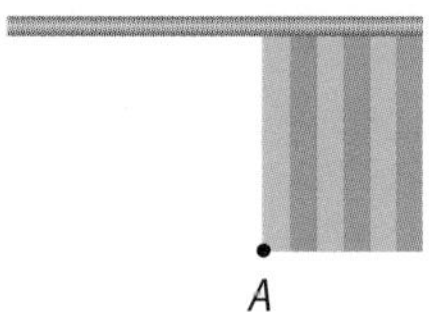	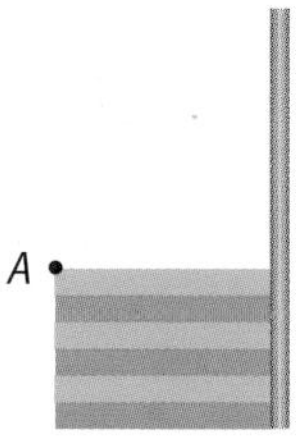	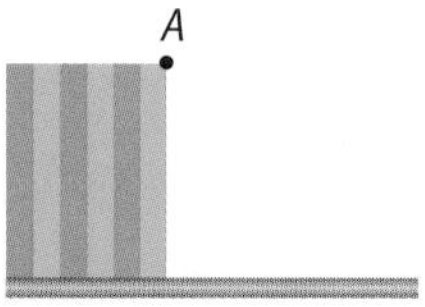

Example 1

Tutor

Draw the figure at the right after a 90° clockwise rotation about point *C*.

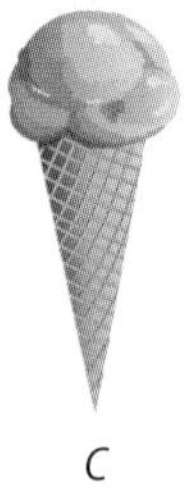

Since point *C* is the center of rotation, it remains in the same position. The figure moves one quarter turn clockwise.

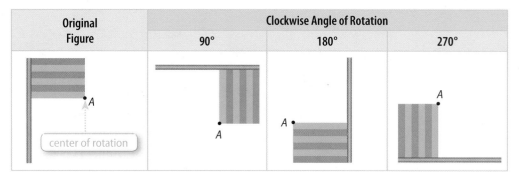

Got It? Do these problems to find out.

Draw the figure in Example 1 after each rotation.

1a. 270° clockwise rotation about point *C*.

1b. 180° counterclockwise about point *C*.

Example 2

Tools Tutor

Quadrilateral *WXYZ* has vertices *W*(−4, −1), *X*(−2, 0), *Y*(−1, −3), and *Z*(−2, −4). Graph the figure and its image after a clockwise rotation of 180° about vertex *X*. Then give the coordinates of the vertices for quadrilateral *W′X′Y′Z′*.

Step 1 Graph the original figure. Then graph vertex *W′* after a 180° rotation about vertex *X*. Note that $m\angle WXW' = 180°$ and $WX = XW'$.

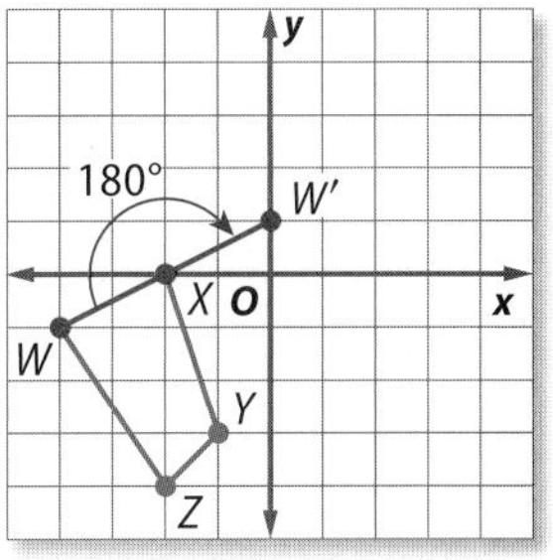

Step 2 Graph the remaining vertices after 180° rotations about vertex *X*. Connect the vertices to form quadrilateral *W′X′Y′Z′*.

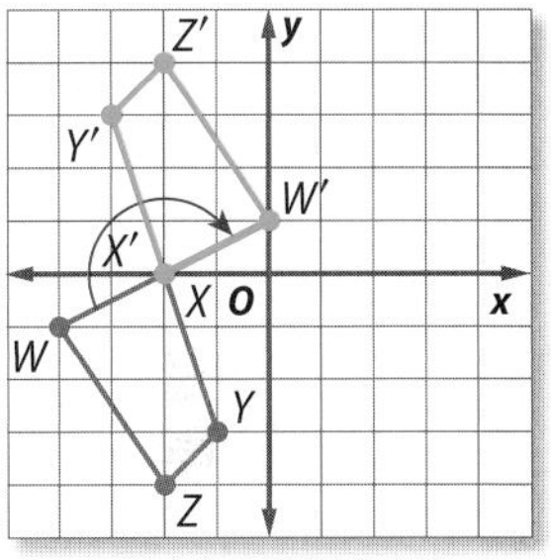

So, the coordinates of the vertices of quadrilateral *W′X′Y′Z′* are *W′*(0, 1), *X′*(−2, 0), *Y′*(−3, 3), and *Z′*(−2, 4).

Got It? **Do this problem to find out.**

2. Graph quadrilateral *WXYZ* and its image after a counterclockwise rotation of 270° about vertex *Y*. Then give the coordinates of the vertices for quadrilateral *W′X′Y′Z′*.

Vocabulary Link
Rotate
Everyday Use to take turns
Math Use to turn around a fixed point

The diagrams below show three clockwise rotations of a figure about the origin.

90° Rotation

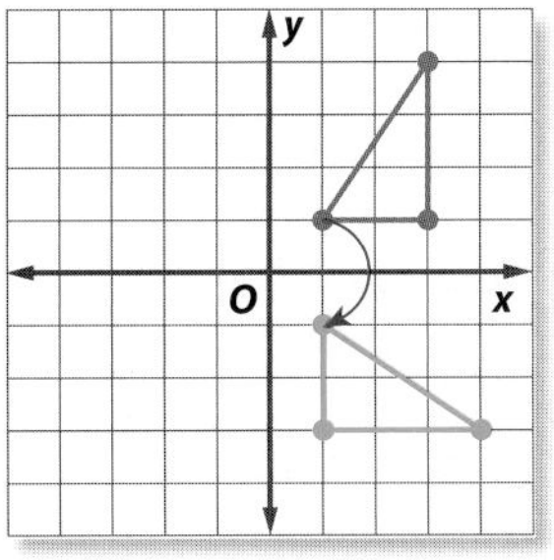

180° Rotation

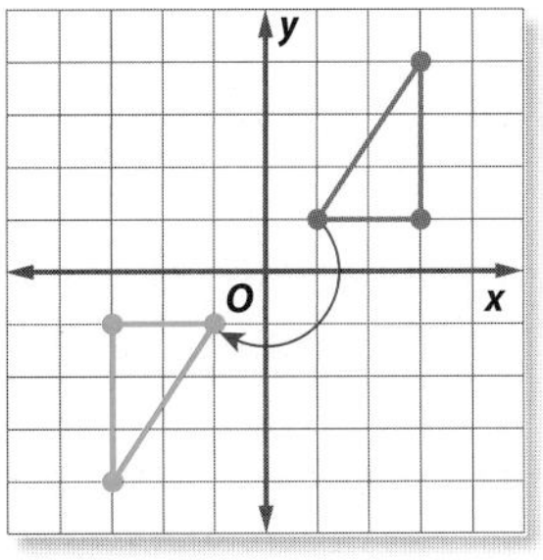

270° Rotation

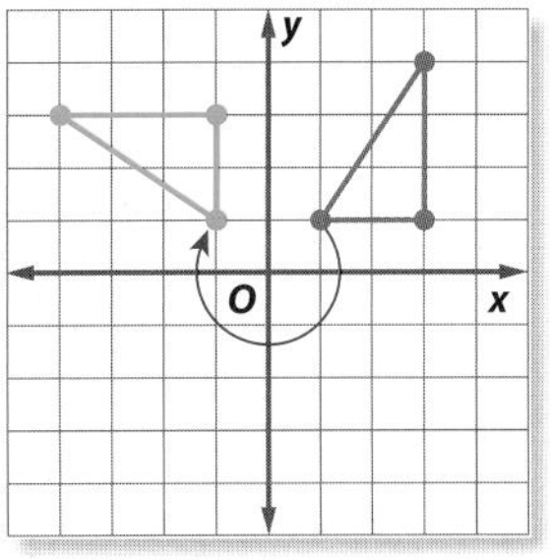

The following diagrams show three counterclockwise rotations about the origin.

270° Rotation

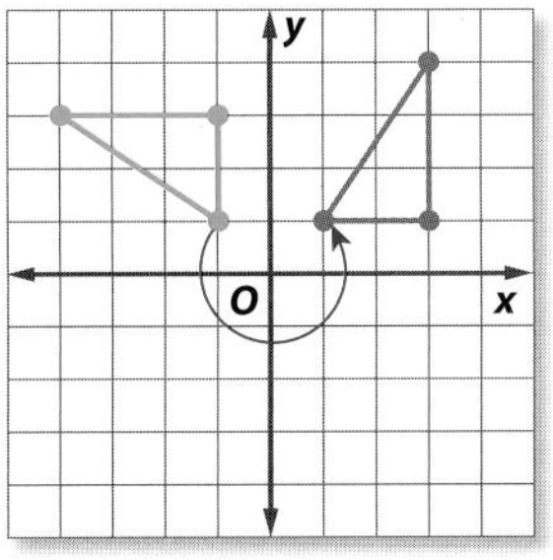

180° Rotation

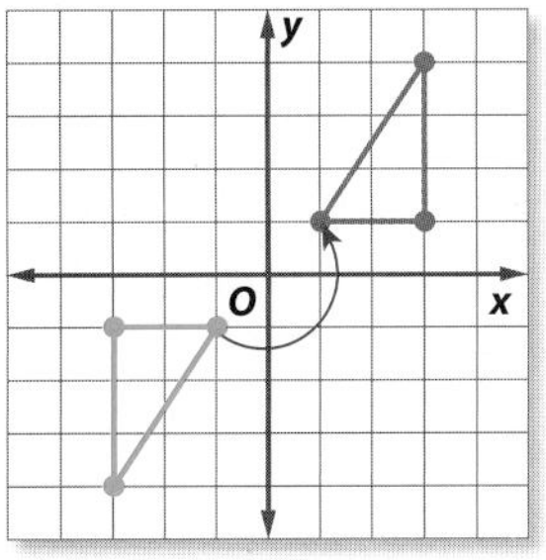

90° Rotation

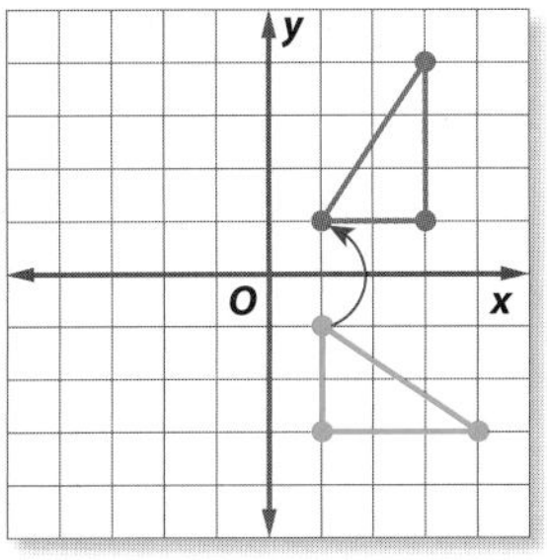

Example 3

Watch Out!
In Example 3, the center of rotation is not on the figure. It is at the origin. Notice that angle *C′OC* is 360° – 270° or 90°. This is the same measure angles *A′OA* and *B′OB* will have.

A triangle has vertices *A*(2, −5), *B*(4, −4), and *C*(2, −1). Graph the triangle and its image after a rotation of 270° clockwise about the origin.

Step 1 Graph △*ABC* on a coordinate plane. A 270° degree rotation is the same as three 90° rotations or $\frac{3}{4}$ of a complete circle. Then graph vertex *C′* after a 270° clockwise rotation about the origin.

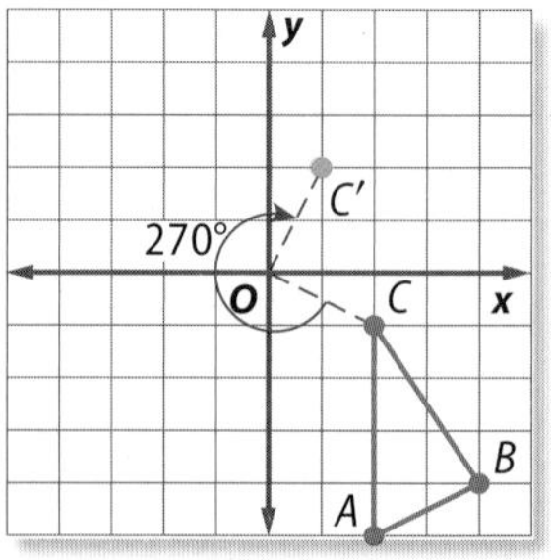

Step 2 Graph the remaining vertices after 270° rotations about the origin. Then connect the vertices to form △*A′B′C′*.

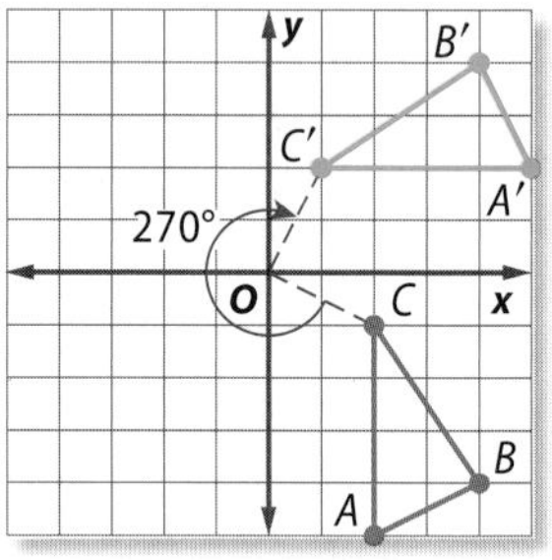

So, the coordinates of the vertices of △*A′B′C′* are *A′*(5, 2), *B′*(4, 4), and *C′*(1, 2).

Got It? **Do this problem to find out.**

3. A figure has vertices *J*(3, 2), *K*(6, 4), *M*(6, 7), and *N*(3, 7). Graph the figure and its image after a rotation of 90° clockwise about the origin.

Rotational Symmetry

If a figure can be rotated less than 360° about its center so that the image matches the original figure, the figure has **rotational symmetry**.

Example 4

Real World

Rotational Symmetry
One complete turn of a figure measures 360° because there are 360° in a circle. A figure that matches itself only after a 360° turn does not have rotational symmetry.

Determine whether the snowflake shown has rotational symmetry. If it does, describe the angle of rotation.

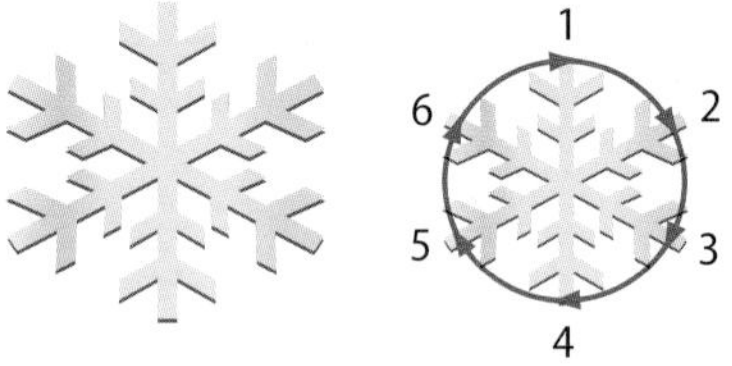

The snowflake can match itself in six positions.

The pattern repeats in 6 even intervals.

So, the angle of rotation is 360° ÷ 6 or 60°.

Got It? **Do this problem to find out.**

Look at the flower at the right.

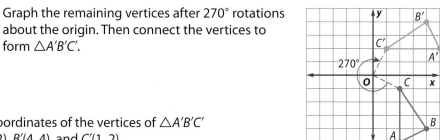

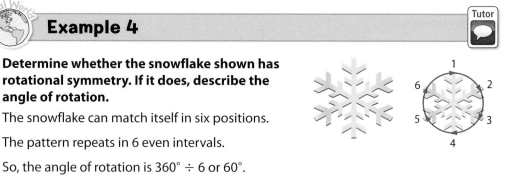

4. Does it have rotational symmetry? If it does, describe the angle of rotation.

Guided Practice

Check

1. Draw the figure at the right after a 270° clockwise rotation about point *A*. (Example 1)

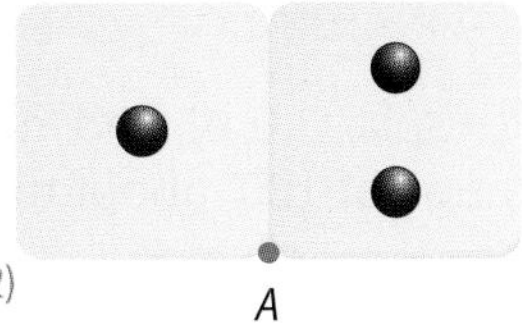

2. Triangle *JKL* has vertices *J*(1, 4), *K*(1, 1), and *L*(5, 1). (Example 2)
   - **a.** Graph the figure and its image after a clockwise rotation of 270° about vertex *J*.
   - **b.** Give the coordinates of the vertices for triangle *J′K′L′*.

3. A figure has vertices *D*(1, 1), *F*(2, 3), *G*(5, 3), and *H*(5, 1). (Example 3)
   - **a.** Graph the figure and its image after a rotation of 180° about the origin.
   - **b.** Give the coordinates of the vertices for figure *D′F′G′H′*.

4. Determine if the blades of the windmill shown at the right have rotational symmetry. If they do, describe the angle of rotation. (Example 4)

Independent Practice

eHelp Go online for Step-by-Step Solutions

Draw each figure after the rotation described. (Example 1)

5. 90° clockwise rotation about point *R*

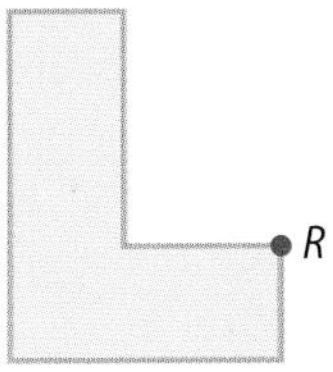

6. 180° clockwise rotation about point *S*

7. Triangle *GHJ* is shown at the right. (Example 2)
   - **a.** Graph the figure after a clockwise rotation of 90°about vertex *G*.
   - **b.** Give the coordinates of the vertices for △*G′H′J′*.

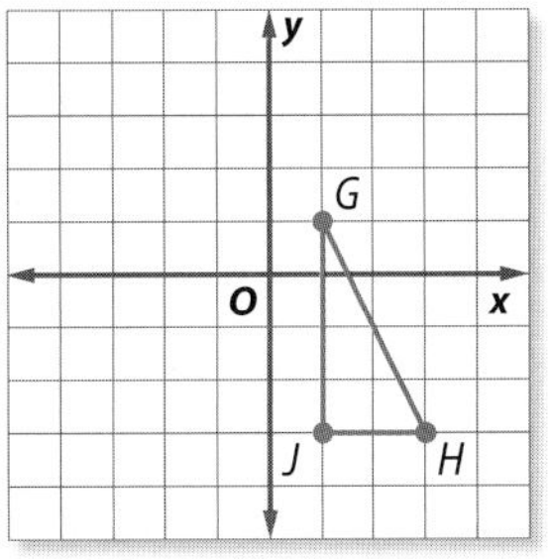

8. Trapezoid *ABCD* has vertices *A*(−3, 1), *B*(−3, 4), *C*(1, 4), and *D*(−1, 1). (Example 3)
   - **a.** Graph the figure and its image after a clockwise rotation of 180° about vertex *D*.
   - **b.** Give the coordinates of the vertices for trapezoid *A′B′C′D′*.

9. A figure has vertices $K(1, -1)$, $L(3, -4)$, $M(1, -5)$, and $N(-1, -4)$. Graph the figure and its image after a rotation of 270° counterclockwise about the origin. (Example 3)

10. A triangle has vertices $P(-3, 1)$, $Q(0, 4)$, and $R(1, -1)$. Graph the triangle and its image after a rotation of 180° about the origin. (Example 3)

11. Describe the rotational symmetry of the star in Exercise 6. (Example 4)

Determine whether each synchronized swimming formation has rotational symmetry. If it does, describe the angle of rotation. (Example 4)

12. 

13. 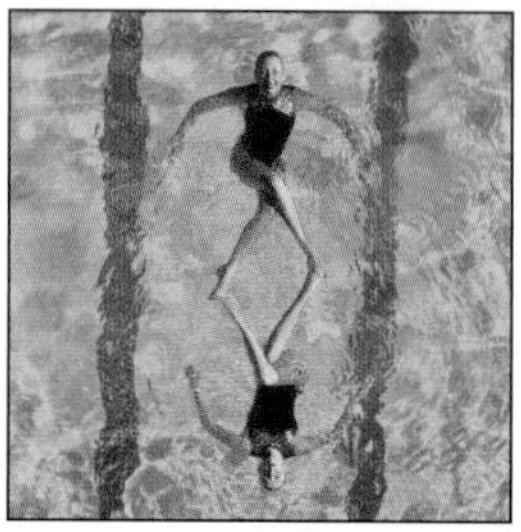

14. 

15. **CCSS Model with Mathematics** Pat is placing the tile at the right in different ways on a wall. Describe three rotations that would change the appearance of the tile. Use drawings to show what the tile looks like after each rotation.

16. **CCSS Multiple Representations** A figure is graphed on the coordinate plane. One of its vertices has coordinates (x, y).

   a. **Symbols** Write the coordinates of the vertex after each rotation.
      - 90° clockwise about the origin
      - 180° about the origin
      - 270° clockwise about the origin

   b. **Analysis** Suppose the vertex of the figure is located in Quadrant II. If the figure is rotated 180° about the origin, in which quadrant will the corresponding vertex of the image be located?

H.O.T. Problems Higher Order Thinking

17. **CCSS Identify Structure** Sketch a figure that has rotational symmetry. Describe the angle of rotation.

18. **CCSS Reason Inductively** Name a clockwise rotation that is equivalent to a rotation of 90° counterclockwise about the origin.

19. **CCSS Persevere with Problems** Triangle RST is rotated 270° clockwise around the origin. Its rotated image has vertices $R'(-5, 1)$, $S'(2, 0)$, and $T'(-3, -6)$. What are the coordinates of triangle RST?

20. **CCSS Construct an Argument** Will a geometric figure and its rotated image *sometimes*, *always*, or *never* have the same perimeter? Explain your reasoning.

21. **Building on the Essential Question** Describe the similarities and differences between reflections and rotations.

Standardized Test Practice

22. Triangle XYZ was rotated about the origin to $\triangle X'Y'Z'$. Which of the following describes the rotation?

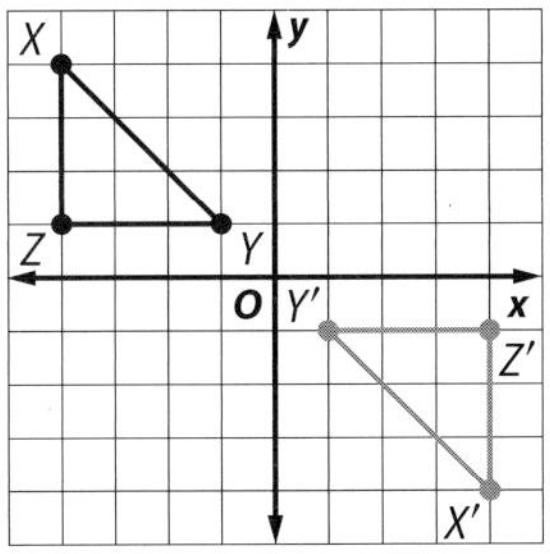

A 90° clockwise about the origin

B 90° counterclockwise about the origin

C 180° clockwise about the origin

D 270° clockwise about the origin

23. Triangle ABC has vertices $A(1, 2)$, $B(-1, -1)$, and $C(2, 0)$. If the triangle is rotated clockwise 90° about the origin, which of the following would be the vertices of $\triangle A'B'C'$?

F $A'(2, 1)$, $B'(-1, -1)$, $C'(0, 2)$

G $A'(-1, 2)$, $B'(1, 1)$, $C'(-2, 0)$

H $A'(-2, 1)$, $B'(1, -1)$, $C'(0, 2)$

J $A'(2, -1)$, $B'(-1, 1)$, $C'(0, -2)$

24. **Short Response** Figure $ABCD$ is shown. Graph the figure after a clockwise rotation of 270° about the origin.

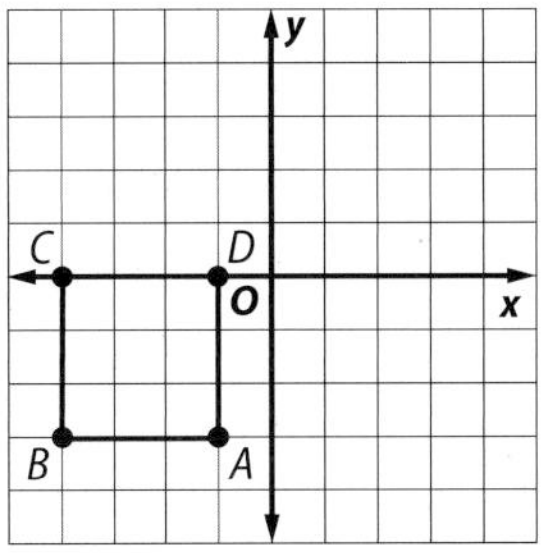

25. Which of the following is true about the image shown below?

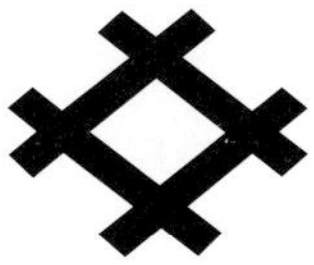

A It does not have rotational symmetry.

B The angle of rotation is 90°.

C The angle of rotation is 180°.

D The angle of rotation is 270°.

Common Core Review

26. If $m\angle B = 17°$ and $\angle A$ and $\angle B$ are complementary, find $m\angle A$. **7.G.5**

27. If $m\angle B = 17°$ and $\angle A$ and $\angle B$ are supplementary, find $m\angle A$. **7.G.5**

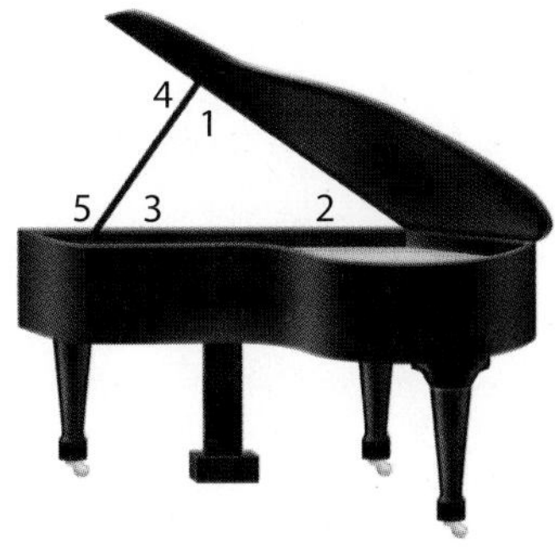

For Exercises 28–30, use the diagram at the right. 7.G.5, 8.G.5

28. Find the measure of $\angle 1$ if the measure of $\angle 4$ is 90°.

29. Find the measure of $\angle 3$ if the measure of $\angle 5$ is 145°.

30. Suppose the measure of $\angle 4 = 95°$ and the measure of $\angle 5 = 135°$. Find the measure of angles 1, 2, and 3.

Solve each equation. 8.EE.7

31. $4x + 157 = 243$

32. $351 = 2x + 89$

33. $6(x - 12) = 138$

34. $73 = 0.7x - 4$

35. $8x - 6 = 10x + 2$

36. $-51 = 3(x + 2)$

37. $5x - 3 = 3x + 1$

38. $-3.1x = -24.8$

39. $8x - 10 = 2(4x - 5)$

Lesson 11-6

Congruence and Transformations

Interactive Study Guide

See pages 257–258 for:
- Getting Started
- Real-World Link
- Notes

Essential Question

How can you determine congruence and similarity?

Common Core State Standards

Content Standards
8.G.2

Mathematical Practices
1, 3, 4

What You'll Learn

- Identify congruency by using transformations.
- Identify transformations.

Real-World Link

Fabrics Interior designers often use fabrics with repeatable patterns. These patterns are created by transforming geometric shapes.

Identify Congruency

Two figures are congruent if the second can be obtained from the first by a series of rotations, reflections, and/or translations.

Example 1

Tutor

Determine if the two figures are congruent by using transformations.

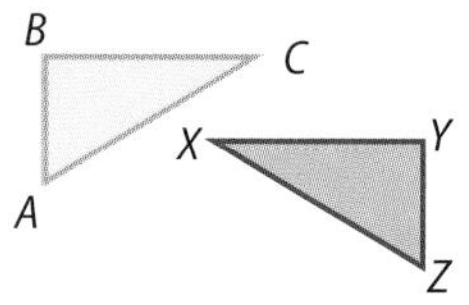

Step 1 Reflect $\triangle ABC$ over a vertical line. Label the vertices of the image A', B', and C'.

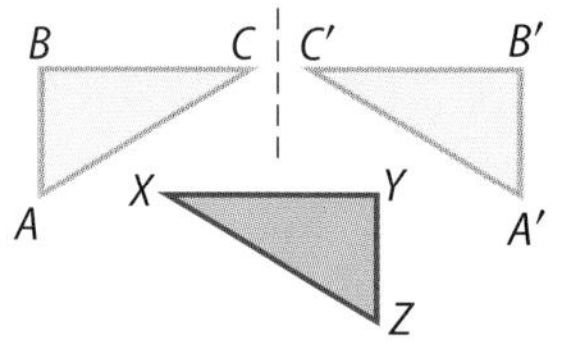

Step 2 Translate $\triangle A'B'C'$ until all sides and angles match $\triangle ZYX$.

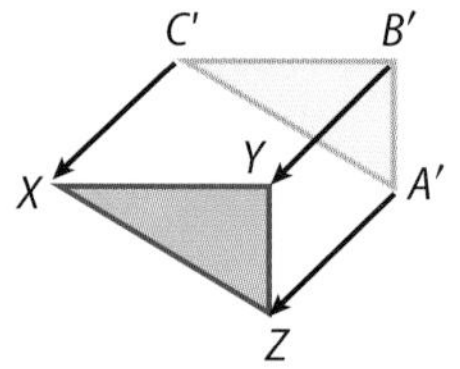

So, the two triangles are congruent because a reflection followed by a translation will map $\triangle ABC$ onto $\triangle ZYX$.

Got It? Do this problem to find out.

1. Determine if the two figures are congruent by using transformations. Explain your reasoning.

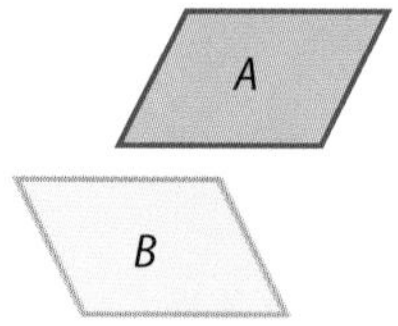

Example 2

Tutor

Determine if the two figures are congruent by using transformations.

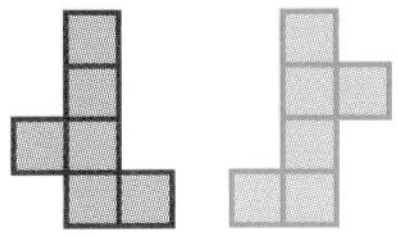

Reflect the red figure over a vertical line.

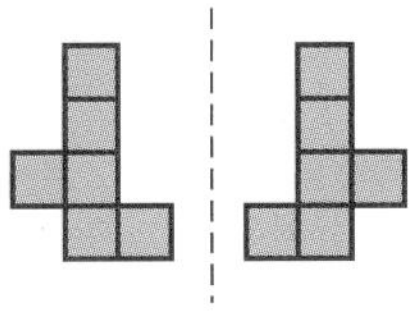

Even if the reflected figure is translated up and over, it will not match the green figure exactly. The two figures are not congruent.

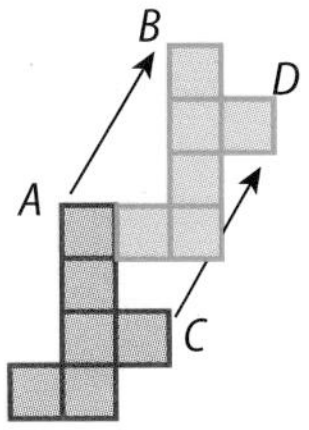

> **Transformations**
> Translations, reflections and rotations are called *isometries*. In an isometry, the distance between two points in an image is the same as the distance in the preimage.
> **iso** / metry
> **same** distance

Got It? Do this problem to find out.

2. Determine if the two figures are congruent by using transformations. Explain your reasoning.

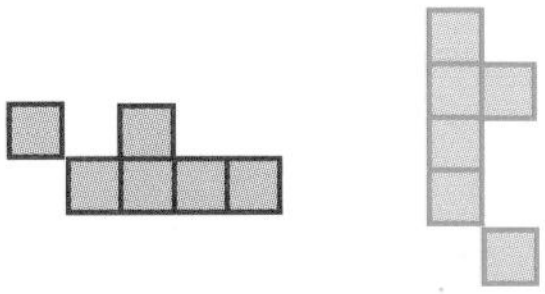

Identify Transformations

If you have two congruent figures, you can determine the transformation, or series of transformations, that maps one figure onto the other by analyzing the orientation or relative position of the figures.

Translation	Reflection	Rotation
• length is the same • orientation is the same	• length is the same • orientation is reversed	• length is the same • orientation is changed
Notice the segments are facing the same way. 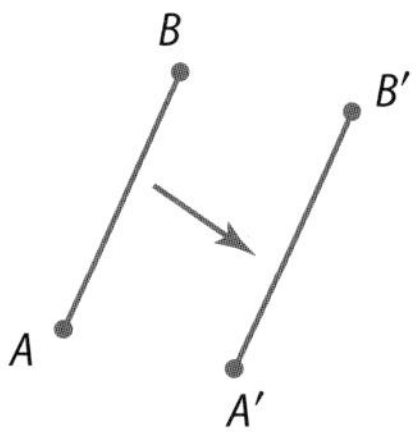	Notice the segments are facing the opposite way. 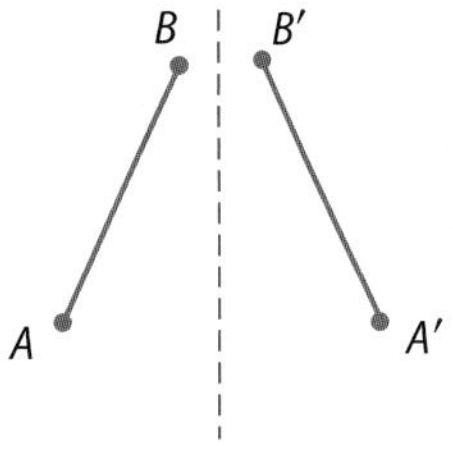	Notice the segments are facing a different way. 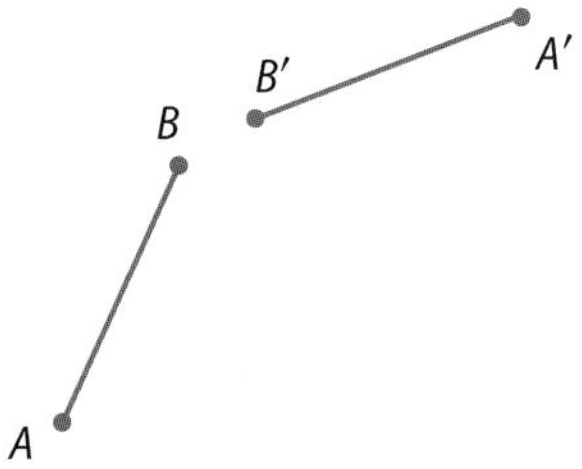

Example 3

Ms. Martinez created the logo shown. What transformations could she have used if the letter "d" is the preimage and the letter "p" is the image? Are the two figures congruent?

> **Watch Out!**
> There may be more than one combination of transformations that maps one figure onto the other. For example, a series of two reflections over intersecting lines is the same as a rotation.

Step 1 Start with the preimage. Determine which transformation will change the orientation of the letter.

Step 2 Rotations or reflections change orientation. Rotate the letter "d" 180° about point *A*.

Step 3 Translate the new image up.

So, Ms. Martinez could have used a rotation and a translation to create the logo. The letters are congruent because images produced by a rotation and translation have the same shape and size.

CHECK Trace the letter "d" with tracing paper. Rotate the letter 180° around Point *A*. Slide it up to line up with the letter "p." The letters are the same shape and size. They are congruent. ✓

Got It? **Do this problem to find out.**

3. What transformations could be used if the letter "W" is the preimage and the letter "M" is the image in the logo shown?

Guided Practice

Determine if the two figures are congruent by using transformations. Explain your reasoning. (Examples 1 and 2)

1.

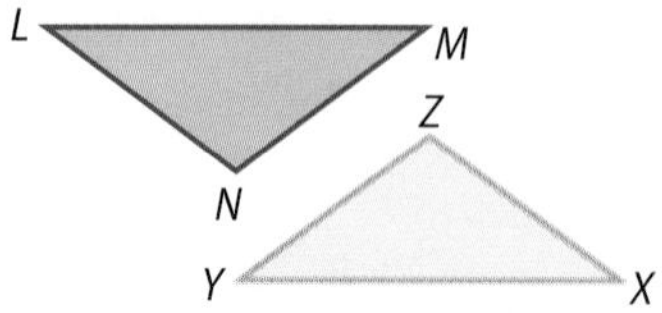

2.

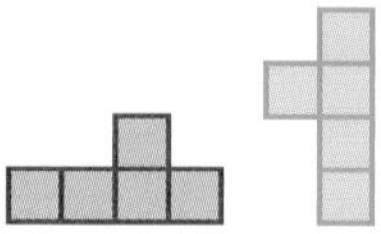

3. The Boyd Box Company uses the logo shown. What transformations could be used if the top, red trapezoid is the preimage and the bottom, blue trapezoid is the image? (Example 3)

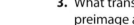

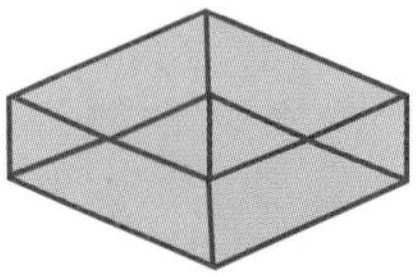

Independent Practice

Go online for Step-by-Step Solutions | eHelp

Determine if the two figures are congruent by using transformations. Explain your reasoning. (Examples 1 and 2)

4.

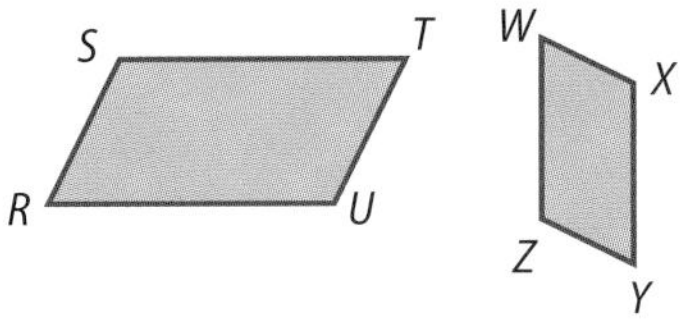

5.

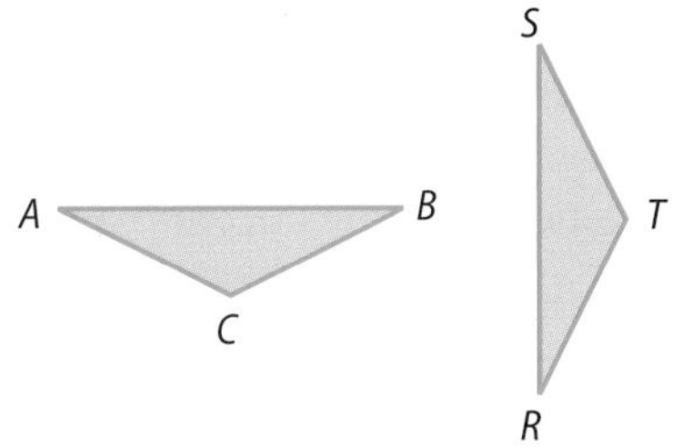

6.

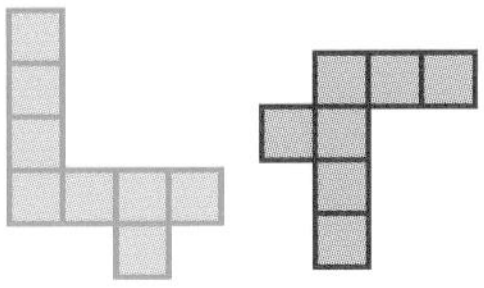

7.

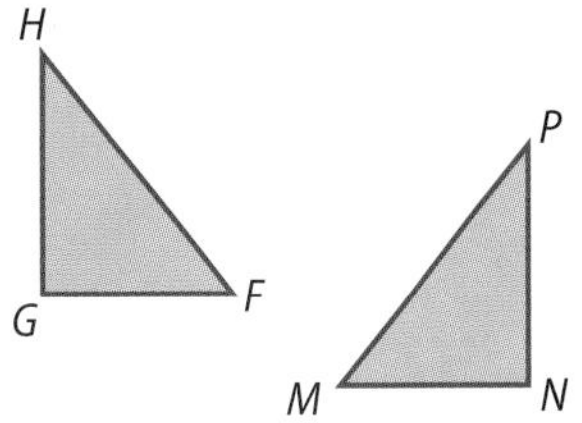

8. Nilda purchased some custom printed stationery with her initials. What transformations could be used if the letter "Z" is the preimage and the letter "N" is the image in the design shown? (Example 3)

9. Simon is illustrating a graphic novel for a friend. He is using the two thought bubbles shown. What transformations did he use if Figure A is the preimage and Figure B is the image? (Example 3)

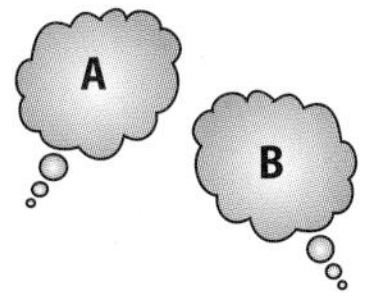

10. CCSS **Justify Conclusions** One way to identify congruent polygons is to prove their matching sides have the same measure for both polygons. Square *CDEF* has vertices at (1, 5), (1, 1), (5, 1), and (5, 5).

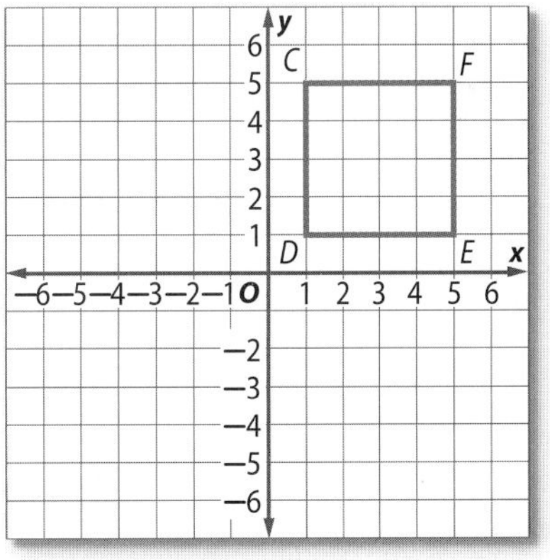

a. Find the lengths of segments *CD*, *DE*, *EF*, and *FC*.

b. Reflect square *CDEF* over the *x*-axis, then translate it 2 units left. Label the vertices of the image *C′D′E′F′*.

c. Find the lengths of segments *C′D′*, *D′E′*, *E′F′*, and *F′C′*.

d. Are the two squares congruent? Justify your response.

Find the lengths of the sides of the preimage with the given vertices and the image after the transformations are performed. Then determine if the two figures are congruent.

11. preimage: (−2, −1), (−2, 2), (1, 2), (1, −1) transformations: translate 1 unit right, then reflect over the *x*-axis

12. preimage: (0, 1), (0, 5), (3, 5), (3, 1) transformations: translate 3 units left and 2 units down, then reflect over the *y*-axis

H.O.T. Problems Higher Order Thinking

13. CCSS **Model with Mathematics** Create a fabric design using a series of transformations that produce congruent figures. Exchange designs with a classmate and determine what transformations were used to create the design.

14. CCSS **Persevere with Problems** Angle *ABC* has points $A(-3, 4)$, $B(-2, 1)$ and $C(2, 2)$. Find the coordinates of the image of the angle after a 90° clockwise rotation about the origin, a translation of 2 units up, and a reflection over the *y*-axis.

15. CCSS **Persevere with Problems** Line segment $\overline{XY}$ has endpoints at $X(3, 1)$ and $Y(-2, 0)$. After a series of transformations, its image has endpoints at $X'(0, 1)$ and $Y'(5, 0)$. Find the transformations that map $\overline{XY}$ onto $\overline{X'Y'}$.

16. **Building on the Essential Question** Explain why rotations, reflections, and translations create congruent images.

Standardized Test Practice

17. **Short Response** Gregory is creating a mosaic for art class. He started by using triangular tiles as shown.

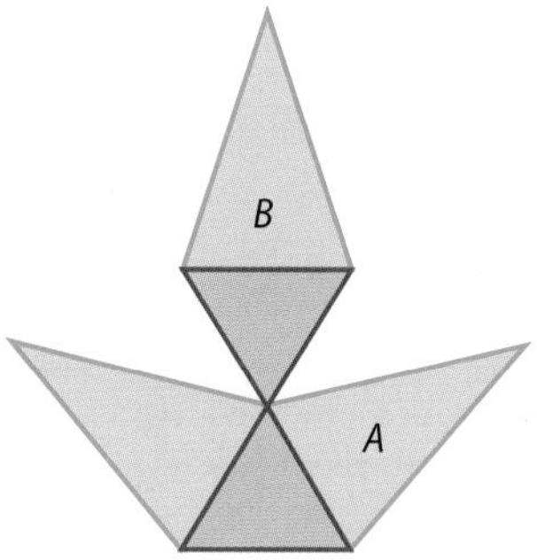

What possible transformations did Gregory use if Figure *A* is the preimage and Figure *B* is the image?

18. Triangle *MNO* is congruent to triangle *RST*.

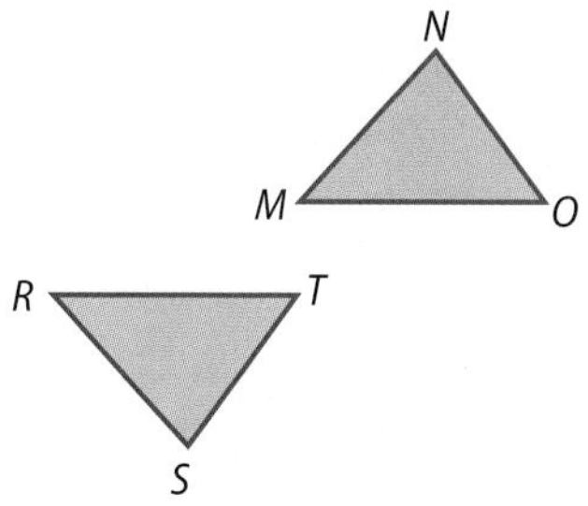

Which series of transformations maps $\triangle MNO$ onto $\triangle RST$?

A 90° clockwise rotation about *M* then reflection

B translation then enlargement

C 90° clockwise rotation about *M* then translation

D reflection then translation

Common Core Review

19. **Use the figure at the right.** 7.EE.1

a. Write and simplify a polynomial expression for the perimeter of the figure in meters.

b. Find the perimeter of the figure if $x = 5$.

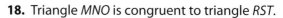

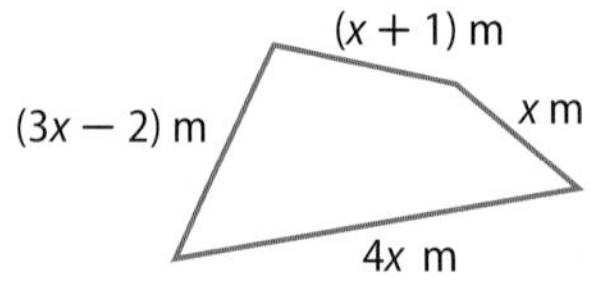

Subtract. Use models if needed. 7.EE.1

20. $(3x + 2) - (x - 10)$

21. $(6x - 1) - (3x - 2)$

22. $(4x + 1) - (3x - 3)$

23. $(5x - 2) - (x + 11)$

24. $(5x - 5) - (-5x + 5)$

25. $(4 + 6x) - (4 - 6x)$

Lesson 11-7

Dilations on the Coordinate Plane

ISG Interactive Study Guide

See pages 259–260 for:
- Getting Started
- Real-World Link
- Notes

Essential Question

How can you determine congruence and similarity?

CCSS Common Core State Standards

Content Standards
8.G.3

Mathematical Practices
1, 3, 4, 5, 6

Vocabulary

dilation

Math Symbols

$(x, y) \rightarrow (kx, ky)$ means the dilation of (x, y) has a scale factor of k

What You'll Learn

- Graph dilations on a coordinate plane.
- Find the scale factor of a dilation.

Real-World Link

Photography Whether you print your own photographs or purchase them from someone else, you can get them printed in just about any size you want. You can also enlarge or reduce the size of your photographs by using digital editing software.

Dilations

A **dilation** is a transformation that enlarges or reduces a figure by a scale factor.

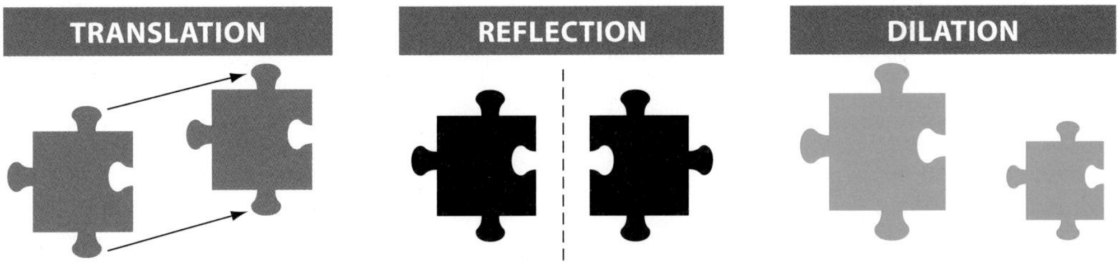

When the center of a dilation is the origin, you can find the coordinates of the image by multiplying each coordinate of the figure by the scale factor. Use the notation $(x, y) \rightarrow (kx, ky)$ to describe the dilation.

Example 1

Tools Tutor

A figure has vertices $J(2, 4)$, $K(2, 6)$, $M(8, 6)$, and $N(8, 2)$. Graph the figure and the image of the polygon after a dilation with a scale factor of $\frac{1}{2}$.

The dilation is $(x, y) \rightarrow \left(\frac{1}{2}x, \frac{1}{2}y\right)$. Multiply the coordinates of each vertex by $\frac{1}{2}$.

Then graph both figures on the same coordinate plane.

$J(2, 4) \rightarrow J'\left(\frac{1}{2} \cdot 2, \frac{1}{2} \cdot 4\right) \rightarrow J'(1, 2)$

$K(2, 6) \rightarrow K'\left(\frac{1}{2} \cdot 2, \frac{1}{2} \cdot 6\right) \rightarrow K'(1, 3)$

$M(8, 6) \rightarrow M'\left(\frac{1}{2} \cdot 8, \frac{1}{2} \cdot 6\right) \rightarrow M'(4, 3)$

$N(8, 2) \rightarrow N'\left(\frac{1}{2} \cdot 8, \frac{1}{2} \cdot 2\right) \rightarrow N'(4, 1)$

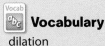

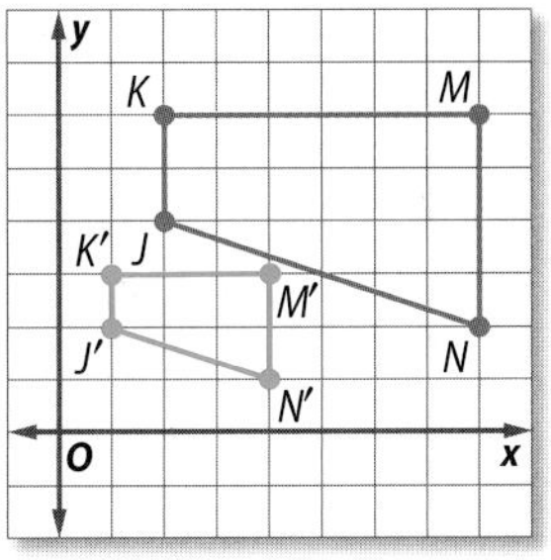

Got It? Do this problem to find out.

1. A figure has vertices $R(-3, 6)$, $S(3, 12)$, and $T(3, 3)$. Graph the figure and the image of the figure after a dilation with a scale factor of $\frac{1}{3}$.

J. Hardy/PhotoAlto

Example 2

Tools | Tutor

Dilate

Everyday Use to expand or widen

Math Use to enlarge or reduce by a scale factor

A triangle has vertices $A(1, -2)$, $B(0, 2)$, and $C(-2, -1)$. Find the coordinates of the triangle after a dilation with a scale factor of 2.

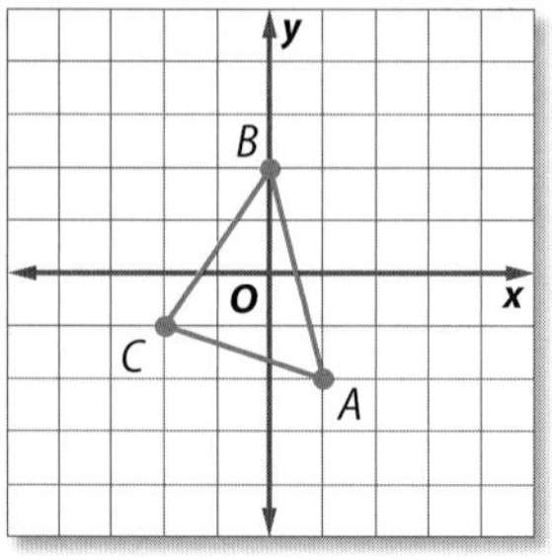

The scale factor is 2, so the dilation is $(x, y) \rightarrow (2x, 2y)$. To dilate the figure, multiply the coordinates of each vertex by 2.

$A(1, -2) \rightarrow (2 \cdot 1, 2 \cdot -2) \rightarrow A'(2, -4)$

$B(0, 2) \rightarrow (2 \cdot 0, 2 \cdot 2) \rightarrow B'(0, 4)$

$C(-2, -1) \rightarrow (2 \cdot -2, 2 \cdot -1) \rightarrow C'(-4, -2)$

The coordinates of triangle ABC after a dilation with a scale factor of 2 are $A'(2, -4)$, $B'(0, 4)$, and $C'(-4, -2)$

Draw a Diagram

Draw a diagram when the problem situation involves spatial reasoning or geometric figures. In this example, drawing a diagram helps you check that the coordinates are correct.

CHECK Graph the triangle and its image on the coordinate plane.

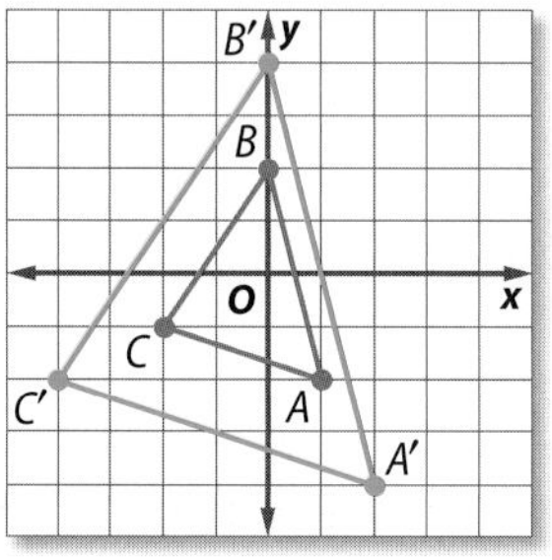

Got It? **Do this problem to find out.**

2. A figure has vertices $X(-3, 6)$, $Y(3, 0)$, and $Z(3, 3)$. Find the coordinates of the figure after a dilation with a scale factor of 3.

Key Concept Find a Scale Factor

A dilation with a scale factor of k will be:

- an enlargement if $k > 1$,
- a reduction if $0 < k < 1$,
- the same as the original figure if $k = 1$.

If a figure has been enlarged, its scale factor is greater than 1. If a figure has been reduced, its scale factor is between 0 and 1. When the scale factor is 1, the size and shape of the image do not change. To determine the exact scale factor, use a ratio.

The sides of the image of a figure being dilated are proportional to the corresponding sides of the original figure. To determine the scale factor, write a ratio comparing a side length of the image to the corresponding side length of the original figure.

Example 3

Tutor

A drawing of a school mascot is to be enlarged as shown at the right so it can be painted on the gymnasium doors. What is the scale factor of the dilation?

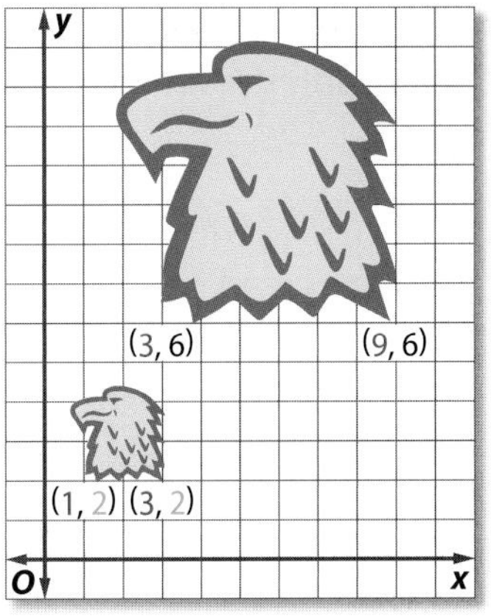

Write a ratio comparing the lengths of the sides of the two images. Subtract the *x*-coordinates to find the lengths.

$$\frac{\text{length of dilation}}{\text{length of original}} = \frac{9 - 3}{3 - 1}$$
$$= \frac{6}{2}$$
$$= 3$$

Similarity
Dilations produce images that are similar (same shape, but not same size).

So, the scale factor of the dilation is 3.

Check for Reasonableness The figure has been enlarged, so $k > 1$. Since $3 > 1$, the answer is reasonable. ✓

Got It? **Do this problem to find out.**

3. Megan wants to reduce an 8-by-10-inch photo to a 2-by-$2\frac{1}{2}$-inch photo. What is the scale factor of the dilation?

Guided Practice

Check

Find the vertices of each figure after a dilation with the given scale factor *k*. Then graph the image. (Examples 1 and 2)

1. $k = \frac{1}{2}$

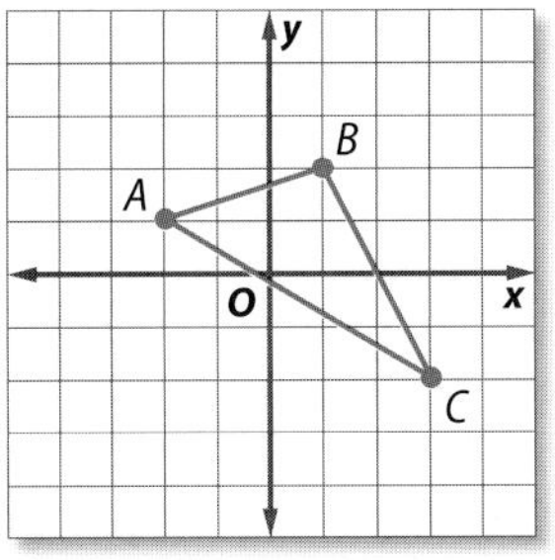

2. $k = \frac{1}{4}$

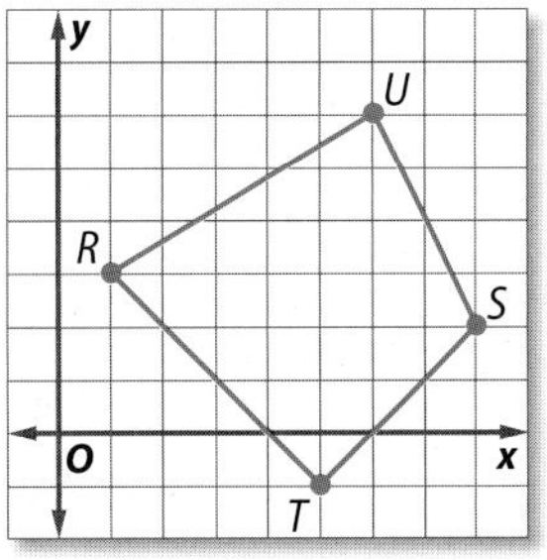

3. A square has vertices $M(0, 0)$, $N(3, -3)$, $O(0, -6)$ and $P(-3, -3)$. Find the coordinates of the square after a dilation with a scale factor of 2.5. (Example 2)

4. Jorge is using a photocopier to reduce a poster that is 8 inches by 14 inches to 5 inches by $8\frac{3}{4}$ inches. What is the scale factor of the dilation? (Example 3)

5. Karen uses grid paper to help her enlarge her painting from 14 inches by 22 inches to 35 inches by 55 inches. What is the scale factor of the dilation? (Example 3)

Independent Practice

eHelp Go online for Step-by-Step Solutions

Find the vertices of each figure after a dilation with the given scale factor *k*. Then graph the image. (Examples 1 and 2)

6. $k = 4$

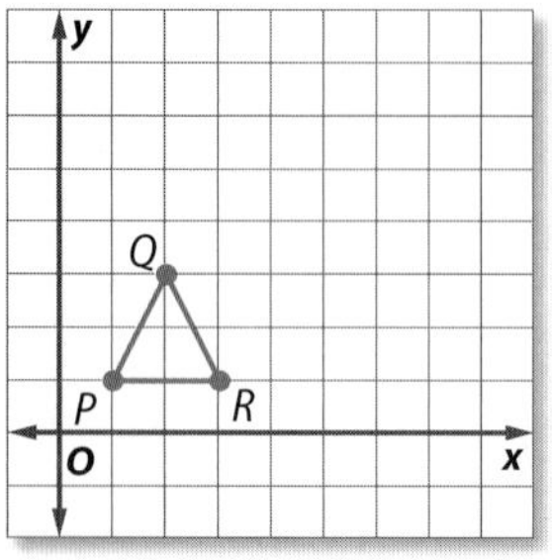

7. $k = 1.5$

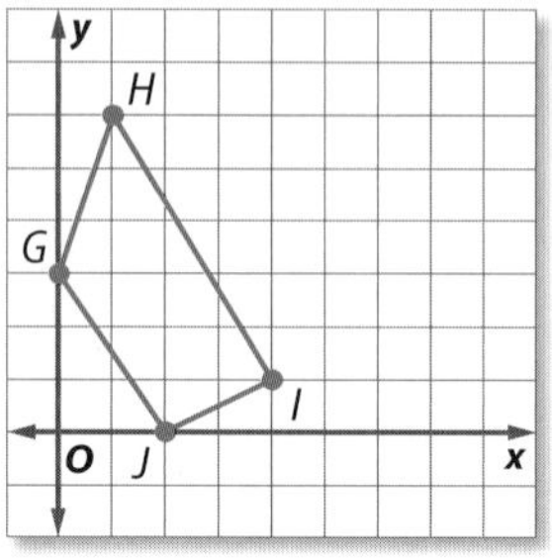

8. $k = \frac{3}{4}$

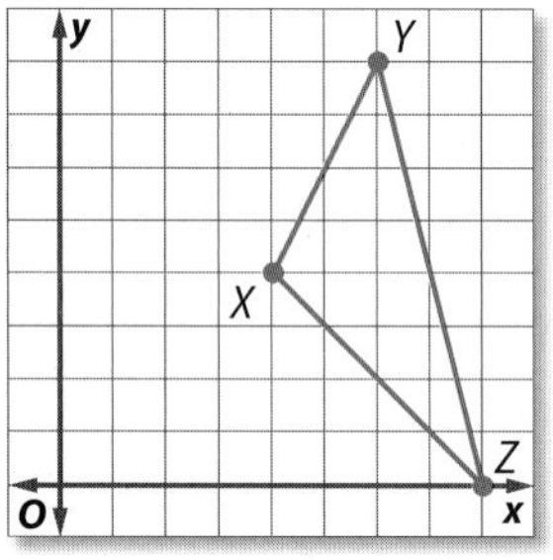

9. $k = \frac{1}{3}$

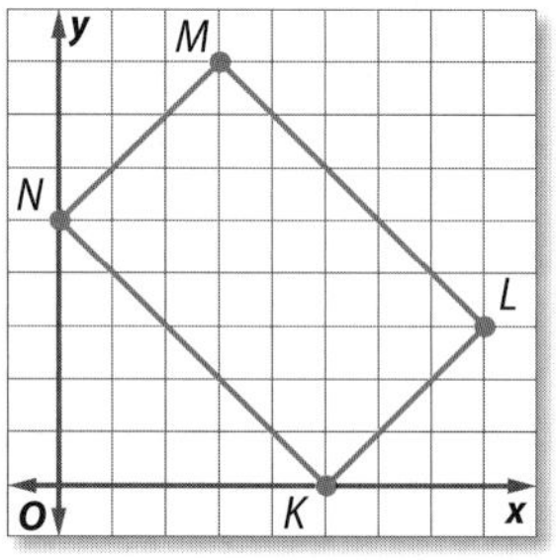

Find the vertices of each figure after a dilation with the given scale factor *k*. Then graph the original image and the dilation. (Examples 1 and 2)

10. $G(-1, 1), H(2, 1), J(3, -2), K(-2, -2); k = 2$

11. $W(0, 0), X(5, -5), Y(5, 10); k = \frac{2}{5}$

12. $R(-1, 2), S(1, 4), T(1, 1); k = 3$

13. $A(-3, -5), B(0, 2), C(3, -1), D(0, -4); k = \frac{1}{2}$

14. STEM In a microscope, the image of a 0.16-millimeter paramecium appears to be 32 millimeters long. What is the scale factor of the dilation? (Example 3)

15. During an eye exam, an optometrist dilates her patient's pupils to 7 millimeters. If the diameter of the pupil before dilation was 4 millimeters, what is the scale factor of the dilation? (Example 3)

16. CCSS **Be Precise** Jung is editing a digital photograph that is 640 pixels wide and 480 pixels high on his computer monitor.

   a. If Jung zooms in on the photo on his monitor 150%, what are the dimensions of the image?

   b. Suppose that Jung is going to use the photograph in a design and wants the image to be 32 pixels wide. What scale factor should he use?

   c. Jung resizes the photograph so it is 600 pixels high. What scale factor did he use?

17 Quadrilateral $A'B'C'D'$ is a dilation of quadrilateral $ABCD$. Find the scale factor of the dilation and classify it as an enlargement or a reduction.

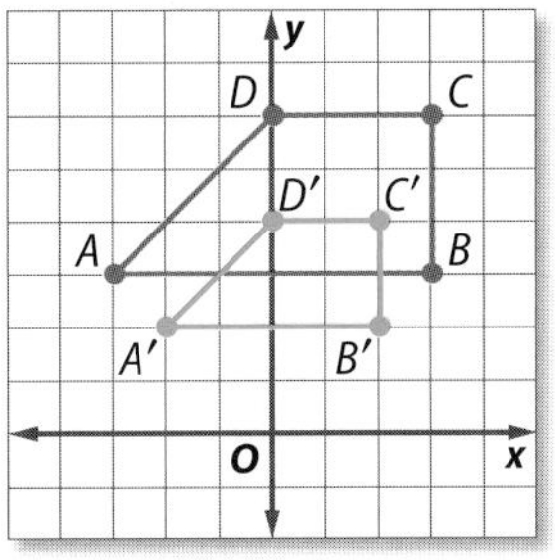

18. Triangle $L'M'N'$ is a dilation of triangle LMN. Find the scale factor of the dilation and classify it as an enlargement or a reduction.

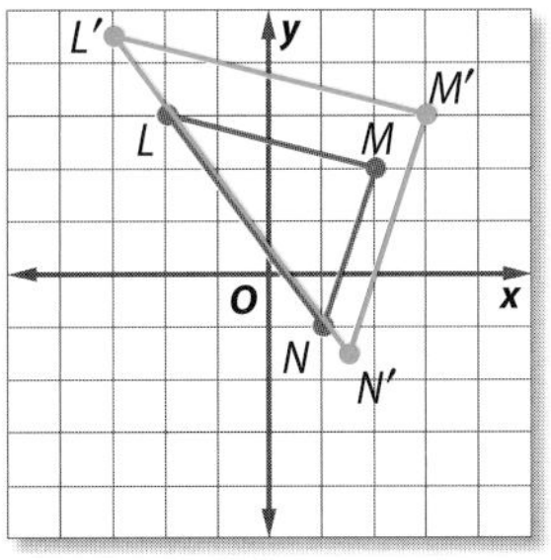

19. Felicia wants to project a 2-inch by 2-inch slide onto a wall to create an image 128 inches by 128 inches. If the slide projector makes the dimensions of the image twice as large for each yard that it is moved away from the wall, how far away should Felicia place the projector?

H.O.T. Problems Higher Order Thinking

20. CCSS **Use Math Tools** Draw a triangle on grid paper. Then draw the image of the triangle after it is moved 5 units right and then dilated by a scale factor of $\frac{1}{3}$.

21. CCSS **Persevere with Problems** Suppose a figure $ABCD$ is dilated by a scale factor of $\frac{1}{2}$ and then reflected over the x-axis and the y-axis. The final image is shown on the graph at the right. Graph the original image and list the coordinates of points A, B, C, and D.

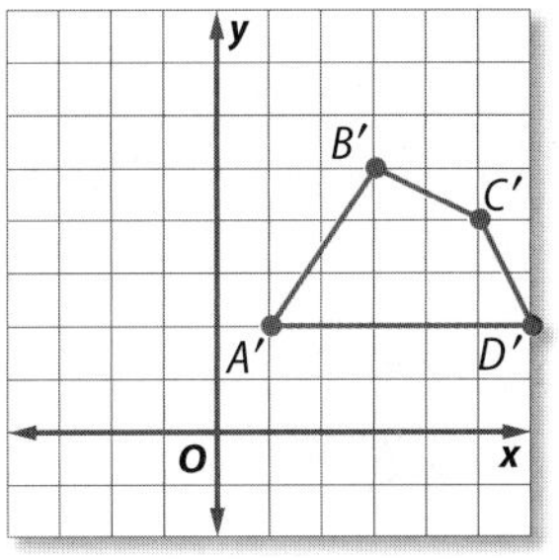

22. CCSS **Reason Inductively** A triangle has coordinates $A(0, 5)$, $B(0, 0)$ and $C(10, 5)$. Find the coordinates of the final image after each set of dilations.

a. $k = 2$ followed by $k = \frac{2}{5}$

b. $k = \frac{2}{5}$ followed by $k = 2$

c. $k = \frac{1}{4}$ followed by $k = 2$

d. $k = 2$ followed by $k = \frac{1}{4}$

23. CCSS **Reason Inductively** Refer to Exercise 22. Does the order in which you perform multiple dilations *always*, *sometimes*, or *never* result in the same image? Explain.

24. CCSS **Which One Doesn't Belong?** Which pair of points does not represent a dilation with center at the origin? Explain.

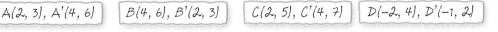

$A(2, 3)$, $A'(4, 6)$	$B(4, 6)$, $B'(2, 3)$	$C(2, 5)$, $C'(4, 7)$	$D(-2, 4)$, $D'(-1, 2)$

25. **Building on the Essential Question** A triangle has one vertex at point $(3, 6)$. It is dilated with a center at the origin by a scale factor of 3. The resulting image is then dilated with a scale factor of $\frac{1}{3}$. What are the coordinates of that vertex after both dilations? Explain your reasoning.

Standardized Test Practice

26. Quadrilateral *PQRS* was dilated to form quadrilateral *P′Q′R′S′*.

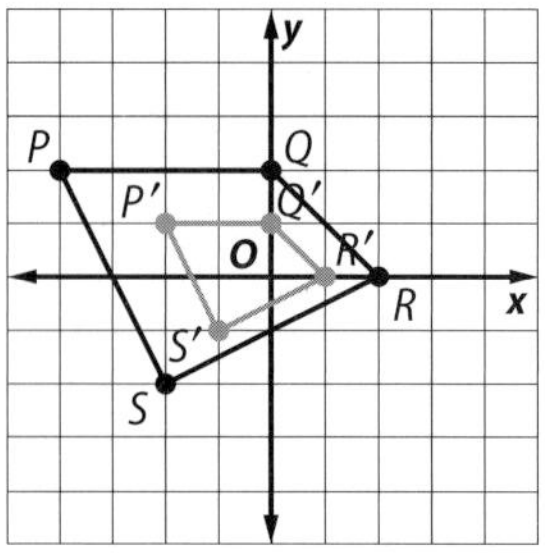

Which number *best* represents the scale factor used to dilate quadrilateral *PQRS* to quadrilateral *P′Q′R′S′*?

A -2

B $-\frac{1}{2}$

C $\frac{1}{2}$

D 2

27. A triangle has vertices $(1, -1)$, $(-2, 6)$, and $(4, 1)$. If the triangle undergoes a dilation with a scale factor of 3, what will be the vertices of the image?

F $(3, -3), (-6, 18), (12, 3)$

G $(3, 3), (6, 18), (12, 3)$

H $(3, 3), (-6, 18), (12, -3)$

J $(3, -3), (-6, -18), (-12, 3)$

28. Let $G(6, -8)$ be a point on triangle *FGH*. What are the coordinates of *G′* if the triangle is dilated by a scale factor of $\frac{1}{2}$?

A $(3, 4)$

B $(3, -4)$

C $(-3, 4)$

D $(-3, -4)$

29. **Short Response** Point $P(-10, 5)$ is on rectangle *PQRS*. If the coordinates of *P* after a dilation are $(-4, 2)$, what is the scale factor of the dilation?

Common Core Review

30. Triangle *RSK* is similar to $\triangle RUV$. What is the value of *x*, rounded to the nearest tenth? 7.RP.2

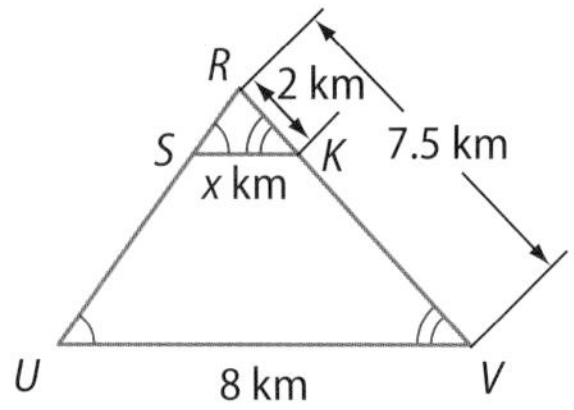

On a map of South Carolina, the scale is 1 inch = 20 miles. Find the actual distance for each map distance. 7.G.1

	From	To	Map Distance
31.	Columbia	Florence	4 inches
32.	Myrtle Beach	Greenville	12 inches

33. **Financial Literacy** The table at the right shows the exchange rates for certain countries compared to the U.S. dollar on a given day. 7.RP.3

Country	Rate per $1 (U.S.)
Great Britain	0.480 pounds
South Africa	6.568 rand

a. What is the cost of an item in U.S. dollars if it costs 14.99 in British pounds?

b. Find the cost of an item in South African rand if it costs 12.50 in U.S. dollars.

34. CCSS **Reason Abstractly** Write a vocabulary term that completes the analogy. 8.G.3

Slide is to translation as __?__ *is to reflection.*

35. Identify the transformation below as a *translation* or a *reflection*. Then describe the transformation. 8.G.3

A triangle has vertices $A(2, -2)$, $B(2, 10)$, and $C(6, -2)$.
The image's vertices are $A'(0, -3)$, $B'(0, 9)$, and $C'(4, -3)$.

Lesson 11-8

Similarity and Transformations

ISG Interactive Study Guide

See pages 261–262 for:
- Getting Started
- Real-World Link
- Notes

Essential Question

How can you determine congruency and similarity?

CCSS Common Core State Standards

Content Standards
8.G.4

Mathematical Practices
1, 3, 4, 6, 7

What You'll Learn

- Use a series of transformations to identify similar figures.
- Use a scale factor to create similar figures.

Real-World Link

Art Artists use grids to transfer their work to a larger area. A grid ensures that what an artist draws or paints retains the correct proportions when it is enlarged.

Identify Similarity

Recall that a dilation changes the size of a figure by a scale factor. Since the size of a figure is changed, the image and the preimage are not congruent. Two figures are similar if the second can be obtained from the first by a sequence of transformations and dilations.

Example 1

Tutor

Determine if the two triangles are similar by using transformations. Explain your reasoning.

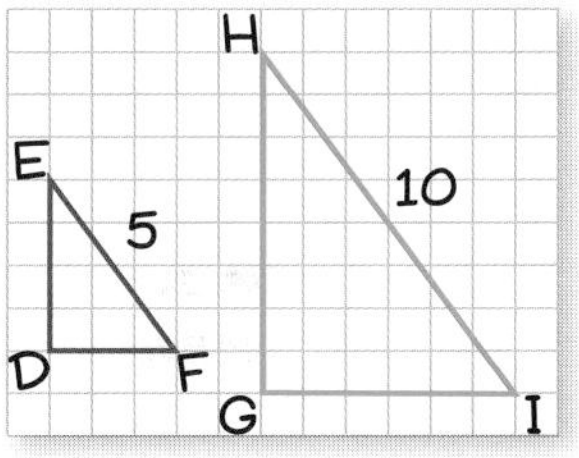

Since the orientation of the figures is the same, one of the transformations is a translation.

Step 1 Translate $\triangle DEF$ down 1 unit and 5 units to the right so D maps onto G.

Step 2 Write ratios comparing the lengths of each side.

$$\frac{HG}{ED} = \frac{8}{4} \text{ or } \frac{2}{1}; \frac{GI}{DF} = \frac{6}{3} \text{ or } \frac{2}{1}; \frac{IH}{FE} = \frac{10}{5} \text{ or } \frac{2}{1}$$

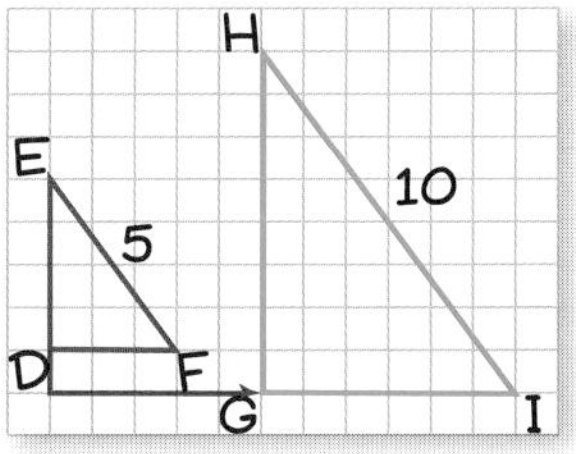

Since the ratios are equal, $\triangle HGI$ is the dilated image of $\triangle EDF$. So, the two triangles are similar because a translation and a dilation maps $\triangle EDF$ onto $\triangle HGI$.

Got It? **Do this problem to find out.**

1. Determine if the two triangles are similar by using transformations. Explain your reasoning.

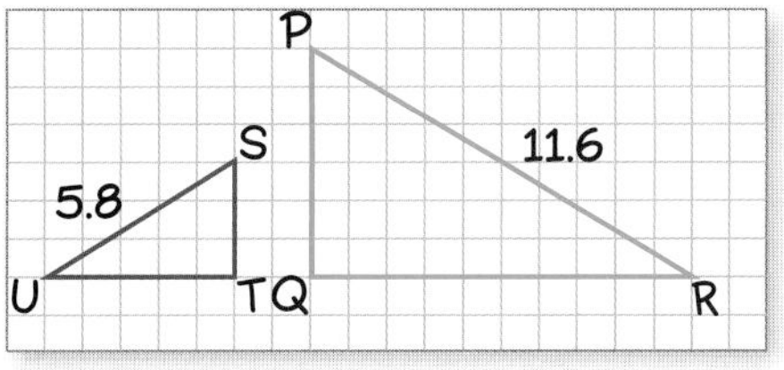

Tetra Images/Getty Images

Example 2

Determine if the two rectangles are similar by using transformations. Explain your reasoning.

The orientation of the figures is different, so one of the transformations is a rotation.

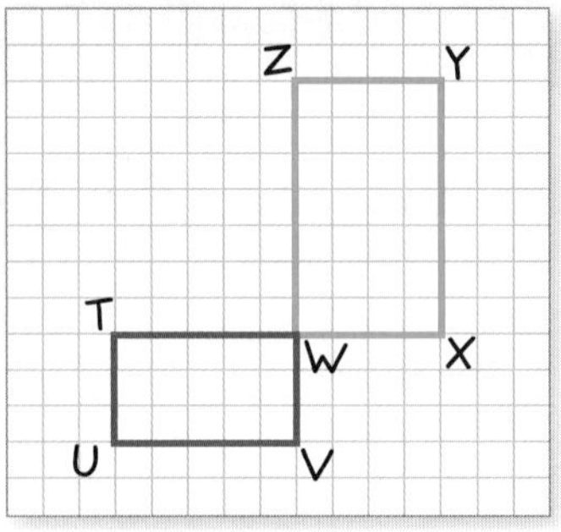

Step 1 Rotate rectangle *VWTU* 90° clockwise about *W* so that it is oriented the same way as rectangle *WXYZ*.

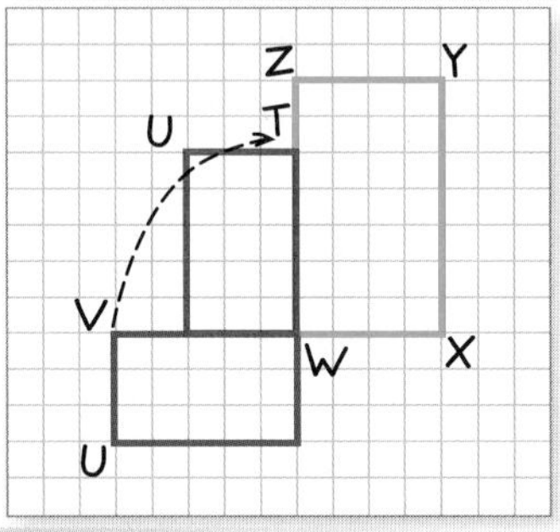

Step 2 Write ratios comparing the lengths of each side.

$\frac{WT}{XY} = \frac{5}{7}; \frac{TU}{YZ} = \frac{3}{4}; \frac{UV}{ZW} = \frac{5}{7}; \frac{VW}{WX} = \frac{3}{4}$

The ratios are not equal. So, the two rectangles are not similar since a dilation did not occur.

Got It? **Do this problem to find out.**

2. Determine if the two figures are similar by using transformations. Explain your reasoning.

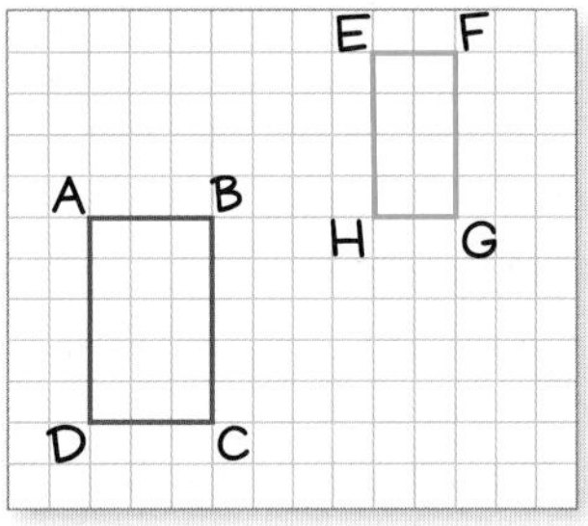

Use the Scale Factor

Scale Factor
If the scale factor is less than 1, the dilation is a reduction. If the scale factor is greater than 1, the dilation is an enlargement.

Similar figures have the same shape but may have different sizes. The sizes of the two figures are related to the scale factor of the dilation. You can find the dimensions of the image when given the scale factor and the dimensions of the original figure.

Example 3

Triangle *ABC* has sides that are 12 feet, 16 feet, and 20 feet in length. It is related to $\triangle CDE$ by a scale factor of 0.5. What are the lengths of the sides of $\triangle CDE$?

Multiply each dimension of $\triangle ABC$ by the scale factor 0.5.

12 ft × 0.5 = 6 ft   16 ft × 0.5 = 8 ft   20 ft × 0.5 = 10 ft

So, the lengths of the sides of triangle *CDE* are 6 feet, 8 feet, and 10 feet.

Got It? **Do this problem to find out.**

3. The sides of a square are 32 inches long. It is related to a smaller square by a scale factor of 0.25. How long are the sides of the smaller square?

Real World Example 4

Tutor

Ken enlarges the photo shown by a scale factor of 2 for his Web page. Then he enlarges the Web page photo by a scale factor of 1.5 to print. What are the dimensions of the print? Are the enlarged photos similar to the original?

3 in.

2 in.

Multiply each dimension of the original photo by 2 to find the dimensions of the Web page photo.

2 in. $\times$ 2 = 4 in. 3 in. $\times$ 2 = 6 in.

So, the Web page photo will be 4 inches by 6 inches. Multiply the dimensions of that photo by 1.5 to find the dimensions of the print.

4 in. $\times$ 1.5 = 6 in. 6 in. $\times$ 1.5 = 9 in.

The printed photo will be 6 inches by 9 inches. All three photos are similar since each enlargement was the result of a dilation.

Got It? Do this problem to find out.

4. An art show offers different sized prints of the same painting. The smallest print measures 24 centimeters by 30 centimeters. A printer enlarges the original by a scale factor of 1.5, and then enlarges the second image by a scale factor of 3. What are the dimensions of the largest print? Are both of the enlarged prints similar to the original?

Guided Practice

Check

Determine if the two figures are similar by using transformations. Explain your reasoning. (Examples 1 and 2)

1. 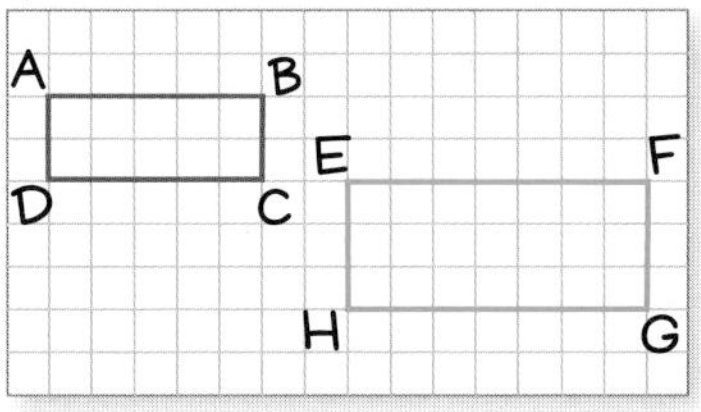

2. 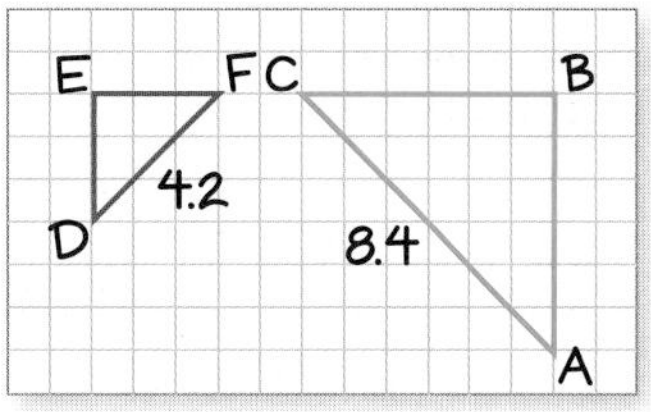

3. The sides of a rectangle are 15 ft and 25 ft in length. It is related to a smaller rectangle by a scale factor of 0.2. How long are the sides of the smaller rectangle? (Example 3)

4. A T-shirt iron-on measures 2 inches by 1 inch. It is enlarged by a scale factor of 3 for the back of the shirt. The second iron-on is enlarged by a scale factor of 2 for the front of the shirt. What are the dimensions of the largest iron-on? Are both of the enlarged iron-ons similar to the original? (Example 4)

Independent Practice

Go online for Step-by-Step Solutions

eHelp

Determine if the two figures are similar by using transformations. Explain your reasoning. (Examples 1 and 2)

5.

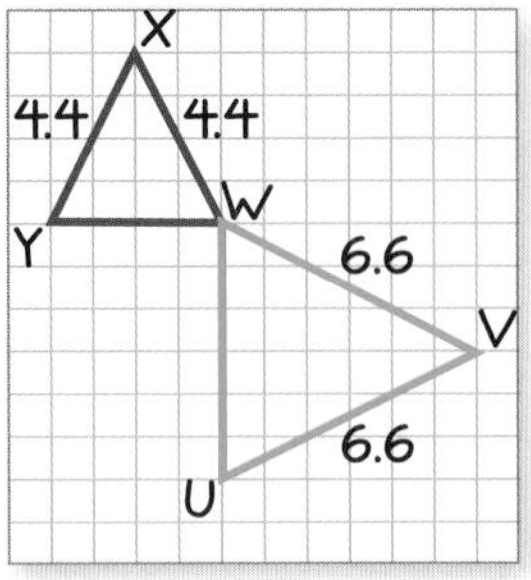

6.

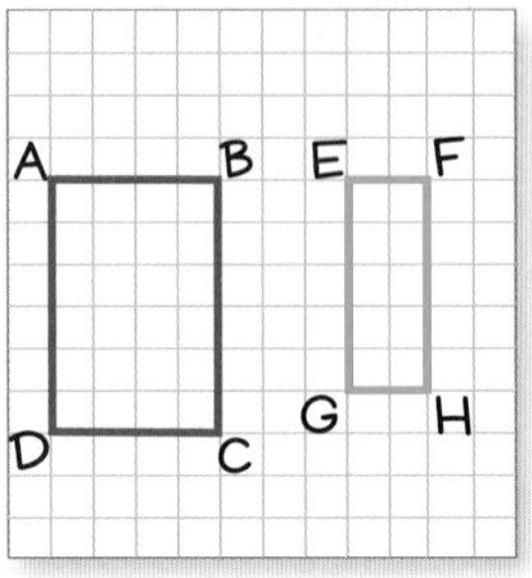

7.

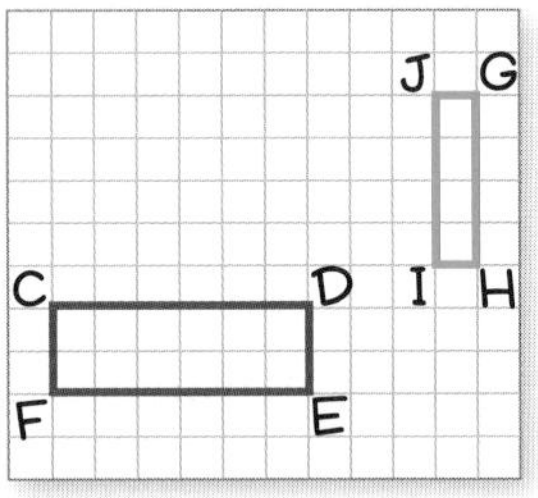

8.

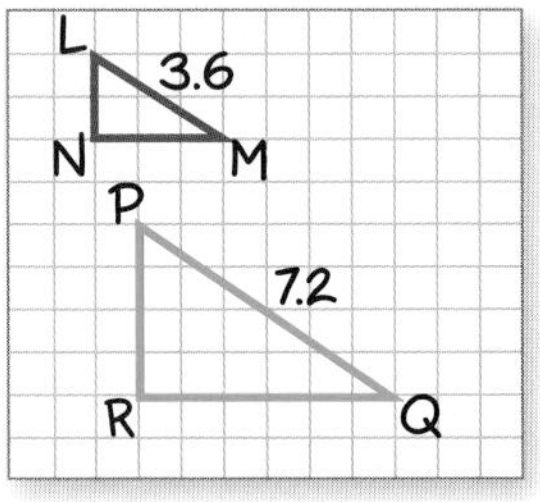

9. The sides of a regular hexagon are 36 units long. It is related to a smaller hexagon by a scale factor of 0.75. How long are the sides of the smaller hexagon? (Example 3)

10. The sides of a figure are 1.2 feet, 3.4 feet, 5 feet, and 8 feet. It is related to a smaller figure by a scale factor of 0.3. How long are the sides of the smaller figure? (Example 3)

11. CCSS **Be Precise** Felisa is creating a scrapbook of her family. A photo of her grandmother measures 3 inches by 5 inches. She enlarges it by a scale factor of 1.5 to place in the scrapbook. Then she enlarges the second photo by a scale factor of 1.5 to place on the cover of the scrapbook. What are the dimensions of the photo for the cover of the scrapbook? Are all of the photos similar? (Example 4)

12. Shannon is making three different sizes of blankets from the same material. The first measures 2.5 feet by 2 feet. She wants to enlarge it by a scale factor of 2 to make the second blanket. Then she will enlarge the second blanket by a scale factor of 1.5 to make the third blanket. What are the dimensions of the third blanket? Are all of the blankets similar? (Example 4)

Each preimage and image are similar. Describe a sequence of transformations that maps the preimage onto the image.

13.

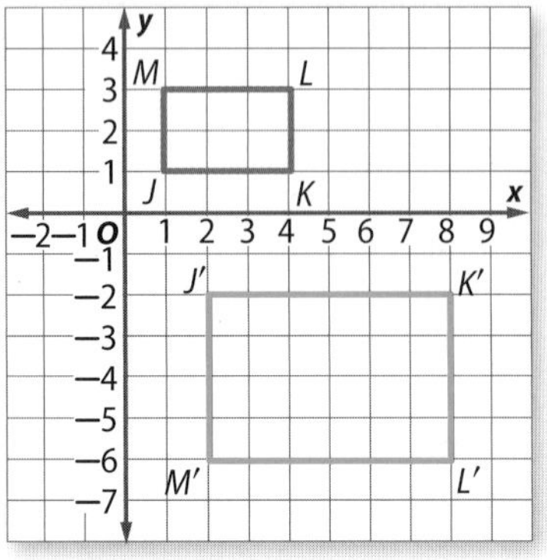

14.

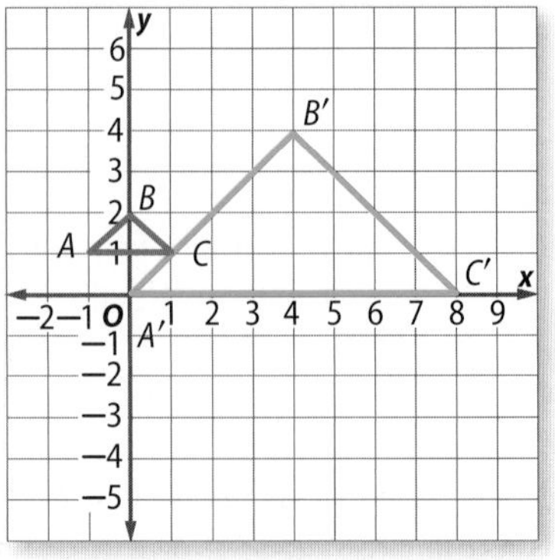

15 In the figure shown, $\triangle A'B'C'$ is the image of $\triangle ABC$ after a dilation followed by a translation.

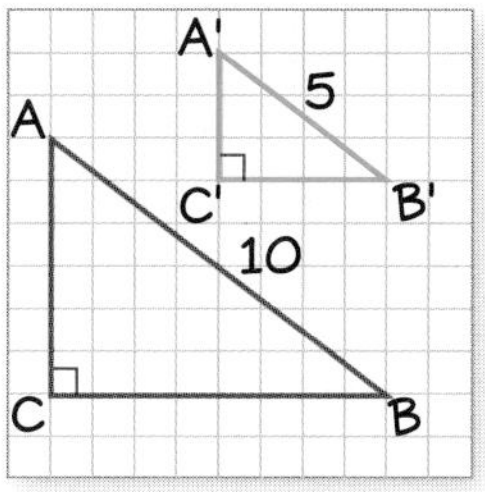

a. Find the lengths of segments *AC* and *CB*.

b. Find the lengths of segments *A′C′* and *C′B′*.

c. Write and simplify the following ratios:

$\frac{AC}{A'C'}, \frac{CB}{C'B'},$ and $\frac{AB}{A'B'}$.

d. What do you notice about the ratios?

16. In the figure shown, trapezoid *RSTU* has vertices *R*(1, 3), *S*(4, 3), *T*(3, 1), and *U*(2, 1).

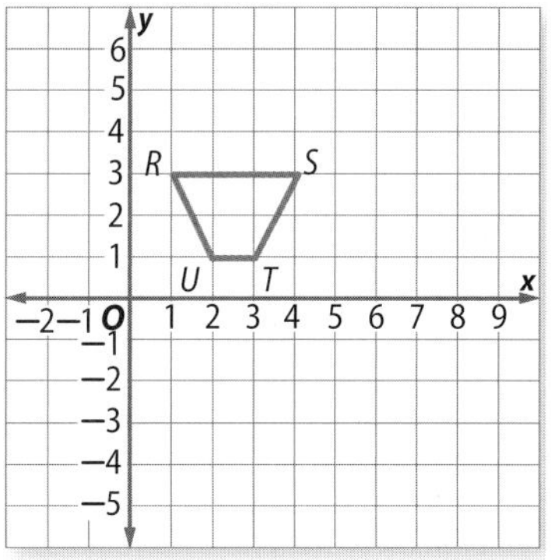

a. Draw *RSTU* on a coordinate grid. Then draw the image of the trapezoid after a translation of 2 units down followed by a dilation with a scale factor of 2. Label the vertices *ABCD*.

b. On a different coordinate grid, draw the image of *RSTU* after a dilation with a scale factor of 2, followed by a translation of 2 units down. Label the vertices *EFGH*.

c. Which figures are similar? Which figures are congruent?

d. Are *ABCD* and *EFGH* in the same location? If they are not, what transformation would map *ABCD* onto *EFGH*?

H.O.T. Problems Higher Order Thinking

17. CCSS **Identify Structure** Using at least one dilation, describe a series of transformations where the image is congruent to the preimage.

18. CCSS **Persevere with Problems** The image of $\triangle DEF$ after two transformations has vertices at *D′*(3, 3), *E′*(6, 3) and *F′*(3, −6).

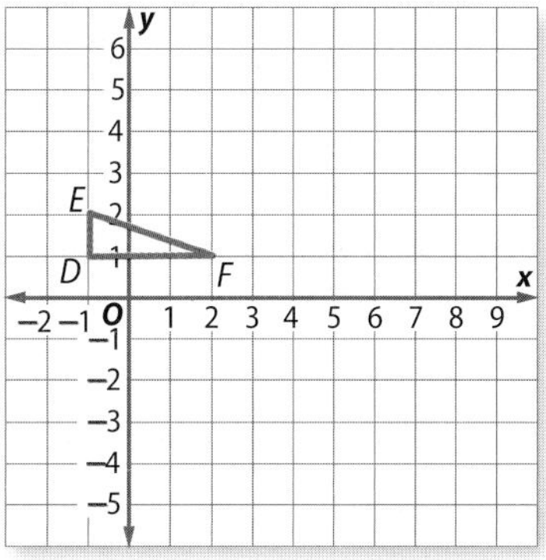

a. If the two triangles are similar, determine what two transformations map $\triangle DEF$ onto $\triangle D'E'F'$.

b. Suppose $\triangle D'E'F'$ is translated up ten units. What are the coordinates of the vertices of the image?

19. CCSS **Reason Inductively** *True* or *false*. If a dilation is in a composition of transformations, the image is similar to the original figure regardless of the order in which you perform the transformations Explain your reasoning.

20. **Building on the Essential Question** Explain the difference between using transformations to create similar and congruent figures.

Standardized Test Practice

21. Triangle *DEF* is the image of $\triangle ABC$ after a sequence of transformations.

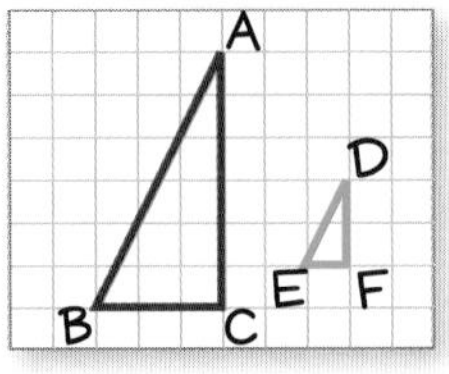

What is the scale factor of the dilation in the sequence?

A 3

B $\frac{1}{3}$

C $-\frac{1}{3}$

D -3

22. Which transformation produces similar figures that are enlargements or reductions?

F translation

G rotation

H reflection

J dilation

23. Trapezoid *ABCD* is similar to trapezoid *PQRS*.

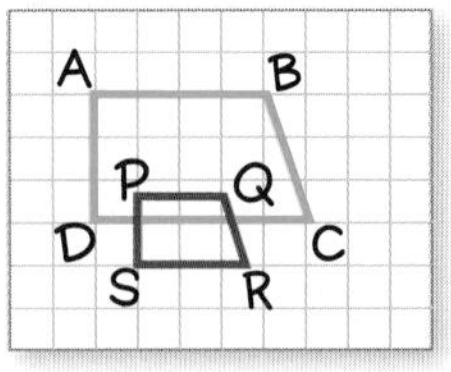

Which series of transformations maps *ABCD* onto *PQRS*?

A rotation then a dilation

B reflection then a dilation

C translation then a dilation

D two dilations

24. **Short Response** Figure B is produced after Figure A is reflected over the *y*-axis and then dilated by a scale factor of $\frac{3}{4}$. What is the ratio comparing the lengths of the sides of Figure A to Figure B?

Common Core Review

Find the discount to the nearest cent. 7.RP.3

25. \$85 cell phone, 20% off

26. \$489 desk, 15% off

27. 25% off of a \$74 baseball glove

28. 25% off of a \$65 pair of roller blades

29. Seri has a coupon for 20% off her entire purchase at a craft store. She buys \$35 acrylic paints that cost \$35. If the sales tax rate is 5%. How much will Seri pay for the acrylic paints?

30. A pair of sunglasses is on sale for 40% off of the original price of \$27. If the tax rate is 5%, what will be the cost for the sunglasses?

For Exercises 31–34, use the diagram at the right. 7.G.5, 8.G.5

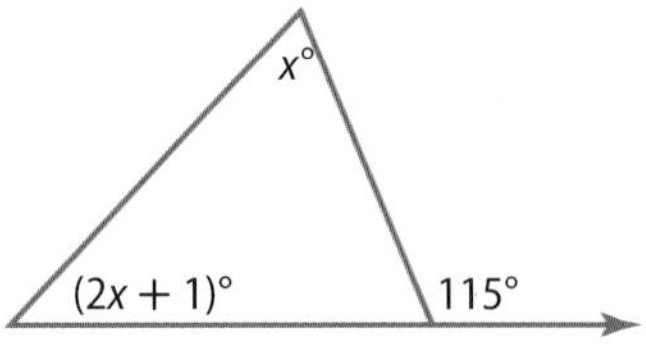

31. What is the measure of the third angle in the triangle?

32. What is the value of *x*?

33. What is the value of $(2x + 1)$?

34. Suppose the exterior angle of the triangle shown had a measure of 118°. Find the measure of the three interior angles of the triangle.

35. Angles *ABC* and *DBE* are vertical angles, with $m\angle ABC = 7x°$ and $m\angle DBE = (3x + 8)°$. 7.G.5

a. Find the value of *x*.

b. Find $m\angle DBE$.

Check

Chapter Review

ISG Interactive Study Guide
See pages 263–266 for:
- Vocabulary Check
- Key Concept Check
- Problem Solving
- Reflect

Lesson-by-Lesson Review

Lesson 11-1 Angle and Line Relationships (pp. 494–500)

In the figure below, $m \parallel n$ and r is a transversal. If $m\angle 4 = 112°$, find the measure of each angle. Explain your reasoning.

1. $\angle 6$
2. $\angle 2$
3. $\angle 8$
4. $\angle 1$

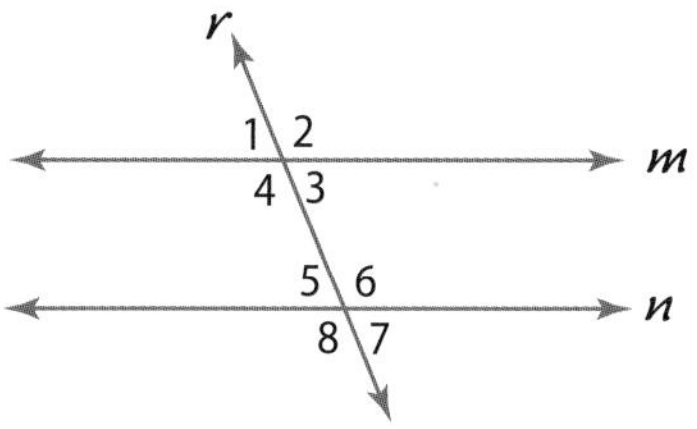

5. A door can swing open 180°. The door is open at an angle of 99°. What is the measure of the angle between the door and the doorjamb?

Example 1

In the figure below, $a \parallel b$ and g is a transversal. If $m\angle 2 = 61°$, find $m\angle 7$ and $m\angle 4$. Explain your reasoning.

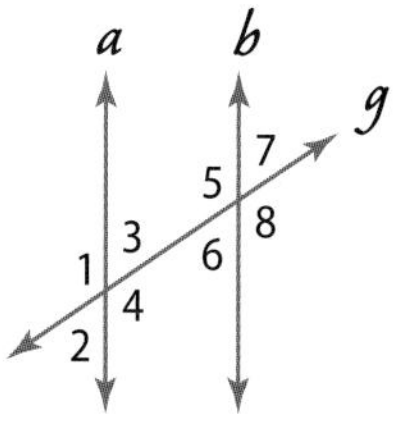

Since $\angle 2$ and $\angle 7$ are alternate exterior angles, they are congruent. So, $m\angle 7 = 61°$.

Since $\angle 2$ and $\angle 4$ are supplementary, the sum of their measures is 180°.

$m\angle \mathbf{2} + m\angle 4 = 180°$	Supplementary angles
$\mathbf{61°} + m\angle 4 = 180°$	Replace m∠2 with 61°.
$m\angle 4 = 119°$	Subtract 61° from each side.

So, $m\angle 4 = 119°$.

Lesson 11-2 Triangles (pp. 503–508)

Find the value of x in each triangle. Then classify each triangle by its angles and by its sides.

6. 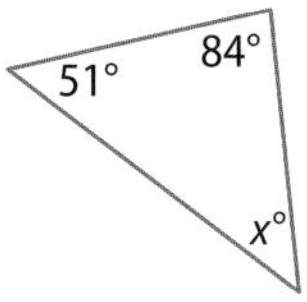

7. 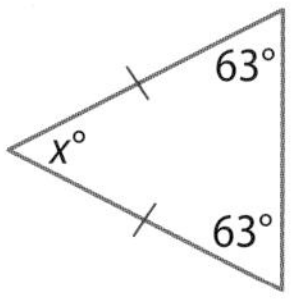

8. 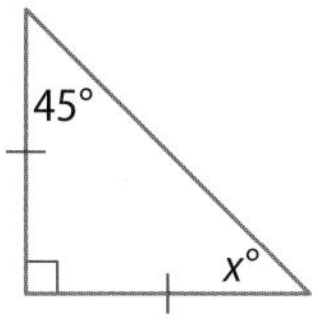

9. 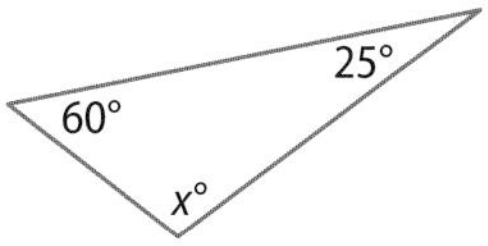

10. Classify the yield sign by its angles and by its sides.

Example 2

Find the value of x in the triangle. Then classify the triangle by its angles and by its sides.

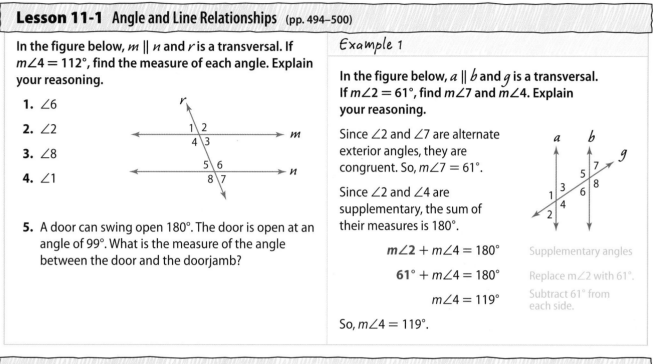

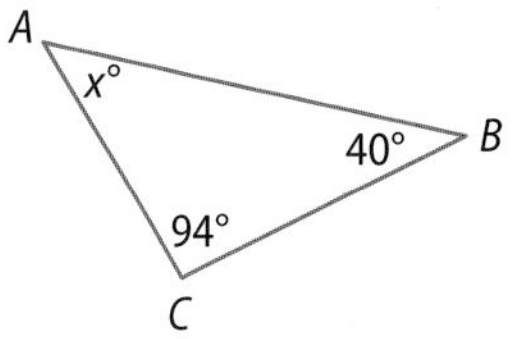

Step 1 **Find the missing angle measure.**

$x + 40 + 94 = 180$	Write an equation.
$x + 134 = 180$	Simplify.
$x + 134 - \mathbf{134} = 180 - \mathbf{134}$	Subtract 134 from each side.
$x = 46$	Simplify.

Step 2 **Classify the triangle.**

Angles: The triangle has an obtuse angle.

Sides: The triangle has no congruent sides.

So, the triangle is obtuse scalene.

Lesson 11-3 Polygons (pp. 513–518)

Determine whether the figure is a polygon. If it is, classify the polygon and state whether it is regular. If it is not a polygon, explain why.

11. 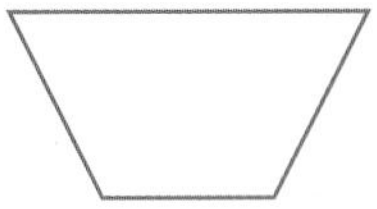

12. 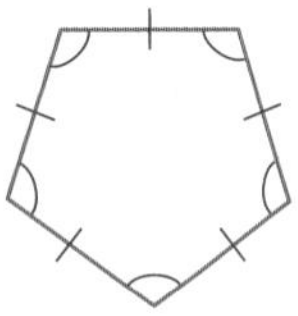

Find the measure of an interior angle of each regular polygon.

13. hexagon

14. 18-gon

15. What is the measure of each interior angle of the stop sign?

Example 3

Find the measure of one interior angle in a regular heptagon.

Step 1 **Find the sum of the measures of the angles. A heptagon has 7 sides. So, $n = 7$.**

$(n - 2)180 = (7 - 2)180$ Replace n with 7.

$= 5(180)$ or $900°$ Simplify.

The sum of the measures of the interior angles is 900°.

Step 2 **Divide the sum by 7 to find the measure of one angle.**

$900 \div 7 \approx 128.6$

So, the measure of one interior angle in a heptagon is about 128.6°.

Lesson 11-4 Translations and Reflections on the Coordinate Plane (pp. 521–526)

16. The vertices of figure $ABCD$ are $A(1, -3)$, $B(4, -3)$, $C(1, -1)$, and $D(-2, -1)$. Find the vertices after a reflection over the x-axis.

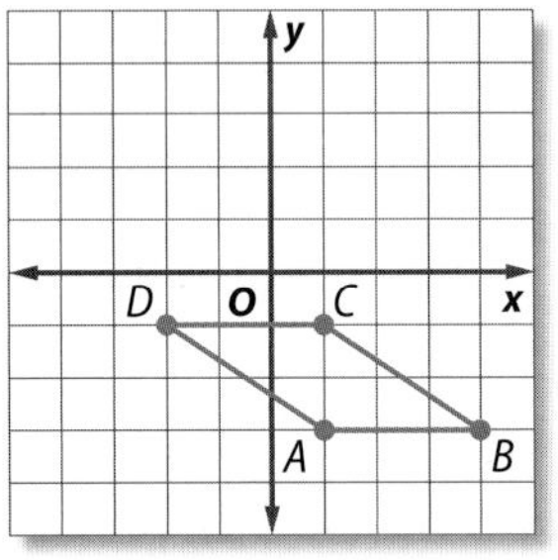

17. A triangle has vertices $N(6, 3)$, $P(3, 9)$, and $Q(9, 6)$. The triangle is translated 2 units right and 2 units down. Graph the figure and its image.

18. What type of transformation is used when moving up an escalator?

Example 4

The vertices of $\triangle JKL$ are $J(1, 2)$, $K(3, 2)$, and $L(1, -1)$. Find the vertices of the image after a translation 3 units left and 2 units up. Then find the vertices of the image after $\triangle JKL$ is reflected over the x-axis.

Translation This translation can be written as $(-3, 2)$.

original		translation		image
$J(1, 2)$	+	$(-3, 2)$	→	$J'(-2, 4)$
$K(3, 2)$	+	$(-3, 2)$	→	$K'(0, 4)$
$L(1, -1)$	+	$(-3, 2)$	→	$L'(-2, 1)$

Reflection Use the same x-coordinate and replace the y-coordinate with its opposite.

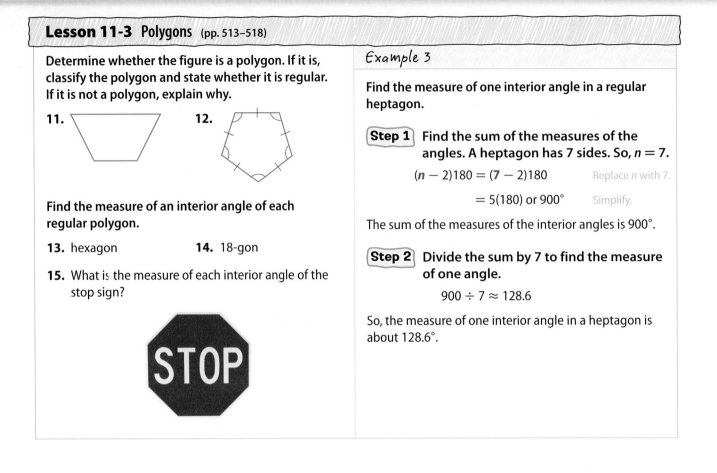

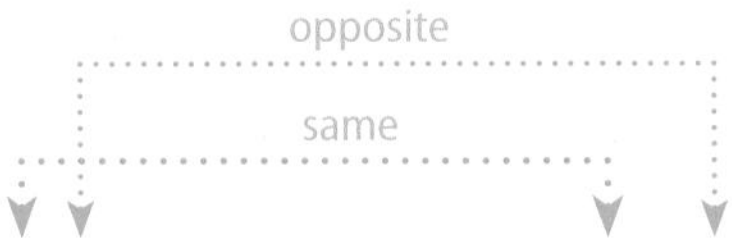

$J(1, 2)$	→	$J'(1, -2)$
$K(3, 2)$	→	$K'(3, -2)$
$L(1, -1)$	→	$L'(1, 1)$

Lesson 11-5 Rotations on the Coordinate Plane (pp. 528–533)

19. Triangle *ABC* has vertices *A* (2, 0), *B* (4, −1), and *C* (1, −3). Graph the figure and its image after a clockwise rotation of 180° about vertex *A*. Give the coordinates of the vertices for triangle *A′B′C′*.

Graph each figure and its image after a clockwise rotation about the origin.

20. triangle *GHJ* with vertices *G* (0, −1), *H* (3, 3), and *J* (2, −3); 270° clockwise rotation

21. quadrilateral *NPQR* with vertices *N* (1, 1), *P* (2, 3), *Q* (4, 2), and *R* (4, −2); 90° clockwise rotation

22. Determine whether the shape of the sign shown at the right has rotational symmetry. If it does, describe the angle of rotation.

Example 5

Triangle *TVW* has vertices *T*(−1, 0), *V*(0, 3), and *W*(2, 2). Graph the figure and its image after a clockwise rotation of 270° about vertex *T*. Give the coordinates of the vertices for triangle *T′V′W′*.

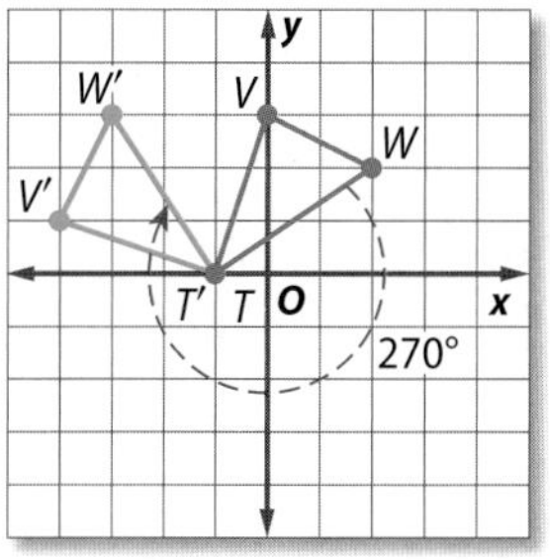

The coordinates of the vertices are *T′* (−1, 0), *V′* (−4, 1), and *W′* (−3, 3).

Lesson 11-6 Congruence and Transformations (pp. 534–538)

23. Determine if the two figures are congruent by using transformations. Explain your reasoning.

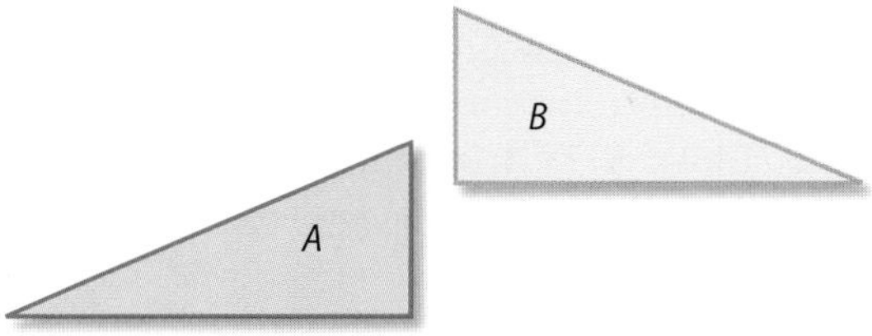

24. Helen made a logo for her company using her initials. What transformation could be used if the red "H" is the preimage and the blue "H" is the image in the design shown?

25. Given two congruent figures, how you can determine the transformation, or series of transformations, that maps one figure onto the other?

Example 6

Determine if the two figures are congruent by using transformations.

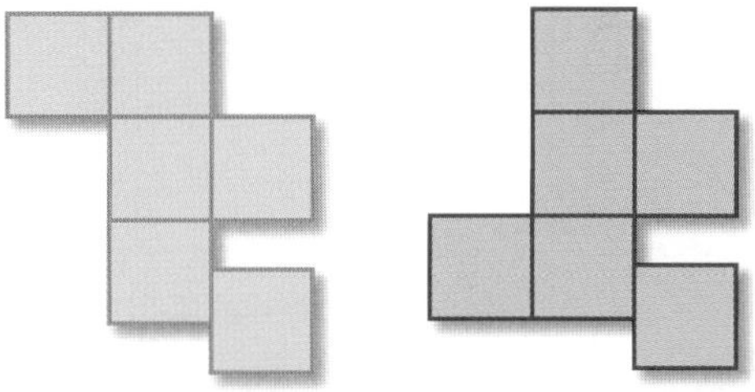

Reflect the green figure over a horizontal line. Then translate the image right.

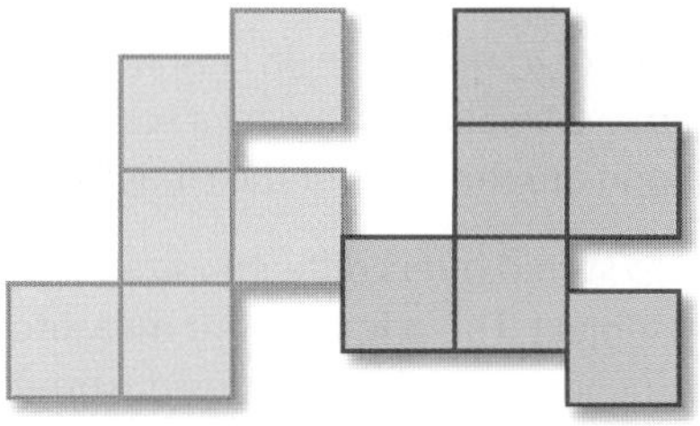

The green figure does not match the red figure exactly. The two figures are not congruent.

Lesson 11-7 Dilations on the Coordinate Plane (pp. 539–544)

Find the vertices of each figure after a dilation with the given scale factor k, centered at the origin. Then graph the original image and the dilation.

26. $W(-3, 0), X(2, 6), Y(6, 2), Z(2, -2); k = \frac{2}{3}$

27. $Q(-2, 3), R(1, 2), S(3, -1), T(-2, -2); k = 3$

28. $L(-4, -2), M(-2, 4), N(4, 0); k = \frac{1}{4}$

29. $F(1, 3), G(3, 4), H(2, 1); k = 2.5$

30. Percy is using a photocopier to increase the dimensions of his 5-inch by 7-inch photograph.

- **a.** What is the scale factor if he enlarges the photograph to 7.5 inches by 10.5 inches?
- **b.** What is the scale factor if he enlarges the photograph to 10 inches by 14 inches?
- **c.** What are the new dimensions if he enlarges the photograph by a scale factor of 2.2?

Example 7

A triangle has vertices $A(-2, -1)$, $B(1, 1)$, and $C(3, -3)$. Find the coordinates of the triangle after a dilation centered at the origin with a scale factor of 2.

To dilate the triangle, multiply the coordinates of each vertex by 2.

$A(-2, -1) \rightarrow A'(-2 \cdot 2, -1 \cdot 2) \rightarrow A'(-4, -2)$

$B(1, 1) \rightarrow B'(1 \cdot 2, 1 \cdot 2) \rightarrow B'(2, 2)$

$C(3, -3) \rightarrow C'(3 \cdot 2, -3 \cdot 2) \rightarrow C'(6, -6)$

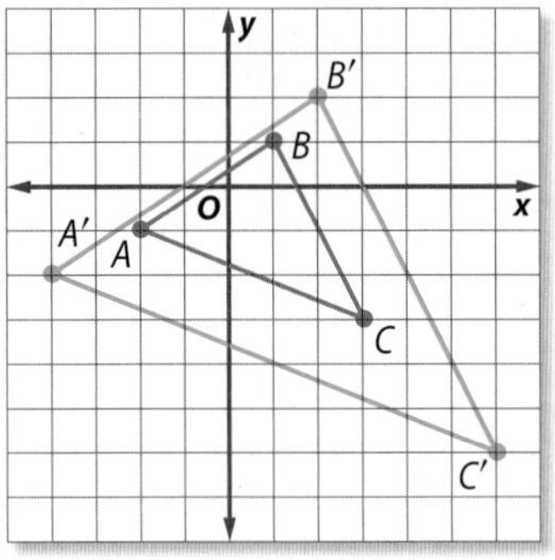

Lesson 11-8 Similarity and Transformations (pp. 545–550)

31. Determine if the two figures are similar by using transformations. Explain your reasoning.

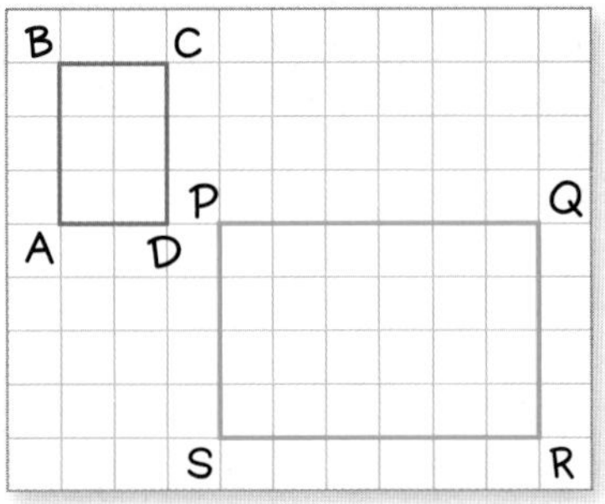

32. Triangle ABC has sides that are 20 centimeters, 38 centimeters, and 50 centimeters in length. It is related to $\triangle CDE$ by a scale factor of 0.5. What are the lengths of the sides of $\triangle CDE$?

33. The sides of a square are 48 centimeters long. It is related to a smaller square by a scale factor of 0.75. How long are the sides of the smaller square?

34. A photo studio offers different-sized prints of the same subject. The original print measures 8 inches by 10 inches. A printer enlarges the original by a scale factor of 1.5, and then enlarges the second image by a scale factor of 3. What are the dimensions of the largest print?

Example 8

Determine if the two figures are similar by using transformations.

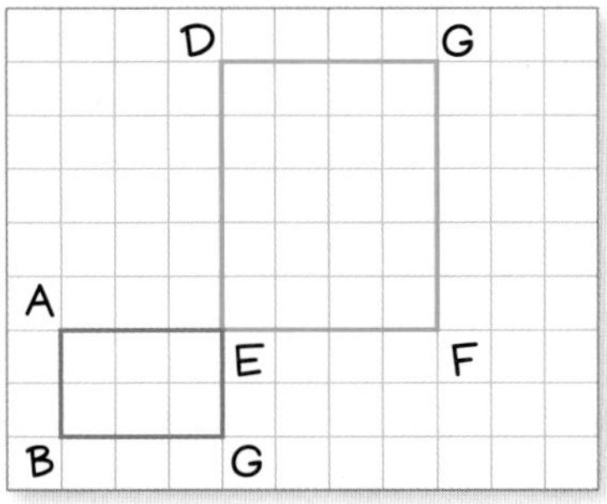

Write the ratios comparing the lengths of the sides.

$\frac{DE}{BG} = \frac{5}{3}$ $\quad \frac{EF}{GE} = \frac{4}{2}$ or 2

$\frac{GF}{AE} = \frac{5}{3}$ $\quad \frac{DG}{BA} = \frac{4}{2}$ or 2

The ratios of the sides are not all equal. So, the two rectangles are not similar since a dilation did not occur.

Chapter 12

Volume and Surface Area

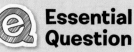

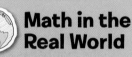

Essential Question

How are two-dimensional figures used to solve problems involving three-dimensional figures?

Common Core State Standards

Content Standards
7.G.3, 7.G.4, 7.G.6, 8.G.9

Mathematical Practices
1, 2, 3, 4, 6, 7, 8

Math in the Real World

Aquariums The main tank at the Okinawa Churaumi Aquarium in Japan has a volume of 7500 cubic meters. The tank holds almost 2 million gallons of water, making it a perfect home for whale sharks, manta rays, and giant schools of fish.

Interactive Study Guide
See pages 267–270 for:

- Chapter Preview
- Are You Ready?
- Foldable Study Organizer

Inquiry Lab

Circles

Inquiry HOW are the circumference and area of a circle related to its radius?

CCSS Content Standards
7.G.4
Mathematical Practices
1, 3, 6

Packaging Hanna has several cans in her pantry. The top of each can is a circle. She wonders if there is a relationship between the distance across each can through its center, or its *diameter*, and the distance around the top of the can, or the *circumference*.

Investigation 1

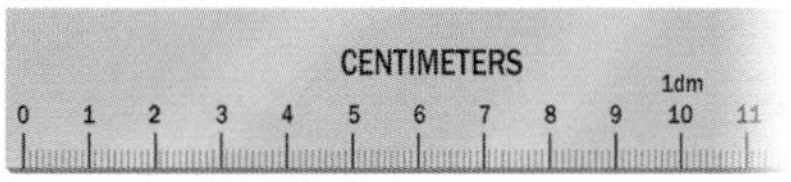

Step 1 Measure the circumference and diameter of the top of some round cans.

Step 2 For each can, find the ratio of the circumference C to the diameter d. Write the number as a decimal and round to the nearest hundredth.

	Can 1	Can 2	Can 3	Can 4
circumference (cm)	23.9	31.9	25.9	51.9
diameter (cm)	7.6	10.2	8.3	16.5
$\frac{\text{circumference } (C)}{\text{diameter } (d)}$	3.14	3.13	3.12	3.15

Collaborate

Work with a partner.

1. Repeat Investigation 1 using four different circular objects.
2. Describe the ratio $\frac{C}{d}$ for the values you calculated.

Analyze

3. **CCSS Make a Conjecture** For any circle, the ratio of the circumference to the diameter is a constant number known as *pi* (π). What do you think is the value of π?

CCSS Be Precise For any circle, $\frac{C}{d} = \pi$. Write formulas for each of the following.

4. Find the circumference C given the diameter d.
5. Find the diameter d given the circumference C.
6. The radius of a circle is one-half its diameter. Write a formula to find the diameter d given the radius r.
7. Write a formula to find the circumference C given the radius r.

Investigation 2

The radius of a circular pizza is r inches. You can use this information to find its area.

Step 1 A circular pizza can be divided into slices of equal size, as shown.

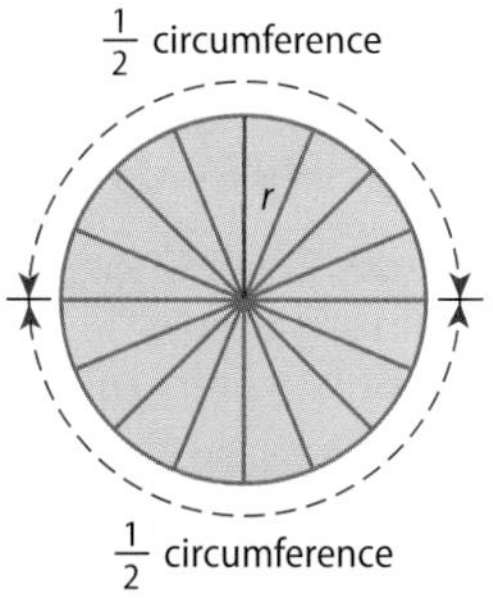

Step 2 Arrange half the slices pointing downward and half pointing upward to form a figure that is close to a parallelogram.

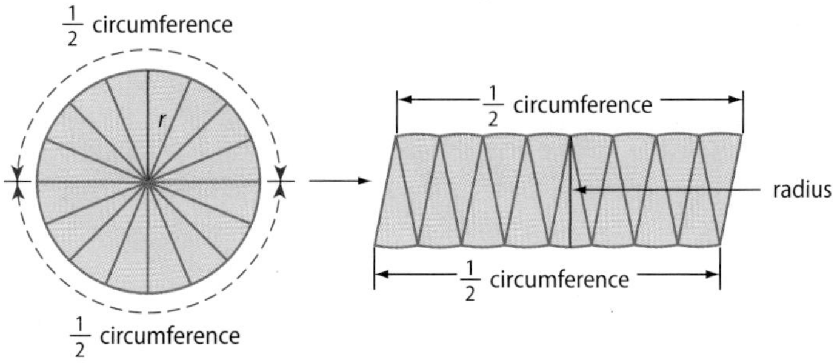

Step 3 The area of the circle is approximately equal to the area of the parallelogram. Find the area of the parallelogram.

$A = bh$	Area of a parallelogram
$A = \left(\frac{1}{2}C\right) \cdot r$	The base equals half the circumference of the circle; the height equals the radius of the circle.
$A = \left(\frac{1}{2} \cdot 2\pi r\right) \cdot r$	Replace C with $2\pi r$.
$A = \pi r \cdot r$	Simplify.
$A = \pi r^2$	Use exponential notation.

So, the area of a circle with radius r is πr^2.

Collaborate

Work with a partner. Find the area of each circle. Use 3.14 for π.

8.

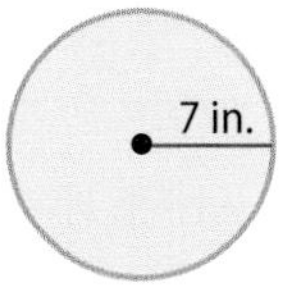

9.

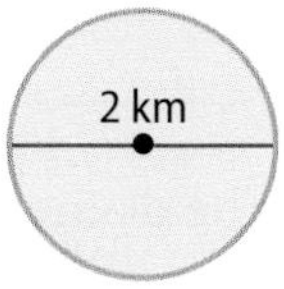

10.

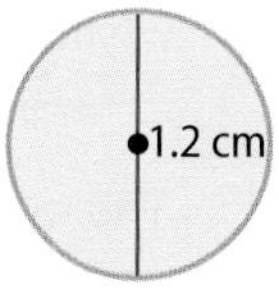

Analyze

11. CCSS **Model with Mathematics** Find a real-world example of a circle. Measure the diameter of the circle and calculate the circle's circumference and area.

12. CCSS **Persevere with Problems** Find a formula for the area of a circle given its circumference.

Reflect

13. Inquiry HOW are the circumference and area of a circle related to its radius?

Lesson 12-1

Circles and Circumference

Interactive Study Guide

See pages 271–272 for:
- Getting Started
- Vocabulary Start-Up
- Notes

Essential Question

How are two-dimensional figures used to solve problems involving three-dimensional figures?

Common Core State Standards

Content Standards
7.G.4

Mathematical Practices
1, 2, 3, 4

Vocabulary

circle
center
diameter
radius
circumference
π (pi)

What You'll Learn

- Find the circumference of circles.
- Solve problems involving circumference.

Real-World Link

Bicycles Some bicycle wheels have wide tires. Others have smooth tires or feature fewer spokes. The type of wheel on a bike depends on how the bike is used—such as for racing, tricks, or touring.

Circumference of Circles

A **circle** is the set of all points in a plane that are the same distance from a given point in the plane.

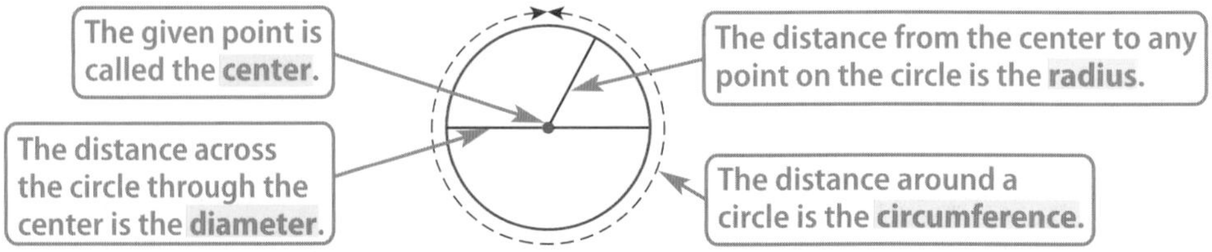

The given point is called the **center**.
The distance across the circle through the center is the **diameter**.
The distance from the center to any point on the circle is the **radius**.
The distance around a circle is the **circumference**.

Key Concept Circumference of a Circle

Watch

Words The circumference C of a circle is equal to its diameter times π, or 2 times its radius times π.

Symbols $C = \pi d$ or $C = 2\pi r$

Model

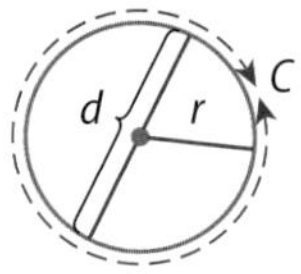

d
r
C

The ratio of the circumference to the diameter is equal to 3.1415926… which is represented by the Greek letter **π (pi)**. So, $\frac{C}{d} = \pi$.

Example 1

Tools Tutor

Find the circumference of the circle. Round to the nearest tenth.

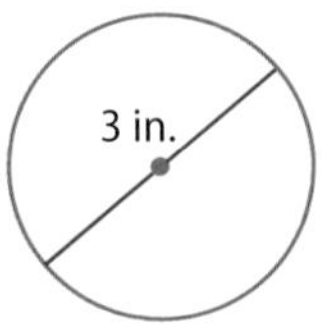

3 in.

$C = \pi \boldsymbol{d}$	Circumference of a circle
$= \pi \cdot \mathbf{3}$	Replace d with 3.
$= 3\pi$	Simplify. This is the *exact* circumference.
3 [×] [2nd] [π] [ENTER] 9.424777961	Use a calculator.

The circumference is about 9.4 inches.

Got It? Do these problems to find out.

1a. diameter = 3.75 ft

1b. diameter = 5.1 mm

Example 2

Tools Tutor

Find the circumference of the circle. Round to the nearest tenth.

$C = 2\pi \mathbf{r}$	Circumference of a circle
$= 2 \cdot \pi \cdot \mathbf{5.3}$	Replace *r* with 5.3.
≈ 33.3	Simplify. Use a calculator.

The circumference is about 33.3 centimeters.

Estimation
If an exact answer is needed, leave the answer in terms of π. The exact answer to Example 2 is 10.6π. If an estimate is sufficient, use a calculator to find a decimal approximation. Round to the indicated place value.

Got It? **Do this problem to find out.**

2. Find the circumference of the circle with a radius of 7 millimeters. Round to the nearest tenth.

Use Circumference to Solve Problems

You can use the circumference of a circle to find the diameter or the radius.

Real World Example 3

Tutor

Bernard works at a community center that has a circular swimming pool with a circumference of 40 meters. He would like to use a rope to divide the pool down the center. What should be the length of the rope?

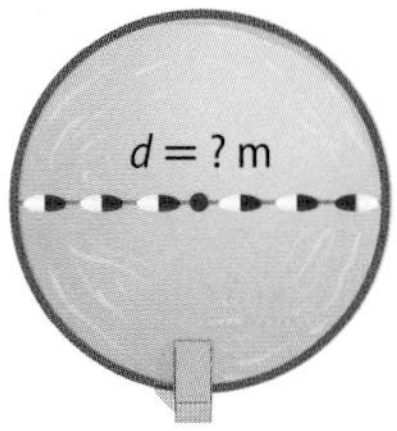

The length of the rope is the diameter of the pool. Use the formula for circumference to find the diameter.

$\mathbf{C} = \pi d$	Circumference of a circle
$\mathbf{40} = \pi \cdot d$	Replace *C* with 40.
$\frac{40}{\pi} = d$	Divide each side by π.
$12.7 \approx d$	Simplify. Use a calculator.

So, the length of the rope should be about 12.7 meters.

Check the reasonableness of the solution by replacing *d* with 12.7 in $C = \pi d$.

$C = \pi \mathbf{d}$	Circumference of a circle
$= \pi \cdot \mathbf{12.7}$	Replace *d* with 12.7.
≈ 39.9	Simplify. Use a calculator.

Since this circumference is close to the original circumference, 40 meters, the solution is reasonable.

Got It? **Do these problems to find out.**

3a. A Ferris wheel has a circumference of 226 meters. What is the radius of the Ferris wheel?

3b. A CD has a diameter of 120 millimeters. A Universal Media Disc (UMD) has a diameter of 60 millimeters. Compare the circumferences of the discs.

Guided Practice

Find the circumference of each circle. Round to the nearest tenth. (Examples 1 and 2)

1.

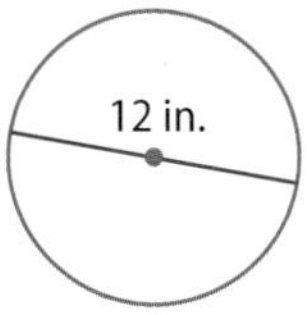

2.

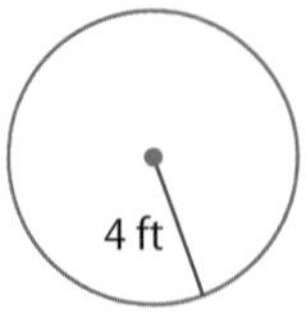

3.

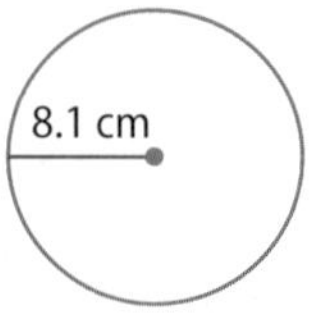

4. diameter = 10.5 centimeters

5. radius = 3.6 kilometers

6. An ancient *timber circle* discovered in the United Kingdom was thought to have been built more than 4000 years ago. Known as "Seahenge," this ancient circle has a circumference of about 21.3 meters. What is the radius of the circle to the nearest tenth? (Example 3)

Independent Practice

Go online for Step-by-Step Solutions

Find the circumference of each circle. Round to the nearest tenth. (Examples 1 and 2)

7

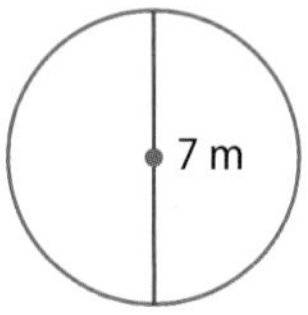

8.

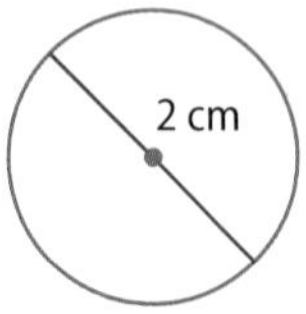

9.

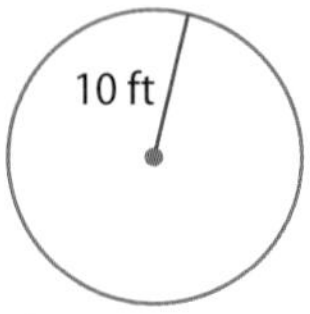

10.

11.

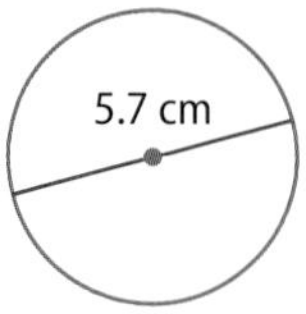

12.

13. diameter = 9.4 meters

14. radius = 13.7 millimeters

15. radius = $11\frac{1}{4}$ feet

16. diameter = $14\frac{3}{5}$ yards

17. Compare the circumferences of the two sand dollars shown below. (Example 3)

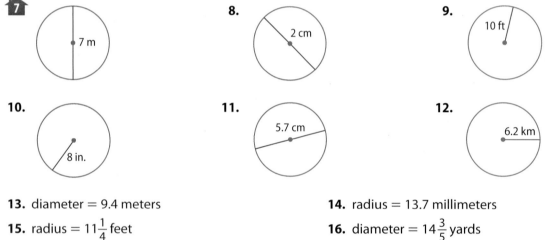

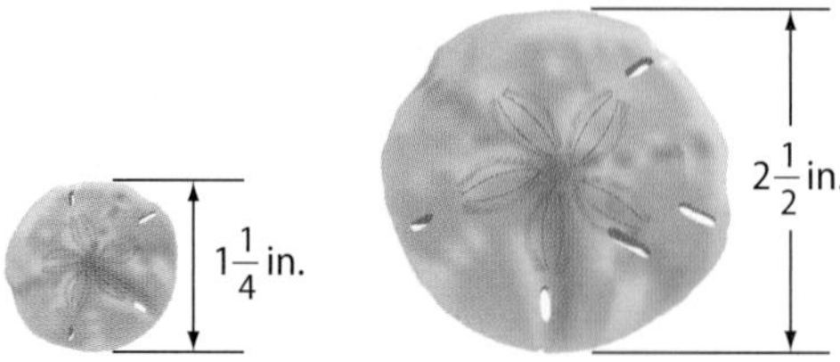

18. The circumference of the moon is about 6790 miles. What is the distance to the center of the moon in kilometers? Round to the nearest kilometer. (*Hint*: 1 mile ≈ 1.6 kilometers) (Example 3)

19. The tower at Philadelphia City Hall contains four clocks that have a radius of about 3.96 meters. Find how far the minute hand travels after each number of rotations around the clock face. Round to the nearest hundredth.

a. 2 rotations

b. $\frac{1}{2}$ rotation

c. $5\frac{3}{4}$ rotations

20. The world's largest carousel in Spring Green, Wisconsin, has a diameter of 80 feet. How far does a rider travel on an outside horse after 10 revolutions? Round to the nearest foot.

21 A circular fountain at a park has a radius of 4 feet. The mayor wants to build a fountain that is quadruple the radius of the current fountain. Find the circumference of the new fountain. Round to the nearest tenth.

22. The world's largest rideable motorcycle travels one mile after about 272.5 rotations of a tire.

a. To the nearest tenth, how many feet does the motorcycle travel after one rotation of a tire? What does this measure represent? (*Hint:* 1 mile = 5280 feet)

b. To the nearest tenth, how many feet tall are the tires?

23. CCSS **Multiple Representations** In this problem, you will investigate the relationship between the radius and circumference of a circle.

Radius (in.)	Circumference (in.)
1	■
2	■
3	■
4	■
5	■

a. Table Make a table of values like the one at the right. Find the circumference of a circle having each radius. Round to the nearest tenth.

b. Graph Use your table to graph the circumference *C* of a circle as a function of the radius *r*.

c. Analyze Describe the slope of the graph. How is the slope related to the formula for finding circumference?

24. STEM The Beaverhead Crater is a large impact crater in the United States. If the circumference of the crater is about 188.5 kilometers, what is the diameter of the crater?

H.O.T. Problems Higher Order Thinking

25. CCSS **Model with Mathematics** Find a circular object in your home. Measure the diameter and use that value to calculate the circumference. Use a tape measure to check your calculation.

26. CCSS **Reason Abstractly** A *variable* is a quantity with a value that changes. In the formula for the circumference of a circle, identify any variables.

27. CCSS **Persevere with Problems** Three congruent circles are inside a rectangle as shown at the right. Which is greater, the length of the rectangle ℓ, or the circumference of one circle? Explain your reasoning.

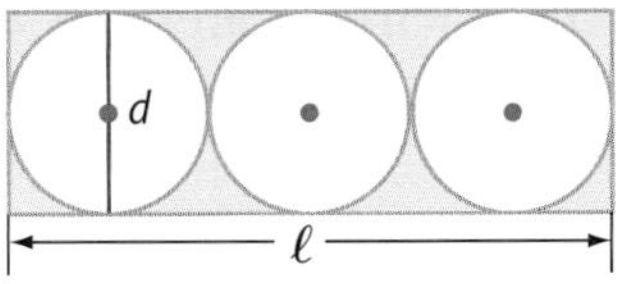

28. CCSS **Justify Conclusions** Explain why the formula $C = \pi d$ can be used to calculate the length of a circular bracelet. If the bracelet is opened up and laid out flat, can the same formula be used to find the length? Explain.

29. **Building on the Essential Question** Describe the relationship between circumference and radius. How does the circumference change if the radius is increased? How does the radius change if the circumference is decreased?

Standardized Test Practice

30. A plate has a radius of 5 inches. Which equation could be used to find the circumference of the plate in inches?

A $C = 2(10\pi)$

B $C = 5\pi$

C $C = 10\pi$

D $C = 2\pi$

31. A bicycle tire has a diameter of 24 inches. Find the circumference of the tire to the nearest tenth of an inch.

F 75.4 in.

G 83.5 in.

H 87.9 in.

J 91.5 in.

32. A planter has a circumference of 37.6 inches. Which measure is *closest* to the diameter of the planter?

A 10 in.

B 11 in.

C 12 in.

D 14 in.

33. **Short Response** The circle shown below has a diameter of 15 feet.

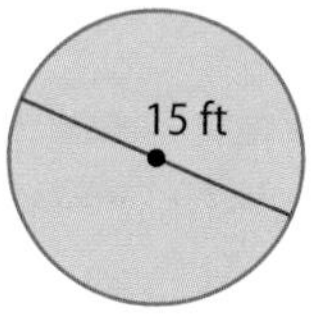

What is the circumference to the nearest foot?

Common Core Review

Find the area of each figure described. 6.G.1

34. parallelogram: base, 10 yards; height, 5 yards

35. trapezoid: height, 3 feet; bases, 15 feet and 20 feet

36. right triangle: height, 6 centimeters; base, 5 centimeters

37. parallelogram: height, 9 inches; base, 10 inches

Find the sum of the measures of the interior angles of each polygon. 8.G.5

38. triangle

39. hexagon

40. pentagon

41. For every order submitted, an online bookstore charges a \$5 shipping fee plus a charge of \$2 per pound on the weight of the items being shipped. The total shipping charges y can be represented by $y = 2x + 5$, where x represents the weight of the order in pounds. Graph the equation. 8.F.3

Solve each inequality. 7.EE.4b

42. $b - 14 > 23$

43. $30 \geq 5n$

44. $\frac{p}{-3} < 9$

45. $9x + 3 \leq 21$

46. $6 - 4m > 14$

47. $2c - 7 > 5c + 14$

48. It costs \$15 per person to take a tour of an underground cave. 7.EE.4a

a. Write an equation to determine the total cost for any number of people to take the tour.

b. What is the total cost if 12 people take the tour?

Lesson 12-2

Area of Circles

ISG Interactive Study Guide

See pages 273–274 for:
- Getting Started
- Real-World Link
- Notes

Essential Question

How are two-dimensional figures used to solve problems involving three-dimensional figures?

CCSS Common Core State Standards

Content Standards
7.G.4

Mathematical Practices
1, 3, 4, 6, 7

What You'll Learn

- Find areas of circles.
- Use areas of circles to solve problems.

Real-World Link

Disc Golf Disc golf is one of today's fastest-growing sports. The putting "green" in disc golf is the circular area that is about 33 feet from the disc catcher.

Key Concept Area of a Circle

Words The area A of a circle in square units is $A = \pi r^2$, where r is the length of the radius.

Symbols $A = \pi r^2$

Model

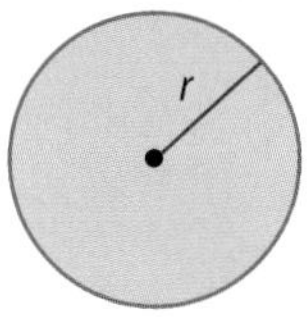

You can find the area of a circle given its radius or diameter.

Example 1

Tools Tutor

Find the area of each circle. Round to the nearest tenth.

a.

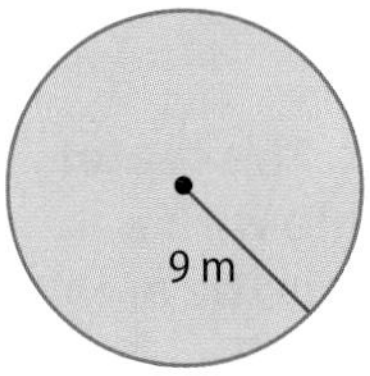

Estimate $A \approx 3 \cdot 9^2$ or 243

$A = \pi r^2$ — Area of a circle

$= \pi \cdot 9^2$ — Replace r with 9.

$= \pi \cdot 81$ — Evaluate 9^2.

≈ 254.5 — Use a calculator.

Check $254.5 \approx 243$ ✔

The area is about 254.5 m^2.

b.

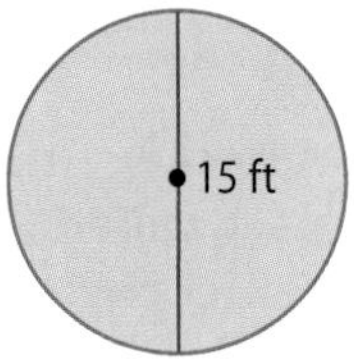

Estimate $A \approx 3 \cdot 8^2$ or 192

$A = \pi r^2$ — Area of a circle

$= \pi \cdot 7.5^2$ — $r = \frac{15}{2}$ or 7.5

$= \pi \cdot 56.25$ — Evaluate 7.5^2.

≈ 176.7 — Use a calculator.

Check $176.7 \approx 192$ ✔

The area is about 176.7 ft^2.

Got It? Do these problems to find out.

1a.

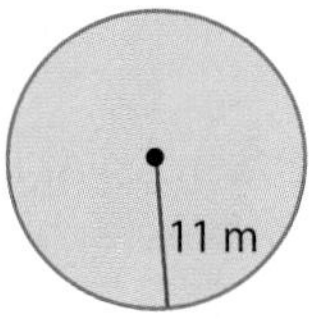

1b.

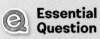

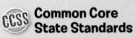

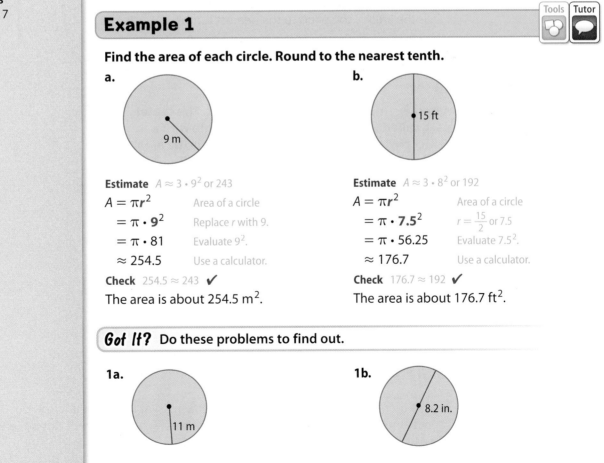

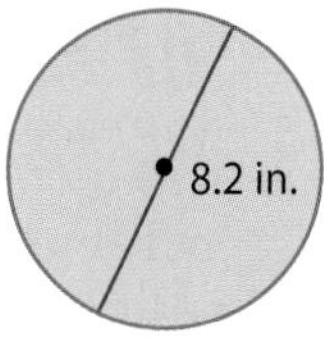

Use Area of Circles to Solve Problems

You can calculate the area of a circle to help you solve real-world problems.

Example 2

Tutor

Estimation

To estimate the area of a circle, multiply the square of the radius by 3.

A wireless fence transmitter at the back door of a house allows a dog to roam freely within a radius of 30 feet, as shown. What is the area of the space the dog has to roam?

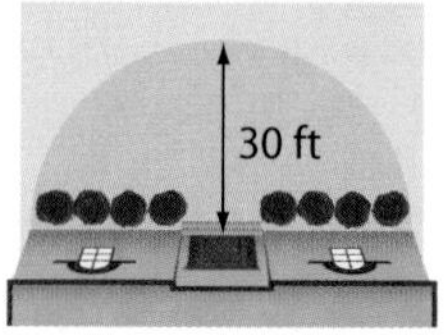

The region is a *semicircle*, or half of a circle. So, the area of a semicircle is half the area of a circle.

$A = \frac{1}{2}\pi r^2$ — Find the area of $\frac{1}{2}$ of the complete circle.

$= \frac{1}{2}\pi(30)^2$ — Replace r with 30.

$= 450\pi$ — Simplify.

≈ 1413.7 — Use a calculator.

So, the dog can roam about 1413.7 square feet.

Got It? Do these problems to find out.

2a. The tabletop shown has a radius of 1.5 feet. What is the area of the tabletop to the nearest tenth?

2b. Mr. Williams is building a patio in the shape of a semicircle. The diameter of the patio is 14 feet. What is the area of the patio to the nearest tenth?

Real World

Example 3

Tutor

A hockey rink is divided into three parts. The center part, called the *neutral zone*, is a rectangle with a center circle, as shown at the right. What is the area of the neutral zone around the center circle?

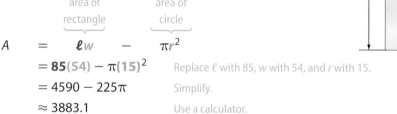

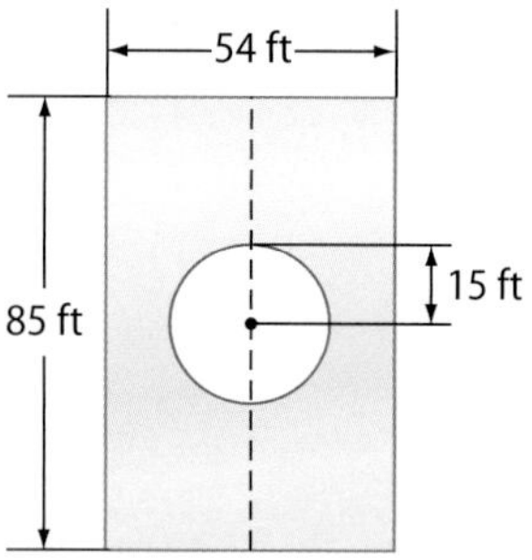

Find the area of the rectangle minus the area of the circle.

$A = \underbrace{\ell w}_{\text{area of rectangle}} - \underbrace{\pi r^2}_{\text{area of circle}}$

$= 85(54) - \pi(15)^2$ — Replace ℓ with 85, w with 54, and r with 15.

$= 4590 - 225\pi$ — Simplify.

≈ 3883.1 — Use a calculator.

So, the area of the neutral zone around the circle is about 3883.1 square feet.

Got It? Do this problem to find out.

3. A *face-off circle* on a hockey rink is 30 feet across. At its center is a red spot 2 feet in diameter. What is the area of the face-off circle that is *not* red? Round to the nearest tenth.

Guided Practice

Check

Find the area of each circle. Round to the nearest tenth. (Example 1)

1.
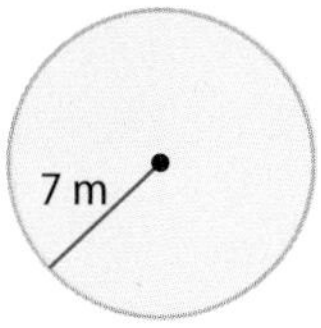

2.
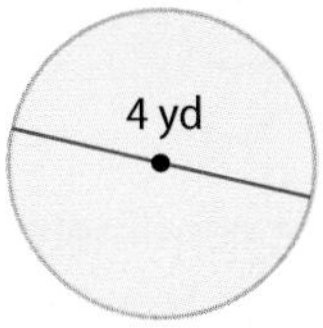

3.
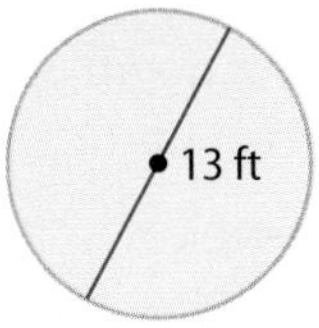

4. radius = 3.6 kilometers

5. diameter = 10.5 centimeters

6. A motion detector at the corner of a building can detect motion outside within a radius of 20 feet as shown. Within what area can it detect motion? Round to the nearest tenth. (Example 2)

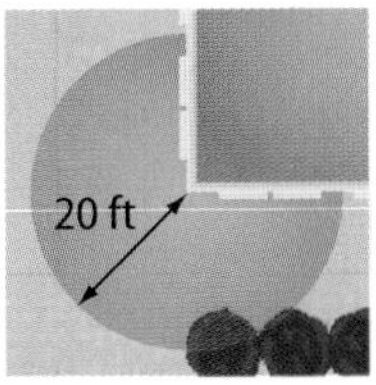

7. May is making doughnuts. First she cuts out a circle of dough with a diameter of 8 centimeters. Then she cuts a hole in the middle with a diameter of 3 centimeters. What is the area of the top of the doughnut? Round to the nearest tenth. (Example 3)

Independent Practice

Go online for Step-by-Step Solutions eHelp

Find the area of each circle. Round to the nearest tenth. (Example 1)

8.
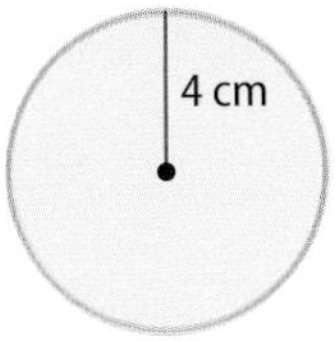

9.
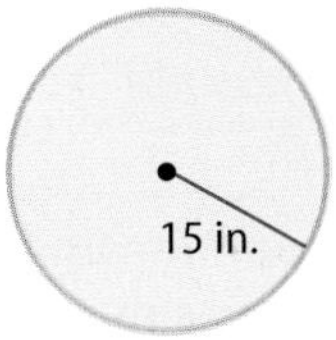

10.
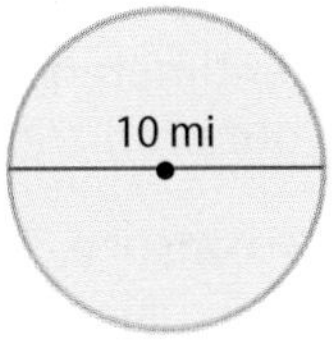

11.
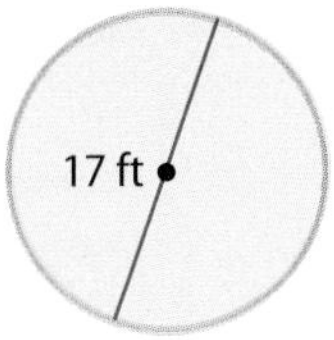

12.
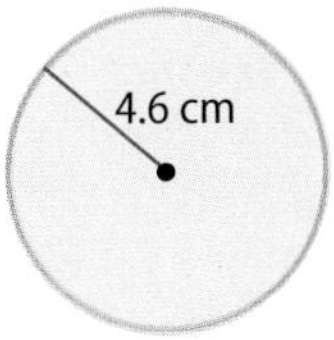

13.
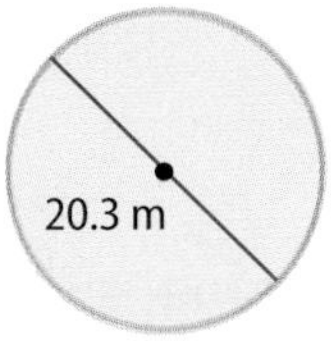

14. radius = 9.6 feet

15. diameter = 24.8 meters

16. diameter = $11\frac{1}{2}$ yards

17. radius = $3\frac{2}{3}$ miles

18. Each shelf of a shelving unit is a quarter circle with a radius of 32 centimeters. What is the area of each shelf? Round to the nearest tenth. (Example 2)

19. Lauren has a sprinkler positioned in her lawn that directs a 12-foot spray in a circular pattern. About how much of the lawn does the sprinkler water if there is a rectangular flower bed 3 feet by 6 feet that is also in the path of the spray? Round to the nearest tenth. (Example 3)

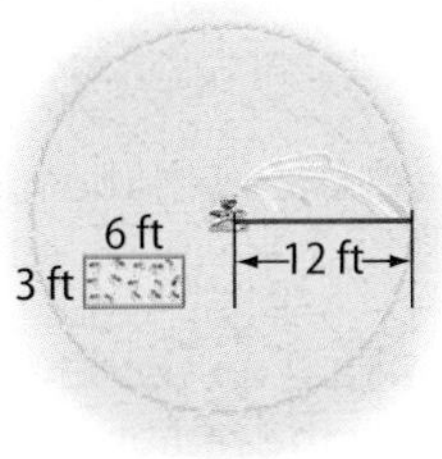

20. What is the area of the CD shown at the right? Round to the nearest tenth. (Example 3)

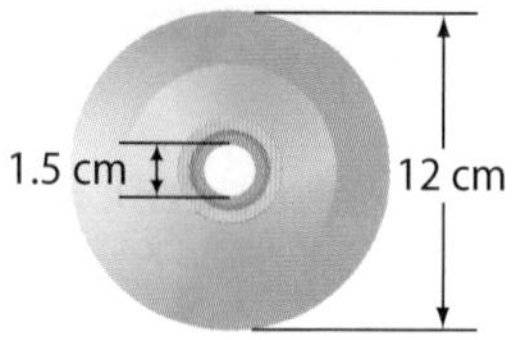

21. **STEM** The trunk of the General Sherman Tree in Sequoia National Park has a circumference of 102.6 feet. If the tree were cut down at the base, what would be the area of the cross section?

22. What is the diameter of a circle if its area is 35.6 square centimeters? Round to the nearest tenth.

23. Find the radius of a circle if its area is 50 square inches. Round to the nearest inch.

Find the distance around and the area of each figure. Round to the nearest tenth.

24. semicircle

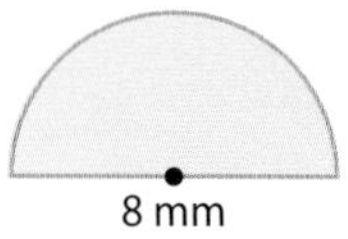

25. semicircle

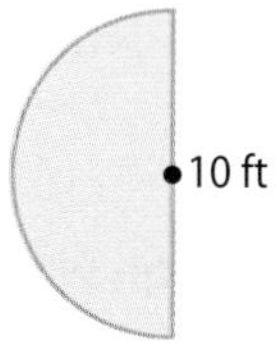

26. quarter circle

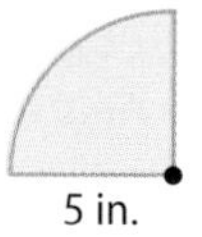

27. **CCSS Multiple Representations** In this problem, you will investigate the area of a circle as the radius changes.

   a. **Table** Make a table like the one at the right. Find the area of each circle to the nearest tenth.

   b. **Analyze** Describe how the area of a circle changes when the radius is doubled.

   c. **Logic** Predict the area of a circle that has a radius of 96 centimeters. Explain your reasoning. Then verify your prediction by finding the area.

Radius (cm)	Area (cm^2)
3	■
6	■
12	■
24	■
48	■

28. **CCSS Identify Structure** The circular radio signal from transmitter A has three times the radius of the circular signal from transmitter B. How many times greater is the area of the signal from transmitter A than from transmitter B? Explain your reasoning.

H.O.T. Problems Higher Order Thinking

29. **CCSS Model with Mathematics** Draw and label a circle that has an area between 800 square centimeters and 820 square centimeters. Label the length of the radius and state the area of the circle to the nearest tenth.

30. **CCSS Construct an Argument** Describe the difference between the circumference and area of a circle and explain how the formulas for circumference and area of a circle are related.

31. **CCSS Persevere with Problems** The radius of circle *B* is 2.5 times the radius of circle *A*. If the area of circle *A* is 8 square yards, what is the area of circle *B*?

32. **CCSS Be Precise** If the measures of the area and circumference of a circle have the same numerical values, what is the radius of the circle? Explain.

33. **Building on the Essential Question** Describe how you can find the area of a circle given the radius, diameter, or circumference.

Standardized Test Practice

34. Find the area of a circle with a diameter of 22 millimeters. Round to the nearest tenth.

A 380.1 mm^2
B 319.5 mm^2
C 189.9 mm^2
D 69.1 mm^2

35. A sprinkler is set to cover the area shown. Find the area of the grass being watered if the sprinkler reaches a distance of 10 feet.

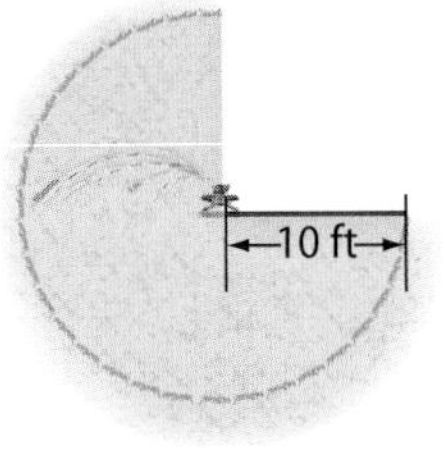

F 47.1 ft^2 **H** 235.6 ft^2
G 157.1 ft^2 **J** 314.2 ft^2

36. The Blackwells have a circular pool with a radius of 10 feet. They want to install a 3–foot sidewalk around the pool. What will be the area of the walkway?

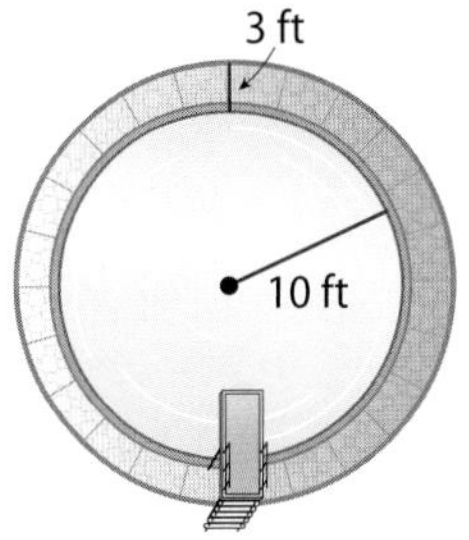

A 216.8 ft^2 **C** 314.2 ft^2
B 285.9 ft^2 **D** 442.2 ft^2

37. **Short Response** The area of a circle is 327.6 square centimeters.

a. Write an algebraic equation in terms of *A* that could be used to find the radius of the circle.

b. Find the radius to the nearest tenth.

Common Core Review

Find the circumference of each circle. Round to the nearest tenth. 7.G.4

38. radius: 8 in. **39.** radius: 12.5 ft **40.** diameter: 21 cm

Find the area of each figure. 6.G.1

41.

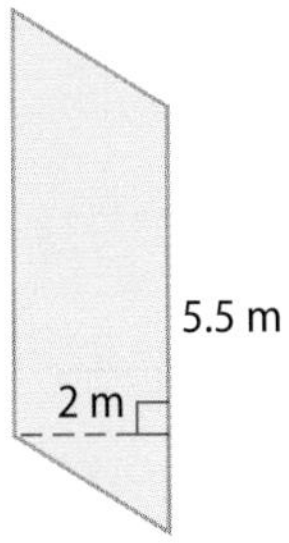

42.

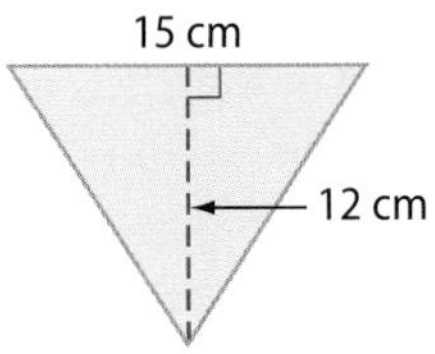

43.

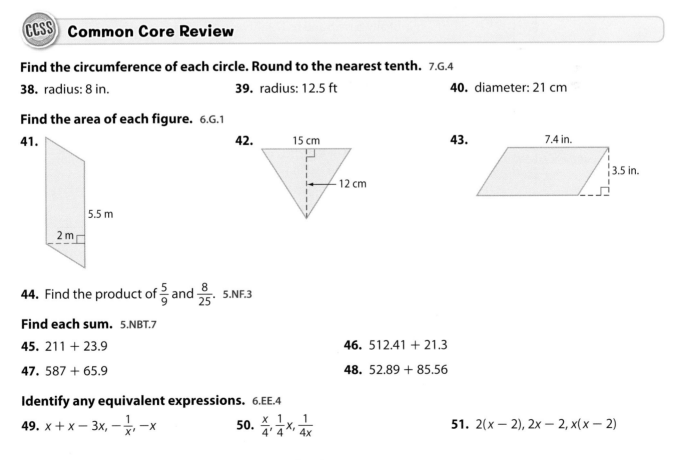

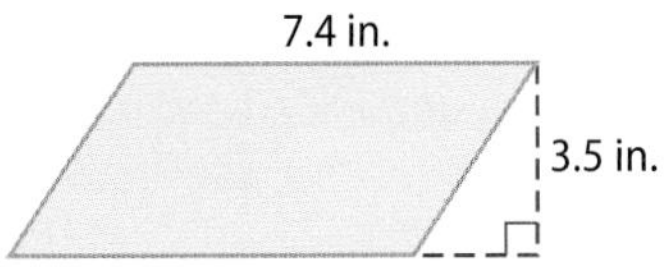

44. Find the product of $\frac{5}{9}$ and $\frac{8}{25}$. 5.NF.3

Find each sum. 5.NBT.7

45. 211 + 23.9 **46.** 512.41 + 21.3

47. 587 + 65.9 **48.** 52.89 + 85.56

Identify any equivalent expressions. 6.EE.4

49. $x + x - 3x, -\frac{1}{x}, -x$ **50.** $\frac{x}{4}, \frac{1}{4}x, \frac{1}{4x}$ **51.** $2(x - 2), 2x - 2, x(x - 2)$

Lesson 12-3

Area of Composite Figures

ISG Interactive Study Guide

See pages 275–276 for:
- Getting Started
- Real-World Link
- Notes

Essential Question

How are two-dimensional figures used to solve problems involving three-dimensional figures?

CCSS Common Core State Standards

Content Standards
7.G.6

Mathematical Practices
1, 2, 3, 4

Vocabulary

composite figure

What You'll Learn

- Find the area of composite figures.
- Solve problems involving the area of composite figures.

Real-World Link

Pools Above-ground and in-ground swimming pools come in a variety of sizes and shapes. Sometimes the shape is made up of different polygons.

Key Concept Area Formulas

Shape	Words	Formula
Parallelogram	The area *A* of a parallelogram is the product of any base *b* and the height *h*.	$A = bh$
Triangle	The area *A* of a triangle is half the product of any base *b* and its height *h*.	$A = \frac{1}{2}bh$
Trapezoid	The area *A* of a trapezoid is half the product of the height *h* and the sum of the bases, b_1 and b_2.	$A = \frac{1}{2}h(b_1 + b_2)$
Circle	The area *A* of a circle is equal to π times the square of the radius *r*.	$A = \pi r^2$

A **composite figure** is made up of two or more shapes. To find the area of a composite figure, decompose the figure into shapes with areas you know how to find. Then find the sum of those areas.

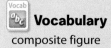

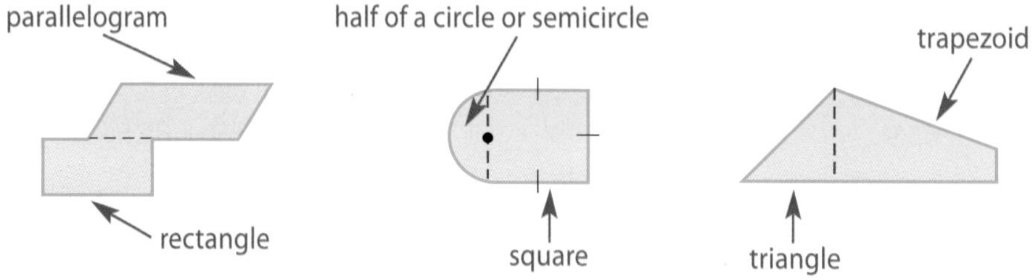

Example 1

Tools Tutor

Find the area of the composite figure.

Separate the figure into a rectangle and a triangle. Find the sum of the areas of the figures.

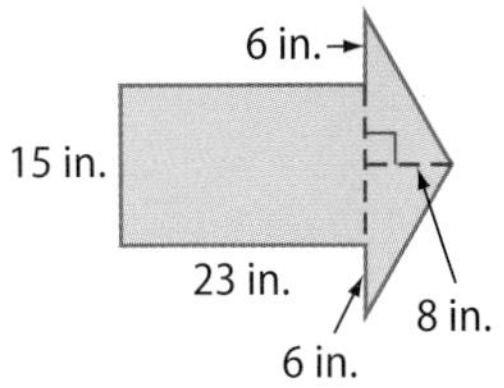

$A = \ell w$ — Area of a rectangle

$= 15 \cdot 23$ — $\ell = 15$ and $w = 23$

$= 345$ — Simplify.

$A = \frac{1}{2}bh$ — Area of a triangle

$= \frac{1}{2} \cdot 27 \cdot 8$ — $b = 6 + 15 + 6$ or 27 and $h = 8$

$= 108$ — Simplify.

The area of the composite figure is 345 + 108 or 453 square inches.

Solve a Simpler Problem

Sometimes it is necessary to break a problem into parts. You can use colored pencils to shade each separate figure that makes up the composite figure a different color. Find the area of each part, then combine the solutions of the parts.

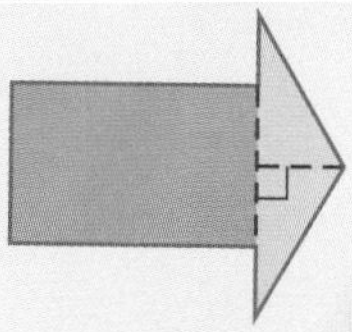

Got It? Do this problem to find out.

1. Find the area of the composite figure. Round to the nearest tenth if necessary.

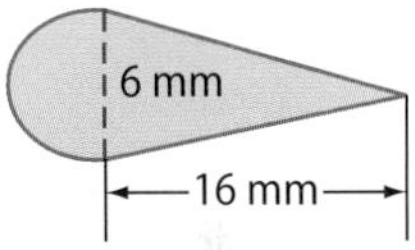

Solve Problems Involving Area

Often the first step in a multi-step problem is to find the area of a composite figure.

Real World Example 2

Tutor

A ceiling tile covers 3 square feet. How many tiles are needed to cover the octagonal ceiling shown at the right?

First, find the area of the ceiling by separating the figure into a rectangle and two trapezoids. Then, find the number of ceiling tiles.

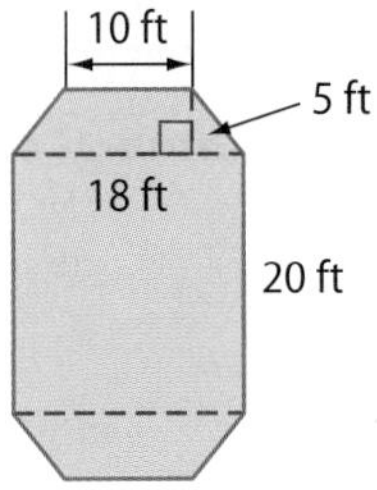

$A = \frac{1}{2}h(b_1 + b_2)$ — Area of a trapezoid

$= \frac{1}{2}(5)(10 + 18)$ — $h = 5$, $b_1 = 10$, $b_2 = 18$

$= 70$ — Simplify.

The area of both trapezoids is $70 \cdot 2$ or 140 square feet.

$A = \ell w$ — Area of a rectangle

$= 18 \cdot 20$ — $\ell = 18$, $w = 20$

$= 360$ — Simplify.

The total area of the ceiling is 360 + 140 or 500 square feet.

Each ceiling tile covers 3 square feet, so the total number of tiles needed is $500 \div 3 \approx 166.7$, or about 167 tiles.

Estimate Area

One way to estimate the area of a composite figure is to draw a rectangle around the figure, as shown below. The area of the figure should be slightly less than the area of the rectangle.

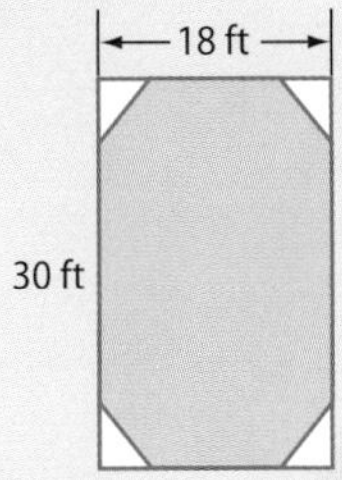

Got It? Do this problem to find out.

2. The L-shaped counter shown is to be replaced with a new countertop that costs \$52 per square foot. What will be the cost of the new countertop?

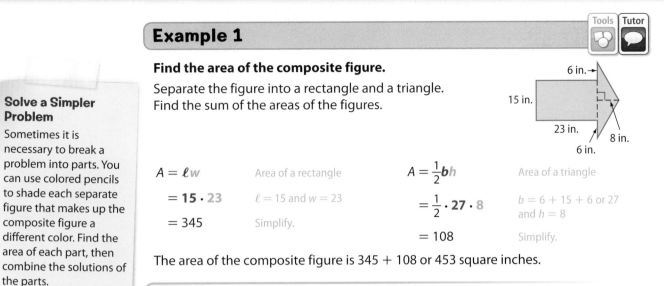

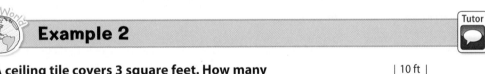

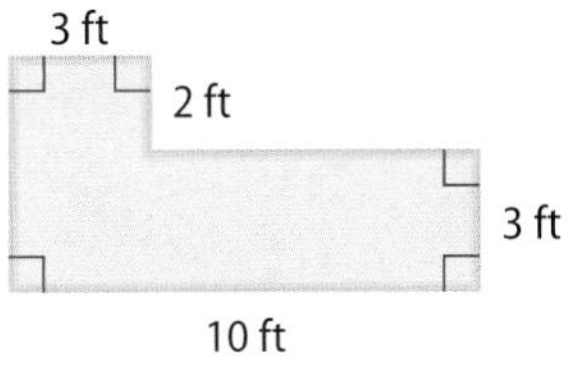

Guided Practice

Check

Find the area of each figure. Round to the nearest tenth if necessary. (Example 1)

1.

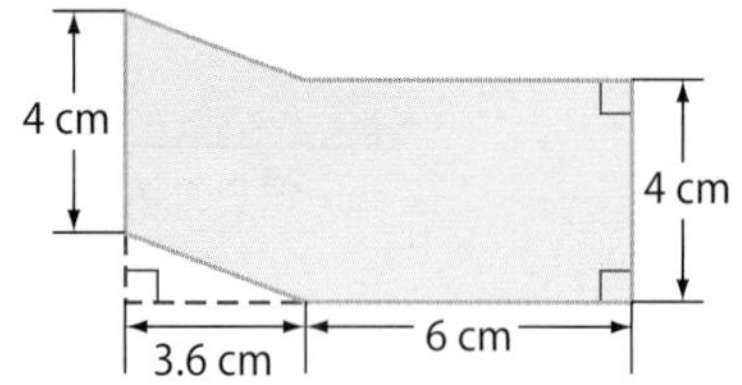

2.

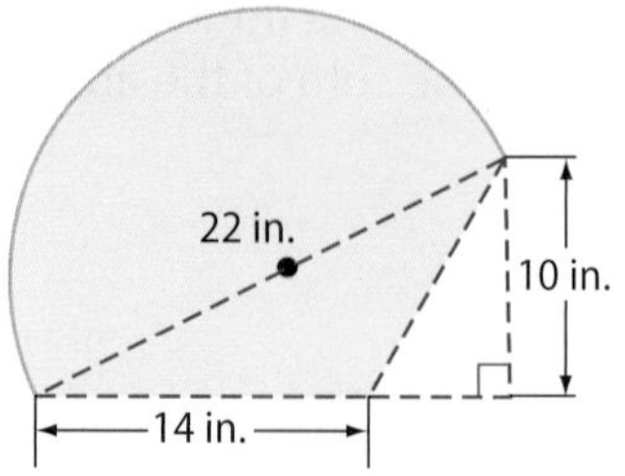

3.

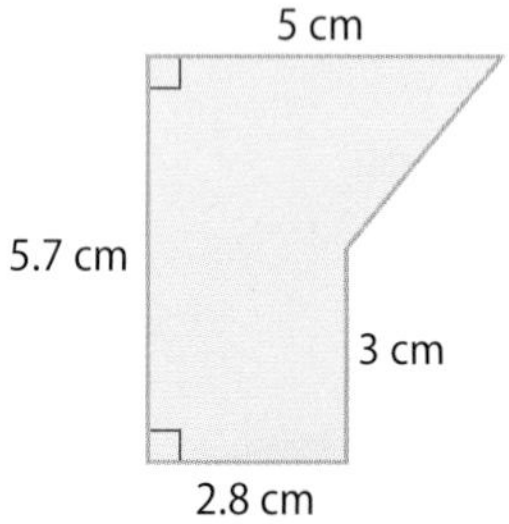

4.

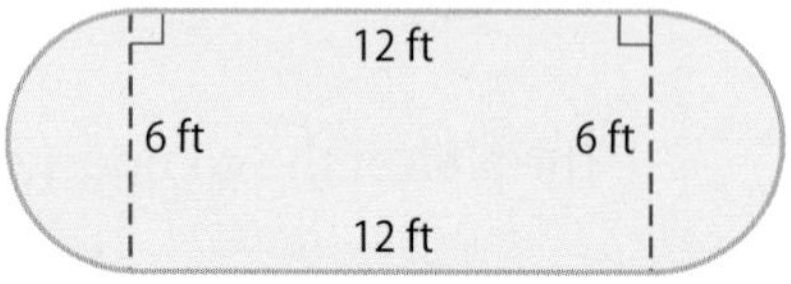

5. The Jamesons hired a landscaper to brick their walkway. One case of bricks costs $25 and covers 6 square feet. (Example 2)

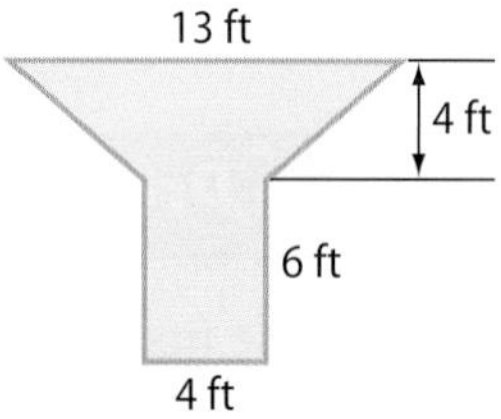

a. What is the area of the walkway?

b. How many cases are needed to cover the walkway with bricks? What would be the cost?

Independent Practice

eHelp

Go online for Step-by-Step Solutions

Find the area of each figure. Round to the nearest tenth if necessary. (Example 1)

6.

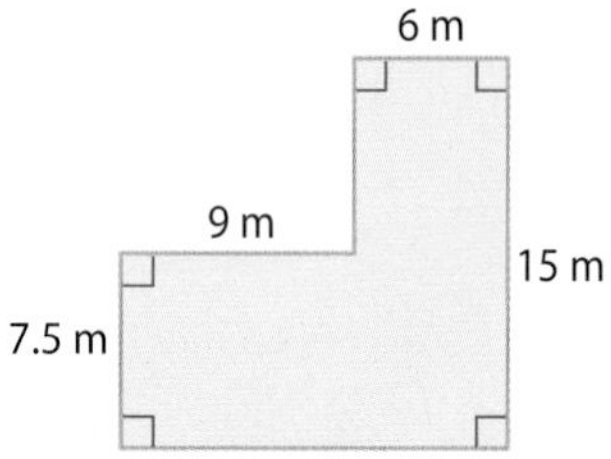

7

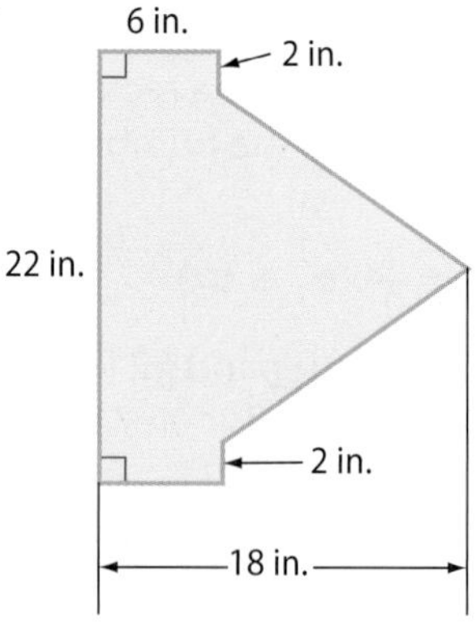

8.

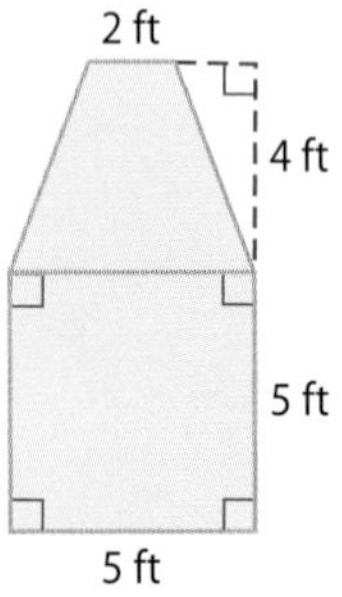

9.

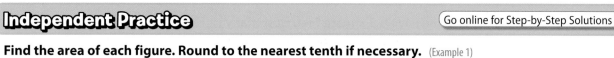

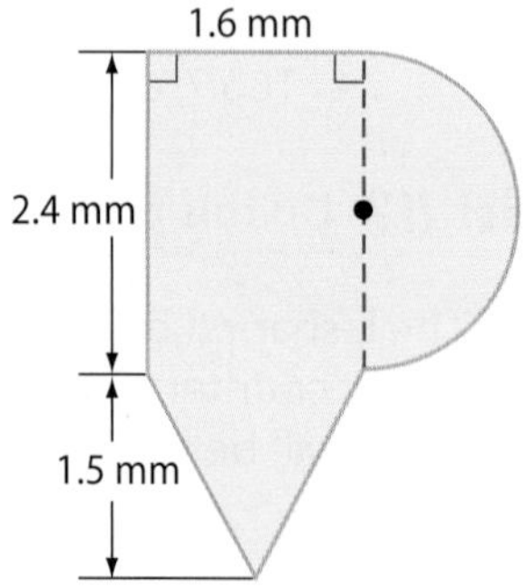

Find the area of each figure. Round to the nearest tenth if necessary.

10.

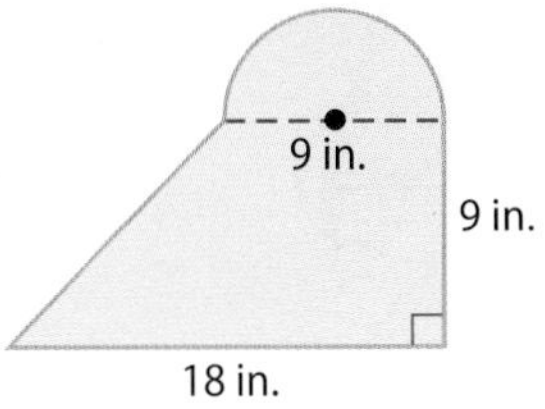

11.

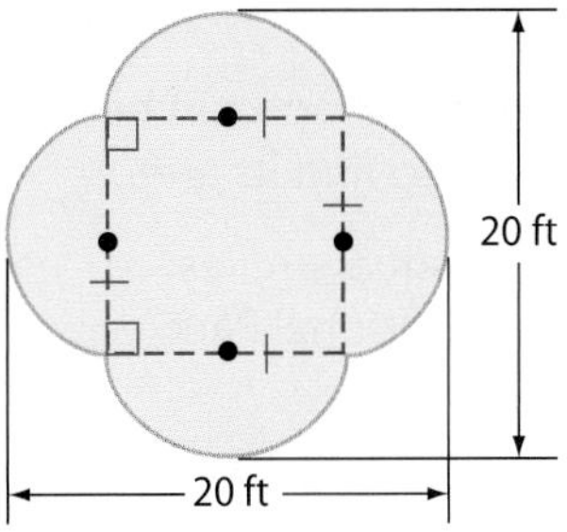

12. Refer to the swimming pool shown at the right. (Example 2)

a. What is the area of the pool's floor?

b. A pool cover costs \$1.70 per square foot. How much would a cover for the pool cost?

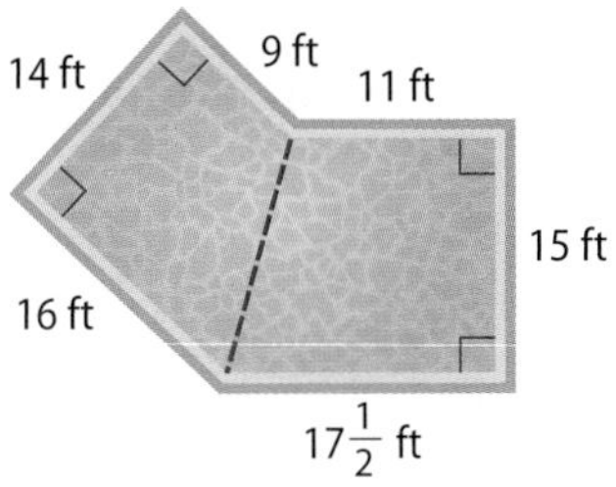

13. **Financial Literacy** Mr. Reyes wants to carpet his family room. (Example 2)

a. What is the area of the space to be carpeted?

b. If carpet costs \$2.25 per square foot, how much would it cost to carpet Mr. Reyes' family room if there is no leftover carpet?

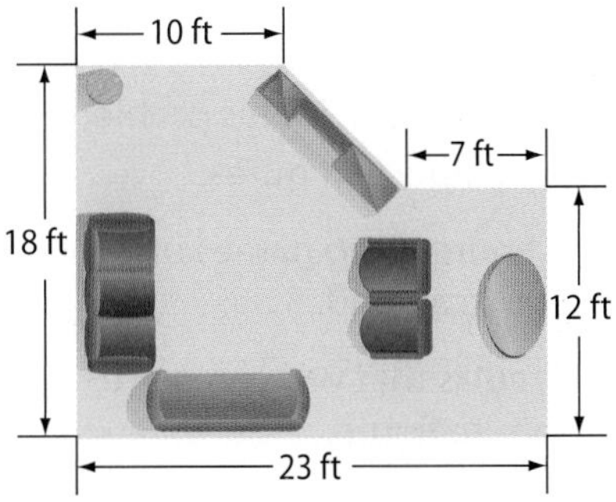

14. What is the area, to the nearest tenth, of a figure that is formed using a rectangle 10 feet long and 7 feet wide and a semicircle with a diameter of 7 feet?

15. A figure is formed using a semicircle and a triangle that has a base of 8 inches and a height of 12 inches. Find the area of each figure to the nearest tenth.

a. The diameter of the semicircle equals the base of the triangle.

b. The diameter of the semicircle equals the height of the triangle.

16. A rectangular room is 15 feet long and 12 feet wide. It has a semicircle on each 12-foot side. What is the area of the room? Round to the nearest tenth.

17 Find the area of the part of the shuffleboard court shown.

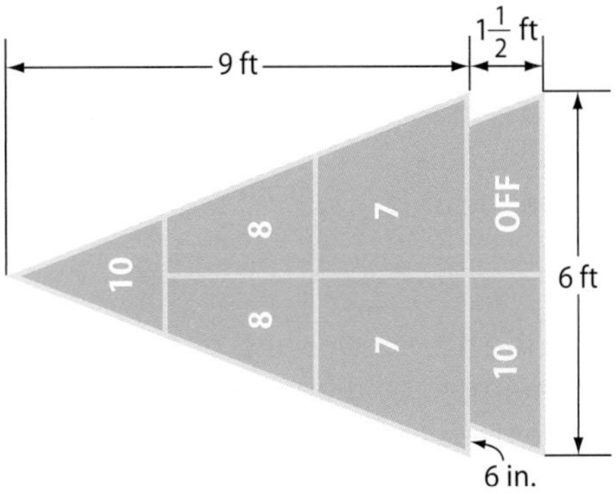

18. A diagram of the Denali National Park Wilderness area in Alaska is shown at the right.

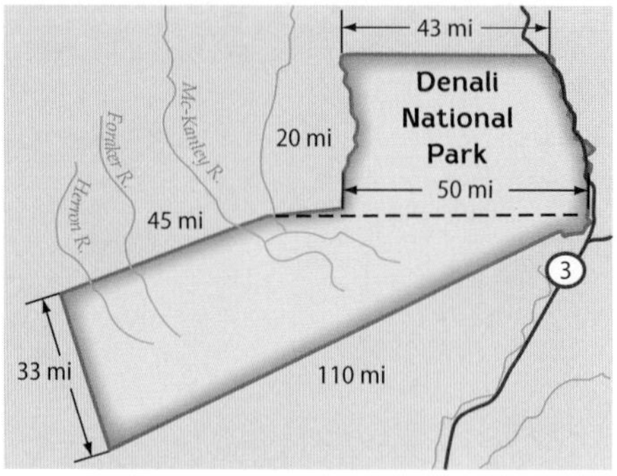

  a. Describe how the irregular shape can be separated into polygons to find its area.

  b. Estimate the number of square miles contained in the Denali National Park Wilderness area.

  c. Estimate the perimeter of the Wilderness area.

Find the area of each shaded region. Round to the nearest tenth, if necessary.

19.

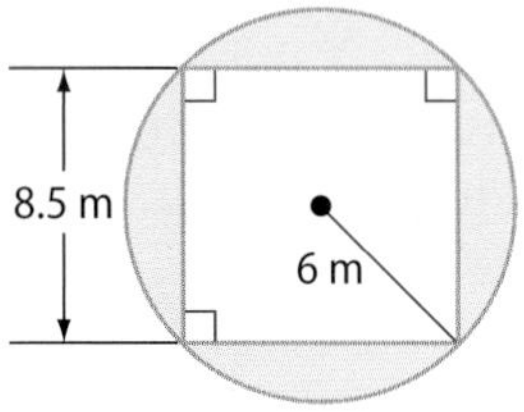

20.

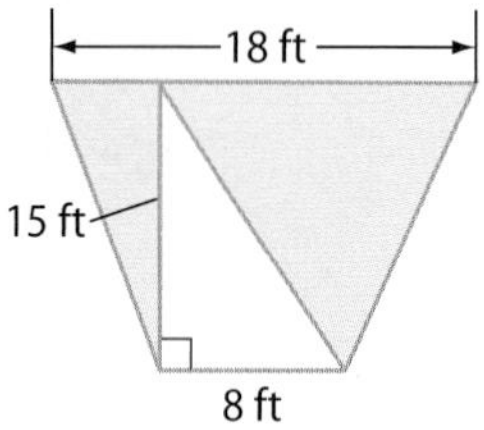

21

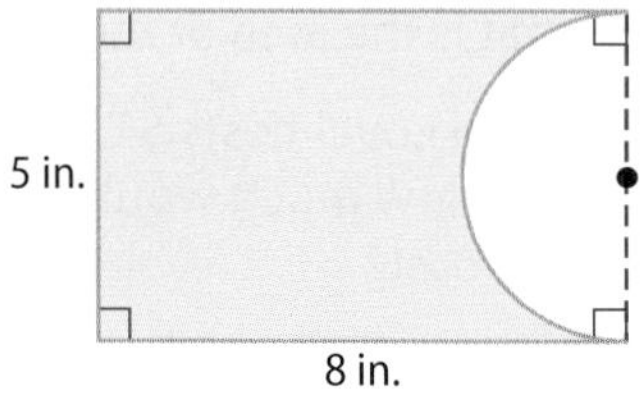

22. **CCSS Multiple Representations** In this problem, you will investigate areas of composite figures.

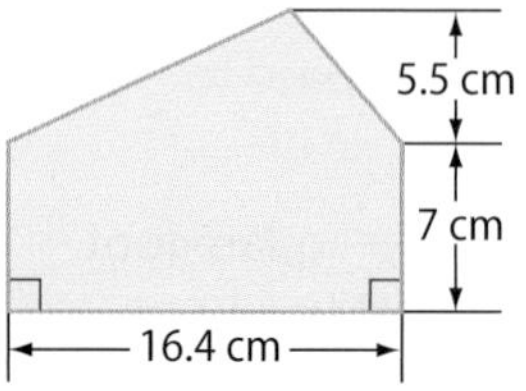

  a. **Words** Describe the figure using names of geometric figures whose areas you can find.

  b. **Symbols** List the formulas that you can use to find the area of the composite figure.

  c. **Analyze** Make a conjecture about how the area of the composite figure changes if each dimension given is doubled. Then test your conjecture by doubling the dimensions and finding the area.

H.O.T. Problems Higher Order Thinking

23. **CCSS Model with Mathematics** Describe real-life composite figures whose areas can be found by separating the figures into geometric figures that have area formulas.

24. **CCSS Persevere with Problems** Reena created the flower at the right by placing semicircles around a regular pentagon. Using the measurements shown, find the area of the flower. Round to the nearest tenth.

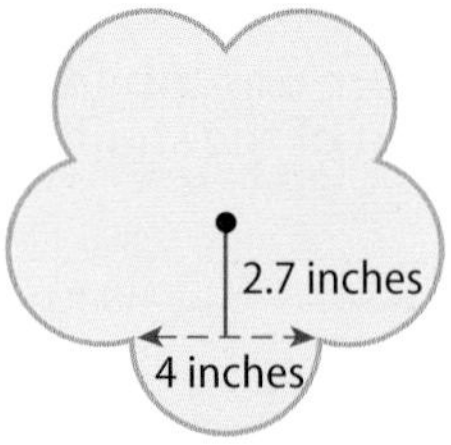

25. **CCSS Reason Abstractly** Suppose a playground is in the shape of a composite figure that has a curved side that is not a semicircle. Describe how you could estimate the area of the playground.

26. **Building on the Essential Question** Describe how circles and polygons help you find the area of a composite figure. Give an example by drawing and labeling a composite figure. Describe the figures that can be used to find the area and then find the area.

Standardized Test Practice

27. Short Response In the diagram, a patio that is 4 feet wide surrounds a swimming pool. What is the area of the patio in square feet? Round to the nearest tenth.

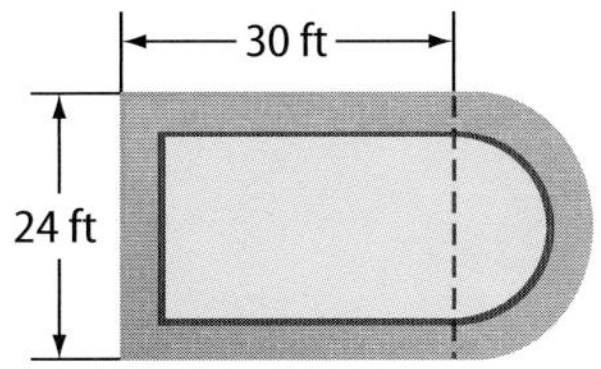

28. Find the area of the shaded region. Use 3.14 for π. Round to the nearest hundredth.

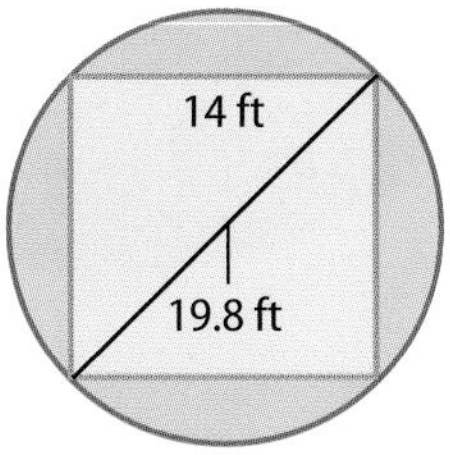

A 392.04 ft^2 **C** 111.75 ft^2
B 503.75 ft^2 **D** 258.17 ft^2

29. The Lin family is buying a cover for their swimming pool shown below. The cover costs \$3.19 per square foot. How much will the cover cost?

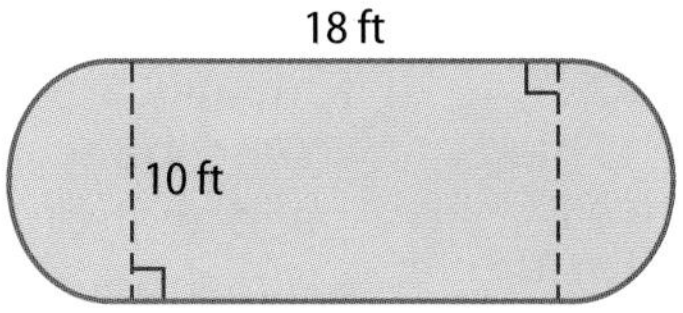

F \$219.27 **H** \$699.47
G \$258.54 **J** \$824.74

30. Find the difference between the areas of the two bedrooms (not including the closet).

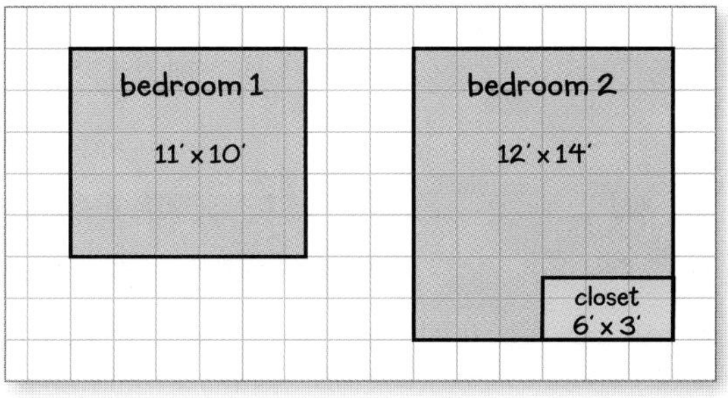

A 27 ft^2 **C** 58 ft^2
B 40 ft^2 **D** 168 ft^2

CCSS Common Core Review

Find the circumference and area of each circle. Round to the nearest tenth. 7.G.4

31. radius 6 centimeters
32. diameter 14 inches
33. diameter $9\frac{1}{2}$ feet
34. radius 9 feet
35. radius 20 inches
36. diameter 8 millimeters

Find the area of each figure described. 6.G.1

37. triangle: base, 9 inches; height, 6 inches

38. trapezoid: height, 3 centimeters; bases, 4 centimeters, 8 centimeters

39. Jeremy can run $3\frac{1}{3}$ miles in 25 minutes. How long would it take him to run 5 miles at this same rate? 7.RP.2c

Find the percent of change. 7.RP.3

40. The cost of a monthly pass for students on the Baxterville Bus Line increased from \$30 to \$33. What was the percent increase in the cost?

41. Membership in the school orchestra fell from 120 members to 102 members this year. What was the percent decrease in membership?

Lesson 12-4

Three-Dimensional Figures

ISG Interactive Study Guide

See pages 277–278 for:
- Getting Started
- Vocabulary Start-Up
- Notes

Essential Question

How are two-dimensional figures used to solve problems involving three-dimensional figures?

CCSS Common Core State Standards

Content Standards
7.G.3

Mathematical Practices
1, 3, 4

Vocabulary

plane
solids
polyhedron
edge
vertex
face
skew lines
prism
bases
pyramid
cylinder
cone
cross section

What You'll Learn

- Identify three-dimensional figures.
- Describe and draw vertical, horizontal, and angled cross sections of three-dimensional figures.

Real-World Link

Puzzles Some three-dimensional jigsaw puzzles contain over 3000 foam pieces! These pieces are used to create everything from historical landmarks to everyday objects that are probably found in your home.

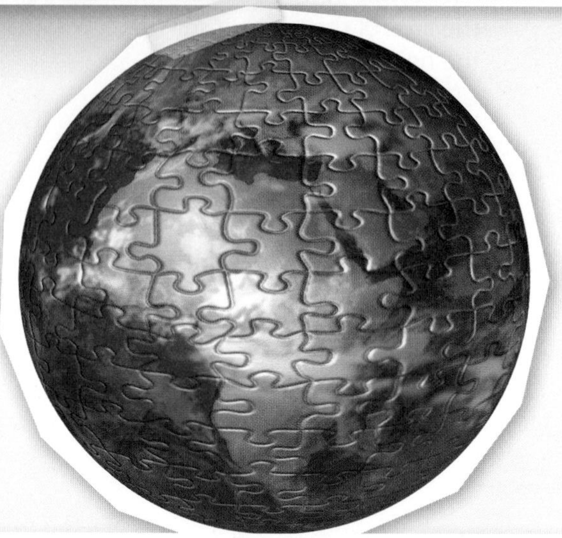

Identify Three-Dimensional Figures

A two-dimensional figure has two dimensions—length and width. A **plane** is a two-dimensional flat surface that extends in all directions. There are different ways that planes may be related in space.

Intersect in a Line

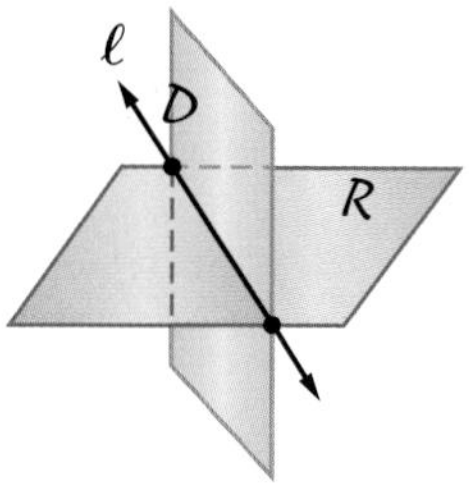

Intersect in a Point

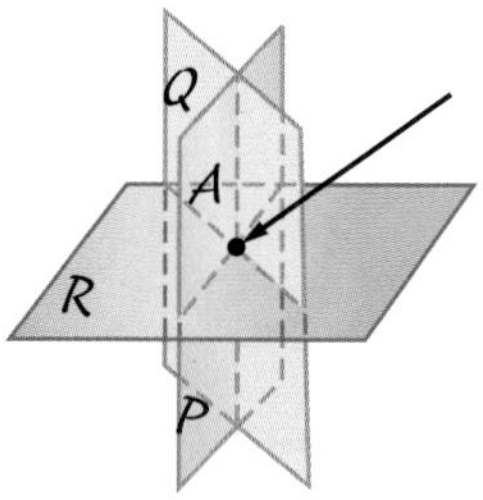

No Intersection

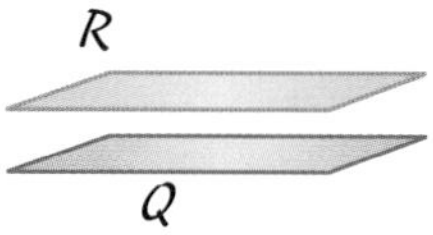

These are called parallel planes.

A three-dimensional figure has three dimensions—length, width, and depth (or height). Intersecting planes can form three-dimensional figures or **solids**. A **polyhedron** is a three-dimensional figure with faces that are polygons.

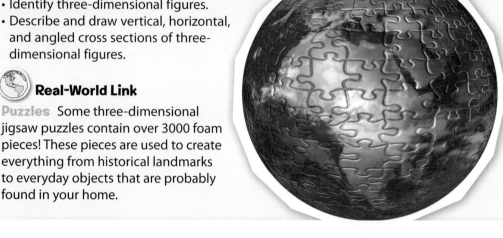

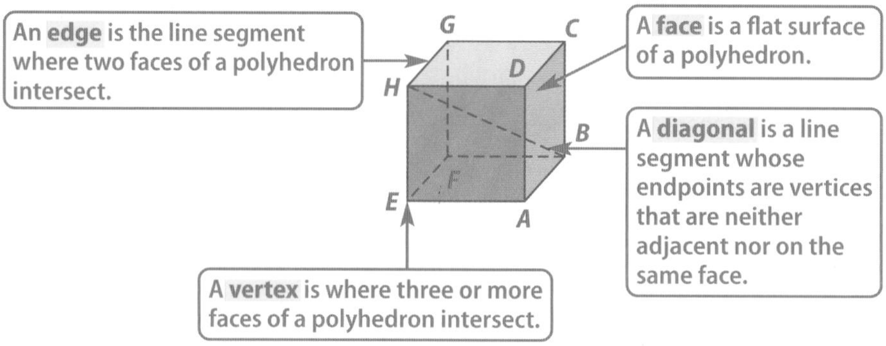

An edge is the line segment where two faces of a polyhedron intersect.
A face is a flat surface of a polyhedron.
A diagonal is a line segment whose endpoints are vertices that are neither adjacent nor on the same face.
A vertex is where three or more faces of a polyhedron intersect.
G
C
D
H
B
F
E
A

Notice in the figure above, $\overline{GC}$ and $\overline{DA}$ do not intersect. These segments are not parallel because they do not lie in the same plane. Lines that do not intersect and do not lie in the same plane are called **skew lines**.

Watch Out!
In a *rectangular* prism, the bases do not have to be the top and bottom. Any two parallel rectangles are bases. In a *triangular* pyramid, any face is a base.

Key Concept Polyhedrons

Polyhedron	triangular prism	rectangular prism	triangular pyramid	rectangular pyramid
Number of Bases	2	2	1	1
Polygon Base	triangle	rectangle	triangle	rectangle
Figure	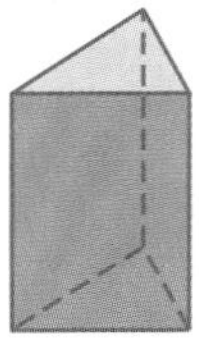	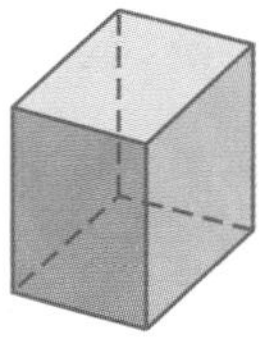	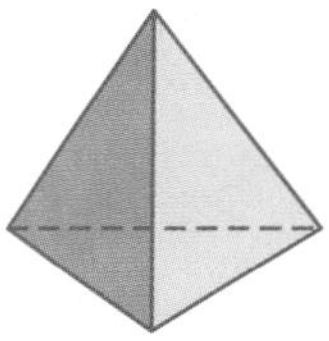	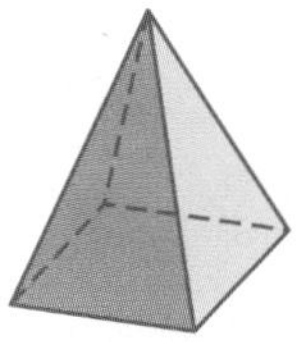

A **prism** is a polyhedron with two parallel, congruent faces called **bases** that are polygons. A **pyramid** is a polyhedron with one base that is a polygon and three or more triangular faces that meet at a common vertex. Prisms and pyramids are named by the shape of their bases.

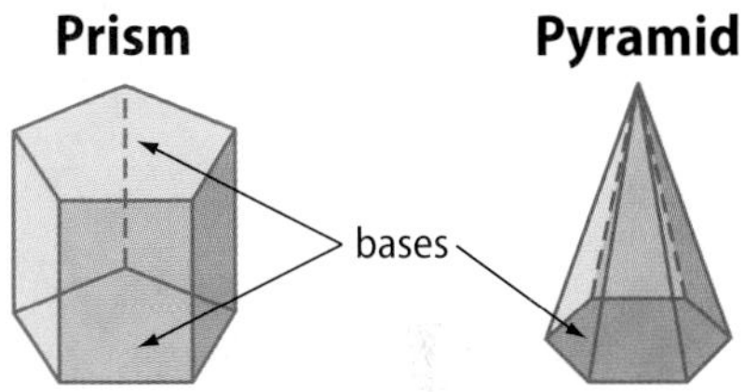

There are solids that are *not* polyhedrons. A **cylinder** is a three-dimensional figure with two parallel congruent circular bases connected by a curved surface. A **cone** is a three-dimensional figure with one circular base connected by a curved surface to a single vertex.

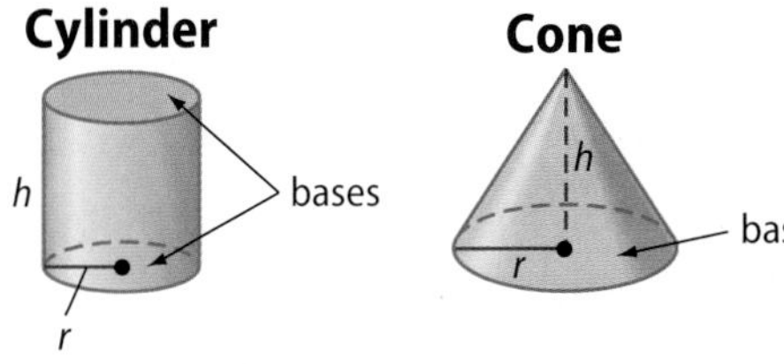

Vocabulary Link
Cone
Everyday Use Lupe had an ice cream *cone* for dessert.
Math Use A solid with one base that is a circle connected to a vertex by a curved side is a *cone*.

Example 1

Tutor

Identify the figure. Name the bases, faces, edges, and vertices. Then identify a pair of skew lines.

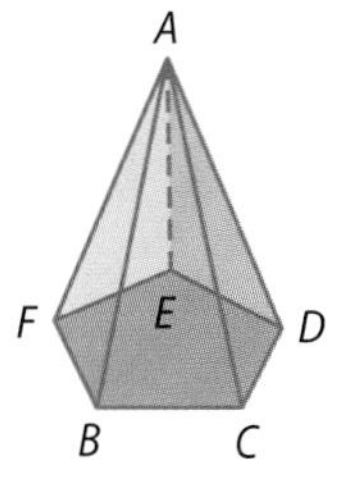

This figure has one pentagonal base, *BCDEF*, so it is a pentagonal pyramid.

faces: *ABF, AFE, AED, ADC, ACB, BCDEF*

edges: $\overline{AB}, \overline{AC}, \overline{AD}, \overline{AE}, \overline{AF}, \overline{BC}, \overline{CD}, \overline{DE}, \overline{EF}, \overline{FB}$

vertices: *A, B, C, D, E, F*

skew lines: $\overline{AB}$ and $\overline{CD}$

Got It? Do these problems to find out.

1a.

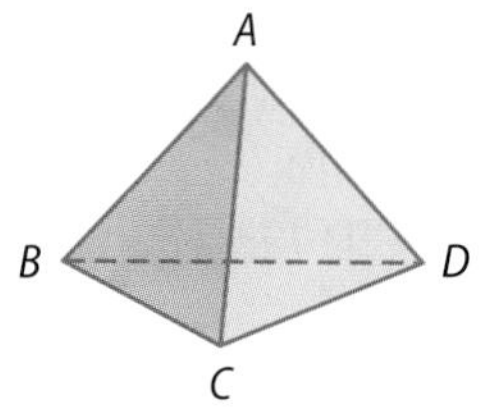

1b.

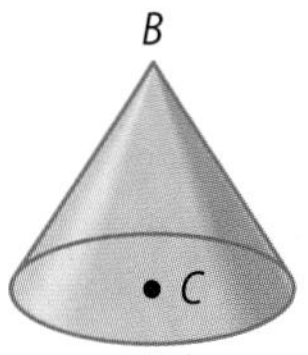

Cross Sections

Interesting shapes can occur when a plane intersects, or slices, a three-dimensional figure. The intersection of the figure and the plane is called a **cross section** of the figure.

Example 2

Tutor

Draw and describe the shape resulting from the following vertical, angled, and horizontal cross sections of a cone.

> **Parabola**
> The shape that results from the angled slice through a cone is called a *parabola*.

Vertical Slice

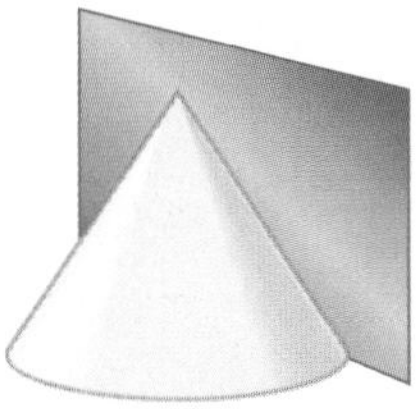

The cross section is a triangle.

Angled Slice

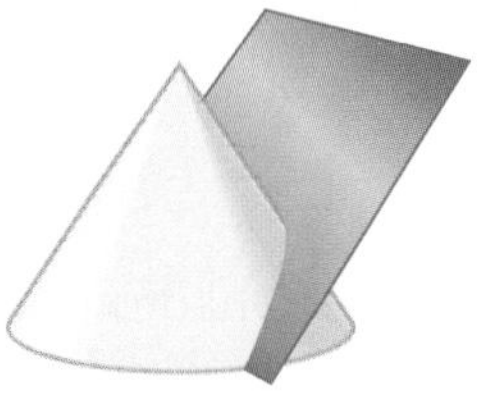

The cross section is a parabola.

Horizontal Slice

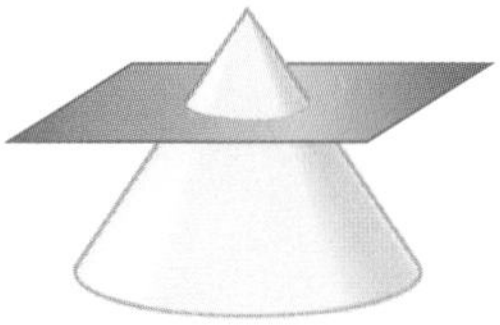

The cross section is a circle.

Got It? Do this problem to find out.

2. Draw and describe the shape resulting from a vertical, angled, and horizontal cross section of a triangular pyramid.

Example 3

Real World

Tutor

A fluorite crystal is shown at the right. Draw the top view and side view. Then draw and describe the shape resulting from a vertical cross section of the figure.

The crystal is a rectangular prism with two square pyramids attached. The vertical cross sections will look similar to the side view.

Top View

Side View

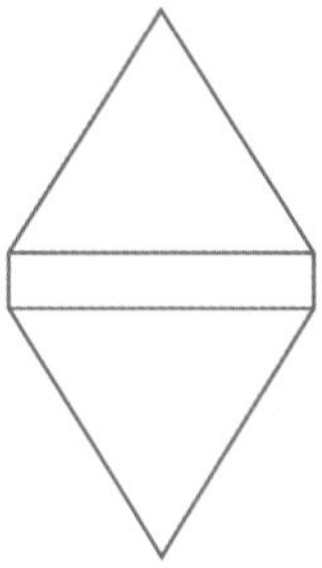

Vertical Cross Section

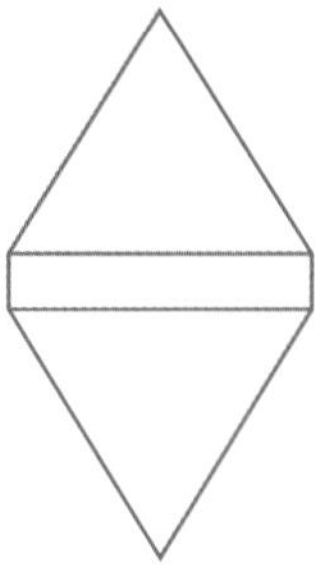

Got It? Do this problem to find out.

3. Lemar and his brother went camping for the weekend. They set up the tent at the right. Draw the top view and side view. Then draw and describe the shape resulting from a vertical cross section of the figure.

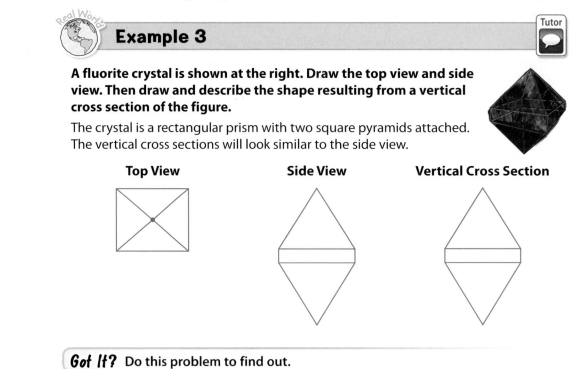

Doug Sherman/Geofile

Guided Practice

Check

Identify each figure. Name the bases, faces, edges, and vertices. Then identify a pair of skew lines. (Example 1)

1.

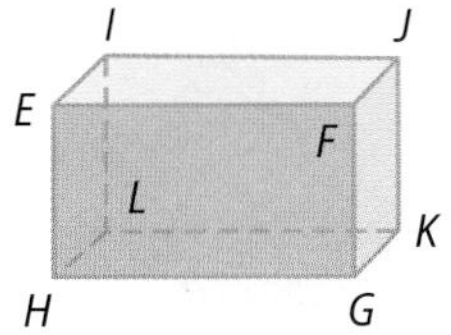

2.

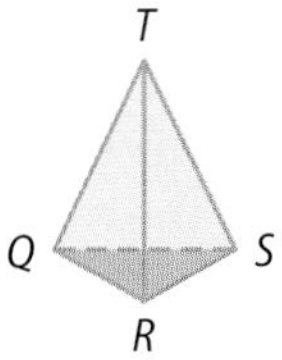

Draw and describe the shape resulting from each cross section. (Example 2)

3.

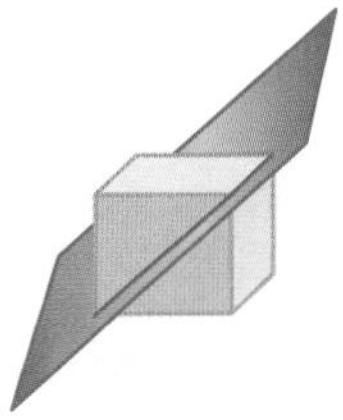

4.

5.

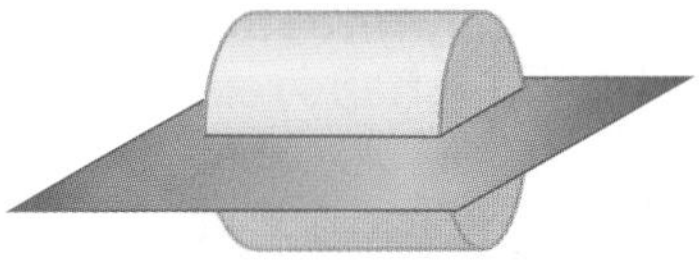

6. A glass paperweight in the shape of a square pyramid is placed on a desk. Draw the top view and side view. Then draw and describe the shape resulting from a vertical cross section of the figure. (Example 3)

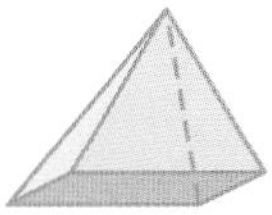

Independent Practice

eHelp Go online for Step-by-Step Solutions

Identify each figure. Name the bases, faces, edges, and vertices. Then identify a pair of skew lines. (Example 1)

7.

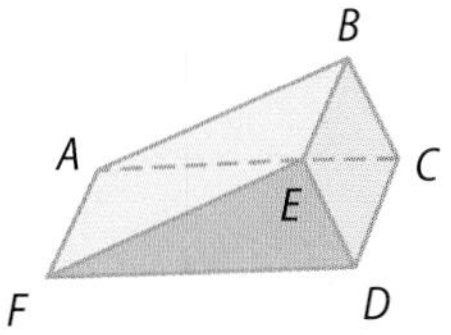

8.

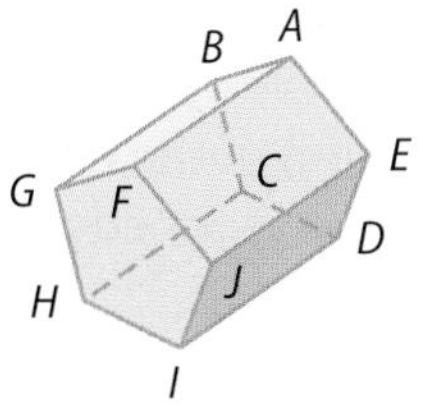

9

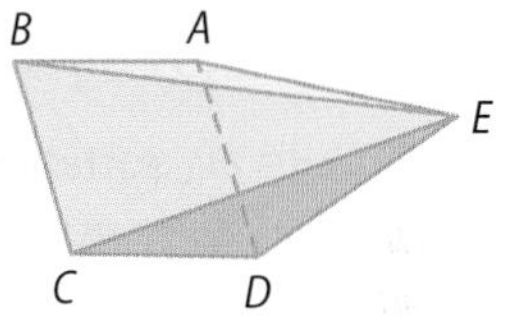

10.

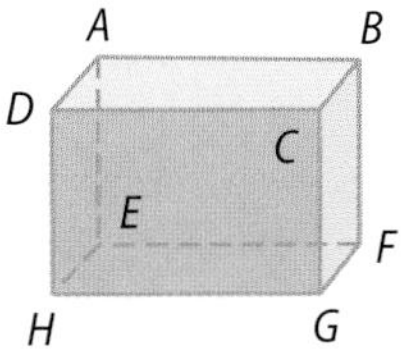

11.

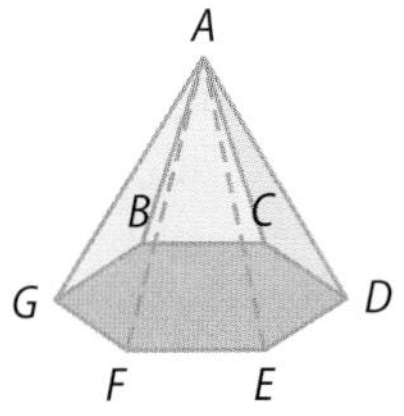

12.

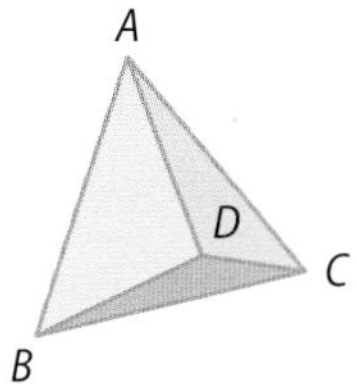

Draw and describe the shape resulting from each cross section. (Example 2)

13.

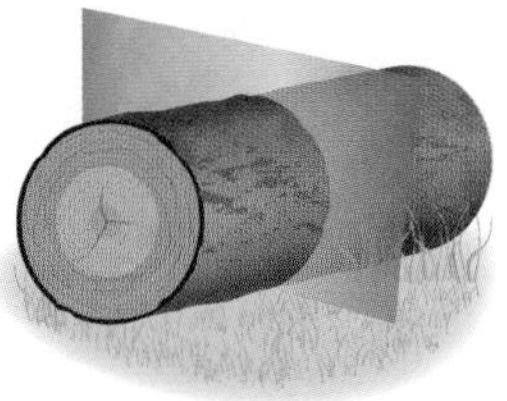

14.

15.

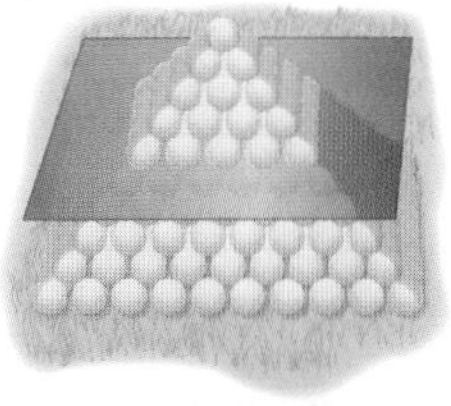

16. For an art project, Jorge is using the box shown at the right. Draw the top view and side view. Then draw and describe the shape resulting from an angled cross section of the figure. (Example 3)

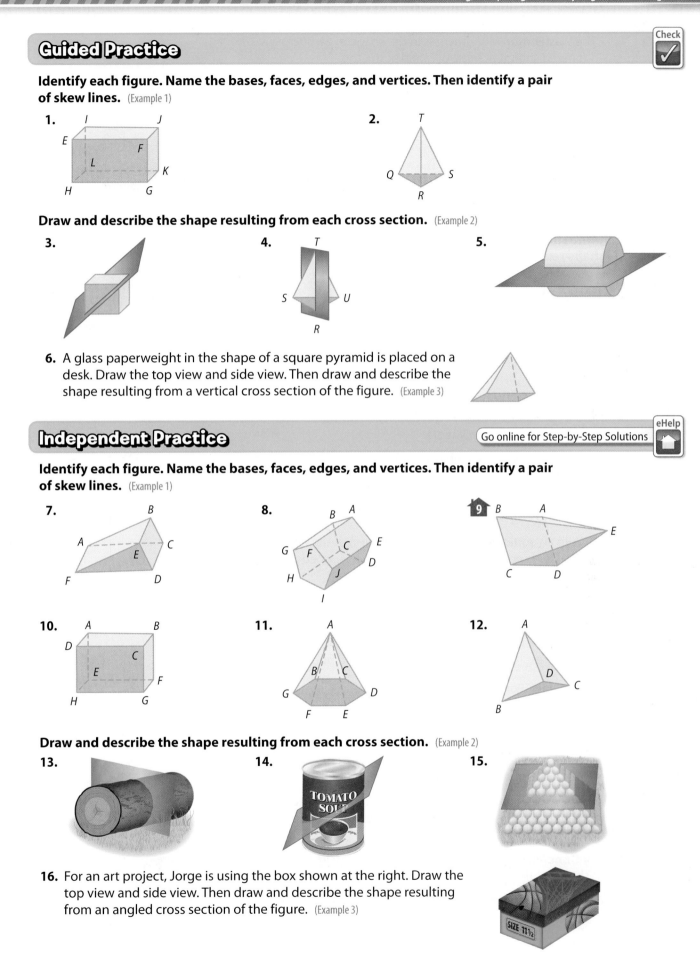

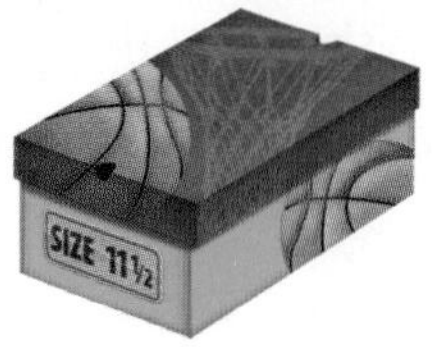

17. **STEM** The Washington Monument is in the shape of an *obelisk*, a four-sided tapered shaft that has a pyramid at the top.

a. Sketch the pyramid-shaped top of the monument and label the vertices.

b. Identify the bases, faces, and edges.

c. Describe the shapes that result from vertical, angled, and horizontal cross sections of the pyramid top of the monument.

d. Draw and describe the shape that would result from an angled cross section through the top and base of the monument.

For each sculpture shown below, draw the top view and side view. Then draw and describe the shape resulting from a vertical cross section of the figure.

18. 

19. 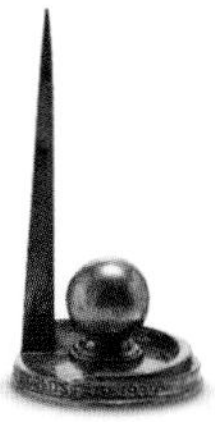

20. 

21 CCSS **Multiple Representations** In this problem, you will investigate Euler's Formula on polyhedra.

Name	Triangular Pyramid	Square Pyramid	Pentagonal Pyramid
Vertices	4	5	■
Faces	■	■	6
Edges	■	8	■

a. Table Draw each figure. Then, copy and complete the table shown.

b. Analyze What do you notice about the number of vertices, faces, and edges?

c. Symbols Write an equation that compares the sum of the number of vertices V and the number of faces F to the number of edges E.

H.O.T. Problems Higher Order Thinking

22. CCSS **Model with Mathematics** Choose a solid object from your home. Draw and describe the shape resulting from a vertical, angled, and horizontal cross section of it.

CCSS **Reason Inductively For Exercises 23–25, determine whether each statement is *always*, *sometimes*, or *never* true.**

23. The bases of a cylinder have different radii.

24. Two planes intersect in a single point.

25. Three planes do not intersect in a point.

26. CCSS **Persevere with Problems** A triangular pyramid has 6 edges. A square pyramid has 8 edges. Write a formula that gives the number of edges e for a pyramid with an n-sided base.

27. CCSS **Justify Conclusions** Explain how you would classify the polyhedron shown at the right.

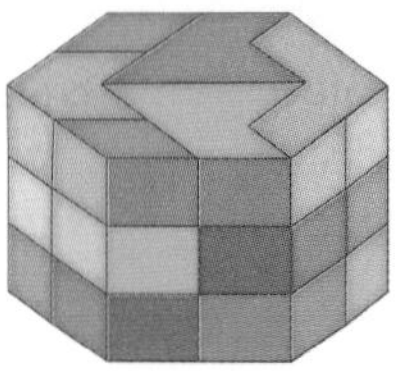

28. **Building on the Essential Question** Are cylinders polyhedrons? Explain.

Standardized Test Practice

29. Which of the following is *not* considered an edge of the triangular prism?

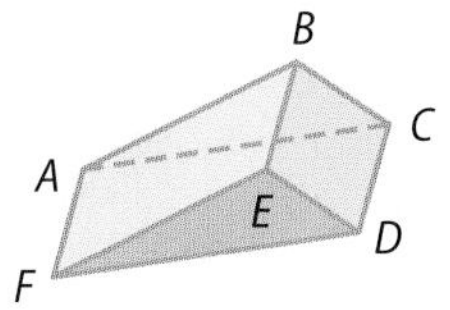

A $\overline{AF}$

B $\overline{BE}$

C $\overline{DE}$

D $\overline{AE}$

30. What three-dimensional figure has one vertex?

F cone

G cylinder

H triangular prism

J triangular pyramid

31. **Short Response** Draw and describe the shape resulting from a vertical cross section of the figure shown.

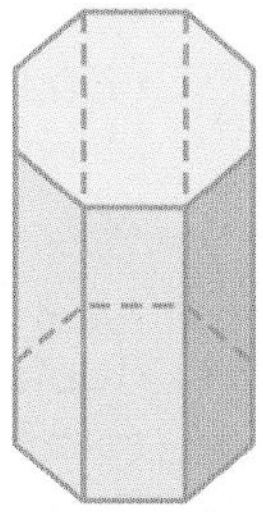

32. Which of the following real-world objects resembles a rectangular prism?

A bowl

B box

C soup can

D stop sign

Common Core Review

Find the area of each figure. Round to the nearest tenth if necessary. 7.G.6

33.

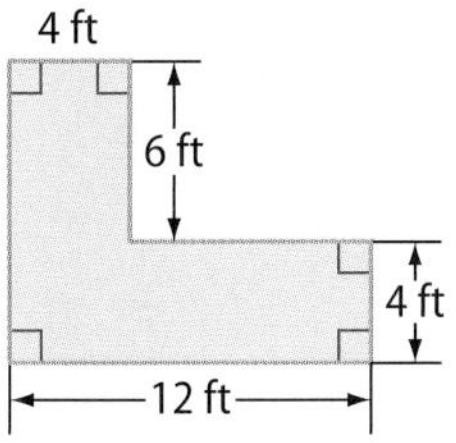

34.

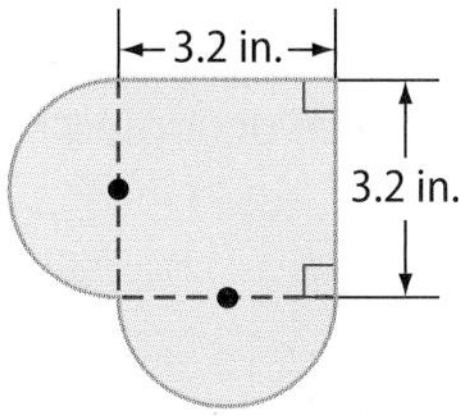

35.

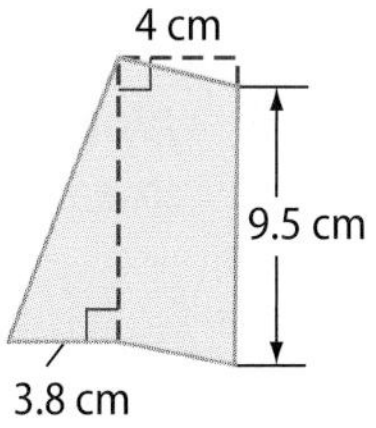

36. During a football game, a marching band can be heard within a radius of 1.7 miles. What is the area in which people can hear the band? Round to the nearest tenth. 7.G.4

Twenty identical cards are numbered from 1 to 20, one number to a card. A card is selected without looking. Find each probability. Express as a fraction and a percent. 7.SP.7

37. $P(9)$

38. $P(\text{even})$

39. $P(25)$

40. $P(\text{greater than } 13)$

41. $P(1, 3, \text{or } 5)$

42. $P(\text{prime})$

43. The experimental probability of getting 3 on a toss of a number cube is $\frac{2}{11}$. If a number other than 3 turned up 45 times, how many times was the cube tossed? 7.SP.7a

Lesson 12-5
Volume of Prisms

ISG Interactive Study Guide

See pages 279–280 for:
- Getting Started
- Real-World Link
- Notes

Essential Question

How are two-dimensional figures used to solve problems involving three-dimensional figures?

Common Core State Standards

Content Standards
7.G.6

Mathematical Practices
1, 3, 4

Vocabulary
volume

What You'll Learn
- Find volumes of prisms.
- Find volumes of composite figures.

Real-World Link

Coolers Do you enjoy camping? Coolers will keep food cold for days in the hot sun. There are many sizes to choose from: 16-quart personal coolers, 100-quart coolers with built-in wheels, and everything in between!

Key Concept: Volume of a Prism

Words The volume V of a prism is the area of the base B times the height h.

Symbols $V = Bh$

Models

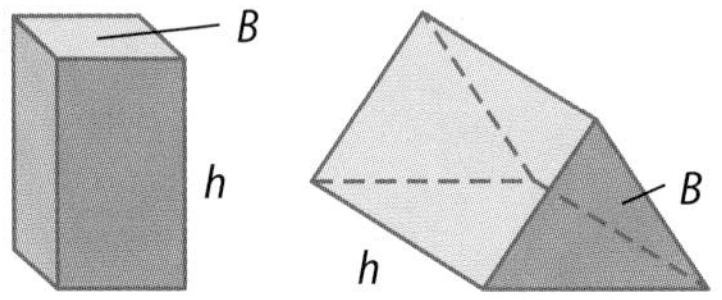

Volume is the measure of space occupied by a three-dimensional region. It is measured in cubic units.

Example 1

Tools Tutor

Find the volume of the rectangular prism.

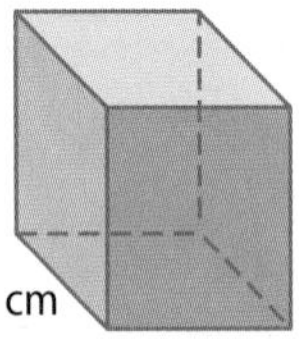

$V = Bh$ — Write the formula for volume of a prism.

$= (\ell w)h$ — The base is a rectangle, so $B = \ell w$.

$= (4.2 \cdot 3) \cdot 5$ or 63 — Replace ℓ with 4.2 cm, w with 3 cm, and h with 5 cm. Then simplify.

The volume is 63 cubic centimeters.

Got It? Do this problem to find out.

1. Find the volume of a rectangular prism with a length of 10 feet, a width of 13 feet, and a height of 21 feet.

Example 2

Tools Tutor

Watch Out! Be sure not to confuse the height of the triangle with the height of the prism.

Find the volume of the triangular prism.

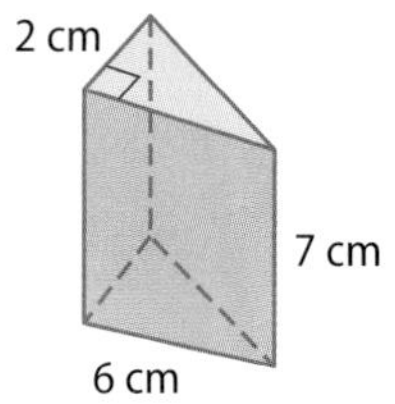

$V = Bh$ — Write the formula for volume of a prism.

$= \left(\frac{1}{2} \cdot 6 \cdot 2\right)h$ — The base is a triangle, so $B = \frac{1}{2}bh$, $b = 6$ and $h = 2$.

$= 6 \cdot 7$ — The height of the prism is 7 cm.

$= 42$ — Simplify.

The volume is 42 cubic centimeters.

Got It? Do these problems to find out.

Find the volume of each triangular prism.

2a.

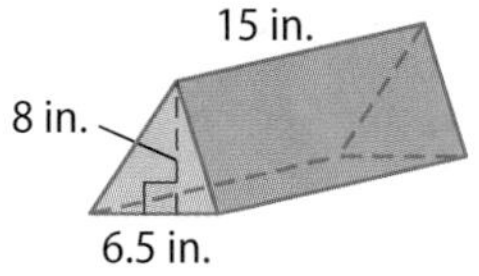

2b.

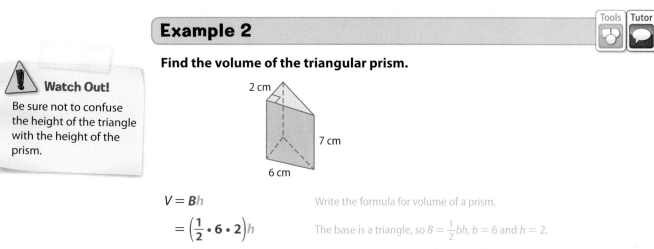

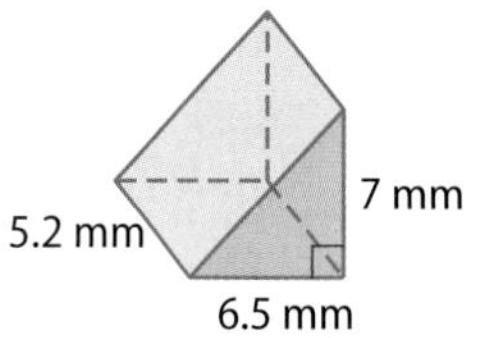

Example 3

Real World Tutor

STEM A room air conditioner can cool a room with a volume of 1600 cubic feet. If a room has a height of 8 feet and a length of 16 feet, what is the maximum width of the room?

Estimate $1600 \div (10 \times 16) = 10$

$V = Bh$ — Write the formula for the volume of a prism.

$V = \ell wh$ — Replace B with ℓw.

$1600 = 16 \cdot w \cdot 8$ — Replace V with 1600, ℓ with 16, and h with 8.

$1600 = 16 \cdot 8 \cdot w$ — Commutative Property

$1600 = 128w$ — Simplify.

$12.5 = w$ — Divide each side by 128.

The maximum width of the room is 12.5 feet.

Check for Reasonableness $12.5 \approx 10$ ✓

Got It? Do this problem to find out.

3. A children's rectangular pool holds 17.5 cubic feet of water. What is the width of the pool if its length is 3.5 feet and its height is 1 foot?

Volumes of Composite Figures

You can find the volume of composite three-dimensional figures by breaking them into smaller pieces.

Example 4

Tutor

Find the volume of the ramp.

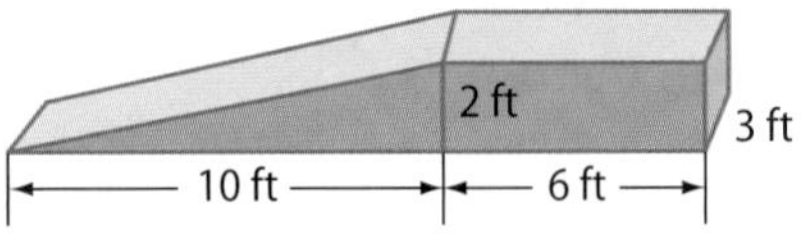

The solid is made up of a triangular prism and a rectangular prism. The volume of the solid is the sum of both volumes.

Triangular Prism	**Rectangular Prism**
$V = Bh$	$V = \ell wh$
$V = \frac{1}{2} \cdot 10 \cdot 2 \cdot 3$	$V = 6 \cdot 3 \cdot 2$
$V = 30$	$V = 36$

The volume of the figure is $30 + 36$ or 66 cubic feet.

Got It? **Do this problem to find out.**

4. Find the volume of the figure at the right.

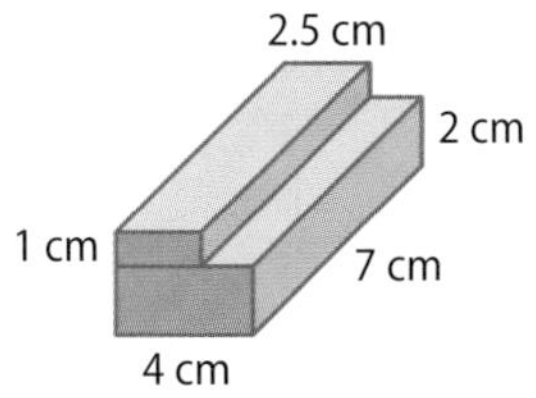

Guided Practice

Check

Find the volume of each figure. (Examples 1 and 2)

1.

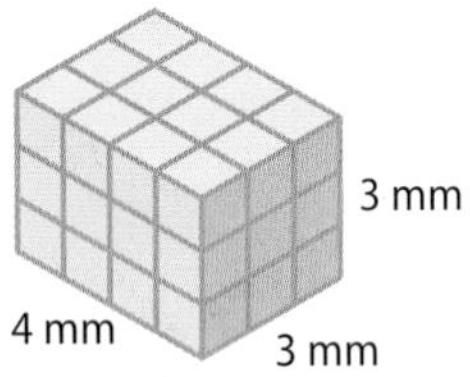

2.

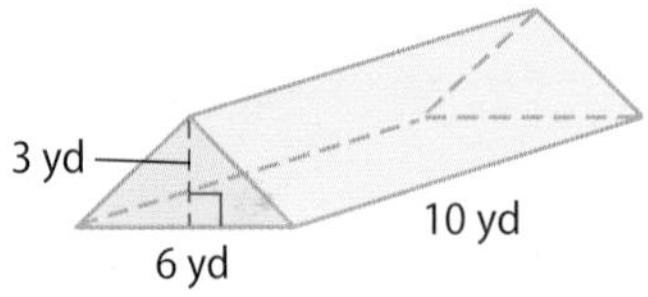

3.

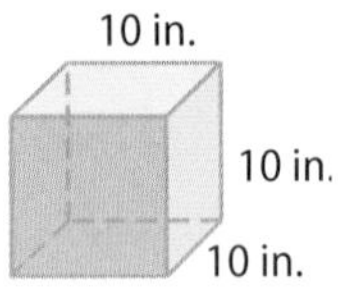

4. A window box has a length of 8.5 inches and a height of 9 inches. If the volume of the box is 2295 cubic inches, what is the width of the box? (Example 3)

5. Find the volume of the figure at the right. (Example 4)

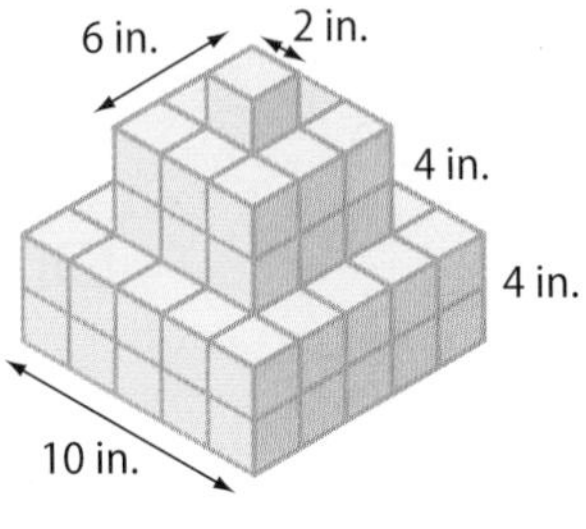

Independent Practice

Go online for Step-by-Step Solutions eHelp

Find the volume of each figure. (Examples 1 and 2)

6.

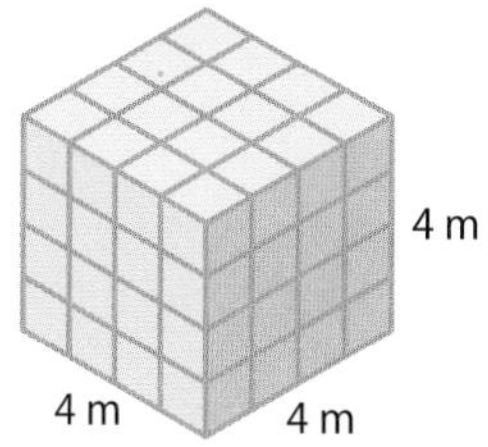

7.

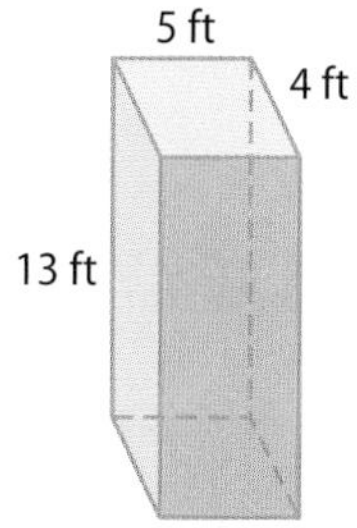

8.

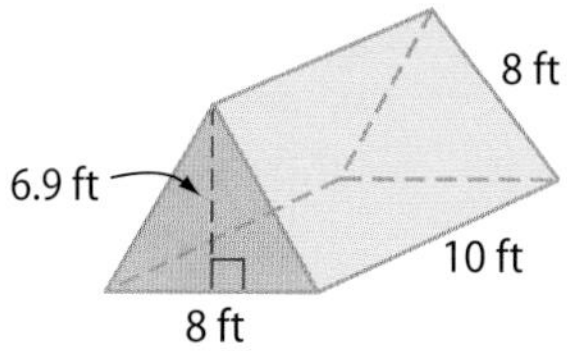

9.

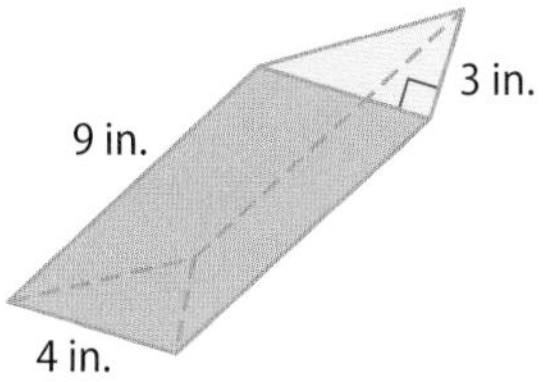

10.

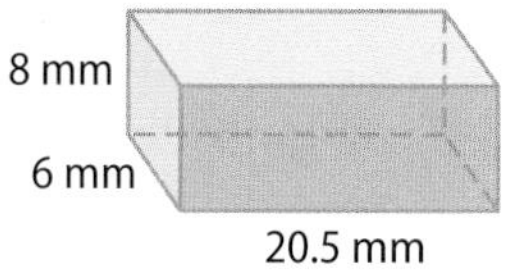

11.

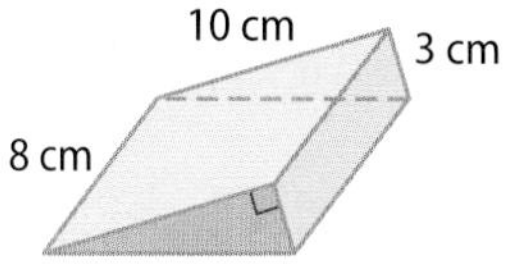

12. triangular prism: base of triangle 6.2 yards, height of triangle 20 yards, height of prism 14 yards

13. rectangular prism: height 4 inches, width $1\frac{1}{2}$ inches, length $\frac{1}{4}$ inch

14. Find the length of a rectangular prism with a width of 4 feet, a height of 6 feet, and a volume of 84 cubic feet. (Example 3)

15. Find the height of a triangular prism with a base length of 10 yards, a base height of 20 yards, and a volume of 600 cubic yards. (Example 3)

16. Josh and his mom are building the dog house shown below. Find the volume of the dog house. (Example 4)

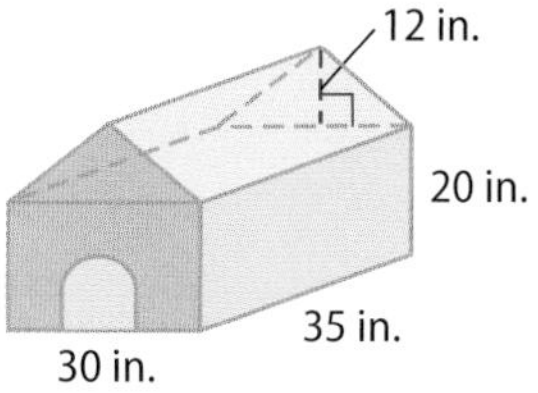

17. Morgan is building a model of some steps with 6-inch foam blocks. What is the total volume of the blocks? (Example 4)

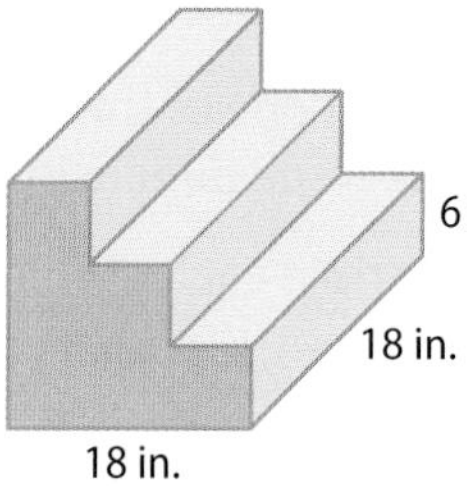

18. Jill is mailing a candle that is in the shape of a triangular prism as shown. She put the candle in a rectangular box that measures 3 inches by 5 inches by 7 inches and places foam pieces around the candle. Find the volume of the foam pieces needed to fill the space between the candle and the box.

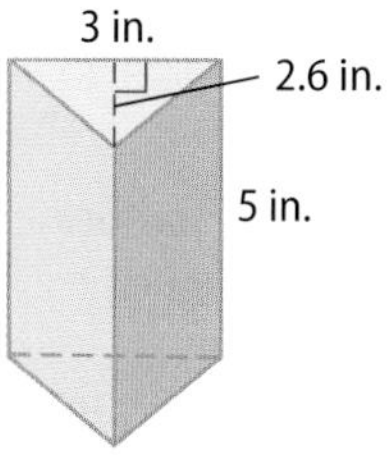

19. The height of a triangular prism is 8 feet, and it has a volume of 200 cubic feet. If the base has a length of 5 feet, what is the height of the base?

20. Ben wants to buy enough potting soil to fill a window box that is 42 inches long, 8 inches wide, and 6 inches high. If one bag of potting soil contains 576 cubic inches, how many bags should he buy?

Find the volume of each figure.

21.

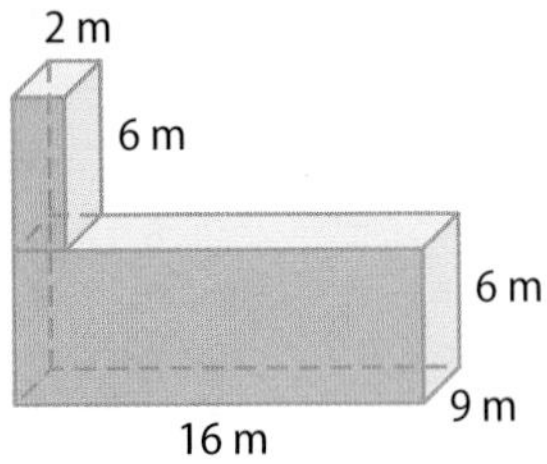

22.

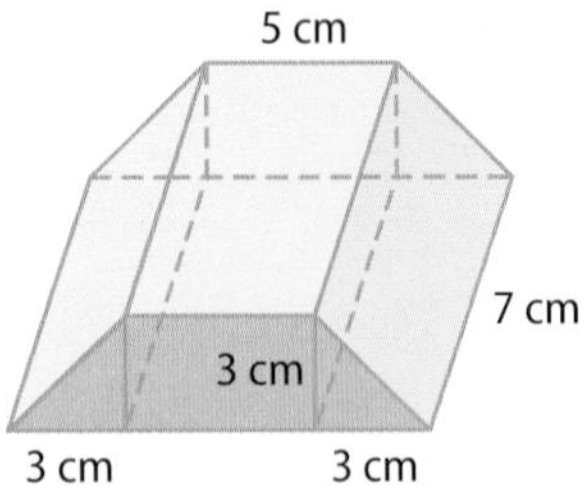

23.

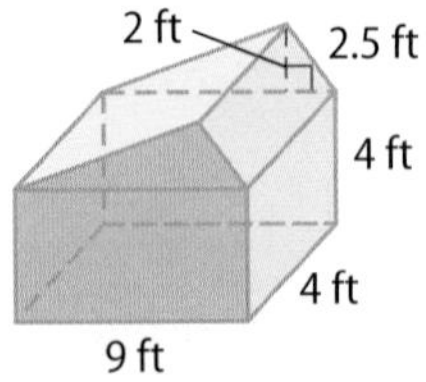

24.

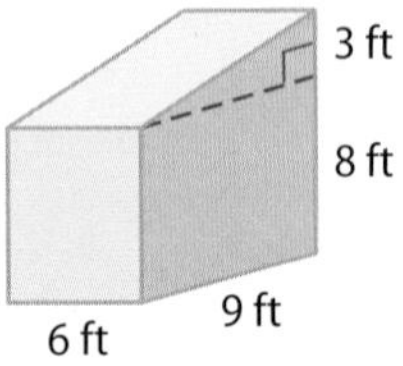

25 A chocolate bar is in the shape of a trapezoidal prism as shown at the right. Find the volume of the chocolate bar.

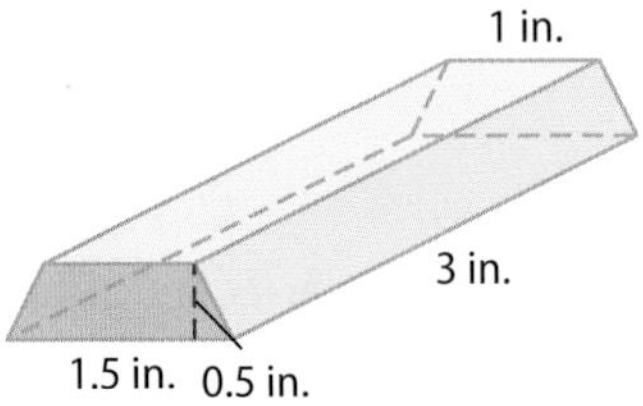

26. Financial Literacy A movie theater sells different sizes of popcorn as shown in the table. If the containers are rectangular prisms, find the ratio of cost to the volume of each bag of popcorn. Which size of popcorn is the best buy?

Popcorn Sizes Available

Size	Length (in.)	Width (in.)	Height (in.)	Price ($)
Small	5	4	8	4.50
Medium	7	5	10	5.75
Large	10	6	12	6.50

27. CCSS **Make a Conjecture** Refer to the figure at the right.

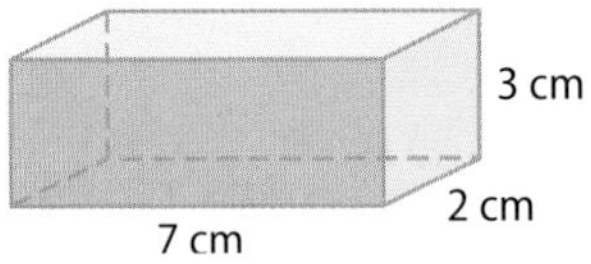

a. What is the volume of the figure?

b. How does the volume change if one of the dimensions is doubled? two dimensions? three dimensions? Explain.

c. Repeat the above steps and triple each dimension. What do you notice?

d. Without calculating, say how the volume of the prism changes if each of the dimensions is multiplied by 6.

H.O.T. Problems Higher Order Thinking

28. CCSS **Justify Conclusions** Without calculating, compare the volumes of the prisms shown. Explain your reasoning to a classmate.

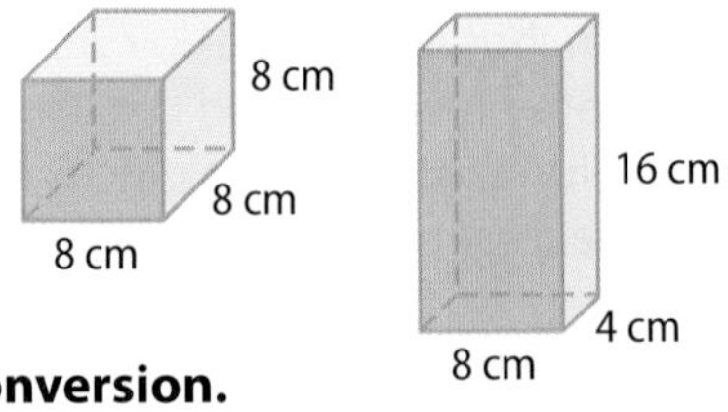

29. CCSS **Persevere with Problems** Find the dimensions of any triangular prism that has a volume of 44 cubic inches.

CCSS **Persevere with Problems Use dimensional analysis to make each conversion.**

30. $5\text{ yd}^3 = \blacksquare\text{ ft}^3$

31. $945\text{ ft}^3 = \blacksquare\text{ yd}^3$

32. $2\text{ m}^3 = \blacksquare\text{ cm}^3$

33. **Building on the Essential Question** Explain how doubling the length, width, and height of a box changes the volume of the box.

Standardized Test Practice

34. Mr. Toshio is filling a 20-foot by 35-foot garden framed by two levels of bricks with topsoil. If the topsoil costs $9 per cubic foot, what other information is needed to find *s*, the cost of the soil?

A the area of the garden
B the perimeter of the garden
C the price per cubic yard of soil
D the height of the bricks

35. Short Response How many centimeters tall is a rectangular prism with a length of 8 centimeters, width of 10 centimeters, and a volume of 960 cubic centimeters?

36. What is the volume of the prism below?

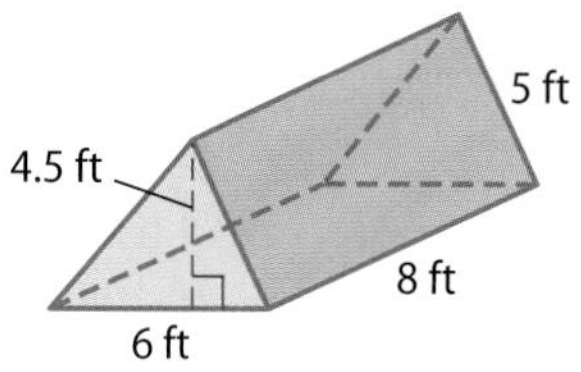

F 67.5 ft^3 **H** 108 ft^3
G 90 ft^3 **J** 216 ft^3

37. Which of the following is the best estimate for the volume of a shoe box with sides that measure 15.75 inches, 9.25 inches, and 8 inches?

A 10 in^3 **C** 1000 in^3
B 100 in^3 **D** $10{,}000\text{ in}^3$

Common Core Review

Find the area of each figure. Round to the nearest tenth. 7.G.6

38.

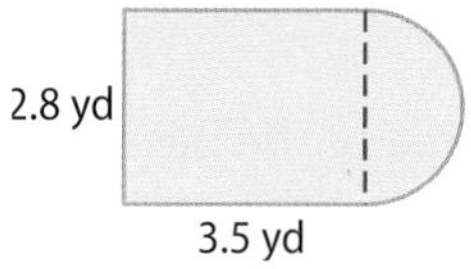

39.

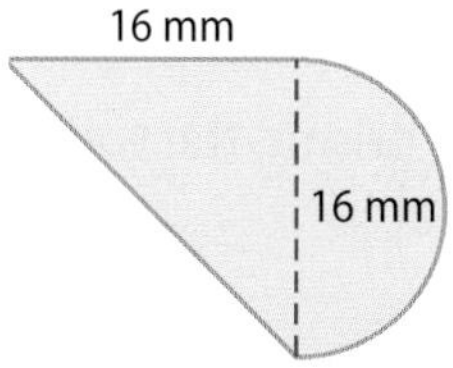

40. Tomás wants to spend less than $100 for a new soccer ball and shoes. The ball costs $24. Write and solve an inequality that gives the amount that Tomás can spend on shoes. 7.EE.4

Find each sum or difference. Write in simplest form. 7.NS.1

41. $3\frac{7}{10} + 8\frac{3}{4}$

42. $-1\frac{1}{6} + 6\frac{5}{9}$

43. $9\frac{1}{5} + \left(-10\frac{1}{2}\right)$

44. $\frac{1}{6} - \frac{3}{4}$

45. $3\frac{4}{5} - 1\frac{1}{3}$

46. $3\frac{4}{9} - \left(-2\frac{2}{3}\right)$

47. Kristin bought a scarf on sale for 25% off the original price of $40. The sales tax on the scarf was 7.5%. What was the total price of the scarf, including tax? 7.RP.3

48. A shirt is on sale for 40% off the original price of $35. What is the sale price? 7.RP.3

Find the area of each circle. Round to the nearest tenth. 7.G.4

49.

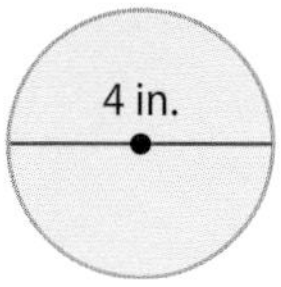

50.

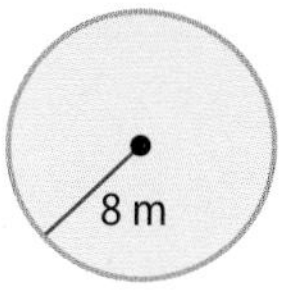

51.

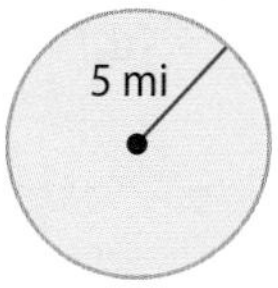

Lesson 12-6

Volume of Cylinders

ISG Interactive Study Guide

See pages 281–282 for:
- Getting Started
- Real-World Link
- Notes

Essential Question

How are two-dimensional figures used to solve problems involving three-dimensional figures?

CCSS Common Core State Standards

Content Standards
8.G.9

Mathematical Practices
1, 2, 3, 4

What You'll Learn

- Find the volumes of circular cylinders.
- Find the volumes of composite figures involving circular cylinders.

Real-World Link

Music Have you ever played music by hitting an ordinary water glass with a spoon? If you want to play a low note, add just a little water to the glass. If you want to play a higher note, add more water.

Key Concept Volume of a Cylinder

		Model
Words	The volume V of a circular cylinder with radius r is the area of the base B times the height h.	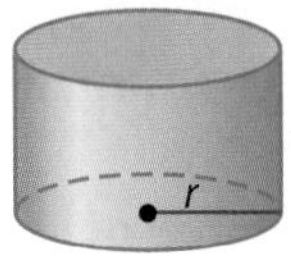
Symbols	$V = Bh$, where $B = \pi r^2$ or $V = \pi r^2 h$	

As with prisms, the volume of a cylinder is the product of the area of the base and the height.

Example 1

Tutor

Find the volume of each cylinder. Round to the nearest tenth.

a. radius of base 3 inches, height 12 inches

$V = Bh$ — Volume of a cylinder

$V = \pi r^2 h$ — Replace B with πr^2.

$= \pi \cdot 3^2 \cdot 12$ — Replace r with 3 and h with 12.

$\approx 339.3 \text{ in}^3$ — Use a calculator.

b. diameter of base 14 centimeters, height 20 centimeters

$V = Bh$ — Volume of a cylinder

$V = \pi r^2 h$ — Replace B with πr^2.

$= \pi \cdot 7^2 \cdot 20$ — Replace r with 7 and h with 20.

$\approx 3078.8 \text{ cm}^3$ — Use a calculator.

Got It? Do these problems to find out.

1a.

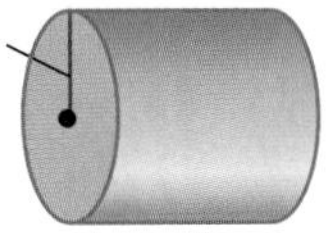

1b.

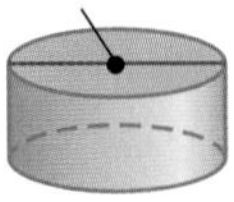

FancyVeerSet21/Alamy

Calculating with
When performing calculations with pi, wait until the final step to use the π key on a calculator or any approximation. Your answer will be more precise.

Example 2

Tools Tutor

The volume of the cylinder is 618 cubic meters. Find the height of the cylinder. Round to the nearest tenth.

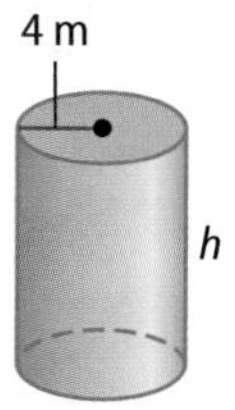

$V = Bh$	Volume of a cylinder
$V = \pi r^2 h$	Replace B with πr^2.
$618 = \pi \cdot 4^2 \cdot h$	Replace V with 618 and r with 4.
$618 = 16\pi h$	Simplify.
$12.3 \approx h$	Divide each side by 16π. Round to the nearest tenth.

The height of the cylinder is about 12.3 meters.

Got It? **Do this problem to find out.**

2. Find the height of a cylinder with a diameter of 10 yards and a volume of 549.5 cubic yards. Round to the nearest tenth.

Volumes of Composite Figures

You can find the volume of a composite figure that includes cylinders.

Example 3

Real World Tutor

An art museum is placing a sculpture on the stone pedestal shown at the right. Find the volume of the pedestal.

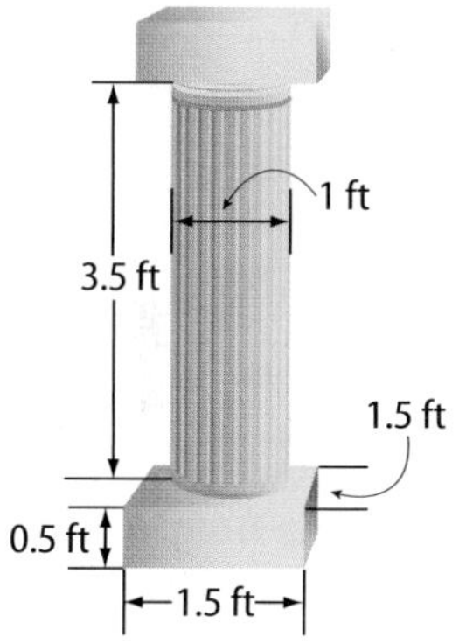

Step 1 Find the volume of the prisms.

$V = Bh$	Volume of a prism
$= 1.5 \cdot 1.5 \cdot 0.5$	The length and width are each 1.5 feet. The height is 0.5 foot.
$= 1.125 \text{ ft}^3$	Simplify.

So, the volume of the two square bases is $2 \cdot 1.125$ or 2.25 ft^3.

Step 2 Find the volume of the cylinder.

$V = \pi r^2 h$	Volume of a cylinder
$= \pi \cdot (0.5)^2 \cdot 3.5$	Replace r with 0.5 and h with 3.5.
$\approx 2.75 \text{ ft}^3$	Use a calculator.

Step 3 Find the volume of the composite figure.

$2.25 + 2.75 = 5$	Add the volumes.

So, the total volume of the pedestal is about 5 ft^3.

Got It? **Do this problem to find out.**

3. Find the volume of the plastic building brick shown at the right. Round to the nearest tenth.

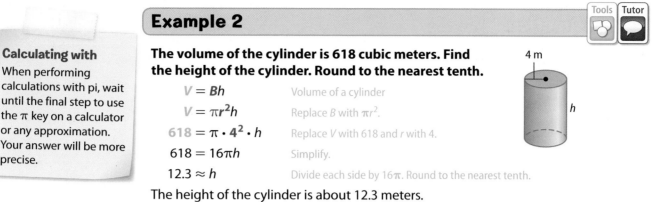

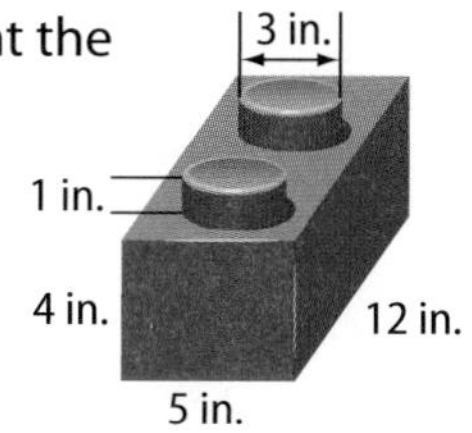

Guided Practice

Check

Find the volume of each cylinder. Round to the nearest tenth. (Example 1)

1. 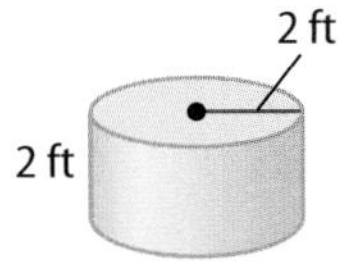

2. diameter of base: 33.2 mm
height: 60 mm

Find the height of each cylinder. Round to the nearest tenth. (Example 2)

3. volume = 283 in^3

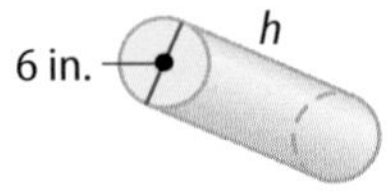

4. volume: 5700 m^3
diameter of base: 22 m

5. An oak peg like the one shown at the right was used in a toy truck. Find the volume of the peg. Round to the nearest tenth. (Example 3)

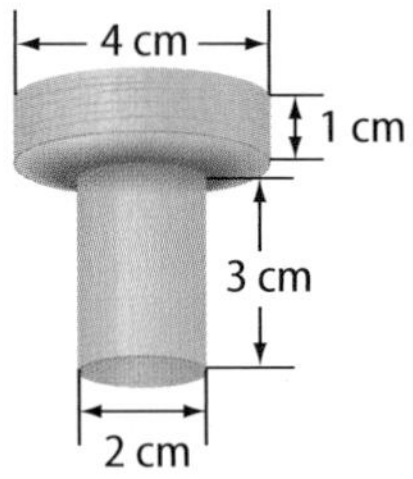

Independent Practice

Go online for Step-by-Step Solutions

eHelp

Find the volume of each figure. Round to the nearest tenth. (Example 1)

6. 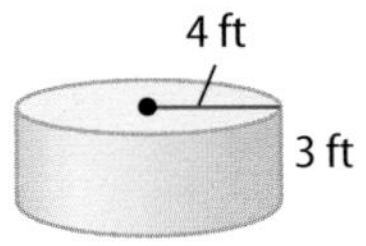

7. 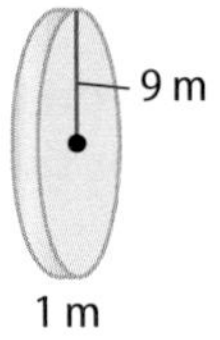

8. 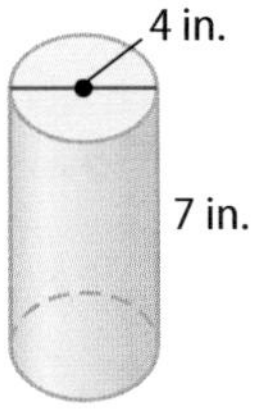

9. radius: 2.2 cm
height: 3 cm

10. diameter: 5 yd
height: 11 yd

11. diameter: 4.6 m
height: 6.1 m

Find the height of each cylinder. Round to the nearest tenth. (Example 2)

12. volume: 41.5 yd^3

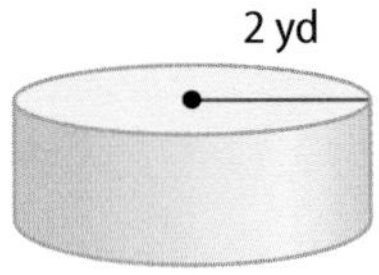

13. volume: 9.7 m^3
diameter: 1.5 m

14. A scented candle is in the shape of a cylinder that is 8 inches tall. The diameter of the candle is 3.5 inches. Find the volume of the candle. Round to the nearest tenth.

15. A soup can is 5.5 inches tall and 4 inches in diameter. What is the volume of the soup can?

Find the volume of each figure. Round to the nearest tenth. (Example 3)

16. 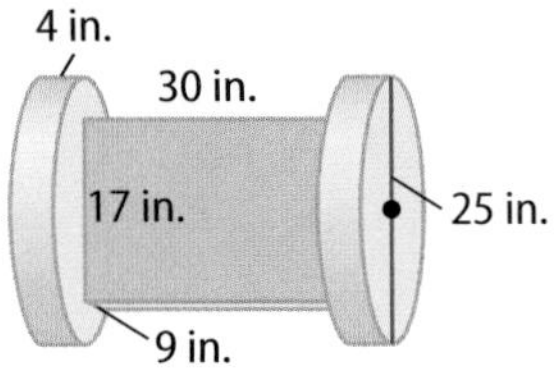

17. 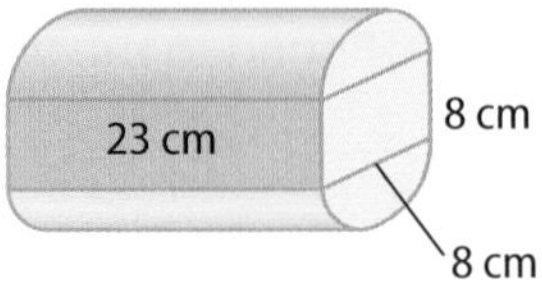

18. Find the volume of the mailbox below. Round to the nearest tenth.

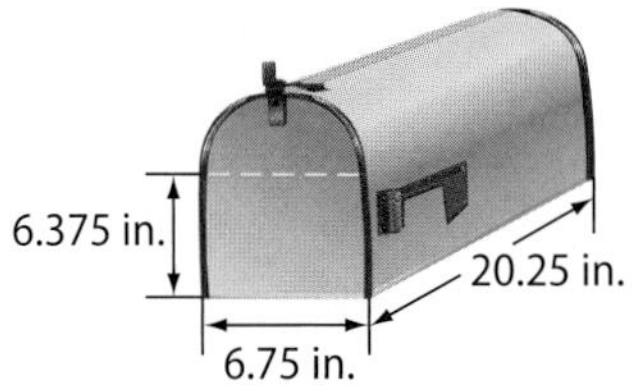

19 A roll of paper towels has the dimensions shown. Find the volume of the roll. Round to the nearest tenth.

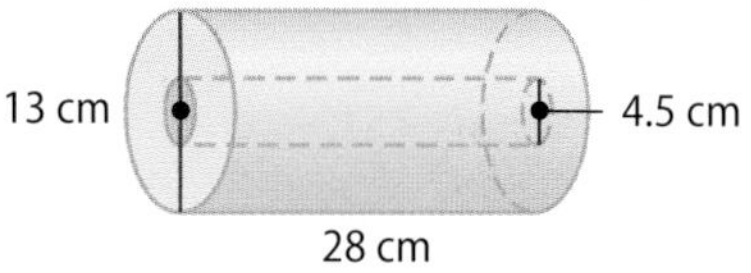

20. CCSS **Multiple Representations** In this problem, you will examine how changing the size of a cylinder affects its volume.

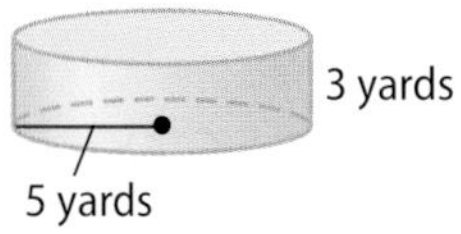

a. Table Copy and complete the table for the cylinder shown. Round to the nearest tenth, if necessary.

Cylinder	Radius	Height	Volume (yd^3)
Original	5	3	■
Multiply radius by 3	$5 \cdot 3 = $ ■	3	■
Multiply height by 3	5	$3 \cdot 3 = $ ■	■
Multiply both by 3	■	■	■

b. Analyze Compare the original volume to the other volumes.

c. Symbols Write an equation to find the volume of a cylinder after a dilation *d*.

21. STEM A cylindrical rain barrel is 38 inches tall and has a diameter of 28 inches. If the volume of one gallon of water is 231 cubic inches, how many gallons of water will the rain barrel hold? Round to the nearest tenth.

22. Kamilah's uncle, a fire captain, said the diameters of fire hoses range from 1.5 inches to 6 inches and the hoses are 50 feet long. Find the approximate minimum and maximum volumes of a fire hose in cubic feet.

H.O.T. Problems Higher Order Thinking

23. CCSS **Model with Mathematics** Write two real-world examples where you would want to change the dimensions of cylinders, but maintain the volumes.

24. CCSS **Persevere with Problems** Two equal-sized sheets of paper are rolled along the length and along the width, as shown. Which cylinder do you think has the greater volume? Explain.

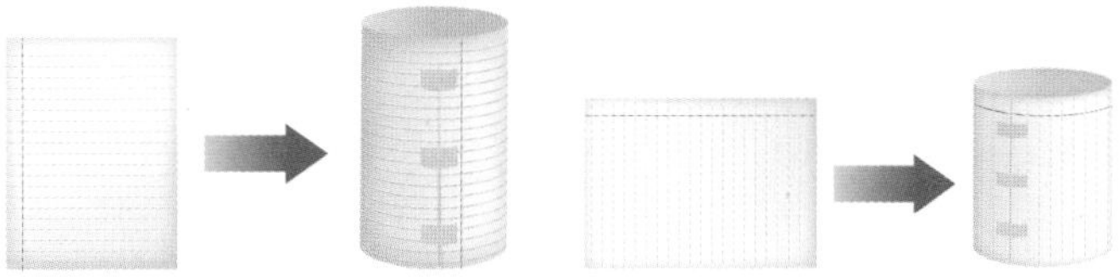

25. CCSS **Reason Abstractly** Find the ratios of the volume of cylinder A to cylinder B.

a. Cylinder A has the same radius but twice the height of cylinder B.

b. Cylinder A has the same height but twice the radius of cylinder B.

26. **Building on the Essential Question** Explain how the formula for the volume of a cylinder is similar to the formula for the volume of a rectangular prism.

Standardized Test Practice

27. Find the maximum amount of water that can fill the trough shown.

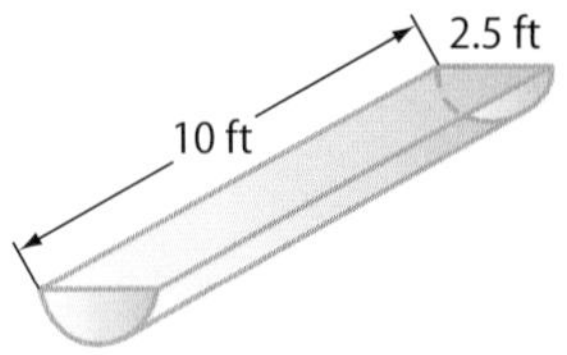

A $20.5\ ft^3$ C $48\ ft^3$
B $24.5\ ft^3$ D $49\ ft^3$

28. What is the volume of a cylinder with a radius of 8 inches and a height of 1 foot? Round to the nearest tenth.

F $25.1\ in^3$ H $301.6\ in^3$
G $201.1\ in^3$ J $2412.7\ in^3$

29. **Short Response** The smallest canister in a set of cylindrical canisters has a height of 11 inches and a radius of 1.5 inches. Round each answer to the nearest tenth.

a. Find the volume of the smallest canister.

b. The medium canister has the same height as the smallest but the radius is tripled. What is the volume of the medium canister?

30. A cylindrical diesel tank is 1.25 meters high and has a radius of 0.60 meter. If the tank can only be filled to an 85% capacity to allow for expansion of the fuel, what is the maximum volume of fuel? Round to the nearest hundredth.

A $4.71\ m^3$ C $1.41\ m^3$
B $2.01\ m^3$ D $1.20\ m^3$

Common Core Review

Find the volume of each figure. 7.G.6

31.

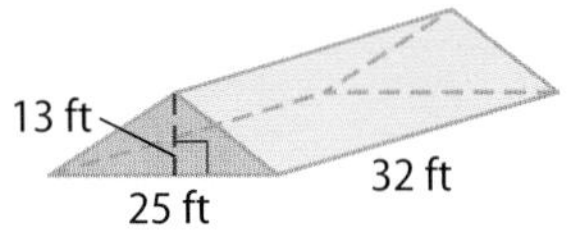

32.

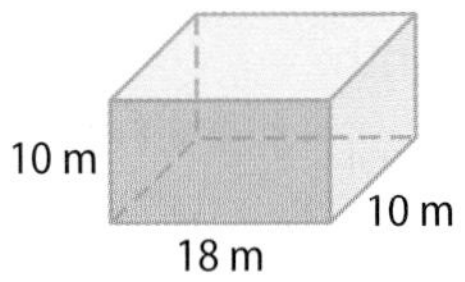

33.

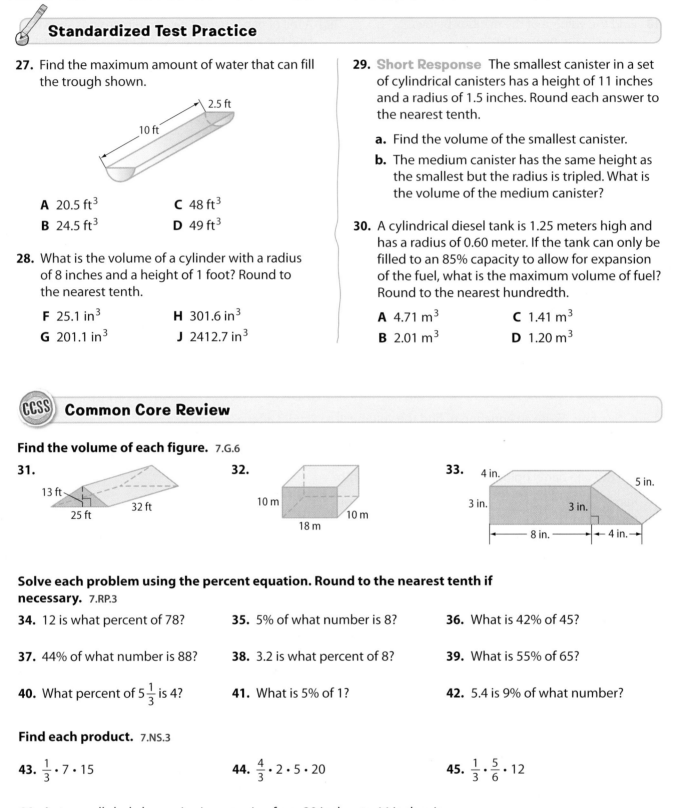

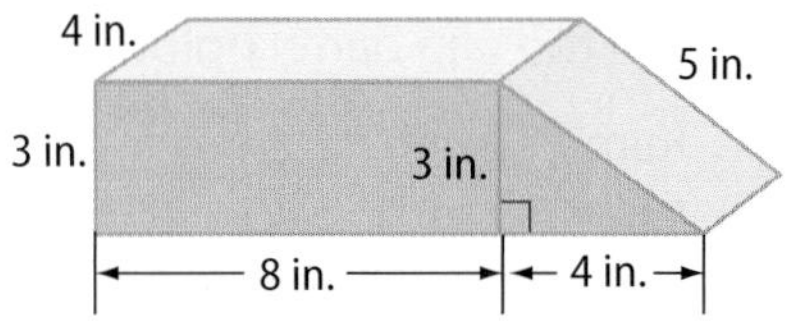

Solve each problem using the percent equation. Round to the nearest tenth if necessary. 7.RP.3

34. 12 is what percent of 78?

35. 5% of what number is 8?

36. What is 42% of 45?

37. 44% of what number is 88?

38. 3.2 is what percent of 8?

39. What is 55% of 65?

40. What percent of $5\frac{1}{3}$ is 4?

41. What is 5% of 1?

42. 5.4 is 9% of what number?

Find each product. 7.NS.3

43. $\frac{1}{3} \cdot 7 \cdot 15$

44. $\frac{4}{3} \cdot 2 \cdot 5 \cdot 20$

45. $\frac{1}{3} \cdot \frac{5}{6} \cdot 12$

46. A store sells hula hoops in sizes ranging from 30 inches to 44 inches in diameter. How much greater is the circumference of the largest hula hoop than the smallest? Round to the nearest inch. 7.G.4

47. A submarine descended 1120 feet in 4 minutes. It ascended 1545 feet in 3 minutes. How much faster was the ascent rate of the submarine than the descent rate? 7.NS.2b

ISG Interactive Study Guide

See page 283 for:
• Mid-Chapter Check

21ST CENTURY CAREER in Architecture

Space Architect

Do you like building things? Are you an excellent problem solver? If so, you have what it takes to be a space architect. Space architects use principles from architecture, design, engineering, and science to create places for people to live and work in outer space. Their designs include transfer vehicles, lunar habitats, and Martian greenhouses. Because of the limitations, space architecture must be very efficient and functional. Every square inch of surface and every cubic inch of space must have a purpose.

College & Career READINESS

Explore college and careers at ccr.mcgraw-hill.com

Is This the Career for You?

Are you interested in a career as a space architect? Take some of the following courses in high school.

- Aerospace Technology
- Calculus
- Geometry
- Introductory Space Planning
- Intro to CAD

Find out how math relates to a career in architecture.

Interactive Study Guide
See page 284 for:
- Problem Solving
- Career Project

Inquiry Lab

Volume of Pyramids and Cones

Inquiry **HOW are the volumes of pyramids and cones related to the volumes of prisms and cylinders?**

CCSS **Content Standards**
7.G.6, 8.G.9
Mathematical Practices
1, 2, 3

Snacks A company sells sunflower seeds in two different packages. One package is a cube with edges that are 2 inches long. The other package is a square pyramid with a base and height that are the same as those of the cube. How is the volume of the pyramid-shaped package related to the volume of the cube-shaped package?

Investigation 1

Make a model of the two packages and then compare the volumes.

Step 1 Draw and cut out five 2-inch squares. Then tape them together as shown.

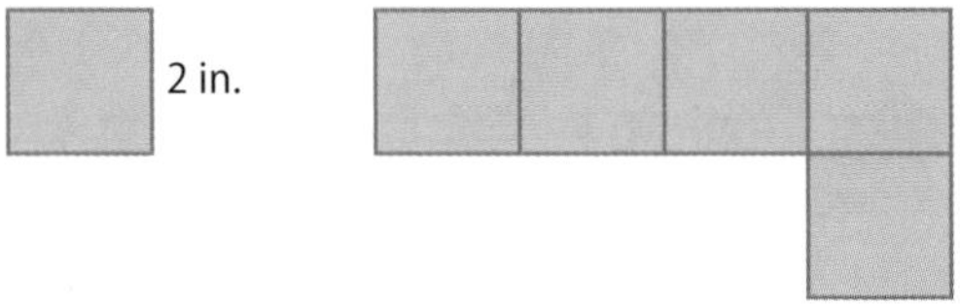

Step 2 Fold and tape to form a cube with an open top.

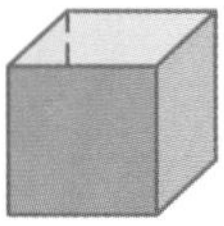

Step 3 Draw and cut out four isosceles triangles with the measurements shown. Then tape them together.

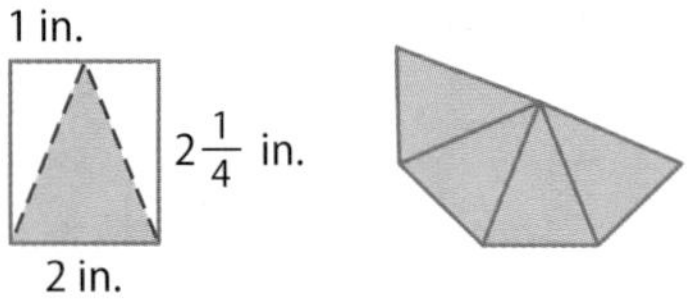

Step 4 Fold and tape to form an open square pyramid.

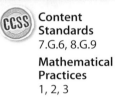

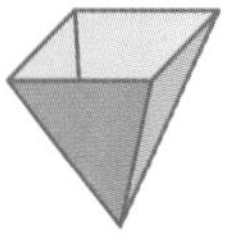

Collaborate

Work with a partner.

1. Compare the base areas and the heights of the rectangular prism (cube) and the pyramid.

2. Fill the pyramid with rice, sliding a ruler across the top to level the amount. Pour the rice into the cube. Repeat the process until the prism is filled. How many times did you fill the pyramid in order to fill the cube?

3. What fraction of the prism volume does one pyramid fill?

4. Repeat the activity using measurements for the cube and the pyramid that are twice the measurements in Investigation 1. How are the two volumes related?

5. CCSS **Reason Inductively** The pyramid shown has the same base area and the same height as the rectangular prism.

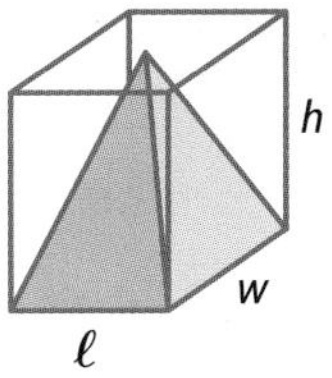

a. Write a formula for the volume of the rectangular prism.

b. Based on your results from the investigation, what formula do you think gives the volume of the pyramid?

Analyze

Find the volume of each pyramid.

6.

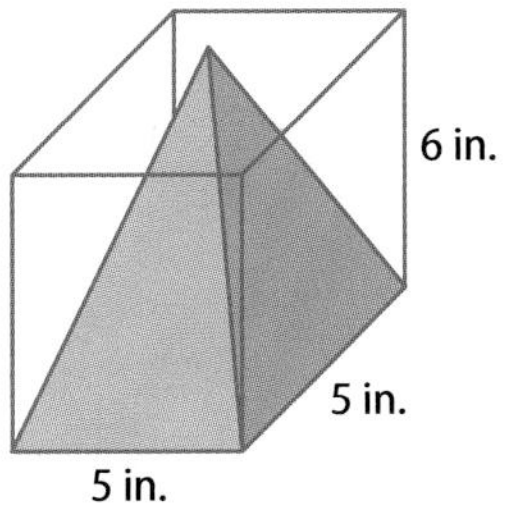

7.

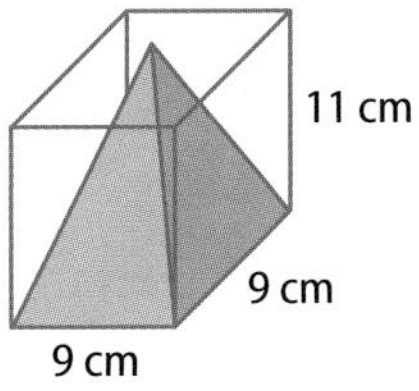

8. CCSS **Justify Conclusions** A movie theater offers popcorn in two different containers, as shown below. The price of popcorn in the rectangular prism is $4.95. What do you think would be a reasonable price for the popcorn in the square pyramid? Why?

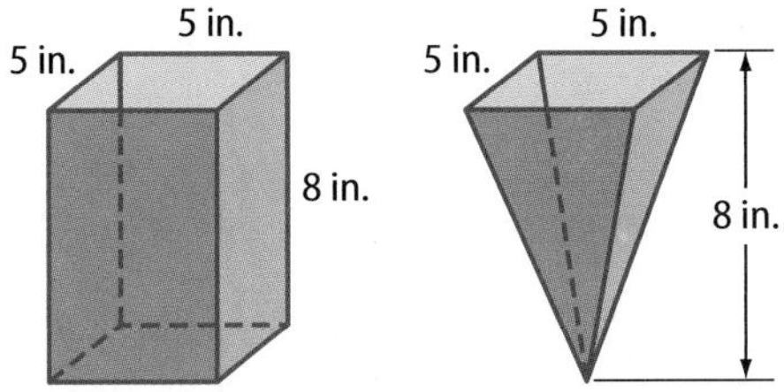

There is a relationship between the volume of a cone and the volume of a cylinder with the same height and radius. Use a container with a cylindrical interior, such as a cup, mug, or can, for the following investigation.

Investigation 2

Follow these steps to make a cone-shaped container that has the same height and radius as a cylindrical container.

Step 1 Roll a piece of paper into a cone so the point rests on the bottom of the cylinder. Be sure that the paper is at or above the top of the cylinder all the way around. Tape the cone along its seam.

Step 2 Trim the excess paper from the top of the cone.

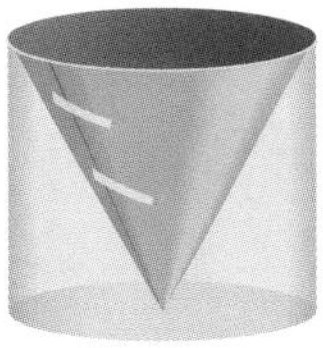

Collaborate

Work with a partner.

9. Compare the base areas and the heights of the cylinder and the cone.

10. Fill the cone with rice, sliding a ruler across the top to level the amount. Pour the rice into the cylinder. Repeat until the cylinder is filled. How many times did you fill the cone in order to fill the cylinder?

11. What fraction of the cylinder volume does one cone fill?

12. **Make a Conjecture** The cone shown in the figure has the same base area and the same height as the cylinder.

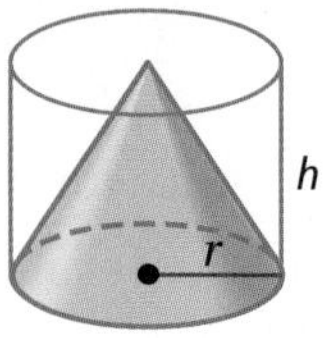

   a. Write a formula for the volume of the cylinder.

   b. What formula do you think gives the volume of the cone?

Analyze

Find the volume of each cone. Write your answers in terms of π.

13.

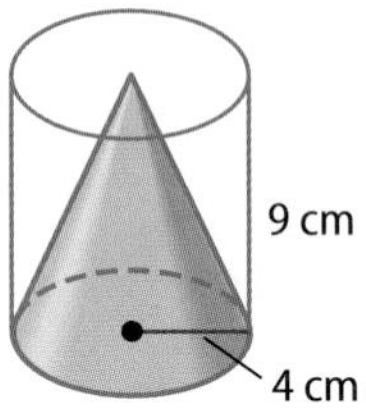

14.

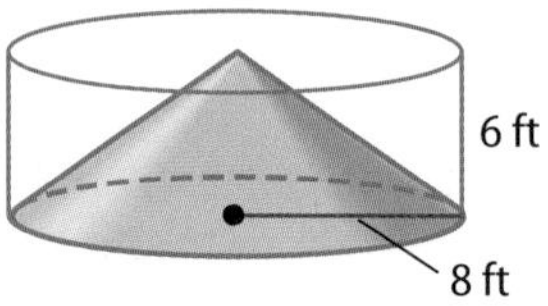

15. A hanging basket for flowers is a cone with an open top. The diameter of the cone is 12 inches and the height of the cone is 6 inches.

   a. Assuming the basket is filled completely, what volume of soil can the basket hold? Round to the nearest cubic inch.

   b. A different hanging basket is a cylinder with the same diameter and same height as the cone-shaped basket. What volume of soil can this basket hold?

16. Each of the three-dimensional figures shown below has a height of 4 inches. How can you determine which figure has the greatest volume without calculating the volume of every figure?

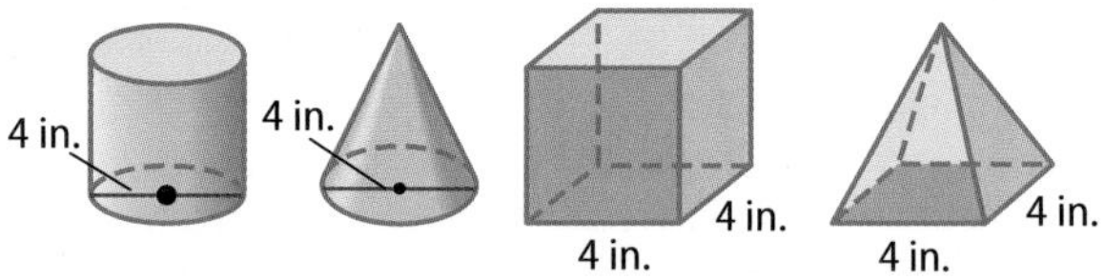

17. **Reason Abstractly** Complete the following analogy. *The volume of a(n) _?_ is related to the volume of a cylinder with the same base area and height in the same way that the volume of a square pyramid is related to the volume of a rectangular prism with the same base area and height.*

18. **Reason Abstractly** The figures below show three shapes of cardboard containers. The cylinder and the cube have the same volume but not the same height. The cone has the same base area and the same height as the cylinder. Can you make a comparison between the volume of the cone and the volume of the cube? Explain.

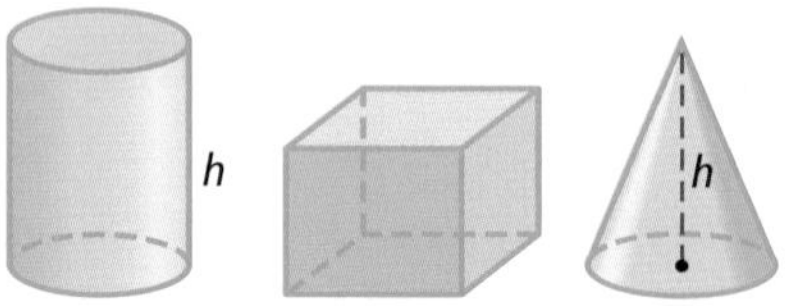

Reflect

19. **Inquiry** HOW are the volumes of pyramids and cones related to the volumes of prisms and cylinders?

Lesson 12-7

Volume of Pyramids, Cones, and Spheres

Interactive Study Guide

See pages 285–286 for:
- Getting Started
- Real-World Link
- Notes

Essential Question

How are two-dimensional figures used to solve problems involving three-dimensional figures?

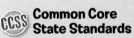

Common Core State Standards

Content Standards
8.G.9

Mathematical Practices
1, 3, 4, 7,

Vocabulary

sphere

What You'll Learn

- Find the volumes of pyramids and cones.
- Find the volumes of spheres.

Real-World Link

Basketball Basketballs can be made out of leather, synthetic materials, or rubber, and they come in different sizes. The diameter of a regulation basketball is usually half the diameter of the basket.

Key Concept Volume of a Pyramid

		Model
Words	The volume V of a pyramid is one-third the area of the base B times the height h.	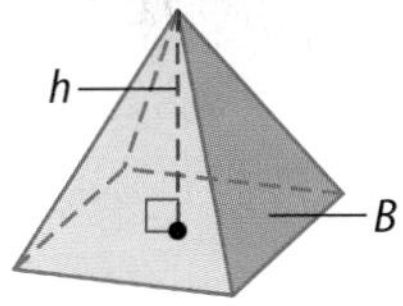
Symbols	$V = \frac{1}{3}Bh$	

A pyramid has one-third the volume of a prism with the same base and height. The height of a pyramid is the perpendicular distance from the vertex to the base.

Example 1

Tools Tutor

Find the volume of the pyramid. Round to the nearest tenth, if necessary.

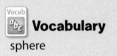

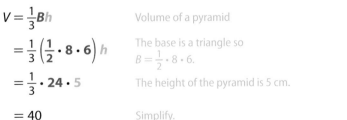

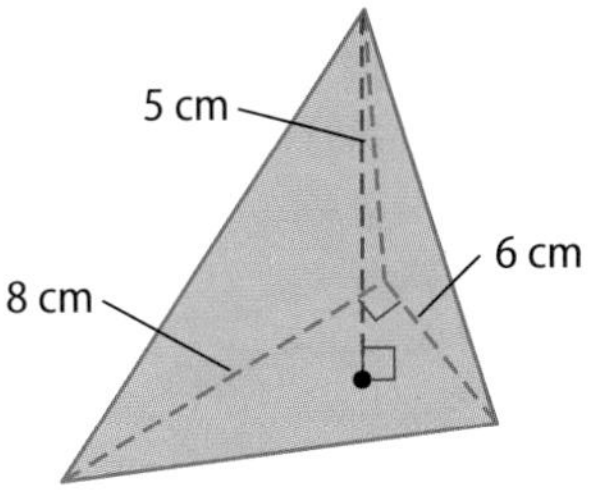

$V = \frac{1}{3}Bh$ — Volume of a pyramid

$= \frac{1}{3}\left(\frac{1}{2} \cdot 8 \cdot 6\right)h$ — The base is a triangle so $B = \frac{1}{2} \cdot 8 \cdot 6$.

$= \frac{1}{3} \cdot 24 \cdot 5$ — The height of the pyramid is 5 cm.

$= 40$ — Simplify.

The volume of the pyramid is 40 cubic centimeters.

Got It? Do these problems to find out.

1a. Find the volume of a pyramid with a base area of 90 square feet and a height of 12 feet.

1b. Find the volume of a pyramid with a base area of 25 square centimeters and a height of 3 centimeters.

Key Concept Volume of a Cone

Words	The volume V of a cone with radius r is one third the area of the base πr^2 times the height h.	**Model** 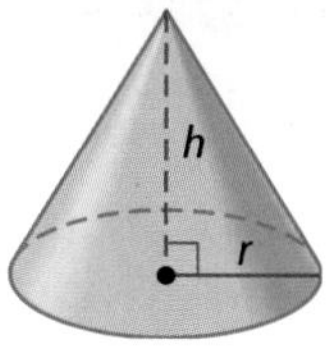
Symbols	$V = \frac{1}{3}Bh$, where $B = \pi r^2$ or $V = \frac{1}{3}\pi r^2 h$	

The volumes of a cone and a cylinder are related in the same way as the volumes of a pyramid and a prism are related.

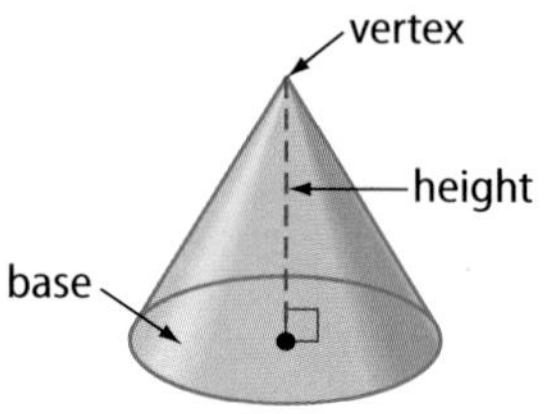

Prism	$V = Bh$
Pyramid	$V = \frac{1}{3}Bh$

Cylinder	$V = \pi r^2 h$
Cone	$V = \frac{1}{3}\pi r^2 h$

The volume of a cone is $\frac{1}{3}$ the volume of a cylinder with the same base and height.

Estimating
Since $\frac{1}{3} \cdot \pi$ is about 1, you can use estimation to determine whether your solution is reasonable.
$V \approx r^2 h$
$\approx 5^2 \cdot 7$
$\approx 25 \cdot 7$ or 175

Example 2

Tools Tutor

Find the volume of the cone. Round to the nearest tenth.

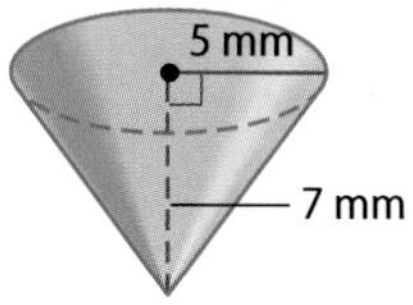

$V = \frac{1}{3}\pi \mathbf{r}^2 \mathbf{h}$ Volume of a cone

$= \frac{1}{3} \cdot \pi \cdot \mathbf{5}^2 \cdot \mathbf{7}$ Replace r with 5 and h with 7.

$\approx 183.3 \text{ mm}^3$ Simplify. Round to the nearest tenth.

Got It? **Do these problems to find out.**

2a. radius 6 ft, height 20 ft

2b. radius $1\frac{1}{2}$ yd, height 9 yd

Key Concept Volume of a Sphere

Words	The volume V of a sphere with radius r is four-thirds times π times the radius cubed.	**Model** 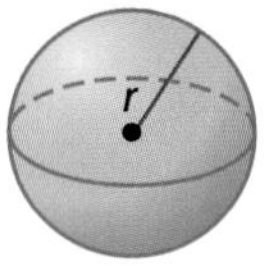
Symbols	$V = \frac{4}{3}\pi r^3$	

A **sphere** is a set of points in space that are a given distance r from the center. Suppose a sphere with radius r is placed inside a cylinder with the same radius r and height $2r$. The volume of the sphere is $\frac{2}{3}$ of the volume of the cylinder. The volume of the cylinder is shown below.

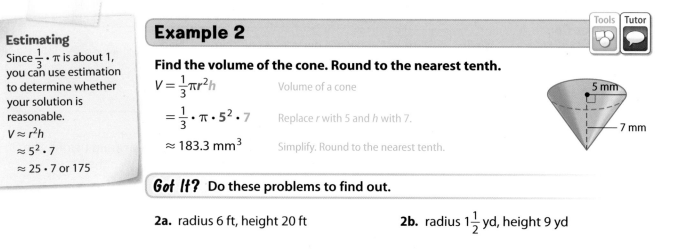

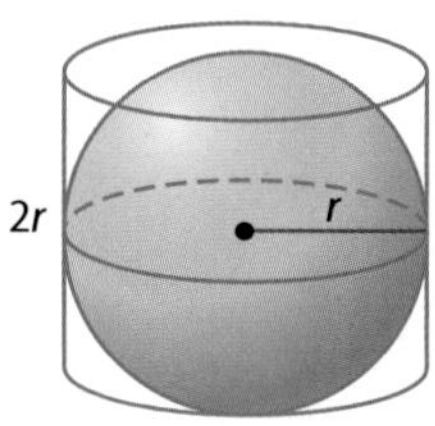

$V = \pi r^2 \mathbf{h}$ Volume of a cylinder

$= \pi r^2 (\mathbf{2r})$ Replace h with $2r$.

$= 2\pi r^3$ Simplify.

Since the sphere is $\frac{2}{3}$ the size of the cylinder, you can find the volume of the sphere.

$V = \left(\frac{2}{3}\right)2\pi r^3$ The sphere is $\frac{2}{3}$ the size of the cylinder.

$= \frac{4}{3}\pi r^3$ Simplify.

Tools Tutor

Example 3

Find the volume of the sphere. Round to the nearest tenth.

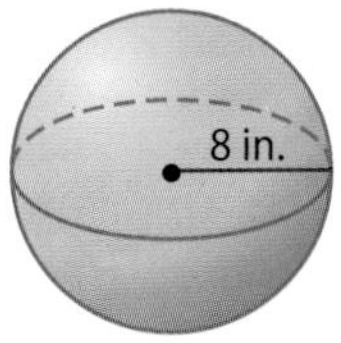

$V = \frac{4}{3}\pi \mathbf{r}^3$ Write the formula for the volume of a sphere.

$= \frac{4}{3} \cdot \pi \cdot \mathbf{8}^3$ Replace *r* with 8.

$\approx 2144.7 \text{ in}^3$ Simplify.

Got It? **Do this problem to find out.**

3. Find the volume of a sphere with a radius of 10 feet. Round to the nearest tenth.

Real World Tutor

Example 4

A spherical scoop of ice cream with a diameter of 6.3 centimeters is placed in a bowl. Find the volume of the ice cream. Then find how long it would take the ice cream to melt if it melts at a rate of 2.1 cubic centimeters every minute.

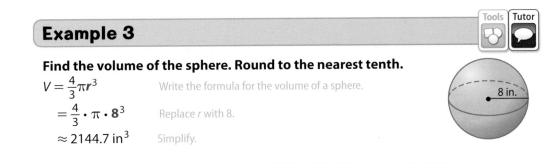

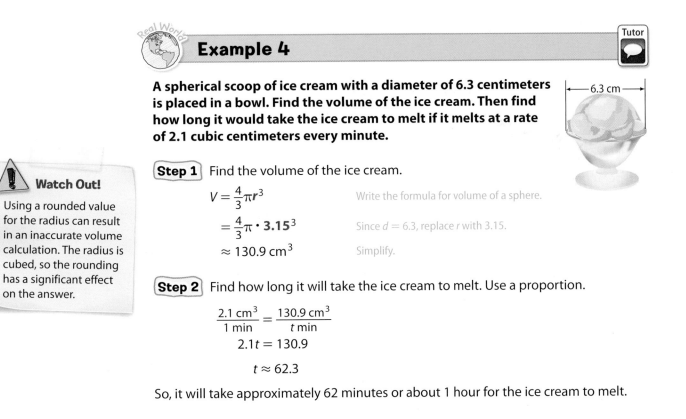

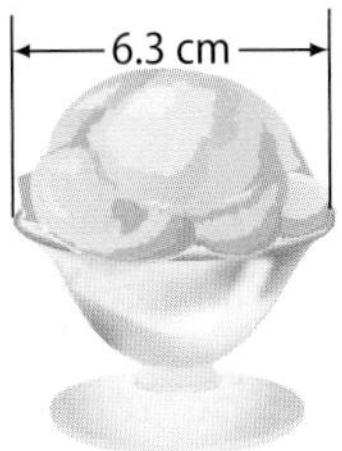

Step 1 Find the volume of the ice cream.

$V = \frac{4}{3}\pi \mathbf{r}^3$ Write the formula for volume of a sphere.

$= \frac{4}{3}\pi \cdot \mathbf{3.15}^3$ Since $d = 6.3$, replace *r* with 3.15.

$\approx 130.9 \text{ cm}^3$ Simplify.

Step 2 Find how long it will take the ice cream to melt. Use a proportion.

$$\frac{2.1 \text{ cm}^3}{1 \text{ min}} = \frac{130.9 \text{ cm}^3}{t \text{ min}}$$

$$2.1t = 130.9$$

$$t \approx 62.3$$

So, it will take approximately 62 minutes or about 1 hour for the ice cream to melt.

Watch Out!

Using a rounded value for the radius can result in an inaccurate volume calculation. The radius is cubed, so the rounding has a significant effect on the answer.

Got It? **Do this problem to find out.**

4. A beachball has a diameter of 18 inches. Find the volume of the ball. Then find how long it will take a hand pump to inflate the ball if it inflates at a rate of 325 cubic inches per minute. Round to the nearest tenth.

Guided Practice

Check

Find the volume of each figure. Round to the nearest tenth, if necessary. (Examples 1-3)

1.

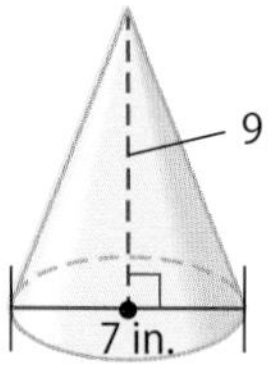

2.

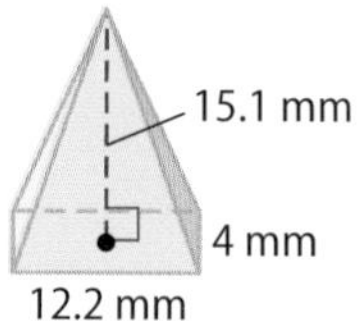

3.

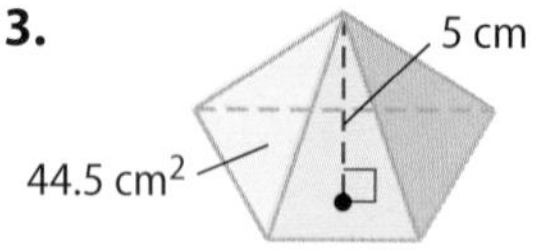

4.

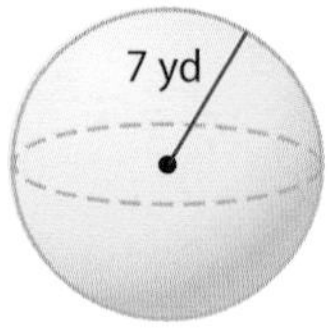

5. Amber purchased a necklace that contained an 8 millimeter diameter round pearl. Find the volume of the pearl to the nearest tenth. (Example 4)

Independent Practice

Go online for Step-by-Step Solutions

eHelp

Find the volume of each figure. Round to the nearest tenth, if necessary. (Examples 1-3)

6.

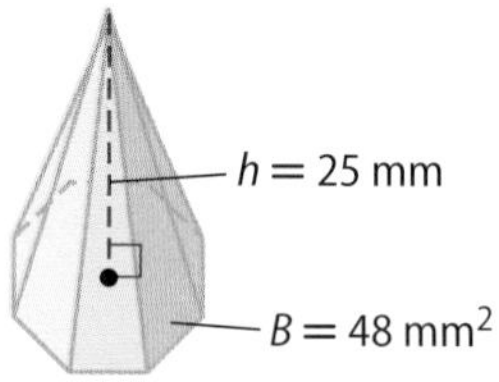

7.

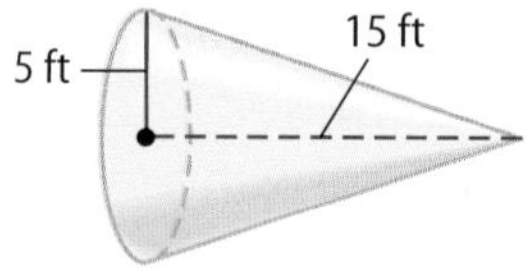

8.

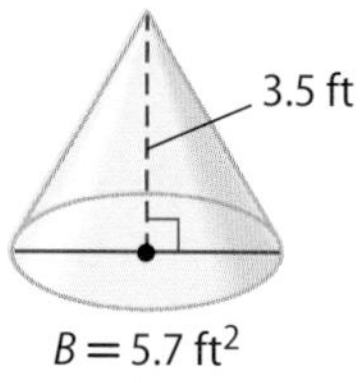

9.

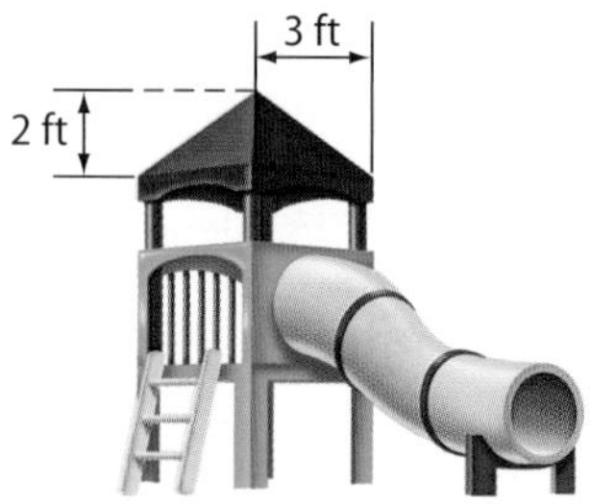

10.

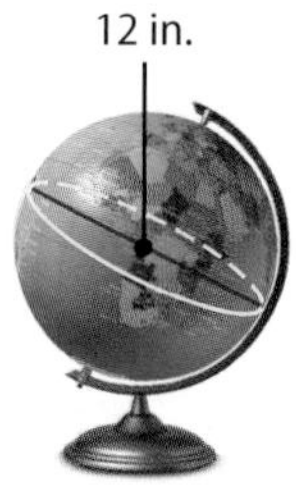

11.

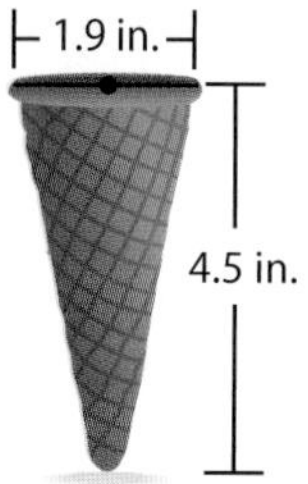

12. pentagonal pyramid: base area 52 in^2, height 12 in.

13. rectangular pyramid: length 3.5 feet, width 2 feet, height 4.5 feet

14. cone: diameter 6.7 mm, height 2.1 mm

15. sphere: radius 7.2 km

16. sphere: diameter 1.8 mm

17 **STEM** A cone-shaped icicle is 2.5 feet long and has a diameter of 1.5 feet. (Example 4)

a. Find the amount of ice in the icicle to the nearest tenth of a cubic foot.

b. At a certain temperature, the icicle melts at a rate of 0.1 cubic foot every 5 minutes. How long will it take for the icicle to melt?

Find the height of each figure. Round to the nearest tenth, if necessary.

18. square pyramid: volume 873.18 m^3, length 12.6 m

19. cone: volume 306.464 ft^3, diameter 8 ft

Find the volume of each figure. Round to the nearest tenth, if necessary.

20.

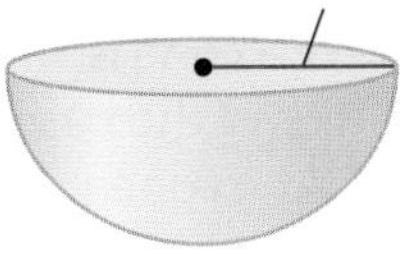

21.

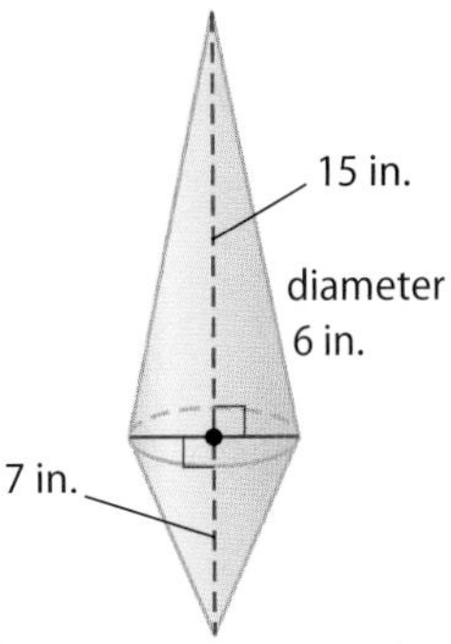

22.

23. **STEM** The solid silver bead shown at the right is made of a cylinder and two cones. If pure silver's mass is 10.5 grams per cubic centimeter, find the approximate mass in grams of the bead. Round to the nearest gram.

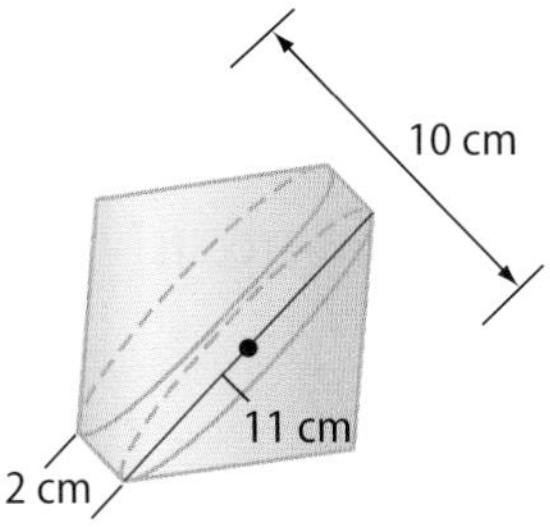

24. The volume of a mini-basketball is about 230 cubic inches. What is its radius? Round to the nearest inch.

25. **CCSS Persevere with Problems** Three golf balls are packaged in a box as shown below. Round each answer to the nearest tenth.

a. If each ball is 4.3 centimeters in diameter, find the volume of the empty space in the box.

b. What percent of the box is full?

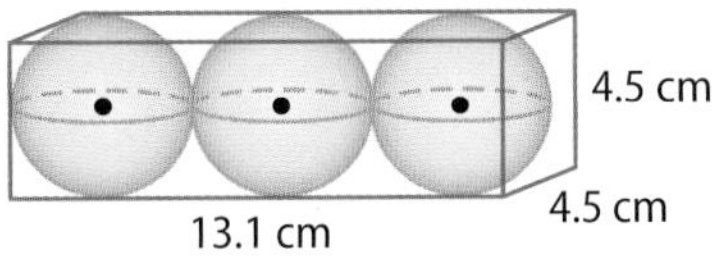

H.O.T. Problems Higher Order Thinking

26. **CCSS Identify Structure** Name the dimensions of two different cones whose volume is between 45 cubic inches and 55 cubic inches.

27. **CCSS Persevere with Problems** A cone contains 93 cubic units of water. All the water is poured equally into three congruent cylindrical containers to a level of 4.5 units. What is the diameter of each container? Round to the nearest tenth.

28. **CCSS Which One Doesn't Belong?** Find the expression that does not represent the height of a three-dimensional figure. Explain your reasoning.

$\frac{3V}{B}$	$\frac{Vr^2}{3\pi}$	$\frac{V}{B}$

29. **CCSS Identify Structure** *True* or *false*? The volumes of a rectangular-based pyramid and a cone with the same heights and equal base areas are equal. Explain your reasoning to a classmate.

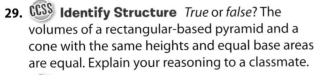

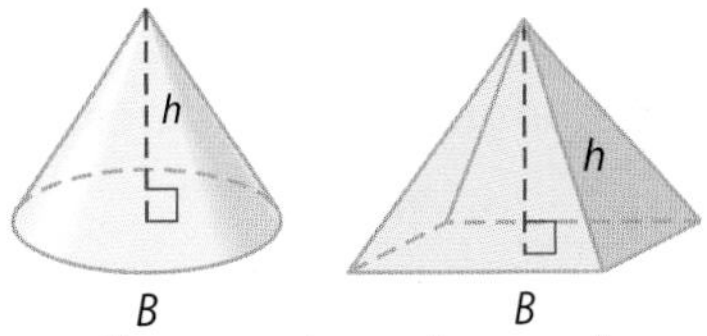

30. **CCSS Persevere with Problems** Which has a greater effect on the volume of a cone: changing the radius or changing the height? Explain.

31. **Building on the Essential Question** Write a real-world example where you would want to change a cylinder to a cone. Describe how the dimensions of the cone would be affected if the volumes remain the same.

Standardized Test Practice

32. What is the volume of a square pyramid with a height of 12 centimeters and base edges of 4 centimeters?

A 48 cm^3 **C** 96 cm^3

B 64 cm^3 **D** 192 cm^3

33. Find the volume of the cone. Use 3.14 for π. Round to the nearest hundredth.

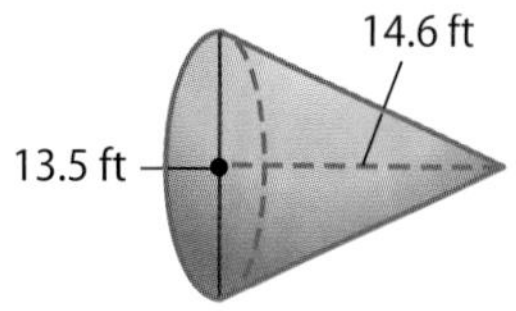

F 8355.07 ft^3 **H** 696.26 ft^3

G 2785.02 ft^3 **J** 618.89 ft^3

34. **Short Response** The volume of a cone is 3768 cubic millimeters. The radius of the base is 5 millimeters. Find the height of the cone.

35. A sphere has a radius of 5 meters. Which of the following is closest to the volume of the sphere in cubic meters?

A 65.4 m^3 **C** 130.8 m^3

B 104.7 m^3 **D** 523.3 m^3

CCSS Common Core Review

Find the volume of each figure. Round to the nearest tenth, if necessary. 7.G.6

36.

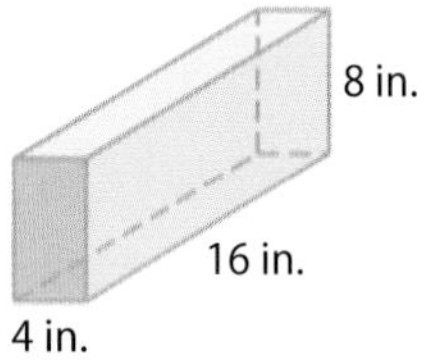

37.

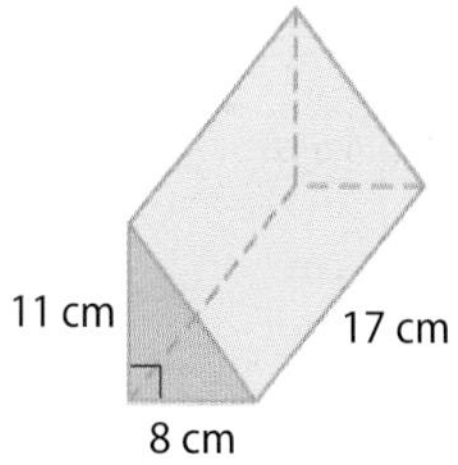

38.

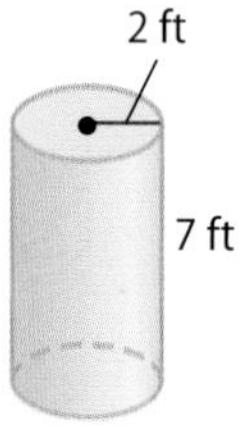

39. An elevator picks up a passenger 268 feet above ground and the door opens at the parking garage 48 feet below ground. How far did the elevator travel? **7.NS.1d**

40. Copy and complete the table. Use the results to write four solutions of $y = x + 5$. Write the solutions as ordered pairs. **6.EE.9**

x	$x + 5$	y
-3	$-3 + 5$	■
-1	■	■
0	■	■
1	■	■

41. Manuel has $15 to buy snack mix that costs $5.75 per pound. Write and solve an equation to find the amount of snack mix he can buy if he spends all $15. Round to the nearest tenth. **6.EE.7**

Find the area of each figure. Round to the nearest tenth, if necessary. 6.G.1, 7.G.4

42.

43.

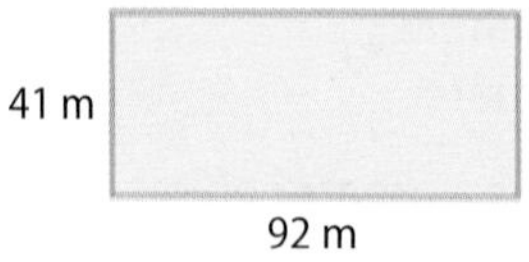

44.

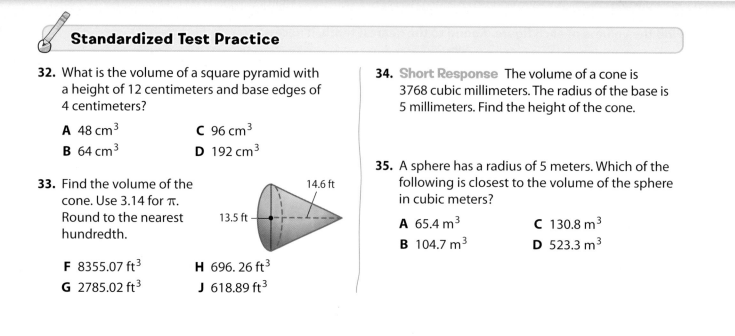

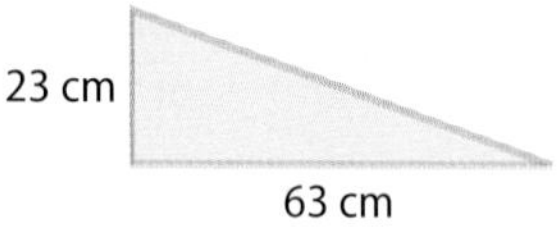

45. Find P(not prime) when you roll a 6-sided number cube. Then describe the likelihood of the event. Write *impossible, unlikely, equally likely, likely,* or *certain.* **7.SP.5**

Inquiry Lab

Surface Area of Prisms

Inquiry HOW can you develop a formula for the surface area of a rectangular prism?

Content Standards
7.G.6

Mathematical Practices
1, 3, 6

Crafts Isaac has a cardboard box like the one shown at the right. As part of a crafts project, he plans to cover the outside of the box with decorative paper. He wants to know how much paper he will need.

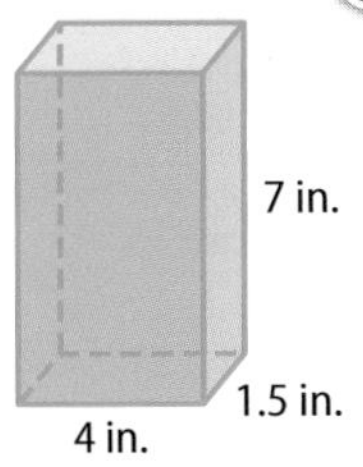

Investigation 1

Nets are two-dimensional patterns of three-dimensional figures. When you construct a net, you are decomposing the three-dimensional figure into separate shapes. You can use a net to find the area of each surface of a three-dimensional figure such as the rectangular prism shown above.

Step 1 To help you make a net for the rectangular prism, find an empty box with a tuck-in lid. Use a marker to label the top, bottom, front, back, left side, and right side.

Step 2 Open the lid. Cut each of the 4 vertical edges. Open the box and lay it flat.

Step 3 Measure each face. Copy the table and record the results. Then find and record the area of each face.

Face	Length (in.)	Width (in.)	Area (in^2)
top	■	■	■
bottom	■	■	■
front	■	■	■
back	■	■	■
left	■	■	■
right	■	■	■

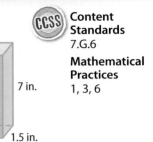

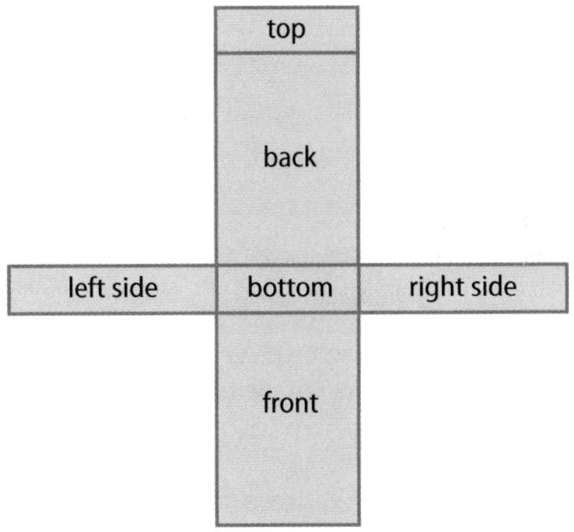

Step 4 Find the sum of the areas of the faces. This is the *surface area* of the prism.

Collaborate

Work with a partner.

1. Repeat Investigation 1 to find the amount of paper Isaac will need to cover his box shown above.
2. Find the perimeter of the top or bottom base of Isaac's box above.
3. Multiply the perimeter of the base by the height of the box.
4. Add the product from Exercise 3 to the sum of the areas of the two bases.

5. Compare your answer from Exercise 4 to the surface area you found in Exercise 1.

6. What do you observe about the areas of opposite faces in the rectangular prism?

Find the surface area of the rectangular prism that corresponds to each net.

7.

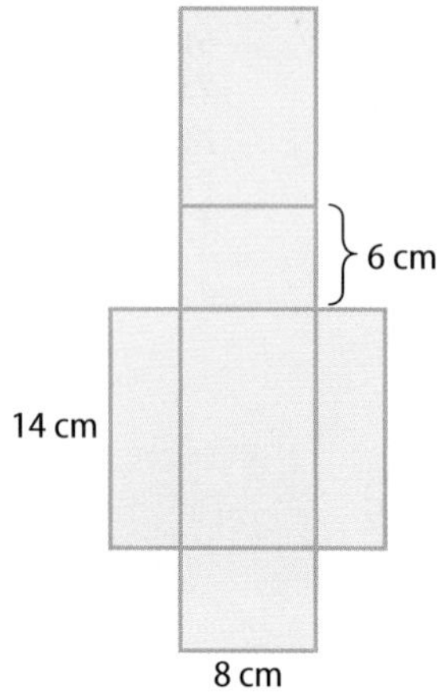

8.

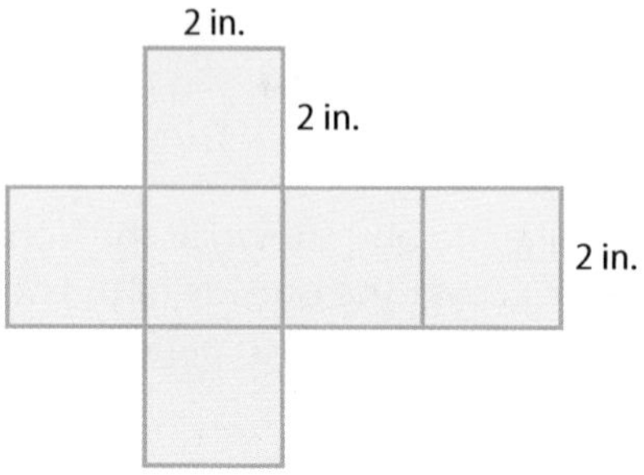

9.

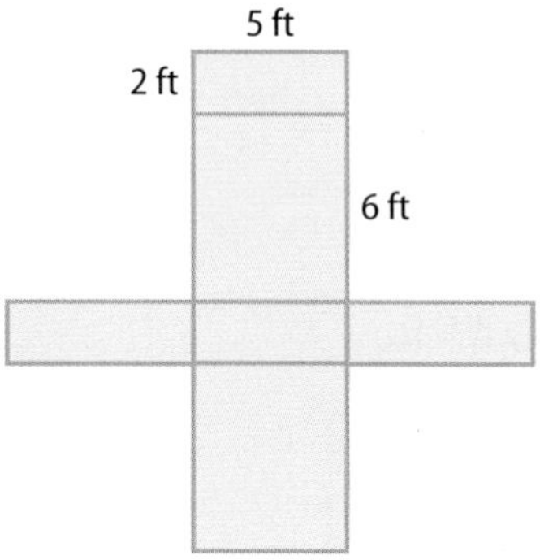

10.

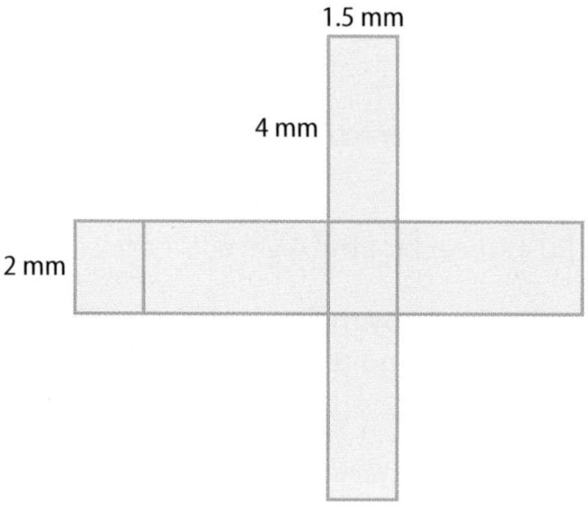

Analyze

11. Is it possible to find the surface area of a rectangular prism without finding the area of every face of the prism? Explain.

12. CCSS **Be Precise** Write a formula for the surface area *S* of a rectangular prism. Be sure to specify what the variables stand for.

13. Chandra wants to find the surface area of the rectangular prism at the right.

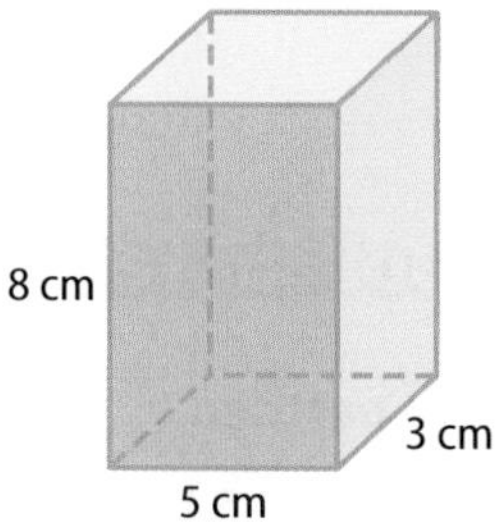

   a. Draw and label two different nets that Chandra could use to find the surface area.

   b. Use each net to find the surface area.

   c. Use the formula from Exercise 8 to find the surface area. Do you get the same result as in part b?

Reflect

14. Inquiry HOW can you develop a formula for the surface area of a rectangular prism?

Lesson 12-8

Surface Area of Prisms

ISG Interactive Study Guide

See pages 287–288 for:
- Getting Started
- Real-World Link
- Notes

Essential Question

How are two-dimensional figures used to solve problems involving three-dimensional figures?

CCSS Common Core State Standards

Content Standards
7.G.6

Mathematical Practices
1, 2, 3, 4, 7

Vocabulary

lateral faces
lateral area
surface area

What You'll Learn

- Find lateral area and surface area of prisms.
- Find surface area of real-world objects shaped like prisms.

Real-World Link

Gift Wrap Gift wrap is often non-recyclable because it is laminated, dyed, or decorated with non-paper items such as plastic or glitter. To reduce landfill waste, make your own wrapping paper from paper bags, or use the wrapping paper you have more than once!

Prisms

If you open up a box or prism and lay it flat, the result is a *net*. Every prism has two congruent parallel bases. Faces that are *not* bases are called **lateral faces**.

The **lateral area** is the sum of the areas of the lateral faces. The **surface area** is the sum of the lateral area plus the area of the bases. In the figures below, the lateral faces are shown in blue. The bases are shown in yellow.

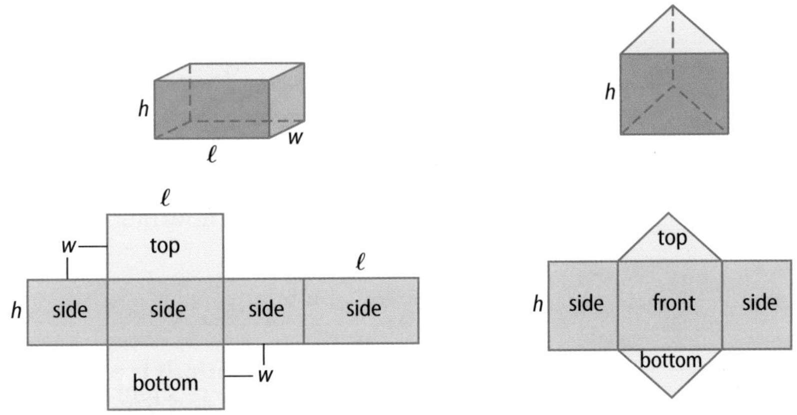

Key Concept

	Lateral Area of Prisms		Surface Area of Prisms
Words	The lateral area L of a prism is the perimeter of the base P times the height h.	Words	The surface area S of a prism is the lateral area L plus the area of the two bases $2B$.
Symbols	$L = Ph$	Symbols	$S = L + 2B$ or $S = Ph + 2B$
Model	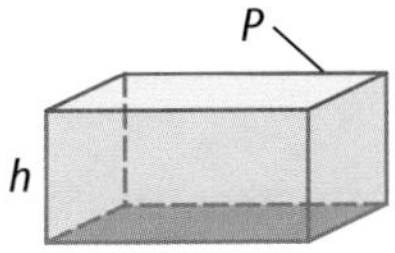	Model	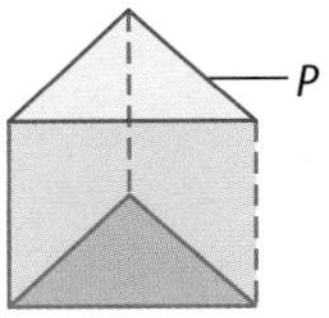

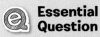

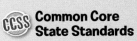

Example 1

Tools Tutor

Find the lateral and surface area of each prism.

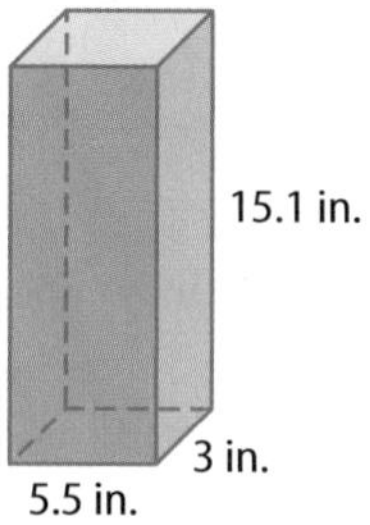

a. Find the lateral area.

$L = Ph$

$= (2\ell + 2w)h$

$= (2 \cdot 5.5 + 2 \cdot 3)15.1$

$= 256.7 \text{ in}^2$

Find the surface area.

$S = L + 2B$

$= L + 2\ell w$

$= 256.7 \text{ in}^2 + 2 \cdot 5.5 \text{ in.} \cdot 3 \text{ in.}$

$= 256.7 + 33 \text{ or } 289.7 \text{ in}^2$

Lateral Area
The lateral area of a rectangular prism may vary, depending on which faces you choose for the bases.

b. Find the lateral area. The lateral area is made up of faces that are *not* parallel.

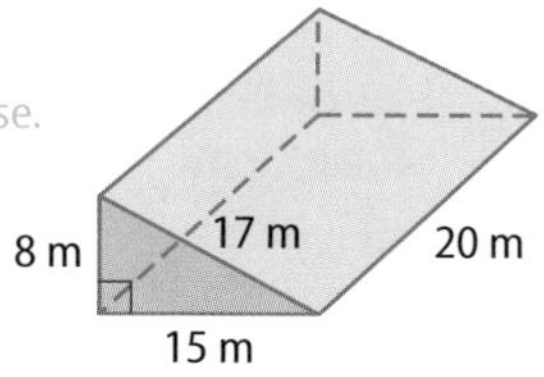

$L = Ph$	Formula for the lateral area
$= (8 + 15 + 17)20$	P is the perimeter of the triangular base.
$= 40 \cdot 20 \text{ or } 800 \text{ m}^2$	Simplify.

Find the surface area.

$S = L + 2B$	Formula for the surface area
$= 800 + 2\left(\frac{1}{2} \cdot 15 \cdot 8\right)$	$B = \frac{1}{2}bh$
$= 800 + 120 \text{ or } 920 \text{ m}^2$	Simplify.

Nets
Another way to find the surface area of any prism is to draw its net on grid paper and find the area of each face.

Got It? Do these problems to find out.

1a.

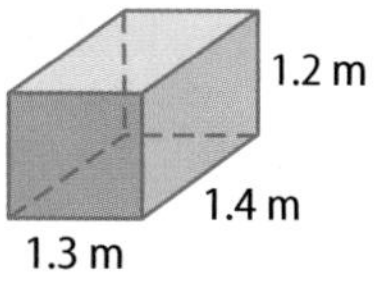

1b.

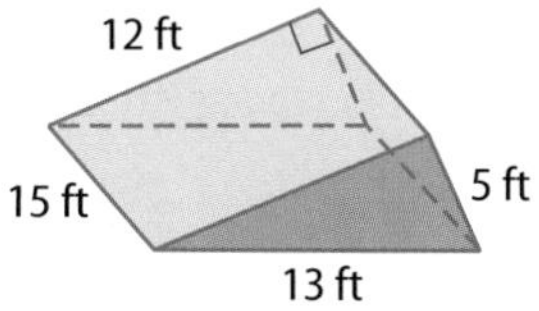

Applying Surface Area

Use surface area formulas to find how much to buy when covering objects.

Real World Example 2

Tools Tutor

Lucia is covering a box with fabric. It is a rectangular prism and measures 10 inches wide, 14 inches long, and 5 inches high. Find the amount of fabric she needs.

Find the lateral area.

$L = Ph$

$= (2 \cdot 14 + 2 \cdot 10)5$

$= (28 + 20)5$

$= 240$

Find the surface area.

$S = L + 2B$

$= 240 + 2 \cdot 14 \cdot 10$

$= 240 + 280$

$= 520 \text{ in}^2$

She will need 520 square inches of fabric.

Got It? Do this problem to find out.

2. Marc is painting a skateboard ramp like the one shown. Find the surface area of the ramp. Then determine how much paint he needs if 1 quart covers 100 square feet.

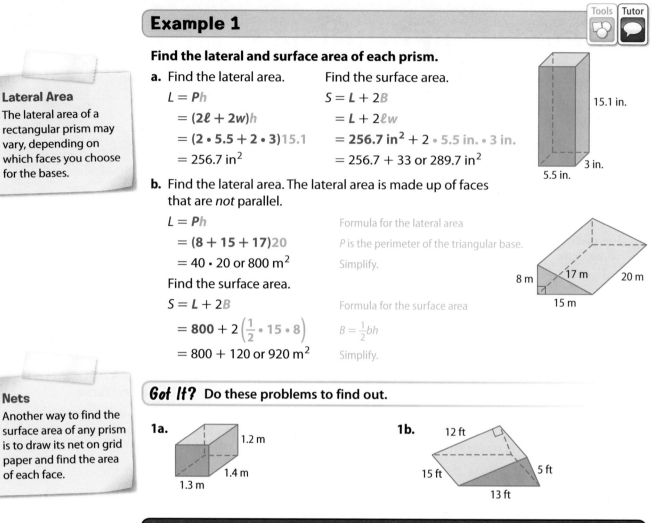

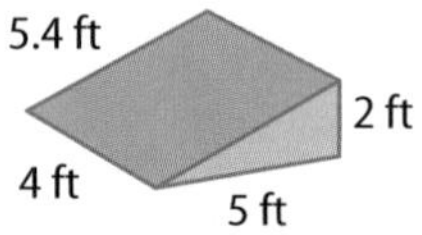

Guided Practice

Check

Find the lateral and surface area of each prism. (Example 1)

1.

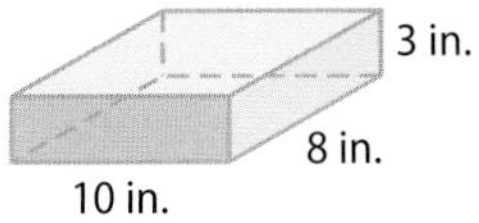

2.

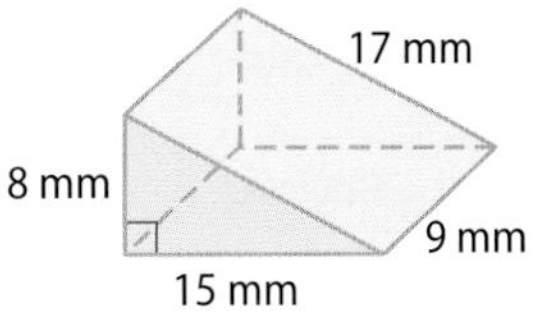

3. rectangular prism: length $5\frac{1}{2}$ yards, width $9\frac{1}{2}$ yards, height 12 yards

4. Find the amount of cardboard needed to make a box for a single slice of pizza. The box is in the shape of a triangular prism as shown. (Example 2)

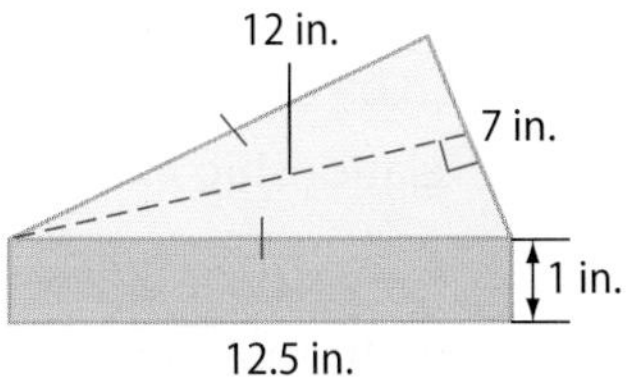

Independent Practice

Go online for Step-by-Step Solutions

eHelp

Find the lateral and surface area of each prism. (Example 1)

5.

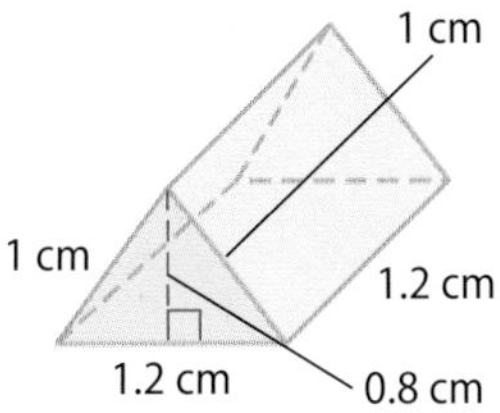

6.

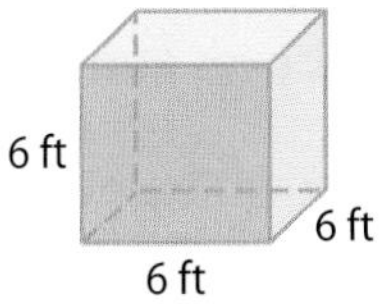

7.

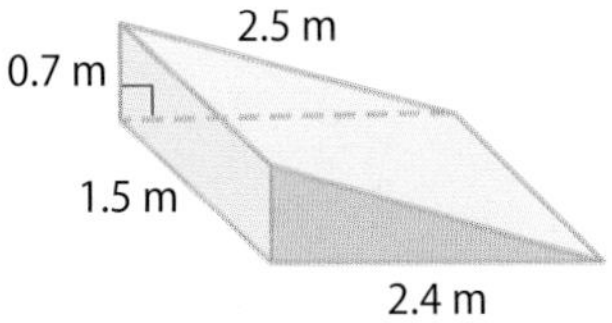

8.

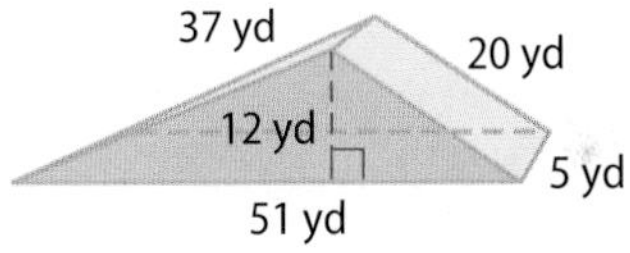

9 rectangular prism: length 2.2 centimeters, width 1.4 centimeters, height 9 centimeters

10. rectangular prism: length $6\frac{1}{2}$ inches, width $8\frac{3}{4}$ inches, height 10 inches

11. Find the amount of paper used to cover the juice box at the right. (Example 1)

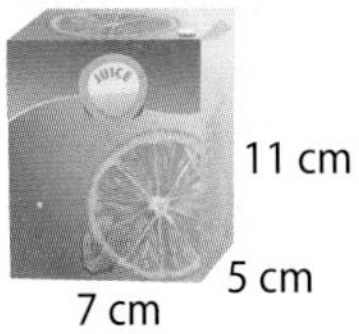

12. Hinto is planning to paint the walls of a bedroom that is 20 feet long, 15 feet wide, and 8 feet high. If Hinto has 1 gallon of paint that covers 400 square feet, how many additional gallons of paint does he need? (Example 2)

Find the surface area of each prism.

13.

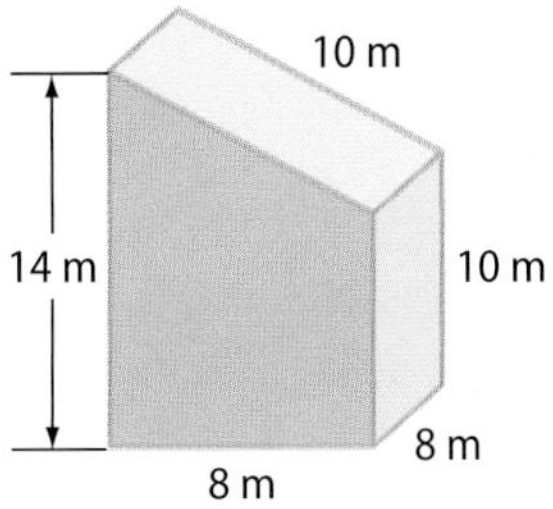

14.

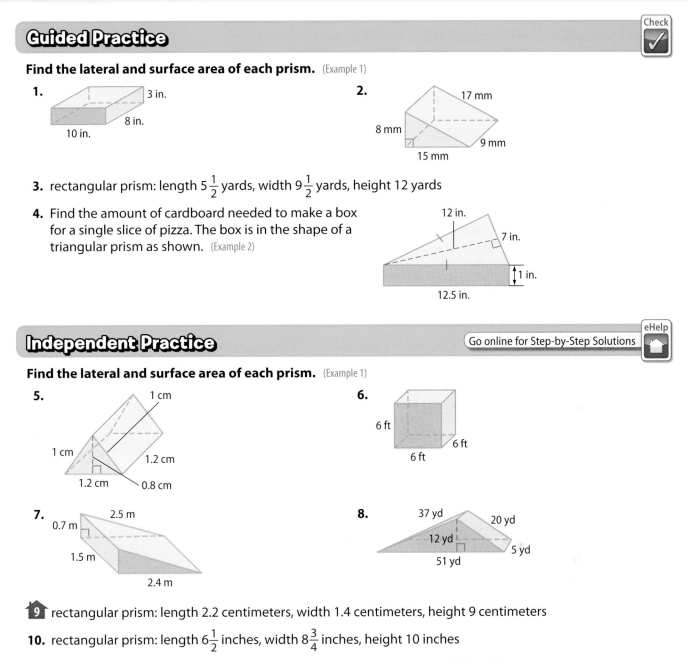

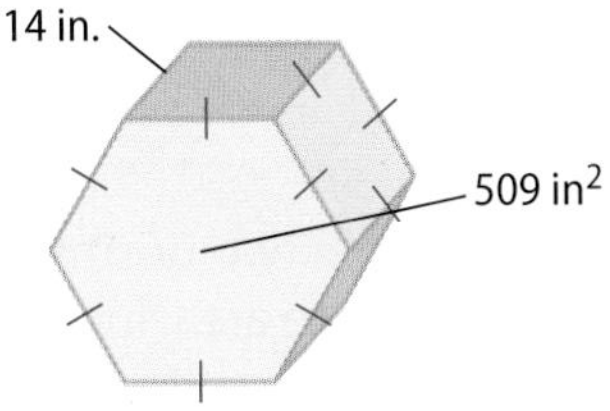

15. **CCSS Multiple Representations** In this problem, you will explore how dilations affect surface area. Use the cube at the right.

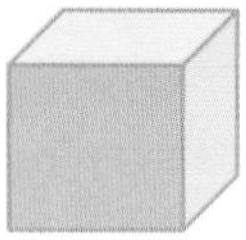

a. **Numbers** Find the surface area of the cube.

b. **Table** Copy and complete the table shown.

Scale Factor of Dilation	Original	× 2	× 3
Length of Side	■	■	■
Surface Area	■	■	■

c. **Analyze** Compare the original surface area to the surface area after a dilation by a scale factor of 2, then by a scale factor of 3.

d. **Words** Does this same relationship exist with any rectangular prism? Explain your reasoning.

16. A shipping box in the shape of a rectangular prism can hold 576 cubic inches of material. The length of the box is 12 inches and the width of the box is 8 inches.

a. Find the surface area of the box.

b. Another box with the same volume has dimensions of 16 inches by 9 inches by 4 inches. Which box has the greater surface area? Explain.

c. Predict the shape of a rectangular prism with the same volume but with a greater surface area than either of the boxes above. Then, test your prediction.

17. **Financial Literacy** The boxes shown below are puzzle boxes that are made from wood. If the wood to make the boxes costs $1.30 per square inch, which box would cost more to make? Explain.

Box 1

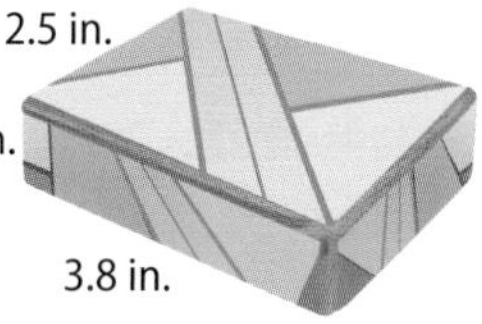

Box 2

H.O.T. Problems Higher Order Thinking

18. **CCSS Reason Abstractly** Name the dimensions of a cube with a surface area between 10 and 20 square inches.

19. **CCSS Persevere with Problems** A box manufacturer wants to make a box that has a volume of 216 cm^3 and uses the least amount of cardboard possible. Find the dimensions of that box.

20. **CCSS Find the Error** Serena calculated the surface area of the cube shown. Find her error and correct it.

$2 \cdot 2 \cdot 2 = 8\ cm^3$

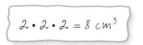

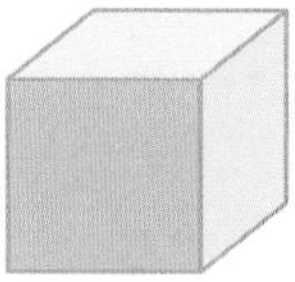

21. **CCSS Identify Structure** Sketch two prisms such that one has a greater volume and the other has a greater surface area. Include real-world units.

22. **Building on the Essential Question** Explain the difference between surface area and volume.

Standardized Test Practice

23. A rectangular cardboard box has the same volume as another rectangular box that measures 6 inches by 14 inches by 20 inches, but with less surface area. Which size box would not meet those requirements?

A 7 in. by 10 in. by 24 in.
B 7 in. by 12 in. by 20 in.
C 10 in. by 12 in. by 14 in.
D 5 in. by 16 in. by 21 in.

24. Short Response The edge lengths of a rectangular prism are tripled. What is the ratio, written as a fraction, of the surface area of the original prism to the surface area of the larger prism?

25. Find the surface area of the figure below to the nearest whole number.

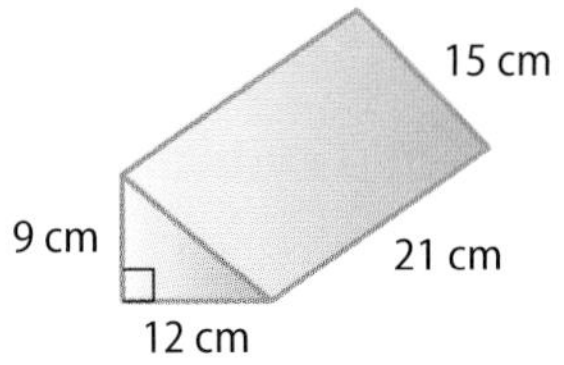

F 1065 cm^2 **H** 777 cm^2
G 945 cm^2 **J** 864 cm^2

26. How much cardboard is needed to make a rectangular box of cereal that measures 10 inches by 8 inches by 2 inches?

A 116 in^2 **C** 232 in^2
B 160 in^2 **D** 248 in^2

Common Core Review

Find the volume of each figure. Round to the nearest tenth. 8.G.9

27.
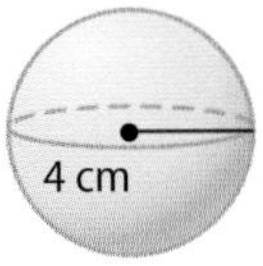

28.
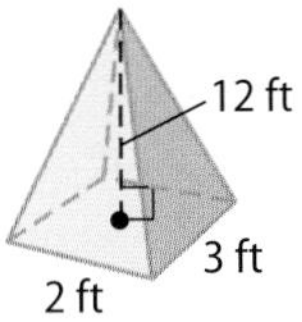

29.
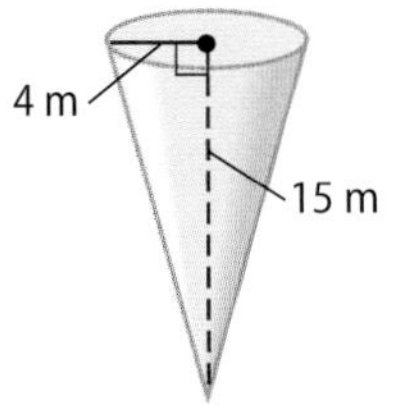

30. A water trough as shown at the right is used to provide water for farm animals. Find the volume of the trough. Round to the nearest tenth. 8.G.9

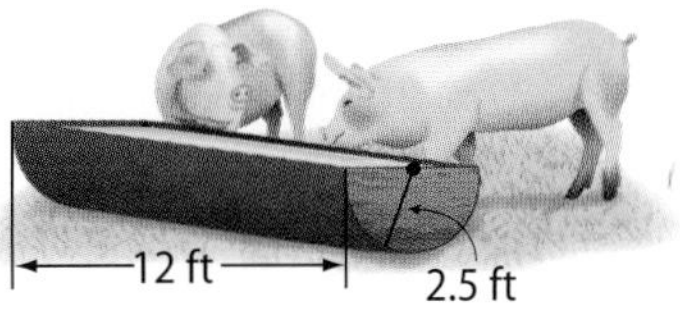

Evaluate each expression if $a = 3$, $b = 5$, and $c = 4$. 8.EE.1

31. b^4

32. c^3

33. $2(3c + 7)^2$

34. $4(2a - c^3)^2$

35. $5(2a + b^3)$

36. $a^0 + b^2 + c^3$

Find the area of each figure. Round to the nearest tenth. 7.G.4, 7.G.6

37.
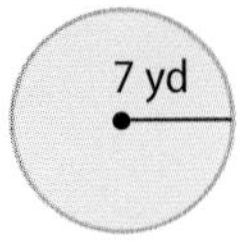

38.
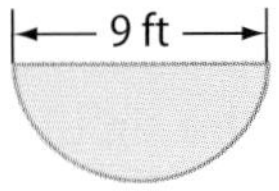

39.
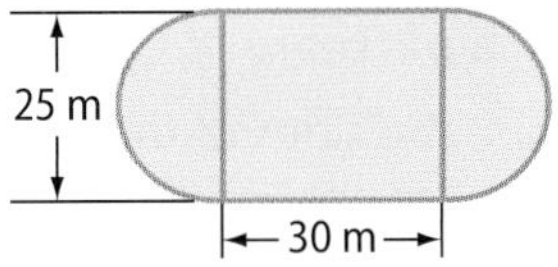

40. An ounce of prevention is said to be worth a pound of cure. 6.RP.3d

a. How many ounces of cure is 2.5 ounces of prevention worth?

b. How many ounces of prevention are worth 72 ounces of cure?

Inquiry Lab

Surface Area of Cylinders

Inquiry **HOW can you use nets to develop a formula for the surface area of a cylinder?**

CCSS Content Standards
7.G.4
Mathematical Practices
1, 3, 6

Manufacturing A cereal company plans to package its oatmeal crisps in a cylindrical cardboard container. The container has the dimensions shown at the right. How much cardboard is needed to manufacture the container?

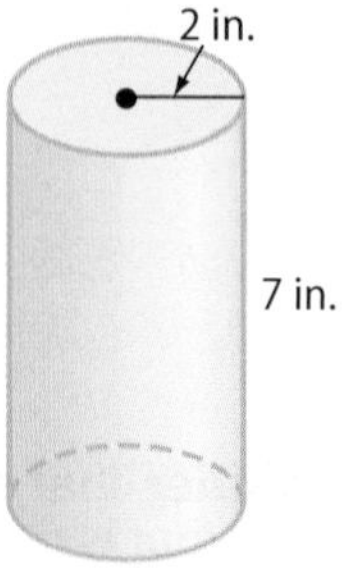

Investigation

You can make a net to help you find the surface area of the cylinder.

Step 1 Use an empty cylinder-shaped container that has a lid. Outline the bases on blank paper and cut them out.

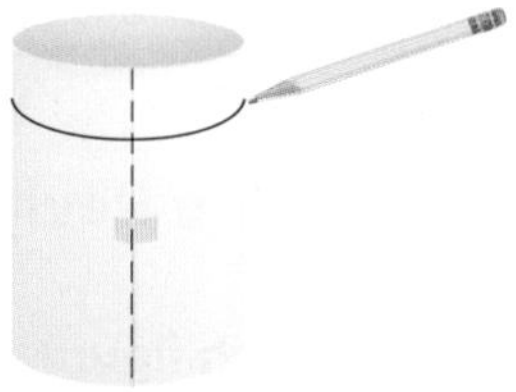

Step 2 Wrap paper around the curved surface of the cylinder and tape it in place. Draw a line from the bottom of the cylinder to the top along the edge of the paper. Draw a line around the top base of the cylinder.

Step 3 Unroll the paper. Cut the paper along the marked line.

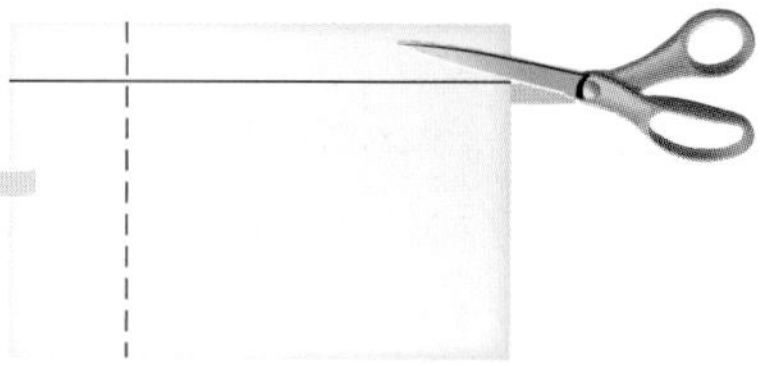

Step 4 Tape the bases and the side together so that they can be re-folded to make the original cylinder. This figure is the net of the cylinder.

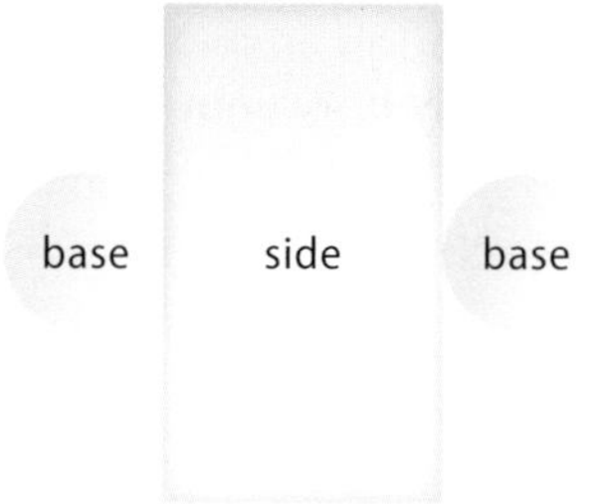

Collaborate

Work with a partner.

1. Measure the sides of the rectangle and the diameter of the circular bases. Find the area of each shape. Then find the sum of these areas.
2. Find the circumerence of the circular base. How does this measurement compare to one of the side lengths of the rectangle?
3. Multiply the circumference by the height of the container. What does this product represent?
4. Add the product from Exercise 3 to the sum of the areas of the two circular bases.

5. How does your answer from Exercise 4 compare to the surface area you found in the Investigation?

6. Use the steps described in Exercises 3 and 4 to find the amount of cardboard needed for the cereal package described at the beginning. Round to the nearest tenth.

Analyze

7. CCSS **Be Precise** Write a formula for the surface area S of a cylinder based on the Investigation. Be sure to specify what the variables stand for.

Find the surface area of the cylinder that corresponds to each net. Round to the nearest hundredth.

8.

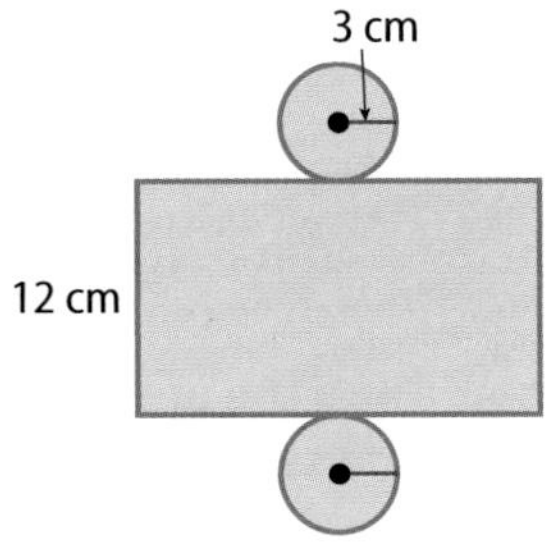

9.

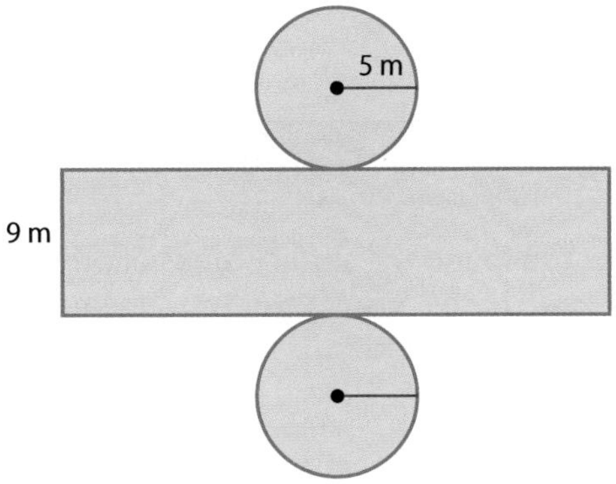

10. CCSS **Reason Inductively** How does the formula for the surface area of a cylinder change given the following?

   a. The top of the cylinder is open.

   b. Both the top and bottom of the cylinder are open.

11. CCSS **Justify Conclusions** Can the net shown below be the net for a cylinder? Explain why or why not.

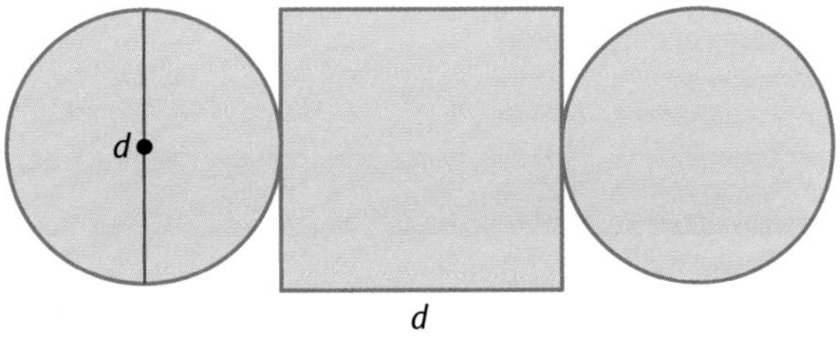

12. CCSS **Justify Conclusions** You are given two of the following measurements for a cylinder: the radius of the base, the height of the cylinder, and the circumference of the base. Which measurement *must* be included for you to be able to find the cylinder's surface area? Explain your reasoning.

Reflect

13. Inquiry HOW can you use nets to develop a formula for the surface area of a cylinder?

Lesson 12-9

Surface Area of Cylinders

ISG Interactive Study Guide

See pages 289–290 for:
- Getting Started
- Real-World Link
- Notes

Essential Question

How are two-dimensional figures used to solve problems involving three-dimensional figures?

CCSS Common Core State Standards

Content Standards
7.G.4

Mathematical Practices
1, 3, 4, 7

What You'll Learn

- Find lateral and surface areas of cylinders.
- Compare surface areas of cylinders.

Real-World Link

Food Many different kinds of food are packaged in cylinder-shaped containers. You can use what you know about circles to determine the amount of material used to make these containers.

Surface Area of Cylinders

You found the surface areas of prisms by adding the lateral area and the area of the two bases, $S = L + 2B$. You can find surface areas of cylinders in the same way. If you unroll a cylinder, its net is a rectangle (lateral area) and two circles (bases).

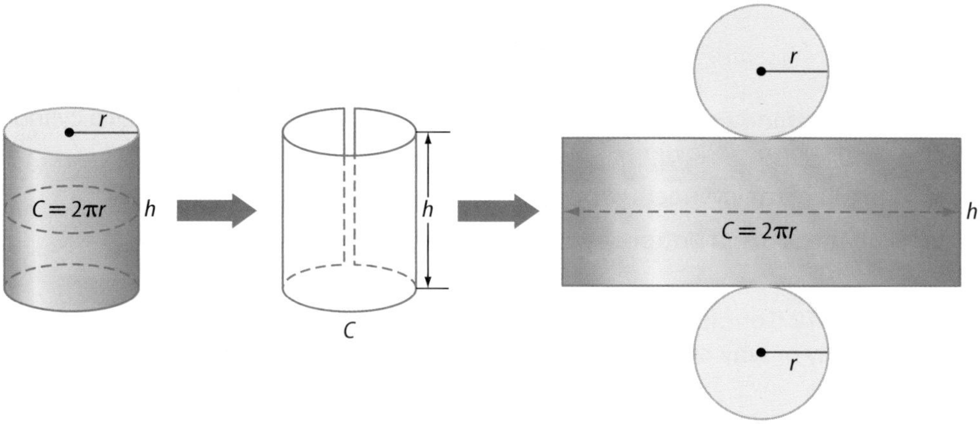

Key Concept

Watch

Lateral Area of Cylinders	
Words	The lateral area L of a cylinder with radius r and height h is the circumference of the base $(2\pi r)$ times the height h.
Symbols	$L = 2\pi rh$
Surface Area of Cylinders	
Words	The surface area S of a cylinder is the lateral area L plus the area of the two bases $(2\pi r^2)$.
Symbols	$S = L + 2B$ or $S = 2\pi rh + 2\pi r^2$

Model

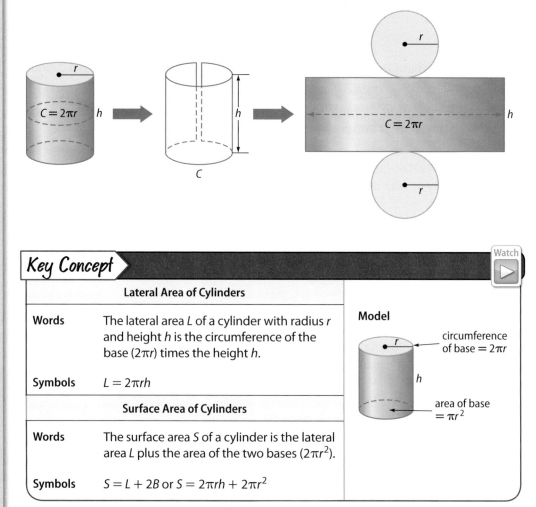

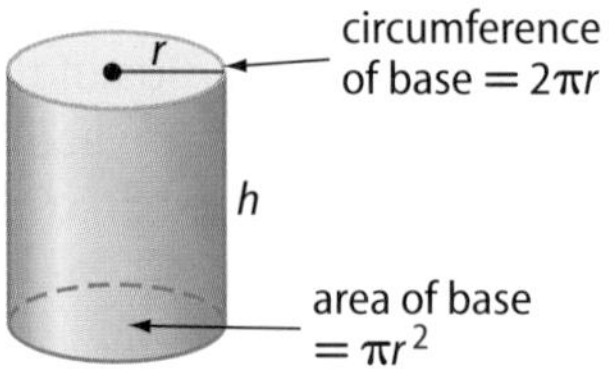

Example 1

Tools Tutor

Find the lateral area and the surface area of each cylinder.

a. Lateral Area

$L = 2\pi rh$
$= 2 \cdot \pi \cdot \mathbf{6} \cdot 12$
$= 144\pi \text{ m}^2$ exact
$\approx 452.4 \text{ m}^2$ approximate

Surface Area

$S = L + 2\pi r^2$
$= \mathbf{144\pi} + 2\pi(6)^2$
$= 216\pi \text{ m}^2$ exact
$\approx 678.6 \text{ m}^2$ approximate

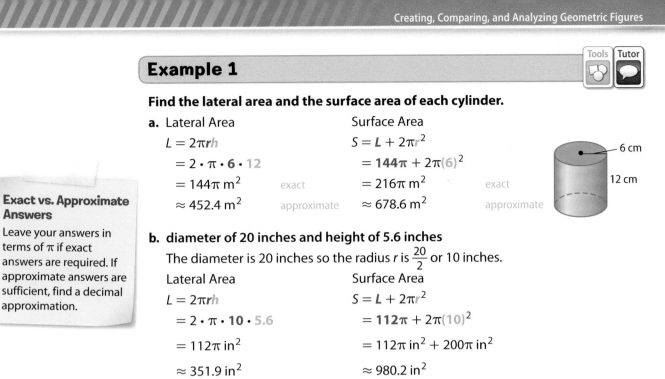

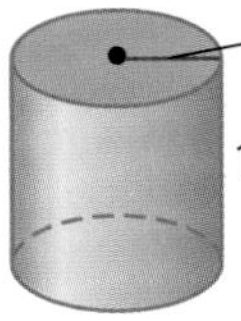

Exact vs. Approximate Answers

Leave your answers in terms of π if exact answers are required. If approximate answers are sufficient, find a decimal approximation.

b. diameter of 20 inches and height of 5.6 inches

The diameter is 20 inches so the radius r is $\frac{20}{2}$ or 10 inches.

Lateral Area

$L = 2\pi rh$
$= 2 \cdot \pi \cdot \mathbf{10} \cdot 5.6$
$= 112\pi \text{ in}^2$
$\approx 351.9 \text{ in}^2$

Surface Area

$S = L + 2\pi r^2$
$= \mathbf{112\pi} + 2\pi(10)^2$
$= 112\pi \text{ in}^2 + 200\pi \text{ in}^2$
$\approx 980.2 \text{ in}^2$

Got It? Do this problem to find out.

1. Find the lateral area and surface area of a cylinder with a radius of 7 centimeters and a height of 12 centimeters. Round to the nearest tenth.

Compare Surface Areas

Changing one dimension of a cylinder does not necessarily mean that surface areas change in the same way.

Example 2

Tutor

Isabel is making two cylindrical candles, each with a radius of 1.5 inches. One candle is 4 inches tall and the other is 8 inches tall. Is the surface area of the larger candle twice that of the smaller candle? Explain.

Find the surface areas of both candles.

Surface Area of Candle A

$S = L + 2\pi r^2$
$= 2\pi rh + 2\pi r^2$
$= 2\pi \cdot \mathbf{1.5} \cdot 4 + 2\pi \cdot (\mathbf{1.5})^2$
$= 12\pi + 4.5\pi$ or 16.5π

Surface Area of Candle B

$S = L + 2\pi r^2$
$= 2\pi rh + 2\pi r^2$
$= 2\pi \cdot \mathbf{1.5} \cdot 8 + 2\pi \cdot (\mathbf{1.5})^2$
$= 24\pi + 4.5\pi$ or 28.5π

The surface area of candle B is *not* twice the surface area of candle A. The lateral area of candle B is twice the lateral area of candle A, but the base areas are equal.

Got It? Do this problem to find out.

2. Which can has a greater surface area: a tuna fish can with diameter 8 centimeters and height 4 centimeters or a soup can with a diameter 4 centimeters and height 8 centimeters?

Guided Practice

Check

Find the lateral and surface area of each cylinder. Round to the nearest tenth. (Example 1)

1.

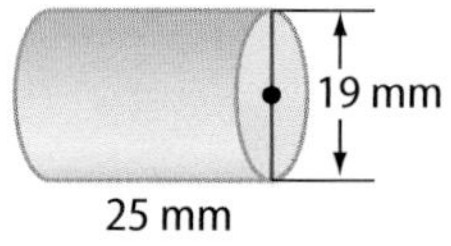

2.

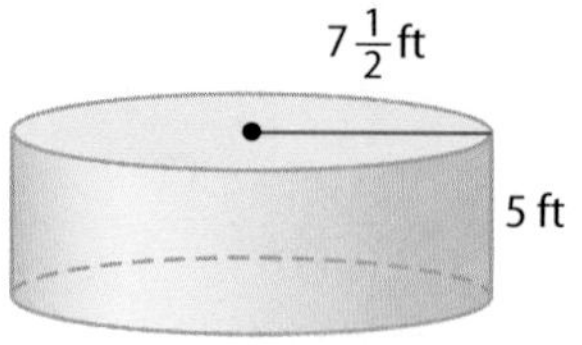

3. diameter of 4.2 meters and a height of 7.8 meters

4. A can of frozen orange juice has the dimensions shown. A soup can has a height of 5.5 inches and radius of 2 inches. Is more paper needed to make the label on the orange juice can or the soup can? Explain. (Example 2)

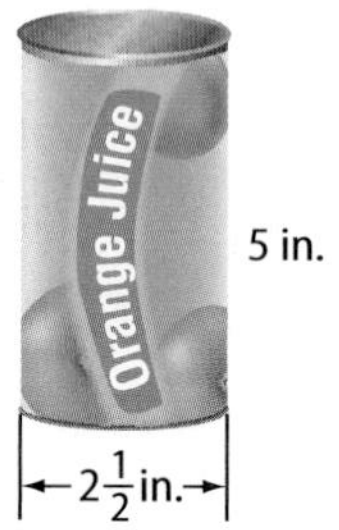

Independent Practice

eHelp

Go online for Step-by-Step Solutions

Find the lateral and surface area of each cylinder. Round to the nearest tenth. (Example 1)

5.

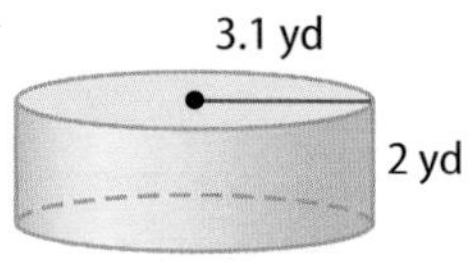

6.

7.

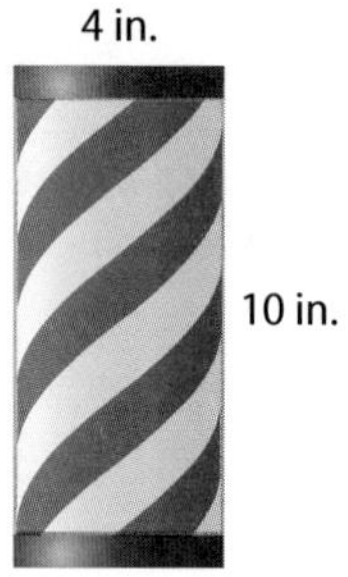

8.

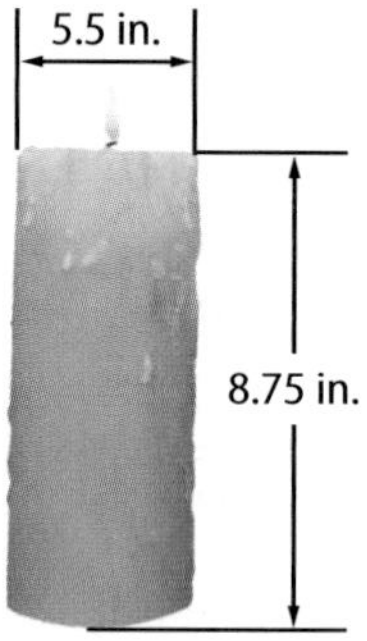

9. radius of $3\frac{2}{3}$ feet and a height of $5\frac{1}{4}$ feet

10. diameter of 9 meters and a height of 7.3 meters

11. The mailing box and mailing tube shown at the right have about the same volume. Which one has a smaller surface area? Explain. (Example 2)

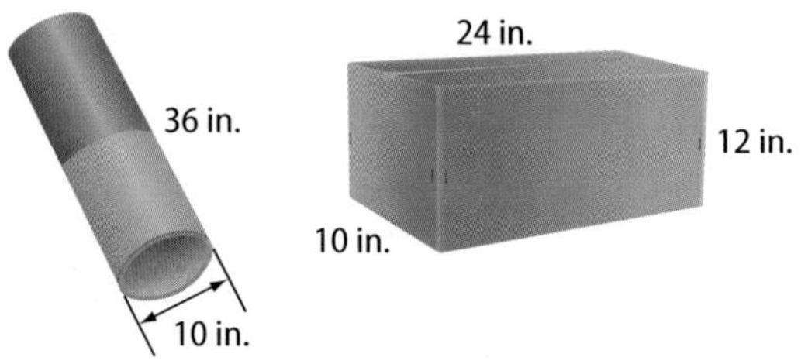

12. Taryn is painting pillars. One pillar is 0.75 meter tall and 0.15 meter in diameter. Another pillar is 0.30 meter in diameter and 0.38 meter tall. Which pillar needs more paint?

13. Find the lateral area (exterior and interior) of the pipe shown. Round to the nearest tenth of a square inch.

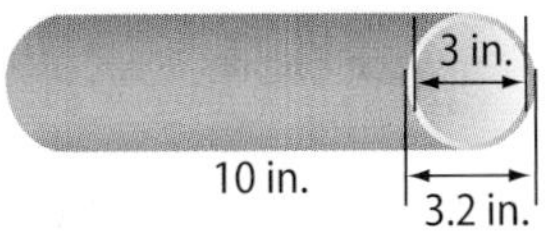

Find the surface area of each figure. Round to the nearest tenth.

14.

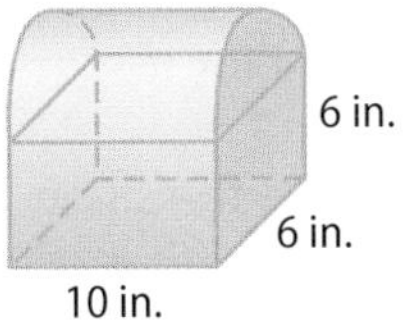

15.

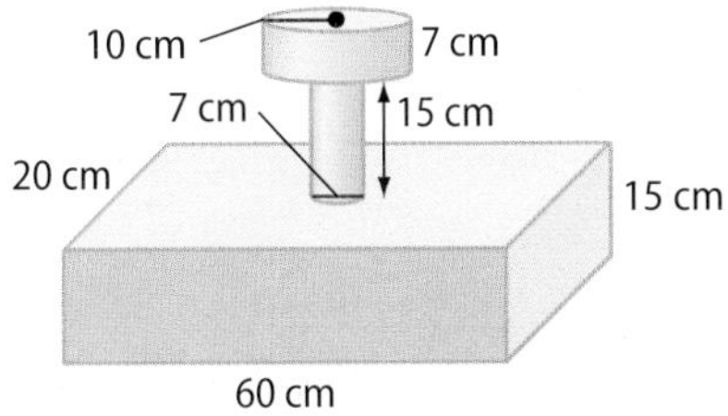

16. A penny's mass is exactly half the mass of a nickel. Using the information in the table, determine the ratio between their surface areas. Round to the nearest tenth. Write as a fraction in simplest form.

Coin	Penny	Nickel
Mass	2.5 g	5.0 g
Diameter	19.05 mm	21.21 mm
Thickness	1.55 mm	1.95 mm

17. Financial Literacy The Student Council is planning a movie night. They will sell popcorn in one of the open-top containers shown. The cost depends on the amount of cardboard used to make each container. Which container should they buy? Use volume and surface area measurements to explain your choice.

18. Find the height of a cylinder if the surface area is 402 square centimeters and the radius is 4 centimeters. Round to the nearest tenth.

H.O.T. Problems Higher Order Thinking

19. CCSS **Identify Structure** Create a prism that has approximately the same surface area as the cylinder shown at the right.

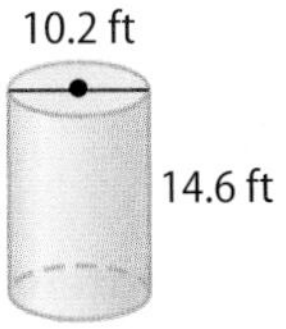

20. CCSS **Persevere with Problems** A cylinder has a radius of 10 centimeters and a height of 6 centimeters. Without calculating, explain whether multiplying the radius or the height by a scale factor of 0.5 would have a greater effect on the surface area of the cylinder.

21. **Building on the Essential Question** Both ice cubes below have a volume of about 22.5 cubic centimeters. Which ice cube would you expect to melt faster? Find the total surface areas to explain your reasoning.

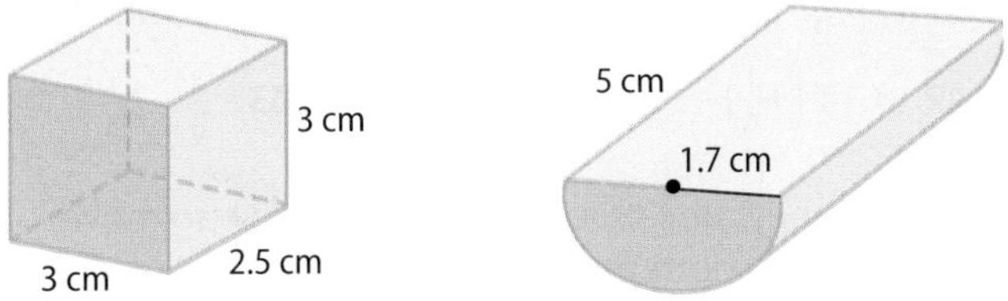

Standardized Test Practice

22. Find the amount of paper needed for the label on the can. Use 3.14 for π and round to the nearest tenth.

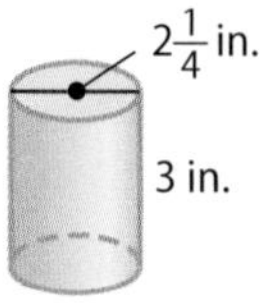

A 21.2 in^2

B 25.5 in^2

C 29.1 in^2

D 42.4 in^2

23. A cylinder has a surface area of 5652 square millimeters. If the diameter of the cylinder is 30 millimeters, what is the height? Use 3.14 for π.

F 10 mm

G 45 mm

H 59 mm

J 105 mm

24. **Short Response** What is the surface area in square meters of a cylinder with a radius of 3.4 meters and a height of 2.8 meters? Round to the nearest hundredth. Use 3.14 for π.

25. A cylindrical plastic bar is to have the same surface area as a metal bar with a radius of 1 inch and a height of 4 inches. Which of the following dimensions meet these requirements? Use 3.14 for π.

A radius: 4 in., height: 2 in.

B radius: 2 in., height: 2 in.

C radius: 2 in., height: 4 in.

D radius: 2 in., height: 0.5 in.

CCSS Common Core Review

Find the lateral and surface area of each figure. 7.G.6

26.

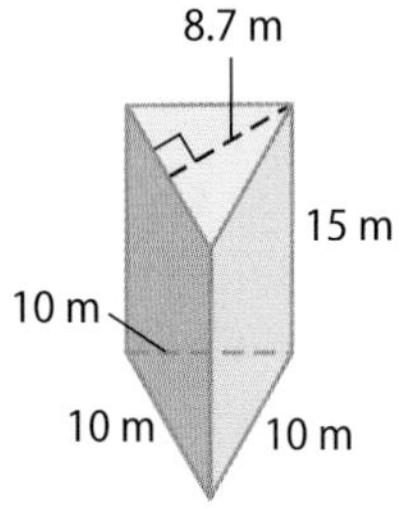

27.

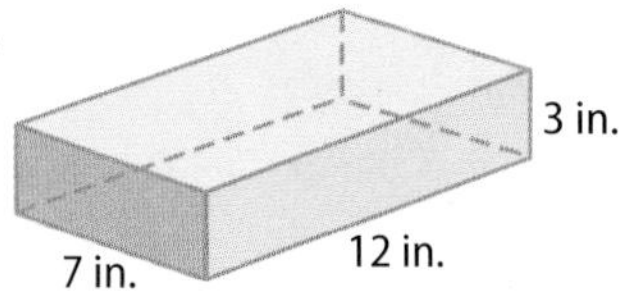

28.

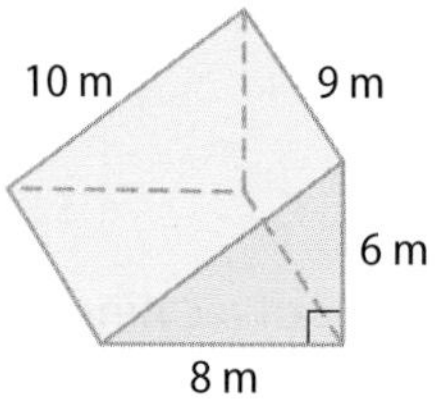

29. The Pyramid of Cestius in Rome, Italy, is 27 meters high and has a square base 22 meters on a side. What is its volume? 8.G.9

For each angle, find the measure of the supplementary angle. 7.G.5

30. 50° **31.** 90° **32.** 43° **33.** 114°

Find the simple interest earned to the nearest cent. 7.RP.3

34. \$1500 at 7.5% for 5 years

35. \$750 at 12.25% for 10 years

36. \$625 at 5.75% for 8 years

37. \$10,150 at 4.5% for 20 years

Find each product. 7.NS.2a, 8.EE.1

38. 2.45^2

39. $4 \cdot \frac{1}{2} \cdot 3 \cdot 8$

40. $3.14(5.4)^2$

41. $7.5 \cdot \frac{1}{2} \cdot 15\frac{2}{3}$

42. $3.14\left(1\frac{1}{4}\right)^2$

43. $\frac{2}{5} \cdot 2\frac{1}{3}$

44. A game requires you to toss a 6-sided number cube twice to determine how to move on a game board. What is the probability that both numbers are at least 5? 7.SP.8a

Lesson 12-10

Surface Area of Pyramids and Cones

ISG Interactive Study Guide

See pages 291–292 for:
- Getting Started
- Real-World Link
- Notes

Essential Question

How are two-dimensional figures used to solve problems involving three-dimensional figures?

CCSS Common Core State Standards

Content Standards
7.G.6

Mathematical Practices
1, 3, 4, 7

Vocab Vocabulary

regular pyramid
slant height

What You'll Learn

- Find lateral areas and surface areas of pyramids.
- Find lateral areas and surface areas of cones.

Real-World Link

Ice Cream One popular story about the invention of the ice cream cone states that it was invented by accident at the 1904 World's Fair. An ice cream vendor ran out of paper dishes, so instead, he used rolled waffles from the booth that was next to his.

Surface Areas of Pyramids

A **regular pyramid** is a pyramid with a base that is a regular polygon. The lateral faces of a regular pyramid are congruent isosceles triangles that intersect at the vertex. The altitude or height of each lateral face is called the **slant height** of the pyramid.

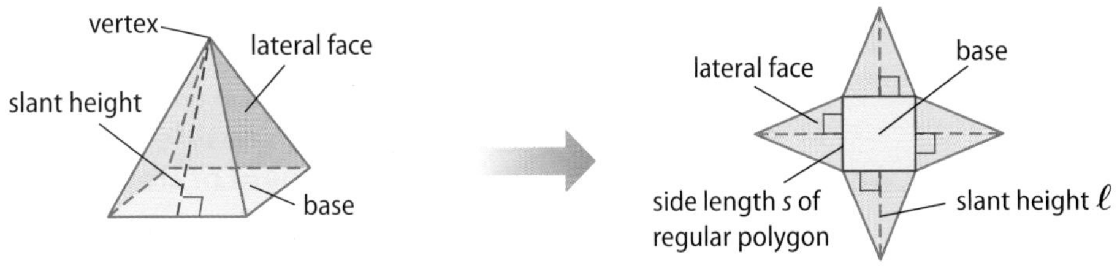

The lateral area of a pyramid is the sum of the areas of its lateral faces, which are all triangles.

$$L = 4\left(\frac{1}{2}s\ell\right) \quad \rightarrow \quad L = \frac{1}{2}(\mathbf{4s})\ell \quad \rightarrow \quad L = \frac{1}{2}\mathbf{P}\ell$$

The total surface area is the lateral surface area plus the area of the base.

Key Concept — Lateral and Surface Areas of Pyramids

	Lateral Area of Pyramids
Words	The lateral area L of a regular pyramid is half the perimeter P of the base times the slant height ℓ.
Symbols	$L = \frac{1}{2}P\ell$
	Surface Area of Pyramids
Words	The total surface area S of a regular pyramid is the lateral area L plus the area of the base B.
Symbols	$S = L + B$ or $S = \frac{1}{2}P\ell + B$

Model

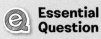

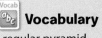

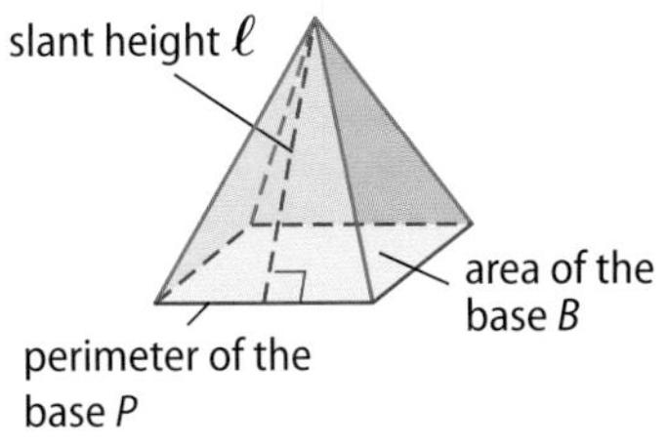

Example 1

Find the lateral and total surface area of the regular pentagonal pyramid.

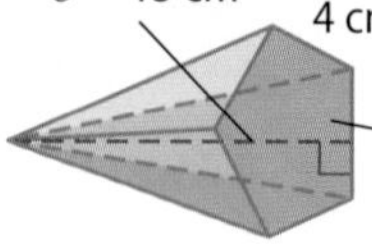

> **Height of a Pyramid and Slant Height**
> The height of a pyramid is the perpendicular distance from the top point to the base of the pyramid. The slant height of the pyramid is the altitude of one of the triangles that forms the sides of the pyramid.

Find the lateral area.

$L = \frac{1}{2}P\ell$	Write the formula.
$= \frac{1}{2}(4 \cdot 5)15$	Replace P with $4 \cdot 5$ and ℓ with 15.
$= 150 \text{ cm}^2$	Simplify.

Find the surface area.

$S = L + B$	Write the formula.
$= 150 + 27.5$	Replace L with 150 and B with 27.5.
$= 177.5 \text{ cm}^2$	Simplify.

The lateral surface area is 150 cm^2 and the total surface area is 177.5 cm^2.

Got It? Do this problem to find out.

1. Find the lateral and surface area of a square pyramid with a base side length of 6 centimeters and a slant height of 18.4 centimeters.

Real World Example 2

In 1483, Leonardo da Vinci sketched a design for a parachute in the shape of a square pyramid. If the length of each side of the base is 12 yards and the slant height is 13.4 yards, find the amount of cloth needed to make the parachute.

Sketch and label the pyramid.

Since the parachute is open on the bottom, to find the amount of cloth needed for the parachute, only find the lateral area of the pyramid.

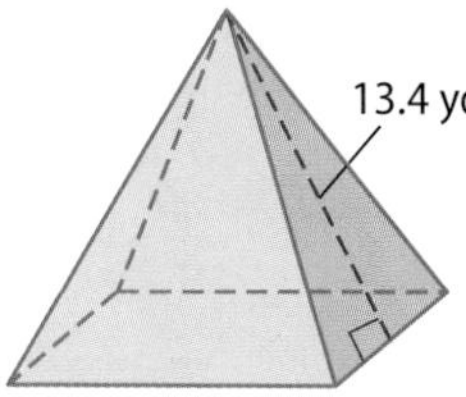

$L = \frac{1}{2}P\ell$	Write the formula for the lateral area.
$= \frac{1}{2}(12 \cdot 4)13.4$	Replace P with $12 \cdot 4$ and ℓ with 13.4.
$= 321.6 \text{ yd}^2$	Simplify.

So, 321.6 square yards of cloth are needed for the parachute.

Got It? Do these problems to find out.

2a. The pyramid at the Louvre Museum is a square pyramid with a slant height of 27 meters and base side lengths of 35 meters. Find the amount of glass on the pyramid.

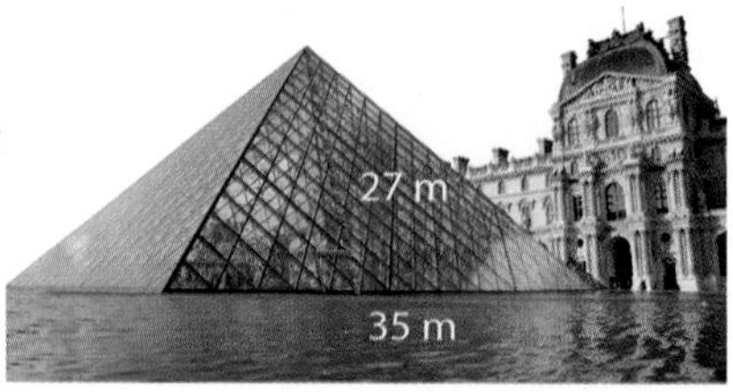

Andrew Ward/Life File/Getty Images

2b. A gift box for party favors is a sealed square pyramid with base side lengths of 3 inches and a slant height of 6 inches. How much paper is used on the gift box? Round to the nearest tenth.

Surface Area of Cones

You can also find surface areas of cones. The net of a cone shows the regions that make up the cone.

Model of Cone

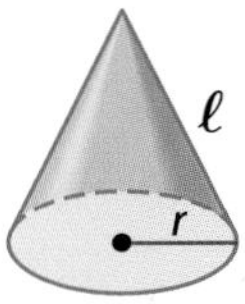

Net of Cone

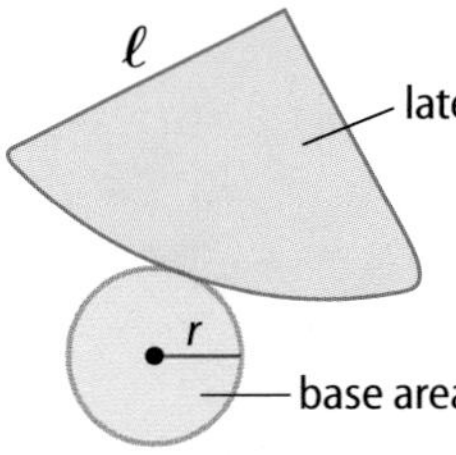

Lateral Area of a Cone

Sometimes you may be asked to find the surface area of an open-ended cone (for example, a cone-shaped hat or an ice-cream cone). Since these cones have no base, use the lateral area.

The lateral area of a cone with slant height ℓ is one-half the circumference of the base, $2\pi r$, times ℓ. So, $L = \frac{1}{2} \cdot 2\pi r \cdot \ell$ or $L = \pi r\ell$. The base of the cone is a circle with area πr^2.

Key Concept Lateral and Surface Areas of Cones

Lateral Area of Cones	
Words	The lateral area L of a cone is π times the radius times the slant height ℓ.
Symbols	$L = \pi r\ell$
Surface Area of Cones	
Words	The surface area S of a cone with slant height ℓ and radius r is the lateral area plus the area of the base.
Symbols	$S = L + \pi r^2$

Model

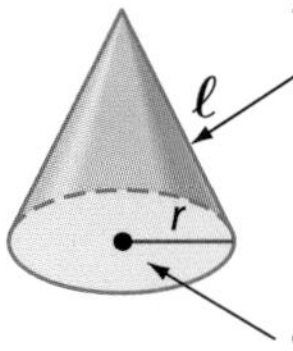

Example 3

Tools Tutor

Find the lateral and surface area of the cone. Round to the nearest tenth.

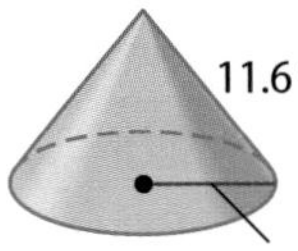

Find the lateral area.

$L = \pi r\ell$ — Lateral area of a cone

$= \pi(7.3)(11.6)$ — Replace r with 7.3 and ℓ with 11.6.

≈ 266.0 — Simplify.

Find the surface area.

$S = L + \pi r^2$ — Surface area of a cone

$= 266 + \pi(7.3)^2$ — Replace r with 7.3.

$\approx 433.4 \text{ ft}^2$ — Simplify.

The surface area of the cone is about 433.4 square feet.

Got It? Do these problems to find out.

3. Find the lateral and surface area of a cone with a radius of $4\frac{1}{4}$ yards and a slant height of 7 yards. Round to the nearest tenth.

Guided Practice

Find the lateral and surface area of each figure. Round to the nearest tenth. (Examples 1 and 3)

1.

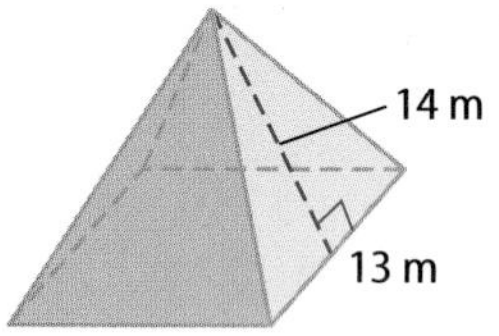

2.

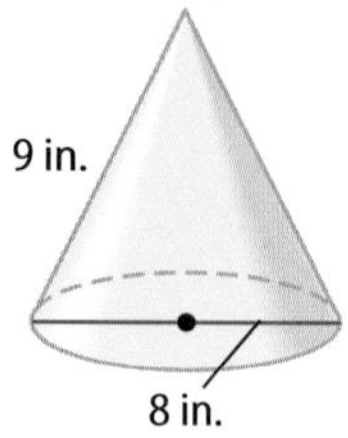

3.

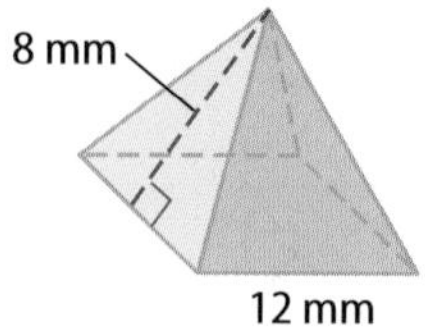

4. The Luxor Hotel in Las Vegas, Nevada, is a square pyramid made from glass with base lengths of 646 feet and a slant height of 476.3 feet. Find the surface area of the glass on the Luxor. (Example 2)

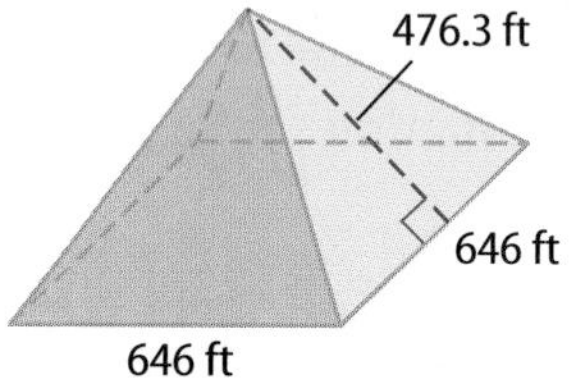

Independent Practice

Go online for Step-by-Step Solutions

Find the lateral and surface area of each figure. Round to the nearest tenth. (Example 1)

5

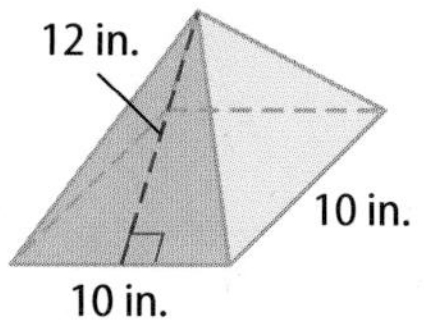

6.

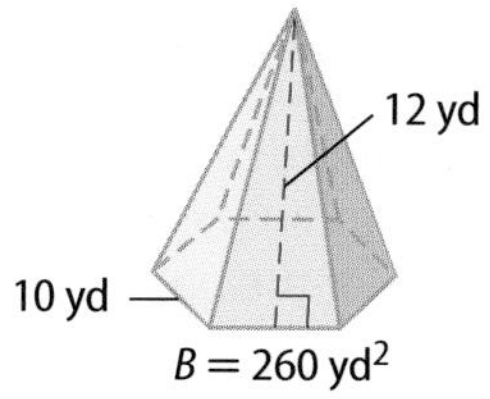

7.

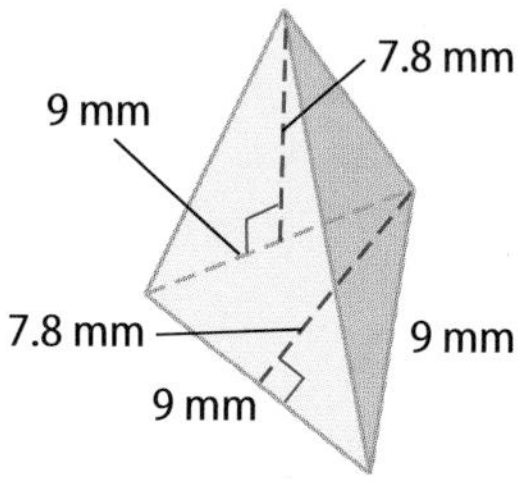

8. triangular pyramid: base side length 6 in., base area $15\frac{3}{5}$ in^2, slant height 8 in.

9. Brianne is making a necklace using the cone-shaped bead shown at the right. Find the approximate surface area of the silver bead. Include the bottom of the bead. (Example 2)

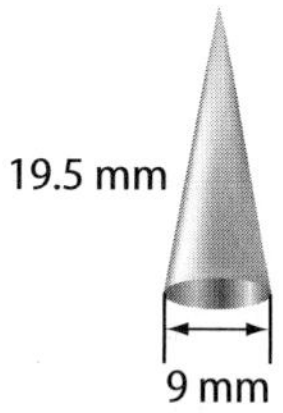

10. The Transamerica Building in San Francisco is shaped like a square pyramid. It has a slant height of about 856 feet, and each side of its base is 145 feet. Find the lateral area of the building. (Example 2)

Find the lateral and surface area of each figure. Round to the nearest tenth. (Example 3)

11.

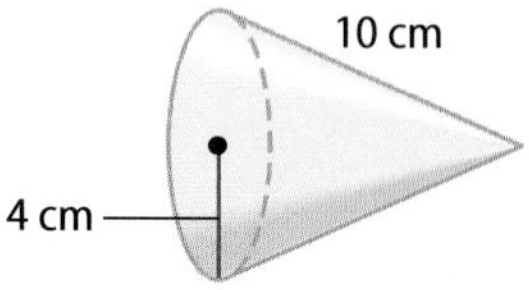

12.

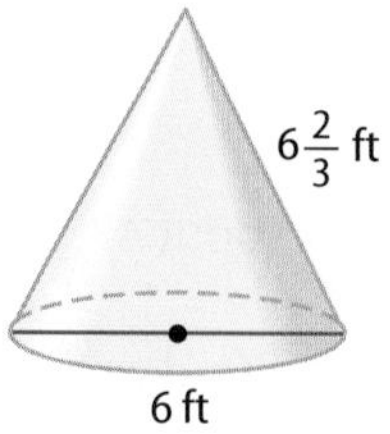

13.

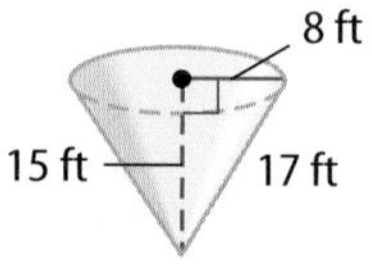

14. cone: diameter 19 cm, slant height 30 cm

Find the surface area of each figure. Round to the nearest tenth.

15.

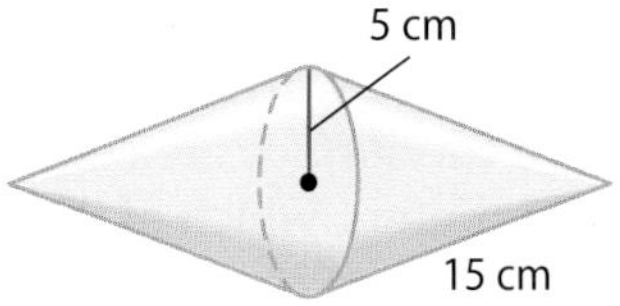

16.

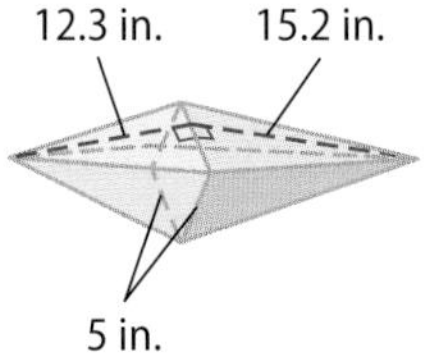

17 Adrienne is making costumes for the school play. She needs to make eight conical medieval hats. She wants each hat to have a slant height of 18 inches with a base circumference of 22 inches. How much material will she use to make the hats? Round to the nearest tenth.

Draw the figure represented by each net. Then find the lateral and total surface area of each figure.

18.

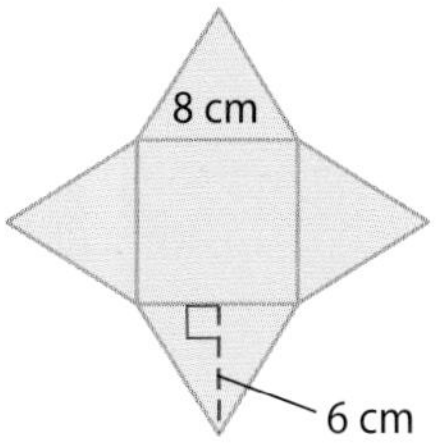

19.

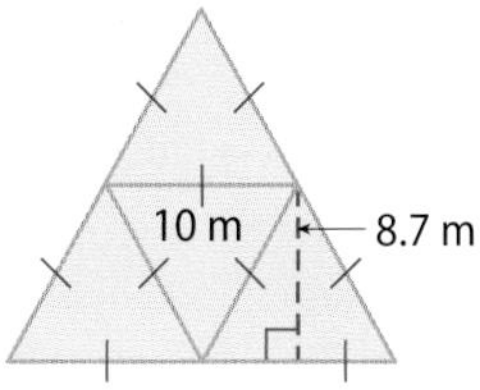

20.

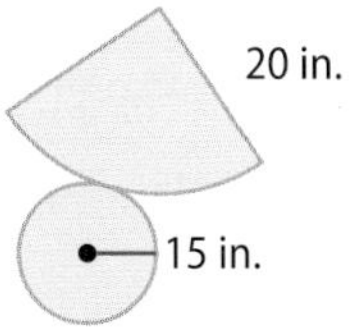

21. A can of water repellant will seal a canvas that is 50 feet by 60 feet. Delsin wants to make a tent in the shape of a cone with a diameter of 30 feet. Find the slant height of the largest tent he can make with a floor that will use only one can of water repellant to seal the outside surface. Round to the nearest tenth.

22. CCSS **Multiple Representations** In this problem, you will examine surface areas of pyramids.

a. **Table** Find the surface area of a square pyramid with a base side length of 1 and a slant height of 10. Then complete the table.

Side Length	1	2	3	4
Slant Height	10	20	30	40
Surface Area	■	■	■	■

b. **Analyze** What happens to the surface area if the base length and slant height are doubled, tripled, or multiplied by 4?

c. **Words** Predict the surface area of a pyramid with the side length of 5 and slant height of 50. Then check your prediction.

H.O.T. Problems Higher Order Thinking

23. CCSS **Identify Structure** Draw a cone with a surface area that is between 100 and 150 square units.

24. CCSS **Justify Conclusions** Which has a greater surface area: a square pyramid with a base of x units and a slant height of ℓ units or a cone with a diameter of x units and a slant height of ℓ units? Explain your reasoning.

25. CCSS **Persevere with Problems** The dimensions of the prism shown are increased by a scale factor and the total surface area of the new prism is 240,000 square meters. What was the scale factor used to create the new prism?

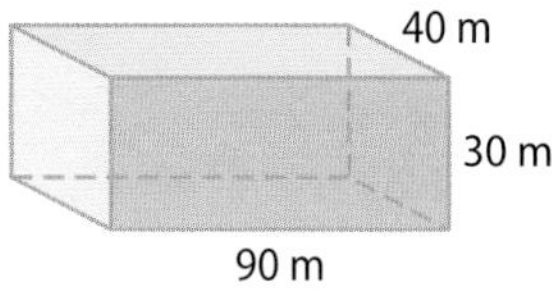

26. **Building on the Essential Question** Explain how to find the surface area of a cone if you are given the diameter and the slant height of the cone.

Standardized Test Practice

27. Find the surface area of a cone with a radius of 6 centimeters and slant height of 15 centimeters. Round to the nearest tenth.

A 141.3 cm^2

B 282.7 cm^2

C 395.8 cm^2

D 565.2 cm^2

28. Find the area of the ice cream cone covered by the wrapper.

F 9.4 in^2

G 11.2 in^2

H 15.5 in^2

J 20.0 in^2

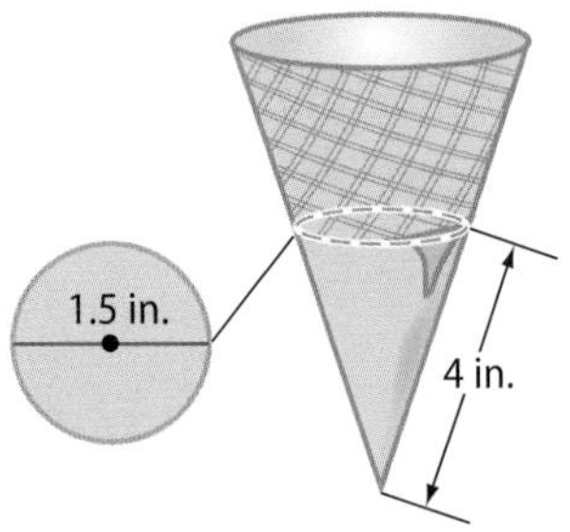

29. A square pyramid has a base with sides measuring 7 meters. If the surface area of the pyramid is 189 square meters, which of the following is the slant height of the pyramid?

A 8 m

B 10 m

C 14 m

D 16 m

30. **Short Response** A square pyramid has a surface area of 39 square feet. After a dilation, the surface area is 351 square feet. What was the scale factor of the dilation?

Common Core Review

For exercises 31–33, find the lateral and surface area of each three-dimensional figure. Round to the nearest tenth. 7.G.4, 7.G.6

31. cylinder: radius 6 inches, height $10\frac{1}{4}$ inches

32. cylinder: diameter 5.7 meters, height 2.3 meters

33. equilateral triangular prism: base height 10.4 feet, base length 12 feet, prism height $3\frac{1}{3}$ feet

34. Find the volume of an ice cream cone that has a radius of 28 mm and a height of 117 mm. 8.G.9

35. Mrs. Morales used the parallelogram at the right as a pattern for a paving stone for her sidewalk. If $m\angle 1$ is 130°, find $m\angle 2$. 8.G.5

36. Find the discount for a $45 shirt that is on sale for 20% off. 7.RP.3

37. A skateboard with a regular price of $80 is on sale for 15% off. If Xavier buys the skateboard on sale, then pays 5% sales tax, what is his total bill? 7.RP.3

Solve each proportion. 7.RP.3

38. $\frac{x}{4} = \frac{3}{5.6}$

39. $\frac{2}{3} = \frac{14}{n}$

40. $\frac{n}{20} = \frac{15}{50}$

41. $\frac{14}{32} = \frac{x}{8}$

42. $\frac{3}{2.2} = \frac{7.5}{y}$

43. $\frac{30}{14} = \frac{m}{1.54}$

44. Find the mean and the mean absolute deviation of the set of points scored by a player in four games. 7.SP.4

Points	12	16	30	42

Check

Chapter Review

ISG Interactive Study Guide
See pages 293–296 for:
- Vocabulary Check
- Key Concept Check
- Problem Solving
- Reflect

Lesson-by-Lesson Review

Lesson 12-1 Circles and Circumference (pp. 558–562)

Find the circumference of each circle. Round to the nearest tenth.

1.

2.

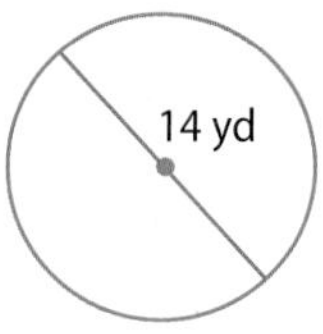

3. diameter $= 5\frac{1}{3}$ ft

4. radius $= 13.5$ mm

5. A Ferris wheel has a diameter of 250 feet. What is the distance, to the nearest whole foot, that riders travel if they stay on it for 15 rotations?

Example 1

Find the circumference of the circle. Round to the nearest tenth.

$C = 2\pi r$ — Circumference of a circle

$= 2 \cdot \pi \cdot \mathbf{9.2}$ — Replace *r* with 9.2.

≈ 57.8 — Simplify.

So, the circumference of the circle is about 57.8 meters.

Lesson 12-2 Area of Circles (pp. 563–567)

Find the area of each circle. Round to the nearest tenth.

6.

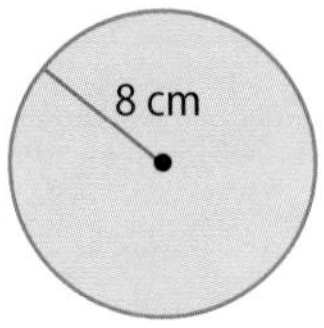

7.

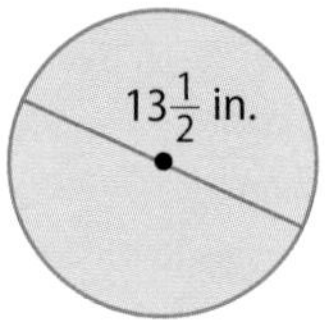

8. A pie has a diameter of 9 inches. Find the area of the top crust. Round to the nearest tenth.

Example 2

Find the area of the circle. Round to the nearest tenth.

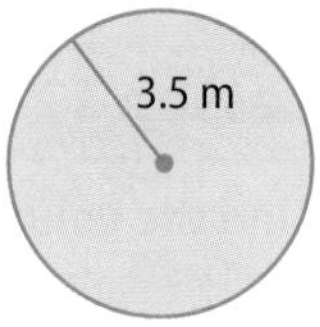

$A = \pi r^2$ — Area of a circle

$= \pi \cdot \mathbf{3.5}^2$ — Replace *r* with 3.5.

≈ 38.5 — Simplify.

So, the area of the circle is about 38.5 square meters.

Lesson 12-3 Area of Composite Figures (pp. 568–573)

9. Find the area of the composite figure.

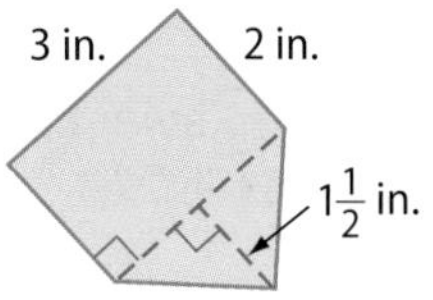

10. The floor plan of a new museum is shown. What is the area of the museum rounded to the nearest square foot?

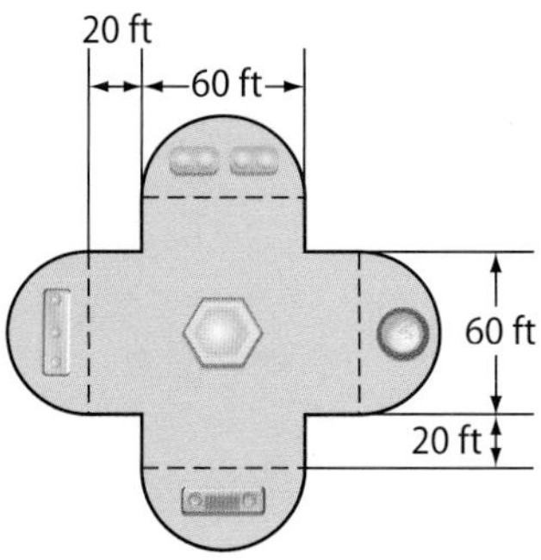

Example 3

Find the area of the composite figure.

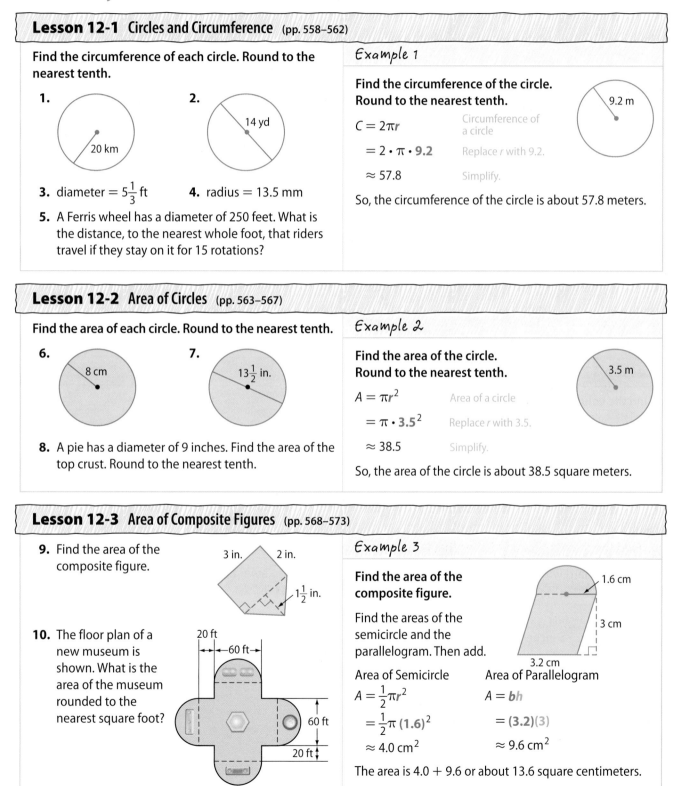

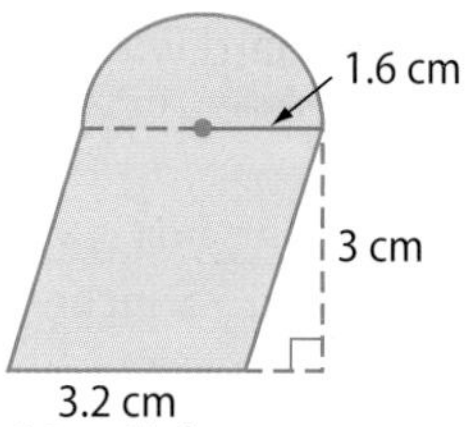

Find the areas of the semicircle and the parallelogram. Then add.

Area of Semicircle	Area of Parallelogram
$A = \frac{1}{2}\pi r^2$	$A = bh$
$= \frac{1}{2}\pi\,(\mathbf{1.6})^2$	$= (\mathbf{3.2})(3)$
$\approx 4.0 \text{ cm}^2$	$\approx 9.6 \text{ cm}^2$

The area is 4.0 + 9.6 or about 13.6 square centimeters.

Lesson 12-4 Three-Dimensional Figures (pp. 574–579)

Identify each figure. Name the bases, faces, edges, and vertices.

11.

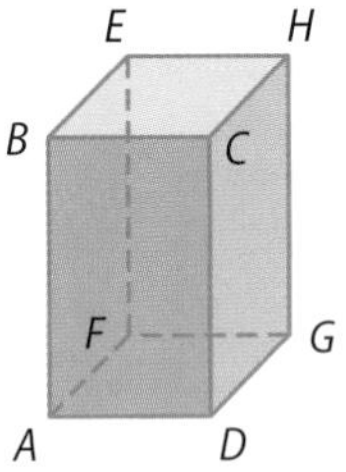

12.

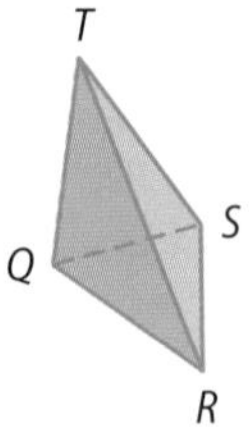

13.

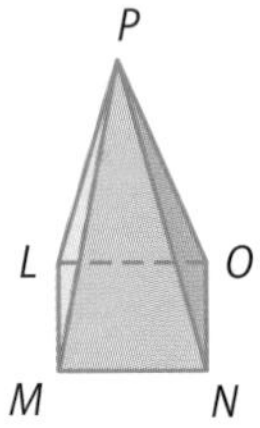

14.

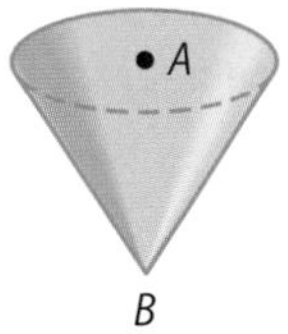

15. Draw the top view and side view of the drum. Then draw and describe the shape resulting from a vertical cross section of the figure.

Example 4

Identify the figure. Name the bases, faces, edges, and vertices.

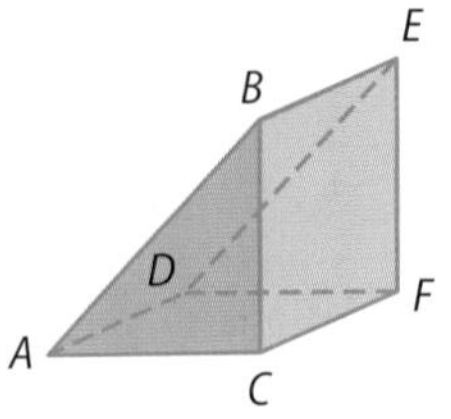

There are two congruent triangular bases, so the solid is a triangular prism.

bases: *ABC, DEF*

faces: *ABED, BCFE, ACFD, ABC, DEF*

edges: $\overline{AB}, \overline{BC}, \overline{AC}, \overline{DE}, \overline{EF}, \overline{DF}, \overline{AD}, \overline{BE}, \overline{CF}$

vertices: *A, B, C, D, E, F*

Lesson 12-5 Volume of Prisms (pp. 580–585)

Find the volume of each prism.

16.

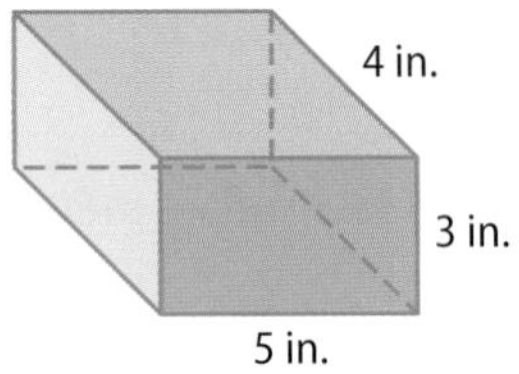

17.

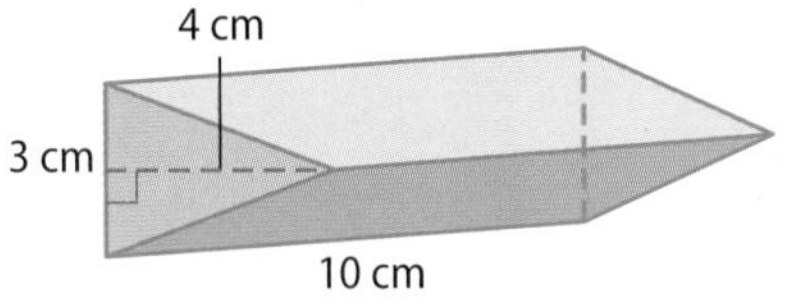

18. A shipping box is 11 inches long, 8.5 inches wide, and 5.5 inches high. What is the volume of the box?

19. Sandra is filling a keepsake storage box that is 40.5 centimeters long, 28 centimeters wide, and 17 centimeters high. What is the volume of the box?

Example 5

Find the volume of the rectangular prism.

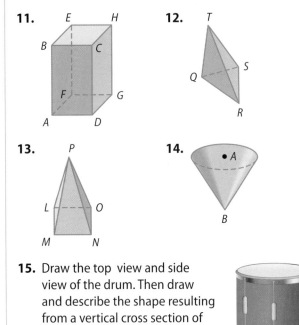

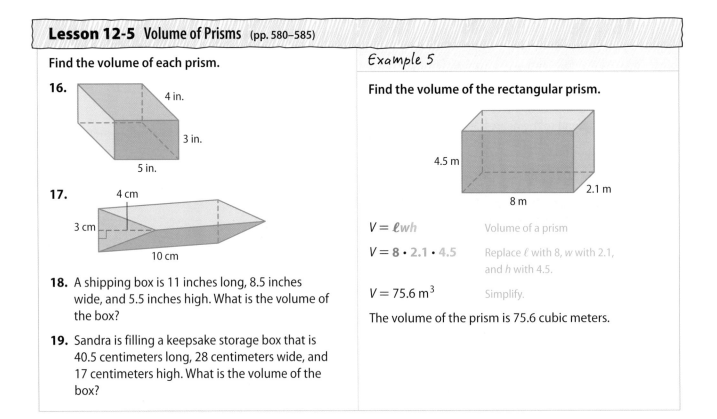

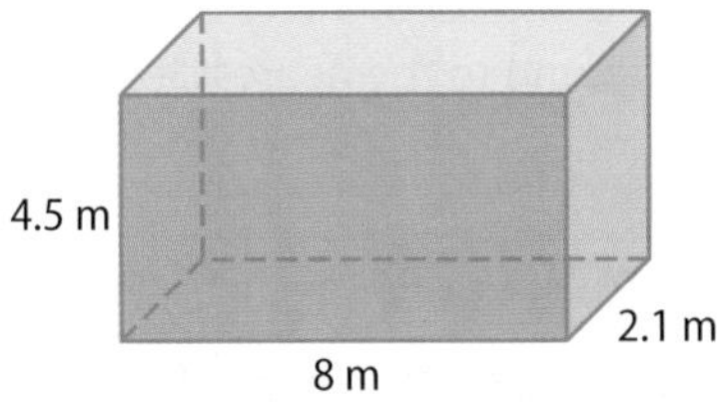

$V = \ell wh$	Volume of a prism
$V = 8 \cdot 2.1 \cdot 4.5$	Replace ℓ with 8, w with 2.1, and h with 4.5.
$V = 75.6\ \text{m}^3$	Simplify.

The volume of the prism is 75.6 cubic meters.

Lesson 12-6 Volume of Cylinders (pp. 586–590)

Find the volume of each cylinder. Round to the nearest tenth, if necessary.

20.
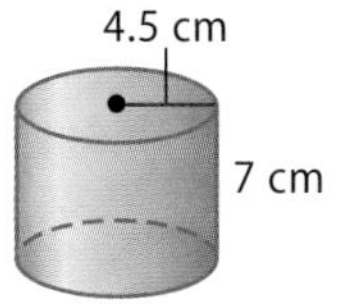

21.
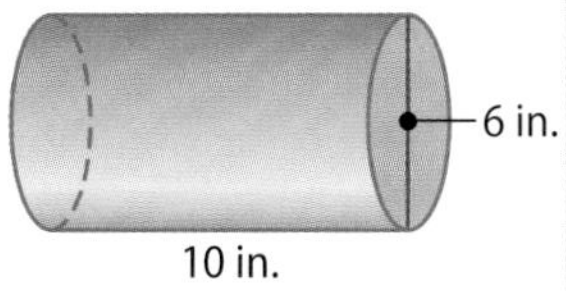

22. A 12-ounce can of soda measures $4\frac{3}{4}$ inches high with a radius of $1\frac{1}{8}$ inches. Find the amount of soda that can fit in the can. Round to the nearest tenth.

Example 6

Find the volume of the cylinder. Round to the nearest tenth, if necessary.

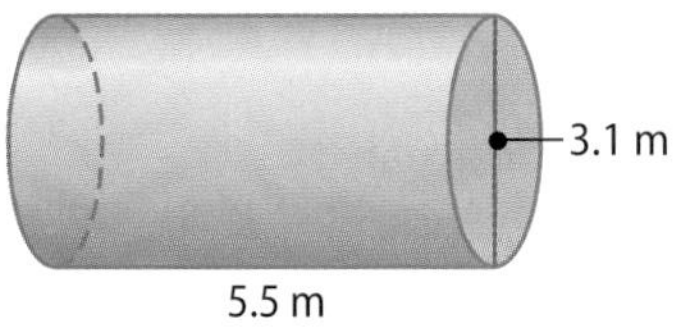

$V = \pi r^2 h$	Volume of a cylinder
$V = \pi(1.55)^2(5.5)$	Replace *r* with 1.55 and *h* with 5.5.
$V \approx 41.5 \text{ m}^3$	Simplify.

Lesson 12-7 Volume of Pyramids, Cones, and Spheres (pp. 595–600)

Find the volume of each figure. Round to the nearest tenth, if necessary.

23.
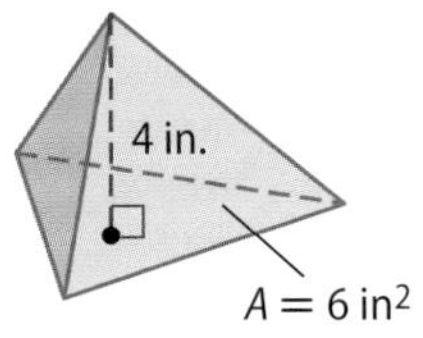

24.
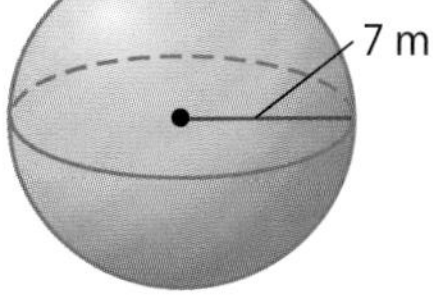

25. Mr. Owens built a conical storage shed with a base 14 feet in diameter and a height of 11 feet. What is the volume of the shed?

Example 7

Find the volume of the cone. Round to the nearest tenth, if necessary.

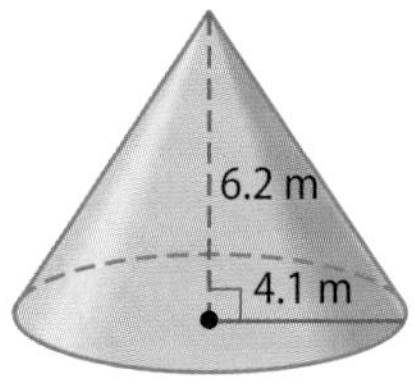

$V = \frac{1}{3}\pi r^2 h$	Volume of a cone
$= \frac{1}{3}\pi(4.1)^2(6.2)$	Replace *r* with 4.1 and *h* with 6.2.
$\approx 109.1 \text{ m}^3$	Simplify.

Lesson 12-8 Surface Area of Prisms (pp. 603–607)

Find the lateral and surface area of each prism. Round to the nearest tenth, if necessary.

26.
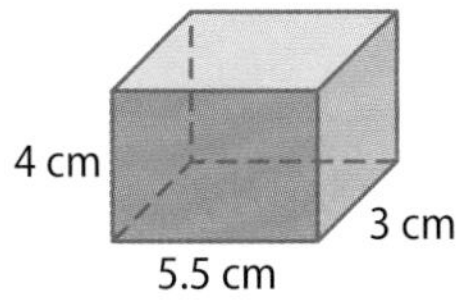

27.
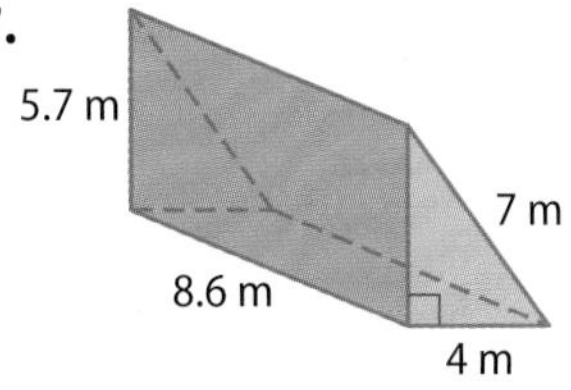

28. Sarah is wrapping a gift that is 12 inches long, 6 inches wide, and 4 inches high. How many square inches of paper are needed to cover the gift?

Example 8

Find the lateral and surface area of the prism.

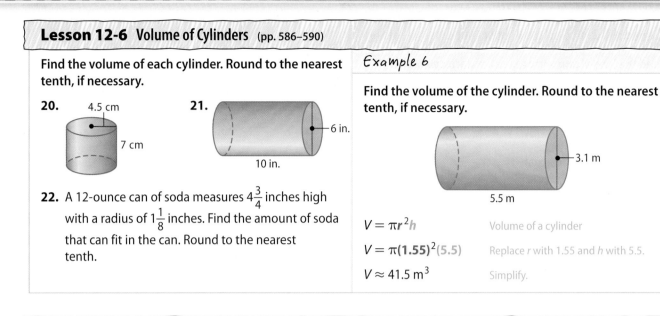

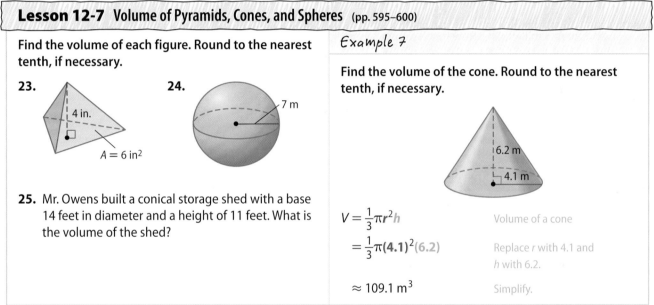

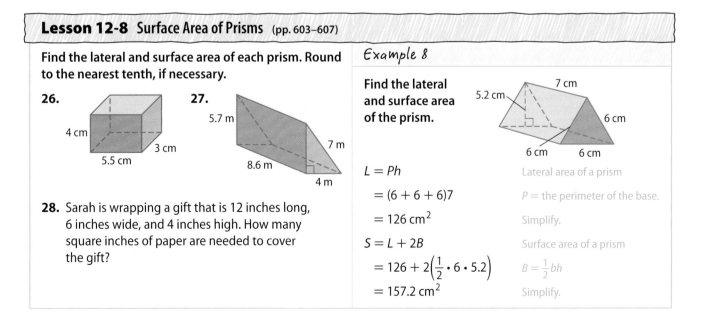

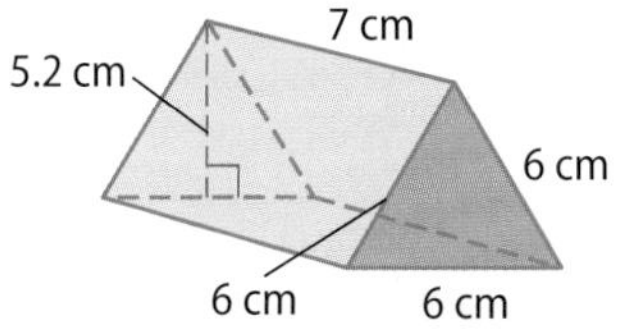

$L = Ph$	Lateral area of a prism
$= (6 + 6 + 6)7$	$P =$ the perimeter of the base.
$= 126 \text{ cm}^2$	Simplify.
$S = L + 2B$	Surface area of a prism
$= 126 + 2\left(\frac{1}{2} \cdot 6 \cdot 5.2\right)$	$B = \frac{1}{2}bh$
$= 157.2 \text{ cm}^2$	Simplify.

Lesson 12-9 Surface Area of Cylinders (pp. 610–614)

Find the lateral and surface area of each cylinder. Round to the nearest tenth, if necessary.

29.

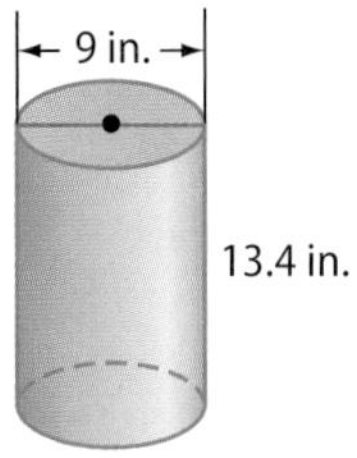

30.

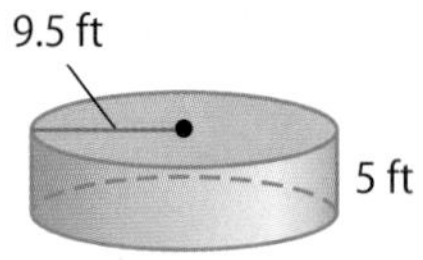

31.

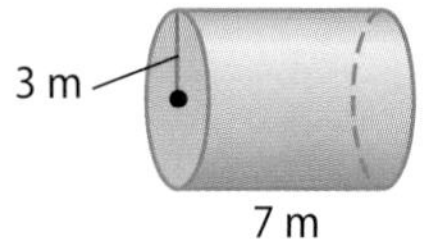

32.

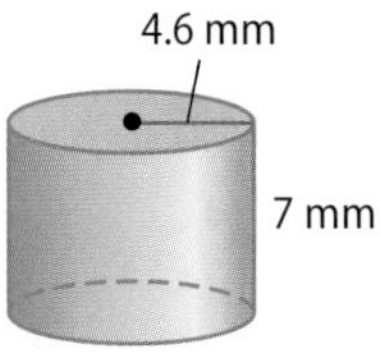

33. A cable is covered by rubber sheathing and has a diameter of 3 millimeters. How much rubber sheathing is there in 100 centimeters of cable?

Example 9

Find the lateral and surface areas of the cylinder below. Round to the nearest tenth, if necessary.

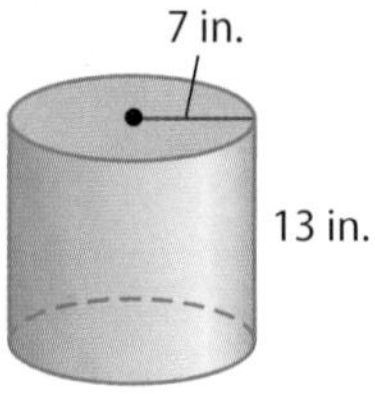

Lateral Area

$L = 2\pi \boldsymbol{r} h$ — Lateral area of a cylinder

$= 2 \cdot \pi \cdot \mathbf{7} \cdot 13$ — Replace *r* with 7 and *h* with 13.

$\approx 571.8 \text{ in}^2$ — Simplify.

Surface Area

$S = \boldsymbol{L} + 2\pi r^2$ — Surface area of a cylinder

$= \mathbf{571.8} + 2\pi(7)^2$ — Replace *L* with 571.8 and *r* with 7.

$\approx 879.7 \text{ in}^2$ — Simplify.

Lesson 12-10 Surface Area of Pyramids and Cones (pp. 615–620)

Find the lateral and surface areas of each figure. Round to the nearest tenth, if necessary.

34.

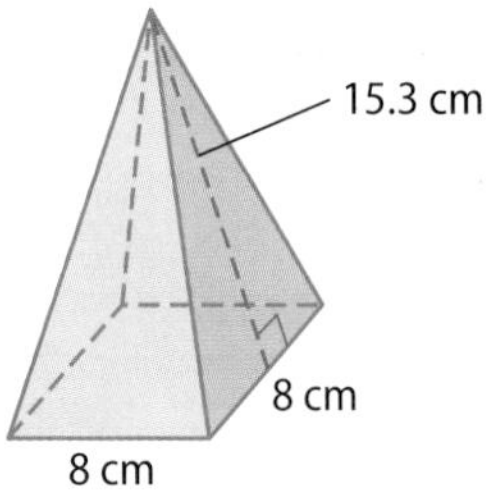

35.

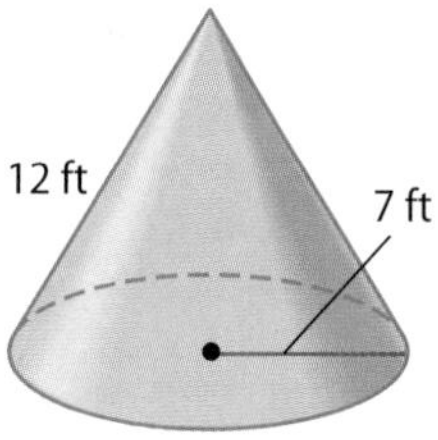

36.

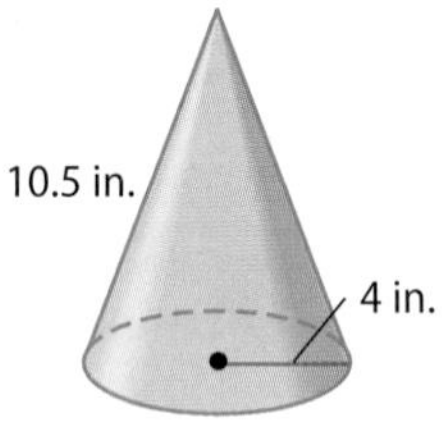

37.

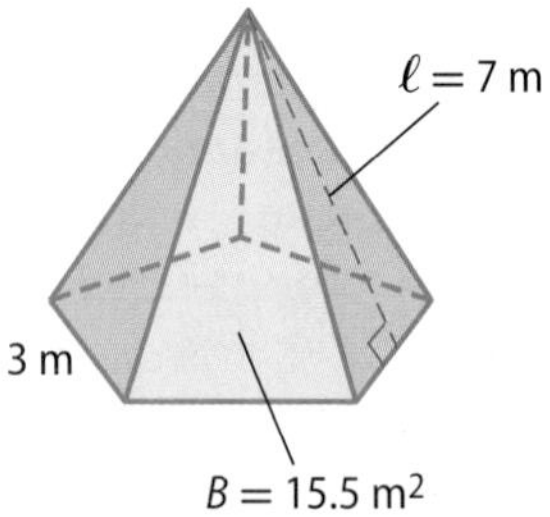

38. A pyramid-shaped roof has a slant height of 18 feet and its square base is 55 feet wide. How many square feet of roofing material is needed to cover the roof?

Example 10

Find the lateral and surface area of the square pyramid.

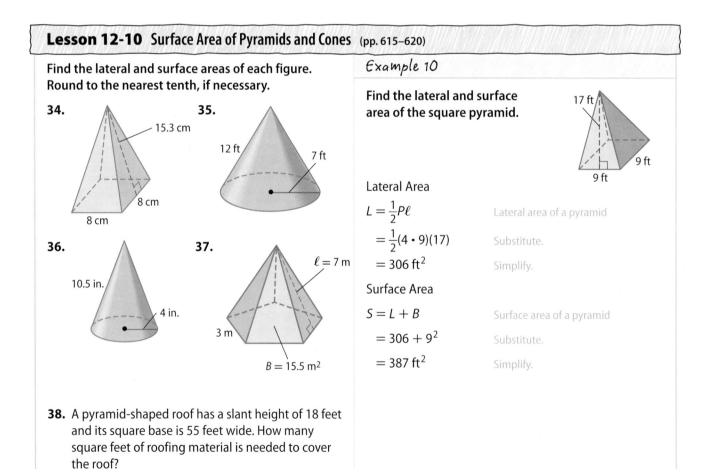

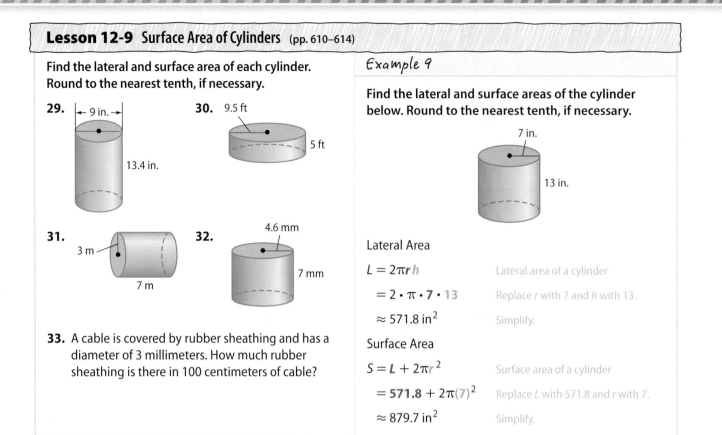

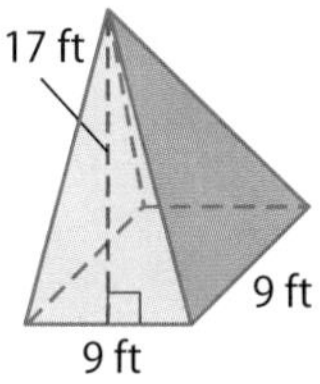

Lateral Area

$L = \frac{1}{2}P\ell$ — Lateral area of a pyramid

$= \frac{1}{2}(4 \cdot 9)(17)$ — Substitute.

$= 306 \text{ ft}^2$ — Simplify.

Surface Area

$S = L + B$ — Surface area of a pyramid

$= 306 + 9^2$ — Substitute.

$= 387 \text{ ft}^2$ — Simplify.

Glossary/Glosario

Go online for the eGlossary. Vocab

The eGlossary contains words and definitions in the following 13 languages:

Arabic
Bengali
Brazilian Portuguese
Chinese
English
Haitian Creole
Hmong
Korean
Russian
Spanish
Tagalog
Urdu
Vietnamese

English

Aa

absolute value The distance a number is from zero on the number line.

additive inverses An integer and its opposite.

adjacent angles Two angles that have the same vertex, share a common side, and do not overlap.

algebra A branch of mathematics dealing with symbols.

algebraic expression An expression that contains sums and/or products of variables and numbers.

alternate exterior angles Nonadjacent exterior angles found on opposite sides of the transversal. In the figure below, ∠1 and ∠7, ∠2 and ∠8 are alternate exterior angles.

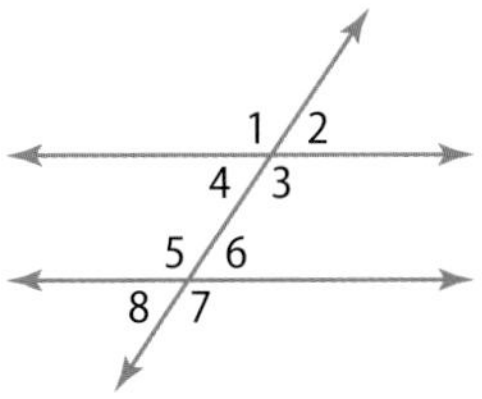

alternate interior angles Nonadjacent interior angles found on opposite sides of the transversal. In the figure above, ∠4 and ∠6, ∠3 and ∠5 are alternate interior angles.

Associative Property The way in which numbers are grouped when added or multiplied does not change the sum or product.

Español

valor absoluto Distancia que un número dista de cero en la recta numérica.

inverso aditivo Un entero y su opuesto.

ángulos adyacentes Dos ángulos que poseen el mismo vértice, comparten un lado y no se traslapan.

álgebra Rama de las matemáticas que tiene que ver con signos.

expresión algebraica Expresión que contiene sumas y/o productos de números y variables.

ángulos alternos externos Ángulos exteriores no adyacentes que se encuentran en lados opuestos de una transversal. En la siguiente figura, ∠1 y ∠7, ∠2 y ∠8 son ángulos alternos externos.

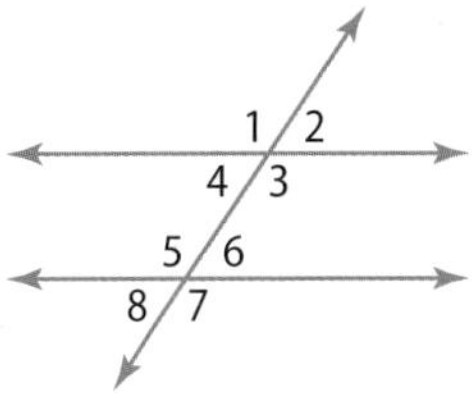

ángulos alternos internos Ángulos interiores no adyacentes que se encuentran en lados opuestos de una transversal. En la figura, ∠4 y ∠6, ∠3 y ∠5 son ángulos alternos internos.

Propiedad asociativa La forma en que se suman o multiplican tres números no altera su suma o producto.

Bb

bar notation In repeating decimals, the line or bar placed over the digits that repeat. For example, $2.\overline{63}$ indicates the digits 63 repeat.

base In 2^4, the base is 2. The base is used as a factor as many times as given by the exponent (4). That is, $2^4 = 2 \times 2 \times 2 \times 2$.

bases The bases of a prism are any two parallel congruent faces.

biased sample A sample that is not representative of a population.

box plot A diagram that divides a set of data into four parts using the median and quartiles. A box is drawn around the quartile values and whiskers extend from each quartile to the extreme data points.

notación de barra En decimales periódicos, la línea o barra que se escribe encima de los dígitos que se repiten. Por ejemplo, en $2.\overline{63}$ la barra encima del 63 indica que los dígitos 63 se repiten.

base En 2^4, la base es 2. La base se usa como factor las veces que indique el exponente (4). Es decir, $2^4 = 2 \times 2 \times 2 \times 2$.

bases Las bases de un prisma son cualquier par de caras paralelas y congruentes.

muestra sesgada Una muestra que no es representativa de una población.

diagrama de caja Diagrama que divide un conjunto de datos en cuatro partes usando la mediana y los cuartiles. Se dibuja una caja alrededor de los cuartiles y se extienden patillas de cada uno de ellos a los valores extremos.

Cc

center The given point from which all points on the circle are the same distance.

center of rotation A fixed point around which shapes move in a circular motion to a new position.

circle The set of all points in a plane that are the same distance from a given point called the center.

circumference The distance around a circle.

coefficient The numerical part of a term that contains a variable.

Commutative Property The order in which numbers are added or multiplied does not change the sum.

complement One of two parts of a probability making a whole.

complementary angles Two angles are complementary if the sum of their measures is 90°.

complex fraction A fraction with a numerator and/or denominator that is also a fraction.

centro Punto dado del cual equidistan todos los puntos de un círculo.

centro de rotación Punto fijo alrededor del cual una figura gira con un movimiento circular hasta alcanzar una nueva posición.

círculo Conjunto de todos los puntos de un plano que están a la misma distancia de un punto dado llamado centro.

circunferencia Longitud del contorno de un círculo.

coeficiente Parte numérica de un término que contiene una variable.

Propiedad conmutativa El orden en que se suman o multiplican los números no altera la suma o el producto.

complemento Una de dos partes de una probabilidad que forman un todo.

ángulos complementarios Dos ángulos son complementarios si la suma de sus medidas es 90°.

fracción compleja Fracción cuyo numerador y/o denominador también es una fracción compleja.

composite figure A figure that is made up of two or more shapes.

compound event Two or more simple events.

compound interest Interest paid on the initial principal and on interest earned in the past.

cone A three-dimensional figure with one circular base. A curved surface connects the base and vertex.

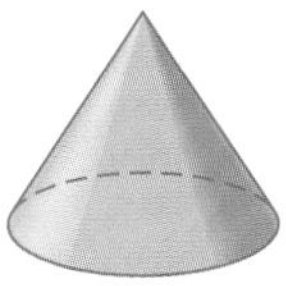

congruent Line segments that have the same length, or angles that have the same measure, or figures that have the same size and shape.

constant A term without a variable.

constant of proportionality A constant ratio or unit rate of a proportion.

constant of variation The slope, or rate of change, in the equation $y = kx$, represented by k.

constant rate of change The rate of change between any two data points in a linear relationship is the same or constant.

convenience sample A sample which includes members of the population that are easily accessed.

coordinate A number that corresponds with a point on a number line.

coordinate plane Another name for the coordinate system.

coordinate system A coordinate system is formed by the intersection of two number lines that meet at right angles at their zero points, also called a coordinate plane.

corresponding angles Angles that have the same position on two different parallel lines cut by a transversal. In the figure, $\angle 1$ and $\angle 5$, $\angle 2$ and $\angle 6$, $\angle 3$ and $\angle 7$, $\angle 4$ and $\angle 8$ are corresponding angles.

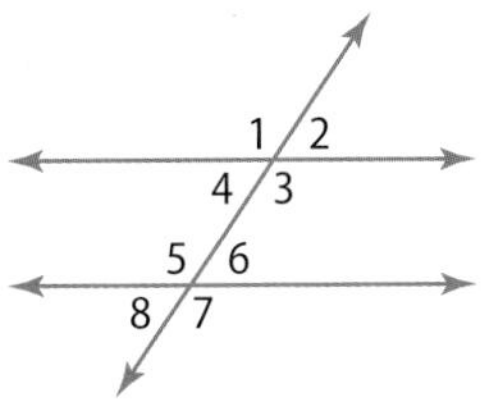

figura compuesta Figura formada de dos o más figuras.

evento compuesto Dos o más eventos simples.

interés compuesto Interés que se paga sobre el capital inicial y sobre el interés que se haya ganado en el pasado.

cono Figura tridimensional con una base circular, la cual posee una superficie curva que une la base con el vértice.

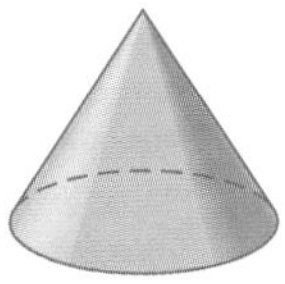

congruentes Segmentos de recta que tienen la misma longitud, o ángulos que tienen la misma medida, o figuras que tienen la misma forma y tamaño.

constante Término sin variables.

constante de proporcionalidad La razón constante o tasa unitaria de una proporción.

constante de variación La pendiente, o razón de cambio, en la ecuación $y = kx$, representada por k.

razón constante de cambio La razón de cambio entre dos puntos cualesquiera en una relación lineal permanece constante o igual.

muestra de conveniencia Muestra que incluye miembros de una población fácilmente accesibles.

coordenada Número que corresponde a un punto en la recta numérica.

plano de coordenadas Otro nombre para el sistema de coordenadas.

sistema de coordenadas Un sistema de coordenadas se forma de la intersección de dos rectas numéricas perpendiculares que se intersecan en sus puntos cero. También llamado plano de coordenadas.

ángulos correspondientes Ángulos que tienen la misma posición en dos rectas paralelas distintas cortadas por una transversal. En la figura, $\angle 1$ y $\angle 5$, $\angle 2$ y $\angle 6$, $\angle 3$ y $\angle 7$, $\angle 4$ y $\angle 8$ son ángulos correspondientes.

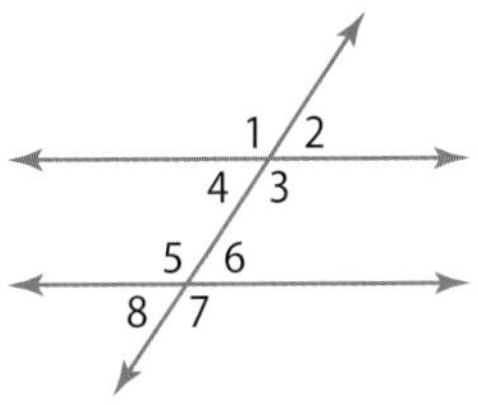

corresponding parts Parts of congruent or similar figures that match.

partes correspondientes Partes de figuras congruentes o semejantes que se corresponden mutuamente.

counterexample An example that shows a conjecture is not true.

contraejemplo Ejemplo que muestra que una conjetura no es verdadera.

cross products If $\frac{a}{c} = \frac{b}{d}$, then ad = bc. If $ad = bc$, then $\frac{a}{c} = \frac{b}{d}$.

productos cruzados Si $\frac{a}{c} = \frac{b}{d}$, entonces $ad = bc$. Si $ad = bc$, entonces $\frac{a}{c} = \frac{b}{d}$.

cross section The intersection of a solid and a plane.

sección transversal Intersección de un sólido con up plano.

cube root One of three equal factors of a number. The cube root of 8 is 2 since $2^3 = 8$.

raíz cúbica Uno de los tres factores iguales de un número. La raíz cúbica de 8 es 2 porque $2^3 = 8$.

cylinder A solid that has two parallel, congruent bases (usually circular) connected with a curved side.

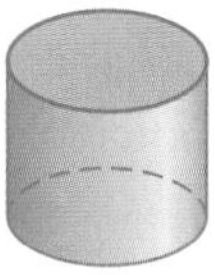

cilindro Sólido que posee dos bases congruentes paralelas (por lo general circulares) unidas por un lado curvo.

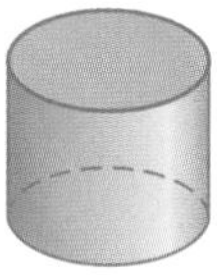

Dd

deductive reasoning The process of using facts, properties, or rules to justify reasoning or reach valid conclusions.

razonamiento deductivo Proceso de usar hechos, propiedades o reglas para justificar un razonamiento o para sacar conclusiones válidas.

defining a variable Choosing a variable and a quantity for the variable to represent in an equation.

definir una variable Seleccionar una variable y una cantidad para la variable para representarla en una ecuación.

dependent variable The variable in a relation with a value that depends on the value of the independent variable.

variable dependiente La variable en una relación cuyo valor depende del valor de la variable independiente.

diagonal A line segment that connects two nonconsecutive vertices.

diagonal Segmento de recta que une dos vértices no consecutivos de an polígono.

diameter The distance across a circle through its center.

diámetro La distancia a través de un círculo pasando por el centro.

dilation A transformation that alters the size of a figure but not its shape.

homotecia Transformación que altera el tamaño de una figura, pero no su forma.

dimensional analysis The process of including units of measurement when computing.

análisis dimensional Proceso que incorpora las unidades de medida al hacer cálculos.

direct variation A special type of linear equation that describes rate of change. A relationship such that as x increases in value, y increases or decreases at a constant rate.

variación directa Tipo especial de ecuación lineal que describe una razón de cambio. Relación en la que si x aumenta de valor, y aumenta o disminuye a una razón constante.

discount The amount by which the regular price of an item is reduced.

descuento Cantidad que se reduce del precio normal de un artículo.

Distributive Property To multiply a sum by a number, multiply each addend by the number outside the parentheses.

$$a(b + c) = ab + ac$$

Propiedad distributiva Para multiplicar una suma por un número, multiplica cada sumando por el número fuera del paréntesis.

$$a(b + c) = ab + ac$$

domain The domain of a relation is the set of all x-coordinates from each pair.

dominio El dominio de una relación es el conjunto de coordenadas x de todos los pares.

double box plot Two box plots graphed on the same number line.

diagrama de doble caja Dos diagramas de caja graficados en la misma recta numérica.

Ee

edge Where two planes intersect in a line.

arista Recta en donde se intersecan dos planos.

equation A mathematical sentence that contains an equals sign (=).

ecuación Enunciado matemático que contiene el signo de igualdad (=).

equivalent equations Two or more equations with the same solution. For example, $x + 4 = 7$ and $x = 3$ are equivalent equations.

ecuaciones equivalentes Dos o más ecuaciones con las mismas soluciones. Por ejemplo, $x + 4 = 7$ *y* $x = 3$ son ecuaciones equivalentes.

equivalent expressions Expressions that have the same value.

expresiones equivalentes Expresiones que tienen el mismo valor.

evaluate Find the numerical value of an expression.

evaluar Calcular el valor numérico de una expresión.

experimental probability What actually occurs in a probability experiment.

probabilidad experimental Lo que realmente sucede en un experimento probabilístico.

exponent In 2^4, the exponent is 4. The exponent tells how many times the base, 2, is used as a factor. So, $2^4 = 2 \times 2 \times 2 \times 2$.

exponente En 2^4, el exponente es 4. El exponente indica cuántas veces se usa la base, 2, como factor. Así, $2^4 = 2 \times 2 \times 2 \times 2$.

exterior angle An angle formed by one side of a triangle and the extension of the adjacent side.

ángulo exterior Ángulo formado por uno de los lados de un triángulo y la extensión del lado adyacente.

Ff

face A flat surface, the side or base of a prism.

cara Superficie plana, el lado o la base de un prisma.

factor To write a number as a product of its factors.

factorizar Escribir un número como el producto de sus propios factores.

factored form A monomial expressed as a product of prime numbers and variables where no variable has an exponent greater than 1.

forma reducida Monomio escrito como el producto de números primos y variables en el que ninguna variable tiene un exponente mayor que 1.

first quartile For a data set with median M, the first quartile is the median of the data values less than M.

plan de cuatro pasos Plan para resolver problemas que consiste en los pasos Comprender, Planear, Resolver y Verificar

four step plan A plan for problem solving consisting of the steps Understand, Plan, Solve, and Check.

primer cuartil Para un conjunto de datos con la mediana M, el premir cuartil es la mediana de los valores menores que M.

function A function is a special relation in which each element of the domain is paired with exactly one element in the range.

función Una función es una relación especial en que a cada elemento del dominio le corresponde un único elemento del rango.

function notation A way to name a function that is defined by an equation. In function notation, the equation $y = 3x - 8$ is written as $f(x) = 3x - 8$.

notación funcional Una manera de nombrar una función definida por una ecuación. En notación funcional, la ecuación $y = 3x - 8$ se escribe $f(x) = 3x - 8$.

function rule The operation performed on the input of a function.

regla de función Operación que se efectúa en el valor de entrada de una función.

function table A table organizing the input, rule, and output of a function.

tabla de funciones Tabla que organiza las entradas, la regla y las salidas de una función.

Gg

graph A dot at the point that corresponds to an ordered pair on a coordinate plane.

gráfica Punto que corresponde a un par ordenado en un plano de coordenadas.

guess, check, and edit A problem-solving strategy in which you make a reasonable guess, check it in the problem, and use the results to improve your guess until you find the solution.

adivinar, verificar, y editar Estrategia para resolver problemas en la cual se hace una conjetura, se comprueba en el problema y se usa el resultado para perfeccionar la conjetura hasta hallar la solución.

Ii

identity An equation that is true for every value of the variable.

identidad Ecuación que es verdadera para cada valor de la variable.

image Every corresponding point on a figure after its transformation.

imagen Todo punto de una figura después de su transformación.

independent variable The variable in a function with a value that is subject to choice.

variable independiente La variable de una función sujeta a elección.

indirect measurement Using the properties of similar triangles to find measurements that are difficult to measure directly.

medición indirecta Uso de las propiedades de triángulos semejantes para hacer mediciones que son difíciles de realizar directamente.

inductive reasoning Making a conjecture based on a pattern of examples or past events.

razoznamiento inductivo Hacer una conjetura con base en un patrón de ejemplos o sucesos pasados.

inequality A mathematical sentence that contains $<, >, \neq, \leq$, or $\geq$.

desigualdad Enunciado matemático que contiene $<, >, \neq, \leq$, o $\geq$.

integer The whole numbers and their opposites.

$\ldots, -3, -2, -1, 0, 1, 2, 3, \ldots$

enteros Los números enteros y sus opuestos.

$\ldots, -3, -2, -1, 0, 1, 2, 3, \ldots$

interest The amount of money paid or earned for the use of money.

interés Cantidad de dinero que se paga o se gana por el uso del dinero.

interior angle An angle inside a polygon.

ángulo interno Ángulo ubicado dentro de un polígono.

interquartile range The range of the middle half of a set of data. It is the difference between the first quartile and the third quartile.

amplitud intercuartílica Amplitud de la mitad central de un conjunto de datos. Es la diferencia entre el cuartil superior y el inferior.

inverse operation Operation that undoes another, such as addition and subtraction.

operaciones inversas Operaciones que se anulan mutuamente, como la adición y la sustracción.

irrational number A number that cannot be expressed as $\frac{a}{b}$, where a and b are integers and b does not equal 0.

número irracional Número que no puede escribirse como $\frac{a}{b}$, donde a y b son enteros y b no es igual a 0.

LI

lateral area The sum of the areas of the lateral faces of a solid.

área lateral Suma de las áreas de las caras laterales de un sólido.

lateral faces The lateral faces of a prism, cylinder, pyramid, or cone are all the surfaces of the figure except the base or bases.

caras laterales Las caras laterales de un prisma, cilindro, pirámide o cono son todas las superficies de la figura, excluyendo la base o las bases.

like fractions Fractions that have the same denominator.

fracciones semejantes Fracciones con el mismo denominador.

like terms Expressions that contain the same variables to the same power, such as $2n$ and $5n$ or $6xy^2$ and $4xy^2$.

términos semejantes Expresiones que tienen las mismas variables elevadas a la misma potencia, como $2n$ y $5n$ ó $6xy^2$ y $4xy^2$.

line of symmetry Each half of a figure is a mirror image of the other half when a line of symmetry is drawn.

eje de simetría Cuando se traza un eje de simetría, cada mitad de una figura es una imagen especular de la otra mitad.

line segment Part of a line containing two endpoints and all the points between them.

segmento de recta Parte de una recta que contiene dos extremos y todos los puntos entre éstos.

linear equation An equation in which the variables appear in separate terms and neither variable contains an exponent other than 1. The graph of a linear equation is a straight line.

ecuación lineal Ecuación en que las variables aparecen en términos separados y en la cual ninguna de ellas tiene un exponente distinto de 1. La gráfica de una ecuación lineal es una recta.

linear expression An algebraic expression in which the variable is raised to the first power.

expresión lineal Expresión algebraica en la que la variable se eleva a la primera potencia.

linear function A function for which the graph is a straight line.

función lineal Función cuya gráfica es una recta.

linear relationship Relationship that has straight line graphs.

relación lineal Relación que al ser graficada forma una línea recta.

look for a pattern A problem-solving strategy in which you analyze the first few terms in a pattern and identify a rule that is used to go from one number to the next.

buscar un patrón Estrategia de resolución de problemas en la cual se analizan los primeros términos de un patrón para identificar la regla que se usa para pasar de un número al siguiente.

Mm

make a table A problem-solving strategy in which you use a table to organize information in an understandable way.

hacer una tabla Estrategia de resolución de problemas en la cual se usa una tabla para organizar información de una manera clara.

markup The amount the price of an item is increased above the price that the store paid for an item.

margen de utilidad Cantidad de aumento en el precio de un artículo por encima del precio que paga la tienda por dicho artículo.

mean absolute deviation The average distance between each data value and the mean.

desviación media absoluta Distancia promedio entre cada dato y la media.

measures of center For a list of numerical data, numbers that can represent the center of the data.

medidas de centro Números que pueden representar todo el conjunto de datos en una lista de datos numéricos.

measures of variability Used to describe the distribution of statistical data.

medidas de variación Se usan para describir la distribución de datos estadísticos.

monomial An expression that is a number, a variable, or a product of numbers and/or variables.

monomio Expresión que es un número, una variable y/o un producto de números y variables.

multiplicative inverse Two numbers whose product is 1.

inversos multiplicativos Dos números cuyo producto es igual a uno.

Nn

negative exponents Negative exponents are the result of repeated division. For $a \neq 0$ and any whole number n, $a^{-n} = \frac{1}{a^n}$.

exponentes negativo Los exponentes negativos son el resultado de la división repetida. Para $a \neq 0$ y cualquier número entero n, $a^{-n} = \frac{1}{a^n}$.

negative number A number less than zero.

número negativo Número menor que cero.

nonproportional A relationship in which two ratios are not equal.

relación no proporcional Relación en la que dos razones no son iguales.

null set or empty set A set with no elements shown by the symbol { } or Ø.

conjunto vacío Conjunto que carece de elementos y que se denota con el símbolo { } o Ø.

numerical expressions A combination of numbers and operations such as addition, subtraction, multiplication, and division.

expresiones numéricas Combinación de números y operaciones, como suma, resta, multiplicación y división.

Oo

opposites Two numbers with the same absolute value but different signs.

opuestos Dos números que tienen el mismo valor absoluto, pero que tienen distintos signos.

order of operations The rules to follow when more than one operation is used in an expression.

1. Do all operations within grouping symbols first; start with the innermost grouping symbols.
2. Evaluate all powers before other operations.
3. Multiply and divide in order from left to right.
4. Add and subtract in order from left to right.

orden de las operaciones Reglas a seguir cuando se usa más de una operación en una expresión.

1. Primero ejecuta todas las operaciones dentro de los símbolos de agrupamiento
2. Evalúa todas las potencias antes que las otras aperaciones.
3. Multiplica y divide en orden de izquierda a derecha.
4. Suma y resta en orden de izquierda a derecha.

ordered pair A pair of numbers used to locate any point on a coordinate plane.

par ordenado Par de números que se usa para ubicar cualquier punto en un plano de coordenadas.

origin The point at which the number lines intersect in a coordinate system.

origen Punto de intersección de las rectas numéricas de un sistema de coordenadas.

outcome Possible result of a probability experiment.

resultado Resultados posibles de un experimento probabilístico.

outliers Data that are more than 1.5 times the interquartile range beyond the quartiles.

valores atípicos Datos que distan de los cuartiles más de 1.5 veces la amplitud intercuartílica.

Pp

parallel lines Two lines in the same plane that do not intersect.

rectas paralelas Dos rectas en el mismo plano que no se intersecan.

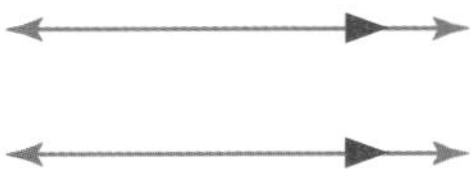

percent equation An equivalent form of the percent proportion, where % is written as a decimal.

Part = Percent × Whole

ecuación porcentual Forma equivalente a la proporción porcentual en la cual el % se escribe como decimal.

Parte = Por ciento × Entero

percent error A measure of the difference between an estimate, prediction, or measurement and the actual

error porcentua Medida de la diferencia entre una estimación, predicción o medida y el valor real.

percent of change The ratio of the increase or decrease of an amount to the original amount.

porcentaje de cambio Razón del aumento o disminución de una cantidad a la cantidad original.

percent of decrease The ratio of an amount of decrease to the previous amount, expressed as a percent. A negative percent of change.

porcentaje de disminución Razón de la cantidad de disminución a la cantidad original, escrita como por ciento. Un por ciento de cambio negativo.

percent of increase The ratio of an amount of increase to the original amount, expressed as a percent.

porcentaje de aumento Razón de la cantidad de aumento a la cantidad original, escrita como por ciento.

percent proportion
$$\frac{\text{part}}{\text{whole}} = \frac{\text{percent}}{100} \text{ or } \frac{a}{b} = \frac{p}{100}$$

perfect cube A number that is the cube of an integer. 1, 8, and 27 are examples of perfect cubes.

perfect square Rational number whose square root is a whole number. 25 is a perfect square because $\sqrt{25} = 5$.

perpendicular lines Lines that intersect to form a right angle.

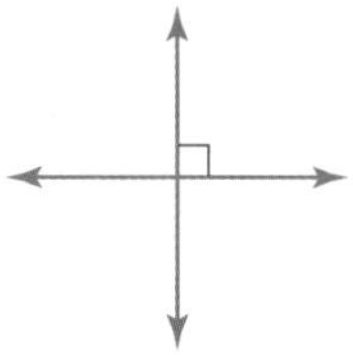

pi, π The ratio of the circumference of a circle to the diameter of the circle. Approximations for π are 3.14 and $\frac{22}{7}$.

plane A two-dimensional flat surface that extends in all directions and contains at least three noncollinear points.

polygon A simple closed figure formed by three or more straight line segments.

polyhedron A solid with flat surfaces that are polygons.

population A larger group used in statistical analysis.

positive number Any number that is greater than zero.

power A number that is expressed using an exponent.

principal The amount of money in an account.

prism A polyhedron that has two parallel, congruent bases in the shape of polygons.

rectangular prism

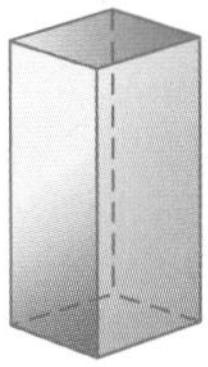

triangular prism

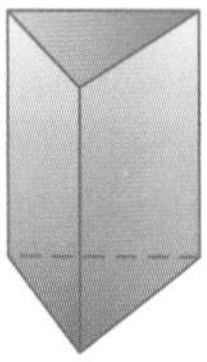

proporción porcentual
$$\frac{\text{parte}}{\text{todo}} = \frac{\text{por ciento}}{100} \text{ o } \frac{a}{b} = \frac{p}{100}$$

cubo perfecto Número que es el cubo de un entero. 1, 8 y 27 son ejemplos de cubos perfectos.

cuadrados perfectos Números racionales cuyas raíces cuadradas son números racionales. 25 es un cuadrado perfecto porque $\sqrt{25} = 5$.

rectas perpendiculares Rectas que se intersecan formando un ángulo recto.

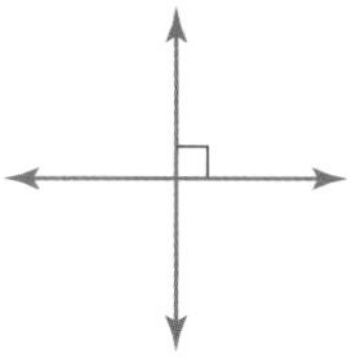

pi, π Razón de la circunferencia de un círculo al diámetro del mismo. 3.14 y $\frac{22}{7}$ son aproximaciones de π.

plano Superficie plana bidimensional que se extiende en todas direcciones y que contiene por lo menos tres puntos no colineales.

polígono Figura cerrada simple formada por tres o más segmentos de recta.

poliedro Sólido con superficies planas que son polígonos.

población Grupo grande que se utiliza en análisis estadísticos.

número positive Todo número mayor que cero.

potencia Número que puede escribirse usando un exponente.

capital Cantidad de dinero en una cuenta.

prisma Poliedro que posee dos bases congruentes paralelas en forma de polígonos.

prisma rectangular

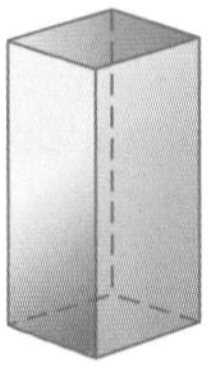

prisma triangular

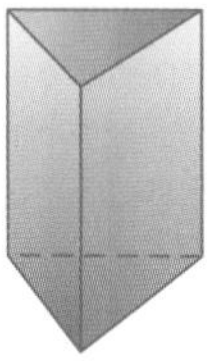

probability The ratio of the number of ways a certain event can occur to the number of possible outcomes.

$$P(\text{event}) = \frac{\text{number of favorable outcomes}}{\text{number of possible outcomes}}$$

probabilidad La razón del número de maneras en que puede ocurrir el evento al número de resultados posibles.

$$P(\text{evento}) = \frac{\text{número de resultados favorables}}{\text{número de resultados posibles}}$$

properties Statements that are true for any number.

propiedades Enunciados que son verdaderos para cualquier número.

proportion A statement of equality of two or more ratios.

proporción Enunciado de la igualdad de dos o más razones.

proportional The ratios of related terms are equal.

proporcional Relación en la que la razón entre los términos relacionados permanece igual.

pyramid A polyhedron that has a polygon for a base and triangles for sides.

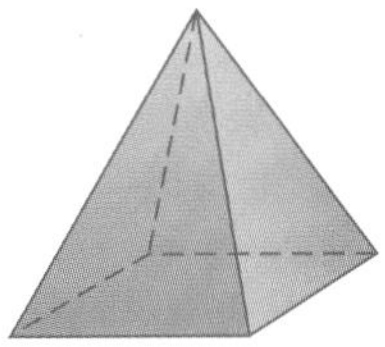

pirámide Poliedro cuya base es un polígono y cuyos lados son triángulos.

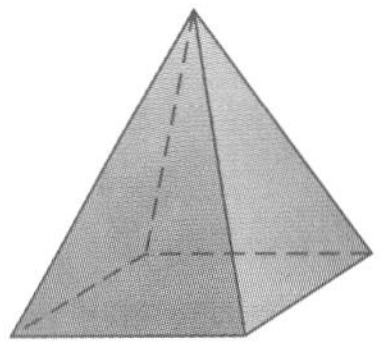

Qq

quadrants The four regions into which the x-axis and y-axis separate the coordinate plane.

cuadrantes Las cuatro regiones en que los ejes x y y dividen el plano de coordenadas.

quartiles The values that divide a set of data into four equal parts.

cuartiles Valores que dividen un conjunto de datos en cuatro partes iguales.

Rr

radical sign The symbol $\sqrt{\ }$ used to indicate a nonnegative square root.

signo radical El símbolo $\sqrt{\ }$ que se usa para indicar la raíz cuadrada no negativa.

radius The distance from the center to any point on the circle.

radio Distancia del centro a cualquier punto de un círculo.

random Outcomes occur at random if each outcome is equally likely to occur.

aleatorio Los resultados son aleatorios si todos son equiprobables.

range The range of a relation is the set of all y-coordinates from each ordered pair.

rango El rango de una relación es el conjunto de coordenadas y de todos los pares.

range A measure of variation that is the difference between the least and greatest values in a set of data.

amplitud Medida de variación que es la diferencia entre los valores máximo y mínimo de un conjunto de datos.

rate A ratio of two measurements having different units.

tasa Razón de dos medidas que tienen unidades distintas.

rate of change A change in one quantity with respect to another quantity.

tasa de cambio Cambio de una cantidad con respecto a otra.

rational numbers The set of numbers that can be written as a fraction in the form $\frac{a}{b}$, where a and b are integers and $b \neq 0$.

números racionales Conjunto de números que puede escribirse como una fracción de la forma $\frac{a}{b}$ donde a y b son enteros y $b \neq 0$.

real numbers The set of rational numbers together with the set of irrational numbers.

números reales El conjunto de los números racionales junto con el de los números irracionales.

reciprocals The multiplicative inverse of a number.

recíprocos El inverso multiplicativo de un número.

reflection A transformation where a figure is flipped over a line. Also called a flip.

reflexión Transformación en que una figura se voltea a través de una recta.

regular polygon A polygon that has all sides congruent and all angles congruent.

polígono regular Polígono con todos los lados y todos los ángulos congruentes.

regular pyramid A pyramid whose base is a regular polygon.

pirámide regular Pirámide cuya base es un polígono regular.

relation A set of ordered pairs.

relación Conjunto de pares ordenados.

repeating decimal A decimal whose digits repeat in groups of one or more without end. Examples are 0.181818... and 0.8333....

decimal periódico Decimal cuyos dígitos se repiten en grupos de uno o más. 0.181818... y 0.8333... son ejemplos de este tipo de decimales.

rotation A transformation where a figure is turned around a fixed point. Also called a turn.

rotación Transformación en que una figura se hace girar alrededor de un punto fijo. También se llama vuelta.

rotational symmetry A figure has rotational symmetry if it can be turned less than 360° about its center and still look like the original.

simetría rotacional Una figura posee simetría rotacional si se puede girar menos de 360° en torno a su centro sin que esto cambia su apariencia con respecto a la figura original.

Ss

sample A subgroup or subset of a population used to represent the whole population.

muestra Subgrupo o subconjunto de una población que se usa para representarla.

sample space The set of all possible outcomes.

espacio muestral Conjunto de todos los resultados posibles.

scale The relationship between the measurements on a drawing or model and the measurements of the real object.

escala Relación entre las medidas de un dibujo o modelo y las medidas de la figura verdadera.

scale drawing A drawing that is used to represent an object that is too large or too small to be drawn at actual size.

dibujo a escala Dibujo que se usa para representar una figura que es demasiado grande o pequeña como para ser dibujada de tamaño natural.

scale factor The ratio of a length on a scale drawing or model to the corresponding length on the real object.

factor de escala Razón de la longitud en un dibujo a escala o modelo a la longitud correspondiente en la figura verdadera.

scale model A model used to represent an object that is too large or too small to be built at actual size.

modelo a escala Modelo que se usa para representar una figura que es demasiado grande o pequeña como para ser construida de tamaño natural.

scientific notation A number in scientific notation is expressed as $a \times 10^n$, where $1 \leq a < 10$ and n is an integer. For example, $5{,}000{,}000 = 5.0 \times 10^6$.

notación científica Un número en notación científica se escribe como $a \times 10^n$, donde $1 \leq a < 10$ y n es un entero. Por ejemplo, $5{,}000{,}000 = 5.0 \times 10^6$.

selling price The amount a customer pays for an item.

precio de venta Cantidad de dinero que paga un consumidor por un artículo.

similar figures Figures that have the same shape but not necessarily the same size.

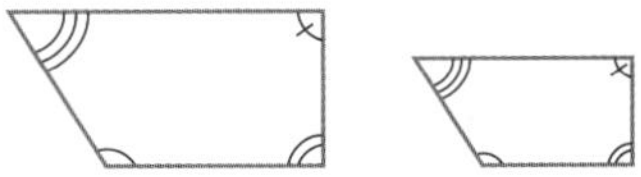

figuras semejantes Figuras que tienen la misma forma, pero no necesariamente el mismo tamaño.

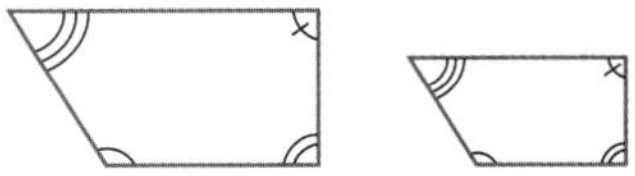

simple event One outcome or a collection of outcomes.

evento simple Resultado o colección de resultados.

simple interest The amount of money paid or earned for the use of money.

$I = prt$ (Interest = principal × rate × time)

interés simple Cantidad que se paga o que se gana por usar el dinero.

$I = crt$ (Interés = capital × rédito × tiempo)

simple random sample A sample where each item or person in the population is as likely to be chosen as any other.

muestra aleatoria simple Muestra de una población que tiene la misma probabilidad de escogerse que cualquier otra.

simplest form An algebraic expression in simplest form has no like terms and no parentheses.

forma reducida Una expresión algebraica reducida no tiene ni términos semejantes ni paréntesis.

simplify To write an expression in a simpler form.

reducir Escribir una expresión en forma más simple.

simplifying the expression To use distribution to combine like terms.

reducir la expresión Usar la distribución para combinar términos semejantes.

simulation A way of modeling a problem situation or event that would be too difficult or impractical to actually perform.

simulación Forma para representar una situación de problema o un suceso que sería muy difícil o poco práctico de realizar en la práctica.

skew lines Lines that do not intersect and are not in the same plane.

rectas alabeadas Rectas que no se intersecan y que no son coplanares.

slant height The length of the altitude of a lateral face of a regular pyramid or cone.

altura oblicua En una pirámide regular o un cono, la longitud de la altura de una cara lateral.

slope The ratio of the rise, or vertical change, to the run, or horizontal change. The slope describes the steepness of a line.

$$\text{slope} = \frac{\text{rise}}{\text{run}}$$

pendiente Razón de la elevación o cambio vertical al desplazamiento o cambio horizontal. La pendiente describe la inclinación de una recta.

$$\text{pendiente} = \frac{\text{elevación}}{\text{desplazamiento}}$$

slope-intercept form A linear equation in the form $y = mx + b$, where m is the slope and b is the y-intercept.

forma pendiente-intersección Una ecuación lineal de la forma $y = mx + b$, donde m es la pendiente y b es la intersección y.

solids Three-dimensional figures.

sólidos Figuras tridimensionales.

solution A value for the variable that makes an equation true. For $x + 7 = 19$, the solution is 12.

solución Valor de la variable que hace verdadera la ecuación. Para $x + 7 = 19$, la solución es 12.

sphere The set of all points in space that are a given distance, *r*, from the center.

esfera El conjunto de todos los puntos en el espacio que se hallan a una distancia *r* del centro.

square root One of the two equal factors of a number. The square root of 25 is 5 since $5^2 = 25$.

raíz cuadrada Uno de los dos factores iguales de un número. La raíz cuadrada de 25 es 5 porque $5^2 = 25$.

standard form A number is in standard form when it does not contain exponents. The standard form for seven hundred thirty-nine is 739.

forma estándar Un número está en forma estándar si no contiene exponentes. Por ejemplo, la forma estándar de setecientos treinta y nueve es 739.

statistics The branch of mathematics that deals with collecting, organizing, and interpreting data.

estadística Rama de las matemáticas cuyo objetivo primordial es la recopilación, organización e interpretación de datos.

substitution Use algebraic methods to find an exact solution of a system of equations.

sustitución Usa métodos algebraicos para hallar una solución exacta a un sistema de ecuaciones.

supplementary angles Two angles are supplementary if the sum of their measures is 180º.

suplementarios ángulos Dos ángulos son suplementarios si sus medidas suman 180º.

surface area The sum of the areas of all the surfaces (faces) of a 3-dimensional figure.

área total Suma de las áreas de todas las superficies (caras) de una figura tridimensional.

system of equations A set of two or more equations with the same variables.

sistema de ecuaciones Conjunto de dos o más ecuaciones con las mismas variables.

systematic random sample A sampling method in which the items or people are selected according to a specific time or item interval.

muestra aleatoria sistemática Muestra en que los elementos de las muestra se escogen según un intervalo de tiempo o elemento específico.

Tt

terminating decimal A repeating decimal which has a repeating digit of 0.

decimal finito Decimal cuyos dígitos terminan.

tessellation A pattern formed by repeating figures that fit together without gaps or overlaps.

teselado Patrón formado por figuras repetidas que no se transiapan y que no dejan espacios entre sí.

theoretical probability What should occur in a probability experiment.

probabilidad teórica Lo que debería ocurrir en un experimento probabilístico.

third quartile For a data set with median *M*, the third quartile is the median of the data values greater than *M*.

tercer cuartil Para un conjunto de datos con la mediana *M*, el tercer cuartil es la mediana de los valores mayores que *M*.

transformation A movement of a geometric figure.

transformación Desplazamiento de una figura geométrica.

translation A transformation where a figure is slid from one position to another without being turned. Also called a slide.

translación Transformación en que una figura se desliza sin girar, de una posición a otra. También se llama deslizamiento.

transversal A line that intersects two parallel lines to form eight angles.

tree diagram A diagram used to show the total number of possible outcomes.

triangle A figure having three sides.

two-step equation An equation that contains two operations.

transversal Recta que interseca dos rectas paralelas formando ocho ángulos.

diagrama de árbol Diagrama que se usa para mostrar el número total de resultados posibles.

triángulo Figura de tres lados.

ecuación de dos pasos Ecuación que contiene dos operaciones.

Uu

unbiased sample A random sample that is representative of a larger sample.

uniform probability model A model in which each outcome has an equal probability of occurring.

unit rate A rate simplified so that it has a denominator of 1.

unlike fractions Fractions with different denominators.

muestra no sesgada Muestra aleatoria que es representativa de una muestra más grande.

modelo de probabilidad uniforme Modelo en el cual cada resultado tiene la misma probabilidad de ocurrir.

razón unitaria Razón reducida que tiene denominador igual a 1.

fracciones con distinto denominador Fracciones cuyos denominadores son diferentes.

Vv

variable A placeholder for any value.

vertex A vertex of a polygon is a point where two sides of the polygon intersect.

vertex Where three or more planes intersect in a point.

vertical angles Two pairs of opposite angles formed by two intersecting lines. The angles formed are congruent. In the figure, the vertical angles are $\angle 1$ and $\angle 3$, $\angle 2$ and $\angle 4$.

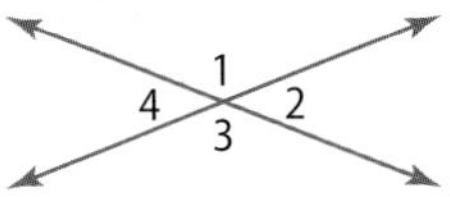

vertical line test If any vertical line drawn on the graph of a relation passes through no more than one point on the graph for each value of x in the domain, then the relation is a function.

volume The measure of space occupied by a solid region.

variable Marcador de posición para cualquier valor.

vértice El vértice de un polígono es un punto en que se intersecan dos lados del mismo.

vértice Punto en que se intersecan tres o más planos.

ángulos opuestos por el vértice Dos pares de ángulos opuestos formados por dos rectas que se intersecan. Los ángulos que resultan son congruentes. En la figura, los ángulos opuestos por el vértice son $\angle 1$ y $\angle 3$, $\angle 2$ y $\angle 4$.

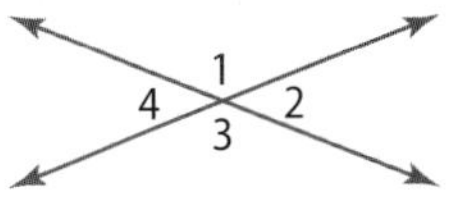

prueba de la recta vertical Si todas las rectas verticales trazadas en la gráfica de una relación no pasan por más un punto para cada valor de x en el dominio, entonces la relación es una función.

volumen Medida del espacio que ocupa un sólido.

voluntary response sample A sample which involves only those who want to participate in the sampling.

muestra de respuesta voluntaria Muestra que involucra sólo aquellos que quieren participar en el muestreo

Ww

work backward A problem-solving strategy in which you start with the end result and undo each step that led to the end result.

trabajar hacia atrás Estrategia de resolución de problemas en la cual se empieza a trabajar a partir del resultado final deshaciendo cada paso que condujo al resultado final.

Xx

***x*-axis** The horizontal number line which helps to form the coordinate system.

eje *x* Recta numérica horizontal que forma parte de un sistema de coordenadas.

***x*-coordinate** The first number of an ordered pair.

coordenada *x* El primer número de un par ordenado.

***x*-intercept** The *x*-coordinate of a point where a graph crosses the *x*-axis.

intersección *x* La coordenada *x* de un punto en que una gráfica interseca el eje *x*.

Yy

***y*-axis** The vertical number line which helps to form the coordinate system.

eje *y* Recta numérica vertical que forma parte de un sistema de coordenadas.

***y*-coordinate** The second number of an ordered pair.

coordenada *y* El segundo número en un par ordenado.

***y*-intercept** The *y*-coordinate of a point where a graph crosses the *y*-axis.

intersección *y* La coordenada *y* de un punto en que una gráfica interseca el eje *y*.

Zz

zero pair A positive tile paired with a negative tile.

par nulo Ficha positiva apareada con una negativa.

Selected Answers

Go online for Step-by-Step Solutions. eHelp

Chapter 1 The Language of Algebra

Pages 4–5 Lesson 1-1

1. Sample answer: \$27 **3a.** 1752 Calories **3b.** 66 Calories **5** \$80 **7.** 127 mi **9.** Sample answer: About how many times greater is Sonia's high score than Brian's high score? Answer: about 3 times greater **11.** Sample answer: If none of the items cost more than \$2, then the greatest possible total cost would be $\$2 \times 6 = \$12$, which is less than the actual total cost. So, at least one of the items must cost more than \$2. **13.** D **15.** C **17.** 70 cm^2 **19.** 7 **21.** 1 **23.** 3

Pages 8–10 Lesson 1-2

1. 6×8 **3.** 26 **5.** 33 **7.** 36 **9.** 5
11. $75 + 6(25)$; \$225;

Number of Passengers	Expression	Cost (\$)
25	75 + 6(25)	225
30	75 + 6(30)	255
35	75 + 6(35)	285
40	75 + 6(40)	315

13. $8 + 4$ **15.** 9×3 **17.** $15 - 10$ **19.** 10 **21** 62 **23.** 76 **25.** 16 **27.** 28 **29.** 2
31 $8 + 6(0.75)$; \$12.50;

Lines	Expression	Total Cost (\$)
6	8 + 6(0.75)	12.50
10	8 + 10(0.75)	15.50
14	8 + 14(0.75)	18.50
18	8 + 18(0.75)	21.50

33a.

Term Number	Number of Toothpicks
1	4
2	6
3	8

33b. The number of toothpicks is two more than twice the term number. **35.** Sample answer: $(6 \times 3) + 2$; $(8 \times 5) - 20$ **37.** Sample answer: Without the order of operations, any expression that contains more than one operation could have many different values. For example, when following the order of operations, the expression $12 + 4 \times 4$ equals $12 + 16$ or 28. When you perform the operations from left to right, the expression $12 + 4 \times 4$ equals 16×4 or 64. **39.** H **41.** Sample answer: 5(\$15) + 4(\$20) **43.** 3.95 **45.** 33.8 **47.** 6.1 **49.** 0

Pages 16–18 Lesson 1-3

1. $c - 4$ **3.** $10h$ **5.** 25 **7.** 16 **9.** 12 **11a.** $16p$ **11b.** 80 fl oz **13** $24 \div s$ **15.** $12n$ **17.** $10n - 4$ **19.** 4 **21.** 34 **23.** 18 **25.** 6 **27.** 18 **29.** 72 **31.** $\frac{w}{231}$ **33.** 13 **35.** 50 **37.** 58 **39** $42b$; $100 \cdot 42 \cdot 6$ or 25,200 lb **41a.** There are four cups in one quart. **41b.** $c \div 4$ **41c.** 25 qt **43.** Sample answer: John wrote 5 and n in the incorrect order. He should subtract 5 from the number: $n - 5$. **45.** Addition then multiplication; sample answer: Order of operations tells us to evaluate what is inside the parentheses first. Since the addition is inside the parentheses, it is first. Then you multiply the value of a by the value of $x + y$. **47.** H **49.** \$35 **51.** 3 **53.** 24 **55.** 41 **57.** 4 **59.** 41 points **61.** 13 points **63.** Knights, Huskies, Wildcats **65.** 35 **67.** 8 **69.** 12 **71.** 36

Pages 21–24 Lesson 1-4

1. no; $10 - 6 \neq 6 - 10$ **3.** Identity (×) **5.** Identity (+) **7.** Associative (×) **9.** $11 + k$ **11.** $16 + m$ **13.** $33z$ **15.** no; $3 \times 5 = 15$ **17.** no; $3 \times 2 = 6$ **19.** Identity (×) **21.** Associative (×) **23.** Associative (+) **25.** Identity (×) **27** $70 + p$ **29.** $r + 56$ **31.** $35g$ **33.** $75b$ **35.** $20 + p$ **37.** $8x$ **39.** Yes; the order in which she does her homework doesn't matter as long as it all gets done. **41.** $114b$ **43** $(2n + 5) + 6n$; $8n + 5$ **45.** $4(11 \cdot 5n)$; $220n$ **47.** $4n + (2n - 7)$; $6n - 7$ **49a.** $5a + 61.25$ **49b.** \$79 **49c.** $(2a + 24.50) + (2a + 19.50)$ **51.** false; $15 + (4 \cdot 6) = 39$ and $(15 + 4) \cdot 6 = 114$; Since $39 \neq 114$, the statement is false. **53a.** No; $2 - 3 = -1$ and -1 is not a whole number. **53b.** No; $1 + 1 = 2$ and 2 is not a member of the set. **53c.** The Closure Property for Multiplication states that because the product of two whole numbers is also a whole number, the set of whole numbers is closed under multiplication. **53d.** Yes; $0 \cdot 0 = 0$, $0 \cdot 1 = 0$, $1 \cdot 0 = 0$, and $1 \cdot 1 = 1$. **55.** D **57.** B **59.** $s + 200$ **61.** $h - 6$ **63a.** 71°F **63b.** 62°F **65.** 45 **67.** 24 **69.** 210 **71.** 45 **73.** 0.07 m **75.** 0.3 L **77.** 12,000 g **79.** 54 ft

Pages 28–30 Lesson 1-5

1.

Figure Number	1	2	3	4	5
Perimeter	6	8	10	12	14

28

3. 45 cards **5.** 15.6 ft **7** 21 ways **9.** \$16,200 **11.** 10 **13.** \$119 **15.** 12 yr and 4 yr **17.** No; Sample answer: Tyler starts with an odd number of cards and then adds an even number of cards each week, so the total number of cards in his collection will always be an odd number. Since 504 is even, he will never have exactly 504 cards. **19.** Sample answer: If the problem involves a sequence of numbers that grows regularly, then look for a pattern. If the problem involves counting all possible outcomes or combinations, then make a table. If the problem involves finding a starting condition, then work backward. **21.** G **23.** 32 **25.** 14 **27.** 25 **29.** 10 **31.** 9 **33.** 3 **35.** 90

Pages 34–36 Lesson 1-6

1–4.

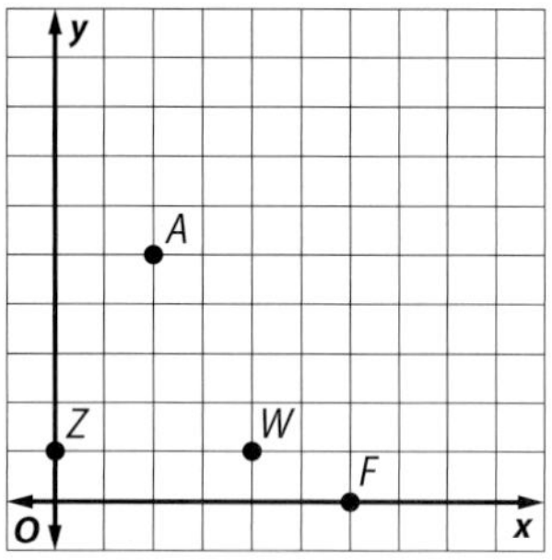

5. (3, 4) **7.** (5, 2)

9.

x	*y*
3	4
1	5
4	2

D = {1, 3, 4}, R = {2, 4, 5}

11a.

x	*y*	*(x, y)*
1	2	(1, 2)
2	4	(2, 4)
3	6	(3, 6)
4	8	(4, 8)

11b. **Pints in Quarts**

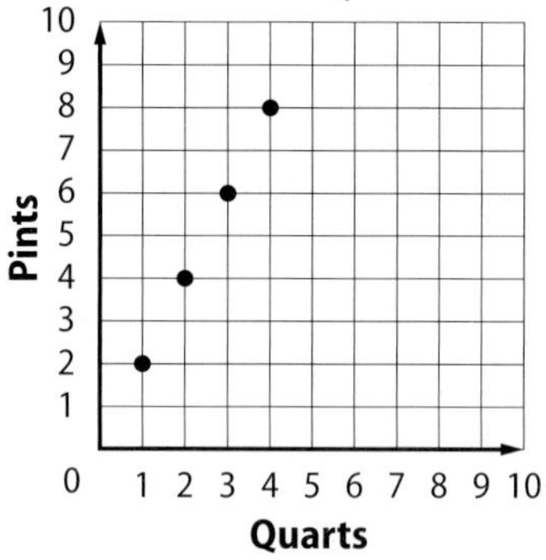

The points appear to lie in a line.

12–19.

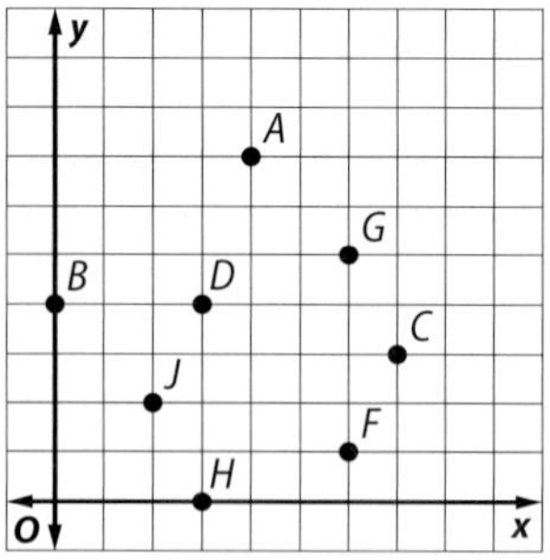

21. (6, 4) **23.** (5, 2) **25.** (7, 2) **27.** (4, 5)

29.

x	*y*
0	2
2	2
4	1
3	5

D = {0, 2, 3, 4}, R = {1, 2, 5}

31.

x	*y*
5	1
3	7
4	8
5	7

D = {3, 4, 5}, R = {1, 7, 8}

33a.

x	*y*	*(x, y)*
1	3	(1, 3)
2	6	(2, 6)
4	12	(4, 12)
6	18	(6, 18)

33b. **Hiking**

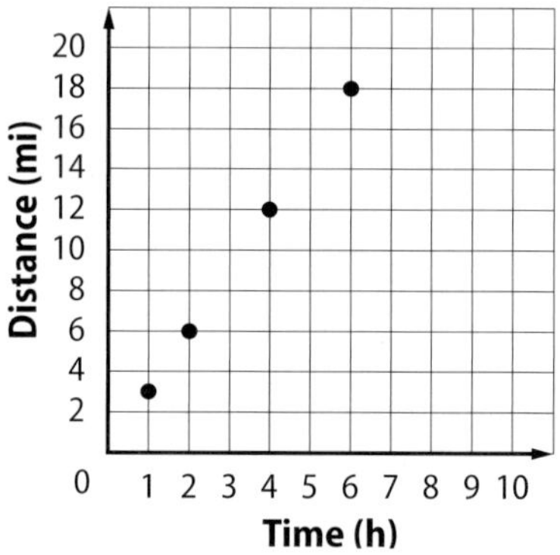

35 a. **Mr. Maloney's Students**

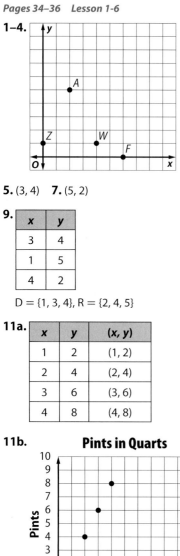

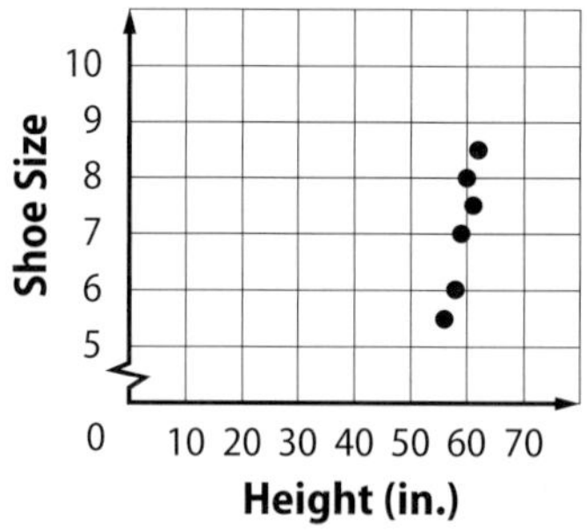

35 b. Sample answer: The graph is not linear like the other graph.

37. ordered pair **39.** Exercise 33; Sample answer: By connecting the points on the graph in Exercise 33, you could determine how far Aaron will have hiked at any time other than a whole number of hours. In Exercise 32, the points do not need to be connected because you would not need to know how much a portion of a pizza would cost. **41.** C **43.** A **45.** Commutative (×) **47.** Identity (+) **49.** no; $(100 \div 10) \div 2 \neq 100 \div (10 \div 2)$ **51a.** $3s + 16$ **51b.** $91 **53.** 7 **55.** 16 **57.** 14 **59.** 2 **61.** 10 **63.** 7

Pages 39–41 Lesson 1-7

1.

Number of Touchdowns	Number of Points
x	*y*
1	6
2	12
5	30
7	42

D: {1, 2, 5, 7}; R: {6, 12, 30, 42}

3a. Sample answer: $z = 16p$

3b.

p	16*p*	*z*
5	16(5)	80
8	16(8)	128
11	16(11)	176
13	16(13)	208

3c.

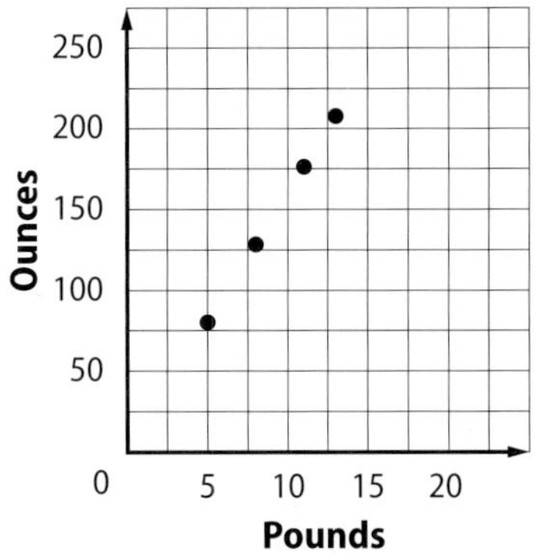

5.

Weight of Cat (lb)	Weight of Dog (lb)
x	*y*
3	7
6	10
9	13
12	16

D: {3, 6, 9, 12}; R: {7, 10, 13, 16}

7.

Ben's Cards	Casey's Cards
x	*y*
3	7
7	23
11	39
15	55

D: {3, 7, 11, 15}; R: {7, 23, 39, 55}

9 a.

m	*g*
1	180
2	165
3	150
4	135
5	120
6	105

9 b. Sample answer: $g = 195 - 15m$

9 c. D: {1, 2, 3, 4, 5, 6}; R: {105, 120, 135, 150, 165, 180}

11a.

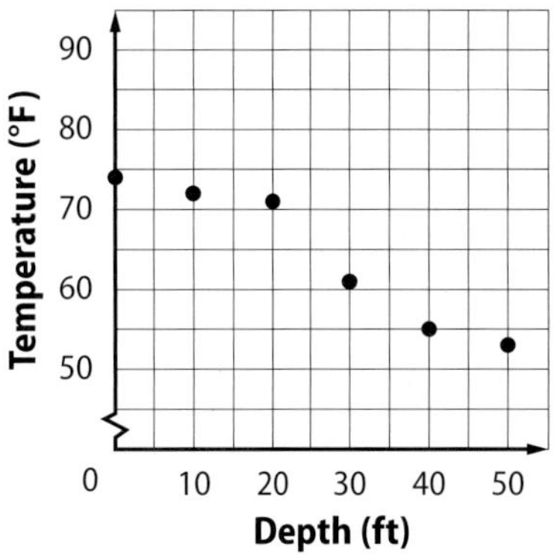

11b. No; the change in temperature is not constant, so one equation cannot be used to find any temperature value.
11c. D: {0, 10, 20, 30, 40, 50}; R: {53, 55, 61, 71, 72, 74}
13. Sample answer: The cost of each used CD is $4 at an electronics store. In the equation $y = 4x$, x represents the number of CDs and y represents the total cost of the CDs.
15. Sample answer: If a newspaper wanted to show the daily temperatures for a month, a table would be appropriate. If a reporter was writing about how the daily temperature was rising at a constant rate, an equation would be appropriate.
17. C **19.** B **21.** (7, 3) **23.** (6, 6) **25.** (3, 4) **27.** $26 + p$ **29.** $48c$ **31.** $75s$ **33.** 7 **35.** 33 **37.** 17 **39.** 235 **41.** 12 **43.** 12

Pages 42–44 Chapter Review

1. about 63 mph **3.** 12 dinner combinations **5.** 19 + 13 **7.** 45 **9.** 30 **11.** 20 **13.** $130 **15.** 20 **17.** 33 **19.** 32 **21.** 1 **23.** $\frac{x}{36}$ **25.** Associative (×) **27.** 12 + v **29.** 9 + x **31.** 25x **33.** 2 quarters, 4 dimes, 3 nickels, 5 pennies; or 4 quarters, 10 pennies **35.** 74 and 76

37.

x	1	2	3	4
y	4	8	12	16

D = {1, 2, 3, 4}; R = {4, 8, 12, 16}

39.

x	1	2	3	4
y	1	4	9	16

D = {1, 2, 3, 4}; R = {1, 4, 9, 16}

41a. $s = v + 5$

41b.

Number of Videos	Number of Songs
v	s
2	7
4	9
6	11
8	13

41c.

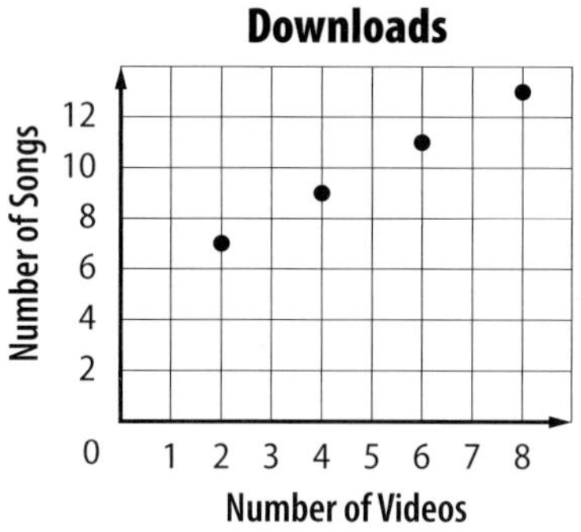

Chapter 2 Operations with Integers

Pages 49–51 Lesson 2-1

1. −500; +500 or 500; a deposit of $500 **3.** 2 > −5; −5 < 2 **5.** 1 > −1; −1 < 1 **7.** < **9.** −80, −48, −45, −39, −34, −27, −23, −2, 12 **11.** 17 **13.** 9 **15.** 18 **17.** −200; +200 or 200; 200 feet above sea level **19.** 0 **21.** 14 > −8; −8 < 14 **23.** 0 > −4; −4 < 0 **25.** |−30| > |−27|; |−27| < |−30| **27.** > **29.** < **31.** > **33.** = **35** −4, −3, −2, −1, +1, +2, +3, +4, +5, +6 **37.** 17 **39.** −15 **41** 5 **43.** 26 **45.** 7 **47.** 21 **49.** 23

51. Movie F; the absolute value of −8 is 8, and 8 is greater than the absolute value of any other change in the table. **53.** −272, −249, −201, −157, −101 **55.** Saturn; −218 < −162 **57.** 5, 0, −5, −13 **59.** 62, 20, −28, −35, −59 **61.** 232, 88, −72, −83, −94, −165 **63.** Always; the absolute value of a number and its opposite are equal. **65.** Sometimes; the expressions are equal when $x = 0$. **67.** 1 **69.** D **71.** −|10|, −|9|, −4, 0, |−5|, 7; Sample answer: Simplify the expressions that involve absolute values, then list the integers in order from least to greatest. **73.** Commutative (+) **75.** Associative (×) **77.** Associative (+) **79.** 2b − 9 **81.** 3 × (g − 2) **83** 195 **85.** 306 **87.** 15,328 R4 **89.** 84 cm^2 **91.** 15 ft^2

Pages 58–60 Lesson 2-2

1. −11 **3.** −7 **5.** 6 + (−10) = x; −4 yd; The team lost a total of 4 yards. **7.** 14 **9.** −3 **11.** −20 **13.** 10 **15.** 2 **17.** −25 **19.** −4 **21.** 11 **23.** 450 + (−160) = d; $290; Ty's account balance is now $290. **25.** −1 **27.** 6 **29.** −1 **31** −11 **33.** −103 + 68 = x; −35 ft; The shark is at a current depth of 35 feet below the surface. **35 a.** Rock: 34%, Rap/Hip Hop: 12%, Pop: 7%, Country: 13% **35 b.** 0%; The sum of the additive inverses, +1 and −1, and +2 and −2, is 0, so the percent of change is 0%. **37** 9 **39.** Sample answer: The temperature was −5°C at noon and then the temperature increased by 4°C over the next hour. What was the temperature at 1:00? **41.** False; sample answer: If $n = -2$, then −(−2) is positive. **43.** Multiplicative Property of 0, Additive Inverse Property **45.** B **47.** H **49.** −54; 54°F; 54°F above zero **51.** (6, 8) **53.** (5, 3) **55.** (2, 1) **57.** Multiplicative Property of 0 **59.** 3w + 1t **61.** 22 **63.** 1 **65.** 4 **67.** 1 **69.** 42 **71.** 137

Pages 65–67 Lesson 2-3

1. −2 **3.** 31 **5.** 6 units **7.** 10 units **9.** −1 **11.** −5 **13.** 25 **15.** −7 **17.** 15 **19** −1 **21.** 10 **23.** −60 **25.** 6 units **27.** 22 units **29.** 20 units **31.** 90 points

33. 3 m;

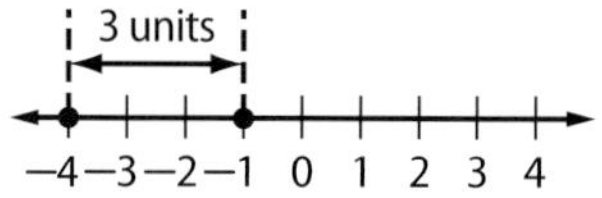

35. −214

37 a.

Date	May 3	May 4	May 5	May 6	May 7
Closing Price	$33.30	$30.59	$31.04	$31.97	$30.15
Change	–	−$2.71	$0.45	$0.93	−$1.82

37 b. $3.64 **39.** Sample answer: 4 − (−7); 11 **41.** False; 2 −(−2) = 4 and (−2) − 2 = −4 **43.** C **45.** 5765 ft **47.** 4 + (−5) = y; −1 yd; The team lost 1 yard in two plays. **49.** > **51a.** 6x

51b.

Number of Tickets (x)	$6x$	Cost ($)
2	6(2)	12
4	6(4)	24
5	6(5)	30
7	6(7)	42

51c.

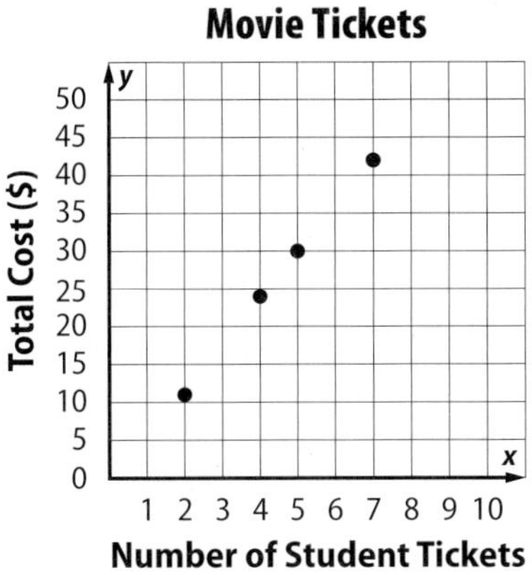

53. $r - 5$ **55.** $7 + n \div 8$ **57.** $39 + s$ **59.** $56p$ **61.** $60b$ **63.** 96 in. **65.** 3 gal

Pages 74–76 Lesson 2-4

1. −42 **3.** 120 **5.** −8(5); −$40 ; Mr. Heppner's account balance is $40 less after 5 days. **7.** $-21ab$ **9.** −88 **11.** −27 **13.** 75 **15.** 49 **17.** −336 **19.** −2(6); −12°F; The temperature dropped 12°F after 6 hours. **21.** $-30m$ **23.** $81mn$ **25.** $-36ab$ **27.** $-54efg$ **29.** −100 **31.** −72 **33.** $780 - 65w$

Week	$780 - 65w$	Amount Owed ($)
2	780 − 65(2)	650
4	780 − 65(4)	520
6	780 − 65(6)	390
8	780 − 65(8)	260

It will take Diego 12 weeks to repay his father.

35. < **37.** > **39.** Team 1 answered 12 questions correctly, 3 questions incorrectly, and passed on 1 question, so Team 1 earned 12(5) + 3(−8) + 1(−2) or 34 points. Team 2 answered 13 questions correctly, 2 questions incorrectly, and passed on 7 questions, so Team 2 earned 13(5) + 2(−8) + 7(−2) or 35 points. Since 34 < 35, Team 2 won.

41a.

Minutes	$-15x + 600$	Altitude (ft)
0	−15(0) + 600	600
5	−15(5) + 600	525
10	−15(10) + 600	450
15	−15(15) + 600	375

41b. the time the balloon began descending

41c.

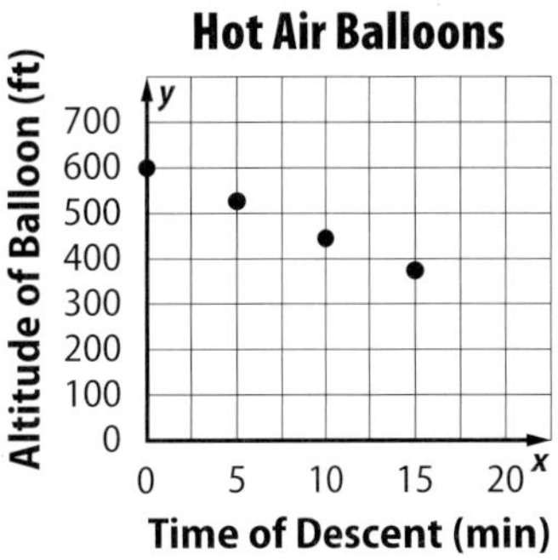

41d. 40 minutes; Sample answer: If you continue the pattern in the table, for every 10 minutes, the balloon drops 150 feet. $150 \times 4 = 600$, so $10 \cdot 4 = 40$ minutes. **43.** −9 and 9 **45.** 0; Multiplicative Property of Zero **47.** true **49.** A **51.** A **53.** 14,776 ft **55.** −16 **57.** −6 **59.** −28 **61.** −2 **63.** −10,924; 10,924 or +10,924; The meaning of the opposite is 10,924 feet above sea level.

Pages 80–82 Lesson 2-5

1. −4 **3.** $8.\overline{6}$ **5.** −16 **7.** −9 **9.** −4 **11.** −7 **13.** −3 **15.** 18 **17.** −15 **19.** 14 **21.** −110 **23.** 5 **25.** 14 **27.** −36 **29.** $42; The mean amount for the last 5 transactions was $42. **31.** −$300; Sample answer: Every month, the company spends $300 more than they earn. **33.** < **35.** −748 **37.** −13 **39.** $-40 \div 8$; −5°F; Each hour the temperature dropped an average of 5°F. **41a.** −244.8 m **41b.** −234.8 m; The mean would be 10 meters higher. **43.** Sample answer: $-110 \div 5 = -22$; Jessie's total score in a word game is −110 points after 5 rounds. What is Jessie's mean score per round? **45.** 4, −1; Divide the previous term by −4. **47.** Sample answer: The Associative Property is not true for the division of integers because the way the integers are grouped affects the solution. $[24 \div (-6)] \div 2 = -2$; $24 \div [(-6 \div 2)] = -8$; The Commutative Property is not true for the division of integers because the order of the integers affects the solution. $-2 \div 10 = -0.2$; $10 \div (-2) = -5$ **49.** F **51.** J **53.** −72 **55.** −88 **57.** 108 **59.** −5 **61.** −7 **63.** 3 **65.** −$7 **67.** (5, 1) **69.** (5, 4)

Pages 85–87 Lesson 2-6

1. (−5, 2) **3.** (5, 2)

5–8.

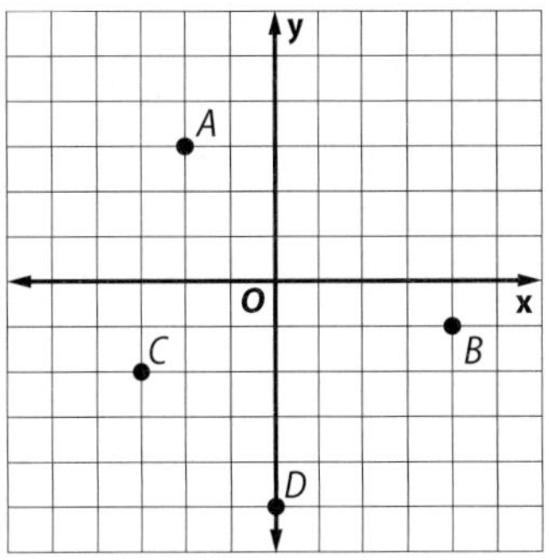

9.

x − y = 4		
x	**y**	**(x, y)**
6	2	(6, 2)
5	1	(5, 1)
4	0	(4, 0)
−1	−5	(−1, −5)
−2	−6	(−2, −6)

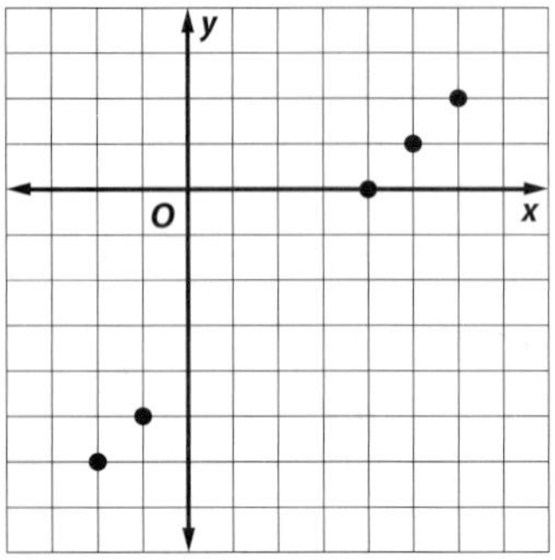

The points are along a diagonal line that crosses the *y*-axis at *y* = −4 and the *x*-axis at *x* = 4.

11. (−3, 1) **13.** (3, 5) **15.** (−4, −3) **17.** (5, −3) **19.** (0, 3)

21–31.

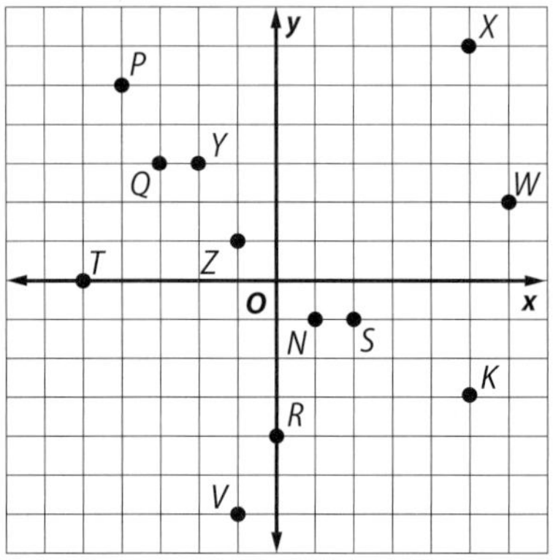

21. II **23.** I **25.** IV **27.** none **29.** II **31.** IV

33. Sample answer:

x − y = 10		
x	**y**	**(x, y)**
30	20	(30, 20)
20	10	(20, 10)
10	0	(10, 0)
−20	−30	(−20, −30)
−30	−40	(−30, −40)

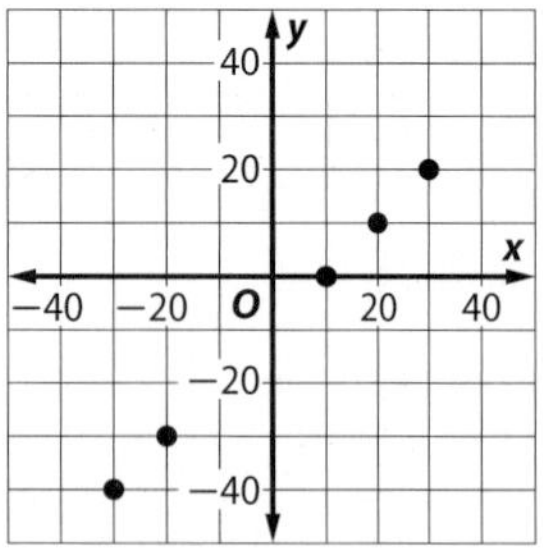

The points on the graph are in a line that slants downward to the left. The line crosses the *x*-axis at *x* = 10.

35. IV **37.** IV

39

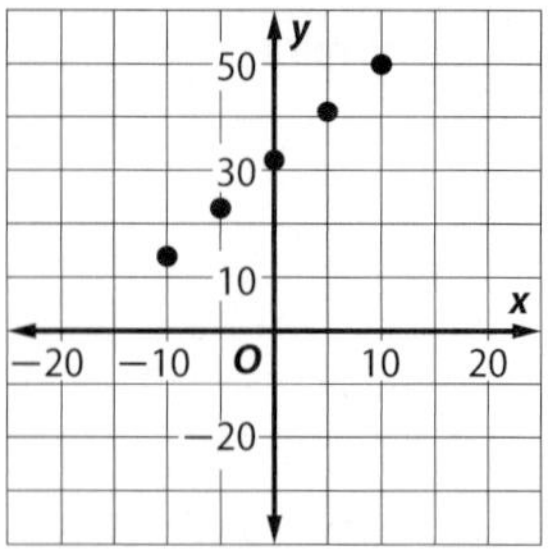

41.

y = x − 4		
x	**y**	**(x, y)**
−2	−6	(−2, −6)
−1	−5	(−1, −5)
0	−4	(0, −4)
1	−3	(1, −3)
2	−2	(2, −2)

43–46.

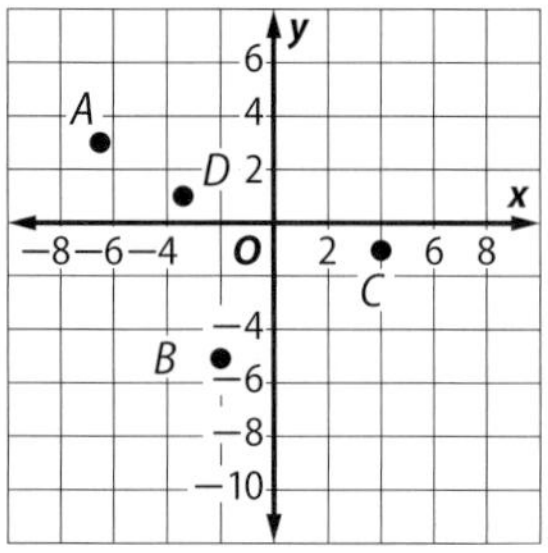

47. Sample answer: (−3, 1)
49a. Sample answer: Never; both coordinates are positive.
49b. Sample answer: Sometimes; both (−2, 0) and (2, 0) lie on the *x*-axis. **51.** C **53.** D **55.** 3 **57.** −50 **59.** −11 s
61. 6 **63.** −70 **65.** 2 **67.** 7 **69.** 6 **71.** −65 **73.** −1

Pages 88–90 Chapter Review

1. −18 > −20; −20 < −18 **3.** > **5.** < **7.** 16 **9.** 34
11. −2 **13.** −10 **15.** −5 **17.** 9 **19.** 7 **21.** −6 **23.** 3
25. −7 **27.** 3 **29.** −2 **31.** −14 **33.** 10 **35.** 0
37. 288°F **39.** −30 **41.** 48 **43.** −140 **45.** −84
47. −48 **49.** 60 **51.** −288 **53.** −12 **55.** 4
57. −6 **59.** 8 **61.** −2 **63.** −80 **65.** 13.4 s
67–76.

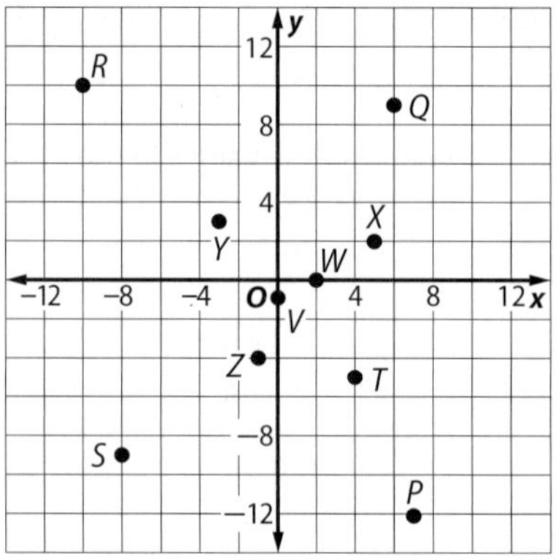

;

67. IV **69.** II **71.** IV **73.** I **75.** none **77.** Andy: I, Owen: II, Cheryl: IV, Denyce: III, and Fabio: IV

Chapter 3 Operations with Rational Numbers

Pages 97–100 Lesson 3-1

1. 0.6 **3.** −0.15 **5.** $-0.\overline{6}$ **7.** 0.8 **9.** $<$ **11.** $>$ **13.** $<$ **15.** test 1 **17.** 0.35 **19.** −0.1875 **21.** 0.36 **23.** −0.4375 **25.** $0.7\overline{3}$ **27.** $-0.\overline{2}$ **29.** 0.929 **31.** $<$ **33.** $=$ **35.** $<$ **37.** $<$ **39.** more than; $\frac{1}{4} = 0.25$ and $0.28 > 0.25$ **41.** $<$ **43.** $>$ **45.** $>$ **47.** $\frac{3}{32}, \frac{5}{16}, \frac{3}{8}, \frac{1}{2}, \frac{3}{4}$ **49.** $2\frac{3}{5}, 2\frac{2}{3}, 2.67$ **51.** $\frac{1}{13}, \frac{2}{25}, 0.089$ **53.** $0.\overline{9}$ **55.** $-10.3\overline{4}$ **57a.** Sample answer: *A*: $\frac{7}{9}$; *B*: $\frac{9}{10}$; *C*: $1\frac{1}{5}$; *D*: $1\frac{3}{10}$; *E*: $1\frac{3}{5}$ **57b.** Sample answer: $\frac{7}{9} < \frac{9}{10}$ **59.** Sample answer: fractional form: customary measurement; decimal form: stock price **61.** $\frac{1}{3} = 0.\overline{3}, \frac{1}{6} = 0.1\overline{6}, \frac{1}{7} = 0.\overline{142857}, \frac{1}{9} = 0.\overline{1}$ **63.** $0.\overline{1}$, $0.\overline{23}$, and $0.\overline{75}$; Sample answer: When the denominator is 9 or 99, the numerator of the fraction is the repeating part of the decimal. **65.** C **67.** B **69.** 240 **71.** 9 **73.** 24 m below the surface **75.** (−4, −3) **77.** (0, −3) **79.** (3, 4) **81.** 23 **83.** 87 **85.** thirty-four hundredths **87.** three-tenths

Pages 103–106 Lesson 3-2

1. $\frac{15}{4}$ **3.** $-\frac{7}{4}$ **5.** $-3\frac{85}{99}$ **7.** $2\frac{27}{50}$ **9.** rational **11.** $\frac{11}{6}$ **13.** $-\frac{87}{8}$ **15.** $3\frac{5}{8}$ **17.** $-5\frac{9}{25}$ **19.** $-1\frac{3}{10}$ **21.** $\frac{253}{500}$ **23.** $-2\frac{5}{9}$ **25.** $\frac{16}{99}$ **27.** $-\frac{1}{11}$ **29.** integer, rational **31.** rational **33.** irrational **35.** Yes; $\frac{5}{8} = 0.625$ and $0.625 > 0.6$, so the bead will fit. **37.** $>$ **39.** $<$ **41.** $>$ **43.** $\frac{652}{999}$ **45.** $\frac{163}{225}$ **47.** $9\frac{241}{990}$ **49a.** 3.1415927;

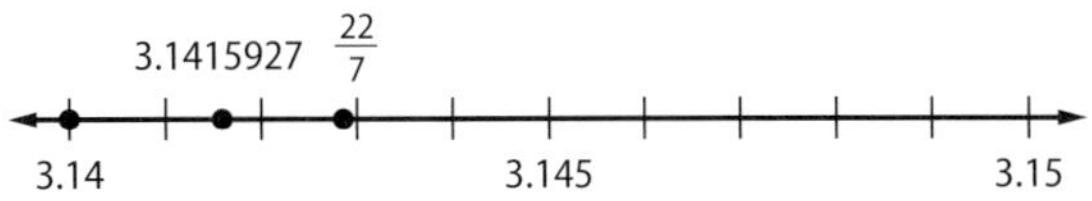

49b. $3.14 < \pi < \frac{22}{7}$ **49c.** Sample answer: If the diameter is a multiple of 7, use $\frac{22}{7}$. Otherwise, use 3.14. **51.** $-3.\overline{42}, -3.4, 3\frac{4}{11}, 3.38$ **53.** $-1.95, -1\frac{13}{14}, -1.9, -1\frac{9}{11}$ **55.** $\frac{7}{9}; \frac{5}{8}; \frac{5}{8} < \frac{5}{7} < \frac{7}{9}$ **57.** Sample answer: Since $0.\overline{76} = 0.76767676...$ and $0.76 = 0.76000000...$, comparing the values in the thousandths place shows that $0.\overline{76}$ is greater than 0.76. **59a.** true; Sample answer: Integers may be written as a fraction with a denominator of 1. Therefore, they belong to the set of rational numbers. **59b.** true; Sample answer: All whole numbers and their opposites belong to the set of integers. **59c.** false; Sample answer: $\frac{1}{2}$ is not an integer. **59d.** true; Sample answer: All natural numbers are rational because they may be written as a fraction with a denominator of 1. **61.** B **63.** C **65.** −0.625 **67.** −0.2 **69a.** Ava: −5; Brennan: −3; Denny: −6; Jose: 0; Hao: −7 **69b.** Jose, Brennan, Ava, Denny, Hao **71.** 1.5 lb **73.** 6050 m **75.** 0 **77.** −280 **79.** −2 **81.** 0

Pages 109–112 Lesson 3-3

1. $\frac{7}{16}$ **3.** $-\frac{1}{8}$ **5.** 5 **7.** $-\frac{98}{75}$ or $-1\frac{23}{75}$ **9.** $-\frac{2}{5}$ **11.** $-\frac{28}{5}$ or $-5\frac{3}{5}$ **13.** 140 towns **15.** $\frac{3}{32}$ **17.** $\frac{8}{27}$ **19.** $\frac{1}{9}$ **21.** $\frac{2}{39}$ **23.** $-\frac{2}{3}$ **25.** −1 **27.** $\frac{5}{2}$ or $2\frac{1}{2}$ **29.** $-\frac{8}{5}$ or $-1\frac{3}{5}$ **31.** −10 **33.** 87 lb **35.** $\frac{1}{21}$ **37.** −3 **39.** −2 **41.** 5 bags **43.** 12 **45.** 10 **47.** 4

51 a.

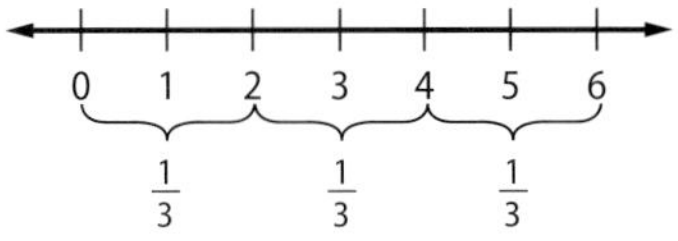

So, $\frac{2}{3}$ of 6 is 4.

51 b.

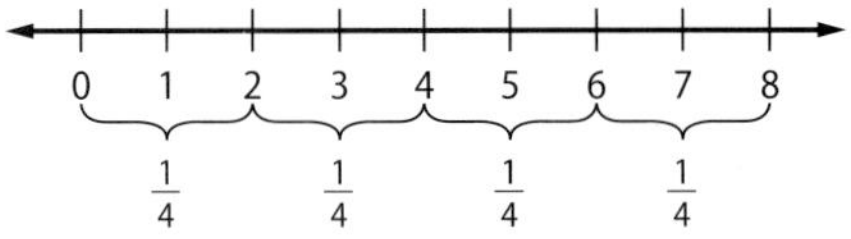

So, $\frac{3}{4}$ of 8 is 6.

51 c.

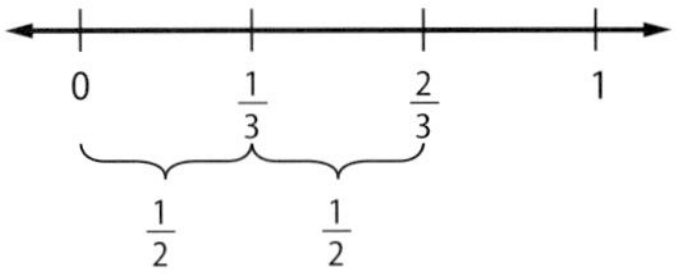

So, $\frac{1}{2}$ of $\frac{2}{3}$ is $\frac{1}{3}$.

51 d.

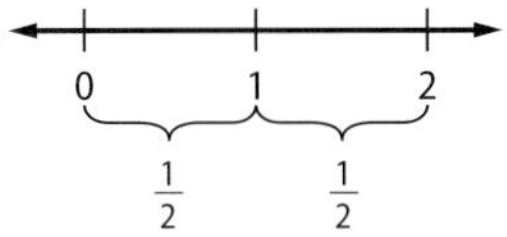

So, $\frac{1}{2}$ of 2 is 1.

51 e.

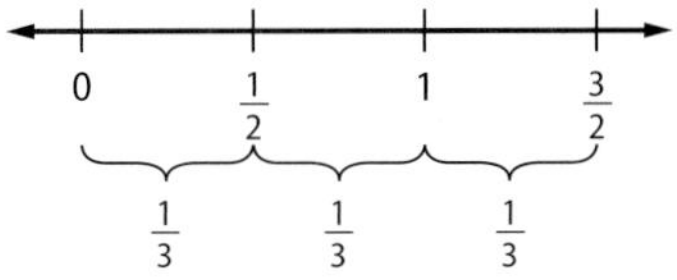

So, $\frac{2}{3}$ of $\frac{3}{2}$ is 1.

51 f.

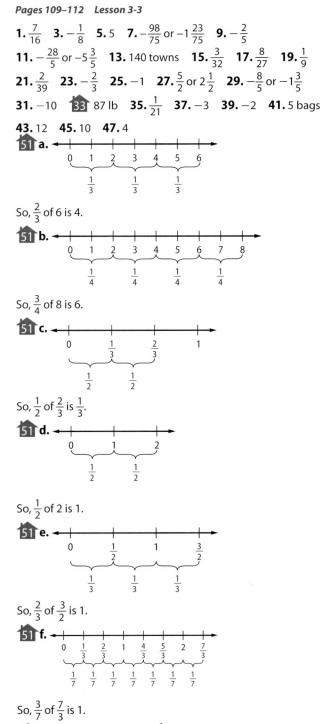

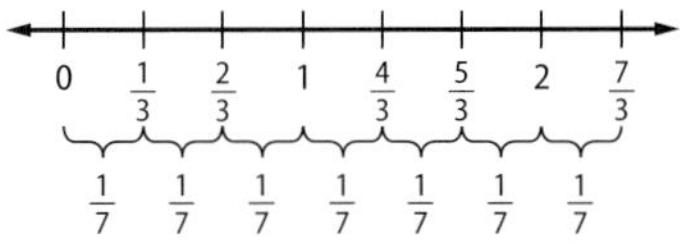

So, $\frac{3}{7}$ of $\frac{7}{3}$ is 1.

51 g. Each product is 1. **51 h.** 1 **53.** She did not write the mixed numbers as fractions before multiplying. $-4\frac{1}{6} \cdot 2\frac{2}{9} = -\frac{25}{\cancel{6}_{3}} \cdot \frac{\cancel{20}^{10}}{9} = -\frac{250}{27}$ or $-9\frac{7}{27}$ **55.** $-\frac{5}{6}$ **57.** $\frac{2}{3}$ **59.** $-\frac{3}{5}$ **61.** Sample answer: 20; $16\frac{4}{5}$; If you have mixed numbers and you round up both of them, you will overestimate the product. Your estimate may be closer to the actual product if you round up one mixed number and round down the other one. **63.** G **65.** $6\frac{2}{3}$ ft; $2\frac{7}{9}$ ft^2 **67.** $\frac{43}{200}$ **69.** $-\frac{1}{3}$ **71.** $-2\frac{5}{99}$ **73.** $<$ **75.** $<$ **77.** $=$ **79.** −70 **81.** −168 **83.** 250 **85.** 32 **87.** 4 **89.** 16 **91.** 2 **93.** 3 **95.** 21

Pages 117–119 Lesson 3-4

1. $\frac{7}{6}$ **3.** $-\frac{1}{63}$ **5.** $-\frac{1}{101}$ **7.** $-\frac{9}{10}$ **9.** $-\frac{2}{9}$ **11.** $\frac{3}{5}$ **13.** 8 squares **15.** $\frac{np}{18}$ **17.** $-\frac{5}{4}$ **19.** $\frac{19}{10}$ **21.** $\frac{1}{19}$ **23.** $\frac{3}{17}$ **25.** $-\frac{9}{2}$ **27.** $-\frac{5}{8}$ **29.** $\frac{1}{54}$ **31.** $-\frac{5}{3}$ or $-1\frac{2}{3}$ **33.** $\frac{57}{28}$ or $2\frac{1}{28}$ **35.** 12 costumes **37.** $\frac{1}{14}$ **39.** $\frac{15}{2}$ or $7\frac{1}{2}$ **41.** 90 mph **43.** $-\frac{3}{25}$ **45.** $-\frac{3}{4}$ **47.** Sample answer: The quotient decreases; $\frac{3}{4} \div \frac{1}{8} = 6$. **49.** Sample answer: $\frac{1}{2} \div \frac{3}{4}$ is $\frac{2}{3}$, and $\frac{2}{3}$ is not a whole number. **51.** always **53.** C **55.** A **57.** $\frac{9}{8}$ or $1\frac{1}{8}$ **59.** $-\frac{3}{22}$ **61.** 1 **63.** $\frac{7}{250}$ mi^2 **65.** −72 **67.** 48 **69.** −68 **71.** $3n$ **73.** $-5n$ **75.** $12 - \frac{6}{3}$ **77.** $15 - (3 + 5)$

Pages 123–125 Lesson 3-5

1. $1\frac{1}{3}$ **3.** $-\frac{1}{2}$ **5.** 10 **7.** $\frac{5}{7}$ **9.** $-\frac{6}{7}$ **11.** $\frac{10}{11}$ unit **13.** $-\frac{1}{3}$ **15.** $\frac{3}{7}$ **17.** $\frac{103}{9}$ or $11\frac{4}{9}$ **19.** $-\frac{2}{11}$ **21.** $-\frac{7}{10}$ **23.** $-\frac{66}{5}$ or $-13\frac{1}{5}$ **25.** $\frac{7}{9}$ **27.** $\frac{1}{9}$ **29.** $4\frac{1}{2}$ c **31.** $\frac{1}{5}$ unit **33.** $\frac{3}{8}$ unit **35.** $-2\frac{8}{9}$ **37a.** $1\frac{1}{8}$ lb; $3\frac{5}{8}$ lb **37b.** $23\frac{5}{8}$ lb **39.** Sample answer: $\frac{1}{3} - 1 = -\frac{2}{3}$ **41.** No; he did not add the fraction part of the mixed numbers correctly. **43.** Sample answer: A recipe requires $\frac{3}{4}$ cup of milk and $1\frac{1}{4}$ cups of water. How much liquid is in the recipe?; 2 cups **45.** J **47.** $-6\frac{1}{3}$ **49.** −10 **51.** $\frac{2}{35}$ **53.** $\frac{5}{9}$ **55.** 17-yard line **57.** 60 **59.** 45 **61.** 40 **63.** 17 **65.** 0 **67.** −6 **69.** 11

Pages 128–131 Lesson 3-6

1. $\frac{2}{3}$ **3.** $\frac{33}{56}$ **5.** $-9\frac{5}{6}$ **7.** $-1\frac{1}{36}$ **9.** $\frac{1}{24}$ **11.** $8\frac{1}{12}$ **13.** $\frac{7}{12}$ c **15.** $\frac{5}{21}$ **17.** $-1\frac{8}{33}$ **19.** $-\frac{23}{42}$ **21.** $-\frac{11}{30}$ **23.** $\frac{1}{5}$ **25.** $12\frac{3}{10}$ **27.** $4\frac{1}{2}$ **29.** $-11\frac{7}{9}$ **31.** $4\frac{19}{48}$ **33.** $-17\frac{3}{14}$ **35.** $\frac{3}{4}$ c

37a.

Perimeter of 20	
Length (ℓ)	Width (w)
8	2
7	3
6	4
5	5

37b. (8, 2), (7, 3), (6, 4), (5, 5);

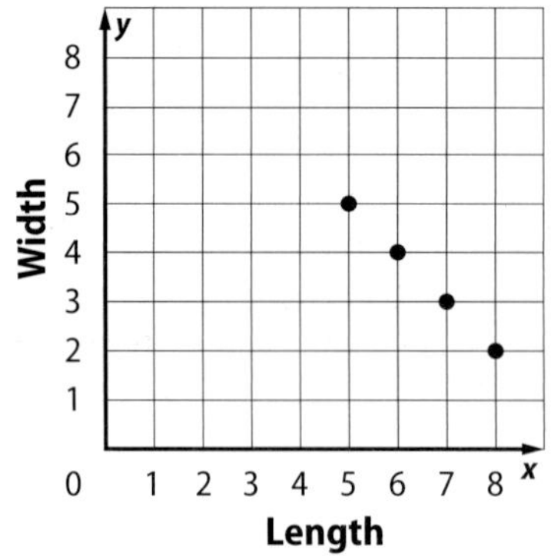

37c. $4\frac{1}{2}$ in. **41.** $-14\frac{5}{8}$ **43.** $-59\frac{1}{6}$ **45.** $36\frac{67}{99}$ **47.** Sample answer: $\frac{2}{3} - \frac{5}{8} = \frac{1}{24}$ **49.** Sample answer: Each of the fractions in the sum $\frac{3}{4} + \frac{1}{8} + \frac{1}{2}$ uses a different unit (parts of a whole). So the addends must be rewritten in the same unit before they can be added. **51.** Sample answer: Fill the $\frac{1}{2}$ cup. From the $\frac{1}{2}$ cup, fill the $\frac{1}{3}$ cup. $\frac{1}{6}$ cup will be left in the $\frac{1}{2}$ cup because $\frac{1}{2} - \frac{1}{3} = \frac{1}{6}$. **53.** D **55.** H **57.** $14\frac{1}{2}$ **59.** $\frac{3}{4}$ **61.** $\frac{2}{3}$ **63.** 6 **65.** $\frac{39}{1{,}000{,}000}$ **67.** 50 **69.** 2.3 ft

Pages 132–134 Chapter Review

1. 0.3 **3.** $-0.8\overline{3}$ **5.** 0.625 **7.** < **9.** > **11.** < **13.** 5.3125 in. **15.** $2\frac{2}{25}$ **17.** $\frac{7}{8}$ **19.** $\frac{1}{9}$ **21.** $\frac{5}{9}$ **23.** integer, rational **25.** irrational **27.** $1\frac{2}{3}$ h **29.** $\frac{3}{20}$ **31.** $3\frac{1}{3}$ **33.** $2\frac{5}{6}$ **35.** 17 **37.** $6\frac{1}{4}$ in. **39.** $-\frac{1}{16}$ **41.** $\frac{5}{19}$ **43.** −11 **45.** $-2\frac{11}{12}$ **47.** $\frac{2}{3}$ **49.** $\frac{24}{25}$ **51.** $4b$ **53.** $\frac{15a}{7}$ **55.** 8 days **57.** $\frac{2}{5}$ **59.** $\frac{5}{7}$ **61.** 11 **63.** $5\frac{1}{2}$ mi **65.** $1\frac{1}{2}$ ft **67.** $\frac{7}{15}$ **69.** $-\frac{16}{21}$ **71.** $10\frac{14}{15}$ **73.** $-4\frac{2}{9}$ **75.** $6\frac{1}{12}$ c **77.** $18\frac{1}{4}$ m

Chapter 4 Powers and Roots

Pages 138–140 Lesson 4-1

1. 2^6 **3.** $\left(-\frac{1}{4}\right)^3$ **5.** $(y - 3)^3$ **7.** 16 cm **9.** 29 **11.** 28.25 **13.** 11^4 **15.** $(-8)^6$ **17.** $\left(-\frac{1}{5}\right)^4$ **19.** $(ab)^4$ or a^4b^4 **21.** $21m^2n^4$ **23.** $(n - 5)^3$ **25a.** 30,000 mi **25b.** 1215 volcanoes **27.** 81 **29.** 28 **31.** 4.25 **33.** 21.625 **35.** 23 **37.** 243 **39a.** 64,000 ft^2; 63,000 ft^2; 36,400 ft^2 **39b.** 36,400 ft^2; 63,000 ft^2; 64,000 ft^2 **39c.** 1000 ft^2 **41.** 1331 **43.** 625 **45.** 512 **47.** 2430 **49.** 1372 **51.** 820.125 **53.** > **55.** < **57.** > **59.** Sample answer: Jin drove 6^2 miles to visit a friend. He drove 4^3 times as far to visit his uncle. How far did he drive to visit his uncle? 2304 mi **61.** 10^8; Sample answer: $10^7 = 10{,}000{,}000$ and $10^8 = 100{,}000{,}000$. 100,000,000 is much closer to 230,000,000 than 10,000,000 is. **63.** Using exponents is a more efficient way to describe and compare numbers. **65.** 512 **67.** B **69.** 30 **71.** 0 **73.** −20 **75.** Commutative (×) **77.** $4\frac{3}{8}$ in. **79.** −150; +150 or 150; 150 feet above sea level **81.** 4 **83.** 2 **85.** 3 **87.** 4 **89.** −7

Pages 143–146 Lesson 4-2

1. $\frac{1}{6^2}$ **3.** $\frac{1}{x^5}$ **5.** 2^{-6} **7.** 3^{-2} **9.** 10^{-3} **11.** $-\frac{1}{64}$ **13.** $\frac{1}{2}$ **15.** $\frac{1}{7^1}$ or $\frac{1}{7}$ **17.** $\frac{1}{(-5)^4}$ **19.** $\frac{1}{k^8}$ **21.** $\frac{1}{r^{20}}$ **23.** 10^{-3} **25.** 6^{-5} **27.** 7^{-2} **29.** 5^{-3} **31.** 10^{-8} **33.** $\frac{1}{144}$ **35.** $-\frac{1}{6}$ **37.** $\frac{1}{64}$ **39.** $-\frac{2}{9}$ **41a.** 10^3 or 1000 times **41b.** 10^4 or 10,000 **41c.** 10^8 or 100,000,000 times **43a.** 10^{-4} **43b.** 10^6 mm^3 **43c.** 10^3 cm^3 **45.** $\frac{1}{10^{15}}$ or $\frac{1}{1{,}000{,}000{,}000{,}000{,}000}$; 10^{-15}

47. $\frac{1}{25} = \frac{1}{5 \cdot 5} = \frac{1}{5^2} = 5^{-2}$ **49a.** $2^{-2} = \frac{1}{4}, (-2)^{-2} = \frac{1}{4}$, $(-2)^2 = 4, 2^2 = 4; 2^{-2} = (-2)^{-2}$ and $(-2)^2 = 2^2$
49b. $2^{-3} = \frac{1}{8}, (-2)^{-3} = -\frac{1}{8}, (-2)^3 = -8, 2^3 = 8$; None of the expressions are equal. **49c.** Sample answer: When you square either a positive or a negative value, the answer is positive. When you cube a positive value, you get a positive, and when you cube a negative value, you get a negative. **49d.** x is an even number **49e.** x is an even number **51.** Sample answer: If $n = 3, \frac{1}{2^n} = \frac{1}{2^3}$ or $\frac{1}{8}$. If $n = 4, \frac{1}{2^n} = \frac{1}{2^4}$ or $\frac{1}{16}$. So, as the value of n increases, the value of $\frac{1}{2^n}$ decreases. **53.** C **55.** D
57. $-509, -505, -435, -410$ **59.** $\frac{1}{9}$ **61.** $\frac{1}{2}$ **63.** $6\frac{13}{15}$
65. -48 **67.** -60 **69.** 15^5 **71.** $36x^2y^4$ **73.** 0.025
75. 0.038 **77.** 0.0045

Pages 150–152 Lesson 4-3

1. 2^{10} **3.** 5^3 **5.** x^{16} **7.** $\frac{1}{m^2}$ **9.** 4^2 **11.** 6^{12} **13.** r^4
15. $\frac{1}{c^9}$ **17.** 2^3 or 8 times **19.** 5^8 **21.** a^9 **23.** 4^1 or 4
25. $\frac{1}{c}$ **27.** $\frac{40}{x^6}$ **29.** $-4m$ **31.** 5^8 **33.** a **35.** $\frac{1}{8^{11}}$ **37.** $\frac{1}{b}$
39. $(-1.5)^5$ **41.** r^{14} **43a.** 10^5 or 100,000 times **43b.** 10^6 or 1,000,000 times **45** 5^4 or 625 lb **47.** $\frac{484}{1}$; Sample answer: For every 484 red blood cells, there is one white blood cell.
49. -2 **51.** 12 **53.** -3 **55.** $8a^3b^{10}$ **57.** n^6 **59.** $21x^2y^6$
61. Sample answer: x^7, x^2 **63.** Sample answer: By the Quotient of Powers Property, $\frac{a^n}{a^n} = a^{n-n}$ or a^0 for $a \neq 0$. Since $\frac{a^n}{a^n} = 1$, then $a^0 = 1$. So any nonzero number raised to the zero power must equal 1. **65.** Sample answer: Write the numbers as powers with the same base. Then divide by subtracting the exponents. **67.** F **69.** $A = \frac{1}{2}bh = \frac{1}{2}(4x^3)(3x^5) = \frac{1}{2} \cdot 4 \cdot 3 \cdot x^3 \cdot x^5 = \frac{1}{2} \cdot 4 \cdot 3 \cdot x^8 = 6x^8$
71. -70 **73.** Chicago: 20^2; St. Louis: 15^2; Nashville: 5^3; Evansville: 10^2; Paducah: 5^2 **75.** $|24 - (-6)|$; 30 inches
77. $\frac{1}{w^6}$ **79.** $-\frac{1}{8}$ **81.** $-\frac{1}{21}$ **83.** $\frac{1}{15}$ **85.** -142

Pages 155–158 Lesson 4-4

1. 4160 **3.** 107,500 **5.** 1.35×10^5 **7.** Sample answer: 7×10^{-6} kg **9.** Sample answer: 2×10^7 cm **11.** 3.7×10^{-2}, 3.4×10^2, 3.5×10^2, 400 **13.** 0.00015 **15.** 0.00951
17. 792.4 **19.** 171,000,000,000 **21** 3.2×10^7
23. 9.18×10^{-4} **25.** 6.752×10^{-3} **27.** Sample answer: 8×10^{-23} lb **29.** Sample answer: 4×10^5 km **31.** 219 mg
33. 3.024×10^2, 2805, 2.81×10^4, 3.2×10^4, 2.08×10^5
35. 9.05×10^{-6}, 905,000, 9.5×10^6, 9,562,301
37a. 0.0000125 cm **37b.** 1.25×10^{-6} m^3 **39.** = **41.** <
43. < **45.** Sample answer: about 2.41×10^6 km
47 Sample answer: about 1.24×10^8 **49.** googol: 1×10^{100}; centillion: 1×10^{303} **51.** 6×10^2; $3 \times 10^2 = 300$ and $2 \times 300 = 600$ or 6×10^2 **53.** B **55.** 1×10^{-4} lb
57. penicillin; 10^5 times greater **59.** n^8 **61.** 3^{13}
63. $[8(6n)] \div 2$; $24n$ **65.** $\frac{1}{3^2}$ **67.** $\frac{1}{(-5)^1}$ **69.** $\frac{1}{5^3}$ **71.** 4^5
73. $2^3 \cdot 3^2$ **75.** $10^2 \cdot 5^4$

Pages 162–165 Lesson 4-5

1. about 1.3×10^{-4} lb **3.** 7.982003×10^7 **5.** 9.85×10^6
7. 2.2×10^{-2} **9.** about 3.27×10^6 mi **11.** 1.88328×10^{12}
13. 2.698×10^1 **15** 7×10^{13} **17.** 1.25×10^7
19. 5.86205×10^5 **21.** 2.144×10^8 **23.** 2.36602×10^2
25. 8.327×10^7 **27.** about 1.75×10^8 times greater
29. 9.7×10^7 km **31** 8×10^1 downloads
33. 1.256×10^{-5} cm^2 **35.** 1×10^{-1} **37.** 6.54×10^{11}
39. 8.88×10^{-25} **41.** 9.0249×10^{12} **43.** Sample answers: $(2.15 \times 10^{-3}) + (2.5 \times 10^{-4})$; $(2.56 \times 10^{-3}) - (1.6 \times 10^{-4})$
45. about 4×10^{89} adults **47.** A **49.** B **51.** -58; $+58$ or 58
53. -4500; $+4500$ or 4500 **55.** $+68$ or 68; -68 **57.** -5
59. -2 **61.** 760 **63.** true **65.** true **67.** true

Pages 171–173 Lesson 4-6

1. 4 **3.** ± 9 **5.** -7 **7.** 90 ft **9.** 13 **11.** -7 **13.** 3
15. -10 **17.** 3 **19.** -12 **21.** ± 1 **23** 6 **25.** -4
27. ± 12 **29.** -12 **31.** 6 **33.** 8 **35.** -4 **37** 14.7 cm; 58.8 cm **39.** 1331 **41a.** 246 **41b.** 811 **41c.** 732 **43.** C
45. H **47.** \$2.00 **49.** \$392.50 **51.** Additive Identity Property **53.** Associative Property (+) **55.** Multiplicative Identity Property **57.** Zero Property of Multiplication
59. -64 **61.** -8

Pages 177–179 Lesson 4-7

1. natural, whole, integer, rational **3.** irrational **5.** > **7.** >
9. 10.15, $10\frac{1}{5}$, $\sqrt{110}$, $10.\overline{5}$ **11.** 8.6, -8.6 **13.** 61.3 ft
15. rational **17.** irrational **19.** rational **21.** integer, rational **23.** natural, whole, integer, rational **25.** rational
27. > **29.** = **31.** < **33.** $\frac{15}{2}$, $\sqrt{64}$, $8.\overline{14}$, $8\frac{1}{7}$
35. $-\frac{31}{6}$, $-\sqrt{26}$, -5, $-\frac{5}{6}$ **37** 11.4, -11.4 **39.** 9, -9
41. 3 **43** 1.9 s **45.** Sometimes; sample answer: $\frac{4}{9}$ can be written as $0.\overline{4}$, but $\frac{1}{2}$ is written as 0.5. **47.** Always; sample answer: All whole numbers are integers. **49.** Rational; $\sqrt{49}$ is rational. **51.** Irrational; π is irrational. **53.** Sample answer: 6.4; $\sqrt{40} \approx 6.32$ **55.** False; $\sqrt{16}$ is rational. **57.** $\sqrt{50}$; It is irrational and the other numbers are rational. **59.** D
61. C **63.** 9 **65.** -7 **67.** ± 14 **69.** 0.000308
71. 849,500 **73.** 2^5 **75.** 16 **77.** 48 **79.** x^6 **81.** $(-2, 3)$
83. $(2, -5)$ **85.** $(-1, -7)$ **87.** 1408 ft

Pages 180–182 Chapter Review

1. 6^5 **3.** x^3 **5.** 243 **7.** -64 **9.** 10 **11.** -248
13. 32 teeth **15.** $\frac{1}{9^4}$ **17.** $\frac{1}{m^5}$ **19.** $\frac{1}{(-4)^3}$ **21.** 6^{-3}
23. 5^{-3} **25.** 4^{-2} or 2^{-4} **27.** 10^{-3} **29.** $(-7)^5$ **31.** x^9
33. $-30a$ **35.** k^6 **37.** 5820 **39.** 0.00011 **41.** 3.79×10^2
43. 4.7×10^4 **45.** 1,988,920,000,000,000 exagrams
47. 1.5×10^{11} **49.** 3.5715×10^7 **51.** 7.1×10^3 lb
53. -5 **55.** 9 **57.** -7 **59.** 7 **61.** natural, whole, integer, rational **63.** irrational **65.** > **67.** > **69.** $m = 8$
71. 14.9 ft

Chapter 5 Ratio, Proportion, and Similar Figures

Pages 185–188 Lesson 5-1

1. $\frac{3}{4}$ **3.** $\frac{3}{7}$ **5.** $\frac{5}{8}$; For every 8 students, 5 participate in sports. **7.** $\frac{16}{3}$ **9.** $\frac{8}{1}$ **11.** $\frac{5}{9}$ **13.** $\frac{3}{4}$ **15.** $\frac{32}{1}$ **17.** $\frac{2}{3}$; For every 3 tables, 2 of them are booths. **19.** $\frac{1}{8}$ **21.** $\frac{36}{7}$ **23a.** $\frac{7}{4}$ **23b.** $\frac{3}{4}$ **23c.** $\frac{11}{15}$ **23d.** $\frac{11}{4}$ **25a.** $\frac{40}{3}$ **25b.** horse, cow, cat, hamster **25c.** Horse; Sample answer: The ratio is approximately 27,273 to 1. **27.** No; sample answer: An 18-pound turkey should be cooked for $4\frac{1}{2}$ hours. **29.** $\frac{27}{9} = \frac{45}{15}$ **31.** $\frac{6}{48} < \frac{14}{88}$ **33.** Sample answers: number of girls to boys in a class; number of apples to bananas in a fruit basket; a sale price of 3 for \$8 **35.** 10 **37.** Always; sample answer: No matter what is used to measure the dimensions of the table, the actual dimensions will never change, so the ratio will always be the same. **39.** C **41.** B **43.** $-12.\overline{3}$ **45.** $2\ 8\overline{3}$ **47.** 0.37 **49.** $30\frac{12}{25}$ **51.** (2, −1) **53.** (3, 4) **55.** (4, −4) **57.** \$1.44 **59.** \$0.28 **61.** 14 **63.** 19 **65.** −15

Pages 191–193 Lesson 5-2

1. \$24 per day **3.** 21.1 points per game **5.** 4.3 gal/min **7.** Mr. Nut; sample answer: Barrel costs \$0.34 per ounce, Mr. Nut costs \$0.32 per ounce, and Chip's costs \$0.35 per ounce. **9.** 26 students per class **11.** 59 mph **13.** \$77 per ticket **15.** 1 pizza for \$6.50 **17.** 22 mi/gal **19.** Party Time; sample answer: The Party Planner and Party Time both charge \$0.25 per plate. **21.** $1\frac{11}{16}$ c **23.** Jenny **25a.** Alicia: 8 ft/s; Jermaine: 6 ft/s **25b.** Alicia: 660 s or 11 min; Jermaine: 880 s or 14 min 40 s **25c.** His line would be just below Jermaine's because he ran at a slower speed than Jermaine. **27.** 0.035 gram of salt for every gram of water **29.** situation b; sample answer: $\frac{120\text{ mi}}{2\text{ h}} = \frac{60\text{ mi}}{1\text{ h}}, \frac{120\text{ mi}}{3\text{ h}} = \frac{40\text{ mi}}{1\text{ h}}$ **31.** \$6.40; Sample answer: The unit rate for the 96-oz container is \$0.05 per ounce. So, 128 ounces would cost \$0.05 × 128 or \$6.40. **33.** B **35.** Large; it has the least cost per ounce. **37.** $\frac{31}{15}$ **39.** $\frac{4}{1}$ **41.** Sample answer: 6(\$20) + 4(\$100) **43.** 8 **45.** 144 **47.** 0.04 **49.** 3500 **51.** $-1\frac{1}{8}$ **53.** $-8\frac{11}{30}$ **55.** $2\frac{51}{80}$

Pages 196–199 Lesson 5-3

1. 24 **3.** $1\frac{1}{3}$ **5.** 22 **7.** $\frac{5}{6}$ page per min **9.** $\frac{1}{15}$ **11.** $\frac{21}{200}$ **13.** $1\frac{1}{2}$ **15.** 20 **17.** $\frac{4}{27}$ **19.** $\frac{1}{10}$ **21.** 2 **23.** $\frac{1}{10}$ **25.** 7 mph **27.** 23 cookies **29.** $\frac{9}{16}$ **31.** $\frac{2}{15}$ **33.** $\frac{31}{400}$ **35a.** 8 costumes; $\frac{3}{4}$ yd **35b.** \$3; Sample answer: Divide the total cost by the number of yards purchased; $44.25 \div 4\frac{3}{4} = 3$. **37.** about 140 soft shells **39.** $\frac{11}{200}$ **41.** $1\frac{3}{5}$ **43.** $1\frac{2}{3}$ **45.** $1\frac{16}{17}$ **47.** Sample answer: $\frac{\frac{1}{32}}{\frac{1}{8}}, \frac{\frac{1}{2}}{2}, \frac{\frac{1}{16}}{\frac{1}{4}}$ **49.** Sample answer: Ratios can be written as fractions. If one of the numbers in a ratio is a fraction, then the ratio will be a complex fraction.

51. H **53.** 3^7 **55.** $\left(\frac{3}{4}\right)^2$ **57.** $(x + 1)^3$ **59.** 8×10^4 **61.** 5.4×10^{-3} **63.** 9.8×10^{-9} **65.** 114-ounce bottle; Sample answer: The 114-ounce bottle costs about \$0.03 per ounce and the 44-ounce bottle costs about \$0.04 per ounce. **67.** isosceles trapezoid **69.** < **71.** =

Pages 203–205 Lesson 5-4

1. 28,800 acres **3.** 20.32 **5.** 425.25 **7.** 4.26 **9.** 138.72 g/min **11.** 1 billion cans per week **13.** 80.67 ft/s **15.** 40.64 **17.** 7.31 **19.** 1.85 **21.** 4.58 **23.** 3250.68 mph **25.** 203.2 **27.** 2188 **29.** 6.02 **31a.** 2.60 m/s **31b.** 12.27 mph **33.** 7 yd/min, 500 m/h, 6 in./s **35.** 18 lb/min, 500 kg/h, 5 oz/s **37.** 72.79 L per week **39.** < **41.** 118.5 euros **43.** 89.41 dollars **45.** Sample answer: Macha needs 120 square feet of carpeting for her bedroom. How many square yards does she need?; $13\frac{1}{3}$ yd^2 **47.** 500 ft/min; Sample answer: All of the other rates are equal to 60 mi/h. **49.** Identity Property; the conversion factor is equal to 1. **51.** C **53.** A **55.** \$45.75 per ticket **57.** 24.2 mi/gal **59.** 12 cans for \$4.25; 6-pack unit cost ≈ \$0.367 per can; 12-pack unit cost ≈ \$0.354 per can **61.** $\frac{1}{4}$ **63.** $\frac{2}{5}$ **65.** $-32y$ **67.** $45m$ **69.** $15r$ **71.** 120 **73.** −126 **75.** −28 **77.** 0.8 **79.** 0.375

Pages 208–210 Lesson 5-5

1. No; the rates are not equal. **3.** 3.19; $c = 3.19g$; \$59.02 **5.** No; the rates are not equal. **7.** Yes; 7; each rate is equal to 7. **9.** 18; $p = 18\ell$; \$126

11.

Number of Rides	1	2	3	4	5
Cost (\$)	5.50	7	8.50	10	11.50

13a. Yes; sample answer:

Hot Dog Packages	1	2	3	4
Hot Dogs	8	16	24	32

In the table, the ratio of hot dogs to hot dog packages is $\frac{8}{1}, \frac{16}{2}, \frac{24}{3}$, and $\frac{32}{4}$. Since these ratios all equal 8, the relationship is proportional. The constant of proportionality is 8.

13b. Yes; sample answer:

Hot Dogs	8	16	24	32
Hot Dog Buns	10	20	30	40

In the table, the ratio of hot dog buns to hot dogs is $\frac{10}{8}, \frac{20}{16}, \frac{30}{24}$, and $\frac{40}{32}$. Since these ratios all equal $\frac{5}{4}$, the relationship is proportional. The constant of proportionality is $\frac{5}{4}$.
15. Sample answer: At Store A, there are always 2 red flowers for every 8 pink flowers in a bouquet. At Store B, there are always 3 more pink flowers than red flowers in a bouquet. The bouquet for Store A is a proportional relationship, while the bouquet for Store B is nonproportional. Store A: $r = 0.25p$ for r red flowers and p pink flowers. Store B: $r = p - 3$ for r red flowers and p pink flowers. **17b.** Sample answer: The ratio of length to width is about equal to 1.618 to 1. **17c.** Sample answer: Pyramid of Khufu in Giza, Egypt; the Taj Mahal in India; the Lincoln Memorial in Washington, D.C. **19.** A **21.** D

23. 10.16 **25.** 681 **27.** 23.3 mi per gal
29. 52.6 mi per day **31.** $1\frac{5}{6}$ ft per min **33.** 151°F; 142°F **35.** $-27 \div 9$; -3°F; The temperature dropped 3°F on average each hour.

Pages 215–217 Lesson 5-6

1.

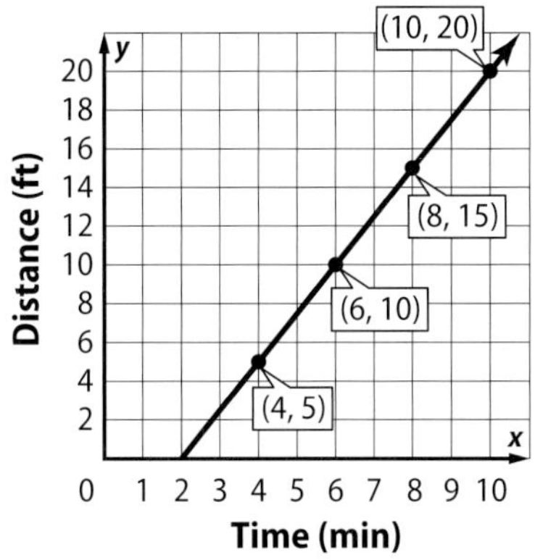

Nonproportional; the line does not pass through the origin.

3.

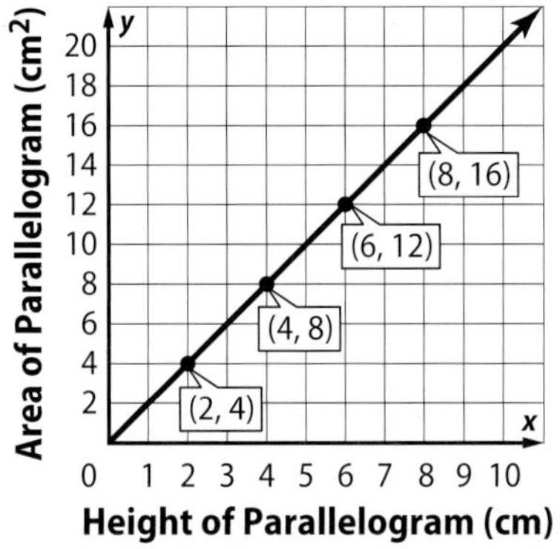

Proportional; the line passes through the origin and is a straight line.

5

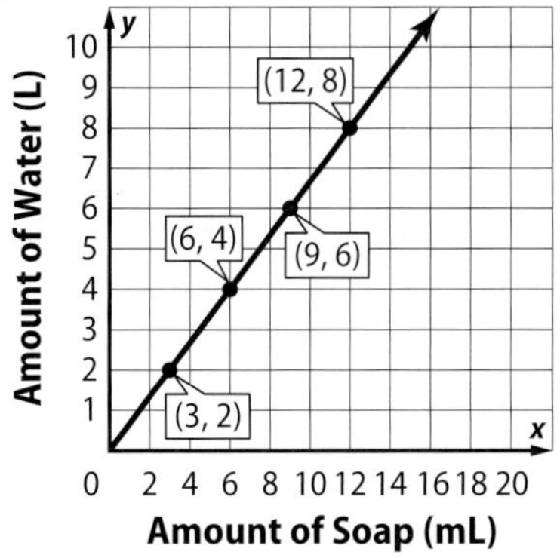

Proportional; the line passes through the origin and is a straight line.

7.

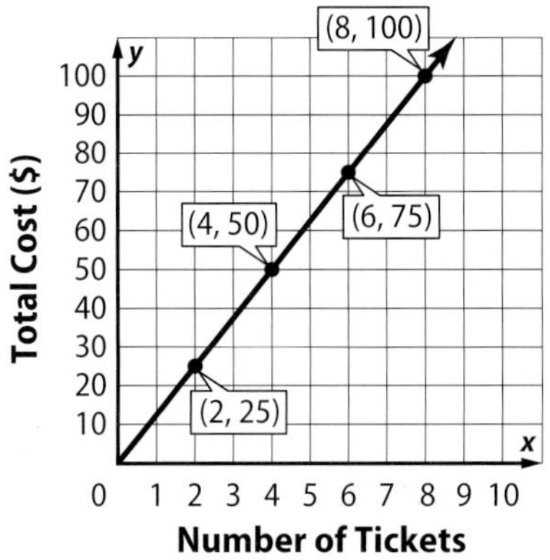

Proportional; the line passes through the origin and is a straight line.

9a. 4; for every 1 unit increase in side length, the perimeter increases by 4 units. **9b.** The perimeter of a square with side length of 0 units has a perimeter of 0 units; the perimeter of a square with side length of 1 unit is 4 units; the perimeter of a square with side length of 3 units is 12 units.

11 \$27.50;

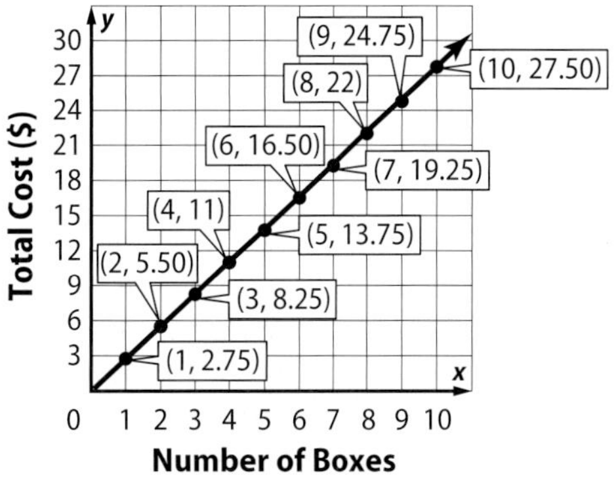

Sample answer: The graph of the relationship is a straight line that passes through the origin and the point (1, 2.75), so the unit rate is 2.75. So, 10 boxes of cereal cost $10 \times \$2.75 = \$27.50$.

13.

Length of Ribbon (yd)	5	7.5	15	25
Total Cost ($)	2	3	6	10

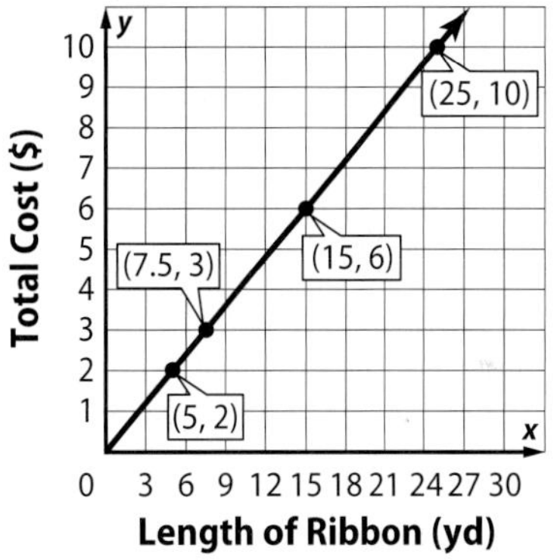

15.

Boxes of Cat Food	3	6	9	15
Weight (lb)	8	16	24	40

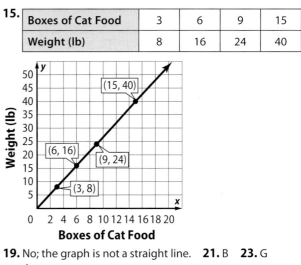

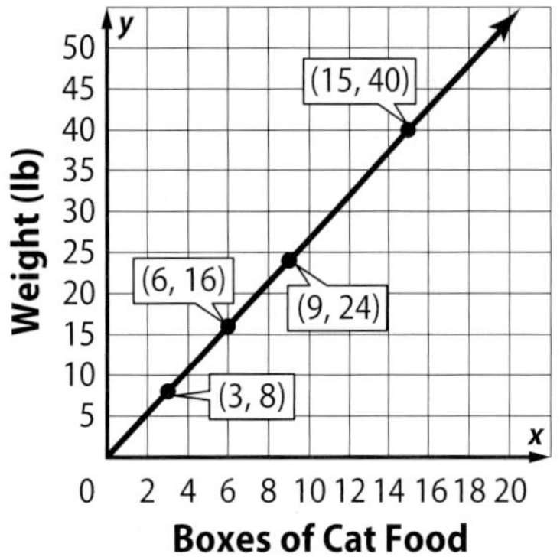

19. No; the graph is not a straight line. **21.** B **23.** G
25. $\frac{4}{3}$ **27.** 0 **29.** 0 **31.** -6 **33.** 10
35. 1.83696×10^{-10} **37.** 7.5×10^{-2}, 7.1×10^{3}, 7.01×10^{4}, 7.15×10^{5}

Pages 221–223 Lesson 5-7

1. 24 **3.** 34 **5.** 29 **7.** 27 in. **9** 36 **11.** 24 **13.** 12 **15.** 14 **17.** 0.96 **19.** 23 **21.** 28 gal **23.** 927.8 ft; 1391.8 ft **25a.** 10 oz **25b.** 27 oz **25c.** $190\frac{2}{3}$ oz **27.** $\frac{s}{0.54} = \frac{4.55}{1.89}$; 1.3 in. **29.** $\frac{20}{4} = \frac{b}{20}$; 100 boxes **31.** 0.8 **33.** 15 **35.** 7 **37 a.** ribbon: $c = 0.89n$; fleece: $c = 4.75n$; satin fabric: $c = 3.45n$; quilting fabric: $c = 2.99n$

37 b.

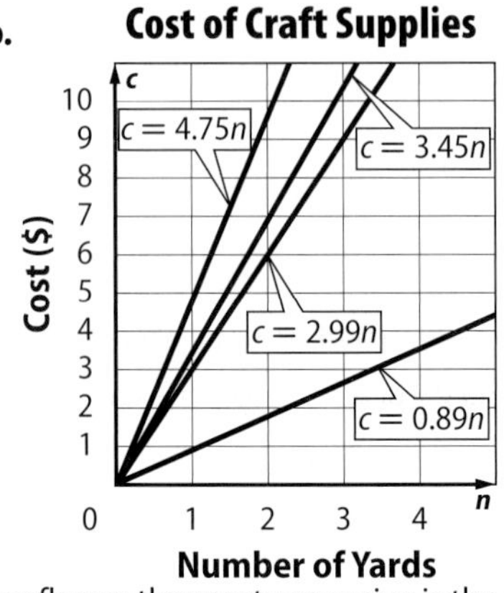

Ribbon; fleece; the most expensive is the one with the steepest line. The least expensive is the one with the line that is the least steep.

37 c. \$0.45; \$51.97 **39a.** ±6 **39b.** ±4 **39c.** ±10 **39d.** ±12 **41.** Sample answer: The product of the length and width is constant. The length is not proportional to the width. For example, if the area is 36 square units:

w	3	6	9
ℓ	12	6	4
Ratio	1:4	1:1	9:4

From the table, you can see that the ratios of width to length are not equal.

43. A **45.** D **47.** yes; $\frac{1}{6}$ **49.** 12.7 **51.** 10.36 **53.** $-\frac{8}{27}$ **55.** $-\frac{9}{20}$ **57.** $-8 < 6$; $6 > -8$

Pages 227–229 Lesson 5-8

1. 1 in. = 2 ft

3.

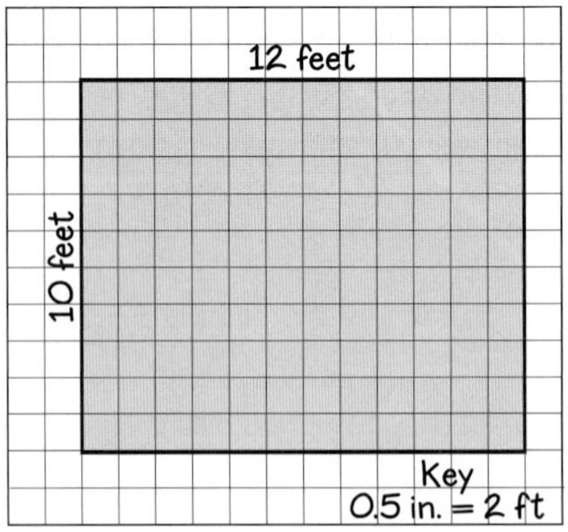

(not shown to actual size)

5 1 in. = 3 ft **7a.** 15 ft by 18 ft **7b.** 30 ft by 9 ft **7c.** 9 ft by 18 ft

9.

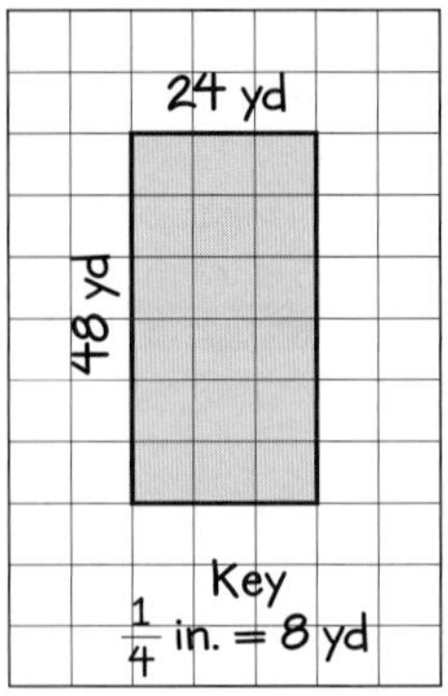

(not shown to actual size)

11 $\frac{1}{50}$ **13.** $\frac{1}{72}$ **15.** $\frac{16}{1}$

17a. Sample answer: $\frac{1}{4}$ in. = 3.6×10^7 mi

17b. Sample answer:

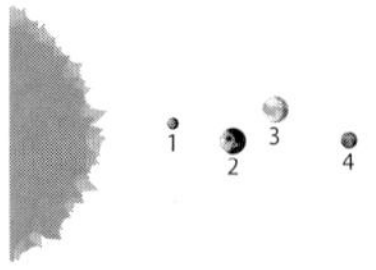

21. Always; sample answer: A scale factor of $\frac{3}{1}$ means that 3 units of the drawing is equal to 1 unit of the object, so the scale drawing or model will be larger than the actual object. **23.** Sample answer: The model is larger than the actual insect. The scale factor is $\frac{2.5}{1}$, which means the model is 2.5 times as large as the insect. **25.** B **27.** H **29.** 4 **31.** 16 **33.** 1.4 **35.** Pattern A: 0, 4, 8, 12, 16, 20, 24, 28, 32, 36; Pattern B: 0, 8, 16, 24, 32, 40, 48, 56, 64, 72 **35a.** Each term in pattern A is half its corresponding term in pattern B. **35b.** Adding 4 is the same as adding half of 8. **37.** $\frac{1 \text{ m}}{3.28 \text{ ft}} = \frac{110 \text{ m}}{x \text{ ft}}$; about 360.8 ft **39.** Multiplicative Identity Property **41.** Multiplicative Property of Zero **43.** Associative Property of Multiplication **45.** 0

Pages 234–237 Lesson 5-9

1. $3\frac{1}{3}$ cm **3.** 2 units **5** 15 km **7.** 2.125 cm **9.** 27 in.

11.

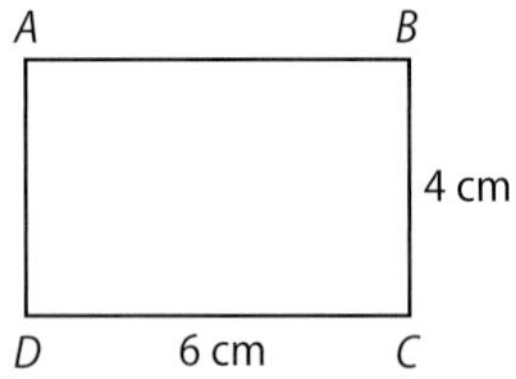

13. 102 m **15 a.** $a + b + c$ **15 b.** ad, bd, cd **15 c.** $ad + bd + cd$ **15 d.** $d(a + b + c)$; Sample answer: The scale factor times the sum of the measures of the sides of $\triangle ABC$ will give the perimeter of $\triangle XYZ$. **15 e.** 3 in. + 4 in. + 5 in. = 12 in.; $12 \times 2 = 24$; The perimeter of $\triangle XYZ$ is 24 inches. **15 f.** Sample answer: If the figures are similar, then the perimeters are also similar. So, the perimeters are proportional. **17.** Sometimes; sample answer: Even though all rectangles have congruent corresponding angles, the sides of one rectangle are not always proportional to the sides of another rectangle. **19.** Tony wrote the proportion incorrectly. He should have solved the problem like this: $\frac{16}{12} = \frac{x}{18}$; $x = 24$ ft

21. Triangle *B* is the original triangle. Since the scale factor is less than 1, the original triangle is being reduced, which means that the scaled triangle will be smaller. The measures of the sides of triangle *A* are less than the measures of the sides of triangle *B*, so triangle *B* must be the original triangle. **23.** J **25.** 72
27. 16 bags **29.** 4 **31.** 4 **33.** $\frac{4}{25}$ **35.** $\frac{1}{9}$ **37.** 4^7 **39.** x

Pages 239–242 Lesson 5-10

1. 9.8 ft **3.** 45 ft **5.** 120 yd **7** 2 ft
9a. Sample answer:

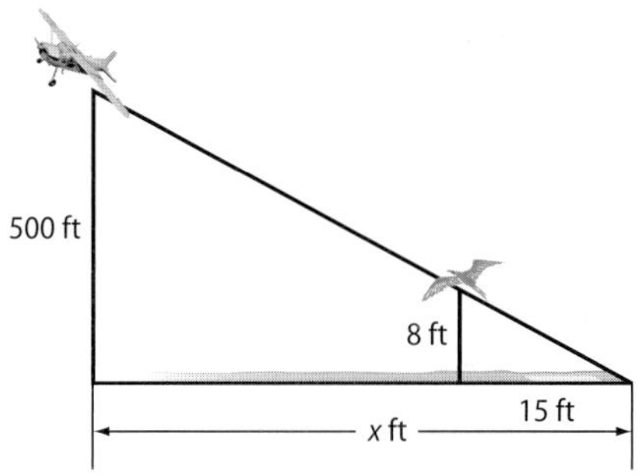

9b. $\frac{500}{8} = \frac{x}{15}$; 937.5 ft **11** 12 ft **13.** $\frac{37.5}{1.5} = \frac{150}{x}$; 6 ft
17. false; Sample answer: You also need to know if the angles are congruent. **19.** D **21.** 10 km **23.** 13.5 **25.** 0.625
27. $0.\overline{6}$ **29.** $\frac{1}{16}$ **31.** $\frac{10}{9}$ or $1\frac{1}{9}$ **33.** 9 **35.** no real solution

Pages 243–246 Chapter Review

1. $\frac{5}{12}$ **3.** $\frac{45}{1}$ **5.** $\frac{3}{4}$; For every 4 times at bat, Jean got 3 hits or in $\frac{3}{4}$ of his at bats, he got a hit. **7.** 80 meters per minute
9. 0.6 mile per minute **11.** 15 **13.** $3\frac{3}{10}$ mph **15.** $\frac{13}{200}$
17. 17.78 **19.** 739.35 **21.** 8.51 **23.** 10.24 **25.** 13.12
27. 318.75 mi **29.** 5.61 m/s **31.** Yes; 8; each rate is equal to 8.
33. 0.25; $c = 0.25r$, 11 rings cost \$2.75; 20 rings cost \$5
35.

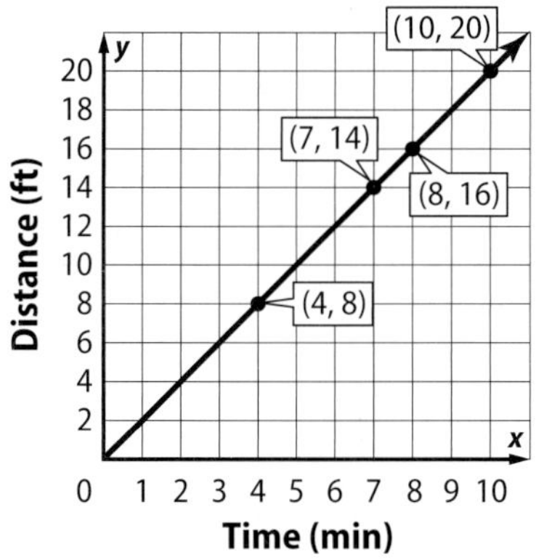

Proportional; the line passes through the origin and is a straight line.
37a. 3; there are 3 squirrels for every tree. **37b.** The point (0, 0) represents the number of squirrels, 0, when there are no trees. The point (3, 9) represents the number of squirrels, 9, when there are 3 trees. The point (5, 15) represents the number of squirrels, 15, when there are 5 trees. **39.** 7.2 **41.** 7
43. 120 ft **45.** 150 ft **47.** 7.2 in **49** 3.5 cm **51.** 31 in.
53. 1.2 m

Chapter 6 Percents

Pages 252–255 Lesson 6-1

1. 25% **3.** 41.44 **5.** 400% **7.** 22.95 **9.** 20 **11.** 39.2%
13. 250 **15** 42% **17.** 36% **19.** 72 **21.** 286 **23.** 30
25. 15 **27.** 20% **29** **a.** 973 people **29** **b.** 766 people
31a. The percent is increased by a factor of 2 and the whole is halved. Each answer is 2. **31b.** 32% of 6.25 = 2 **33.** 300%
35. 4.1 **37.** 2.2 **39a.** 12.5% **39b.** Greater than; the new percent increases from 30.9% to 32%. **41a.** $\frac{45 + x}{120 + x} = \frac{40}{100}$; 5 cell phones **41b.** 125 cell phones **43.** 5% of 80, 25% of 80, 25% of 160; If the percent is the same but the whole is greater, then the part is greater. If the whole is the same but the percent is greater, then the part is greater. **45.** 3
47. Always; sample answer: To find x% of y, solve $\frac{x}{100} = \frac{n}{y}$ for n. So, $n = \frac{xy}{100}$. To find y% of x, solve $\frac{y}{100} = \frac{n}{x}$ for n. So, $n = \frac{xy}{100}$. **49.** C **51.** H **53.** daily newspaper **55** $1\frac{1}{5}$
57. $\frac{5}{6}$ **59.** $\frac{2}{3}$ **61.** $-\frac{3}{10}$ **63.** $\frac{11}{36}$ **65.** 3 **67.** 19
69. 62.0 mi/h **71.** 1.03×10^7

Pages 258–260 Lesson 6-2

1. 12 **3.** 3.7 **5.** 0.72 **7.** 160 **9.** 25 **11.** Sample answer: 7; 11% is about 10% or $\frac{1}{10}$; $\frac{1}{10} \cdot 70 = 7$ **13.** Sample answer: 8; 19 is about 20 and 40% is $\frac{2}{5}$; $\frac{2}{5} \cdot 20 = 8$ **15.** Sample answer: 2; 598 is about 600 and 1% of 600 is 6; $\frac{1}{3} \cdot 6 = 2$ **17.** Sample answer: 90; 359 is about 360; 24% is about 25% or $\frac{1}{4}$; $\frac{1}{4} \cdot 360 = 90$
19. Sample answer: 32; 81 is about 80, $37\frac{1}{2}$% is about 40% or $\frac{2}{5}$; $\frac{2}{5} \cdot 80 = 32$ **21.** 43 seeds **23.** 10 **25.** 72 **27.** 6 **29.** 28 **31.** 25.9 **33.** 4 **35.** Sample answer: 3; 16% is about 15% or $\frac{3}{20}$; $\frac{3}{20} \cdot 20 = 3$ **37.** Sample answer: 44; 46% is about 50% or $\frac{1}{2}$; $\frac{1}{2} \cdot 88 = 44$ **39.** Sample answer: 2.5; 507 is about 500 and 1% of 500 is 5; $\frac{1}{2} \cdot 5 = 2.5$ **41** Sample answer: 45; 148% is about 150%, 100% of 30 is 30 and 50% of 30 is 15; $30 + 15 = 45$ **43.** Sample answer: 6; 801 is about 800 and 1% of 800 is 8; $\frac{3}{4} \cdot 8 = 6$ **45.** Sample answer: 60; 117% is about 120%, 100% of 50 is 50 and 20% of 50 is 10; $50 + 10 = 60$
47. \$6 **49.** Sample answer: 600 students **51** Sample answer: 180 twelfth graders **53.** Sample answer: $x = 400$, $y = 100$; since 40% is 4 times 10%, x must be 4 times y.
55. Tobie found 60% of 400. Janine found 62.5% of 400.
57. D **59.** A **61.** movies; Sample answer: $\frac{3}{8} = 37.5\%$ and $\frac{3}{5} = 60\%$. Since 60% is greater than 37.5% and 35%, movies are the favorite activity. **63.** 4 **65.** $\frac{1}{6}$ **67.** 0.20 **69.** 0.04

Pages 264–266 Lesson 6-3

1. 30 **3.** $33\frac{1}{3}$% **5.** 275 **7.** 114 boxes **9** 10.24 **11.** 20%
13. 64 **15.** 78 **17.** \$78 **19.** 39.9 **21.** 26% **23.** \$2500
25 **a.** $c = 1.2 \cdot 120$; 144 cars; 24 cars

25b.

Percent Increase	Number of Cars
5%	126
15%	138
25%	150
35%	162

27a. $s = 1.06 \cdot 500$; $530

27b.

x	y
2010	$500.00
2011	$530.00
2012	$561.80
2013	$595.51
2014	$631.24
2015	$669.11

27c. Sample answer: No; the base amount is different each year so the part changes even though the percent remains constant. **29.** 96 **31.** Sample answer: The percent is greater than 100% because otherwise the part would be less than or equal to the whole. **33.** No; Suppose an item costs $100. A 10% discount would be a discount of $10, so the discounted price would be $90. Adding a 10% sales tax adds $9. So, the final price is $90 + $9 or $99, not $100. **35.** B **37.** A **39.** 48 **41.** 9 **43.** chocolate pieces: 2 c; peanuts: 3 c **45.** 26.4 **47.** 19.2 **49.** $(xy)^5$ or x^5y^5 **51.** $3(z - 8)^4$ **53.** 45

Pages 272–274 Lesson 6-4

1. −20%; decrease **3.** about −30.8%; decrease **5.** $6.\overline{6}$% **7.** 85.7%; increase **9.** −9.8%; decrease **11.** −12.5%; decrease **13.** 41.4%; increase **15.** 13.2%; increase **17.** $11.\overline{1}$% **19.** 37.5% **21** 2.34% **23.** 3088 calls **25.** −11.7% **27.** 62.9 s; 62.9 < 74 **29.** B; It increased by 22% and restaurant A increased by about 3%.

31a.

City	Population 2000	Population 2006	Amount of Change	% of change
Raleigh, NC	276,093	356,321	80,228	29%
Columbia, SC	116,278	119,961	3683	3%
Frankfort, KY	27,741	27,077	−664	−2%
Columbus, OH	711,470	733,203	21,733	3%

31b. Sample answer: The percents of change were the same but the amount of change for Columbus was much greater than the amount of change for Columbia. The percents were the same because the original amounts for each city were very different. **33.** Sample answer: actual length 20 cm, calculated length 19 cm **35.** 245 mm, 255 mm **37.** C **39.** C **41.** 8.64 **43.** 33.375 **45.** about 90 Cal **47.** $\frac{13}{30}$ **49.** $5\frac{1}{20}$ **51.** 0.2 **53.** 0.9 **55.** −10

Pages 277–280 Lesson 6-5

1. 37.50 **3.** $26.40 **5.** $46.08 **7.** $60.50 **9.** $19.28 **11.** $35.78 **13.** $77.00 **15.** $46.24 **17.** $341.70 **19.** $19.50 **21.** $44.80 **23.** $19.95 **25.** $45.00

27.

Marked-down Price	$2.99	$4.99	$6.99	$8.99	$10.99
Original Price	$3.74	$6.24	$8.74	$11.24	$13.74

29. $317.36 **31.** markup; 25.6% **33.** discount; −50.4% **35.** markup; 8.4% **37.** No; Lassen charges $240.80, and Pineapple charges $246.43. **39.** No; the shirts are marked up to $15 and then marked down to $7.50. **41.** Sample answer: A bookshelf is marked down to $10; the regular price is $200. **43a.** False; sample answer: An item costs $25 and you want to mark it up 125%. Multiply $25 by 125% or 1.25. The new price is $25 + $31.25 or $56.25. **43b.** true **45.** Sample answer: An Internet service provider offers a plan for $40 per month. If the plan is 45% off this month, what is the cost of the plan?; $22 **47.** F **49.** 100% **51.** $\frac{47}{10}$ **53.** $-\frac{37}{4}$ **55.** $-\frac{103}{100}$ **57.** 9 **59.** 5 **61.** −22 **63.** add 2; 24, 26 **65.** divide by 4; 4, 1 **67.** 81 **69.** −25

Pages 283–285 Lesson 6-6

7–9, 17–21, 27–29. Answers may vary due to rounding. **1.** $567 **3.** $117.81 **5.** 3.23% **7.** $643.71 **9.** $3512.55 **11.** $193.75 **13.** $242.40 **15.** $919.80 **17.** $4264.86 **19.** $683.88 **21.** $15,700.17 **23.** 42 months **25.** Option A; Sample answer: Option A earned $281.25 in interest and Option B earned $273.91. **27.** $15,496.72 **29.** $1123.16 **31.** $137.77; $39.68 **33.** Sample answer: $2000 at 1%. Using the simple interest formula $I = \$2000 \cdot 0.01 \cdot 4$ or $80. **35.** No; Sabino did not convert the time to years. If he multiplies $2500 \cdot 0.0575 \cdot 1.5$, the interest is $215.63. **37.** Sample answer: With simple interest, the amount of money earned will be the same each year because it is always applied to the initial amount. With compound interest, the amount of interest will increase each year because it is being applied to the new total after the interest is added each year. **39.** J **41.** Plan B, $16.29 **43.** 3% **45a.** $\frac{1}{4} \times 7$ or 1.75 billion **45b.** $\frac{1}{5} \times 7$ or 1.4 billion **47.** $\frac{5a}{c}$ **49a.** $m = 8d - 1$ **49b.** 71 min **51.** 1.47 **53.** 23.1%

Pages 288–290 Chapter Review

1. 20% **3.** 35 **5.** 86.4 mi **7.** 36 CDs **9.** 43 students **11.** 22 **13.** 1.67 **15.** 54; 62% is about 60% or $\frac{3}{5}$, $\frac{3}{5} \cdot 90 = 54$ **17.** 325; 100% of 250 is 250 and 30% of 250 is 75, 250 + 75 = 325. **19.** about 29 **21.** 4 **23.** 150% **25.** 35% **27.** 27 **29.** No; the jersey is now 75% off the original price. If the jersey was originally $100, it is $50 after the first markdown. After the manager takes 50% off of $50, the jersey is $25, or 75% off the original price. **31.** 34.5%; increase **33.** $44.\overline{4}$% **35.** 60% **37.** $204 **39.** $235 **41.** $35.81 **43.** $299 **45.** $251.56 **47.** $3464.21 **49.** 6.6%

Chapter 7 Algebraic Expressions

Pages 294–296 Lesson 7-1

1. $7 \cdot 9 + 7 \cdot 3$; 84 **3.** $7 \cdot 2.2 + 8 \cdot 2.2$; 33 **5.** \$6.50; $5(\$1 + \$0.30) = 5(\$1) + 5(\$0.30)$ **7.** $5p + 20$ **9.** $9.5a - 95$ **11.** $4.5 \cdot 16 + 4.5 \cdot 8$; 108 **13.** $12.3 \cdot 9 + 12.3 \cdot 4$; 159.9 **15.** $\frac{5}{8} \cdot 20 - \frac{5}{8} \cdot 4$; 10 **17.** \$6.50(3 + 5), \$6.50(3) + \$6.50(5); \$52 **19** $\frac{4}{5}t - 12$ **21.** $12b + 48$ **23.** $-\frac{1}{2}n - 2$ **25.** 176 **27.** 779 **29.** 3000 **31.** $5 \cdot 4 + 5 \cdot \frac{1}{5}$; 21 **33.** $6 \cdot 4 + 6 \cdot \frac{2}{3}$; 28 **35** $9 \cdot 2 + 9 \cdot \frac{1}{3}$; 21 yd **37.** Sample answer: Julia did not distribute 3 to the second term inside the parentheses; the correct answer is $3x + 6$. **39.** Sample answer: You can break up $2\frac{1}{2}$ to be $2 + \frac{1}{2}$ and $4\frac{1}{2}$ to be $4 + \frac{1}{2}$. Then set up the multiplication expression $2\left(4 + \frac{1}{2}\right) + \frac{1}{2}\left(4 + \frac{1}{2}\right)$. Distribute: $2 \cdot 4 + 2 \cdot \frac{1}{2} + \frac{1}{2} \cdot 4 + \frac{1}{2} \cdot \frac{1}{2}$. Add: $8 + 1 + 2 + \frac{1}{4} = 11\frac{1}{4}$. **41.** G **43.** $c = d(45 - 10)$ **45.** $-3\frac{1}{6}$ **47.** $-\frac{362}{63}$ or $-5\frac{47}{63}$ **49.** 38 ft **51.** 108 **53.** −12

Pages 301–304 Lesson 7-2

1. terms: $-2a$, $3a$, $5b$; like terms: $-2a$, $3a$; coefficients: −2, 3, 5; constants: none **3.** terms: mn, $4m$, $6n$, $2mn$; like terms: mn, $2mn$; coefficients: 1, 4, 6, 2; constants: none **5.** terms: $3x$, $4x$, $5y$; like terms: $3x$, $4x$; coefficients: 3, 4, 5; constants: none **7.** $8x + 3$ **9.** $2x - 4$ **11.** $m + 5$ **13.** $2x + 7$ **15.** terms: $3a$, 2, $3a$, 7; like terms: $3a$, $3a$; coefficients: 3, 3; constants: 2, 7 **17.** terms: $3c$, $4d$, $5c$, 8; like terms: $3c$, $5c$; coefficients: 3, 4, 5; constant: 8 **19.** terms: $4x$, $4y$, $4z$, 4; like terms: none; coefficients: 4, 4, 4; constant: 4 **21.** $7a$ **23.** $-4m + 5$ **25.** $11p + 8$ **27.** $-3a - 6b$ **29.** $6x + 30$ **31.** $-18 + 3r$ **33** $2y - 5$ **35.** $7x + 6$ **37.** $5x - 2y$ **39.** $-17m - 8n + y$ **41.** $-\frac{3}{4}m + \frac{3}{2}n$ **43.** $\frac{22}{15}a + \frac{14}{15}b$

45a. $7 + 4x + (-4) + (-2x) + 3$;

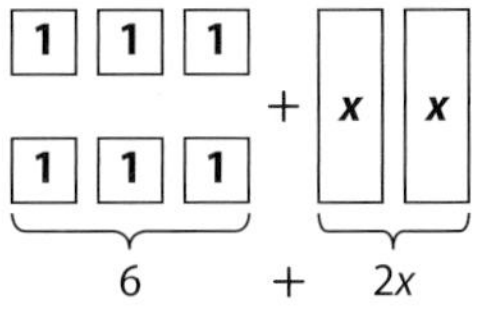

45b. $-8 + (-3x) + 2 + (-5x) + 2x + 4$;

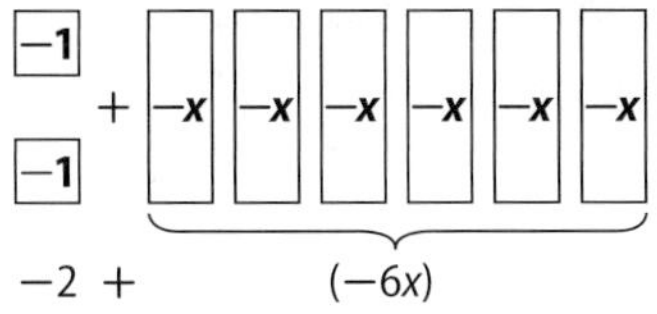

47. $6c - 8$

49

$16 \cdot (-31) + 16 \cdot 32$	$= 16(-31 + 32)$	Distributive Property
	$= 16(1)$	Simplify.
	$= 16$	Multiplicative Identity

51.

$24 \cdot (-15) + 36 \cdot 15$	$= (-24 + 36)15$	Distributive Property
	$= (12)15$	Simplify.
	$= (10 + 2)15$	Distributive Property
	$= 150 + 30$ or 180	Simplify.

53. $4x + 2$ **55.** $5x + xy + 2y + 10$ **57.** $-6x - 12$; The other expressions are equivalent to $-6x + 12$. **59.** Sample answer: The friend multiplied 5 by +2 instead of −2.

$$4x - 2(x + 5) = 4x - 2x - 10$$
$$= 2x - 10$$

61. G **63.** $9a - 9b - 6$ **65.** $-5a + 30$ **67.** 4(\$7 + \$3), 4(\$7) + 4(\$3); \$40 **69.** $-6 < -2$; $-2 > -6$ **71.** $0 > -9$; $-9 < 0$ **73.** $|15| < |18|$; $|18| > |15|$ **75.** 23 **77.** 3 **79.** 74 **81.** −72 **83.** −56

Pages 307–308 Lesson 7-3

1. $3x + 8$ **3.** $-x + 2$ **5a.** $6x + 3$ **5b.** 27 units **7** $-6x + 9$ **9.** $-4x + 9$ **11.** $2x$ **13.** $18x$; 14.4 in. **15 a.** $55x + 12$ **15 b.** $68x + 15$ **15 c.** $123x + 27$ **17.** $a + 2b - 2$ **19.** $5y - 2$ **21.** Sample answer: First, combine like terms. Then you can get rid of zero pairs and simplify. **23.** F **25.** D **27.** $\frac{1}{2}$ **29.** $\frac{1}{5}$ **31.** $\frac{1}{4}$ **33.** ±7 **35.** 4 **37.** −8 **39.** \$109.14 **41.** $-3x + 15$ **43.** $\frac{2}{9}z - 6$ **45.** $-\frac{1}{16}m + \frac{3}{4}$ **47.** $4 - 11x$

Pages 312–313 Lesson 7-4

1. $3x + 4$ **3.** $11x - 5$ **5a.** $x + 1.55$ **5b.** \$3.55 **7.** $-3x + 7$ **9** $2x + 9$ **11.** $-x + 2$ **13** \$6.20 **15.** Sample answer: $5x + 4$ and $x + 3$ **17.** Sample answer: The rule to add the inverse when subtracting integers is applied to each term in the linear expression that is being subtracted. **19.** G **21.** $x - 3$ **23.** 40% **25.** 48.75 **27.** 70 **29.** $8w + 72$ **31.** $4b + 8$ **33.** 45% **35.** −7 **37.** −4 **39.** −356 **41.** −47 **43.** 63

Pages 318–320 Lesson 7-5

1. 2 **3.** $5a$ **5.** $14s$ **7.** $27s$ **9.** $11mn$ **11.** $3(2 + x)$ **13.** cannot be factored **15.** $2(7x - 8)$ **17.** $8(3 + 4x)$ **19.** $5(5x + 24)$ **21.** 24 **23** $20x$ **25.** $6rs$ **27.** $36k$ **29.** $25xy$ **31** $5(x + 1)$ **33.** cannot be factored **35.** $8(4 + 3x)$ **37.** $6(3x + 1)$ **39.** 4 units by $(x - 2)$ units **41.** $5(x + 4)$ units^2 **43.** $4(5x + 19)$ units^2 **45** $(x + 2)$ dollars **47.** Sample answer: $\frac{1}{2}(x + 8)$ **49.** Sample answer: $\frac{3}{4}(x - 32)$ **51.** Sample answer: $20m$ and $12mn$ **53.** Sample answer: The Distributive Property shows that $a(b + c)$ and $ab + ac$ are equal. The GCF is the number that is distributed to each factor inside the parentheses. **55.** B **57.** 2^{10} **59.** $12x^7$ **61.** $\frac{16}{s}$ **63.** 21 mi; $6(1.5 + 2) = 6 \cdot 1.5 + 6 \cdot 2$ **65.** $x - 2$ **67.** $\frac{1}{8}x - 5$

Pages 321–322 Chapter Review

1. $\frac{7}{8} \cdot 8 + \frac{7}{8} \cdot 5$; $11\frac{3}{8}$ **3.** $7y + 21$ **5.** $-\frac{2}{3}b + 6$ **7.** \$30; $5(2.50 + 3.50)$; $5 \cdot 2.50 + 5 \cdot 3.50$ **9.** $11a$ **11.** $5m + 3$ **13.** $7a + 18$ **15.** $5x - 5$ **17.** $-2x + 7$ **19.** $13x - 9$ **21.** $-2x - 4$ **23.** $6x + 2$ **25.** $4(x + 3)$ **27.** $3(x - 5)$ **29.** $2(7x + 5)$ **31.** cannot be factored **33.** $(5x + 3)$ cm **35.** 2 in. by $(4x - 1)$ in.

Chapter 8 Equations and Inequalities

Pages 326–329 Lesson 8-1

1. -50 **3.** 12 **5.** -36 **7.** -8.5 **9.** $\frac{1}{15}$ **11.** $\frac{1}{6}x = 54$; 324 lb **13.** -120 **15** -2.4 **17.** 16.5 **19.** 392 **21.** -64 **23.** $\frac{1}{3}$ **25.** -80 **27.** $\frac{5}{6}x = 420$; 504 strawberries **29.** -4 **31.** -12 **33.** -64 **35.** $\frac{2}{3}$ **37.** 0.9 or $\frac{9}{10}$ **39.** $-\frac{7}{27}$ **41.** 70 bpm **43.** Sample answer: The value of m is $\frac{10}{9} \cdot 5\frac{1}{10}$, which is approximately $1 \cdot 5$, so the solution should be close to 5.

45 **a.** $d = 52.5t$

45 **b.**

Time (days)	1	2	3	4	5	6
Distance (miles)	52.5	105	157.5	210	262.5	315

45 **c.**

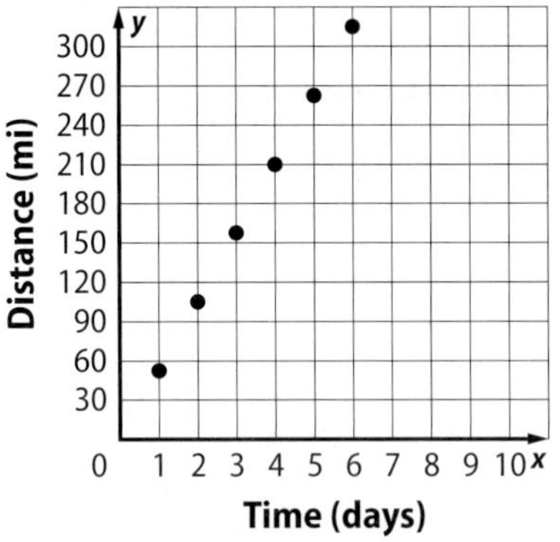

45 **d.** about 9 days **45** **e.** 40 days **47.** Sample answer: A shirt was on sale for half off. If the sale price was \$15.60, find the original price; \$31.20. **49.** Yes; the solution is 0 when $q = 0$. In this case, $x = \frac{0}{p} = 0$. **51.** Sample answer: Yes; he can multiply each side of the equation by $\frac{1}{3}$ instead of dividing by 3. **53.** J **55.** $2s = 64.98$; \$32.49 **57.** -3 **59.** -0.7 **61.** -8 **63.** $-\frac{1}{2}$ **65.** $5t + 15$ **67.** $12p + 4$ **69.** $-4x - 5$ **71.** $13x + 6$ **73.** $10y + 80$ **75.** Gabriel: 20; Elias: 10 **77.** $-\frac{1}{5}$ **79.** 67 **81.** 216 **83.** -95 **85.** -10 **87.** 1

Pages 335–338 Lesson 8-2

1. 4 **3.** -4 **5.** 1 **7.** -8 **9.** 30 **11.** -32 **13.** -3 **15.** 5 **17.** -8 **19.** 10 **21.** 4 **23.** 18 **25.** 30 weeks **27.** 5 **29** -4 **31.** -6 **33.** -18 **35.** 76 **37.** 11 **39.** 8 **41.** 30 **43.** -11 **45.** 0.3 **47.** 1.6 **49.** -2 **51.** 15 months **53.** 8 s **55.** 5 **57.** 1.3 **59.** $\frac{5}{12}$ **61.** 30.4 **63** 6 girls **65.** 15.1 **67.** -14.5 **69.** Sample answer: You spent \$7 at the bookstore and bought lunch for 2 days. You spent a total of \$15. How much was lunch? \$4 **71.** Sample answer: $\frac{1}{5}x + 4 = 6$. By the Subtraction Property of Equality, $\frac{1}{5}x = 2$. By the Multiplication Property of Equality, $5 \cdot \frac{1}{5}x = 5 \cdot 2$. So, $x = 10$. **73.** Sample answer: The problems both involve multiplying by 3 and adding 5. They are different because $3(2) + 5$ is an expression, while $3x + 5 = 11$ is an equation in which you solve by subtracting 5 then dividing by 3. **75.** G **77.** 19 **79.** 56 **81.** -32 **83.** 3.83 **85.** $2x + 40$ **87.** -40 **89.** 18 **91.** -560 **93.** 62 **95.** -14 **97.** 3

Pages 341–343 Lesson 8-3

1. $\frac{n}{3} - 8 = 16$ **3.** $\frac{2}{3}t - 12 = 98$ **5.** $2x - 14 = 96$; 41 cars **7** $\frac{1}{2}x + 18 = 8$ **9.** $3x - 3 = 48$ **11.** $2g + 37 = 497$; 230 field goals **13.** $\frac{x}{3} + 8.50 = 13.25$; \$14.25 **15.** $6d + 0.90 = 2.40$; \$0.25 **17** $24 = 2n + 8$; 8 action movies

19a. $47 + 15w$

Number of Weeks	Amount of Savings (\$)
1	62
2	77
3	92
4	107

19b.

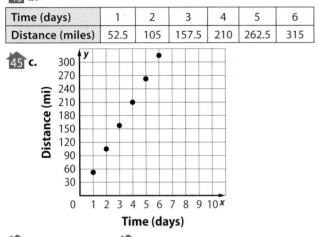

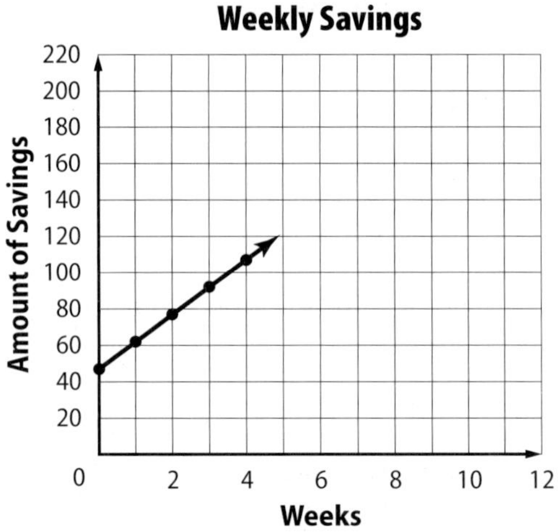

Extend the line so that it goes through the point that has 212 as its y-value and find the corresponding x-value.
19c. $47 + 15w = 212$; 11 weeks **19d.** The line graph will be large and depending upon the scale used, may be more difficult to determine an exact value. Solving an equation is quick and ends with an exact answer. **21.** 8, 10, 12 **23.** 24 yrs old **25.** D **27.** C **29.** -22 **31.** -10 **33.** 4 **35.** 10 **37.** 7.84 **39.** -24 **41.** $n(t + h + s)$ **43.** 5×10^8 **45.** $8y - 16$ **47.** $18 + 2p$

Pages 349–351 Lesson 8-4

1. 12 **3.** -2 **5.** 40 **7.** 45 **9.** 0 **11.** 6 **13.** $24(f + 4) = 216$; \$5 **15.** 3 **17.** -11 **19.** 17 **21.** -9.6 **23.** $10\frac{1}{3}$ **25.** -5.1 **27.** $5(d + 1.5) = 71.25$; 12.75 mi **29.** $85(7 + m) = 1955$; 16 months **31.** $20.4 = 6(x - 1.2)$; 4.6 **33.** Sample answer: You can estimate the solution of the equation by solving $8(x + 3) = 32$, which has the solution $x = 1$. So, the solution of the given equation should be close to 1. Therefore, the student's solution must be incorrect. **35.** -15 **37.** -5.2 **39.** $\frac{8}{9}$ **41.** 4 **43a.** $6(4.95 + s) = 50.70$; \$3.50 **43b.** $4(10.15 + x) = 76$; \$8.85 **45.** She did not apply the Distributive Property correctly. The correct solution is $p = -\frac{2}{3}$. **47.** $3(2j - 4) = 114$; 21 **49.** B **51.** H **53.** 50%; increase **55.** -7.4%; decrease **57.** 57.9%; increase **59.** $8.50 + 3p$ **61.** -14 **63.** -36 **65.** -28 **67.** 60 **69.** -4

Pages 358–360 Lesson 8-5

1. 3 **3.** −10 **5.** −27 **7.** $50 + 1.99m = 3.99m$; 25 DVDs **9.** 1 **11.** −0.75 **13.** −0.5 **15.** −80 **17.** $5 + 0.5s = s$; 10 songs **19.** 3.25 **21.** $\frac{4}{5}$ **23.** 0 **25.** 57 text messages **27.** FL: 1350 mi; TX: 367 mi **29.** Sample answer: Use the information in the figure to write and solve an equation; $w = 55$ ft, $w + 40 = 95$ ft, $w + 45 = 100$ ft **31.** $t = 1.8t + 32$, $-32 = 0.8t$, $-40 = t$. At −40°, the temperatures are the same. **33.** Sample answer: The annual membership fee for a movie club is $54, which allows you to buy tickets for $3.50 each. If the local movie theater charges $8.00 for tickets, determine how many movie tickets you would need to buy through the club for the cost to equal that of buying tickets at the regular price. Answer: 12 **35.** G **37.** C **39.** $22.50 + 17.50x = 83.75$; 3.5 hours **41.** $7x$ **43.** $11y$ **45.** $2s + 7$ **47.** $y - 2$ **49.** $2x + 4$ **51.** 53 **53.** 24

Pages 364–366 Lesson 8-6

1. $x \leq 45$ **3.** $f > 8000$ **5.** true

7. 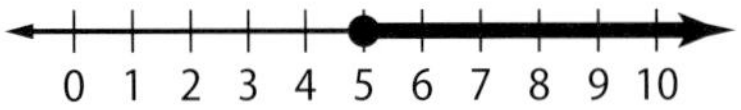

9. 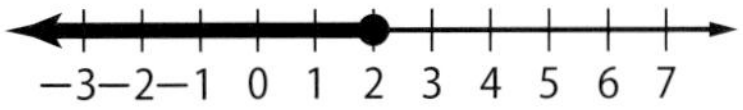

11. $y > 5$ **13.** $c < 2$ **15.** $s \leq 50$ **17.** $l \leq 15$
19. false **21.** true **23.** true

25. 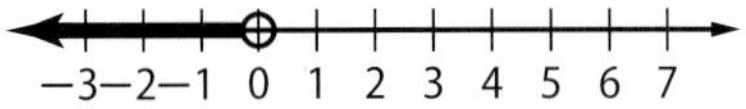

27. 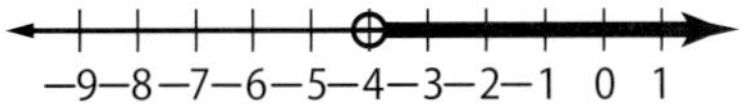

29. 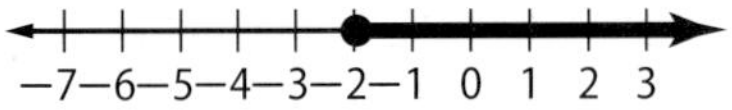

31. $x \leq -1$ **33.** $x \geq -10$ **35.** $x < 9$ **37.** $s \geq 1500$

41. Sample answer: The Wilson family spent more than $20.50 on groceries. **43.** Since the symbol is <, the circle should be open. **45.** A **47.** J **49.** −12 **51.** 2 **53.** 12 **55.** −2
57. $234 + 45.50w = 780$; 12 weeks **59.** $\frac{x}{4} + 2 = 14.75$; $51
61. 63 cm^2 **63.** 18 m^2

Pages 370–373 Lesson 8-7

1. $y \leq 5$ **3.** $x > -4$

5. $f < 4$ 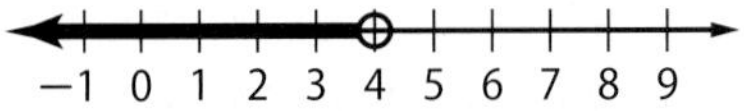

7. $q \geq -2$ 

9. $g > -27$ 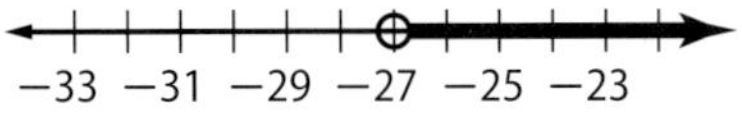

11. $a < 22$ **13.** $y \leq 10.7$ **15.** $p > 26$ **17.** $n \leq 1.5$
19. $c \leq 24.1$ **21.** $b > 15\frac{3}{4}$ **23.** $2 + 0.75y \leq 10$; at most 10 games

25. $x > -9$ 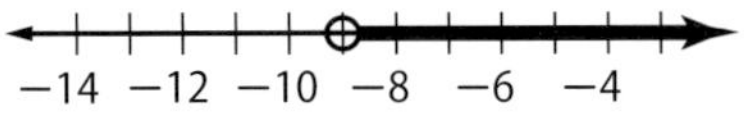

27. $n > 7$ 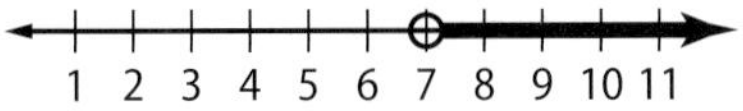

29. $r \leq -4.5$ 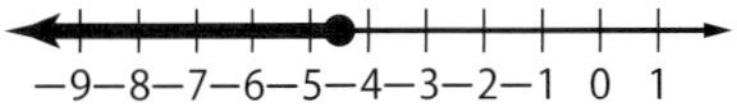

31. $y > 18$ 

33. $b \leq 2$ 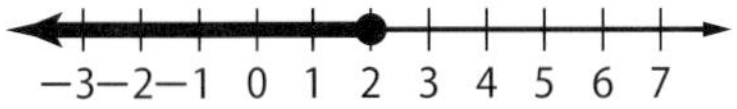

35. $c \leq -\frac{9}{5}$ or $-1\frac{4}{5}$ 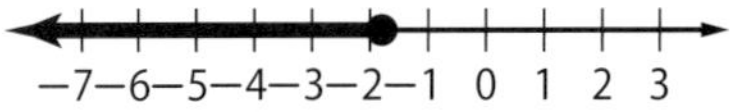

37. at most $2\frac{1}{4}$ h or 2 h 15 min **39.** $a < 3.9$ **41.** $c \leq -21$
43. $r \leq 3$ **45.** $t \leq 8.5$ **47.** $g \geq \frac{3}{2}$ **49.** $\frac{n}{-3} > \frac{5}{6}$; $n < -\frac{5}{2}$
51. $n - 15 \leq -8$; $n \leq 7$ **53.** $28 \geq 7n$; $n \leq 4$

55a. 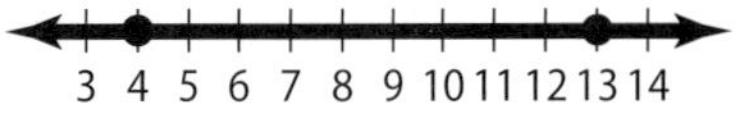

55b. Yes; it represents the solutions that satisfy both inequalities. **55c.** $4 \leq b \leq 13$

55d. 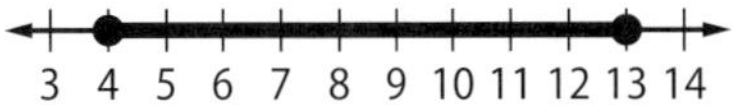

57. 

59a. $99.2 + t > 101$; $t > 1.8$; more than 1.8°F **59b.** $98.6 - t < 95$; $t > 3.6$; more than 3.6°F lower **61.** Sample answer: $x + (-3.60) \leq 15.00$; $x \leq 18.60$. After a discount of $3.60 was applied to an item, the new price was at most $15. Find the original price of the item. **63.** false; If $x = -2$, then $2x = -4$, $-2 > -4$. **65.** Sample answer: To solve an inequality involving multiplication, divide each side by the same number. To solve an inequality involving division, multiply each side by the same number. If the number is positive, keep the inequality symbol. If the number is negative, reverse the symbol.
67. J **69.** F **71.** $c \leq 8500$ **73.** $s < 625$ **75.** $c > 200$
77. 3 **79.** −1 **81.** −3 **83.** 7.5 **85.** 13 **87.** −13

Pages 377–379 Lesson 8-8

1. 6 **3.** 11 **5.** −2 **7.** null set; no solution
9. identity; all numbers

11. $k \leq 12$ 

13. $g < 2$ 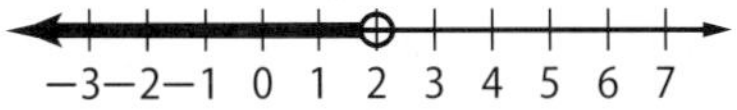

15. $z < -2$ 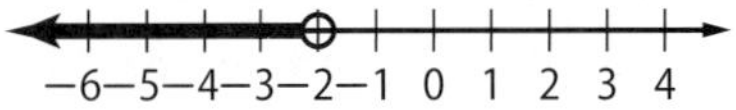

17. $8 **19.** 2 **21.** 4 **23.** null set; no solution
25. null set; no solution **27.** identity; all numbers

29. $h \geq -2$ 

31. $s < -5$ 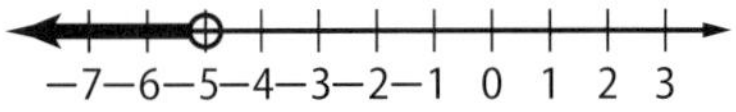

33. $h \leq -7$ 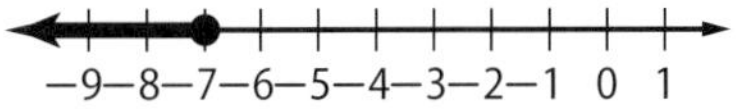

35. $t < 10$ 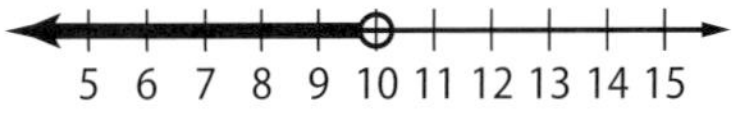

37. He can buy at most 6 granola bars and 6 magazines.
39. $s \geq 79$ **41.** 2.8 **43.** $n \geq 13.2$ **45.** $x < 84$ **47.** 4
49. 4.73 mi

51. 2

$5(2f - 1) = 3(f + 3)$	Write the equation.
$10f - 5 = 3f + 9$	Distributive Property
$10f - 5 - 3f = 3f + 9 - 3f$	Subtraction Property of Equality
$7f - 5 = 9$	Simplify.
$7f - 5 + 5 = 9 + 5$	Addition Property of Equality
$\frac{7f}{7} = \frac{14}{7}$	Division Property of Equality
$f = 2$	Simplify.

53. $p > -1$

$p > \frac{2}{3}\left(p - \frac{1}{2}\right)$	Write the inequality.
$p > \frac{2}{3}p - \frac{1}{3}$	Distributive Property
$p - \frac{2}{3}p > \frac{2}{3}p - \frac{1}{3} - \frac{2}{3}p$	Subtraction Property of Inequality
$\frac{1}{3}p > -\frac{1}{3}$	Simplify.
$\frac{1}{3}p \cdot 3 > -\frac{1}{3} \cdot 3$	Multiplication Property of Inequality
$p > -1$	Simplify.

55. Sample answer: You can add $6x$ to each side of the inequality before you solve.

57a. Identity; All numbers

$5x - 6 \geq 3(x - 2) + 2x$	Write the inequality.
$5x - 6 \geq 3x - 6 + 2x$	Distributive Property
$5x - 6 \geq 5x - 6$	Simplify.
$5x - 5x - 6 \geq 5x - 5x - 6$	Subtraction Property of Inequality
$-6 \geq -6$	Simplify.

57b. Null set; No solution

$12p + 17 \leq 3(4p - 8)$	Write the inequality.
$12p + 17 \leq 12p - 24$	Distributive Property
$12p + 17 - 17 \leq 12p - 24 - 17$	Subtraction Property of Inequality
$12p \leq 12p - 41$	Simplify.
$12p - 12p \leq 12p - 41 - 12p$	Subtraction Property of Inequality
$0 \leq -41$	Simplify.

59. C **61.** J

63. $t \geq -16$

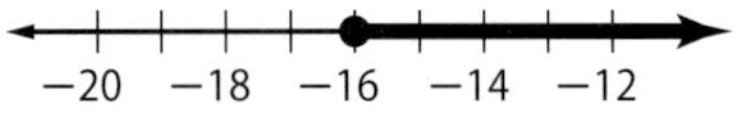

65. $a \leq 7$

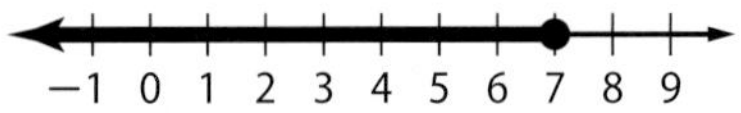

67. $w > \frac{1}{4}$

69. $m \leq 60$ **71.** 0.2 **73.** 7.3 **75.** −3.75 **77a.** −$753
77b. expenses > income **79.** 3.6 **81.** $7.40

Pages 380–382 Chapter Review

1. 2 **3.** −1.1 **5.** −64 **7.** $4.8x = 336$; 70 granola bars
9. 3 **11.** 9 **13.** $-\frac{1}{7}$ **15.** 35 books **17.** $2n - 6 = -22$; −8
19. 13 **21.** 35 **23.** $6\frac{1}{3}$ **25.** $4(p - 1.15) = 8.40$; $3.25
27. −6 **29.** 2 **31.** 4 **33.** $20 + 5m = 30 + 3m$; 5 months
35. $j \leq 15$ **37.** true **39.** true **41.** false
43. $a \leq 18$

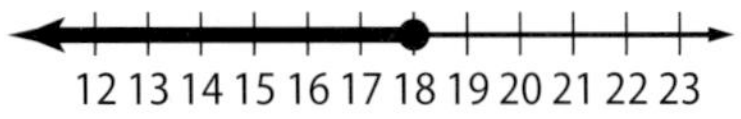

45. $z \geq 8$

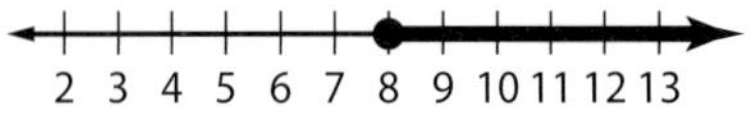

47. $x > \frac{13}{5}$

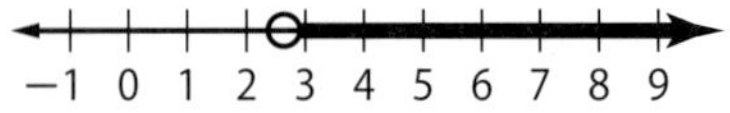

49. $20x + 30 \leq 150$; $x \leq 6$ **51.** 6 **53.** identity; all numbers
55. $g \leq 3$ **57.** $c \geq 3$ **59.** at least 20 cars

Chapter 9 Linear Functions

Pages 387–389 Lesson 9-1

1. This relation is a function; each element of the domain is paired with exactly one element of the range. **3.** The graph is not a function; it does not pass the vertical line test. At least one input value has more than one output value. **5.** 14
7. 44 **9a.** Since the total cost depends on the number of people, the total cost c is the dependent variable and the number of people p is the independent variable; $c(p) = 125 + 15p$.
9b. 6 people **11.** This is not a function because 24 is paired with two range values, 16 and 17. **13.** This is not a function; the domain value 3 is paired with four range values, 1, 3, 7, and 9. **15.** This is a function; each element of the domain is paired with exactly one element of the range.
17. This graph is a function; it passes the vertical line test.
19 18 **21.** −54 **23.** 74 **25.** −86 **27 a.** Since the total cost depends on the number of miles, the total cost c is the dependent variable and the number of miles m is the independent variable; $c(m) = 50 + 0.55m$. **27 b.** 142 mi
29. Sample answer: This graph shows the distance d in miles that sound travels in t seconds. This represents a function because the vertical line test shows that for each value of t, a vertical line passes through no more than one point on the graph.

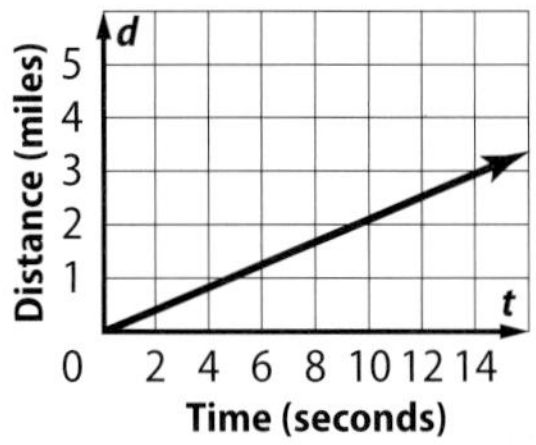

31. 101 **33.** −982 **35.** A **37.** $c(s) = 15 + 1.75s$; $41.25
39. $R \geq 55$ **41.** $L \geq 50\%$ **43.** $y = -5$ **45.** $x = -3$ **47.** $z = 0$
49. $\frac{6}{7}$ **51.** $-\frac{7}{20}$ **53.** $\frac{8}{9}$ **55.** $\frac{27}{25}$ or $1\frac{2}{25}$ **57.** 40 **59.** 52

Pages 393–395 Lesson 9-2

1.

x	$y = x + 7$	y
−1	$y = -1 + 7$	6
0	$y = 0 + 7$	7
1	$y = 1 + 7$	8
2	$y = 2 + 7$	9

(−1, 6), (0, 7), (1, 8), (2, 9)

3. Sample answer: (−1, 4), (0, 5), (1, 6), (2, 7) **5.** Sample answer: (−1, 3), (0, 6), (1, 9), (2, 12) **7.** Sample answer: (1, 10) means that she earns $10 for working 1 hour; (2, 20) means that she earns $20 for working 2 hours.

9.

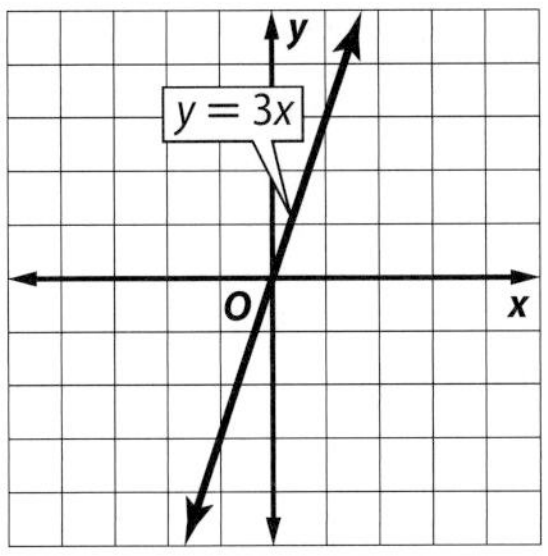

11.

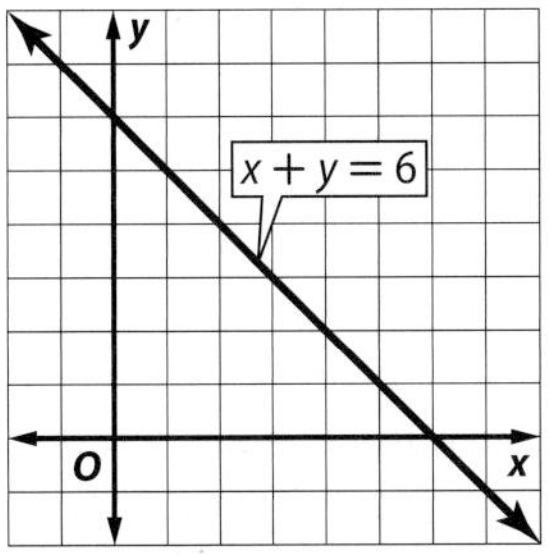

13.

x	$y = x - 2$	y
−1	$y = -1 - 2$	−3
0	$y = 0 - 2$	−2
1	$y = 1 - 2$	−1
2	$y = 2 - 2$	0

(−1, −3), (0, −2), (1, −1), (2, 0)

15.

x	$y = 5x + 1$	y
−2	$y = 5(-2) + 1$	−9
−1	$y = 5(-1) + 1$	−4
0	$y = 5(0) + 1$	1
1	$y = 5(1) + 1$	6

(−2, −9), (−1, −4), (0, 1), (1, 6)

17 Sample answer: (−1, −8), (0, 0), (1, 8), (2, 16) **19.** Sample answer: (−1, 6), (0, 7), (1, 8), (2, 9) **21.** Sample answer: (−1, 3), (0, 5), (1, 7), (2, 9) **23.** Sample answer: (−1, −2), (0, −3), (1, −4), (2, −5) **25.** Sample answer: (1, 6.3) means that a circle with a radius of 1 unit has a circumference of about 6.3 units; (2, 12.6) means that a circle with a radius of 2 units has a circumference of about 12.6 units.

27.

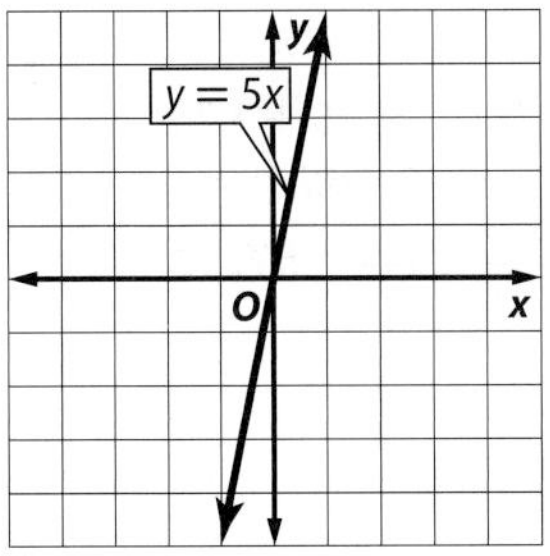

29.

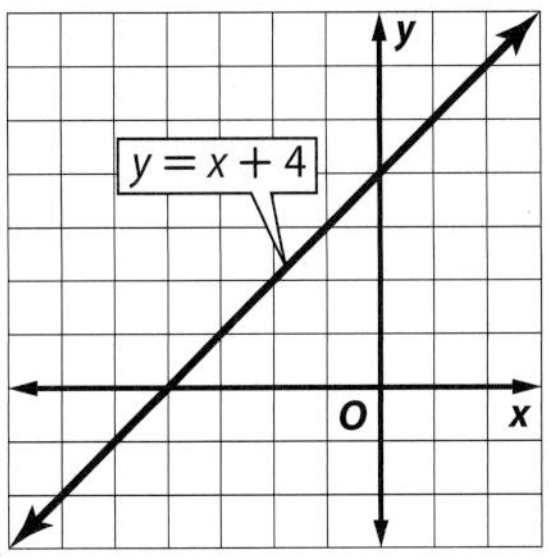

31.

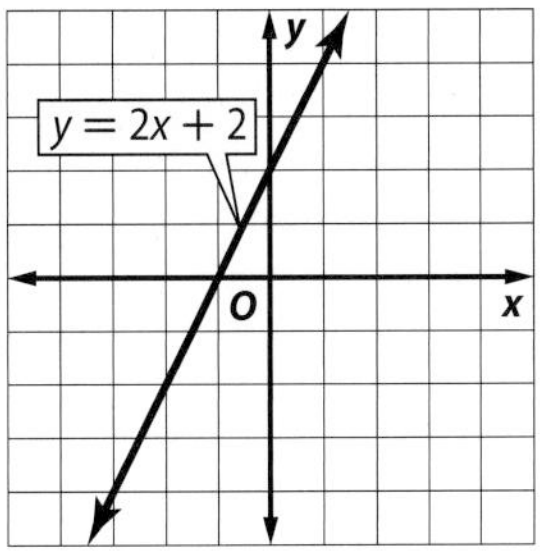

33.

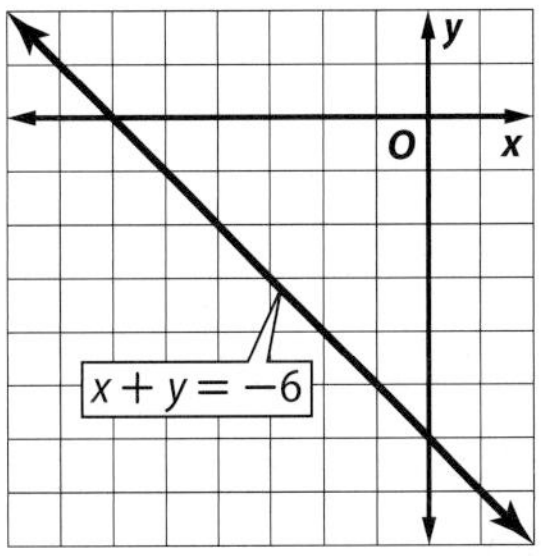

35 a. $x = 8$; $y = 8$

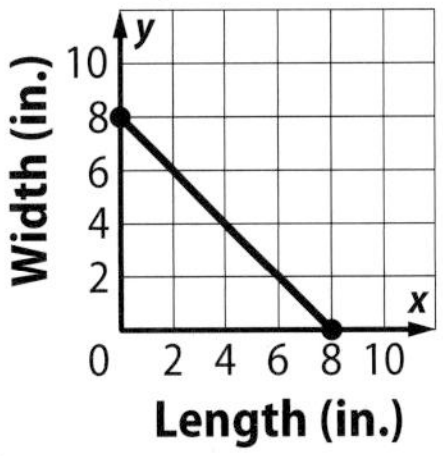

35 b. x-intercept: (8, 0), y-intercept:(0, 8); the length and width must each be less than 8, otherwise the other dimension would be 0 or negative.

37a.

Time (min) x	$y = 0.5x$	Distance (mi) y
5	$y = 0.5(5)$	2.5
10	$y = 0.5(10)$	5
15	$y = 0.5(15)$	7.5
20	$y = 0.5(20)$	10

37b. No; the output values increase by 2.5.

37c. **Distance a Whale Swims**

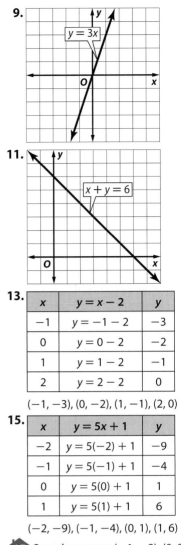

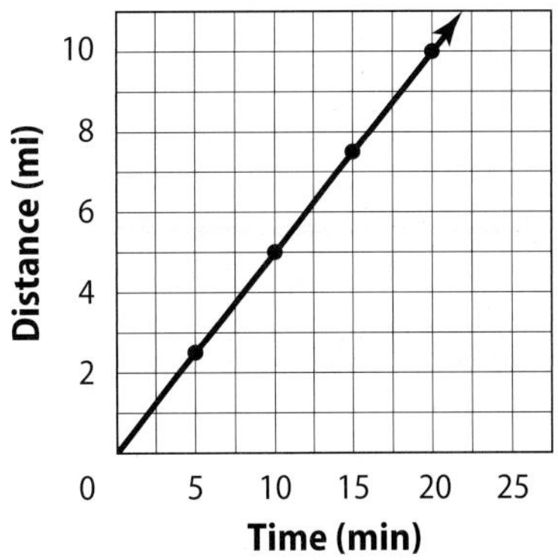

37d. $y = 0.5x$; The miles y that a whale swims is one-half or 0.5 of the time x spent swimming, so $y = 0.5x$. **39.** Sample answer: $x = 1$ **41.** Sample answer: Linear equations use variables to show the relationship between the domain values and the range values of a function. Functions can be represented using a table, a graph, a verbal description, or an equation. **43.** D **45.** B **47.** 105 calendars **49.** $7(x + 8)$ units^2 **51.** $-4\frac{1}{2}$ or −4.5 **53.** $\frac{6}{25}$ or 0.24 **55.** $y - 10$ **57.** $11x + 7$ **59.** $-15z - 8$

Pages 399–402 Lesson 9-3

1. \$2.40 per item **3.** $\frac{1}{2}$ in./wk; The plant grows $\frac{1}{2}$ in. per week. **5.** −4 **7.** \$0.25 per photo **9** −0.5 cm per s **11.** 15 miles per hour; The distance a bicyclist travels increases 15 miles for each hour of travel. **13.** $\frac{1}{3}$ **15.** 0 **17** $\frac{6}{7}$ **19.** 3 **21.** 0

23a. −2

23b.

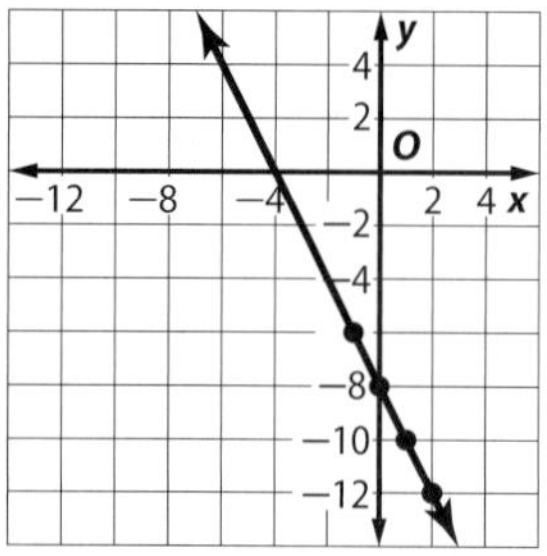

23c. The point is where the line intersects the *y*-axis.

25. Sample answer: (0, 2) and (4, 4)

27. Sample answer:

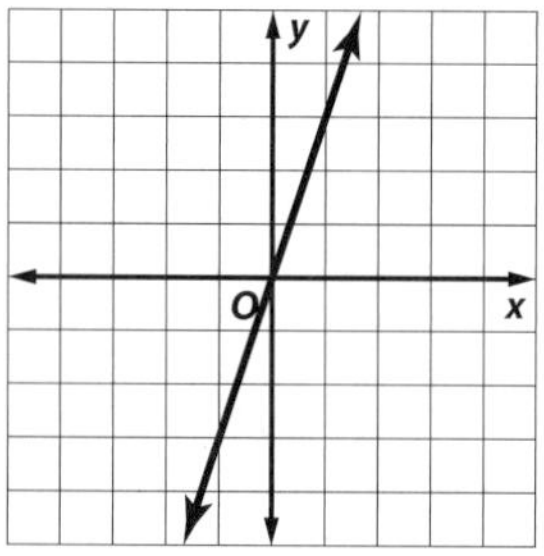

rate of change: $\frac{3}{1}$

29. Sample answer: sequence B; Since the common difference is greater, its terms increase at a faster rate and the points form a steeper line.

31. Sample answer:

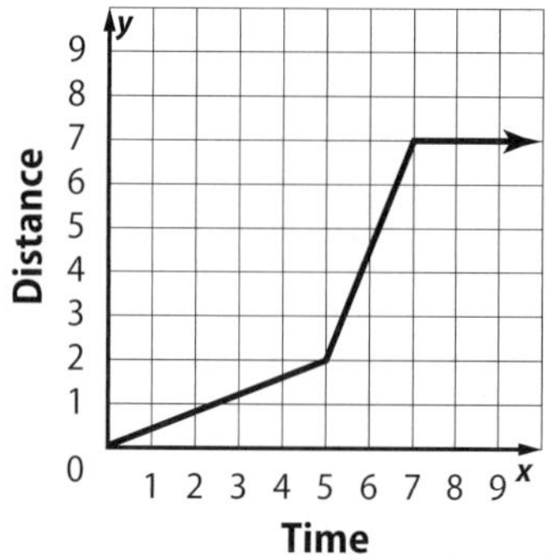

33. D **35.** C **37.** Sample answer: (−1, −12), (0, 0), (1, 12), (2, 24) **39.** Sample answer: (−1, 8), (0, 5), (1, 2), (2, −1) **41.** 75 **43.** 105 **45.** 71 **47.** 20 **49a.** Illinois: 6114 mi^2; Kentucky: 19,506 mi^2; Michigan: 25,391 mi^2; New York: 26,487 mi^2; Ohio: 11,179 mi^2 **49b.** New York

Pages 408–410 Lesson 9-4

1a. This is not a direct variation since the graph does not pass through the origin, and the ratio of centimeters to seconds is not constant. **1b.** This is a direct variation since the graph passes through the origin, and the ratio of centimeters to seconds is constant. **3a.** $y = 0.25x$ **3b.** 5 in.

5 This is a direct variation since the graph passes through the origin and the ratio of the number of weights to the total weight is constant. **7.** The constant of variation is 26.2. This means that Conrad runs 26.2 miles in each marathon.

9 **a.** $y = 0.4x$ **9** **b.** 360 psi **11.** Erica; Sample answer: The unit rate for her pay is \$10.15 per hour, which is more than \$9.20 per hour. **13a.** false; Sample answer: the graph does not pass through the origin. **13b.** true; Sample answer: the rate of change is constant. **15.** Sample answer: Compare the equation to $y = mx$, where m is the constant of variation. $y = 20x$, where 20 is the constant of variation. $y = 20x + 1$ does not have a constant of variation. **17.** H **19.** $y = 12.75x$ **21.** 0 **23.** −1 **25.** 170

Pages 414–417 Lesson 9-5

1. 2; 6 **3.** −7; 0

5.

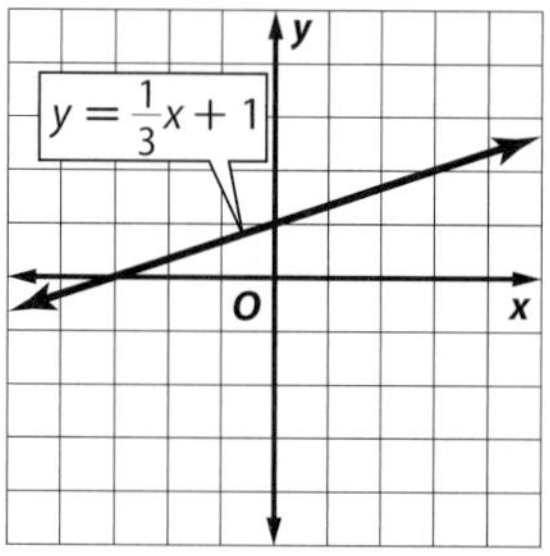

7.

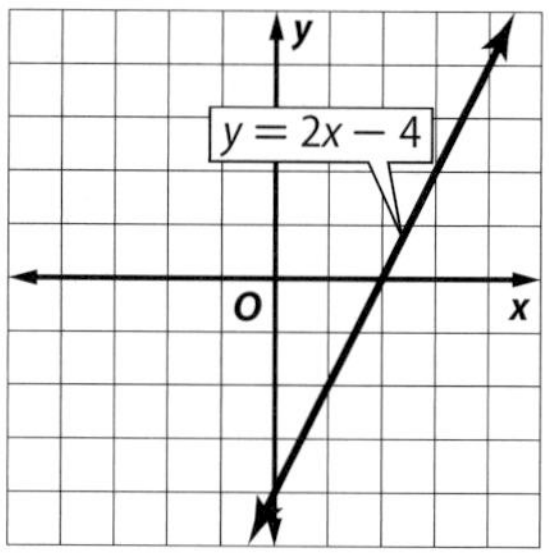

9a.

Flying a Kite

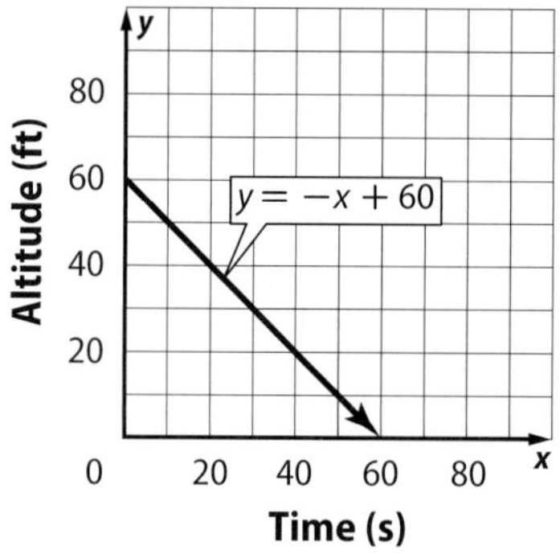

9b. The *y*-intercept 60 represents the initial height of the kite. The slope −1 represents the descent of 1 foot per second.

11. $-\frac{5}{2}$; −2 **13.** −9; 0 **15.** −4; 0 **17.** $\frac{1}{2}$; 6

19.

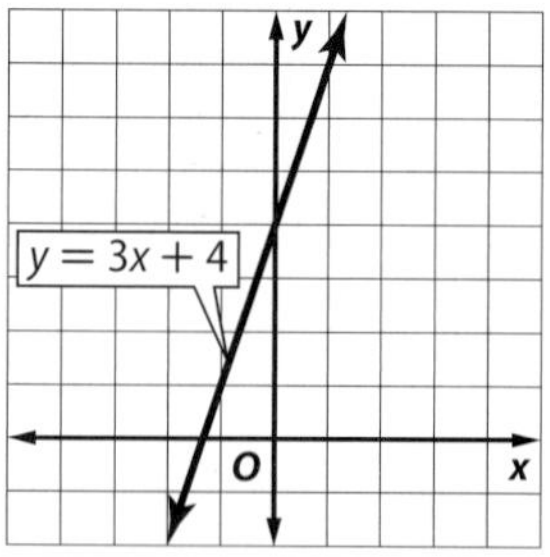

21.

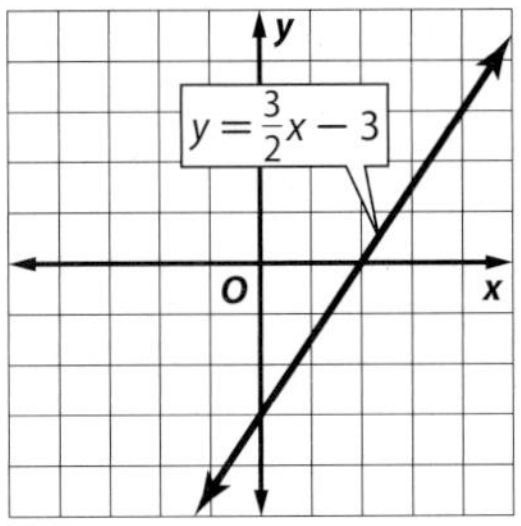

23.

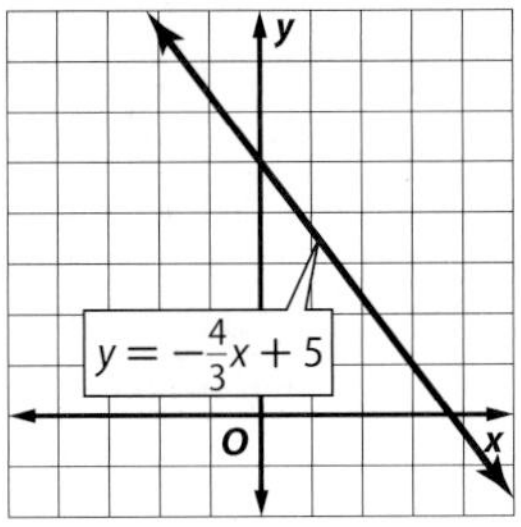

25 a.

A Flying Albatross

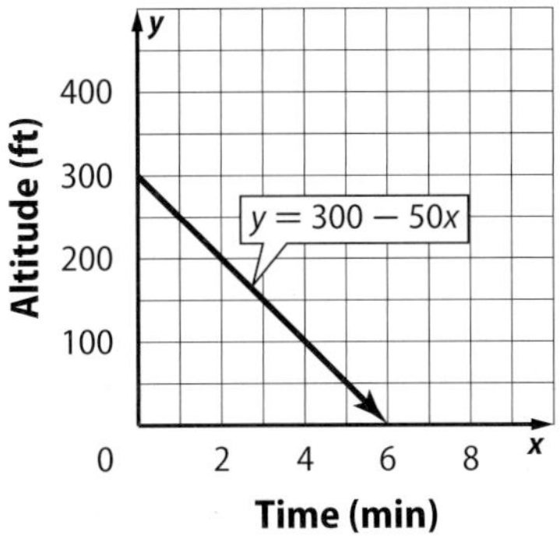

25 b. 300 is the y-intercept, which represents the original height. The slope is -50, which represents descending at 50 ft/min. **27a.** $y = 15x + 18$; $y = 15x + 12$

27b.

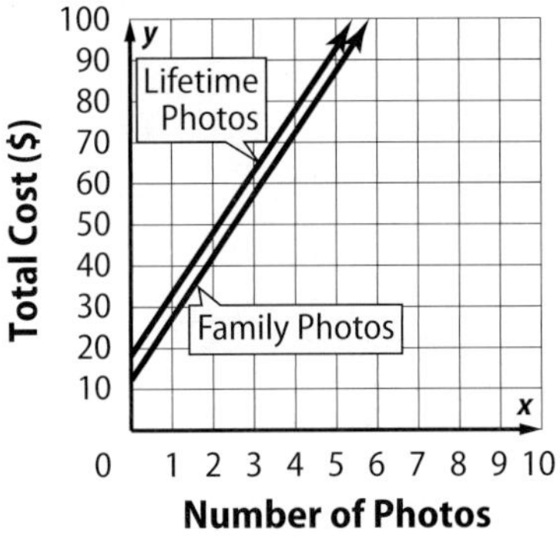

27c. No; the lines are parallel and parallel lines do not intersect. **27d.** Each line has a slope of 15.

29.

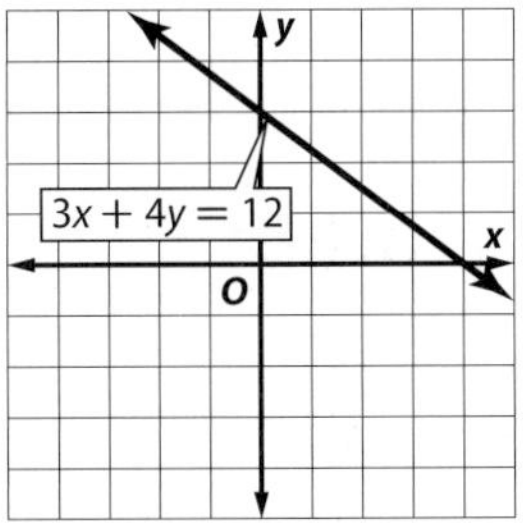

31.

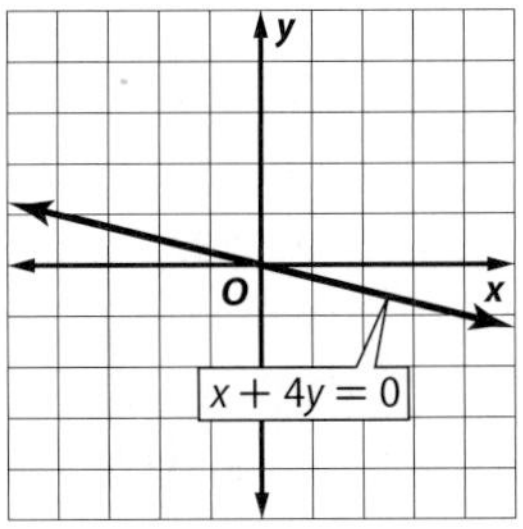

33 2 **35.** $y = 2x + 4$ **37.** Sample answer: A photographer charges a \$50 sitting fee to to take family portraits and then \$15 for each portrait purchased. The total cost y can be represented by the equation $y = 50 + 15x$, where x represents the number of portraits purchased. A family has \$300 to spend on the portraits. How many portraits can they purchase? Answer: 16; Graphs may vary. **39.** The graph becomes less steep. **41.** Sample answer: Graph the y-intercept. Then use the slope to locate a second point on the line. Draw a line through the two points. **43.** J **45.** The y-intercept 35 is the initial cost of the pass and the slope 25 is the cost per time of skiing. **47.** \$2.80/lb; The birdseed costs \$2.80 per pound.

49. $\frac{5}{12}$ **51.** -22 **53.** 2 **55.** 32 **57.** $\approx 4 \times 10^{-7}$ cm **59.** $\approx 2 \times 10^{7}$ mi **61.** $\approx 1 \times 10^{-43}$ mg

Pages 422–424 Lesson 9-6

1. (2, −2)

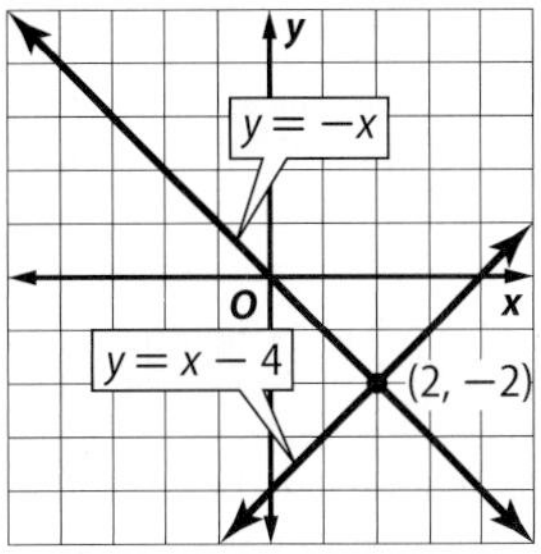

3. infinitely many solutions

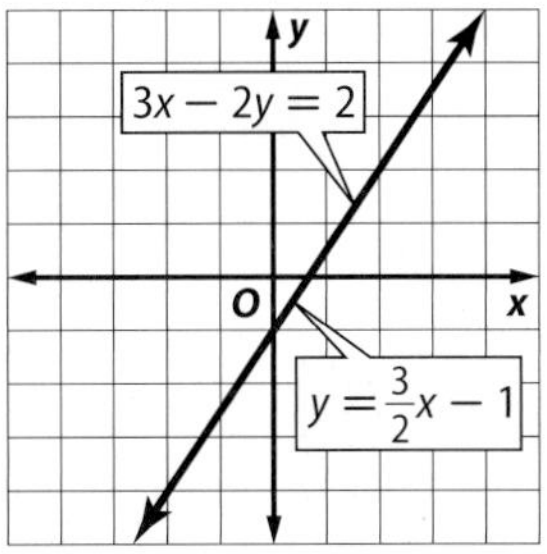

5. (1, 2)

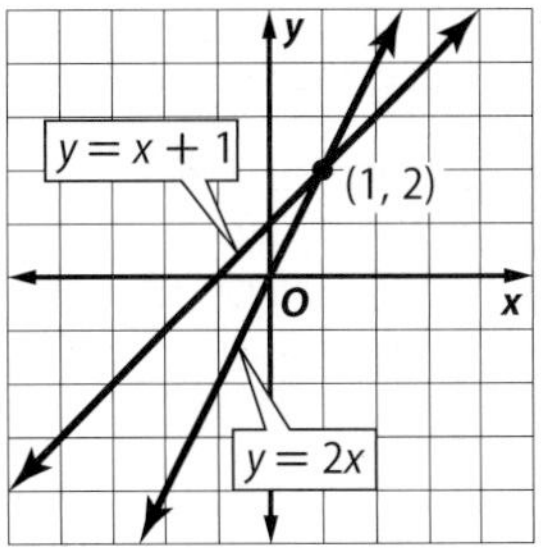

7 (3, −2)

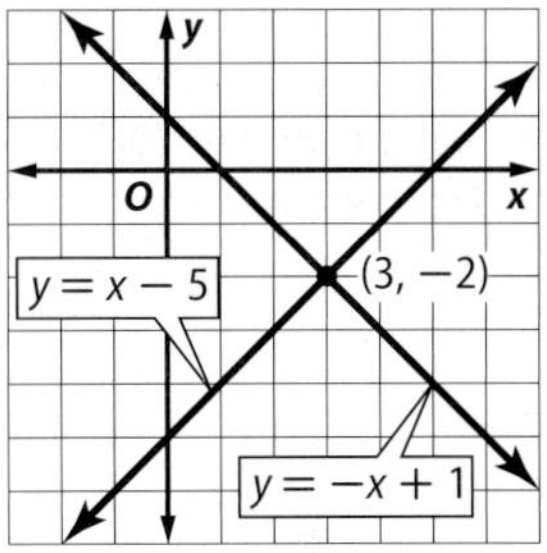

9. no solution

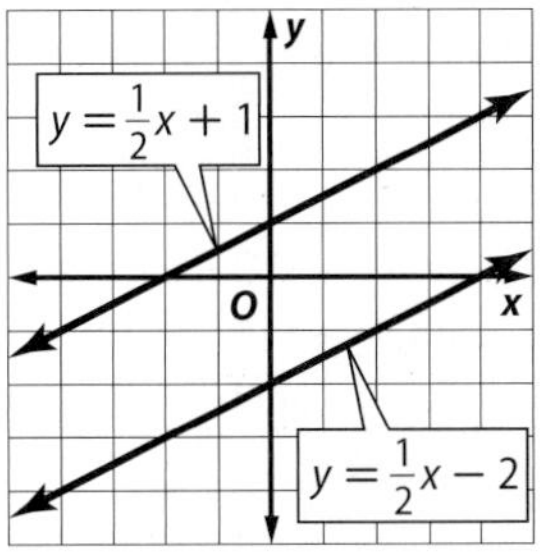

11a. $y = 5x + 50$ and $y = -5x + 90$

11b.

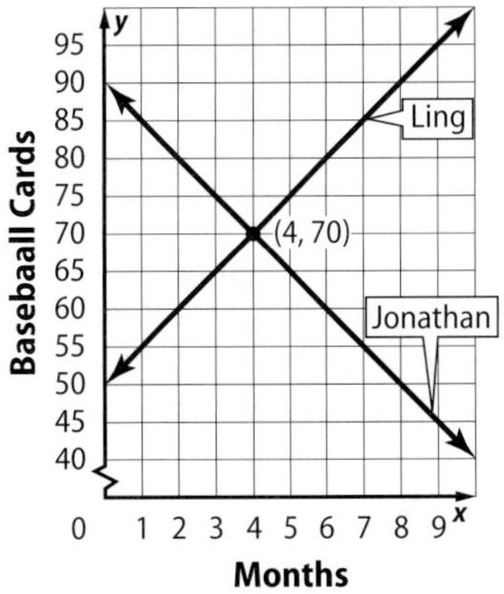

The solution (4, 70) means that after 4 months, they will have the same number of cards (70).

13. (1, 3)

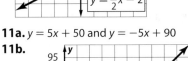

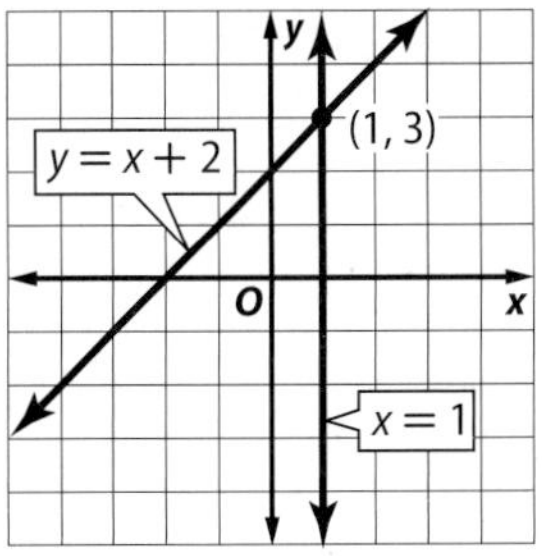

15. (4, 5)

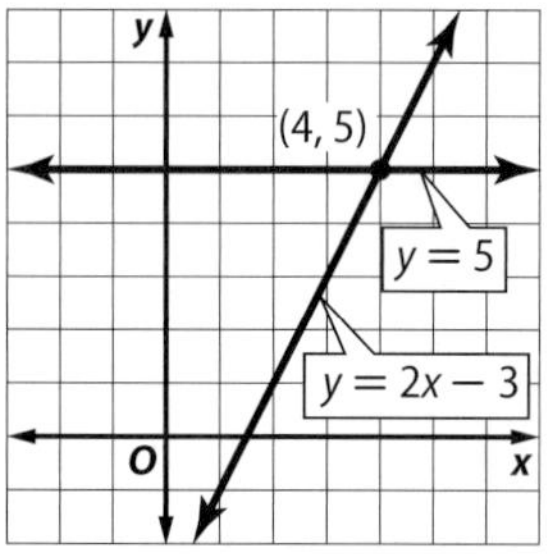

17. (−3, 5)

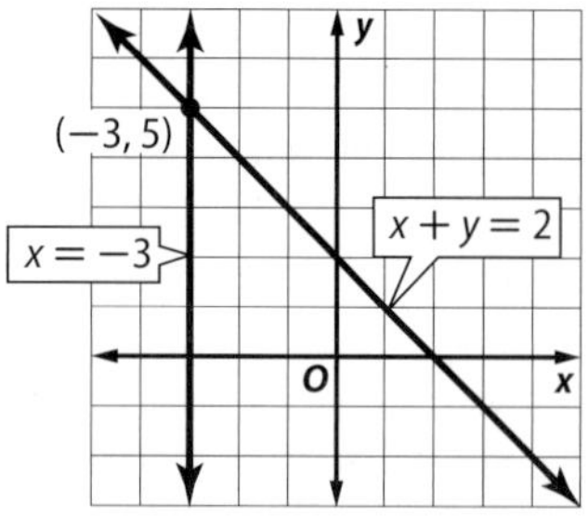

19 a. $2x + y = 8$ and $3x + 2y = 13$ **19 b.** (3, 2); Bagels cost \$3 each and orange juice costs \$2 per can. **21.** Sample answer: If a system of equations has 1 solution, the graphs are intersecting lines. If the system has no solution, the graphs are parallel lines. If the system has infinitely many solutions, then the graphs are the same line. **23.** Sample answer: Two lines can be parallel and never intersect, intersect in exactly one point, or be the same line. Therefore a system of linear equations cannot have exactly two solutions. **25.** G **27.** 12 **29.** 56 **31.** \$10.80 **33.** \$36.75 **35.** 25 ft/s **37.** 8 h **39.** This is not a function because 3 in the domain value is paired with two range values, 3 and 1. **41.** 6561 **43.** 125

Pages 427–429 Lesson 9-7

1. (−6, 2) **3.** (7, −1) **5.** no solution **7.** $x + y = 6$; $2x + 3y = 13$; (5, 1); 5 \$2 cards, 1 \$3 card **9.** (−2, 6) **11** (5, 16) **13.** infinitely many solutions **15.** no solution **17.** (−3, 9) **19.** no solution **21** $x + 2y = 30$; $x + y = 19$; (8, 11); One number is 8, the other number is 11. **23.** $x + y = 15$; $2x = 3y$; (9, 6); The two digits are 9 and 6, so the number is 96. **25.** Sample answer: Since the solution is (1, p), $x = 1$. Replace x with 1 in the system of equations. Solve for y in the first equation. Since $y = 2$, $p = 2$. Replace x with 1 and y with 2 in the second equation. Then solve for m; $m = -3$. **27.** B **29.** A **31.** $x = 15$ **33.** all numbers **35.** $x = -10$ **37.** $\$20x + \$600 \geq \$1000$; $x \geq 20$ **39.** 9 **41.** 7 **43.** ±5 **45.** 3 **47.** −8 **49.** $x = 10$ **51.** $x = -3$ **53.** $x = -2$

Pages 430–432 Chapter Review

1. No; the domain value 4 is paired with 2 range values, 1 and 2. **3.** Sample answer: (−1, 5), (0, 0),(1, −5), (2, −10) **5.** Sample answer: (−1, 8), (0, 9),(1, 10), (2, 11)

7.

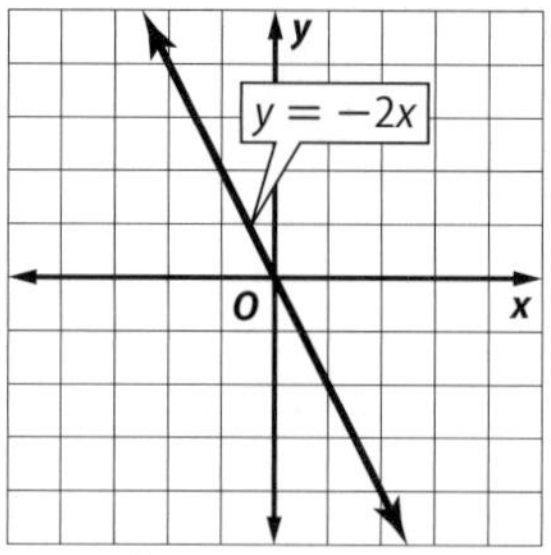

9. Sample answer: (0, 4) means she can buy 0 small smoothies and 4 large smoothies with \$12; (6, 1) means she can buy 6 small smoothies and 1 large smoothie with \$12.

11. $\frac{1}{2}$ **13.** $\frac{1}{6}$ **15.** \$63 **17.** 4; 7 **19.** −5; 0 **21.** −8; −7

23.

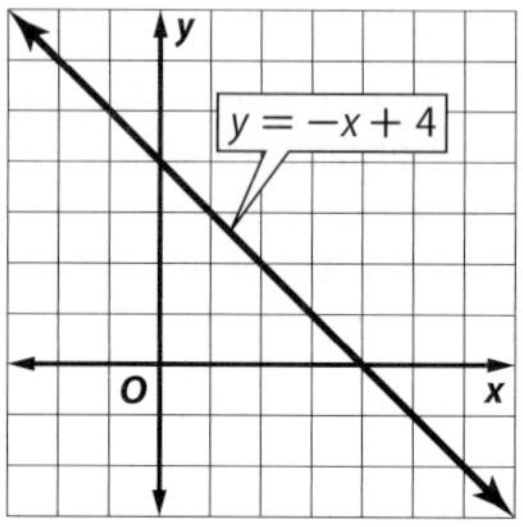

25.

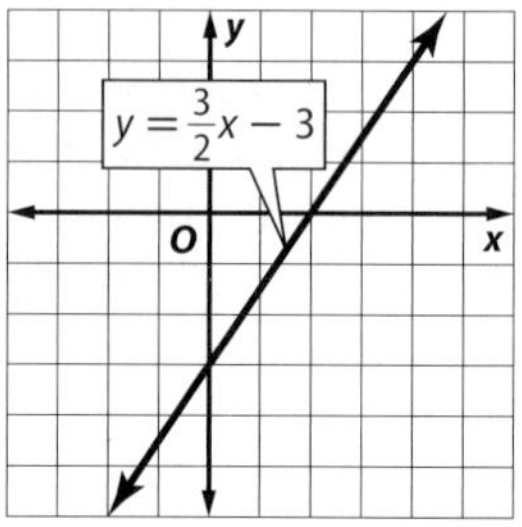

27. Slope: 2; y-intercept: 7; the slope 2 represents the ascent in ft per second. The y-intercept 7 represents the initial altitude in ft before the balloon is released.

29. (−2, 2)

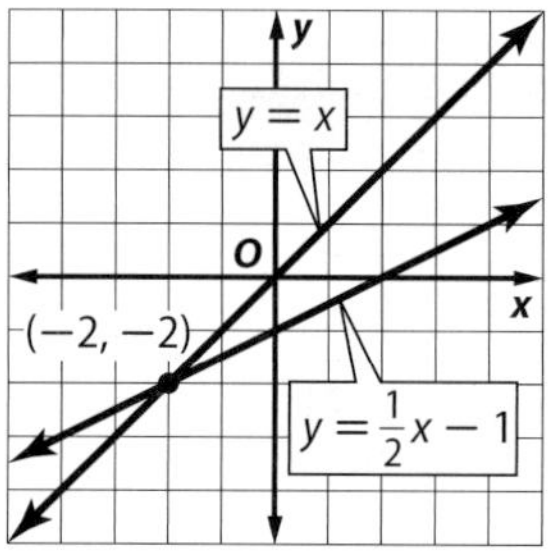

31. (−1, −1)

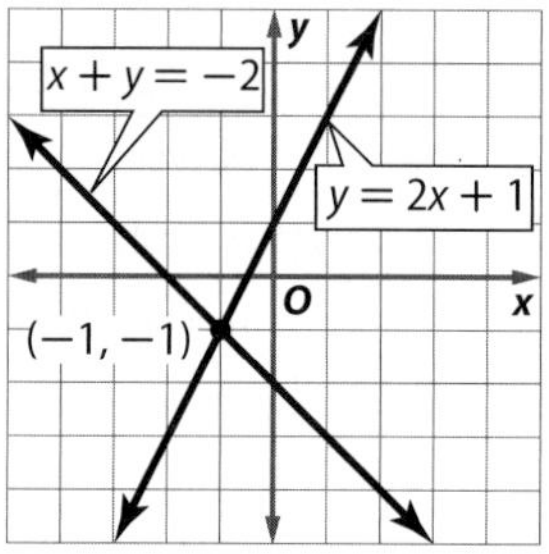

33. Sample answer: $x + y = 9$, $x - y = 1$; 5 and 4; $x = 5$; $y = 4$ **35.** (5, 4) **37.** (0, 2) **39.** (7, 2) **41.** no solution **43.** $3x - y = 11$; $x + y = 1$; (3, −2); One number is 3; the other number is −2.

Chapter 10 Statistics and Probability

Pages 437–439 Lesson 10-1

1. mean: 157.6 mph; median: 157 mph; no mode **3.** The mean height for 1991 to 1995 is 6.09 m and the mean height for 2007 to 2011 is 5.97 m. The median height for 1991 to 1995 is 6.10 and the median height for 2007 to 2011 is 5.95. Using either mean or median, the best heights for 1991 to 1995 are greater than the best heights for 2007 to 2011. **5.** mean; Sample answer: The mean shows the longest amount of time. **7** $245; $242.50; $200 **9.** Sample answer: mean; there are no extreme values; 855.4 wins. **11** Median; sample answer: The median is the greatest value. Since he wants to show a high test score, he would use the median. **13.** Mode; sample answer: More students have 2 siblings than any other number. It is the best representation of the data.
17. Sometimes; for example, the data set: $3, $5, $5, $5, and $7; the mean is $5, the median is $5, and the mode is $5.

19. Sample answer: The median home price would be useful because it is not affected by the cost of the very expensive homes. The cost of half the homes in the county would be greater than the median cost and half would be less.
21. Sample answer: The median or the mean best represents the data. The data are all close together. The median is 371 passes and the mean is 374.6 passes.

Completed Passes in the NFL	
Player	**Number of Passes**
Matt Schaub	396
Peyton Manning	393
Tom Brady	371
Drew Brees	363
Aaron Rodgers	350

23. J **25.** 140 **27.** 2($13.86) ÷ 12; Each member paid $2.31.
29. 0.5 **31.** $-\frac{1}{4}$ **33.** 66 mL/s **35.** 0.10 errors/game

Pages 444–446 Lesson 10-2

1. R: 145; M: 115; Q_1: 90; Q_3: 145; IR: 55; none **3a.** The fruits' range is 30. The vegetables' range is 20. So, the number of Calories in fruits varies more than the number of Calories in vegetables. **3b.** 80; With the outliers, the mean increased by 57.5 − 54.3, or 3.2; the median increased by 55–50, or 5; and the mode did not change. **5.** R: 1.05; M: 0.555; Q_1: 0.455; Q_3: 0.63; IR: 0.175; 1.25 **7** R: 15; M: 21.5; Q_1: 18; Q_3: 25; IR: 7; no outlier **9a.** The range for 2011 is 21 points, and the range for 2012 is 26 points. So, her points scored per game varied more in 2012 than in 2011. **9b.** yes; 2012
9c. Without the outlier, the range is 12 points. The inclusion of the outlier makes the range 26 points. So, the range is increased by 14 points. **11 a.** Antelope **11 b.** Antelope: R: 63; M: 58; Q_1: 33.5; Q_3: 75.5; IR: 42; Augusta: R: 52; M: 55.5; Q_1: 37.5; Q_3: 72.5; IR: 35

11 c. Sample answer: The median average high temperature of Antelope is only slightly greater than the median average high temperature of Augusta. The interquartile range for Antelope is 42°, while the interquartile range for Augusta is 35°. **11 d.** The appropriate measure of center to describe the average temperature is the mean or median, since they are roughly the same for Augusta. The mean is about 55.17; median: 55.5; mode: none.
11 e. Sample answer: Antelope has a greater range, or spread, of temperatures than Augusta. Both cities have about the same mean average temperature. Augusta has a mean of 55.17 and Antelope has a mean of 54.92. The median for Antelope is only slighter greater than median for Augusta. Since the third quartile and the first quartile for Antelope are higher and lower, respectively, than those for Augusta, it follows that Augusta's high temperatures do not fluctuate as much as Antelope's. **13.** Sample answer: 1, 2, 3, 4, 5, 6, 7, 10, 20, 21, 22, 23, 24, 25, 26; 8, 8, 8, 8, 9, 9, 9, 10, 11, 11, 11, 12, 12, 13, 14
15. False; sample answer: The interquartile range deals with the middle values of a data set. **17.** Sample answer: An outlier is a value that is either much larger or much smaller than the median. Therefore, an outlier greatly increases or decreases the range. It does not affect the interquartile range.
19. F **21.** R: 68; M: 36.5; Q_1: 30; Q_3: 53; IR: 23; no outlier
23. 4.3; 4.2; 4.1 and 4.2 **25.** 6.7, 6.8, 6.9, 7.0, 7.8, 8.7 **27.** 65
29. $\frac{5}{12}$ **31.** (−5, 6) **33.** (14, −3)

Pages 449–451 Lesson 10-3

1. 9.8; The average distance between each data value and the mean number of petals is 9.8. **3.** 3.5; The average distance between each data value and the mean weight is 3.5 oz. **5** 5; 2.6 **7.** January; the mean absolute deviation for January is 8, which is greater than 2.5, the mean absolute deviation for July. **9 a.** 2 yd **9 b.** 6 ft **9 c.** The mean absolute deviation in feet is 3 times as great as the mean absolute deviation in yards. **11.** Sample answer: trips from home of 7 mi and 17 mi **13a.** The mean absolute deviation stays at $12. **13b.** The mean absolute deviation doubles from $12 to $24. **15.** B **17.** C **19.** The range before the sale is $15 and the range during the sale is $13. So, the prices varied more before the sale than during the sale. **21a.** 70% **21b.** 7% **21c.** The percent error in part a is greater.
23. 0, −0.5 **25.** $-\frac{2}{3}$, 20

Pages 457–460 Lesson 10-4

1. Sample answer: Second period's data has a mean of 16 points with a mean absolute deviation of about 0.8 points. Fifth period's data has a mean of 17 points with a mean absolute deviation of about 1.4 points. Fifth period's scores centered around a greater value. However, the variation is also greater, which means the scores were more spread out.
3. Sample answer: The times at Lucy's Steakhouse have a median of 20 minutes with an interquartile range of 20 minutes. The times at Gary's Grill have a median of 15 minutes with an interquartile range of 10 minutes. In general, a customer will wait longer at Lucy's Steakhouse.

5 this season; Sample answer: Both seasons' scores have a median of 20 points, but last season's scores have an interquartile range of 15 points while this season's interquartile range is 10 points. So the football team's performance was more consistent this season.
7. The median number of visitors to Summer Lake is less than the median number of visitors to Canyon Overlook.
9 Canyon Overlook; Sample answer: During half of the days, there were at least 80 visitors to the park. On any given day, there were no more than 80 visitors to Summer Lake.
11. Sample answer: The data shown in the histograms are shown only as intervals. Specific values are not shown.
13. double box plot **15.** B **17.** 5 **19.** 4 **21.** 0
23. The median rating of 3 stars is greater than the mean rating of about 2.88 stars. **25.** $-6\frac{1}{2}$ **27.** identity; all numbers

Pages 464–467 Lesson 10-5

1. Biased, convenience sample; all the students on a school bus might come from the same neighborhood/economic status.
3. Unbiased, stratified random sample; the shoppers are first divided into nonoverlapping groups and then 1 male and 1 female are selected randomly from each group.
5. Unbiased, simple random sample; the students are randomly selected. **7** Biased, voluntary response survey; those teenagers who are interested in participating in the survey are part of the sample. **9.** This sampling method is not valid because it will include teenagers who do not attend the school. So, the results cannot lead to a valid prediction. **11.** Yes; this is a systematic random survey because the sample is selected according to an interval; 1225 people. **13.** No; only 80 out of 300 customers agree. From this random survey, you can predict that only about 27% of the customers would like an international movie section, so the store should not add such an area. **15a.** Sample answer: They could survey everyone in the school, including the students and staff. **15b.** Sample answer: What types of activities should we offer at the carnival? How many and what types of games should we offer? Would you attend the carnival? **15c.** Sample answer: The council could use the results to determine the games and activities that are most popular in the school population. **17.** Sample answer: Survey your classmates about their favorite sport. **19.** Yes; sample answer: If questions are asked in a neutral tone, then a more accurate answer can be expected. However, if the person asking the questions changes their tone of voice it can persuade someone to give an inaccurate response. **21.** B **23.** J
25. $(8x + 10)$ in. **27.** 146.9%; increase **29.** 23.9
31. $50(5t - 2)$ **33.** $\frac{7}{8}$ **35.** $-\frac{1}{24}$

Pages 472–474 Lesson 10-6

1. $\frac{1}{10}$ or 10%; unlikely **3.** $\frac{2}{5}$ or 40%; unlikely **5.** $\frac{2}{5}$ or 40%; unlikely **7.** $\frac{1}{2}$ or 50%; equally likely **9.** $\frac{9}{10}$ or 90%; likely **11.** $\frac{5}{6}$ or $83.\overline{3}$%; likely **13** $\frac{7}{20}$ or 35%; unlikely **15.** $\frac{2}{5}$ or 40%; unlikely **17.** $\frac{3}{10}$ or 30%; unlikely **19.** 0 or 0%; impossible **21.** $\frac{1}{10}$ or 10%; unlikely **23.** 1 or 100%; certain **25.** $\frac{1}{2}$ or 50%; equally likely **27.** $\frac{3}{5}$ or 60%; likely **29.** 1 or 100%; certain **31.** $\frac{19}{20}$ or 95%; likely **33.** 62.6% **35.** $\frac{37}{40}$ or 92.5% **37** $\frac{4}{13}$ **39.** $\frac{1}{2}$; $P(\text{red or white}) = \frac{1}{10} + \frac{2}{5}$ or $\frac{1}{2}$, so $P(\text{blue}) = 1 - \frac{1}{2}$ or $\frac{1}{2}$ **41.** Sample answer: choosing the correct answer on a 4-choice multiple-choice test by random guessing **43.** 10 girls **45.** D **47.** $\frac{1}{5}$ or 20% **49a.** The graph shows the number of states that have a certain number of roller coasters; most states have less than 10 roller coasters. **49b.** to show a break in the vertical scale **49c.** The graph shows that 30 states have anywhere from 0 to 9 roller coasters, but it does not indicate how many states have none.
51. (5, 9) **53.** (3, 0) **55.** (−1, −2) **57.** (1, 4) **59.** (6, 6)

Pages 478–481 Lesson 10-7

1. $\frac{1}{5}$ or 20%; Sample answer: The theoretical probability of rolling a 4 is $\frac{1}{6}$ or about 17%. So, rolling a 4 in the experiment occurred slightly more often than expected. **3.** 150
5a. $\frac{3}{10}$ or 30% **5b.** $\frac{1}{4}$ or 25% **5c.** The theoretical probability of spinning a 3 is $\frac{1}{4}$ or 25%. This is the same as the experimental probability. **7a.** 68 **7b.** 108
9a. $\frac{9}{40}$ or 22.5%; This is the experimental probability that an adult prefers to pay bills online. **9b.** 910 **11** 11,200
13. The player needs to make $\frac{5}{7}$ or about 71% of the free throws. Since the player made 0.76 or 76% of the free throws during the past season, the player is likely to win the tournament.
15. 48 **17.** Sample answer: The theoretical probability is $\frac{7}{35}$ or 20%. So, she would expect to select a navy pair of socks 4 out of 20 times. Since she selected a navy pair 6 times, her experimental probability of $\frac{6}{20}$ or 30% exceeded the theoretical probability.
19. $\frac{19}{50}$ or 38% **21.** A **23.** $\frac{5}{6}$ or $83.\overline{3}$%; likely **25.** $\frac{1}{3}$ or $33.\overline{3}$%; unlikely **27.** $\frac{1}{2}$ or 50%; equally likely **29.** $\frac{2}{3}$ or $66.\overline{6}$%; likely
31. 1 or 100%; certain **33.** all numbers **35.** no solution
37. no solution

Pages 484–486 Lesson 10-8

1. 6 **3.** 36 **5.** $\frac{1}{64}$ **7.** 12 **9.** 6 **11** 8 **13.** $\frac{1}{8}$ **15a.** $\frac{1}{10}$ **15b.** $\frac{1}{2}$ **17a.** $\frac{7}{16}$ **17b.** $\frac{1}{4}$ **19.** He should have multiplied rather than added. The correct answer is 80 outfits.
21. 5^x **23.** A **25.** C **27.** No; this is a biased sample. The students who participate in band are more likely to enjoy music, so more of them may say music is their favorite class, than compared to the entire school population. **29a.** 15; 23

29b. Chicken; whereas chicken sandwiches have 8–20 grams of fat, burgers have 10–36 grams of fat **29c.** Chicken: R: 12; M: 15; Q_1: 13; Q_3: 18.5; IR: 5.5; outliers: none; Burgers: R: 26; M: 23; Q_1: 17; Q_3: 31.5; IR: 14.5; outliers: none **31.** $3x + 7$ **33.** $6x - 6$ **35.** $-9x - 3$

Pages 490–492 Chapter Review

1. 22.4 students; 22 students; 22 students **3.** 2.1 in.; 2 in.; 1.5 in., 2 in., 2.5 in. **5.** R: 4; M: 33; Q_1: 32; Q_3: 34; IR: 2; none **7.** R: 5; M: 7; Q_1: 5.5; Q_3: 9; IR: 3.5; none **9.** 21.25 **11.** 6 **13.** 6 **15.** The median number of floors on Main St. is 40 with an interquartile range of 10. The median number of floors on Grand Ave. is 18 with an interquartile range of 5. Main St. has more variability and centers around a higher number of floors than Grand Ave. **17.** No; the sample is biased. Diners at a steakhouse are more likely to choose steak as their favorite meal than randomly-chosen diners. **19.** $\frac{3}{5}$ or 60%; likely **21.** 1 or 100%; certain **23.** $\frac{31}{100}$ or 31% **25.** 99 **27.** $\frac{1}{12}$ **29.** $\frac{1}{3}$

Chapter 11 Congruence, Similarity, and Transformations

Pages 497–500 Lesson 11-1

1. supplementary angles; 34 **3.** supplementary angles; 75 **5.** 128°; ∠1 and ∠8 are alternate exterior angles, so they are congruent. **7a.** The angles are alternate interior angles. **7b.** 117 **9.** $x = 20.4$, $m\angle Q = 139.2°$, $m\angle R = 40.8°$ **11.** vertical angles; 152 **13.** supplementary angles; 121.1 **15.** 52.6°; ∠6 and ∠3 are alternate interior angles, so they are congruent. **17.** 127.4°; ∠6 and ∠2 are supplementary angles, so the sum of their measures is 180°. **19.** 52.6°; ∠6 and ∠4 are vertical angles, so they are congruent. **21.** 75° **23.** $x = 24$, $m\angle M = 27°$, $m\angle P = 63°$ **25 a.** $2x + 6x = 180$; 22.5 **25 b.** 45°; 135° **27.** $x = 5$, $m\angle 4 = 75°$, $m\angle 2 = 75°$ **29.** $m\angle B = 35°$, $m\angle C = 35°$ **31a.** $x + 2x = 90$; 30 **31b.** 30°; 60° **33 a.** $m\angle WXY = 60°$, $m\angle YXZ = 120°$ **33 b.** Sample answer: 6:00, 15 s **33 c.** 150° **35.** Sample answer:

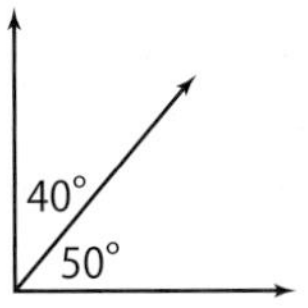

37. The sum of their measures is 180°. The interior angles on the same side of a transversal are supplementary. **39.** C **41.** A **43.** 98 **45.** 9.3%; increase **47.** $a < -6$ **49.** 96 **51.** −27 **53.** 65 **55.** 85 **57.** 30

Pages 506–508 Lesson 11-2

1. 45; acute scalene **3.** 45; right isosceles **5.** 30° **7** 30; obtuse isosceles **9.** 75; right scalene **11.** 70; acute isosceles **13** 12°, 72°, 96° **15.** 90°, 60°, 30° **17.** 85° **19.** $x = 47$ **21.** $x = 53$ **23** 53°, 55°, 72° **25.** scalene **27.** isosceles **29.** Miguel should have said that an equilateral triangle is never an obtuse triangle since all of the angles measure 60°. **31.** The sum of the angles must equal 90° because 180° − 90° = 90°. **33.** D **35.** C **37.** 9, −9 **39.** 9.2, −9.2 **41.** −7 **43.** 6 **45.** 4:00 P.M. **47.** 9 **49.** 289

Pages 516–518 Lesson 11-3

1. The figure is not a polygon because it is an open figure. Two of the sides are not connected. **3.** The figure has 5 sides that only intersect at their endpoints. It is a pentagon. **5.** 154.3° **7.** yes **9.** The figure has 8 sides that only intersect at their endpoints. It is an octagon. **11.** The figure is not a polygon because it has sides that cross each other. **13.** The figure is not a polygon because it has a curved side. **15.** 1620° **17.** 3960° **19** 120° **21.** No; each interior angle of a regular 12-gon measures 150°, and 360° is not evenly divisible by 150°. **23.** No; each interior angle of a regular 20-gon measures 162°, and 360° is not evenly divisible by 162°. **25.** 15-gon **27.** 30-gon

29.

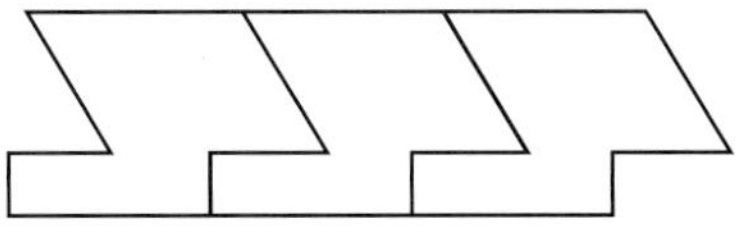

31.

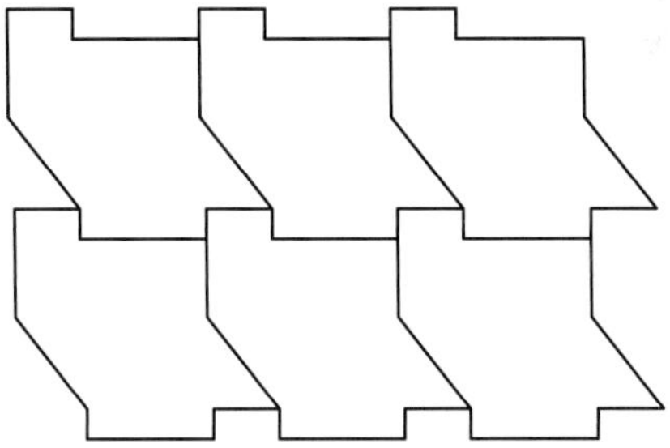

33. 120° **35.** 24° **37.** It increases by 180°

39.

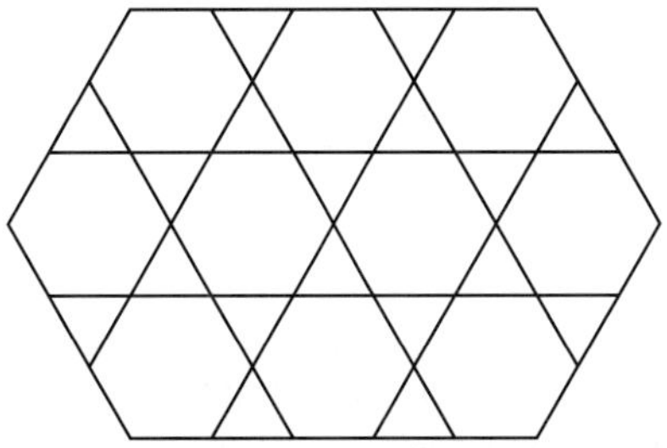

41. B **43.** A **45.** $x = 36$ **47.** 10 bags **49.** $y = 9$ **51.** $m = 8$ **53.** −3 **55.** 2

Pages 524–526 Lesson 11-4

1. $M'(1, 3)$, $N'(3, 4)$, $P'(1, 7)$ **3** $X'(4, -1)$, $Y'(2, -5)$, $Z'(6, -5)$

5.

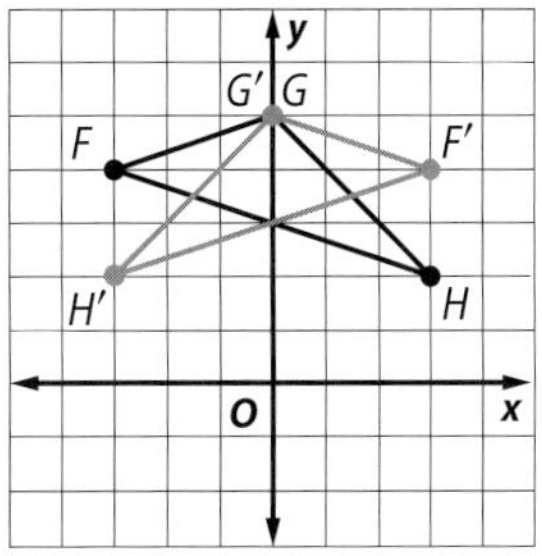

7. 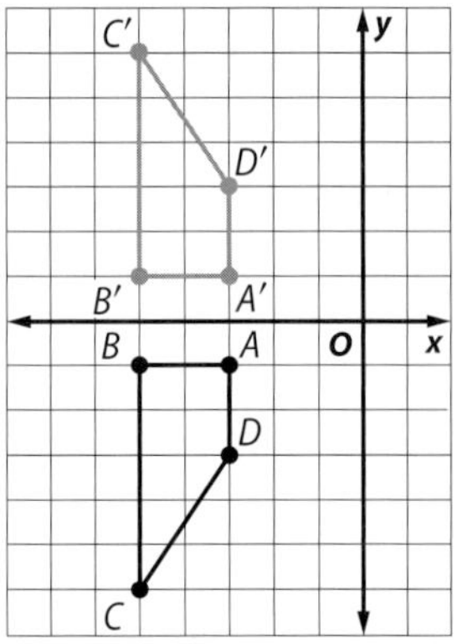

9 5 units left and 3 units up   11 reflection over the x-axis
13. It is a translation 5 units left and 4 units down.
15. $S'(-14, 2)$, $T'(0, 9)$; The translation is 6 units left and 2 units up.
17. 

; y-axis

19. No; for example, the vertices of $\triangle ABC$ are $A(1, 3)$, $B(7, 3)$, and $C(1, 6)$. After a reflection over the x-axis and a reflection over the y-axis, the image's vertices are $A'(-1, -3)$, $B'(-7, -3)$, and $C'(-1, -6)$. No single reflection or translation maps $\triangle ABC$ to $\triangle A'B'C'$. **21.** No single transformation is equivalent to the original two. The new figure is a reflection of the original but moved up and over to the right. **23.** D **25.** H **27.** $(-4, 4)$ **29.** $(-3, -2)$ **31.** 49 points **33.** $3n$ **35.** $3(x + 9)$ **37.** cannot be factored **39.** cannot be factored

Pages 531–533 Lesson 11-5

1. 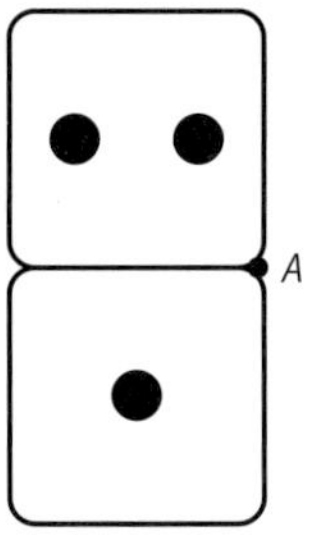

3a. 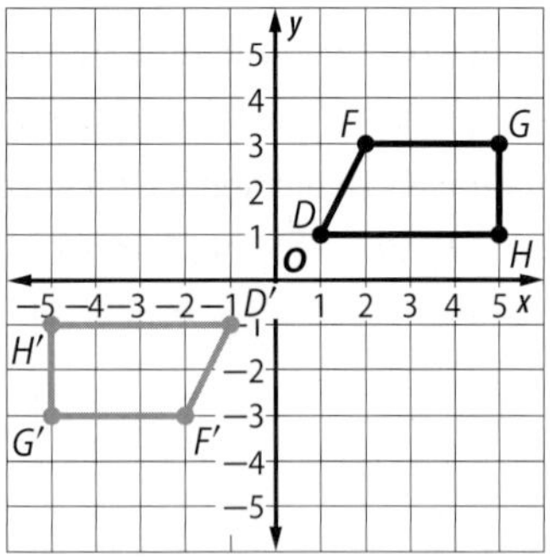

3b. $D'(-1, -1)$, $F'(-2, -3)$, $G'(-5, -3)$, $H'(-5, -1)$
5. 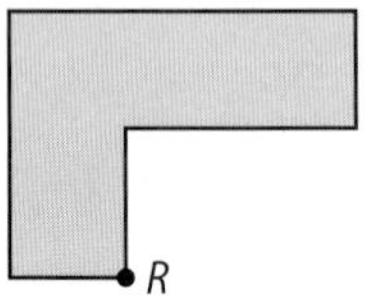

7 **a.** 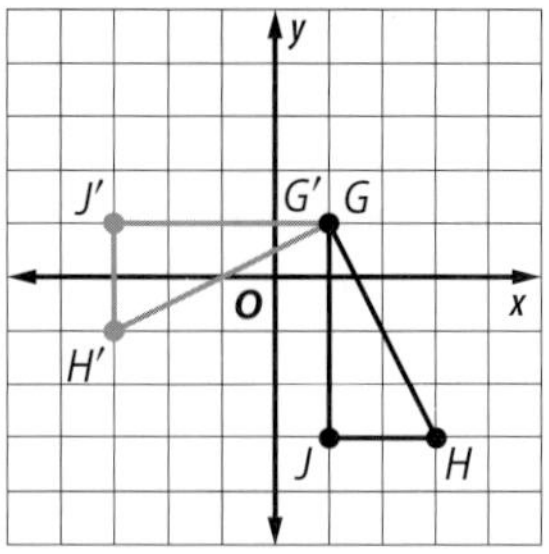

7 **b.** $G'(1, 1)$, $H'(-3, -1)$, $J'(-3, 1)$
9. 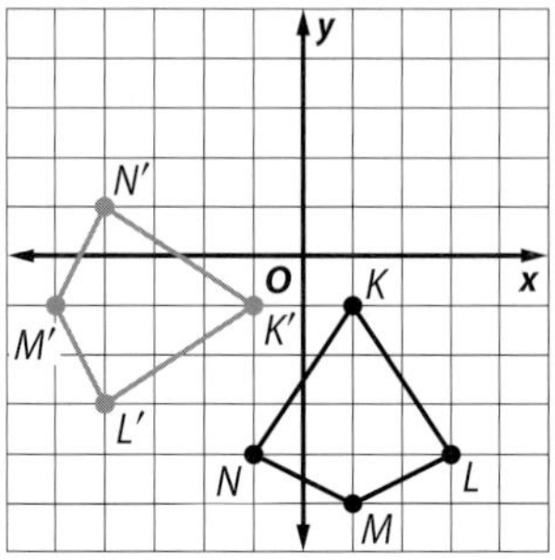

11. 72° **13.** yes; 180°
15 90° clockwise 
270° clockwise 
180° 

17. Sample answer: 90° 

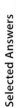

19. $R(1, 5)$, $S(0, -2)$, $T(-6, 3)$ **21.** With both transformations, the original figure and the image are congruent. In a reflection, a figure is reflected over a line. In a rotation, a figure is rotated about a point. **23.** J **25.** C **27.** 163° **29.** 35° **31.** 21.5 **33.** 35 **35.** −4 **37.** 2 **39.** infinitely many solutions

Pages 536–538 Lesson 11-6

1. The two triangles are congruent because a reflection or a rotation followed by a translation will map $\triangle LMN$ onto $\triangle XYZ$. **3.** Sample answer: a rotation of 180° about the center of the logo 5 The two triangles are congruent because a counterclockwise rotation of 90° followed by a translation will map $\triangle ABC$ onto $\triangle RST$. **7.** The two triangles are congruent because a reflection followed by a translation will map $\triangle FGH$ onto $\triangle MNP$. 9 Sample answer: a reflection followed by a translation **11.** 3 units, 3 units, 3 units, 3 units; 3 units, 3 units, 3 units, 3 units; yes **15.** Sample answer: a reflection over the y-axis followed by a translation of 3 units to the right **17.** Sample answer: rotation and translation **19a.** $9x - 1$ **19b.** 44 units **21.** $3x + 1$ **23.** $4x - 13$ **25.** $12x$

Pages 541–544 Lesson 11-7

1. $A'(-1, 0.5)$, $B'(0.5, 1)$, $C'(1.5, -1)$

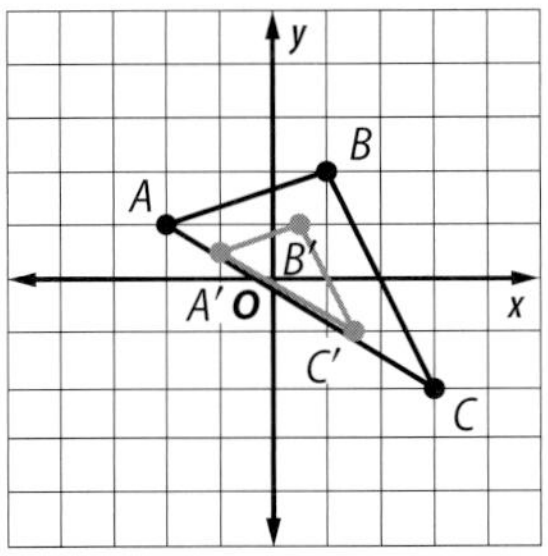

3. $M'(0, 0)$, $N'(7.5, -7.5)$, $O'(0, -15)$, $P'(-7.5, -7.5)$ **5.** 2.5

7. $G'(0, 4.5)$, $H'(1.5, 9)$, $I'(6, 1.5)$, $J'(3, 0)$

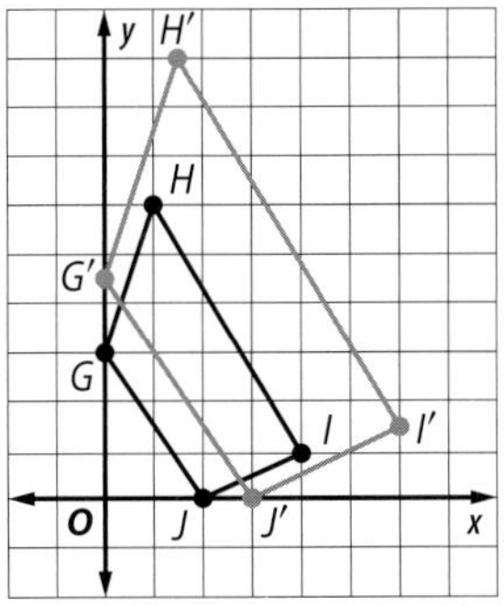

9. $K'\left(1\frac{2}{3}, 0\right)$, $L'\left(2\frac{2}{3}, 1\right)$, $M'\left(1, 2\frac{2}{3}\right)$, $N'\left(0, 1\frac{2}{3}\right)$

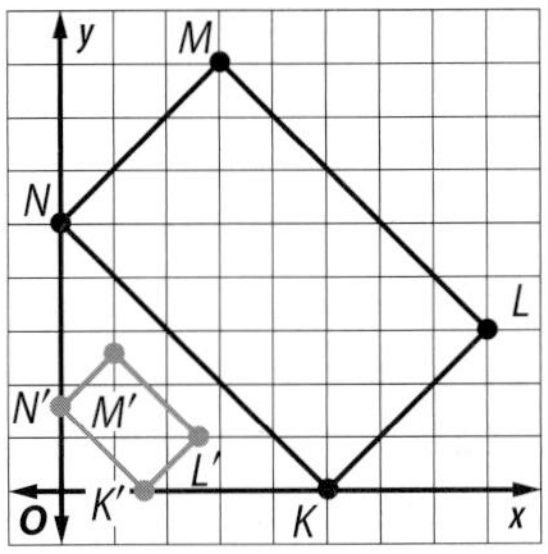

11. $W'(0, 0)$, $X'(2, -2)$, $Y'(2, 4)$

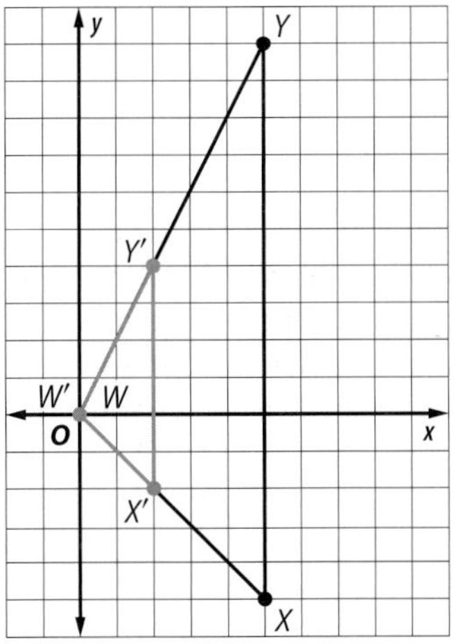

13. $A'(-1.5, -2.5)$, $B'(0, 1)$, $C'(1.5, -0.5)$, $D'(0, -2)$

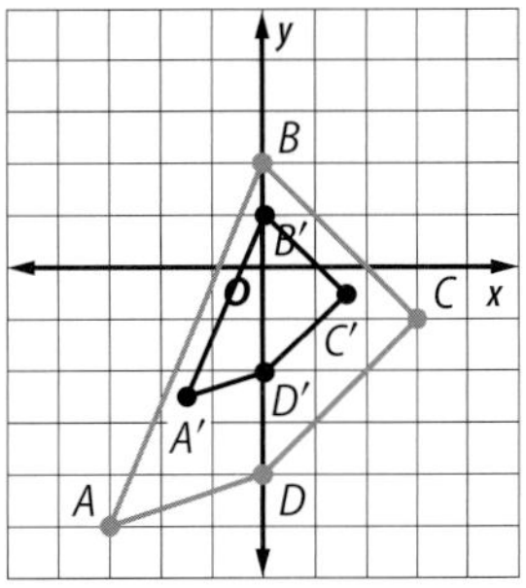

15. $\frac{7}{4}$ **17.** scale factor: $\frac{2}{3}$; reduction **19.** 6 yd

21. Sample answer: $A(-2, -4)$, $B(-6, -10)$, $C(-10, -8)$, $D(-12, -4)$

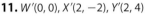

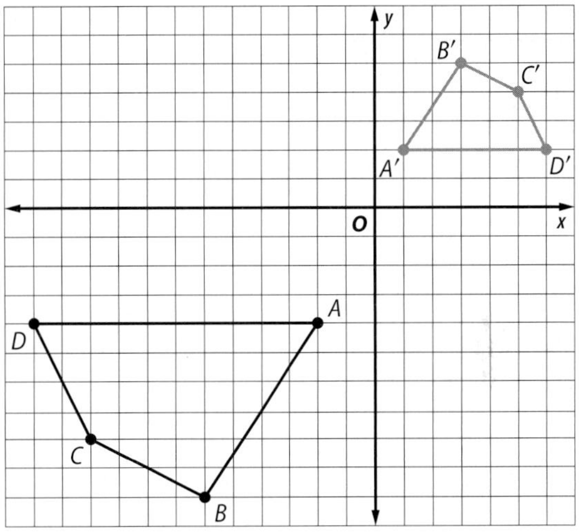

23. always; Sample answer: In examples a–d, both sets of two dilations have the same image. **25.** (3, 6); Sample answer: Multiplying by 3 and then by $\frac{1}{3}$ is the same as multiplying by 1. When multiplied by 1, the point will remain the same. **27.** F **29.** 0.4 **31.** 80 mi **33a.** \$31.23 **33b.** 82.1 rand **35.** The transformation is a translation. The triangle is translated left 2 units and down 1 unit.

Pages 547–550 Lesson 11-8

1. no; Sample answer: The ratios a of the side lengths are not equal for all of the sides; $\frac{EH}{AD} = \frac{3}{2}$, while $\frac{EF}{AB} = \frac{7}{5}$. **3.** 3 ft and 5 ft **5.** yes; Sample answer: A rotation and a dilation with a scale factor of $\frac{3}{2}$ maps $\triangle XYW$ onto $\triangle VUW$. **7.** no; Sample answer: The ratios of the side lengths are not equal for all of the sides; $\frac{JG}{FC} = \frac{1}{2}$, while $\frac{GH}{CD} = \frac{4}{6}$. **9.** 27 units **11.** 6.75 in. by 11.25 in.; yes **13.** Sample answer: reflection over the *x*-axis followed by a dilation with a scale factor of 2
15a. $AC = 6$ units, $CB = 8$ units **15b.** $A'C' = 3$ units, $C'B' = 4$ units **15c.** $\frac{2}{1}$; $\frac{2}{1}$; $\frac{2}{1}$ **15d.** Sample answer: The ratios are equal. **19.** true; Sample answer: The size and shape of the image are not affected by the order of the compositions. **21.** B **23.** C **25.** \$17 **27.** \$18.50 **29.** \$29.40 **31.** 65° **33.** 77 **35a.** 2 **35b.** 14°

Pages 551–554 Chapter Review

1. 112°; ∠6 and ∠4 are alternate interior angles, so they are congruent. **3.** 112°; ∠8 and ∠4 are corresponding angles, so they are congruent. **5.** 81° **7.** 54; acute isosceles **9.** 95; obtuse scalene **11.** The figure has 4 sides that only intersect at their endpoints. It is a quadrilateral that is not regular. **13.** 120° **15.** 135°

17.

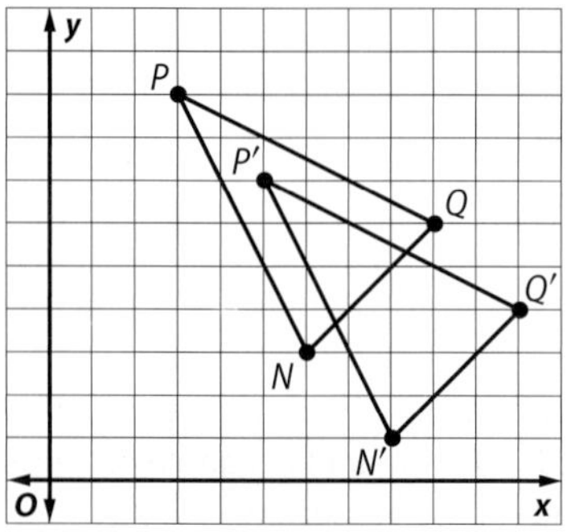

19. $A'(2, 0)$, $B'(0, 1)$, $C'(3, 3)$

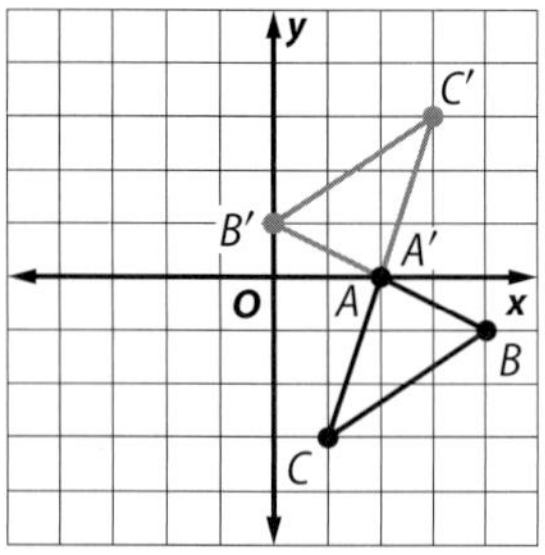

21.

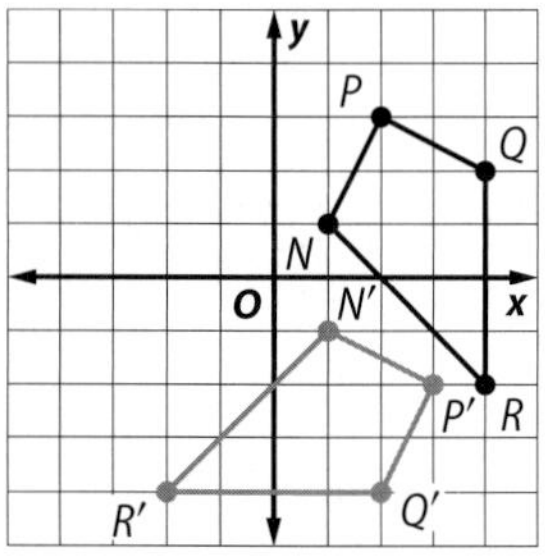

23. The figures are congruent because a reflection followed by a translation will map figure *A* onto figure *B*. **25.** Sample answer: Analyze the orientation or relative position of the figures.

27. $Q'(-6, 9)$, $R'(3, 6)$, $S'(9, -3)$, $T'(-6, -6)$

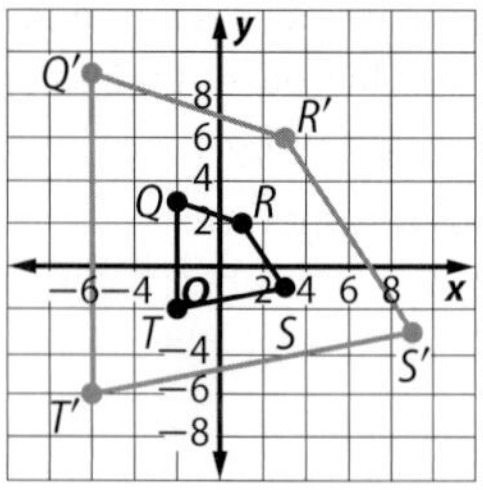

29. $F'(2.5, 7.5)$, $G'(7.5, 10)$, $H'(5, 2.5)$

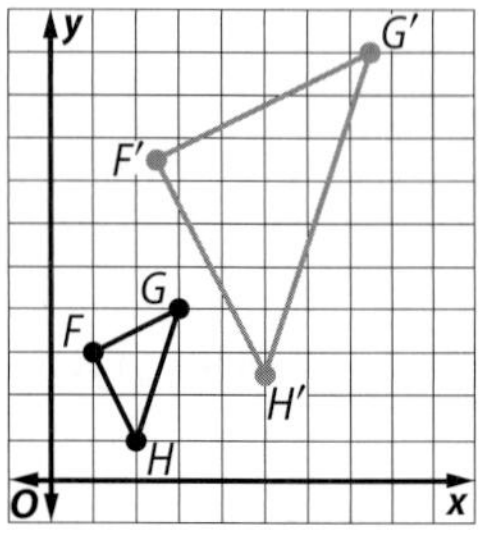

31. yes; Sample answer: Since the orientation of the figures is different, one of the transformations is a rotation. The ratios of the side lengths are equal. So, the two triangles are similar because a rotation and a dilation map rectangle *ABCD* to rectangle *PQRS*. **33.** 36 cm

Chapter 12 Volume and Surface Area

Pages 560–562 Lesson 12-1

1. 37.7 in. **3.** 50.9 cm **5.** 22.6 km **7** 22.0 m **9.** 62.8 ft **11.** 17.9 cm **13.** 29.5 m **15.** 70.7 ft **17.** Sample answer: Small sand dollar: $C \approx 3.9$ in., large sand dollar: $C \approx 7.9$ in.; the circumference of the small sand dollar is half the circumference of the large sand dollar. **19a.** 49.76 m **19b.** 12.44 m **19c.** 143.07 m **21** 100.5 ft

23a.

Radius (in.)	Circumference (in.)
1	6.3
2	12.6
3	18.8
4	25.1
5	31.4

23b. Sample answer:

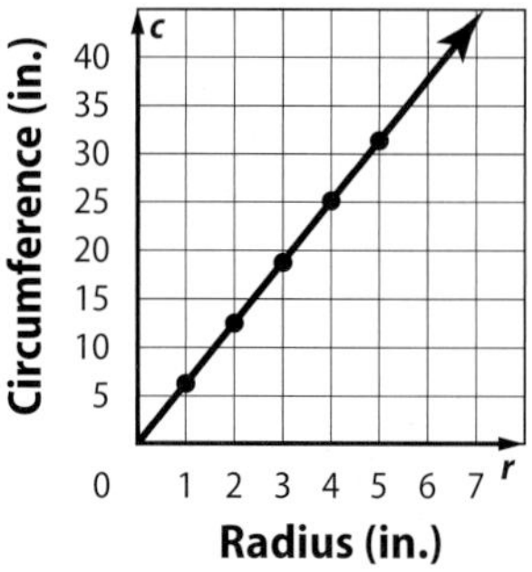

23c. 2π; Since the formula for the circumference of a circle, $C = 2\pi r$, is in the form of $y = mx$, 2π is the slope. **25.** Sample answer: A glass with a diameter of 7 centimeters has a circumference of about 22.0 centimeters. **27.** Circumference of one circle; ℓ equals $3d$ and the circumference of one circle is approximately $3.14d$ **29.** Circumference is $2\pi r$ or about 6.3 times the radius. The circumference increases as the radius increases. The radius decreases as the circumference decreases. **31.** F **33.** 47 ft **35.** 52.5 ft^2 **37.** 90 in^2 **39.** 720°

41.

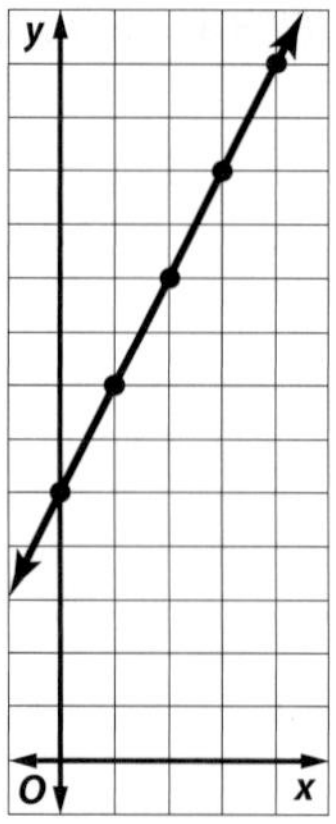

43. $n \leq 6$ **45.** $x \leq 2$ **47.** $c < -7$

Pages 565–567 Lesson 12-2

1. 153.9 m^2 **3.** 132.7 ft^2 **5.** 86.6 cm^2 **7.** 43.2 cm^2 **9.** 706.9 in^2 **11.** 227.0 ft^2 **13.** 323.7 m^2 **15.** 483.1 m^2 **17.** 42.2 mi^2 **19** 434.4 ft^2 **21.** 837.7 ft^2 **23.** 4 in. **25** 25.7 ft; 39.3 ft^2

27a.

Radius (cm)	Area (cm^2)
3	28.3
6	113.1
12	452.4
24	1809.6
48	7238.2

27b. The area is multiplied by 4. **27c.** Sample answer: Since $96 = 48 \cdot 2$, the area should be $4 \cdot 7238.2$ or about 28,952.8 cm^2; actual area ≈ 28,952.9 cm^2.

29. Sample answer:

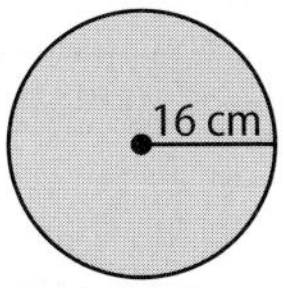

804.2 cm^2

31. 50 yd^2 **33.** If you know the radius, substitute the value for r in $A = \pi r^2$. If you know the diameter, first divide by 2 to find the radius. Then substitute the value for r in $A = \pi r^2$. If you know the circumference, substitute the value for C in $C = 2\pi r$ and solve for r to find the radius. Then substitute the value for r in $A = \pi r^2$. **35.** H **37a.** $r = \sqrt{\frac{A}{\pi}}$ **37b.** 10.2 cm **39.** 78.5 ft **41.** 11 m^2 **43.** 25.9 in^2 **45.** 234.9 **47.** 652.9 **49.** $x + x - 3x$ and $-x$ **51.** no equivalent expressions

Pages 570–573 Lesson 12-3

1. 38.4 cm^2 **3.** 18.9 cm^2 **5a.** 58 ft^2 **5b.** 10 cases; $250 **7** 240 in^2 **9.** 7.3 mm^2 **11.** 257.1 ft^2 **13a.** 354 ft^2 **13b.** $796.50 **15a.** 73.1 in^2 **15b.** 104.5 in^2 **17** 35.25 ft^2 **19.** 40.8 m^2 **21** 30.2 in^2 **23.** Sample answer: states, parks, shopping malls **25.** Sample answer: Use polygons to approximate the shape of the curved side of the playground. **27.** 429.7 ft^2 **29.** J **31.** 37.7 cm; 113.1 cm^2 **33.** 29.8 ft; 70.9 ft^2 **35.** 125.7 in.; 1256.6 in^2 **37.** 27 in^2 **39.** 37 min 30 s **41.** 15%

Pages 577–579 Lesson 12-4

1. rectangular prism; bases: *EIJF* and *HLKG*, *EILH* and *FGKJ*, *IJKL* and *EFGH*; faces: *EFGH*, *EFJI*, *GHLK*, *JFGK*, *EHLI*, *IJKL*; edges: $\overline{EF}$, $\overline{FJ}$, $\overline{JI}$, $\overline{IE}$, $\overline{HG}$, $\overline{GK}$, $\overline{KL}$, $\overline{LH}$, $\overline{EH}$, $\overline{IL}$, $\overline{FG}$, $\overline{JK}$; vertices: *E*, *F*, *G*, *H*, *I*, *J*, *K*, *L*; Sample answer: $\overline{TR}$ and $\overline{QS}$

3. 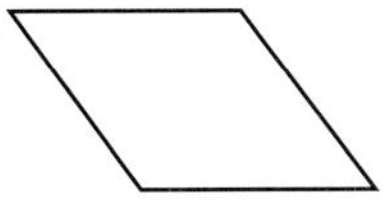 parallelogram

5. 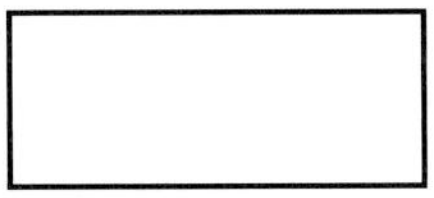 rectangle

7. triangular prism, bases: *DEF*, *ABC*; faces: *ABEF*, *BCDE*, *ACDF*, *FED*, *ABC*; edges: $\overline{AB}$, $\overline{AC}$, $\overline{AF}$, $\overline{BC}$, $\overline{BE}$, $\overline{DF}$, $\overline{CD}$, $\overline{ED}$, $\overline{EF}$; vertices: *A*, *B*, *C*, *D*, *E*, *F*; Sample answer: $\overline{AB}$ and $\overline{CD}$ **9** rectangular pyramid; base: *ABCD*; faces: *EAB*, *EBC*, *EDC*, *EDA*, *ABCD*; edges: $\overline{EA}$, $\overline{EB}$, $\overline{EC}$, $\overline{ED}$, $\overline{AB}$, $\overline{BC}$, $\overline{CD}$, $\overline{DA}$; vertices: *A*, *B*, *C*, *D*, *E*; Sample answer: $\overline{AB}$ and $\overline{ED}$ **11.** hexagonal pyramid; base: *BCDEFG*; faces: *ABC*, *ACD*, *ADE*, *AEF*, *AFG*, *AGB*, *BCDEFG*; edges: $\overline{AB}$, $\overline{AC}$, $\overline{AD}$, $\overline{AE}$, $\overline{AF}$, $\overline{AG}$, $\overline{BC}$, $\overline{CD}$, $\overline{DE}$, $\overline{EF}$, $\overline{FG}$, $\overline{GB}$; vertices: *A*, *B*, *C*, *D*, *E*, *F*, *G*; Sample answer: $\overline{AF}$ and $\overline{ED}$

13. 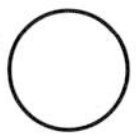 circle **15.**  square

17a.

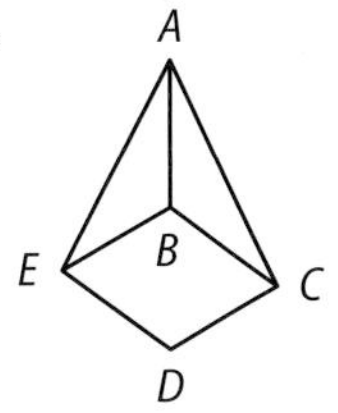

17b. base: *BCDE*; faces *ABC*, *ABE*, *ACD*, *ADE*, *BCDE*; edges: $\overline{AB}$, $\overline{AC}$, $\overline{AD}$, $\overline{AE}$, $\overline{BC}$, $\overline{CD}$, $\overline{DE}$, $\overline{EB}$; **17c.** triangle, triangle or trapezoid, square **17d.** top: triangle, base: trapezoid

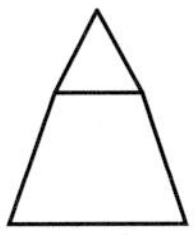

19.

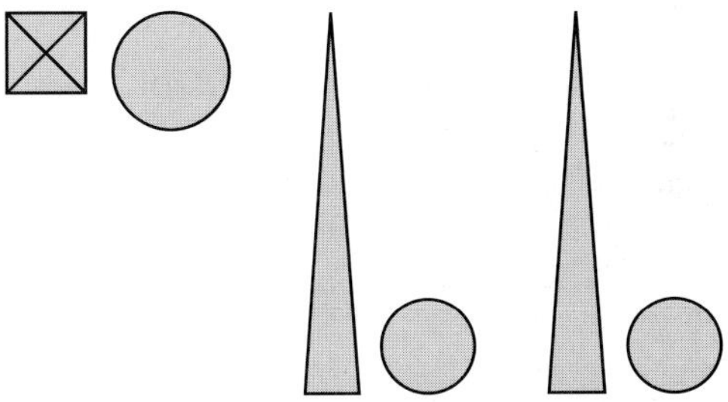

triangle and circle

21 a.

Name	Triangular Pyramid	Square Pyramid	Pentagonal Pyramid
Figure	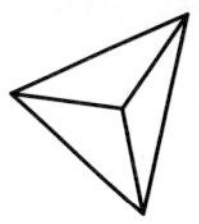	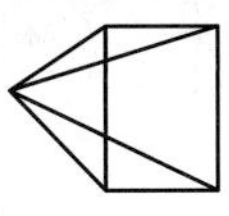	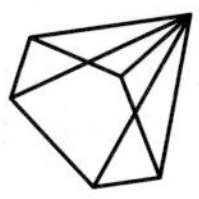
Vertices	4	5	6
Faces	4	5	6
Edges	6	8	10

21 b. Sample answer: The number of vertices is equal to the number of faces. The number of edges increases by 2 as the number of vertices of the pyramid increases by 1.

21 c. $V + F = E + 2$ **23.** never **25.** sometimes **27.** Sample answer: Examine the base of the object; it is a hexagon. Since it has two parallel, congruent bases, it is also a prism. The figure is a hexagonal prism. **29.** D

31. Sample answer:

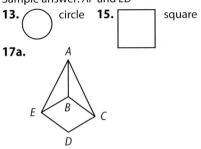

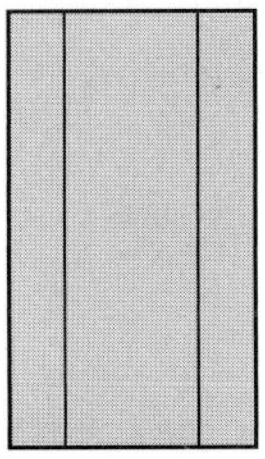

The shape resulting from a vertical cross section is a rectangle. **33.** 72 ft^2 **35.** 56.1 cm^2 **37.** $\frac{1}{20}$ or 5% **39.** 0 or 0% **41.** $\frac{3}{20}$ or 15% **43.** 55 times

Pages 582–585 Lesson 12-5

1. 36 mm^3 **3.** 1000 in^3 **5.** 552 in^3 **7.** 260 ft^3 **9.** 54 in^3 **11.** 120 cm^3 **13.** $1\frac{1}{2}$ in^3 **15.** 6 yd **17.** 3888 in^3 **19.** 10 ft **21.** 972 m^3 **23.** 180 ft^3 **25.** 1.875 in^3 **27a.** 42 cm^3 **27b.** The original volume is multiplied by 2 because $V = w\ell(2h) = 2w\ell h$; The original volume is multiplied by 2^2 or 4 because $V = (2w\ell)(2h) = 4w\ell h$; The original volume is multiplied by 2^3 or 8 because $V = (2w)(2\ell)(2h) = 8w\ell h$. **27c.** When one dimension is tripled, the volume is multiplied by 3. When two dimensions are tripled, the original volume is multiplied by 3^2. When three dimensions are tripled, the original volume is multiplied by 3^3. **27d.** The volume is multiplied by 216. **29.** Sample answer: base length 4 in., base height 2 in., prism height 11 in. **31.** 35 **33.** The volume is multiplied by 2^3 or 8. **35.** 12 **37.** C **39.** 228.5 mm^2 **41.** $12\frac{9}{20}$ **43.** $-1\frac{3}{10}$ **45.** $2\frac{7}{15}$ **47.** \$32.25 **49.** 12.6 in^2 **51.** 78.5 mi^2

Pages 588–590 Lesson 12-6

1. 25.1 ft^3 **3.** 10.0 in. **5.** 22.0 cm^3 **7.** 254.5 m^3 **9.** 45.6 cm^3 **11.** 101.4 m^3 **13.** 5.5 m **15.** 69.1 in^3 **17.** 2628.1 cm^3 **19.** 3271.2 cm^3 **21.** 101.3 gal **23.** Sample answer: Changing the packaging size of a drink; changing the size of a garbage can to fit in a narrower but taller space. **25a.** 2:1 **25b.** 4:1 **27.** B **29a.** 77.8 in^3 **29b.** 699.8 in^3 **31.** 5200 ft^3 **33.** 120 in^3 **35.** 160 **37.** 200 **39.** 35.75 **41.** 0.05 **43.** 35 **45.** $3\frac{1}{3}$ **47.** 235 ft/min

Pages 598–600 Lesson 12-7

1. 115.5 in^3 **3.** 74.2 cm^3 **5.** 268.1 mm^3 **7.** 392.7 ft^3 **9.** 6 ft^3 **11.** 4.3 in^3 **13.** 10.5 ft^3 **15.** 1563.5 km^3 **17a.** 1.5 ft^3 **17b.** about 75 min or 1 h 15 min **19.** 18.3 ft **21.** 207.4 in^3 **23.** 4657 g **25a.** 140.4 cm^3 **25b.** 47.1% **27.** 3.0 units **29.** True; sample answer: Both volumes are $\frac{1}{3}Bh$. **31.** Changing the shape of packaging; if the volumes remained the same and they have congruent bases, the cone's height would be 3 times the height of the cylinder. **33.** H **35.** D **37.** 748 cm^3 **39.** 316 ft **41.** $5.75p = 15$; 2.6 lb **43.** 3772 m^2 **45.** 0.5; equally likely

Pages 605–607 Lesson 12-8

1. Sample answer: 108 in^2; 268 in^2 **3.** Sample answer: 360 yd^2; 464.5 yd^2 **5.** 3.84 cm^2; 4.8 cm^2 **7.** 8.4 m^2; 10.08 m^2 **9.** Sample answer: 64.8 cm^2; 70.96 cm^2 **11.** 334 in^2 **13.** 528 m^2 **15a.** 150 in^2

15b.

Scale Factor of Dilation	Original	× 2	× 3
Length of Side	5	10	15
Surface Area	150	600	1350

15c. Sample answer: The surface area after a dilation of scale factor of 2 is four times the original; with a scale factor of 3, it is 9 times the original. **15d.** Yes. The formula for surface area is $S = Ph + 2B$. After doubling the side lengths, the new surface area is $S = (2P)(2h) + 2(4B)$, which simplifies to $S = 4(Ph + 2B)$. **17.** Box 1 would be more expensive to make. The surface area of Box 1 is 36.64 in^2, which would cost \$47.63 to make. The surface area of Box 2 is 34.56 in^2, which would cost \$44.93 to make. **19.** The box is a cube that measures 6 cm on each side. **21.** Sample answer: Prism A with dimensions 3 by 3 by 3 and Prism B with dimensions 10 by 2 by 1. Prism A has the greater volume while Prism B has the greater surface area. **23.** D 25. J **27.** 268.1 cm^3 **29.** 251.3 m^3 **31.** 625 **33.** 722 **35.** 655 **37.** 153.9 yd^2 **39.** 1240.9 m^2

Pages 612–614 Lesson 12-9

1. 1492.3 mm^2; 2059.3 mm^2 **3.** 102.9 m^2; 130.6 m^2 **5.** 39.0 yd^2; 99.3 yd^2 **7.** 125.7 in^2; 150.8 in^2 **9.** 121.0 ft^2; 205.4 ft^2 **11.** tube; Sample answer: The box has a surface area of 1296 in^2 and the tube has a surface area of about 1288.1 in^2. **13.** 194.7 in^2 **15.** 6121.0 cm^2 **17.** The cylindrical popcorn container because it uses less material (surface area) and also holds about the same amount of popcorn (volume). Rectangular Prism: $V = 275$ in^3, $S = 237.5$ in^2; Cylinder: $V = 274.3$ in^3, $S = 211.1$ in^2 **21.** The ice cube in the shape of a half cylinder would melt at a faster rate because it has a greater total surface area exposed to the air. **23.** G **25.** D **27.** Sample answer: 114 in^2; 282 in^2 **29.** 4356 m^3 **31.** 90° **33.** 66° **35.** \$918.75 **37.** \$9135 **39.** 48 **41.** 58.75 **43.** $\frac{14}{15}$

Pages 618–620 Lesson 12-10

1. 364 m^2; 533 m^2 **3.** 192 mm^2; 336 mm^2 **5.** 240 in^2; 340 in^2 **7.** 105.3 mm^2; 140.4 mm^2 **9.** 339.3 mm^2 **11.** 125.7 cm^2; 175.9 cm^2 **13.** 427.3 ft^2; 628.3 ft^2 **15.** 471.2 cm^2 **17.** Sample answer: 1583.2 in^2

19.  130.5 m^2; 174 m^2

21. 48.7 ft **25.** 4 **27.** C **29.** B **31.** 386.4 in^2; 612.6 in^2 **33.** 120 ft^2; 244.8 ft^2 **35.** 50° **37.** \$71.40 **39.** 21 **41.** 3.5 **43.** 3.3

Pages 621–624 Chapter Review

1. 125.7 km **3.** 16.8 ft **5.** about 11,781 ft **7.** 143.1 in^2 **9.** 8.25 in^2 **11.** rectangular prism; Sample answer: bases: *AFGD, BEHC*; faces: *ABCD, EFGH, BEFA, ADGF, BCHE, CHGD*; edges: $\overline{AB}, \overline{BC}, \overline{CD}, \overline{AD}, \overline{EF}, \overline{EH}, \overline{FG}, \overline{GH}, \overline{EB}, \overline{HC}, \overline{AF}, \overline{GD}$; vertices: *A, B, C, D, E, F, G, H* **13.** square pyramid; base: *LMNO*; faces: *LPM, LPO, NPO, NPM, LMNO*; edges: $\overline{LM}, \overline{MN}, \overline{NO}, \overline{OL}, \overline{LP}, \overline{MP}, \overline{NP}, \overline{OP}$; vertices: *L, M, N, O, P*

15. Top View Side View Cross Section

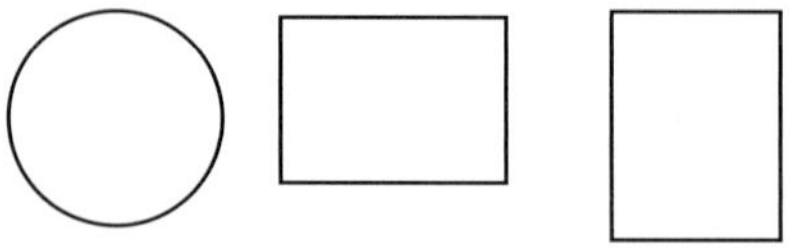

The vertical cross section of the drum is a rectangle. **17.** 60 cm^3 **19.** 19,278 cm^3 **21.** 282.7 in^3 **23.** 8 in^3 **25.** 564.4 ft^3 **27.** 143.6 m^2; 166.4 m^2 **29.** 378.9 in^2; 506.1 in^2 **31.** 131.9 m^2; 188.5 m^2 **33.** 94.2 cm^2 **35.** 263.9 ft^2; 417.8 ft^2 **37.** 52.5 m^2; 68 m^2

Index

Dd

Ee

Ff

Gg

Hh

Ii

Kk

Ll

Mm

Nn

Oo

Pp

Qq

Rr

Ss

Tt

Uu

Vv

Ww

Xx

Yy

Zz

Mathematics Reference Sheet

Formulas

Perimeter	square	$P = 4s$	
	rectangle	$P = 2\ell + 2w$ or $P = 2(\ell + w)$	
Circumference	circle	$C = 2\pi r$ or $C = \pi d$	
Area	square	$A = s^2$	
	rectangle	$A = \ell w$	
	parallelogram	$A = bh$	
	triangle	$A = \frac{1}{2}bh$	
	trapezoid	$A = \frac{1}{2}h(b_1 + b_2)$	
	circle	$A = \pi r^2$	
Volume	cube	$V = s^3$	
	rectangular prism	$V = Bh$ or $V = \ell wh$	
	triangular prism	$V = Bh$	
	cylinder	$V = Bh$ or $V = \pi r^2 h$	
	pyramid	$V = \frac{1}{3}Bh$	
	cone	$V = \frac{1}{3}Bh$ or $V = \frac{1}{3}\pi r^2 h$	
	sphere	$V = \frac{4}{3}\pi r^3$	
Surface Area	cube	$V = 6s^2$	
	prism	**Lateral Area:** $L = Ph$	**Surface Area:** $S = L + 2B$ or $S = Ph + 2B$
	cylinder	**Lateral Area:** $L = 2\pi rh$	**Surface Area:** $S = L + 2B$ or $S = 2\pi rh + 2\pi r^2$
	pyramid	**Lateral Area:** $L = \frac{1}{2}P\ell$	**Surface Area:** $S = L + B$ or $S = \frac{1}{2}P\ell + B$
	cone	**Lateral Area:** $L = \pi r\ell$	**Surface Area:** $S = L + B$ or $S = \pi r\ell + \pi r^2$
Temperature	Fahrenheit to Celsius	$C = \frac{5}{9}(F - 32)$	
	Celsius to Fahrenheit	$F = \frac{9}{5}C + 32$	
Slope	line	$m = \frac{\text{rise}}{\text{run}}$ or $m = \frac{\text{change in } y}{\text{change in } x}$ or $m = \frac{y_2 - y_1}{x_2 - x_1}$	

Measures

Length	1 kilometer (km) = 1000 meters (m) 1 meter = 100 centimeters (cm) 1 centimeter = 10 millimeters (mm)	1 foot (ft) = 12 inches (in.) 1 yard (yd) = 3 feet or 36 inches 1 mile (mi) = 1760 yards or 5280 feet
Volume and Capacity	1 kiloliter (kL) = 1000 liters (L) 1 liter = 1000 milliliters (mL)	1 cup (c) = 8 fluid ounces (fl oz) 1 pint (pt) = 2 cups 1 quart (qt) = 2 pints 1 gallon (gal) = 4 quarts
Weight and Mass	1 kilogram (kg) = 1000 grams (g) 1 gram = 1000 milligrams (mg) 1 metric ton (t) = 1000 kilograms	1 pound (lb) = 16 ounces (oz) 1 ton (T) = 2000 pounds
Time	1 minute (min) = 60 seconds (s) 1 hour (h) = 60 minutes 1 day (d) = 24 hours	1 week (wk) = 7 days 1 year (yr) = 12 months (mo) or 52 weeks or 365 days 1 leap year = 366 days
Metric to Customary	1 meter ≈ 39.37 inches 1 kilometer ≈ 0.62 mile 1 centimeter ≈ 0.39 inch	1 kilogram ≈ 2.2 pounds 1 gram ≈ 0.035 ounce 1 liter ≈ 1.057 quarts

Symbols

Number and Operations

$a \cdot b$ $a \times b$ ab $a(b)$	*a* times *b*	$\neq$	is not equal to	$\approx$	is approximately equal to
		$>$	is greater than	$\pm$	plus or minus
		$<$	is less than	%	percent
		$\geq$	is greater than or equal to	$a : b$	the ratio of *a* to *b*, or $\frac{a}{b}$
		$\leq$	is less than or equal to	$0.7\overline{5}$	repeating decimal 0.7555…

Algebra and Functions

$-a$	opposite or additive inverse of *a*	a^{-n}	$\frac{1}{a^n}$	$\sqrt{x}$	principal (positive) square root of *x*
a^n	*a* to the nth power	$\lvert x \rvert$	absolute value of *x*	$f(x)$	function, *f* of *x*

Geometry and Measurement

$\cong$	is congruent to	AB	the length of $\overline{AB}$	$\triangle ABC$	triangle *ABC*
$\sim$	is similar to	∟	right angle	O	origin
$^\circ$	degree(s)	$\parallel$	is parallel to	π	pi $\approx \left(3.14 \text{ or } \frac{22}{7}\right)$
$\overleftrightarrow{AB}$	line *AB*	$\perp$	is perpendicular to	(a, b)	ordered pair with *x*-coordinate *a* and *y*-coordinate *b*
$\overrightarrow{AB}$	ray *AB*	$\angle A$	angle *A*		
$\overline{AB}$	line segment *AB*	$m\angle A$	measure of angle *A*		

Probability and Statistics

$P(A)$	probability of event *A*